Forensic Science
Second Edition

HANDBOOK OF ANALYTICAL SEPARATIONS

Series Editor: ROGER M. SMITH

Forensic Science

Second Edition

Edited by

M.J. BOGUSZ

Department of Pathology and Laboratory Medicine,
King Faisal Specialist Hospital and Research Centre,
Riyadh, Saudi Arabia

ELSEVIER

Amsterdam – Boston – Heidelberg – London – New York – Oxford
Paris – San Diego – San Francisco – Singapore – Sydney – Tokyo

Elsevier
Radarweg 29, PO Box 211, 1000 AE Amsterdam, The Netherlands
Linacre House, Jordan Hill, Oxford OX2 8DP, UK

First edition 2008

Notice
No responsibility is assumed by the publisher for any injury and/or damage to persons
or property as a matter of products liability, negligence or otherwise, or from any use
or operation of any methods, products, instructions or ideas contained in the material
herein. Because of rapid advances in the medical sciences, in particular, independent
verification of diagnoses and drug dosages should be made

Library of Congress Cataloging-in-Publication Data
A catalog record for this book is available from the Library of Congress

British Library Cataloguing in Publication Data
A catalogue record for this book is available from the British Library

ISBN: 978-0-444-52214-6
ISSN: 1567-7192

For information on all Elsevier publications
visit our website at books.elsevier.com

Printed and bound in The Netherlands

08 09 10 11 12   10 9 8 7 6 5 4 3 2 1

Preface to the second edition

This book is dedicated to late Dr. Irving Sunshine,
The Grandfather of modern forensic toxicology,
and teacher of many generations of us.

The methods and procedures used in forensic sciences depend on the development of techniques having the highest possible reliability. Because of the importance and consequences of their expert evidence and their great sense of responsibility, forensic scientists often adopt a critical and cautious attitude to new methods. Nevertheless, this professional caution does not prevent the forensic society from rapidly adopting all the new techniques that have been proved to be reliable scientific tools in various disciplines. In the six years that have elapsed since the publication of the first edition of this volume, several such techniques have been widely implemented in forensic practice, the skyrocketing popularity of liquid chromatography-mass spectrometry being the best example. Therefore, the need for this new edition was recognized by the publisher, and it is hoped that it will be accepted by prospective readers.

Recent years have also seen an enormous growth in public interest in forensic methods and disciplines, almost certainly stimulated by some spectacular criminal cases. With this interest has also come an increase in expectations as can be seen in various television serials and novels, which show infallible forensic experts who solve all possible problems within hours, with the help of very complicated but also very easy methods. Another important factor that has given rise to this increasing interest in forensic sciences is the worldwide spread of the most dangerous forms of organized crime, such as international terrorism and drug trafficking. The general public feels that forensic experts are in the forefront of the fight against these plagues of the modern world.

The general outline of this book has not changed: it is mainly devoted to the analysis of the most important groups of drugs and poisons. However, some new important fields have emerged in recent years. Therefore, in the part of the volume devoted to actual and emerging problems of forensic toxicology, new topics are dealt with, such as the analytical aspects of alcohol markers, toxicology of herbal remedies, and pharmacogenomics applied to forensic toxicology. The previously existing fields of application such as drugs and driving, the analysis of unconventional matrices, doping analysis, and quality assurance have again been covered, as in the previous edition.

Special emphasis in this edition has been placed on toxicological screening. Analytical screening procedures are among the basic strategies of forensic toxicology, and are presented in detail in six chapters, covering the most important chromatographic methods, such as GC, GC-MS, HPLC, LC-MS, CE, and LC-ICP-MS. Additionally, in each chapter dealing with particular groups of drugs, preliminary methods (mainly immunoassays) are presented along with separation methods. This might look out of place in a volume devoted to analytical separations, but in forensic practice it is impossible to separate preliminary methods like immunoassays, from the confirmative ones like chromatographic procedures.

In the part of the book devoted to forensic chemistry, the analysis of explosives, of chemical warfare agents, of fire debris in arson cases and writing media is presented.

The book concludes with chapters dealing with forensic genetics applied to the identification of individuals and biological traces.

The choice of topics and the highlighting of particular problems is always a matter of subjective preference by the person responsible, in this case the editor. He must accept the criticism that is unavoidably caused by omitting some important themes, but it must be kept in mind that forensic science is such a multi-faceted discipline that it is not possible to cover everything in a single volume, however large.

The authors of the chapters are the pioneers in their fields, both as scientists and as case experts. Their wisdom, knowledge, experience, and very active collaboration has made this work possible. As an editor, I feel very happy and honored to have had the opportunity of working with such a distinguished team.

I would like to express my thanks to Roger M. Smith, Series Editor, and to Elsevier Science Publishing, for entrusting me with this volume again, and for all their support during the entire preparation process.

Maciej J. Bogusz

Preface to the first edition

"Work is always play when you follow your heart, but after the first burst of ideas it became a game." Michael Hawley, Cambridge, MA, 1997.

The term "forensic" has a double meaning. In the most popular sense, forensic science is understood as a particular scientific discipline (medicine, toxicology or chemistry) that is applied for the needs of civil or criminal law. Therefore, an obvious synonym of forensic medicine is legal medicine. In a broader and deeper sense, all "forensic" disciplines belong to *forum*, i.e. are subjected–more than any other scientific activities–to public debate and public control. Forensic experts are obliged to explain the smallest details of the methods used, to substantiate the choice of the applied technique and to give their unbiased conclusions–all under the critical and often mistrustful look of the servants of Justice, as well as the general public, including the media. The final result of the work of forensic scientists–expert evidence–exerts a direct influence on the fate of a given individual. This burden is a most important stimulus, which determines the way of thinking and acting in forensic sciences.

The purpose of this volume is to present critical, up-to-date information on the separation methods applied in various disciplines of forensic science. The book cannot and should not replace a scientific paper in regard to the depth and coverage of a specific problem. It should, however, present the relevant problems *in statu nascendi*, showing their development, potential importance and future perspectives. This book was written by forensic scientists not only for forensic scientists, but also for other colleagues interested in the particular analytical aspects of substances or materials involved.

The general structure of this volume corresponds to the most important forensic disciplines that apply various separation methods, i.e. forensic toxicology, chemistry and serohematology. The chapters devoted to forensic toxicology are focused on particular groups of illicit and therapeutic drugs and other substances of forensic interest. Also, some specific problems of forensic toxicology, such as drugs and driving, doping control, quality assurance, chiral separations, use of alternative matrices and general analytical strategy, are separately discussed. In the part devoted to forensic chemistry, the most important problems of the analysis of explosives, arson accelerants and writing media are individually covered. The part concerning forensic identification of individuals and biological traces presents all of the relevant separation techniques applied to endogenous macromolecules.

The focus of this volume is the responsibility of the editor. The choice of presented topics was not always obvious, taking into account the multifaceted world of forensic sciences. Nevertheless, it is sincerely hoped that the most relevant problems are covered adequately and in proportion. The individual chapters were written by scientists who not only carry out important scientific activity in their field but are also known as very experienced forensic experts. Therefore, this volume gives not only the information about the application of separation methods in forensics, but also reflects the forensic community itself. Forensic scientists are usually passionate about their work, trying to create new methods or to adapt existing methods to their own needs. In doing so, they tend always to put down a personal signature on their work. It was said once that forensic toxicologists would rather share a toothbrush, than their analytical method. I believe that this is caused not, or not only, by personal ambition, but by the consciousness that the final results of the work may be of enormous significance to other person. The analysis report may disrupt a person's professional and family life, may ruin the good name of an olympic champion or may simply send somebody to jail for a lifetime. Therefore, the name of the play of forensic scientist may be: perfective tracking, but the game must be called: professional responsibility.

It is my very pleasant duty to express my sincere thanks to the Series Editor, Roger M. Smith, who entrusted me this volume, and all co-authors for their creative and timely collaboration.

I wish to address my thanks to Ms. Reina Bolt of Elsevier Science for her kind and forgiving assistance during the whole editing procedure.

Maciej J. Bogusz
April 2000

List of contributors

Rolf E. Aderjan, Institute of Legal Medicine and Traffic Medicine, University Hospital — Ruprecht-Karls University of Heidelberg 69115 Heidelberg, Voßstr. 2, Germany

Valery N. Aginsky, Riley, Welch & Aginsky, Forensic Documents Examinations, Inc., P.O. Box 80225, Lansing, MI48908, USA

Mohammed Al-Tufail, Department of Pathology and Laboratory Medicine, King Faisal Specialist Hospital and Research Center, P.O. Box 3354, MBC10, 11211, Riyadh, Saudi Arabia

Vanesa Álvarez-Iglesias, Unidad de Xenética, Instituto de Medicina Legal, Facultad de Medicina, Universidad de Santiago de Compostela, 15782, Santiago de Compostela, Spain

Olof Beck, Department of Medicine, Division of Clinical Pharmacology, Karolinska Institute and Karolinska University Hospital, SE-171 76 Stockholm, Sweden

Gert De Boeck, Section of Toxicology, Federal Public Service Justice, National Institute of Criminalistics and Criminology (N.I.C.C.), Vilvoordsesteenweg 100, 1120 Brussels, Belgium

Maciej J. Bogusz, Department of Pathology and Laboratory Medicine, King Faisal Specialist Hospital and Research Center, P.O. Box 3354, MBC10, 11211, Riyadh, Saudi Arabia

Federica Bortolotti, Department of Medicine and Public Health, Section of Forensic Medicine, University of Verona, 37134 Verona, Italy

María Brión, Grupo de Medicina Xenómica, CeGen-Institute of Legal Medicine, University of Santiago de Compostela, 15782, Santiago de Compostela, Spain

Angel Carracedo, Unidad de Xenética, Instituto de Medicina Legal, Facultad de Medicina, Universidad de Santiago de Compostela, 15782, Santiago de Compostela, Spain; Centro Nacional de Xenoripado (CeGen), Grupo de Medicina xenómico, Hospital clínico Universitario, 15706, Galicia, Spain

Joseph A. Caruso, Department of Chemistry, University of Cincinnati, Cincinnati, OH 45221-0172, USA

María Cerezo, Unidad de Xenética, Instituto de Medicina Legal, Facultad de Medicina, Universidad de Santiago de Compostela, 15782, Santiago de Compostela, Spain

John T. Cody, Air Force Drug Testing Laboratory, 2730 Louis Bauer Drive, Brooks City-Base, TX 78235-5132, USA

Edward J. Cone, ConeChem Research, LLC, 441 Fairtree Drive, Severna Park, MD 21146, USA

Jean M. H. Conemans, Central Department for Pharmacy, 's-Hertogenbosch Hospital, 5232 JL 's-Hertogenbosch, The Netherlands

Paul A. D'Agostino, DRDC Suffield, P.O. Box 4000 Station Main, Medicine Hat, AB, Canada, TIA 8K6

William D. Darwin, National Institute on Drug Abuse, Chemistry and Drug Metabolism Section, 5500 Nathan Shock Drive, Baltimore, MD 21224, USA

Julia A. Dolan, Bureau of Alcohol, Tobacco, Firearms and Explosives, Forensic Science Laboratory – Washington, Ammendale, Maryland, USA

Mahmoud A. ElSohly, ElSohly Laboratories, Incorporated (ELI), 5 Industrial Park Drive, Oxford, MS 38655, USA; National Center for Development of Natural Products, School of Pharmacy, University of Mississippi, University, MS 38677, USA

Ameriga Fanigliulo, Department of Medicine and Public Health, Section of Forensic Medicine, University of Verona, 37134 Verona, Italy

Merja Gergov, Department of Forensic Medicine, University of Helsinki, P.O. Box 40, FI-00014, Helsinki, Finland

Iva Gomes, IPATIMUP, Institute of Molecular Pathology and Immunology of the University of Porto, 4200-465 Porto, Portugal; Grupo de Medicina Xenómica, CeGen-Institute of Legal Medicine, University of Santiago de Compostela, 15782 Santiago de Compostela, Spain

Waseem Gul, ElSohly Laboratories, Incorporated (ELI), 5 Industrial Park Drive, Oxford, MS 38655, USA; National Center for Development of Natural Products, School of Pharmacy, University of Mississippi, University, MS 38677, USA

Leonor Gusmão, IPATIMUP, Institute of Molecular Pathology and Immunology of the University of Porto, 4200-465 Porto, Portugal

Ricardo Gutiérrez Gallego, Department of Experimental and Health Sciences, Universitat Pompeu Fabra and Pharmacology Research Unit, Institut Municipal d'Investigació Mèdica IMIM-Hospital del Mar, Dr. Aiguader 88, 08003 Barcelona, Spain

Anders Helander, Department of Clinical Neuroscience, Karolinska Institute and Karolinska University Hospital, SE-171 76 Stockholm, Sweden

Jessica Jennings Smith, Office of the Chief Medical Examiner, 200 South Adams Street, Wilmington, DE 19801, USA

Rebecca Jufer Phipps, Office of the Chief Medical Examiner, 111 Penn Street, Baltimore, MD 21201, USA

Kevin M. Kubachka, Department of Chemistry, University of Cincinnati, Cincinnati, OH 45221-0172, USA

Thomas Kraemer, Institute of Legal Medicine, Forensic Toxicology, Saarland University, D-66421 Homburg, Germany

Marleen Laloup, Section of Toxicology, Federal Public Service Justice, National Institute of Criminalistics and Criminology (N.I.C.C.), Vilvoordsesteenweg 100, 1120 Brussels, Belgium

Willy E. Lambert, Laboratorium voor Toxicologie, Universiteit Gent, Harelbekestraat 72 9000 Gent, Belgium

María Victoria Lareu, Unidad de Xenética, Instituto de Medicina Legal, Facultad de Medicina, Universidad de Santiago de Compostela, 15782, Santiago de Compostela, Spain

José Marcos, Department of Experimental and the Health Sciences, Universitat Pompeu Fabra, Dr. Aiguader 88, 08003 Barcelona, Spain

Hans H. Maurer, Department of Experimental and Clinical Toxicology, Institute of Experimental and Clinical Pharmacology and Toxicology, Saarland University, D-66421 Homburg, Germany

Ana Mosquera, Unidad de Xenética, Instituto de Medicina Legal, Facultad de Medicina, Universidad de Santiago de Compostela, 15782, Santiago de Compostela, Spain

Nuria Naverán, Unidad de Xenética, Instituto de Medicina Legal, Facultad de Medicina, Universidad de Santiago de Compostela, 15782, Santiago de Compostela, Spain

Ilkka Ojanperä, Department of Forensic Medicine, University of Helsinki, P.O. Box 40, FI-00014, Finland

Jennifer Pascali, Department of Medicine and Public Health, Section of Forensic Medicine, University of Verona, 37134 Verona, Italy

Chris Phillips, Grupo de Medicina Xenómica, Hospital Clínico Universitario, 15706, Galicia, Spain

Fritz Pragst, Institute of Legal Medicine, University Hospital Charité, Hittorfstr. 18, 14195 Berlin, Germany

Elke Raes, Department of Clinical Biology, Microbiology and Immunology, 185 De Pintelaan, Ghent University, Ghent, Belgium

Ilpo Rasanen, Department of Forensic Medicine, University of Helsinki, P.O. Box 40, FI-00014, Finland

Douglas D. Richardson, Department of Chemistry, University of Cincinnati, Cincinnati, OH 45221-0172, USA

Antonio Salas, Unidad de Xenética, Instituto de Medicina Legal, Facultad de Medicina, Universidad de Santiago de Compostela, 15782, Santiago de Compostela, Spain; Grupo de Medicina Xenómica, Hospital Clínico Universitario, 15706, Galicia, Spain

Maissa Salem, Department of Analytical Chemistry, Faculty of Pharmacy, University of Cairo, Kasr El-Aini, Cairo, Egypt

Nele Samyn, Section of Toxicology, Federal Public Service Justice, National Institute of Criminalistics and Criminology (N.I.C.C.), Vilvoordsesteenweg 100, 1120 Brussels, Belgium

Jordi Segura, Department of Experimental and Health Sciences, Universitat Pompeu Fabra and Pharmacology Research Unit, Institut Municipal d'Investigació Mèdica IMIM-Hospital del Mar, Dr. Aiguader 88, 08003 Barcelona, Spain

Beatriz Solorino, Centro Nacional de Xenotipado (CeGen), Hospital Clínico Universitario, 15706, Galicia, Spain

Franco Tagliaro, Department of Medicine and Public Health, Section of Forensic Medicine, University of Verona, 37134 Verona, Italy

Donald R. A. Uges, Laboratory for Clinical and Forensic Toxicology, University Medical Center, Groningen, P.O. Box 30.001, 9700 RB, The Netherlands

Jet C. Van De Steene, Laboratorium voor Toxicologie, Universiteit Gent, Harelbekestraat 729000 Gent, Belgium

Rosa Ventura, Department of Experimental and Health Sciences, Universitat Pompeu Fabra and Pharmacology Research Unit, Institut Municipal d'Investigació Mèdica IMIM-Hospital del Mar, Dr. Aiguader 88, 08003 Barcelona, Spain

Alain Verstraete, Department of Clinical Biology, Microbiology and Immunology, 185 De Pintelaan, Ghent University, Ghent, Belgium; Laboratory of Clinical Biology - Toxicology, Ghent University Hospital, 185 De Pintelaan, Ghent, Belgium

Robert Wennig, Toxicology Laboratory, Laboratoire National de Santé, Université du Luxembourg, Luxembourg

Steven H. Y. Wong, Department of Pathology, Medical College of Wisconsin, USA Toxicology Department, Milwaukee County Medical Examiner's office Milwaukee, WI, USA

Jehuda Yinon, Department of Environmental Science, Weizmann Institute of Science, Rehovot 76100, Israel

Contents

Chapter 3.   Amphetamines

Chapter 4.   Hallucinogens

Chapter 10. Mushroom toxins

Ilkka Ojanperä

PART 2: SCREENING PROCEDURES USED IN FORENSIC TOXICOLOGY

Chapter 14.   Forensic screening with liquid chromatography-mass spectrometry

Merja Gergov   491

Chapter 15.   Forensic toxicological screening with capillary electrophoresis and related techniques

Ameriga Fanigliulo, Federica Bortolotti,
Jennifer Pascali and Franco Tagliaro   513

PART 3: ACTUAL AND EMERGING PROBLEMS OF FORENSIC TOXICOLOGY

Chapter 17.   Analytical markers of acute and chronic alcohol consumption
Anders Helander and Olof Beck

Chapter 18. Toxicological aspects of herbal remedies
Maciej J. Bogusz and Mohammed Al-Tufail   589

Chapter 19. Drugs and driving
Elke Raes, Alain Verstraete and Robert Wennig . . .   611

PART 5: FORENSIC IDENTIFICATION OF INDIVIDUALS AND BIOLOGICAL TRACES

PART 1:
COMPOUNDS OF IMPORTANCE IN FORENSIC TOXICOLOGY

M.J. Bogusz (Ed.). Forensic Science
Handbook of Analytical Separations, Vol. 6

CHAPTER 1

Opioids: methods of forensic analysis

Maciej J. Bogusz

*Department of Pathology and Laboratory Medicine, King Faisal Specialist Hospital and Research Center,
P.O. Box 3354, MBC 10, 11211, Riyadh, Saudi Arabia*

1.1 INTRODUCTION

In this chapter, the use of separation methods for the isolation, identification, and quantitative analysis of natural and synthetic opiates is reviewed. Strictly speaking, the term "opiate" refers specifically to the products derived from the opium poppy. The review focuses on morphine derivatives and synthetic or semisynthetic opiates, showing agonistic action at opioid receptors OP_1 (δ), OP_2 (κ) or OP_3 (μ). The action of opiates on opioid receptors has been reviewed elsewhere [1,2]. The present overview focuses on forensic analytical applications, devoted mainly to biological samples. These applications are divided into several sub-chapters, covering:

– preliminary methods for opioid detection in non-biological and biological samples,
– isolation of opioids from different biological matrices,
– analysis of opium poppy constituents in plant material and in body fluids,
– separation and detection of heroin, its congeners and its specific metabolites in illicit drug preparations and in body fluids,
– analysis of morphine and other natural and synthetic opiates in body fluids and organs.

In each sub-chapter the relevant separation techniques: thin layer chromatography (TLC), gas chromatography (GC), high-performance liquid chromatography (HPLC), and capillary electrophoresis (CE), combined with various detection methods, are reviewed in turn.

1.2 PRELIMINARY METHODS FOR OPIATE DETECTION

Preliminary tests play a dual role in forensic toxicology. First of all their use fulfills the main condition of forensic analysis, i.e. the application of two independent

References pp. 62–72

methods for positive results whenever possible. The second purpose of these tests is to exclude samples that definetly do not contain any opiates. Since a negative result from a preliminary test is usually decisive, there is no room for false-negative results. Therefore, preliminary tests should show broad group specificity and possibly high sensitivity, whereas an absolute specificity is not required. An unequivocal identification and quantitation is usually carried out in the confirmatory step of analysis.

1.2.1 Methods used for street drugs

Preliminary testing in field conditions is mainly performed by law enforcement officers (police, prison, or customs officers). The testing devices are simple and robust, and usually based on well-known color reactions. The main task of these tests is to select suspicious samples or materials for possible further examination with confirmatory methods.

Narcopouch® (ODV Inc., Paris, ME, USA) is a battery of color tests for the detection of opiates, amphetamines, cocaine, barbiturates, cannabinoids, and lysergic acid diethylamide (LSD) in street samples. This test uses color reactions with several reagents, e.g. Marquis, Meyer's, Mecke, Ehrlich's, Fast Blue B, and Koppanyi. The whole procedure is performed in a plastic pouch by visual inspection. The Herosol® (Mistral Detection Ltd., Jerusalem, Israel) field kit consists of a spray reagent and special test paper. A suspected surface (e.g. skin) is wiped with the paper, which is then sprayed with Herosol. A violet color indicates the presence of heroin. A similar heroin test Detect Now™ is supplied by Test Medical Symptoms@Home, Inc., and is marketed via the internet as a simple test for parents who want to check their children for drug use. The NIK® (Public Safety Inc., Armor Holding, Jacksonville, FL, USA) narcotic field test consists of individual ampoule tests for the main groups of drugs of abuse, among them opiates/amphetamines, and heroin/opium. Drug Wipe and Drug Wipe II (Securetec AG, Germany) are immunochemical tests designed for the detection of drugs of abuse on different surfaces, e.g. luggage, passports, currency, and also on the skin or the tongue. Therefore, these tests may be used for the detection of drugs in sweat or saliva. The detection limit for opiates is 25 ng of morphine equivalent. With a portable reader a colorimetric quantitation may be performed.

1.2.2 Methods used for biological fluids

Preliminary methods used for biological fluids may be divided according to different criteria. From the technical point of view, it is obvious to divide these methods into on-site and laboratory tests. These two groups of tests are discussed below. However, preliminary tests may be used not only for forensic or preventive purposes, e.g. in employee screening, but also as a diagnostic procedure in suspected acute poisoning. In on-site tests applied in a clinical emergency ward, the confirmation analysis sometimes is not of primary importance. In the case of a suspected heroin overdose, the positive result of a preliminary opiate test is an indication for the administration of an opiate antagonist, e.g. naloxone, instead of waiting for the results of the confirmatory analysis. This practice is not limited to health professionals; the distribution of naloxone for administration in addicts' home by their companions or family members is a novel approach, which is being tested in the USA, Germany, and the UK [3].

1.2.2.1 On-site tests

There is a multitude of socially critical situations that demand full sobriety and an unaffected psychomotorical condition. On-site tests are widely used among very different social groups, including automobile drivers, incarcerated criminals, the military, athletes, employees of the oil industry, and others. The particular value of the on-site test is in the testing of mobile groups located in remote areas. The most important is the ability to use non-invasive sampling. For this reason, testing of saliva or sweat instead of urine or blood has become particularly attractive. It should be added that on-site tests might also be very valuable in monitoring some therapeutic drugs. For these reasons, such tests carry an alternative name of "point-of-care tests". Jenkins and Goldberger published a comprehensive review of on-site drug testing, comprising all aspects of this technique [4]. Table 1.1 shows some popular on-site testing devices used for opiate detection. All these tests use the immunoassay principle and are capable of detecting the whole panel of drugs of abuse, including amphetamines, benzodiazepine, cannabinoids, cocaine, and opiates.

Crouch *et al.* [5] performed a field evaluation of five on-site drug-testing devices: AccuSign, Rapid Drug Screen, TesT-Cup-5, TesTstik, and Triage. Four hundred urine samples were collected and tested at two sites. All positive results were confirmed using gas chromatography-mass spectrometry (GC-MS; for morphine and codeine) or liquid chromatography-mass spectrometry (LC-MS; for hydrocodone and hydromorphone). One false-negative result was observed. The false-positive rate was below 0.25% for all devices. Gronholm and Lillsunde [6] evaluated eight on-site devices for urine and oral fluid assay of opiates and other drugs. In the case of opiates, the accuracy ranged from 94 to 98% for both matrices. In a multicenter evaluation of

TABLE 1.1
ON-SITE TESTS USED FOR OPIATE DETECTION

Name	Manufacturer	Calibrator	Cutoff (ng/ml)	Matrix
Cozart RapiScan	www.cozart.co.uk	Morphine	10	Saliva
ONTRAK TesTcup	www.rochediagnostics.com	Morphine	300	Urine
ONTRAK TesTStik	www.rochediagnostics.com	Morphine	300	Urine
Syva RapidTest	www.dadebehring.com	Morphine	300	Urine
Triage DOA	www.biosite.com	Morphine	300	Urine
Rapid Drug Screen	www.bioscaninc.com	Morphine	300	Urine
AccuSign	www.pbmc.com	Morphine	300	Urine
Instant-View™Morphine	www.alcopro.com	Morphine	300	Urine

References pp. 62–72

the immunochromatographic on-site urine test Frontline® published by Wennig *et al.*
[7], the sensitivity and specificity for opiates was above 99% for all centers involved.
Buchan *et al.* [8] evaluated the accuracy and specificity of four on-site kits for urine
testing. Opiates were correctly detected in 100% of cases by all the kits.

Barrett *et al.* [9] evaluated an ORALscreen System for the on-site analysis of
drugs in oral fluids and found very good agreement with laboratory-based urine
screening test results for 2–3 days following drug use.

1.2.2.2 Laboratory tests

Laboratory tests are used in situations when the number of examined samples is
quite high, e.g. drug screening of employees or military personnel. These tests are
also based on the immunoassay principle, and usually comprise not only opiates, but
also a whole battery of tests, including amphetamines, cocaine, benzodiazepines,
barbiturates, cannabinoids, and methadone.

For the immunoassays, the selectivities of all the available tests are similar. All react
with a broad range of opiates, including morphine and its glucuronides, codeine, and
semisynthetic opiates, such as dihydrocodeine (DHC) or hydrocodone. One exception
is the CEDIA® 6-AM assay (developed by Microgenics, Fremont, CA, USA), which
is selective for 6-monoacetylmorphine (the primary metabolite of heroin). George and
Parmar [10] analyzed 1100 urine specimens with the CEDIA® 6-AM and failed to
confirm 21 out of 282 positive specimens, which had been identified using GC-MS.
The performance of the Microgenics CEDIA® 6-AM assay was also assessed by
Holler *et al.* [11]. A total of 37,713 urine samples from service members on active duty
were analyzed by both the CEDIA® 6-AM reagent and the Roche Abuscreen ON-
LINE opiate reagent. Three samples screened positive for 6-AM at the cutoff of 10 ng/
ml; one of the three samples was confirmed positive by GC-MS above the cutoff of
10 ng/ml; the two remaining samples were not confirmed at the limit of detection
(LOD) of 2.1 ng/ml. Additionally, 87 human urine samples known to contain 6-AM by
GC-MS were re-analyzed using the CEDIA® 6-AM assay and gave positive results.

Several authors compared different laboratory-screening systems under real-life
conditions. Cone *et al.* [12] compared the sensitivity, specificity, and accuracy of
fluorescence polarization immunoassay (FPIA), enzyme multiplied immunoassay
technique (EMIT) and radioimmunoassay (RIA) for opiate detection in urine. In
all cases, the apparent sensitivities of the assays were higher than the cutoff required
by government organizations. However, the pattern of sensitivity and selectivity of
each assay was different. Armbruster *et al.* [13] compared the performance of analyzers based on different detection principles, including EMIT, kinetic interaction
of microparticles in solution (KIMS), FPIA, and RIA. For the opiates, EMIT gave
3% of non-confirmed positive results, whereas the other tests gave no false results.
Smith *et al.* [14] tested four commercial enzyme immunoassays, using a 300 µg/l
cutoff, and two immunoassays using a 2000 µg/l cutoff. The study was carried out
on 920 urine samples taken from 11 volunteers receiving different intravenous
or inhalatory doses of heroin. The specificity and sensitivity of the assays were
different, but morphine was detectable in urine for at least 12 h after heroin

administration. Cone *et al.* [15] conducted a study in order to establish the cutoff concentrations of drugs of abuse in oral fluids for workplace testing. The Intercept immunoassay followed by GC-MS-MS was applied. Intercept^{TM} is a laboratory-based system, where oral fluid samples are collected on-site with an adsorbent device and are then analyzed in the laboratory. 3908 out of 77,218 oral fluid specimens (5.06%) were positive for different drugs. In the case of opiates, a very high (66.7%) prevalence of 6-AM confirmations was observed, suggesting a high usefulness of oral fluid testing. Cheever *et al.* [16] compared the selectivity of two enzyme immunoassays, FPIA, KIMS, and two enzyme-linked immunosorbent assays (ELISAs) for the cross-reactivity of l-α-acetylmethadol (LAAM) and methadol (a common metabolite of LAAM and methadone) in a methadone immunoassay. Both compounds showed a high cross-reactivity with most immunoassays, indicating the need for chromatographic confirmation of results.

Some preliminary tests were adapted for postmortem samples. Moore *et al.* [17] applied ELISA for drug abuse screening in postmortem blood and tissue homogenates. The morphine assay was very specific for free morphine, but less sensitive than class opiate screening. The latter assay was recommended for screening postmortem specimens. Kemp *et al.* [18] evaluated a commercial ELISA for the opiate/benzodiazepine screening of postmortem blood samples. Ninety positive and forty negative specimens were verified with GC-MS. At the cutoff of 20 µg/l morphine equivalent, the sensitivity was 95% and the specificity 92% versus GC-MS.

Several studies were performed to check and compare the selectivity of various immunoassays used for opiate detection. Kerrigan *et al.* [19] compared the analytical performance of two ELISA tests for the detection of opiates and five other drugs of abuse in blood and urine. Fifteen out of the 855 samples gave discordant results for opiates. The number of false positives was one and three, respectively. Schütz *et al.* [20] studied possible interference from the therapeutic use of apomorphine with the CEDIA and FPIA immunoassays for opiates. Apomorphine (as the commercial preparation Ixense) is widely used in the treatment of erectile disorders. No false-positive results were observed using the recommended cutoff values for urine.

The broad use and importance of drug testing has created an illegal market for procedures and products that promise to "beat the test". Besides the dilution of urine through excessive drinking or sample substitution, several manufacturers offer different kits and reagents, which may be added to urine to avoid the detection of drugs. As countermeasures to adulteration, the following steps can be applied: measuring the temperature, specific gravity, pH and creatinine content of urine, as well as using specific chemical tests for the detection of chemical adulterants. Cody *et al.* [21,22] examined the influence of the "Stealth" adulterant on the detectability of morphine or codeine in urine samples. "Stealth" consists of peroxidase and peroxide and is advertised as being undetectable by adulteration tests. It was demonstrated that samples with low concentrations of morphine and codeine (2.5 mg/l urine) gave a negative response both in immunochemical and GC-MS examination, while the typical urine parameters remained unchanged. Microgenics has developed a special assay named "Sample Check" which detects any possible interference with the CEDIA® assays caused by sample adulteration. This assay replaced a complex panel of adulteration assays.

1.3 ISOLATION OF OPIATES FROM BIOSAMPLES

1.3.1 Solvent extraction

Solvent extraction has been widely used for the isolation of opioids from non-biological, plant, and human samples for decades. As a new development, liquid-phase microextraction (LPME) described by Rasmussen *et al.* [23] and Ugland *et al.* [24] should be mentioned. This technique was applied by Ho *et al.* [25,26] for isolation of methadone.

1.3.2 Solid-phase extraction

Solid-phase column extraction methods have been frequently used for opiate isolation from biological material. In this section, only those studies will be reviewed that deal directly with the assessment of solid-phase extraction (SPE) as an isolation method in toxicology or with the comparison of different SPE materials. Usually, the studies involved were not limited only to opiates, but included other drugs of forensic or clinical toxicological interest.

The optimization of extraction conditions taking into account all three interacting factors, analyte, sorbent, and eluent, has been the subject of numerous studies [27–31]. SPE in a disk format [32,33] consumed about 10–20 times less solvent than the classical column cartridge extraction. Disk-format SPE has been applied for the isolation of opiates by Degel [34]. De Zeeuw *et al.* [35] tested the efficiency of SPE in a disk format for the broad-spectrum isolation of drugs from urine, using selective elution of acidic/neutral and basic drugs. All the drugs were detected with GC-FID. The disk procedure allowed a 60% reduction in the elution volumes and processing time in comparison with the standard SPE method.

1.3.2.1 Reversed-phase SPE

The first SPE methods for the isolation of morphine and its metabolites (normorphine, morphine glucuronides) from serum and urine were published by Svensson *et al.* [36,37]. Bouquillon *et al.* [38] applied C_{18} cartridges for the isolation of morphine and hydromorphone from plasma. The applicability of C_8, polytetrafluoro-ethylene (PTFE)-based extraction disks (Empore^{TM}) for the isolation of different acidic and basic drugs (including codeine) from urine was tested by Ensing *et al.* [39]. The recovery of codeine averaged 76%, using methanol elution. Several comparative studies, utilizing different SPE cartridges, were performed [34,40–43]. These studies indicated substantial differences in extraction recoveries between different products.

Recent years have seen the widespread application of automated SPE systems applicable to cartridges and 96-well plates, which are offered by several manufactures, e.g. Gilson Inc. (www.gilson.com), Waters Inc. (www.waters.com), Caliper (www.caliperls.com), or Varian Inc. (www.varianinc.com) among others. These systems allow the handling of a multitude of samples in a short time and are particularly suitable for drug testing in a workplace or military environment.

1.3.2.2 Mixed-phase SPE

Mixed-phase (reversed-phase – cation exchange) SPE Bond Elut Certify™ cartridges were used for the isolation of morphine, codeine, hydrocodone, hydromorphone, and oxycodone from urine after β-glucuronidase hydrolysis [44]. The recovery of all the drugs, determined by GC-MS, was independent of the pH of the urine and exceeded 80%. Bond Elut Certify™ columns were also applied to the extraction of morphine from whole blood [45]. Several methods of sample pretreatment were tested and the elution procedure was optimized. The recovery was over 70%. The method was extended to the isolation of a range of acidic, neutral, and basic drugs from whole blood [46] using differential elution. Capillary gas chromatography-nitrogen-phosphorus detection (GC-NPD) and GC-MS were used for the separation. Bogusz *et al.* [47] examined four commercially available types of mixed-phase SPE cartridges, using morphine, codeine, and 6-AM as test compounds; HPLC with amperometric detection and GC-MS (ion trap) as the analytical methods. All the extracts were chromatographically pure with both the detection methods. A distinct variability in the extraction recoveries was observed, not only among various products, but also among various batches of the same brand. This study showed that there is a need and room for improvement in the quality of SPE cartridges. Weinmann *et al.* [48] developed a method for the simultaneous isolation of morphine, codeine, benzoylecgonine, and amphetamine from 0.1 ml serum, using Chromabond Drug mixed-phase columns. The extracted drugs were determined by GC-MS (selected ion monitoring, SIM) after derivatization with pentafluoropropionyl anhydride (PFPA). The limit of quantitation (LOQ) was 1 µg/l for morphine and codeine.

1.3.3 Solid-phase microextraction

Solid-phase microextraction (SPME) was introduced by Pawliszyn's research team in the 1990s [49] as a universal, solvent-free isolation technique, which is particularly suitable for volatile and thermally stable compounds. In the case of opioids, this technique was used for the isolation of methadone and its metabolites [50–52] and methadone with pethidine [53]. Staerk and Kulpmann [54] applied headspace SPME at a high temperature (200°C) combined with simultaneous derivatization for the isolation of drugs of abuse from urine. In full-scan GC-MS, the LODs for opiates and methadone were 100 and 200 µg/l, respectively. The simultaneous extraction of methadone, 2-ethylidene-1,5-dimethyl-3,3-diphenylpyrrolidine (EDDP), amphetamine, cocaines, and cannabinoids from saliva using SPME was developed by Fucci *et al.* [55]. The drugs were determined with GC-MS.

1.3.4 Supercritical fluid extraction

Supercritical fluid extraction (SFE), introduced in the 1980s, together with supercritical fluid chromatography (SFC), promised a new quality in the isolation of forensically relevant compounds. However, these hopes were not fulfilled. According to a bitter statement by Georges Guiochon: "Unlike Cinderella, SFC was invited three times to the ball, never made it, and probably won't dance" [56]. Nevertheless, some applications of SFE for opiate analysis have been published. SFE with supercritical CO_2 modified with methanol–TEA–water was applied to the isolation of morphine, 6-AM, and codeine from hair, with recoveries ranging from 53 to 96% [57]. Brewer *et al.* [58] applied SFE with CO_2 modified with 10% methanol for the isolation of morphine, codeine, and benzoylecgonine from human hair. The procedure was faster and gave higher recoveries than the conventional acid hydrolysis. GC-MS was used for the detection. SFE was applied by Allen *et al.* [59] and Scott and Oliver [60] to the isolation of morphine and 6-AM from blood and vitreous humor and clean extracts were reported. In the reviews of the applications of SFE and SFC in forensic samples, some opioids were mentioned [61,62].

1.4 *PAPAVER SOMNIFERUM* AS A SOURCE OF OPIATES

1.4.1 Investigation of the plant material

The studies of the composition and alkaloid content in the *Papaver* plant and in poppy seeds are of forensic relevance, since the plant material has often been used for illegal, home-baked morphine preparations. In many countries, the cultivation of poppy plants as well as the distribution and any use of poppy seeds of any kind is strictly forbidden. Among the multitude of publications concerning *Papaver* plant, only some have been included here on the basis of their forensic relevance.

1.4.1.1 Thin layer chromatography

Circular multi-layer overpressured layer chromatography (ML-OPLC), followed by HPLC-UV was used for the determination of the morphine and thebaine content in poppy capsules [63]. Popa *et al.* [64] isolated morphine and codeine from poppy capsules with solvent or SPE and subjected the extracts to TLC-UV densitometric examination at 275 nm after elution with ethyl acetate:toluene:methanol:ammonia (68:18:10:5, v/v).

1.4.1.2 Gas chromatography

Paul *et al.* [65] investigated which alkaloids may be helpful in differentiating between heroin and poppy seed consumption. Poppy seeds, originating from India and the Netherlands, were subjected to solvent extraction. Part of the extracts was acetylated with acetic anhydride/pyridine. Both acetylated and underivatized extracts were analyzed by GC-MS in SIM mode. Morphine, codeine, thebaine, papaverine, and noscapine were identified in the extracts at different concentrations (Fig. 1.1).

As well as poppy seed extracts, Mexican and Southwest Asian heroin samples were analyzed, which contained heroin, 6-AM, 6-acetylcodeine, and papaverine, but not thebaine and noscapine. The authors postulated that the detection of urinary noscapine, papaverine, or thebaine could be used to differentiate poppy seed consumption from illicit heroin use. However, noscapine may occur in illicit heroin when

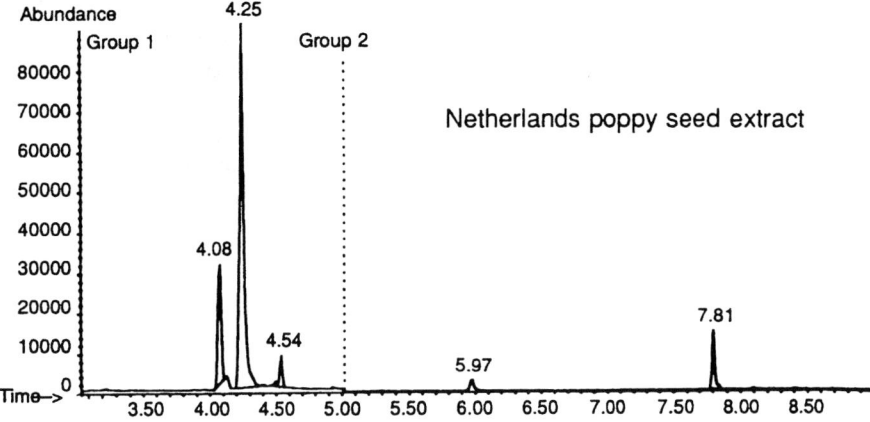

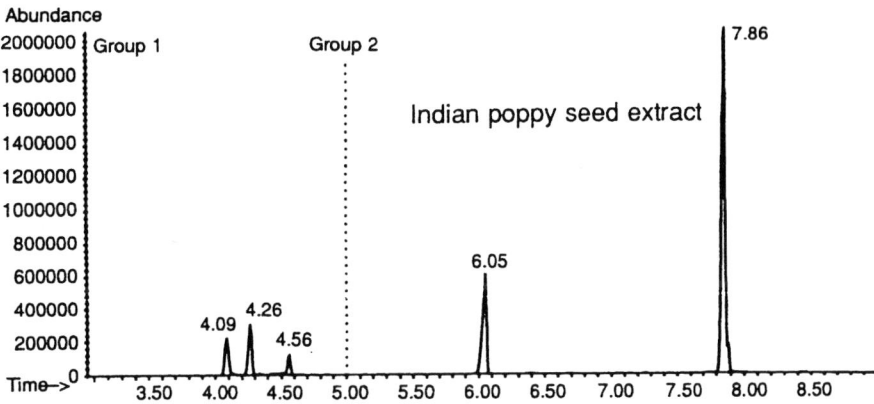

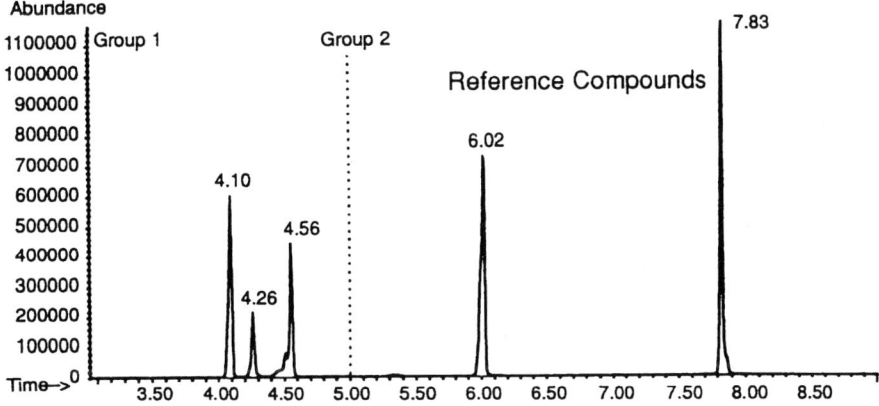

Fig. 1.1. Selected ion chromatogram of opium alkaloids derived from a range of seeds of *P. somniferum*. Following retention times of compounds are given: codeine, 4.10; morphine, 4.26; thebaine, 4.56; papaverine, 6.02; and narcotine, 7.83. From Paul *et al.* [65] with permission of G. Thieme Verlag.

References pp. 62–72

a particular production process is applied. In a study by Huizer [66] of 220 illicit heroin samples, the noscapine content ranged from 13 to 21%. Al-Amri *et al.* [67,68] studied the applicability of detecting reticuline and neopine in urine as markers of opium use. Reticuline (a precursor of opium alkaloids) was detected and characterized as its trimethylsilyl ethers, acetyl esters, and methyl ethers by GC-EI-MS and GC-CI-MS in opium and the urine of opium users. Also neopine, a minor opium alkaloid and an isomer of codeine (also known as β-codeine), has been detected in both the urine of opium users and pharmaceutical codeine users. Detection of these compounds in urine may help to differentiate between opium and heroin use.

1.4.1.3 Liquid chromatography

Supercritical fluid chromatography with carbon dioxide on packed aminopropyl-bonded or straight silica columns has been applied to the separation of opium alkaloids extracted from poppy straw [69]. Methanol, water and triethylamine were used as modifiers. The alkaloids were separated within 2 to 10 min and were detected with a diode array detector (DAD). Krenn *et al.* [70] analyzed poppy straw and opium by HPLC after sonication in 2.5% acetic acid. The filtered extract was adjusted to pH 9.0 and re-extracted with dichloromethane-isopropanol using Extrelut columns. The HPLC analysis was carried out on a C_{18} column with UV detection (280 nm). The method was used to investigate the alkaloid content of 24 samples of gum opium and 80 samples of poppy straw of different origins.

1.4.1.4 Capillary electrophoresis

Crude morphine preparations, poppy straw extracts, and opium, containing morphine, codeine, thebaine, papaverine, noscapine, narceine, oripavine, cryptopine, and salutaridine were examined by micellar electrokinetic capillary chromatography (MEKC) with UV detection at 254 nm [71]. The drugs were separated on an uncoated fused silica capillary in less than 10 min. Lurie *et al.* [72] developed a CE method with dynamically coated capillaries for the analysis of the major opium alkaloids in opium. The results obtained for morphine, papaverine, codeine, noscapine, and thebaine in opium gum and opium latex samples showed good agreement with values obtained by HPLC. CE provided better resolution and was faster than HPLC. Reproducible separations for over 500 samples have been obtained on a single capillary. The CE conditions were also applicable to the analysis of LSD.

1.4.2 Morphine and other opiates in body fluids after ingestion of poppy seeds

Poppy seeds are commonly used in a number of cakes and pastries. These seeds may contain considerable amounts of morphine or codeine. It was therefore of forensic importance to assess whether, and to what extent, the intake of poppy seed-containing products is associated with the elimination from the body of measurable amounts of psychoactive opiates. Since the alleged poppy seed cake ingestion was often being used as an explanation in the cases of positive opiates in urine (the "poppy seed defence"), it is important to differentiate between opiates originating from poppy seeds and from illicit heroin.

Earlier studies on this topic were discussed in detail in the previous edition of this book [73]. Generally, these studies showed that the consumption of poppy seeds or bakery products containing poppy seeds might be associated with the urinary excretion of morphine and codeine in considerable concentrations. Additionally, no substance was found whose presence in urine might unequivocally confirm the consumption of poppy seeds [74–83].

Therefore, persons subjected to drug testing should avoid products containing poppy seeds. In a more recent study, Moeller *et al.* [84] administered poppy seed products (containing 50 mg morphine/kg poppy seeds) to five volunteers. All the on-site tests on urine were enzyme immunologically positive for opiates and were positive to morphine by GC/MS. All the blood samples were negative to morphine by enzyme immunoassay (EIA) and to free morphine by GC/MS. However, after hydrolysis, morphine was detected by GC/MS in all cases. Since German legislation applies a threshold of 10 ng/ml of free morphine in blood as proof of morphine/heroin use, the consumers of poppy products should not expect any legal consequences if blood sample are properly analyzed. Lewis *et al.* [85] determined thebaine, 6-AM morphine, codeine, DHC, oxycodone, hydrocodone, and hydromorphone in postmortem specimens originating from the victims of civil aviation accidents. The samples were subjected to automated SPE and analyzed with GC-MS after silanization. Thebaine was included in this study as a putative indicator of poppy seed ingestion. Rohrig and Moore [86] investigated the detectability of morphine and codeine in oral fluids after the ingestion of poppy seeds and poppy seed bagels, using GC-MS. Morphine concentrations higher than the suggested cutoff concentrations (40 ng/ml for oral fluid, 300 ng/ml for urine) were detected in oral fluids up to 1 h and up to 8 h in urine. This study demonstrated that the ingestion of poppy seed products might cause positive opiate results in oral fluids.

The "poppy-cake dilemma" is also relevant in high-performance sports, since morphine is included on the IOC's list of banned substances at a level exceeding 1 mg/l urine. Thevis *et al.* [87] analyzed eight commercially available samples of baking mixtures with poppy seeds for the presence of morphine using GC-MS. One selected batch was used for baking a typical cake, which was given to nine volunteers. The morphine concentration in the urine in many samples was higher than 1 mg/l, and reached 10 mg/l. The authors confirmed the warning concerning the use of poppy-seed containing products by athletes. Not only poppy seeds, but also several herbal teas present on the market, may be sources of morphine, since they contain parts from the plant *P. somniferum.* Therefore, it was important to verify whether the consumption of such beverages may lead to the elimination from the body of morphine at the relevant concentrations. Van Thuyne *et al.* [88] applied two sorts of herbal tea containing *Papaveris fructus* to five male volunteers. Morphine was detected in the urine of all the volunteers after the consumption of two 120-ml cups of tea. Maximum morphine concentrations were 4.3 and 7.4 mg/l, respectively. Therefore, athletes should be warned against the use of herbal teas containing parts of poppy plants, as well as against the use of food products containing poppy seeds.

The problem of morphine excretion after the ingestion of poppy seeds is relevant also in animal sport. Kollias-Baker and Sams [89] applied 1, 5, and 10-g doses of poppy seeds to four horses and analyzed their plasma and urine for morphine. Morphine was detectable in plasma for at least 4 h and in urine for up to 24 h after administration of poppy seeds. No behavioral changes were noted in the animals.

1.5 HEROIN AND ASSOCIATED ILLICIT OPIATE FORMULATIONS

1.5.1 Investigation of illicit preparations (street drugs). Profiling

1.5.1.1 Thin layer chromatography

Nair *et al.* [90] assessed the separating power of 35 TLC systems reported in the literature for opiate analysis. The developing system consisting of chloroform–*n*-hexane–triethylamine (9:9:4) was capable of separating eight opiates and five potential adulterants, with a LOD of 0.1 μg. Several TLC systems were studied by Huizer [66]. The best results were obtained with the systems chloroform–cyclohexane–diethylamine (8:10:3) and toluene–diethylamine (85:15) (Table 1.2).

1.5.1.2 Gas chromatography

GC-FID was used for the determination of illicit heroin constituents and adulterants useful for profiling [91,92]. Neumann [93] derivatized street heroin samples with *N*-methyl-*N*-trimethylsilyltrifluoroacetamide (MSTFA). This provided a very

TABLE 1.2
TLC RF-VALUES OF HEROIN AND SOME IMPURITIES AND ADULTERANTS IN TOLUENE-DIETHYLAMINE (85:15) DEVELOPING SYSTEM (FROM REF. [66], WITH PERMISSION OF THE AUTHOR)

Compound	Rf	Compound	Rf
Dipyrone	00	Caffeine	38
Piracetam	01	6MAM	39
Acetylsalicylic acid	01	Strychnine	44
Phenolphtaleine	02	Papaverine	55
Nicotinamine	04	Heroin	58
Paracetamol	05	Acetylcodeine	61
Phenobarbital	05	Aminophenazone	64
Morphine	07	Quinine ethylcarbonate	66
Barbital	11	Noscapine	70
Acetylprocaine	18	Lidocaine	71
Quinine	20	Methaqualone	73
Phenacetin	22	Acetylthebaol	75
Phenazone	31	Cocaine	80
Codeine	33	*N*-Phenyl-2-naphtylamine	84
Procaine	35		

good separation of all the compounds. Neumann [94] also presented data on the trends in the occurrence of adulterants (e.g. caffeine, paracetamol, procaine, phenobarbital) that were most frequently encountered in illicit heroin from 1986 to 1992.

Kaa [95] described changes in illicit heroin content and adulterant profiles in Denmark during the period from 1981 to 1992 and observed similar trends to those in Germany described by Neumann [94]. 6-AM in illicit drugs may originate not only from heroin as its deacetylation product, but also from partially acetylated morphine. Therefore, illicit drug samples that contain only 6-AM, without traces of heroin, cannot be classified as illicit heroin [96]. Myors *et al.* [97] assessed different GC parameters useful for the profiling of Southeast Asian heroin. From the library of 649 impurities detected by GC-MS, 18 parameters were selected, which were applied to the identification of the origin of the samples. The European approach to heroin impurity profiling was presented by Stromberg *et al.* [98], who established a gas chromatographic profiling system, harmonized for laboratories in Sweden, Germany, and the Netherlands. Sixteen chromatographic parameters were used for the identification of Southwest Asian heroin, which is prevalent on the European drug market. The study demonstrated the high interlaboratory variability in the parameters, limiting the usefulness of a common database. The best option is still the use of an in-house database for identification. Sharma *et al.* [99] quantified the constituents of illicit heroin specimens seized from different regions of eastern India. GC-MS and HP-TLC were used for the determination of heroin, morphine, 6-AM, and acetylcodeine. Unfortunately, the origin of the specimens could not be identified due to the unavailability of reference samples. Swiss and Australian authors have developed artificial neural networks to validate illicit drug classification using the profiling method [100]. This method established links between samples using a combination of principal component analysis and calculation of a correlation value between samples. Heroin samples were analyzed by GC to separate the major alkaloids. Statistical analysis was then performed on 3371 samples. About 20 "chemical classes" have been identified. A model study showed that in 96% of cases the neural network attributed the seizure to the right "chemical class".

Brenneisen *et al.* [101] studied the identity of pyrolysis products of heroin, which are generated after the heating of street heroin for inhalation ("chasing the dragon"). Heroin samples were heated on aluminum foil at 250–400°C and analyzed by GC-MS. Seventy-two pyrolysis products were detected, and half of these could be identified.

Mannitol hexaacetate (MHA) has been detected by GC-MS in some brown illicit drug seizures in United Arab Emirates [102]. Diacetylmorphine was completely lacking. The presence of MHA as a genuine ingredient of the drug seizures rather than a storage- or an analytical artifact has been verified. MHA was probably formed as a result of the addition of mannitol, as a diluent, before the acetylation step in the processing of the heroin preparation. This early dilution in the production and distribution chain of the illicit drug may be highly indicative of a specific production process for heroin in a trafficking organization.

Volatile compounds occluded in heroin preparations, may be helpful in sample recognition. Cartier *et al.* [103] identified traces of 16 different solvents in 41 illicit

and basics drugs of abuse. Anostos *et al.* [119] used CE on an uncoated fused silica capillary for the determination of carbohydrates in heroin drug seizures.

1.5.1.5 Multi-method approach:

Huizer [66] identified components present in crude opium, purified morphine, and illicit heroin using TLC, HPLC, and GC. The procedures used for the illicit isolation of morphine from opium (the lime method and the ammonia method) may be recognized on the basis of the percentage composition of crude morphine. Also, during the acetylation step, various characteristic impurities may be formed. Straight-phase HPLC of illicit heroin provided general information concerning the composition of the sample, and capillary GC-FID of silylated heroin samples according to Neumann and Gloger [120] demonstrated distinct differences between each production batch of illicit heroin – even originating from the same production unit (Fig. 1.3).

   Chiarotti *et al.* [121] presented a multi-method approach to the comparative analysis of illicit heroin samples, using headspace GC on Porapak Q column for volatiles, HPLC for sugar diluents, AAS for trace metals, and GC-MS(IT – ion trap) for organic constituents. A combination of HPLC-DAD and GC-NPD has been used for the analysis of illicit heroin and cocaine samples [122]. The alkaloids and adulterants were identified through retention parameters and UV spectra. Besacier *et al.* [123] performed the analysis of illicit heroin in three steps. In the first step, all major and

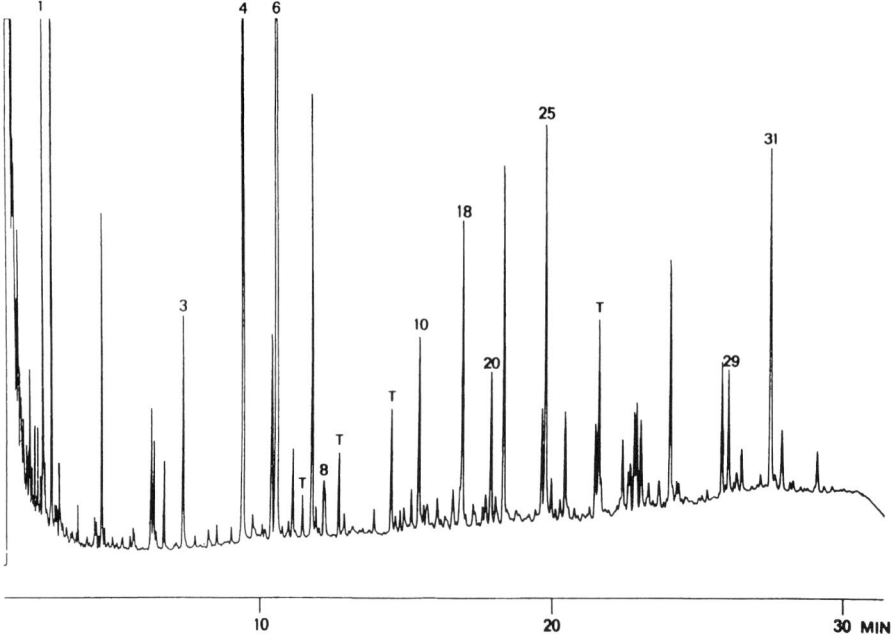

Fig. 1.3. Capillary GC profile of South-West Asian heroin sample, treated according to the method of Neumann and Gloger [120]. From Huizer [66] with permission of the author.

minor heroin constituents were identified and quantified using GC-FID. In the second step, the GC-FID analysis of impurities was carried out . In the third and last step, the isotope ratios $^{13}C/^{12}C$ were measured using GC-isotope ratio mass spectrometer. This procedure, according to the authors, enabled the batch identification of a given sample with a high degree of certainty. Bora *et al.* [124] measured the levels of ten elements in 44 illicit heroin samples originating from Southeast Anatolia, using ICP-OES and AAS. The observed profiles were useful in determining the source and trafficking routes of the heroin.

Kala and Lechowicz [125] analyzed Polish substitutes for heroin, the so-called kompot or makiwara, which are produced from macerated poppy straw or capsules subjected to extraction and acetylation, using HPLC-DAD and GC-MS (ion trap) and found very large variations between batches. As main constituents, morphine, codeine, and acetylcodeine, were found. 3-MAM, 6-AM, and heroin occurred usually in lower concentrations.

In Western Europe, particularly in the Netherlands, heroin is mostly taken in smokable form [126]. Therefore, smokable heroin preparations were considered for use in the maintenance of chronic treatment-resistant heroin addicts. Klous *et al.* [127] conducted a study on the thermal stability and recovery of diacetylmorphine base which had been subjected to controlled volatilization at temperatures from 50 to 300 °C. Caffeine was added to the mixture since this drug is often present in street heroin preparations. Controlled heating was carried out using differential scanning calorimetry, the recovery of drugs and of degradation product 6-AM was measured with HPLC-UV. It was demonstrated that around 41% of the diacetylmorphine and around 100% of the caffeine was recovered after volatilization. The addition of caffeine increased the recovery of diacetylmorphine and reduced its degradation. Dams *et al.* [128] reviewed the analytical procedures used for the quantification of minor components of heroin seizures, such as the impurities related to the origin and manufacture. By combining these data, complex characterizations, i.e. impurity profiles, chemical signatures or fingerprints, can be obtained and used for comparative analysis. Such analysis can be used for tactical (batch-to-batch comparison) and strategic (origin determination) intelligence purposes.

1.5.2 Heroin metabolites in biological matrices

Heroin is usually self-administered intravenously. In the last decade, however, a growing preference for other routes of administration has been observed, including smoking or intranasal administration ("snorting"). This has been caused by several factors, including the fear of HIV, the possibility of administration without leaving external marks on the body and the decrease in the price of street heroin. Irrespective of the administration route, heroin is rapidly deacetylated to 6-monoacetyl-morphine (6-AM). The half-life of heroin in blood after intravenous injection was estimated at 2–8 min [129,130], after smoking at 3–5 min [131], after intranasal or intramuscular administration at 5–6 min [132,133]. 6-AM is deacetylated at a somewhat slower rate than morphine; the half-life after intravenous administration was

6–38 min, and 5, 11, and 12 min after smoking, intranasal, and intramuscular administration, respectively. The half-life of morphine was estimated at ca. 30 min after heroin smoking, and at 60–180 min after administration by other routes. Fig. 1.4 shows the main steps of heroin biotransformation.

The extremely short plasma half-life of heroin indicates that this drug might be detectable in blood only under experimental conditions, when the sample is taken almost immediately after administration, or in the case of a very massive heroin overdose (e.g. in the "body packer syndrome"). 6-AM, a specific heroin metabolite, may be detected in the blood of living subjects a short time after heroin intake. On the other hand, 6-AM (about 0.5% of the heroin dose) and some unchanged heroin are eliminated in the urine and may be detected for several hours [134]. 6-AM is the only known specific metabolite of heroin.

1.5.2.1 Urine

1.5.2.1.1 Gas chromatography

Fehn and Megges [135] introduced the GC-MS determination of 6-AM in urine as a specific heroin metabolite in forensic toxicological practice. They isolated 6-AM with SPE C_{18} cartridges and determined it by GC-MS after PFPA derivatization. A LOD of 2 µg/l was reported. In subsequent studies, several authors reported the presence of 6-AM in urine of heroin addicts. Usually, solvent extraction or SPE followed by GC-MS(SIM) was applied [136–139].

In recent years, the differentiation between the intake of pure DAM and illicit heroin has become relevant since the introduction of heroin prescription programs in some countries, including Switzerland, Great Britain, Germany, and the Netherlands. One of the basic requirements of these programs is that the participants must not use any illicit drugs, particularly illicit heroin. In illicit heroin not only is present, but also several other opiates, including 6-AM, acetylcodeine, codeine, papaverine, and noscapine as well as a number of adulterants. It must be stressed that only acetylcodeine (AC) may be regarded as specific marker of illicit heroin use. AC is produced from codeine during the acetylation of opium. Its content in illicit heroin ranges from 2 to 7% [140]. A method for the simultaneous determination of acetylcodeine, 6-AM, morphine, codeine, and norcodeine by GC-MS was described by O'Neal and Poklis [141]. Examination of 69 morphine/codeine-positive urine samples revealed AC in 6 cases, whereas 6-AM was detected in 13 cases. The concentrations of AC were much lower that those of 6-AM. In the second study [142], O'Neal and Poklis analyzed 100 morphine-positive urine samples and found AC in 37 samples at concentrations ranging from 2 to 290 µg/l (median, 11 µg/l). 6-AM was also present in these samples at concentrations ranging from 49 to 12600 µg/l (median, 740 µg/l). Codeine – a possible metabolite of AC – was found in all urine samples. The authors concluded that 6-AM was a much more sensitive marker of illicit heroin use than AC. On the other hand, AC may play a very important role as a specific indicator of illicit street heroin use. Staub *et al.* [143] detected AC in over 85% and 6-AM in over 94% of 71 urine samples obtained from consumers of illegal heroin. Urine samples taken from 532 participants in a heroin maintenance program

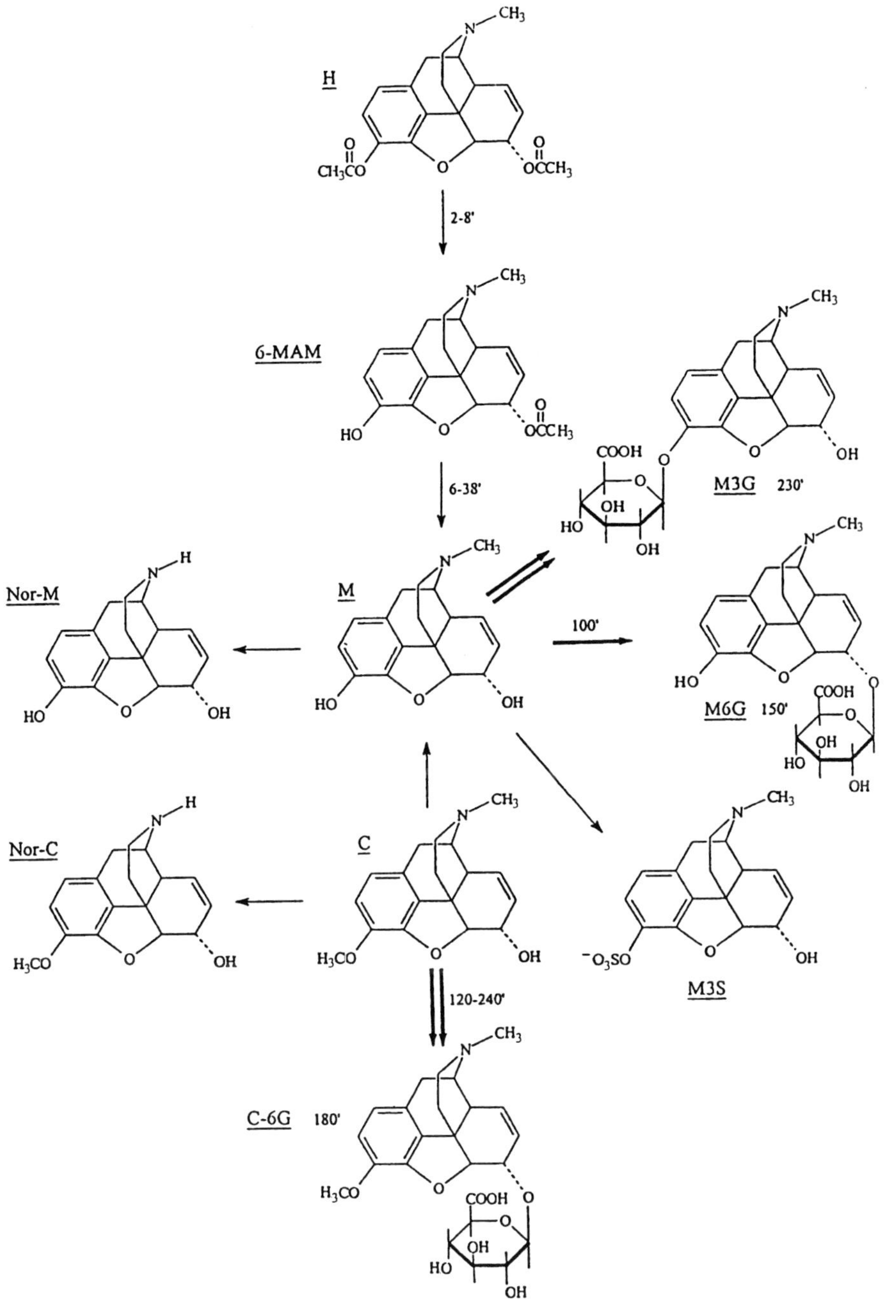

Fig. 1.4. Metabolic pathways of heroin and codeine. From Bogusz [73] with permission of Elsevier.

in the UK were subjected to GC-MS analysis for the putative markers of street heroin abuse [144]. Among the morphine-positive samples, 61% were positive for at least one of the codeine, meconine, and putative papaverine and noscapine metabolites. The detection of urinary noscapine and papaverine metabolites was recommended as an indication of street heroin abuse.

Brenneisen and Hasler [145] studied the pharmacokinetics of acetylcodeine, administered intravenously to healthy volunteers. The peak urine concentration appeared at 2 h, and the detection window in urine was 8 h. SPE followed by GC-MS was applied to the determination of acetylcodeine and its metabolite codeine in urine. In selected cases, papaverine and noscapine were also measured. In a study of 105 participants in a heroin maintenance program, 15 urine samples were positive for acetylcodeine, and 8 for acetylcodeine, papaverine, and noscapine.

1.5.2.1.2 Liquid chromatography

In earlier studies reversed-phase HPLC with electrochemical or UV detection was used for the determination of 6-AM and other heroin metabolites in urine after solvent extraction or SPE [146–148]. Low and Taylor [149] used a straight phase HPLC for the analysis of 6-AM, heroin, morphine, codeine, DHC, and pholcodine in urine extracts. UV detection at 280 nm was applied and the LODs varied from 4 to 20 µg/l. An HPLC method for the simultaneous determination of 17 opium alkaloids in urine and blood was published by Dams et al. [150]. The drugs were isolated with a cation exchange SPE, separated on a "high-speed" phenyl column (53 mm × 7 mm) within 12 min, and detected with DAD and fluorescence detectors. LODs in the range between 2.5 and 9.7 µg/l were observed. Fernandez et al. [151] developed an HPLC-DAD method for the determination of morphine, codeine, and 6-AM, as well as cocaine and methadone and their metabolites in urine. For each analyte group, the optimal signal wavelength was chosen. The drugs were isolated from urine with Bond Elut Certify cartridges and separated on RP-8 column in an ACN-phosphate buffer gradient. The LODs were in the range 0.1 mg/l.

In the past decade, several LC-MS or LC-MS-MS procedures for the determination of heroin metabolites in urine have been developed. Bogusz et al. [152] determined putative street heroin markers in 25 morphine-positive urine samples, in order to differentiate between the administration of illicit heroin and prescription diamorphine. LC-APCI (atmospheric pressure chemical ionization)-MS (positive ions) was applied after SPE (Fig. 1.5). Codeine-6-glucuronide (C6G) was found in all samples, codeine in 24, noscapine in 22, 6-AM in 16, papaverine in 14, DAM in 12, and AC in 4 samples. A very similar study was published later by Musshoff et al. [153], who determined morphine, 6-AM, morphine-3-glucuronide (M3G), morphine-6-glucuronide (M6G), acetylcodeine, codeine, noscapine, and papaverine in urine samples using the LC-ESI-MS-MS technique. In urine samples originating from the consumers of street heroin, acetylcodeine, codeine, noscapine, and papaverine were frequently detected. Katagi et al. [154] developed an automatic method for the determination of heroin, 6-AM, and morphine as well as AC, codeine, and DHC in urine. Urine samples were applied to a cation exchange column, and after column switching, the drugs were eluted and separated on an analytical cation

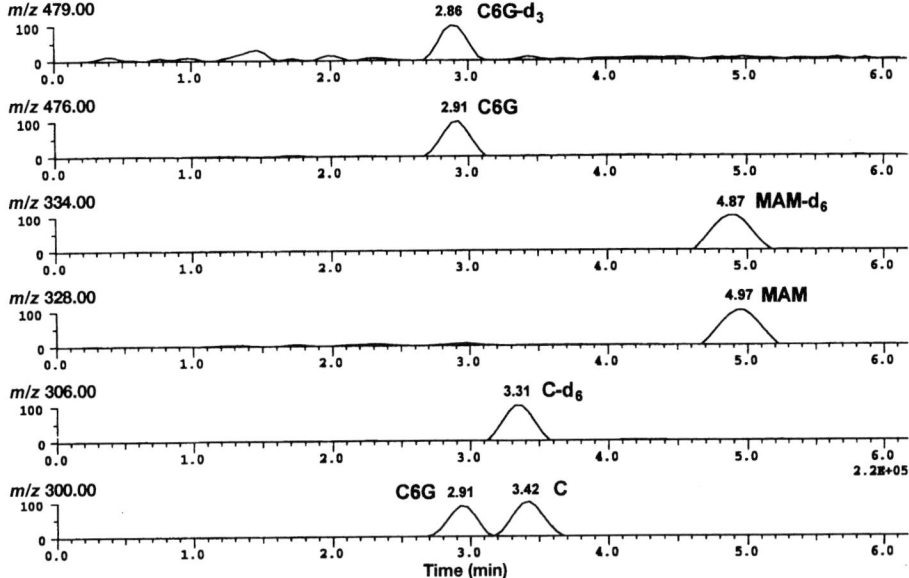

Fig. 1.5. LC-APCI-MS ion chromatogram of urine extract from heroin consumer, showing the presence of codeine, C6G, and 6-AM. From Bogusz *et al.* [152] with permission of Preston Publications.

exchange column in ACN–ammonium acetate (70:30). The detection was carried out by ESI-MS in full scan or SIM mode. Protonated quasi-molecular ions or ACN adducts were monitored. The LODs ranged from 2 to 30 μg/l in full scan mode and from 0.1 to 3 μg/l in SIM.

Klous *et al.* [155] presented a novel strategy for detecting the use of illicit heroin by addicts receiving prescribed diacetylmorphine. A deuterated analogue of heroin was added (1:20) to pharmaceutical, smokable heroin (a powder mixture of 75% w/w diacetylmorphine base and 25% w/w caffeine). Plasma and urine samples were collected from nine male patients who had used pharmaceutical, smokable heroin during a 4-day stay in a closed clinical research unit. The samples were analyzed by LC-ESI-MS-MS. The ratios of the deuterated and undeuterated diacetylmorphine and 6-acetylmorphine in plasma and urine were calculated from the peak areas of these substances in the respective chromatograms. A 6-AM/6-AM-d3 ratio in urine above 32.8 was considered indicative of co-use of illicit heroin. The ratio was detectable in urine for 4–9.5 h after use of the pharmaceutical, smokable heroin. The addition of stable, isotopically labeled heroin to pharmaceutical, smokable heroin was considered to be a feasible strategy for the detection of use of street heroin by the participants in heroin-assisted treatment.

1.5.2.1.3 Capillary electrophoresis
Taylor *et al.* [156] described a CE method for the separation of heroin, 6-AM, morphine, codeine, DHC, and pholcodine. The method was applied to the determination

of pholcodine, DHC, and morphine in urine extracts. LODs of 10 µg/l were reported. Wu and Tsai [157] applied CE to the determination of morphine and M3G in urine. The specimens were only filtrated, acidified to pH 2–3 and centrifuged. The LODs were 0.2 and 0.5 mg/l for morphine and M3G, respectively, using UV detection. The same research group [158] developed a CE method for the detection of morphine in urine using MS (IT) detection. A LOD of 10 µg/l was achieved. Wey and Thormann [159] used CE coupled with mass spectrometric detection for the determination of morphine, M3G, 6-AM, codeine, C6G, DHC, methadone, and EDDP in urine. Solvent extraction and SPE on a mixed-mode polymer phase was applied. Detection limits for the free opioids were at the level of 100–200 ng/ml, using 2 ml urine sample. Much improved (ppb) sensitivity was obtained by infusing the extract directly into the source of the MS system. However, with this method it was impossible to distinguish between M3G and M6G. Alnajjar *et al.* [160] developed a CE method with native fluorescence detection of normorphine, morphine, 6-AM, and codeine. An excitation wavelength of 245 nm with a cut-off emission filter of 320 nm was used. The detection limits were in the range of 200 ng/ml. For a highly sensitive analysis, LIF was applied using a two-step precolumn derivatization procedure. Drugs extracted from urine were first subjected to an *N*-demethylation reaction involving the use of 1-chloroethyl chloroformate and then derivatized using fluorescein isothiocyanate isomer and analyzed by CE coupled to a LIF detector. The estimated instrumental detection limits of the fluorescein isothiocyanate (FITC) derivatives were in the range 50–100 pg/ml, detection with excitation and emission wavelengths of 488 and 520 nm, respectively. A novel multi-target antibody to morphine and derivatives was developed and coupled with CNBr-activated Sepharose 4B to form a multi-target immunoaffinity column for the determination of morphine, codeine, acetylcodeine, 6-monoacetylmorphine, and M3G in the urine of heroin abusers [161]. The analytes were extracted from the urine of drug addicts and separated using CE using β-cyclodextrin. The UV detection was carried out at 214 nm. The recovery ranged from 91 to 105% and the detection limit was between 10 and 20 ng/ml.

1.5.2.2 Blood

1.5.2.2.1 Gas chromatography

Heroin metabolites in blood have been usually determined simultaneously with codeine, which also appears after street heroin abuse. Schuberth and Schuberth [162] published a GC-MS method for the determination of 6-AM, morphine, and codeine in blood. The blood samples were subjected to methanol precipitation, SPE on C_{18} cartridges and derivatization with PFPA. Musshoff and Daldrup [163] modified this procedure by using ACN for blood precipitation and changing the SPE procedure. Blood samples were also subjected to acid hydrolysis in order to measure the total amounts of opiates. High purity extracts were reported; the LOD was below 1 µg/l. Wasels and Belleville [164] presented an overview of GC-MS and derivatization procedures for the identification of 6-AM, morphine, and codeine. Wang *et al.* [165] published a method for the simultaneous determination of heroin and its metabolites 6-AM, morphine, and normorphine as well as cocaine and its metabolites in hair, plasma, saliva, and urine. The drugs were extracted from biosamples with SPE cartridges and derivatized with bis-(trimethylsilyltrifluoroacetamide) (BSTFA)/ trimethylchlorosilane (TMCS) before GC-MS (SIM) analysis. Heroin, 6-AM, and morphine levels were monitored in saliva after the experimental administration of intranasal heroin. Goldberger *et al.* [166] developed a GC-MS method for the determination of heroin, 6-AM, and morphine in body fluids and organs of 21 victims of heroin overdose. The samples were extracted with SPE cartridges and partially derivatized with *N*-methyl-bis-trifluoroacetamide (MBTFA) (for 6-AM and morphine). Heroin was determined without derivatization. 6-AM was detected in all 21 urine samples and in 14 blood samples. Heroin was present in 17 urine samples, but not in blood. The authors used the concentration ratios of the drugs for the evaluation of the rapidity of death. Moeller and Mueller [167] determined 6-AM in serum, urine, and hair of heroin users by GC-MS. The drug was isolated with SPE and derivatized with PFPA. 6-AM was detected in 19 of 25 analyzed opiate-positive urine samples, the concentration ranging from 1 to 9950 µg/l. In five serum samples 6-AM levels of 2–9 µg/l were observed. Guillot *et al.* [168] developed a GC-MS method for the determination of heroin, 6-AM, and morphine in postmortem blood, urine, and vitreous humor. The drugs were isolated by alkaline solvent extraction and were subjected to propionylation in the presence of 4-dimethylaminopyridine. The quantitation limits were 2 µg/l for morphine and 6-AM and 5 µg/l for heroin. GC-MS-EI and HPLC with electrochemical detection were applied in a case of fatal oral heroin poisoning. The concentrations of heroin, 6-AM, and morphine in blood were 109, 168, and 1140 µg/l, respectively [169].

Gas chromatographic methods for the determination of heroin, its metabolites, and associated compounds in body fluids are summarized in Table 1.3.

1.5.2.2.2 Liquid chromatography

The HPLC methods used for opiate agonists since 1999 were reviewed by Bogusz [73] and Pichini *et al.* [174]. The advent of LC-MS in the 1990s brought very important progress in the determination of opiates and its metabolites in biological fluids. LC-MS is the only analytical technique that allows the specific detection of the parent opiates and all the polar metabolites without derivatization and without acidic or enzymatic cleavage.

Zuccaro *et al.* [175] developed a LC-ESI-MS method for the simultaneous determination of heroin, 6-AM, morphine, M3G, and M6G in serum. The drugs were extracted with SPE C_2 cartridges and separated on a straight phase silica column in a methanol–ACN–formic acid mobile phase. The authors used a silica column in order to separate all substances in one run under isocratic conditions. The LOD for heroin was 0.5 µg/l, for 6-AM 4 µg/l. The method was applied to a pharmacokinetic study of heroin-treated mice. Bogusz *et al.* [176] used LC-APCI-MS for the determination of heroin metabolites (6-AM, M3G, M6G, and morphine) in blood, cerebrospinal fluid, vitreous humor, and urine of heroin victims. The drugs were extracted with C_{18} cartridges; the LOD for 6-AM was 0.5 µg/l. This procedure was extended to the LC-APCI-MS method for the determination of 6-AM, M3G, M6G, morphine, codeine, and C6G, using deuterated ISs for each compound [177], and was applied to routine casework [178].

TABLE 1.3
GAS CHROMATOGRAPHIC METHODS FOR HEROIN, 6-MAM, MORPHINE, CODEINE, AND METABOLITES

Drug	Sample	Isolation	Derivatization	Column, conditions	Detection	LOD (µg/l)	Ref.
6-MAM	Urine	SPE C_{18}	PFPA	OV-1, 230°	EI-MS (SIM)	2	[135]
6-MAM	Urine	SPE or l/l	Propionylation	DB 5, 130–250°	EI-MS (SIM)	0.8	[136]
6-MAM	Urine	l/l alkaline	Propionylation	RSL 200, 146–246°	EI-MS (SIM)	n.s.	[137]
6-MAM	Urine	SPE Certify	TFA	HP-1, 150–300°	EI-MS (SIM)	n.s.	[138]
AC, 6-MAM, M, C, NC	Urine	SPE	Propionylation	HP-1, 170–280°	EI-MS (SIM)	0.5	[141]
6-MAM, M, C	Blood	SPE C_{18}	PFPA	DB 5, 150–256°	EI-MS (SIM)	0.5	[162]
6-MAM, M, C, DHC	Blood	SPE C_{18}	PFPA	OV 1, 150–220°	EI-MS (SIM)	1	[163]
Heroin, 6-MAM	Serum, saliva, urine, hair	SPE	BSTFA/TMCS	HP 1, 70–250°	EI-MS (SIM)	1	[165]
Heroin, 6-MAM, M	Body fluids, organs	SPE	MBTFA	Rtx-5, 150–290°	EI-MS (SIM)	1	[166]
6-MAM	Serum, urine hair	SPE C_{18}	PFPA	n.s.	EI-MS (SIM)	n.s.	[167]
M	Blood	l/l pH 9	PFPA	DB-5, 100–300°C	EI-MS-MS	1	[170]
M, C	Blood	SPE C_{18}	PFPA	CP-Sil5, 200–300°C	NCI-MS (SIM)	2–5	[171]
M, C, NM	Plasma	l/l pH 9.5	HBFA	HP-1, 100–257°C	NCI-MS (SIM)	1 pg on col.	[172]
M	Plasma	l/l pH 9	PFPA	HP-5MS, 150–250°C	EI-MS (SIM)	0.2	[173]

Abbreviations: M = morphine, C = codeine, AC = acetylcodeine, NC = norcodeine, NM = normorphine, DHC = dihydrocodeine, n.s. = not stated, on col. = on-column l/l = liquid/liquid.

LC-MS methods for the determination of morphine, codeine, and their metabolites are summarized in Table 1.4.

Some authors developed LC-MS methods for the simultaneous determination of opiates and other drugs. Cailleux *et al.* [185] extracted opiate agonists (morphine, 6-AM, codeine, norcodeine, pholcodine, codethyline) as well as nalorphine and cocaine and its metabolites (benzoylecgonine, ecgonine methyl ester, cocaethylene, and anhydromethylecgonine) from blood, plasma or urine with chloroform/isopropanol (95:5) at pH 9. The drugs were separated on an octyl column in ACN–ammonium formate–formic acid. Protonated molecular ions and one fragment for each substance were monitored using ESI-MS-MS. The quantitation was carried out using deuterated ISs. The limits of quantitation were 10 µg/l for opiates and 5 µg/l for cocaines and were higher than these reported after SPE. A similar approach was taken by Rook *et al.* [186] who isolated heroin, 6-AM, morphine, M3G, M6G, acetylcodeine, codeine, cocaine, BE, methadone, and EMDP from human plasma with mixed mode SPE. The drugs were separated on a Zorbax Bonus reversed-phase column and were detected with tandem MS in multiple reaction monitoring (MRM) mode, using one transition for each compound (Fig. 1.6). This assay was developed for pharmacokinetic studies with prescribed heroin.

Concheiro *et al.* [187] published a LC-ESI-MS(SIM) method for the simultaneous determination of morphine, 6-AM, amphetamine, methamphetamine, 3, 4-methylenedioxyamphetamine (MDA), 3, 4-methylenedioxymethamphetamine (MDMA), 3, 4-methylenedioxyethylamphetamine (MDEA), *N*-methyl-2, 3-methylenedioxyphenylbutan-2-amine (MDBD), cocaine, and benzoylecgonine in 0.2 ml plasma. The drugs were extracted by SPE on OASIS® HLB cartridges and were separated on an Atlantis C18 column within 17 min. The method was applied to 156 cases of road traffic fatalities.

1.5.2.3 Alternative matrices

The use of alternative samples may bring several advantages in the forensic analysis for drugs of abuse and also for opiates. The analyses of sweat or oral fluids have the advantage of non-invasive collection, while the analysis of hair may expand the detection window to months after the exposure. Most of the alternative samples can be collected by police officers and not just by medical personnel. A review of the application of unconventional samples and alternative matrices was carried out by Kintz and Samyn [188], and is covered in Chapter 21 of this book.

Pichini *et al.* [189] developed a HPLC method with UV detection at 254 nm for the determination of heroin, 6-AM, morphine, and codeine in human hair. Hair specimens were subjected to acid hydrolysis and SPE. HPLC was carried out in ACN-phosphate buffer at pH 2.1. The LODs for 6-AM, morphine, and codeine were 0.5 ng/mg hair, for heroin 5 ng/g. Samyn *et al.* [190] performed a feasibility study on alternative samples under real-life conditions. Oral fluid, sweat wipes, blood, and urine samples were obtained from 180 drivers in Belgium, who failed the field sobriety test during a police check. Mostly, cannabinoids, amphetamines, and cocaine were detected, while the number of opiate positives was lower. The sampling of sweat

TABLE 1.4
LIQUID CHROMATOGRAPHIC METHODS FOR HEROIN, 6-MAM, MORPHINE, CODEINE, AND METABOLITES

Drug	Sample	Isolation	Column, elution conditions	Detection	LOD (µg/l)	Ref.
M, M3G, M6G, 6-MAM, Cod, C6G	Serum, urine	SPE	C18, ACN-HCOONH$_4$ isocr.	APCI-Q, SIM	0.1–10	[176]
M, M3G, M6G	Plasma	SPE C$_{18}$	ODS, ACN-HCOOH	ESI-Q	0.8–5	[179]
M, M3G, M6G	Plasma	SPE	Silica, ACN-HCOOH isocr.	ESI-QQQ, MRM	0.5–1.0	[180]
M, M3G, M6G, NorM	Serum, urine	SPE	C18, ACN-HCOOH grad.	ESI-QQQ, MRM	0.3–2.5	[181]
M, M3G, M6G, 6-MAM, Cod, C6G	Serum	SPE	C18, ACN-HCOONH$_4$ grad.	ESI-Q, SIM	0.5–5.0	[182]
M, M3G, M6G	Serum	SPE	C18, ACN-HCOONH$_4$ isocr.	ESI-QQQ, MRM	1.0–5.0	[183]
M, M3G, M6G	Plasma	SPE	C18, ACN-HCOOH isocr.	ESI-QQQ, MRM	0.25–0.5	[184]
M, M3G, M6G, Cod, Her, Meth, EMDP	Plasma	SPE	C18, ACN- HCOONH$_4$ grad.	ESI-QQQ, MRM	5 LOQ	[186]
M, M3G, M6G	Plasma	SPE 96 well	C18, ACN-MeOH-HCOOH isocr.	ESI-Q, SIM	0.5–5.0	[225]
M, M3G, M6G, 6-MAM, Cod, C6G	Urine	SPE	C18, ACN-MeOH-CH$_3$COONH$_4$ grad.	ESI-IT, MRM	10	[226]

Abbreviations: M = morphine, M3G = morphine-3-glucuronide, M6G = morphine-6-glucuronide, Cod = codeine, C6G = codeine-6-glucuronide, Her = heroin, Meth = methadone, EMDP = 2-ethyl-5-methyl-3,3-diphenyl-pyrroline, APCI = atmospheric pressure chemical ionization, ESI = electrospray ionization, Q = single stage quadrupole, QQQ = triple stage quadrupole, SIM = selected ion monitoring, MRM = multiple reaction monitoring, IT = ion trap.

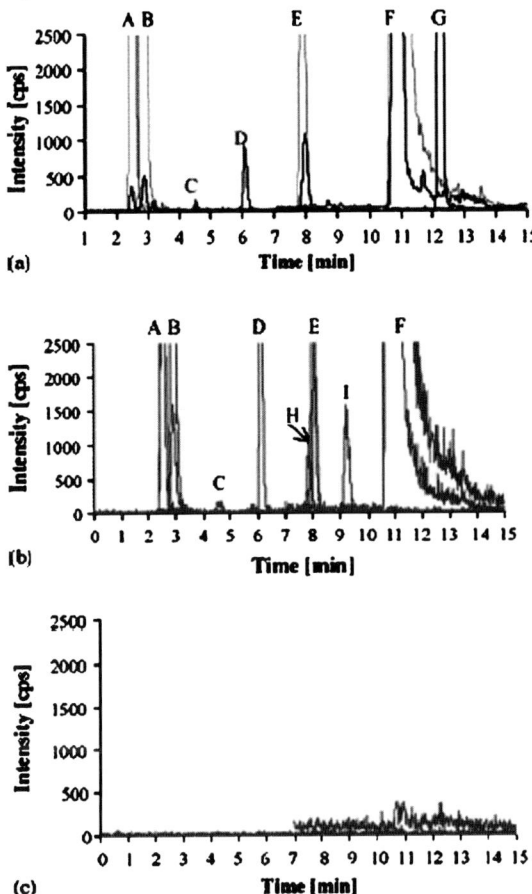

Fig. 1.6. Total ion chromatogram of all analytes at LOQ level in plasma extract (a), chromatogram of patient plasma extract (b), and chromatogram of blank plasma extract (c). A = M3G and M3G-d3, B = morphine and morphine-d3, C = M6G, D = 6-AM, E = heroin and heroin-d6, F = methadone and methadone-d9, G = EMDP, H = cocaine, I = BE. From Rook *et al.* [186] with permission of Elsevier.

appeared to be simpler in field conditions than saliva. Barnes *et al.* [191] performed a semiquantitative analysis of opiates in oral fluid, using a modified Cozart Microplate Opiate EIA Oral Fluid Kit, followed by GC-MS for the confirmation and quantitation of codeine, norcodeine, morphine, and normorphine. Specimens (1406) of oral fluid were collected from 19 subjects prior to and up to 72 h following controlled administration of oral codeine. The results indicated that the Opiate ELISA efficiently detected oral codeine use, according to official UK and SAMSHA criteria.

Several authors have used alternative samples to assess prenatal exposure to opiates.

Vinner *et al.* [192] applied GC-MS for the detection of opiates, together with other drugs (cannabis, cocaine, amphetamine, LSD, and benzodiazepines) in neonatal hair. Potential combined dependences and/or substitutive therapeutics (methadone or buprenorphine (BP)) were also assessed in 17 mother/neonate couples. Gestational opiate exposure profiles were correlated with the observed withdrawal syndromes. Neonatal hair analysis could contribute to assessing *in utero* exposure to opiates, particularly when results in urine and meconium are negative or when these matrices are not available. Montgomery *et al.* [193] assessed the agreement of testing for fetal exposure to illicit drugs in 118 paired specimens of meconium vs. umbilical cord tissue obtained from pregnancies where there was a high suspicion of illicit drug use by the mothers. Each specimen was tested for amphetamines, opiates, cocaine, cannabinoids, and phencyclidine using immunoassays. The agreement of the drug screening results between cord and meconium was above 90% for all drugs tested. Umbilical cord tissue performed as well as meconium in assessing fetal drug exposure. The results of studies using the cord may have a more rapid return for the clinician, because waiting for meconium to be passed sometimes requires several days.

1.5.3 Morphine as heroin metabolite or therapeutic drug

1.5.3.1 Morphine and its metabolites in biosamples taken from living subjects

1.5.3.1.1 Thin layer chromatography
In earlier publications, TLC was applied for the detection of heroin metabolites and other opioids in urine, usually after solvent extraction or SPE [194–197]. The LOD was in the range 0.5–1 mg/l.

1.5.3.1.2 Gas chromatography
GC-MS was recommended as a confirmation method for opiate identification in urine drug screening [198]. The need to handle large numbers of urine samples in the shortest possible time brought several logistic and analytical problems. The main concern was focused on sample pretreatment procedures, including the optimization of urine hydrolysis and the derivatization of opiates.

The effective hydrolysis of conjugates is critical for all further steps in opiates determination with GC-MS. Zezulak *et al.* [199] stressed the diversity of commercially available β-glucuronidase preparations, which may originate from snail (*Helix Pomatia*), beef liver, limpets (*Patella vulgata*) or bacteria. Each enzyme preparation showed a different specific activity and pH optimum. An enzyme of bacterial origin (β-glucuronidase from *Escherichia coli*, EC 3.2.1.31) was preferred for the GC-MS analysis of opiates. Lin *et al.* [200] compared the performance of three acid hydrolysis and four enzymatic hydrolysis procedures used for total morphine and codeine measurements in urine. The drugs were measured by GC-FID and GC-MS-ITD (ion trap detection) after SPE. Acid hydrolysis with 6.5 M HCl and the addition of bisulfite appeared as the method of choice.

C6G, M3G, and M6G are often present together in urine samples, therefore Hackett *et al.* [201] examined the hydrolysis procedures for these metabolites. β-Glucuronidase obtained from *E. coli* and *H. pomatia* as well as acid hydrolysis using different concentrations of hydrochloric acid were used. Samples were extracted with SPE, derivatized and quantified by GC-MS in SIM mode. C6G was much more resistant to hydrolysis than M3G and M6G. The optimized hydrolysis method using 50% HCl for 1.5 h at 120 °C gave reproducible results that approached the spiked concentration.

Several studies were devoted to the assessment of different derivatization reagents, including: acetic acid anhydride, trifluoroacetic acid (TFA), PFPA, hexafluorobutyric acid (HFBA), or BSTFA/TMSCl (trimethylsilyl chloride) for opioids analyzed by GC-MS [202–204]. Wasels and Belleville [164] reviewed the GC-MS procedures used for the identification of 6-AM, morphine, and codeine. All the relevant steps were scrutinized in this review, including extraction, hydrolysis of conjugates, and derivatization methods. It was concluded that SPE had the advantage of decreasing the background noise and that it is gradually replacing solvent extraction. The possibility of confusing morphine and codeine with hydromorphone and hydrocodone was studied by Fenton *et al.* [205]. Chemical reduction with sodium borohydride and subsequent trimethylsilylation resulted in a better separation of the compounds and improved the quantitation of morphine in the presence of hydromorphone. Brooks and Smiths [206] applied mild acetylation of urine samples under aqueous conditions with subsequent solvent extraction. Under these conditions, only morphine and hydromorphone were converted into their respective 3-monoacetates and virtually no interference from hydrocodone and hydromorphone with codeine and morphine was observed. Broussard *et al.* [207] prevented the interference of keto-opiates (hydromorphone, oxymorphone, hydrocodone, and oxycodone) in morphine and codeine determinations by the addition of hydroxylamine before silylation to form oxime derivatives. The keto-opiates could then be separated from morphine and codeine. Rettinger *et al.* [208] evaluated a number of derivatization methods as well as contributions to deuterated ISs from unlabeled drugs. PFPA derivatives of morphine, codeine, hydromorphone, hydrocodone, and oxycodone showed the best resolution. The use of higher labeled standards (D6 instead of D3) improved quantitation at the low and high ends of the curve due to a diminished contribution of labeled compounds to the target drug ions and *vice versa*. Bogusz [209] raised the problem of a contribution of non-deuterated morphine to the D3-labeled standard and postulated the use of highly deuterated compounds as ISs for LC-API-MS.

Contrary to urine examination, the determination of morphine in blood or plasma may give important information concerning the acute influence at a given time. The methods applied for blood are usually devoted to the determination of both free and conjugated fractions of drug. Phillips *et al.* [170] applied GC-EI-MS-MS for the determination of free morphine in blood after ethyl acetate extraction and PFPA derivatization. The possible interference of codeine and 6-AM was studied. A LOD of 1 μg/l was observed for morphine. Schmitt *et al.* [171] developed a GC-CI-MS (negative and positive ions) method for the determination of PFPA derivatives of free morphine and codeine in blood samples. Negative chemical ionization (NCI) appeared more sensitive and was applied in forensic practice. Cone and Darwin [210] reviewed the GC-MS methods for the simultaneous determination of morphine and related opiates, including heroin, 6-AM, codeine and others in biological fluids. A growing number of methods for the simultaneous determination of the different drugs and metabolites were observed. Watson *et al.* [172] determined free and total morphine, codeine, and normorphine in plasma. Unconjugated drugs were isolated by solvent extraction and derivatized with HFBA before GC-MS-NCI determination. For enzymatic hydrolysis, several sources of enzyme were tested and the *E. coli* glucuronidase was found to be the most effective. The LOD was estimated at 0.25 µg/l. Fryirs *et al.* [173] isolated free morphine from plasma with an organic solvent and derivatized with PFPA. GC-MS-EI (SIM) analysis was performed, and the LOD was 0.2 µg/l.

Gunnar *et al.* [211] developed a procedure for the simultaneous identification, screening, and quantitation of 30 drugs of abuse using 0.25 ml of human oral fluid. After sequential SPE elution, an optimized derivatization procedure was performed. Amphetamine-type stimulant drugs were acylated with HFBA, benzodiazepines and Δ-9-THC were silylated with *N*-methyl-*N*-(*tert*-butyldimethylsilyl)trifluoroacetamide, whereas BE, codeine, ethylmorphine, 6-monoacetylmorphine, morphine, pholcodine, BP, and norbuprenorphine were silylated with *N*-methyl-*N*-(trimethylsilyl)trifluoro-acetamide. All the derivatives were simultaneously determined with GC-EI-MS.

1.5.3.1.3 *Liquid chromatography*

The application of HPLC methods to the analysis of morphine and its metabolites was stimulated by the recognition of the role of M6G as an active morphine metabolite [212,213]. Moreover, it was demonstrated that the M6G receptor might be a major site of heroin action [214]. Svensson [37] first developed a procedure for the determination of morphine glucuronides in biofluids. This method was based on SPE with C_{18} cartridges and subsequent HPLC separation with UV or electrochemical (coulometric) detection. The problem of the different detectability of morphine glucuronides was solved by using HPLC with coulometric or fluorescence detection [215–221]. The introduction of LC-MS in analytical toxicology brought new possibilities for the determination of morphine metabolites and rendered all previous methods obsolete. These techniques have also been discussed in Section 1.5.2.2.2. Tyrefors *et al.* [179] determined morphine, M3G, and M6G in human serum with ESI-LC-MS with external standardization. Bogusz *et al.* [176,177,222] applied LC-APCI-MS for the determination of morphine, M3G, M6G, as well as other opiates, to the analysis of blood and urine samples. Isotope dilution was applied for quantitation.

Zheng *et al.* [223] used an ESI-LC-MS-MS system for the determination of morphine and glucuronides isolated from plasma samples from rats. Using plasma samples of 100 µl, detection limits of 3.8–12 µg/l were achieved. Shou *et al.* [224] applied LC-ESI-MS-MS for the rapid, high throughput analysis of morphine, M3G, and M6G in plasma. The compounds were isolated with SPE in 96-well format and separated on a silica column 50 mm × 3 mm column at flow rate of 1.5–9.9 ml/min. Chromatographic run times ranged from 0.6 to 2 min; the LODs ranged from 0.5 to 10 µg/l. Whittington and Kharash [225] applied SPE in a 96-well plate format, followed by LC-ESI-MS-MS, for the determination of morphine, M3G, and M6G in 0.5 ml of human plasma (Fig. 1.7). Murphy and Huestis [226] developed an LC-ESI-MS-MS procedure for the determination of morphine, codeine, M3G, M6G, and C6G in 0.2 ml human urine, using SPE cartridges for drug isolation.

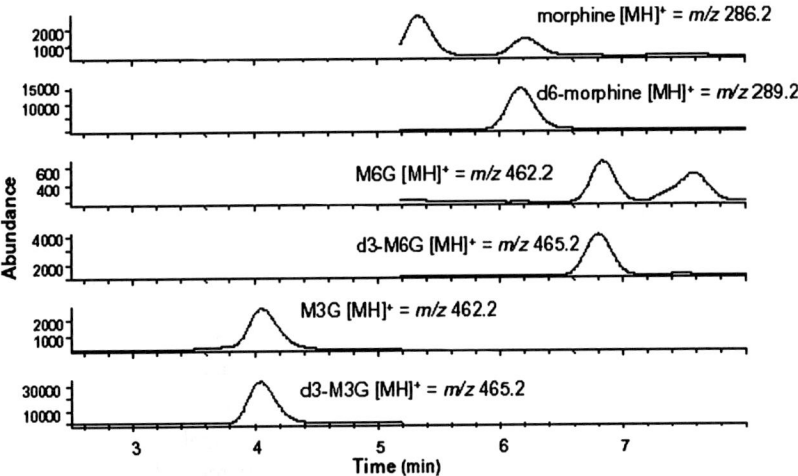

Fig. 1.7. LC-ESI-MS-MS of morphine and morphine glucuronides extracted from serum at the concentrations of 0.5 ng/ml for morphine and M6G, and 2 ng/ml for M3G. From Whittington *et al.* [225] with permission of Elsevier.

A review of HPLC methods used for direct detection of drug glucuronides in biological matrices was given by Kaushik and LaCourse [227].

1.5.3.2 Morphine and its metabolites in autopsy material after morphine or heroin overdose

The purposes of morphine determination in forensic autopsy samples are different from those in living subjects. The following points may be mentioned here:

- The measured concentration of drug should be helpful in explaning of the fatal outcome.
- A differentiation between heroin, morphine, or codeine intake should be made
- Analytical results may give some indications concerning the rapidity of death after assumed heroin administration.
- The analysis should include other groups of compounds, since heroin victims usually abuse several other drugs, including cocaine or benzodiazepines.

All the above-mentioned points dictate the need to apply a particular analytical strategy, i.e. the use of a method that is universal in regard to the kind of biosample and to the substances detected. Usually, not only morphine, but also codeine, 6-MAM, acetylcodeine, morphine, and codeine glucuronides should be determined, as well as other non-opioid drugs. The methods that fulfill these requirements have been reviewed in Section 1.5.2. Mass spectrometric detection, coupled with GC or HPLC, appeared to be the most versatile approach. Aderjan *et al.* [228] applied the method of Glare *et al.* [220] for the determination of morphine and its glucuronides in autopsy blood samples taken from heroin victims. The molecular ratios were helpful in differentiating between rapid and protracted death. A similar observation was made by Bogusz *et al.* [177,178] who used LC-APCI-MS for the determination of morphine and its glucuronides in autopsy blood. Kerrigan *et al.* [229] reported a case of a 44-year-old male with end-stage pancreatic cancer, who was receiving a morphine infusion for pain control *via* a single subclavian intravenous catheter. Comprehensive toxicology on autopsy samples indicated that morphine was the only drug present, in extraordinarily high concentrations. The free morphine concentrations in heart blood, vitreous fluid, brain, liver, stomach contents, and urine were 96 mg/l, 52 mg/l, 26 mg/kg, 88 mg/kg, 82 mg/l, and 976 mg/l, respectively. The total morphine concentrations in heart blood, vitreous fluid, brain, liver, and stomach contents were 421 mg/l, 238 mg/l, 65 mg/kg, 256 mg/kg, and 325 mg/l, respectively. Records indicated that the infusion pump might have continued to deliver the drug for 15–45 min following death.

Advanced postmortem biochemical processes may distinctly influence the analytical strategy, due to the limited availability of some materials, e.g. blood or urine. This problem has stimulated research into some alternative matrices useful for forensic toxicological assessment. Cengiz *et al.* [230] determined morphine in postmortem rabbit bone marrow and in blood. The aim of this study was to predict how long after death a buried body could be analyzed for opiates in soft tissues and to show the accessibility and suitability of bone marrow as a useful toxicological specimen from buried bodies. Morphine in doses of 0.3–1.1 mg/g was administered to nine albino rabbits. One hour after the injections, the rabbits were sacrificed and buried after the collection of blood, urine, and bone marrow samples. Femur bone marrow specimens were collected on the 7 and 14 days after burial. Morphine was determined with CEDIA® immunoassay. The morphine doses correlated well with the drug concentrations in blood and bone marrow. The morphine concentrations in the bone marrow at 7 and 14 day postmortem decreased consistently. It was concluded that bone marrow could be a useful alternative specimen in postmortem cases.

A particular application of opiate analysis is the determination of drugs in carnivorous fly larvae infesting the decayed corpse. Goff *et al.* [231] demonstrated the presence of morphine in the larvae of the flesh fly feeding on the tissues of a rabbit previously injected with heroin. An interesting observation was that the larvae feeding on these tissues developed more rapidly than those feeding on tissues from controls. Introna *et al.* [232] observed a positive radioimmunoassay reaction on opiates in fly larvae fed on opiate-positive liver specimens. Kintz *et al.* [233] determined morphine and codeine in the blood and bile of a putrefied cadaver and the fly larvae found on the corpse. The larvae were washed, homogenized in saline, and subjected to solvent extraction after enzymatic hydrolysis with β-glucuronidase. The extract was derivatized with BSTFA/TMCS and examined with GC-MS (ion trap). French authors [234–237] performed systematic experimental studies on the usefulness of necrophagous larvae *Coleoptera* and *Diptera* for the postmortem diagnosis of opiate poisoning. Rabbits were given morphine in dosages corresponding to human overdose and sacrificed. Eggs of flies were planted in the eyes, nostrils, and mouth of the carcasses and the larvae were analyzed for morphine at different stages of development. Radioimmunoassay and immunohistochemistry were applied as detection methods. Morphine was detected in all the larvae; however, a correlation between the dosage and morphine levels was not found. Campobasso *et al.* [238] investigated the correlation between the concentrations of drugs in human tissues and fly larvae feeding on these tissues. Samples of liver were taken from 18 cases in which preliminary toxicological screening indicated the presence of drugs. Blowfly larvae (*Diptera: Calliphoridae*) were fed on these samples and subsequently analyzed for drug content. All drugs detected in human tissues (opiates, cocaine, barbiturates, antidepressants) were also detected in insect specimens using immunoassay and GC-MS. Comparisons of the drug concentrations between those in the human tissues and blowfly larvae showed different patterns of distribution that may be attributed to differences in physiology. The results confirm the reliability of entomological specimens for qualitative analyses, although quantitative extrapolations are unreliable.

1.6 OTHER OPIATE AGONISTS

1.6.1 Codeine and dihydrocodeine

Several authors have developed methods for the simultaneous determination of codeine and its metabolites together with other opiates, particularly morphine. These studies have been reviewed in Sections 1.4 and 1.5, devoted to opiates and heroin. In this section, only the studies devoted solely to codeine, DHC and their metabolites will be reviewed.

DHC is a semisynthetic opiate, which was used at first as an analgesic and antitussive drug. Since the late 1980s DHC has been extensively used in Germany in the treatment of heroin addicts, and in consequence a number of fatal poisonings were observed [239]. DHC possesses a primary addiction potential and may be abused [240]. In the human body DHC undergoes *N*-demethylation to nor-DHC and *O*-demethylation to very toxic dihydromorphine (DHM). All these drugs are conjugated to the corresponding glucuronides [241] (Fig. 1.8).

1.6.1.1 Gas chromatography

Seno *et al.* [242] determined underivatized codeine and DHC in plasma and urine using GC with surface ionization detection (SID) after SPE. The LOD was estimated at 2.5 μg/l for both drugs. Hofmann *et al.* [243] determined DHC and DHM in serum by NCI-GC-MS-MS after derivatization with PFPA. Codeine and morphine were used as ISs. The limits of quantitation were 2 μg/l for DHC and 0.04 μg/l for DHM, respectively. Kintz *et al.* [244] found codeine in blood at a concentration of 22.1 mg/l in fatal mixed intoxication with ethanol (1.25 g/l). The distribution of drug in the organs was also studied. Sachs *et al.* [245], who examined hair samples of heroin abusers, frequently observed the presence of DHC using GC-MS after derivatization with HFBA. Wilkins *et al.* [246] determined codeine and morphine as codeine metabolites in human hair by PCI-GC-MS. The limits of detection for both drugs were 10 pg on-column. This allowed codeine to be detected in hair for at least 8 weeks after a single oral dose of 120 mg. The same group analyzed codeine and

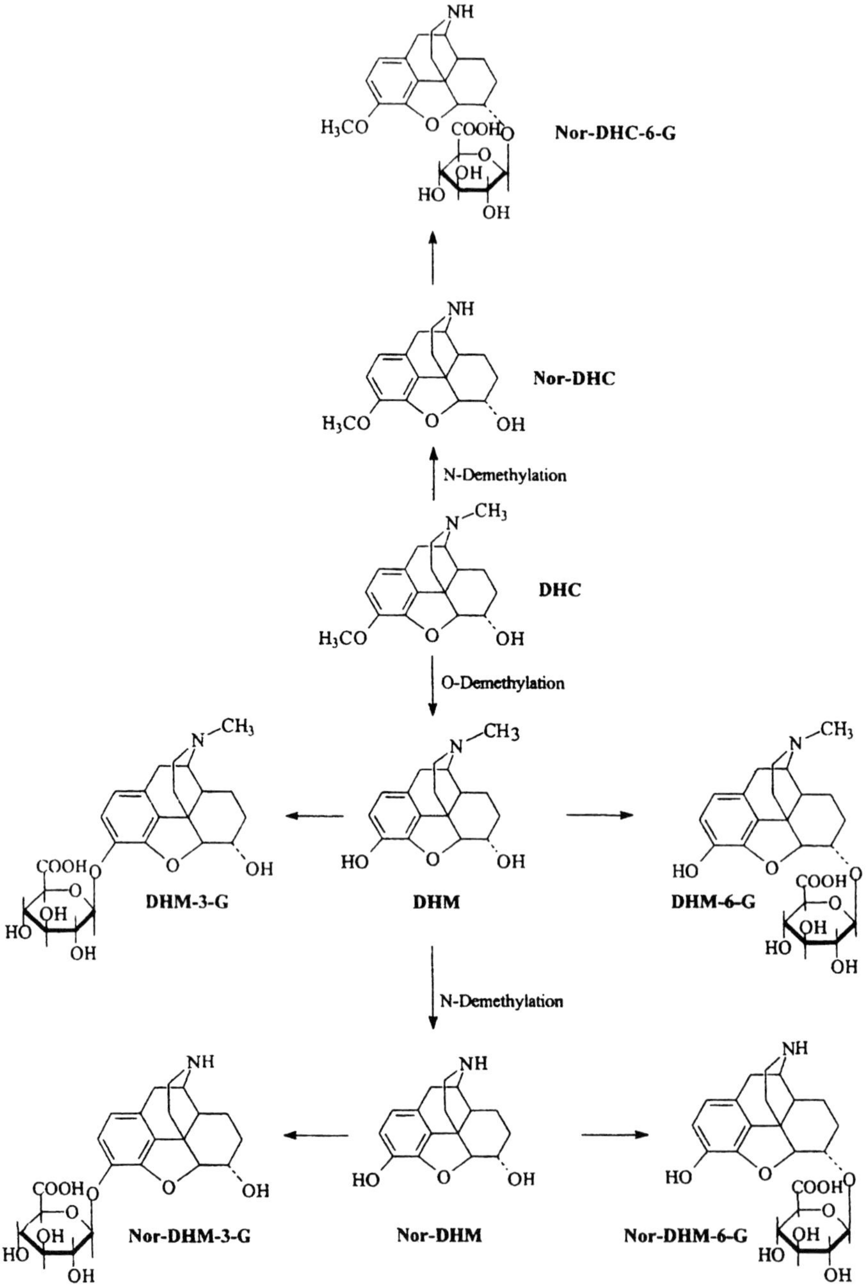

Fig. 1.8. Metabolic pathways of DHC. From Bogusz [178] with permission of Elsevier.

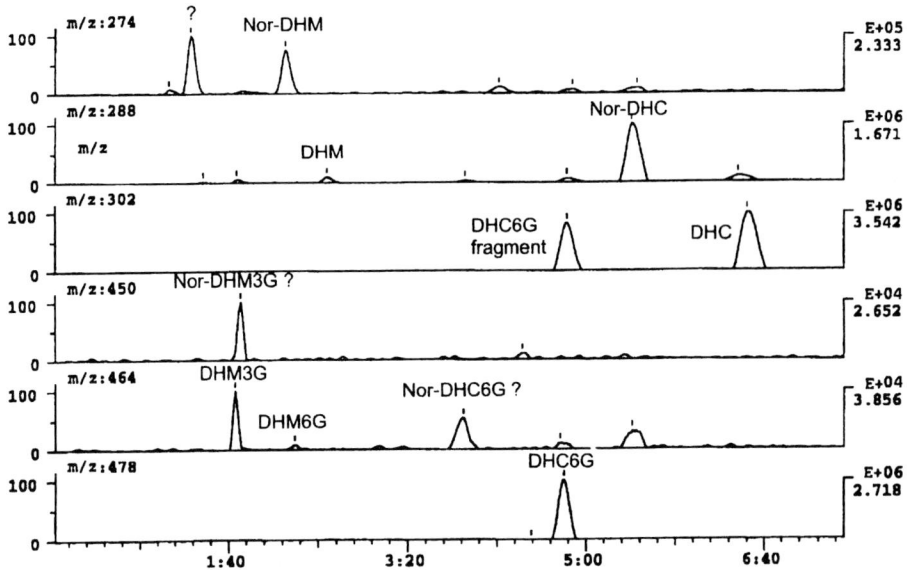

Fig. 1.9. LC-APCI-MS chromatogram of urine extract after administration of 10 mg DHC orally. DHM, dihydromorphine; G, glucuronide. From Bogusz [178] with permission of Elsevier.

morphine in rat hair after a long-term, chronic application of codeine, using an iontrap GC-MS [247]. The kinetics of the drug incorporation into hair was followed. The excretion of DHC metabolites in urine was studied by Balikova *et al.* [248] who applied GC/MS after SPE and cleavage of conjugates.

1.6.1.2 Liquid chromatography

In earlier procedures, codeine and its metabolites, norcodeine, morphine, and normorphine as well as their corresponding glucuronides (C6G, M3G, M6G) were determined in plasma and urine samples by HPLC with electrochemical or fluorimetric detection [249–252]. The LOD for codeine was around 5 µg/l. Lafolie *et al.* [253] determined codeine and metabolites (C6G, M3G, M6G, morphine) in the plasma of 13 volunteers after an experimental intake of 25 and 50 mg of codeine. HPLC with electrochemical and UV detection was used. Large interindividual variability of the peak concentrations of analytes was observed and the need for careful interpretation of the results was stressed.

DHC metabolites: DHM and nor-DHC, formed in liver microsomal incubates, were determined by HPLC with UV detection after alkaline solvent extraction [254]. The distribution of DHC and its metabolites: DHM, DHM3G, nor-DHC, and DHC6G in different blood vessels and organs was examined in fatal DHC intoxication cases [239]. HPLC with fluorescence detection was used. The authors stressed the role of DHM as an active, toxic metabolite of DHC. Bogusz [178] identified metabolites of DHC: Nor-DHC, DHM, DHC6G, DHM3G, DHM6G, Nor-DHM3G, in the extract of 1 ml urine after administration of 10 mg DHC orally. LC-APCI-MS was used (Fig. 1.9).

1.6.1.3 Capillary electrophoresis

Hufschmid *et al.* [255] determined urinary DHC and DHM by MEKC. Both urine extracts and non-pretreated urine samples were analyzed. The method applied appeared to be valuable for metabolic studies. Wey *et al.* [256] presented a CE-MS (IT) procedure for the determination of codeine, DHC, and their glucuronides. The metabolites were detected in urine samples after oral administration of 7 mg codeine or 25 mg DHC.

1.6.2 Buprenorphine

Buprenorphine (Bp), an oripavine derivative, was obtained from thebaine and displays partial agonist and antagonist opioid activity [257]. The drug was initially used as a potent analgesic (marketed under the commercial name Temgesic or Buprenex). Further studies demonstrated its applicability in the treatment of heroin addiction [258], and were presented in a monograph edited by Kintz and Marquet [259]. Sublingual BP tablets prescribed for addiction therapy were used on intravenous heroin addicts [260].

1.6.2.1 Immunoassays

The wide use of BP as an alternative in the therapeutic treatment of heroin abusers dictated a need for a rapid and sensitive test, such as an immunoassay, essential for therapy monitoring. Debrabandere *et al.* [261] first described an immunoassay for the detection of BP in urine samples. The assay was used for the detection of BP in urine specimens of persons suspected of Temgesic abuse. In recent years, commercial BP ELISA immunoassays were developed by Microgenics (for urine), and by International Diagnostic System (IDS), Inc. (for plasma). Cirimele *et al.* [262] evaluated the Singlestep ELISA urine BP assay from Microgenics. The immunoassay showed no cross-reactivity with other opiates and opioids. A low cross-reactivity (3% at 1 ng/ml) was observed at low concentrations of norbuprenorphine. On the basis of the parallel determination of 76 urine samples for BP with LC-ESI-MS and ELISA, an optimum cutoff concentration of 2 µg/l was determined for the immunoassay. None of the potential urine adulterants (hypochloride 50 ml/l, sodium nitrite 50 g/l, liquid soap 50 ml/l, and sodium chloride 50 g/l) was able to cause a false-negative response in the immunoassay. Böttcher and Beck [263] evaluated a new immunochemical test based on CEDIA® technology for use in clinical urine drug testing. The method was compared with an existing ELISA method and a GC-MS method on urine specimens from patients in heroin substitution treatment. The agreement in the qualitative results with an existing ELISA method was 96.8%. The sensitivity of CEDIA® at a concentration of 5 µg/l was 99.5%. A false-positive response was discovered in patients receiving DHC. Cirimele *et al.* [264] published an assessment of the serum BP assay developed by International Diagnostic Systems Corporation (IDS). The authors investigated the applicability of this assay for other biological matrices such as urine, blood, and hair specimens. Low concentrations of BP were detected with the ELISA test and were confirmed by LC-MS (0.3 µg/l in urine, 0.2 µg/l in blood, and 40 pg/mg in hair). This immunoassay had no cross-reactivity with DHC, or other opiates and opioids. Cross-reactivity (1%) was measured for a norbuprenorphine concentration of 50 ng/ml. The ELISA method produced false-positive results in <21% of the cases, but no false-negative results were observed with the immunological test. Four potential adulterants (hypochloride 50 ml/l, sodium nitrite 50 g/l, liquid soap 50 ml/l, and sodium chloride 50 g/l) did not cause a false-negative response in the immunoassay. De Giovanni *et al.* [265] evaluated the BP serum assay from IDS for the qualitative determination of BP in other matrices, such as urine, saliva, and hair. The specimens of urine, plasma, saliva, and hair were collected from 18 heroin addicts treated with BP for at least 1 year. All urine samples were positive; 1 serum sample, 6 saliva samples, and 12 hair samples showed negative results. All positive results were confirmed with GC-MS at a LOD of 0.2 ng/ml. The authors concluded that the serum assay could be applicable to saliva.

1.6.2.2 Gas chromatography

Everhard *et al.* [266] modified a GC-ECD method of BP determination, developed initially by Cone *et al.* [267]. The method was used for pharmacokinetic studies and the bioavailability parameters were given. The stability of BP and morphine was assessed in spiked blood samples [268]. The drugs were determined by GC-MS (SIM) after silylation. Both drugs remained unchanged at $-20°C$, morphine was very stable at 4°C and 25°C (90% after 12 months storage), and BP was stable at 4°C and 25°C (80 and 70% after 12 months storage). Kuhlman *et al.* [269,270] developed a NCI-MS-MS method for pharmacokinetic applications. BP, norbuprenorphine (NBP), and ISs (BP-D$_4$, norcodeine) were derivatized with HFBA. BP, due to its analgesic and euphorizing properties, may be abused as doping substance in sport. Lisi *et al.* [271] developed a GC-MS method for the detection of BP and NBP in urine. Urine was hydrolyzed with β-glucuronidase and subjected to extractive alkylation with hexane-iodomethane. The derivatives of BP and NBP were determined by GC-MS (SIM), using BP-D$_4$ as IS. The cyclic artifacts of BP and NBP, as observed by Cone *et al.* [272], were not formed. BP and NBP were easily detected in urine taken 42.5 h after a sublingual dose of 0.2 mg Temgesic.

1.6.2.3 Liquid chromatography

Debrabandere *et al.* [273] described an HPLC method with electrochemical detection for the detection of BP and NBP in urine samples after alkaline toluene extraction. Detection limits of 0.2 µg/l for BP and 0.15 µg/l for NBP were reported. This method was successfully applied by Kintz *et al.* [274,275] for the examination of hair samples taken from BP addicts and from heroin abusers treated with BP. The authors also tried LC-MS (ESI and PBI), using instruments of an earlier generation. In a particle beam interface, the BP molecule was thermally destroyed to many small fragments, and the sensitivity of an old electrospray interface was not high enough to detect the drug in a hair extract. In 1997, several LC-MS methods for the determination of BP and NBP were published. The main advantage in comparison with GC-MS was simpler sample pretreatment due to the omission of the derivatization step. Hoja *et al.*

[276] determined BP and NBP in whole blood by LC-ESI-MS after β-glucuronidase hydrolysis, acetone precipitation, and Extrelut (toluene-ether) extraction. The LOQ was 0.1 ng/ml for both analytes. Tracqui *et al.* [277] applied LC-ESI-MS to the determination of BP and NBP in blood, urine, and hair samples. A solvent extraction with a chloroform–isopropanol–heptane mixture at pH 8.4 was applied. The mass spectra of BP, NBP, and IS (BP-D$_4$) exhibited only protonated molecular peaks. The sensitivity was comparable with other ESI-MS methods. Moody *et al.* [278] developed an LC-ESI-MS-MS method for BP determination in plasma and compared it with an existing GC-PCI-MS method. The LC-MS-MS method appeared more sensitive (LOQ, 0.1 ng/ml) than GC-MS (LOQ, 0.5 ng/ml) and could demonstrate the presence of drug up to 96 h after administration. The mass spectrum of BP observed by Hoja *et al.* [276] and Moody *et al.* [278] was very similar, showing the protonated molecular ion as the base peak ion and small fragments at m/z 414 and 396, respectively. The protonated molecule of BP remained stable up to a collision energy of 20 V and at higher energies was shattered to many product ions of low intensity. In a later study, Moody *et al.* [279] applied LC-ESI-MS-MS procedure for the determination of BP and NBP in human plasma. The transitions m/z 468 to 396 for BP and m/z 414 to 101 for NBP were monitored, and a LOQ of 0.1 µg/l was achieved for both compounds.

Gaulier *et al.* [280] reported the suicidal poisoning of 25-year-old male heroin addict with high dose of BP. BP and NBP were determined in body fluids and organs with LC-ESI-MS after deproteinization and SPE. In the gastric contents, only BP was found at a concentration of 899 mg/l. The following concentrations were found in selected matrices: in blood, BP 3.3 mg/l, NBP 0.4 mg/l; in bile, BP 2035 mg/l, NBP 536 mg/l; in brain, BP 6.4 mg/l, NBP 3.9 mg/l. Besides BP and NBP, 7-aminoflunitrazepam was found in the blood, urine, and gastric contents. Polettini and Huestis [281] developed a LC-ESI-MS-MS method for the determination of BP, NBP, and BP glucuronide (BPG) in human plasma. SPE with C$_{18}$ cartridges and gradient elution was used. For BP, NBP as well as for the deuterated analogues used as ISs, the protonated molecule ions were monitored, for BPG the protonated molecule and BP aglycone. The LOQ was 0.1 µg/l for all compounds. On the base of the transition m/z 590→414, nor-buprenorphine glucuronide (NBPG) was also tentatively detected. The reference standard for this compound was not available. The authors stated that the useful fragmentation of the BP molecule was not possible; after increasing of fragmentation energy this compound dissipated to very small particles. Later, when a reference standard of NBPG became available, Murphy and Huestis [282] published a study on the determination of BP, NBP, BPG, and NBPG in plasma with LC-ESI-MS-MS. BP-D$_4$ and NPB-D$_3$ were used as ISs. The following transitions were monitored: for BP, 468.4 to 396.5 and 414.3; for NPB, 414.3 to 326.1 and 340.3. For glucuronides, both aglycone product ions (BP and NBP) were monitored (Fig. 1.10).

Ceccato *et al.* [283] determined BP and NBP in human plasma by LC-APCI-MS-MS. Automated SPE on C8 disposable extraction cartridges was used to isolate the compounds from the biological matrix. The separation was obtained on a C-18 column using a mobile phase consisting of methanol and 50 mM ammonium acetate solution (50:50, v/v). Clonazepam was used as the IS. The MS/MS ion transitions monitored were m/z 468→468, 414→414 and 316→270 for BP, NBP, and IS, respectively. The limits of quantification were around 10 pg/ml for BP and 50 pg/ml for NBP. Kronstrand *et al.* [284] applied direct injection of urine followed by LC-MS-MS detection for the analysis of BP, NBP, and their glucuronides. The compounds were separated on a phenyl column with a gradient of ACN and ammonium formate buffer. The following transitions were monitored: 468→468, 414, and 396 for BP, 414→101 for NBP, 644→468 and 590 for BPG, and 590→414 for NBPG, respectively. Some ion suppression was observed in the directly injected samples. The application of SPE before analysis allowed a reduction in the LOQ to 1 ng/ml, as compared with 20 ng/ml for direct injection. The method was applied to the analysis of urine samples obtained from patients receiving BP as well as from the abusers. The concentrations of BP and NBP in patients ranged between 31 and 1080 ng/l and 48–2050 ng/ml, respectively. In suspected abusers, the ranges were 2.3–796 ng/ml and 5.0–2580 ng/ml.

Grimm *et al.* [285] studied the transfer of BP and NBP into the breast milk of lactating women receiving BP maintenance therapy. The drugs were isolated by SPE and LLE and subjected to LC-ESI-MS-MS analysis. The LOQ were 0.18 ng/ml for

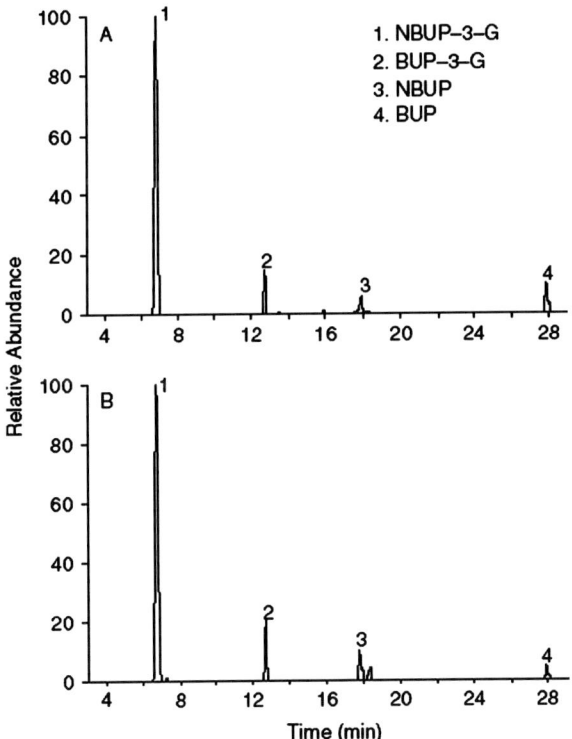

Fig. 1.10. Reconstructed SRM chromatograms of plasma extracts from two participants of BP maintenance program. Concentrations of NBUP-3-G, BUP-3-G, NBU, and BUP were (A) 9.8, 0.7, 1.3, and 1.4, and (B) 9.7, 1.0, 1.8, and 12.2 ng/ml. From Murphy *et al.* [282] with permission of John Wiley Sons, Ltd.

TABLE 1.5
GAS CHROMATOGRAPHIC METHODS FOR SYNTHETIC OPIOIDS

Drug	Sample	Isolation	Derivatization	Column, conditions	Detection	LOD (µg/l)	Ref.
BP	Blood	Extrelut + SCX	Silylation	CPSil-5, 180–300°	PCI-MS (SIM)	1 pg on column	[268]
BP	Plasma	l/l pH 9.1	HFBA	HP 1, 150–325o	ECD	0.1 BP	[266]
BP, NBP	Plasma	SPE	HFBA	DB-5, 125–300°	NCI-MS-MS	0.15 BP 0.016 NBP	[270]
BP, NBP	Urine	l/l alkaline	Methylation	HP 2, 247–310°	EI-MS (SIM)	0.2 both	[271]
Meth	Plasma, urine, CSF	l/l pH	–	SE-52	NPD	0.5 LOQ	[298]
Meth, EDDP	Urine	l/l alkaline	–	DB-5, 190°	EI-MS (SIM)	50	[297]
Meth, EDDP, EMDP	Hair	l/l alkaline	–	DB-5, 80–280°	PCI-MS-ITD	0.5 ng/mg	[300]
Meth, EDDP, EMDP	Plasma, urine, liver	SPE	–	HP-1, 80–280°	PCI-MS (SIM)		[299]
Tramadol	Blood	l/l pH 9	–	EC-5, 80–295°	EI-MS (SIM)	10	[324]
Tramadol	Plasma	SPE C18	–	HP-5	EI-MS (SIM)	1	[323]
Tramadol	Plasma	SPME	–		EI-MS (SIM)	0.2	[331]

Abbreviations: BP = buprenorphine,   NBP = norbuprenorphine,   Meth = methadone,   EDDP = 2-ethylidene-1,5-dimethyl-3,3-diphenylpyrrolidine,
EMDP = 2-ethyl-5-methyl-3,3-diphenylpyrroline, l/l = liquid/liquid.

TABLE 1.6
LIQUID CHROMATOGRAPHIC METHODS FOR SYNTHETIC OPIOIDS

Drug	Sample	Isolation	Column, elution conditions	Detection	LOD (µg/l)	Ref.
BP	Plasma	l/l pH 10.5	C8, H_2O-MeOH-ACN-HCOOH	ESI-QQQ, MRM	0.1 LOQ	[278]
BP, NBP	Blood	Extrelut pH 9	C18, ACN-NH₄COOH	ESI-Q (SIM)	0.1 LOQ BP, NBP	[275]
BP, NBP	Blood, urine, hair	l/l pH 8.4	C18, ACN-NH₄COOH	ESI-Q (SIM)	0.1 BP, 0.05 NBP	[277]
BP, NBP	Hair	l/l pH 8.5	CN, ACN-phosphate buffer	EC, ESI	0.02 ng/mg BP, 0.01 NBP	[275]
BP, NBP, BUG	Plasma	SPE	C18, ACN-HCOONH₄ grad.	ESI-QQQ, MRM	0.1	[281]
BP, NBP, BUG, NBPG	Plasma	SPE	C18, ACN-NH₄COOCH₃	ESI-IT, MRM	0.3	[282]
Meth, EDDP	Urine, meconium	l/l pH 9	C18, ACN-phosphate buffer +TEA	DAD 204 nm	76M, 127 EDDP	[305]
R/S-Meth, R/S-EDDP	Hair	SPE C18	Chiral-AGP, PropOH-NH₄COOH	ESI-Q (SIM)	0.2 M, 0.1 EDDP	[312]
R/S-Meth	Serum	SPE mixed	Chiral-AGP	UV 205 nm		[310]
R/S-Meth	Serum	l/l	Chiral-AGP + CN, ACN-phosphate buffer	UV 200 nm	1.5 LOQ	[309]
R/S-Meth	Plasma	l/l	Chiral-AGP	UV 215 mm	2.5 LOQ	[307]
R/S-Meth	Plasma	l/l	Chiral-AGP	UV 212 nm		[306]
R/S-Meth, R/S-EDDP	Serum	Ultrafiltration, l/l	Chiral-AGP	ESI-QQQ	0.3	[318]
(+)/(−)-Tramadol	Plasma	SPE C2	Chiralcel OD-R, ACN-phosphate buffer	Fluorimetry	0.5	[339]
(+)/(−)-Tramadol, ODT	Plasma	SPE	Chiralpak AD	APCI-QQQ		[340]
Fentanyl	Plasma	SPE 96 plate	Silica, ACN-TFA isocr.	ESI-QQQ, MRM	0.05	[383]
Sufentanil	Serum	L/l	C18, ACN-TFA isocr.	ESI-IT	0.005, 0.01	[386]
Sufentanil	Plasma	SPE	Si HILIC, ACN-HCOOH grad	ESI-QQQ	0.00025	[388]
Ketobemidone, Nor-K	Urine	SPE	C8, ACN-HCOOH grad.	ESI-Q, SIM	25	[391]

Abbreviations: BP = buprenorphine, NBP = norbuprenorphine, BUG = buprenorphine glucuronide, NBPG = norbuprenorphine glucuronide, Meth = methadone, EDDP = 2-ethylidene-1,5-dimethyl-3,3-diphenylpyrrolidine, Nor-K = nor-ketobemidone, EC = electrochemical detection, Q = single stage quadrupole, QQQ = triple stage quadrupole, SIM = selected ion monitoring, MRM = multiple reaction monitoring, ACN = acetonitrile, TFA = trifluoroacetic acid, l/l = liquid/liquid.

BP and 0.20 ng/ml for NBP, respectively, using a sample volume of 0.5 ml milk. The BP and NBP concentrations determined in ten random breast milk samples collected over four successive days from a lactating woman during BP maintenance therapy ranged from 1.0 to 14.7 and 0.6 to 6.3 ng/ml, respectively. The drug exposure of the infant was considered to be low.

Tables 1.5 and 1.6 show selected gas- and liquid-chromatographic methods applied for synthetic opiates.

1.6.3 Methadone

Methadone, a morphine substitute synthesized in Germany during World War II, initially found limited application due to its very long elimination half-life and subsequent accumulation. These properties drew the attention of Dole, who first applied methadone as a heroin substitute in the therapy of addicts [286]. In the past 20 years, due to the international proliferation of methadone maintenance programs, this drug has become the most widely used opioid agonist [287]. This has dictated the need for methadone monitoring in body fluids, in order to control the compliance and to prevent toxicity. The wide availability of methadone is associated with its illicit use, and with a growing number of drug-associated death cases, particularly among heroin addicts [288,289].

1.6.3.1 Immunoassays

The immunoassays for methadone are always included in a standard preliminary screening package for urine testing on drugs of abuse. Some specific studies concerning methadone and its metabolites have been published in recent years. Chikhi-Chorfi *et al.* [290] developed antibodies selective for (*R*)-methadone (levomethadone) and for racemic (*R-S*)-methadone. Both antibodies showed low (0.5%) cross-reactivity with the EDDP metabolite and no cross-reactivity with other opioids. An ELISA procedure has been developed for the determination of both forms of methadone in the serum of opiate addicts under maintenance treatment.

Standard methadone immunoassays do not cross-react with a prevalent metabolite EDDP. This may be seen as a drawback, since some drug addicts, who are supposed to ingest methadone, spike the urine sample taken for control analysis. In such samples, high levels of methadone are detected, but not EDDP metabolite. Microgenics Corp. have developed a selective CEDIA® EDDP assay. George *et al.* [291] used this assay for the screening of 1381 urine specimens, in parallel with a standard methadone EMIT immunoassay. 39% samples were positive by the methadone assay, and 46% were positive for EDDP. In seven cases, only high methadone concentrations were found, with negative results for EDDP. These urine specimens originated most probably from "spikers", i.e. subjects who added methadone to urine. LAAM was approved as a substitute for methadone. This drug, as well as methadol – a common metabolite of LAAM and methadone – showed very high cross-reactivity with all methadone immunoassays [16].

Methadone immunoassays have been applied to alternative samples. A Cozart RapiScan test was developed for the detection of methadone in saliva. The results obtained with this assay were in agreement with the results of GC-MS confirmation [292,293]. Cooper *et al.* [294] evaluated the performance of the Cozart RapiScan immunoassay in a group of 198 addicts treated with methadone and in 40 volunteer donors who were not drug users. Oral fluid specimens were analyzed in the laboratory by immunoassay and by GC-MS. A total of 103 samples were confirmed positive for methadone. The cutoff of 30 ng/ml in diluted oral fluid was applied. ElSohly *et al.* [295] used EMIT urine immunoassay for the detection of methadone in 50 meconium samples and compared the results with GC-MS. All EMIT results were negative. In GC-MS analysis, four samples contained low concentrations of methadone and high concentrations of EDDP. The authors suggested the use of immunoassays directed to EDDP (e.g. EDDP-CEDIA®) for the detection of prenatal exposure to methadone.

1.6.3.2 Gas chromatography

GC or GC-MS assays used for methadone determination usually include EDDP [296–298]. In GC-PCI-MS procedure, Alburges *et al.* [299] determined methadone, EDDP, and 2-ethyl-5-methyl-3,3-diphenyl-pyrroline (EMDP) in human plasma, urine, and liver microsomes. The protonated molecules of drugs and their tri-deuterated analogues, used as ISs, were monitored. A LOQ of 10 µg/l was reported. The method was applied to the determination of methadone in the body fluids of 33 patients under going methadone treatment. Methadone was found in all plasma samples, EDDP in 15 plasma samples, whereas EMDP was detectable in small concentrations in some urine samples. Wilkins *et al.* [300] determined methadone, EDDP, and EMDP in hair samples by GC-MS-IT. Cooper and Oliver [301] optimized a mixed-mode SPE column extraction for the isolation of methadone, EDDP and EMDP from whole blood. Clean extracts and high recoveries were reported, using GC-MS-SIM as the detection technique. The detection limits for all substances were 5 µg/l. Sporkert and Pragst [50] applied automatic headspace-SPME combined with GC-MS (SIM) for the determination of methadone, EDDP, and EMDP in human hair. The LODs were 0.03 and 0.05 ng/mg for methadone and metabolites, respectively. The method was applied to the analysis of 26 drug fatalities, and in 19 cases positive results were observed. Lachenmeier *et al.* [302] isolated methadone from hair with headspace solid-phase dynamic extraction with subsequent determination by GC-MS-MS. The method was faster than conventional methods of hair analysis and more robust than SPME. Methadone and EDDP was determined in human saliva and plasma using SPME and GC-MS [51,52]. The LODs were 40 ng/ml and 8 ng/ml for methadone and EDDP, respectively. The comparison with solvent extraction showed shorter procedure times and better recovery for SPME.

Some reports demonstrated that the wide use of methadone is accompanied by its use as an infanticidial poison. Couper *et al.* [303] presented the case of an infant fatality involving methadone. A mother allegedly found her child unresponsive in a crib. The infant was taken to a hospital and was pronounced dead. Methadone was detected in subclavian blood with GC-MS at a concentration of 670 ng/ml. The cause of death was determined to be "methadone intoxication", and the manner of death was "homicide". Kintz *et al.* [304] reported two cases of the criminal use of methadone. In the first case, a 14-month-old girl was found dead at home. Toxicological analysis for methadone was carried out by GC-MS and revealed 1071 and 148 ng/ml for methadone and EDDP, respectively. Hair (6 cm) tested positive at 1.91 and 0.82 ng/mg for methadone and EDDP, respectively. In the second case, a 5-month-old girl in a coma was taken to a hospital pediatric unit. An antemortem blood analysis revealed methadone exposure (142 ng/ml), and the baby was declared dead 12 days after admission. Hair analysis (5 cm) by segmentation was positive for methadone in the range 1.0 (root) to 21.3 ng/mg (end). The death in both cases was attributed to accidental asphyxia in situations where the mothers used methadone to sedate the children.

1.6.3.3 Liquid chromatography

HPLC-DAD was applied to the determination of methadone and EDDP in meconium of neonates delivered by methadone-using mothers [305].

Methadone contains a chiral carbon atom and exists in two enantiomeric forms: (*S*)-(+)methadone and the 25–50 times more potent (*R*)-(-)methadone, known also as levomethadone. In the methadone maintenance therapy of heroin addicts, both levomethadone and racemic forms are applied. It is of pharmacokinetic importance to separate the methadone enantiomers and, hence, several stereoselective HPLC methods were developed for this purpose. The procedures were published in the 1990s based on UV detection [306–311]. Usually a serial coupling of RP and chiral column was used. These studies showed differences in the bioavailability and elimination of the two chiral forms of methadone.

The first LC-ESI-MS method for the enantioselective separation of methadone and EDDP was published by Kintz *et al.* [312]. Deuterated analogues of all compounds involved were used for quantification. The method was applied to the analysis of hair samples originating from subjects receiving a racemic drug. Both enantiomers were detected and the data collected suggested the predominance of the *R*-enantiomer in hair, which was in contrast to previous observations on serum [306,309]. Ortelli *et al.* [313] applied LC-MS for the enantioselective determination of methadone in saliva and serum. The method was applied to the analysis of samples taken from heroin addicts participating in a methadone maintenance program. The results of total methadone determination showed poor correlation between saliva and serum, while the enantiomeric ratios of drug correlated very well. Dale *et al.* [314] applied LC-MS to the study of the pharmacokinetics of methadone and EDDP in healthy volunteers after nasal, intravenous, and oral administration.

Several methods have been published for the rapid, automated determination of methadone, its enantiomers and metabolites using LC-MS or LC-MS-MS. Souverain *et al.* [315] developed a method for on-line extraction, chiral separation, and MS determination of methadone from 50 μl in plasma. Liang *et al.* [316] separated methadone enantiomers on a Chiral-AGP column in a high-throughput

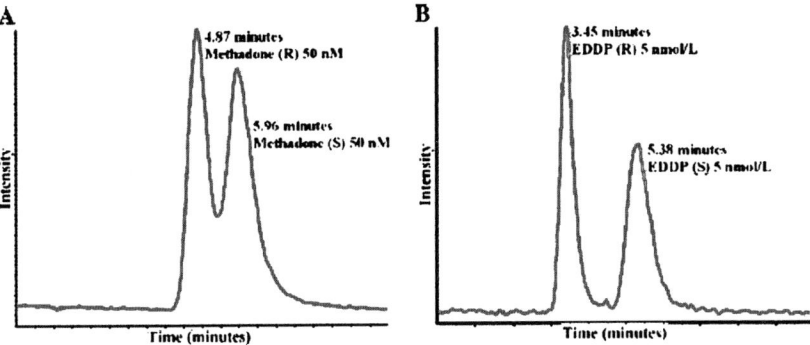

Fig. 1.11. Representative LC-MS chromatogram of methadone (A) and EDDP (B) enantiomers. From Etter *et al.* [318] with permission of Elsevier.

LC-MS-MS method. The stability of the enantiomers was studied under different conditions. On-line SPE and stereoselective LC-MS determination of methadone and EDDP enantiomers in plasma was presented by Whittington *et al.* [317]. Etter *et al.* [318] measured free and protein-bound *R*- and *S*-enantiomers of methadone and EDDP in serum. To determine the free fraction, serum samples were filtered using ultrafiltration membranes with a molecular weight cut-off of 10,000 Da and extracted using liquid–liquid extraction. The extract was evaporated, reconstituted, and analysed by LC-MS-MS. The total analyte was determined by extracting unfiltered samples. Enantiomeric separations were carried out by chiral chromatography. The LC conditions resulted in the baseline separation of *R*- and *S*-EDDP, and 85% resolution of methadone enantiomers. The total instrument run-time was 10 min (Fig. 1.11).

1.6.3.4 Capillary electrophoresis

Molteni *et al.* [319] investigated the possibility of methadone determination in urine by CE. The drug and its metabolite could be easily determined by cationic capillary zone electrophoresis but the application of MEKC was not successful. LOQ of 20 µg/l was achieved. Thormann *et al.* [320] developed two CE methods for the detection of methadone and EDDP in urine; an electrokinetic capillary-based immunoassay as a screening procedure, and a combination of CE with ESI-MS-MS for confirmation. Esteban *et al.* [321] developed a CE method to detect the interactions between methadone and anti-retroviral compounds. The enantiomers of methadone and EDDP were resolved within 4 min using a chiral electrophoresis mixture. The *R*-methadone plasma concentration decreased in a patient following the commencement of the anti-retroviral therapy, and returned to the previous higher levels after the progressive dose increased. Monitoring of *R*-methadone plasma levels appeared useful for the dose adjustment of methadone. A stereoselective method for the simultaneous determination of methadone, EDDP, and EMDP by CE was developed by Kelly *et al.* [322]. Five β-cyclodextrin background electrolyte additives were evaluated for resolution efficiency. Baseline resolution of the methadone enantiomers was achieved within 4 min.

1.6.4 Tramadol

Tramadol is a centrally acting analgesic introduced in the late 1970s as a weak μ-opioid receptor agonist and was widely prescribed as an alternative to the opiates. Tramadol causes less respiratory depression than morphine at recommended doses and has been widely prescribed in patients with mild pain. In the 1990s, the drug found its way onto the drug abuse scene, as reported below.

1.6.4.1 Gas chromatography

Merslavic and Zupancic-Kraj [323] published a GC-MS method for tramadol determination in plasma, using SPE on C_{18} cartridges. Goeringer *et al.* [324] determined tramadol and its metabolites *N*-desmethyltramadol (NDT) and *O*-desmethyltramadol (ODT) in blood from drug-related deaths and drug-impaired drivers by GC-MS after alkaline butyl chloride extraction. Artifactual formation of NDT in the injection port of GC was reported. In all cases a number of other relevant drugs were found. This study showed that every case of suspected tramadol intoxication must be very carefully scrutinized in regard to the role of coexisting substances. In a monointoxication with tramadol, a concentration of 13 mg/l was found. A GC-MS method was used, but the details were not given. Also, ODT was identified but not quantified [325]. Levine *et al.* [326] reported four cases in which tramadol was found, but death was attributed to other causes, including coronary disease, drowning or a gun shot wound. Tramadol, NDT, and ODT were extracted with *n*-butyl chloride in alkaline conditions and identified by GC-EI-MS. Quantitative determination of tramadol was performed by GC-NPD in body fluids and organs and the distribution data were presented. The authors stressed that urine is the specimen of choice for identifying tramadol use. In contrast to the finding of Sticht *et al.* [327], no evidence of sequestration of the drug in the liver or kidney was found, which was consistent with the reported volume of distribution of 3 l/kg. Gambaro *et al.* [328] determined tramadol in plasma by GC-MS using nefopam hydrochloride as IS. Plasma samples drawn from subjects in the postoperative period, who had been treated with two different initial intravenous bolus of tramadol (50 and 100 mg) followed by tramadol at the same infusion rate (12 mg/h) were analysed. GC-MS was used in scan mode for qualitative analysis and in SIM mode for quantitation, selecting the ion m/z 58 for tramadol and m/z 179 for IS. The LOD was 10 ng/ml, and the limit of quantification was 40 ng/ml. Leis *et al.* [329] used the *N*-ethyl analogue of tramadol as the IS for the quantitative measurement of tramadol in human plasma. The underivatized drug was analyzed by GC-MS (SIM) after solvent extraction. The ions m/z 58 and m/z 73 were used for tramadol and the IS, respectively. The calibration curve was linear in the range 5–640 ng/ml plasma. Data on solution stability, long- and short-term stability of tramadol in plasma samples, freeze-thaw-stability, as well as inter- and intra-day precision and accuracy were presented. The method has been applied to the pharmacokinetic profiling of tramadol. A headspace solid-phase microextraction (HS-SPME) combined with GC-MS was developed for the determination of tramadol in plasma by

Sha *et al.* [330]. The HS-SPME procedure was performed with a poly-dimethylsiloxane/divinylbenzene (PDMS/DVB) fiber, using 0.5 ml of plasma mixed with 0.5 ml of 0.1 M NaOH at 100 °C, with stirring for 30 min. The calibration curve showed linearity in the range 1–400 ng/ml. The detection limit for tramadol in plasma was 0.2 ng/ml. The proposed method was applied for the determination of tramadol in human plasma samples from 10 healthy volunteers after a single oral administration. Hadidi *et al.* [331] determined tramadol in hair using SPE and GC-MS. The LOD was 0.5 ng/mg, and the average recovery was 90.75%. The calibration curve was linear over the concentration range 0.5–5.0 ng/mg hair. The method was tested on 11 hair samples taken from patients using tramadol as prescribed by their physician along with other different drugs for treating chronic illnesses. Tramadol was detected in all hair samples at a concentration of 0.176–16.3 ng/mg.

1.6.4.2 Liquid chromatography

Sticht *et al.* [327] described the distribution of tramadol in a drug associated death case. The following tramadol concentrations were found postmortem (mg/kg): in peripheral blood 5.6, in heart blood 15.1, in heart muscle 14.9, in brain 14.7, in lung 23.2, in liver 20.0. Tramadol was determined by HPLC on octyl column, the metabolites were not analyzed. Nobilis *et al.* [332] developed a HPLC method with fluorescence detection for the pharmacokinetic study of two commercial tramadol preparations. The drug was extracted with *t*-butyl methyl ether in alkaline conditions and separated on an RP-18 column. The LOQ was 17 µg/l. In a later study [333] Nobilis *et al.* described simultaneous HPLC determination of tramadol and ODT in human plasma, using N^1,N^1-dimethylsulfanilamide as the IS. The analytical procedure involved solvent extraction, separation on a LiChrospher 60 RP-selectB column in a mobile phase consisting of ACN and 0.01 M phosphate buffer, pH 2.8 (3:7). The whole analysis lasted 19 min. Fluorescence detection λ(ex) 202 nm/λ (em) 296 nm for tramadol and its metabolite, λ (ex) 264 nm/λ (em) 344 nm for N^1,N^1-dimethylsulfanilamide was used. The method was applied to pharmacokinetic studies of tramadol in human volunteers.

A LC-MS-MS method for the simultaneous determination of tramadol *N*-oxide and several of its major metabolites in the plasma of rats and dogs was published by Juzwin *et al.* [334]. SPE was used for the isolation followed by reversed-phase liquid chromatography coupled with tandem mass spectrometric detection in the positive ionization mode. The assay was linear for all analytes over concentrations ranging from approximately 6 to 2000 ng/ml. The overall recovery of the analytes ranged from approximately 40 to 64% in rat plasma and 53 to 75% in dog plasma. This assay has proven to be sensitive, specific, and reproducible, and it has been implemented in preclinical pharmacokinetic studies.

Tramadol metabolites in urine were identified after a single oral administration of 100 mg of drug to three male volunteers [335]. Unchanged tramadol and a total of 23 metabolites, consisting of 11 Phase I metabolites (M1-11) and 12 conjugates (seven glucuronides, five sulphates), were detected, characterized, and tentatively identified in urine on the basis of MS and MS-MS data. Zhao *et al.* [336] determined tramadol and ODT in human plasma and amniotic fluid by LC-APCI-MS-MS after solvent extraction. Diphenhydramine was used as IS for quantitation. The distribution of tramadol and ODT in maternity and fetus were studied. The calibration curves for tramadol and ODT were linear in the range from 8.0 to 800.0 ng/ml and 1.0 to 400.0 ng/ml for plasma and amniotic fluid, respectively. The method was applied to the measurement of tramadol and ODT concentrations in maternal vein, umbilical vein, umbilical artery, and amniotic fluid.

Tramadol possesses two stereogenic centers and is normally used in therapy as the racemate of the *trans*-isomer, which is more active than the *cis*-isomer. Also, the (+)-*trans*-tramadol is about 10-fold more potent than the (−)-*trans*-tramadol [337]. Interindividual differences of the enantiomeric ratios of tramadol, NDT, and ODT in urine were studied by Elsing and Blaschke [338] using Chiralpak AD and Chiralcel OD columns. Ceccato *et al.* [339] developed a HPLC method for the determination of the enantiomers of *trans*-tramadol and its *O*-desmethylated metabolite in plasma, using automatic SPE and chiral liquid chromatography with UV (220 nm) and fluorometric detection. The influence of the SPE sorbent, elution conditions and type of chiral column on the detectability of the substances was studied. The LOD of 0.5 ng/ml for both enantiomers was reported. In the later study, Ceccato *et al.* [340] developed a LC-APCI-MS-MS method for the determination of enantiomers of tramadol and ODT in human plasma. The compounds were isolated with automated SPE on disposable extraction cartridges (C2) using an ASPEC system. The enantiomeric separation of tramadol and ODT was achieved on a Chiralpak AD column. The mobile phase was isohexane–ethanol–diethylamine (97:3:0.1, v/v). MS–MS analysis was carried out in the positive ion mode, using selected reaction monitoring. The MS–MS ion transitions monitored were $264 \rightarrow 58$ for tramadol, $250 \rightarrow 58$ for ODT, and $278 \rightarrow 58$ for ethyltramadol, used as IS. The recoveries were around 90% for both T and ODT. The method was also selective for other metabolites, NDT and *N,O*-desmethyltramadol (NODT). The method was used to investigate plasma concentration of enantiomers of T and ODT in a pharmacokinetic study.

1.6.4.3 Capillary electrophoresis

Soetebeer *et al.* [341] developed a CE method for the direct determination of tramadol in human urine without extraction or preconcentration. Laser-induced native fluorescence with a frequency doubled argon ion laser at an excitation wavelength of 257 nm was used as the detection method. This detection method was about 1,000-fold more sensitive compared to UV detection. In the next study, Soetebeer *et al.* [342] applied CE with laser-induced native fluorescence detection for the chiral separation of directly applied urine samples. Carboxymethyl-β-cyclodextrin and methyl-β-cyclodextrin were used for the separation of enantiomers. Furthermore, ODT glucuronide was determined from the urine samples and the ratio of the diastereomers was determined. After oral administration of 150 mg tramadol hydrochloride to a healthy volunteer, about 11.4% of the dose was excreted as 1*S*,2*S*-tramadol, 16.4% as 1*R*,2*R*-tramadol, and 23.7% as ODT-glucuronide. The amount of 1*S*,2*S*-ODT-glucuronide was more than 3-fold higher than 1*R*,2*R*-ODT-glucuronide. The data showed the occurrence of a stereoselective metabolism of tramadol. Rudaz *et al.* [343] applied CE-ESI-MS to the simultaneous enantiosep-aration of tramadol and its main phase-I metabolites. The partial filling technique was efficient at avoiding MS contamination by the chiral selector. The procedure was applied to the stereoselective analysis of tramadol and its main metabolites in plasma after a simple liquid–liquid extraction.

CE with UV detection was used by Lehtonen *et al.* [344] to separate tramadol and its five phase-I and three phase-II metabolites. After purification and 5-fold con-centration of the sample (SPE with Oasis MCX cartridges), the parent drug and its metabolites were detected. Diastereomeric separation of tramadol glucuronides in *in vitro* samples was achieved. Both separations showed that glucuronidation *in vitro* produces glucuronide diastereomers in different amounts. The authentic urine sample was also analyzed by a micellar method, but unambiguous identification of the glucuronide diastereomers was not achieved owing to many interferences.

1.6.5 Keto-opioids

Semisynthetic 6-keto-opioids (hydrocodone, hydromorphone, oxycodone, and ox-ymorphone) are widely used as analgesics and antitussive drugs. These compounds have achieved popularity as abused drugs in some European countries [345]. Sees *et al.* [346] examined the non-medical use of OxyContin (controlled release oxyco-done HCl) in the United States, based on data from the 1999–2001 Substance Abuse and Mental Health Services Administration National Household Survey on Drug Abuse. Reported non-medical OxyContin use in the United States increased from 0.1% in 1999 to 0.4% in 2001. Non-medical OxyContin users showed a pattern of more serious drug abuse: they used multiple drugs, used needles for drug injection, and had higher rates of abuse and dependence. Wolf *et al.* [347] reviewed 172 fatal cases involving the use of oxycodone in Palm Beach County. Eighteen cases were attributed to oxycodone toxicity, 117 to combined drug toxicity, 23 to trauma, 9 to natural causes, and 5 to another drug or drugs. The postmortem blood concentra-tions of oxycodone overlapped among the groups. This study confirmed that deaths in which oxycodone is a factor are most commonly cases of combined drug toxicity.

1.6.5.1 Immunoassays

In recent years, several manufacturers have developed immunoassays for the de-tection of oxycodone and oxymorphone in urine or blood. Microgenics Inc. devel-oped the DRI Oxycodone Assay for urine. This assay was evaluated by two groups of authors. Abadie *et al.* [348] used the DRI immunoassay to determine oxycodone results in a total of 148 urine samples from four different sample groups. GC-MS was subsequently used to confirm the presence or absence of oxycodone (or its primary metabolite, noroxycodone). The new DRI immunoassay was used to eval-uate 17,069 urine samples to estimate oxycodone misuse profiles during a 4-month period. The sensitivity and specificity of the new oxycodone immunoassay were 97.7 and 100%, respectively, at the cutoff concentration of 300 ng/ml. The assay linearity was 1,250 ng/ml, and the sensitivity was 10 ng/ml. Backer *et al.* [349] evaluated the performance of the DRI Oxycodone enzyme immunoassay for the detection of oxycodone and its primary metabolite, oxymorphone, in urine, by testing 1523 consecutive urine specimens collected from pain management patients. All specimens were tested with the DRI-Oxy assay at a cutoff of 100 ng/ml and then screened by GC-MS for opiates, including oxycodone and oxymorphone. 435 specimens yielded positive results by the DRI-Oxy assay. In 433 specimens, this result was confirmed by GC-MS. Of the 433 positive specimens, 189 contained other opiates including codeine, hydrocodone, hydromorphone, and morphine. These other opiates were also present in 54% of the oxycodone negative specimens. The DRI-Oxy assay demonstrated no cross-reactivity for codeine, > 75 mg/l, hydrocodone, > 75 mg/l, hydromorphone, > 12 mg/l, and morphine, > 163 mg/l. Spiehler *et al.* [350] investigated the accuracy of screening postmortem whole blood for oxycodone with Neogen Oxycodone/Oxymorphone ELISA (Neogen Corporation, Lexington, KY, USA) using the ratio of the oxycodone immunoassay response to the response for the specimen obtained with a general opiate-class immunoassay. Fifty-eight specimens, which were negative for opiates and 158 postmortem whole blood specimens positive for opiates including 66 specimens known to contain oxycodone were assayed. The sensitivity of the ELISA response ratio for the presence of oxycodone at a response ratio cutoff of 2.0 was $89.4 \pm 3.8\%$ and the specificity was $88.1 \pm 3.2\%$.

1.6.5.2 Gas chromatography

Most reports concerning 6-keto-opiates coped with the problem the of chromatographic differentiation of these drugs from morphine or codeine. These papers, in which GC procedures were used, were discussed earlier in the section devoted to morphine analysis [205,206,215]. Cone and Darwin in their review of opiate analysis [210] also discussed the application of GC-MS methods for keto-opiates. A gas chromatographic method for oxycodone was developed by Kapil *et al.* [351]. The drug and IS (hydrocodone) were extracted from plasma with toluene-isopropanol and quantified with a nitrogen detector. A LOQ of 1.8 μg/l was reported. Moore *et al.* [352] reported the detection of hydrocodone in meconium samples in two cases. The drug was isolated with methyl *t*-butyl ether in alkaline conditions, trimethylsilylated and analyzed with GC-MS (SIM). In one case, hydromorphone (hydrocodone metabolite) and codeine were also found. In the Czech Republic, an illicit hydrocodone preparation called "Brown" has been abused for 20 years. Beside hydrocodone, "Brown" contains codeine as the precursor and DHC as the by-product. Balikova and Maresova [240] described a case of a fatal overdose of "Brown" together with ethylmorphine and morphine. The drugs were determined in autopsy blood with an ion trap GC-MS after SPE. The following concentrations of unconjugated drugs were found: hydrocodone 15.9 mg/l, hydromorphone 11.88 mg/l, ethylmorphine 15.60 mg/l, morphine 12.15 mg/l, DHC 2.26 mg/l, codeine 0.5 mg/l, and norcodeine 0.14 mg/l. Jones *et al.* [353] determined codeine, morphine, hydrocodone, hydromorphone, 6-acetylmorphine, and oxycodone in the hair and oral fluids of addicts by GC-MS after derivatization with methoxyamine. The use of this derivatization reagent prevented the formation of multiple derivatives that may originate from the keto- or enol form of keto-opioids, as observed for silyl derivatives.

A procedure for the determination of oxycodone in meconium using direct ELISA microplate technology followed by EI-GC-MS was described by Le *et al.* [354]. Oxycodone may cause complications in neonates after maternal drug abuse. The cross-reactivity of oxycodone to the morphine antibody was only 5–6% in standard EMIT opiate assay. A positive screening value would require a high concentration of drug to be present, so a protocol for the detection of oxycodone in meconium using a direct ELISA microplate immunoassay followed by GC-MS was developed. Meatherall [355] presented the simultaneous analysis of seven opiates, codeine, morphine, 6-acetylmorphine, hydrocodone, hydromorphone, oxycodone, and oxymorphone, in blood samples by GC-MS. A combined pretreatment procedure, consisting of an acetonitrile precipitation, alkaline extraction, acidic back extraction, and alkaline re-extraction was applied. The residue was consecutively derivatized with methoxyamine and propionic anhydride in pyridine. After a purification step, the extracts were analyzed by full scan EI-GC-MS. The method was linear to at least 2000 ng/ml. For each opiate, the LOQ was 10 ng/ml, and the LOD was 2 ng/ml.

1.6.5.3 Liquid chromatography

Bouquillon *et al.* [38] developed a HPLC method for the simultaneous determination of hydromorphone and morphine in plasma. A coulometric detection was used. The limits of quantitation (2.5 ng/ml for hydromorphone and 1.2 ng/ml for morphine) were sufficient for pharmacokinetic studies. Wright *et al.* [356] synthesized hydromorphone-3-glucuronide (H3G) from hydromorphone, using rat liver microsomes. The crude product was purified by semi-preparative HPLC with UV detection. H3G evoked similar behavioral effects in rats as morphine or M3G or normorphine-3-glucuronide. Chen *et al.* [357] published LC-ESI-MS-MS procedures for the determination of hydrocodone and hydromorphone in plasma. The drugs and deuterated analogues were extracted with solvent and separated from glucuronides using a 50 mm × 2 mm silica column and a mobile phase consisting of ACN–water–formic acid (80:20:1). The LOQ was 0.1 µg/l. A LC-MS procedure for the determination of oxycodone, oxymorphone, and noroxycodone in Ringer solution, rat plasma and rat brain tissue by liquid chromatography/mass spectrometry was described by Bostrom *et al.* [358]. Deuterated analogs of the substances were used as ISs. Samples in Ringer solution were analyzed by direct injection, for plasma an ACN precipitation was applied, for brain tissue a C_{18} SPE was used. The LOQ for rat plasma was 0.5 ng/ml and the methods were linear in the range 0.5–250 ng/ml for all substances. The LOQ of oxycodone was 20 ng/g brain, and for oxymorphone and noroxycodone 4 ng/g brain. Ammonium acetate (5 mM) in 45% ACN was used as the mobile phase, and a SB-CN column was used for separation. The total run time of all methods was 9 min. Edinboro *et al.* [359] developed a method for the direct analysis of 10 opiate compounds in urine using ESI-LC-MS-MS. The opiates included were M3G, M6G, morphine, norcodeine, codeine, 6MAM, oxycodone, oxymorphone, hydromorphone, and hydrocodone. Urine samples were directly injected into LC-MS-MS after centrifugation. The separation and detection of all compounds was accomplished within 6 min. Eighty-nine urine samples previously analyzed by GC-MS were re-analyzed by the LC-MS-MS method. The procedure showed a high degree of agreement with GC-MS, both for positive and negative samples. The LC-MS-MS method identified 19 samples with additional opiates in the positive samples. Edwards and Smith [360] used LC-ESI-MS-MS for the simultaneous determination of morphine, M3G, oxycodone, and noroxycodone, in 50 µl samples of rat serum. Deuterated analogues of each compound were used as ISs. The samples were precipitated with ACN, which was removed from the supernatant by centrifugal evaporation before analysis. Cheremina *et al.* [361] determined oxycodone and its metabolite noroxycodone in human plasma with HPLC-UV, using solvent extraction and detection at 205 nm. Codeine was used as the IS, the quantitation range was 2–100 ng/ml.

1.6.5.4 Capillary electrophoresis

Wey and Thormann [362] determined oxycodone and its metabolites oxymorphone and noroxymorphone in urine with CE-MS (IT) and CE-UV. The existence of glucuronidated 2nd phase metabolites was postulated. The group of Thormann [363] identified phase-I and phase-II metabolites of oxycodone in human urine, using CE-ESI-MSn (IT). Several phase-I metabolites (such as oxymorphone, noroxycodone, noroxymorphone, 6-oxycodol, nor-6-oxycodol, oxycodone-*N*-oxide, and 6-oxycodol-*N*-oxide) and phase-II conjugates with glucuronic acid of several of these compounds could be detected in alkaline solid-phase extracts of urine collected from a patient undergoing therapy with a daily ingestion of 240–320 mg of oxycodone hydrochloride.

1.6.6 Fentanyl and related drugs (sufentanil, alfentanil, remifentanil)

Fentanyl and its structural analogs are very potent, specific µ-receptor agonists of synthetic origin. Beside therapeutical applications as an analgesic, fentanyl appeared in the 1970s in the illicit drug market. The methyl- or fluoroderivatives of fentanyl, sold as "super heroin" or "China White" turned out to be particularly dangerous and several reports of drug-associated death cases were published [364]. In the late 1990s, fentanyl-related death cases were also reported in Europe. Kronstrand *et al.* [365] registered nine fentanyl-associated fatalities that occurred among drug-addicts. The street samples involved in these cases contained fentanyl as an additive in low-concentration amphetamine powders with caffeine, phenazone, and sugar as cutting agents. Fentanyl concentrations ranged from 0.5 to 17 ng/g blood, and from 5 to 160 ng/ml urine. Other drugs found were amphetamine, ethanol, and benzodiazepines (five cases). According to Kintz *et al.* [366] fentanyl, alfentanil, and sufentanil are the most commonly abused drugs among anesthesiologists. The authors described five such cases; hair analysis appeared as the method of choice.

1.6.6.1 Immunoassays

ELISA and RIA methods for the detection of fentanyl, alfentanil, sufentanil, and carfentanil in equine urine were published by Tobin *et al.* [367]. The assay was applied to animal doping control. The possible of use of this assay for human drug abuse monitoring was suggested. Käferstein and Sticht [368] evaluated three fentanyl enzyme immunoassays (from COZART, STC, and DIAGNOSTIX) for use with serum samples from forensic and clinical cases. The cutoff for all assays could be set at 0.5 ng/ml. The presence of the typical drugs of abuse, e.g. heroin, methadone, cocaine, cannabinoids, and amphetamines including the derivatives of methylenedioxyamphetamine, did not generate false-positive results. No cross-reactivity was also observed at toxic levels of benzodiazepines and paracetamol and therapeutic levels of barbiturates, phenothiazines, antidepressants, and analgesics. Nowadays, fentanyl assays are mostly based on ELISA technique. Neogen Corp. (www.neogen.com) developed ELISAs for fentanyl and sufentanil in plasma. Bio-Quant Inc. (www.bio-quant.com) manufactures ELISA kits for the detection of fentanyl in blood, serum, plasma, and urine. Other companies delivering ELISA kits for fentanyl may be found at www.biocompare.com.

1.6.6.2 Gas chromatography

Watts and Caplan [369] used dual column GC with nitrogen sensitive and mass spectrometric detectors for the determination of fentanyl in whole blood. Two capillary columns of different polarities (5 and 50% phenyl methyl silicone) were used. Several related substances (sufentanil, carfentanil, lofentanil, and alfentanil) were also examined. The LOD for the nitrogen detector was found at 0.1 μg/l, and for MS detection at 0.05 μg/l. Ohta *et al.* [370] discriminated between fentanyl and its 24 analogues using GC, GC-MS, and condensed-phase IR. Esposito and Winek [371] used GC-MS-EI for the identification of 3-methylfentanyl in street samples. Szeitz *et al.* [372] developed a GC-MS assay of fentanyl, suitable for pharmacokinetic studies of the transdermally administered drug in a postoperative swine. Sufentanil was used as an IS. Quantitation in the SIM mode was possible down to 0.05 μg/l. A similar method was described by Fryirs *et al.* [373], who determined fentanyl with GC-MS (SIM) in plasma. A LOD of 0.02 μg/l was observed. Fentanyl and sufentanil were determined in hair specimens of tumor patients receiving these drugs percutaneously or intravenously. The assay was performed with GC-PCI-MS-MS [374]. The introduction of trandermal patches containing fentanyl (Duragesic) was associated with numerous reports of acute intoxication. Postmortem distribution of fentanyl after Duragesic administration was studied by Anderson and Muto [375]. Fentanyl was extracted with alkaline butyl chloride and determined by GC-MS. The distribution data of fentanyl in heart and femoral blood, urine, vitreous humor, bile, and organs were given. Kuhlman *et al.* [376] reported that fentanyl-associated death cases have increased from 3 in 2000 to 12 in 2002 in southwestern Virginia. Of the 23 cases, 19 were attributed to fentanyl misuse or abuse of fentanyl transdermal patches. Fentanyl was identified using a SPE basic drug screen in blood and/or urine followed by full scan gas EI-GC-MS. Quantitation was performed using GC-MS-SIM. The method was linear from 1 to 50 ng/ml with a LOD of 1ng/ml. Fentanyl blood concentrations ranged from 2 to 48 ng/ml with a mean concentration of 18 ng/ml. Two cases of a fatal intravenous injection of the content from fentanyl patches were reported in Norway by Lilleng *et al.* [377]. Confirmatory analysis of fentanyl and morphine was performed by GC-MS. In the first case, the toxicological analysis revealed fentanyl (2.7 ng/ml), morphine (31.4 ng/ml), and ethanol (1.1 g/l) in post-mortem blood. In the second case, the analysis revealed fentanyl (13.8 ng/ml), 7-aminoclonazepam (57.1 ng/ml), and sertralin (91.9 ng/ml) in postmortem blood. Police investigations revealed that both the deceased had bought the patches from the same source. Tharp *et al.* [378] presented four fatal cases of fentanyl abuse, in which the drug was extracted from transdermal patches and injected intravenously. In all of these cases, needles, syringes, and fentanyl patches were recovered at the scene. All reported deaths were attributed to fentanyl intoxication, with blood concentrations ranging from 5 to 27 ng/ml. Excretion of fentanyl and its major metabolite norfent-anyl in chronic pain patients treated with the Duragesic transdermal patches was studied by Poklis and Backer [379]. These patches are available in four sizes releasing 25, 50, 75, and 100 µg/h fentanyl, respectively. Five hundred and forty six random urine specimens were collected from chronic pain patients wearing different trans-dermal patches. Urine specimens were collected from hours after application to several days later after continuous fentanyl release. Fentanyl and norfentanyl were isolated by SPE and quantified by GC-MS-SIM. The LODs and LOQs for both drugs were 3 ng/ml. The results obtained showed a wide interindividual variability of con-centrations of fentanyl and its metabolite. The values obtained during therapeutic use far exceeded concentrations previously reported in fatal poisoning. Urine norfentanyl concentrations were 3–4 times higher than those of fentanyl.

Paradis *et al.* [380] isolated sufentanil from plasma with SPME. Sufentanil and fentanyl (IS) were extracted from plasma with a 65-micron polydimethylsiloxane-divinylbenzene (PDMS-DVB) fiber for 30 min using salting out agents under basic conditions. The drugs were determined by GC-MS. The calibration curve was linear over a concentration range of 6–50 ng/ml. The limit of quantification was 6 ng/ml for a plasma volume of 1 ml. The SPME method was regarded as useful for a rapid extraction of sufentanil from human plasma. A GC-MS procedure was developed for the determination of fentanyl, sufentanil, and alfentanil and their major nor-metabolites in urine of potentially exposed opioid production workers [381]. The drugs were isolated with SPE and derivatized with PFPA. LODs of 2.5 pg fentanyl/ml, 2.5 pg sufentanil/ml, and 7.5 pg alfentanil/ml urine were achieved. Urine samples were found to be stable for at least 2 months when stored at –30 °C. The analytical procedure has been successfully applied to the biological monitoring survey of fentanyl exposed production workers. The same research group [382] published a GC-MS procedure for the determination of tentanyl, sufentanil, and alfentanil in air and surface contamination wipes. The analytical and sampling procedures have been applied in an explorative field study.

1.6.6.3 Liquid chromatography

Shou *et al.* [383] determined fentanyl in plasma, using automated 96-well SPE, straight-phase chromatography, and ESI-MS-MS. The LOQ was 50 ng/ml plasma, based on 0.25 ml sample volume. Simultaneous determination of fentanyl and norfentanyl with LC-MS was described by two groups. Koch *et al.* [384] extracted drugs from plasma with toluene. Chromatography was performed using a Zirchrom-PBD (50 mm × 2.1 mm) column with a mobile phase of ACN-ammonium acetate (10 mM), citrate (0.1 mM, pH 4.4) (45:55, v/v) with a flow rate of 0.3 ml/min. MS detection was performed using ESI in the positive mode. The LOQ for fentanyl was 25 pg/ml and norfentanyl was 50 pg/ml. Huynh *et al.* [385] determined fentanyl and norfentanyl in urine after solvent extraction. The drugs were separated on a reversed-phase YMC Pro C18-column followed by ESI-MS-MS detection in positive ion mode. The methods have been applied for the determination of fentanyl in plasma and fentanyl/norfentanyl in urine samples. A LC-ESI-MS-MS method for therapeutic drug monitoring of sufentanil was published by Martens-Lobenhoffer [386]. The drug was extracted from serum with toluene–isopropanol (10:1) and separated on an ODS column. The LOQ of 10 ng/l was achieved. Palleschi *et al.* [387] described a LC-ESI-MS-MS method for the quantification of sufentanil in human plasma. Fentanyl was used as the IS. SPE on C18 cartridges was used for drug isolation. Chromatographic separation of the analytes was obtained using an RP-18 column. Three transitions were monitored during MRM analysis. The LOQ for sufentanil in human plasma samples was 0.3 ng ml. The method was used for sufentanil determination in maternal plasma samples collected after epidural administration of a single sufentanil dose to women in labor, and at birth in arterial and venous umbilical cord plasma samples from the newborns. Schmidt *et al.* [388] described a LC-ESI-MS-MS procedure for the determination of sufentanil in the plasma of parturients and umbilical plasma of their neonates following patient-controlled epidural analgesia. The drug and its IS (d_5-sufentanil) were extracted by SPE and separated on straight phase HILIC column in ACN–formic acid–ammonium acetate gradient. LOQ of 0.25 pg/ml was achieved.

Alfentanil is an intravenous narcotic analgesic with a short duration of action. The concentrations of this drug were measured in plasma and tissues of experimental animals by GC-MS in a pharmacokinetic study [389].

Another member of the fentanyl group, remifentanil, is an analgesic which has considerable abuse potential in racing horses. Lehner *et al.* [390] studied the metabolism of remifentanil after intravenous administration of 5 mg of this drug to a horse. A major metabolite of remifentanil was identified.

Since January 2005, the list of prohibited substances established by the World Anti-Doping Agency prohibits the use of the opioid agent fentanyl as well as its related drugs in professional and amateur sports. A LC-ESI-MS-MS screening and confirmation method was developed that enables the identification of fentanyl, alfentanil, remifentanil, and sufentanil as well as their *N*-dealkylated or de-esterified metabolites utilizing SPE of a 2 ml urine. The LODs for all drugs were at 0.5 ng/ml. The mass spectrometric behavior of fentanyl after electrospray ionization and collision-induced dissociation was studied by the synthesis and analysis of structurally related compounds, and dissociation pathways were proposed allowing the characterization of target analytes and corresponding metabolites [391].

1.6.7 Ketobemidone

Ketobemidone is a synthetic opioid agonist and narcotic analgesic, which is frequently abused, particularly in Scandinavian countries. Breindahl *et al.* [392] developed a LC-ESI-MS method for the determination of ketobemidone and its demethylated metabolite in urine. Mixed-bed SPE cartridges were used for the isolation with a recovery over 90%. Protonated quasi-molecular ions for both substances as well as three fragments for ketobemidone were monitored. The LOD was 25 µg/l. Sundstrom *et al.* [393] applied ESI-MS-MS for the determination of ketobemidone, its five phase-I metabolites as well as glucuronides of ketobemidone and norketobemidone in human urine. The same group used ESI-qTOF-MS besides LC-MS-MS for the determination of glucuronides of ketobemidone, nor-, and hydroxymethoxyketobemidone in urine [394]. The accuracy of the mass measurement was better than 2 ppm. Lampinen *et al.* [395] presented a LC-MS-MS method for the determination of ketobemidone in human plasma. Solvent extraction and SPE of plasma samples were compared. Both methods showed good precision and accuracy. Ketobemidone could be quantified at 0.43 nM, with a relative standard deviation of 17.5% ($n = 19$) using solvent extraction and 18.6% ($n = 10$) using SPE. This level was an order of magnitude lower than earlier reported quantification limits. Quantitative data from plasma samples analyzed by LC-MS-MS were in good agreement with those obtained by gas chromatography with chemical ionization mass spectrometry (GC-CI/MS). This indicated that LC-MS-MS is a good alternative method to GC-MS as it was more sensitive and the time-consuming derivatization could be avoided.

1.6.8 Butorphanol, dextrometorphan

Andraus and Siquera [396] determined butorphanol in the urine of racehorses with ELISA immunoassay kits followed by GC-MS. After intramuscular application of 8 mg Torbugesic to the horse, the detection window in urine was up to 104 h with ELISA, and up to 24 h with GC-MS. A LC-MS-MS assay for butorphanol in human plasma was described by Boulton *et al.* [397]. The cyclopropyl analogue of butorphanol, was employed as an IS. The drugs were isolated from plasma by solvent extraction and separated on a Partisil RP8 (5 µm) column with a mobile phase consisting of methanol–water–formic acid (90:10:0.1) at a flow-rate of 0.3 ml/min. The standard curve was linear from 13.7 to 1374 pg/ml. The LOQ was 13.7 pg/ml. Butorphanol in plasma was stable over 3 freeze/thaw cycles and at room temperature for 1 day. Butorphanol was measured in plasma in two healthy subjects for 24 h following a 1 mg intranasal dose.

Wu *et al.* [398] described a GC method for the determination of dextrometorphan (DEX) and its main metabolite dextrorphan (DOR) in human urine. The drugs were subjected to solvent extraction and detected with GC/FID on an HP-1 (17 m × 0.22 mm) column. The method was used for the phenotyping of a Chinese population. A SPME procedure, followed by GC-MS, was developed for the determination of DEX and DOR in human plasma [399]. Three different polymers were synthesized as coated fibers using sol-gel methodologies. DEX was converted to its acetyl-derivative prior to extraction and subsequent determination. The porosities of the coated fibers were examined by SEM. The effects of different parameters, such as the fiber coating type, extraction mode, agitation method, sample volume, extraction time, and desorption condition, were investigated and optimized. The LODs were 10 and 15 pg/ml for DEX and DOR, respectively. Linear ranges were obtained from 30 pg/ml to 2 µg/ml for DEX and from 50 pg/ml to 2 µg/ml for DOR.

Several LC-MS methods were published for the determination of dextrometor-phan. Lutz *et al.* [400] developed a LC-MS-MS method for the determination of DEX and DOR in human saliva and urine, with a LOQ of 0.27 ng/ml. The method was applied to the study of DEX metabolism. Additionally, four additional metabolites of DEX as well as two glucuronides were measured in urine. The study, carried out on 170 subjects, demonstrated clear differences in the metabolic profiles of "extensive" and "poor metabolizers". Arellano *et al.* [401] applied LC-MS to the determination of DEX and its three metabolites in tissue culture medium in the study of drug metabolism in rat intestine. The LOQs were in the range of 3–6 pg per 10 µl of injection volume. A simultaneous determination of DEX and DOR in urine, based on SPE after acid hydrolysis and LC-MS-MS was published by Constanzer *et al.* [402]. The drugs and the IS (levallorphan) were detected in the positive ion-ization mode using MRM. The following transitions were monitored: m/z 272→215, 258→201, and 284→201 for DEX, DOR, and IS, respectively. The analytes were separated on a RP column (50 mm × 2.0 mm) using two mobile phases consisting of MeOH and water containing 0.1% TFA (pH 3.0). The LOQ was 250 ng/ml for both compounds from 1 ml of urine. The need for the careful assessment of the selectivity of the LC-MS-MS assay in the presence of metabolites and the assessment of the matrix effect were emphasized. A high-throughput LC-MS-MS method, which combined on-line sample extraction with turbulent flow chromatography with a monolithic column separation, has been developed for direct injection analysis of DEX and DOR in human plasma [403]. A strategy of assessing and reducing the matrix suppression effect on the on-line extraction LC/MS/MS has been discussed. The total run time with a baseline separation of the two analytes was < 1.5 min. Another fast, on-line SPE coupled with LC-MS-MS for the determination of DEX, DOR, and guaifenesin in human plasma was described by Kuhlenbeck *et al.* [404]. A Prospekt-2 system coupled to LC-MS was used. Using stable-isotope-labeled ISs for each analyte, the Prospekt-2 on-line methodology was evaluated for sensitivity, matrix suppression, and other parameters. The LOQ for the on-line SPE procedure for DEX, DOR, and guaifenesin was 0.05, 0.05, and 5.0 ng/ml, respectively, using a 0.1 ml sample volume. The linear range for DEX and DOR was 0.05–50 ng/ml. Accuracy ranged from 90 to 112% for all three analytes, while the precision, as measured by the %RSD, ranged from 1.5 to 16.0%

1.7 CONCLUDING SUMMARY

Immunoassays are the most important techniques used for the preliminary testing of opiate agonists. The trend towards the use of immunoassays for alternative samples, available in a non-invasive way, e.g. saliva or sweat, is observed. On-site tests, used by law enforcement officers, are gaining more and more popularity.

TLC is still in use for the preliminary detection of opiates in plant material, in street drugs, and in urine. The method has the advantage of simplicity, speed, and low cost and is therefore preferred in modestly equipped laboratories. Positive results, however, always need a confirmative analysis with mass spectrometric detection.

SPE is gradually replacing solvent extraction procedures for the isolation of opiate agonists and their metabolites from biological samples. The advantages of this work-up technique include: a broader polarity spectrum of isolated substances and rather pure extracts. So far, SPE is mainly used in a column format; the use of 96-well plate format is growing, particularly in on-line combination with LC-MS. SPE in disk format seems to have been abandoned. The tendency to automation of SPE is observed with the main application being to large sets of samples, e.g. in workplace or military drug testing.

For identification purposes of unknown opiates of low or middle polarity, GC coupled with a full-scan electron impact mass spectrometry is a most important tool. This method has been usually used for the confirmation of the results of presumptive immunochemical tests.

Unequivocal identification and quantitative analysis of defined opiates may be performed by gas or liquid chromatography, coupled with a mass spectrometer. The past decade has brought a breakthrough in the development of the liquid chromatography-atmospheric pressure ionization mass spectrometry. This technique, utilizing electrospray or atmospheric pressure chemical ionization sources and single quadrupole, triple quadrupole, or ion trap mass analyzer, has shown distinct advantages over GC-MS in regard to the spectrum of detectable drugs and simplicity of sample preparation. LC-MS-MS, used in the MRM mode, has became the main method for substance detection and quantitation in opiate analysis. This trend will be certainly continued in the future.

As well as mass spectrometry, other detection modes used with GC (nitrogen-selective or electron capture detection) or with liquid chromatography (diode array detection, electrochemical or fluorimetric detection) are still being successfully used for dedicated purposes in opiate analysis. The advantages of these techniques being lower costs and sometimes very high sensitivity. Due to their lower selectivity these methods are particularly valuable for analysis of less complicated matrices, including illicit drug specimens or pharmaceutical preparations.

CE, in combination with UV (DAD) or MS-detection, combines the most important features of gas and liquid chromatography; it offers separation efficiency comparable with capillary GC, and is applicable to polar and thermally unstable compounds, which may be analyzed with HPLC. However, the technique is successfully used only in some centers and obviously requires much skill and experience. At the moment, it cannot compete with LC–MS.

1.8 ABBREVIATIONS

| 6-AM | 6-Acetylmorphine |
| ACN | Acetonitrile |

APCI	Atmospheric pressure chemical ionization
BE	Benzoylecgonine
BSTFA	bis-(Trimethylsilyltrifluoroacetamide)
CE	Capillary electrophoresis
CE-MS	Capillary electrophoresis-mass spectrometry
CI	Chemical ionization
CID	Collision-induced dissociation
CSF	Cerebrospinal fluid
CZE	Capillary zone electrophoresis
DAD	Diode array detector
DAM	Diacetylmorphine
DEA	Diethylamine
DHC	Dihydrocodeine
DMOA	Dimethyloctylamine
EDDP	2-Ethylidene-1,5-dimethyl-3,3-diphenylpyrrolidine
EMDP	2-Ethyl-5-methyl-3,3-diphenylpyrroline
EMIT	Enzyme multiplied immunoassay technique
EI	Electron impact ionization
EIA	Enzyme immunoassay
ESI	Electrospray ionization
FAB	Fast atom bombardment
FID	Flame ionization detector
FPIA	Fluorescence polarization immunoassay
FITC	Fluorescein isothiocyanate
G6PDH	Glucose-6-phosphodehydrogenase
GC	Gas chromatography
HFBA	Hexafluorobutyric acid
HPLC	High-performance liquid chromatography
IDS	International Diagnostic Systems Corporation
ITD	Ion trap detection
KIMS	Kinetic interaction of microparticles in solution
LC–MS	Liquid chromatography-mass spectrometry
LOD	Limit of detection
LOQ	Limit of quantitation
LSD	Lysergic acid diethylamide
M3G	Morphine-3-glucuronide
M6G	Morphine-6-glucuronide
MBTFA	*N*-Methyl-bis-trifluoroacetamide
MDA	3,4-Methylenedioxyamphetamine
MDBD	*N*-Methyl-2,3-methylenedioxyphenylbutan-2-amine
MDEA	3,4-Methylenedioxyethylamphetamine
MDMA	3,4-Methylenedioxymethamphetamine
MEKC	Micellar electrokinetic capillary chromatography
MeOH	Methanol
MSTFA	*N*-Methyl-*N*-trimethylsilyltrifluoroacetamide

References pp. 62–72

MRM	Multiple reaction monitoring
NCI	Negative chemical ionization
NPD	Nitrogen-phosphorus detection
ODS	Octadecylsilica
PBI	Particle beam ionization
PCI	Positive chemical ionization
PFPA	Pentafluoropropionyl anhydride
PTFE	Polytetrafluoroethylene
POD	Peroxidase
RIA	Radioimmunoassay
RP	Reversed phase
SFE	Supercritical fluid extraction
SIM	Selected ion monitoring
SPE	Solid-phase extraction
SPME	Solid-phase microextraction
TEA	Triethyleneamine
TFA	Trifluoroacetic acid
TLC	Thin-layer chromatography
TMCS	Trimethylchlorosilane
TMS	Trimethylsilyl
TSP	Thermospray ionization
QTOF	Quadrupole-time-of flight

1.9 REFERENCES

1   B.N. Dhawan, F. Cesselin, R. Raghubir, T. Reisine, P.B. Bradley, P.S. Porthogese and M. Hamon, Pharmacol. Rev., 48 (1996) 567.
2   G.J. Kilpatrick and T.W. Smith, Med. Res. Reviews, 25 (2005) 521.
3   K.A. Sporer, Brit. Med. J., 326 (2003) 442.
4   A.J. Jenkins and B.A. Goldberger (Eds.), On-site drug testing. Humana Press, Totowa, New Jersey, 2002.
5   D.J. Crouch, R.K. Hersch, R.F. Cook, J.F. Frank and J.M. Walsh, J. Anal. Toxicol., 26 (2002) 493.
6   M. Gronholm and P. Lillsunde, Forensic Sci. Int., 121 (2001) 37.
7   R. Wennig, M. Moeller, J.M. Haguenoer, A. Marocchi, F. Zoppi, B.L. Smith, R. de la Torre, C.A. Carstensen, A. Goerlach-Graw, J. Schaeffler and R. Leibberger, J. Anal. Toxicol., 22 (1998) 148.
8   B. Buchan, J.M. Walsh and P.E. Leaverton, J. Forensic Sci., 43 (1998) 395.
9   C. Barrett, C. Good and C. Moore, Forensic Sci. Int., 122 (2001) 163.
10 S. George and S. Parmar, J. Anal. Toxicol., 26 (2002) 233.
11 J.M. Holler, T.Z. Bosy, K.L. Klette, R. Wiegand, J. Jemionek and A. Jacobs, J. Anal. Toxicol., 28 (2004) 489.
12 E.J. Cone, S. Dickerson, B.D. Paul and J.M. Mitchell, J. Anal. Toxicol., 16 (1992) 72.
13 D.A. Armbruster, R.H. Scharzhoff, E.C. Hubster and M.K. Liserio, Clin. Chem., 39 (1993) 2137.
14 M.L. Smith, E.T. Shimomura, J. Summers, B.D. Paul, D. Nichols, R. Shippee, A.J. Jenkins, W.D. Darwin and E.J. Cone, J. Anal. Toxicol., 24 (2000) 522.
15 E.J. Cone, L. Presley, M. Lehrer, W. Seiter, M. Smith, K.W. Kardos, D. Fritch, S. Salamone and R.S. Niedbala, J. Anal. Toxicol., 26 (2002) 541.

16   M.L. Cheever, G.A. Armendariz and D.E. Moody, J. Anal. Toxicol., 23 (1999) 500.
17   K. Moore, C. Werner, R.M. Zannelli, B. Levine and M.L. Smith, Forensic Sci. Int., 106 (1999) 93.
18   P. Kemp, G. Sneed, T. Kupiec and V. Spiehler, J. Anal. Toxicol., 26 (2002) 504.
19   S. Kerrigan and W.H. Phillips, Clin. Chem., 47 (2001) 540.
20   H. Schütz, F. Erdmann, M. Risse and G. Weiler, Arzneimittelforschung, 52 (2002) 716.
21   J.T. Cody and S. Valtier, J. Anal. Toxicol., 25 (2001) 466.
22   J.T. Cody, S. Valtier and J. Kuhlman, J. Anal. Toxicol., 25 (2001) 572.
23   K.E. Rasmussen, S. Pedersen-Bjergaard, M. Krogh, H.G. Ugl and T. Grønhaug, J. Chromatogr. A, 873 (2000) 3.
24   H.G. Ugland, M. Krogh and K.E. Rasmussen, J. Chromatogr. A, 749 (2000) 85.
25   T.S. Ho, S. Pedersen-Bjergaard and K.E. Rasmussen, J. Chromatogr. A, 963 (2002) 3.
26   T.S. Ho, S. Pedersen-Bjergaard and K.E Rasmussen, Analyst, 127 (2002) 608.
27   V. Marko, L. Soltes and K. Radova, J. Chromatogr. Sci., 28 (1990) 403.
28   L. Soltes, Biomed. Chromatogr., 6 (1992) 43.
29   A. Gelencser, G. Kiss, Z. Krivacsy, Z. Varga-Puchony and J. Hlavay, J. Chromatogr. A, 693 (1995) 217.
30   J. Scheurer and C.M. Moore, J. Anal. Toxicol., 16 (1992) 264.
31   E.M. Thurman; M.S. Mills, Solid phase extraction: principles and practice, John Wiley & Sons, New York, 1998.
32   D.L. King, M.J. Gabor, P.A. Martel and C.M. O'Donnell, Clin. Chem., 35 (1989) 163.
33   D.D. Blevins and D.O. Hall, LC-GC Int. (1998), Suppl., September 17, pp. 16–21.
34   F. Degel, Clin. Biochem., 29 (1996) 529.
35   R.A. De Zeeuw, J. Wijsbeek and J.P. Franke, J. Anal. Toxicol., 24 (2000) 97.
36   J.O. Svensson, A. Rane, J. Säwe and F. Sjöqvist, J. Chromatogr., 230 (1982) 427.
37   J.O. Svensson, J. Chromatogr., 375 (1986) 174.
38   A.I. Bouquillon, D. Freeman and D.E. Moulin, J. Chromatogr., 577 (1992) 354.
39   K. Ensing, J.P. Franke, A. Temmink, X.H. Chen and R.A. de Zeeuw, J. Forensic Sci., 37 (1992) 460.
40   M.E. Soares, V. Seabra, M. de Lourdes and A.M. Bastos, J. Liq. Chromatogr., 15 (1992) 1533.
41   I. Papadoyannis, A. Zotou, V. Samanidou, G. Theodoridis and F. Zougrou, J. Liq. Chromatogr., 16 (1993) 3017.
42   G. Theodoridis, I. Papadoyannis, H. Tsoukali-Papadopoulou and G. Vasilikiotis, J. Liq. Chromatogr., 18 (1995) 1973.
43   A. Geier, D. Bergemann and L. von Meyer, Int. J. Legal Med., 109 (1996) 80.
44   W. Huang, W. Andollo and W.L. Hearn, J. Anal. Toxicol., 16 (1992) 307.
45   X.H. Chen, A.L.C. Hommerson, P.G.M. Zweipfenning, J.P. Franke, C.W. Harmen-Boverhof, K. Ensing and R.A. de Zeeuw, J. Forensic Sci., 38 (1993) 668.
46   P.G.M. Zweipfenning, A.H. Wilderink, P. Horsthuis, J.P. Franke and R.A. de Zeeuw, J. Chromatogr. A, 674 (1994) 87.
47   M.J. Bogusz, R.D. Maier, K.H. Schiwy-Bochat and U. Kohls, J. Chromatogr. B, 683 (1996) 177.
48   W. Weinmann, M. Renz, C. Pelz, P. Brauchle, S. Vogt and S. Pollak, Blutalkohol, 35 (1998) 195.
49   H. Lord and J. Pawliszyn, J. Chromatogr. A, 885 (2000) 153.
50   F. Sporkert and F. Pragst, J. Chromatogr. B, 746 (2000) 255.
51   A.M. Bermejo, R. Seara, A.C. Dos Santos Lucas, M.J. Tabernero, P. Fernandez and R. Marsili, J. Anal. Toxicol., 24 (2000) 66.
52   A.C. Dos Santos Lucas, A. Bermejo, P. Fernandez and M.J. Tabernero, J. Anal. Toxicol., 24 (2000) 93.
53   S.W. Myung, S. Kim, J.H. Park, M. Kim, J.C. Lee and T.J. Kim, Analyst, 124 (1999) 1283.
54   U. Staerk and W.R. Kulpmann, J. Chromatogr. B, 745 (2000) 399.
55   N. Fucci, N. De Giovanni and M. Chiarotti, Forensic Sci. Int., 134 (2003) 40.
56   G. Guiochon, Int. Lab., 29 (1999) 13C.
57   V. Cirimele, P. Kintz, R. Majdalani and P. Mangin, J. Chromatogr. B, 673 (1995) 173.
58   W.E. Brewer, R.C. Galipo, K.W. Sellers and S.L. Morgan, Anal. Chem., 73 (2001) 2371.

59   D.L. Allen, K.S. Scott and J.S. Oliver, J. Anal. Toxicol., 23 (1999) 16.
60   K.S. Scott and J.S. Oliver, Med. Sci. Law., 39 (1999) 77.
61   C. Staub, Forensic Sci. Int., 84 (1997) 295.
62   C. Radcliffe, K. Maguire and B. Lockwood, J. Biochem. Biophys. Methods, 43 (2000) 261.
63   Z. Fater, Z. Samu, M. Szatmary and S. Nyiredy, Acta Pharm. Hung., 67 (1997) 211.
64   D.S. Popa, R. Oprean, E. Curea and N. Preda, J. Pharm. Biomed. Anal., 18 (1998) 645.
65   B.D. Paul, C. Dreka, E.S. Knight and M.L. Smith, Planta Med., 62 (1996) 544.
66   H. Huizer, Analytical Studies on Illicit Heroin, Ph.D.Thesis, University of Leiden, The Nether-
      lands, 1988.
67   A.M. Al-Amri, R.M. Smith, B.M. El-Haj and M.H. Juma'a, Forensic Sci. Int., 140 (2004) 175.
68   A.M. Al-Amri, R.M. Smith and B.M. El-Haj, Anal. Bioanal. Chem, 382 (2005) 830.
69   J.L. Janicot, M. Caude and R. Rosset, J. Chromatogr., 437 (1988) 351.
70   L. Krenn, S. Glantschnig and U. Sorgner, Chromatographia, 47 (1998) 21.
71   V.C. Trenerry, R.J. Wells and J. Robertson, J. Chromatogr. A, 718 (1995) 217.
72   I.S. Lurie, S. Panicker, P.A. Hays, A.D. Garcia and B.L. Geer, J. Chromatogr. A, 984 (2003) 109.
73   M.J. Bogusz, Opioid agonists, in: M.J. Bogusz (Ed.) R.M. Smith (Series Ed.), Forensic Science,
      Handbook of Analytical Separations, Vol. 2. Elsevier Sciences, Amsterdam, 2000, pp. 3–65.
74   K. Bjerver, J. Johnsson and J. Schuberth, J. Pharm. Pharmacol., 34 (1982) 798.
75   G. Fritschi and W.R. Prescott Jr, Forensic Sci. Int., 27 (1985) 111.
76   R.E. Struempler, J. Anal. Toxicol., 11 (1987) 97.
77   A.B. Zebelman, B.L. Troyer, G.L. Randall and J.D. Batjer, J. Anal. Toxicol., 11 (1987) 131.
78   H.N. ElSohly, M.A. ElSohly and D.F. Stanford, J. Anal. Toxicol, 14 (1990) 308.
79   C.M. Selavka, J. Forensic Sci., 36 (1991) 685.
80   K.D. Meneely, J. Forensic Sci., 37 (1992) 1158.
81   M.G. Pelders and J.J.W. Ros, J. Forensic Sci., 41 (1996) 209.
82   G. Casella, A.H.B. Wu, B.R. Shaw and D.W. Hill, J. Anal. Toxicol., 21 (1997) 376.
83   C. Meadway, S. George and R. Braithwaite, Forensic Sci. Int., 96 (1998) 29.
84   M.R. Moeller, K. Hammer and O. Engel, Forensic Sci. Int., 143 (2004) 183.
85   R.J. Lewis, R.D. Johnson and R.A. Hattrup, J. Chromatogr. B, 822 (2005) 137.
86   T.P. Rohrig and C. Moore, J. Anal. Toxicol., 27 (2003) 449.
87   M. Thevis, G. Opfermann and W. Schänzer, J. Anal. Toxicol., 27 (2003) 53.
88   W. Van Thuyne, P. Van Eenoo and F.T. Delbeke, J. Chromatogr. B, 785 (2003) 254.
89   C. Kollias-Baker and R. Sams, J. Anal. Toxicol., 26 (2002) 81.
90   N.K. Nair, V. Navaratnam and V. Rajananda, J. Chromatogr., 366 (1986) 363.
91   A. Sperling, J. Chromatogr., 538 (1991) 269.
92   C. Barnfield, S. Burns, D.L. Byrom and A.V. Kemmenoe, Forensic Sci. Int., 39 (1988) 107.
93   H. Neumann, Forensic Sci. Int., 44 (1990) 85.
94   H. Neumann, Forensic Sci. Int., 69 (1994) 7.
95   E. Kaa, Forensic Sci. Int., 64 (1994) 171.
96   A. Sibley, Forensic Sci. Int., 77 (1996) 159.
97   R.B. Myors, P.T. Crisp, S.V. Skopec and R.J. Wells, Analyst, 126 (2001) 679.
98   L. Stromberg, L. Lundberg, H. Neumann, B. Bobon, H. Huizer and N.W. van der Stelt, Forensic
      Sci. Int., 114 (2000) 67.
99   S.P. Sharma, B.C. Purkait and S.C. Lahiri, Forensic Sci. Int., 152 (2005) 235.
100   P. Esseiva, F. Anglada, L. Dujourdy, F. Taroni, P. Margot, E. Du Pasquier, M. Dawson, C. Roux
      and P. Doble, Talanta, 67 (2005) 360.
101   R. Brenneisen, F. Hasler and D. Wursch, J. Anal. Toxicol., 26 (2002) 561.
102   B.M. El-Haj, A.M. Al-Amri and H.S. Ali, Forensic Sci. Int., 145 (2004) 41.
103   J. Cartier, O. Gueniat and M.D. Cole, Sci. Justice, 37 (1997) 175.
104   I.S. Lurie and S.M. Carr, J. Liquid Chromatogr., 9 (1986) 2485.
105   I.S. Lurie and K. Mc Guiness, J. Liquid Chromatogr., 10 (1987) 2189.
106   P.A. Hays and I.S. Lurie, J. Liquid Chromatogr., 14 (1991) 3513.
107   A. Johnston and L.A. King, Forensic Sci. Int., 95 (1998) 47.

108   R. Dams, T. Benijst, W. Gunther, W. Lambert and A. De Leenheer, Anal. Chem., 74 (2002) 3206.
109   M.M.K. Reddy, P. Ghosh, S.N. Rasool, R.K. Sarin and R.B. Sashidhar, J. Chromatogr. A, 1088 (2005) 158.
110   F. Tagliaro and F.P. Smith, Trends Anal. Chem., 15 (1996) 513.
111   F.v. Heeren and W. Thormann, Electrophoresis, 18 (1997) 2415.
112   I.S. Lurie, K.C. Chan, T.K. Spratley, J.F. Casale and H.J. Issaq, J. Chromatogr. B, 669 (1995) 3.
113   I.S. Lurie, Int. Lab. (1996) March, 21–28
114   I.S. Lurie, J. Chromatogr. A, 780 (1997) 265.
115   F. Tagliaro, S. Turina and F.P. Smith, Forensic Sci. Int., 77 (1996) 211.
116   I.S. Lurie, D.S. Anex, Y. Fintschenko and W.Y. Choi, J. Chromatogr. A, 924 (2001) 421.
117   I.S. Lurie, P.A. Hays, A.E. Garcia and S. Panicker, J. Chromatogr. A, 1034 (2004) 227.
118   I.S. Lurie, P.A. Hays and K. Parker, Electrophoresis, 25 (2004) 1580.
119   N. Anostos, N.W. Barrett, S.W. Lewis, J.R. Pearson and K.P. Kirkbride, J. Forensic Sci., 50 (2005) 1039.
120   H. Neumann and M. Gloger, Chromatographia, 16 (1982) 261.
121   M. Chiarotti, N. Fucci and C. Furnari, Forensic Sci. Int., 50 (1991) 47.
122   A.F. Hernandez, A. Pla, J. Moliz, F. Gil, M.C. Gonzalvo and E. Villanueva, J. Forensic Sci., 37 (1992) 1276.
123   F. Besacier, H. Chaudron-Thozet, M. Rousseau-Tsangaris, J. Girard and A. Lamotte, Forensic Sci. Int., 85 (1997) 113.
124   T. Bora, M. Merdivan and C. Hamamci, J. Forensic Sci., 47 (2002) 959.
125   M. Kala and W. Lechowicz, In: Proceedings of the XXXV TIAFT Meeting, Centre of Behavioural and Forensic Toxicology, University of Padova, 1997, pp. 521–535.
126   V.M. Hendriks, W. Van den Brink, P. Blanken, I.J. Bosman and J.M. Van Ree, Eur. Neuropsychopharmacol., 11 (2001) 241.
127   M.G. Klous, G.M. Bronner, B. Nuijen, J.M. van Ree and J.H. Beijnen, J. Pharm. Biomed. Anal., 39 (2005) 944.
128   R. Dams, T. Benijst, W.E. Lambert, D.L. Massart and A.P. De Leenheer, Forensic Sci. Int., 123 (2001) 81.
129   J.G. Umans, T.S.K. Chiu, R.A. Lipman, M.F. Schulz, S.U. Shin and C.E. Inturrisi, J. Chromatogr., 233 (1982) 213.
130   C.E. Inturrisi, M.B. Bax, K.M. Foley, K. Schutz, S.U. Shin and R.W. Houde, New Engl. J. Med., 310 (1984) 1213.
131   A.J. Jenkins, R.M. Keenan, J.E. Heningfield and E.J. Cone, J. Anal. Toxicol., 18 (1994) 317.
132   E.J. Cone, B.A. Holicky, T.M. Grant, W.D. Darwin and B.A. Goldberger, J. Anal. Toxicol., 17 (1993) 327.
133   G. Skopp, B. Ganssmann, E.J. Cone and R. Aderjan, J. Anal. Toxicol., 21 (1997) 105.
134   E.J. Cone, P. Welch, J.M. Mitchell and B.D. Paul, J. Anal. Toxicol., 15 (1991) 1.
135   J. Fehn and G. Megges, J. Anal. Toxicol, 9 (1985) 134.
136   B.D. Paul, J.M. Mitchell, L.D. Mell and J. Irving, J. Anal. Toxicol., 13 (1989) 2.
137   R.W. Romberg and V.E. Brown, J. Anal. Toxicol., 14 (1990) 58.
138   D.C. Fuller and W.H. Anderson, J. Anal. Toxicol., 16 (1992) 315.
139   C. Meadway, S. George and R. Braithwaite, Forensic Sci. Int., 127 (2002) 136.
140   W.H. Soine, Med. Res. Rev., 6 (1986) 41.
141   C.L. O'Neal and A. Poklis, J. Anal. Toxicol., 21 (1997) 427.
142   C.L. O'Neal and A. Poklis, Forensic Sci. Int., 95 (1998) 1.
143   C. Staub, M. Marset, A. Mino and P. Mangin, Clin. Chem., 47 (2001) 301.
144   N. McLachlan-Troup, G.W. Taylor and B.C. Trathen, Addict. Biol., 6 (2001) 223.
145   R. Brenneisen and F. Hasler, J. Forensic Sci., 47 (2002) 885.
146   H.J. Derks, K. Van Twillert, D.P. Pereboom-De Fauw, G. Zomer and J.G. Loeber, J. Chromatogr., 370 (1986) 173.
147   H. Hanisch and L.v. Meyer, J. Anal. Toxicol., 17 (1993) 48.
148   J. Gerostamoulos, K. Crump, I. McIntyre and O.H. Drummer, J. Chromatogr., 617 (1993) 152.

149   A.S. Low and R.B. Taylor, J. Chromatogr. B, 663 (1995) 225.

150   R. Dams, T. Benijst, W.E. Lambert and A.P. De Leenheer, J. Chromatogr. B, 773 (2002) 53.

151   P. Fernandez, C. Vasquez, L. Morales and A.M. Bermejo, J. Appl. Toxicol., 25 (2005) 200.

152   M.J. Bogusz, R.D. Maier, M. Erkens and U. Kohls, J. Anal. Toxicol., 25 (2001) 431.

153   F. Musshoff, J. Trafkowski and B. Madea, J. Chromatogr. B, 811 (2004) 47.

154   M. Katagi, M. Nishikawa, M. Tatsuno, A. Miki and H. Tsushihashi, J. Chromatogr. B, 751 (2001) 177.

155   M.G. Klous, E.J. Rook, M.J.X. Hillebrand, W. van den Brink, J. van Ree and J. Beijnen, J. Anal. Toxicol., 29 (2005) 564.

156   R.B. Taylor, A.S. Low and R.G. Reid, J. Chromatogr. B, 675 (1996) 213.

157   W.S. Wu and J.L. Tsai, Biomed. Chromatogr., 13 (1999) 216.

158   J.L. Tsai, W.S. Wu and H.H. Lee, Electrophoresis, 21 (2000) 1580.

159   A.B. Wey and W. Thormann, J. Chromatogr. A, 916 (2001) 225.

160   A. Alnajjar, J.A. Butcher and B. McCord, Electrophoresis, 25 (2004) 1592.

161   X.-H. Qi, J.-Q. Mi, X.-X. Zhang and W.-B. Chang, Anal. Chim. Acta, 551 (2005) 115.

162   J. Schuberth and J. Schuberth, J. Chromatogr., 490 (1989) 444.

163   F. Musshoff and T. Daldrup, Int. J. Leg. Med., 106 (1993) 107.

164   R. Wasels and F. Belleville, J. Chromatogr. A, 674 (1994) 225.

165   W.L. Wang, W.D. Darwin and E.J. Cone, J. Chromatogr. B, 660 (1994) 279.

166   B.A. Goldberger, E.J. Cone, T.M. Grant, Y.H. Caplan, B.S. Levine and J.E. Smialek, J. Anal. Toxicol., 18 (1994) 22.

167   M.R. Moeller and C. Mueller, Forensic Sci. Int., 70 (1995) 125.

168   J.G. Guillot, M. Lefebvre and J.P. Weber, J. Anal. Toxicol., 21 (1997) 127.

169   P.P. Rop, M. Fornaris, T. Salmon, J. Burle and M. Bresson, J. Anal. Toxicol., 21 (1997) 232.

170   W.H. Phillips, K. Ota and N.A. Wade, J. Anal. Toxicol., 13 (1989) 268.

171   G. Schmitt, M. Bogusz, R. Aderjan and C. Meyer, Z. Rechtsmed., 103 (1990) 513.

172   D.G. Watson, Q. Su, J.M. Midley, E. Doyle and N.S. Morton, J. Pharm. Biomed. Anal., 13 (1995) 27.

173   B. Fryirs, M. Dawson and L.E. Mather, J. Chromatogr. B, 693 (1997) 51.

174   S. Pichini, I. Altieri, M. Pellegrini, P. Zuccaro and R. Pacifici, Mass Spectrom. Rev., 18 (1999) 119.

175   P. Zuccaro, R. Ricciarello, S. Pichini, R. Pacifici, I. Altieri, M. Pellegrini and G. D'Ascenzo, J. Anal. Toxicol., 21 (1997) 268.

176   M.J. Bogusz, R.D. Maier and S. Driessen, J. Anal. Toxicol., 21 (1997) 346.

177   M.J. Bogusz, R.D. Maier, M. Erkens and S. Driessen, J. Chromatogr. B, 703 (1997) 115.

178   M.J. Bogusz, J. Chromatogr. B, 748 (2000) 3.

179   N. Tyrefors, B. Hyllbrant, L. Ekman, M. Johansson and L. Langström, J. Chromatogr. A, 729 (1996) 279.

180   W. Naidong, J.W. Lee, X. Jiang, M. Wehlin, J.D. Hulse and P.P. Lin, J. Chromatogr. B, 735 (1999) 255.

181   G. Schanzle, S. Li, G. Mikus and U. Hofmann, J. Chromatogr. B, 721 (1999) 55.

182   A. Dienes-Nagy, L. Rivier, G. Giroud, M. Augsburger and P. Mangin, J. Chromatogr. A, 854 (1999) 109.

183   M. Blanchet, G. Bru, M. Guerret, M. Bromet-Petit and N. Bromet, J. Chromatogr. A, 854 (1999) 93–108.

184   M.H. Slawson, D.J. Crouch, D.M. Andrenyak, D.E. Rollins, J.K. Lu and P.L. Bailey, J. Anal. Toxicol., 23 (1999) 468.

185   A. Cailleux, A. Le Bouil, B. Auger, G. Bonsergent, A. Turcant and P. Allain, J. Anal. Toxicol., 23 (1999) 620–624.

186   E.J. Rook, M.J.X. Hillebrand, H. Rosing, J.M. van Ree and J.H. Beijnen, J. Chromatogr. B, 824 (2005) 213.

187   M. Concheiro, A. de Castro, O. Quintela, M. Lopez-Rivadulla and A. Cruz, J. Chromatogr. B, 832 (2006) 81.

188   P. Kintz and N. Samyn, Unconventional samples and alternative matrices. in: M.J. Bogusz (Ed.), R.M. Smith (Series Ed.), Forensic science, handbook of analytical separations, Vol. 2. Elsevier Sciences, Amsterdam, 2000, pp. 459–488.

189 S. Pichini, I. Altieri, M. Pellegrini, R. Pacifici and P. Zuccaro, J. Liq. Chromatogr. Rel. Technol., 22 (1999) 873.
190 N. Samyn, G. De Boeck and A. Verstraete, J. Forensic Sci., 47 (2002) 1380.
191 A.J. Barnes, I. Kim, R. Schepers, E.T. Moolchan, L. Wilson, G. Cooper and C.H. Reid, C. Hand and M. A. Huestis,. J. Anal. Toxicol, 27 (2003) 402.
192 E. Vinner, J. Vignau, D. Thibault, X. Codaccioni, C. Brassart, L. Humbert and M. L'Hermitte, Forensic Sci. Int., 133 (2003) 57.
193 D. Montgomery, C. Plate, S.C. Alder, M. Jones, J. Jones and R.D. Christensen, J. Perinatol., 26 (2006) 11.
194 K. Wolff, M.J.S. Anderson and A.W. Hay, Ann. Clin. Biochem., 27 (1990) 482.
195 D.J. Dietzen, J. Koening and J. Turk, J. Anal. Toxicol., 19 (1995) 299.
196 J. Vecerkova, Soud. Lek., 42 (1997) 32.
197 R. Jain, R. Ray, B.M. Tripathi and C. Singh, Indian J. Pharmacol., 28 (1996) 220.
198 R. De la Torre, J. Segura, R. de Zeeuw and J. Williams, Ann. Clin. Biochem., 34 (1997) 339.
199 M. Zezulak, J.J. Snyder and S.B. Needleman, J. Forensic Sci., 38 (1993) 1275.
200 Z. Lin, P. Lafolie and O. Beck, J. Anal. Toxicol., 18 (1994) 129.
201 P.L. Hackett, L.J. Dusci, K.F. Kenneth and G.M. Chiswell, Therap. Drug Monit., 24 (2002) 652.
202 B.D. Paul, L.D. Mell, J.M. Mitchell, J. Irving and A.J. Novak, J. Anal. Toxicol., 9 (1985) 222.
203 B.H. Chen, E.H. Taylor and A.A. Pappas, J. Anal. Toxicol., 14 (1990) 12.
204 G.F. Grinstead, J. Anal. Toxicol., 15 (1991) 293.
205 J. Fenton, J. Mummert and M. Childers, J. Anal. Toxicol., 18 (1994) 159.
206 K.E. Brooks and N.B. Smiths, J. Anal. Toxicol., 20 (1996) 269.
207 L.A. Broussard, L.C. Presley, T. Pittman, R. Clouette and G.H. Wimbish, Clin. Chem., 43 (1997) 1029.
208 M.M. Rettinger, C.J. Jones, M.A. Re, J.B. Rettinger and A.S. Zisman, Comparision of derivatizing reagents and internal standards for the analysis of opiates, presented at the SOFT-TIAFT Meeting, Albuquerque, NM 1998.
209 M.J. Bogusz, J. Anal. Toxicol., 21 (1997) 246.
210 E.J. Cone and W.D. Darwin, J. Chromatogr., 580 (1992) 43.
211 T. Gunnar, K. Ariniemi and P. Lillsunde, J. Mass Spectrom., 40 (2005) 739.
212 R.J. Osborne, S.P. Joel, D. Trew and M.L. Slevin, Clin. Pharmacol. Ther., 47 (1990) 12.
213 R.T. Penson, S.P. Joel, K. Bakhshi, S.J. Clark, R.M. Langford and M.L. Slevin, Clin. Pharmacol. Ther., 68 (2000) 667.
214 G.C. Rossi, G.P. Brown, L. Leventhal, K. Yang and G.W. Pasternak, Neurosci. Lett., 216 (1996) 1.
215 M. Barberi-Heyob, J.R. Merlin, I. Krakowski, C. Kettani, E. Collin and P. Poulain, Bull. Cancer, 78 (1991) 1063.
216 J.L. Mason, S.P. Ashmore and A.R. Aitkenhead, J. Chromatogr. B, 570 (1991) 191.
217 R.K. Portenoy, E. Khan, M. Layman, J. Lapin, M.G. Malkin, K.M. Foley, H.T. Thaler, D.J. Cerbone and C.E. Inturrisi, Neurology, 41 (1991) 1457.
218 P. Joel, R.J. Osborne and M.L. Slevin, J. Chromatogr., 430 (1988) 394.
219 Y. Rothsteyn and B. Weingarten, Ther. Drug Monit., 18 (1996) 179.
220 P.A. Glare, T.D. Walsh and C.E. Pippenger, Ther. Drug Monit., 13 (1991) 226.
221 C.E. Hartley, M. Green, M. Quinn and M.I. Levene, Biomed. Chromatogr., 7 (1993) 34.
222 M.J. Bogusz, R.D. Maier, K.D. Krüger and U. Kohls, J. Anal. Toxicol., 22 (1998) 549.
223 M. Zheng, K.M. McErlane and M.C. Ong, J. Pharmaceut. Biomed. Anal., 16 (1988) 971.
224 W.Z. Shou, Y.L. Chen, A. Eerkes, Y.Q. Tang, L. Magis, X. Jiang and W. Naidong, Rapid Comm. Mass Spectrom., 16 (2002) 1613.
225 D. Whittington and E.D. Kharash, J. Chromatogr. B, 796 (2003) 95.
226 C.M. Murphy and M. Huestis, J. Mass Spectrom., 40 (2005) 1412.
227 R. Kaushik and W.R. LaCourse, Anal. Chim. Acta, 556 (2006) 255.
228 R. Aderjan, S. Hofmann, G. Schmitt and G. Skopp, J. Anal. Toxicol., 19 (1995) 163.
229 S. Kerrigan, D. Honey and G. Baker, J. Anal. Toxicol., 28 (2004) 529.

230   S. Cengiz, Ö. Ulukan, I. Ates and H. Tugcu, Forensic Sci. Int., 156 (2006) 91.
231   M.L. Goff, W.A. Brown, K.A. Hewadikaram and A.I. Omori, J. Forensic. Sci., 36 (1991) 537.
232   F. Introna, C. Lo Dico, Y.H. Caplan and J.E. Smialek, J. Forensic Sci., 35 (1990) 118.
233   P. Kintz, V. Cirimele, Y. Edel, C. Jamey and P. Mangin, J. Forensic Sci., 39 (1994) 1497.
234   V. Hedouin, B. Bourel, L. Martin-Bouyer, A. Becart, G. Tournel, M. Devaux and D. Gosset, J. Forensic Sci., 44 (1999) 351.
235   V. Hedouin, B. Bourel, A. Becart, G. Tournel, M. Devaux, M.L. Goff and D. Gosset, J. Forensic Sci., 46 (2001) 12.
236   B. Bourel, L. Fleurisse, V. Hedouin, J.C. Cailliez, C. Creusy, D. Gosset and M.L. Goff, J. Forensic Sci., 46 (2001) 596.
237   B. Bourel, G. Tournel, V. Hedouin, M.L. Goff and D. Gosset, J. Forensic Sci., 46 (2001) 600.
238   C.P. Campobasso, M. Gherardi, M. Caligara, L. Sironi and F. Introna, Int. J. Legal Med., 118 (2004) 210.
239   G. Skopp, K. Klinder, L. Potsch, G. Zimmer, R. Lutz, R. Aderjan and R. Mattern, Forensic Sci. Int., 95 (1998) 99.
240   M. Balikova and V. Maresova, Forensic Sci. Int., 94 (1998) 201.
241   R. Aderjan and G. Skopp, Ther. Drug Monit., 20 (1998) 561.
242   H. Seno, H. Hattori, S. Kurono, T. Yamada, T. Kumazawa, A. Ishii and O. Suzuki, J. Chromatogr. B, 673 (1995) 189.
243   U. Hofmann, M.F. Fromm, S. Sohnson and G. Mikus, J. Chromatogr. B., 663 (1995) 59.
244   P. Kintz, A. Tracqui and P. Mangin, Int. J. Legal Med., 104 (1991) 177.
245   H. Sachs, R. Denk and I. Raff, Int. J. Legal Med., 105 (1993) 247.
246   D. Wilkins, D.E. Rollins, J. Seaman, H. Haughey, G. Krueger and R.L. Foltz, J. Anal. Toxicol., 19 (1995) 269.
247   S.P. Gygi, D.G. Wilkins and D.E. Rollins, J. Anal. Toxicol., 19 (1995) 387.
248   M. Balikova, V. Maresova and V. Habrdova, J. Chromatogr. B, 752 (2001) 179.
249   C.P. Verwey-Van Wissen, P.M. Koopman-Kimenai and T.B. Vree, J. Chromatogr., 570 (1991) 309.
250   S.S. Mohammed, M. Butschkau and H. Derendorf, J. Liquid Chromatogr., 16 (1993) 2325.
251   J.O. Svensson, Q.Y. Yue and J. Säwe, J. Chromatogr. B, 674 (1995) 49.
252   H. He, S.D. Shay, Y. Caraco, M. Wood and A.J. Wood, J. Chromatogr. B, 708 (1998) 185.
253   P. Lafolie, O. Beck, Z. Lin, F. Albertioni and L. Boreus, J. Anal. Toxicol., 20 (1996) 541.
254   L.C. Kirkwood, R.L. Nation and A.A. Somogyi, J. Chromatogr. B, 701 (1997) 129.
255   E. Hufschmid, R. Theurillat, C.H. Wilder-Smith and W. Thormann, J. Chromatogr. B, 678 (1996) 43.
256   A.B. Wey, J. Caslavska and W. Thormann, J. Chromatogr. A, 895 (2000) 133.
257   S.L. Walsh, K.L. Preston, G.E. Bigelow and M.L. Stitzer, J. Pharmacol. Exp. Ther., 274 (1995) 361.
258   L. Amass, J.B. Kamien and S.K. Mikulich, Drug Alcohol Depend., 58 (2000) 143.
259   P. Kintz and P. Marquet (Eds.), Buprenorphine therapy of opiate addiction. Humana Press, Totowa, USA, 2002.
260   G. Vidal-Trecan, I. Vareson, N. Nabet and A. Boisonnas, Drug Alcohol Depend., 69 (2003) 175.
261   L. Debrabandere, M. Van Bouven and P. Daenens, J. Forensic sci., 40 (1995) 250.
262   V. Cirimele, P. Kintz, S. Lohner and B. Ludes, J. Anal. Toxicol., 27 (2003) 103.
263   M. Böttcher and O. Beck, J. Anal. Toxicol., 29 (2005) 769.
264   V. Cirimele, S. Etienne, M. Villain, B. Ludes and P. Kintz, Forensic Sci. Int., 143 (2004) 153.
265   N. De Giovanni, N. Fucci, S. Scarlata and G. Donzelli, Clin. Chem. Lab. Med, 43 (2005) 1377.
266   E.T. Everhard, P. Cheung, P. Schwonek, K. Zabel, E.C. Tisdale, P. Jacob, J. Mendelson and R.T. Jones, Clin. Chem., 43 (1997) 2292.
267   E.J. Cone, C.W. Gorodetzky, D. Yousefnejad and W.D. Darwin, J. Chromatogr., 335 (1985) 291.
268   K.A. Hadidi and J.S. Oliver, Int. J. Med. Leg., 111 (1998) 165.
269   J.J. Kuhlman, J. Magluilo Jr., E.J. Cone and B. Levine, J. Anal. Toxicol., 20 (1996) 229.
270   J.J. Kuhlman, S. Lalani, J. Magluilo Jr., B. Levine, W.D. Darwin, R.E. Johnson and E.J. Cone, J. Anal. Toxicol., 20 (1996) 369.

271 A.M. Lisi, R. Kazlauskas and G.J. Trout, J. Chromatogr. B, 692 (1997) 67.

272 E.J. Cone, C.W. Gorodetzky, W.D. Darwin and W.F. Bunchwald, J. Pharm. Sci., 73 (1984) 243.

273 L. Debrabandere, M. Van Boven and P. Daenens, J. Forensic Sci., 37 (1992) 82.

274 P. Kintz, J. Anal. Toxicol., 17 (1993) 443.

275 P. Kintz, A. Tracqui and P. Mangin, J. Forensic Sci. Soc., 34 (1994) 95.

276 H. Hoja, P. Marquet, B. Verneuil, H. Lofti, J.L. Dupuy and G. Lachatre, J. Anal. Toxicol., 21 (1997) 160.

277 A. Tracqui, P. Kintz and P. Mangin, J. Forensic Sci., 42 (1997) 111.

278 D.E. Moody, J.D. Laycock, A.C. Spanbauer, D.J. Crouch, R.L. Foltz, J.L. Josephs, L. Amass and W.K. Bickel, J. Anal. Toxicol., 21 (1997) 406.

279 D.E. Moody, M.H. Slawson, E.C. Strain, J.D. Laycock, A.C. Spanbauer and R.L. Foltz, Anal. Biochem., 306 (2002) 31.

280 J.M. Gaulier, P. Marquet, E. Lacassie, J.L. Dupuy and G. Lachatre, J. Forensic Sci., 45 (2000) 226.

281 A. Polettini and M.A. Huestis, J. Chromatogr. B, 754 (2001) 447.

282 C.M. Murphy and M.A. Huestis, J. Mass Spectrom., 40 (2005) 70.

283 A. Ceccato, R. Klinkenberg, P. Hubert and B. Streel, J. Pharm. Biomed. Anal., 32 (2003) 619.

284 R. Kronstrand, T.G. Selden and M. Josefsson, J. Anal. Toxicol., 27 (2003) 464.

285 D. Grimm, E. Pauly, J. Poschl, O. Linderkamp and G. Skopp, Ther. Drug Monit., 27 (2005) 526.

286 V.P. Dole, . In: A. Tagliamonte, L. Maremmani (Eds.), Methadone maintenance. Comes of age, Drug addiction and related clinical problems, Springer Verlag, Heidelberg–Wien–New York, 1995, pp. 45–63.

287 R.G. Newman, . In: A. Tagliamonte, L. Maremmani (Eds.), The pharmacological rational for methadone treatment of narcotic addiction, Drug addiction and related clinical problems, Springer Verlag, Heidelberg-Wien-New York, 1995, pp. 109–136.

288 R. La Harpe and O. Fryc, Arch. Kriminol., 196 (1995) 24.

289 A. Heinemann, J. Ribbat, K. Püschel, S. Iwersen and A. Schmoldt, Rechtsmedizin, 8 (1998) 55.

290 N. Chikhi-Chorfi, H. Galons, C. Pham-Huy, M. Thevenin, J.M. Warnet and J.R. Claude, Chirality, 13 (2001) 187.

291 S. George, S. Parmar, C. Meadway and R.A. Braithwaite, Ann. Clin. Biochem., 37 (2000) 350.

292 L. Moore, J. Wicks, V. Spiehler and R. Holgate, J. Anal. Toxicol., 25 (2001) 520.

293 N. De Giovanni, N. Fucci, M. Chiarotti and S. Scarlata, J. Chromatogr. B, 773 (2002) 1.

294 G. Cooper, L. Wilson, C. Reid, D. Baldwin, C.H and and V. Spiehler, J. Forensic Sci., 50 (2005) 928.

295 M. ElSohly, S. Feng and T.P. Murphy, J. Anal. Toxicol., 25 (2001) 40.

296 P. Kintz, P. Mangin, A.A. Lugniert and A.J. Chaumont, J. Toxicol. Clin. Exp., 10 (1990) 15.

297 L.D. Baugh, R.H. Liu and A.S. Walia, J. Forensic Sci., 36 (1991) 548.

298 N. Schmidt, R. Sittl, K. Brune and G. Geisslinger, Pharm. Res., 10 (1993) 441.

299 M.E. Alburges, W. Huang, R.L. Foltz and D.E. Moody, J. Anal. Toxicol., 20 (1996) 362.

300 D.G. Wilkins, P.R. Nasagawa, S.P. Gygi, R.L. Foltz and D.E. Rollins, J. Anal. Toxicol., 20 (1996) 355.

301 G.A.A. Cooper and J.S. Oliver, J. Anal. Toxicol., 22 (1998) 389.

302 D.W. Lachenmeier, L. Kroener, F. Musshoff and B. Madea, Rapid Commun. Mass Spectrom., 17 (2003) 472.

303 F.J. Couper, K. Chopra and M.L. Pierre-Louis, Forensic Sci. Int., 153 (2005) 71.

304 P. Kintz, M. Villain, V. Dumestre-Toulet, B. Capolaghi and V. Cirimele, Ther. Drug Monit., 27 (2005) 741.

305 L.M. Stolk, S.M. Coenradie, B.J. Smit and H.L. van As, J. Anal. Toxicol., 21 (1997) 154.

306 O. Beck, L.O. Boreus, P. Lafolie and G. Jacobson, J. Chromatogr., 570 (1991) 198.

307 N. Schmidt, K. Brune and G. Geisslinger, J. Chromatogr., 583 (1992) 195.

308 R.L. Norris, P.J. Ravenscroft and S.M. Pond, J. Chromatogr. B, 661 (1994) 346.

309 K. Kristensen, H.R. Angelo and T. Blemmer, J. Chromatogr. A, 666 (1994) 283.

310 S. Rudaz and J.L. Veuthey, J. Pharm. Biomed. Anal., 14 (1996) 1271.

311 H.R. Angelo, O. Beck and K. Kristensen, J. Chromatogr. B, 724 (1999) 35.

312 P. Kintz, H.P. Eser, A. Tracqui, M. Moeller, V. Cirimele and P. Mangin, J. Forensic Sci., 42 (1997) 291.

313 D. Ortelli, S. Rudaz, A.F. Chevalley, J.J. Deglon, L. Balant and J.L. Veuthey, J. Chromatogr. A, 871 (2000) 163.

314 O. Dale, C. Hoffer, P. Sheffels and E.D. Kharasch, Clin. Pharmacol. Ther., 72 (2002) 536.

315 S. Souverain, C. Eap, J.L. Veuthey and S. Rudaz, Clin. Chem. Lab. Med., 41 (2003) 1615.

316 H.R. Liang, R.L. Foltz, M. Meng and O.P. Bennett, J. Chromatogr. B, 806 (2004) 191.

317 D. Whittington, P. Sheffels and E.D. Kharash, J. Chromatogr. B, 809 (2004) 313.

318 M.L. Etter, S. George, K. Graybiel, J. Eichhorst and D.C. Lehotay, Clin. Biochem., 38 (2005) 1095.

319 S. Molteni, J. Caslavska, D. Allemann and W. Thormann, J. Chromatogr. B, 658 (1994) 355.

320 W. Thormann, M. Lanz, J. Caslavska, P. Siegenthaler and R. Portmann, Electrophoresis, 19 (1998) 57.

321 J. Esteban, M. de la Cruz Pellin, C. Gimeno, J. Barril, E. Mora, J. Gimenez and E. Vilanova, Toxicol. Lett., 15 (2004) 243.

322 T. Kelly, P. Doble and M. Dawson, Electrophoresis, 24 (2003) 2106.

323 C. Merslavic and L. Zupancic-Kraj, J. Chromatogr. B, 693 (1997) 222.

324 K.E. Goeringer, B.K. Logan and G.D. Christian, J. Anal. Toxicol., 21 (1997) 529.

325 K.J. Lusthof and P.G.M. Zweipfenning, J. Anal. Toxicol., 22 (1998) 260.

326 B. Levine, V. Ramcharitar and J.E. Smialek, J. Anal. Toxicol., 21 (1997) 43.

327 G. Sticht, P. Schmidt and H. Käferstein, Rechtsmedizin, 7 (1997) 127.

328 V. Gambaro, C. Benvenuti, L. De Ferrari, L. Dell'Acqua and F. Fare, Farmaco, 58 (2003) 947.

329 H.J. Leis, G. Fauler and W. Windischhofer, J. Chromatogr. B, 804 (2004) 369.

330 Y.F. Sha, S. Shen and G.L. Duan, J. Pharm. Biomed. Anal., 37 (2005) 143.

331 K.A. Hadidi, J.K. Almasad, T. Al-Nsour and S. Abu-Ragheib, Forensic Sci. Int., 135 (2003) 129.

332 M. Nobilis, J. Pastera, P. Anzenbacher, D. Svoboda, J. Kopecky and F. Perlik, J. Chromatogr. B, 681 (1996) 177.

333 M. Nobilis, J. Kopecky, J. Kvetina, J. Chladek, Z. Svoboda, V. Vorisek, F. Perlik, M. Pour and J. Kunes, J. Chromatogr. A, 949 (2002) 11.

334 S.J. Juzwin, D.C. Wang, N.J. Anderson and F.A. Wong, J. Pharm. Biomed. Anal., 22(2) (2000) 469.

335 W.N. Wu, L.A. McKown and S. Liao, Xenobiotica, 32 (2002) 411.

336 L.M. Zhao, X.Y. Chen, J.J. Cui, M. Sunita and D.F. Zhong, Yao Xue Xue Bao, 39 (2004) 458.

337 E. Frankus, E. Friderichs, S.M. Kim and G. Osterloh, Arzneim. Forsch., 28 (1978) 114.

338 B. Elsing and G. Blaschke, J. Chromatogr., 612 (1993) 223.

339 A. Ceccato, P. Chiap, P. Hubert and J. Crommen, J. Chromatogr. B, 698 (1997) 161.

340 A. Ceccato, F. Vanderbist, J.Y. Pabst and B. Streel, J. Chromatogr. B, 748 (2000) 65.

341 U.B. Soetebeer, M.O. Schierenberg, H. Schulz, G. Grunefeld, P. Andresen and G. Blaschke, J. Chromatogr. B, 745 (2000) 271.

342 U.B. Soetebeer, M.O. Schierenberg, H. Schulz, P. Andresen and G. Blaschke, J. Chromatogr. B, 765 (2001) 3.

343 S. Rudaz, S. Cherkaoui, P. Dayer, S. Falani and J.L. Veuthey, J. Chromatogr. A, 868 (2000) 295.

344 P. Lehtonen, H. Siren, I. Ojanpera and R. Kostiainen, J. Chromatogr. A, 1041 (2004) 227.

345 J. Vecerkova, Criminalistics, 25 (1992) 216.

346 K.L. Sees, M.E. DiMarino, N.K. Ruediger, C.T. Sweeney, S. Shiffman and J. Pain Palliat, Care Pharmacother., 19 (2005) 13.

347 B.C. Wolf, W.A. Lavezzi, L.M. Sullivan and L.M. Flanagan, J. Forensic Sci., 50 (2005) 192.

348 J.M. Abadie, K.H. Allison, D.A. Black, J. Garbin, A.J. Saxon and D.D. Bankson, J. Anal. Toxicol., 29 (2005) 825.

349 R.C. Backer, J.R. Monforte and A. Poklis, J. Anal. Toxicol., 29 (2005) 675.

350 V.R. Spiehler, L. DeCicco, J.R. McCutcheon, T. Kupiec and P. Kemp, J. Forensic Sci., 40 (2004) 621.

351 R.P. Kapil, P.K. Padovani, S.Y. King and G.N. Lam, J. Chromatogr., 577 (1992) 283.

352 C.M. Moore, D. Deitermann, D. Lewis and J. Leikin, J. Anal. Toxicol., 10 (1995) 514.

353 J. Jones, K. Tomlinson and C. Moore, J. Anal. Toxicol., 26 (2002) 171.

354 N.L. Le, A.E. Reiter, K. Tomlinson, J. Jones and C. Moore, J. Anal. Toxicol., 29 (2005) 54.

355 R. Meatherall, J. Anal. Toxicol., 29 (2005) 301.

356 A.W. Wright, M.L. Nocente and M.T. Smith, Life Sci., 63 (1998) 401.

357 Y.L. Chen, G.D. Hanson, X. Jiang and W. Naidong, J. Chromatogr. B, 769 (2002) 55.

358 E. Bostrom, B. Jansson, M. Hammarlund-Udenaes and U.S. Simmonsson, Rapid Commun. Mass Spectrom., 18 (2004) 2565.

359 L.E. Edinboro, R.C. Backer and A. Poklis, J. Anal. Toxicol., 29 (2005) 704.

360 S.E. Edwards and M.T. Smith, J. Chromatogr. B, 814 (2005) 241.

361 O. Cheremina, I. Bachmakov, A. Neubert, K. Brune, F. Maertin and B. Hinz, Biomed. Chromatogr. (2005) online.

362 A.B. Wey and W. Thormann, J. Chromatogr. B, 770 (2002) 191.

363 A. Baldacci, J. Caslavska, A.B. Wey and W. Thormann, J. Chromatogr. A, 1051 (2004) 273.

364 R.C. Baselt, Disposition of toxic drugs and chemicals in man, (5th ed). Chemical Toxicology Institute, Foster City, California, 2000, pp. 353–356.

365 R. Kronstrand, H. Druid, P. Holmgren and J. Rajs, Forensic Sci. Int., 88 (1997) 185.

366 P. Kintz, M. Villain, V. Dumestre and V. Cirimele, Forensic Sci. Int., 153 (2005) 153.

367 T. Tobin, S. Kwiatkowski, D.S. Watt, H.H. Tai, C.L. Tai, W.E. Woods, J.P. Goodman, D.G. Taylor, T.J. Weckman and J.M. Yang, Res. Commun. Chem. Pathol. Pharmacol., 63 (1989) 129.

368 H. Käferstein and G. Sticht, Forensic Sci. Int., 113 (2000) 353.

369 V. Watts and Y. Caplan, J. Anal. Toxicol., 12 (1988) 246.

370 H. Ohta, S. Suzuki and K. Ogasawara, J. Anal. Toxicol., 23 (1999) 280.

371 F.M. Esposito and C.L. Winek, J. Forensic Sci., 26 (1991) 86.

372 A. Szeitz, K.W. Riggs and C. Harvey-Clark, J. Chromatogr. B, 675 (1996) 33.

373 B. Fryirs, A. Woodhouse, J.L. Huang, M. Dawson and L.E. Mather, J. Chromatogr. B, 688 (1997) 79.

374 H. Sachs, M. Uhls, G. Hege-Scheuning and E. Schneider, Int. J. Leg. Med., 109 (1996) 213.

375 D.T. Anderson and J.J. Muto, J. Anal. Toxicol., 24 (2000) 627.

376 J.J. Kuhlman Jr., R. McCauley, T.J. Valouch and G.S. Behonick, J. Anal. Toxicol., 27 (2003) 499.

377 P.K. Lilleng, L.I. Mehlum, L. Bachs and I. Morild, J. Forensic Sci., 49 (2004) 1364.

378 A.M. Tharp, R.E. Winecker and D.C. Winston, Am. J. Forensic Med. Pathol., 25 (2004) 178.

379 A. Poklis and R. Backer, J. Anal. Toxicol., 28 (2004) 422.

380 C. Paradis, C. Dufresne, M. Bolon and R. Boulieu, Ther. Drug Monit., 24 (2002) 768.

381 N.F. Van Nimmen, K.L. Poels and H.A. Veulemans, J. Chromatogr. B, 804 (2004) 375–387.

382 N.F. Vam Nimmen and H.A. Veulemans, J. Chromatogr. A, 1035 (2004) 249.

383 W.Z. Shou, X. Jiang, B.D. Beato and W. Naidong, Rapid Commun. Mass Spectrom., 15 (2001) 466.

384 D.E. Koch, R. Isaza, J.W. Carpenter and R.P. Hunter, J. Pharm. Biomed. Anal., 34 (2004) 577.

385 N.H. Huynh, N. Tyrefors, L. Ekman and M. Johansson, J. Pharm. Biomed. Anal., 37 (2005) 1095.

386 J. Martens-Lobenhoffer, J. Chromatogr. B., 769 (2002) 227.

387 L. Palleschi, L. Lucentini, E. Ferretti, F. Anastasi, M. Amoroso and G. Draisci, J. Pharm. Biomed. Anal., 32 (2003) 329.

388 R. Schmidt, D.H. Bremerich and G. Geisslinger, J. Chromatogr. B, 836 (2006) 98.

389 S.R. Edwards, C.F. Minto and L.E. Mather, Br. J. Anesth., 88 (2002) 94.

390 A.F. Lehner, P. Almeida, J. Jacobs, J.D. Harkins, W. Karpiesiuk, W.E. Woods, L. Dirikolu, J.M. Bosken, W.G. Carter, J. Boyle, C. Holtz, T. Heller, C. Nattrass, M. Fisher and T. Tobin, J. Anal. Toxicol., 24 (2000) 309.

391 M. Thevis, H. Geyer, D. Bahr and W. Schanzer, Eur. J. Mass Spectrom., 11 (2005) 419.

392 T. Breindahl and K. Andreasen, J. Chromatogr. B, 736 (1999) 103.

393 I. Sundstrom, U. Bondesson and M. Hedel and, J. Chromatogr. B, 763 (2001) 121.

394 I. Sundstrom, M. Hedel and, U. Bondesson and P.E. Andren, J. Mass Spectrom., 37 (2002) 414.

395   M. Lampinen, U. Bondesson, E. Fredriksson and M. Hedel and, J. Chromatogr. B, 789 (2003) 347.

396   M.H. Andraus and M.E. Siquera, J. Chromatogr. B, 704 (1997) 143.

397   D.W. Boulton, G.F. Duncan and N.N. Vachharajani, J. Chromatogr. B, 775 (2002) 57.

398   Y.J. Wu, Y.Y. Cheng, C.S. Zeng and M.M. Ma, J. Chromatogr. B, 784 (2003) 219.

399   H. Bagheri, A. Es-haghi and M.R. Rouini, J. Chromatogr. B, 818 (2005) 147.

400   U. Lutz, W. Volkel, R.W. Lutz and W.K. Lutz, J. Chromatogr. B, 813 (2004) 217.

401   C. Arellano, C. Philibert, E.N. Dane e Yakan, C. Vachoux, O. Lacombe, J. Woodley and G. Houin, J. Chromatogr. B, 819 (2005) 105.

402   M.L. Constanzer, C.M. Chavez-Eng, I. Fu, E.J. Woolf and B.K. Matuszewski, J. Chromatogr. B, 816 (2005) 297.

403   S. Shou, H. Zhou, M. Larson, D.L. Miller, D. Mao, X. Jiang and W. Naidong, Rapid Commun. Mass Spectrom., 19 (2005) 2144.

404   D.L. Kuhlenbeck, T.H. Eichold, S.H. Hoke, T.R. Baker, R. Mensen and K.R. Wehmeyer, Eur. J. Mass Spectrom., 11 (2005) 199.

M.J. Bogusz (Ed.). Forensic Science
Handbook of Analytical Separations, Vol. 6
73

CHAPTER 2

Current methods for the separation and analysis of cocaine analytes

Rebecca Jufer Phipps[1], Jessica Jennings Smith[2], William D. Darwin[3] and Edward J. Cone[4]

[1] *Office of the Chief Medical Examiner, 111 Penn Street, Baltimore, MD 21201, USA*
[2] *Office of the Chief Medical Examiner, 200 South Adams Street, Wilmington, DE 19801, USA*
[3] *National Institute on Drug Abuse, Chemistry and Drug Metabolism Section, 5500 Nathan Shock Drive, Baltimore, MD 21224, USA*
[4] *ConeChem Research, LLC, 441 Fairtree Drive, Severna Park, MD 21146, USA*

ABSTRACT

The continuing need for sensitive and specific analytical methods for the detection and quantitation of cocaine is reflected by the number of publications that continue to be devoted to this topic. This review focuses on immunoassay-screening methods as well as chromatographic methods that were reported over approximately the last two decades for the determination of cocaine analytes in various biological specimens. Illicit cocaine analysis is addressed briefly in the introduction. The reviewed methods are summarized in tables to provide additional information on each assay. Solid-phase extraction was the most frequently applied technique to isolate cocaine analytes from biological matrices. Also, it was no surprise that gas chromatography-mass spectrometry operated in the positive ion–electron impact ionization mode was the most widely reported instrument used for the detection of cocaine analytes. However, other analytical methodologies, such as liquid chromatography-mass spectrometry, are becoming more important for the analysis of cocaine with the growing interest in identifying and quantitating multiple cocaine analytes with varying physiochemical properties.

References pp. 120–125

2.1 INTRODUCTION

2.1.1 Historical cocaine use

The primary source of cocaine is the *Erythroxylum coca* plant that grows abundantly on the eastern slopes of the Andes Mountains in Peru and Bolivia. In addition, there are other varieties of cocaine-containing plants that are cultivated in Columbia and on the desert coast of Peru. It has been reported that cocaine use occurred prior to the time of the Inca period (AD 800–1000), and possibly as early as 3000 years ago. The earliest archeological evidence of cocaine consumption comes from the pre-Incan tribes in Peru. This evidence includes remnants of coca leaves that have been recovered from tombs in Bolivia and Peru, dating back to about AD 600. During this period, it is thought that coca leaves were chewed primarily to enhance physical performance and to decrease the need for food and rest. Coca leaves were frequently chewed with lime to increase the amount of un-ionized cocaine, thereby increasing the efficiency of cocaine extraction from the leaf. Additionally, there is evidence of the early use of cocaine as a local anesthetic during trepanination procedures (a crude medical procedure involving the removal of a circular piece of cranium).

In the late 1800s, coca leaf extract was marketed in a variety of forms for numerous purposes ranging from enhancement of athletic performance to treatment of depression and morphine addiction. The most well-known of these extracts were those marketed by Mariani, including a wine, an elixir, pastilles and a tea. However, this type of cocaine use was banned in the United States by the passage of a federal anti-cocaine law, the Harrison Act, in 1914. The Harrison Act, as well as 46 local state laws passed prior to it, gave pharmacists and physicians regulatory power over the distribution of cocaine. The restrictions on cocaine use increased further when the Harrison Act and the Narcotic Drugs Import and Export Act of 1914 were amended in 1919 and 1922, respectively.

Illicit cocaine use grew slowly for many years, and then increased at a rapid pace in the 1970s. An all-time high of 5.7 million users in the United States was reached in 1985, according to data collected for the National Household Survey on Drug Abuse. Thereafter, there was some decline in use, but recent surveys on drug use have indicated that cocaine use has stabilized. The 2004 National Survey on Drug Use and Health reported that an estimated 2.0 million people were current cocaine users, compared with only 166,000 current heroin users. The widespread use of cocaine has created significant public health problems and a continuing need for sensitive and specific analytical methodologies for the identification and quantitation of cocaine analytes.

2.1.2 Illicit cocaine analysis

The analysis of cocaine samples became increasingly complex as illicit street sources of cocaine emerged in different forms. Most of the cocaine seized today is contaminated with various manufacturing by-products, intentionally added adulterants or diluents, naturally occurring alkaloids and products resulting from the chemical breakdown of cocaine. Common adulterants and diluents that have been identified in illicit cocaine samples included lidocaine, benzocaine, procaine, mannitol, lactose, dextrose and sucrose. Other impurities that have been identified in illicit cocaine samples included pseudococaine, anhydroecgonine, anhydroecgonine methyl ester, *trans*-cinnamic acid, ecgonine methyl ester, ecgonine, tropacocaine, benzoylecgonine, norcocaine, beta-truxinic acid, alpha-truxillic acid, *cis*- and *trans*-cinnamoyl ecgonine methyl ester, *N*-formylnorcocaine, *cis*- and *trans*- cinnamoylcocaine, truxillines, *N*-benzoylnorecgonine methyl ester, norecgonine methyl ester, hydroxy-cocaines, nortropacocaine, *N*-formylnorecgonine methyl ester, 3′,4′,5′-trimethoxy-cocaine, 3′,4′,5′-trimethoxytropacocaine, 3′,4′,5′-trimethoxy-*cis*-cinnamoylcocaine and 3′,4′,5′-trimethoxy-*trans*-cinnamoylcocaine [1–6]. The structures of the most commonly identified of these substances are illustrated in Fig. 2.1. It is important to identify the contaminants present in illicit cocaine seizures because such information can be valuable in determining the geographical origin of the seizure as well as identifying potentially harmful adulterants present in the cocaine mixture.

There has been substantial interest in developing "fingerprinting" and comparative analysis methods for the analysis of illicit cocaine samples. Several different techniques have been employed. Two selected techniques are summarized in Table 2.1. A simple and rapid method reported by Janzen *et al.* [7] focussed on the identification of four impurities in an illicit cocaine mixture, tropacocaine, norcocaine and *cis*- and *trans*- cinnamoylcocaine. The area ratios of each component were calculated with respect to cocaine and then compiled in a computerized database. The authors concluded that this method would serve as a useful tool to establish commonality of origin of illicit cocaine samples provided that an adequate database is developed. In addition, this method utilized instruments that are available in most forensic laboratories and could be performed on a routine basis. Casale *et al.* [2] have reported a comprehensive method for the analysis of illicit cocaine samples that is capable of detecting 14 coca-related impurities. With this method, unadulterated illicit cocaine samples are analyzed by gas chromatography with flame ionization detection (GC-FID) following derivatization with *N,O*-bis(trimethylsilyl) acetamide (BSA). The authors also indicated that this method could be applied to adulterated samples containing cocaine hydrochloride or cocaine base. This technique was referred to as CISPA (chromatographic impurities signature profile analyses). The authors performed studies to ensure that artifactual production of analytes during the analysis procedure did not occur. Since this technique measured numerous compounds, it provided a powerful comparison method for illicit cocaine samples. However, its most practical application to illicit cocaine sample comparison would be in combination with a computerized database with searching capabilities.

Recent research has also focused on the use of carbon (δ^{13}C) and nitrogen (δ^{15}N) isotope ratios for the determination of the region of origin of illicit cocaine. This method of analysis has proven successful for determination of illicit cocaine origin, particularly when combined with the alkaloid content of the sample. Ehleringer *et al.* [8] accurately predicted the region of origin of 200 illicit cocaine samples by combining isotope ratio analysis with truxilline and trimethoxycocaine content.

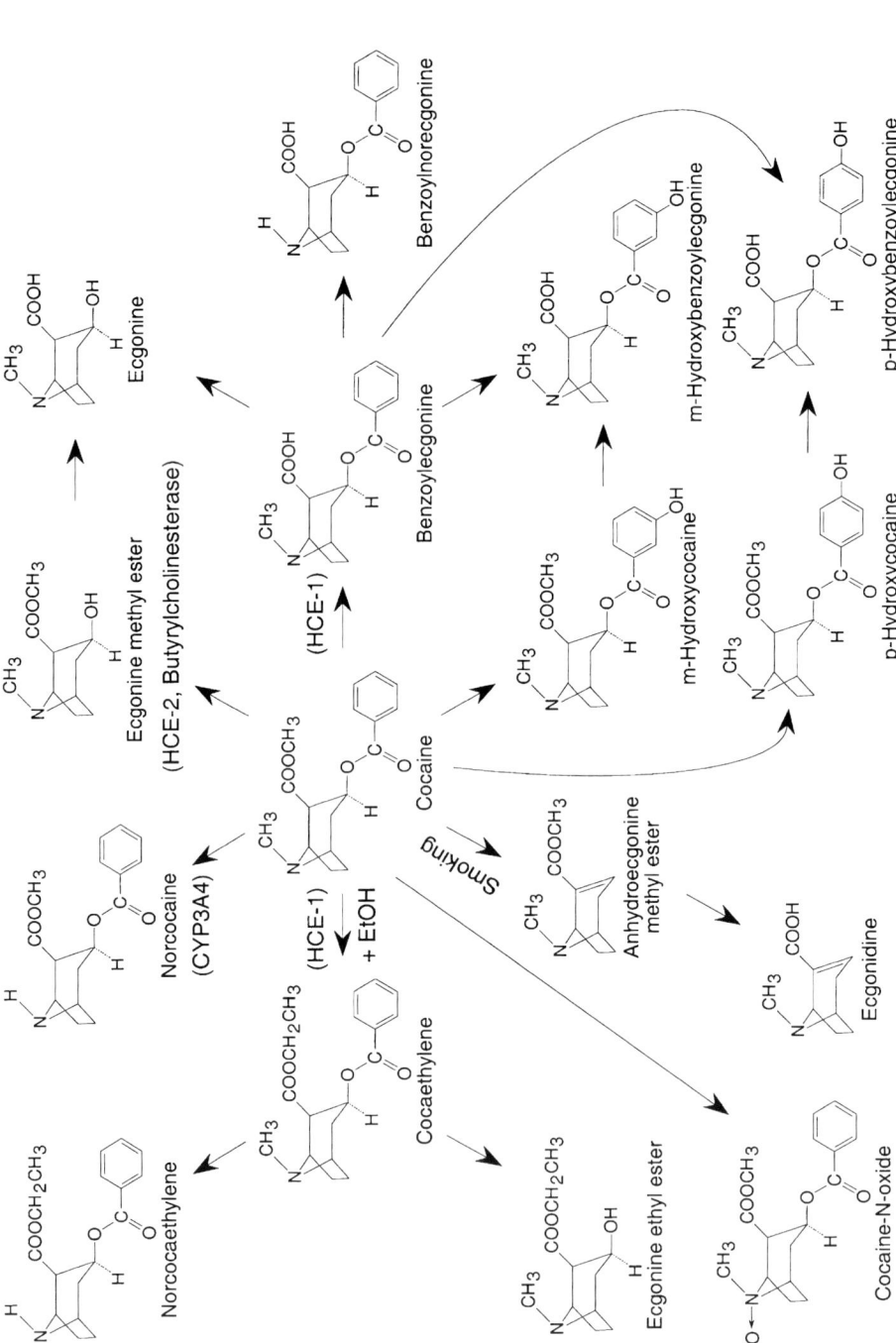

Fig. 2.1. Structures of selected alkaloids in coca leaf and alkaloidal impurities in illicit cocaine: (1) *N*-benzoylnorecgonine methyl ester; (2) pseudoecgonine; (3) pseudococaine; (4) *N*-formylnorcocaine; (5) tropacocaine; (6) anhydroecgonine methyl ester; (7) benzoylecgonine; (8) norcocaine; (9) *trans*-cinnamoylcocaine; (10) *cis*-cinnamoylcocaine.

TABLE 2.1
METHODS FOR THE ANALYSIS OF ILLICIT COCAINE SAMPLES

Analytes Detected	Sample Type	Sample Preparation	Internal Standard(s)	Brief Description	Ref.
Cocaine, tropacocaine, norcocaine and *cis*- and *trans*-cinnamoylcocaine	Illicit cocaine powder	0.5 g of sample was ground and dissolved in 2 mL ethanol	Bupivacaine	COC quantitation: 1 μL of sample was injected into a GC equipped with a DB-1 methyl silicone column (12.5 m, 0.32 mm, 0.25 μm) and a flame ionization detector isothermal detection was employed (230°C) Sample profiling: 1 μL of sample was injected into a GC equipped with an HP-1 cross-linked methyl silicone column (12.5 m, 0.20 mm, 0.5 μm) and a nitrogen–phosphorus detector. Temperature ramping was employed (120°C for 2 min, 6°C/min to 320°C, hold for 5 min; injector temp.: 215°C; detector temp.: 325°C)	[7]
Benzoic acid, anhydroecgonine methyl ester, anhydroecgonine, *trans*-cinnamic acid, ecgonine methyl ester, ecgonine, tropacocaine, cocaine, benzoylecgonine,	Illicit cocaine powder	4–5 mg sample of unadulterated cocaine was mixed with IS in chloroform and derivatized with BSA	*p*-fluorococaine	5 μL of sample was injected into a GC equipped with a DB-1701-coated capillary column (30 m, 0.25 mm, 0.25 μ) and a flame ionization detector. Temperature ramping was employed (180°C for 1 min, 4°C/min to 200°C, 6°C/min to 275°C, hold for 11.5 min; injector temp.: 230°C; detector temp.: 280°C)	[2]

References pp. 120–125

TABLE 2.1
CONTINUED

Analytes Detected	Sample Type	Sample Preparation	Internal Standard(s)	Brief Description	Ref.
norcocaine, beta-truxinic acid, alpha-truxillic acid, *trans*-cinnamoyl ecgonine methyl ester, *N*-formyl cocaine and *cis*-cinnamoylecgonine methyl ester					
Volatile organic compounds (25) present as residual solvents	Illicit cocaine powder	15–30 mg sample was mixed with IS in aqueous 22% sodium sulfate	Acetone-d$_6$ Isopropanol-d$_8$ 2-chloro-2-methylpropane-d$_9$ n-hexane-d$_{14}$ Toluene-d$_8$	2 mL of headspace was injected (8:1 split ratio) into a GC equipped with a DB-1624-coated capillary column (75 m, 0.53 mm, 3 μ) and an ion trap mass spectrometric detector. Temperature ramping was employed (35°C for 14 min, 7°C/min to 210°C, hold for 3 min; injector temp.: 180°C; detector temp.: 200°C)	[9]
Volatile organic compounds (32) present as residual solvents	Illicit cocaine powder	100 mg sample was mixed with aqueous saturated sodium chloride	None	SPME fiber was thermally desorbed onto GC injector for 1 min at 250°C. The analytes were separated on an HP1301 capillary column (60 m, 0.25 mm, 1 μ) and detected with an electron impact mass spectrometer. Temperature ramping was employed (50°C for 1 min, 30°C/min to 80°C, 10°C/min to 150°C, 20°C/min to 270°C, hold for 1 min)	[10]

An additional technique that has been applied to comparative cocaine analysis is the determination of residual solvents in illicit cocaine samples. Morello and Meyers [9] developed a method for the analysis of 25 residual solvents by headspace gas chromatography-mass spectrometry (GC-MS). Their method utilized five deuterated internal standards and provided qualitative and quantitative data. This study also examined the effects of adulterants on the determination of residual solvents in illicit cocaine. The authors concluded that adulterated samples containing sugars, starches and inorganics could be analyzed by this method with minimal interference. However, the presence of neutral or basic ion-pair salts as adulterants warrants a more conservative approach; only samples with less than 20% total ion-pair salt are amenable to analysis. Chiarotti *et al.* [10] analyzed residual solvents with a method based on GC-MS following solid-phase microextraction. This method identified 32 volatile organic compounds in 47 cocaine exhibits. Through a statistical analysis of their residual solvent composition, 47 cocaine exhibits were divided into 4 distinct classes.

2.1.3 Metabolism of cocaine in humans

In humans, cocaine (COC) is extensively metabolized by both enzymatic and non-enzymatic pathways. Cocaine's primary metabolites include benzoylecgonine (BZE) and ecgonine methyl ester (EME) while norcocaine (NCOC), benzoylnorecgonine (BNE), *m*- and *p*-Hydroxycocaine (HOCOC) and *m*- and *p*-Hydroxybenzoylecgonine (HOBZE) are usually detected at considerably lower concentrations. In addition, cocaine *N*-oxide (CNO) has recently been identified and quantitated in a single meconium sample [11]. It was present at a concentration comparable to COC. Subsequent meconium analyses detected CNO in 12 of 21 meconium samples from drug-positive meconium samples [12]. CNO has also been identified and quantitated in human plasma; however, plasma CNO concentrations never exceeded 3% of the cocaine concentration [13]. Other unique analytes that may serve as biomarkers for co-ingestion of alcohol include cocaethylene (CE), ecgonine ethyl ester (EEE) and norcocaethylene (NCE). Anhydroecgonine methyl ester (AEME) results from cocaine pyrolysis and can be used as a marker for crack cocaine use. Depending on the biological matrix examined, the amount of parent drug (cocaine) varies. The major analytes detected in various biological matrices are summarized in Table 2.2. Although it is infrequently tested for, reports have indicated that ecgonine may also be present in appreciable amounts following cocaine use [14,15]. A metabolic scheme for cocaine is illustrated in Fig. 2.2.

Clinical studies of cocaine administration have demonstrated that BZE is the primary metabolite detected in plasma and urine following cocaine administration by the intranasal, intravenous, oral and smoked routes [16,17]. The formation of BZE results from the cleavage of the methyl ester bond of cocaine through both enzymatic and non-enzymatic pathways [18–20].

Another major metabolite of cocaine, EME, is produced by the enzymatic hydrolysis of the benzoyl ester of cocaine. Both liver and plasma cholinesterases have

TABLE 2.2
CONSIDERATIONS FOR BIOLOGICAL SPECIMEN ANALYSIS

Biological Matrix	Primary Analytes Detected	Ref.
Adipose tissue	COC, BZE, CE	[31]
Brain	COC, BZE	[32,33]
Breast milk	COC	[34]
Hair	COC, BZE	[35,36]
Liver	COC, BZE	[33]
Meconium	COC, BZE, EME, *m*-HOBZE, *p*-HOBZE	[29,37–39]
Nails	COC, BZE	[40–42]
Plasma	BZE ($t_{1/2} = 2.6$–5.1 h)	[16]
Saliva	COC, BZE	[43]
Sebum	COC	[44]
Semen	COC, BZE	[45]
Skin	COC, CE	[31]
Sweat	COC, BZE, EME	[46]
Teeth	COC, BZE	[47]
Urine	BZE ($t_{1/2} = 7.5$ h)	[24]
	EME ($t_{1/2} = 3.6$ h)	
Vitreous humor	BZE	[48–50]

been reported to carry out this conversion [21–23]. EME is generally detected only at low concentrations in plasma following cocaine administration by the intranasal, intravenous and smoked routes [16]. However, following controlled oral administration, EME was present at plasma concentrations up to four times that of cocaine [17]. In urine, EME is usually present at substantial concentrations, sometimes at concentrations greater than BZE [24,25].

Minor metabolites of cocaine include NCOC, BNE, *m*- and *p*-HOCOC and *m*- and *p*-HOBZE. NCOC is produced enzymatically by the human cytochrome P-450 3A4 and has been recognized as a precursor of hepatotoxic cocaine metabolites [26–28]. NCOC has been detected at low concentrations in postmortem blood specimens [29]. However, NCOC was present in plasma at substantial concentrations following oral administration [17]. Both *m*- and *p*-HOCOC have been reported to be produced by hepatic microsomes in mice, rats and guinea pigs [30]. In addition, *p*-HOCOC has been shown to have comparable pharmacological activity to cocaine when administered to mice [30].

There are several unique metabolites that are formed when ethanol is consumed during the period of cocaine use, including CE, EEE and NCE. The presence of these metabolites can serve as a marker for concomitant ethanol and cocaine use. The liver carboxylesterase that forms BZE is also responsible for the formation of CE in the presence of cocaine and ethanol [22,23]. In the presence of ethanol, the transesterification of COC to CE is about 3.5 times faster than hydrolysis to BZE. Another cocaine analyte, AEME, is a cocaine pyrolysis product that can serve as a marker for the smoked route of administration. Ecgonidine (ECGD), a metabolite of AEME, has recently gained attention as another marker for the smoked route of administration that is less likely to have resulted from environmental exposure [51].

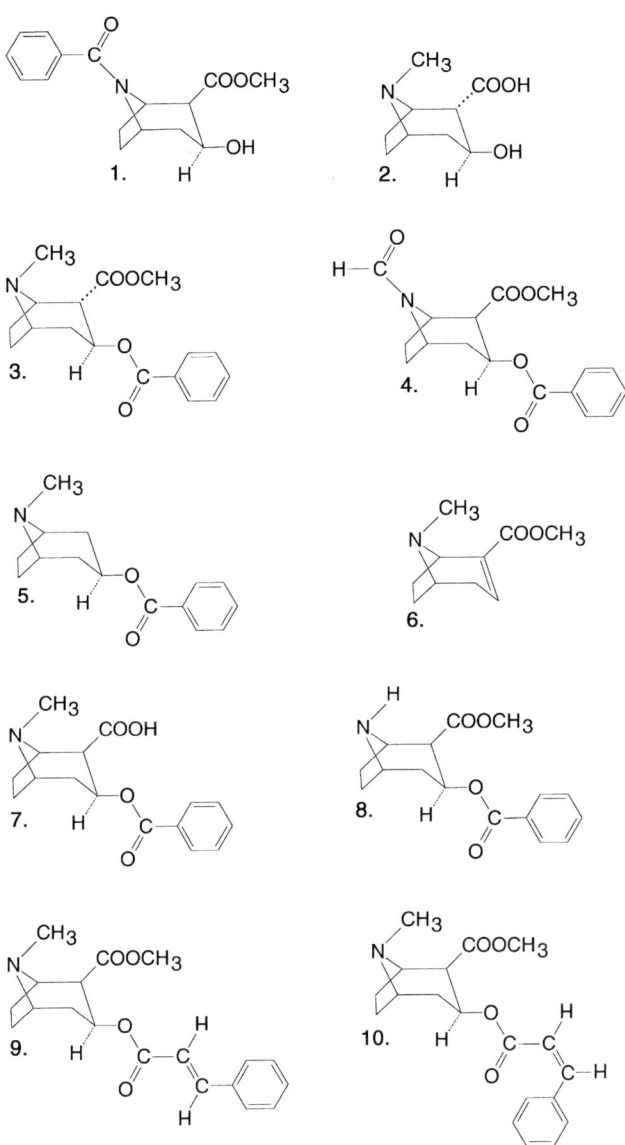

Fig. 2.2. Biotransformation and thermal degradation products of cocaine. Enzymes responsible for metabolic conversions are indicated in parentheses. HCE-1, human liver carboxylesterase-1; HCE-2, human liver carboxylesterase-2.

Analysis of postmortem specimens for ECGD revealed that ECGD was present in postmortem blood, urine and liver at higher concentrations than AEME, indicating that it may be a more useful marker for smoked cocaine use [52]. A unique cocaine metabolite that is formed after the co-abuse of smoked cocaine and ethanol, anhydroecgonine ethyl ester (AEEE), has recently been identified in postmortem urine [53].

2.1.4 Biological specimen considerations

The analysis of cocaine in biological specimens is complicated by the instability of cocaine in various biological matrices. The benzoyl and methyl esters of cocaine can be readily hydrolyzed *via* enzymatic or chemical hydrolysis. Cocaine is particularly vulnerable to hydrolysis in cholinesterase-containing specimens, including blood and plasma. The stability of COC and BZE in whole blood has recently been investigated [54]. In this study, blood was collected from living individuals who were under the influence of cocaine. Blood was collected into vacutainer tubes containing sodium fluoride (0.25%) and potassium oxalate. The blood was analyzed for COC and BZE at the time of collection, stored at ambient temperature and then reanalyzed after three months, six months, one year or two years. Each blood specimen was subjected to only one reanalysis. Cocaine was the least stable of the two analytes, with no detectable cocaine present after the initial analysis at the time of collection. BZE was detected in some specimens after three months, six months and one year at markedly decreased concentrations, but was not detected in any specimen reanalyzed after two years.

Another study found that cocaine is most stable in a refrigerated or frozen blood specimen when the specimen pH is adjusted to 5 and a cholinesterase inhibitor is added [55]. In this study, COC hydrolysis to BZE was shown to be pH dependent (increased pH = increased hydrolysis), while the hydrolysis to EME was shown to be dependent on cholinesterase activity. It was concluded that effective preservation of cocaine blood specimens was attainable by acidification of the specimen in combination with the addition of a cholinesterase inhibitor. However, another study has indicated that cocaine remains fairly stable under refrigerated or frozen conditions (without acidification or inhibitor) for at least 48 h [56].

Fandino *et al.* [57] examined the stability of AEME in human plasma. The authors found that 50% of AEME was hydrolyzed to ECGD in plasma stored at room temperature for five days. However, the stability of AEME was greatly improved by lowering pH to 5, adding an esterase inhibitor and refrigeration. A study examining the stability of AEME and ECGD in sheep plasma reported similar findings [58].

The stability of cocaine and metabolites in urine specimens may not be as critical an issue as it is with blood or plasma specimens. A recent study evaluated the stability of 10 cocaine analytes in frozen urine over a six-month period [15]. The results indicated that most cocaine analytes were stable under these conditions with the exception of BNE, which had a coefficient of variation of 34%. Another study assessed the stability of BZE in frozen urine specimens and found no significant changes in the BZE concentration over a period of 45 days [59]. A separate study evaluated concentrations of EME and COC in urine under refrigerated and frozen conditions over a period of six months [60]. The authors observed that EME concentrations remained stable under both conditions although COC concentrations decreased substantially.

2.1.5 Postmortem stability issues

Concerns arise regarding the stability of cocaine when postmortem specimens are collected. The postmortem interval may be exceedingly long, and the autopsy, laboratory analysis and storage conditions may result in cocaine hydrolysis (chemical and enzymatic). This process is particularly evident in unpreserved blood where the majority of cocaine can disappear as a result of hydrolysis. In addition to chemical and enzymatic degradation, concentrations are also affected by the postmortem release of cocaine analytes that have been sequestered in tissues. Several studies have demonstrated that postmortem cocaine release may contribute to the lack of agreement between cocaine concentrations in blood collected from different sites [61,62]. However, the postmortem redistribution pattern of cocaine and metabolites does not appear to occur in a predictable manner [62]. Consequently, it is important to carefully document the site of collection of all specimens. Sometimes, conventional postmortem specimens are unavailable for cocaine analysis. In these cases, alternative specimens may be analyzed. Cocaine and BZE have been detected in insect larvae, beetle feces, decomposed skeletal muscle, bloody decomposition fluid and mummified tissue [63,64].

Occasionally, the postmortem toxicologist is tasked with the analysis of embalmed or formalin-fixed tissues. Cingolani *et al.* [65] reported on the detection of cocaine and BZE in fixed liver tissue and formalin solution. Briefly, their study involved the analysis of fresh liver tissue, formalin-fixed liver tissue and the formalin solutions in which the liver tissues were fixed. They included postmortem liver specimens from four cocaine-positive cases. The fixed liver specimens and formalin solutions were analyzed after four weeks of preservation and compared to the results obtained from the fresh liver analysis. The results indicated that 12% of the BZE present in the fresh tissues was recovered in the fixed tissues, while 84% was recovered in the formalin solutions. In addition, cocaine was detected in one fresh liver specimen but was not detected in the fixed tissue or formalin solution. The authors concluded that BZE shows good stability in formalin solutions and that the double evaluation of BZE in fixed tissues and formalin solutions may provide a more accurate indication of the original quantity of cocaine and BZE present in the same tissues before preservation.

2.2 CURRENT METHODS FOR THE DETERMINATION OF COCAINE

2.2.1 Immunoassay-screening methods

Immunoassays are commonly used for drug-screening purposes due to their simplicity, sensitivity and rapidity. These screening methods are typically instrument based, have the capability to analyze many samples in a short period of time and require little or no sample preparation. The general mechanism of these assays is based on the formation of an antibody–antigen complex and each assay targets a specific analyte. Immunoassays for cocaine detection typically use BZE as the target analyte, although considerable cross-reactivity to other cocaine analytes is observed with some assays. The types of immunoassays that will be discussed include Enzyme-Multiplied Immunoassay Technique® (EMIT®), radioimmunoassay (RIA), fluorescence polarization immunoassay (FPIA), kinetic interaction of microparticles in solution (KIMS), Cloned Enzyme Donor Immunoassay (CEDIA®), enzyme-linked immunosorbent assay (ELISA) and various on-site testing devices. A brief description of each immunoassay will be provided; additional details are included in a chapter by Smith [66]. Immunoassays yield qualitative and semi-quantitative results. Thus, a specific confirmatory technique such as GC-MS or high-performance liquid chromatography (HPLC) should be utilized for quantitative analysis.

2.2.1.1 Enzyme-Multiplied Immunoassay Technique®

EMIT®, a trademark method of Dade Behring, Inc. (Deerfield, IL), is a homogenous enzyme assay that utilizes an antigen linked to the enzyme glucose-6-phosphate dehydrogenase. This enzyme oxidizes glucose-6-phosphate to guconolactone-6-phosphate and reduces the cofactor nicotinamide adenine dinucleotide (NAD) to NADH. The enzyme activity is determined by spectrophotometrically measuring the amount of NADH produced. When antibodies specific to the particular drug of interest are mixed with the specimen, they will bind to drug in the specimen in addition to the enzyme-labeled antigen. When the enzyme-labeled antigen binds to antibody, the enzyme becomes less active. The amount of NADH produced is directly related to the amount of drug present in the specimen. Advantages of EMIT® include the long shelf lives of kits and the ability to measure bound drug without separating from unbound drug. A notable disadvantage, however, is the possibility of interferences from cross-reacting components in the matrix.

In a study by Luzzi *et al.* [67], the objective was to evaluate the performance of immunoassays for drugs of abuse below established Substance Abuse and Mental Health Services Administration (SAMHSA) cutoff concentrations. The EMIT® reagents were run on a Hitachi 717 using a single-calibration point. The authors reported that a coefficient of variance (CV) of <20% was achieved at a cutoff concentration of 60 µg/L BZE for the EMIT® cocaine metabolite assay.

ElSohly *et al.* [39] used the EMIT® assay to analyze meconium samples. The authors evaluated several sample preparation methods for meconium screening and determined that methanolic extraction was most effective. The cutoff concentration used in this study was 200 ng/g BZE. The GC-MS BZE confirmation rate of positive specimens was 67%, which led the authors to recommend that additional cocaine metabolites should be included in the GC-MS confirmation procedure for meconium.

Contreras *et al.* [68] analyzed pericardial fluid using both semi-quantitative EMIT® and GC-MS. They found consistently low results for cocaine analysis *via* EMIT® compared to GC-MS. They suggested the lower EMIT® result could be due to matrix interferences since pericardial fluid contains higher concentrations of sodium chloride than urine (the specimen for which EMIT® is designed). The cocaine concentrations in pericardial fluid were approximately two-times higher than those in blood while BZE concentrations were about 50% higher in pericardial fluid relative to blood.

2.2.1.2 Radioimmunoassay

Radioimmunoassay is a heterogeneous assay that employs radiolabeled drug (typically ^{125}I). There are two different methods of RIA that are commonly employed for drug detection in biological matrices, double-antibody RIA and coated-tube RIA. With double-antibody RIA, a second antibody is added to facilitate precipitation of the bound primary antibody. Once the primary/secondary antibody–antigen complex precipitates, the unbound labeled drug can be easily removed. With coated-tube RIA, the primary antibody is coated on the inside of each tube. The unbound labeled drug can be easily removed by pouring off the supernatant. The samples from each RIA method are analyzed in a gamma counter to determine the counts per minute, which is inversely proportional to the amount of drug present in the original specimen. Radioimmunoassays are both sensitive and specific but require special handling and disposal of radioactive waste. Additionally, shelf life is limited to approximately two months due to radioactive decay.

Swartz *et al.* [69] used an RIA method for the detection of illicit substances in hair of 203 individuals with schizophrenia. The drug use of these subjects was assessed with self-reported illicit substance use, hair analysis and urine analysis. The hair samples were screened by RIA; no confirmatory testing was performed. The authors found that 40 subjects tested positive for cocaine by hair analysis, while only 13 subjects had a positive urine test and only 13 self-reported substance use. The authors concluded that RIA of hair appears to be a promising method for improving the detection of illicit substance use among persons with schizophrenia.

2.2.1.3 Fluorescence polarization immunoassay

FPIA involves a competitive binding reaction between antibody, drug and a tracer, which is typically fluorescein-labeled drug. The antibody can bind both the drug in the specimen and in the tracer. If the drug concentration in the specimen is high, more of the tracer will remain unbound. However, if the drug concentration in the specimen is low, more of the tracer will bind to the antibody. If the tracer is bound to antibody, it will not rotate freely and there will be little loss of polarization. If the tracer is unbound, the polarization is reduced. Consequently, the amount of drug present in the specimen is inversely proportional to the intensity of the polarized light. Abbott Laboratories is a major commercial supplier of FPIA screening assays that run on their analyzers (TDx®, ADx® and AxSYM®).

Cone *et al.* [70] reported a method for enhancing the detection of positive results in urine specimens with the TDx® cocaine metabolite assay. The authors developed the zero-threshold criteria method to detect specimens that contained sufficient BZE to confirm by GC-MS but produced a screening result below the cutoff concentration. This methodology was based on an evaluation of the immunoassay response rates of the negative specimens. Application of this method resulted in a 50% increase in the detection of specimens positive for BZE. Schilling *et al.* [71] compared the ADx® system with two on-site testing devices, the ONTRAK kit from Roche Diagnostics and the EZ-SCREEN kit from Environmental Diagnostics. The authors concluded that the ADx® assay had the best sensitivity and specificity of the three when used for detecting cocaine and opiates in 345 urine specimens.

Luzzi *et al.* [67] evaluated the performance of the TDx® cocaine metabolite assay below the established SAMHSA cutoff concentration. The TDx® reagents were run on an Abbott AxSYM® using a six-point calibration curve. The authors reported that a coefficient of variance of <20% was achieved at a cutoff concentration of 72 µg/L BZE for the TDx® cocaine metabolite assay.

2.2.1.4 Kinetic interaction of microparticles in solution

Roche Diagnostics patented the KIMS method, which is sold as Abuscreen® OnLine automated assays. This assay is based on the interaction of antibody, drug–microparticle conjugate and free drug in the specimen. The drug present in a specimen competes with drug–microparticle conjugate for antibody-binding sites. When a specimen contains no drug, the drug-microparticle conjugate binds to the antibody and forms particle aggregates that scatter transmitted light, resulting in an increase in absorbance. When a specimen contains the drug of interest, less of the antibody is available to bind the drug-microparticle conjugate, preventing aggregate formation. This results in a decrease of light absorbance proportional to specimen drug concentration. At a urine cutoff concentration of 300 ng/mL BZE, the Abuscreen® On-Line Cocaine Metabolite-Qualitative assay shows low cross-reactivity to other cocaine analytes (<5% for COC, EME and ECG).

Lu and Taylor [72] compared OnLine KIMS and EMIT® II in two different laboratories and found a concordance of 97–99% for cocaine analysis results from the two assays. The urine specimens analyzed in this study were collected from participants in the Arrestee Drug Abuse Monitoring program. The screen results were confirmed by GC-MS. The EMIT® II assay correctly identified seven negative specimens that screened positive by the KIMS assay. The authors concluded that the EMIT® II assay has greater specificity in detecting true cocaine negatives and is more sensitive in detecting true cocaine positives.

Feldman *et al.* [73] evaluated the Roche Diagnostics ONLINE® DAT II, an improved type of KIMS assay. This assay modified the original KIMS assay by coupling the antibody to the microparticle surface and by keeping a multivalent drug-conjugate in solution. Improvements in reagent shelf lives, stability of the standard curve and selectivity for the cocaine assay were observed when comparing ONLINE® II to OnLine I and CEDIA® assays.

2.2.1.5 Cloned Enzyme Donor Immunoassay®

CEDIA® is a competitive, homogenous assay marketed by Microgenics Corporation. In this technique, the bacterial enzyme β-galactosidase has been genetically engineered into two inactive fragments – the enzyme acceptor (EA) and the enzyme donor (ED), which can reassemble to form active enzyme. The reassociated enzyme cleaves a substrate, producing a color change that is read spectrophotometrically.

The ED fragment is linked to the drug and cannot reassemble to form active enzyme when it is bound to antibody. When there is no drug present in the sample, the antibody binds to the enzyme–drug conjugate on the inactive fragment, preventing the formation of active enzyme. When drug is present in the sample, it competitively binds to antibody, leaving the enzyme–drug conjugate free to form active enzyme. The concentration of drug in the specimen is directly proportional to the amount of active enzyme formed and the resultant absorbance change. At a urine cutoff concentration of 300 ng/mL BZE, the CEDIA® assay shows good cross-reactivity for COC and CE but low cross-reactivity for EME and ECG.

Kidwell *et al.* [74] analyzed urine, PharmChek™ sweat patches and skin swabs to assess cocaine use of 10 subjects over a four-week period. Three different immunoassay techniques were investigated in this study – Microgenics CEDIA®, Cozart® ELISA and OraSure ELISA. Both the OraSure and Cozart® assays produced acceptable results that correlated well. The Microgenics CEDIA® assay showed sufficient sensitivity for the matrices tested but had poor precision, which the authors suspected was affected by running the assay in manual mode rather than using an automated analyzer. Comparison of the results of the matrices analyzed suggested that the most reliable method for detecting cocaine use was daily urinalysis.

In a study by Chronister *et al.* [75], the CEDIA® DAU cocaine assay was used for the detection of BZE in vitreous humor specimens. The vitreous humor specimens were analyzed directly, with no pretreatment. Of the 392 vitreous humor specimens tested, 23 were positive on the CEDIA® assay, which used a cutoff concentration of 100 ng/mL. Only one presumptive positive sample was not confirmed by GC-MS, and this sample had a screen value near the cutoff. The authors also reported that analysis of vitreous humor specimens for cocaine improved the detection rate of cocaine analytes by 0.7% during the 14-month study.

Hattab *et al.* [76] evaluated modified cutoff concentrations with the CEDIA® cocaine metabolite immunoassay to determine if lower threshold concentrations would improve the detection of drug use or exposure in pregnant women and newborns. The threshold calibrator for cocaine was lowered by diluting 300 ng/mL threshold calibrator 1:3 with saline solution. The authors analyzed 911 urine specimens that initially screened negative for cocaine metabolite with the modified threshold assay. The lower threshold resulted in the detection of an additional 38 positive specimens, 28 of which were confirmed by GC-MS.

2.2.1.6 Enzyme-linked immunosorbent assay

ELISA is a competitive, heterogeneous enzyme immunoassay. There are different variations of this technique and kits are available from a number of manufacturers. In the kits from OraSure Technologies, Inc. (Bethlehem, PA), specimen and enzyme-conjugated drug are added to microtiter plates, which are coated with drug-specific antibodies. The samples are incubated and free drug in the specimen and drug–enzyme conjugate compete for antibody-binding sites in the plates. Following incubation, the plates are washed to remove any unbound drug. Enzyme activity is reduced when the enzyme–drug conjugate is bound to the antibodies coated on the microtiter plate. Addition of a substrate solution produces a colored product in the presence of active enzyme. After a second incubation, the enzyme reaction is stopped and the absorbance is read. The amount of color produced is inversely proportional to the amount of drug present in the original specimen. While ELISA immunoassays tend to be more expensive than most homogenous assays, this method is advantageous in many ways. ELISA not only has excellent sensitivity, but it is also less subject to matrix effects, is easily automated and the kits have relatively long shelf lives.

Kerrigan and Phillips [77] compared kits from two different ELISA vendors, Immunalysis Corp. and STC Diagnostics, Inc. (presently OraSure Technologies, Inc.) for the analysis of whole blood and urine. The STC kit required 500 μL blood and 200 μL urine for analysis, whereas 10 μL blood and urine were required for the Immunalysis assays. Evaluation of the analytical performance for kits from both vendors indicated greater sensitivity for the Immunalysis assay and improved precision and lot-to-lot reproducibility for the STC assay. The total CV for the STC BZE kit was 7.5% compared to 23% for Immunalysis. The LODs for blood and urine for STC were 5 and 38 μg/L BZE compared to 5 and 19 μg/L for Immunalysis.

Niedbala *et al.* [78] reported cross-reactivities and precision data for the STC presently OraSure Technologies, Inc. Cocaine Metabolite MICRO-PLATE EIA kit for the analysis of urine and oral fluid. The STC assay was 100% cross-reactive to BZE, 12.9% to COC and 13.8% to CE. Nine adulterants including food products and household chemicals as well as numerous commonly encountered drugs were reported not to interfere with the screening assay. When the EIA screening results were compared with GC-MS results, the oral fluid assay was slightly less sensitive in the detection of cocaine metabolites compared to the urine assay.

Lachenmeier *et al.* [79] analyzed hair samples for the presence of cocaine using OraSure oral fluid ELISA kits. The oral fluid kit was selected because of its high cross-reactivity with parent drug, which is why an oral fluid kit is actually better suited for hair analysis compared to urine immunoassays. Hair specimens required washing for 5 min with 5 mL deionized water, followed by 5 mL of petroleum benzene and then 5 mL methylene chloride. Methanol (4 mL) was then added to 50 mg of hair, and the extraction was performed by ultrasonication for 5 h at 50 °C. For ELISA analysis, 1 mL of the methanol was evaporated to dryness and reconstituted in 250 μL buffer solution (pH 7.4). A cutoff of 5 ng/mL was used for cocaine. When compared with GC-MS confirmation results, the sensitivity of the ELISA assay for the detection of cocaine in hair was 100% while the specificity was only 66%.

Moore *et al.* [80] evaluated the STC (presently OraSure Technologies, Inc.) Cocaine 5-plate Saliva Micro-Plate kit for the detection of cocaine in hair specimens. Specimens were also analyzed by FPIA and GC-MS. Sample preparation in this study involved adding 3 mL 0.1 M hydrochloric acid to powdered hair samples followed by overnight incubation at 60 °C. After decanting the acid, 200 μL ethanol, 2 mL deionized water and 200 μL 12 N sodium hydroxide were added, followed by 30 min of incubation at 80 °C. Hair samples were allowed to cool, and then 1 mL glacial acetic acid was added dropwise. Next the samples were centrifuged, and the supernatant was combined with the acid and the hair was discarded. Lastly, 9 mL deionized water was added to each extract. The authors concluded that the STC Micro-Plate EIA is a valid alternative to other screening assays for the detection of cocaine in hair. In addition, the authors reported that the EIA shows increased sensitivity, specificity and efficiency relative to the FPIA assay.

Spiehler *et al.* [81] evaluated the performance of the Neogen™ Cocaine/BZE microtiter plate ELISA assay for the screening of postmortem blood. Blood specimens were analyzed at 1:5 and 1:50 dilutions. The optimal cutoff concentrations were determined to be 5 ng/mL BZE for the 1:5 dilution and 50 ng/mL BZE for the 1:50 dilution. The authors reported that the specimens from decomposed bodies produced non-zero results, but none were above the 5 ng/mL cutoff at a 1:5 dilution. The sensitivities and specificities were all >93% at both cutoff concentrations.

The sensitivity, specificity and efficiency of the Cozart® Microplate EIA Cocaine Oral Fluid kit at proposed oral fluid screening and cutoff concentrations were determined by Kim *et al.* [82]. GC-MS was employed as a reference method. SAMHSA's proposed cutoffs for oral fluid testing are 20 µg/L BZE for screening and 8 µg/L cocaine and/or BZE for confirmation. At these cutoffs, sensitivity was 92.2%, specificity was 84.7% and efficiency was 88.8%. Similar results were obtained using the United Kingdom's oral fluid cutoff concentrations of 30 µg/L for screening and 15 µg/L cocaine, BZE and/or EME for confirmation.

In a validation study by Cooper *et al.* [83], the specificity and sensitivity of the Cozart® Microplate EIA kit for cocaine in oral fluid were 95.7% and 100%, respectively (using a screening cutoff of 30 ng/mL BZE and a confirmation cutoff of 15 ng/mL BZE). Potential adulterants, such as ethanol, chewing gum, coffee and water, were analyzed for interference with the assay performance, and none were found to alter the results.

2.2.1.7 On-site testing kits

On-site testing kits, also known as point-of-collection and point-of-care (POC) devices, are widely used for their rapid determination of drug-free urine specimens. These devices are generally based on immunochromatography where a target analyte migrates along a test strip. The analyte competes for antibody at fixed locations along the strip, and a colored reaction line indicates that analyte has been detected in the specimen [84]. In addition to the rapid results of on-site testing kits, there is no need for special collection facilities or highly trained testing personnel. However, since on-site devices are considered screening tests, a confirmatory test should be completed for all positive results.

The Cozart® RapiScan Oral Fluid Drug Testing System was evaluated for cocaine detection by Kolbrich *et al.* [85]. The authors compared this on-site testing device to the Cozart® Microplate ELISA system and found similar performance between the two assays. In addition, the RapiScan device showed suitable sensitivity and specificity at the proposed SAMHSA oral fluid cutoff concentrations. Fucci *et al.* [86] evaluated the RapiScan system for the detection of drugs in 146 vitreous humor samples. Since vitreous humor is of limited volume and typically contains low concentrations of drugs, the screening technique used to analyze this matrix must be sensitive and require little specimen volume. Due to the high viscosity of vitreous humor, the samples were centrifuged prior to analysis, but no other pre-treatment was necessary. Seventeen percent of the vitreous humor samples screened positive for cocaine. All positive results and 20% of the negative results were con-firmed by GC-MS. No false negative or false positive cocaine results were observed.

Moody *et al.* [84] compared two on-site urine-testing devices, the Instant-View 5-Panel Test Card and OnTrak TesTcup Pro 5 with GC-MS analysis. For all three tests, 300 ng/mL BZE was used as the cutoff. Analysis of precision samples fortified near the cutoff determined an accuracy of 82.1% for the Instant-View device and an accuracy of 90.7% for the TesTcup device. Yacoubian Jr. *et al.* [87] reported a comparison study of OnTrak TesTcup Pro 5 to urinalysis *via* EMIT®. Urine sam-ples from 136 male arrestees were analyzed in this study. Using the EMIT® result as a reference, the sensitivity for cocaine was 86%, the specificity was 99% and the percent agreement of results between the two assays was 96%.

2.2.2 Sample preparation

Biological specimen analysis most often requires a sample preparation and concen-tration step prior to analysis. Frequently, this is accomplished with liquid–liquid extraction (LLE) or solid-phase extraction (SPE). The SPE extraction phases that are most commonly reported for cocaine analysis include the nonpolar C_8 and C_{18} phases, strong cation exchange phases and specialty mixed-mode phases that com-bine a nonpolar phase with a strong cation exchange phase. The mixed-mode phases are frequently employed for the isolation of cocaine and several metabolites.

The short half-life and extensive metabolism of cocaine necessitates the inclusion of metabolites in assays to provide sensitive and extended detection. However, ex-traction of multiple cocaine analytes from a biological matrix can be challenging due to their different physiochemical properties, and the most efficient extraction tech-nique is dependent upon the analytes of interest. The cocaine analytes that are most readily isolated from biological fluids by LLE are the weak base analytes, including COC and NCOC, since they will extract into an organic solvent at a basic pH. However, isolation of cocaine analytes that are amphoteric, including BZE, BNE and ECG by LLE is more complicated. An efficient LLE requires careful regulation of the pH and selection of a solvent that has sufficient polarity to extract these polar analytes. The properties of selected cocaine analytes are displayed in Table 2.3. A liquid–liquid extraction can generally be optimized for several components, but for isolation of multiple cocaine analytes from biological matrices, mixed-mode SPE appears to be the preferred technique. The mixed-mode SPE columns are especially useful for the isolation of multiple cocaine analytes since the two phases can provide excellent recovery for both nonpolar and polar analytes in a single extraction pro-cedure. The advantages and disadvantages of LLE and SPE are summarized in Table 2.4.

Several published methods have utilized LLE schemes to extract cocaine and analytes [88–92]. The solvents employed varied, generally depending on the analytes

TABLE 2.3
SELECTED PROPERTIES OF COCAINE AND ANALYTES

Analyte	Chemical Properties
Cocaine	Weak base ($pK_a = 8.6$)
Benzoylecgonine	Amphoteric ($pK_a = 2.25$ and 11.2)
Ecgonine	Amphoteric ($pK_a = 2.8$ and 11.1)
Ecgonine methyl ester	Weak bases
Norcocaine	
Cocaethylene	
Ecgonine ethyl ester	
Anhydroecgonine methyl ester	
Norcocaethylene	
Benzoylnorecgonine	Amphoteric
Ecgonidine	
m-Hydroxycocaine	Behave similarly to parent compound (COC or BZE)
p-Hydroxycocaine	
m-Hydroxybenzoylecgonine	
p-Hydroxybenzoylecgonine	

TABLE 2.4
ADVANTAGES AND DISADVANTAGES OF LLE AND SPE

Technique	Advantages	Disadvantages
Liquid–Liquid Extraction	• Increased sample contact with solvent may be beneficial for solid specimens such as tissues or hair • Relatively inexpensive	• Often requires considerable volume of solvent • Conditions typically are not optimal for multiple drugs
Solid-phase extraction	• Minimal solvent volume required • Suitable for multiple drugs with differing chemical properties • Less time required for extraction • Easily automated • Cleaner extracts are often produced • Results are usually very reproducible	• Lot-to-lot variability of SPE columns possible • Columns can be expensive

of interest. LLE has been applied quite successfully for the determination of COC, NCOC and CE [88–90]. However, when BZE and BNE are incorporated into the assay, it becomes difficult to optimize the LLE scheme for all analytes. Ma *et al.* [90] used a chloroform:ethanol mixture to extract COC, BZE, NCOC, CE, NCE and BNE from serum. They indicated that the recovery of BZE and BNE was dependent on the chloroform:ethanol ratio, since both analytes are insoluble in chloroform. While maximizing the recoveries of COC and NCOC, their method attained a recovery of only 35–40% for BNE. Lampert *et al.* [93] experienced a similar situation while attempting to isolate COC, NCOC, BZE and BNE by LLE. Sandberg *et al.* [94] attempted to use LLE for COC, BZE, NCOC and BNE but selected to use SPE as a result of the decreased recovery and sensitivity and increased variability experienced with the solvent extraction. In addition, Clauwaert *et al.* [92] evaluated 20 different LLE schemes for the analysis of COC, BZE and CE and compared the results to those obtained following SPE. They found that SPE provided increased extraction recoveries and superior chromatogram quality. The authors also found that sonication and dilution of blood specimens alleviated the problem of clogged SPE columns.

The preferred extraction technique for the analysis of cocaine and analytes by HPLC and GC appears to be SPE with a mixed mode column. The majority of the referenced methods using this type of SPE column reported extraction recoveries of at least 70%. The analytes extracted by these methods included COC, BZE, EME, NCOC, BNE, *m*- and *p*-HOCOC, *m*- and *p*-HOBZE, CE, EEE, NCE and AEME. Virag *et al.* [95] reported a method for the analysis of COC, BZE, EME and NCOC using a mixed-mode SPE column. The authors indicated that BNE could readily be incorporated into this method if the elution solvent volume was increased from 6 mL to 10 mL. Another SPE scheme utilized a strong cation exchange column followed by a C_{18} column [96]. The authors incorporated two SPE columns into their assay because they found that no single SPE extraction phase would cleanly and efficiently extract COC, BZE, NCOC and BNE from serum.

Ecgonine and ecgonidine are very polar cocaine analytes that are difficult to extract efficiently in combination with other cocaine analytes. In the most commonly employed SPE methods, ECG and ECGD are not retained during sample loading at pH close to 6. Nishikawa *et al.* [97] isolated COC, BZE, EME, NCOC and ECG with two SPE cartridges (Bond Elut Certify and Bond Elut SCX). The Bond Elut Certify column retained COC, BZE, NCOC, but only a small fraction of EME and ECG, so the sample was collected after passing through this column and then applied to the SCX column for EME and ECG extraction. Moreover, the authors were able to improve ECG recovery by the addition of ethylenediaminetetraacetic acid (EDTA) to the sample prior to its application to the SCX column. Paul *et al.* [15] developed a method for the analysis of 14 cocaine analytes in blood and urine. ECG, ECGD and norecgonidine were incorporated into this procedure that included two solid-phase extractions on mixed-mode columns (C_8 and benzene sulfonic acid). Briefly, the sample was adjusted to pH 6.0 and loaded onto the SPE columns. ECG, ECGD and norecgonidine, which were not retained, were collected, readjusted to pH 2–3 and passed through another SPE column. The authors reported that the acidic conditions of the second extraction allowed the columns to adsorb the polar compounds more efficiently. In addition, a polar elution solvent (methanol, 2-propanol and ammonia) was necessary to elute ECG, ECGD and norecgonidine from the columns; the more commonly used elution solvent (methylene chloride, 2-propanol and ammonia) gave poor recovery. The overall recovery for ECG was 29% and recoveries for ECGD and norecgonidine were greater than 83%. This method provides good recovery and limits of detection (LODs) for multiple cocaine analytes; however, it may be difficult to adapt to routine analysis since it involves a dual solid-phase extraction as well as several derivatization schemes and instrumental parameter settings for different cocaine analyte groups.

Simultaneous extraction of ECG and ECGD with other cocaine analytes has produced mixed results. Cardona *et al.* [98] developed a procedure for the analysis of ECG and 10 additional cocaine analytes in blood, urine and muscle. This procedure involved extraction with a Bond Elut® Certify SPE column, hydrochloride salt formation, derivatization with PFPA/PFPOH and GC-MS analysis. Ecgonine was extracted with a recovery of 15%, resulting in an LOD of 640 ng/mL and a limit of quantitation (LOQ) of 800 ng/mL. Lewis *et al.* [99] published a similar method. This method also employed extraction with a Bond Elut® Certify SPE column, derivatization with PFPA/PFPOH and GC-MS analysis for the analysis of ECG, ECGD and 11 other cocaine analytes. The extraction recoveries of ECG and ECGD ranged from 0.7 to 4.1%; however, the authors were able to achieve an LOQ of 25 ng/mL for ECG and 12.5 ng/mL for ECGD.

2.2.3 Thin layer chromatographic procedures

Thin layer chromatography (TLC) is a method of separation of the components of a mixture by elution on a stationary phase composed of finely divided particles bonded on a plate. The mobile phase is a liquid phase that moves through the stationary phase by capillary action. The most frequently employed stationary phase for cocaine analysis is silica gel. Various mobile phases have been used in combination with silica gel plates for cocaine analyses. The parameters for the detection of cocaine with recently reported TLC methods are summarized in Table 2.5. In addition, a commercially available thin layer chromatography kit, Toxi-Lab®, is produced by the Ansys Corporation. The Toxi-Lab® A system can be used to detect cocaine and BZE in urine. The LODs for these two analytes with the Toxi-Lab® A system are 1.0 μg/mL for cocaine and 1.0 μg/mL or 250 ng/mL for BZE.

TLC is a suitable analytical technique for several sample types, including drug standards, illicit drug mixtures and biological specimens. Recent methods have primarily focused on detection of COC, BZE and CE [100–102]. Comparison of a high-performance TLC analysis to HPLC analysis for the detection of cocaine, BZE and CE in urine indicated that the high-performance TLC method was a suitable analytical method only when the three analytes were present in high concentrations [103]. Advantages of the TLC methods include minimal sample preparation (generally a single-step LLE or SPE) and a short analysis time, making them useful as a screening technique. However, they are of limited quantitative use because of their high LODs.

2.2.4 Liquid chromatographic procedures

Since the early 1970s, liquid chromatography has developed as the foremost separation method for organic substances. Because the mobile phase is a liquid, the requirement for vaporization is eliminated, and therefore, liquid chromatography can separate a much broader range of substances than gas chromatography. Species that have been successfully resolved include inorganic ions, amino acids, drugs,

TABLE 2.5
TLC METHODS FOR THE DETECTION OF COCAINE AND ANALYTES

Analytes Detected	Specification Type	Sample Preparation	Stationary Phase	Mobile Phase	Visualization Technique	Limit of Detection (LOD)	Ref.
COC and CE	Urine	SPE – non-ionic resin column	Silica gel	Hexane:toluene:diethylamine (65:20:5)	Iodoplatinate spray	5 μg/L	[101]
COC	Urine	BZE methylated to COC: SPE – Chem Elut	Kieselgel 60 F$_{254}$ 5554	EtOAc:MeOH:ammonia:water (43:5:0.5:1.5)	Dragendorff followed by iodoplatinate	500 μg/L	[100]
BZE (COC, EME and ECG visible under TLC conditions)	Urine	SPE – Clean Screen DAU	Silica gel	EtOAc:MeOH:dichloromethane:ammonium hydroxide (3:3:1:0.6)	Ludy Tenger's reagent followed by 20% sulfuric acid	Approximately 0.28 mg/L	[104]

sugars, oligonucleotides and proteins. Both analytical-scale liquid chromatography with samples at the picogram-to-milligram level and preparative-scale liquid chromatography at the tens-of-grams level are available.

Similar to other chromatographic techniques, liquid chromatographic separation is based on the relative amounts of each solute distributed between a moving solvent stream, called the mobile phase, and a contiguous stationary phase. Kinetic molecular motion continuously exchanges solute molecules between the two phases. The separation occurs because the different species are transported at different rates in the direction of solvent flow. The driving force for solute migration is the moving solvent, and the resistive force is the solute affinity for the stationary phase; the combination of these forces, as manipulated by the analyst, produces the separation.

Very small particles with a thin film of stationary phase placed in small-diameter columns enable liquid chromatography to achieve the resolving power of gas chromatography. The development of the technique now termed HPLC was dependent on the development of pumps that would deliver a steady stream of liquid at high pressure to the column to force the liquid through the narrow interstitial channels of the packed columns at reasonable rates and detectors that would sense the small sample sizes analyzed. After significant improvements from classical liquid–solid chromatography, now HPLC is conducted with porous particles as small as 3–5 μm in diameter, and liquid pumps are used to drive the liquid through the particle-filled column. High resolution and fast separations are achieved since the small particles provide good efficiency with fast mobile phase velocities (1 cm/s or higher). This technique is also important in purification, and separated substances can be automatically collected after the column using a fraction collector.

The most common liquid–solid chromatography technique for drug analysis is reversed-phase chromatography, in which the liquid mobile phase is an aqueous fraction combined with an organic solvent such as methanol or acetonitrile and the stationary phase surface is nonpolar or hydrocarbon-like. In contrast to normal phase chromatography, where the adsorbent surface is polar, in reverse-phase chromatography, the elution of substances from the column is in the order of decreasing polarity. In addition, separation is dependent on the nonpolar properties of the substances.

Some forms of chromatography can detect substances present at picogram (10^{-12} gram) levels, thus making the method a superior trace analytical technique extensively used in the detection of chlorinated pesticides in biological materials and the environment, in forensic science and in the detection of both therapeutic and abused drugs. Its resolving power is unmatched among separation methods.

Isocratic mobile phases are sometimes employed, but the general elution problem encountered in liquid chromatography involves samples that contain both weakly and strongly retained solutes. This is handled in a manner analogous to the temperature programming used in gas chromatography. With the process of gradient elution, the concentration of well-retained solutes in the mobile phase is increased by constantly changing the composition, and hence the polarity, of the mobile phase during the separation. Gradient elution is often applied to cocaine analysis since the polarities of some cocaine analytes vary substantially.

Liquid chromatographic detectors sense the solute in the mobile phase as they emerge from the column. There are various types of detectors routinely used for the liquid chromatographic analysis of drugs and their metabolites, including the ultraviolet spectrophotometer (and photodiode array), electrochemical, fluorescence and mass spectrometric detectors. The detectors most frequently employed for cocaine analysis are the ultraviolet spectrophotometer and the mass spectrometer.

2.2.4.1 *Liquid chromatography with ultraviolet spectrophotometry or photodiode array detection*

The majority of HPLC-UV methods (Table 2.6) for the detection of cocaine and analytes have employed various extraction techniques to prepare specimens for analysis, primarily SPE. However, several methods have been developed that involve direct injection of biological specimens [48,105]. One direct injection method used a precolumn (CH-8 Lichrospher) to concentrate analytes from the injected specimen. This method was applied to vitreous humor (VH), plasma and cerebrospinal fluid (CSF). The authors commented that although this technique minimized specimen preparation, a major disadvantage was the decreased lifetime of the preconcentration column, about 50–60 injections for VH and CSF and only 10–15 injections for plasma [48]. Another direct injection method utilized a cyano precolumn to concentrate cocaine analytes in urine [105]. The cyano column was found to provide superior recovery and cleaner extracts when preliminary evaluations were completed with C_8, C_{18} and cyano SPE columns. Applying this method, the authors were able to inject 50 urine specimens before a significant rise in back pressure occurred. Another method that minimized sample preparation included only centrifugation, evaporation and filtration steps [106]. The simplified sample preparation produced good recoveries of COC, BZE, tropacocaine and benzoic acid from serum, hepatic cytosol and microsomes. Also, this paper provided a means of assessing cocaine metabolism in cell cultures, a tool that may be useful for the elucidation of additional metabolic pathways for cocaine.

Most of the HPLC methods that employed ultraviolet spectrophotometry (UV) or photodiode array detection did not include EME or ECG because these analytes do not have a chromophore that enables sensitive UV detection. To include EME in their assay, Virag *et al.* [95] derivatized EME to *p*-fluorococaine. This provided reproducible and sensitive detection of EME.

The most common type of HPLC column used for cocaine analysis was the reverse phase C_{18} column. Kim and Bornheim [106] reported that the use of a double-endcapped C_{18} column eliminated that need for mobile phase chemical modifiers. Other authors made modifications to the C_{18} column including the addition of a precolumn filter, a C_{18} precolumn or a C_8 precolumn [90,95,107,108]. Both of the direct injection techniques employed a C_8 column, one with a C_8 precolumn and the other with a cyanopropyl precolumn [48,105]. Other column types used included cyanopropyl, C_6 and tandem cyanopropyl and silica [89,91,93,95]. Williams *et al.* [89] used a cyanopropyl column to separate COC, CE and NCOC without the use of an ion-pairing agent or column heating. The use of tandem

TABLE 2.6
HPLC-UV METHODS FOR THE DETECTION OF COCAINE AND ANALYTES

Analytes Detected	Specification Type	Sample Preparation	Internal Standard(s)	Column Type	Mobile Phase	Detector Details	Performance Characteristics	Ref.
COC and CE	Serum	Specimens preserved with sodium fluoride; LLE – solvent: hexane	Lidocaine	Supelcosil LC-CN cyanopropyl (250 × 4.6 mm, 5 μm)	ACN:(pH 7.4) phosphate buffer (38:62)	UV absorbance monitored at 230 nm	Recovery (Rec): 95–100% Linearity (Lin): 25–3200 ng/mL Accuracy (Acc): 88–92%	[89]
COC	Plasma	Specimen preserved with sodium fluoride; LLE – solvents: hexane with 2% isoamyl alcohol; back extract into 0.1 M HCl	n-propyl benzoylecgonine	Spherisorb C6 (250 mm, 5 μm)	ACN:(50 mM, pH 3.0) phosphate buffer containing hexanesulfonic acid (1.88 g/L) (1:2)	UV absorbance monitored at 235 nm 0.02 or 0.01 absorbance units full scale (a.u.f.s.)	Rec: average 71.4% LOD: 5μg/L Lin: 12–500 μg/L; 500–5000 μg/L Coefficient of variance (CV): 8.6%	[91]
COC, BZE and CE	Blood, serum	Specimens preserved with 2%NaF SPE – Bond Elut Certify Various L-L schemes also evaluated, but SPE was superior	2'-methylbenzoyl-ecgonine and 2'-methylcocaine	Hypersil BDS C18 (150 × 4.6 mm; 5 μm)	Eluent A: 0.045 M ammonium acetate in water:MeOH:ACN (80:10:10) Eluent B: 0.045 ammonium acetate in MeOH:ACN:water (40:40:20) Gradient: linear 100% A to 47.2% A over 19 min; 2 min hold	Photodiode array collected spectra every 21 ms over 221–400 nm at a bandwidth of 4 nm; 236 nm used for chromatogram construction	Rec: 88–95% LOD: 0.02 μg/mL Acc: 99–104% CV: 1–5%	[92]
COC and BZE	Plasma	SPE – Extrelut	Bupivacaine	Bondapak C18 (30 × 3.9 mm; 10 μm) with a guard column (5 × 6 mm)	0.1 M ammonium acetate:ACN:MeOH (40:30:30)	Photodiode array detector collected absorbance spectra over 200–350 nm for peak at 230 nm	Rec: 70% Lin: 50–500 ng/ mL	[107]

TABLE 2.6
CONTINUED

Analytes Detected	Specification Type	Internal Standard(s)	Sample Preparation	Column Type	Mobile Phase	Detector Details	Performance Characteristics	Ref.
COC, BZE, BNE and NCOC	Serum	Tolazoline	Specimens preserved with sodium fluoride; SPE – Bond Elut SCX followed by C18 (Analytichem Int.)	Tandem columns: cyanopropyl and silica (100 × 4.6 mm, 5 µm) with 2 precolumns (15 × 4.6 mm, 7 µm); one precolumn was placed before the injector, one prior to the analytical columns	ACN:(6.25 mM, pH 2.9) phosphate buffer (80:20)	UV absorbance monitored at 228 nm 0.001 a.u.f.s.	Rec: 75–86% LOD: 1 ng/mL Lin: 0–200 ng/mL Acc: 98–115% RSD: 0.7–13.8%	[93]
COC, BZE, EME and NCOC	Plasma	tropacocaine	Specimens preserved with saturated sodium fluoride SPE – Bond Elut Certify reconstituted extract was split; one portion analyzed for COC, BZE, NCOC; remaining portion derivatized with p-fluorobenzoyl chloride for EME analysis	COC, BZE, NCOC: Nucleosil C$_{18}$ (250 × 4.6 mm, 5 µm) with a Brownlee C$_8$ guard cartridge (30 × 4.6 mm, 5 µm); EME: tandem columns, Bakerbond Cyanopropyl (250 × 4.6 mm, 5 µm) and Microsorb Silica (150 × 4.6 mm, 5 µm)	COC, BZE, NCOC: 0.05 M citric acid: (0.1 M, pH 3.0) dibasic sodium phosphate (4:1) with 18% ACN and 0.3% triethylamine EME: (0.01 M, pH 3.0) monobasic sodium phosphate buffer:ACN (70:30)	COC, BZE, NCOC: UV absorbance monitored at 235 nm 0.01 a.u.f.s. EME: UV absorbance monitored at 235 nm 0.005 a.u.f.s.	Rec: 78–86% LOD (ng/mL) COC: 75 BZE: 35 EME: 90 NCOC: 80 Lin: 0.3–5 µg/mL CV: 1.8–11.0%	[95]
COC, BZE, BNE and NCOC (CE and NCE with a longer run time)	Plasma	3-isobutyl-1-methylxanthine	LLE – solvents: chloroform:MeOH (87.5:12.5)	Brownlee C$_{18}$ (100 × 2.1 mm, 5 µm) with a 2 µm Rheodyne precolumn filter	MeOH:ACN:(25.8mM, pH 2.2) sodium acetate buffer (12.5:10:77.5) containing 1.29×10^{-4} M tetrabutylammonium phosphate	UV absorbance was monitored at 235 nm	Rec: 69–82% (35–40% for BNE) LOD: 2.5 ng/mL Lin: 0.2–1 µg/mL CV: 1.2–10.9%	[90]
COC and BZE	Urine	Mepivacaine	None – direct injection	C$_8$ (150 × 4.6 mm, 5 µm) with a precolumn (Resolve CN Guard Pak precolumn)	0.025 M monobasic potassium phosphate:ACN:diethylamine (88:10:2) adjusted to pH 3 with o-phosphoric acid	UV absorbance monitored at 230 nm	Rec: near 100% (from guard column) LOD: 5 ng/mL Lin: 0–50 µg/mL	[105]

Analytes	Matrix	Extraction	IS	Column	Mobile phase	Detection	Results	Ref
COC and BZE	Vitreous humor	None: applied directly to precolumn through a glass wool filter	None – use of an IS did not improve precision or accuracy of method	CH-8 Lichrospher (250 × 4.6 mm) with a precolumn (3 cm, 40 μm)	(0.05 M, pH 3) phosphate buffer:ACN (70:30)	Photodiode array detector employed	Rec: 90% LOD: 0.01 μg/mL Lin: 0.02–1.0 μg/mL CV: <5%	[48]
COC, BZE, BNE and NCOC	Meconium	SPE – Clean Screen DAU	Lidocaine	Microsorb C$_{18}$ (100 × 4.6 mm, 3 μm)	(0.01 M, pH 2.0) phosphate buffer:ACN (87:13) with 58 μL tetrabutylammonium hydroxide	UV absorbance was monitored at 233 nm	Rec: 71–90% LOD: 0.05 μg/g Lin: 0.05–5 μg/g Acc: 85–93% CV: 3.1–6.9%	[109]
COC, BZE and benzoic acid	Serum, Hepatic cytosol, Microsomes	Centrifuge sample; evaporate supernatant; reconstitute, filter and inject	Tropacocaine	Altima C$_{18}$ (250 × 4.6 mm; 5 μm)	ACN:0.05 M pH 3 potassium phosphate buffer (67:33)	UV absorbance monitored at 230 nm	Rec: 82–96% LOD: 0.3 nmol/mL Lin: 1.5–1500 nmol/mL	[106]
COC and NCOC	Plasma, Hepatic cell cultures	LLE – solvents: diethyl ether and 0.1% TMAHS	None	Spherisorb RP-18 (100 × 4.6 mm, 5 μm)	ACN:0.1% aq. TMAHS (60:40)	UV absorbance monitored at 230 nm 0.02 a.u.f.s.	Rec: 100% LOD: 2 ng/mL Lin: 20–2000 ng/mL RSD: 0.9–8.3%	[88]
COC and BZE	Plasma, Urine	SPE – Bond Elut Certify	Methaqualone	Lichrospher RP18 (125 × 4 mm, 5 μm) with a guard column (4 × 4 mm)	MeOH:(0.02 M, pH 7) phosphate buffer (70:30) a flow rate gradient was employed	UV absorbance was monitored at 235 nm	Rec: 77–97% LOD: 5 ng/mL (urine) 12.5 ng/mL (plasma) Lin: 0.1–20 μg/mL CV: <5%	[108]
COC, BZE, BNE and NCOC	Plasma, urine, amniotic fluid	SPE – Chromprep PRP-1	Lidocaine	RP-C$_{18}$ (100 × 3.2 mm, 3 μm)	(0.01 M, pH 2.1) phosphate buffer with 0.0002 M TBA-OH:ACN (94:6)	UV absorbance monitored at 233 nm 0.002 a.u.f.s.	Rec: 85–92% LOD: 35 ng/mL (BZE, BNE) 75 ng/mL (COC, NCOC) Lin: 35–2260 ng/mL Acc: 96% CV: <7%	[94]

cyanopropyl and silica columns provided good resolution of *p*-fluorococaine (EME derivative) and COC in addition to eliminating the need for sample purification following derivatization [95]. In addition, a cyanopropyl column has been used in tandem with a silica column to shift the retention of various interfering substances without altering the relative retention of COC, BZE, NCOC and BNE [93].

Generally, the mobile phases employed were composed of a substantial aqueous fraction. Lampert and Stewart [93] investigated the effects of making modifications to the mobile phase used for separation on a silica column. They observed that even small changes in the ionic strength of the mobile phase influenced the retention and separation of cocaine analytes. The authors also observed that organic modifier concentration did not significantly influence analyte retention, but it did affect peak shape (a high concentration of modifier generally produced better peak shapes).

Several of the referenced HPLC methods used lidocaine or tropacocaine as an internal standard. While these internal standards may have worked well for the reported applications, one must make appropriate modifications when attempting to apply these methods to analysis of different sample types that may already contain lidocaine and tropacocaine. In summary, HPLC-UV can provide sensitive and robust methods for the analysis of cocaine and analytes with instrumentation that is economically available to most laboratories. HPLC methods also produce results that are comparable to those obtained with GC analysis [91,110].

2.2.4.2 Liquid chromatography with mass spectrometric detection

A growing number of methods have been reported for the analysis of cocaine and analytes by HPLC-MS or HPLC-MS-MS. These methods are summarized in Table 2.7. Most of these methods utilized SPE for sample preparation [97,111–113]. A unique application was reported by Sosnoff *et al.* [114] who developed a method of extracting BZE from dried blood spots that were collected from newborns. The purpose of the study was to generate epidemiological data on the prevalence of cocaine use during pregnancy. The extraction was accomplished with a simple elution of the blood spot into 2 mM aqueous ammonium acetate, deproteination with methanol and evaporation. The recovery of this extraction was not reported.

The mobile phases for HPLC-MS (-MS) consisted of methanol and acetonitrile in combination with volatile buffers, primarily ammonium acetate. Nishikawa *et al.* [97] evaluated various mobile phase compositions and observed that a higher concentration of ammonium acetate yielded faster elution of cocaine and its metabolites, particularly EME and ECG on a GS-320 H column (for aqueous steric exclusion chromatography). They also commented that the spectra of cocaine and analytes showed little variation with different mobile phase compositions. Similar to HPLC-UV methods, column phases were typically C_{18} or C_8 for the HPLC-MS methods. Nishikawa *et al.* [97] utilized a GS-320 H column after unsuccessfully attempting separation of cocaine and metabolites on an L-column ODS and ODP-50. However, their assay included EME, ECG and NCOC, analytes that were not common to all HPLC-MS methods. Nine of the reviewed LC-MS methods were capable of analyzing multiple drugs of abuse [111,115–123]. These methods

TABLE 2.7
HPLC-MS(-MS) METHODS FOR THE DETECTION OF COCAINE AND ANALYTES

Analytes Detected	Specification Type	Sample Preparation	Internal Standard(s)	Column Type	Mobile Phase	Detector Details	Performance Characteristics	Ref.
COC, BZE, EME and ECG	Plasma	SPE – Bond Elut Certify® and Bond Elut SCX®	D₃COC, D₃BZE, D₃EME, D₃ECG	Zorbax® Eclipse XDB-C8 (150 × 2.1 mm, 5μm)	Eluent A: 20mM ammonium acetate, pH 3.6 Eluent B: MeOH:ACN (1:1) Gradient: 99% eluent A for 2 min, followed by an decrease to 20% eluent A over 1 min and held for the duration of the 11 min run	Turbo ionspray interface API 365 triple quadrupole MS MRM mode	Rec: >82% (ECG = 40%) LOD: 2.8–4.4 ng/mL LOQ: 10.0–15.1 ng/mL Lin: 10/15 – 1000 ng/mL	[124]
BZE	Dried blood spots	Overnight elution of BZE from spot with 2mM ammonium acetate, protein removal with MeOH, filtration and evaporation	D₃BZE	2 Perkin-Elmer C₁₈ columns in series (30 × 4.6 mm, 3 μm)	25 mM ammonium acetate in MeOH:water (50:50)	APCI interface Triple quadrupole MS MRM with argon	LOD: 2 ng/mL Lin: 0–166 ng/mL	[114]
COC and BZE with other drug classes	Urine	SPE – Sep-pak C₁₈	None	L-column ODS (150 × 4.6 mm)	100% 100mM ammonium acetate for 1 min followed by a linear gradient to ACN:100 mM ammonium acetate (40:60) over 20 min	Thermospray interface: vaporizer temperature decreased linearly from 170–150°C with the change in mobile phase composition	Data indicated is for all analytes assayed Rec: 88–99% LOD: (ng/mL) SCAN: 50–400 SIM: 2–40 CV: 4.5–9.5%	[111]
COC, BZE, EME, NCOC, BNE, *m*-HOBZE, *p*-HOBZE, *m*-HOCOC, *p*-HOCOC, ECG, AEME, CE, NCE and EEE with other drugs of abuse	Urine	Centrifugation followed by direct injection	D₈COC, D₈BZE, D₃EME, D₃ECG, D₃NCOC, D₃CE	Synergi Polar RP (150 × 2.0 mm, 4 μm)	Eluent A: 10mM ammonium formate in water, 0.001% formic acid (pH 4.5) Eluent B: ACN Gradient elution	Orthogonal APCI source LCQ Deca ion trap MS SRM	Rec: 36–49% LOQ: 10–100 ng/mL Lin: 10/50 – 2500/10,000 ng/mL	[116]

TABLE 2.7
CONTINUED

Analytes Detected	Specification Type	Sample Preparation	Internal Standard(s)	Column Type	Mobile Phase	Detector Details	Performance Characteristics	Ref.
COC, BZE, EME, NCOC and ECG	Urine	SPE – Bond Elut Certify and Bond Elut SCX	None	Asahipak GS-320H packed column (250 × 7.6 mm)	10mM ammonium acetate:ACN (90:10) for 2 min followed by 20 mM ammonium acetate:ACN (50:50) for 30 min	APCI interface vaporizer temp: 350°C desolvation temp: 390°C drift voltage: 105 focus voltage: 120	Rec: 87–95% ECG- 40% LOD: (ng injected) SCAN: 40–80 SIM: 1–16 Lin: 20–100ng injected	[97]
COC and BZE with other drugs of abuse	Oral fluid	Oral fluid collected on pad and placed into 15% MeOH in 25 mM ammonium acetate; sample centrifuged, internal standard added and injected on chromatographic system	D₃COC, D₃BZE	HyPURITY C8 (50 × 3 mm, 5 μm)	Eluent A: 25 mM ammonium acetate in 5% MeOH Eluent B: MeOH with 0.05% formic acid and 2% propan-2-ol Gradient: 100% eluent A for 1 min, decreased linearly to 5% eluent A over 3 min. held for 1.5 min	Electrospray interface API 3000 tandem MS MRM mode	NA	[117]
COC and BZE with other drugs of abuse	Oral fluid	SPE – Bond Elut Certify®	2′-methylcocaine	Hypersil BDS (100 × 2.1 mm, 3 μm)	MeOH:Water with 10mM ammonium formate, pH 5 Gradient:linear increase from 6% to 41.2% methanol within 20 min	Electrospray interface (Z-spray®) Micromass quadruple time of flight MS	Rec: >89% LOD: <0.3 ng/mL LOQ: 2 ng/mL	[118]
COC, BZE and CE	Oral fluid	SPE – mixed mode IST™confirm HCX	2′-methyl-cocaine, 2′-methylbenzoylecgonine	Hypersil BDS C₁₈ (100 × 2.1 mm, 3 μm)	Eluent A: 0.05 M ammonium acetate in water:MeOH:ACN (82.5:8.75:8.75) Eluent B: 0.05 M ammonium acetate in water:MeOH:ACN (17.6:41.2:41.2) Gradient: 100% eluent A for 10 min, decreased linearly to 50% eluent A within 17 min, held for 1 min	Electrospray interface (Z-spray®) Micromass quadruple time of flight MS	Rec: >85% LOD: 1 μg/L LOQ: 10 μg/L Lin: 10–1000 μg/L	[125]

Analytes	Matrix	Extraction	Internal standard	Column	Eluent/Gradient	Detection	Performance characteristics	Ref.
COC and BZE with other drugs of abuse	Oral fluid	SPE – Oasis® MCX	D_3COC, D_8BZE	Xterra MS C_{18} (150 × 2.1 mm, 3.5 µm)	10mM ammonium bicarbonate (pH 10) and MeOH. Gradient: 30% MeOH for 3 min, increased to 50% MeOH over 1 min, linearly increased to 75% MeOH from 4 to 12 min, increased to 90% MeOH at 12 min, held for 1 min	Electrospray interface (Z-spray®) Micromass Quattro Ultima tandem MS MRM mode	Rec: >85% LOD: 0.2–0.5 µg/L LOQ: 2 µg/L Lin: 2–200 µg/L	[119]
COC and BZE with other drugs of abuse	Hair	Hair was incubated in mobile phase at 37ºC for 18 hrs; aliquot removed and injected onto chromatographic system	D_3COC	Zorbax® Eclipse XDB-C8 (150 × 2.1 mm, 5 µm)	Eluent A: ACN:MeOH:20mM formate buffer, pH 3.0 (10:10:80) Eluent B: ACN:MeOH:20mM formate buffer, pH 3.0 (35:35:30) Gradient: 100% eluent A linearly decreased to 35% eluent A from 0.5 to 7 min	Electrospray interface Sciex API 2000 MS-MS MRM mode	Rec: >77% LOD: 3–5 pg/mg LOQ: 10–16 pg/mg	[120]
COC, BZE and CE	Hair	Hair was cut and washed, then digested overnight with 0.1 M HCl and extracted by SPE – IST Isolute Confirm HCX	2'-methylbenzoylecgonine, 2'-methylcocaine	Hypersil BDS C_{18} (125 × 2.1 mm, 3 µm)	Eluent A: 0.045 M aqueous ammonium acetate:MeOH:ACN (84:8:8) Eluent B: MeOH:ACN:water (42:42:16) Gradient: 100% eluent A for 7 min, followed by a linear gradient to 65% eluent A over 27 min with a final hold of 3 min	Fluorescence detection: Excitation 1 = 242 nm; emission 1 = 315 nm MS detection: electrospray interface Triple quadrupole VG Quattro II Full scan and SRM data obtained	Performance characteristics based on Fluorescene detection Rec: 88–92% LOD: 0.025 ng/mg LOQ: 0.1 ng/mg Lin: 0.1–20 ng/mg Acc: 98–101% CV: 3–13%	[113]
COC, BZE, CE, m-HOBZE and p-HOBZE with opiates	Meconium	Methanolic extraction followed by SPE – Bond Elut Certify®	Nalorphine	Zorbax® Eclipse XDB-C8 (150 × 4.6 mm)	1% acetic acid: CAN Gradient: 97% 1% acetic acid linearly decreased to 73% in 8 min; returning to initial conditions in 11 min; 25 min run	Electrospray interface Agilent G1946D MSD SIM mode	Rec: >60% LOD: 0.0004–0.0015 µg/g LOQ: 0.001–0.003 µg/g	[121,122]

TABLE 2.7
CONTINUED

Analytes Detected	Specification Type	Sample Preparation	Internal Standard(s)	Column Type	Mobile Phase	Detector Details	Performance Characteristics	Ref.
COC, BZE, EME, NCOC, BNE, CNO, m-HOBZE, p-HOBZE, m-HOCOC, p-HOCOC, ECG, AEME, ECGD, CE, NCE and EEE	Meconium	Methanolic extraction followed by SPE – Bond Elut Certify®	D₃COC	Zorbax® Eclipse XDB-C8 (150 × 2.1 mm, 5 μm)	Eluent A: 20 mM ammonium acetate (pH 2.7) Eluent B: MeOH:ACN (1:1) Gradient elution	Electrospray interface (Z-spray®) Micromass Quattro II triple quadrupole MS MRM mode	Rec: 36–49% LOQ: 1–5 ng/g Lin: 0.005–5 μg/g	[12]
COC and CNO	Meconium	SPE – Silica cartridges	D₃COC	Eclipse® XDB-C-8 (150 × 2.1 mm, 5 μm) Column temp.: 40°C	Eluent A: 20 mM ammonium acetate(pH 2.7) Eluent B: MeOH:ACN (50:50) Multi-step gradient from 100%A to 100%B over 23 min	UV- 230 nm MS- Micromass Quattro II triple quadrupole MS with an ESI ion source	Lin: 0.01–2 μg/mL	[11]
COC, BZE and EME with other drugs of abuse	Serum, blood, urine, vitreous humor, cerebrospinal fluid, bile	SPE – Bond Elut C₁₈	D₃COC, D₃BZE, D₃EME	LC-Super-spher RP-18 (125 × 3 mm, 4 μm)	ACN:50 mM ammonium formate buffer (for COC, BZE.15:85): (for EME. 5:95)	APCI interface SSQ 7000 single quadrupole MS SIM mode	LOD: 0.2 μg/mL (COC:0.5 μg/mL) LOQ: 2 × LOD Lin: 1–200 μg/L	[115]
COC, BZE, EME and CE with opiates	Blood, plasma, urine	Single-step LLE – Chloroform:Isopropanol (95:5, v/v)	D₃BZE, D₃EME	Spheris-orb RP 8S (100 × 2.1 mm, 5 μm)	Water:ACN (50:50) containing 0.1% formic acid and 2 mM ammonium formate	Ionspray interface Sciex API 300 triple quadrupole MS MRM mode	Rec: >85% LOQ: 5 ng/mL Lin: up to 1000 ng/mL	[123]

were suitable for routine application. A unique method for the detection and quantitation of CNO was reported by Wang and Bartlett [11]. CNO is a thermally labile metabolite of cocaine, so high-performance liquid chromatography-electron ionization mass spectrometry (HPLC-EIMS) was the preferred technique. The authors also reported that CNO is converted to COC in the injection port when analyzed by GC-MS. Consequently, CNO may contribute to cocaine concentrations measured by GC-MS.

2.2.5 Gas chromatographic procedures

Cocaine analysis is often performed by gas–liquid chromatography. In this technique, a gaseous fluid serves as the mobile phase, or carrier gas. Frequently used carrier gases, including helium, hydrogen and nitrogen, have very weak intermolecular interactions with solutes. A liquid stationary phase provides retention through solution forces. Thus, all interactions responsible for the selective retention of solutes occur in the stationary phase. A wide variety of liquid stationary phases have been employed for a diversity of chromatographic separations, including those of drugs and their metabolites.

Mixtures of solutes that have wide boiling point or polarity ranges, or a large variety of functional groups, pose a particular problem. At low-column-operating temperatures, the solutes with high volatility appear early on the chromatogram as well-resolved peaks. Solutes with low volatility progress slowly through the column, with ample opportunity for peak broadening. These solutes may appear as very low, broad peaks that could be overlooked. An increase in column temperature increases the concentration of the solutes in the gas phase. However, this impacts upon the solutes of high volatility, which are now spending most of their time in the gaseous mobile phase and migrate rapidly through the column to appear as unresolved peaks. The succeeding less-volatile solutes are adequately resolved. A simple solution is to employ temperature programming during the course of the separation. The well-resolved, highly volatile solutes are removed from the column at the lower temperatures before the solvents of low volatility leave the origin at the column inlet. Temperature programming is an essential technique for the GC analysis of multiple drugs and their metabolites. In addition, chemical derivatization of the solutes can be used to modify their chromatographic behavior. Various derivatization schemes for cocaine analytes are described below.

Gas chromatographic detectors sense the gaseous solutes in the mobile phase as they emerge from the column. There are several types of detectors routinely used for the GC analysis of drugs and their metabolites, including flame ionization, nitrogen–phosphorus, electron capture and mass spectrometric detectors. In the reviewed methods, the most commonly employed detectors were the nitrogen phosphorus detector (NPD) and the mass spectrometer.

In recently reported methods, extraction for GC analysis was performed with LLE or SPE. Garside *et al.* [126] reported a single-step LLE for the extraction of COC from urine. The extraction procedure extracted only nonpolar analytes and provided sensitive detection of cocaine. Generally, extraction of cocaine and metabolites was accomplished with SPE. Jennison *et al.* [127] completed a 12-month study of an SPE system for the extraction of BZE from urine. They found that reliable results could be obtained for this period across several analysts and across several column lots. They also reported that SPE produced much cleaner extracts than LLE. Washing and hydrolysis steps were indicated for some methods that analyzed hair and nails [35,40,41,49,128–131]. Cirimele *et al.* [130] evaluated several decontamination and hydrolysis methods for this type of specimen. They reported that methylene chloride was the most effective wash solvent, removing over 60% of cocaine contamination from hair. The authors also determined that higher drug recoveries were obtained with acid and enzymatic hydrolysis compared to methanolic extraction. Nail analysis was accomplished with either methanolic extraction or phosphate buffer extraction followed by SPE [40,41]. Engelhart *et al.* [41] found that soaking nails in phosphate buffer for an extended period solubilized the nail matrix. It was established that COC was stable under these conditions. Hernandez *et al.* [32] developed an extraction method for human brain specimens. Brain tissue was prepared as a 1:1 homogenate in water. Some specimens were treated with lipase prior to SPE. The results indicated that the lipase digestion did not alter drug recovery or prevent column blockage.

The gas chromatographic analysis of cocaine analytes often requires derivatization because of the polar nature of several cocaine metabolites. The most frequently employed derivatizing agents included BSTFA, MTBSTFA, MSTFA, dipropylacetal, diazopropane and PFPA. Thompson *et al.* [133] reported a unique approach to derivatization with MTBSTFA and BSTFA. A microwave oven at high power was used to prepare derivatives, resulting in a five-fold reduction in the time required for derivatization. The microwave derivatization produced the same yield of derivative as conventional heating and a good correlation was observed between the two techniques. Isenschmid *et al.* [134] developed a derivatization method that produced unique high molecular weight derivatives for BZE and EME. However, this procedure required the analysis of two extracts for each specimen, as separate derivatization reactions were used for each analyte. A modification of this derivatization technique was employed by Smirnow and Logan [135]. A pre-extraction derivatization scheme involved deproteination, propylation of organic acids and primary and secondary amines and modification of organic alcohols to *p*-nitrobenzoyl esters. After these chemical reactions, the analytes of interest could be extracted by LLE. This method was especially useful for the analysis of ECG, which is very difficult to isolate using standard LLE or SPE. Paul *et al.* [136] developed a rapid procedure for the propylation of BZE. The resulting derivative produced more abundant ions for GC-MS analysis and was stable for at least five days. Crouch *et al.* [137] compared the derivatization of cocaine analytes with MTBSFA and BSTFA. They reported that the t-butyldimethylsilyl derivatives of BZE and EME produced higher molecular weight ions and were stable for up to two weeks, while the trimethylsilyl derivatives were quite susceptible to hydrolysis.

Predominant GC column phases included 100% dimethylpolysiloxane (similar columns: HP-1, DB-1, Rtx-1, OV-1, SPB-1, SE-30, CP-Sil 5CB, RSL-150, MTX-1,

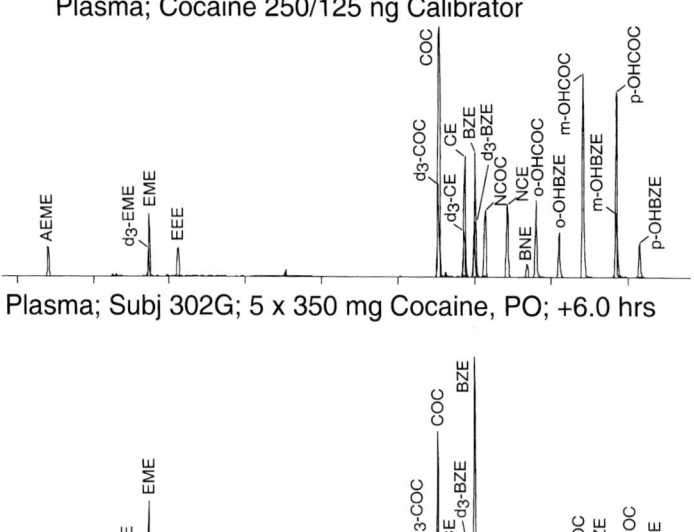

Fig. 2.3. SIM chromatograms of a calibrator and a plasma sample from a clinical study. The calibrator contained 250 ng of COC, BZE and EME and 125 ng of all other analytes. The plasma sample was collected 6 h following the administration of five doses of oral cocaine and contained EME, COC, BZE, NCOC, BNE, *m*- and *p*-HOCOC and *m*- and *p*-HOBZE.

BP-1, 007-1 and MDN-1) and 95% dimethylpolysiloxane with 5% diphenyl (similar columns: HP-5, DB-5, Rtx-5, PTE-5, CP-Sil 8CB, SE-54, Mtx-5, OV-5, SE-52, GC-5, 007-2, RSL-200, MDN-5 and BP-5). Some manufacturers produce columns containing these phases but with increased specifications for retention indices and capacity factor. Most of the time, adequate resolution of cocaine analytes can be achieved on these phases. As an illustration of a typical separation, Fig. 2.3 displays a GC-MS SIM chromatogram in which 15 cocaine analytes were separated. The column phase employed in this separation was a 100% dimethylpolysiloxane. This separation was accomplished with the method reported by Darwin *et al.* [138].

2.2.5.1 Gas chromatography with nitrogen phosphorus detection

Although a few cocaine analysis methods employing an electron capture detector or a flame ionization detector (FID) have recently been reported, their focus has primarily been the analysis of illicit cocaine samples [2,6,138]. The NPD is frequently employed for the gas chromatographic analysis of cocaine analytes in biological specimens. Table 2.8 provides a summary of cocaine analysis methods that have

TABLE 2.8
GC-NPD METHODS FOR THE DETECTION OF COCAINE AND ANALYTES

Analytes Detected	Specification Type	Sample Preparation	Internal Standard(s)	Column Type	GC Temperature Program	Performance Characteristics	Ref.
COC and BZE	Plasma	SPE – Bond Elut Certify Derivatization: conversion to butyl ester	Benzoylecgonine *n*-propyl ester	HP Ultra 2 (12 m. 0.2 mm. 0.33 μm) Stationary Phase: 5% diphenyl. 95% dimethylpoly-siloxane	Injector: 250°C Detector: 300°C Oven: 200°C for 2 min, ramp to 280°C at 30°C/min, hold to final time of 11 min	Rec: 70% (COC) LOD: 1 ng/mL RSD: 2–9%	[139]
COC	Serum, plasma, blood	LLE – solvents: single step in an ice bath	Maprotiline	3% SP2250 packed on Supelcoport (80–100 mesh) (2.4 m. 2 mm)	Injector: 300°C Detector: 300°C Oven: 270°C isothermal	LOD: 0.02 mg/L Lin: 0.05–10 mg/L CV: 0.7–4%	[140]
EME	Urine	LLE – solvents: chloroform:isobutanol (98:2)	Mexiletine	HP 530 μm megabore column (10 m. 0.53 mm) 50% phenylmethyl-silicone	Injector: 270°C Detector: 300°C Oven: 150°C held for 3 min, then ramped to 280°C at 70°C/min with a final hold time of 2 min	Rec: 96–101% LOD: <0.1 μg/mL Lin: 0–4 μg Precision: 1.4–14%	[141]
COC, BZE and EME	Urine	SPE – Bond Elut Certify Derivatization: Carboxy groups derivatized to ethyl esters followed by derivatization of Hydroxy and phenolic groups with MSTFA	Levallorphan	HP 5% phenyl-methyl silicone (25 m. 0.2 mm. 0.33 μm)	Injector: 280°C Detector: 290°C Oven: 150°C ramped to 280°C at 10°C/min	Rec: 84–88%, 41.5% (EME) LOD: 100 ng/mL: 250 ng/mL (EME) Lin: 100–2000 ng/mL: 250–2500 ng/mL (EME) RSD: 3–10%	[142]
COC and CE	Blood, liver, neurological tissue (tissue specimens prepared as 1:4 homogenates)	LLE – solvents: three step with hexane wash of acid extract	Propylbenzoyl-egonine	DB-17 (15 m. 0.53 mm. 0.1 μm) Stationary Phase: 50: phenylmethyl-silicone	Injector: 250°C Detector: 300°C Oven: 150°C ramped to 255°C at 8°C/min, with a final hold time of 3 min	Rec: 84–87% LOD: 0.02 mg/L Lin: 0.05–10 mg/LCV: 3.3–5.6%	[143]

coupled a GC with an NPD. Overall, the reviewed GC-NPD methods were quite sensitive, with COC LODs ranging from 1 to 100 ng/mL. The LODs for BZE and CE were comparable to COC, while the LODs for EME were slightly higher. The matrices analyzed included serum, plasma, blood, urine, liver and neurological tissue. GC-NPD methods are valuable for the screening of specimens for cocaine analytes since they are moderately priced and provide sensitive results.

2.2.5.2 Gas chromatography methods with mass spectrometric detection

GC-MS was by far the most extensively reported analytical technique for the detection and quantitation of cocaine analytes. Table 2.9 summarizes these methods. Selected-ion monitoring in the electron impact ionization mode was most often used, though some methods employed positive or negative chemical ionization, ion-trap MS and tandem MS. One advantage of chemical ionization methods is that they can be useful for determining molecular weights of novel cocaine metabolites. Also, the ion trap methods can provide increased sensitivity with MS-MS capabilities. The matrices analyzed by GC-MS included plasma, blood, urine, saliva, meconium, hair, nails, brain, amniotic fluid, umbilical cord tissue and vernix caseosa. Limits of detection for cocaine and analytes in biological fluids by GC-MS were often less than 10 ng/mL.

Although GC-MS (-MS) equipment is fairly expensive in comparison to some of the other methodologies discussed, it is regarded as the "gold-standard" of drug testing methodologies, as it provides spectral identification and sensitive and selective detection of drugs and their metabolites.

2.2.6 Direct mass spectrometric analysis

Several methods have been reported that involve direct injection of an extract into a MS detector without prior chromatographic separation. These methods are displayed in Table 2.10. One tandem MS method employed the technique of flow injection analysis [112]. The specimen was extracted by SPE and then injected directly into the ion-spray source. This method was validated for the detection of multiple drugs of abuse and produced quantitation results comparable to GC-MS. Its advantages included a short analysis time (approximately 3 min) and sensitive and selective detection. Traldi *et al.* [131] developed a method for the analysis of hair extracts by direct injection into an ion trap in the MS-MS mode. The method detected COC easily; however, the analysis of metabolites was difficult since the detection of BZE, EME and ECG was complicated by the presence of interfering ions. Papa *et al.* [155] reported a direct exposure probe (DEP) MS method for the confirmation of COC and BZE. Urine was lyophilized and reconstituted in ethyl acetate and methanol, then analyzed by DEP negative ionization chemical ionization MS. The authors compared DEP-EI spectra to DEP-CI spectra of spiked urine and found that the CI spectra were much cleaner and more closely matched a pure standard. Sensitive detection of about 1 ng of pure standard was accomplished by

TABLE 2.9
GC-MS(-MS) METHODS FOR THE DETECTION OF COCAINE AND ANALYTES

Analytes Detected	Specification Type	Sample Preparation	Internal Standard(s)	Column Type	GC Temperature Program	Detector Details	Performance Characteristics	Ref.
COC, BZE, EME, NCOC, CE and ECG	Blood	Deproteination with MeOH ACN, extractive alkylation, extraction into n-butyl chloride, back extraction for clean-up	D_8COC, D_3BZE, D_3EME, D_3ECG	BP-5 (30 m, 0.32 μm)	Oven: 80 C ramped to 295 C at 15 C/ min, with a final hold time of 8 min	HP 5970 SIM mode	Lin: 10–1000 ng/mL CV: <15%	[135]
COC and BZE	Blood	Extracted with Amberlite XAD-2 Derivatized with diazopropane	D_3COC, D_3BZE	DB-5 (30 m × 0.25 mm, 0.1 μm) with a deactivated precolumn	Injector: 270 C Detector: 290 C Oven: 100 C for 2 min, ramped to 290 C at 18 C/min	HP 5970 SIM mode	Rec: >84% LOD: 0.00025 mg/dL CV: ~3%	[144]
COC, BZE, EME and ECG	Urine	SPE – SCX Derivatization with diazopropane to form propyl esters	D_3BZE, D_3ECG	BP-5 (30 m, 0.32 μm)	Oven: 100 C ramped to 295 C over 15 min with a final hold time of 7 min	HP 5970 MSD SIM mode	Rec: COC: 95% BZE: 85% EME: <50% ECG: 65% LOQ: 0.5 μg/mL Lin: 1–100 μg/mL CV: <7% (EME and ECG)	[145]
COC, BZE, EME and CE	Urine	SPE – Clean Screen DAU Derivatized with BSTFA or MTBSTFA Microwave was used for rapid derivatization	D_3COC, D_3BZE, D_3EME, D_3CE	HP Ultra-2	Injector: 250 C Oven: 120 C for 1 min, ramped to 310 C at 18 C/min, with a final hold time of 5 min	HP 5970 MSD SIM mode	Rec: COC – 75% BZE: 90% EME: 50% CE: 50–75% LOD: 100 ng/mL Lin: 250–5000 ng/mL CV: <8.6%	[133]
COC, BZE	Urine	SPE – Mini Bed Amberlite XAD-2 columns Derivatized with MSTFA	D_3COC, D_3BZE	HP Ultra-1 (12 m × 0.2 mm, 0.33 μm)	Injector: 250 C Detector: 280 C Oven: 200 C isothermal	HP 5970A MSD SIM mode	Rec: 75–80% LOD: 50 ng/mL Lin: 50–4000 ng/mL	[146]
COC	Urine	LLE – solvent: petroleum ether	D_3COC	Rtx-5 (15 m × 0.25 mm, 0.1 μm)	Injector: 250 C Detector: 290 C Oven: 150 C for 0.5 min, ramped to 250 C at 25 C/min, with a final hold of 2.5 min	HP 5972 MSD SIM mode	Rec: 40–59% LOD: 5 ng/mL Lin: 25–4000 ng/mL Acc: 98.6–101.7% CV: 1.9–6.6%	[126]

Analyte	Matrix	Extraction/derivatization	IS	Column	GC conditions	Instrument/detection	Performance	Ref.
BZE	Urine	SPE – Clean Screen Dau 303 Derivatization with BSTFA	D_3BZE	SB-5 (12.5 m × 0.2 mm, 0.25 µm)	Injector: 280°C Detector: 280°C Oven: 170°C for 0.5 min, ramped to 300°C at 25°C/min, with a final hold time of 0.5 min	HP 5970 SIM mode	Rec: ~70% CV: <10.4%	[127]
BZE	Urine	SPE – Clean Screen DAU Derivatized with DMF-dipropylacetal or DMF-diisopropylacetal in pyridine	D_3BZE	DB-5 (15 m × 0.25 mm)	Injector: 280°C Detector: 280°C Oven: 180°C for 1 min, ramped to 270°C at 30°C/min with a 2 min hold, then to 320°C at 40°C/min with a final hold of 1 min	HP5972 MSD SIM mode	Rec: 90% yield for deriv. rxn Lin: 10–8000 ng/mL CV: <8.3%	[136]
COC, BZE, EME	Oral Fluid	Oral fluid was extracted in a Toxitube A® Derivatized with BSTFA/TMCS	D_3COC, D_3BZE, D_3EME	DB-5 (12 m × 0.2 mm, 0.33 µm)	Injector: 250°C Detector: 320°C Oven: 90°C for 1 min, ramped to 180°C at 20°C/min, then to 240°C at 5°C/min and to 290°C at 30°C/min	HP 5973 MSD PCI/SIM mode	Rec: 96–102% LOD: 0.2–2.2 ng/mL LOQ: 0.8–7.4 ng/mL Lin: to 1000 ng/mL CV: 2.7–6.7%	[147]
COC, BZE, EME, CE and opiates	Hair	Methylene chloride wash, sample pulverized and incubated with 0.1 N HCl, 1 N NaOH, MeOH, or an enzyme solution followed by LLE – chloroform:isopropanol: n-heptane (50:17:33), back extraction into 0.2 N HCl and re-extraction into chloroform Derivatized with BSTFA	D_3COC, D_3BZE, D_3EME D_3CE	HP5-MS (30 m × 0.25 mm)	Injector: 250°C Detector: 320°C Oven: 60°C for 1 min, ramped to 290°C at 30°C/min with a final hold of 6 min	HP 5971 MSD SIM mode	No performance data – results compared extraction methods with 19 hair samples collected from drug abusers	[130]
COC, BZE, EME, CE and opiates	Hair	Phosphate buffer wash followed by two washes with dichloromethane, then hair is pulverized and extracted with 0.1 M HCl followed by SPE – Isolute Derivatized with BSTFA	D_3COC, D_3BZE, D_3EME, D_3CE	CP SIL 8 CB (25 m × 0.25 mm, 0.25 µm)	Injector: 280°C Detector: 300°C Oven: 50°C for 2 min, ramped to 310°C at 15°C/min, with a final hold of 4.67 min	HP 5972 MSD Scan mode	Rec: >86% (52% for EME) LOD: 0.12–0.28 ng/mg RSD: <9% (15.7% for EME)	[49]

TABLE 2.9
CONTINUED

Analytes Detected	Specification Type	Sample Preparation	Internal Standard(s)	Column Type	GC Temperature Program	Detector Details	Performance Characteristics	Ref.
COC, BZE, CE, EME and NCOC	Hair	Hair was incubated overnight in 0.1 N HCl followed by extraction by SPE – Clean Screen DAU Derivatized with HFIP and TFAA	D3COC, D3BZE, D3EME, D3NCOC	DB-5 MS (15 m × 0.25 mm, 0.25 μm)	Injector: 240 C Detector: 300 C Oven: 130 C ramped to 270 C at 25 C/min, then to 280 C at 6 C/min then to 300 C at 35 C/min	Finnegan MAT TSQ 7000 Tandem MS PCI mode Reagent gas – ammonia	Rec: 47–110% LOD: 0.01 ng/mg LOQ: 0.05 ng/mg Lin: 0.01–0.5 ng/mg CV: <8%	[148]
COC, BZE, EME, NCOC, CE and NCE	Hair	Methanolic wash, extraction with 0.05 M sulfuric acid followed by SPE – Clean Screen DAU Derivatized with BSTFA	D3COC, D3BZE, D3EME	HP-1 (20 m × 0.2 mm, 0.33μm)	Injector: 250 C Detector: 280 C Oven: 120 C for 1 min, ramped to 220 C at 35 C/min with a 0.25 min hold, then to 250 C at 10 C/min with a 0.25 min hold, then to 260 C at 3.5 C/min	HP 5970B SIM mode	Rec: ~90% LOD: 0.1 ng/mg (50 mg sample) Lin: 12.5–1000 ng (for 10–100 mg samples)	[35]
COC, BZE and EME	Hair	Hair was washed with 1% SDS, water and MeOH, then digested in enzyme solution followed by extraction by SPE – Bond Elut Certify Derivatized with MTBSTFA	Difluoro-cocaine	DB-5 (15 m × 0.25 mm, 0.1 μm)	Injector: 260 C Detector: 260 C Oven: 100 C for 1 min, ramped to 300 C at 25 C/min, with a final hold of 2 min	Finnigan ITS-40 ion trap MS Full scan (m/z 80–440) Reagent gas – isobutene	Rec: 90–127% LOQ: 0.1 ng/mg Lin: 0.1–100 ng/mg CV: 10–28%	[132]
COC, EME, EME and CE	Hair	Hair was washed with dichloromethane, water and MeOH, pulverized, then digested in 0.1 N HCl followed by extraction by SPE – HCX Isolute	D3COC, D3EME	DB-5 MS (15 m × 0.25 mm, 0.25 μm)	Injector: 75°C for 1 min, ramped to 280 C at 50 C/min Oven: 75°C for 1 min, ramped to 170 C at 15 C/min, then to 210 C at 5 C/min and to 310 C at 30 C/min	Varian Saturn 2000 ion trap MS PCI mode Reagent gas– isobutene	Rec: 96–102% LOD: 0.005–0.05 ng/ mg LOQ: 0.05–0.10 ng/ mg Lin: to 5.0 ng/mg	[149]

Analytes	Matrix	Sample preparation	Internal standard	Column	GC conditions	MS detection	Results	Ref.
COC, BZE, EME	Hair	Hair washed with water and acetone, hydrolyzed in enzyme solution twice, then extracted by SPE – Chromabond C$_{18}$ Derivatized with PFPA/PFPOH	D$_3$COC, D$_3$BZE, D$_3$EME	HP Ultra-2 (12 m × 0.2 mm, 0.33 μm)	Injector: 260°C Detector: 280°C Oven: 70°C for 3 min, ramped to 180°C at 15°C/min, then to 240°C at 5°C/min and to 300°C at 30°C/min with a final hold of 5 min	HP 5971A MSD SIM mode	Rec: COC – 90% BZE: 75% EME: 50% LOD: 0.1 ng/mg (1 ng/mg for EME) Lin: 0–25 ng/mg (0–50 ng/mg for COC) CV: <14%	[129]
COC, BZE and NCOC	Nails	Nails were washed with MeOH, then incubated in MeOH for 16 h followed by extraction by SPE – Bond Elut Certify Derivatized with PFPA and PFP	D$_3$COC, D$_3$BZE, D$_3$NCOC	HP-5 MS (30 m × 0.25 mm, 0.25 μm)	Injector: 240°C Detector: 300°C Oven: 130°C ramped to 270°C at 25°C/min, then to 280°C at 6°C/min then to 300°C at 35°C/min	HP 5972 MSD SIM mode	Rec: 47–110% LOD: 3–3.5 ng/mg CV: 1–11%	[150]
COC, BZE, EME, NCOC, BNE, CE, EEE and AEME	Finger-nails, toenails	Methanolic wash and extraction followed by SPE – Clean Screen DAU Derivatized with BSTFA	D$_3$COC, D$_3$BZE, D$_3$EME, D$_3$CE	HP-5MS (30 m × 0.25 mm, 0.25 μm)		HP 5972 SIM mode	LOD: 0.1 ng/mg (0.25 ng/mg for EEE and BNE) Lin: 0.1–10 ng/mg	[40]
COC, BZE, NCOC, CE and opiates	Toenails	Methanolic wash, extracted into phosphate buffer, followed by SPE – Clean Screen DAU Derivatized with MSTFA	D$_3$BZE	DB-5 (30 m × 0.25 mm, 0.25 μm)	Injector: 200°C Oven: 150°C for 2 min, ramped to 200°C at 50°C/min with a 1 min hold, then to 260°C at 10°C/min with a final hold of 4 min	HP 5970 MSD SIM mode	Rec: >93% for COC, BZE LOQ: 0.3 ng on column Lin: 0.3–60 ng on column	[41]
COC, BZE and CE	Brain tissue	Brain was homogenized in Tissue-Tearer® (some specimens were incubated with lipase), followed by SPE – Clean Screen DAU Derivatized with hexafluoro-2-propanol and PFPA	D$_3$COC, D$_3$BZE, D$_3$CE	DB5-MS (15 m × 0.25 mm, 0.25 μm)	Injector: 105°C for 0.25 min, ramped to 290°C at 180°C/min, with a final hold of 11.7 min Detector: 280°C Oven: 115°C for 1.3 min, ramped to 290°C at 15°C/min with a final hold of 2 min	Finnigan ITS-40 ion trap GC-MS Full Scan mode (m/z 75–450)	Rec: 54% (60% for COC) LOD: 25 ng/g (1 g sample) Lin: 50–10,000 ng/g CV: <7%	[32]

TABLE 2.9
CONTINUED

Analytes Detected	Specification Type	Sample Preparation	Internal Standard(s)	Column Type	GC Temperature Program	Detector Details	Performance Characteristics	Ref.
COC, NCOC, BZE, CE, NCE, AEME, EME and EEE with other drugs of abuse	Skin	Skin was homogenized in MeOH, sonicated, evaporated and reconstituted in buffer followed by extraction by SPE – Clean Screen DAU Derivatized with MTBSTFA and BSTFA	D$_3$COC, D$_3$BZE, D$_3$EME, D$_3$CE, D$_3$NCOC	HP-5 MS (12 × m × 0.2 mm, 0.33 μm)	Injector: 250°C Oven: 70°C for 1.5 min, ramped to 170°C at 23°C/min, then to 310°C at 21°C/min with a 5 min final hold	Agilent 5973 MSD PCI mode Reagent gas – methane	Rec: 93–104% LOD: 1.25–5.0 ng/biopsy Lin: to 100 ng/biopsy CV: <6%	[151]
COC, BZE, CE and m-HOBZE	Vernix caseosa	SPE – Isolute HCX Derivatization with MTBSTFA	D$_3$COC, D$_3$BZE, D$_3$CE	DB5-MS (25 × m × 0.2 mm, 0.33 μm)	Injector: 270°C Detector: 310°C Oven: 100°C for 1 min, ramped to 230°C at 30°C/min, then to 249°C at 3°C/min, and to 310°C at 30°C/min with a final hold time of 5.3 min	HP5971A MSD SIM mode	Rec: >80%	[152]
COC, BZE and EME	Blood, Urine	SPE – Chem Elut Eluent split – one portion derivatized with DMF and N,N-DMF dipropylacetal for BZE analysis; remaining portion derivatized with 4-fluorobenzoylchloride for EME analysis	D$_3$COC, D$_3$BZE, D$_3$EME	HP-1 (12 m)	Injector: 250°C Detector: 250°C Oven: 100°C for 1 min, ramped to 260°C at 30°C/min	HP 5870 MSD SIM mode	Lin: 0.05–4 mg/L (0.025–0.2 mg/L for COC) CV: <3%	[134]
COC, BZE, EME, NCOC, BNE, CE, EEE, NCE, AEME and opiates	Plasma, urine, saliva, hair	Hair – Methanolic wash and extraction prior to SPE All samples: SPE – Clean Screen DAU Derivatized with BSTFA	D$_3$COC, D$_3$BZE, D$_3$EME, D$_3$CE	HP-1 (12 × m × 0.2 mm, 0.33 μm)	Injector: 250°C Detector: 280°C Oven: 70°C for 1 min, ramped to 220°C at 35°C/min with a 0.25 min hold, then to 250°C at 10°C/min with a final hold of 3 min	HP 5970 MSD SIM mode	LOD$_{hair}$: 0.1 ng/mg (0.5 ng/mg for BNE) LOD$_{PL,SALIVA}$: 1 ng/mL (5 ng/mL for BNE)	[128]

Analytes	Matrix	Extraction/Derivatization	Internal standards	Column	GC conditions	Instrument	Results	Ref.
COC, EME, BZE and CE	Blood, plasma, urine, tissue	SPE – Clean Screen DAU Derivatized with MTBSTFA	D$_3$COC, D$_3$BZE, D$_3$EME	DB-5 (15 × m × 0.32 mm, 1 μm)	Injector: 250°C Detector: 250°C Oven: 115°C for 1.5 min, ramped to 280°C at 20°C/min with a final hold of 1 min	Finnigan MAT 4500 PCI – Reagent gas methane: ammonia (4:1) Ion source: 130°C	Rec: >80% Lin: 2.5–2000 μg/L Acc: most within 8.6% CV: <12%	[137]
COC, BZE, EME and CE	Blood, plasma, meconium	Extracted analytes into MeOH, then extracted MeOH by SPE – Bond Elut Certify Derivatized with PFPA and PFP	D$_3$COC, D$_3$BZE, D$_3$EME, D$_3$CE	HP Ultra-2 (12 m × 0.2 mm, 0.33 μm)	Injector: 260°C Detector: 280°C Oven: 145°C for 4.5 min, ramped to 210°C at 60°C/min with an 8 min hold, then to 280°C at 60°C/min with a final hold of 0.25 min	HP 5970 MSD SIM mode	Rec: >81% at 500 ng/g (58–78% at lower concentrations) LOD: 6 ng/g – higher for BL and PL (23 ng/g for EME) Lin: 0–1000 ng/g and 0–1000 ng/mL	[153]
COC, BZE, EME, NCOC, CE, EEE and *m*-HOBZE	Amniotic fluid, umbilical cord tissue	SPE – Clean Screen® Derivatized with MTBSTFA	Bupivacaine	HP-5 MS (30 m × 0.25 mm, 0.25 μm)	Injector: 275°C Detector: 290°C Oven: 90°C for 1 min, ramped to 220°C at 30°C/min with a 0.5 min hold, then to 330°C at 20°C/min with a final hold of 1 min	HP 5972A MSD SIM mode	Rec: 76–120% LOD: 5–50 ng/mL and 2.5–25 ng/g LOQ: 10–50 ng/mL and 2.5–50 ng/g Lin: 25–750 ng/mL or ng/g CV: 6–20%	[154]

TABLE 2.10
METHODS FOR THE DIRECT MASS SPECTROMETRIC ANALYSIS OF COCAINE AND ANALYTES

Analytes Detected	Specification Type	Sample Preparation	Internal Standard(s)	Injection Solvent	Detector Details	Performance Characteristics	Ref.
COC and BZE with other drugs of abuse	Urine	Urine specimen lyophilized and reconstituted in EtOAc:MeOH (3:1)	None	None	Finnigan 4510 GC/MS operated in NICI mode Reagent gas: ammonia	LOD: 1 ng	[155]
COC and BZE	Urine	None – direct injection	None		Bioprobe® continuous flow fast atom bombardment fitted to a TSQ 70 MS/MS	Data obtained from injection of standard solutions LOD: 100 ng/mL (1000 ng/mL for BZE) Lin: linear below 10 ng injected	[156]
COC, BZE, EME, ECG and morphine	Hair	Wash hair with ether and 0.01 M HCl, digest with 0.5 M HCl and extract with Toxitubes® A	None	MeOH	Finnigan MAT ion trap MS	LOD: 5–30 ppb technique for qualitative ID in hair	[131]
COC, BZE, ECG and phencyclidine	Hair	Hair washed in pentane and placed in solid-probe cup	p-methyl phencyclidine	Sample introduced in solid-probe cup	Finnigan-MAT TSQ 70 Daughter mode or parent mode	NA	[157]
BZE with other drug classes	Serum and urine	SPE – Chromabond mixed mode	D₃BZE	Flow injection analysis ACN:10 mM ammonium acetate (1:1)	API 300 triple quadrupole with an ionspray source MRM	Rec: 86% LOD: <2 ng/mL LOQ: <4 ng/mL	[112]

this method. A direct analysis method for confirmation of hair samples was reported by Kidwell [157]. Following a pentane wash, the hair sample was placed in a solid-probe cup and analyzed for COC, BZE and ECG by tandem MS. Seifert *et al.* [156] detected COC and BZE in unextracted urine with combined fast atom bombardment and tandem MS. Although characteristic spectra were obtained, the authors noted that reproducibility was decreased after repeated injections, probably as a result of the precipitation of salts and/or proteins in the capillary tubing of the inlet.

2.2.7 Other analytical methods

Recently-emerging techniques for the analysis of cocaine analytes include supercritical fluid extraction (SFE) and capillary electrophoresis (CEL), including capillary zone electrophoresis (CZE) and micellar electrokinetic capillary chromatography (MECC). Morrison *et al.* [158] investigated matrix and modifier effects on the SFE of COC and BZE. When they applied SFE to hair, COC recovery was limited by its desorption from the hair-binding sites rather than its solubility. BZE recovery was low under all conditions studied. Cirimele *et al.* [159] reported the isolation of COC from hair using a SFE procedure that was also capable of extracting opiates and cannabinoids.

Both CZE and MECC separate analytes based on electrophoretic principles. However, because MECC also incorporates separation mechanisms based on reversed phase chromatography, the two techniques can be used to provide complimentary results [160]. Cocaine analytes and other drugs of abuse have been qualitatively and quantitatively analyzed in illicit drug samples and biological specimens by capillary electrophoresis [160–168]. Detection methods interfaced with CEL systems included UV absorption (UV and diode array detectors), electrospray ionization MS and time-of-flight MS. These methods were employed for the analysis of various sample types, including hair and illicit cocaine mixtures [161,164]. Lurie *et al.* [161] developed CEL-photodiode array and CEL-laser-induced fluorescence methods for the detection of eight isomeric truxillines present in illicit cocaine mixtures. Sensitive LODs were achieved with CEL methods, sometimes as low as the femtomole range [163]. In addition, quantitative results from CEL have been found to be comparable to those from GC and LC methods [166,167]. This technique has certainly demonstrated its potential for the forensic analysis of cocaine.

2.3 CONCLUSIONS AND PERSPECTIVES

Various methods have been presented for the analysis of cocaine and analytes. A summary of the most frequently applied analytical techniques is presented in Table 2.11. The development of more sensitive and specific analytical methods has resulted in the recent identification and quantitation of several cocaine metabolites, including the hydroxycocaines, the hydroxybenzoylecgonines and cocaine N-oxide. Also, this has led to a focus on the testing of alternative matrices, including hair,

MS-MS	Tandem mass spectrometry
MSTFA	*N*-methyl-*N*-trimethylsilyltrifluoroacetamide
MTBSTFA	*N*-methyl-*N*-(tert-butyldimethylsilyl)-trifluoroacetamide
NA	Not available
NAD	Nicotinamide adenine dinucleotide
NADH	The reduced form of NAD
NCE	Norcocaethylene
NCOC	Norcocaine
NICI	Negative impact chemical ionization
NPD	Nitrogen phosphorus detector
PCI	Positive chemical ionization
PDA	Photo diode array
PFP	Pentafluoro-1-propanol
PFPA	Pentafluoropropionic anhydride
PFPOH	Pentafluoropropanol
POC	Point-of-collection/point-of-care
RIA	Radioimmunoassay
Rec	Recovery
RSD	Relative standard deviation
SAMHSA	Substance Abuse and Mental Health Services Administration
SFE	Supercritical fluid extraction
SIM	Selected ion monitoring
SPE	Solid-phase extraction
TLC	Thin layer chromatography
TMAHS	Tetramethylammonium hydrogen sulfate
UV	Ultraviolet spectrophotometry
VH	Vitreous humor

2.5 REFERENCES

1   J.F. Casale and J.M. Moore, J. Forensic Sci., 6 (1994) 1537.
2   J.F. Casale and R.W. Waggoner Jr., J. Forensic Sci., 36 (1991) 1312.
3   J.G. Ensing, C. Racamy and R.A. de Zeeuw, J. Forensic Sci., 37 (1992) 446.
4   J.M. Moore and D.A. Cooper, J. Forensic Sci., 38 (1993) 1286.
5   M.J. LeBelle, B. Dawson, G. Lauriault and C. Savard, Analyst, 116 (1991) 1063.
6   J.F. Casale and J.M. Moore, J. Forensic Sci., 39 (1994) 462.
7   K.E. Janzen, L. Walter and A.R. Fernando, J. Forensic Sci., 37 (1992) 436.
8   J.R. Ehleringer, J.F. Casale, M.J. Lott and V.L. Ford, Nature, 408 (2000) 311.
9   D.R. Morello and R.P. Meyers, J. Forensic Sci., 40 (1995) 957.
10 M. Chiarotti, R. Marsili and A. Moreda-Pineiro, J. Chromatogr. B, Analyt. Technol. Biomed. Life Sci., 772 (2002) 249.
11 P.P. Wang and M.G. Bartlett, J. Anal. Toxicol., 23 (1999) 62.
12 Y. Xia, P. Wang, M.G. Bartlett, H.M. Solomon and K.L. Busch, Anal. Chem., 72 (2000) 764.
13 S.N. Lin, S.L. Walsh, D.E. Moody and R.L. Foltz, Anal. Chem., 75 (2003) 4335.

14   C.L. Hornbeck, K.M. Barton and R.J. Czarny, J. Anal. Toxicol., 19 (1995) 133.
15   B.D. Paul, S. Lalani, T. Bosy, A.J. Jacobs and M.A. Huestis, Biomed. Chromatogr., 19 (2005) 677.
16   E.J. Cone, J. Anal. Toxicol., 19 (1995) 459.
17   R.A. Jufer, S.L. Walsh and E.J. Cone, J. Anal. Toxicol., 22 (1998) 435.
18   R.C. Baselt, J. Chromatogr., 268 (1983) 502.
19   V.D. Gupta, Int. J. Pharm., 10 (1982) 249.
20   M.R. Brzezinski, T.L. Abraham, C.L. Stone, R.A. Dean and W.F. Bosron, Biochem. Pharmacol., 48 (1994) 1747.
21   T. Inaba, D.J. Stewart and W. Kalow, Clin. Pharmacol. Ther., 23 (1978) 547.
22   R.A. Dean, C.D. Christian, R.H. Sample and W.F. Bosron, FASEB J., 5 (1991) 2735.
23   E.V. Pindel, N.Y. Kedishvili, T.L. Abraham, M.R. Brzezinski, J. Zhang, R.A. Dean and W.F. Bosron, J. Biol. Chem., 272 (1997) 14769.
24   J. Ambre, J. Anal. Toxicol., 9 (1985) 241.
25   V. Ramcharitar, B. Levine and J.E. Smialek, J. Forensic Sci., 40 (1995) 99.
26   B.W. LeDuc, P.R. Sinclair, L. Shuster, J.F. Sinclair, J.E. Evans and D.J. Greenblatt, Pharmacology, 46 (1993) 294.
27   U.A. Boelsterli and C. Goldlin, Arch. Toxicol., 65 (1991) 351.
28   F.M. Ndikum-Moffor, T.R. Schoeb and S.M. Roberts, J. Pharmacol. Exp. Ther., 284 (1998) 413.
29   D.M. Chinn, D.J. Crouch, M.A. Peat, B.S. Finkle and T.A. Jennison, J. Anal. Toxicol., 4 (1980) 37.
30   K. Watanabe, Y. Hida, T. Matsunaga, I. Yamamoto and H. Yoshimura, Biol. Pharm. Bull., 16(10) (1993) 1041.
31   J.A. Levisky, D.L. Bowerman, W.W. Jenkins and S.B. Karch, Forensic Sci. Int., 110 (2000) 35.
32   A. Hernandez, W. Andollo and W.L. Hearn, Forensic Sci. Int., 65 (1994) 149.
33   V.R. Spiehler and D. Reed, J. Forensic Sci., 30 (1985) 1003.
34   R.E. Winecker, B.A. Goldberger, I.R. Tebbett, M. Behnke, F.D. Eyler, J.L. Karlix, K. Wobie, M. Conlon, D. Phillips and R.L. Bertholf, J. Forensic Sci., 46 (2001) 1221.
35   E.J. Cone, D. Yousefnejad, W.D. Darwin and T. Maquire, J. Anal. Toxicol., 15 (1991) 250.
36   A.M.B. Barrera and S.S. Rossi, Forensic Sci. Int., 70 (1995) 203.
37   D.E. Lewis, C.M. Moore and J.B. Leikin, Clin Toxicol., 32 (1994) 697.
38   B.W. Steele, E.S. Bandstra, N.C. Wu, G.W. Hime and W.L. Hearn, J. Anal. Toxicol., 17 (1993) 348.
39   M.A. ElSohly, D.F. Stanford, T.P. Murphy, B.M. Lester, L.L. Wright, V.L. Smeriglio, J. Verter, C.R. Bauer, S. Shankaran, H.S. Bada and H.C. Walls, J. Anal. Toxicol., 23 (1999) 436.
40   D. Garside, J.D. Ropero-Miller, B.A. Goldberger, W.F. Hamilton and W.R. Maples, J. Forensic Sci., 43 (1998) 974.
41   D.A. Engelhart, E.S. Lavins and C.A. Sutheimer, J. Anal. Toxicol., 22 (1998) 314.
42   D.A. Engelhart and A.J. Jenkins, J. Anal. Toxicol., 26 (2002) 489.
43   W. Schramm, R.H. Smith, P.A. Craig and D.A. Kidwell, J. Anal. Toxicol., 16 (1992) 1.
44   R.E. Joseph Jr., J.M. Oyler, A.T. Wstadik, C. Ohuoha and E.J. Cone, J. Anal. Toxicol., 21 (1997) 6.
45   E.J. Cone, K. Kato and M. Hillsgrove, J. Anal. Toxicol., 20 (1996) 139.
46   E.J. Cone, M.J. Hillsgrove, A.J. Jenkins, R.M. Keenan and W.D. Darwin, J. Anal. Toxicol., 18 (1994) 298.
47   M. Pellegrini, A. Casa, E. Marchei, R. Pacifici, R. Mayne, V. Barbero, O. Garcia-Algar and S. Pichini, J. Pharm. Biomed. Anal., 40 (2006) 662.
48   B.K. Logan and D.T. Stafford, J. Forensic Sci., 35 (1990) 1303.
49   Y. Gaillard and G. Pepin, Forensic Sci. Int., 86 (1997) 49.
50   P.E. McKinney, S. Phillips, H.F. Gomex, J. Brent, M. MacIntyre and W.A. Watson, J. Forensic Sci., 40 (1995) 102.
51   B.D. Paul, L.K. McWhorter and M.L. Smith, Detection of Urinary ecgonidine as an indicator of active smoking of cocaine, Presented at the Society of Forensic Toxicologist Annual meeting, Albuquerque, NM, 1998.

52   E.T. Shimomura, G.D. Hodge and B.D. Paul, Clin. Chem., 47 (2001) 1040.

53   A.L. Myers, H.E. Williams, J.C. Kraner and P.S. Callery, J. Forensic Sci., 50 (2005) 1481.

54   S.N. Giorgi and J.E. Meeker, J. Anal. Toxicol., 19 (1995) 392.

55   D.S. Isenschmid, B.S. Levine and Y.H. Caplan, J. Anal. Toxicol., 13 (1989) 250.

56   W.C. Brogan III, P.M. Kemp, R.O. Bost, D.B. Glamann, R.A. Lange and L.D. Hillis, J. Anal. Toxicol., 16 (1992) 152.

57   A.S. Fandino, S.W. Toennes and G.F. Kauert, J. Anal. Toxicol., 26 (2002) 567.

58   K.B. Scheidweiler, J. Shojaie, M.A. Plessinger, R.W. Wood and T.C. Kwong, Clin. Chem., 46 (2000) 1787.

59   B.D. Paul, R.M. McKinley, J.K. Walsh Jr., T.S. Jamir and M.R. Past, J. Anal. Toxicol., 17 (1993) 378.

60   B. Levine, V. Ramcharitar and J.E. Smialek, J. Forensic Sci., 41 (1996) 126.

61   W.L. Hearn, E.E. Keran, H. Wei and G. Hime, J. Forensic Sci., 36 (1991) 673.

62   B.K. Logan and D. Smirnow, J. Anal. Toxicol., 21 (1997) 23.

63   K.B. Nolte, R.D. Pinder and W.D. Lord, J. Forensic Sci., 37 (1992) 1179.

64   D.T. Manhoff, I. Hood, F. Caputo, J. Perry, S. Rosen and H.G. Mirchandani, J. Forensic Sci., 36 (1991) 1732.

65   M. Cingolani, M. Cippitelli, R. Froldi, V. Gambaro and G. Tassoni, J. Anal. Toxicol., 28 (2004) 16.

66   M.L. Smith, Immunoassay. In: B. Levine (Ed.), Principles of forensic toxicology, AACC Press, Washington, DC, 2003, pp. 117–138.

67   V.I. Luzzi, A.N. Saunders, J.W. Koenig, J. Turk, S.F. Lo, U.C. Garg and D.J. Dietzen, Clin. Chem., 50 (2004) 717.

68   M.T. Contreras, A.F. Hernandez, M. Gonzalez, S. Gonzalez, R. Ventura, A. Pla, J.L. Valverde, J. Segura and R.D. Torre, Forensic Sci. Int., 164 (2006) 168.

69   M.S. Swartz, J.W. Swanson and M.J. Hannon, Psychiatr. Serv., 54 (2003) 891.

70   E.J. Cone, A.H. Sampson-Cone, W.D. Darwin, M.A. Huestis and J.M. Oyler, J. Anal. Toxicol., 27 (2003) 386.

71   R.F. Schilling, B. Bidassie and N. El Bassel, J. Psychoactive Drugs, 31 (1999) 305.

72   N.T. Lu and B.G. Taylor, Forensic Sci. Int., 157 (2006) 106.

73   M. Feldman, D. Kuntz, K. Botelho, D.C. Ananias, M. Gnezda, D.K. Hoch, S.L. Jordan, S. Rashid and Y. Zhao, J. Anal. Toxicol., 28 (2004) 593.

74   D.A. Kidwell, J.D. Kidwell, F. Shinohara, C. Harper, K. Roarty, K. Bernadt, R.A. McCaulley and F.P. Smith, Forensic Sci. Int., 133 (2003) 63.

75   C.W. Chronister, J.C. Walrath and B.A. Goldberger, J. Anal. Toxicol., 25 (2001) 621.

76   E.M. Hattab, B.A. Goldberger, L.M. Johannsen, P.W. Kindland, F. Ticino, C.W. Chronister and R.L. Bertholf, Ann. Clin. Lab. Sci., 30 (2000) 85.

77   S. Kerrigan and J.W. Phillips Jr., Clin. Chem., 47 (2001) 540.

78   R.S. Niedbala, K. Kardos, T. Fries, A. Cannon and A. Davis, J. Anal. Toxicol., 25 (2001) 62.

79   K. Lachenmeier, F. Musshoff and B. Madea, Forensic Sci. Int., 159 (2006) 189.

80   C. Moore, D. Deitermann, D. Lewis, B. Feeley and R.S. Niedbala, J. Forensic Sci., 44 (1999) 609.

81   V. Spiehler, D.S. Isenschmid, P. Matthews, P. Kemp and T. Kupiec, J. Anal. Toxicol., 27 (2003) 587.

82   I. Kim, A.J. Barnes, R. Schepers, E.T. Moolchan, L. Wilson, G. Cooper, C. Reid, C. Hand and M.A. Huestis, Clin. Chem., 49 (2003) 1498.

83   G. Cooper, L. Wilson, C. Reid, D. Baldwin, C. Hand and V. Spieher, J. Anal. Toxicol., 28 (2004) 498.

84   D.E. Moody, W.B. Fang, D.M. Andrenyak, K.M. Monti and C. Jones, J. Anal. Toxicol., 30 (2006) 50.

85   E.A. Kolbrich, I. Kim, A.J. Barnes, E.T. Moolchan, L. Wilson, G.A. Cooper, C. Reid, D. Baldwin, C.W. Hand and M.A. Huestis, J. Anal. Toxicol., 27 (2003) 407.

86   N. Fucci, N. De Giovanni, F. De Giorgio, R. Liddi and M. Chiarotti, Forensic Sci. Int., 156 (2006) 102.

87   G.S. Yacoubian Jr., E.D. Wish and J.D. Choyka, J. Psychoactive Drugs, 34 (2002) 325.
88   P. Bouis, G. Taccard and U.A. Boelsterli, J. Chromatogr., 526 (1990) 447.
89   C.L. Williams, S.C. Laizure, R.B. Parker and J.J. Lima, J. Chromatogr. B, Biomed. Appl., 681 (1996) 271.
90   F. Ma, J. Zhang and C.E. Lau, J. Chromatogr. B, Biomed. Sci. Appl., 693 (1997) 307.
91   P. Jatlow and H. Nadim, Clin. Chem., 36 (1990) 1436.
92   K.M. Clauwaert, J.F. Van Bocxlaer, W.E. Lambert and A.P. De Leenheer, J. Chromatogr. Sci., 35 (1997) 321.
93   B.M. Lampert and J.T. Stewart, J. Chromatogr., 495 (1989) 153.
94   J.A. Sandberg and G.D. Olsen, J. Chromatogr., 525 (1990) 113.
95   L. Virag, B. Mets and S. Jamdar, J. Chromatogr. B, Biomed. Appl., 681 (1996) 263.
96   J.L. Katz, P. Terry and J.M. Witkin, Life Sci., 50 (1992) 1351.
97   M. Nishikawa, K. Nakajima, M. Tatsuno, F. Kasuya, K. Igarashi, M. Fukui and H. Tsuchihashi, Forensic Sci. Int., 66 (1994) 149.
98   P.S. Cardona, A.K. Chaturvedi, J.W. Soper and D.V. Canfield, Forensic Sci. Int., 157 (2006) 46.
99   R.J. Lewis, R.D. Johnson, M.K. Angier and R.M. Ritter, J. Chromatogr. B, Analyt. Technol. Biomed. Life Sci., 806 (2004) 141.
100 P. Lillsunde and T. Korte, J. Anal. Toxicol., 15 (1991) 71.
101 D.N. Bailey, Am. J. Clin. Pathol., 101 (1994) 342.
102 P.D. Whitter and P.L. Cary, J. Anal. Toxicol., 10 (1986) 68.
103 L. Antonilli, C. Suriano, M.C. Grassi and P. Nencini, J. Chromatogr. B, Biomed. Sci. Appl., 751 (2001) 19.
104 M.J. Kogan, D.J. Pierson, M.M. Durkin and N.J. Willson, J. Chromatogr., 490 (1989) 236.
105 A.K. Larsen Jr. and I.R. Tebbett, J. Forensic Sci., 37 (1992) 636.
106 K.Y. Kim and L.M. Bornheim, J. Chromatogr. Sci., 35 (1997) 287.
107 C. Rerat, M. Sauvain, P.P. Rop, E. Ruiz, M. Bresson and A. Viala, J. Ethnopharmacol., 56 (1997) 173.
108 P. Fernandez, N. Lafuente, A.M. Bermejo, M. Lopez-Rivadulla and A. Cruz, J. Anal. Toxicol., 20 (1996) 224.
109 L.J. Murphey, G.D. Olsen and R.J. Konkol, J. Chromatogr., 613 (1993) 330.
110 D.L. Phillips, I.R. Tebbett and R.L. Bertholf, J. Anal. Toxicol., 20 (1996) 305.
111 M. Tatsuno, M. Nishikawa, M. Katagi and H. Tsuchihashi, J. Anal. Toxicol., 20 (1996) 281.
112 W. Weinmann and M. Svoboda, J. Anal. Toxicol., 22 (1998) 319.
113 K.M. Clauwaert, J.F. Van Bocxlaer, W.E. Lambert, E.G. Van den Eeckhout, F. Lemiere, E.L. Esmans and A.P. De Leenheer, Anal. Chem., 70 (1998) 2336.
114 C.S. Sosnoff, Q. Ann, J.T. Bernert Jr., M.K. Powell, B.B. Miller, L.O. Henderson, W.H. Hannon and E.J. Sampson, J. Anal. Toxicol., 20 (1996) 179.
115 M.J. Bogusz, R.D. Maier, K.D. Kruger and U. Kohls, J. Anal. Toxicol., 22 (1998) 549.
116 R. Dams, C.M. Murphy, W.E. Lambert and M.A. Huestis, Rapid Commun. Mass Spectrom., 17 (2003) 1665.
117 K.R. Allen, R. Azad, H.P. Field and D.K. Blake, Ann. Clin. Biochem., 42 (2005) 277.
118 K.A. Mortier, K.E. Maudens, W.E. Lambert, K.M. Clauwaert, J.F. Van Bocxlaer, D.L. Deforce, C.H. Van Peteghem and A.P. De Leenheer, J. Chromatogr. B, Analyt. Technol. Biomed. Life Sci., 779 (2002) 321.
119 M. Wood, M. Laloup, M.M. Ramirez Fernandez, K.M. Jenkins, M.S. Young, J.G. Ramaekers, G. De Boeck and N. Samyn, Forensic Sci. Int., 150 (2005) 227.
120 R. Kronstrand, I. Nystrom, J. Strandberg and H. Druid, Forensic Sci. Int., 145 (2004) 183.
121 S. Pichini, R. Pacifici, M. Pellegrini, E. Marchei, E. Perez-Alarcon, C. Puig, O. Vall and O. Garcia-Algar, J. Chromatogr. B, Analyt. Technol. Biomed. Life Sci., 794 (2003) 281.
122 S. Pichini, E. Marchei, R. Pacifici, M. Pellegrini, J. Lozano and O. Garcia-Algar, J. Chromatogr. B, Analyt. Technol. Biomed. Life Sci., 820 (2005) 151.
123 A. Cailleux, A. Le Bouil, B. Auger, G. Bonsergent, A. Turcant and P. Allain, J. Anal. Toxicol., 23 (1999) 620.

124 A. Klingmann, G. Skopp and R. Aderjan, J. Anal. Toxicol., 25 (2001) 425.
125 K. Clauwaert, T. Decaestecker, K. Mortier, W. Lambert, D. Deforce, C. Van Peteghem and J. Van Bocxlaer, J. Anal. Toxicol., 28 (2004) 655.
126 D. Garside, B.A. Goldberger, K.L. Preston and E.J. Cone, J. Chromatogr., 692 (1997) 61.
127 T.A. Jennison, C.W. Jones, E. Wozniak and F.M. Urry, J. Chromatogr. Sci., 32 (1994) 126.
128 W.L. Wang, W.D. Darwin and E.J. Cone, J. Chromatogr., 660 (1994) 279.
129 M.R. Moller, P. Fey and S. Rimbach, J. Anal. Toxicol., 16 (1992) 291.
130 V. Cirimele, P. Kintz and P. Mangin, Comparison of different extraction procedures for drugs in hair of drug addicts. In: E.J. Cone, M.J. Welch and M.B. Grigson Babecki (Eds.), Hair testing for drugs of abuse: international research on standards and technology(NIH-Pub. No. 95-3727). National Institute on Drug Abuse, Rockville, MD, 1995, pp. 277–288
131 P. Traldi, D. Favretto and F. Tagliaro, Forensic Sci. Int., 63 (1993) 239.
132 M.R. Harkey, G.L. Henderson and C. Zhou, J. Anal. Toxicol., 15 (1991) 260.
133 W.C. Thompson and A. Dasgupta, Am. J. Clin. Pathol., 104 (1995) 187.
134 D.S. Isenschmid, B.S. Levine and Y.H. Caplan, J. Anal. Toxicol., 12 (1988) 242.
135 D. Smirnow and B.K. Logan, J. Anal. Toxicol., 20 (1996) 463.
136 B.D. Paul, C. Dreka, J.L. Summers and M.L. Smith, J. Anal. Toxicol., 20 (1996) 506.
137 D.J. Crouch, M.E. Alburges, A.C. Spanbauer, D.E. Rollins and D.E. Moody, J. Anal. Toxicol., 19 (1995) 352.
138 W.D. Darwin, J. Oyler and E. Cone. Simultaneous EC/MS Assay for cocaine, codeine, 6-Acetylmorphine and Metabolite in Human Biological Specimer, Presented at the Society of Forensic Toxicologists Annual Meeting, Snowbird, UT, 1997.
139 L. Virag, S. Jamdar, C.R. Chao and H.O. Morishima, J. Chromatogr. B, Biomed. Appl., 658 (1994) 135.
140 S. Dawling, E.G. Essex, N. Ward and B. Widdop, Ann. Clin. Biochem., 27 (1990) 478.
141 J. Vasiliades, J. Anal. Toxicol., 13 (1989) 127.
142 J. Ortuno, T.R. de la, J. Segura and J. Cami, J. Pharm. Biomed. Anal., 8 (1990) 911.
143 G.W. Hime, W.L. Hearn, S. Rose and J. Cofino, J. Anal. Toxicol., 15 (1991) 241.
144 M.R. Corburt and E.M. Koves, J. Forensic Sci., 39 (1994) 136.
145 K.L. Peterson, B.K. Logan and G.D. Christian, Forensic Sci. Int., 73 (1995) 183.
146 R.W. Taylor, N.C. Jain and M.P. George, J. Anal. Toxicol., 11 (1987) 233.
147 P. Campora, A.M. Bermejo, M.J. Tabernero and P. Fernandez, J. Anal. Toxicol., 27 (2003) 270.
148 J.A. Bourland, E.F. Hayes, R.C. Kelly, S.A. Sweeney and M.M. Hatab, J. Anal. Toxicol., 24 (2000) 489.
149 E. Cognard, S. Rudaz, S. Bouchonnet and C. Staub, J. Chromatogr. B, Analyt. Technol. Biomed. Life Sci., 826 (2005) 17.
150 S. Valente-Campos, M. Yonamine, R.L. Moraes Moreau and O.A. Silva, Forensic Sci. Int., 159 (2006) 218.
151 W. Yang, A.J. Barnes, M.G. Ripple, D.R. Fowler, E.J. Cone, E.T. Moolchan, H. Chung and M.A. Huestis, J. Chromatogr. B, Analyt. Technol. Biomed. Life Sci., 833 (2006) 210.
152 C. Moore, D. Dempsey, D. Deitermann, D. Lewis and J. Leikin, J. Anal. Toxicol., 20 (1996) 509.
153 G.M. Abusada, I.K. Abukhalaf, D.D. Alford, I. Vinzon-Bautista, A.K. Pramanik, N.A. Ansari, J.E. Manno and B.R. Manno, J. Anal. Toxicol., 17 (1993) 353.
154 R.E. Winecker, B.A. Goldberger, I. Tebbett, M. Behnke, F.D. Eyler, M. Conlon, K. Wobie, J. Karlix and R.L. Bertholf, J. Anal. Toxicol., 21 (1997) 97.
155 V.M. Papa, P.S. Ng, E.F. Robbins and D.W. Ou, Biomed. Environ. Mass Spectrom., 16 (1988) 263.
156 W.E. Seifert Jr., A. Ballatore and R.M. Caprioli, Rapid Commun. Mass Spectrom., 3 (1989) 117.
157 D.A. Kidwell, J. Forensic Sci., 38 (1993) 272.
158 J.F. Morrison, S.N. Chesler, W.J. Yoo and C.M. Selavka, Anal. Chem., 70 (1998) 163.
159 V. Cirimele, P. Kintz, R. Majdalani and P. Mangin, J. Chromatogr. B, Biomed. Appl., 673 (1995) 173.
160 F. Tagliaro, F.P. Smith, S. Turrina, V. Equisetto and M. Marigo, J. Chromatogr. A, 735 (1996) 227.

161 I.S. Lurie, P.A. Hays, J.F. Casale, J.M. Moore, D.M. Castell, K.C. Chan and H.J. Issaq, Electrophoresis, 19 (1998) 51.

162 I.S. Lurie, J.M. Moore, D.A. Cooper and T.C. Kram, J. Chromatogr., 405 (1987) 273.

163 I.M. Lazar, G. Naisbitt and M.L. Lee, Analyst, 123 (1998) 1449.

164 F. Tagliaro, W.F. Smyth, S. Turrina, Z. Deyl and M. Marigo, Forensic Sci. Int., 70 (1995) 93.

165 F. Tagliaro, Z. Deyl and M. Marigo, Capillary electrophoresis: a novel tool for toxicological investigation: its potential in the analysis of body fluids and hair. In: E.J. Cone, M.J. Welch and M.B. Grigson Babecki (Eds.), Hair testing for drugs of abuse: international research on standards and technology (NIH-Pub. No. 95-3727). National Institute on Drug Abuse, Rockville, MD, 1995, pp. 225–247.

166 J.A. Walker, H.L. Marche, N. Newby and E.J. Bechtold, J. Forensic Sci., 41 (1996) 824.

167 V.C. Trenerry, J. Robertson and R.J. Wells, Electrophoresis, 15 (1994) 103.

168 I.S. Lurie, J. Chromatogr. A, 780 (1997) 265.

M.J. Bogusz (Ed.). Forensic Science
Handbook of Analytical Separations, Vol. 6
© 2008 Elsevier B.V. All rights reserved.

CHAPTER 3

Amphetamines ☆

John T. Cody

Air Force Drug Testing Laboratory, 2730 Louis Bauer Drive, Brooks City-Base, TX 78235-5132, USA

3.1 INTRODUCTION

Amphetamine and methamphetamine are powerful central nervous system (CNS) stimulants that have been in use since the early 1900s. These drugs have a single asymmetric center and therefore exist as two enantiomers, each of which has different pharmacological activities. The structures of amphetamine and methamphetamine enantiomers are shown in Fig. 3.1. Over the years, substitutions have been made to the molecules to alter their activity. The alterations may have been to increase CNS activity, or to decrease side effects associated with the native drug.

A group of related drugs, often referred to as 'designer amphetamines', includes several methylenedioxy analogues which, although thought by some to have useful medicinal effects, are currently not recognized to have any legitimate medical use and are therefore included in Schedule I of controlled substances. They are often found on the illicit market as an abused drug. These compounds include 3,4-methylenedioxyamphetamine (MDA), 3,4-methylenedioxymethamphetamine (MDMA) and 3,4-methylenedioxyethylamphetamine (MDEA). MDA, the first of these to appear, was widely used but had a number of undesirable side effects. MDMA soon followed and was favored because of its lack of significant side effects and more pleasurable experiences. The structures of MDA, MDMA and MDEA as well as some of their deuterated analogs are shown in Fig. 3.2.

Historically, a variety of other amphetamine related compounds have been abused. These include 2-methoxy, 4-hydroxy, 2,5-dimethoxyamphetamine (DMA), 4-bromo-2,5-dimethoxyamphetamine (DOB), 4-bromo-2,5-dimethoxy-β-phenethylamine (BDMPEA), 3,4,5-trimethoxyamphetamine (TMA), 3,4-methylenedioxyamphetamine (MDA), *N,N*-dimethyl-3,4-methylenedioxy, *N*-hydroxy-3,4-methylenedioxy

---

☆The views expressed in this article are those of the author and do not reflect the official policy of the Department of Defense or other Departments of the US Government

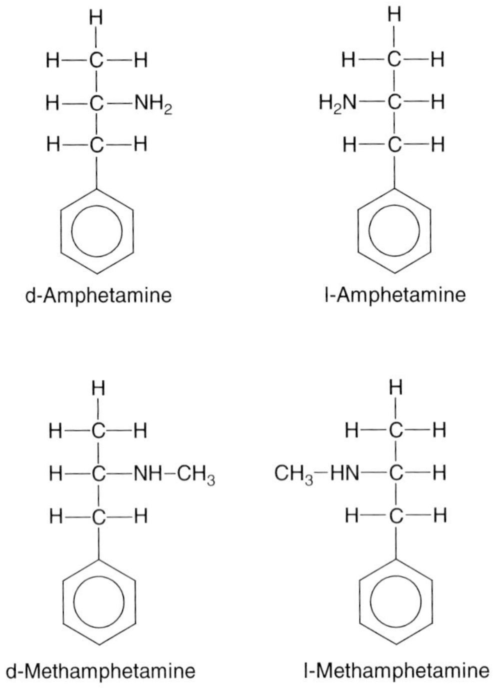

Fig. 3.1. Structures of amphetamine and methamphetamine enantiomers.

(N-OH-MDA), 2,5-dimethoxy-4-ethylamphetamine (DOE) and 2,5-dimethoxy-4-methylamphetamine (DOM). Analysis of these compounds has been described by various authors and most are the same methods used for identification of amphetamines and therefore are not described in detail in this chapter. A few examples will be included, but only those that relate to the drugs otherwise included in this chapter.

N-Substitution of amphetamine or methamphetamine has been used in the preparation of a number of drugs that have been developed for a variety of reasons. Many are the result of attempts to provide the anorexic effect of amphetamine and methamphetamine while decreasing the undesirable side effects and rapid tolerance seen with those drugs. Some of these drugs are metabolized to methamphetamine and/or amphetamine by the body. The structures of these amphetamine/methamphetamine precursor drugs are shown in Fig. 3.3.

There are numerous references describing analysis of amphetamines using a variety of extraction, derivatization and instrumental methods. Many of these procedures, though developed many years ago, are still effective and remain in wide use. Recently, developed procedures focus more on the elimination of interferences, shortening analysis time, decreasing solvent use (and its purchase and related disposal costs), detection with new detectors, isolation from alternative matrices (hair, meconium, sweat, etc.) or procedures that take advantage of the sensitivity of newer analytical

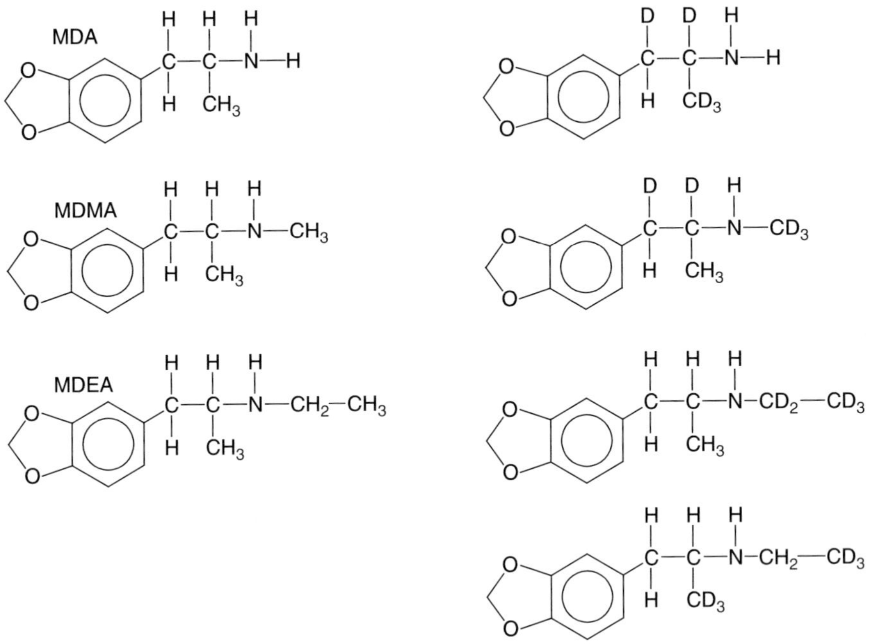

Fig. 3.2. Structures of MDA, MDMA, MDEA, d_5-MDA, d_5-MDMA and d_5-MDEA.

instruments and techniques. The references used in this review are primarily from the 1990s with a few exceptions for the description of initial methods or important developments.

3.2 EXTRACTION

Preparation of samples for analysis involves varying degrees of time and effort. The most common methods are liquid–liquid extraction and solid-phase extraction. Most procedures utilize one or the other of these two methods. Additionally, several other procedures have been proposed that do not involve extraction using one of these traditional methods. These include direct analysis of samples in the biological matrix, most often seen with liquid chromatography (LC) procedures, and solid phase and solvent microextraction. Solid-phase microextraction (SPME) is a relatively new methodology in the extraction of compounds from biological origin and therefore not widely used in the analytical community, but has recently been employed for analysis of amphetamines and its use has increased dramatically since the first writing of this chapter. These methodologies are discussed later in this chapter.

   The analysis of amphetamine, methamphetamine and related parent compounds requires no hydrolysis since they are not conjugated by the body. Analysis of most of their metabolites, however, does require hydrolysis. Hydrolysis of conjugates of metabolites of amphetamines is typically accomplished using either acid or

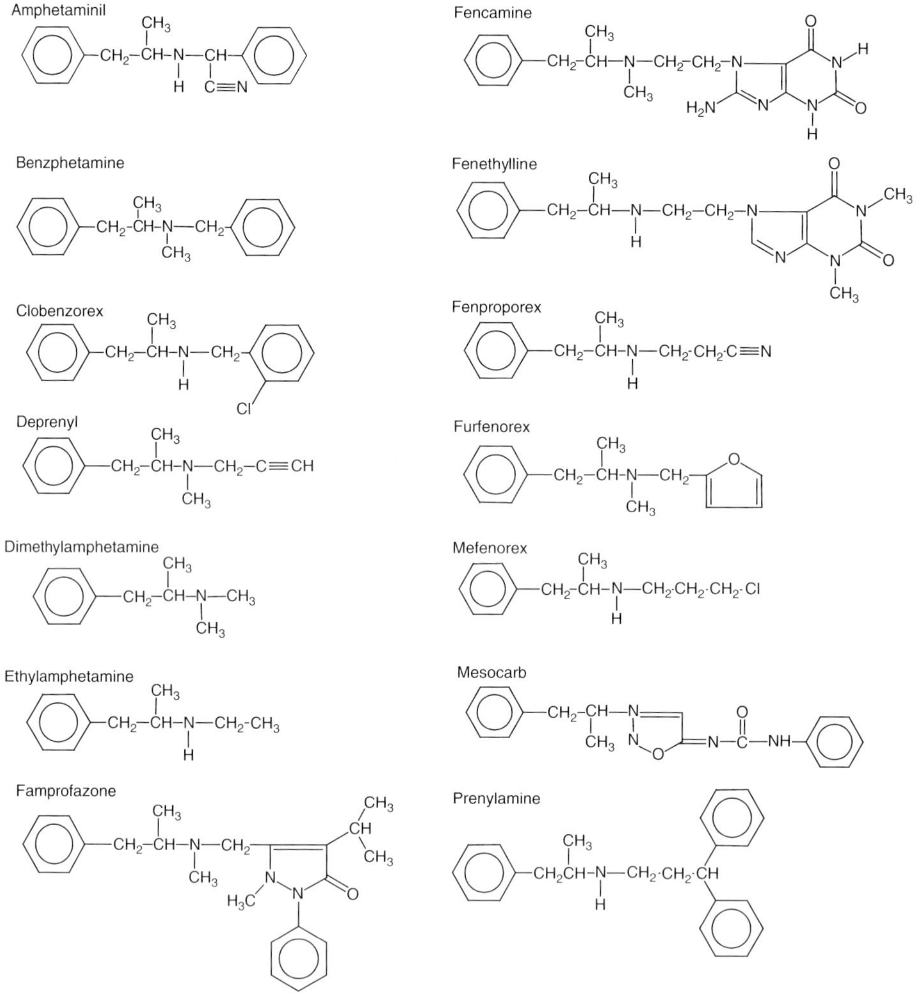

Fig. 3.3. Structures of amphetamine and methamphetamine precursor drugs.

enzymatic procedures. Generally, enzyme hydrolysis gives a cleaner sample than seen with acid hydrolysis, although this is not universally true. Acid hydrolysis procedures generally require less time than their enzymatic counterparts. Although the amphetamines do not require hydrolysis, some methods are designed to isolate and identify a large number of drugs including the amphetamines. As a result, these methods involve hydrolysis of the sample prior to extraction to release analytes of interest, which are normally conjugated. Fortunately, the amphetamines are generally stable to these conditions.

Depending on the target analytes, pretreatment of a sample may be accomplished to avoid interference from some amphetamine related compounds. Compounds such as ephedrine, pseudoephedrine and phenylpropanolamine are commonly found in samples, despite the fact they have been removed from open sale, and often in very high concentrations. These compounds are readily extracted from samples using the same procedures designed to extract amphetamines and as a result, they may cause interference. Although in many cases, the selective nature of the mass spectrometry (MS) avoids this problem, the amount of material sometimes seen in samples may make it impossible to properly identify amphetamine and/or methamphetamine if present. To eliminate that interference, periodate can be used to destroy the hydroxy containing compounds. Several methods have been described for periodate oxidation of these drugs including use of the reagent at room [1,2] and elevated [3,4] temperatures. At very high concentrations of these hydroxy containing compounds, periodate treatment may not completely eliminate all of the compound from the sample. In such cases, however, even when some of the compound remains after periodate treatment, its concentration is sufficiently decreased and thus causes no interference. Paul *et al.* described demethylation of methamphetamine when periodate oxidation was carried out at pH values of 9.1 or higher. They determined that at pH levels below that level the degradation did not occur. The periodate reaction with ephedrine, pseudoephedrine and phenylpropanolamine was shown to take place at pH values above 5.2, causing the authors to recommend adjustment of pH above 5.2 and below 9.1 [4].

Pyrolysis products of smoked methamphetamine were extracted by placing a C-8 column in line with a pump that drew in the mainstream smoke. Sidestream smoke was collected by drawing smoke from the area of the burning end of the cigarette. The components were eluted off the cartridge with 3 ml of methanol and the dried extract was reconstituted in a small amount of methanol then analyzed. Analysis in this case was accomplished using a high-resolution double focusing gas chromatography-mass spectrometry-mass spectrometry (GC-MS-MS) instrument. A standard cigarette control showed the products that came from tobacco smoke. Those results were then compared with the products from the analysis of the smoked methamphetamine study. This resulted in the identification of a number of different pyrolysis products of methamphetamine [5].

An extraction process commonly used with solid materials called supercritical fluid extraction has been applied to the extraction of amphetamines from biological samples. Supercritical fluid extraction reduces the time needed for the extraction and concentration of organic compounds. The solvent strength of a supercritical fluid is directly related to its density, which can be modified by changing pressure and temperature. Use of supercritical fluids that are gases at room temperature makes sample concentration relatively simple. Supercritical fluid extraction techniques followed by GC-MS analysis were applied to extract MDA, MDMA and MDEA from hair. Supercritical fluid extraction conditions were 3800 psi with a flow-rate of 2 ml/min and temperature 70°C. The modifier was chloroform:isopropyl alcohol (90:10, v/v) pumped at 10% in 90% CO in dynamic extraction mode for 30 min. Extracts were derivatized using pentafluoroproprionic anhydride (PFPA):ethyl acetate (1:1, v/v) followed by analysis by GC-MS using mephentermine as the internal standard [6]. Another supercritical fluid extraction procedure for the determination of amphetamine and methamphetamine in urine was described by Wang *et al.* [7]. The procedure called for simultaneous supercritical fluid extraction and derivatization rather than the more common stepwise process citing the finding that the extraction efficiencies of 95% (RSD = 3.8%) and 89% (RSD = 4.0%) for amphetamine and methamphetamine, respectively were superior to those for the stepwise process.

A procedure designed to use only a small volume of urine described the application of 0.1 mL urine sample to an extraction column (0.2 g of Extrelut and sodium carbonate (3:1, w/w) in a pasture pipette; 10 mm I.D.) for 20 min followed by the addition of propylchloroformate in ethyl acetate to the column for at least 10 min. The derivatized drug was then eluted by 0.2 mL of ethyl acetate, which was then injected into the GC [8]. Recoveries of amphetamine and methamphetamine from urine were 100 and 102%, respectively. The linearity of the assay was reported to be 0.50–50 µg/mL.

3.2.1 Liquid–Liquid Extraction

Liquid–liquid extraction is the oldest yet still widely used method for the extraction of drugs, including amphetamines, from biological samples. The amphetamines are basic drugs with pKa values of approximately 10. As a result, liquid–liquid extraction methods commonly employ an organic solvent to extract the drugs from their aqueous environment. In order for that to happen effectively, the amine group is neutralized by raising the pH of the matrix. Once the charge on the amine group is gone, the amphetamines become highly soluble in organic solvents and are readily extracted. Methods using these basic procedures extract both basic and neutral compounds. Some procedures further purify the initial extract by back extracting the basic drugs into an aqueous solvent. This is usually accomplished with an acid solution that causes the amine to become positively charged thus making it more soluble in the aqueous phase. Neutral compounds remain in the organic layer and are discarded thereby cleaning up the sample extract. The basic drugs are then extracted from the aqueous layer by increasing the pH as before and extracting into an organic solvent. A simple example of this principle is the extraction of amphetamines from 2 mL of urine to which internal standard has been added along with sodium hydroxide to make the sample basic. 1-Chlorobutane (5 mL) is added and the tube shaken then the solvent transferred to another tube. To this dilute sulfuric acid solution is added and the tubes shaken to back extract the amphetamines. After discarding the solvent layer, which contains neutral compounds, sodium hydroxide is added and the basic drugs extracted into 5 mL of 1-chlorobutane. This extract is then dried under a stream of nitrogen, derivatized with heptafluorobutyric anhydride (HFBA) and analyzed by GC-MS [9]. This extraction method proved successful for the analysis of a variety of amphetamine and amphetamine related drugs (amphetamine, methamphetamine, MDA, MDMA, MDEA, and a number of related precursor drugs).

Urine samples make up the majority of assays conducted for the detection of amphetamines. In many respects, urine samples are relatively easy to work with. Extraction of other samples, such as blood, plasma, tissue, etc., often involves additional steps. A common first step with blood and plasma, for example, is to eliminate the protein in the sample. This can be accomplished by a number of different methods, including salting out the protein, cold precipitation, use of acetonitrile or some other agent to precipitate the proteins. Tissue samples are typically homogenized, followed by removal of the solid matter prior to extraction. After such initial steps, extraction procedures typically are the same regardless of the initial matrix. Recent developments in solid-phase materials have allowed extraction of blood and plasma samples without the necessity of any prior steps being taken. In addition, some liquid–liquid extraction procedures involve no special treatment and extract directly from the matrix. An example of this is a simple, rapid procedure for the extraction of amphetamine and methamphetamine from blood as described by Gjerde *et al.* [10]. This report described the extraction of drugs using 2 mL of cyclohexane after addition of internal standard and base. Shaking the tubes for 10 min resulted in extraction of amphetamine and methamphetamine from the blood providing good linearity and low detection limits.

A number of reports have been made regarding the extraction of amphetamines from less commonly used samples (sweat, hair, saliva, meconium, etc.). Suzuki *et al.* reported extraction of amphetamine and methamphetamine from filter paper or gauze which had been used for the collection of sweat. The extracts were derivatized and analyzed by GC-MS. Fay *et al.* described analysis of sweat collected with a commercially available 'Band-Aid' like patch (PharmChek) using acetate buffer and methanol to elute the drugs from the patch. The drugs were then extracted using isoamyl alcohol:hexane after the solution was made alkaline with sodium carbonate. Clean-up was further enhanced by back extraction into acid followed by final extraction into 1-chlorobutane. The extracts were then derivatized with carbethoxyhexafluorobutyryl chloride (CB) and analyzed by GC-MS. Immunoassay analysis was also accomplished on the initial patch eluate [11]. Analysis of saliva samples has been described by several investigators [12–17]. Extraction of amphetamines from meconium has been described using initial homogenization with HCl and liquid–liquid extraction of the supernatant using heptane:methylene chloride:ethylene dichloride:isopropanol (50:17:17:16), derivatization then analysis by GC-MS. This method gave a detection limit of 1 ng/g [18].

Several methods for the analysis of amphetamines from hair have been described. A serious concern in the analysis of hair samples is the potential for external contamination of the hair with the drug. This raises the concern that any detected drug(s) might be the result of contamination rather than from use of the drug. Measurement of metabolites rather than the actual drug can be helpful, but in many cases, the metabolites are not incorporated into hair. Analysis of drugs in hair generally follows a series of steps, decontamination, homogenization, extraction, clean-up, derivatization and analysis. Procedures used include acid or base hydrolysis, methanol sonication or enzyme hydrolysis. A comparison of these techniques by Kintz and Cirimele [19] reported alkaline hydrolysis to provide the best recovery. Most procedures for the identification of drugs in hair use GC-MS. Procedures have been published for the analysis of amphetamine, methamphetamine, MDA, MDMA and MDEA in hair [20–29] along with several reviews [25,30–32].

3.2.2 Solid-Phase

Solid-phase extraction has, over the last few years, greatly increased in usage. Advancements in the materials used in the production of sorbents are the primary reason for the increased use. Columns containing a variety of different sorbent materials can easily be found and refinements have led to consistency from cartridge to cartridge, a problem seen with early solid-phase extraction procedures. Another commonly encountered deficiency in early solid-phase extraction materials was the tendency for the column to get plugged with the sample matrix, particularly when used with blood, plasma, serum or tissue homogenates. This limited the utility of solid-phase extraction, but to a large extent, this problem has been corrected. Currently available cartridges allow for consistent extraction from a variety of biological matrices. In addition, automation of solid-phase extraction techniques is much easier and more practical to accomplish than for liquid–liquid. A number of different automated techniques have been described using solid-phase extraction and the number of those applications continue to increase. Automation, more than increased effectiveness, is the driving force increasing the use of solid-phase extraction due to its overall impact on lab operations.

Most manufacturers of solid-phase extraction columns provide recommend extraction protocols for many of the common drugs including procedures for the amphetamines. Typically, these procedures provide acceptable results and are thus often used by laboratories without modification.

A combination of liquid–liquid and solid-phase extraction using the Extrelut column was described for the extraction of amphetamine and methamphetamine from skeletonized remains [33]. This method extracted material from the marrow of the bone. Due to the nature of the material, a combined liquid–liquid and solid-phase extraction process was used. The extracts were derivatized with trifluoroacetic anhydride (TFA) and analyzed by chemical ionization MS. A procedure designed to identify many different drugs, blood, tissue and urine from postmortem analysis was analyzed for amphetamines including amphetamine, methamphetamine, MDA, MDMA, MDEA and the p-methoxy derivatives of amphetamine and methamphetamine has been described. Blood (2 mL) was adjusted to pH 9.0 with a borate buffer, vortex mixed and poured into an extraction column (Extrelut silica). After 10 min, the sample was eluted with methylene chloride:isopropanol (9:1). Tissue samples were homogenized to which 2 mL of phosphoric acid and 10 mL of saturated ammonium chloride solution was added and the mixture was then heated for 10 min. After cooling, the material was filtered then extracted in the same manner as described for blood. Samples were analyzed using GC-NPD or, after derivatization with pentafluoropropionic anhydride and hexafluoroisopropanol, were analyzed by GC-MS. Use of ethylamphetamine and methylenedioxypropylamphetamine as internal standards allowed analysis by either GC-NPD or GC-MS [34].

A procedure designed to use only a small volume of urine described the application of 0.1 mL urine sample to an extraction column (0.2 g of Extrelut and sodium carbonate (3:1, w/w) in a Pasteur pipette; 10 mm i.d. × 100 mm) for 20 min followed by the addition of propylchloroformate in ethyl acetate to the column for at least

10 min. The derivatized drug was then eluted by 0.2 mL of ethyl acetate, which was then injected into the GC [8]. Recoveries of amphetamine and methamphetamine from urine were 100 and 102%, respectively. The linearity of the assay was reported to be 0.50–50 µg/mL.

3.3 DERIVATIZATION

Derivatization of amphetamines is not required for their analysis by most chromatographic techniques. The use of derivatives, however, is common practice with these drugs, particularly with gas chromatography (GC) and to a lesser extent with LC. Detection of amphetamines using GC detectors such as flame ionization detection (FID) is readily accomplished. Owing to the presence of the amine nitrogen, a nitrogen-phosphorus detector (NPD) can detect the amphetamines without derivatization. In fact, they can provide a sensitive and selective identification of amphetamines because of the selectivity afforded nitrogen containing amine group. Generally however, amphetamines tend to show tailing peaks when chromatographed through most GC columns if not derivatized. Derivatization dramatically improves the chromatographic behavior of the amphetamines, which lessens the potential for interference and enhances the ability of instrument software integrators to accurately determine peak areas for quantitative analysis. As a result, most GC procedures for the analysis of amphetamines incorporate a derivatization step.

Liquid chromatographic analysis of amphetamines can also be accomplished without derivatization. The chromatographic properties of amphetamines on LC columns is generally better behaved than seen with GC. However, even in LC procedures, derivatization of the amphetamines can enhance their detectability and therefore the sensitivity of the method. For example, amphetamines do absorb in the ultraviolet (UV) but the absorbance is not very strong. Derivatization with a chromophore can give the drug a much stronger absorbance than the original drug. Some derivatives can also impart properties amenable to fluorescence or electrochemical detection which are generally more specific and sensitive than UV.

Amphetamine and methamphetamine each exist as one of two possible enantiomers that cannot be separated using typical (achiral) analytical chromatographic columns. Asymmetric derivatizing reagents can be used to generate diastereomers which are readily separated. Additionally, although less often, chiral columns are used for the separation of enantiomers. Capillary electrophoresis is increasingly being used for separation of enantiomers. Identification of which enantiomers are present and their proportions can be a helpful tool in interpretation of analytical results.

3.3.1 Achiral Derivatization

The most common derivatization techniques for the analysis of amphetamines are designed to assist with identification and accurate quantitation. Although the amphetamines have an asymmetric center, the enantiomers are not separated on typical chromatographic columns unless derivatized with a chiral derivatizing reagent. Such reagents are described in the next section of this chapter. Most common derivatizing reagents used in the analysis of amphetamines improve their chromatographic behavior. In addition, depending on the reagent used, they can facilitate selective ionization. For example, electron capture negative ionization MS does not work well with amphetamines unless they have been derivatized with an electronegative reagent.

It is well known that the electron ionization mass spectrum of amphetamine and methamphetamine are quite unremarkable. The spectra are dominated by a single ion that could come from many different biological compounds. Derivatization with most reagents result in a more complex, and therefore more unique, mass spectrum. See Fig. 3.4 for an example of the mass spectra of amphetamine and methamphetamine and several common derivatives. Some derivatives add significant mass to the molecule and its fragments. This can be an advantage since, generally speaking, the higher the mass, the more unique the ion thus making it less likely to suffer from interference. Although this is an advantage in some procedures, in others designed to isolate and identify many different compounds, large derivatives may have some disadvantages. Masses associated with derivatized amphetamine and related compounds tend to be well within the operating mass range of all commonly used mass spectrometers.

The most common derivatization schemes for amphetamines involve silylation, acylation and alkylation. The general schemes for these reactions are detailed in several excellent reference books and review articles [35–38] and therefore will not be described here.

A plethora of reports describe the use of TFA, PFPA, HFBA and acetyl derivatives. Each of these derivatizing reagents has its own adherents and each can provide excellent analytical results. These derivatives are typically made by incubation of the sample extract with the derivatizing reagent at elevated temperatures. Most procedures call for reaction with these reagents at 50–70°C for 15–30 min. A few derivatives do not require elevated temperatures and can be used at room temperature. For example, heptafluorobutyryl chloride will react with the amine nitrogen at room temperature. Dasgupta and Spies [39] described derivatization of amphetamine and methamphetamine using 2,2,2-trichloroethyl chloroformate for GC electron ionization and chemical ionization mass spectral analysis. The derivatization was conducted at room temperature for 10 min. The performance of this derivative was shown to be comparable to a method using PFPA.

The process of derivatization has been shown to be hastened the use of microwave radiation. Several procedures have used this technique to form derivatives more rapidly. One report involved the evaluation of microwave-assisted derivatization of amphetamine, methamphetamine and MDMA [40]. Trifluoroacetyl, pentafluoropropyl, heptafluorobutyryl and perfluorooctanoyl derivatives were made for each of the drugs. The study showed rapid formation of the trifluoroacetyl, pentafluoropropyl, heptafluorobutyryl derivatives of all three drugs in much less time than conventional heating methods requiring only 45 s, 1 and 6 min, respectively for each of the three derivatizing reagents. However, the perfluorooctanoyl derivatization of methamphetamine and MDMA did not perform well with this method.

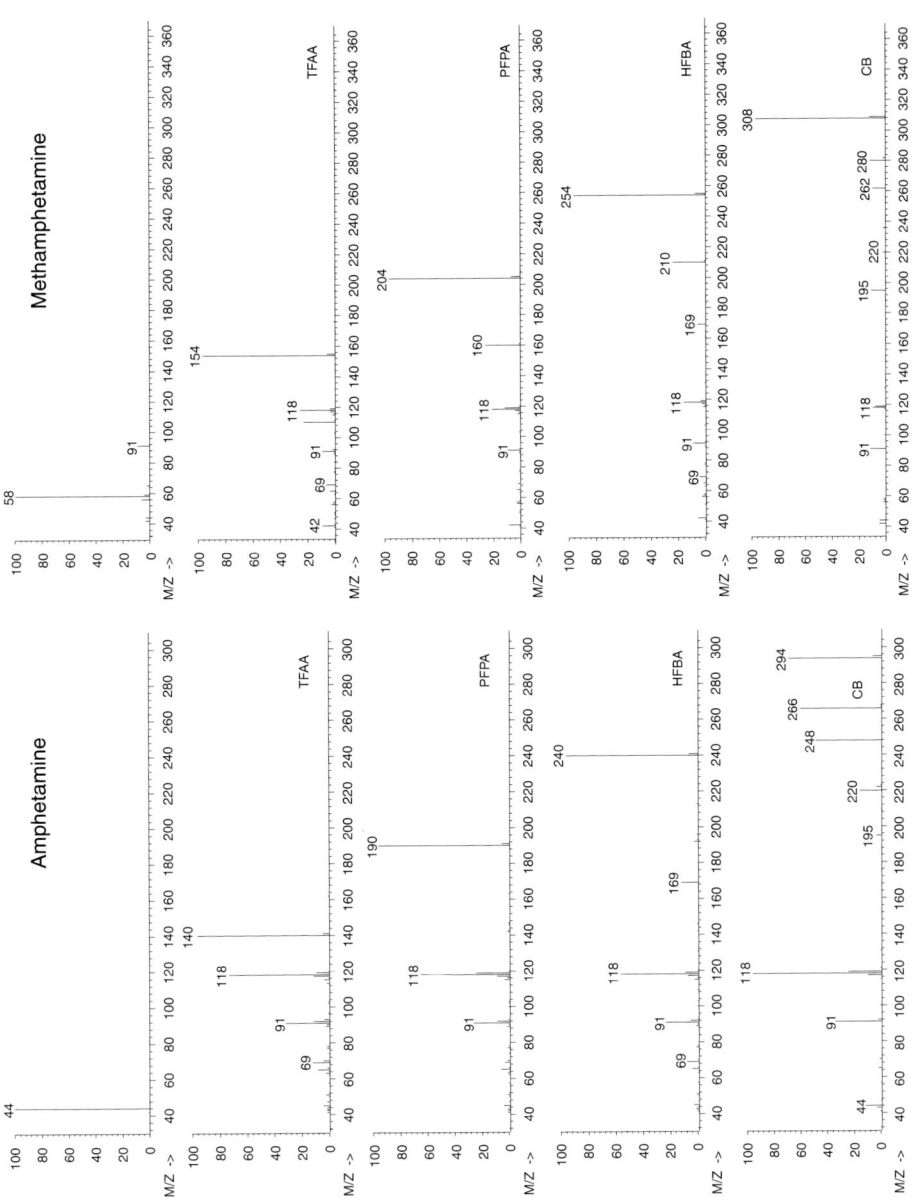

Fig. 3.4. Mass spectra of several derivatives of amphetamine and methamphetamine.

Mixed derivatization is also used for analysis of amphetamines and related compounds. This process typically involves the derivatization with one reagent followed by derivatization with a second. In these cases the first reagent will react with both amines and alcohols, while the second will displace the first reagent from one of those groups which then leads to one derivative on the amine and another on the HO-group [41].

Another method involves derivatization during the extraction process. This extractive derivatization of drugs saves the time required for the concentration of the extract and a separate derivatization step. This method has been successfully employed by several investigators [42,43]. Solid-phase derivatization is another option. This is a process where the derivatizing reagent is fixed on a support. The sample is exposed to this and then the derivatized molecules are released for analysis. Several such procedures have been described [44–46]. Still another derivatization method involves injection of the drug and derivatizing reagent into the injection port of the GC. This process has been used with amphetamines for a number of years for both quantitative [47] and enantiomer [48] analysis. The process is simple in that the same syringe contains the drug extract and the derivatizing reagent. This has been shown to be an effective and rapid method for analysis. One such method used TFA that was injected into the injection port with the extract. The derivatization reaction took place rapidly in the high temperature of the injection port. Unfortunately, the reaction with anhydrides also produces the corresponding acid as a by-product. The acid degrades the GC column and requires higher levels of maintenance and earlier column replacement than encountered with off line derivatization procedures.

Yamada *et al.* described an interesting procedure for the identification of amphetamine and methamphetamine on solid-phase extraction cartridges [49]. The procedure involved derivatization of the drugs using dansyl chloride assessing fluorescence on the cartridge to determine the presence of the amphetamines. Following extraction, the drugs are analyzed using one of several analytical techniques. Yamada *et al.* [50] described the use of GC-MS and high-performance liquid chromatography (HPLC) with UV detection. Derivatization and extraction was accomplished simultaneously by addition of the derivatizing reagent and sample to the solid-phase extraction column. Analysis of both amphetamine and methamphetamine gave acceptable results using this procedure. Comparison to HPLC using UV detection showed good agreement (correlation; $r = 0.95$) between the methods. Another method detailed in several publications using an alternative derivatization method is described by Nishida *et al.* [51,52]. The procedure called for the adsorption of amphetamine and methamphetamine on the surface of Extrelut. The bound drugs were then derivatized with propylchloroformate. Extraction efficiencies from spiked blood were 89.7 and 90.3% for amphetamine and methamphetamine, respectively. Within and between day relative standard deviations were <5%. The linear range was reported to be 12.5–2000 ng/g.

Chang *et al.* [53] described the analysis of a wide variety of deuterated isotopomers of amphetamine and methamphetamine. They also evaluated trimethylsilyl-, trichloroacetyl- and pentafluoropropionyl-derivatization of the drugs to evaluate which were suitable based upon their production of sufficiently high mass free of interference by the contribution from the other component of the pair, or so called "cross-contribution". The authors stated, the data showed, that derivatization methods play a significant role in deciding which deuterated analogue provides the most suitable ion pairs that cause the least cross-contribution. This lead to the suggestion that the most suitable internal standard varies with the derivative used for the analysis.

In addition to the derivatives described above, others that have successfully been used include pentafluorobenzolyl chloride [54], propylchloroformate [43], acetic anhydride [55–59], trifluoroacetylation and trimethylsilylation [41], perfluorooctanoyl chloride [10], *N*-methyl-*N*-*t*-butyldimethylsilyl trifluoroacetamide (MTBSTFA) [60] and CB [61,62]. The CB derivative has been used successfully in several laboratories, but the fragment ions monitored are the result of losses from the derivative itself and do not represent different fragments of the drug molecule, something that should be kept in mind by those using this procedure [62]. A GC-MS procedure for detection of cathinone (Khat) and methcathinone in urine using 4-carboethoxyhexa-fluorobutyryl as the derivatizing reagent was described by Paul and Cole [63]. The procedure used amphetamine-d6 and methamphetamine-d9 as internal standards for cathinone and methcathinone, respectively. The assay was linear from 25–5000 ng/mL for cathinone and 12.5–5000 ng/mL for methcathinone with recoveries of 86 and 78%, respectively.

Simple acetylation has some advantages, particularly when implementing a technique that is designed to analyze a large variety of compounds. The acetyl group improves the mass spectral characteristics, chromatography and has a large library of mass spectra available for identification purposes [64].

Derivatization procedures for use with LC methods provide both better chromatography and detection of amphetamines. A number of different derivatizing reagents are used with LC procedures including 3,5-dinitrobenzoyl chloride (DNB) [65] that was used with solid-phase extraction disks where the extraction and derivatization was accomplished simultaneously. The authors analyzed amphetamine, methamphetamine and several related compounds on both conventional and chiral columns. This same reagent was also used in another study [66]. Other reagents include phenylisothiocyanate [67], 4-(*N*,*N*-dimethylaminosulphonyl)-7-fluoro-2,1,3-benzoxadiazole (DBD-F) [68], fluorescein-4-isothiocyanate [69], 1,2-naphthoquinone-4-sulphonate [70–74], fluorenylmethylchloroformate-l-prolyl chloride (FMOC) [75] and dansyl derivatization [29,76–79] have also been used with LC analysis of amphetamines.

SPME is an extraction method using a unique system for the extraction of analytes of interest on to fibers that are designed to isolate analytes from the biological matrix. The process is detailed elsewhere in this chapter but a few examples of these procedures are mentioned here because they involve derivatization of the analytes as a part of the process, something not done in many SPME methods. A few examples include derivatization in the injection port by injecting heptafluorobutyric anhydride followed by insertion of the extraction fiber which allowed desorption and derivatization of the analytes in a procedure reminiscent of other on-column derivatization procedures [80]. Another procedure involved headspace derivatization using heptafluorobutyric anhydride ethyl acetate solution in an oil bath. Interestingly, this procedure significantly increased the sensitivity but had a negative impact by increasing the variability of the assay [81]. Huang *et al.* [82] used two derivatizing reagents, heptafluorobutyric anhydride and heptafluorobutyric chloride, to obtain their desired effect. The reagents were combined together in the vial along with the sample, KOH and NaCl. By combining the two derivatizing reagents and placing them in an insert they were able to achieve the high sensitivity they sought from the procedure.

3.3.2 Chiral Derivatization

Chiral derivatization of the amphetamines is designed to convert the enantiomers, which co-chromatograph, to diastereomers that can be separated using common GC columns. There are chiral GC columns designed to directly separate the enantiomers and do so without chiral derivatization. These columns have been used with success, but have several drawbacks. Generally, they are more expensive than achiral columns and need to be replaced more frequently than their achiral counterparts. They also tend to have lower tolerance to temperatures often encountered in GC procedures, and therefore tend to break down more readily. Since they are specific purpose columns, they must be dedicated to that function and cannot be used for the general analysis of samples.

There are a number of different chiral derivatizing reagents available, but the most commonly used is trifluoroacetyl-l-prolyl chloride (l-TPC). The derivatization reaction can be carried out at room temperature for 15 min [83]. An alternative procedure is to derivatize at 85–90°C for 10 min [84]. See Fig. 3.5 for an example of enantiomeric separation of amphetamine and related compounds using this reagent. Another method for the formation of the l-TPC derivatives of amphetamine and methamphetamine is on-column [48], or more precisely a reaction in the injection port. This method involved drawing 3 μl of urine extract into a 10 μl syringe, followed by 3 μl of l-TPC reagent, then injecting the contents of the syringe into the injection port set at 250°C. The resulting diastereomers are readily separated by a number of different GC temperature programs and achiral GC stationary phases. Co-injection of analytes and derivatizing reagent in these publications required manual injection and thus lacked the ability to be automated, which severely limited the use of this technique.

In an extensive study of different procedures, Maurer *et al.* [85] evaluated various GC and LC procedures. Standard electron ionization and chemical ionization procedures were compared relative to the usefulness of the information they provide. Similarly the authors described the use of atmospheric pressure ionization LC-MS for the analysis of various compounds of clinical and forensic interest. Enantiomers of amphetamine and methamphetamine were analyzed using negative ion chemical ionization of the *S*-heptafluorobutyrylprolylchloride derivative of the drug. GC conditions were as follows: splitless injection; column, 5% phenyl methyl siloxane (HP-5 MS, 30 m × 0.25 mm i.d.); injection port and transfer line temperatures,

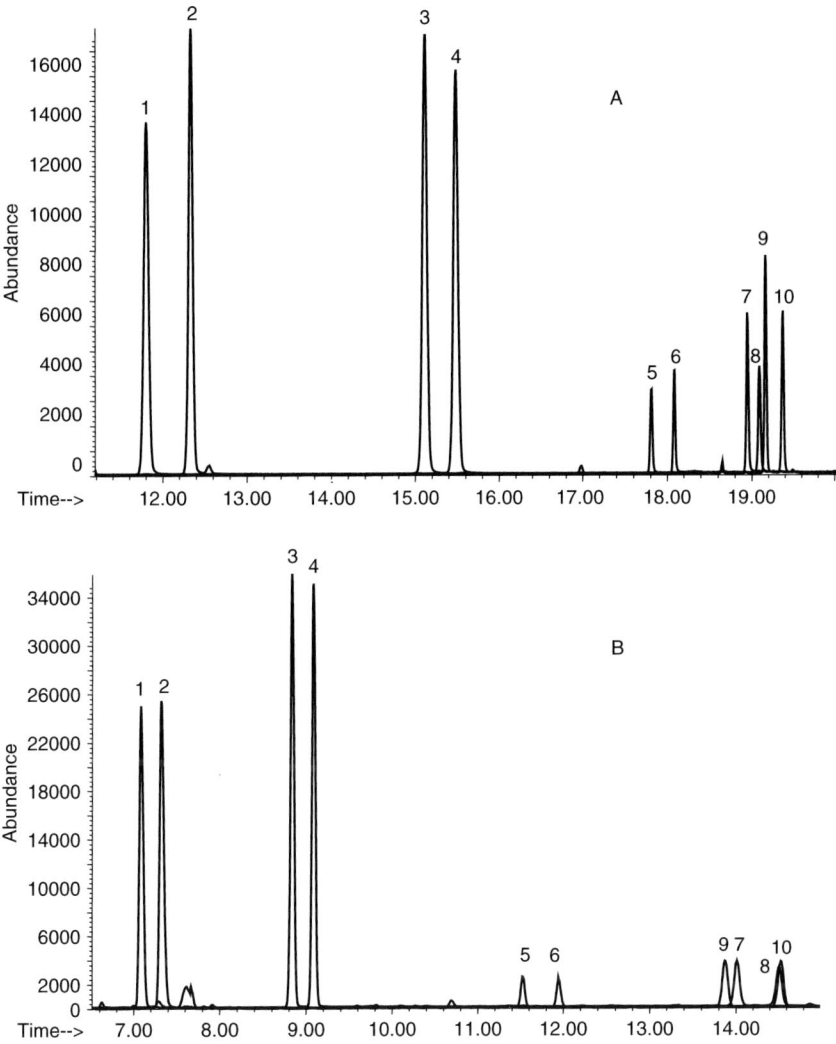

Fig. 3.5. Chromatography of enantiomers on HP-1 and DB-17 columns, peaks are identified by number: *l*-amphetamine (peak 1), *d*-amphetamine (peak 2), *l*-methamphetamine (peak 3), *d*-methamphetamine (peak 4), *l*-MDA (peak 5), *d*-MDA (peak 6), *l*-MDMA (peak 7), *d*-MDMA (peak 8), *l*-MDEA (peak 9), *d*-MDEA (peak 10).

280°C, carrier gas, helium; flow rate, 1 mL/min; column temperature, 100–180°C at 30°C/min, to 230°C at 5°C/min, and to 310°C at 30°C/min. Often chemical ionization does not yield fragment ions but in the case of *S*-heptafluorobutyrylprolylchloride derivatized amphetamine and methamphetamine, there are fragment ions that can be monitored. Ions monitored were m/z 388, 368, 428 for amphetamine, m/z 399, 379, 439 for amphetamine-D11 (internal standard), m/z 402, 382, 442 for

References pp. 165–174

methamphetamine, m/z 407, 387, 447 for methamphetamine-D5 (internal standard) [85]. Peters *et al.* described negative ion chemical ionization gas chromatographic-mass spectrometric analysis for the determination of amphetamine and methamphetamine enantiomers in plasma or serum. The enantiomers were derivatized with *S*-heptafluorobutyrylprolyl chloride and separated using a GC method. The method was linear from 5 to 250 μg/L with extraction efficiencies of 88.9–98.6% using a simple solid-phase extraction. GC conditions were as follows: 5% phenyl methyl siloxane column (HP-5 MS; 30 m × 0.25 mm i.d.); injection port temperature, 280°C; column temperature, 100°C increased to 180°C at 30°C/min, to 230°C at 5°C/min, and to 310°C at 30°C/min, transfer line, 280°C; negative ion chemical ionization (NICI), methane (2 mL/min); source temperature, 150°C. Ions monitored were: m/z 399, 379 and 439 for amphetamine-d11, and m/z 388, 368 and 428 for amphetamine; (Electron Multiplier Voltage (EMV) increased by 400 V) m/z 407, 387 and 447 for methamphetamine-d5 and m/z 402, 382 and 442 for methamphetamine. The increased sensitivity afforded by NICI allowed the use of a sample size of 0.2 mL [86,87].

Enantiomer separation of amphetamine, methamphetamine, MDA, MDMA and MDEA enantiomers using (*S*)-(+)-α-methoxy-α-(trifluoromethy)phenylacetyl chloride (MTPA) was described as an alternative to using the more common (*S*)-(−)-trifluoroacetylprolyl chloride (l-TPC) [88,89]. The TPC is prone to suffer from racemization under certain conditions, an issue for accurate determination of the proportion of an enantiomer. MTPA is stable to such changes and offers a potential advantage as a result. The authors described the use of three ions for each analyte and two for each internal standard with the exception of MDEA for which only a single ion was described because there was interference seen with other candidate ions making the procedure incapable of meeting the standard of three ions for the identification of an analyte. The authors note this to be an uncommon analyte that, if encountered, could acceptably be analyzed using an alternative procedure. Although the procedure identified three m/z values to monitor for methamphetamine, two of the ions at m/z 274 and 275 represent a fragment and its isotope peak, generally not considered acceptable for qualifier ion identification of an unknown. Its ability to correctly determine the proportion of d- and l-enantiomers, however, is not hampered by this limitation.

Wang *et al.* [90] described the analysis of amphetamine and methamphetamine using both chiral and achiral derivatization following a simple liquid–liquid extraction. Chiral derivatization was accomplished using *N*-trifluoroacetyl-1-prolyl chloride giving resolutions of 2.2 and 2.0 for amphetamine and methamphetamine, respectively. The ions monitored were m/z 240, 126, 96 and 237, 118, 91 for d8-amphetamine and amphetamine, respectively using 240 and 237 as quantifying ions. Methamphetamine and d8-methamphetamine were monitored at m/z 251, 118, 91 and 258, 122, 92, respectively and using 251 and 258 for quantitation. Achiral derivatization was accomplished using pentafluoropropionylation using PFPA. The ions monitored were m/z 193, 126, 96 for d8-amphetamine and m/z 190, 118, 91 for amphetamine using 190 and 193 for quantitation. Methamphetamine and d8-methamphetamine were monitored at m/z 204, 160, 118 and 211, 163, 122, respectively and using 251 and 258 for quantitation. Monitoring three unique fragments of the TPC derivative make this assay capable of forensically acceptably identification and quantitation of the enantiomers of both amphetamine and methamphetamine in a single assay. Quantitative comparison between PFPA and TPC derivatization showed the assays to be comparable quantitatively. The method had limits of detection and quantitation of 40 and 45 ng/mL, respectively.

Chiral analysis of amphetamine and methamphetamine by LC was accomplished with pre-column derivatization using Marfey's reagent (1-fluoro-2,4-dinitrophenyl-5-l-aniline amide) [91]. This reagent was compared with another chiral reagent (-)-1-(9-fluorenyl)ethyl chloroformate (FLEC) and both produced comparable results. Both reagents are detected by their fluorescence. Other derivatives utilized include (-)-α-methoxy-α-(trifluoromethyl)phenylacetyl chloride (MTPA) [54,88,89, 92–95], *R*-(+)-1-phenylethylisocyanate (PEIC), 2,3,4-tri-*O*-acetyl-α-d-arabinopyranosyl isothiocyanate (AITC) [93], 2,3,4,6-tetra-*O*-acetyl-β-d-glucopyranosyl isothiocyanate (GITC) [93,96], 4-nitrophenylsulfonyl-l-prolyl chloride (NPSP) [97], FLEC [71,98–101] and FMOC [102]. Enantiomeric separation has also been accomplished by using *o*-phthaldialdehyde and an optically active thiol [103,104].

3.4 GAS CHROMATOGRAPHY

GC analysis of amphetamines can involve the use of FID, nitrogen-phosphorus detector (NPD), electron capture detection (ECD) and MS. See Fig. 3.6 for an example of the chromatography of amphetamine and several related compounds. However, the drugs are typically derivatized because the spectrum of the amphetamines, particularly amphetamine and methamphetamine, lacks uniqueness (see Fig. 3.4).

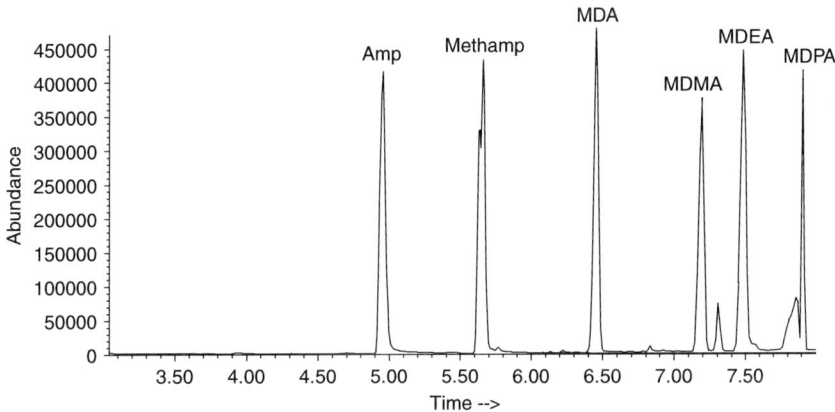

Fig. 3.6. Chromatography of amphetamine, methamphetamine, MDA, MDMA, MDEA, MDPA.

3.4.1 Achiral Gas Chromatography

Many procedures have been published for the analysis of amphetamine and methamphetamine by GC-MS and others which involve the analysis of these two drugs together with other related drugs including their metabolites or methylenedioxy analogues. Analysis of the methylenedioxy analogues generally requires no substantial modification of procedures designed for amphetamine and methamphetamine except to extend run times, etc. to accommodate those compounds.

The majority of procedures described in the recent literature involve MS as the GC detector. With the advances made in the instrumentation improving the ease of use of the instruments and decreasing costs, most laboratories now have MS capabilities. In the forensic analysis of drugs, the mass spectrometer is important for unequivocal results. Several procedures, however, are described for the use of other detectors. Often these procedures are for rapid screening of compounds and are then followed by mass spectral analysis for confirmation.

The use of an NPD has some advantages due to the selective nature of the detector. Since amphetamines are nitrogen containing compounds, many other potentially interfering compounds that are co-extracted are not detected and therefore do not interfere with their analysis. One procedure using NPD involved the analysis of amphetamine, methamphetamine, ephedrine, norephedrine, and related compounds using methyl chloroformate as the derivatizing reagent. The procedure gave good results, but it was found that the excess reagent from this derivatization process needed to be removed to avoid rapid deterioration of the detector [105]. In a method used for screening a large number of drugs from urine, samples were extracted, derivatized and analyzed by NPD and GC-MS [106]. Ortuno *et al.* [107] reported a procedure using NPD for the analysis of MDMA and several metabolites (MDA, 4-hydroxy-3-methoxymethamphetamine, 4-hydroxy-3-methoxyamphetamine) in plasma and urine. The plasma samples were analyzed in splitless mode on a 5% phenyl-methylsilicone column (HP Ultra-2) with temperatures from 70°C for 2 min to 100°C at 30°C/min then to 200°C at 20°C/min, and finally to 280°C at 25°C/min. Urine extracts were analyzed using a temperature program from 100°C to 280°C at 15°C/min. This assay used N-Methyl-bis-trifluoroacetamide (MBTFA) as the derivatizing reagent and methylenedioxypropylamphetamine as the internal standard. The assay proved to have low ng/mL detection and quantitation limits. Jenkins *et al.* [108] similarly used methylenedioxypropylamphetamine as the internal standard for a procedure that used NPD as a GC detector and also used LC with UV detection as well as LC-MS in the analysis of MDMA and metabolites. The use of methylenedioxypropylamphetamine, since it separates well from the other analytes, allowed the option of using either NPD, LC with UV detection or LC-MS.

ECD of 1 mL plasma extracts following liquid–liquid extraction and pentafluorobenzenesulfonylation was reported by Asghar *et al.* [109]. The method gave excellent sensitivity with a linear range of 1–50 ng/mL. The method used fused-silica capillary column 5% phenylmethylsilicone (25 m × 0.32 mm i.d.; Hewlett-Packard, Palo Alto, CA, USA). Helium (2 mL/min) was used as carrier gas, with methane:argon (5:95) as the makeup gas at a flow rate of 35 mL/min. Injection port and detector temperatures were 280 and 325°C, respectively. The oven was set at 105°C for 0.50 min to 300°C at 5°C/min to a final temperature of 300°C and held for 15 min. Analyte identification was confirmed by GC-MS analysis using the same GC conditions as the electron capture detector.

MS as the detector is the most common method used in the analysis of amphetamines, therefore, most of the recent literature utilizes this technique. Procedures for the analysis of amphetamines using GC-MS most commonly use deuterated isotopomers as internal standards. The ability of the MS to select individual ions of the internal standard from that of the drug of interest allows use of these compounds despite the fact they typically co-elute. The similarity of the deuterated isotopomer to the native drug is such that the accuracy and precision of these procedures is unparalleled compared to using any other compound as an internal standard. However, deuterated internal standards must be selected wisely. Depending on the assay and fragments monitored, some are not viable because they have ions in common with the native drug. For example, a report describing the enantiomeric analysis of the methylenedioxyamphetamines (MDA, MDMA, MDEA) evaluated the use of MDEA-d$_5$ and MDEA-d$_6$ as the internal standard for the separation of MDEA enantiomers. For purposes of enantiomer separation, both of these internal standards proved equally acceptable. For quantitative and identification purposes using HFB derivatives, the MDEA-d$_5$ had a common ion with MDEA making it a poor choice [83]. Other investigators have evaluated internal standards for the purpose of analysis of amphetamine finding most, but not all, acceptable [9,110]. Many assays for amphetamines use perfluoroacetylation, using reagents such as TFA, PFPA and HFBA for derivatization in addition to others described earlier in this chapter. One method using HFBA has been described for amphetamine and methamphetamine [9]. This method used liquid–liquid extraction followed by derivatization and analysis by electron ionization MS. Because of the potential for interferences, the method was described on several different columns. With minor modification, this method was used to analyze the methylenedioxy analogues MDA, MDMA and MDEA, and for several precursor drugs [2,111–115] and amphetamine from several studies of amphetamine excretion [116,117].

A method for the analysis of amphetamine, methamphetamine, MDA, MDMA, MDEA, ephedrine, pseudoephedrine together with a number of other drugs has been described [118]. This method used solid-phase extraction following enzyme hydrolysis of the urine samples. The extract was derivatized using propionic anhydride with pyridine. The derivatized extract was chromatographed on an HP-5 column with an initial temperature of 85°C for 0.7 min to 285°C at 14°C/min. The MS was set to scan from m/z 40 to 500 and gave a detection limit of 100 ng/mL for the amphetamines.

A method by Dasgupta and Spies [39] used 2,2,2-trichloroethyl chloroformate as the derivatizing reagent and N-propyl amphetamine as the internal standard and showed weak molecular ion peaks by electron ionization but intense peaks at m/z 218, 220 and 222. Chemical ionization showed strong (M + 1) at m/z 310 and 312 and intense peaks at m/z 274 and 276. The method showed good reproducibility with relative standard deviations of 1,000 ng/mL: 4.8 and 3.6% (within-run) and 5.3 and 6.7% (between-run) for amphetamine and methamphetamine, respectively.

The linear range was 250–5,000 ng/mL with a detection limit of 100 ng/mL in scan mode. Results using this method compared well with PFPA derivatization.

A study described the excretion of amphetamine and methamphetamine following controlled administration of methamphetamine to healthy volunteers [119]. The analytical procedure used an HP-1 column with a temperature program from 80°C for 2 min to 200°C at 40°C/min. Urine samples (1–4 mL) were extracted and derivatized with PFPA. Quantitation was accomplished using deuterated internal standards for each analyte and monitored using m/z 194, 123, 122 for d_5-amphetamine, m/z 190, 118, 91 for amphetamine, m/z 208, 163, 120 for d_5-methamphetamine, and m/z 204, 160, 118 for methamphetamine using the m/z 194:190 and m/z 208:204 ion pairs for quantitation. The limit of quantitation for both analytes was determined to be 27.5 ng/mL and a detection limit of 1.7 ng/mL for both using 4 mL urine samples. To keep the concentrations of analyte within the linear range of the assay, sample volumes were adjusted.

A GC-MS method for analysis of MDEA, MDMA, and its metabolites, 4-hydroxy-3-methoxyamphetamine (HMA), MDA and 4-hydroxy-3-methoxyamphetamine (HMMA) from 1 mL of urine has been described [120]. After acid hydrolysis, the samples were extracted using solid-phase extraction and derivatized with heptafluorobutyric acid anhydride. Limits of quantification were 25 ng/mL for MDEA, MDMA, and its metabolites. The assay was linear to 5,000 ng/mL for MDEA, MDMA, HMA, MDA and HMMA. The extraction efficiencies were >85.5% for all analytes.

Kim *et al.* [121] described a GC high resolution MS method for the analysis of methamphetamine and amphetamine in hair. The procedure involved decontamination of hair, acid hydrolysis and extraction followed by GC-MS analysis. The limits of detection were 9 pg/mg for methamphetamine and 21 pg/mg for amphetamine using a 30 mg hair sample. By using high resolution MS, limits of detection were reported to be 2.4–4.4 times lower than those seen with low resolution MS.

A procedure developed to identify the presence of amphetamine, methamphetamine, methylenedioxyamphetamine, methylenedioxymethylamphetamine, methylenedioxyethylamphetamine and methylenedioxyphenyl-*N*-methyl-2-butanamine in urine samples by positive chemical ionization GC-MS used methanol as the reagent gas [122]. The method used a simple liquid extraction procedure followed by the GC-MS analysis using an ion trap monitoring the protonated molecular ions at m/z 136, 150, 180, 194, 208. The authors indicted the extraction efficiency was between 62 and 66% for amphetamine and MDA and between 73 and 85% for the other compounds. Some may question the use of only a single ion for the analysis but one must consider the selectivity coming from the lack of fragmentation and molecular weight information for identification.

3.4.2 Chiral Gas Chromatography

Separation of the enantiomers of amphetamine, methamphetamine, MDA, MDMA and MDEA by GC using different stationary phases has been described [83].

Separation of the l-TPC derivatives was described on DB-1 (cross-linked methyl silicone) and DB-17 (50% phenyl methyl silicone) equivalent GC columns. GC conditions for the DB-17 column were an initial oven temperature at 120–210°C at 30°C/min, then to 260°C at 6°C/min and held for 1 min. Using the DB-1 column the conditions were 130°C to 190°C at 4°C/min, then to 250°C at 25°C/min and held for 2 min. In both cases, the injection port and interface temperatures were 270°C. Ions monitored were m/z 237 for amphetamine and MDA, m/z 241 for amphetamine-D_5 and MDA-D_5, m/z 251 for methamphetamine and MDMA, m/z 255 for metham-phetamine-D_5 and MDMA-D_5, m/z 265 for MDEA and m/z 270 for MDEA-D_5. The method could be used to separate and identify the enantiomers of each of the analytes at concentrations ranging from 5 to 10,000 ng/mL across the full range (0–100%) of each enantiomer. All enantiomer peaks were easily baseline separated on the DB-1 column but on the DB-17, the d-enantiomers of MDMA and MDEA were not resolved. While this would cause a problem with many GC detectors, the mass spectrometer was easily able to selectively monitor the unique ions for each of the analytes and distinguish them from one another. In addition, it is unlikely to find both MDMA and MDEA abused at the same time, therefore the likelihood of a sample having both of the compounds in them would be low. In the description of the analysis of MDEA enantiomers, Paul *et al.* [88] indicated an issue with common ion interference that prevented the use of more than a single ion for that analyte despite the use of a different derivatizing reagent and GC conditions.

Other investigators have also reported enantiomer separation using chiral prolyl derivatives [123–127].

Fallon *et al.* [92] reported the enantiomeric disposition of MDMA and met-abolites following administration of MDMA (40 mg) to eight healthy volunteers followed by collection of urine and plasma samples. Following extraction, the drugs were derivatized using MTPA, and chromatographed on a DB-17 column at 50°C for 2 min to 250°C at 25°C/min then to 290°C at 2°C/min using NPD for detection. Plasma samples were extracted, derivatized and chromatographed on a DB-1 equivalent (HP ultra 1) column at 100°C for 3 min to 285°C at 15°C/min and held for 5 min and analyzed by MS. Ions at m/z 119, 139, 162, 189, 260 for amphetamine, 135, 162, 189, 260 for MDA, 135, 162, 189, 260 for MDMA were used to detect the compounds of interest. Quantitation was accomplished using m/z 162 for the drugs and m/z 148 for the internal standard methoxyphenamine. The assay showed close agreement between actual and measured enantiomer composition at three different concentrations and four different ratios.

MTPA was used by a number of other authors (as noted earlier) for the analysis of amphetamine and related compounds. Using a DB-5MS (20 m × 0.18 mm i.d.; J&W Scientific Rancho Cordova, CA, USA) and GC conditions of: injector port temperature 140°C. The initial oven temperature was set to 140°C for 0.5 min then to 215°C at 15°C/min then to 285°C at 35°C/min and held for 1 min. The transfer line temperature was maintained at 280°C. This method was used for the separation of amphetamine and methamphetamine enantiomers [89]. Methamphetamine enantiomers were separated by 0.07 min at a retention time of >7 min, putting the enantiomers within 1% of one another, a common requirement for acceptable retention time variation within an analytical run. Since both enantiomers elute close together, it is important to carefully evaluate which enantiomer peak is being evaluated by the instrument data system. The procedure gave relative standard deviations, at three different concentrations (500, 2000 and 4000 ng/mL) of 0.4–7.2% for amphetamine and from 0.5 to 4.0% for methamphetamine. The assay, using a $1/X$ weighted regression, gave a linear range of 25–10,000 ng/mL. A similar procedure was described for the analysis of MTPA derivatized amphetamine, methamphetamine, MDA, MDMA and MDEA on a 5% phenyl polysiloxane capillary column (15 m × 0.25 mm i.d.; J&W Scientific Rancho Cordova) using the following GC conditions: Oven temperature was increased from 140°C for 0.5 min to 215°C at 15°C/min for 1.5 min to 285°C at 35°C/min where it was held for 1.0 min. The injector and transfer line temperatures were at 160°C and 260°C, respectively [88]. The assay gave a linear range of 25–5,000 ng/mL for MDEA and 25–10,000 ng/mL for amphetamine, methamphetamine, MDA, MDMA. Variation at the cutoff concentration (500 ng/mL) was within $\pm 11\%$ for all analytes.

Peters *et al.* [86] described a negative ion chemical ionization GC-MS assay for determination of amphetamine and methamphetamine enantiomers in plasma or serum. The enantiomers were derivatized with *S*-heptafluorobutyrylprolyl chloride and separated using a GC. The method was linear from 5 to 250 µg/L with extraction efficiencies of 88.9–98.6% using a simple solid-phase extraction. GC condition were as follows 5% phenyl methyl siloxane column (HP-5MS; 30 m × 0.25 mm i.d.); injection port temperature, 280°C; column temperature, 100 °C increased to 180°C at 30°C/min, to 230°C at 5°C/min, and to 310°C at 30°C/min, transfer line, 280°C; NICI, methane (2 mL/min); source temperature, 150°C. Ions monitored were: m/z 399, 379 and 439 for amphetamine-$_{d11}$, and m/z 388, 368, and 428 for amphetamine; (EMV increased by 400 V) m/z 407, 387, and 447 for methamphetamine-$_{d5}$ and m/z 402, 382, and 442 for methamphetamine. The increased sensitivity afforded by negative ion chemical ionization allowed the use of a sample size of 0.2 mL. In a subsequent publication describing the metabolic profile of MDMA following administration of the drug in a controlled study by the same author, a thorough validation of the assay was also delineated [87]. In another procedure using NICI Leis *et al.* [128] also described a method for the quantitative analysis of amphetamine enantiomers in human plasma by GC/negative ion chemical ionization MS. The method was linear within a range 0.006–50 ng/mL. Following a simple liquid extraction of a 1 mL of plasma and derivatization with (*S*)-heptafluorobutyrylprolyl chloride GC conditions used were: SGE-BPX5 fused-silica capillary column (15 m × 0.25 mm i.d.; ThermoQuest), injector 280°C, column temperature was 100°C for 1 min, followed by an increase of 40°C/min to 180°C, 5°C/min from 180–195°C, 40°C/min to 310°C for 2 min, transfer line 315°C. Chemical ionization was accomplished with methane with an emission current of 300 mA. Ions monitored for this pharmacokinetic study were m/z 368.1 and 373.1 for amphetamine and internal standard, respectively.

Analysis of MDMA and related regioisomers has been described by several authors, helping to ensure these closely related compounds can be readily differentiated [129–132]. The combination of the mass spectral characteristics and particularly the chromatographic behavior of these analytes are described by both groups relative to their importance for the proper identification of these compounds. Optimization showed that very slow program rates gave the best separation on non-polar stationary phases but the run time upwards of 85 min was impractical. The use of narrow bore capillary columns with the same phases improved the analysis time to approximately half an hour due to their enhanced resolving power. Optimization of a DB-35MS, a polar stationary phase column, allowed resolution of 10 side chain and ring regioisomers of MDMA in approximately 4.5 min [131].

3.5 LIQUID CHROMATOGRAPHY

LC, in the form of HPLC, is an important analytical tool. It can utilize a wide variety of detectors with varying degrees of specificity. One advantage of LC is that the extracts need not be as clean as those typically used for GC or GC-MS analysis, thereby allowing simpler extraction procedures. Underivatized compounds have been shown to give reasonably good results on LC columns. The amphetamines, which typically show tailing peaks on GC when not derivatized, have better chromatographic behavior in LC. In addition, thermal labile compounds can be analyzed by LC without concern over thermal degradation that would limit the use of GC. In addition, evaluation of metabolic products, including conjugated drugs can be accomplished without the requirement for hydrolysis. Detectors used for the analysis of amphetamines include UV, photodiode array, fluorescence, electrochemical and MS.

3.5.1 Achiral Liquid Chromatography

A number of investigators have described methods for the analysis of amphetamines using LC without derivatization. As previously mentioned, amphetamines do not have native fluorescence and are not strong UV absorbers. Nor do the amphetamines have a strongly electronegative character that would lend itself to electrochemical detection. In any case, several investigators have successfully used LC without derivatization [67,133–138]. Many LC procedures do utilize derivatives to enhance the analysis of amphetamines. Derivatization not only tends to improve chromatography, but the sensitivity can be greatly increased by using a derivatizing reagent with strong UV, fluorescent or electrochemical characteristics that will provide a strong signal at the detector.

Bogusz *et al.* [67] described a method for analysis of amphetamine, methamphetamine, MDA, MDMA, MDEA and eight other sympathomimetic amines by HPLC with UV detection at 250 nm and with photo diode array (PDA) detection. The procedure also reported mass spectral detection using atmospheric pressure chemical ionization (APCI). Limits of detection for the UV detectors were 50–100 ng/mL using the PDA and 10–30 ng/mL using the single wavelength at 250 nm. Selected ion monitoring of the drugs and their deuterated isotopomers was, as expected, very selective and sensitive. Detection limits of 1 ng/mL for methamphetamine, MDMA and MDEA and 5 ng/mL for amphetamine and MDA were seen. The use of deuterated isotopomers provides generally superior quantitative results, however, caution in the selection of the amount of deuterium is important to ensure accurate quantitative analysis particularly with LC/MS techniques [139].

Weinmann and Svoboda described a procedure for the analysis of drugs from urine and serum [140]. The procedure used solid-phase extraction of the samples and direct injection of the extract into the LC–MS interface without the use of a column. This method is based on the selective ability of tandem MS to separate the analytes of interest. The instrument was set up to isolate the parent ions of the drugs of interest, which included amphetamine and methamphetamine, and their deuterated analogues with subsequent collisional dissociation to product ions that were then detected. This process is very selective and sensitive with applicability to a variety of different analytes. There are several advantages to a process such as this including the speed at which the analysis can be accomplished. Not depending on chromatographic resolution of the analytes eliminates significant amounts of time from this procedure.

Al-Dirbashi and co-workers described achiral and chiral high-performance liquid chromatographic procedures for the determination of methamphetamine and amphetamine in urine [141]. A small (10 μL) sample was extracted and derivatized with 4-(4,5-diphenyl-1*H*-imidazol-2-yl)-benzoyl chloride. The extract was chromatographed using Tris-HCl buffer:acetonitrile (45:55, v/v) at 0.2 mL/min for achiral analysis. Chiral analysis was accomplished using a semi-micro OD-RH column with sodium hexafluorophosphate:acetonitrile (44:56, v/v) at a flow rate of 0.1 mL/min as the mobile phase. Both methods gave comparable results with the highest relative standard deviation observed at 9.6%. This procedure was used to analyze hair from known users. It proved to give acceptable results with resolution between enantiomers of 3.4 and 1.1, respectively for amphetamine and methamphetamine [142].

Another method for analysis of MDMA, MDEA and MDA in whole blood, serum, vitreous humor and urine using fluorescent detection was described by Clauwaert *et al.* [143]. The method was linear over the range of 2–1,000 ng/mL for whole blood, serum and vitreous humor, and 0.1–5 μg/mL for urine. Relative standard deviations of 2.5–19% were seen and the limits of detection and quantification were 0.8 and 2 ng/mL, respectively, for whole blood, serum and vitreous humor, and 2.5 ng/mL and 0.1 μg/mL for urine.

Concheiro *et al.* [144] described a procedure for analysis of MDMA, MDA, MDEA and MBDB in oral fluid. HPLC with fluorescence detection allowed good separation of the analytes in an isocratic run of only 10 min. The method showed limits of detection and quantitation of 2 and 10 ng/mL, respectively for all analytes. The linear range was described from 10 to 250 ng/mL. Unlike amphetamine and methamphetamine, which have little native fluorescence, the ring-substituted amphetamines exhibit substantial native fluorescence, with excitation and emission maxima at 285 and 320 nm, respectively.

Jenkins *et al.* [108] developed a procedure using NPD as a GC detector and also used LC with UV detection as well as LC-MS in the analysis of MDMA and metabolites. The LC-MS method gave better sensitivity and, although NPD gave excellent selectivity, it required considerably greater effort to clean up the sample to meet that seen with the LC-MS. The method covered the range of 0.10–20 µg/mL and gave limits of quantitation of 0.1 µg/mL for MDMA and MDA and 0.04 µg/mL for HMMA.

The ability of LC to chromatograph a wide variety of analytes of differing polarities and compounds that are thermal labile allows for the identification of conjugates directly. Glucuronide and sulfate conjugates of 4-hydroxymethamphetamine can be analyzed in the same analytical runs as amphetamine and methamphetamine. An LC-MS-MS procedure for determination of methamphetamine and its metabolites, amphetamine, 4-hydroxymethamphetamine, glucuronide conjugated 4-hydroxymethamphetamine and sulfate conjugated 4-hydroxymethamphetamine in urine has been described [145,146]. After deproteinization of urine samples with methanol, LC-MS through a C-18 column with a gradient elution program provided determination of these analytes within 20 min.

Other methods include studies by Meyer *et al.* [147], Nakashima *et al.* [148], Kaddoumi *et al.* [149], da Costa *et al.* [150] who used a sodium dodecyl sulfate (SDS) ion-pairing reagent to improve separation and efficiency, Kaddoumi *et al.* [149], Wada *et al.* [151], Muller and Windberg [152] who examined the drug in tablets, Mc Fadden *et al.* [153], who examined the drug in tablets using monolithic (Chromolith RP18e) columns which showed that monolithic columns gave greater efficiency at higher flow rates compared to particulate columns without loss of peak resolution.

3.5.2 Chiral Liquid Chromatography

Al-Dirbashi *et al.* described use of 4-(4,5-diphenyl-1*H*-imidazol-2-yl)benzoyl chloride (DIB-Cl). This derivative has strong fluorescent properties and was used for the separation of enantiomers of amphetamine, methamphetamine and 4-hydroxymethamphetamine. Derivatization of the extract was accomplished at room temperature for 10 min. The derivatized extract was chromatographed through an ODS column and the compounds detected by fluorescent detection (excitation 330 nm, emission 440 nm). The detection capability of the assay was comparable to or better than most published procedures. The investigators also evaluated a chiral LC column (Chiralcel OD-R), which did not separate the d-enantiomers of amphetamine and methamphetamine [154,155]. In an earlier study [156], l-TPC derivatives of amphetamine and methamphetamine were chromatographed through a *N*-3,5-(dinitrobenzoyl)phenylglycine chiral column. The d- and l-enantiomers of methamphetamine were not completely separated by this method. Katagi *et al.* [137] described a method for the HPLC and LC/MS analysis of amphetamine, methamphetamine and p-hydroxymethamphetamine enantiomers from urine. After solid-phase extraction, the extracts were separated through a β-cyclodextrin (CD) phenylcarbamate-bonded silica column. Elution was accomplished using acetonitrile:methanol:potassium phosphate buffer (50 mM) (10:30:60) with a flow rate of 1 mL/min. The assay was linear 200–20,000 ng/mL with relative standard deviations of 1.67–2.35% at 2,000 ng/mL. The detection limits were 50 ng/mL for amphetamine enantiomers and d-methamphetamine and 100 ng/mL for l-methamphetamine. Samples were also analyzed using thermospray

LC-MS. The authors used acetonitrile:methanol:ammonium acetate (100 mM) (10:30:60) as the solvent system. Detection limits for the LC-MS analysis were 10–20 ng/mL using scan mode and 0.5–1 ng/mL using selected ion monitoring. Herraez-Hernandez *et al.* [71] described pre-column derivatization of drugs. Three different derivatives were evaluated by this method including 1,2-naphthoquinone-4-sulfonate, *o*-phthaldialdehyde and 9-fluorenylmethyl chloroformate using column switching. The derivatization was accomplished in a pre-column packed with un-modified ODS stationary phase into which the derivatives were injected followed by separation on an analytical column.

Direct enantiomer separation has been reported by a number of investigators. Aboul-Enein and Serignese described a direct, isocratic method the enantiomeric resolution of cathinone, amphetamine, norephedrine and norphenylephrine on an S-18-crown-6-ether chiral stationary phase [157]. Use of chiralcel OB and OJ columns in series was used by Nagai and Kamiyama [158] for the enantiomeric separation of amphetamine, methamphetamine and hydroxy metabolites of samples of the drug confiscated on the street and from the urine of users. Other direct separation methods have also been described [137,159–161].

A procedure for the analysis of the enantiomeric composition of amphetamine and methamphetamine from human hair samples using a 0.46 cm × 15.0 cm column with a chiral stationary phase having 3,5-dimethylphenylcarbamoylated-β-CD as the chiral selector (Cyclobond I 2000 DMP; Advanced Separation Technologies, Astec, Whippany, NJ, USA) has been described [162]. The mobile phase was 1% triethylammonium acetate/acetonitrile (65:35) at a flow rate of 1.0 mL/min. Separation of amphetamine enantiomers was easily achieved but baseline resolution of methamphetamine enantiomers was not using the chiral column as described. The authors combined the separation afforded by an achiral C-18 column in series with the chiral column, which then provided baseline resolution of methamphetamine as well as amphetamine.

Another alternative to chiral derivatization option is the use of polarimetry to identify enantiomers. In a study of the stereoselective disposition of methylenedioxy analogues, MDA, MDMA and MDEA were analyzed by HPLC using two different detectors. A UV detector was used to identify the drugs and polarimetry was used to identify the optically active forms of these drugs [163].

3.6 ALTERNATIVE TECHNIQUES

Traditional GC and LC analysis of amphetamines is widely used throughout the world. Several alternative techniques are available and some have been available for many years (immunoassays, thin-layer chromatography (TLC), etc.) while others are more recent. Capillary electrophoresis is an alternative separation technique with significant capabilities owing to its tremendous resolving power and ability to work with very small sample sizes. While liquid–liquid and solid-phase extraction still account for the vast majority of isolation techniques, SPME is being implemented in more and more analytical procedures, including the analysis of amphetamines.

These techniques are described in this chapter for a general background into the procedures and description of some uses and potential applications and not as a thorough reviews of these analytical procedures.

Another technique that has been used with amphetamines is supercritical fluid chromatography. This technique is not widely used as yet for the analysis of drugs, but has some interesting potential. Several papers have described the use of this technique for amphetamine analysis. One of the advantages of supercritical fluid chromatography is that it typically does not require derivatization of the compounds. In the case of amphetamines however, derivatization was shown to be important, thus eliminating one of the common advantages of the procedure. The equipment necessary to carry out this procedure is not found in many laboratories at this time, therefore its use is not widespread. Descriptions of several procedures using supercritical fluid chromatography for the analysis of amphetamines have been published [164,165] and a review of the technique has also been written [166].

3.6.1 Thin-Layer Chromatography

TLC is a method that has been used in the analysis of drugs for many, many years. Its use has been superseded in recent years by other techniques that give more definitive results. Nonetheless, TLC is still used and has some advantages over other screening techniques. For example, compared to immunoassay, it can detect a drug and its metabolites, rather than being targeted to a single compound as are most immunoassays. In addition, detection of multiple drugs within a single sample is possible as is the ability to detect various drugs and/or metabolites for which immunoassay tests are currently not available. Compared to other methods that allow for rapid screening of samples for multiple compounds (i.e. GC, LC, etc.), TLC is relatively inexpensive with regard to the capital investment required for the equipment used for other methods. In addition, the ability to have a second method for the identification and confirmation of a drug has significant advantages in a forensic environment.

One commonly used TLC assay system is Toxi-Lab (ANSYS, Inc., Irvine, CA, USA). This system involves methods for the identification of over 300 different drugs/drug metabolites through an extraction, chromatography and visualization process. The visualization process involves four stages, which allow for the identification of the drugs based on their migration (Rf) and color under the four conditions used for visualization. Despite the resolving power of this combined chromatographic and visualization process, the sympathomimetic amines pose a challenge to the system. Differentiation of compounds, such as amphetamine, methamphetamine, ephedrine, pseudoephedrine, phenylpropanolamine and phentermine requires a separate procedure using different solvents to adequately separate these compounds.

Another TLC system is Drug-Skreen II (Eppendorf-Brinkman, Inc., Westbury, NY, USA). This process involves alkaline extraction and separation on silica-coated plates using sequential spraying to visualize different drugs. The amphetamines are detected under UV light.

Kato and Ogamo [167] used citric acid/acetic anhydride reagent for detection of abused tertiary amino drugs, such as dimethylamphetamine, by TLC. The method is reported to be 2.5–15 times greater than conventional detection with Dragendorff reagent. The same group also developed a method for the detection of *p*-hydroxymethamphetamine in urine. The method uses liquid–liquid extraction with acetonitrile and solid-phase extraction by TLC with oxidation using potassium hexacyanoferrate(III) and sodium hydroxide to detect the fluorophor of *p*-hydroxymethamphetamine. The detection limit is reported to be 10 ng ($n = 3$), in the range compatible with samples from methamphetamine abusers [168]. Maresova *et al.* noted a spot on their TLC plate, which was later characterized as 6-Cl-MDMA [169].

In addition to these systems, there are a myriad of other TLC procedures used by analysts to detect amphetamines. The variety of these procedures are beyond the scope of this chapter but can be found in the early literature on drug analysis.

3.6.2 Capillary Electrophoresis

Capillary electrophoresis, while not a new method, has seen rapid growth in the analysis of drugs over the past few years. Early descriptions of the technique were provided some years ago [170]. Weinbeger and Laurie [171] described a method for the analysis of drugs of forensic interest and other laboratories have produced a number of analytical papers dealing with this technique. Capillary electrophoresis offers some significant advantage over other commonly available techniques. These include analysis of analytes from a variety of matrices requiring little sample preparation and often no derivatization. The technique is commonly used with only a few nanoliters of sample or sample extract on-column. As a result, the total sample volume required for analysis is often <0.1 mL. More recently, techniques have been developed to concentrate samples so that there may be substantial concentration of the sample during the analytical procedure thus increasing the overall sensitivity of the method. Several reviews of the technique have been written within the last few years, and readers are referred to those papers for further information [172–183].

Much of the recent literature regarding capillary electrophoresis and the analysis of amphetamines describes chiral analysis of drugs [184–195]. Chiral analysis of drugs is an important area and adds significant specific information for the interpretation of results, particularly in the forensic analysis of amphetamines as discussed elsewhere in this chapter.

Some specific examples of the use of this technique for the analysis of amphetamines and related compounds include the following. Kuroda *et al.* [196] used both capillary zone electrophoresis (CZE) and micellar electrokinetic capillary chromatography (MECC) to detect amphetamine, methamphetamine, 2-phenethylamine, 4-hydroxyamphetamine and 4-hydroxymethamphetamine using 1-phenethylamine as an internal standard when analyzing urine samples. Using a 50 µl sample, the drugs were separated within 15 min with detection limits in the low fmol/injection range using a UV detector. Derivatizing the sample extract with 4-fluoro-7-nitro-2,1,3-benzoxadiazole (NBD-F), yielded a fluorescent compound which was detected using laser-induced fluorescence (LIF). This method lowered the detection limits to the low to mid attomole/injection range although it required 45 min to separate the analytes using MECC. Capillary electrophoresis using LIF detection of amphetamines has also been described by Choi *et al.* [197].

A method has been described for the analysis of the enantiomers of amphetamine, methamphetamine, MDA, MDMA, MDEA and ephedrine with β-CD as the chiral agent. In this case, an uncoated capillary (45 cm × 50 μm i.d.) was used with a potential of 10 kV with pH 2.5 phosphate buffer. Detection was by monitoring at 200 nm or scanning from 190 to 400 nm [198]. Wallenborg *et al.* [194] demonstrated separation of norephedrine, ephedrine, cathinone, pseudoephedrine, methcathinone, amphetamine and methamphetamine using micellar electrokinetic capillary chromatography (MEKC) and LIF detection. The authors used neutral and negatively charged CDs with and without the addition of an organic modifier and/or SDS for chiral separation of the enantiomers. The authors reported the best results were obtained using a highly sulfated gamma-CD combined with a low concentration of SDS. To obtain complete separation of a mixture of (+/−)-norephedrine, (+/−)-ephedrine, (+/−)-pseudoephedrine, (+/−)-methcathinone, (+/−)-amphetamine and (+/−)-methamphetamine it was necessary to add a small amount (1.5 mM) of SDS to the separation buffer. Chiral separation was achieved in 7 min using an S-folded separation channel, 8 kV separation voltage and a buffer consisting of 50 mM phosphate (pH 7.35), 10 mM highly sulfated gamma-CD and 1.5 mM SDS. Chinaka *et al.* [193] described a method for the chiral analysis of methamphetamine, amphetamine, dimethylamphetamine, ephedrine, norephedrine, methylephedrine, MDMA, MDA and 3,4-methylenedioxy-*N*-ethylamphetamine from urine using a mixture of β-CD and heptakis(2,6-di-*O*-methyl)-β-CD. The method gave a linear range of 0.2–500 μg/ml for ephedrine and methamphetamine with detection limits of all enantiomers were 0.1 μg/mL, capability that is suitable for routine analysis. Pizarro *et al.* [194] characterized the identification of (*S*)-3,4-methylenedioxymethamphetamine, (*S*)-4-hydroxy-3-methoxymethamphetamine and (*S*)-3,4-dihydroxymethamphetamine using enantiomeric enrichment to assign unequivocal identification of the analytes. Another capillary electrophoresis method using MS for the detection method was designed for the enantiomer analysis methamphetamine, amphetamine, dimethylamphetamine and *p*-hydroxymethamphetamine in urine [188]. As with other methods, the chiral selector was a mixture of 3 mM β-CD and 10 mM heptakis(2,6-di-*O*-methyl)-β-CD. The detection limits were 0.03 μg/mL for the enantiomers of amphetamine and methamphetamine, and 0.05 μg/mL for the para-hydroxylated metabolite of methamphetamine using selected ion monitoring. Iio *et al.* [185] described a similar procedure using capillary electrophoresis/MS for the enantiomer analysis of methamphetamine, amphetamine, dimethylamphetamine, ephedrine, norephedrine and methylephedrine in urine. The separation was completed within 25 min with detection limits of 0.01 μg/mL for the enantiomers of methamphetamine, amphetamine, dimethylamphetamine, ephedrine, and methylephedrine, and 0.02 μg/mL for the enantiomers of norephedrine. The linear range was in the range of 0.05–10 μg/mL. In a study designed to evaluate the use of both aqueous and non-aqueous buffers in together with β-CD for the chiral separation of (*R*)- and (*S*)-3,4-methylenedioxymethamphetamine and 3,4-methylenedioxyam-phetamine, *N,N*-dimethyl-3,4-methylenedioxyamphetamine, 1-(1,3-benzodioxol)-5-yl-2-butylamine (BDB) and *N*-methyl-1-(1,3-benzodioxol-5-yl)-2-butylamine (MBDB). Huang *et al.* [184] described procedures in CZE and MEKC modes which allowed them to characterize the elution order of each of the analytes.

Achrial methods described by a number of investigators successfully identified and quantitated amphetamine and related compounds. Non-aqueous capillary electrophoresis was used in a number of applications for the analysis of various illicit drugs including amphetamines. Backofen *et al.* [199] described a method for the analysis of amphetamine, methamphetamine, MDA, MDMA, mescaline, cocaine and benzoylecgonine using an acetonitrile-based buffer solution containing 10 mM sodium acetate and 1 M acetic acid. The authors employed electrochemical detection using a Pt microdisk electrode. Chung *et al.* [200] also used non-aqueous capillary electrophoresis with fluorescence detection using a cryogenic molecular fluorescence technique at 77 K. Fang *et al.* [201] also using non-aqueous capillary electrophoresis and fluorescence spectroscopy described a procedure for analysis of MDMA in less than 5 min. This procedure used liquid–liquid of MDMA from urine and gave a detection limit (*S/N* = 3) of 50 ng/mL with no derivatization. An evaluation of detection techniques was conducted comparing LIF detection of MDMA, among other compounds, with a XeHg lamp-based detection system. The study showed that the XeHg lamp gave results comparable to those found using LIF [202].

Other methods for the chiral analysis of amphetamine, methamphetamine and related amines and/or their metabolites, have also been described [203–207]. Several of these describe evaluation of critical parameters such as temperature, buffer concentration, pH, chiral separator and applied voltage [203,205,207].

3.6.3 Solid-Phase Microextraction

SPME is a technique that has been utilized for a number of different procedures in recent years. Although available since the early 1990s, this technique has made substantial strides in the last few years to offer a viable method for the sample analysis. SPME utilizes a solventless extraction procedure combining several steps in the normal processing of samples. The detection of volatile amphetamines from the headspace of a sample has been described by several investigators [208,209]. While somewhat effective, these methods sampled only the headspace with no ability to concentrate the analyte of interest. This principle together with the ability to bind the drug of interest to a fiber, thus concentrating the compound on the fiber, is used in headspace SPME.

In 1990, SPME was described by Arthur and Pawliszyn [210]. Recent developments are summarized in a number of general reviews [211–213]. The technique involves use of a fused silica fiber typically coated with a sorbent material such as polydimethylsiloxane or polyacrylate. Fibers are exposed to the drugs of interest by exposure of the fiber to the headspace above the sample or by immersion of the fiber into the sample itself. Amphetamines have been analyzed using both the headspace and immersion techniques. The technique has been described in a variety of recent publications. Automation of the technique hampers the ability to use in a number of different situations but it is becoming more frequently used in analytical procedures. One consideration of analytical methods is sample-processing time. Often samples take upwards of 30–80 min for their analysis. One must consider this in terms of total analysis time, however, when comparing to the more traditional off line extraction and derivatization followed by instrumental analysis that this represents total analysis time for a sample.

The SPME of amphetamine and methamphetamine from hair was reported after base hydrolysis [214]. The method involved direct immersion of the polydimethylsiloxane fiber into the sample and was linear from 4 to 200 ng/mg. Amphetamine and methamphetamine were extracted from whole blood by exposing the polydimethylsiloxane fiber in the headspace at 80°C for 5 min. The method was linear from 10 to 200 ng/mL with a detection limit of 10 ng/mL from 0.5 mL of whole blood [215]. Battu *et al.* [216] described a procedure for the analysis of amphetamine and methamphetamine along with 19 other drugs using SPME followed by GC-MS analysis. Centini *et al.* [217] described a procedure for the analysis of amphetamine, methamphetamine, methylenedioxyamphetamine and methylenedioxymethamphetamine in 1 mL of urine heated to 75°C for 30 min followed by exposure of the fiber to the headspace for 15 min. The method was evaluated with samples from 100 to 2,000 ng/mL of each analyte and showed the method to be a viable alternative to more traditional techniques. Another method using SPME headspace was described for urine samples at 80°C for 5 min followed by analysis with GC and GC-MS. The detection limit was reported to be 100 ng/mL and a linear range of 200–100,000 ng/mL [218]. Use of the immersion technique was described using 1 mL of urine at 65°C for 30 min using a polydimethylsiloxane/divinylbenzene fiber. The method gave a detection limit of 10 ng/mL with a linear range of 12–200 ng/mL by GC-NPD [219].

A method for the analysis of derivatized amphetamine and methamphetamine in postmortem blood was developed using headspace SPME and GC-MS. The process involved the combination of blood, deuterated methamphetamine (internal standard), tri-*N*-propylamine and pentafluorobenzyl bromide in a vial. The vial was heated at 90°C and stirred for 30 min then the extraction fiber was inserted into the headspace for an additional 30 min while stirring continued. The linearity was reported to be 0.5–1000 ng/G for both drugs [220]. Similarly, a SPME procedure for the analysis of amphetamine, methamphetamine, MDA and MDMA from urine followed by GC-MS analysis has been described. Extraction was accomplished by exposing a polydimethylsiloxane fiber in a vial, previously heated to 100°C for 20 min, for 10 min. The analytes were derivatized by exposing the fiber to TFA for 20 min in the headspace of another vial heated at 60°C for 20 min. The derivatized analytes were desorbed in the GC injection port for 5 min. The assay showed linearity from 50 to 1000 ng/mL with limits of quantitation of 10 ng/mL for amphetamine and methamphetamine and 20 ng/mL for MDA and MDMA. Extraction efficiencies were determined to be 71.89–103.24% [221]. Another headspace SPME procedure was described for the analysis of amphetamine, methamphetamine and fenfluramine in whole blood. The procedure involved exposure of the extraction fiber to the headspace of a vial heated to 70°C for 15 min. Derivatization was accomplished in the injection port by injecting heptafluorobutyric anhydride followed by insertion of the extraction fiber which allowed desorption and derivatization of the analytes. Analysis time was approximately 30 min per sample. The assay showed linearity from 10 to 1,000 ng/G. The limits of detection were 5.0 ng/G for fenfluramine and methamphetamine, and 10 ng/G for amphetamine [80].

Lee *et al.* described a SPME procedure that evaluated the impact of derivatization on the analysis of amphetamine and methamphetamine in serum. Headspace derivatization was accomplished using 6 μL 20% heptafluorobutyric anhydride ethyl acetate solution (v/v) in an oil bath (270°C) for 10 s. The precision was approximately 7% without derivatization and approximately 17% with headspace derivatization making derivatization an unappealing option. Derivatization did have a significant impact on sensitivity, however, giving approximately an order of magnitude better performance than seen with the underivatized analytes [81]. Two derivatizing reagents were used in a headspace SPME procedure designed for the analysis of amphetamine and methamphetamine [82]. Both heptafluorobutyric anhydride and heptafluorobutyric chloride were combined together in the vial along with the sample with added deuterated internal standard and KOH and NaCl. The SPME device was a 100 μm polydimethylsiloxane fiber that was inserted into the vial and exposed to the headspace. The vial was heated and stirred at 100 °C and 600 rpm for 20 min for evaporation/adsorption/derivatization to complete. Some of the compounds that went into the headspace bound directly to the fiber and were then derivatized or they reacted with the derivatizing reagent and then bound to the fiber. This process overcomes the problem of using these derivatizing reagents where there is high moisture content, a problem for analysis of aqueous samples. GC-MS analysis gave limits of detection of 0.3 ng/ml and the limits of quantitation of 1.0 ng/ml for both drugs using 1 mL of urine. Koster *et al.* [222] used amphetamine as a model compound to improve the detectability and extractability of drugs from biological samples using SPME. The procedure used direct immersion of a polydimethylsiloxane-coated fiber into buffered urine. On-fiber derivatization was performed simultaneously or after the extraction using pentafluorobenzoyl chloride and analyzed by GC using electron capture or MS as the detector. The investigators found the derivatizing reagent interacted with the fiber and matrix compounds, requiring a reagent loading time of 5 min to obtain a linear range from 250 pg/mL to 15 ng/mL [222].

A SPME method designed for the analysis of a variety of analytes including tetrahydrocannabinol (THC), amphetamine, methamphetamine, cocaine and ethanol in saliva using GC-MS analysis has also been described. The extraction involved two distinct consecutive SPMEs. THC was extracted by submersing the polydimethylsiloxane fiber in the vial for 20 min followed by alkalinization of the sample to facilitate the subsequent SPME of amphetamine, methamphetamine and cocaine. The amphetamines were derivatized by adding 2 μL of butylchloroformate to the sample. The procedure gave limits of quantitation of 5 ng/mL and 0.5 ng/mL for amphetamine and methamphetamine, respectively with relative standard deviations of

<8% for all analytes tested [223]. Vu [224] used SPME GC-MS to determine likely candidates for use in a pseudomethamphetamine for the making training aids for drug detecting canines.

SPME has improved over the last few years with modifications of pH, stirring, temperature, addition of salts and selection of sorbents. These alterations have increased the recovery of the analytes of interest and decreased the time required for sample analysis. The significant advantage of this technique is the simplicity of the process, allowing analysis with little analyst intervention. It also eliminates the need for solvents that are both expensive to use and dispose of as well as posing a health hazard. The drawback of this technique is the time involved in analysis of samples. Although the technique requires little analyst intervention, the time for the analysis of each sample would limit the current methods use in high volume laboratories. Advances in this methodology and related technology and automation in the future promise to enhance the utility of SPME.

3.7 INTERPRETATION OF ANALYTICAL RESULTS

Analysis of samples for the detection and quantitation of amphetamines is an important and demanding task. Another critical part of this process, once the analytical results are obtained, is to interpret those results. The purpose of this interpretation can range from helping a clinician evaluate the results from a sick patient to interpretation of results from a forensic sample where issues such as pharmacological effect, time since dose, amount ingested, potential impact on behavior or performance, contribution to any morbidity or mortality can have significant impact. Interpretation of results is discussed here for illustration purposes since a thorough description is beyond the scope of this chapter. The topic is discussed to help demonstrate the importance of having sound analytical data and a substantial understanding of the behavior of the drugs in the body. Without this information, results by themselves have little value or meaning.

Unlike many other compounds, amphetamines do not follow an orderly and easily predictable excretion profile. Interpretation of all urine drug results must be tempered with the understanding that the degree of hydration can have a significant influence on the concentration of the drug in the urine. Some drugs lend themselves to the 'correction' of concentration based on the concentration of creatinine. Unfortunately the excretion of amphetamines, like many basic drugs, is strongly influenced by urine pH in addition to dilution. As a result, while useful, knowing the concentration of the drug is limited in interpretation of the amount ingested or time since ingestion.

Analysis of enantiomer composition can, in some cases, provide significant additional information for interpretation. The d-enantiomer is the most centrally active form of amphetamine and methamphetamine. Knowing the enantiomer(s) present can be helpful in the interpretation of results. It can help to determine what the source of the drug was, and to some extend, assist in determination of how long since the drug was taken.

References pp. 165–174

3.7.1 Metabolic Profile

The metabolism of amphetamine and methamphetamine was initially studied many years ago [225–228] including the difference in metabolism of the enantiomers [116,117,229–234]. These studies also clearly demonstrated the fact that the excretion of these drugs was dependent on the pH of the urine [12,228,235–238]. Controlling urine pH allows for relatively straightforward evaluation of the excretion of these drugs. In the normal physiological condition, however, urine pH varies considerably and the impact on the excretion of these drugs is dramatically affected.

In addition to the effect of pH on the excretion of these drugs, the stereochemistry of the molecules has a strong influence on their metabolism. There is no indication the enantiomers of amphetamine or methamphetamine are excreted at different rates. There is ample evidence, however, that they are metabolized at substantially different rates. The d-enantiomers of both amphetamine and methamphetamine are metabolized more rapidly than the l-enantiomer. For this reason, the proportion of the d- to the l-enantiomer changes with time. Initially, the amount of both enantiomers is nearly equal, followed by a greater amount of the l-enantiomer over time. As a result, knowledge of the enantiomer proportion can be helpful in the interpretation of analytical data. As an example, interpretation of amphetamine enantiomer results was described in several studies using mixed salt amphetamine preparation containing a ratio of 3:1 (d:l) enantiomers of amphetamine. One study described the excretion profile following the administration of a single dose [117] and another described the excretion following administration of multiple doses of the same drug formulation [116]. Both studies described results that could easily be differentiated from those expected from administration of d-amphetamine alone. It was also possible to assess the possibility of the results being from racemic amphetamine. The racemic drug yields results that begin with essentially the same proportion of each enantiomer. Shortly thereafter the l-amphetamine predominates due to its lower rate of metabolism compared to the d-antipod. Administration of the mixture described in these studies showed the d-enantiomer to predominate as long as the samples remained positive ($\geq 500\,\mathrm{ng/mL}$ total drug concentration). Amphetamine enantiomer ratios have also been used to monitor compliance with drug treatment regimens. Evaluation of the enantiomers allows for determination of the source of amphetamine used by the participants whether it is from the treatment program, an illicit source or both [234,239].

3.7.2 Source Differentiation

Interpretation of the source of the amphetamine and/or methamphetamine is a difficult task. There is much, however, that can be determined by evaluation of the drug, its metabolites, potential precursors and impurities. Pharmaceutical amphetamine and methamphetamine are high purity drugs that can often be separated from illicitly produced drugs that typically are not purified to the high degree seen in the legitimate pharmaceutical industry. Moore *et al.* [240–242], describe the profiling of contaminants found in the illicit preparation of methamphetamine as a means of identification of the source of the drug as not being legitimate. Obviously, those drugs that contain amphetamine would not give rise to methamphetamine, which can eliminate several potential sources for the drug. Enantiomer analysis can also be a very useful tool in the evaluation of the source of the drug. Prescription drugs can be characterized by their enantiomer composition. While this does not allow for the demonstration that a drug was from a legitimate medical source, inconsistent findings can easily exclude a legitimate source. For example, the presence of d-methamphetamine in a sample, which is alleged to be the result of the use of a Vicks® inhaler, would disprove the assertion without question since the inhaler contains the l-enantiomer. Likewise, presence of l-methamphetamine in a sample, which is allegedly the result of administration of desoxyn® (contains only d-methamphetamine), would clearly demonstrate the legitimate medication was not the source of the drug found.

Another potential source for amphetamine and/or methamphetamine are precursor drugs. Precursor in this case refers to compounds that are metabolically converted to methamphetamine and/or amphetamine by the body. These drugs have been described and reviewed in several publications [64,243–245]. Many of these drugs are available by prescription, in which case, a valid medical prescription would help to resolve the issue, although it must be remembered that abuse of a prescription drug is still abuse. Some of these drugs, depending on the country, are available over-the-counter. Drugs that are metabolic precursors and their literature references to amphetamine or methamphetamine include: amphetaminil [246], benzphetamine [71,112,247–259], clobenzorex [2,57,115,260–264], deprenyl (selegiline) [187,265–284], dimethylamphetamine [248,249, 285–294], ethylamphetamine [49,229–231,249,295–300], famprofazone [301–312], fencamine [313], fenethylline [49,314–319], fenproporex [111,113,232,320–326], furfenorex [249,250,327,328], mefenorex [299,320,329–331], mesocarb [332–336] and prenylamine [337–344].

Several of these drugs have been studied to identify the enantiomeric composition of the metabolically produced (meth)amphetamine. Examples of some of these are described here to illustrate the information that can help identify the involvement of these drugs. Deprenyl is an example of these drugs, which has received attention in the analytical literature [187,195,280,281,283,345]. The fact that the stereochemistry of the methamphetamine portion of the molecule is the l-enantiomer makes its analysis and interpretation relatively easy compared to others in the group. While several procedures have been published describing the identification and quantitation of unique metabolites, the need for such procedures in routine analysis is small due to the unique enantiomeric form of the drug. For most purposes, simply identifying the form of the methamphetamine and amphetamine present to be the l-enantiomer is all that is necessary.

Famprofazone metabolism has been studied extensively and a number of metabolites have been described that can be used in the interpretation of analytical results [305–311,346–348]. Although unique metabolites have been found, the amphetamine and methamphetamine could be detected in urine for a period after these unique metabolites were no longer detectable. In addition to the metabolites of famprofazone, several studies describing the enantioselective metabolism and resultant proportion of the amphetamine and methamphetamine enantiomers have been described [301,303,304,310,311]. These studies revealed that both enantiomers of amphetamine and methamphetamine were seen in samples after administration of famprofazone. Long only manufactured in Switzerland, famprofazone is used in a medication now made in Taiwan making the drug much more easily obtainable than before.

Fenproporex is another example of a precursor drug that has been evaluated for differentiation from other sources. Analysis of the drug has been described in a number of different reports [25,41,111,322,323,349–351] including enantiomer analysis [114,232]. It was shown that the administration of fenproporex produced both d- and l-amphetamine. In addition, it was shown with both single dose and multidose (one capsule daily for 7 days) administration of fenproporex, when the concentration of amphetamine was greater than or equal to 500 ng/mL, fenproporex itself was also detected [111,113] although caution should be applied as the concentration approaches 500 ng/mL. Kraemer *et al.* [326] described 14 metabolites, including amphetamine, of fenproporex. The procedure used acid hydrolysis followed by extraction at pH 8–9 and acetylation before analysis by GC-MS. The same group evaluated the metabolism of fenproporex in an animal model and *in vitro* with human liver microsomes to determine the impact of metabolic enzymes in the process [324]. This extensive study compliments previous investigations, noted above, of fenproporex metabolism and excretion that allow interpretation of analytical results.

Benzphetamine has also been studied by a number of investigators. The presence of methamphetamine and amphetamine following the administration of benzphetamine is well established [64,112,247,250,255–257,352]. The fact that the enantiomeric composition of the drug is d-only is also well established and the d-enantiomer of amphetamine and methamphetamine are found in samples following the administration of this drug. The parent drug, however, is typically not detected in urine. As a result, looking for the parent drug to assess the possible involvement of benzphetamine, unlike that seen with fenproporex, is not a viable option. However, it has been shown that the proportion of amphetamine to methamphetamine is much different than expected with use/abuse of methamphetamine and this information can be useful in interpretation. In every case, the amount of amphetamine was far in excess of what would be seen from administration of methamphetamine [112,247]. In several subjects, the amount of amphetamine actually exceeded the amount of methamphetamine [112]. In a recent study, employing LC-electrospray ionization MS Sato *et al.* [258] described the identification and quantitation of benzphetamine, *N*-benzylamphetamine, *p*-hydroxybenzphetamine, *p*-hydroxy-*N*-benzylamphetamine, methamphetamine and amphetamine using a mixed mode solid-phase extraction giving limits of quantitation (signal-to-noise ratio 10) of 700 pg/mL, 300 pg/mL, 500 pg/mL, 1.4 ng/mL, 6 ng/mL and 10 ng/mL, respectively. Deuterium-labeled isotopomers, used as internal standards for quantitative analysis, were chromatographically separated from the corresponding unlabeled compound. The method used an alkaline-resistant HPLC column (Develosil ODS-UG-5, 150 mm × 31.0 mm i.d.; Nomura, Seto, Japan) to accommodate the mobile-phase system that was composed of acetonitrile and

10 mM ammonium acetate buffer (pH 9.0) with a stepwise gradient elution was carried out as follows: 5–35% (5–11 min) to 55% (16–22 min) to 80% (26–45 min) at 80–100 µL/min. This procedure allows for characterization of benzphetamine and metabolites, which add significant utility to the interpretation of analytical results. While conducted using an animal model, the study suggests use of unique metabolites may be viable for human sample evaluation as well.

Another anorexic drug that has been evaluated as a precursor for amphetamine is clobenzorex. The amphetamine produced from metabolism of clobenzorex is the d-enantiomer [2]. The parent could be detected for some time following administration of the drug, but some samples were shown to contain considerable amounts of amphetamine (≥ 500 ng/mL) while no parent drug could be detected. Detection of the parent together with concentrations of the amphetamine can be very useful in determining the potential for the positive result being the result of the use of this drug. When the parent drug is no longer detected, however, metabolites hold valuable information. Qualitative identification of several metabolites of this drug, one of which was detected in at least one subject as long as amphetamine has been described [57]. Further studies of this drug conducted by Valtier and Cody described the analysis of samples following controlled administration of clobenzorex [115,264]. The studies characterize a hydroxy metabolite of clobenzorex which can be used to identify its use compared to use of amphetamine, an important consideration because the parent compound cannot be detected for very long following administration of the drug [115,264]. The authors found 4-hydroxyclobenzorex could be detected for up to 91.5 h following administration of a single 30-mg dose of clobenzorex and up to 152 h post last dose in a multidose study. The metabolite was extracted using liquid–liquid extraction following acid hydrolysis the derivatized and analyzed along with amphetamine by GC-MS monitoring ions at m/z 125, 330 and 364 for 4-hydroxyclobenzorex and its 3-Cl regioisomer, which the authors used as an internal standard.

3.8 CONCLUSION

Amphetamines are chemically a simple group of compounds but are, in many ways, a difficult group to deal with. Some amphetamines have a variety of legitimate uses, but have a high abuse potential and are therefore scheduled drugs where the medical use is strictly controlled. The high abuse potential is borne out by the amount of these drugs that are abused throughout the world. Not only are amphetamine and methamphetamine widely abused but some of their methylenedioxy analogues are as well. The use of these drugs has also risen in the past decade and they are a serious drug related issue in many parts of the world.

Analysis of these drugs can be accomplished using a variety of analytical methods. Use of a number of analytical procedures involving different instruments, extraction procedures, derivatives, etc. shows the diversity of the analytical methodology that has been brought to bare on the analysis of these drugs. Method development to improve sensitivity, increase speed of analysis, elimination of interference has

References pp. 165–174

continued over the many years these drugs have been around and has lead to procedures for many different biological matrices (sweat, saliva, meconium, etc.) in addition to more traditional substances (blood, plasma, urine, tissue, etc.). The continued use and abuse of the amphetamines and rapid development of analytical procedures promises to make this an active area for many years to come.

3.9 ABBREVIATIONS

AITC	2,3,4-tri-*O*-acetyl-α-d-arabinopyranosyl isothiocyanate
APCI	Atmospheric pressure chemical ionization
BDMPEA	4-Bromo-2,5-dimethoxy-β-phenethylamine
CB	Carbethoxyhexafluorobutyryl chloride
CD	Cyclodextrin
CI	Chemical ionization
CZE	Capillary zone electrophoresis
DB-1	Cross-linked methyl silicone
DB-5	5% Phenyl methyl silicone
DB-17	50% Phenyl methyl silicone
DBD-F	4-(*N*,*N*-dimethylaminosulphonyl)-7-fluoro-2,1,3-benzoxadiazole
DIB-Cl	4-(4,5-diphenyl-1*H*-imidazol-2-yl)benzoyl chloride
DMA	2,5-Dimethoxyamphetamine
DNB	3,5-dinitrobenzoyl chloride
DOB	4-Bromo-2,5-dimethoxyamphetamine
DOE	2,5-Dimethoxy-4-ethylamphetamine
DOM	2,5-Dimethoxy-4-methylamphetamine
EC	Electrochemical detection
ECD	Electron capture detection
EI	Electron ionization
FID	Flame ionization detection
FLEC	(-)-1-(9-Fluorenyl)ethyl chloroformate
FMOC	Fluorenylmethylchloroformate-l-prolyl chloride
FMOC-Cl	Fluorenylmethylchloroformate
GC	Gas chromatography
GC-MS	Gas chromatography-mass spectrometry
GC-MS-MS	Gas chromatography-mass spectrometry-mass spectrometry
GITC	2,3,4,6-tetra-*O*-acetyl-β-D-glucopyranosyl isothiocyanate
HFBA	Heptafluorobutyric anhydride
l-HPC	Heptafluorobutyryl-l-prolyl chloride
HPLC	High-performance liquid chromatography
LC	Liquid chromatography
LC-MS	Liquid chromatography-mass spectrometry
LC-MS-MS	Liquid chromatography-mass spectrometry-mass spectrometry

LIF	Laser-induced fluorescence
MDA	3,4-Methylenedioxyamphetamine
MDEA	3,4-Methylenedioxyethylamphetamine
MDMA	3,4-Methylenedioxymethamphetamine
MECC	Micellar electrokinetic capillary chromatography
MS	Mass spectrometry
MS-MS	Mass spectrometry-mass spectrometry
MTBSTFA	*N*-methyl-*N*-*t*-butyldimethylsilyl trifluoroacetamide
MTPA	(-)-α-Methoxy-α-(trifluoromethyl)phenylacetyl chloride
NBD-F	4-Fluoro-7-nitro-2,1,3-benzoxadiazole
N-OH-MDA	*N*-hydroxy-3,4-methylenedioxy
NPD	Nitrogen-phosphorus detector
NPSP	4-Nitrophenylsulfonyl-l-prolyl chloride
PDA	Photo diode array
PEIC	*R*-(+)-1-phenylethylisocyanate
PFPA	Pentafluoroproprionic anhydride
l-PPC	Pentafluoropropionyl-l-prolyl chloride
psi	Pounds per square inch
SDS	Sodium dodecyl sulfate
SIM	Selected ion monitoring
SPME	Solid-phase microextraction
TFA	Trifluoroacetic anhydride
TLC	Thin-layer chromatography
TMA	3,4,5-Trimethoxyamphetamine
TMS	Trimethylsilyl
l-TPC	Trifluoroacetyl-l-prolyl chloride
UV	Ultraviolet

3.10 REFERENCES

1 M.A. ElSohly, D.F. Stanford, D. Sherman, H. Shah, D. Bernot and C.E. Turner, J. Anal. Toxicol., 16 (1992) 109.
2 S. Valtier and J.T. Cody, J. Forensic Sci., 44 (1999) 17.
3 C.L. Hornbeck, J.E. Carrig and R.J. Czarny, J. Anal. Toxicol., 17 (1993) 257.
4 B.D. Paul, M.R. Past, R.M. McKinley, J.D. Foreman, L.K. McWhorter and J.J. Snyder, J. Anal. Toxicol., 18 (1994) 331.
5 M.-R. Lee, J. Jeng, W.-S. Hsiang and B.-H. Hwang, J. Anal. Toxicol., 23 (1999) 41.
6 D.L. Allen and J.S. Oliver, Forensic Sci. Int., 107 (2000) 191.
7 S.M. Wang, Y.S. Giang and Y.C. Ling, J. Chromatogr. B, Biomed. Sci. Appl., 759 (2001) 17.
8 M. Nishida, A. Namera, M. Yashiki and K. Kimura, Forensic Sci. Int., 143 (2004) 163.
9 S. Valtier and J.T. Cody, J. Anal. Toxicol., 19 (1995) 375.
10 H. Gjerde, I. Hasvold, G. Pettersen and A.S. Christophersen, J. Anal. Toxicol., 17 (1993) 65.
11 J. Fay, R. Fogerson, D. Schoendorfer, R.S. Niedbala and V. Spiehler, J. Anal. Toxicol., 20 (1996) 398.
12 S.H. Wan, S.B. Matin and D.L. Azarnoff, Clin. Pharmacol. Ther., 23 (1978) 585.
13 F.P. Smith, Forensic Sci. Int., 17 (1981) 225.
14 S. Suzuki, T. Inoue, H. Hori and S. Inayama, J. Anal. Toxicol., 13 (1989) 176.

15   T. Inoue and S. Seta, Forensic Sci. Rev., 4 (1992) 89.
16   A. J. Jenkins, in: B. Levine (Ed.), Principles of forensic toxicology, AACC Press, Washington DC, 1999, p. 31.
17   D.A. Kidwell, J.C. Holland and S. Athanaselis, J. Chromatogr. B, Biomed. Appl., 713 (1998) 111.
18   K.T. Nakamura, E.L. Ayau, C.F. Uyehara, C.L. Eisenhauer, L.M. Iwamoto and D.E. Lewis, Dev. Pharmacol. Ther., 19 (1992) 183.
19   P. Kintz and V. Cirimele, Forensic Sc. Int., 84 (1997) 151.
20   R. Kikura, Y. Nakahara, T. Mieczkowski and F. Tagliaro, Forensic Sci. Int., 84 (1997) 165.
21   J. Rohrich and G. Kauert, Forensic Sci. Int., 84 (1997) 179.
22   P. Kintz, V. Cirimele, A. Tracqui and P. Mangin, J. Chromatogr. B, Biomed. Appl., 670 (1995) 162.
23   I. Koide, O. Noguchi, K. Okada, A. Yokoyama, H. Oda, S. Yamamoto and H. Kataoka, J. Chromatogr. B, Biomed. Sci. Appl., 707 (1998) 99.
24   A. Miki, T. Keller, P. Regenscheit, R. Dirnhofer, M. Tatsuno, M. Katagi, M. Nishikawa and H. Tsuchihashi, J. Chromatogr. B, Biomed. Appl., 692 (1997) 319.
25   Y. Nakahara, Forensic Sci. Int., 70 (1995) 135.
26   Y. Nakahara, K. Takahashi, M. Shimamine and Y. Takeda, J. Forensic Sci., 36 (1991) 70.
27   Y. Nakahara, K. Takahashi, Y. Takeda, K. Konuma, S. Fukui and T. Tokui, Forensic Sci. Int., 46 (1990) 243.
28   F. Tagliaro, Z. De Battisti, A. Groppi, Y. Nakahara, D. Scarcella, R. Valentini and M. Marigo, J. Chromatogr. B, Biomed. Sci. Appl., 723 (1999) 195.
29   N. Takayama, S. Tanaka and K. Hayakawa, Biomed. Chromatogr., 11 (1997) 25.
30   T.A. Brettell, K. Inman, N. Rudin and R. Saferstein, Anal. Chem., 71 (1999) 235R.
31   M.R. Moeller, J. Chromatogr., 580 (1992) 125.
32   H. Sachs and P. Kintz, J. Chromatogr. B, Biomed. Appl., 713 (1998) 147.
33   T. Kojima, I. Okamoto, T. Miyazaki, F. Chikasue, M. Yashiki and K. Nakamura, Forensic Sci. Int., 31 (1986) 93.
34   C. Tamayo-Lora, T. Tena and A. Rodriguez, Forensic Sci. Int., 85 (1997) 149.
35   J.M. Halket and K. Blau (Eds.), 2nd ed., Handbook for derivatives for chromatography, Wiley, New York, 1993.
36   D.R. Knapp, Handbook of analytical derivatization reactions, Wiley, New York, 1979.
37   D.R. Knapp, Methods Enzymol., 193 (1990) 314.
38   G.B. Baker, R.T. Coutts and A. Holt, J. Pharmacol. Toxicol. Methods, 31 (1994) 143.
39   A. Dasgupta and J. Spies, Am. J. Clin. Path., 109 (1998) 527.
40   W.C. Thompson and A. Dasgupta, Clin. Chem., 40 (1994) 1703.
41   A. Solans, M. Carnicero, R. de la Torre and J. Segura, J. Anal. Toxicol., 19 (1995) 104.
42   K. Hara, S. Kashimura, Y. Hieda and M. Kageura, J. Anal. Toxicol., 21 (1997) 54.
43   R. Meatherall, J. Anal. Toxicol., 19 (1995) 316.
44   A.J. Bourque and I.S. Krull, J. Chromatogr., 537 (1991) 123.
45   C.X. Gao, D. Schmalzing and I.S. Krull, Biomed. Chromatogr., 5 (1991) 23.
46   M.E. Szulc and I.S. Krull, Biomed. Chromatogr., 6 (1992) 269.
47   G.A. Eiceman, C.S. Leasure and S.L. Selim, J. Chromatogr. Sci., 22 (1984) 509.
48   R.L. Fitzgerald, J.M. Ramos Jr., S.C. Bogema and A. Poklis, J. Anal. Toxicol., 12 (1988) 255.
49   H. Yamada, S. Ikeda-Wada and K. Oguri, Biol. Pharm. Bull., 21 (1998) 778.
50   H. Yamada, A. Yamahara, S. Yasuda, M. Abe, K. Oguri, S. Fukushima and S. Ikeda-Wada, J. Anal. Toxicol., 26 (2002) 17.
51   M. Nishida, A. Namera, M. Yashiki and T. Kojima, Forensic Sci. Int., 125 (2002) 156.
52   M. Nishida, A. Namera, M. Yashiki and T. Kojima, J. Chromatogr. B, 789 (2003) 65.
53   D.L. Lin, W.T. Chang, T.L. Kuo and R.H. Liu, J. Anal. Toxicol., 24 (2000) 275.
54   H.-S. Shin and M. Donike, Anal. Chem., 68 (1996) 3015.
55   H.K. Ensslin, K.A. Kovar and H.H. Maurer, J. Chromatogr. B, Biomed. Appl., 683 (1996) 189.
56   H.H. Maurer, J. Chromatogr., 580 (1992) 3.
57   H.H. Maurer, T. Kraemer, O. Ledvinka, C.J. Schmitt and A.A. Weber, J. Chromatogr. B, Biomed. Appl., 689 (1997) 81.

58 T. Kraemer, I. Vernaleken and H.H. Maurer, J. Chromatogr. B, Biomed. Appl., 702 (1997) 93.
59 H.H. Maurer, Ther. Drug Monit., 18 (1996) 465.
60 R. Melgar and R.C. Kelly, J. Anal. Toxicol., 17 (1993) 399.
61 E.M. Thurman, M.J. Pedersen, R.L. Stout and T. Martin, J. Anal. Toxicol., 16 (1992) 19.
62 R.J. Czarny and C.L. Hornbeck, J. Anal. Toxicol., 13 (1989) 257.
63 B.D. Paul and K.A. Cole, J. Anal. Toxicol., 25 (2001) 525.
64 T. Kraemer and H.H. Maurer, J. Chromatogr. B, Biomed. Appl., 713 (1998) 163.
65 P. Campins Falco, C. Molins Legua, A. Sevillano Cabeza and R. Porras Serrano, Analyst, 122 (1997) 673.
66 R. Herraez-Hernandez, P. Campins-Falco and A. Sevillano-Cabeza, J. Chromatogr. Sci., 35 (1997) 169.
67 M.J. Bogusz, M. Kala and R.D. Maier, J. Anal. Toxicol., 21 (1997) 59.
68 K. Nakashima, K. Suetsugu, K. Yoshida, S. Akiyama, S. Uzu and K. Imai, Biomed. Chromatogr., 6 (1992) 149.
69 O. Al-Dirbashi, N. Kuroda, S. Akiyama and K. Nakashima, J. Chromatogr. B, Biomed. Sci. Appl., 695 (1997) 251.
70 P. Campins-Falco, A. Sevillano-Cabeza, C. Molins-Legua and M. Kohlmann, J. Chromatogr. B, Biomed. Appl., 687 (1996) 239.
71 R. Herraez-Hernandez, P. Campins-Falco and A. Sevillano-Cabeza, Anal. Chem., 68 (1996) 734.
72 C. Molins Legua, P. Campins Falco and A. Sevillano Cabeza, J. Chromatogr. B, Biomed. Appl., 672 (1995) 81.
73 P. Campins Falco, C. Molins Legua, R. Herraez Hernandez and A. Sevillano Cabeza, J. Chromatogr. B, Biomed. Appl., 663 (1995) 235.
74 L. Tedeschi, G. Frison, F. Castagna, R. Giorgetti and S.D. Ferrara, Int. J. Legal Med., 105 (1993) 265.
75 G. Maeder, M. Pelletier and W. Haerdi, J. Chromatogr., 593 (1992) 9.
76 T.K. Wang and M.S. Fuh, J. Chromatogr. B, Biomed. Appl., 686 (1996) 285.
77 K. Hayakawa, Y. Miyoshi, H. Kurimoto, Y. Matsushima, N. Takayama, S. Tanaka and M. Miyazaki, Biol. Pharm. Bull., 16 (1993) 817.
78 K. Hayakawa, N. Imaizumi, H. Ishikura, E. Minogawa, N. Takayama, H. Kobayashi and M. Miyazaki, J. Chromatogr., 515 (1990) 459.
79 C. Molins-Legua, P. Campins-Falco and A. Sevillano-Cabeza, Analyst, 123 (1998) 2871.
80 A. Namera, M. Yashiki, J. Liu, K. Okajima, K. Hara, T. Imamura and T. Kojima, Forensic Sci. Int., 109 (2000) 215.
81 M.R. Lee, Y.S. Song, B.H. Hwang and C.C. Chou, J. Chromatogr. A, 896 (2000) 265.
82 M.K. Huang, C. Liu and S.D. Huang, Analyst, 127 (2002) 1203.
83 D. Hensley and J.T. Cody, J. Anal. Toxicol., 23 (1999) 518.
84 B.J.A. Cooke, J. Anal. Toxicol., 18 (1994) 49.
85 H.H. Maurer, T. Kraemer, C. Kratzsch, F.T. Peters and A.A. Weber, Ther. Drug Monit., 24 (2002) 117.
86 F.T. Peters, T. Kraemer and H.H. Maurer, Clin. Chem., 48 (2002) 1472.
87 F.T. Peters, N. Samyn, C.T. Lamers, W.J. Riedel, T. Kraemer, G. de Boeck and H.H. Maurer, Clin. Chem., 51 (2005) 1811.
88 B.D. Paul, J. Jemionek, D. Lesser, A. Jacobs and D.A. Searles, J. Anal. Toxicol., 28 (2004) 449.
89 J.M. Holler, S.P. Vorce, T.Z. Bosy and A. Jacobs, J. Anal. Toxicol., 29 (2005) 652.
90 S.M. Wang, T.C. Wang and Y.S. Giang, J. Chromatogr. B, Analyt. Technol. Biomed. Life Sci., 816 (2005) 131.
91 B.S. Foster, D.D. Gilbert, A. Hutchaleelaha and M. Mayersohn, J. Anal. Toxicol., 22 (1998) 265.
92 J.K. Fallon, A.T. Kicman, J.A. Henry, P.J. Milligan, D.A. Cowan and A.J. Hutt, Clin. Chem., 45 (1999) 1058.
93 K.J. Miller, J. Gal and M.M. Ames, J. Chromatogr., 307 (1984) 335.
94 L.B. Rasmussen, K.H. Olsen and S.S. Johansen, J. Chromatogr. B, 842 (2006) 136.

95   N. Pizarro, A. Llebaria, S. Cano, J. Joglar, M. Farre, J. Segura and R. de la Torre, Rapid Commun. Mass Spectrom., 17 (2003) 330.

96   F.T. Noggle Jr. and C.R. Clark, J. Forensic Sci., 31 (1986) 732.

97   J.M. Barksdale and C.R. Clark, J. Chromatogr. Sci., 23 (1985) 176.

98   J. Sukbuntherng, A. Hutchaleelaha, H.H. Chow and M. Mayersohn, J. Anal. Toxicol., 19 (1995) 139.

99   A. Hutchaleelaha, A. Walters, H.H. Chow and M. Mayersohn, J. Chromatogr. B, Biomed. Appl., 658 (1994) 103.

100 R. La Croix, E. Pianezzola and M. Strolin Benedetti, J. Chromatogr. B, Biomed. Appl., 656 (1994) 251.

101 A.J. Bourque and I.S. Krull, Biomed. Chromatogr., 8 (1994) 53.

102 C.X. Gao and I.S. Krull, J. Pharm. Biomed. Anal., 7 (1989) 1183.

103 D.M. Desai and J. Gal, J. Chromatogr., 629 (1993) 215.

104 H. Spahn-Langguth, G. Hahn, E. Mutschler, W. Mohrke and P. Langguth, J. Chromatogr., 584 (1992) 229.

105 J. Jonsson, R. Kronstrand and M. Hatanpaa, J. Forensic Sci., 41 (1996) 148.

106 D.S. Lho, H.S. Shin, B.K. Kang and J. Park, J. Anal. Toxicol., 14 (1990) 73.

107 J. Ortuno, N. Pizarro, M. Farre, M. Mas, J. Segura, J. Cami, R. Brenneisen and R. de la Torre, J. Chromatogr. B, Biomed. Sci. Appl., 723 (1999) 221.

108 K.M. Jenkins, M.S. Young, C.R. Mallet and A.A. Elian, J. Anal. Toxicol., 28 (2004) 50.

109 S.J. Asghar, G.B. Baker, G.A. Rauw and P.H. Silverstone, J. Pharmacol. Toxicol. Methods, 46 (2002) 111.

110 Y.-S. Ho, R.H. Liu, A.W. Nichols and S.D. Kumar, J. Forensic Sci., 35 (1990) 123.

111 J.T. Cody and S. Valtier, J. Anal. Toxicol., 20 (1996) 425.

112 J.T. Cody and S. Valtier, J. Anal. Toxicol., 22 (1998) 299.

113 J.T. Cody, S. Valtier and S. Stillman, J. Anal. Toxicol., 23 (1999) 187.

114 J.T. Cody and S. Valtier, J. Anal. Toxicol., 21 (1997) 84.

115 J.T. Cody and S. Valtier, J. Anal. Toxicol., 25 (2001) 158.

116 J.T. Cody, S. Valtier and S.L. Nelson, J. Anal. Toxicol., 28 (2004) 563.

117 J.T. Cody, S. Valtier and S.L. Nelson, J. Anal. Toxicol., 27 (2003) 485.

118 J.H. Galloway, M. Ashford, I.D. Marsh, M. Holden and A.R. Forrest, J. Clin Pathol., 51 (1998) 326.

119 J.L. Valentine, G.L. Kearns, C. Sparks, L.G. Letzig, C.R. Valentine, S.A. Shappell, D.F. Neri and C.A. DeJohn, J. Anal. Toxicol., 19 (1995) 581.

120 S.O. Pirnay, T.T. Abraham and M.A. Huestis, Clin. Chem., 52 (2006) 1728.

121 J.Y. Kim, S.I. Suh, M.K. In and B.C. Chung, J. Anal. Toxicol., 29 (2005) 370.

122 M. Pellegrini, F. Rosati, R. Pacifici, R. Zuccaro, F.S. Romolo and A. Lopez, J. Chromatogr. B, Analyt. Technol. Biomed. Life Sci., 769 (2002) 243.

123 K.M. Hegadoren, G.B. Baker and R.T. Coutts, Res. Commun. Subst. Abuse, 14 (1993) 67.

124 R.L. Fitzgerald, R.V. Blanke, R.A. Glennon, M.Y. Yousif, J.A. Rosecrans and A. Poklis, J. Chromatogr., 490 (1989) 59.

125 K.A. Moore, A. Mozayani, M.F. Fierro and A. Poklis, Forensic Sci. Int., 83 (1996) 111.

126 H.K. Lim, Z. Su and R.L. Foltz, Biol. Mass Spectrom., 22 (1993) 403.

127 D. De Boer, L.P. Tan, P. Gorter, R.M.A. Van de Wal, J.J. Kettenes-van den Bosch, E.A. De Bruijn and R.A.A. Maes, J. Mass Spectrom., 32 (1997) 1236.

128 H.J. Leis, G.N. Rechberger, G. Fauler and W. Windischhofer, Rapid Commun. Mass Spectrom., 17 (2003) 569.

129 J.T. Cody and S. Valtier, J. Anal. Toxicol., 26 (2002) 537.

130 L. Aalberg, J. DeRuiter, F.T. Noggle, E. Sippola and C.R. Clark, J. Chromatogr. Sci., 38 (2000) 329.

131 L. Aalberg, J. DeRuiter, E. Sippola and C.R. Clark, J. Chromatogr. Sci., 42 (2004) 293.

132 L. Aalberg, C.R. Clark and J. Deruiter, J. Chromatogr. Sci., 42 (2004) 464.

133 H.J. Helmlin, K. Bracher, D. Bourquin, D. Vonlanthen and R. Brenneisen, J. Anal. Toxicol., 20 (1996) 432.

134   E.R. Garrett, K. Seyda and P. Marroum, Acta. Pharm. Nord., 3 (1991) 9.
135   R.E. Michel, A.B. Rege and W.J. George, J. Neurosci. Methods, 50 (1993) 61.
136   R. Hartley, M. Green, M. Quinn and M.I. Levene, Arch. Dis. Child., 69 (1993) 55.
137   M. Katagi, H. Nishioka, K. Nakajima, H. Tsuchihashi, H. Fujima, H. Wada, K. Nakamura and K. Makino, J. Chromatogr. B, Biomed. Appl., 676 (1996) 35.
138   N.Y. Li, Y. Li and E.M. Sellers, Eur. J. Drug Metab. Pharmacokinet., 22 (1997) 427.
139   M.J. Bogusz, J. Anal. Toxicol., 21 (1997) 246.
140   W. Weinmann and M. Svoboda, J. Anal. Toxicol., 22 (1998) 319.
141   O.Y. Al-Dirbashi, M. Wada, N. Kuroda, M. Takahashi and K. Nakashima, J. Forensic Sci., 45 (2000) 708.
142   O.Y. Al-Dirbashi, N. Kuroda, M. Wada, M. Takahashi and K. Nakashima, Biomed. Chromatogr., 14 (2000) 293.
143   K.M. Clauwaert, J.F. Van Bocxlaer, E.A. De Letter, S. Van Calenbergh, W.E. Lambert and A.P. De Leenheer, Clin. Chem., 46 (2000) 1968.
144   M. Concheiro, A. de Castro, O. Quintela, M. Lopez-Rivadulla and A. Cruz, Forensic Sci. Int., 150 (2005) 221.
145   N. Shima, H.T. Kamata, M. Katagi and H. Tsuchihashi, Xenobiotica, 36 (2006) 259.
146   N. Shima, H. Tsutsumi, T. Kamata, M. Nishikawa, M. Katagi, A. Miki and H. Tsuchihashi, J. Chromatogr. B, Analyt. Technol. Biomed. Life Sci., 830 (2006) 64.
147   A. Meyer, A. Mayerhofer, K.A. Kovar and W.J. Schmidt, Neurosci. Lett., 330 (2002) 193.
148   K. Nakashima, A. Kaddoumi, Y. Ishida, T. Itoh and K. Taki, Biomed. Chromatogr., 17 (2003) 471.
149   A. Kaddoumi, R. Kikura-Hanajiri and K. Nakashima, Biomed. Chromatogr., 18 (2004) 202.
150   J.L. da Costa and A.A. da Matta Chasin, J. Chromatogr. B, Analyt. Technol. Biomed. Life Sci., 811 (2004) 41.
151   M. Wada, S. Nakamura, M. Tomita, M.N. Nakashima and K. Nakashima, Luminescence, 20 (2005) 210.
152   I.B. Muller and C.N. Windberg, J. Chromatogr. Sci., 43 (2005) 434.
153   K. Mc Fadden, J. Gillespie, B. Carney and D. O'Driscoll, J. Chromatogr. A, 1120 (2006) 54.
154   O. Al-Dirbashi, N. Kuroda, F. Menichini, S. Noda, M. Minemoto and K. Nakashima, Analyst, 123 (1998) 2333.
155   O. Al-Dirbashi, J. Qvarnstrom, K. Irgum and K. Nakashima, J. Chromatogr. B, Biomed Sci. Appl., 712 (1998) 105.
156   S.M. Hayes, R.H. Liu, W.S. Tsang, M.G. Legendre, R.J. Berni, D.J. Pillion, S. Barnes and M.H. Ho, J. Chromatogr., 398 (1987) 239.
157   H.Y. Aboul-Enein and V. Serignese, Biomed. Chromatogr., 11 (1997) 7.
158   T. Nagai and S. Kamiyama, J. Anal. Toxicol., 15 (1991) 299.
159   T. Nagai, S. Kamiyama and K. Matsushima, J. Anal. Toxicol., 19 (1995) 225.
160   T. Nagai and S. Kamiyama, J. Chromatogr., 525 (1990) 203.
161   Y. Makino, A. Suzuki, T. Ogawa and O. Shirota, J. Chromatogr. B, Biomed. Sci. Appl., 29 (1999) 97.
162   K.W. Phinney and L.C. Sander, Anal. Bioanal. Chem., 378 (2004) 144.
163   K. Matsushima, T. Nagai and S. Kamiyama, J. Anal. Toxicol., 22 (1998) 33.
164   J.L. Veuthey and W. Haerdi, J. Chromatogr., 515 (1990) 385.
165   A.A. Descombes, J.L. Veuthey and W. Haerdi, Anal. Chem., 339 (1991) 480.
166   Y. McAvoy, B. Backstrom, K. Janhunen, A. Stewart and M.D. Cole, Forensic Sci. Int., 99 (1999) 107.
167   N. Kato and A. Ogamo, Sci. Justice, 41 (2001) 239.
168   N. Kato, H. Kubo and H. Homma, Anal. Sci., 21 (2005) 1117.
169   V. Maresova, J. Hampl, Z. Chundela, F. Zrcek, M. Polasek and J. Chadt, J. Anal. Toxicol., 29 (2005) 353.
170   F.E.P. Mikkers, F.M. Everaerts and T.P.E.M. Verheggen, J. Chromatogr. A, 713 (1979) 11.
171   R. Weinbeger and I.S. Lurie, Anal. Chem., 63 (1991) 823.
172   F. Tagliaro, S. Turrina, P. Pisi, F.P. Smith and M. Marigo, J. Chromatogr. B, Biomed Appl., 713 (1998) 27.
173   D.K. Lloyd, J. Chromatogr., 735 (1996) 29.

174 W. Thormann, Ther. Drug Monit., 24 (2002) 222.
175 W.F. Smyth, J. Chromatogr. B, Analyt. Technol. Biomed. Life Sci., 824 (2005) 1.
176 N. Anastos, N.W. Barnett and S.W. Lewis, Talanta, 67 (2005) 269.
177 N.P. Lemos, F. Bortolotti, G. Manetto, R.A. Anderson, F. Cittadini and F. Tagliaro, Sci. Justice, 41 (2001) 203.
178 A. Van Eeckhaut and Y. Michotte, Electrophoresis, 27 (2006) 2880.
179 K. Stulik, V. Pacakova, J. Suchankova and P. Coufal, J. Chromatogr. B, Analyt. Technol. Biomed. Life Sci., 841 (2006) 79.
180 F. Svec, J. Chromatogr. B, Analyt. Technol. Biomed. Life Sci., 841 (2006) 52.
181 U. Holzgrabe, D. Brinz, S. Kopec, C. Weber and Y. Bitar, Electrophoresis, 27 (2006) 2283.
182 A.S. Ptolemy and P. Britz-McKibbin, J. Chromatogr. A, 1106 (2006) 7.
183 F. Tagliaro and F. Bortolotti, Electrophoresis, 27 (2006) 231.
184 Y.S. Huang, C.C. Tsai, J.T. Liu and C.H. Lin, Electrophoresis, 26 (2005) 3904.
185 R. Iio, S. Chinaka, N. Takayama and K. Hayakawa, Anal. Sci., 21 (2005) 15.
186 E. Szoko, T. Tabi, T. Borbas, B. Dalmadi, K. Tihanyi and K. Magyar, Electrophoresis, 25 (2004) 2866.
187 T. Tabi, K. Magyar and E. Szoko, Electrophoresis, 24 (2003) 2665.
188 R. Iio, S. Chinaka, S. Tanaka, N. Takayama and K. Hayakawa, Analyst, 128 (2003) 646.
189 A.S. Liau, J.T. Liu, L.C. Lin, Y.C. Chiu, Y.R. Shu, C.C. Tsai and C.H. Lin, Forensic Sci. Int., 134 (2003) 17.
190 Y.S. Huang, J.T. Liu, L.C. Lin and C.H. Lin, Electrophoresis, 24 (2003) 1097.
191 W.C. Cheng, W.M. Lee, M.F. Chan, P. Tsui and K.L. Dao, J. Forensic Sci., 47 (2002) 1248.
192 N. Pizarro, R. de la Torre, M. Farre, J. Segura, A. Llebaria and J. Joglar, Bioorg. Med. Chem., 10 (2002) 1085.
193 S. Chinaka, S. Tanaka, N. Takayama, K. Komai, T. Ohshima and K. Ueda, J. Chromatogr. B, Biomed. Sci. Appl., 749 (2000) 111.
194 S.R. Wallenborg, I.S. Lurie, D.W. Arnold and C.G. Bailey, Electrophoresis, 21 (2000) 3257.
195 E.M. Kim, H.S. Chung, K.J. Lee and H.J. Kim, J. Anal. Toxicol., 24 (2000) 238.
196 N. Kuroda, R. Nomura, O. al-Dirbashi, S. Akiyama and K. Nakashima, J. Chromatogr., 798 (1998) 325.
197 J. Choi, C. Kim and M.J. Choi, J. Chromatogr. B, Biomed. Appl., 705 (1998) 277.
198 F. Tagliaro, G. Manetto, S. Bellini, D. Scarcella, F.P. Smith and M. Marigo, Electrophoresis, 19 (1998) 42.
199 U. Backofen, F.M. Matysik, W. Hoffmann and C.E. Lunte, Fresen. J. Anal. Chem., 367 (2000) 359.
200 Y.L. Chung, J.T. Liu and C.H. Lin, J. Chromatogr. B, Biomed. Sci. Appl., 759 (2001) 219.
201 C. Fang, Y.-L. Chung, J.-T. Liu and C.-H. Lin, Forensic Sci. Int., 125 (2002) 142.
202 J. Caslavska and W. Thormann, Electrophoresis, 25 (2004) 1623.
203 D. Scarcella, F. Tagliaro, S. Turrina, G. Manetto, Y. Nakahara, F.P. Smith and M. Marigo, Forensic Sci. Int., 89 (1997) 33.
204 M. Lanz, R. Brenneisen and W. Thormann, Electrophoresis, 18 (1997) 1035.
205 E. Varesio, J.Y. Gauvrit, R. Longeray, P. Lanteri and J.L. Veuthey, Electrophoresis, 18 (1997) 931.
206 J. Sevcik, Z. Stransky, B.A. Ingelse and K. Lemr, J. Pharm. Biomed. Anal., 14 (1996) 1089.
207 A. Guttman, Electrophoresis, 16 (1995) 1900.
208 H. Tsuchihashi, K. Nakajima, M. Nishikawa, K. Shiomi and S. Takahashi, J. Chromatogr., 467 (1989) 227.
209 D. Martinez and M.P. Gimenez, Hum. Toxicol., 2 (1983) 391.
210 C.L. Arthur and J. Pawliszyn, Anal. Chem, 62 (1990) 2145.
211 C. Dietz, J. Sanz and C. Camara, J. Chromatogr. A, 1103 (2006) 183.
212 G. Theodoridis and G.J. de Jong, Adv. Chromatogr., 43 (2005) 231.
213 G. Vas and K. Vekey, J. Mass Spectrom., 39 (2004) 233.
214 I. Koide, A. Yokoyama, O. Noguchi, K. Okada and H. Oda, Jap. J. Forensic Toxicol., 14 (1996) 142.

215   N. Nagasawa, M. Yashiki, Y. Iwasaki, K. Hara and T. Kojima, Forensic Sci. Int., 78 (1996) 95.

216   C. Battu, P. Marquet, A.L. Fauconnet, E. Lacassie and G. Lachatre, J. Chromatogr. Sci., 36 (1998) 1.

217   F. Centini, A. Masti and I. Barni Comparini, Forensic Sci. Int., 83 (1996) 161.

218   M. Yashiki, T. Kojima, T. Miyazaki, N. Nagasawa, Y. Iwasaki and K. Hara, Forensic Sci. Int., 76 (1995) 169.

219   A. Ishii, H. Seno, T. Kumazawa, M. Nishikawa, K. Watanabe, H. Hattori and O. Suzuki, Jap. J. Forensic Toxicol., 14 (1996) 228.

220   K. Okajima, A. Namera, M. Yashiki, I. Tsukue and T. Kojima, Forensic Sci. Int., 116 (2001) 15.

221   C. Jurado, M.P. Gimenez, T. Soriano, M. Menendez and M. Repetto, J. Anal. Toxicol., 24 (2000) 11.

222   E.H. Koster, C.H. Bruins and G.J. de Jong, Analyst, 127 (2002) 598.

223   M. Yonamine, N. Tawil, R.L. Moreau and O. Alves Silva, J. Chromatogr. B, 789 (2003) 73.

224   D.T. Vu, J. Forensic Sci., 46 (2001) 1014.

225   J. Caldwell, L.G. Dring and R.T. Williams, Biochem. J., 129 (1972) 11.

226   A.H. Beckett and M. Rowland, J. Pharm. Pharmacol., 17 (1965) 109S.

227   A.H. Beckett and M. Rowland, J. Pharm. Pharmacol., 17 (1965) 628.

228   A.H. Beckett and M. Rowland, Nature, 206 (1965) 1260.

229   A.H. Beckett and E.V.B. Shenoy, J. Pharm. Pharmacol., 25 (1973) 793.

230   A.H. Beckett and K. Haya, Xenobiotica, 8 (1978) 85.

231   A.H. Beckett, L.G. Brookes and E.V.B. Shenoy, J. Pharm. Pharmacol., 21 (1969) 151S.

232   A.H. Beckett, E.V.B. Shenoy and J.A. Salmon, J. Pharm. Pharmacol., 24 (1972) 194.

233   S.J. Tulloch, Y. Zhang, A. McLean and K.N. Wolf, Pharmacotherapy, 22 (2002) 1405.

234   S. George and R.A. Braithwaite, J. Anal. Toxicol., 24 (2000) 223.

235   A.H. Beckett, M. Rowland and P. Turner, Lancet, 1 (1965) 303.

236   T. B. Vree and P.T. Henderson. In: J. Caldwell (Ed.), Amphetamines and related stimulants: Chemical, biological, clinical, and sociological aspects, CRC Press, Boca Raton, FL, 1980, p. 47.

237   A.H. Beckett and M. Rowland, Nature, 204 (1964) 1203.

238   J.M. Davis, I.J. Kopin, L. Lemberger and J. Axelrod, Ann. NY Acad. Sci., 179 (1971) 493.

239   I. Nystrom, T. Trygg, P. Woxler, J. Ahlner and R. Kronstrand, J. Anal. Toxicol., 29 (2005) 682.

240   K.A. Moore and A. Poklis, J. Anal. Toxicol., 19 (1995) 549.

241   K.A. Moore, W.H. Soine and A. Poklis, J. Anal. Toxicol., 19 (1995) 542.

242   K.A. Moore, A. Ismaiel and A. Poklis, J. Anal. Toxicol., 20 (1996) 89.

243   J.T. Cody, in: R.H. Liu and B.A. Goldberger (Eds.), Handbook of workplace drug testing, AACC Press, Washington, DC, 1995, p. 239.

244   J.T. Cody, Forensic Sci. Rev., 5 (1993) 109.

245   J.T. Cody, J. Occup. Environ. Med., 44 (2002) 435.

246   G. Remberg, J. Marsel, G. Doring and G. Spiteller, Arch. Toxicol., 29 (1972) 153.

247   R.D. Budd and N.C. Jain, J. Anal. Toxicol., 2 (1978) 241.

248   Y. Nakahara, R. Kikura and K. Takahashi, Life Sci., 63 (1998) 883.

249   A.H. Beckett, G.T. Tucker and A.C. Moffat, J. Pharm. Pharmacol., 19 (1967) 273.

250   J. Marsel, G. Doring, G. Remberg and G. Spiteller, Zeitschrift Fur Rechtsmedizin – J. Legal Med., 70 (1972) 245.

251   J.P. Brooks, M. Phillips, D.T. Stafford and J.S. Bell, Am. J. Forensic Med. Pathol., 3 (1982) 245.

252   A. Fujinami, T. Miyazawa and Y. Kobayashi, Ann. Clin. Biochem., 35 (1998) 775.

253   A. Fujinami, T. Miyazawa, N. Tagawa and Y. Kobayashi, Biol. Pharm. Bull., 21 (1998) 1207.

254   M. Spatzenegger and W. Jaeger, Drug Metab. Rev., 27 (1995) 397.

255   T. Inoue and S. Suzuki, Xenobiotica, 16 (1986) 691.

256   T. Inoue, S. Suzuki and T. Niwaguchi, Xenobiotica, 13 (1983) 241.

257   T. Niwaguchi, T. Inoue and S. Suzuki, Xenobiotica, 12 (1982) 617.

258   M. Sato, T. Mitsui and H. Nagase, J. Chromatogr. B, Biomed. Sci. Appl., 751 (2001) 277.

259   M.V. Bach, R.T. Coutts and G.B. Baker, Xenobiotica, 30 (2000) 297.

260   B. Glasson, A. Benakis and M. Thomasset, Arzneimittel-Forschung, 21 (1971) 1985.

261   K.L. Baden, S. Valtier and J.T. Cody, J. Anal. Toxicol., 23 (1999) 511.

262   J.A. Tarver, J. Anal. Toxicol., 18 (1994) 183.

263   S. Valtier and J.T. Cody. In: V. Spiehler (Ed.), Proceedings of the 1998 joint SOFT/TIAFT international meeting, SOFT/TIAFT, Newport Beach, CA, 1999, p. 358.

264   S. Valtier and J.T. Cody, J. Anal. Toxicol., 24 (2000) 606.

265   G.P. Reynolds, P. Riederer, M. Sandler, K. Jellinger and D. Seemann, J. Neural Transm., 43 (1978) 271.

266   T.C. Kupiec and A.K. Chaturvedi, J. Forensic Sci., 44 (1999) 222.

267   W.P. Melega, A.K. Cho, D. Schmitz, R. Kuczenski and D.S. Segal, J. Pharmacol. Exp. Ther., 288 (1999) 752.

268   J. Lengyel, K. Magyar, I. Hollosi, T. Bartok, M. Bathori, H. Kalasz and S. Furst, J. Chromatogr. A, 762 (1997) 321.

269   H.J. Mascher, C. Kikuta, A. Millendorfer, H. Schiel and G. Ludwig, Int. J. Clin. Pharmacol. Ther., 35 (1997) 9.

270   M. Hasegawa, K. Matsubara, S. Fukushima, C. Maseda, T. Uezono and K. Kimura, Forensic Sci. Int., 101 (1999) 95.

271   G.P. Reynolds, J.D. Elsworth, K. Blau, M. Sandler, A.J. Lees and G.M. Stern, Br. J. Clin. Pharmacol., 6 (1978) 542.

272   F. Karoum, L.W. Chuang, T. Eisler, D.B. Calne, M.R. Liebowitz, F.M. Quitkin, D.F. Klein and R.J. Wyatt, Neurology, 32 (1982) 503.

273   M.L.J. Reimer, O.A. Mamer, A.P. Zavitsanos, A.W. Siddiqui and D. Dadgar, Biol. Mass Spectrom., 22 (1993) 235.

274   S.R. Philips, J. Pharm. Pharmacol., 33 (1981) 739.

275   J.D. Elsworth, M. Sandler, A.J. Lees, C. Ward and G.M. Stern, J. Neural Transm., 54 (1982) 105.

276   E.H. Heinonen, V. Myllyla, K. Sotaniemi, R. Lamintausta, J.S. Salonen, M. Anttila, M. Savijarvi, M. Kotila and U.K. Rinne, Acta Neurol. Scand. – Suppl., 126 (1989) 93.

277   J.E. Meeker and P.C. Reynolds, J. Anal. Toxicol., 14 (1990) 330.

278   Z. Tarjanyi, H. Kalasz, G. Szebeni, I. Hollosi, M. Bathori and S. Furst, J. Pharm. Biomed. Anal., 17 (1998) 725.

279   K. Nishida, S. Itoh, N. Inoue, K. Kudo and N. Ikeda, J. Anal. Toxicol., 30 (2006) 232.

280   R. Kronstrand, J. Ahlner, N. Dizdar and G. Larson, J. Anal. Toxicol., 27 (2003) 135.

281   M. Katagi, M. Tatsuno, H. Tsutsumi, A. Miki, T. Kamata, H. Nishioka, K. Nakajima, M. Nishikawa and H. Tsuchihashi, Xenobiotica, 32 (2002) 823.

282   K.T. Kivisto, J.S. Wang, J.T. Backman, L. Nyman, P. Taavitsainen, M. Anttila and P.J. Neuvonen, Eur. J. Clin. Pharmacol., 57 (2001) 37.

283   R. Kronstrand, M.C. Andersson, J. Ahlner and G. Larson, J. Anal. Toxicol., 25 (2001) 594.

284   A. Clarke, F. Brewer, E.S. Johnson, N. Mallard, F. Hartig, S. Taylor and T.H. Corn, J. Neural Transm., 110 (2003) 1241.

285   S.W. Myung, H.K. Min, S. Kim, M. Kim, J.B. Cho and T.J. Kim, J. Chromatogr. B, Biomed. Appl., 716 (1998) 359.

286   K. Takahashi, A. Ishigami, M. Shimamine, M. Uchiyama, T. Ochiai, K. Sekita, Y. Kawasaki, T. Furuya and M. Tobe, Eisei Shikenjo Hokoku – Bulletin of National Institute of Hygienic Sciences, 105 (1987) 1.

287   H. Blume, Arzneimittel-Forschung, 31 (1981) 805.

288   T. Inoue and S. Suzuki, Xenobiotica, 17 (1987) 965.

289   J.M. Witkin, G.A. Ricaurte and J.L. Katz, J. Pharmacol. Exp. Ther., 253 (1990) 466.

290   D.M. Shakleya, J.C. Kraner, J.A. Kaplan, P.M. Gannett and P.S. Callery, Forensic Sci. Int., 157 (2006) 87.

291   M. Sato, M. Hida and H. Nagase, Forensic Sci. Int., 128 (2002) 146.

292   M. Sato, M. Hida and H. Nagase, J. Anal. Toxicol., 25 (2001) 304.

293   M. Katagi, M. Tatsuno, A. Miki, M. Nishikawa and H. Tsuchihashi, J. Anal. Toxicol., 24 (2000) 354.

294   R. Kikura, Y. Nakahara and S. Kojima, J. Chromatogr. B, Biomed. Sci. Appl., 741 (2000) 163.

295   T. Nagai, H. Kanaya, K. Matsushima and S. Kamiyama, J. Anal. Toxicol., 21 (1997) 112.

296   K. Matsushima, T. Nagai, H. Kanaya, Y. Kato, M. Takahashi and S. Kamiyama, Nippon Hoigaku Zasshi – Japanese J. Legal Med., 52 (1998) 19.

297   A.H. Beckett and K. Haya, J. Pharm. Pharmacol., 29 (1977) 89.
298   F.T. Delbeke and M. Debackere, Arzneimittel-Forschung, 36 (1986) 1413.
299   R.T. Williams, J. Caldwell and L.G. Dring, in: E. Usdin and S.H. Snyder (Eds.), Frontiers in catecholamine research, Pergamon Press, New York, 1973, p. 927.
300   Y. Makino, T. Higuchi, S. Ohta and M. Hirobe, Forensic Sci. Int., 41 (1989) 83.
301   J.T. Cody, Forensic Sci. Int., 80 (1996) 189.
302   F. Musshoff and T. Kraemer, Int. J. Legal Med., 111 (1998) 305.
303   M. Neugebauer, A. Khedr, N. El-Rabbat, M. El-Kommos and G. Saleh, Biomed. Chromatogr., 11 (1997) 356.
304   H.S. Shin, Chirality, 9 (1997) 52.
305   H.S. Shin, B.B. Park, S.N. Choi, J.J. Oh, C.P. Hong and H. Ryu, J. Anal. Toxicol., 22 (1998) 55.
306   Y. Yoo, H. Chung and H. Choi, J. Anal. Toxicol., 18 (1994) 265.
307   R. Mrongovius, M. Neugebauer and G. Rucker, Eur. J. Med. Chem., 19 (1984) 161.
308   M. Neugebauer, J. Pharm. Biomed Anal., 2 (1984) 53.
309   E.S. Oh, S.K. Hong and G.I. Kang, Xenobiotica, 22 (1992) 377.
310   A.T. Rodriguez, S. Valtier and J.T. Cody, J. Anal. Toxicol., 28 (2004) 432.
311   B. Greenhill, S. Valtier and J.T. Cody, J. Anal. Toxicol., 27 (2003) 479.
312   B. Greenhill, Metabolic profile of the drug Famprofazone, Masters Thesis, Graduate College in Biomedical Sciences (Clinical Laboratory Sciences), University of Texas Health Sciences Center, San Antonio, 2002.
313   J. Mallol, L. Pitarch, R. Coronas and A. Pons Jr., Arzneimittel-Forschung, 24 (1974) 1301.
314   S. Goenechea and H. Brzezinka, Arch. Kriminol., 173 (1984) 97.
315   R. Kikura and Y. Nakahara, J. Anal. Toxicol., 21 (1997) 291.
316   R. Iffland, Arch. Kriminol., 169 (1982) 81.
317   B. Nickel, G. Niebch, G. Peter, A. von Schlichtegroll and U. Tibes, Drug Alcohol Depend., 17 (1986) 235.
318   G. Kristen, A. Schaefer and A. von Schlichtegroll, Drug Alcohol Depend., 17 (1986) 259.
319   T. Ellison, L. Levy, J. Bolger and R. Okun, Eur. J. Pharmacol., 13 (1970) 123.
320   A.J. Nazarali, G.B. Baker, R.T. Coutts and F.M. Pasutto, Prog. Neuropsychopharmacol. Biol. Psychiatry, 7 (1983) 813.
321   M.J. Berry, R.H. Poyser and M.I. Robertson, J. Pharm. Pharmacol., 23 (1971) 140.
322   G. Tognoni, P.L. Morselli and S. Garattini, Eur. J. Pharmacol., 20 (1972) 125.
323   R.B. Sznelwar, Eur. J. Toxicol. Environ. Hyg., 8 (1975) 5.
324   T. Kraemer, T. Pflugmann, M. Bossmann, N.M. Kneller, F.T. Peters, L.D. Paul, D. Springer, R.F. Staack and H.H. Maurer, Biochem. Pharmacol., 68 (2004) 947.
325   R.R. Bell, S.B. Crookham, W.A. Dunn, K.M. Grates and T.M. Reiber, J. Anal. Toxicol., 25 (2001) 652.
326   T. Kraemer, G.A. Theis, A.A. Weber and H.H. Maurer, J. Chromatogr. B, Biomed. Sci. Appl., 738 (2000) 107.
327   T. Inoue, T. Yasuda, S. Suzuki, T. Kishi and T. Niwaguchi, Xenobiotica, 16 (2007) 109.
328   J.R. Boissier, J. Hirtz, C. Dumont and A. Gerardin, Ann. Pharm. Fr., 26 (1968) 215.
329   J. Engel, G. Kristen, A. Schaefer and A. von Schlichtegroll, Drug Alcohol Depend., 17 (1986) 229.
330   T. Kraemer, I. Vernaleken and H.H. Maurer, J. Chromatogr. B, Biomed. Appl., 702 (1997) 93.
331   J.E. Blum, Arzneimittel-Forschung, 19 (1969) 748.
332   M. Polgar, L. Vereczkey, L. Szporny, G. Czira, J. Tamas, E. Gacs-Baitz and S. Holly, Xenobiotica, 9 (1979) 511.
333   H. Pyo, S.-J. Park, J. Park, J.K. Yoo and B. Yoon, J. Chromatogr. B, Biomed. Appl., 687 (1996) 261.
334   S.B. Seredenin and I.V. Rybina, Farmakol. Toksikol., 48 (1985) 79.
335   A.V. Shpak, S.A. Appolonova and V.A. Semenov, J. Chromatogr. Sci., 43 (2005) 11.
336   S.A. Appolonova, A.V. Shpak and V.A. Semenov, J. Chromatogr. B, 800 (2004) 281.
337   D. Palm, F. Hansjoachim and H. Grobecker, Life Sci., 8 (1969) 247.

338 H.J. Dengler, M. Eichelbaum and M. Schomerus. In: U.S. von Euler, C. Bartorelli, A. Berreta, H.J. Dengler and A. Giotti (Eds.), Clinico-pharmacological and therapeutical aspects of prenylamine, Casa Editrice, Milan, 1970, p. 26.

339 Y. Gietl, H. Spahn, H. Knauf and E. Mutschler, Eur. J. Clin. Pharmacol., 38 (1990) 587.

340 Y. Gietl, H. Spahn and E. Mutschler, J. Chromatogr., 426 (1988) 304.

341 Y. Gietl, H. Spahn and E. Mutschler, Arzneimittel-Forschung, 39 (1989) 853.

342 W.D. Paar, D. Brockmeier, M. Hirzebruch, E.K. Schmidt, G.E. von Unruh and H.J. Dengler, Arzneimittel-Forschung, 40 (1990) 657.

343 G. Remberg, M. Eichelbaum, G. Spiteller and H.J. Dengler, Biomed. Mass Spectrom., 4 (1977) 297.

344 T. Kraemer, S.K. Roditis, F.T. Peters and H.H. Maurer, J. Anal. Toxicol., 27 (2003) 68.

345 M.H. Slawson, J.L. Taccogno, R.L. Foltz and D.E. Moody, J. Anal. Toxicol., 26 (2002) 430.

346 H.S. Shin, J.S. Park, P.B. Park and S.J. Yun, J. Chromatogr. B, Biomed. Appl., 661 (1994) 255.

347 H.-S. Shin, B.-B. Park, S.N. Choi, J.J. Oh, C.P. Hong and H. Ryu, J. Anal. Toxicol., 22 (1998) 55.

348 H.-S. Shin and J. Park, Korean Biochem. J., 26 (1993) 741.

349 S. Valtier and J.T. Cody, J. Anal. Toxicol., 21 (1997) 84.

350 J. Park, S. Park, D. Lho and B. Chung, in: J. Yinon, (Ed.), Forensic applications of mass spectrometry, CRC Press, Boca Raton, 1995, p. 95.

351 Y. Nakahara and R. Kikura, Archives of Toxicology, 70 (1996) 841.

352 R. Kikura and Y. Nakahara, Biol. Pharm. Bull., 18 (1995) 1694.

M.J. Bogusz (Ed.). Forensic Science
Handbook of Analytical Separations, Vol. 6
© 2008 Elsevier B.V. All rights reserved.

CHAPTER 4

Hallucinogens [☆]

John T. Cody

Air Force Drug Testing Laboratory, 2730 Louis Bauer Drive, Brooks City-Base, TX 78235-5132, USA

4.1 INTRODUCTION

Hallucinogen is a term used to describe compounds that alter a person's perception of reality. Typically, a hallucinogen causes the user to have a heightened state of awareness of sensory input (audio, visual, etc.) and diminished control over the experience. Sometimes, the sensation seems to allow the user to be both the participant and observer. Hallucinogens are divided into several categories. One method of characterization of hallucinogens is as nitrogen containing or non-nitrogen containing. Other classification schemes classify them as indolylalkylamines and phenylalkylamines. Regardless of the classification scheme, most of the compounds are plant alkaloids or derivatives.

Some drugs classified as hallucinogens, such as marijuana, are described elsewhere in this book and are not discussed in this chapter. In addition, a variety of amphetamine-related compounds have some hallucinogenic properties. With the exception of the methylenedioxy compounds, most of these drugs are no longer commonly used. Due to the chemical similarity of these compounds to amphetamine and methamphetamine, procedures for the analysis of amphetamines are also viable for these substituted amphetamines with only minor modifications (extending run times, monitoring appropriate mass ranges for mass spectrometry detection, etc.). Therefore, specific procedures for the analysis of these compounds are not described in this chapter.

Several of the above compounds are not considered by some to fit the classical definition of hallucinogen. These include the methylenedioxy analogues of amphetamine and methamphetamine. The effects of these compounds differ from that of a classic hallucinogen and are often classified as entactogens. The hallucinogenic

---

[☆] The views expressed in this article are those of the author and do not reflect the official policy of the Department of Defense or other Departments of the US Government.

References pp. 199–201

properties of these drugs also differ dramatically depending on the enantiomer. Only one enantiomer of MDMA, for example, has hallucinogenic properties while the other enantiomer does not. This fine distinction is lost, however, since on the illicit market these drugs are racemic and therefore exhibit properties from both enantiomers.

While there are many different drugs that could be described, this chapter will focus on the most common examples including lysergic acid diethylamide (LSD), mescaline, psilocybin and phencyclidine (PCP). LSD, mescaline and psilocybin are chemically different structures (see Figs. 4.1–4.4), yet they have similar pharmacological activities. Mescaline, a phenylethylamine derivative, and psilocybin, an indolethylamine derivative, are found in nature. LSD is a synthetic compound but is closely related to naturally occurring ergot alkaloids. These three drugs resemble the neurotransmitters norepinephrine, dopamine and serotonin. PCP will also be described in detail as a part of this chapter. It is abused by itself or in conjunction with other drugs, such as being smoked with marijuana. The continued use of LSD and the difficulty encountered in its routine analysis make it an important topic for discussion. Multiple procedures are available for the analysis of this drug and several of its metabolites. In an extensive review of analytical procedures Moeller and Kraemer [1] described the analysis of samples using a variety of techniques focusing primarily on driving under the influence. The review describes the analysis of

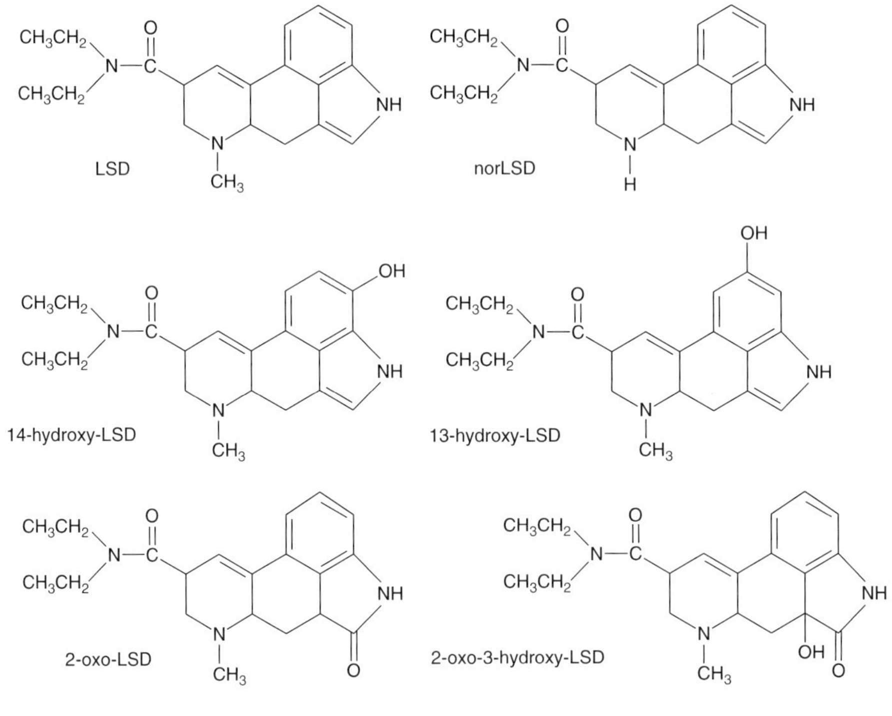

Fig. 4.1. LSD and metabolites.

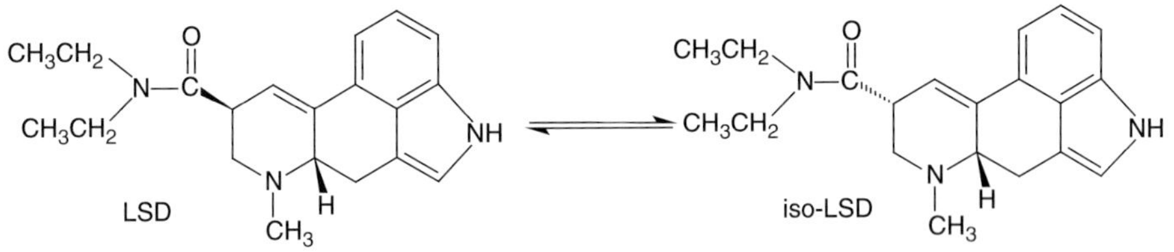

Fig. 4.2. LSD and iso-LSD.

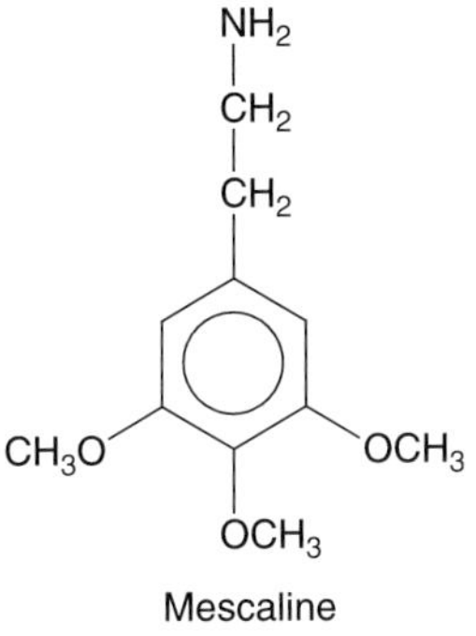

Mescaline

Fig. 4.3. Mescaline.

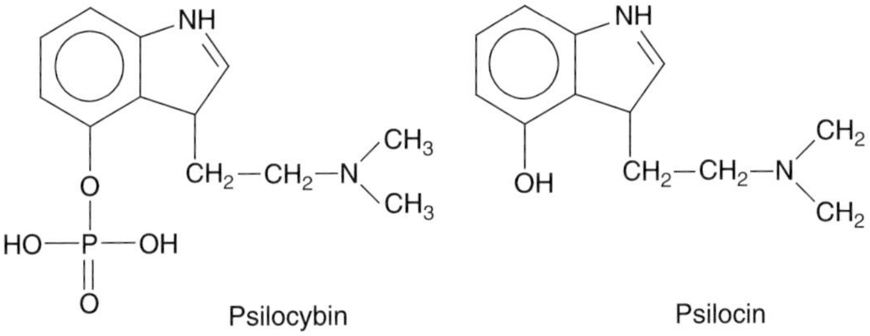

Psilocybin   Psilocin

Fig. 4.4. Psilocybin and psilocin.

numerous drugs including LSD, PCP, psilocybin and psilocin using gas chromatography-mass spectrometry (GC-MS) and liquid chromatography-mass spectrometry (LC-MS) procedures available in the literature.

4.2 LYSERGIC ACID DIETHYLAMIDE (LSD)

LSD is an indolalkylamine whose pharmacologic activity was discovered when its inventor, Albert Hoffman, inadvertently ingested some of the material. LSD is a chemically synthesized compound derived from closely related naturally occurring alkaloids. Lysergic acid is present in a fungus *Claviceps purpurea* that grows on grain plants. Another related naturally occurring compound is lysergic acid amide. This compound is found in the seeds of morning glory plants and the Hawaiian baby wood rose. Recently, there has been an increase in discussion regarding the potential for valid use of drugs such as LSD in medical treatment, a subject generally taboo until recently [2].

Most commonly, LSD is found in the illicit market on blotter paper. It is also found on sugar cubes and as a liquid among other forms. Methods for the analysis of LSD in these forms have included the same analytical tools used for the analysis of LSD from biological matrices.

LSD is a colorless, odorless, tasteless liquid. It is a very potent drug and typically taken in low doses. As a result of the high potency and low doses, detection of LSD in biological matrices poses a significant analytical challenge. Complicating the analytical picture even further is the fact that less than 3% of the LSD ingested is excreted in the urine intact [3]. As a result, LSD is found in urine following use for only a short period of time, generally less than 24 h when using a cutoff level of 200 pg/mL and perhaps as much as several days using detection limits currently available with some analytical procedures. Studies of the metabolism of LSD have identified a number of different metabolites, most of which are also found in low concentrations for short periods of time. One metabolite, 2-oxo-3-hydroxy-LSD, has been shown in several studies to be found in higher concentration than LSD itself in most, but not all, samples analyzed [4–6] (see Fig. 4.1 for the chemical structure of LSD and several of its important metabolites).

4.2.1 Extraction

Extraction of LSD from blotter paper, the most common form found in the illicit market, is a relatively simple process. Commonly, 100 µg of LSD salt are found on examination of blotter paper. Virtually all illicit preparations contain LSD and its isomer iso-LSD (See Fig. 4.2). The proportion of LSD to iso-LSD allows for the determination of source of the drug since the amount of each of these drugs can vary dramatically between batches (10–70% iso-LSD) depending on the source (illicit laboratory). Iso-LSD has no pharmacological activity, therefore sources with high proportions of iso-LSD have less pharmacological activity than those with proportionately higher LSD. In addition to blotter paper, LSD is also sold on gelatin cubes, sugar cubes, microdots (small tablets) and in liquid form in small dropper bottles. Isolation and identification of LSD from sugar cubes and the liquid form has been described by Kilmer [7] using GC-MS and high-performance liquid chromatography (HPLC). Veress described an extensive evaluation of parameters involved with the isolation of LSD from blotter paper. Solvent, time, temperature and extraction method's effects were characterized and analysis using HPLC with UV detection was evaluated [8].

Extraction of LSD from biological matrices is a far more challenging task due to the inherent low concentrations of the drug. Even with the sensitivity associated with selection ion monitoring MS, analysis of LSD is difficult and commonly accomplished using elaborate extraction schemes. Extraction of the drug from blood, plasma and tissue homogenates requires some different handling than that used for urine in order to afford the elimination of cells and proteins is important in these cases. Numerous approaches have been used for this problem over the years, and any of the currently available methods can be employed successfully with LSD. Plasma often only requires dilution with a buffer to prevent plugging of solid-phase extraction cartridges. Methods of cold precipitation, salting out of proteins and addition of solvents such as acetonitrile have also been used as a preliminary step in the extraction of drugs, including LSD, from these sample matrices. Once this is accomplished, the extracts are typically handled the same as urine.

The concentrations found in routine analysis of LSD pose a challenge for benchtop GC-MS systems. As a result, extraction becomes more critical to ensure high recovery along with a very high level of selectivity to the analyte. One method for the extraction of LSD from urine involves a combination of liquid- and solid-phase extractions [9]. Urine samples were made basic with ammonium hydroxide, saturated with sodium chloride and extracted into 1-chlorobutane. Following evaporation, the extract was dissolved in isooctane:methylene chloride:triethylamine (50:50:0.1) and poured through solid-phase extraction columns and eluted with methanol:methylene chloride:triethylamine (0.2:10:0.01). The extract from the column was then dried and reconstituted in 1-chlorobutane and the alkaloids extracted with 3 mL of phosphate buffer. This extract was then washed with 1-chlorobutane followed by making the solution basic with ammonium hydroxide and saturating with sodium chloride. The drugs of interest were then extracted using 1-chlorobutane. This extract was evaporated then reconstituted in ethanol containing triethylamine. The authors felt triethylamine helped to recover the LSD from the glass tube by displacing the LSD that had bound to glass surfaces. The extract was then derivatized with BSTFA and analyzed by GC-MS. This extraction procedure is extensive and laborious. It was the author's belief that to reliably conduct routine analysis of LSD using standard benchtop GC-MS systems required this degree of clean-up to ensure the drugs were efficiently recovered (69%) and potential interferences eliminated from the sample extract.

This method also included an additional step to convert iso-LSD, a common contaminant of illicit LSD, into LSD using ethanolic sodium hydroxide and heat (50°C for 10 min). This process converted iso-LSD into LSD with approximately 98% efficiency, thus effectively combining both of the compounds into one single peak. This allowed assessment of the total amount of LSD, in either isomeric form, in the sample since neither of these drugs is naturally occurring nor have any legitimate source. Although at its surface a reasonable procedure, it has not been adopted by the forensic community.

One procedure for the analysis of LSD in serum involved liquid–liquid extraction. After making the sample basic, 1-chlorobutane was used to extract the LSD. The solvent was then transferred to a silanized glass vial and evaporated under nitrogen. The dried extract was dissolved in methanol then injected into GC using flame ionization detection. Some samples showed interference when using this simple extraction procedure. In such cases, the extraction scheme was expanded to further clean up the extract. This was accomplished by taking the dried extract and redissolving in phosphate buffer and washing with 1-chlorobutane and cyclohexane (1:1). The LSD was then extracted from the buffer using 1-chlorobutane. Following extraction, the extract was derivatized with MSTFA with pyridine and analyzed by electron ionization GC-MS. The extraction efficiency of this method was reported to be 76% for the single-step extraction and 66% for the more extensive procedure. This method was reported to be linear from 100 pg/mL to 10 ng/mL [10].

An automated procedure for the extraction of LSD and nor-LSD from blood, serum, plasma and urine has also been described [11]. This method resulted in recovery of LSD at 95% or above at concentrations of 0.1–5 ng/mL. Within-run precision of the assay was less than 3% and between-run was less than 10%. The reproducibility of nor-LSD by this method was not as good as for LSD and the method was therefore only used for qualitative identification of nor-LSD. Detection limits using liquid chromatography-mass spectrometry-mass spectrometry (LC-MS-MS) following this extraction procedure were 25 pg/mL for both LSD and nor-LSD and quantitative limits were 50 pg/mL for both drugs.

The extraction of LSD and nor-LSD from hair has been reported by Nakahara *et al.* [12]. The hair was first washed using sodium dodecyl sulfate (SDS) and dried in a desiccator. A solution (2 mL) of methanol:5 M HCl (20:1) was added to samples (20 mg) of hair that were placed into an ultrasonicator for 1 h then stored at room temperature for 14 h. The samples were filtered, neutralized with ammonium hydroxide and evaporated. Purification was accomplished by extracting the drug from 0.1 M sodium hydroxide using dichloromethane. The extract was then derivatized and analyzed by GC-MS using both deuterated LSD and lysergic acid methyl propylamide (LAMPA) as internal standards.

Immunoaffinity extraction of compounds found in very low concentrations in biological matrices is an appealing method for several reasons. The specificity of the antibody to the compound of interest allows for isolation of only those compounds that bind to the antibody. The extraction specificity is thus directly related to the specificity of the antibody to the drug of interest. Unlike other methods that extract compounds that share similar physical and chemical properties, this process will extract only those compounds that bind to the antibody. Several reports of using immunoaffinity extraction have been published. Francis and Craston [13] developed a method for immunoaffinity extraction of LSD. Analysis of the extract by HPLC and LC-MS using electrospray ionization gave detection limits of approximately 500 pg/mL. Although a viable technique, the detection limit for this method was higher than would be considered desirable in most cases. (The previous US Department of Defense cutoff for LSD was 200 pg/mL which required a control at not more than 100 pg/mL.) A similar immunoaffinity extraction procedure has been described [14] that employed electrospray ionization LC-MS. The detection limit reported for that method was also 500 pg/mL. Immunoaffinity extraction was also described by Henion and coworkers [15,16]. The process identified LSD and a number of metabolites in a unique on-line extraction process. This method offered a significant advantage since the extraction was on-line but the distinct disadvantage of the antibody not being covalently bonded to the support material. Therefore, when the drug was eluted from the column, so too was the antibody necessitating generation of a new column to analyze each sample. This method was useful for the isolation of the drug of interest and several metabolites from a relatively large volume of urine. It also allowed isolation of metabolites from the samples in sufficient amounts to allow characterization of the metabolic products. An immunochemical method for the isolation of LSD has also been reported by Kerrigan and Brooks [17]. Recoveries of the LSD were reported to be greater than 80%. The method worked well with both blood and urine samples. Use of this method with blood samples gave excellent results. No special preparative procedures were required with blood samples, they were simply diluted with buffer and applied to the resin.

A commercially available immunoaffinity resin is available that has been used for the analysis of LSD in urine samples (LSD ImmunElute, Microgenics, Pleasanton, CA, USA). This allows for the analysis of LSD and several metabolites without requiring investigators to prepare their own immunoaffinity resin for the purpose of extraction.

As seen from the examples above, the extraction of LSD from biological matrices is a complex process and in many ways dependent on the analytical method being used.

4.2.2 Derivatization

Derivatization of LSD is commonly used to enhance detectability when analyzed with gas chromatography (GC) techniques. A variety of different derivatives have been used with this drug and some of its metabolites. Depending on the detector system, the derivative selected can have an important influence. Negative ion chemical ionization methods for the detection of LSD require the use of a derivative with electronegative components to enhance the formation of negative ions. Lim *et al.* [3] used trifluoroimidazole-1,4-dimethylpiperazine to derivatize LSD and used resonance capture negative ion chemical ionization with methane as the reagent gas to identify LSD from urine. This method was later modified to allow detection of the drug from plasma [18]. Low detection limits were easily seen using this procedure in large part due to the conservation of virtually all of the ion current in the molecular anion. The method has a high degree of specificity due to the selective nature of the negative ionization technique. Some individuals, however, prefer to see multiple ions that can be compared with a standard to assist in the positive identification of the drug.

Use of trimethylsilation is a common technique for electron ionization and positive ion chemical ionization. GC analysis of LSD has traditionally used derivatization because of absorptive losses of the drug in the instrument. These losses are lessened by derivatization of the drug. The instrument (injection port and column) is also commonly treated by injection of derivatizing reagent prior to injection of the drug to avoid potential active sites in the injection port, and GC column, from binding the LSD. Another recently described technique to address this issue is the use of on-column injection to avoid the problems associated with the injection port and the losses encountered there. In this way, it was possible to analyze LSD underivatized on GC.

Derivatization of LSD is not required when using liquid chromatography (LC) techniques. The drug and its metabolites are readily separated and show acceptable peak shapes. The absorptive problems seen with GC do not cause comparable problems for LC procedures. LSD exhibits natural fluorescence, therefore fluorescence detectors can be used with acceptable results without the requirement for derivatization. MS analysis with LC is also used effectively particularly when using mass spectrometry-mass spectrometry (MS-MS) procedures.

4.2.3 Gas chromatography

GC-MS has been used for the analysis for LSD for several years with reasonable success. The low levels of the drug found in samples makes the analysis, even with the sensitivity of selected ion monitoring (SIM), a difficult process. Increased specificity and sensitivity have been achieved by use of MS-MS techniques. While new procedures continue to be developed for GC, the majority utilize other separation techniques.

Among recent publications, virtually none describe GC-MS procedures with the exception of a procedure developed to determine the presence and concentration of the 2-oxo-3-hydroxy-LSD [19]. The limit of detection was 0.5 ng/mL and limit of quantitation of 1.0 ng/mL.

Poch *et al.* [20] described the isolation and identification of LSD, iso-LSD, nor-LSD and 2-oxo-3-hydroxy-LSD using a simple liquid–liquid extraction of urine using methylene chloride:isopropanol (95:5). This extract was dried, reconstituted in phosphate buffer and further purified by solid-phase extraction (SPE). Analysis was accomplished by LC-MS-MS using a Finnigan LCQ ion-trap mass spectrometer. Identification of nor-LSD was accomplished by trapping the ion at m/z 310 and dissociating to product ions at m/z 237, 209 and 183 using m/z 237 as the quantitative ion. 2-oxo-3-hydroxy-LSD was identified by isolating the ion at m/z 356 and dissociating to product ions at m/z 338, 265 and 237 using m/z 338 for quantitation. Verstraete and Van de Velde described a method for the analysis of LSD and 2-oxo-3-hydroxy-LSD by electron ionization GC-MS. Their procedure used solid-phase extraction following a procedure recommended for amphetamines followed by derivatization with BSTFA. 2-Oxo-3-hydroxy-LSD was monitored using ions at m/z 499 and 309. This procedure yielded only the di-derivatized 2-oxo-3-hydroxy-LSD when analyzed using traditional injection methods. The authors did observe the mono-derivatized compound when using an on-column injection method. Another procedure reported for this analysis used positive ion chemical ionization gas chromatography-mass spectrometry-mass spectrometry (GC-MS-MS) [4,21]. This method used solid-phase extraction of 4 mL of urine followed by derivatization with BSTFA and 1% trimethylchlorosilane. The authors reported finding both the mono- and di-derivatized 2-oxo-3-hydroxy-LSD. Both derivatives eluted at nearly the same retention time on a DB-5 column but were readily separated by MS-MS analysis by selecting their respective precursor ions. LSD was monitored by following transition of the protonated molecular ion at m/z 396 to the product ion at m/z 295. 2-oxo-3-hydroxy-LSD was determined by monitoring the transition of the precursor ion at m/z 500 to the product ion at m/z 309. This method gave detection limits for LSD and 2-oxo-3-hydroxy-LSD of 10 pg/mL.

Libong *et al.* [22] described a positive chemical ionization GC ion-trap tandem MS method for the detection and quantitation of LSD in whole blood. Two milliliters of whole blood were extracted using SPE. Eight microliters of this extract were then injected using a cold on-column injection technique. The procedure was linear from 0.02–10.0 ng/mL using LAMPA as the internal standard.

Another method for the analysis of LSD in urine utilized ion-trap GC-MS-MS analysis [23]. This procedure allowed the detection of LSD from 5 mL of urine using solid-phase extraction (Clean Screen, Worldwide Monitoring, Bristol, PA, USA). The underivatized extract was injected into a temperature controlled, liquid carbon dioxide-cooled injector on a Varian GC (Varian, Sugarland TX, USA). The injector was held at 85°C for 0.2 min then rapidly raised to 300°C at 180°C/min. The assay used the molecular ion at m/z 323 as the precursor and monitored product ions at m/z 280, 222 and 196 for LSD and the internal standard LAMPA. Quantitation was based on peak areas of m/z 222 for the drug and internal standard. Between-run precision of the assay ranged from 11 to 12% at concentrations ranging from 150 to 450 pg/mL. Within-run precision (at 400 pg/mL) was determined to have a relative standard deviation of approximately 5%.

4.2.4 Liquid chromatography

Using liquid chromatography as the separation method for LSD analysis has been used for many years. The natural fluorescence of LSD allows for reasonably low detection limits. Development of atmospheric pressure ionization techniques have opened a significant opportunity to use this well-established chromatographic technique with the analytical power of MS. Multi-stage MS and hybrid mass spectrometers have given rise to a number of valuable techniques for the analysis of LSD and related compounds.

Some of the first analytical methods utilizing LC coupled with MS for analysis of LSD were described by Henion and coworkers [15,16,24,25]. Forensic analysis of routine samples has been reported by a number of investigators [14,26]. Several different compounds have been used as internal standards for quantitative analysis of LSD including methylsergide, LAMPA and a deuterated isotopomer of LSD. Methylsergide was shown to yield less precise results than use of the deuterated isotopomer of LSD [27]. In experiments evaluating reproducibility of quantitative results between 1.0 and 2.5 ng/mL of LSD, the deuterated internal standard method consistently gave more precise results. This method was linear from 500 pg/mL to 10 ng/mL. Although the lower limit is reasonably high for routine use, it suited the requirements of the author's laboratory. They reported the ability to detect as low as 100 pg/mL using the method if required.

Generally, stable isotopomers are superior to closely chemically related compounds used for internal standards in quantitative analysis. In the case of the commercially available deuterated LSD, the deuterium is located on a portion of the molecule that is lost in one of the major fragmentations of the molecule. As a result there was a problem using the compound as internal standard since the drug itself and the deuterated internal standard gave rise to a common fragment. Since the compounds overlap chromatographically, use of the deuterated internal standard for monitoring of multiple ions, as commonly done in the forensic analysis using SIM, was not a viable option. Multi-stage MS avoids this limitation by its ability to isolate the ion for each individually even though they co-elute. This allows for the use of the ions without the limitation seen with single-stage mass analysis.

Another significant limitation of many analytical procedures that describe the analysis of LSD is that their linear range is too high to be of practical use in the analysis of routine use of LSD. Because the dose is commonly so low, low pg/mL concentrations of the drug are all that exist within less than a day after use of the drug. As a result, procedures that describe lower limits of 1 ng/mL or higher are of little practical use for biological samples except in a situation where the drug may be suspected in an overdose situation. The capability of some of the currently available instrumentation allows routine analysis of low concentrations of LSD in a variety of biological matrices. An LC-MS-MS method was developed for the forensic analysis of LSD, iso-LSD and 2-oxo-3-hydroxy-LSD. Following liquid–liquid extraction of the analytes and deuterated LSD (internal standard) from whole blood or urine with butyl acetate, positive electrospray ionization with a triple quadrupole mass spectrometer was used to identify and quantitate LSD and iso-LSD using multiple reaction monitoring. Two transitions and their ratio were monitored along with retention time for the identification of the analytes. This procedure was linear from 0.01 to 50 μg/kg for LSD and iso-LSD. Despite the tremendous analytical power of tandem mass spectrometry, this procedure demonstrated the necessity for chromatographic separation in order to properly identify the two isomers. Since the isomers have the same molecular weight, separation of the two using MS-MS is not possible. Although not a desirable or intended product of illicit synthesis of LSD, iso-LSD is commonly seen in illicit LSD samples, in many cases even exceeding the concentration of LSD itself. Although 2-oxo-3-hydroxy-LSD was determined by this method, no quantitative parameters were presented [28]. The quantitative determination of 2-oxo-3-hydroxy-LSD was described in a procedure using solid-phase extraction of urine followed by LC-MS analysis. The average extraction efficiency was reported to be 92% with limits of detection and quantitation at 250 pg/mL. The assay was linear from 250 to 30,000 pg/mL. Although the procedure does provide a viable method for analysis of 2-oxo-3-hydroxy-LSD, it does not include analysis of LSD itself in the same analytical run [29].

Two studies looked at the stability of LSD and/or metabolites under various conditions. No significant loss of 2-oxo-3-hydroxy-LSD was seen at normal pH values under refrigeration or frozen conditions for up to 9 and 60 days, respectively.

There was, however, substantial degradation at room (24°C) or elevated (50°C) temperatures. Exposure to fluorescent light did not cause degradation in this study [30]. Sunlight however was shown to cause significant degradation in a study by Skopp *et al.* [31] who observed a drop to only 6% of initial concentration in three days for 2-oxo-3-hydroxy-LSD and 3% for nor-LSD. Unlike Klette, these authors reported a temperature-dependent change in concentration of LSD, 2-oxo-3-hydroxy-LSD and nor-LSD. The compounds were stable at –20°C but not above that temperature although they reported LSD as stable at 4°C and 22°C for three days when protected from light.

Cui *et al.* [32] described the quantitation of LSD using atmospheric pressure matrix-assisted laser desorption/ionization ion-trap mass spectrometry following solid-phase extraction from urine samples. The authors demonstrated that, using their analytical technique, selected reaction monitoring gave superior results compared to SIM with regard to greater linear range and a lower limit of quantitation. In this case the lower limit was 1 ng/mL which, as previously noted, is relatively high for LSD analysis. The authors also reported their results to be comparable to electrospray ionization LC-MS.

Investigators evaluated the potential forensic application of LC-MS using a traditional quadrupole with an ion trap for the detection and quantitation of 2-oxo-3-hydroxy lysergic acid diethylamide [33]. The authors found both procedures to be linear from 0 to 8,000 pg/mL with correlation coefficients greater than 0.99. The limits of detection and quantitation were 400 pg/mL for both. A number ($n = 68$) of human urine samples previously identified as containing LSD were tested by both methods. These samples showed a mean concentration of 2-oxo-3-hydroxy-LSD nearly 16 times higher than the LSD concentration. Overall, both methods produced similar results which lead the authors to suggest either could be used in routine analysis in high-volume drug-testing laboratories.

Bergemann *et al.* [34] describe a method using HPLC with fluorescence detection. The method was able to detect LSD at a level of 50 pg/mL with acceptable precision. LSD was isolated from urine (2 mL) using SPE. Blood was first mixed with distilled water and centrifuged. Two milliliters of the supernatant was then mixed with buffer and the mixture applied to the same SPE cartridges used for analysis of urine samples. Quantitation was accomplished using methylsergide as the internal standard. The procedure had a detection limit of 1 ng/mL, standard deviations of 36 and 37% for within-run and between-run were seen at that concentration. At 10 ng/mL, the values were 9.8 and 13.5%, respectively, a more usable degree of variation. Another method using fluorescence detection used immunoaffinity extraction (LSD ImmunElute) for extraction from serum and hair samples. The limit of detection (S:N = 3) was reported to be < 50 pg for both matrices.

4.2.5 Thin layer chromatography

Thin layer chromatography (TLC) of LSD is generally not sensitive enough to use for identification of the drug in biological samples. A method using high-performance TLC [35] following a single-step alkaline extraction using ether:methylene chloride:isoamyl alcohol (70:30:0.5) accompanied with a wash using ammonium hydroxide allowed the detection of LSD at 0.4 ng/mL with a relative standard deviation at 1 ng/mL of less than 10%. Although this proved to be a viable method for the identification of LSD at high concentrations, the low levels of LSD typically encountered in biological matrices limits the utility of this method.

4.2.6 Capillary electrophoresis and other techniques

Cai and Henion [24] described a procedure for the analysis of LSD and metabolites by several analytical techniques, including capillary electrophoresis (CE). In this case, CE was coupled with MS-MS for LSD and a number of its metabolites. Frost and coworkers described a CE method for the analysis of LSD and nor-LSD using laser-induced fluorescence (LIF) which gave detection limits in the range 100–200 pg/mL which compares favorably with many of the GC and LC procedures currently in use [36,37]. The method involved extraction of LSD from blood followed by fluorescence detection at 435 nm using a HeCd laser with an excitation wavelength of 325 nm in a citrate–acetate buffer. The citrate-acetate buffer was used because of its ability to allow greater fluorescence than seen when using a phosphate buffer with the same samples.

In a description of the analysis of a large number of drugs, including LSD by CE with diode array detection Lurie *et al.* [38] detailed both qualitative and quantitative analyses of seized materials. The technique involves the use of run buffers that contain additives that provide for secondary equilibrium and/or dynamic coating of the capillary. Dynamic coating of the capillary surface was accomplished by rapid flushes of 0.1 M sodium hydroxide, water, buffer containing polycation coating reagent and a buffer containing a polyanionic coating reagent or a micelle coating reagent. Polyanionic coating reagents are used for analysis of moderately basic compounds.

On-line concentration of LSD and iso-LSD in mouse blood was described in a study using capillary electrophoresis/fluorescence spectroscopy with sodium dodecyl sulfate as the surfactant [39]. The procedure used only a 50 μL sample of blood, thus leading to the need for on-line sample concentration. Sweeping micellar electrokinetic chromatography and cation-selective exhaustive injection-sweep-micellar electrokinetic chromatography were used to evaluate optimum conditions. Using a mixture of acetonitrile–methanol–water solution (5:35:60 v/v/v) containing phosphate (50 mM), SDS (100 mM) and Brij-30 (3 mM) provided optimal performance. In the sweeping micellar electrokinetic chromatography mode, for a 65 cm capillary (effective length, 60 cm) the optimum injection length is approximately 27 cm, whereas in cation-selective exhaustive injection-sweep-micellar electrokinetic chromatography mode, the optimum electrokinetic injection time is approximately 1200 s. In comparison with the normal injection used in the CE separation by the micellar electrokinetic chromatography mode, approximately 400 and approximately 100,000-fold improvement (S/N = 3) in detection sensitivity, respectively, can be obtained.

In a study to detect LSD from tablets, Fang *et al.* evaluated sweeping-micellar electrokinetic chromatography [40]. A cryogenic molecular fluorescence experiment performed at 77 K produced on-line spectra that were readily distinguishable and could be unambiguously assigned. Results were compared with GC-MS and reported by the authors to be accurate, sensitive and rapid making it a reliable complementary method to GC-MS for analysis of seized material.

Among the variety of parameters evaluated for the analysis of LSD using capillary chromatography Djordjevic *et al.* described the impact of temperature [41]. By adjusting column temperature and applied electric field, a fast separation using micellar electrokinetic capillary chromatography was developed for the separation of D-lysergic acid diethylamide derivatives. A baseline separation of nine derivatives of D-lysergic acid diethylamide was accomplished in less than 12 min by maintaining a column temperature of 60°C. The authors describe the greater separation power of this condition compared to a column temperature of 20°C while using the same applied electric field.

4.3 MESCALINE

Mescaline is a naturally occurring alkaloid found in the peyote cactus *Lophophora williamsii* (see Fig. 4.3 for the chemical structure of mescaline). The cactus is fairly small and its heads are dome-shaped. The heads are harvested and dried as peyote buttons. The cactus grows in southern Texas and northern Mexico. Although a natural product, mescaline is also synthesized for illicit use.

Mescaline is analyzed from the plant material by extraction of the drug from the buttons using alcohol. One such analysis utilized basic methanol for the extraction with methoxamine as an internal standard. The sample extracts were separated by HPLC on an ODS column and identified using UV detection [42].

A method for the analysis of a variety of psychotropic phenylalkylamines from plant material has also been used for analysis of mescaline from biological fluids [42]. Urine samples were extracted using cation exchange SPE. The analytes of interest were chromatographed through an ODS column using acetonitrile:water:phosphoric acid:hexylamine as the mobile phase with UV detection at 198 or 205 nm. TLC using the ToxiLab system can readily identify the presence of mescaline following the procedures described by the manufacturer.

In a case report describing a death caused by a gunshot wound, mescaline was detected and characterized in the decedent's body. Extraction was *via* a simple liquid–liquid procedure using 1-butyl chloride. Samples were analyzed by GC with a nitrogen–phosphorus detector (NPD). Concentrations ranged from 2.2 mg/kg in brain to 8.2 mg/kg in the liver. Blood and vitreous showed concentrations of 2.95 μg/mL and 2.36 μg/mL, respectively [43].

Mescaline and a series of related compounds were analyzed in blood and plasma by GC-MS using a method developed by Habrdova *et al.* [44]. Samples were analyzed by SIM following solid-phase extraction using mixed-mode SPE cartridges. Analytes were derivatized with Heptafluorobutyric Anhydride (HFBA) prior to analysis. The procedure met generally accepted validation criteria with the exception of 2C-T-2 and 2C-T-7 whose results did not meet acceptable parameters. The assay produced acceptable results ranging from 5 to 500 ng/mL, certainly adequate for routine analysis of these compounds.

4.4 PSILOCYBIN

Psilocybin is a naturally occurring compound found in mushrooms of the genus *Psilocybe* commonly referred to as 'magic mushrooms'. The drug is typically ingested by eating the mushrooms. In addition to psilocybin, the mushroom also contains small amounts of psilocin which is nearly twice as potent as psilocybin. The mushrooms are commonly found in the Pacific Northwest region of the U S, Texas, Hawaii and Florida. Related species of fungus that produce psilocybin are found in various other parts of the world. Although naturally occurring compounds, psilocin and psilocybin are also chemically synthesized [45] (see Fig. 4.4 for the chemical structure of psilocybin and psilocin).

Psilocybin is thermally labile and thus does not lend itself well to the high temperatures encountered in GC. Derivatization of psilocybin helps to stabilize the compound and does allow for its analysis. Other chromatographic techniques, such as TLC, HPLC and CE have also been used for the analysis of these compounds.

4.4.1 Extraction

Psilocin and psilocybin were extracted from the mushroom by grinding the plant material in chloroform. The extract was then derivatized with MSTFA and qualitatively analyzed using ion mobility spectrometry. Quantitative analysis was accomplished using GC-MS [46]. Another method used capillary zone electrophoresis (CZE) for the analysis of psilocybin in mushrooms. Propyl chloroformate was used to derivatize the compound and the extracts were analyzed at pH 11.5. This method was successful for analysis of psilocybin but not for psilocin. Psilocin could be analyzed using CZE when the running buffer was reduced to 7.2. Attempts to use micellar electrokinetic chromatography for these two compounds was not successful [47]. Another method using methanol to extract psilocybin from the plant material has been reported. This method also used CZE at pH 11.5 with barbital as internal standard giving a linear range 0.01–1.0 mg/mL [48]. Methanol was also used in a single extraction step by other investigators who concluded it to be the best for analysis of psilocybin from mushrooms [49]. They also used aqueous alcohol which increased the amount of psilocin recovered from the plant material. This was ultimately discovered to be an anomalous result from the degradation of psilocybin to psilocin. Use of dilute acetic acid solutions was found to be better for this recovery without the degradation problem.

Extraction from plasma has been reported using both liquid–liquid and solid-phase extraction. A study by Lindenblatt *et al.* [50] evaluated both liquid–liquid and solid-phase extraction for the analysis of psilocin in plasma. Both methods gave acceptable results but the study determined SPE to have some advantages over its liquid–liquid counterpart, including greater recovery of the drugs allowing for the use of smaller sample volumes. The solid-phase extraction in this case was in-line making the process easier and less time consuming.

4.4.2 Chromatography

Psilocybin and psilocin have been analyzed by TLC, GC-MS, HPLC as well as CE as described above. HPLC analysis of the compounds from plasma, serum and urine has been described. One of these methods used ascorbic acid to help stabilize the phenolic group and column switching to detect psilocybin and its metabolites psilocin and 4-hydroxyindole-3-acetic acid. Lindenblatt *et al.* [50] also used HPLC with electrochemical detection to detect psilocybin and psilocin, each of which required a different HPLC system.

Another method using HPLC analysis of psilocin and psilocybin in mushroom extracts used chemiluminescence detection. A number of extraction methods for psilocin and psilocybin were investigated, with a simple methanolic extraction found to be most effective. Extracts were separated on a C12 column using a methanol:10 mM ammonium formate (95:5% v/v), pH 3.5 mobile phase. Acidic potassium permanganate and tris(2,2′-bipyridyl)ruthenium(II) reagents were used for the chemiluminescence detection system which gave improved detectability compared to the more common UV absorption at 269 nm. Detection limits were 1.2×10^{-8} and 3.5×10^{-9} mol/L for psilocin and psilocybin, respectively.

Kamata *et al.* [51] described an LC-MS-MS procedure for analysis of psilocin and psilocybin in the plant material. Detection limits ranged from 1 to 25 pg by LC-MS in the SIM mode, and the intra- and inter-day relative standard deviations were estimated to be 4.21–5.93% by LC-MS-MS in the selected reaction monitoring mode.

In a study to determine the optimum hydrolysis conditions for release of psilocin from its conjugate, acid, basic and enzymatic hydrolysis procedures were evaluated [52]. Evaluation of acid and alkaline hydrolysis was conducted using concentrated hydrochloric acid or 10 M potassium hydroxide solution followed by incubation at 50°C for 30 min followed by neutralization with 1.2 M hydrochloric acid or 10 M aqueous potassium hydroxide solution. Enzymes evaluated in the study included β-glucuronidases (EC 3.2.1.31) from bovine liver (Type B-1), *H. pomatia* (Type H-1) and *E. coli* (Type IX-A) (Sigma Co., St. Louis, MO, USA), and that from Ampullaria (Wako, Osaka, Japan). Enzyme solutions were prepared in 0.5 M acetate buffer (25,000 units/mL) immediately prior to use. Results showed neither psilocin nor psilocin glucuronide could be detected following acid hydrolysis. Alkaline hydrolysis resulted in no decrease in the concentration of conjugated psilocin indicating the conjugate is stable under these basic conditions. Enzyme hydrolysis showed β-glucuronidase from *E. coli* (5000 units/mL, pH 6, for 2 h at 37°C) gave 100% conversion of the glucuronide conjugate to the free compound. The other three enzymes gave varying degrees of cleavage but none resulted in 100% hydrolysis. It was also noted that incubation times greater than 2 h showed lower total recoveries presumed to be caused by degradation of the compound. LC-MS-MS conditions were set to the following voltages: capillary, 4.5 kV; cone, 30 V; collision energy, 15 eV. Precursor and product ions were m/z 205 to 160 for psilocin and m/z 381 to 205 for the conjugate.

Huikko *et al.* [53] examined the effect of gas flow parameters on capillary electrophoresis-mass spectrometry (CE-MS) using sheath–liquid interfaces. The effects of nebulizing and drying gas velocity and drying gas temperature on separation and sensitivity were systematically evaluated and showed substantial impact on sample analysis. Nebulizing gas velocity proved to be critical in the optimization of the CE-MS method, since it affected both sensitivity and separation efficiency. Increasing nebulizing gas velocity increased sensitivity but decreased resolution of the compounds studied. Increasing drying gas velocity or temperature did not affect the apparent mobility or separation efficiency. Temperature could be increased to achieve optimal detection sensitivity. The authors studied a slightly different coaxial sheath-liquid CE-MS interface design and found that optimizing the nebulizing gas flow gave better detection sensitivity but had no effect on CE separation efficiency. These different effects for different CE-MS interfaces caused the authors to recommend the cross-sectional dimensions of the fused-silica and steel capillaries, and the gas streamlines should be optimized when CE-MS interfaces are built. In addition, the effect of gas flow on CE separation should also be studied when optimizing CE-MS conditions.

The analysis of psilocin by several investigators showed it was primarily excreted as the glucuronide conjugate rather than the free compound [54,55]. One study designed to determine the pharmacokinetics of psilocybin in humans showed rapid dephosphorylation of psilocybin to psilocin with further metabolism to 4-hydroxytryptophole and 4-hydroxyindole-3-acetic acid. Enzymatic hydrolysis of urine samples using β-glucuronidase from *H. pomatia*, or *E. coli* showed a significant increase in psilocin concentration indicating a significant degree of conjugation. Following extraction and derivatization with MSTFA the procedure gave a limit of quantitation of 10 ng/mL from a 5 mL sample [54].

In a procedure using the REMEDi HS Sticht and Kaferstein [55] confirmed that most psilocin was excreted as the glucuronide conjugate. In one sample, they measured free psilocin at a concentration of 0.23 mg/L while the total in urine was 1.76 mg/L, showing the conjugate to be over seven times the free concentration. The REMEDi HS was unable to detect psilocin in serum due to the low concentration. However, a GC-MS method using trideuterated-morphine as the internal standard following silylation with MSTFA showed free psilocin at 0.018 mg/L while the total was 0.052 mg/L, showing the conjugate to be over three times the free concentration.

LC-MS analysis of a variety of drugs including psilocybin and psilocin using atmospheric pressure chemical ionization was described by Bogusz [56]. The method used a simple SPE(with the exception of LSD) and was able to identify a wide variety of drugs of forensic interest. While this method provided viable results for all analytes, it showed that electrospray ionization yielded greater sensitivity for very polar drugs, such as psilocybin and psilocin, than did atmospheric pressure chemical ionization.

One procedure for the analysis of psilocin and psilocybin in urine stabilized the psilocybin prior to analysis using 350 μL of freshly prepared ascorbic acid solution (94 mg/mL), storage at –78°C over dry-ice then freeze-drying overnight. Sample extracts (methanol from the freeze-dried extract) were analyzed using a Spherisorb RP-8 column (Chemie Brunschwig, Basel, Switzerland) with 46% (v/v) water, containing 0.3 M ammonium acetate buffer (pH 8.3) and 54% (v/v) methanol as mobile phase using EC detection. The assay gave average ($n = 3$) recoveries of 102.7% and relative standard deviations of 4.5%. The limit of quantitation (S/N = 1:5) was 10 μg/L for psilocin [57]. These investigators also found psilocin to be substantially conjugated and hydrolysis was necessary to get the total concentrations of the drug. Samples were hydrolyzed by addition of 60 mg of lyophilized β-glucuronidase (type B-1 bovine liver; 560,000 units/g Sigma, Buchs, Switzerland; 11,200 units/mL of urine) to urine samples and incubation for 5 h at 40°C. This led to a twofold higher psilocin concentrations than was seen without hydrolysis. $18 \pm 7\%$ of unconjugated psilocin was determined to have decomposed during the five hour incubation.

4.5 PHENCYCLIDINE (PCP)

PCP is classified as an anesthetic and was initially developed for that purpose. It was synthesized in the 1950s and was used in several clinical trials as a surgical anesthetic. After being approved for human use, reports of significant untoward side-effects caused it to no longer be used in human surgical procedures. It continued to be used some time in animal surgery, but subsequently that too was also stopped. However, it is generally abused for its hallucinogenic properties.

PCP is synthesized from 1-piperidinocyclohexanecarbonitrile (PCC) and 1-phenylcyclohexylamine (PCH). In most cases, some of the starting material is found in the final product. PCP is metabolized by the body to 4-phenyl-4-piperidinocyclohexanol (PPC), 1-(1-phenylcyclohexyl)-4-hydroxypiperidine (PCHP) and 5-(N-(1'-phenylcyclohexyl)amino)pentanoic acid (PCA) [58–62] (see Fig. 4.5 for the chemical structure of PCP and its metabolites).

PCP has been analyzed using a variety of methods including TLC, GC, GC-MS, etc. Each of these methods has their advantages and disadvantages depending on the specific situation. A comparison of several of these analytical methods has been published which describe the pros and cons of the assays [63].

4.5.1 Extraction

Despite the fact PCP is commonly analyzed in the low ng/mL concentrations, it is a relatively easy drug to isolate and identify. Since this drug has been used for many

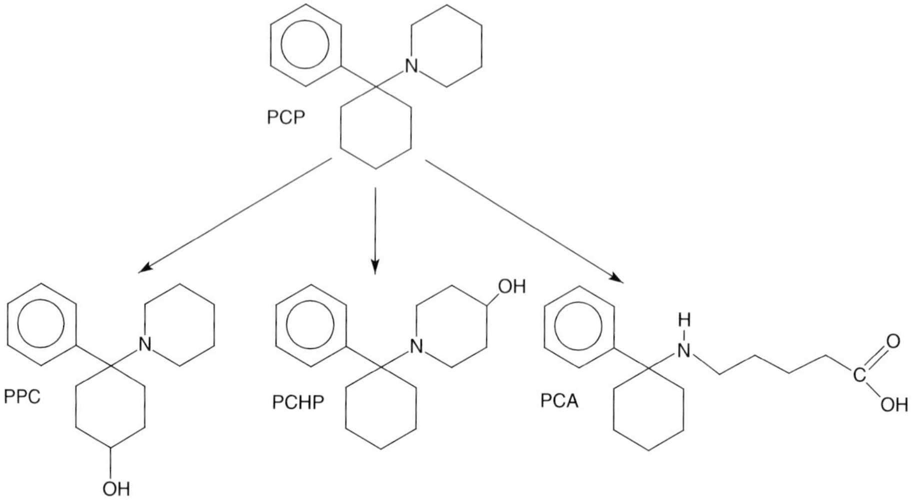

Fig. 4.5. Phencyclidine.

years, a wide variety of methods are available for its analysis. Procedures are also available to analyze its metabolites. Typically, however, only the parent drug is analyzed. Several procedures have been described for both liquid–liquid and solid-phase extraction of PCP along with more unique techniques such as solvent microextraction.

Extraction of PCP from hair has also been described by a number of investigators. Nakahara *et al.* [64] described the analysis of PCP and several of its metabolites in hair. Three different extraction methods were evaluated for this purpose. In all cases, the hair (5 mg) was washed with SDS and dried before analysis. Extractions included: methanol:5 M HCl (20:1), 10% HCl and 2 M sodium hydroxide. After hydrolysis, the digests were extracted using solid-phase extraction, derivatized and analyzed by GC-MS. Extraction of PCP for each of the three methods was nearly the same (within 20%), however, the recovery of metabolites was far greater for the methanol:5 M HCl (20:1) procedure than either of the other two with metabolite recoveries of at least double that seen for the other procedures. Sakamoto *et al.* also reported the use of methanol:5 M HCl (20:1) in a procedure for the isolation of PCP and metabolites from rat hair [65]. Slawson and coworkers used overnight digestion of hair in 1 M sodium hydroxide followed by SPE of the drug from the digest. Using MS-MS, Kidwell described a procedure for the analysis of hair samples for cocaine and PCP using no extraction. The hair samples were washed in pentane, air dried then added to a solids probe cup and the hair directly inserted into the mass spectrometer [66].

Stevenson *et al.* [67] described a solid-phase extraction method for PCP from urine using SPE columns and compared that with a liquid–liquid procedure. The solid-phase procedure was determined to provide a cleaner extract and the recovery was essentially 100% when evaluating spiked samples at 35 ng/mL. The authors found the extraction worked well but they did observe that following extraction, evaporation at temperatures exceeding 30°C caused a decrease in the amount of PCP detected. A liquid–liquid extraction procedure was described using 5 mL of urine that had been made basic by addition of ammonium and sodium hydroxide and extracted with toluene:heptane:isoamyl alcohol (78:20:2). The extracts were then dried and reconstituted in methanol and injected into GC-MS.

Meconium has, in recent years, been used as a sample of choice to determine a history of drug use during pregnancy which exceeds the utility of urine or even hair testing. This matrix does pose some challenges for sample preparation. Moriya *et al.* [68] described an extraction procedure using chloroform:isopropanol (3:1) followed by analysis by immunoassay and GC-MS.

Using a technique initially developed by Jeannot and Cantwell [69] and Casari and Andrews [70] described a method for the analysis of amphetamines (amphetamine, methamphetamine, MDA, MDMA, MDEA) and PCP. Two milliliter urine samples were extracted using a device, consisting of a small piece of porous hollow fiber, containing an organic solvent, in this case 2 µL of chloroform. The needle tip was set so that the needle tip was about 5 mm below the surface of the sample solution. The syringe plunger was depressed to generate the solvent drop and the sample solution was constantly stirred to promote diffusion of the drugs from the urine into the solvent. After 8 min of extraction, the drop was retracted into the needle then the solvent was injected into the GC. The system was operated in the split mode (10:1) with the injection port temperature set at 250°C. The oven was programmed from 140°C to 300°C at 20°C/min. Separation was accomplished using a 30 m 5% phenylmethylsiloxane (320 µm i.d. with 0.25 µm film thickness) with a flow rate of 1.5 mL/min. The detector temperature was left constant at 300°C. Plasma gas flow for the PDHID was 30 mL/min of helium. For the method described in the publication, the authors used the solvent that gave the best results for all analytes. In the case of PCP, it was noted that 1-octanol gave significantly better results, however, it showed interference and poor recovery with the amphetamines analyzed therefore the compromise selection of the solvent. Extraction time was set to 8 min for the analysis. This gave adequate sensitivity even though higher recovery may have come from greater exposure time. Since the GC method had an 8 min cycle time, setting the extraction time to the same, or less, ensured the analysis time was not extended. The PCP portion of the assay gave results linear from 0.015 to 1.0 µg/mL and a detection limit of 0.07 µg/mL (LOD determined by results of samples analyzed including the mean of blank samples and the slope of the calibration curve; see reference for formula).

4.5.2 Gas chromatography

Gas chromatography analysis of samples has been used for many years. Due to the presence of the nitrogen in the molecule, PCP is amenable to analysis using a NPD. Using NPD to detect lower levels of PCP in urine was described by Kandiko *et al.* [71]. The key element in decreasing the detection limit to as low as 15 ng/mL was acetylation of the column's packing material. A more recent method using NPD has been described for the detection of PCP and a number of other non-opiate narcotic analgesics from plasma using a simple alkaline extraction [72]. This method was able to detect concentrations as low as 10 ng/mL on a routine basis. A procedure for the analysis of a variety of basic drugs has been reported by Chen *et al.* [73]. The procedure utilized solid-phase extraction of basic drugs from urine, plasma and blood. Analysis was completed using GC with a cross-linked methyl silicone column. Since this procedure was designed to isolate many different drugs, many other biological compounds were also isolated, yet the selectivity of the NPD allowed for identification of amine-containing drugs without further clean-up.

Mass spectral analysis of PCP using electron ionization gives rise to a number of different prominent ions (m/z 186, 200, 242, 243). Monitoring of three ions with appropriate attention to ion ratios is standard procedure for forensic laboratories. PCP is a bit unique among compounds in that it has two prominent ions that are only one mass unit apart (m/z 242 and 243). Generally it is not appropriate to monitor ions from organic compounds that differ by only a single mass unit because they represent the naturally occurring C-13 isotope of carbon; however, in this case the monitoring of the two ions is fully justified because they are clearly different fragments. This can easily be documented by evaluating the mass spectrum of the deuterated PCP which has a corresponding fragment that differs by two mass units.

Tai *et al.* [74] described a method for the analysis of PCP standard reference material. The analysis encompassed GC-MS using electron ionization on a standard quadrupole instrument, an ion trap mass spectrometer and also by LC-MS. Results of all methods gave results that were in close agreement. The ion trap used a 14% cyanopropylphenyl-methylpolysiloxane (DB-1701) column at 99°C for 0.5 min to 280°C at 40°C/min. The other GC-MS used a 5% phenylmethyl-dimethylpolysiloxane (DB-5) column at 140°C for 0.5 min to 280°C at 30°C/min. Since the PCP was standard reference material placed into certified negative urine, the only ions monitored were at m/z 200 and 205, base peaks for the drug and its deuterated isotopomer.

Another GC-MS method described the analysis of the drug and its deuterated internal standard by injection into a 5% phenylmethyl-dimethylpolysiloxane column at 120°C for 1 min to 190°C at 40°C/min. Ions monitored for PCP were m/z 200, 243 and 186 and m/z 200 and 205 were used for quantitation. The limit of detection with this method was reported to be 0.47 ng/mL with a limit of quantitation of 1.38 ng/mL. The method was shown to be linear up to 1,000 ng/mL.

GC-MS analysis was also used for the analysis of PCP and its amino acid metabolite 5-(*N*-(1′-phenylcyclohexyl)amino)pentanoic acid (PCA) [75]. Following SPE, which the authors determined to be superior to liquid–liquid extraction, the extract was reacted with methanolic HCl at 60°C for 1 h. Water was then added along with ammonium hydroxide and the analytes extracted into hexane. This was evaporated and the extract reconstituted into a small amount of isooctane and analyzed by GC-MS by monitoring the ions at m/z 289, 246, 159 and 294, 251, 164 for the methyl derivative of the metabolite and its penta-deuterated internal standard, respectively. The assay was linear from 10 to 150 ng/mL for the PCA with concentrations as low as 2 ng/mL being detectable. A method using di-fluorophencyclidine as internal standard has also been described [76]. This method used both full scan and SIM for the analysis of PCP. The detection limit for both modes was 0.25 ng/mL. The assay was linear to 500 ng/mL in both cases with lower limits of linearity of 0.5 and 0.32 ng/mL for full scan and SIM, respectively. Keta-mine has also been used as an internal standard for the analysis of PCP. A method reported by Mule and Casella used liquid–liquid extraction and had a limit of quantitation of 10 ng/mL [77].

Analysis of PCP and its metabolites in hair has been described by several inves-tigators. Slawson *et al.* [78] used an ion-trap mass spectrometer to identify and quantitate PCP using deuterated PCP as the internal standard. This assay was linear from 0.1 to 50 ng/mg of hair. Between-run precision was less than 6% when assessed at several different concentrations. Sakamoto *et al.* analyzed PCP and its metabolites PCHP and PPC in rat hair using GC-MS. Following extraction, the metabolites were derivatized with N,O-bis(trimethylsilyl) acetamide. Deuterated isotopomers of each of these analytes were used as internal standards. Ions were monitored at m/z 186, 200, 242 (PCP), m/z 172, 288, 331 (TMS derivative of PCHP) and m/z 200, 254, 331 (TMS derivative of PPC). PCP and metabolites could be detected at 0.1 mg/kg.

Another method describing the analysis of PCP in hair by GC-MS is given by Cairns [79], approximately 10 mg of hair was extracted using a liquid–liquid ex-traction procedure. GC-MS analysis was performed on a 30 M DB5-MS capillary column (J & W, 0.25 mm i.d. 0.25 μm film thickness). Ions monitored for PCP were m/z 200, 242 and 243, and m/z 205 and 246 for its penta-deuterated isotopomer internal standard. Quantitation was based on a single-point calibration based on one standard at the cutoff concentration [79]. The authors strongly emphasized in the report the importance of monitoring sample washes in order to identify the possibility of external contamination as the source of the drug detected.

Gas chromatography (GC)/surface ionization organic mass spectrometry (SIOMS) was used for the determination of PCP in whole blood and urine. Sur-face ionization is a process by which neutral atoms or molecules are converted to ions on an incandescent metal surface. In this case the metal used was a rhenium filament heated by application of approximately 1.6 A. Oxygen gas was supplied to the filament to keep the rhenium surface stable. SIOMS is selective for compounds containing tertiary amino groups; therefore it is applicable for only limited numbers of compounds. However, SIOMS is very advantageous for sensitive determination of such compounds. Spectra obtained by SIOMS differ from the more common EI. Base peaks were at m/z 200 in EI mode and m/z 156 in SI mode. In the SI mode, other significant ions appeared at m/z 242, 156, 180, 206 and 230. In the case of pethidine, the internal standard used for the analysis, virtually only a single peak was seen at m/z 170 in SI mode. The method gave a linear range easily reaching the goal of low level detection and quantitation of PCP in whole blood or urine. The method was more sensitive than typically required for urine; however, intoxication by PCP is often associated with blood concentrations of approximately 5 ng/mL. The detection limit for the method (S/N = 3) was 0.01 ng/mL in both whole blood and urine. The relative standard deviations for both with-in and between day were $< = 10.3\%$.

PCP has been analyzed using methods designed for a variety of different drugs. These methods were developed to afford a single procedure capable of analyzing a large number of different compounds of interest. Having a single procedure provides several advantages including simpler and less expensive training and conduct of the analyses. Typically these general procedures are less specific and less sensitive than their single analyte competitors. Examples of such assays include a procedure developed to analyze samples for the metabolites of cocaine, opiates, cannabinoids, amphetamines and phencyclidine in meconium [80]. Montgomery *et al.* also used a multi-drug assay, designed to assay meconium and umbilical cord tissue for amphetamines, opiates, cocaine, cannabinoids and phencyclidine. Results from the comparison led to the conclusion the umbilical cord tissue provided a more rapid assessment of exposure compared with meconium because it was always more readily available than meconium [81]. Another multi-analyte procedure described the analysis of urine specimens (2 mL) that were diluted with 2 mL 0.1 M phosphate buffer (2 mL, pH 5.0) and then applied to the preconditioned extraction disc. Samples were extracted by washing each column with 1 mL water. Acidic and neutral drugs were eluted with 1 mL ethyl acetate:acetone (1:1); basic drugs were eluted with 1 mL ammoniated ethyl acetate. Eluates were collected separately then evaporated to approximately 0.1 mL. Evaluation of the procedure was accomplished by gas chromatography-flame-ionization detection. Recoveries were 75–100% and relative standard deviations were approximately 5%. The extraction disc was compared to standard SPE-packed bed columns and showed reductions in elution volumes and processing time of approximately 60–65 [82].

4.5.3 Liquid chromatography

HPLC methods have been used for analysis of PCP for many years. One method of analysis involves direct analysis of biological samples without the need for extraction prior to injection on the instrument. The REMEDi is an HPLC system consisting of several columns which are used to preperatively separate the sample components and detect what is present using UV detection. The instrument is generally used for clinical and emergency toxicology and therefore is designed to produce rapid results. Sensitivity of the system is a problem for the analysis of PCP which cannot be determined at levels consistent with most other analytical procedures. Given the setting the instrument is designed to be used, such low levels of the drug would likely have no clinical significance. One application using the REMEDi is described by Baskin and Morgan [83].

As part of the validation of PCP standard reference material, analysis was accomplished by both GC-MS and LC-MS methods [74] as mentioned above. This method used a C-18 column with an isocratic mobile phase of 1 mM heptanesulfonic acid, 20 mM ammonium acetate and 3.5% glacial acetic acid at a flow rate of 1 mL/min. Thermospray ionization was used and the protonated molecular ions at m/z 244 and 249 were monitored for the drug and the deuterated internal standard.

A method for the analysis of phencyclidine by liquid chromatography-tandem mass spectrometry method in small (i.e. 50 μL) volumes of rat serum has been described [84]. Samples were extracted using a mixed-mode SPE then separated isocratically using a narrow-bore (2.1 mm i.d., 3 μm Hypersil phenyl column using pH 2.7 ammonium formate buffer:methanol 40:60 v/v) mobile phase. Detection was accomplished using positive ion electrospray ionization multiple reaction monitoring. Mass spectra were obtained and peaks at m/z 244, 159 and 86 were observed. Tandem mass spectra using m/z 244 as the precursor ion and found ions at m/z 159, 86 and 91. The transition from m/z 244 -> 159 gave optimal PCP sensitivity and precision. Matrix-associated ion suppression was evaluated and found to not be a negative factor in the assay and did not significantly impact the accuracy (100–112%) or precision (relative standard deviation $< = 8\%$). The limit of quantitation was 1 ng/mL from 50 μL serum sample. The method offers the advantage of sufficient sensitivity to analyze very small sample sizes and still provide a viable linear range [84].

4.5.4 Thin layer chromatography

Thin layer chromatography is a method used to evaluate the presence of PCP. One TLC assay system (Toxi-Lab, ANSYS, Inc., Irvine, CA, USA) involves methods for the identification a large number of drugs and drug metabolites. PCP can be identified by this system in stage IV as a brown spot. Occasionally, there is interference with the identification of PCP by other drugs; however, there are alternative solvent systems that can resolve the identification. The most significant drawback to the use of TLC is the lack of sensitivity. When PCP is used by itself by an abuser, the concentration found in urine can be quite high. In cases where the PCP is used in conjunction with another drug, the concentrations are typically much lower. In either case, the detection limit of approximately 500 ng/mL limits the utility of this technique for the analysis of PCP.

Some TLC procedures have been developed that can detect lower levels of PCP from urine. Jain reported a method that could detect PCP at 200 ng/mL which is significantly lower than most other TLC assays [85]. While this procedure is more sensitive than others, it too is still not sensitive enough to evaluate the presence of PCP in samples as are HPLC, GC and GC-MS procedures.

4.5.5 Capillary electrophoresis and other techniques

Chen and Evangelista described a novel method for the immunochemical binding of drugs followed by CE with LIF [86]. This method was rapid (less than 5 min) and could detect PCP at a level of 4 nmole/L. Surface ion detection of PCP was described by Ishii *et al.* [87]. This method extracted PCP from urine and blood using solid-phase extraction (C-18 column). Recovery of PCP was determined to be over 85% from blood and urine using pethidine as the internal standard. The assay was linear from 1.25 to 20 ng/mL and a detection limit of 0.75 ng/mL.

4.6 CONCLUSION

Use of hallucinogens remains a significant problem for a population of drug abusers. These drugs have a long history and their popularity comes and goes with time, but they remain a constant presence in the drug community. LSD in particular has risen in popularity in recent years and the use of low doses of this drug makes its isolation and identification a significant analytical challenge. Most analytical tools commonly used in toxicology labs have been utilized to identify these drugs. All these techniques have their place and are routinely used. The common method used in most forensic work in biological matrices involves confirmation with mass spectrometry. Development of atmospheric pressure ionization has taken LC-MS from a difficult analytical process to one that has increased applicability including analysis of drugs of abuse. Advances in CE holds interest for the future. The resolving power of this analytical technique is impressive; however, development of automation of this analytical process is important for its more wide-spread use.

4.7 ABBREVIATIONS

CE-MS	Capillary electrophoresis-mass spectrometry
CZE	Capillary zone electrophoresis
DB-1	Cross-linked methyl silicone
DB-17	50% Phenyl methyl silicone
DB-5	5% Phenyl methyl silicone
FID	Flame ionization detector
GC	Gas chromatography
GC-MS	Gas chromatography-mass spectrometry
GC-MS-MS	Gas chromatography-mass spectrometry-mass spectrometry
HFBA	Heptafluorobutyric anhydride
HPLC	High-performance liquid chromatography
iso-LSD	Iso-Lysergic acid diethylamide
LAMPA	Lysergic acid methyl propylamide
LC	Liquid chromatography
LC-MS	Liquid chromatography-mass spectrometry
LC-MS-MS	Liquid chromatography-mass spectrometry-mass spectrometry
LIF	Laser-induced fluorescence
LSD	Lysergic acid diethylamide
MDA	3,4-Methylenedioxyamphetamine
MDEA	3,4-Methylenedioxyethylamphetamine

MDMA	3,4-Methylenedioxymethamphetamine
MS	Mass spectrometry
MS-MS	Mass spectrometry-mass spectrometry
Nor-LSD	Nor-Lysergic acid diethylamide
NPD	Nitrogen–phosphorus detector
ODS	Octadecyl silyl
PCA	5-(N-(1'-Phenylcyclohexyl)amino)pentanoic acid
PCC	1-Piperidinocyclohexanecarbonitrile
PCH	1-Phenylcyclohexalamine
PCHP	1-(1-Phenylcyclohexyl)-4-hydroxypiperidine
PCP	Phencyclidine
PPC	4-Phenyl-4-piperidino-cyclohexanol
SIM	Selected ion monitoring
SIOMS	Surface ionization organic mass spectrometry
SPME	Solid-phase microextraction
TLC	Thin layer chromatography
TMS	Trimethylsilyl
UV	Ultraviolet

4.8 REFERENCES

1   M.R. Moeller and T. Kraemer, Ther. Drug Monit., 24 (2002) 210.
2   R.A. Sewell, J.H. Halpern and H.G. Pope, J. Neurol., 66 (2006) 1920.
3   H.K. Lim, D. Andrenyak, P. Francom, R.L. Foltz and R.T. Jones, Anal. Chem., 60 (1988) 1420.
4   S.A. Reuschel, S.E. Percey, S. Liu, D.M. Eades and R.L. Foltz, J. Anal. Toxicol., 23 (1999) 306.
5   G.K. Poch, K.L. Klette, D.A. Hallare, M.G. Manglicmot, R.J. Czarny, L.K. McWhorter and C.J. Anderson, J. Chromatogr. B Biomed. Appl., 724 (1999) 23.
6   A.G. Verstraete and E.J. Van de Velde, Acta Clin. Belg. Suppl., 1 (1999) 94.
7   S.D. Kilmer, J. Forensic Sci., 39 (1994) 860.
8   T. Veress, J. Forensic Sci., 38 (1993) 1105.
9   E.D. Clarkson, D. Lesser and B.D. Paul, Clin. Chem., 44 (1998) 287.
10   F. Musshoff and T. Daldrup, Forensic Sci. Int., 88 (1997) 133.
11   J. de Kanel, W.E. Vickery, B. Waldner, R.M. Monahan and F.X. Diamond, J. Forensic Sci., 43 (1998) 622.
12   Y. Nakahara, R. Kikura, K. Takahashi, R. Foltz and T. Mieczkowski, J. Anal. Toxicol., 20 (1996) 323.
13   J.M. Francis and D.H. Craston, Analyst, 121 (1996) 177.
14   K.S. Webb, P.B. Baker, N.P. Cassells, J.M. Francis, D.E. Johnston, S.L. Lancaster, P.S. Minty, G.D. Reed and S.A. White, J. Forensic Sci., 41 (1996) 938.
15   J. Cai and J. Henion, Anal. Chem., 68 (1996) 72.
16   G.S. Rule and J.D. Henion, J. Chromatogr., 582 (1992) 103.
17   S. Kerrigan and D.E. Brooks, J. Immunol. Methods, 224 (1999) 11.
18   D.I. Papac and R.L. Foltz, J. Anal. Toxicol., 14 (1990) 189.
19   B.T. Burnley and S. George, J. Anal. Toxicol, 27 (2003) 249.
20   G.K. Poch, K.L. Klette, D.A. Hallare, M.G. Manglicmot, R.J. Czarny, L.K. McWhorter and C.J. Anderson, J. Chromatogr. B: Biomedical Applications, 724 (1999) 23.
21   S.A. Reuschel, R.L. Foltz, S.E. Percey, S. Liu and D.M. Eades, Proceedings of the 1998 joint SOFT/TIAFT international meeting, in: V. Spiehler (Ed.), SOFT/TIAFT, Newport Beach, CA, 1999, p. 538.

22   D. Libong, S. Bouchonnet and I. Ricordel, J. Anal. Toxicol., 27 (2003) 24.
23   J.H. Sklerov, K.S. Kalasinsky and C.A. Ehorn, J. Anal. Toxicol., 23 (1999) 474.
24   J. Cai and J. Henion, J. Anal. Toxicol., 20 (1996) 27.
25   K.L. Duffin, T. Wachs and J.D. Henion, Anal. Chem., 64 (1992) 61.
26   S.A. White, T. Catterick, M.E. Harrison, D.E. Johnston, G.D. Reed and K.S. Webb, J. Chromatogr. B Biomed. Appl., 689 (1997) 335.
27   S.A. White, A.S. Kidd and K.S. Webb, J. Forensic Sci., 44 (1999) 375.
28   S.S. Johansen and J.L. Jensen, J. Chromatogr., B, Analyt. Technol. Biomed. Life Sci., 825 (2005) 21.
29   C.K. Horn, K.L. Klette and P.R. Stout, J. Anal. Toxicol., 27 (2003) 459.
30   K.L. Klette, C.K. Horn, P.R. Stout and C.J. Anderson, J. Anal. Toxicol., 26 (2002) 193.
31   G. Skopp, L. Potsch, R. Mattern and R. Aderjan, Clin. Chem., 48 (2002) 1615.
32   M. Cui, M.A. McCooeye, C. Fraser and Z. Mester, Anal. Chem., 76 (2004) 7143.
33   G.K. Poch, K.L. Klette and C. Anderson, J. Anal. Toxicol., 24 (2000) 170.
34   D. Bergemann, A. Geier and L. von Meyer, J. Forensic Sci., 44 (1999) 372.
35   L.M. Blum, E.F. Carenzo and F. Rieders, J. Anal. Toxicol., 14 (1990) 285.
36   M. Frost and H. Kohler, Forensic Sci. Int., 92 (1998) 213.
37   M. Frost, H. Kohler and G. Blaschke, J. Chromatogr., B, Biomed. Appl., 693 (1997) 313.
38   I.S. Lurie, P.A. Hays and K. Parker, Electrophoresis, 25 (2004) 1580.
39   C. Fang, J.T. Liu, S.H. Chou and C.H. Lin, Electrophoresis, 24 (2003) 1031.
40   C. Fang, J.T. Liu and C.H. Lin, Electrophoresis, 24 (2003) 1025.
41   M.N. Djordjevic, F. Fitzpatrick and F. Houdiere, Electrophoresis, 21 (2000) 724.
42   H.J. Helmlin and R. Brenneisen, J. Chromatogr., 593 (1992) 87.
43   J.L. Henry, J. Epley and T.P. Rohrig, J. Anal. Toxicol., 27 (2003) 381.
44   V. Habrdova, F.T. Peters, D.S. Theobald and H.H. Maurer, J. Mass Spectrom., 40 (2005) 785.
45   O. Shirota, W. Hakamata and Y. Goda, J. Nat. Prod., 66 (2003) 885.
46   T. Keller, A. Schneider, P. Regenscheit, R. Dirnhofer, T. Rucker, J. Jaspers and W. Kisser, Forensic Sci. Int., 99 (1999) 93.
47   S. Pedersen-Bjergaard, K.E. Rasmussen and E. Sannes, Electrophoresis, 19 (1998) 27.
48   S. Pedersen-Bjergaard, E. Sannes, K.E. Rasmussen and F. Tonnesen, J. Chromatogr. B, Biomed. Sci. Appl., 694 (1997) 375.
49   J. Gartz, J. Basic Microbiol., 34 (1994) 17.
50   H. Lindenblatt, E. Kramer, P. Holzmann-Erens, E. Gouzoulis-Mayfrank and K.A. Kovar, J. Chromatogr. B, Biomed. Sci. Appl., 709 (1998) 255.
51   T. Kamata, M. Nishikawa, M. Katagi and H. Tsuchihashi, J. Forensic Sci., 50 (2005) 336.
52   T. Kamata, M. Nishikawa, M. Katagi and H. Tsuchihashi, J. Chromatogr. B: Analytical Technologies in the Biomedical & Life Sciences, 796 (2003) 421.
53   K. Huikko, T. Kotiaho and R. Kostiainen, Rapid Commun. Mass Spectrom., 16 (2002) 1562.
54   A.F. Grieshaber, K.A. Moore and B. Levine, J. Forensic Sci., 46 (2001) 627.
55   G. Sticht and H. Kaferstein, Forensic Sci. Int., 13 (2000) 403.
56   M.J. Bogusz, J. Chromatogr. B, Biomed. Appl., 748 (2000) 3.
57   F. Hasler, D. Bourquin, R. Brenneisen and F.X. Vollenweider, J. Pharm. Biomed. Anal., 30 (2002) 331.
58   L.K. Wong and K. Biemann, Clin. Toxicol., 9 (1976) 583.
59   J.K. Baker, J.G. Wohlford, B.J. Bradbury and P.W. Wirth, J. Med. Chem., 4 (1981) 666.
60   C.E. Cook, M. Perez-Reyes, A.R. Jeffcoat and D.R. Brine, Fed. Proc., 42 (1983) 2566.
61   L.S. Cohen, L. Gosenfeld and J. Wilkins Jr., N. Engl. J. Med., 306 (1982) 1427.
62   C.E. Cook, D.R. Brine and A.R. Jeffcoat, Clin. Pharmacol. Ther., 31 (1982) 625.
63   J.K. Fallon, A.T. Kicman, J.A. Henry, P.J. Milligan, D.A. Cowan and A.J. Hutt, Clin. Chem., 45 (1999) 1585.
64   Y. Nakahara, K. Takahashi, T. Sakamoto, A. Tanaka, V.A. Hill and W.A. Baumgartner, J. Anal. Toxicol., 21 (1997) 356.
65   T. Sakamoto, A. Tanaka and Y. Nakahara, J. Anal. Toxicol., 20 (1996) 124.

66 D.A. Kidwell, J. Forensic Sci., 38 (1993) 272.
67 C.C. Stevenson, D.L. Cibull, G.E. Platoff Jr., D.M. Bush and J.A. Gere, J. Anal. Toxicol., 16 (1992) 337.
68 F. Moriya, K.M. Chan, T.T. Noguchi and P.Y. Wu, J. Anal. Toxicol., 18 (1994) 41.
69 M.A. Jeannot and F.F. Cantwell, Anal. Chem., 68 (1996) 2236.
70 C. Casari and A.R. Andrews, Forensic Sci. Int., 120 (2001) 165.
71 C.T. Kandiko, S. Browning, T. Cooper and W.A. Cox, J. Chromatogr., 528 (1990) 208.
72 P. Kintz, A. Tracqui, A.J. Lugnier, P. Mangin and A.A. Chaumont, Methods Find. Exp. Clin. Pharmacol., 12 (1990) 193.
73 X.H. Chen, J.P. Franke, J. Wijsbeek and R.A. de Zeeuw, J. Anal. Toxicol., 18 (1994) 150.
74 S.S.-C. Tai, R.G. Christensen, K. Coakley, P. Ellerbe, T. Long and M.J. Welch, J. Anal. Toxicol., 20 (1996) 43.
75 M.A. ElSohly, T.L. Little Jr., J.M. Mitchell, B.D. Paul, L.D. Mell Jr. and J. Irving, J. Anal. Toxicol., 12 (1988) 180.
76 A.H. Wu, T.A. Onigbinde, S.S. Wong and K.G. Johnson, J. Anal. Toxicol., 16 (1992) 202.
77 S.J. Mule and G.A. Casella, J. Anal. Toxicol., 12 (1988) 102.
78 M.H. Slawson, D.G. Wilkins, R.L. Foltz and D.E. Rollins, J. Anal. Toxicol., 20 (1996) 350.
79 T. Cairns, V. Hill, M. Schaffer and W. Thistle, Forensic Sci. Int., 145 (2004) 97.
80 B.M. Lester, M. ElSohly, L.L. Wright, V.L. Smeriglio, J. Verter, C.R. Bauer, S. Shankaran, H.S. Bada, H.H. Walls, M.A. Huestis, L.P. Finnegan and P.L. Maza, Pediatrics, 107 (2001) 309.
81 D. Montgomery, C. Plate, S.C. Alder, M. Jones, J. Jones and R.D. Christensen, J. Perinat., 26 (2006) 11.
82 R.A. de Zeeuw, J. Wijsbeek and J.P. Franke, J. Anal. Toxicol., 24 (2000) 97.
83 L.B. Baskin and D.L. Morgan, Tex. Med., 93 (1997) 50.
84 H.P. Hendrickson, E.C. Whaley and S.M. Owens, J. Mass Spectrom., 40 (2005) 19.
85 N.C. Jain, R.D. Budd, W.J. Leung and T.C. Sneath, J. Anal. Toxicol., 1 (1977) 77.
86 F.T. Chen and R.A. Evangelista, Clin. Chem., 40 (1994) 1819.
87 A. Ishii, H. Seno, T. Kumazawa, M. Nishikawa, K. Watanabe and O. Suzuki, Int. J. Legal Med., 108 (1996) 244.

M.J. Bogusz (Ed.). Forensic Science
Handbook of Analytical Separations, Vol. 6

CHAPTER 5

Cannabinoids analysis: analytical methods for different biological specimens

Mahmoud A. ElSohly[1,2], Waseem Gul[1,2] and Maissa Salem[3]

[1]*ElSohly Laboratories, Incorporated (ELI), 5 Industrial Park Drive, Oxford, MS 38655, USA*
[2]*National Center for Development of Natural Products, School of Pharmacy, University of Mississippi, University, MS 38677, USA*
[3]*Department of Analytical Chemistry, Faculty of Pharmacy, University of Cairo, Kasr El-Aini, Cairo, Egypt*

5.1 INTRODUCTION

Cannabinoids are a group of compounds unique to the cannabis plant (*Cannabis sativa* L.) of which Δ^9-tetrahydrocannabinol (THC) is the most active component that causes psychedelic activity. They are responsible for most of the pharmacological effects of the plant. These psychoactive constituents are present mainly in the flowering and fruiting tops and leaves of the plant.

Three cannabis preparations are illicitly trafficked: herbal cannabis (marijuana), cannabis resin with fine plant particles (hashish), and cannabis extract (cannabis oil or hash oil) [1].

Herbal cannabis (marijuana) is the most widely used illicit drug in the world [2–4]. It is prepared by collecting the flowering tops and leaves of the female plants and drying them in the air. The dried material may then be compressed into blocks or left as loose herbal material.

Cannabis resin (hashish) is prepared by threshing the herbal material, often against a wall, to separate the fibrous parts of the plant from the resin-producing parts, then compressing them into slabs. Alternatively, the flowering and fruiting tops are rubbed between the palms of the hands, which are then scraped periodically to remove the resin.

Cannabis oil (hashish oil) is an extremely potent preparation. It is prepared from the herbal or resin material by liquid extraction; the extract is often concentrated prior to trafficking and contains up to 60% of the active principle (THC).

The major active constituent of cannabis, Δ^9-tetrahydrocannabinol (THC), was first characterized in 1964 by Gaoni and Mechoulam [5]. To date, 70 cannabinoids have been identified [6].

The development of methods for the determination of cannabinoids is an area of increasing interest and a large number of publications appear every year describing a variety of analytical techniques, which vary in sensitivity, specificity, and instrumentation. Articles providing extensive reviews of the various analytical techniques have also been written [2,7–10]. This chapter will focus mainly on the methods published in the recent past with special emphasis on those methods that appear to be more practical and feasible for routine analysis of these compounds in various types of biological specimens. However, because of the large number of publications, this study is not meant to be exhaustive.

Various types of biological samples can be analyzed for cannabinoids to test for marijuana use.

5.2 ANALYSIS OF CANNABINOIDS IN URINE

Urine appears to be the biological fluid of choice to test for the presence of Δ^9-THC metabolites in the human body. Many THC metabolites are excreted in urine, but the major urinary metabolite is Δ^9-THC-11-oic-acid (THC-COOH), either free or conjugated as glucuronide [11]. Urinalysis has the advantage of being able to detect THC metabolites for a relatively long period of time. These metabolites, being highly lipophilic, are readily distributed to body tissues and are slowly eliminated in the urine [12]. THC metabolites, therefore, persist in urine for several days after smoking a single marijuana cigarette, and three to four weeks may be required for elimination of all metabolites in the case of heavy users [13].

The general approach for the analysis of THC metabolites in urine is to screen the samples by an immunoassay method such as radioimmunoassays (RIA), enzyme immunoassays (EIA) or fluorescence polarization immunoassays (FPIA). The presumed positive samples are then confirmed by another more specific method such as gas chromatography-mass spectrometry (GC-MS).

5.2.1 Immunoassays

Immunoassays are the most widely used screening methods for cannabinoids in urine. These methods are based on developing antibodies specific to the drug to be tested and/or one or more of its major urinary metabolites. EIA and RIA are among the most commonly used methods, although the RIA method has lost favor in the recent years.

5.2.1.1 Radioimmunoassay

RIA methods are very sensitive assays, which have been widely used for many years. However, the assays have the inherent disadvantages of limited stability of radio-labelled compounds and the need for special disposal of radioactive materials and special handling to avoid health hazards [14]. Radiolabelling is usually carried out using either ³H or ¹²⁵I. ¹²⁵I radiotracers are usually preferred since higher specific activity can be obtained, and separation and gamma counting are simpler than the liquid scintillation counting used for ³H tracers.

A simple and sensitive RIA method using ¹²⁵I tracer was described by Law *et al.* [15], which required small sample volume and allowed the detection of cannabinoid metabolites many days after consumption. The sensitivity of the RIA method was then coupled with high performance liquid chromatography (HPLC) and the combined HPLC-RIA method was then used for the analysis of THC metabolites in urine and in blood [16]. Clatworthy *et al.* [17] compared the ¹²⁵I-RIA method of Law *et al.* [15] with another ³H-RIA method and the results obtained were confirmed by GC-MS.

The specificity of the Abuscreen® RIA for cannabinoids, a method which was in commercial use for many years, was assessed by Jones *et al.* [18] who examined 41 cannabinoid and non-cannabinoid phenolic constituents for potential cross-reactivity and found that only cannabinoids of the dibenzopyran type structure cross-react with the antiserum. ElSohly *et al.* [19] tested the specificity of the assay with respect to indole carboxylic acids where none of the compounds tested showed any cross-reactivity. Altunkaya and Smith [20] reported false-positive and false-negative results of RIA for cannabinoids in urine samples. The interfering substance was not identified but the authors suggested that the cause was the presence of proteinaceous material in the urine.

Because of the limitations described above, RIA methods have been largely abandoned and their current use limited to specific research applications.

5.2.1.2 Enzyme immunoassays

EIA are the most commonly used screening methods for the detection of cannabinoids in urine today. EIA methods are rapid, simple, and do not require special precautions for handling and disposal.

Several publications describing the utilization of enzyme multiplied immunoassay techniques (EMIT) for the determination of cannabinoids in urine have been reported [21–23]. The initial screening by EMIT was followed by confirmation either by TLC [24,25], HPLC [26,27], or most commonly by GC-MS [28,29].

Rapid, cost-effective urine testing of a large number of urine samples with the EMIT urine cannabinoid assay was automated through the use of a centrifugal analyzer [26,30–32], Monarch analyzer [33], or a chemistry analyzer [34].

Specificity of the EMIT d.a.u. cannabinoid assay with respect to 162 drugs was studied by Allen and Stiles [35].The presence of nabilone, a synthetic cannabinoid used as an antinauseant, did not affect the results of the assay [36].

An enhanced chemiluminescent EIA for the detection of cannabinoids in urine samples was developed by Sharma *et al.* [14]. The assay is based on the horseradish peroxidase catalyzed oxidation of luminol by H_2O_2 in the presence of *p*-iodophenol under mildly basic conditions. The method is sensitive, simple, and suitable for the automation and routine screening of large numbers of samples.

The use of EMIT assays as a semi-quantitative method is controversial. Standefer and Backer [33] reported that quantitative results were obtained from the EMIT assays, using a multiple-point calibration curve which is updated regularly. However, other authors reported many factors that hinder this quantification including: cross-reactivity of the assay with several chemically related substances, sample carryover from highly concentrated samples, and absorbance changes that reached a plateau near the medium calibrator. Therefore, it was suggested that EMIT immunoassays be used only as a qualitative tool [32,37].

5.2.1.3 Fluorescence polarization immunoassays

Colbert *et al.* [38] developed two fluoroimmunoassays for the detection of cannabinoids in urine. The first was a single-reagent polarization immunoassay, which did not require a sample separation step but lacked sensitivity. The second assay had sensitivity comparable with RIA and could be automated. Stopped flow-FPIA (SF-FPIA) was also used for the determination of drugs of abuse in urine. They were suitable for routine screening programs, being faster and having lower detection limits, and better within- and between-assay precision than conventional FPIA [39].

ElSohly *et al.* [40] evaluated the cross-reactivity of the Abbott TDx® cannabinoid assay against a variety of cannabinoid and non-cannabinoid phenolic compounds. The antiserum was found to cross-react equally to 11-nor-Δ^9-THC-COOH, its glucuronide and to the corresponding Δ^8-isomer. The hydroxylated derivatives of Δ^9-THC and Δ^8-THC and other cannabinoids in general show limited binding potential toward the antibody.

The Abbott AxSYM assay for drugs of abuse was evaluated and compared with the Syva EMIT d.a.u./Roche Cobas Mira S Plus, Abbott TDx and ADx, Syva EMIT d.a.u./Syva ETS Plus, Syva EMIT II/Hitachi 717, and Roche Abuscreen OnLine/Roche Cobas Mira S Plus. The system's advantages, including stability of the calibration curves for 3–4 months, the possibility of providing semi-quantitative results, and the ability to process emergency samples, made it useful for routine analysis of drugs of abuse in urine samples [41].

5.2.1.4 Enzyme-linked immunosorbent assays

Microanalysis of cannabis components and their metabolites was also carried out by enzyme-linked immunosorbent assays (ELISA). The application of the method to the analysis of THC metabolites in plasma and urine was suggested [42].

Fraser *et al.* [43] used ELISA and EIA assays for the screening of urine samples for cannabinoids followed by GC-MS confirmation.

5.2.1.5 Kinetic interaction of microparticles in solution

Another type of immunoassay, which depends on the kinetic interaction of microparticles in solution (KIMS), is the Abuscreen OnLine assay. Hailer *et al.* [44] evaluated the Abuscreen OnLine cannabinoids assay using the COBAS FARA II automatic analyzer where modifications were made in the cutoff definition, calibration curve and reagent volume to obtain maximum sensitivity and reagent economy. The results were compared with the EMIT d.a.u. assay, and the authors concluded that the OnLine cannabinoids assay was a good alternative to EMIT d.a.u. in terms of low detection limits, calibration curve stability, and cost effectiveness.

Armbruster *et al.* [45] compared the Roche OnLine assay, the Syva EMIT II assay, and the Abbott TDx FPIA with the Roche Abuscreen RIA assay. The On-Line assay and the EMIT II were reported to be better than the RIA procedure in terms of time and effort.

Microgenics' cloned enzyme donor immunoassay (CEDIA) and KIMS were evaluated for cannabinoids, amphetamines, barbiturates, benzodiazepines, benzoyl-ecgonine, LSD, methadone, and opiates [46]. Cannabinoids showed 99.3% concordant results, where there was only one negative sample by KIMS (cutoff 50 µg/L) and positive by CEDIA at a cutoff level of 25 µg/L. The CEDIA and KIMS results for all eight drugs were in good agreement (93.3–100%).

Feldman *et al.* [47] developed four OnLine DAT II assays by modifying the original KIMS technology for the evaluation of cocaine, methadone, opiates, and THC for improved performance and enhanced ease of use. These assays are being applied to COBAS INTEGRA and Roche/Hitachi line of analyzers. Cutoffs for THC assay were 20, 50, and 100 ng/mL with 0–100, 0–300, and 0–300 ng/mL dynamic ranges, respectively.

5.2.1.6 On-site testing kits

Many on-site testing kits for the analysis of cannabinoids in urine are now commercially available. Compared with laboratory-based immunoassays, these kits have the advantages of being simple, easily performed, allow rapid access to the test results and do not need costly instrumentation or highly trained personnel. Several authors tested the performance of many of these kits and compared their results with other laboratory-based methods.

Armbruster and Krolak [48] evaluated the Abuscreen *ONTRAK* assay (Roche Diagnostic systems) and compared the results with those obtained using RIA, FPIA, and GC-MS confirmation. Results agreement was observed but the authors criticized the subjective nature of identifying the results and the absence of a positive control in the test kit.

The immunoassay TRIAGE[TM] was applied to the detection of several classes of compounds including cannabinoids in postmortem urine samples [49]. Two difficulties were encountered. The first related to the nature of the postmortem urine samples, which contained significant amounts of sediment that reportedly blocked the nylon membrane, inhibiting complete absorption of the reaction mixture after spotting onto the detection area. This was overcome by removing the excess solution from the detection zone and increasing the amount of wash solution used. The second difficulty was the dependence of the color intensity produced on the drug concentration, making judgment of the results difficult, especially for inexperienced users. Nevertheless, the results obtained showed good agreement with the Abbott

ADx FPIA and when compared to GC-MS, a 95% confirmation rate for cannabinoids was reported.

Jenkins *et al.* [50,51] assessed the validity of the EZ-SCREEN® cannabinoid test and the accuPINCH™ THC test for the analysis of cannabinoids in urine. In both cases, 178 clinical urine samples, 72 urine samples containing known amounts of drug, and 50 drug-free urine samples were randomized and analyzed under blind conditions. The results were interpreted independently by three readers. The EZ-SCREEN® cannabinoid test showed high sensitivity for THC-COOH and low cross-reactivity to THC and 11-OH-THC. The LoD was reported to be much lower than the detection average specified by the manufacturer and that positive results should always be confirmed by GC-MS. The assay was easy to perform, provided rapid results, and could be used for on-site drug testing [50].

The accuPINCH™ THC test is a competitive EIA that is used for the detection of THC-COOH and shows relatively low cross-reactivity with THC and other cannabinoids. The assay was highly affected by sample turbidity, which interfered with color interpretation on the detection disk, but the assay was relatively insensitive to changes in sample temperature [51].

Triage® panel for drugs of abuse is a rapid immunoassay for the simultaneous detection of seven drugs in a single sample [52]. De La Torre *et al.* [53] evaluated the degree of concordance between the Triage® results and those obtained by FPIA and demonstrated that the performance of both assays was comparable and that the results of the assay were independent of the laboratory personnel's skills.

The Bionike One-Step tests for the detection of drugs of abuse in urine are used for on-site testing of amphetamines, methamphetamine, benzodiazepines, cannabinoids, methadone, and opiates. These tests were evaluated, and the results obtained were in good agreement with the EMIT d.a.u. assays [54].

Another simple and rapid test that screens for five different classes of drugs of abuse in urine samples is the Advisor™ drug screening system developed by Parsons *et al.* [55]. The system is composed of a multi-chambered vessel that automatically distributes the liquid reagent into distinct assay channels. Each of them tests for a specific class of drugs of abuse. The results of the tests compared well with other automated immunoassays for drugs of abuse.

Korte *et al.* [56] compared the results obtained with RapiTest THC for the detection of cannabinoids in urine with the results obtained with the EMIT d.a.u. and with gas chromatographic-mass spectrometric methods. The results correlate well together when operating above the cutoff concentrations of the methods. At low drug concentration, the color of the band is faint and inexperienced users may find it difficult to judge the results.

Two separate on-site testing kits for drugs of abuse, the ONTRAK TESTCUP and the Abuscreen ONTRAK, were compared, and the results obtained were further compared with another laboratory-based immunoassay, the Abuscreen OnLine [57]. The ONTRAK TESTCUP tests for three drug classes (benzoylecgonine, THC-COOH, and morphine) simultaneously, while the Abuscreen ONTRAK tests have a separate single kit for each drug class. Both systems agreed with the OnLine assays in identifying drug positive and drug negative samples.

The performance of the Abusign™ Drugs-of-Abuse Slide Tests was evaluated by Ros *et al.* [58]. Inter- and intra-individual agreement was tested by comparing the readings of four persons at different time intervals after incubation. Comparison with the FPIA-ADx method was also done and all the samples were confirmed by GC-MS.

For the Abusign cannabinoids (50 ng/mL) slide test, the method was found to be more sensitive than the FPIA-ADx test, but the specificity was lower. The drawback of this method was that the test results depended on the reader and on the time at which the test was read, especially when the concentration of the drug of abuse was near the cutoff. The authors, therefore, concluded that the test was not suitable for screening of drugs of abuse in situations in which a reliable test result was required. The test may be of value in emergency toxicology when a quick result is needed.

Wennig *et al.* [59] developed and evaluated the one-step dip-and-read immuno-chromatographic FRONTLINE® Rapid Tests for drugs of abuse testing in urine samples. Multicenter evaluation of the rapid tests was performed at six European sites, each following the same protocol, by comparing them with FPIA and EMIT assays. The evaluations showed reliable results for the rapid tests of cannabinoids, cocaine, and opiates as compared with the FPIA and EMIT.

Several publications comparing different types of immunoassays to each other and/or to chromatographic methods appearevery year.

Irving *et al.* [60] analyzed 200 urine specimens with two EIA (EMIT-st and EMIT-d.a.u.) and an RIA (Abuscreen RIA), and those samples found to be positive were further analyzed by gas–liquid chromatography with flame ionization detection (FID), gas–liquid chromatography/mass spectrometry, and an experimental RIA from the Research Triangle Institute. The aim of this study was to evaluate the two EIA by comparing the results with those obtained from other methods. The two assays were found to give 98–94% confirmation rates for positive results when compared with GC-MS. The authors noted that the high cutoff levels established eliminated false positives but allowed a high false-negative rate. Attempts to quantify the results of the RIA were unsuccessful.

Jones *et al.* [61] compared five methods, namely, Abuscreen RIA, EMIT d.a.u., HPLC, GC/electrochemical detection (ECD), and GC-MS, for the analysis of THC-COOH in urine. RIA and the EIA were used as screening procedures, and the other methods were used for confirmation of presumptive positives. Quantitative estimates obtained by the immunoassay procedure were always higher than those obtained by the chromatographic methods, probably because of the cross-reactivity of other THC metabolites with the antisera of both immunoassay procedures. The data obtained from the chromatographic methods were compared, and good correlation coefficients were obtained. The effect of storage of urine samples was studied and found to affect the concentration of THC-COOH.

Another comparative study between six cannabinoid metabolite assays was presented by Frederick *et al.* [62]. These assays were two EIA (EMIT-st and EMIT d.a.u.), two RIA (Abuscreen RIA and Immunalysis), one TLC assay (Toxi-Lab), and a new GC-MS method. The four immunoassays were used for screening purposes because of their simplicity and speed. When low levels of THC-COOH were present, the Immunalysis RIA was recommended, while the EMIT-st and the Abuscreen were useful for screening higher levels of THC-COOH. The Toxi-Lab TLC and the GC-MS methods could both be used for confirmation.

Comparison between the TDx assay and the EMIT-Cobas assay for the detection of cannabinoids in urine from prison inmates was done by Karlsson and Stroem [63]. HPLC was used for confirmation. It was found that high background urine may affect the reliability of the results of the TDx assay, a problem that can be solved by diluting the samples and reanalyzing, or by setting the instrument background to a higher level. Apart from this, the TDx assay was reported to be reliable, with an excellent precision and curve stability. The EMIT-Cobas was reported to be faster, with the time to analyze one carousel being approximately 8 min, compared to 20 min for one TDx carousel. However, it was necessary to run the EMIT calibrators in each carousel because of the lack of curve stability.

Comparative results of five cannabinoid immunoassays were reported by Barnhill and Wells [64]. The five assays were the cannabinoid TLC assay (Toxi-Lab), the Syva EMIT urine cannabinoid assay, the DPC cannabinoids double antibody RIA, the Abuscreen RIA, TDx cannabinoids assay, and the urine THC direct RIA (Immunalysis). In general, the RIA gave a greater proportion of positive results than did the EIA or the FPIA.

Kogan *et al.* [65] compared the results of the Syva EMIT® d.a.u. and the Roche Abuscreen® RIA, which were the most widely used, commercially available immunoassays for detecting cannabinoids in urine. The results of both assays agreed qualitatively; however, there was no correlation between the semi-quantitative values obtained from both methods. The results of the immunoassays were confirmed by a modified bonded-phase adsorption/thin layer chromatography (BPA-TLC) and by GC-MS. The BPA-TLC was based on a visual color reaction between the developed spots and the spraying reagent, Fast Blue RR. It was a simpler noninstrumental technique, easier to interpret than quantitative GC-MS, and could be used successfully when only a qualitative confirmation is needed. However, the technique had limited utility for forensic purposes only.

Comparison of the Abbott FPIA and the Roche RIA for the analysis of 142 urine samples containing THC-COOH with subsequent confirmation by GC-MS was done by Budgett *et al.* [66]. The authors concluded that both immunoassays produced similar results and either of them could be used in a mass-drug-screening laboratory.

Weaver *et al.* [67] correlated the results of three commercial immunoassay kits, Abuscreen®, TDx®, and EMIT® with the concentration of THC-COOH determined by GC-MS. None of the methods studied showed perfect correlation with the results of GC-MS, but a significant correlation still exists. Attempts to select an appropriate cutoff value for each assay based on the derived regression equation were also done.

Another comparative study was conducted by Altunkaya *et al.* [68], who compared the results of four immunoassays, namely, EMIT d.a.u. Cannabinoid 20 (Syva Corp.), DPC cannabinoids RIA (Diagnostic Products Corp.), and the Roche Diagnostics System's Cannabinoids-1 RIA and Cannabinoids-2 RIA assays. The four immunoassays correlated well with GC-MS, but the DPC-RIA was selected by the authors as the method of choice because it provided quantitative results that might be used to calculate the concentration of the extracts to be injected on the GC-MS.

Armbruster *et al.* [45] compared three non-radioisotopic immunoassays with the RIA (Roche Abuscreen) previously used in their laboratory and reported that the RIA tests had several drawbacks including short reagent shelf-life, a need for special handling and disposal of wastes, and the requirement of a fully automated system for analysis. The assays compared were the Syva EMIT II, the Abbott TDx FPIA, and the Roche OnLine. RIA and OnLine assays exhibited equivalent performance, detecting 99% of GC-MS marijuana confirmed samples. The TDx detected 95% of the samples, while the EMIT II assay detected 88%. The EMIT II and the OnLine assays were reported to be better than the RIA procedure in terms of time and effort.

A similar comparative study was conducted by Kintz *et al.* [69], where the results of the EMIT d.a.u., the Abbott ADx FPIA, and the Abuscreen OnLine assays were correlated with the GC-MS method. All methods compared favorably and could be successfully used for the screening of THC-COOH in urine samples. However, there was no correlation between the quantitative results obtained by the immuno-assays and those by GC-MS, possibly due to the presence of different cross-reacting metabolites of THC.

Comparison between six immunoassays (EIA-EMIT and EZ-SCREEN, FPIA-ADx, RIA-Coat-A-Count, LI-Abuscreen ONTRAK, and CBI-Triage) and three chromatographic methods (TLC-Toxi-Lab, HPLC, and HPLC-REMEDI Drug Profiling System) with GC-MS confirmation of the results was done by Ferrara *et al.* [70]. The values of sensitivity, specificity, false-positive and false-negative rates were reported for each technique. Statistical analysis of the results allowed the determi-nation of predictive positive and negative values for each single technique and for combinations of immunochemical and chromatographic techniques. A decision-making process for the determination of the best combination of these techniques was also presented.

Huestis *et al.* [71] studied the detection times of cannabinoids in urine following the administration of a single marijuana cigarette using different commercial can-nabinoid immunoassays (EMIT® d.a.u.™ 100, EMIT d.a.u. 50, EMIT d.a.u. 20, EMIT II 100, EMIT II 50, Abuscreen® OnLine™ and Abuscreen RIA, DRI™, and ADx). The results were compared with GC-MS results at a 15 ng/mL cutoff concentration.

The effect of adulterants in urine samples on RIA and on FPIA was studied [72]. A number of readily accessible chemicals such as sodium chloride, bleach, potassium hydroxide, soap, 2-propanol, and ammonia, were added to test tubes containing urine samples, which were then analyzed by RIA and FPIA. For the THC-COOH RIA, false positives occurred with potassium hydroxide and bleach adulterants, while soap caused false-negative results. No adulterant caused FPIA false positives, but false negatives were observed with bleach.

A comparison was made for five non-instrumental urine drug testing devices (Syva RapidTest d.a.u. 8, Syva RapidCup d.a.u. 5, RocheTestcup 5, Biosite Triage, and

Casco-Nerl microLINE Drug Screen Card), using a challenging clinical specimen set with drug concentrations close to the immunoassay screening cutoffs [73]. Based on GC-MS confirmation cutoffs, the non-instrumental devices demonstrated an overall accuracy of 70% (66–74%) when compared with the Syva ETS analyzer (80%).

A comparison was made between on-site immunoassay drug-testing devices and GC-MS [74]. In this study, 800 people and two devices for oral testing and eight on-site devices for urine were used. Good results were obtained for the urine on-site devices, with accuracies of 83–99% for amphetamines, 97–99% for cannabinoids, 94–98% for opiates, and 90–98% for benzodiazepines. Detection of amphetamines and opiates was possible in oral fluids with the on-site devices, but these devices were not sensitive for the lower levels of benzodiazepines and cannabinoids.

5.2.2 Chromatographic methods

Chromatographic methods can be used for qualitative and quantitative screening and/or confirmation of cannabinoids in biological specimens [2]. For the analysis of urine specimens, these methods focus mainly on the major urinary metabolite, THC-COOH. A preliminary hydrolysis step is often required to analyze the free and the glucuronide forms, which increases the concentration of THC-COOH. Hydrolysis can be done enzymatically, using β-glucuronidase enzyme or with strongly alkaline solutions such as sodium or potassium hydroxides, since the majority of the THC-COOH exists as an ester glucuronide. Unlike immunoassays, chromatographic methods require extensive sample clean up using either liquid–liquid extraction methods or solid-phase extraction (SPE) methods.

5.2.2.1 Thin layer chromatography

Thin layer chromatography (TLC) has been used for the screening and identification of cannabinoids for many years. Immunoassays have almost replaced TLC as a screening method. However, TLC can still be used in developing countries where instrumentation and reagents required by other methods might be lacking. The availability of HPTLC plates, which improves the separation of compounds over that obtained by regular TLC plates and the development of densitometric techniques, which allow in situ determination of the separated compounds on the plate, may increase the use of TLC again. TLC methods have the advantage of being more specific to THC-COOH than immunoassays, which are known to cross-react to many THC metabolites. Several publications reported the use of TLC as either a screening or a confirmatory technique.

Nakamura *et al.* [75] used a TLC procedure previously described by Kaistha and Tadrus [76] as a screening and a clean-up procedure for the isolation of THC-COOH from urine samples. The spot corresponding to THC-COOH was visualized with Fast Blue B and then scraped off the plate and eluted with methanol for further analysis by GC-MS.

Kanter *et al.* [77] developed a sequential TLC method for the isolation and identification of THC-COOH from urine. In this method, the pH of a volume of urine containing 50 mg of creatinine was adjusted to 4.7–6.3; hydrolyzed with β-glucuronidase, extracted with ether, washed with 5% NaHCO$_3$, and then evaporated under nitrogen. The residue was dissolved in dichloromethane, spotted on a silica gel G plate, and chromatographed sequentially with two mobile phases, the first consisting of acetone–chloroform–triethlyamine (80:20:1) and the second consisting of petroleum ether–ether–glacial acetic acid (50:50:1.5). The plate was sprayed with a freshly prepared alkaline solution of Fast Blue B. A magenta red color of R$_f$ approximately of 0.1 or corresponding to that of a reference standard indicated a positive response. The results obtained were compared with those produced by EMIT. Good correlation was obtained for samples having a THC-COOH concentration above the detection limit of the immunoassay technique or for completely negative samples; those samples in the borderline range gave mixed results, which could be explained by the fact that immunoassay measures total cannabinoids while TLC measures THC-COOH only.

Lillsunde and Korte [78] used TLC for preliminary screening of drugs of abuse in urine samples followed by confirmation by GC-MS. For screening of cannabinoids, samples were extracted with *n*-hexane–ethyl acetate (7:1) after alkaline hydrolysis with 10 N KOH. The extract was evaporated and the residue dissolved in 50 μL ethanol and applied onto a TLC plate. *n*-Hexane–1,4-dioxane–methanol (35:10:5) was used as mobile phase, while alkaline solution of Fast Blue B was used as the spraying reagent. THC-COOH was confirmed by GC-MS as its methylated derivative.

Commercially available TLC procedures for the detection of THC-COOH in urine are also available. These include the TOXI-LAB Cannabinoid Screen method, the TOXI-GRAMS MS (THC) and the Toxi.Prep THC metabolites. These methods have been evaluated by many authors [21,22,62,64,79,80].

In the TOXI-LAB procedure, urine samples were hydrolyzed at room temperature with KOH and then extracted with a mixture of ethyl acetate and hexane (1:9). The extracts were concentrated onto discs, and those discs were inserted into a toxigram together with a blank toxi disc and a standard disc containing 350 ng of Δ^8-THC-COOH. The plate was then developed using a mixture of heptane–acetone–glacial acetic acid (70:30:1) and visualized with Fast Blue BB salt. The TOXI-LAB method allowed simultaneous extraction of 10 samples with one control and one standard, using a disposable applicator cartridge. Frederick *et al.* [62] compared the Toxi-Lab cannabinoid screen method with four commercially available immunoassay procedures and a GC-MS method, while Wells *et al.* [64] compared it with five cannabinoid immunoassay systems. Foltz and Sunshine [22] compared it with the EMIT d.a.u. assay and with a reference GC-MS method. Sutheimer *et al.* [21] compared the TLC method with two EIA methods, EMIT-st and EMIT d.a.u. In general, the Toxi-Lab procedure was simple, easy to perform, and required minimal cost and instrumentation. The system did not provide the high throughput capacity of automated EMIT but was much better than conventional TLC [22,79]. The Toxi-Lab assay was reported to be successfully used as a screening method for urine samples or as a confirmatory technique to the immunoassays to minimize the need and cost of the GC-MS confirmation [21].

The TOXI-GRAMS MS (THC) procedure was described by King *et al.* [79]. It consisted of biphasic thin layer chromatograms made of glass-fiber paper impregnated with silica gel and chemically modified alkyl-silica layer along one edge. Urine samples were hydrolyzed at room temperature with KOH, then acidified with glacial acetic acid, and transferred to the cells of a cartridge applicator. The chromatogram was developed in *n*-heptane–acetone–glacial acetic acid (50:50:1) and visualized with Fast Blue BB salt. This method was reported to have the advantages of increased sensitivity and increased specificity. All the THC-COOH extracted from the sample was applied to the plate, minimizing sample loss associated with liquid–liquid extraction, drug adsorption into glass, and transfer of extracts to TLC plate. Therefore, the sensitivity was greatly enhanced and, when compared with the liquid–liquid extraction method of Sutheimer *et al.* [21], showed lesser interferences from co-extracted drugs and urinary artifacts; thus specificity was also increased [79].

The Toxi-Prep (TP) system is a semi-automated system that utilizes the SPE technique for the extraction of THC metabolites from urine. Steinberg *et al.* [80] compared the Toxi.Prep THC metabolites system with the Toxi-Lab cannabinoid screen method for evaluating THC metabolites in urine. In the TP method, urine samples were hydrolyzed, loaded onto a preconditioned column, and the columns were washed with 0.5 mL of 20% acetic acid followed by 0.5 mL hexane. Acid elution reagent (400 μL, hexane–ethyl acetate–glacial acetic acid (70:30:0.1)) were added to each SPE column and allowed to spot directly onto the chromatogram. The chromatograms were developed using heptane–acetone–glacial acetic acid (70:30:1) and visualized by Fast Blue BB salt followed by exposure to diethylamine vapors. The TP system was reported to have many advantages over the Toxi-Lab method including 40% labor reduction by automation of the different steps of extraction, washing, and spotting, leading to cost reduction, the need for less extraction solvent and less urine, and giving cleaner chromatograms, which result in increased sensitivity.

The BPA/TLC method for the determination of THC-COOH in human urine was developed by Kogan *et al.* [81]. In this method, 10 mL urine was hydrolyzed with NaOH, then the pH was adjusted to pH 1–3 and extracted with Bond-Elut THC columns. THC-COOH was eluted with acetone. Methylene chloride was added to the eluate, the mixture was vortexed, and the upper layer removed. The lower layer was then partitioned with hexane to get rid of any remaining water, the hexane was evaporated, and the residue reconstituted with 10 μL acetone and spotted on a TLC plate. The developing system was ethyl acetate–methanol–water-conc. ammonia (12:5:0.5:1) and the spraying reagent was Fast Blue RR. This method could be used as a confirmation method for the EMIT cannabinoid drug screen procedure.

The visualization step was modified by spraying the plate after developing with concentrated ammonium hydroxide then with Fast Blue RR spray [65]. The base intensified the color and made visualization of THC-COOH instantaneous. The authors used the modified method for confirmation of EMIT d.a.u. and Abuscreen RIA urine cannabinoids immunoassays, and the results were compared with GC-MS. The non-instrumental BPA-TLC assay was simpler to perform and interpret than the

GC-MS and could be used for qualitative confirmation of THC-COOH in urine after screening with immunoassays.

Vereby *et al.* [82] applied the method of Kogan *et al.* [65] to the confirmation of 100 urine samples which screened positive for cannabinoids by EMIT d.a.u. Another modification of the method was done by Vu Duc [25] who quantitated the method using scanning densitometry at 485 nm, and used petroleum ether (40–60°C): diethyl-ether–glacial acetic acid (5:5:0.1) as the developing system to obtain better separation of THC-COOH and 11-OH-Δ^9-THC. The author also reported that the thin layer plates could be stored in a freezer, wrapped in aluminum foil, for further analysis by GC-MS. This could be done by scraping the spots corresponding to THC-COOH and eluting with ethyl acetate followed by derivatization with TMS. This was advantageous since two confirmation methods could be applied to a single urine specimen.

High efficiency TLC (HETLC) together with an HPLC technique was used by Black *et al.* [26] for confirmation of EMIT urine cannabinoid assay. The method used for the isolation of THC-COOH from urine samples was that developed by ElSohly *et al.* [83] and consisted of the addition of an internal standard, followed by basic hydrolysis, then extraction on a Bond-Elut-THC column. Elution was done with acetonitrile. For HETLC, the eluant was evaporated and the residue reconstituted with methanol and applied on an HETLC plate. The plate was developed using hexane–acetone–glacial acetic acid as mobile phase, and the spots were visualized using an alkaline solution of Fast Blue B salt as the spraying reagent. The results of HPLC and HETLC were always in agreement, suggesting the use of HPTLC as a confirmatory technique for EMIT.

Another HPTLC procedure for the detection of THC-COOH in urine was described by Meatherall and Garriott [84]. This method involved alkaline hydrolysis of the urine sample followed by extraction of THC-COOH from acidified solution with hexane. The hexane was evaporated and the residue reconstituted with 50 μL of $CHCl_3/CH_3OH$ and spotted onto a Kieselgel 60 HPTLC. Development was done using heptane–butanol–acetic acid (90:9:1) as mobile phase, and visualization was done by sequential dipping of the plate in diethylamine, then in 0.1% Fast Blue BB solution. Cannabinol (CBN) was used as internal standard; although the R_f for THC-COOH and CBN were variable, the RR_f was consistent. Fast Blue B, Fast Blue RR, and Fast Blue BB were tried as visualizing reagents, and no differences in the color intensity were observed. Fast Blue B and RR dissolved more slowly in water and imparted a yellow background to the plate. Moreover, Fast Blue B is a potential carcinogen; therefore, 0.1% solution of Fast Blue BB was chosen for routine use.

A qualitative TLC method for the identification of cannabis metabolites in human urine was described by Haensel and Stroemmer [85]. Quantitation of THC-COOH can be done using densitometry [86].

5.2.2.2 High performance liquid chromatography

Combining the separating power of HPLC with different detectors has led to the development of several methods that can be used for the detection of cannabinoids in urine samples.

HPLC with immunoassay detection is a powerful tool that couples the specificity and the separation power of HPLC and the sensitivity of the RIA. It was first introduced by Twitchett *et al.* [87] and was used for the analysis of LSD in body fluids. The combined technique was then used for the analysis of THC and its metabolites in urine and plasma [88–90]. The coupling of the two techniques overcame the problems of cross-reactivity of the RIA and allowed the use of a sensitive and relatively non-specific antigen in the RIA [90]. For the analysis of urine, hydrolyzed samples were injected onto the HPLC column and a stepped solvent elution program was used. The concentrations of THC, CBN, mono-hydroxylated metabolites, di-hydroxylated metabolites, Δ^9-THC-11-oic acid, Δ^9-THC-11-oic acid ester glucuronide can be quantified in the eluting fraction by RIA. The method cannot be used, however, for routine use because of the low sample throughput. A modified method using single acidic elution instead of the stepped gradient elution [91] and a ^{125}I-RIA method [15] was then used by Law *et al.* [16] for the confirmation of cannabis use by the analysis of blood and urine samples. Peat *et al.* [92] studied the HPLC-IA profiles for the analysis of cannabinoid metabolites in urine samples. The samples were chromatographed on a reverse-phase system using a gradient of acetonitrile in water (pH 3.3). Four different antisera, three different RIA procedures, and one EMIT were used for the detection of the eluting fractions.

An HPLC method with UV detection for the determination of THC-COOH was developed by ElSohly *et al.* [83]. Hydrolyzed urine samples were cleaned up using Bond-Elut®-THC columns, and then injected on a reverse-phase column with acetonitrile–phosphoric acid (50 mM) (65:35) as the mobile phase. The clean-up procedure using Bond-Elut® columns had the advantages of saving time and reagents, and the final eluate was clean and could be injected directly onto the HPLC column without evaporation or derivatization. The HPLC method described was rapid and reproducible and could be used as an alternative to GC. This method was compared with four other previously published methods, namely, RIA, EIA, GC/ECD, and GC-MS [61] and was adopted by Black *et al.* [26] for the confirmation of positive results obtained using the EMIT Urine Cannabinoid assay.

Preliminary sample preparation using SPE methods followed by HPLC analysis with UV detection was also used by many authors [93–96].

Bourquin and Brenneisen [93] used Bond-Elut®-THC-SPE columns for the isolation of THC-COOH, which was analyzed by HPLC on C_8 column using acetonitrile–aqueous phosphoric acid (50 mM) (68.5:31.5) as eluting solvent followed by photodiode-array detection. The method was used to confirm 100 urine samples screened positive by immunoassays.

Parry *et al.* [94] used Supelclean DrugPak-T SPE tubes for the isolation of THC-COOH from urine samples prior to analysis by HPLC or GC and reported absolute and relative recoveries higher than 85% and 92%, respectively. HPLC analysis was then performed using a C_{18} column and 55:45 mixture of acetonitrile and 2% acetic acid in water as the mobile phase followed by UV detection at 280 nm.

Ferrara *et al.* [95] used various types of SPE columns for the isolation of metabolites of drugs of abuse from urine samples. Adsorbex RP8 100-mg columns

(Merck) were used for the isolation of THC-COOH and chromatographic separation was done on C_8 column using 0.05 M phosphoric acid–acetonitrile (35:65, v/v) as the mobile phase.

Bianchi and Donzelli [96] used disposable C_{18} SPE cartridges (100 mg) from Bio-Rad Labs and a reversed-phase column with acetonitrile–phosphate buffer (0.125 M) (55:45) as the mobile phase. The proposed method was reported as being precise, sensitive, and linear over a wide range of concentrations, did not require more than 30 min, and could, therefore, be used for routine analysis of large numbers of samples.

THC-COOH can be determined in urine samples by a combination of liquid chromatography with UV detection and gas chromatography (GC) with electron-capture detection [97]. Delta-8-THC-11-oic acid was used as the internal standard, and the pentafluoropropyl-pentafluoropropionyl derivatives were used for GC. HPLC served as a clean-up step for the GC analysis, leading to an increase in the selectivity and sensitivity of the method. Moreover, the LC step could be used alone for the determination of THC-COOH in high concentrations. However, HPLC remained a sophisticated tool for use in sample clean up; therefore, another procedure was presented by L. Karlsson [98]. The author described a fully automated HPLC system in which hydrolyzed urine samples were directly injected onto a CN pre-column, followed by chromatographic separations on two different columns (CN and C_8) in series by means of a column-switching technique. Two detectors were used: an UV detector after the first column, and an electrochemical detector after the second column. This method was reported to have the advantages of selectivity, low detection limit (2 ng/mL), and minimum sample pre-treatment; however, a long time was needed for each run and the sample throughput was therefore low (two urine samples per hour).

Another HPLC method with EC detection for the determination of THC metabolites in urine was presented by Nakahara *et al.* [99]. The method involved automatic sample extraction with ODS-minicolumns followed by separation of THC, THC-COOH, and 11-OH-THC on a reversed-phase silica C8 column with acetonitrile–methanol–H_2SO_4 (0.02 N) (35:15:50) as the mobile phase. The method was linear in the concentration range of 10–500 ng/mL, and the limit of detection was 0.5 ng/mL.

5.2.2.3 Liquid chromatography/mass spectrometry

Weinmann *et al.* [100] developed a method using automated SPE and LC coupled to tandem mass spectrometry (LC/MS-MS) with negative atmospheric chemical ionization (APCI) for the detection of THC-COOH in urine samples. Prior to SPE, conjugates of THC-COOH were hydrolyzed. No derivatization step was needed and the run time was 6.5 min. Thus, this method reduces the sample preparation step and also provides a shorter analysis time. The LoD and LLoQ were 2.0 and 5.1 ng/mL, respectively. Another method was developed for the detection of THC-COOH and THC-COOH-glucuronide [101]. THC-COOH and THC-COOH-glucuronide were extracted in one step using ethyl acetate–diethylether (1:1, v/v). The generation of molecular ions of THC-COOH (MH$^+$, m/z 345) and THC-COOH-glucuronide (MH$^+$, m/z 521) was achieved using a PE/SCIEX turboionspray source in positive ionization mode. THC-COOH-d$_3$ was used as the internal standard.

5.2.2.4 Gas chromatography

5.2.2.4.1 Gas chromatography/flame ionization detection

Irving *et al.* [60] used gas–liquid chromatography with FID and GC-MS for the confirmation of the positive results of immunoassays. The authors concluded that the GLC/FID method was not sufficiently sensitive, and a more sensitive assay was needed if higher confirmation rates were to be attained.

Parry *et al.* [94] used GC/FID for the analysis of urine samples after extraction using Supelclean DrugPak-T SPE tubes and derivatization with BSTFA.

5.2.2.4.2 Gas chromatography/electrochemical detection

ElSohly *et al.* [102] developed a gas chromatographic/electron-capture detection GC/ECD procedure for the determination of THC-COOH in urine samples. Samples were hydrolyzed with 10 N KOH, shaken with 2 mL hexane–ethyl acetate (7:1), and the organic phase was discarded. The pH of the aqueous phase was adjusted to 2–2.5 and the THC-COOH and CBN-COOH (used as internal standard) were extracted with hexane–ethyl acetate (7:1). Derivatization was done with pentafluorobenzyl bromide (PFBBr) in a biphasic system using benzyl tributylammonium hydroxide as a phase transfer catalyst. Jones *et al.* [61] compared the previously described procedure with four other published methods, namely RIA, EIA, HPLC, and GC-MS. The described procedure was sensitive, accurate, and reproducible and needed only a small volume of urine. Another GC/ECD for the determination of THC-COOH in human urine was presented by Rosenfeld *et al.* [103]. They increased the specificity of the assay by selective derivatization of the phenolic group using PFBBr in pentanol in alkaline medium (0.1 N NaOH), and by purification by chromatography on XAD-2 resin to produce an extract almost free from interference.

Micellar electrokinetic capillary chromatography (MECC) with on-column multiwavelength detection was used for the analysis of THC-COOH in urine samples. This technique required concentrated samples; therefore, the extraction and concentration steps were very important for the analysis. Four different SPE columns, namely, Bond-Elut THC cartridges, Bond-Elut Certify II columns, Clean Screen THC columns, and Bond-Elut Certify columns were investigated. The first two SPE columns provided a simple and clean electropherogram but the recovery of THC-COOH was low. Clean Screen THC and Bond-Elut Certify columns provided a more complex electropherogram but the peak corresponding to THC-COOH was well separated and the extraction efficiency was good (80±10%). Therefore, these columns were used for the confirmation of urine samples screened positive by FPIA [12].

5.2.2.5 Gas chromatography-mass spectrometry

GC-MS is the method of choice for the confirmation of cannabinoids in urine [104]. It has the highest sensitivity and specificity of all the techniques. GC-MS methods are usually used as reference for evaluating other cannabinoid assays [2].

A modified GC-MS procedure for the detection of past and recurrent marijuana use was described by Joern *et al.* [105]. The method, a modification of the methods of Karlsson *et al.* [106] and Foltz *et al.* [107], included preparing the standards in alkaline solution to minimize adsorption onto glass and plastic surfaces and using potassium hydroxide–methanol (1:4) for hydrolysis to obtain a cleaner extract. The internal standard was d_3-THC-COOH, and the derivatizing agents used were pentafluoropropionic anhydride (PFPA) and pentafluoropropanol (PFPOH). The new GC-MS method was reported to be more indicative of recent marijuana use than the EMIT semi-quantitative concentration values. Stout *et al.* [108] used PFPA and PFPOH for derivatization of THC and THC-COOH in the evaluation of the performance of d_3-THC-COOH and d_9-THC-COOH as internal standards. The method utilized a positive pressure manifold anion-exchange polymer-based SPE, which was followed by elution directly into the automated liquid sampling (ALS) vials. The LoD for THC-COOH was 0.875 ng/mL by GC-MS. To answer the question of whether a positive drug test for marijuana was the result of the sole use of Marinol®, ElSohly *et al.* [109] used Tetrahydrocannabivarin-9-carboxylic acid (THCV-COOH) as a marker for marijuana ingestion. THCV is the C3 homolog of THC, commonly found as a companion cannabinoid to THC in the cannabis plant, which is metabolized by human hepatocytes to THCV-COOH.

Needleman *et al.* [110] developed a liquid–liquid extraction method followed by GC-MS for the determination of THC-COOH in urine. The extraction procedure used isobutanol–hexane (1:9) for initial extraction from urine samples followed by back extraction into 0.1 N NaOH. The aqueous layer was again extracted with methylene chloride, which was evaporated to dryness. The sample was derivatized with tetramethylammonium hydroxide–dimethyl sulfoxide (1:1) followed by the addition of iodomethane.

Clouette *et al.* [111] developed a GC-MS with electron ionization mode for the determination of THC-COOH utilizing its *t*-butyldimethylsilyl derivative. Trideuterated THC-COOH was added to the samples followed by alkaline hydrolysis and extraction with hexane–ethyl acetate (7:1.5) from acidic solution. Derivatization was done with MTBSTFA at 110°C for 15 min. The derivative obtained was more stable than the trimethylsilyl derivative and could be used for routine analysis of THC-COOH in urine samples.

Most of the GC-MS procedures developed focused on the determination of THC-COOH as a marker for marijuana use, with little or no attention given to other metabolites. Kemp *et al.* [112,113] developed a GC-MS method for the simultaneous determination of THC and six of its metabolites, namely, 8α-OH-THC, 8β-OH-THC, 11-OH-THC, 8α,11-diOH-THC, 8β,11-diOH-THC, and THC-COOH, in addition to CBN and cannabidiol (CBD). The different steps described in the procedure were optimized to achieve cleaner extracts, maximum recovery of the analytes and adequate chromatographic resolution of the extracted compounds. Therefore, the influence of hydrolysis conditions (base hydrolysis or enzyme hydrolysis, enzyme concentration and incubation time), solvent combinations used for extraction, and type of derivatizing agent were studied. Optimum results were obtained using enzyme hydrolysis with 5000 units of bacterial β-glucuronidase from *Escherichia coli* incubated at pH 6.8 for 16 h [113]. Extraction was done with hexane–ethyl acetate (7:1) and derivatization was done with BSTFA in 1% TMCS.

Szirmai *et al.* [114] described a GC-MS method for the determination of three major acidic metabolites of Δ^1-THC, namely, THC-7-oic acid, 1,4″, 5″-bisnor-Δ^1-THC-7,3″-dioic acid, and 4″-hydroxy-Δ^1-THC-7-oic acid. Five derivatization systems (CH_2N_2-BSTFA, CH_2N_2-MBTFA, BSTFA, TFE-PFPA, and TMAH-methyl iodide) were examined.

All the procedures previously mentioned used liquid–liquid extraction method for the isolation of THC metabolites from urine samples. SPE methods were developed in an attempt to produce cleaner and more concentrated extracts. Comparison between four extraction procedures for the isolation of THC-COOH from urine samples was presented by Congost *et al.* [115]. The procedures presented were two solid–liquid methods and two liquid–liquid methods. The first solid–liquid procedure used octadecylsilane-bonded silica resin while the second procedure used an ion exchange (NH_4^+ Cl^- resin). In one liquid–liquid procedure, the acidified urine samples were extracted with hexane–ethyl acetate (7:1), the organic layer was extracted with alkali, and the solution was acidified and re-extracted with hexane–ethyl acetate (7:1). The other liquid–liquid extraction method involved a one-step extraction with hexane–ethyl acetate (9:1) from alkaline solution. The best results were obtained with the last procedure. The authors also suggested a derivatizing agent consisting of a mixture of *N*-methyl-*N*-trimethylsilyl-trifluoroacetamide (MSTFA), trimethyliodosilane (TMIS), and dithioeritrithol (100:0.2:1, v/v/w) and compared it with MSTFA.

SPE methods are gaining increasing use in sample preparation techniques, and many publications appear each year utilizing and/or evaluating SPE cartridges. Nakamura *et al.* [116] used Sep-PAK cartridges for clean-up of urine samples prior to GC-MS analysis. McCurdy *et al.* [117] used C_{18} bonded-phase adsorption (BPA) columns for the extraction of THC-COOH in evaluating the suitability of the ion-trap detector for the detection of THC-COOH, while Paul *et al.* [118] used cartridges containing strongly basic anion-exchange resin (E.I. Du Pont de Nemours & Co) for the detection of THC-COOH using GC-MS. Supelclean DrugPak-T SPE tubes were evaluated by Parry *et al.* [94], CLEAN SCREEN® reduced solvent volume (RSV) SPE columns were evaluated by O'Dell *et al.* [119], and Empore extraction disk cartridges (C_{18}) were evaluated by Singh and Johnson [120]. The Toxi-lab SPEC extraction discs were used by Wu *et al.* [121] for the extraction and simultaneous elution and derivatization of THC-COOH to produce the trimethylsilyl derivatives.

Quantitative interpretation of the results of chromatographic methods necessitates the use of internal standards like 11-nor-9-carboxy-CBN [61,83,102], CBN [93], oxyphenbutazone [122], and ketoprofen [115]. The most commonly used internal standard is the trideuterated derivative of Δ^9-THC-COOH [105,111,112,119,120]. The trideuterated isomer has the disadvantage of having a fragment in common with the natural metabolite at m/z 316 when using the methyl derivative [123]. This results in distortion of the ion ratio of the internal standard and limits the dynamic range of the analysis. Therefore, ElSohly *et al.* [123,124] developed a new internal standard, hexadeutero-Δ^8-THC-9-COOH, having the advantages of a wider linear dynamic range and having no common ion with THC-COOH using different derivatives. The d_6-THC-COOH was used by Wu *et al.* [121] for the analysis of THC-COOH in urine samples by GC-MS.

A new internal standard, $^2H_{10}$-Δ^1-THC-7-oic acid was evaluated by Szirmai *et al.* [114] and can be used as an alternative to the previous internal standards. Stout *et al.* [108] evaluated the performance of d_3-THC-COOH and d_9-THC-COOH as internal standards. The authors determined that d_9-THC-COOH was the preferred internal standard for their method.

5.3 ANALYSIS OF CANNABINOIDS IN BLOOD

The analysis of cannabinoids in blood is an alternative to urine analysis, where THC and its metabolites can be detected for a relatively short time after intake. Therefore, the detection of THC along with its metabolites indicates the recent use of cannabis and their levels may correlate with an actual state of intoxication.

5.3.1 Immunoassays

Immunoassay methods for screening blood samples for cannabinoids are now widely used. The methods employed are often based on the use of those tests primarily developed for use with urine samples.

5.3.1.1 Enzyme multiplied immunoassay techniques

In 1978, E. L. Slightom [125] first reported the application of homogenous EIA to the analysis of drugs in biological fluids other than urine. This was followed by many attempts to refine the EMIT assays for use with blood samples.

Asselin *et al.* [126] described a simple method for the detection of THC in methanolic extract of blood using EMIT d.a.u. cannabinoids urine assay. This method had the advantage of requiring only 1 mL of whole blood, and it also avoided the lengthy extraction procedure previously used. The results obtained encouraged many authors to use methanolic blood extracts for the detection of cannabinoids [127–129].

Perrigo and Joynt [127] made two modifications in the procedure suggested by Syva in the 3M619 Kit product literature to improve the sensitivity of the assays. These modifications included increasing the amount of the sample in the measurement kit and increasing the flow cell temperature. Coupling the advantages of using the methanolic blood extraction procedure with those of using an automatic analyzer allowed the processing of a large number of samples in a short period of time and at low cost. Moreover, the small volume requirements of the automatic analyzer resulted in a five to ten-fold drug enrichment [129,130].

The addition of *N,N*-dimethylformamide (DMF) to serum, plasma, or blood resulted in a clear, colorless supernatant, which does not cause light scattering or irrelevant absorbance in the spectrophotometric measurements of the EMIT analysis [131].

Another procedure for the extraction of THC metabolites from whole blood was suggested by Lewellen and McCurdy [132]. This procedure involved precipitation of the blood proteins with acetone, followed by evaporation and reconstitution of the residue in a 1:1 ratio of EMIT buffer and methanol.

5.3.1.2 Fluorescence polarization immunoassays

Bogusz *et al.* [133] determined drugs of abuse in whole blood by FPIA (FPIA–Abbott TDx and ADx) after protein precipitation with acetone. The results obtained were compared with the acetone precipitation EMIT d.a.u. method. The authors concluded that FPIA was less influenced by matrix effects and was not affected by the decomposition of blood, which means that it could be utilized to analyze autopsy blood samples.

FPIA was also used for the analysis of blood samples for the presence of cannabinoids, and the confirmation and quantitation of THC, 11-hydroxy-THC, and 11-nor-9-carboxy-THC was done by GC-MS [134].

5.3.1.3 Radioimmunoassays

RIA were also used for the determination of THC and THC-COOH in blood and serum samples [15,135]. Hanson *et al.* [135] compared ^{3}H- and ^{125}I-RIA and GC-MS for the determination of cannabinoids in blood and serum. They concluded that both RIA methods could be used to detect THC and THC-COOH, and that serum was a better specimen than blood in terms of accuracy, sensitivity, reproducibility, and specificity.

Moody *et al.* [136] compared the results obtained for the analysis of cannabinoids by RIA using methanol-extracted blood with those obtained using non-extracted blood. The results of both methods were compared with GC-MS analysis. Both procedures were qualitatively similar, but the methanol extract procedure proved to be superior in providing semi-quantitative results that could be correlated with those obtained by GC-MS.

5.3.1.4 KIMS assays

Moody and Medina [137] used the Roche OnLine® KIMS assay to detect cannabinoids in serum. They modified the KIMS method used by Armbruster *et al.* [45] for the detection of abused drugs in urine. Modifications were made to increase the sensitivity of the assay because drug concentrations in serum are usually lower than in urine. Direct measurement of unextracted sera was not possible. Therefore, extraction of the samples was done by the addition of 7 mL of chloroform–isopropanol

(9:1), the organic phase was then separated, dried, and the residue was reconstituted with ethanol and potassium phosphate (pH 7.4).

5.3.1.5 Enzyme-linked immunosorbent assays

THC metabolites can be detected by ELISA [42,138]. When ELISA procedures were applied to the detection of drugs of abuse in whole blood, they were found to be more sensitive and less time-consuming than the EMIT procedures [138].

5.3.1.6 Cloned enzyme donor immunoassay

Another type of immunoassays used for the analysis of cannabinoids in whole blood is the Microgenics CEDIA DAU. Cagle *et al.* [139] compared the CEDIA DAU assay (EIA) and the Abbott AxSym system (FPIA) for the analysis of whole blood. Protein precipitation with acetone was used for the CEDIA assay, while for the FPIA addition of acetonitrile at a ratio of 1:2 (blood–acetonitrile) was found to give the best results. The results obtained were confirmed by GC-MS, which was found to correlate better with FPIA ($r = 0.75$) than with EIA($r = 0.22$).

5.3.2 Chromatographic methods

5.3.2.1 Thin layer chromatography

Quantitative separation and analysis of THC, CBN, and CBD can be done by separation on silica gel HPTLC plates followed by densitometric scanning of the separated compounds [140]. This procedure, however, uses two extraction steps, initial SPE using C_{18}-Sep-Pak cartridge. The eluate obtained was evaporated, reconstituted with acetone and derivatized with dansyl chloride. The dansyl derivatives were then extracted with diethyl ether. The final extract, almost free of interfering compounds, was then spotted on HPTLC plates and developed using isooctane–ethyl acetate–acetic acid (30:10:1).

5.3.2.2 High performance liquid chromatography

Law *et al.* [16] described a method for the confirmation of cannabis use by the analysis of blood and urine samples by combined HPLC and RIA. This method, which resulted from the modification and improvements of already published methods [88,89,91], coupled the separation power of HPLC and the sensitivity of RIA. It allowed the complete analysis of at least six samples per day and could, therefore, be used for routine toxicological analysis of Δ^9-THC-11-oic acid and its glucuronide derivative in methanol extracts of blood samples.

   HPLC with ECD (HPLC/ECD) was also used for the analysis of plasma samples [99,141]. Both methods utilized a preliminary SPE. Zweipfenning *et al.* [141] used Bond-Elut C_{18} SPE columns for the isolation of THC, followed by HPLC analysis on C_{18} column using tetrahydrofuran–methanol–sodium citrate buffer (0.005 M), pH 7.0 (7.5:68:24.5, v/v) as the mobile phase. Nakahara *et al.* [99] used an automatic extractor equipped with ODS-minicolumn for the extraction of THC and its major metabolites (THC-COOH and 11-OH-THC), followed by analysis on Zorbax C_8 column using a mobile phase composed of acetonitrile–methanol–H_2SO_4 (0.2 N) (35:15:50).

5.3.2.3 Liquid chromatography-mass spectrometry and LC-MS/MS

Guinea pig plasma was analyzed by LC-MS using negative mode electrospray ionization detection for Δ^8-THC and Δ^8-THC-COOH [142]. Yang and Xie [143] used solid-phase microextraction membrane (SPMEM) and detected THC and CBD in blood and brain of injected male mice, and in spiked human urine by using LC-MS. Maralikova and Weinmann [144] used LC-MS/MS for the detection of THC, 11-OH-THC, and THC-COOH in human plasma. Automated silica-based SPE was used for sample clean up. LC-MS/MS was equipped with a turbo ion spray interface and triple quardrupole mass analyzer using positive electrospray ionization and multiple-reaction monitoring. The LoD was 0.2 ng/mL for THC and 11-OH-THC and 1.6 ng/mL for THC-COOH, while LoQ was 0.8 ng/mL for THC and 11-OH-THC and 4.3 ng/mL for THC-COOH.

5.3.2.4 Gas chromatography

GC with electron-capture detector was used for the determination of CBD, the most abundant cannabinoid in hashish and in fiber-type *Cannabis*, in plasma [145]. Tetrahydrocannabidiol was used as the internal standard. Liquid–liquid extraction with hexane-1.5% isoamyl alcohol was used. The extracts were concentrated, washed with NaOH, then with HCl, and evaporated to dryness. The pentafluorobenzyl derivatives were then analyzed by GC using an electron-capture detector.

Another liquid–liquid extraction method for the determination of THC in blood by GC with nitrogen selective detector was proposed by Ritchie *et al.* [146]. The procedure comprised hexane extraction of whole blood, followed by re-extraction into alkaline methanol, and derivatization of THC and the internal standard (Δ^8-THC) using 3-pyridinediazonium chloride solution. The mixture was then acidified and back extracted into hexane. The hexane was evaporated, and the residue was reconstituted with methanol. The phenolic groups of THC and the internal standard were methylated by on-column flash alkylation with TMAH and then injected onto the GC.

A solid support reagent, consisting of PFBBr deposited upon XAD-2 resin, was used to extract and derivatize Δ^9-THC, 11-hydroxy-Δ^9-THC, and 11-nor-9-carboxy-Δ^9–THC from plasma samples. The pentafluorobenzyl derivatives could then be analyzed by GC/ECD or GC-MS/ NICI [147].

5.3.2.5 Gas chromatography-mass spectrometry

GC-MS methods are the most widely used confirmatory techniques for the detection of cannabinoids in whole blood, serum, or plasma. Sample clean up before analysis is necessary and is usually done by liquid–liquid extraction [17,112,135,136, 148–150], or by SPE [151,152].

Derivatization of the samples is also necessary. Hanson *et al.* [135] utilized trimethylphenyl ammonium hydroxide to form the methyl derivative of THC, which was then analyzed by electron-impact selected ion monitoring GC-MS. Garriott *et al.* [148] used trimethylanilinium hydroxide as derivatizing agent for the determination of Δ^9-THC, 11-hydroxy-Δ^9-THC, and 11-nor-Δ^9-THC-9-carboxylic acid in blood. Trifluoroacetic anhydride derivatization procedure was used for the determination of THC in plasma using a GC-MS operated in the negative chemical-ionization mode and retrofitted with a High Energy Dynode detector system [149]. This detector improved the limit of detection of THC in plasma by 6.25fold, over that obtained with the same GC-MS system without the new detector. Moody *et al.* [136] compared RIA and GC-MS for the analysis of forensic blood specimens for cannabinoids. Blood specimens were analyzed by negative ion chemical ionization GC-MS with deuterated internal standards for the trifluoroacetyl derivative of THC and the methyl ester trifluoroacetyl derivative of THC-COOH. Bis(trimethylsilyl) trifluoroacetamide (BSTFA) was used for derivatization of THC-COOH by Clatworthy *et al.* [17] for the development of a GC-MS method for the detection of THC-COOH in blood, and by Kemp *et al.* [112] for the analysis of THC and six metabolites, namely, 8α-hydroxy-Δ^9-THC, 8β-hydroxy-Δ^9-THC, 11-hydroxy-Δ^9-THC, 8α-11-dihydroxy-Δ^9-THC, 8β-11-dihydroxy-Δ^9-THC, and 11-nor-9-carboxy-Δ^9-THC. The method of Kemp *et al.* [112] had also the advantage of being able to detect CBD and CBN in plasma. Simultaneous quantitation of THC and THC-COOH in serum by GC-MS using tetrabutyl-ammonium hydroxide in DMSO was also reported [151]. Trimethylsilyl derivatization was also used for the determination of CBD in plasma utilizing GC/ion-trap mass spectrometry in positive ion chemical ionization mode [153].

The GC-MS-MS method was used to confirm the unusually high levels of THC in two postmortem samples [154]. In this method, electron-impact mass fragmentation of the trimethylsilyl derivatives yielded a full scan mass fragmentation pattern. The most abundant ions are again fragmented to produce another spectrum characteristic of THC.

Chi *et al.* [155] used PFPA in PFPOH derivatization for the analysis of THC in whole blood using GC-MS in electron-impact mode.

An automated SPE method (using Zymark RapidTrace SPE Workstation with a RSV SPE copolymer cartridge) was developed for the simultaneous extraction, confirmation, and quantitation of THC and THC-COOH from whole blood [156]. Quantitation was done by GC-MS using electron ionization mode with selected ion monitoring of 3 ions for each analyte. The LoD for THC and THC-COOH were 1.6 and 0.8 ng/mL while the LoQs were 2 and 1 ng/mL, respectively.

Steinmeyer *et al.* [157] validated a method for the quantification of THC, 11-OH-THC, and THC-COOH in serum. SPE was used to isolate the analytes, which were derivatized by methylation and analyzed in the selected ion mode using GC-MS. The LoD for THC, 11-OH-THC, and THC-COOH were 0.52, 0.49, and 0.65 ng/mL, respectively.

THC concentrations in human plasma from three individuals who smoked marijuana were 151, 266, and 99 ng/mL drawn immediately after the end of smoking while THC-COOH concentrations were 41, 52, and 171 ng/mL [158]. SPE was used for plasma samples, while trifluoroacetic anhydride and hexafluoroisopropanol were used for derivatization. THC and THC-COOH were detected using GC-MS in the negative ion chemical ionization mode with LoQs of 0.5 and 2.5 ng/mL for THC and THC-COOH, respectively.

Schutz *et al.* [159] developed a GC-MS method for the detection of THC, THC-CCOH, 11-OH-THC morphine, codeine, cocaine, benzoylecgonine, methylecgonine, cocaethylene, amphetamine, methamphetamine, 3,4-methylenedioxyamphetamine (MDA), 3,4-methylenedioxymetamphetamine, and *N*-methyl-1-(3,4-methylene-dioxyphenyl)-2-butanamine in small blood samples and blood stains using solid-phase SPE columns and a pipetting robot (Gilson Aspec XL). The LoDs are in the order of 0.15–0.82 ng/50 μL spot (cannabinoids), 1.62–4.10 ng/50 μL spot (amphetamines), 1.67–4.70 ng/50 μL spot (cocaine and derivatives), and 4.53–4.91 ng/50 μL (opiates). A GC-MS method was used for the analysis of THC, 11-OH-THC, THC-COOH, CBD, and CBN in plasma after oral application of small doses of THC and cannabis extract [160]. The LoDs were between 0.15 and 0.29 ng/mL for THC, 11-OH-THC, THC-COOH, CBD, and 1.1 ng/mL for CBN.

A GC-MS method was used for the detection of THC and THC-COOH in whole blood samples [161]. In this method, conventional solvent extraction was followed by a clean up using solid-phase cartridges. The LoD was better than 1 ng/mL with extraction efficiencies greater than 80% for THC and 70% for THC-COOH.

A simple extraction procedure for THC, and its three metabolites (11-OH-THC, THC-COOH, and 8β-11OH-diOH-THC) from urine, plasma, and meconium was developed based on immunoaffinity chromatography [162]. Using the affinity resin prepared by immobilization of THC antibody onto cyanogen bromide-activated Sepharose 4B, THC and its three metabolites were extracted from urine and plasma. The same procedure was used for analysis of meconium with some modifications. After derivatization of the samples, GC-MS was used for analysis in the electron impact ionization (EI) mode with SIM monitoring. The LoDs ranged from 0.5 to 2.5 ng/mL in plasma and urine and from 1.0 to 2.5 ng/g in meconium. The extraction recovery from meconium, however, was lower than that of plasma and urine, ranging from 52 to 72% at 10 ng/g level.

SPE (C_{18}) cartridges were used to extract THC, 11-OH-THC and THC-COOH from serum and their trimethylsilyl derivatives were analyzed by GC-MS-MS system based on an ion trap with external ionization [163]. The quantitation of three analytes was achieved in relation to trideuterated internal standards in dual MS-MS mode. Confirmation of these analytes was done by registering the daughter spectra in full scan mode. The LoDs for THC, 11-OH-THC, and THC-COOH were 0.25, 0.5, and <2.5 μg/L, respectively.

For the identification and quantification of THC in rabbit plasma, two ionization techniques were utilized for GC-MS [164]. EI (TMS derivatized) was used after intravenous administration, while negative chemical ionization (NCI) (TFA derivatized) was used after sublingual administration with deuterated internal standard in both cases. The method was successful in analyzing THC from rabbit plasma.

The method used by Richard *et al.* incorporates *E. coli* β-glucuronidase hydrolysis of plasma samples to cleave glucuronic acid moieties and simultaneous SPE of THC, 11-OH-THC, and THC-COOH [165]. After addition of deuterated analogs for each analyte as internal standards, quantification was done on a bench top positive chemical ionization (PCI) GC-MS. LoDs for THC, 11-OH-THC, and THC-COOH were 0.5, 0.5, and 1.0 ng/mL. Plasma samples were collected from individuals participating in a controlled oral THC administration study and analyzed by this method.

5.4 ANALYSIS OF CANNABINOIDS IN HAIR

Hair is another sample that can be analyzed for the presence of drugs of abuse. Drugs persist in hair for months after consumption; therefore, hair analysis can be used as a tool for detection of drug use in forensic sciences, in traffic, and occupational medicine and in clinical toxicology [166,167]. Balabanova *et al.* [168] was the first author to publish a method for the RIA detection of cannabinoids in hair followed by GC-MS confirmation of Δ^9-THC. However, this paper was subject to criticism because the SIM chromatograms shown in the publication were very poor [169,170]. Since this time, many papers have been published describing the use of GC-MS methods for the detection of cannabinoids in hair samples. THC-COOH was determined in hair by GC-MS after alkaline hydrolysis and extraction from acid solution on Baker C_{18} columns, followed by derivatization with methyl iodide [171] or with PFPA and pentafluoropropionyl alcohol (PFP-OH), with levallorphan as the internal standard [172]. Alternatively, liquid–liquid extraction and deuterated internal standards were used for the determination of THC-COOH in hair [173] and for the determination of THC and THC-COOH in human hair and pubic hair [174]. In both methods, hair samples were first decontaminated with methylene chloride, then pulverized and incubated in NaOH to destroy the protein matrix of the hair. Samples were then extracted with *n*-hexane–ethyl acetate (9:1) after acidification with acetic acid. The organic phase was washed with 1 mL 0.1 N NaOH followed by 1 mL 0.1 N HCl, then evaporated to dryness and derivatized with PFPA and PFP-OH. Young *et al.* developed an analytical method for the evaluation of CBD, CBN and THC level in human hair using GC-MS [175].

Hair samples were washed with isopropanol and, after the addition of deuterated internal standard, the hair samples were incubated in 1.0 M NaOH for 10 min at 95°C. These hydrolyzed (digested) samples were then extracted with *n*-hexane–ethyl acetate (7.5:2.5), evaporated, derivatized, and analyzed by GC-MS. Baptista *et al.* [176] used β-glucuronidase/aryl sulfatase for hydrolysis and found that, for the quantification of THC-COOH, GC-MS-NCI (negative ion chemical ionization mode) using methane gas as reagent gas is more sensitive than the GC-MS-EI method, which may give rise to false negatives.

Cirimele *et al.* [177] proposed a simpler method for the simultaneous identification of THC, CBN, and CBD in hair samples, using THC-d$_3$ as the internal standard. This method is a rapid screening method that does not require derivatization prior to analysis. Jurado *et al.* [178] described a method for the simultaneous quantification of opiates, cocaine, and cannabinoids in hair. In this method, the sample was decontaminated with dichloromethane, then two consecutive hydrolyses were done: the first is an acid hydrolysis followed by organic solvent extraction of opiates and cocaine; this is followed by alkaline hydrolysis and extraction of the cannabinoids with organic solvent after addition of maleic acid. Wilkins *et al.* [179] utilized a liquid–liquid extraction procedure prior to quantitative analysis of THC, 11-OH-THC, and THC-COOH in human hair by GC-MS. The extraction procedure included digestion of the sample with NaOH, followed by extraction with hexane–ethyl acetate (9:1v/v), the organic phase was then further extracted for THC and 11-OH-THC and the aqueous phase was used for THC-COOH. Sabina and Maecello described a method for application of solid-phase microextraction (SPME) to cannabis in hair [180]. Hair samples were washed with petroleum ether, hydrolyzed with NaOH, neutralized, deuterated internal standard was added and directly submitted to SPME. The SPME elute was analyzed by GC-MS. The LoD for both CBN and THC was 0.1 ng/mg while CBD had 0.2 ng/mg LoD.

A GC-MS-MS method was used by Mieczkowski [181] for the confirmation of the presence of THC and THC-COOH in hair samples screened by RIA for cannabinoids. He concluded that although RIA screening of hair samples for cannabinoids is efficient, the results should be confirmed by GC-MS-MS methods.

Sachs and Dressler developed a method for the detection of THC-COOH in hair by GC-MS after HPLC clean up [182]. After the sample was digested with 2 M NaOH at 95°C and the neutralized liquid was extracted with a mixture of *n*-hexane and ethyl acetate, the dried residue was reconstituted in acetonitrile–methanol–sulfuric acid (0.01 M) (49:21:30, v/v/v) and the cannabinoids were separated by HPLC, derivatized and analyzed by GC-MS. The LoD and LoQ for THC-COOH were 0.3 and 1.1 pg/mg, respectively.

Musshoff *et al.* [183] developed a fully automated procedure using alkaline hydrolysis and headspace SPME (HS-SPME) followed by on-fiber derivatization with *N*-methyl-*N*-trimethylsilyl-trifluoroactamide (MSTFA) and detection of cannabinoids by GC-MS. The authors concluded that this automated HS-SPME-GC-MS procedure is substantially faster than the conventional methods of hair analysis. Headspace solid-phase dynamic extraction (HS-SPDE) was also used for detection of cannabinoids in human hair samples [184]. SPDE is a further development of SPME, based on an inside needle capillary absorption trap.

5.5 ANALYSIS OF CANNABINOIDS IN MECONIUM

The analysis of meconium for the presence of drugs of abuse has gained interest in recent years. It is now a widely accepted alternative to infant's blood and maternal urine to detect prenatal exposure to these drugs. Although meconium appears to be a more difficult sample to analyze because of the additional steps required to disrupt the tissues and to extract and clean up the samples, it has the advantages of being easier to collect than blood and urine, and it increases the window of detection to the last months of gestation [185].

Ostrea *et al.* [186–188] were the first authors to publish methods for the screening of drugs of abuse in meconium. The analysis of cannabinoid metabolites in meconium was done by mixing the sample with methanol, allowing it to stand at room temperature for 10 min, then centrifuging and testing the supernatant for cannabinoid metabolites by RIA [188]. The authors analyzed the meconium and urine of 20 infants of drug-dependent mothers for the metabolites of heroin, cocaine, and cannabinoids and concluded that meconium contains more drug metabolites than urine and is therefore more useful in detecting fetal exposure to drugs-of-abuse [188].

Nair *et al.* [189] used the procedure of Ostrea *et al.* [188] for the analysis of 141 meconium samples and also concluded that meconium is a superior sample than urine for the detection of fetal exposure to drugs.

EMIT was also used for the screening of meconium samples for the presence of cocaine, cannabinoids, opiates, and methadone [190]. The method consisted of extracting 0.5–1 g meconium with methanol and evaporating the extract to dryness. The residue was reconstituted with 1 mL methanol and divided into two portions, one used for the EMIT and the other saved for confirmation of the results by GC-MS. Comparison between meconium, maternal urine, and neonatal urine was also done and the authors found that maternal urine is more useful than meconium for the detection of THC metabolites [190].

FPIA followed by HPLC with diode-array detection was also used for the analysis of THC-COOH in meconium samples [191]. The extraction of THC-COOH from meconium samples was done with 5 mL water and one drop of NaOH and the supernatant was assayed by FPIA. For the HPLC method, the aqueous extract was partitioned with hexane–ethyl acetate (80:20), then the organic phase was evaporated and the residue reconstituted with the mobile phase, which is composed of acetonitrile–phosphoric acid (50 mM) (65:35), then injected onto a C_{18} column.

Another method for the determination of THC-COOH in meconium was presented by Moore *et al.* [192]. Extraction of meconium samples was done using acetic acid. Diphenylamine in acetone was then added and the mixture was centrifuged. The supernatant was filtered, evaporated to dryness, and the residue was reconstituted with the appropriate buffer and analyzed by FPIA. Confirmation of the results was done by GC-MS using deuterated internal standards and *N*-methyl-*N*-(*tert*-butyldimethylsilyl) trifluoroacetamide as derivatizing agent.

One problem encountered in the determination of THC-COOH in meconium was the low confirmation rate. Wingert *et al.* [190] failed to confirm any of the positive specimens screened by EMIT, Moore *et al.* [192] reported a 20% confirmation rate for samples analyzed by FPIA and confirmed by GC-MS, while ElSohly *et al.* [193] reported a 26% confirmation rate for samples screened by EMIT and confirmed by GC-MS.

A study of the elimination profile of Δ^9-THC in meconium was therefore conducted by ElSohly and Feng [194]. The authors found that in addition to THC-COOH, two other major metabolites of THC, namely 11-OH- 9-THC and 8β–11-diOH-Δ^9-THC, are found in meconium, mainly as their glucuronides. Enzymatic hydrolysis of meconium samples followed by determination of the three metabolites is therefore necessary to increase the confirmation rate of samples screening positive for cannabinoids by immunoassays.

Coles *et al.* [195] analyzed meconium samples for THC, 9-carboxy-THC, and 11-OH-THC using GC-MS with LoD of 5 ng/g for 9-carboxy-THC and 11-OH-THC with more than 66% recovery at 100 ng/g for both metabolites. A GC-MS method was developed for the analysis of 24 meconium specimens, which showed that 11-OH-THC is an important metabolite in meconium [162].

5.6 OTHER BIOLOGICAL SPECIMENS

Sweat, skin, saliva, and breath are other biological matrices that can be analyzed for the presence of cannabinoids.

Sweat and saliva are easier to collect than urine and blood but drug concentrations are lower and the window of detection is often shorter than urine. Their use may be of value for detecting if someone is driving while intoxicated and for surveying populations for illicit drug use [196].

RIA and mass spectrometry were used for the analysis of methadone, cocaine, THC, benzodiazepine, barbiturates, morphine, and cotinine in porcine sweat and the data obtained indicated depositions of those drugs in axillary hair [197]. The effect of pilocarpine stimulation on the concentration of THC in perspiration samples obtained from THC smokers was also determined [198]. The use of sweat patches for detection of drugs of abuse may be advantageous over urine analysis because the patch can be worn for a week without discomfort and can therefore provide a cumulative estimate of the degree of exposure to drugs for a whole week [199,200].

Skin swabs were also used for the detection of cannabinoids, opiates, and cocaine on the skin of drug abusers using an on-the-spot immunological test and GC-MS [201]. Drug residues on the hands of human subjects were also detected using a sampling method based on aspirating and trapping the drug microparticles on a filter plug followed by ion-mobility spectrometry [202].

Lemos *et al.* [203] evaluated fingernail clippings as analytical specimens for the detection and quantitation of cannabinoids. Detergent, water and methanol washes followed by alkaline hydrolysis and liquid–liquid extraction were used. The mean cannabinoid concentration in fingernail clippings of six known cannabis users was 1.03 ng/mg detected by RIA. When GC-MS was used, the mean THC concentration was 1.44 ng/mg in fingernail clippings of 14 known cannabis users. The average

THC-COOH concentrations in fingernail clippings of three known cannabis users was 19.85 ng/mg by GC-MS when extracted in acidic pH.

The detection of cannabinoids in breath and saliva may be particularly useful in traffic control where a non-invasive and simple method of sample collection is required.

The concentration of THC in breath ranges from 10 to 56 ng/sample taken 15 min after smoking and can be detected for about 1 h later [204]. A breath analyzer consisting of a tube containing Fast Blue Salt B, NaOH, and silica gel and a mouthpiece was developed by Volkmann *et al.* [205]. Consumption of hashish or marijuana can be detected by the color of the indicator changing to red when the person blows into the mouthpiece.

5.6.1 Oral fluid

In saliva, the concentration of THC may reach 1000 ng/mL after the administration of 5–20 mg THC and then fades to 50 ng/mL after 3–4 h [206]. Kircher and Parlar [206] developed an HPLC method for the determination of THC in human saliva. They prepared an immunoaffinity column by covalent immobilization of cannabinoid specific IgG on epoxy-activated silica and utilized it for sample clean up and enrichment. This was followed by the transfer of the cannabinoid fraction to an analytical RP column using a column-switching procedure. The authors were able to separate THC from CBN and CBD and achieved a limit of quantification of 20 ng THC/milliliter, using an UV detector at 220 nm.

The point-of-collection oral fluid drug testing devices Oratect (Branan) and Uplink (OraSure) were evaluated for their ability to detect cannabinoids, amphetamines, cocaine, and opiates [207]. For cannabinoids and cocaine, Drugwipe (Securtec) was also evaluated. The performance of all three devices in THC detection was poor, but Branan and OraSure detected well THC-COOH, amphetamine, methamphetamine, and opiates. Nine saliva specimens were positive for cannabis using the On-site OraLine® IV s.a.t. device, with THC concentrations ranging from 3 to 265 ng/mL and confirmed by GC-MS [208]. One OraLine® device positive was not confirmed by GC-MS, which gave a LoQ of 1 ng/mL.

RIA was used for the analysis of oral fluid specimens, while plasma specimens were analyzed by GC-MS [209]. The similarity in oral fluid and plasma concentrations indicated that there is a physiological link between these specimens. This evidence suggested that during cannabis smoking, THC is deposited in the oral cavity.

Moore *et al.* [210] screened oral fluid specimens by ELISA and confirmed by GC-MS for THC and THC-COOH. Quantisal™ oral fluid collection device was used for the first time by Moore *et al.* [211].

Saliva samples were collected by the EPITOPE system and after an SPME step were analyzed on GC-MS [212]. THC and CBD showed positive results up to 13 h after use. SPME and direct immersion-SPME (DI-SPME) followed by GC-MS were also used for the detection of THC, CBD, CBN, cocaine, EDDP, cocaethylene, amphetamine, methamphetamine, MDMA, MDEA, and MBDB in saliva samples [213].

The effects of adulterants and foodstuffs were investigated using the Oral Fluid drug screen, Oratect, on oral fluid drug tests [214]. This study revealed that common foods, beverages, food ingredients, cosmetics, and hygienic products do not cause false positive results when tested 30 min after their consumption.

5.6.1.1 Chromatographic methods

5.6.1.1.1 Gas chromatography-mass spectrometry-MS

Cone *et al.* [215] did comparative studies of the oral fluid testing using intercept immunoassay and GC-MS-MS confirmation versus urine testing and determined that oral fluid testing produces equivalent results to urine testing.

Oral fluid specimens collected from cannabis-free volunteers but exposed to cannabis smoke were screened by EIA for cannabinoids (cutoff concentration is 3 ng/mL) and tested by GC-MS-MS (LoD and LoQ is 0.75 ng/mL) [216]. This study concluded that the risk of positive oral fluid tests from passive cannabis inhalation is limited to a period of approximately 30 min following exposure.

Cannabinoid Intercept MICRO-PLATE EIA was used for the analysis of oral fluid samples from passive cannabis exposure, while the LoD and LoQ for THC in GC-MS-MS assay was 0.3 and 0.75 ng/mL, respectively [217].

5.6.1.1.2 Liquid chromatography-mass spectrometry and LC/MS-MS

In contrast to existing GC-MS methods, no extensive sample clean up and time-consuming derivatization steps are needed to analyze the samples by LC-MS. LC-MS was used to detect THC in oral fluid samples with a LoD and LoQ of 1.0 and 2.0 ng/mL, respectively [218]. The oral fluid was extracted using Bond-Elut LRC-Certify SPE columns and THC was analyzed by LC-MS [219]. Concheiro *et al.* [220] analyzed THC in oral fluid by using 200 µL of sample and achieved a LoD of 2 ng/mL.

Laloup *et al.* [221] developed a simple and rapid method for the analysis of THC in oral fluid using LC-MS-MS. The use of liquid–liquid extraction by hexane was highly effective and decreased the interferences present in the matrix. XTerra MS C18 column was used for chromatographic separation using 1 mM ammonium formate–methanol (10:90, v/v) as the solvent system isocratically. By using 100 and 500 µL of oral fluid, the LoQs were 0.5 and 0.1 ng/mL, respectively.

5.7 AUTOPSY MATERIALS

Blood and urine are the most widely used autopsy samples. The determination of THC in forensic blood samples [130,132,133,136,146,148,151,154] and postmortem urine samples [49,148] has been discussed above under the analysis of cannabinoids in blood and urine.

Other autopsy materials include human solid tissues such as liver, kidney, brain, spleen, stomach, and intestine. Kudo *et al.* [222] developed a simple and sensitive method that can be used for routine forensic analysis of THC in human solid tissues. Tissue samples were homogenized in acetonitrile, the sample was then centrifuged and the supernatant made alkaline by the addition of NaOH. The alkaline solution was shaken with hexane–ethyl acetate (9:1), the organic phase was then separated and shaken again with 0.1 M HCl. Finally, the organic layer was evaporated, derivatized by methylation and analyzed by GC-MS. The application of the method to samples taken from an autopsied individual allowed the study of the distribution of THC in human tissues. THC was found in all tissues except urine. The highest concentration was found in adipose tissues, then in the lungs and the lowest concentration was in the whole blood and liver.

An HPLC/ECD was developed for the determination of THC in rat brain tissue. Methanol was used for protein precipitation and initial extraction of THC from brain tissues. After evaporation of the methanolic extract, the residue was dissolved in hexane–ethyl acetate (7:3) and the solution washed with 0.05 M H_2SO_4. The organic phase was then evaporated and the residue reconstituted with mobile phase–methanol (25:10) then injected onto a C_{18} column. The internal standard used was 4-dodecylresorcinol and the mobile phase was methanol–acetonitrile–H_2SO_4 (0.01 M) (21:24:55).

5.8 ANALYSIS OF CANNABINOIDS IN CRUDE CANNABIS PREPARATIONS

Crude cannabis preparations include marijuana (the dried leaves and flowering tops of the female plants), hashish (the dried resin with fine plant particles), and hash oil (the concentrated extract of the plant material).

The most commonly used methods of analysis over the last two decades have involved GC with FID (GC-FID), GC-MS, and HPLC. The following summarizes some of the procedures described over the last few years for the analysis of these preparations.

Morita and Ando [223] described a GC-MS procedure for the analysis of the different cannabinoids in hash oil in which 11 compounds were separated and identified. These included Δ^9-THC, CBD, CBC, and CBN, along with some C_3 homologs. The composition of major mass spectral fragments of Δ^9-THC were proposed.

In 1988, Brenneisen and ElSohly [224] described a high-resolution capillary GC-FID and a GC-MS procedure for the identification of the different components of a cannabis extract to establish the chemical profiles (chemical signature) of samples of different geographical origin. The components analyzed included terpenes, alkanes, cannabinoids, and non-cannabinoid phenols. Over 100 different components were identified, and the procedure proved to be of forensic value in tracing the geographical origin of a cannabis sample through its chemical profile. In addition, the separation of free cannabinoids and their carboxylic acid precursors was accomplished by HPLC analysis of the samples using a Beckman Ultrasphere

3 μm ODS column (75 mm × 4.6 mm). More than 40 components were detected using a UV detector in the HPLC tracing.

In 1995, Hida *et al.* [225] reported on the classification of hashish by pyrolysis–GC in the presence of powdered chromium, followed by cluster analysis of the normalized pyrograms (the peaks in each pyrogram were normalized against the highest peak in that pyrogram). The results of the cluster analysis were presented in easily interpreted visual representations known as dendograms. The dendograms were used to compare unknown hashish samples with those of samples from different sources for classification purposes.

A GC-FID procedure for the routine analysis of confiscated marijuana samples and quantitation of several cannabinoids including Δ^9-THC, CBD, CBC, CBN, CBG, and THCV was described by Ross *et al.* [226]. The procedure involved the extraction of a small amount of sample (100 mg) with a methanol–chloroform mixture (99:1) containing the internal standard (4-androstene-3, 17-dione) followed by the direct analysis of the extract on a DB-1 column.

Analysis of neutral cannabinoids by HPLC was reported by Veress *et al.* [227], using two types of bonded-phase columns. An amino-bonded-phase column was used, which allows the extraction of plant material with non-polar solvents followed by direct injection of the extract without pre-separation. The results obtained by the amino-bonded column were compared with those obtained by a reverse-phase method, which required sample clean up using a C_{18}-Sep-Pak cartridge prior to HPLC analysis. The authors concluded that the amino-bonded-phase HPLC procedure was superior to that using the reversed phase for the quantitation of neutral cannabinoids.

Several analytical procedures (TLC, GC-FID with both packed and capillary columns, and HPLC) have been described in detail for the analysis of cannabinoids (neutral and acidic) in different cannabis products (marijuana, hashish, and hashish oil) in a manual prepared by the Division of Narcotic Drugs of the United Nations [228]. The manual is a compilation of methods for sampling and analysis of cannabis products, recommended for use by National Narcotics Laboratories. Bosy and Cole [229,230] used GC-MS for the determination of THC amounts in hemp seed oil. HPLC was used for the determination of THC and THC-COOH in hemp-containing foods [231]. Ross *et al.* [232] analyzed the total THC content of both drug- and fiber-type cannabis seeds by GC-MS.

The quantitation of the individual cannabinoids was accomplished by the use of internal standards, which varied depending on the method and included the use of long-chain hydrocarbons (e.g. *n*-tetradecane or *n*-docosane), steroids (androst-4-ene-3,17-dione and cholestane), and phthalates (dibenzyl phthalate or di-*n*-octyl phthalate).

HPLC was used for the analysis of THC, CBD, and CBN along with their acid precursor (THCA, CBDA, and CBNA), using a reversed-phase column (7 μm particle size) and a mixture of methanol and 0.01 M sulfuric acid (80:20) as the mobile phase [233]. The authors carried out standardized storage conditions with hashish samples along with pure cannabinoids and concluded that the total values of CBD-CBDA, CBN-CBNA, and THC-THCA were important in the judgment of hashish samples.

Hazekamp *et al.* [234] developed a ^{1}H-NMR method for the quantitative analysis of cannabinoids present in *C. sativa* plant material. The distinguishable signals of cannabinoids were in the range of δ 4.0–7.0 in the ^{1}H-NMR spectrum. Anthracene was used as the internal standard. The quantitation of the target compound was performed by calculating the relative ratio of the peak area of selected proton signals of the target compound to the known amount of the internal standard. This method allows the simple and rapid quantitation of cannabinoids without any chromatographic purification with 5 min analysis time.

Elias and Lawrence [235] summarized different instrumental methods used in drug interdiction. These methods used for detecting concealed drugs were categorized into two main techniques based on bulk detection and air sampling. The bulk detection techniques included X-ray imaging, gamma backscattering, thermal neutron activation, and other systems, while the air sampling techniques included acetone vapor detection, mass spectrometry, gas spectrometry, and ion-mobility spectrometry. The authors concluded that these methods have their limitations and pointed to the continued need for other more effective and selective methods.

5.9 CONCLUDING REMARKS

The scientific literature today is rich in methods to analyze (both qualitatively and quantitatively) for the presence of cannabinoids in biological specimens with a variety of techniques. The diversity of the techniques available to the analyst is such that one could carry out the task without the need for adding new instrumentation to a modestly equipped laboratory. Therefore, the objective of this chapter is to provide an overview of the technologies available with reference to such technologies so that the analyst reviewing this information can find it easy to follow and be directed to information pertinent to the problem at hand. It is hoped that this chapter has met this goal and thatreaders will find it a useful and easy reference to the information sought.

5.10 REFERENCES

1   J.W. Fairbairn. In: J.D.P. Graham (Ed.), Cannabis health, Academic, London, England, 1976, p. 3.
2   D.L. King, P.A. Martel and C.M. O'Donnel, Clin. Lab. Med., 7 (1987) 641.
3   M.A. ElSohly; H.N. ElSohly, Cocaine, marijuana, designer drugs: chemistry, pharmacology and behavior, CRC Press, Boca Raton, FL, 1989, p. 145.
4   D.J. Harvey, ISI Atlas Sci.: Pharmacol., 1 (1987) 208.
5   Y. Gaoni and R. Mechoulam, J. Am. Chem. Soc., 86 (1964) 1646.
6   M.A. ElSohly and D. Slade, Life Sci., 78 (2005) 539.
7   D.J. Harvey. In: A.S. Curry (Ed.), Analytical methods in human toxicology, Vol. Part 1. Verlag Chemie, Weinheim, 1985, p. 257.
8   L. Vollner, D. Bieniek and F. Korte, Regul. Toxicol. Pharmacol., 6 (1986) 348.
9   R. Mechoulam, N.K. McCallum and S. Burstein, Chem. Rev., 76 (1976) 75.
10 E. Cook, in, Marijuana, cocaine and traffic safety, Vol. 2, Alcohol Information Service, Div. Brain Information Service, University of California, Los Angeles, CA, 1986, p. 79.

11   P.L. Williams and A.C. Moffat, J. Pharm. Pharmacol., 32 (1980) 445.
12   P. Wernly and W. Thorman, J. Chromatogr., 608 (1992) 251.
13   J.S. Cridland, D. Rottanburg and A.H. Robins, Hum. Toxicol., 2 (1983) 641.
14   J.D. Sharma, G.W. Aherne and V. Marks, Analyst, 114 (1989) 1279.
15   B. Law, P.A. Mason, A.C. Moffat and L.J. King, J. Anal. Toxicol., 8 (1984) 14.
16   B. Law, P.A. Mason, A.C. Moffat and L.J. King, J. Anal. Toxicol., 8 (1984) 19.
17   A.J. Clatworthy, M.C.H. Oon, R.N. Smith and M.J. Whitehouse, Forensic Sci. Int., 46 (1990) 219.
18   A.B. Jones, H.N. ElSohly and M.A. ElSohly, J. Anal. Toxicol., 8 (1984) 252.
19   M.A. ElSohly, A.B. Jones, H.N. ElSohly and D.F. Stanford, J. Anal. Toxicol., 9 (1985) 190.
20   D. Altunkaya and R.N. Smith, Forensic Sci. Int., 47 (1990) 195.
21   C.A. Sutheimer, R. Yarborough, B.R. Hepler and I. Sunshine, J. Anal. Toxicol., 9 (1985) 156.
22   R.L. Foltz and I. Sunshine, J. Anal. Toxicol., 14 (1990) 375.
23   H. Kaeferstein and G. Sticht, Beitr. Gerichtl. Med., 48 (1990) 51.
24   K. Verebey, D. Jukofsky and S.J. Mule, Res. Commun. Subst. Abuse., 6 (1985) 1.
25   T. Vu Duc, J. Anal. Toxicol., 11 (1987) 83.
26   D.L. Black, B.A. Goldberger, D.S. Isenschmid, S.M. White and Y.H. Caplan, J. Anal. Toxicol., 8 (1984) 224.
27   E. Johansson and M.M. Halldin, J. Anal. Toxicol., 14 (1989) 1989.
28   S.J. Mule and G.A. Casella, J. Anal. Toxicol., 12 (1988) 102.
29   R.E. Struempler, G. Nelson and F.M. Urry, J. Anal. Toxicol., 21 (1997) 283.
30   S. Atasoy and N. Ozer, Adli Tip Derg., 1 (1985) 30.
31   F.M.L. Moore and D. Simpson, Med. Lab. Sci., 48 (1991) 76.
32   V.M. Haver, J.L. Romson and S.M.H. Sadrzadeh, J. Anal.Toxicol., 15 (1991) 98.
33   J.C. Standefer and R.C. Backer. Clin. Chem. (Winston-Salem, N. C.), 37 (1991) 733.
34   R.C. Sreenivasam, T.C. Sneath and N.C. Jain, J. Anal.Toxicol., 17 (1993) 370.
35   L.V.J. Allen and M.L. Stiles, J. Anal. Toxicol., 12 (1988) 45.
36   A.D. Fraser and R. Meatherall, J. Anal.Toxicol., 13 (1989) 240.
37   R.C. Baselt, J. Anal. Toxicol., 13 (1989) 1.
38   D.L. Colbert, A.M. Sidki, G. Gallacher and J. Landon. Analyst (London), 112 (1987), 1483.
39   D. Perez-Bendito, A. Gomez-Hens and A. Gaikwad. Clin. Chem. (Washington, DC), 40 (1994) 1489.
40   M.A. ElSohly, A.B. Jones and H.N. ElSohly, J. Anal. Toxicol., 14 (1990) 277.
41   L. von Meyer, E. Haensler, G. Lardet, A. Scholer and W. Sieghart, Eur. J. Clin. Chem., Clin. Biochem., 35 (1997) 133.
42   K. Watanabe, Y. Tateoka, T. Matsunaga, I. Yamamoto, Y. Shoyama and H. Yoshimura, J. Forensic Toxicol., 15 (1997) 118.
43   AD. Fraser, J. Zamecnik, J. Keravel, L. McGrath and J. Wells, Forensic Sci. Int., 121 (2001) 16.
44   M. Hailer, Y. Glienke, I.-M. Schwab and L. von Meyer, J. Anal.Toxicol., 19 (1995) 99.
45   D.A. Armbruster, R.H. Schwarzhoff, E.C. Hubster and M.K. Liserio, Clin. Chem., 39 (1993) 2137.
46   L. Schwettmann, W.R. Kulpamann and C. Vidal, Clin. Chem. Lab. Med., 44 (2006) 479.
47   M. Feldman, D. Kuntz, K. Botelho, D.C. Ananias, M. Gnezda, D.K. Hoch, S.L. Jordan, S. Rashid and Y. Zhao, J. Anal. Toxicol., 28 (2004) 593.
48   D.A. Armbruster and J.M. Krolak, J. Anal. Toxicol., 16 (1992) 172.
49   J. Rohrich, K. Schmidt and H. Bratzke, J. Anal.Toxicol., 18 (1994) 407.
50   A.J. Jenkins, L.C. Mills, W.D. Darwin, M.A. Huestis, E.J. Cone and J.M. Mitchell, J. Anal.Toxicol., 17 (1993) 292.
51   A.J. Jenkins, W.D. Darwin, M.A. Huestis, E.J. Cone and J.M. Mitchell, J. Anal.Toxicol., 19 (1995) 5.
52   K.F. Buechler, S. Moi, B. Noar, D. McGrath, J. Vilella, M. Clancy, A. Shenhav, A. Colleymore, G. Valkirs, T. Lee, J.F. Bruni, M. Walsh, R. Hoffman, F. Ahmusty, M. Nowakowski, J. Buechler, M. Mitchell, D. Boyd, N. Stiso and R. Anderson, Clin. Chem., 38 (1992) 1678.
53   R. De La Torre, A. Domingo-Salvany, R. Badia, G. Gonzalez, D. McFarlane, L. San and M. Torrens. Clin. Chem. (Washington, DC), 42 (1996) 1433.
54   E.R.S. Brown, D.R. Jarvie and D. Simpson, Ann. Clin. Biochem., 34 (1997) 74.

55   R.G. Parsons, R. Kowal, D. LeBlond, V.T. Yue, L. Neagarder, L. Bond, D. Garcia, D. Slater and P. Rogers. Clin. Chem. (Washington, DC), 39 (1993) 1899.

56   T. Korte, J. Pykalainen, P. Lillsunde and T. Seppala, J. Anal. Toxicol., 21 (1997) 49.

57   D.J. Crouch, M.L. Cheever, D.M. Andrenyak, D.J. Kuntz and D.L. Loughmiller, J. Forensic Sci., 43 (1998) 35.

58   J.J.W. Ros, M.G. Pelders and A.C.G. Egberts, J. Anal. Toxicol., 22 (1998) 40.

59   R. Wennig, M.R. Moeller, J.M. Haguenoer, A. Marocchi, F. Zoppi, B.L. Smith, R. De La Torre, C.A. Carstensen, A. Goerlach-Graw, J. Schaeffler and R. Leinberger, J. Anal. Toxicol., 22 (1998) 148.

60   J. Irving, B. Leeb, R.L. Foltz, C.E. Cook, J.T. Bursey and R.E. Willette, J. Anal. Toxicol., 8 (1984) 192.

61   A.B. Jones, H.N. ELSohly, E.S. Arafat and M.A. ElSohly, J. Anal. Toxicol., 8 (1984) 249.

62   D.L. Frederick, J. Green and M.W. Fowler, J. Anal. Toxicol., 9 (1985) 116.

63   L. Karlsson and M. Stroem, J. Anal.Toxicol., 12 (1988) 319.

64   D.J. Wells and M.T. Barnhill. Clin. Chem. (Winston-Salem, NC), 35 (1989) 2241.

65   M.J. Kogan, J.A. Razi, D.J. Pierson and N.J. Willson, J. Forensic Sci., 31 (1986) 494.

66   W.T. Budgett, B. Levine, A. Xu and M.L. Smith, Forensic Sci., 37 (1992) 632.

67   M.L. Weaver, B.K. Gan, E. Allen, L.D. Baugh, F.Y. Liao, R.H. Liu, J.G. Langner, A.S. Walia and L.F. Cook, Forensic Sci. Int., 49 (1991) 43.

68   D. Altunkaya, A.J. Clatworthy, R.N. Smith and I.J. Start, Forensic Sci. Int., 50 (1991) 15.

69   P. Kintz, D. Machart, C. Jamey and P. Mangin, J. Anal. Toxicol., 19 (1995) 304.

70   S.D. Ferrara, L. Tedeschi, G. Frison, G. Brusini, F. Castagna, B. Bernadelli and D. Soregaroli, J. Anal. Toxicol., 18 (1994) 278.

71   M.A. Huestis, J.M. Mitchell and E.J. Cone, J. Anal. Toxicol., 19 (1995) 443.

72   W. Bronner, P. Nyman and D. von Minden, J. Anal. Toxicol., 14 (1990) 368.

73   L.J. Kadehjian, J. Anal. Toxicol., 25 (2001) 670.

74   M. Gronholm and P. Lillsunde, Forensic Sci. Int., 121 (2001) 37.

75   G.R. Nakamura, W.J. Stall, V.A. Folen and R.G. Masters, J. Chromatogr., 264 (1983) 336.

76   K.K. Kaistha and R. Tadrus, J. Chromatogr., 237 (1982) 528.

77   S.L. Kanter, L.E. Hollister and M. Musumeci, J. Chromatogr., 234 (1982) 201.

78   P. Lillsunde and T. Korte, J. Anal. Toxicol., 15 (1991) 71.

79   D.L. King, M.J. Gabor, P.A. Martel and C.M. O'Donnell, Clin. Chem. (Winston-Salem, NC), 35 (1989) 163.

80   D.M. Steinberg, L.J. Sokoll, K.C. Bowles, J.H. Nichols, R. Roberts, S.K. Schultheis and O.d.C. Michael, Clin. Chem. (Washington, DC), 43 (1997) 2099.

81   M.J. Kogan, E. Newman and N.J. Willson, J. Chromatogr., 306 (1984) 441.

82   K. Vereby, S.J. Mule, J. Alrazi and M. Lehrer, J. Anal. Toxicol., 10 (1986) 79.

83   M.A. ElSohly, H.N. ElSohly, A.B. Jones, P.A. Dimson and K.E. Wells, J. Anal. Toxicol., 7 (1983) 262.

84   R.C. Meatherall and J.C. Garriott, J. Anal. Toxicol., 12 (1988) 136.

85   W. Haensel and R. Stroemmer, GIT Fachz. Lab., 32 (1988) 156.

86   W. Haensel and R. Stroemmer, GIT-Suppl. Issue, 3 (1988) 45.

87   P.J. Twitchett, S.M. Fletcher, A.T. Sullivan and A.C. Moffat, J. Chromatogr., 150 (1978) 73.

88   P.L. Williams, A.C. Moffat and L.J. King, J. Chromatogr., 155 (1978) 273.

89   P.L. Williams, A.C. Moffat and L.J. King, J. Chromatogr., 186 (1979) 595.

90   A.C. Moffat, P.L. Williams and L.J. King (Eds.), Combined high performance liquid chromatography and radioimmunoassay method for the analysis of delta-9-tetrahydrocannabinol and its metabolites in plasma and urine, NIDA Res. Monogr., 42 (1982).

91   B. Law, P.L. Williams and A.C. Moffat, Vet. Hum. Toxicol., 21 (1979) 144.

92   M.A. Peat, M.E. Deyman and J.R. Johnson, J. Forensic Sci., 29 (1984) 110.

93   D. Bourquin and R. Brenneisen, Anal. Chim. Acta., 198 (1987) 183.

94   R.C. Parry, L. Nolan, R.E. Shirey, G.D. Wachob and D.J. Gisch, J. Anal. Toxicol., 14 (1990) 39.

95   S.D. Ferrara, L. Tedeschi, G. Frison and F. Castagna, J. Anal. Toxicol., 16 (1992) 217.

96   V. Bianchi and G. Donzelli, J. Chromatogr. B: Biomed. Appl., 675 (1996) 162.
97   L. Karlsson and C. Roos, J. Chromatogr., 306 (1984) 183.
98   L. Karlsson, J. Chromatogr., 417 (1987) 309.
99   Y. Nakahara, H. Sekine and C.E. Cook, J. Anal.Toxicol., 13 (1989) 22.
100   W. Weinmann, M. Goerner, S. Vogt, R. Goerke and S. Pollak, Forensic Sci. Int., 121 (2001) 121.
101   W. Weinmann, S. Vogt, R. Goerke, C. Muller and A. Bromberger, Forensic Sci. Int., 113 (2000) 381.
102   M.A. ElSohly, E.S. Arafat and A.B. Jones, J. Anal. Toxicol., 8 (1984) 7.
103   J.M. Rosenfeld, Y. Moharir and S.D. Sandler, Anal. Chem., 61 (1989) 925.
104   D. Catlin, D. Cowan, M. Donike, D. Fraisse, H. Oftebro and S. Rendic, J. Automat. Chem., 14 (1992) 85.
105   W.A. Joern, J. Anal. Toxicol., 11 (1987) 49.
106   L. Karlsson, J. Jonsson, K. Aberg and C. Roos, J. Anal. Toxicol., 7 (1983) 198.
107   R.L. Foltz (Ed.), Advances in analytical toxicology, Vol. 1 Biomedical Publications, Foster City, CA, 1984.
108   P.R. Stout, C.K. Horn and K.L. Klette, J. Anal. Toxicol., 25 (2001) 550.
109   M.A. ElSohly, H. DeWit, S.R. Wachtel, S. Feng and T.P. Murphy, J. Anal. Toxicol., 25 (2001) 565.
110   S.B. Needleman, K. Goodin and W. Severino, J. Anal. Toxicol., 15 (1991) 179.
111   R. Clouette, M. Jacob, P. Koteel and M. Spain, J. Anal. Toxicol., 17 (1993) 1.
112   P.M. Kemp, I.K. Abukhalaf, J.E. Manno, B.R. Manno, D.D. Alford and G.A. Abusada, J. Anal.Toxicol., 19 (1995) 285.
113   P.M. Kemp, I.K. Abukhalaf, J.E. Manno, B.R. Manno, D.D. Alford, M.E. McWilliams, F.E. Nixon, M.J. Fitzgerald, R.R. Reeves and M.J. Wood, J. Anal. Toxicol., 19 (1995) 292.
114   M. Szirmai, O. Beck, N. Stephansson and M.M. Halldin, J. Anal. Toxicol., 20 (1996) 573.
115   M. Congost, R. De La Torre and J. Segura, Biomed. Environ. Mass Spectrom., 16 (1988) 367.
116   G.R. Nakamura, W.J. Stall, R.G. Masters and V.A. Folen, Anal. Chem., 57 (1985) 1492.
117   H.H. McCurdy, L.J. Lewellen, L.S. Callahan and P.S. Childs, J. Anal. Toxicol., 10 (1986) 175.
118   B.D. Paul, L.D. Mell, J.M. Mitchell, R.M. Mckinley and J. Irving, J. Anal. Toxicol., 11 (1987) 1.
119   L. O'Dell, K. Rymut, G. Chaney, T. Darpino and M. Telepchak, J. Anal. Toxicol., 21 (1997) 433.
120   J. Singh and L. Johnson, J. Anal. Toxicol., 21 (1997) 384.
121   A.H.B. Wu, N. Liu, Y.J. Cho, K.G. Johnson and S.S. Wong, J. Anal. Toxicol., 17 (1993) 215.
122   J.D. Whiting and W.W. Manders, J. Anal. Toxicol., 6 (1982) 49.
123   M.A. ElSohly, D.F. Stanford and T.L. Little Jr., J. Anal. Toxicol., 12 (1988) 54.
124   M.A. ElSohly, T.L. Little and D.F. Stanford, J. Anal. Toxicol., 16 (1992) 188.
125   E.L. Slightom, J. Forensic Sci., 23 (1978) 292.
126   W.M. Asselin, J.M. Leslie and B. Mckinley, J. Anal. Toxicol., 12 (1988) 207.
127   B.J. Perrigo and B.P. Joynt, J. Anal.Toxicol., 13 (1989) 235.
128   H. Gjerde, Forensic Sci. Int., 50 (1991) 121.
129   D.T. Diosi and D.C. Harvey, J. Anal.Toxicol., 17 (1993) 133.
130   H. Gjerde, A.S. Christophersen, B. Skuterud, K. Klemetsen and J. Morland, Forensic Sci. Int., 44 (1990) 179.
131   L.M. Blum, R.A. Klinger and F. Rieders, J. Anal.Toxicol., 13 (1989) 285.
132   L.J. Lewellen and H.H. McCurdy, J. Anal. Toxicol., 12 (1988) 260.
133   M. Bogusz, R. Aderjan, G. Schmitt, E. Nadler and B. Neureither, Forensic Sci. Int., 48 (1990) 27.
134   C.R. Goodall and B.J. Basteyens, J. Anal. Toxicol., 19 (1995) 419.
135   V.W. Hanson, M.H. Buonarati, R.C. Baselt, N.A. Wade, C. Yep, A.A. Biasotti, V.C. Reeve, A.S. Wong and M.W. Orbanowsky, J. Anal. Toxicol., 7 (1983) 96.
136   D.E. Moody, L.F. Rittenhouse and K.M. Monti, J. Anal.Toxicol., 16 (1992) 297.
137   D.E. Moody and A. Medina, Clin. Chem., 41 (1995) 1664.
138   B.J. Perrigo and B.P. Joynt, J. Can. Soc. Forensic Sci., 28 (1995) 261.
139   J.C. Cagle, H.H. McCurdy, Y.M. Pan, K.J. Ayton, W.H. Wall and E.T. Solomons, J. Anal. Toxicol., 21 (1997) 213.

140 G. Alemany, A. Gamundi, M.C. Nicolau and D. Saro, Biomed. Chromatogr., 7 (1993) 273.
141 P.G.M. Zweipfenning, J.A. Lisman, A.Y.N. Van Haren, G.R. Dijkstra and J.J.M. Holthuis, J. Chromatogr., 456 (1988) 83.
142 S. Valiveti, D.C. Hammell, D.C. Earles and A.L. Stinchcomb, J. Pharm. Biomed. Anal., 38 (2005) 112.
143 R. Yang and W. Xie, Forensic Sci. Int., 162 (2006) 135.
144 B. Maralikova and W. Weinmann, J. Mass Spectrum., 39 (2004) 526.
145 A.B. Jones, M.A. ElSohly, J.A. Bedford and C.E. Turner, J. Chromatogr., 226 (1981) 99.
146 L.K. Ritchie, Y.H. Caplan and J. Park, J. Anal. Toxicol., 11 (1987) 205.
147 J.M. Rosenfeld, R.A. McLeod and R.L. Foltz, Anal. Chem., 58 (1986) 716.
148 J.C. Garriott, V.J.M. Di Maio and R.G. Rodriguez, J. Forensic Sci., 31 (1986) 1274.
149 L.M. Shaw, J. Edling-Owens and R. Mattes, Clin. Chem., 37 (1991) 2062.
150 P. Kintz and V. Cirimele, Biomed. Chromatogr., 11 (1997) 371.
151 M.R. Moeller, G. Doerr and S. Warth, J. Forensic Sci., 37 (1992) 969.
152 W.E. Stonebraker, T.C. Lamoreaux, M. Bebault, S.A. Rasmussen, B.R. Jepson and B.K. BecK, Am. Clin. Lab., 17 (1998) 18.
153 P. Consroe, K. Kennedy and K. Schram, Pharmacol. Biochem. Behav., 40 (1991) 517.
154 M. Collins, J. Easson, G. Hansen, A. Hodda and K. Lewis, J. Anal. Toxicol., 21 (1997) 538.
155 M. Hok Chi Chu and O.H. Drummer, J. Anal. Toxicol., 26 (2002) 575.
156 J.A. D'Asaro, J. Anal. Toxicol., 24 (2000) 289.
157 S. Steinmeyer, D. Bregel, S. Warth, T. Kraemer and M.R. Moeller, J. Chromatogr. B: Biomed. Sci. Appl., 772 (2002) 239.
158 W. Huang, D.E. Moody, D.M. Andrenyak, E.K. Smith, R.L. Foltz, M.A. Huestis and J.F. Newton, J. Anal. Toxicol., 25 (2001) 531.
159 H. Schutz, J.C. Gotta, F. Erdmann, M. Risse and G. Weiler, Forensic Sci. Int., 126 (2002) 191.
160 T. Nadulski, F. Sporkert, M. Schnelle, A.M. Stadeimann, P. Roser, T. Schefter and F. Pragst, J. Anal. Toxicol., 29 (2005) 782.
161 P.D. Felgate and A.C. Dinan, J. Anal. Toxicol., 24 (2000) 127.
162 S. Feng, M.A. ElSohly, S. Salamone and M.Y. Salem, J. Anal. Toxicol., 24 (2000) 395.
163 J.P. Weller, M. Wolf and S. Szidat, J. Anal. Toxicol., 24 (2000) 359.
164 J. Mannila, M. Lehtonen, T. Jaervinen and P. Jarho, J. Chromatogr. B: Biomed. Sci. Appl., 810 (2004) 283.
165 R.A. Gustafson, E.T. Moolchan, A. Barnes, B. Levine and M.A. Huestis, J. Chromatogr. B: Biomed. Sci. Appl., 798 (2003) 145.
166 M.R. Moeller, P. Fey and H. Sachs, Forensic Sci. Int., 63 (1993) 43.
167 H. Sachs and P. Kintz, J. Chromatogr. B: Biomed. Sci. Appl., 713 (1998) 147.
168 S. Balabanova, P.G. Arnold, V. Luckow, H. Brunner and H.U. Wolf, Z. Rechtsmed., 102 (1989) 503.
169 H. Kaferstein and G. Sticht, Z. Rechtsmed., 103 (1990) 393.
170 M. Bogusz, Z. Rechtsmed., 103 (1990) 621.
171 H. Sachs and M.R. Moeller, Fresenius Z. Anal. Chem., 334 (1989) 713.
172 M. R. Moeller and P. Fey, Bull. Soc. Sci. Med. (Grand Duche Luxembourg), 127 Suppl., (1990) 460.
173 P. Kintz, V. Cirimele and P. Mangin, J. Forensic Sci., 40 (1995) 619.
174 V. Cirimele, P. Kintz and P. Mangin, Forensic Sci. Int., 70 (1995) 175.
175 J.Y. Kim, S. Sung III, M.K. in, K. Jung and B.C. Chung, Arch. Pharm. Res., 28 (2005) 1091.
176 M.J. Baptista, P. Monsanto, P.V. Monsanto, E.G.P. Marquest, A. Bermejo, S. Avila, A.M. Castanheira, C. Margalho, M. Barroso and D.N. Vieira, Forensic Sci. Int., 128 (2002) 66.
177 V. Cirimele, H. Sachs, P. Kintz and P. Mangin, J. Anal. Toxicol., 20 (1996) 13.
178 C. Jurado, M.P. Gimenez, M. Menendez and M. Repetto, Forensic Sci. Int., 70 (1995) 165.
179 D. Wilkins, H. Haughey, E. Cone, M. Huestis, R. Foltz and D. Rollins, J. Anal. Toxicol., 19 (1995) 483.
180 S. Strano-Rossi and M. Chiarotti, J. Anal. Toxicol., 23 (1999) 7.

181   T. Mieczkowski, Forensic Sci. Int., 70 (1995) 83.
182   H. Sachs and U. Dressler, Forensic Sci. Int., 107 (2000) 239.
183   F. Musshoff, H.P. Junker, D.W. Lachenmeier, L. Kroener and B. Madea, J. Anal. Toxicol., 26 (2002) 554.
184   F. Musshoff, D.W. Lachenmeier, L. Kroener and B. Madea, Forensic Sci. Int., 133 (2003) 32.
185   C. Moore, A. Negrusz and L. Douglas, J. Chromatogr. B: Biomed. Sci. Appl., 713 (1998) 137.
186   E.M.J. Ostrea, S.N. Lynn, R.H. Wayne and J.C. Stryker, Dev. Pharmacol. Ther., 1 (1980) 163.
187   E.M.J. Ostrea, P.M. Parks and M.J. Brady, Clin. Chem., 34 (1988) 2372.
188   E.M.J. Ostrea, M.J. Brady, P.M. Parks, D.C. Asensio and A. Naluz, J. Pediatrics, 1 (1989) 474.
189   P. Nair, B.A. Rothblum and R. Hebel, Clin. Pediatr., 33 (1994) 280.
190   W.E. Wingert, M.S. Feldman, M.H. Kim, L. Noble, I. Hand and J.J. Yoon, J. Forensic Sci., 39 (1994) 150.
191   M. Goosensen, L.M.L. Stolk and B.J. Smith, J. Anal. Toxicol., 19 (1995) 330.
192   C. Moore, D. Lewis, J. Becker and J. Leikin, J. Anal. Toxicol., 20 (1996) 50.
193   M.A. ElSohly, C. Walls, M.B. Lester, C.R. Bauer, S. Shankaran, H. Bada, L. Wright, V. Smeriglio and H. Kraus-Steinrauf, Pediatr. Res., 35 (1994) 83A.
194   M.A. ElSohly and S. Feng, J. Anal.Toxicol., 22 (1998) 329.
195   R. Coles, T.T. Clements, G.J. Nelson, G.A. McMillin and F.M. Urry, J. Anal. Toxicol., 29 (2005) 522.
196   D.A. Kidwell, J.C. Holland and S. Athanaselis, J. Chromatogr. B: Biomed. Sci. Appl., 713 (1998) 111.
197   S. Balabanova and E. Schneider, Beitr. Gerichtl. Med., 48 (1990) 45.
198   S. Balabanova, G. Buehler, H.J. Boschek, H. Schneitler, M. Froehlich and A. Froehlich, Dermatol. Monatsschr., 178 (1992) 357.
199   P. Kintz, Ther. Drug Monit., 18 (1996) 450.
200   P. Kintz, A. Tracqui, P. Mangin and Y. Edel, J. Anal. Toxicol., 20 (1996) 393.
201   G. Skopp, L. Poetsch, G. Zimmer and R. Mattern, Blutalkohol, 34 (1997) 427.
202   A.H. Lawrence, Forensic Sci. Int., 34 (1987) 73.
203   N.P. Lemos, R.A. Anderson and J. Roy Robertson, J. Anal. Toxicol., 23 (1999) 147.
204   J.R. Soares, J.D. Grant and S.J. Gross, NIDA Res. Monogr., 42 (1982) 44.
205   J. Volkmann, S. Ullwer, M. Muehlenberg and M. Topf, Cancolor rapid test for detecting cannabinoids in traffic control. Ger. Offen DE 19,607,646, September 11, 1997.
206   V. Kircher and H. Parlar, J. Chromatogr. B: Biomed. Appl., 677 (1996) 245.
207   D.J. Crouch, J. Michael Walsh, R. Flegel, L. Cangianelli, J. Baudys and R. Atkins, J. Anal. Toxicol., 29 (2005) 244.
208   V. Cirimele, M. Villain, P. Mura, M. Bernard and P. Kintz, Forensic Sci. Int., 161 (2006) 180.
209   M.A. Huestis and E.J. Cone, J. Anal. Toxicol., 28 (2004) 394.
210   C. Moore, C. Coulter, S. Rana, M. Vincent and J. Soares, J. Anal. Toxicol., 30 (2006) 413.
211   C. Moore, C. Coulter, S. Rana, M. Vincent and J. Soares, J. Anal. Toxicol., 30 (2006) 409.
212   N. Fucci, N. De Giovanni and S. Scarlata, Forensic Sci. Int., 119 (2001) 318.
213   N. Fucci, N. De Giovanni and M. Chiarotti, Forensic Sci. Int., 134 (2003) 40.
214   K. Papafotiou, J.D. Carter and C. Stough, Psychopharmacology (Berlin), 180 (2005) 107.
215   E.J. Cone, L. Presley, M. Lehrer, W. Seiter, M. Smith, K.W. Kardos, D. Fritch, S. Salamone and R.S. Niedbala, J. Anal. Toxicol., 26 (2002) 541.
216   S. Niedbala, K. Kardos, S. Salamone, D. Fritch, M. Bronsgeest and E.J. Cone, J. Anal. Toxicol., 28 (2004) 546.
217   R. Sam Niedbala, K.W. Kardos, D.F. Fritch, K.P. Kunsman, K.A. Blum, G.A. Newland, J. Waga, L. Kurtz, M. Bronsgeest and E.J. Cone, J. Anal. Toxicol., 29 (2005) 607.
218   H. Teixeira, P. Proenca, A. Castanheira, S. Santos, M. Lopez-Rivadulla, F. Corte-Real, E.P. Marques and D. Nuno Vieira, Forensic Sci. Int., 146 (2004) S61.
219   H. Teixeira, P. Proenca, A. Verstraete, F. Corte-Real and D. Nuno Vieira, Forensic Sci. Int., 150 (2005) 205.

220 M. Concherio, A. deCastro, O. Quintela, A. Cruz and M. Lopez-Rivadulla, J. Chromatogr. B, 810 (2004) 319.

221 M. Laloup, M.D.M.R. Fernandez, M. Wood, G.D. Boeck, C. Henquet, V. Maess and N. Samyn, J. Chromatogr. A, 1082 (2005) 15.

222 K. Kudo, T. Nagata, K. Kimura, T. Imamura and N. Jitsufuchi, J. Anal. Toxicol., 19 (1995) 87.

223 M. Morita and H. Ando, Kagaku Keisatsu Kenkyusho Hokoku, Hokagaku Hen., 37 (1984) 137.

224 R. Brenneisen and M.A. ElSohly, J. Forensic Sci., 33 (1998) 1385.

225 M. Hida, T. Mitsi, Y. Minami and Y. Fujimura, J. Anal. Appl. Pyrol., 32 (1995) 197.

226 S. Ross, M. Parker, R. Arafat, K. Lovett and M.A. ElSohly, Am. Lab., 16F (1996) 128.

227 T. Veress, J. Szanto and L. Leisztner, in: H. Kalasz and L. Ettre (Eds.), Chromatog. Institute of Forensic Sciences, Budapest, Hungary, 87, 1988, p. 481.

228 United Nations, Division of Narcotic Drugs, Recommended Methods for Testing Cannabinoids, Vienna, Austria, 1987, p. 38.

229 T.Z. Bozy and K.A. Cole, J. Anal. Toxicol., 24 (2000) 562.

230 C. Boes, B. Palavinskas, B. Dusemund, G. Gebhart and W. Blass, Lebensmittelchemie, 54 (2000) 104.

231 O. Zoller, P. Rhyn and B. Zimmerli, J. Chromatogr. A, 872 (2000) 101.

232 S.A. Ross, Z. Mehmedic, T.P. Murphy and M.A. ElSohly, J. Anal. Toxicol., 24 (2000) 715.

233 K. Kavor and H. Linder, Arch. Pharm. (Weinheim), 324 (1991) 329.

234 A. Hazekamp, Y.H. Choi and R. Verpoorte, Chem. Pharm. Bull., 52 (2004) 718.

235 L. Elias, A.H. Lawrence. In: T. Gough (Ed.), The analysis of drugs of abuse, Wiley, NY, 1991, p. 373.

M.J. Bogusz (Ed.). Forensic Science
Handbook of Analytical Separations, Vol. 6
© 2008 Elsevier B.V. All rights reserved.

243

CHAPTER 6

Sedatives and hypnotics

Thomas Kraemer[1] and Hans H. Maurer[2]

[1]*Institute of Legal Medicine, Forensic Toxicology, Saarland University, D-66421 Homburg, Germany;*
[2]*Department of Experimental and Clinical Toxicology, Institute of Experimental and Clinical Pharmacology
and Toxicology, Saarland University, D-66421 Homburg, Germany*

ABSTRACT

In this chapter, procedures for the detection of sedatives and hypnotics in blood and
urine as well as in alternative biomatrices such as oral fluid, sweat and hair are
reported. Other matrices of forensic interest, such as autopsy samples or non-biological samples, are also considered. The sedatives and hypnotics are divided here
in five classes: barbiturates, benzodiazepines, benzodiazepine BZ1 (omega 1) receptor agonists (zopiclone zolpidem, zaleplon and eszopiclone), diphenhydramine, and
other sedatives and hypnotics, including meprobamate, methaqualone, chloral hydrate and clomethiazole. For each class, some chemical and pharmacological information is given. Sample preparation from different biomatrices, autopsy samples
and non-biological matrices is discussed, as well as derivatization procedures for gas
chromatography (GC). In the analysis subsection, GC, high-performance liquid
chromatography and capillary electrophoresis procedures are reported with several
methods for detection. Some figures are provided, showing the structures of important analytes and typical chromatograms and/or UV or mass spectra. The chapter ends with a conclusion and some comments on future perspectives.

6.1 INTRODUCTION

Sedative–hypnotic drugs are one of the largest groups of drugs. In this chapter, they
are classified into barbiturates, benzodiazepines, benzodiazepine BZ1 (omega 1)
receptor agonists (zopiclone zolpidem, zaleplon and eszopiclone), diphenhydramine
and other sedative–hypnotics, including meprobamate, methaqualone, chloral

hydrate and clomethiazole. They are widely used for the treatment of insomnia, anxiety states and convulsive disorders as well as for anesthetic and preanesthetic medication. Because of their central nervous and respiratory depressant effects, they may cause, alone or in combination with other drugs and/or ethanol, severe intoxication for which treatment is necessary. They may mimic brain death [1]. Furthermore, they may impair driving ability and the fitness to work with machines even after therapeutic doses. In particular, barbiturates and benzodiazepines may lead to drug dependence and they are misused by heroin addicts to ease the withdrawal symptoms from heroin or to augment the effects of "weak heroin". Having been introduced to the market as drugs with minimal abuse and dependence potential, zolpidem, zopiclone, zaleplon and eszopiclone have become more and more the target of criticism because of a growing number of dependence case reports in the literature [2–13]. For all these reasons, sedative–hypnotics may be encountered in clinical or forensic toxicological analysis.

6.2 BARBITURATES

6.2.1 Introduction

Barbiturates are still used and misused, but with decreasing tendency. Nevertheless, there are important reasons, why screening for and quantification of barbiturates is necessary in clinical and forensic toxicology. Phenobarbital and its precursor primidone are still used as anticonvulsants for which drug monitoring is necessary. Thiopental is widely used as short time intravenous anesthetic. Thiopental and its metabolite pentobarbital are often to be monitored for decision of brain death [1,14]. The question of penal responsibility of a criminal after ingestion of barbiturates must be answered in the assessment of crimes. Barbiturates may reduce the fitness to drive a car or to work at machines and they may lead to addiction or to severe intoxications. Also in cases of drug-facilitated sexual assault, determination of barbiturates might be necessary [15,16].

6.2.2 Structural features of barbiturates

Barbiturates are 5,5-disubstituted barbituric acid derivatives. In Fig. 6.1, some typical representatives are shown.

In contrast to barbituric acid, which is five times more acidic (pKa 4.0) than acetic acid (pKa 4.75), the 5,5-disubstituted derivatives are only weak acids (pKa about 8). This is of importance for their biological effects and for their analytical behavior. Under physiological conditions, barbituric acid is deprotonated, while 5,5-disubstituted analogues are mainly unionized and can cross the blood-brain barrier. For extraction from aqueous matrices, similar reflections apply: strong acidic pH values are not necessary for sufficient isolation.

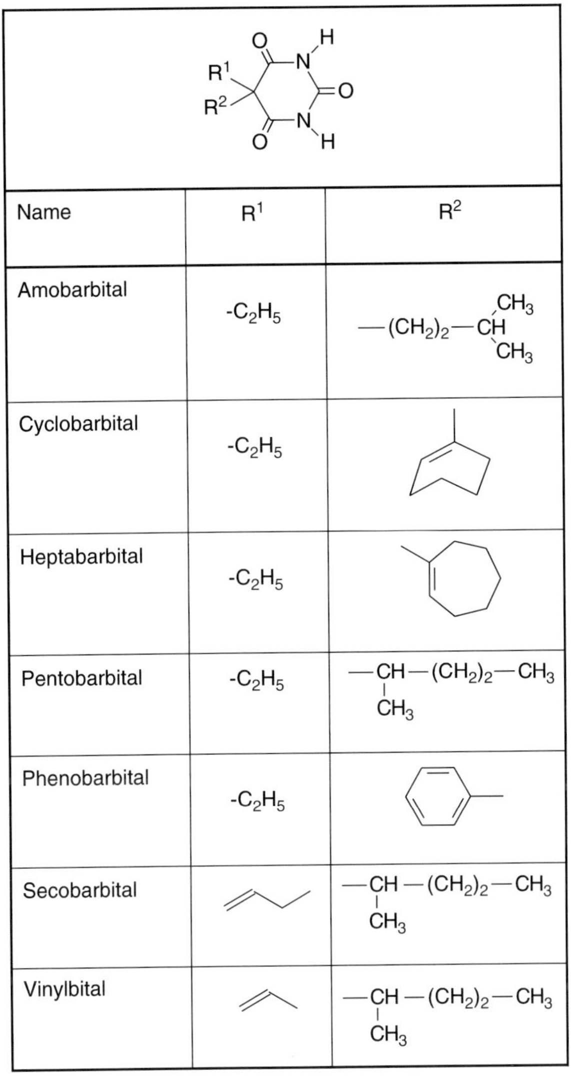

Name	R¹	R²
Amobarbital	$-C_2H_5$	$-(CH_2)_2-CH\big(CH_3\big)CH_3$
Cyclobarbital	$-C_2H_5$	
Heptabarbital	$-C_2H_5$	
Pentobarbital	$-C_2H_5$	$-CH(CH_3)-(CH_2)_2-CH_3$
Phenobarbital	$-C_2H_5$	
Secobarbital		$-CH(CH_3)-(CH_2)_2-CH_3$
Vinylbital		$-CH(CH_3)-(CH_2)_2-CH_3$

Fig. 6.1. Structures of typical barbiturates.

6.2.3 Sample preparation

6.2.3.1 Sample pretreatment and extraction of blood (serum, plasma) or urine

Suitable sample preparation is an important prerequisite for chromatography in biosamples. It involves isolation and, if necessary, cleavage of conjugates and/or derivatization of the barbiturates and their metabolites. Cleavage of conjugates is not necessary for barbiturate screening in contrast to other drug classes such as e.g. benzodiazepines. However, barbiturates can also be analyzed after acid hydrolysis within a general screening procedure [17–19]. Prior to blood, serum or plasma extraction, precipitation of proteins may be useful, which can be achieved using solvents such as acetone, acetonitrile, butyl acetate [1,20–26] or mixtures of methanol, zinc sulfate and ethylene glycol [27]. Deproteinization by adding saturated sodium sulfate solution has further advantages: the organic phase is kept free from water and salting-out effects may improve the extraction rates of liquid–liquid extraction (LLE) [28,29].

Liquid–liquid and solid-phase procedures are used for extraction of barbiturates from biomatrices. Some more exotic procedures such as supercritical fluid extraction are also described [30]. As this technique is not widely used, it is not further discussed here. Use of artificial receptors for analytical purposes was also described. The extraction solvent was enriched with such artificial receptor that works on the basis of molecular recognition. Recoveries of over 90% were determined from human control serum using a volume ratio (organic/serum) as small as 0.5. In the absence of this receptor, the volume ratio had to be > 10 to achieve similar extraction efficiencies [31]. This procedure is interesting from the scientific point of view, but it is not yet used in routine work.

6.2.3.1.1 Liquid–liquid extraction (LLE) procedures

Many different solvents were used for extraction of barbiturates from biological matrices at slightly acidic pH. They include dichloromethane, diethyl ether, toluene, butyl chloride, ethyl acetate, hexane or others. Often mixtures of these solvents were used [32–36]. In some procedures, salting-out effects help improve extraction [17,37]. As all these procedures work quite well with sufficient recoveries, no special recommendation is given here. However, use of chloroform [31,38,39] should be avoided nowadays. Time-consuming back extraction for cleanup [40,41] seems to be not necessary.

6.2.3.1.2 Solid-phase extraction (SPE) and solid-phase microextraction (SPME) procedures

Sample pretreatment for solid-phase extraction (SPE) depends on the sample type: whole blood and tissue (homogenates) need deproteinization and filtration/centrifugation steps before application to the SPE columns, whereas for urine usually a simple dilution step and/or centrifugation is satisfactory. Whatever SPE column is used, the analyst should keep in mind, that there might be large differences from batch-to-batch, and that the same sorbents from different manufacturers also lead to different results [42]. Therefore, use of a suitable internal standard (e.g. deuterated analytes) is recommended.

Many SPE procedures are reported in the literature. Procedures include use of C8, C18, mixed-phase Bond-Elut or Bond-Elut Certify® columns [43–54]. Most of these procedures work quite well, so that the decision, what kind of column should be used for SPE, is often influenced by practical considerations (e.g. column type used in the laboratory for other determinations). However, use of toxic chloroform for elution of retained analytes [47,55,56] should be avoided. In 2006, Cantu et al. [57] described optimization of solid-phase microextraction (SPME) procedures for the determination of tricyclic antidepressants and anticonvulsants (including pheno-barbital). Important factors for the SPME efficiency such as fiber coatings, extraction time, pH, ionic strength, influence of plasma proteins and desorption conditions are discussed [57]. SPE discs allow faster sample processing and smaller volumes of solvents. Application also to barbiturates have already been published [58,59]. A good overview over applications of SPE procedures can be found in several review articles [60–65].

6.2.3.2 Sample pretreatment and extraction of alternative matrices (oral fluid, sweat or hair)

Besides classical biomatrices such as blood and urine, alternative matrices such as oral fluid, sweat and hair have been tested for their usefulness in analytical toxicology. Development of more sensitive analytical equipment was a prerequisite for analysis in these matrices.

The value of drug testing or drug monitoring in alternative matrices should not be overestimated, especially for the following reasons. The general amount of sample is limited. The concentrations of drugs in oral fluid, sweat or hair usually are lower than in urine. The window of detection in sweat and oral fluid is shorter than that in urine. Nevertheless, sweat and oral fluid testing offers advantages over urine in the ease of collection. Hair samples also are easy to collect, and allow detection of chronic or past use of drugs and/or medicaments. However, there still is controversy on how to interpret the results, particularly concerning external contamination, cosmetic treatment or ethnical bias. Principles and kinetics of drug incorporation into hair are still under discussion as well as correlation between blood and hair concentrations [66]. Several reviews on the topic of hair analysis have been published in the past few years [67–71].

6.2.3.2.1 Sample preparation of oral fluid or sweat

For therapeutic drug monitoring, increased usage of oral fluid testing can be observed, whereas sweat was only minimally explored. Both oral fluid and sweat require extraction steps from the collection devices before analysis. Concentrations are lower than e.g. in urine. Therefore, the window of detection is shorter. Dierich and Soyka concluded that urine testing was still more reliable for barbiturates compared to oral fluid testing using immunoassays [72]. Also the triage system did not fulfill requirements for reliable detection in oral fluid [73].

Oral fluid is collected by spitting or by using cotton swabs. Production of saliva can be stimulated by chewing (glass marbles, parafilm, teflone pieces) and/or by giving citric acid in the mouth. Many variables in the oral fluid collection may have pronounced effects on oral fluid–drug concentrations [74]. Oral fluid samples can be extracted in the same way as plasma samples. For separation and detection, gas chromatography (GC) and gas chromatography-mass spectrometry (GC-MS) or high-performance liquid chromatography (HPLC)-UV and liquid chromatography (LC)-MS are suitable.

Analysis of barbiturates in sweat is hardly mentioned in the literature. Therefore, recommendations cannot be given. Common considerations on oral fluid and sweat testing have been reviewed by Kidwell *et al.* [75] and more recently by Verstraete [76].

6.2.3.2.2 Sample preparation of hair

Before extraction, decontamination using solvents (e.g. acetone, dichloromethane, petrolether, water) and homogenization (e.g. in a ball mill) are necessary. Cleavage of conjugates during the extraction procedure is usually performed enzymatically by addition of glucuronidase/arylsulfatase or by simple acid hydrolysis using hydrochloric acid. As already mentioned, barbiturates are not markedly altered during acid hydrolysis. Clean-up steps using SPE, SPME or LLE may help improve sensitivity. Detection of barbiturates is preferably performed by GC-MS after derivatization preferably by methylation [77–79]. Detection techniques of most of these drugs were summarized by Tracqui [80]. Frison *et al.* [15] used GC-MS/MS after SPME of head and pubic hair in a case of sexual assault for determination of thiopental and its metabolite pentobarbital. Saisho *et al.* [81] tested four typical extraction methods for barbiturates, using NH_4OH–methanol–acetone, TFA–methanol–acetone, sodium hydroxide and proteinase K using an animal model. Methanol–acetone–NH_4OH (10:10:1) was the best method concerning high extraction efficiency and low noise.

6.2.3.3 Sample pretreatment and extraction of body tissues and other autopsy material

Extraction of barbiturates from body tissues can be achieved by homogenizing the tissues in phosphate buffer followed by LLE or SPE as described above [33,36]. To improve the extraction rate, digestion steps using e.g. papain, neutrase or collagenase can be useful [33,82]. Sometimes fly larvae have also been used for toxicological analysis to provide data concerning the postmortem interval when poisoning is suspected. After washing, the larvae were dried and homogenized. After LLE drug concentrations were determined by GC-MS in the single-ion monitoring (SIM)-mode. Concentrations of drugs in the larvae were shown to be lower than those in the corresponding feeding tissues [83]. Cingolani *et al.* [84] tested stability of barbiturates in fixed tissues and formalin solutions. Barbiturates showed good stability even in biological specimens subjected to chemical fixation [84]. However, in other studies, not all tested barbiturates were stable under embalming conditions (e.g. phenobarbital) thus making it necessary to analyze for parent barbiturates or their predicted decomposition products [85].

6.2.3.4 Sample pretreatment and extraction of non-biological samples

Pure substances can usually be solved in solvents (e.g. methanol, ethanol, acetone, ethyl acetate, diethyl ether) and can be directly analyzed by GC, LC or capillary electrophoresis (CE) procedures. The same extraction methods, which are suitable for biomatrices can also be applied to extraction of barbiturates from non-biological matrices such as beverages, food or pharmaceutical formulations [86–89].

6.2.3.5 Derivatization for GC and GC-MS

Concentrations of barbiturates in blood and urine are relatively high [17,55]. Therefore, derivatization of barbiturates prior to GC-(MS) analysis is not necessary in most cases. Nevertheless, derivatization can further improve determination of barbiturates [90,91]. Methylation is used most often [17,35,37,46,90,91] and can therefore be recommended. Ethylation [34] or propylation [92] of barbiturates seems not to bring further advantages. Another group of reagents for barbiturate derivatization are the dimethylformamide dialkylacetal compounds. Optimization of this derivatization is described in a paper from Kushnir and Urry [93]. In a series of review articles, Halket and Zaikin, Halket *et al.* and Zaikin and Halket examined all aspects of derivatization [94–99].

6.2.4 Analysis of barbiturates

6.2.4.1 GC and GC-MS procedures

Fused-silica capillary columns are suitable for separation of barbiturates. Low polarity dimethylpolysiloxane types of column work as well as 5% diphenyl methylpolysiloxane types [17,32,55,56,91,92].

Nitrogen-phosphorus-selective detection is sometimes used for detection [32,55,90,91] but the detector of choice is MS, providing good sensitivity and best specificity. Thus, most of the procedures in the literature are GC-MS procedures [1,15,17,34,37,46–50,92,100,101]. Other detectors for GC such as flame ionization detector (FID) or Fourier transformation infrared (FTIR) detector are mentioned in the literature, but they are not of importance for this task.

6.2.4.1.1 Screening and confirmation of barbiturates
The usual strategy for analyzing barbiturates in urine first includes a screening test and second a confirmatory test. Different immunoassays for indication of barbiturates in urine and plasma are commercially available and can be used for screening in order to differentiate between negative and presumptively positive samples. Positive results must be confirmed by a second independent method that is at least as sensitive as the screening test and that provides the highest level of confidence in the result. Without doubt, GC-MS is still the reference method for confirmation of positive screening tests [34,35,47,48,102–104].

Screening and confirmation can be performed in one step using GC-MS in the electron-impact (EI), full-scan mode. Such a procedure, including universal sample preparation for many different drug classes (acid hydrolysis, LLE and acetylation) is described by Maurer [17–19,105]. For barbiturate screening, mass chromatography based on full-scan recording was used. The identity of the peaks in the mass chromatograms was confirmed by computerized comparison of the peaks underlying mass spectra with reference spectra [106,107]. Besides most of the barbiturates, the following groups of medicaments can simultaneously be covered by this procedure: amphetamine derivatives, benzodiazepines, opioids, analgesics, antidepressants, neuroleptics, antiparkinsonians, anticonvulsants, antihistamines, β-blockers, anti-arrhythmics, laxatives, and different designer drug classes.

6.2.4.1.2 Quantification of barbiturates in blood, plasma or serum

GC-MS is also suitable for quantification of barbiturates. However, instead of full-scan mode SIM mode should be used, which provides higher sensitivity and precision. Further prerequisite for precise quantification is the use of suitable internal standards [1,35,46,54,100]. Best results can be achieved with deuterated internal standards, which are commercially available.

6.2.4.2 HPLC and LC-MS procedures

Common C18 or C8 packing materials for LC columns have been most widely used and can be recommended for separation of barbiturates [25–27,44,53,108–117]. Isocratic or gradient elution procedures using acetonitrile or methanol/buffered water mixtures or mixtures of solvents (acetonitrile, hexane, isopropanol) with phosphate or acetate buffers result in sufficient separation. Also mobile phase systems containing ion-pair reagents such as cetrimide were used for barbiturates [118], but they are not of great importance. LC-MS does not allow the use of non-volatile buffers. Column-switching desalting systems can be used to overcome this disadvantage [119]. Of course, it is more useful to use mobile phase systems with volatile buffers [120–122]. Using chiral stationary phases (e.g. Chiralcel), separation of the barbiturate enantiomers can be achieved [123].

6.2.4.2.1 Screening and confirmation of barbiturates

As given above, the usual strategy for analyzing barbiturates in urine first includes a prescreening using immunoassays and second a confirmatory test. Ferrara *et al.* [44] used HPLC-UV for confirmation of positive immunoassays in urine. Diode-array detection (DAD) provides much better specificity than simple UV detection. Screening for drugs of abuse in plasma or serum was described using HPLC-UV-DAD [109]. This assay was capable of detecting and identifying therapeutic and toxic amounts of barbiturates, anticonvulsants, diuretics, non-steroidal anti-inflammatory drugs, sulfonylurea antidiabetic drugs, theophylline and analgesic drugs. Nevertheless, full-scan electron impact GC-MS provides best specificity, which is important, especially in forensic toxicology. With LC-MS becoming more and more a standard technique, it is also used for barbiturate screening and confirmation [120–122].

6.2.4.2.2 Quantification of barbiturates in blood, plasma or serum

For separation and quantification of barbiturates in plasma, several LC procedures were described with different detection modes [22,24,25,27,33,53,108,109,115–117,124,125]. Since therapeutic plasma concentrations of barbiturates are relatively high (> 1 mg/L) requirements on the analytical technique are not very high. Therefore, simple UV detection may be sufficient for determination of barbiturates in biosamples in terms of sensitivity. Requirements on specificity are however not met.

Even if it seems not to be necessary to employ LC-MS for barbiturate analysis, several LC-MS procedures, which also include barbiturate determination, have been published in the past few years [30,119–122,126].

6.2.4.3 Electrokinetic procedures

CE and micellar electrokinetic capillary chromatography (MEKC) are rapidly growing analytical technologies. The great interest raised by them is due to its high efficiency, mass sensitivity, minimum needs of solvent and sample volumes and in particular to the high versatility in terms of separation modes. However, sample matrix components and salts can have deleterious effects on CE separations. For urine, dilution of the sample can help, but this procedure works only when analytes such as barbiturates are present in high concentrations. Extraction of urine before CE was also recommended [127,128]. Ferslew *et al.* [129] used toxi-tubes for LLE of barbiturates from blood vitreous humor and urine before MEKC. Plasma or serum can be analyzed by direct injection techniques [127] or after deproteinization and/or extraction [130]. Deproteinization should not be performed using acids or salts, which would have deleterious effects on CE separations. Solvents such as acetonitrile are suitable. Sample preparation techniques used in CE are often transferred from LC or GC methods. However, mainly the small volumes cause problems to handle, when using devices not specifically tailored for such purpose. Some CE techniques allow sample enrichment during injection, known as sample stacking [39].

For detection of barbiturates after CE separation, UV(DAD) [131–133] and MS [134] have been used. Simple UV detection at 214 nm after MEKC (SDS in phosphate-borate buffer – acetonitrile pH 8.5) gave good results for determination of several barbiturates in plasma and urine after LLE [129]. On-column multiwavelength detection (195 and 320 nm) after MEKC (SDS in phosphate-borate buffer pH 7.8) allowed sufficient sensitivity for determination of barbiturates in plasma [127]. Using different cyclodextrins as chiral separators in CE, determination of barbiturate enantiomers is also possible [135,136]. The status of CE techniques in forensic toxicology has been displayed in detail in the review of Tagliaro and Bortolotti [137]. Today, CE techniques are still not very widely used in forensic toxicology. The hopes that all problems of the method will be solved and CE may be a useful tool in forensic toxicology did not come true. In a more recent review Pucci and Raggi concluded again that electrokinetic chromatographic methods might become a robust and widely accepted method and that their use might continue to grow [138].

6.2.4.4 Magnet resonance spectroscopic procedures

Administration of barbiturates (and other drugs) is clincally monitored by determining serum concentrations of the drugs. The actual drug concentrations in the brain parenchyma are not known. Volume selective proton MR spectroscopy has been used for *in vitro* determination of anticonvulsants under standard clinical conditions using a single-voxel stimulated-echo acquisition mode (STEAM) sequence. The authors expected this technique to be able to monitor anticonvulsants in brain parenchyma *in vivo* [139].

6.3 BENZODIAZEPINES

6.3.1 Introduction

Benzodiazepines are used as tranquilizers, hypnotics, anesthetics, anticonvulsants or muscle relaxants and belong to the most frequently prescribed drugs. They may reduce the fitness to drive a car or to work at machines and they may lead to addiction or severe intoxications, especially in combination with alcohol. Therefore, screening for benzodiazepines is necessary in clinical, forensic and occupational toxicology. Midazolam and other benzodiazepines are often to be monitored for decision of brain death [1,14]. Also in cases of drug-facilitated sexual assault, determination of benzodiazepines might be necessary [11,16,140–149].

6.3.2 Structural features of benzodiazepines

The classical benzodiazepines have a 5-aryl-1,4-diazepine structure, characterized by a benzene ring condensed to the 10- and 11-positions of the 1,4-diazepine ring. The arylsubstituent at position 5 is usually phenyl or a 2-halogenated phenyl ring. Annulation of an imidazole ring in 1,2-position leads to imidazo- or diazolo- benzodiazepines such as midazolam or loprazolam. The primary benzene ring can be replaced with a thienyl ring to give compounds such as brotizolam, clotiazepam and bentazepam. There are also some 1,5-benzodiazepines such as clobazam. Flumazenil is a benzodiazepine receptor antagonist, which is structurally related to the benzodiazepines (see Fig. 6.2b).

During phase I metabolism, benzodiazepines are N-dealkylated, hydroxylated at C-3 and/or hydroxylated at the phenyl ring in C-5. Such metabolism can lead to common metabolites. For example, oxazepam is a common target metabolite in urine for many 1,4-benzodiazepines, such as diazepam, nordazepam, temazepam, clorazepate, chlordiazepoxide, ketazolam, camazepam or medazepam. Other 1,4-benzodiazepines have a nitro-substituent at C-7 (flunitrazepam, clonazepam, nitrazepam etc.). They are metabolized to 7-amino metabolites, which should be target metabolites in urine and postmortem blood specimens.

6.3.3 Sample preparation

6.3.3.1 Sample pretreatment and extraction of blood (serum, plasma) or urine

For successful sample preparation, knowledge is necessary on how the analytes are present in the corresponding biomatrix and on how sample preparation steps can alter the analytes. Like most of the centrally acting drugs, benzodiazepines are

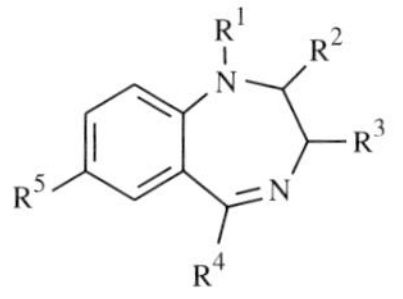

Name	R^1	R^2	R^3	R^4	R^5
Bromazepam	H	=O	H	2'-pyridyl	Br
Camazepam	CH_3	=O	$-O-\overset{O}{\overset{\|}{C}}-N(CH_3)_2$	Phenyl	Cl
Clonazepam	H	=O	H	2-Cl-phenyl	NO_2
Clorazepate	H	=O	COOH	Phenyl	Cl
Diazepam	CH_3	=O	H	Phenyl	Cl
Flunitrazepam	CH_3	=O	H	2-F-phenyl	NO_2
Flurazepam	 $\diagdown\diagup N(C_2H_5)_2$	=O	H	2-F-phenyl	Cl
Halazepam	CF_3CH_2	=O	H	Phenyl	Cl
Lorazepam	H	=O	OH	2-Cl-phenyl	Cl
Lormetazepam	CH_3	=O	OH	2-Cl-phenyl	Cl
Nordazepam	H	=O	H	Phenyl	Cl
Nitrazepam	H	=O	H	Phenyl	NO_2
Oxazepam	H	=O	OH	Phenyl	Cl
Prazepam	 $\diagdown\diagup\triangle$	=O	H	Phenyl	Cl
Quazepam	CF_3CH_2	=S	H	2-F-phenyl	Cl
Temazepam	CH_3	=O	OH	Phenyl	Cl
Tetrazepam	CH_3	=O	H	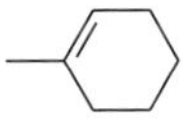 cyclohexenyl	Cl

Fig. 6.2. Structures of (a) 1,4-benzodiazepines, (b) imidazo- and triazolo-benzodiazepines and some other benzodiazepines.

References pp. 274–286

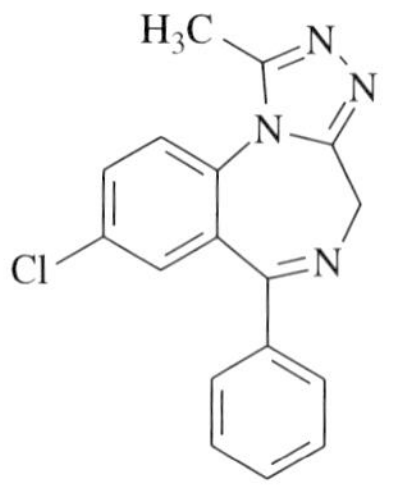

Alprazolam

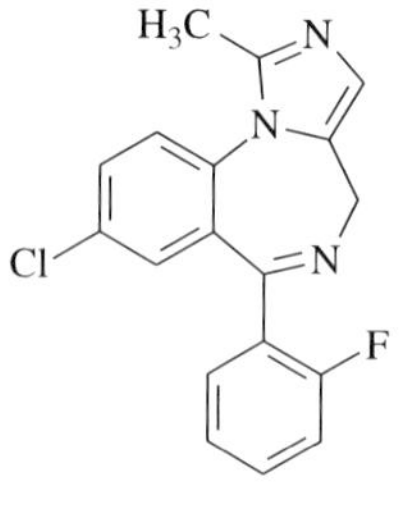

Midazolam

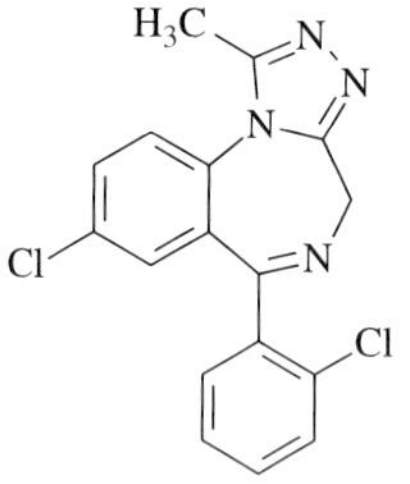

Triazolam

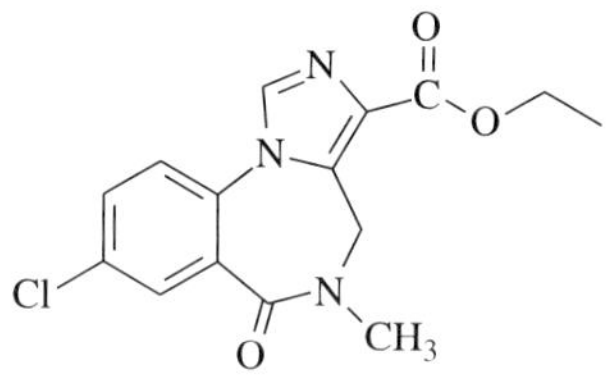

Flumazenil

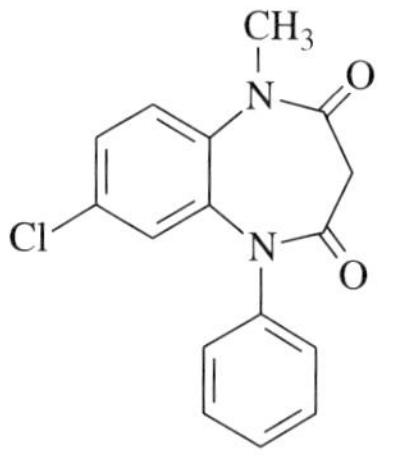

Clobazam

Fig. 6.2 (Continued)

lipophilic and are extensively metabolized. Some benzodiazepines and/or their phase I metabolites can be conjugated with glucuronic or sulfuric acid. Therefore, cleavage of conjugates is necessary before analysis of benzodiazepines, especially in urine. In forensic toxicology, enzymatic hydrolysis using glucuronidase (and arylsulfatase) is recommended [150–163]. Unfortunately, enzymatic hydrolysis is time-consuming and expensive. If results are requested in a short time (e.g. in clinical toxicology) rapid acid hydrolysis using hydrochloric acid is more convenient. However, benzodiazepines are hydrolyzed to benzophenone derivatives [164]. Nevertheless, these benzophenones can be used as target compounds for analysis [18,19,163,165,166]. A compromise of both cleavage techniques is the use of column packed immobilized glucuronidase/arylsulfatase. It combines the advantages of both methods, the speed of acid hydrolysis and the gentle cleavage of enzymatic hydrolysis [167,168].

As described below, liquid–liquid and solid-phase procedures are used for extraction of benzodiazepines from biomatrices. Some more exotic procedures such as supercritical fluid extraction are also described [169,170]. Since this technique is not very widely used it is not further discussed here. This applies also to dialysis as an extraction method [171].

With LC-MS being more and more a standard technique in clinical and forensic toxicological laboratories, the direct determination of benzodiazepine conjugates has also become a routine task [172–174].

6.3.3.1.1 Liquid–liquid extraction (LLE) procedures

For extraction of benzodiazepines from biological matrices many different solvents were used at slightly alkaline pH. They include chlorobutane, dichloromethane, diethyl ether, toluene, butyl chloride, ethyl acetate, butyl acetate, hexane and others. Often mixtures of these solvents were used [1,102,166,175–199]. As all these procedures work quite well with sufficient recoveries, no special recommendation is given here. However, use of toxic chloroform [41,154,200–203] should be avoided, whenever possible. Back extraction as further clean-up [40,41] seems not to bring advantages. Therefore it can be renounced.

Another possibility to get clean extracts is the use of immunoaffinity extraction (IAE) [204–206]. However, antibodies are expensive and saturation effects may impair quantification procedures. Therefore, this powerful method is not very widely used in toxicological laboratories.

6.3.3.1.2 Solid-phase extraction (SPE) procedures

Many SPE procedures for benzodiazepines are reported in the literature. Difficulties in sample pretreatment and in batch-to-batch conformity are already discussed in 6.2.3.1.2. Procedures include use of C2, C8, C18, mixed-mode columns. Even polar cyano phases were tested [11,44,121,150,153,156,207–231]. Most of these procedures work quite well. Using column-switching HPLC, online SPE can also be performed [214,232].

Comparative studies can be found in ref. [221]. A good overview of extraction procedures can be found in different reviews [60,63,65,158,233].

6.3.3.2 Sample pretreatment and extraction of alternative matrices (oral fluid, sweat or hair)

Common aspects of sample pretreatment of alternative matrices are already discussed (cf. 6.2.3.2).

6.3.3.2.1 Sample preparation of oral fluid or sweat

Oral fluid and sweat require common extraction steps from the collection devices before analysis. Concentrations are lower than those e.g. in urine. Therefore, the window of detection is shorter. These considerations show, that these matrices are not the ones of choice [75]. Nevertheless, oral fluid testing has become an interesting alternative for testing acute influence of drugs e.g. roadside drug testing in driving under the influence of drugs cases. Consequently, the number of papers on this topic is growing [73,210,234–236].

Analysis of benzodiazepines in sweat is hardly mentioned in the literature [237]. Common considerations on oral fluid and sweat testing have been reviewed by Kidwell *et al.* [75] and more recently by Verstraete [76].

6.3.3.2.2 Sample preparation of hair

Again, before extraction decontamination using different solvents (e.g. acetone, dichloromethane, petrolether, water) and homogenization (e.g. in a ball mill) is necessary. Cleavage of conjugates during the extraction procedure is usually performed enzymatically by addition of glucuronidase/arylsulfatase or by simple acid hydrolysis using hydrochloric acid. Clean-up steps using SPE or LLE may help improve analysis. Concentrations of benzodiazepines in hair are low and require very sensitive detection techniques. Therefore, detection was preferably performed by GC-MS in the negative ion chemical ionization (NICI) mode after suitable derivatization [77,202,238–240]. Gaillard and Pepin [241] proposed combination of modern diode-array detecors and GC-MS for detection of benzodiazepines and other drugs in hair. Nowadays, with LC-MS/MS having become a routine technique in the clinical and forensic toxicological lab, hair analysis is often done by this technique [141,143,179,184,187,242–245]. In Fig. 6.3, selected reaction monitoring (SRM) chromatograms are shown obtained after analysis of the root segment of the hair of

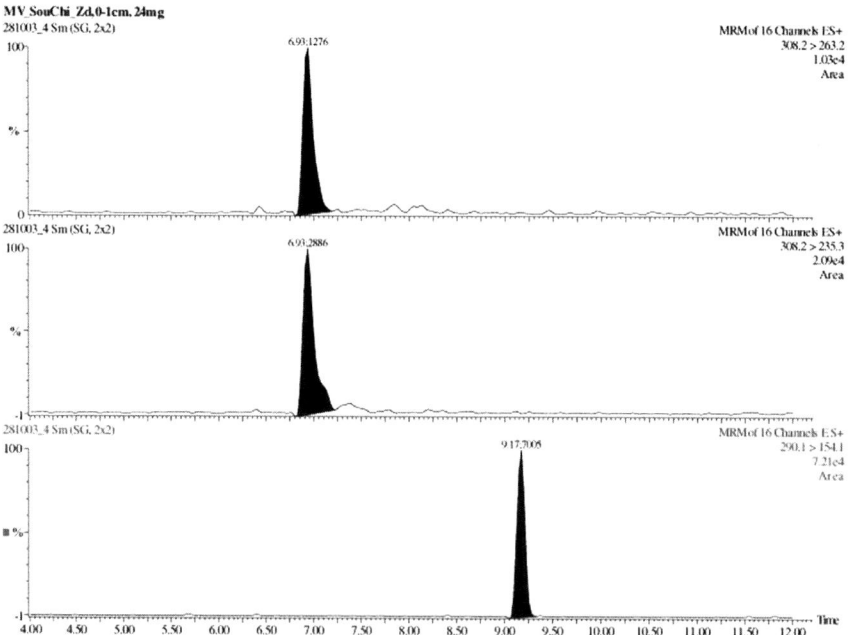

Fig. 6.3. SRM chromatograms obtained after analysis of the root segment of a hair sample of a volunteer who was administered a single dose of 6 mg of bromazepam 1 month before. On the top, the two daughter ions of bromazepam, on the bottom, the daughter ion of the IS. Concentration was 4.7 pg/mg. (Taken from ref. [184].)

a volunteer who was administered a single dose of 6 mg of bromazepam 1 month before (taken from ref. [184]). More details on hair analysis can be found in some recent reviews [67,68,70,71].

6.3.3.3 Sample pretreatment and extraction of body tissues and other autopsy material

Most of the extraction procedures given above also work for postmortem samples after corresponding homogenization and/or enzymatic digestion (cf. 6.3.3.1). Nevertheless, there are some special things to observe in postmortem specimens. Nitrobenzodiazepines are reduced rapidly and almost quantitatively to the corresponding 7-amino metabolites [246]. Therefore, these 7-amino compounds should serve as target compounds in such specimens. Sometimes fly larvae and puparia can also be used for detection of drugs in postmortem cases. A simple homogenization, followed by acetonitrile precipitation can be suitable for pretreatment. Puparia can be pulverized and extracted by ultrasonification in methanol and analyzed by LC-MS/MS [247,248].

6.3.3.4 Sample pretreatment and extraction of non-biological samples

Pure substances can usually be solved in solvents (e.g. methanol, ethanol, acetone, ethyl acetate, diethyl ether) and can be directly analyzed by simple GC or LC procedures. The same extraction methods, which are suitable for biomatrices, can also be applied to extraction of benzodiazepines from non-biological matrices such as beverages, food or pharmaceutical formulations. Because of the high analyte concentrations UV detection for LC methods, capillary electrophoretic procedures or even thin-layer chromatography (TLC) methods are used for their determination [249–259].

6.3.3.5 Derivatization for GC and GC-MS

Most of the published GC-ECD (electron capture detector) procedures for benzodiazepines renounce derivatization, because detection sensitivity seems to be sufficient. Derivatization is recommended for detection of benzodiazepines and/or their metabolites or benzophenones, when containing primary amino or hydroxy groups. Derivatization usually leads to better thermal stability and derivatized benzodiazepines give well defined mass spectra. For derivatization of benzodiazepines, their metabolites or the corresponding benzophenones, trimethylsilylation using BSTFA/TMCS mixtures or acetylation using an acetic anhydride–pyridine mixture are suitable [18,19,102,158,166]. The corresponding mass spectra are included in published databanks [106,107]. They also include mass spectra of underivatized, (trifluoro)acetylated, trimethylsilylated or perfluoroalkyated and/or perfluoroacylated benzodiazepines, their metabolites and benzophenones. For more information on more exotic derivatization procedures such as combined propionylation/propylation the review of Segura *et al.* [260] on derivatization procedures is recommended. More recent information can be found in a series of review articles by Halket and Zaikin, Halket *et al.* and Zaikin and Halket [94–99]. With the benzodiazepines containing halogens or containing groups derivatizable with corresponding electronegative reagents, NICI for GC-MS is suitable for sensitive determination of benzodiazepines [148,176,261–265].

6.3.4 Analysis of benzodiazepines

6.3.4.1 GC and GC-MS procedures

For separation of the benzodiazepines fused-silica capillary columns are suitable. Low polarity dimethylpolysiloxane column types [102,161,197,223] are as suitable as more polar 5% diphenyl methylpolysiloxane types [156,157,162,171,194,266]. For underivatized benzodiazepines cyanopropylphenyl stationary phases can give better separations [267]. Since benzodiazepines contain electronegative substituents (halogen atoms), they are suitable for electron capture detectors which provide good sensitivity (ca. 1ng/mL). Similar sensitivity (ca. 0.1 ng/mL) with higher specificity can be reached using GC-MS in the NICI mode [148,176,197,261,264,268]. Because of the very low concentrations of benzodiazepines in hair and also in sweat, detection is preferably done by GC-MS-NICI [143,148,238,239,269,270].

6.3.4.1.1 Screening and confirmation of benzodiazepines
Different immunoassays for indication of benzodiazepines are commercially available and can be used for screening in urine in order to differentiate between negative and presumptively positive samples. Positive results can be confirmed by GC-MS procedures [102–104,151,156,157,161,166,199,223,271–277] or by LC-MS procedures [234,243]. In Fig. 6.4 overlaid ion chromatograms of blood samples after matrix supported LLE are shown for the simultaneous LC-MS determination of 33 benzodiazepines including metabolites and benzodiazepine-like substances (taken from ref. [191]).

As described in 6.2.4.1.1, screening and confirmation can be performed in one step using GC-MS in the EI full-scan mode. Such a procedure, including universal sample preparation for many different drug classes (acid hydrolysis, LLE and acetylation) is described by Maurer [17–19]. The benzophenones formed during hydrolysis can be detected using mass chromatography based on full-scan recording. The identity of the peaks in the mass chromatograms was confirmed by computerized comparison of the peaks underlying mass spectra with reference spectra [106,107]. Besides most of the benzodiazepines, the following groups of medicaments can simultaneously be covered by this procedure: amphetamine derivatives, barbiturates, opioids, analgesics, antidepressants, neuroleptics, antiparkinsonians, anticonvulsants, antihistamines, β-blockers, antiarrhythmics and laxatives.

6.3.4.1.2 Quantification of benzodiazepines in blood, plasma or serum
While benzodiazepines can easily be screened in urine using GC-MS, GC-MS quantification in blood suffers, e.g. from thermal instability and low volatility of some of the parent compounds [102,158]. Nevertheless, GC procedures were also

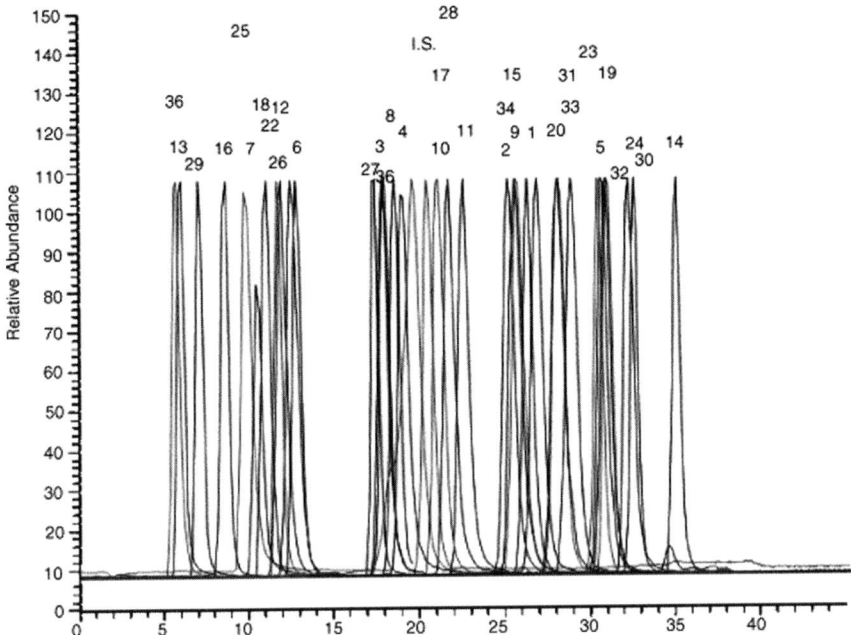

Fig. 6.4. Overlaid ion chromatograms of blood samples after matrix supported LLE for the simultaneous LC-MS determination of 33 benzodiazepines including metabolites and benzodiazepine-like substances: (1) alprazolam; (2) OH-alprazolam; (3) bromazepam; (4) OH-bromazepam; (5) brotizolam; (6) chlordiazepoxide; (7) norchlordiazepoxide; (8) demoxepam; (9) clobazam; (10) desmethylclobazam; (11) clonazepam; (12) acetamidoclonazepam; (13) aminoclonazepam; (14) diazepam; (15) flunitrazepam; (16) 7-aminoflunitrazepam; (17) *N*-desmethylflunitrazepam; (18) flurazepam (19) desalkylflurazepam; (20) OH-ethylflurazepam; (21) internal standard; (22) loprazolam; (23) lorazepam; (24) lormetazepam; (25) desmethylmedazepam; (26) midazolam; (27) 1-OH-midazolam; (28) nitrazepam; (29) acetamidonitrazepam; (30) nordazepam; (31) oxazepam; (32) temazepam; (33) triazolam; (34) OH-triazolam; (35) zolpidem; (36) zopiclone. (Taken from ref. [191].)

used for quantification of benzodiazepines in blood, plasma or serum. Besides MS, nitrogen-phosphorus selective and electron-capture detectors were used [171,267,278,279]. GC-MS in the SIM mode instead of full-scan mode should be used, since it provides higher sensitivity and precision. Further prerequisite for precise quantification is the use of suitable internal standards [224]. Best results can be achieved with deuterated internal standards, which are commercially available [1,268,280–282]. Use of non-deuterated benzodiazepines as internal standards for determination of other benzodiazepines [216,265,278,283,284] cannot be recommended for general purposes, because nowadays every exotic benzodiazepine may appear in every part of the world, thus disturbing useful quantification.

6.3.4.2 HPLC procedures

For separation of benzodiazepines, common C18 packing materials are most widely used [11,141–143,179,180,182,186,195,201,211,221,228,248,252,254,285–292]. C8 packing materials may also be suitable [192,232,293–298]. Isocratic procedures using methanol/unbuffered water mixtures or mixtures of solvents (acetonitrile, hexane, isopropanol) with phosphate or acetate buffers result in sufficient separation. Addition of amines as mobile phase modifiers is described [299], but seems not to be unequivocally necessary. If separation of ionic compounds is desired, mobile phase systems containing ion-pair reagents such as tetramethyl ammonium or tetrabutyl ammonium salts or methane sulfonic acid can be used [44,300]. For LC-MS volatile solvents and buffers such as methanol, acetonitrile, formate-buffers have to be used [141,143,177,179,187,188,244,301].

6.3.4.2.1 Screening and confirmation of benzodiazepines

As given above, the usual strategy for analyzing benzodiazepines in urine first includes a prescreening using immunoassays and second a confirmatory test. Some authors used HPLC-UV for confirmation of positive immunoassays or direct analysis in urine [44,302,303]. DAD provides much better specificity than simple UV detection [198,220,226,228,253,287,289,298,304–310]. Commercial databanks containing UV-spectra of hundreds of drugs and medicaments are available [311–313]. A class-independent drug screen in (postmortem) plasma or serum was described using HPLC- UV-DAD [109]. This assay was capable of detecting and identifying therapeutic and toxic amounts of benzodiazepines, barbiturates, anticonvulsants, diuretics, non-steroidal anti-inflammatory drugs, sulfonylurea antidiabetic drugs, theophylline and analgesic drugs. Nevertheless, MS detection provides best specificity, and with the LC-MS apparatus becoming a lab standard, more and more procedures using this technique have been published [140,150,175,179,184,188,191,214,234,242,244,314–316]. A good overview on multi-analyte procedures using LC-MS or LC-MS/MS can be found in the review of Maurer [317]. Allen proposed to replace immunoassays by LC tandem MS for the routine measurement of drugs of abuse in oral fluid. He concluded, that LC-MS/MS offered a more flexible, specific and sensitive alternative to the screening of oral fluid samples for drugs of abuse than immunoassay techniques. A single sample injection could be suitable for detection of a wide range of drugs and metabolites [318].

6.3.4.2.2 Quantification of benzodiazepines in blood, plasma or serum

For separation and quantification of benzodiazepines in plasma, several LC procedures with UV-DA detection were described [228,231,246,319–325].

LC-MS with its high sensitivity and specificity is very suitable for determination of benzodiazepines in plasma. Consequently, it is most often used for this purpose [173,175,178,179,181,185,188,189,191,200,209,211,212,316,326–329]. In 2005, Maurer reviewed procedures for quantification of drugs in the biosamples blood, plasma, serum, or oral fluid using LC coupled with single-stage or tandem mass spectrometry (LC-MS, LC-MS/MS). Therein, basic information on the procedures has been given in tables. The pros and cons of such LC-MS procedures including sample work-up and ion suppression effects have been critically discussed.

6.3.4.3 CE procedures

Benzodiazepines have far less concentrations in body fluids than barbiturates have. Therefore, determination of benzodiazepines in plasma using CE has hardly been described [330]. Concentrations in beverages used in drug facilitated crimes or in tablets are quite high. CE procedures can be used for their determination [249–251,330]. Baldacci and Thormann [172] were able to analyze lorazepam and its 30-glucuronide in human urine by CE. They found evidence for the formation of two distinct diastereomeric glucuronides [172]. CE can be coupled also to a mass spectrometer (CE-MS or CE-MS/MS) thus reaching good specificity and sensitivity [172,331,332]. In a series of reviews Smyth has given a critical evaluation of CE-MS analytical methods for the detection and determination of small molecular mass drug molecules, such as the 1,4-benzodiazepines. Analytical information on sample concentration techniques, CE separation conditions, recoveries from biological matrices and limits of detection have been given therein [333–335].

6.4 ZOPICLONE, ZOLPIDEM, ZALEPLON AND ESZOPICLONE

6.4.1 Introduction

Zopiclone, zolpidem, zaleplon and eszopiclone are benzodiazepine BZ1 (omega 1) receptor agonists. They have rapid onset of action and short elimination half-life. Unlike benzodiazepines, they have weak myorelaxant and anticonvulsant effects. Nevertheless, they may reduce the fitness to drive a car or to work at machines and they may lead to addiction or severe intoxications, especially in combination with alcohol [3,5,6,10,336–338]. Therefore, screening for these drugs is necessary in clinical, forensic and occupational toxicology. Also in cases of drug-facilitated sexual assault, determination of benzodiazepine BZ1 receptor agonists might be necessary [144,146,242,339–343]. Eszopiclone is discussed to have less addictive properties and has been approved for the long-term treatment of insomnia [344–346].

6.4.2 Structural features of zopiclone, zolpidem, zaleplon and eszopiclone

Zopiclone (*R,S*-6-(5-chloro-2-piridyl)-6,7-dihydro-7-oxo-5*H*-pyrrolo[3,4-b]pyrazin-5-yl 4-methyl-1-piperazinecarboxylate, Fig. 6.5) is a sedative–hypnotic agent possessing a short duration of action and few associated side effects. The maximum plasma level is about 80 ng/mL with a therapeutically active dose of 7.5 mg. It is extensively metabolized *via* three major pathways: decarboxylation, oxidation and demethylation. Zopiclone is unstable in nucleophilic solvents such as methanol or ethanol. Stability seems to be slightly better in isopropanol, acetonitrile and toluene. Eszopiclone is the pure *S*(+)-enantiomer of zopiclone. As it is claimed to be the active enantiomer, dosage is lower (1–3 mg) than that for the racemate. As the physicochemical properties of eszopiclone are identical to that of zopiclone (except

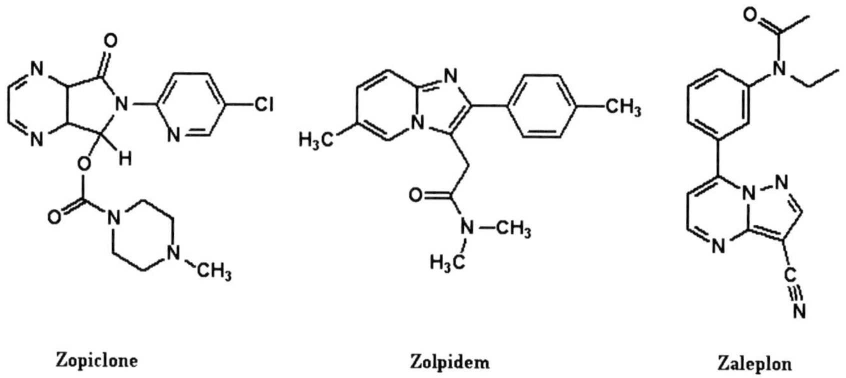

Zopiclone   **Zolpidem**   **Zaleplon**

Fig. 6.5. Structures of zopiclone, zolpidem and zaleplon.

for the rotation direction of the plane of polarized light), analytical procedures for determination of zopiclone should also work for determination of the pure *S*-enantiomer.

Zolpidem (*N,N*,6-trimethyl-2-*p*-tolylimidazo[1,2a]pyridine-3-acetamide; Fig. 6.5) has maximum plasma levels of about 200 ng/mL with a therapeutically active dose of 10 mg. It is metabolized to inactive metabolites *via* oxidation of each of the methyl groups on the phenyl moieties and *via* hydroxylation of the imidazopyridine moiety.

Zaleplon (*N*-[3-(7-cyano-1,5,9-triazabicyclo[4.3.0]nona-2,4,6,8-tetraen-2-yl)phe-nyl]-*N*-ethyl-acetamide; Fig. 6.5) is given in therapeutic doses of 5 or 10 mg. It is extensively metabolized to inactive metabolites mainly *via* oxidation to 5-oxo-zale-plon with <1% of dose excreted unchanged in urine.

6.4.3 Sample preparation

6.4.3.1 Sample pretreatment and extraction of blood (serum, plasma) or urine

Common aspects of sample pretreatment of blood (serum, plasma) or urine are discussed in subsections 6.2.3.1 and 6.3.3.1.

As described below, liquid–liquid and solid-phase procedures are used for extraction of the benzodiazepine BZ1 (omega 1) receptor agonists from biomatrices.

6.4.3.1.1 Liquid–liquid extraction (LLE) procedures

Many different solvents were used for extraction of the benzodiazepine BZ1 (omega 1) receptor agonists from biological matrices at slightly alkaline pH. They include isoamyl alcohol, diethyl ether, toluene, butyl chloride, butyl acetate, ethyl acetate, dichloromethane, hexane or others. Often mixtures of these solvents were used [176,179,184,188,242,278,347–357]. As all these procedures work quite well with sufficient recoveries, no special recommendation is given here.

6.4.3.1.2 Solid-phase extraction (SPE) procedures
Common aspects of SPE procedures, such as sample pretreatment or problems with batch-to-batch conformity are discussed in subsections 6.2.3.1.2 and 6.3.3.1.2.

Many SPE procedures are reported in the literature. Procedures include use of C2, C8, C18 or mixed-mode columns [11,121,210,224,354,358–362]. Most of these procedures work quite well.

6.4.3.2 Sample pretreatment and extraction of alternative matrices (oral fluid, sweat or hair)

Common aspects of sample pretreatment of alternative matrices are already discussed (cf. 6.2.3.2.).

6.4.3.2.1 Sample preparation of oral fluid or sweat
Only a few papers have been published which described the determination of zopiclone, zolpidem or zaleplon in oral fluid. For separation and detection, CE with UV laser-induced fluorescence detection, GC-MS and LC-MS/MS have been used [210,314,363,364]. Detection of these substances in sweat has not been described. Special recommendations cannot be given. Common aspects of analysis in these matrices are discussed in subsection 6.2.3.2.1.

6.4.3.2.2 Sample preparation of hair
Common aspects of sample preparation of hair are given in subsection 6.2.3.2.2. Determination of zopiclone, zolpidem and zaleplon in hair is mostly done by LC-MS/MS methods [144,146,179,184,242,339,341,343].

6.4.3.3 Sample pretreatment and extraction of body tissues and other autopsy material

Extraction of zopiclone and zolpidem from body tissues can be achieved by homogenizing the tissues in phosphate buffer followed by LLE or SPE as described above [33,365–368]. To improve the extraction rate, digestion steps using e.g. papain, neutrase or collagenase can be useful [33,82].

Tissue distribution and potential for postmortem diffusion of zopiclone was investigated by Pounder and Davies [369]. They concluded that zopiclone showed little preferential concentration in solid organs and consequently had relatively stable postmortem blood concentrations, with little drug redistribution artifacts. Postmortem diffusion from gastric drug residue could elevate drug levels in parts of the liver and lung. Corresponding data for zolpidem can be found in refs. [367,368].

6.4.3.4 Sample pretreatment and extraction of non-biological samples

Pure substances can usually be solved in solvents (e.g. methanol, ethanol, acetone, ethyl acetate, diethyl ether) and can be directly analyzed by GC or LC procedures [370,371]. The same extraction methods, which are suitable for biomatrices, can also be applied to extraction of zopiclone or zolpidem from non-biological matrices such as beverages, food or pharmaceutical formulations. For the detection of zopiclone in pharmaceutical tablets an ion-pair reversed-phase high-performance liquid chromatographic method with UV detection was described [372]. Known potential degradation products of zopiclone were separated, allowing simultaneous detection of zopiclone and its degradation products. Kelani described selective potentiometric determination of zolpidem hemitartrate in tablets by using polymeric membrane electrodes [373].

6.4.3.5 Derivatization for GC and GC-MS

Zopiclone is less suitable for GC analysis due to formation of artifacts. The parent compounds of Zopiclone and Zaleplon have no derivatizable groups. The corresponding N-dealkyl metabolites can be acylated using different reagents. Gunnar *et al.* [374] tested different silylating reagents for several hypnotics including zaleplon and zolpidem.

The parent compound zolpidem also has no derivatizable groups. However, the hydroxy metabolite should be derivatized for better chromatographic properties. Trimethylsilylation, acetylation or perfluoroacylation should be suitable [176,210,278,337,351,367,368,374]. Mass spectra of such derivatives can be found in refs. [106,107]. Again, more information on derivatization can be found in a series of review articles by Halket and Zaikin, Halket *et al.* and Zaikin and Halket [94–99].

6.4.4 Analysis of zopiclone, zolpidem, zaleplon and eszopiclone

6.4.4.1 GC and GC-MS procedures

Zopiclone is not an ideal candidate for GC analysis because of its high thermal instability. Using electron-impact and positive chemical ionization MS, Boniface *et al.* [352] elucidated structures of the decomposition products of zopiclone after GC.

Zolpidem is stable under the usual GC conditions. A few GC methods for determination of zolpidem and zaleplon can be found in the literature [176,210,278,337,351,367,368,374].

6.4.4.1.1 Screening and confirmation of zopiclone, zolpidem, zaleplon and eszopiclone
Immunoassays for screening for zopiclone, zolpidem, zaleplon and eszopiclone are not yet commercially available. Mannaert *et al.* [375,376] developed a stereoselective radioimmunoassay for the analysis of zopiclone and its metabolites in urine. The same working group developed a fluorescence polarization immunoassay for the routine detection of N-desmethylzopiclone in urine samples. The reagents were adapted for use on the Vitalab Eclair analyzer. N-fluoresceinthiocarbamyl desmethylzopiclone was synthesized as a fluorescent tracer [377].

Today, screening and confirmation must still be done by other methods. Screening and confirmation can be performed in one step using GC-MS in the EI, full-scan mode. The above described (6.2.4.1.1) universal procedure by Maurer, including universal sample preparation for many different drug classes (acid hydrolysis, LLE and acetylation) [17–19,105] can also be applied to zopiclone (degradation products), zaleplon and zolpidem. For zopiclone zaleplon and zolpidem screening, mass chromatography based on full-scan recording was used. The identity of the peaks in the mass chromatograms was confirmed by computerized comparison of the peaks underlying mass spectra with reference spectra [106,107].

6.4.4.1.2 Quantification of zopiclone, zolpidem, zaleplon and eszopiclone in blood, plasma or serum

As described above, GC is not ideal for analysis of zopiclone. Nevertheless, GC procedures for quantification of zopiclone in plasma are described. Gaillard *et al.* [359] transformed zopiclone to the known decomposition product during a SPE procedure. After SPE (C18), the chromatograms showed only one single peak corresponding to the decomposition product of zopiclone. Prazepam was used as internal standard. Electron capture detection allowed good sensitivity [359].

In a series of papers, Gunnar *et al.* have examined GC-ECD, GC-EI-MS and GC-NICI-MS procedures for (screening and) quantification of several drugs (up to 51) including zopiclone, zaleplon and zolpidem [176,278,374]. In Fig. 6.6, GC-MS selected ion chromatograms of a spiked serum sample after LLE and silylation for screening of 51 drugs including zolpidem are shown (taken from ref. [278]).

6.4.4.2 HPLC and LC-MS procedures

As described above, HPLC is more suitable for determination of zopiclone than GC is. Therefore, most of the published procedures for determination of zopiclone are HPLC procedures. Also for zolpidem and zaleplon, more HPLC procedures are described. Common C18 or C8 packing materials for LC columns have been most widely used and can be recommended for separation of zopiclone, zolpidem or zaleplon [11,121,146,179,184,188,203,314,339,343,347,348,350,352,354,360,363,378–386]. More polar packing materials such as phenyl [387] or cyano propyl sorbents [355] were seldom used. Isocratic or gradient elution procedures using acetonitrile or

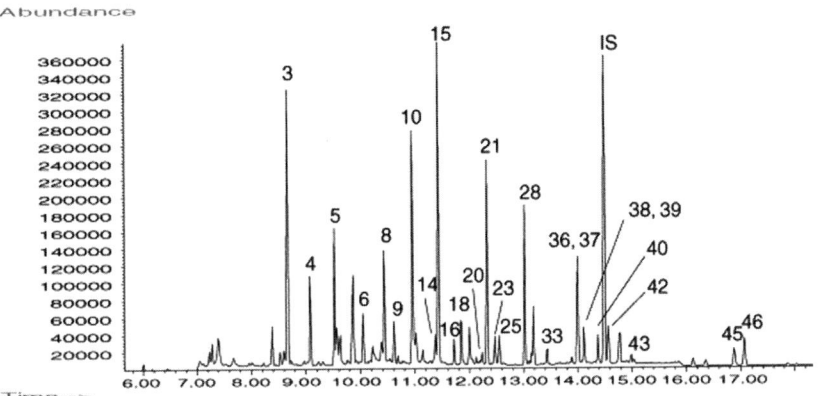

Fig. 6.6. GC-MS selected ion chromatograms of a spiked serum sample after LLE and silylation for screening of 51 drugs including zolpidem. (Taken from ref. [278].)

methanol/buffered water mixtures or mixtures of solvents (acetonitrile, hexane, isopropanol) with phosphate, acetate or formate buffers result in sufficient separation. Also mobile phase systems containing ion-pair reagents were used for determination of zopiclone in tablets [372]. Fluorescence or UV detection is seldom used [347,348,354,380,381]. Nowadays detection is preferably done by MS.

6.4.4.2.1 Screening and confirmation of zopiclone, zolpidem, zaleplon and eszopiclone
As given above, immunoassays for screening purposes are not available. Screening in urine using HPLC on phenyl column with direct injection of diluted urine and fluorescence detection has been described by Ascalone *et al.* [387]. Because of its higher specificity DAD is preferred over simple UV-detection [203,384,386,388]. Lambert *et al.* [384] and also Tracqui *et al.* [389] described the use of HPLC-DAD for systematic toxicological analysis (STA). However, the universality and specificity of GC-MS for STA [18,19,102,105] is not reached.

As already stated, MS detection is preferred also for liquid chromatographic separations. Nordgren *et al.* [362] proposed direct injection of urine and LC-MS/MS with rapid chromatography and atmospheric pressure chemical ionization as screening step. Confirmation has been done by a second LC-MS/MS method [362]. LC-MS has been used in many multi analyte procedures in the past few years [11,121,179,184,188,242,314].

6.4.4.2.2 Quantification of zopiclone, zolpidem, zaleplon and eszopiclone in blood, plasma or serum
For separation and quantification of zopiclone, zolpidem, zaleplon and eszopiclone in plasma, several LC procedures were described with different detection modes mostly using MS detection [179,188,203,347,348,350,352,354,360,378–382,385–387]. Chiral analysis of zopiclone is described by Gebauer *et al.* [380].

6.4.4.3 CE procedures

Determination of zopiclone, zolpidem, zaleplon and eszopiclone in plasma using CE has hardly been described. Using UV laser-induced fluorescence detection, the determination of zolpidem and zaleplon was possible after direct injection of urine on the capillary [361,390]. For detection of zopiclone, a LLE prior to CE was necessary [364]. Some improvement in CE technique still seems to be necessary, before low dosed pharmaceuticals can routinely be determined. As mentioned above, the status of CE techniques in forensic toxicology has been displayed in detail in the review of Tagliaro and Bortolotti [137]. Today, CE techniques are still not very widely used in forensic toxicology. The hopes that all problems of the method will be solved and CE may be a useful tool in forensic toxicology did not come true. In a more recent review, Pucci and Raggi concluded again that electrokinetic chromatographic methods might become a robust and widely accepted method and that their use might continue to grow [138]. In his recent reviews, Smyth supposed that CE-MS should be suitable for determination of zopiclone and zolpidem in biofluids [333,334].

6.5 DIPHENHYDRAMINE

6.5.1 Introduction

Diphenhydramine is clinically used as antihistaminic, antitussive and sedative–hypnotic drug. The main metabolites are demethyl, bis-demethyl diphenhydramine and diphenylmethoxyacetic acid (DPMA). In addition, one- and two-fold hydroxylation of one of the phenyl moieties followed by methylation of one of the hydroxy groups was described [391]. The phase I metabolites can be conjugated with glucuronic or sulfuric acid. Diphenhydramine may lead to severe poisonings and has also been abused in drug facilitated crimes [392].

6.5.2 Structural features of diphenhydramine

Diphenhydramine (2-(diphenyl methoxy)-*N*,*N*-dimethylethylamine) is an ethanolamine derivative, which acts as antagonist at the histamine H_1 receptor with anticholinergic side effects. The structure is given in Fig. 6.7. The ether bond is susceptible to acid hydrolysis. Therefore, the target compounds for urinalysis after acidic cleavage of conjugates are the corresponding carbinols [391].

6.5.3 Sample preparation

6.5.3.1 Sample pretreatment and extraction of blood (serum, plasma) or urine

Liquid–liquid and solid-phase procedures are used for extraction of diphenhydramine from biomatrices.

6.5.3.1.1 Liquid–liquid extraction (LLE) procedures
Many different solvents were used for extraction of diphenhydramine from biological matrices at slightly alkaline pH. They include butyl acetate, diethyl ether, toluene, ethyl acetate, dichloromethane or others. Often mixtures of these solvents were used [389,391,393–398]. As all these procedures work quite well with sufficient recoveries, no special recommendation is given here.

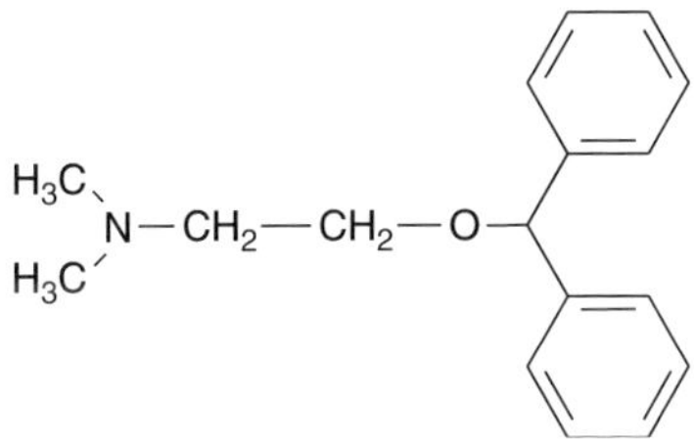

Fig. 6.7. Structure of diphenhydramine.

6.5.3.1.2 Solid-phase extraction (SPE) procedures

Many SPE procedures are reported in the literature [121,399–402]. Procedures include use of C8, C18 or mixed mode columns. Most of these procedures work quite well.

A more exotic extraction method was described by Moore that utilizes a cation-exchange column for extraction of diphenhydramine from urine of greyhounds [403]. The applicability of SPME was tested for diphenhydramine in body fluids or hair [404,405]. Again, a good overview over applications of SPE procedures can be found in several review articles [60–65].

6.5.3.2 Sample pretreatment and extraction of alternative matrices (oral fluid, sweat or hair)

The use of alternative matrices for detection of diphenhydramine is hardly mentioned in the literature. Therefore, no special recommendations can be given. Sporkert and Pragst used headspace solid-phase micro extraction (HS-SPME) for determination of several lipophilic drugs including diphenhydramine [404]. Common aspects of these matrices are discussed above 6.2.3.2.

6.5.3.3 Sample pretreatment and extraction of body tissues and other autopsy material

Only common recommendations can be given, which are already discussed above (cf. 6.2.3.3, 6.3.3.3).

6.5.3.4 Sample pretreatment and extraction of non-biological samples

Pure substances can usually be solved in solvents (e.g. methanol, ethanol, acetone, ethyl acetate, diethyl ether) and can be directly analyzed by GC or LC procedures. The same extraction methods, which are suitable for biomatrices, can also be applied to the extraction of diphenhydramine from non-biological matrices such as beverages, food or pharmaceutical formulations.

Lau and Cheung isolated diphenhydramine from cough-cold syrups using LLE at alkaline pH. Unfortunately, they used the toxic chloroform for this purpose [406,407]. Diphenhydramine is often part of cough-cold formulations. For its determination in such pharmaceuticals different methods including hydrophilic interaction liquid chromatography (HILIC), micellar liquid chromatography, FT-Raman spectroscopy, HPLC, (non-aqueous) CE, native fluorescence flow-through optosensometry, flow injection spectrophotometry, hydrophobic interaction electrokinetic chromatography and even LC-MS have been used [408–419].

6.5.3.5 Derivatization for GC and GC-MS

Diphenhydramine has no derivatizable groups. Therefore, derivatization is not used, if only the parent compound is targeted [396,400–402,405]. However, the demethyl metabolites as well as the hydroxy metabolites and the products of acid hydrolysis (cf. 6.5.2) should be derivatized for better chromatographic properties. Trimethylsilylation, acetylation or perfluoroacylation should be suitable. Derivatization using *N*-methyl-*N*-(*tert*-butyldimethylsilyl) trifluoroacetamide is also described [397]. Mass spectra of diphenhydramine and of its derivatized metabolites and hydrolysis products can be found in refs. [106,107]. For more information on derivatization procedures, the review of Segura *et al.* [260] and for even more detailed information the review series by Halket and Zaikin, Halket *et al.* and Zaikin and Halket are recommended [94–99].

6.5.4 Analysis of diphenhydramine

6.5.4.1 GC and GC-MS procedures

6.5.4.1.1 Screening and confirmation of diphenhydramine
Immunoassays for screening for diphenhydramine are not commercially available. Today, screening and confirmation must still be done by other methods. Screening and confirmation can be performed in one step using GC-MS in the EI, full-scan mode. The above described (6.2.4.1.1.) universal and comprehensive procedure by Maurer, including universal sample preparation for many different drug classes (acid hydrolysis, LLE and acetylation) [18,19,105] can also be applied to alkanolamine antihistamines used as sedative–hypnotics such as diphenhydramine or doxylamine [391]. Other screening procedures, which used GC with NP detection, suffer from less specificity [402,420]. Hasegawa *et al.* [399] used a pipette tip SPE before GC-MS for determination of 10 antihistamine drugs in human plasma.

6.5.4.1.2 Quantification of diphenhydramine in blood, plasma or serum
GC with NPD detection has been used for quantification of diphenhydramine in plasma samples in a pharmacokinetic study [421]. GC-MS is also suitable for quantification of diphenhydramine [399]. However, instead of full-scan mode SIM mode should be used, which provides higher sensitivity and precision. Further prerequisite for precise quantification is the use of suitable internal standards. Best results can be achieved with deuterated internal standards, which are commercially available [396]. Tonn *et al.* [397] quantified DPMA, a major metabolite of diphenhydramine in blood and urine using GC-MS in the single-ion monitoring mode with deuterated DPMA as internal standard.

6.5.4.2 HPLC procedures

6.5.4.2.1 Screening and confirmation of diphenhydramine
Screening in urine using HPLC with simple UV detection suffers from lack of specificity, so that it cannot be recommended. Because of its higher specificity DAD should be preferred. A series of studies on specific difficulties of HPLC-DAD for screening purposes was published by Bogusz *et al.* These authors tested the applicability of base-deactivated reversed-phase columns for systematic toxicological analysis [422], the use of corrected retention indices based on 1-nitroalkane and alkyl arylketone scales for HPLC identification of basic drugs [423,424], the influence of biological matrix on chromatographic behavior [425] and the possibility of

interlaboratory exchange of retention indices and UV spectra of toxicologically relevant substances [426]. Most of the screening procedures described in the literature are HPLC-DAD procedures. Diphenhydramine can easily be detected by these methods [384,389,427,428]. The REMEDI® system is also able to detect diphenhydramine [429–431]. Of course, diphenhydramine can also be screened for and quantified by LC-MS [121].

6.5.4.2.2 Quantification of diphenhydramine in blood, plasma or serum
Diphenhydramine is most often quantified in blood (plasma, serum) using GC-NPD or GC-MS. Nevertheless, several LC procedures were described for separation and quantification of diphenhydramine in plasma [432,433] including LC-MS methods [121].

6.5.4.3 CE procedures

Determination of diphenhydramine in plasma using CE has hardly been described. Some published procedures seem to be more of theoretical value than of practical use, since they are not tested in authentic samples [434,435].

As mentioned above (cf. 6.2.4.3), the current status of CE techniques in forensic toxicology is displayed in detail in the review of Tagliaro and Bortolotti [137]. Today, CE techniques are still not very widely used in forensic toxicology. The hopes that all problems of the method will be solved and CE may be a useful tool in forensic toxicology did not come true. In a more recent review, Pucci and Raggi [138] concluded again that electrokinetic chromatographic methods might become a robust and widely accepted method and that their use might continue to grow.

6.6 OTHER SEDATIVES AND HYPNOTICS

In this chapter, different sedative–hypnotics are summarized. In Fig. 6.8, the structures of meprobamate, methaqualone, chloral hydrate and clomethiazole are given.

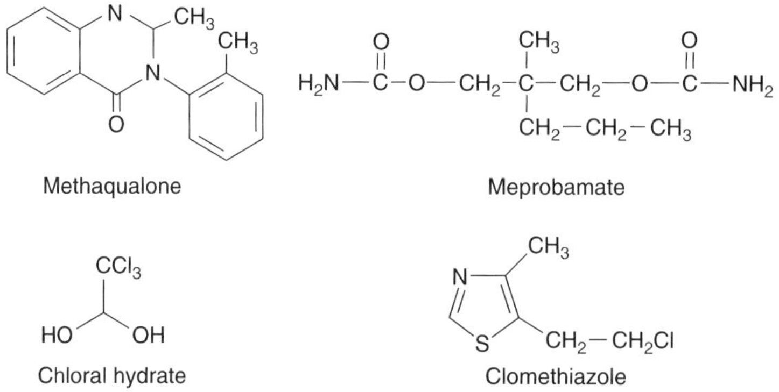

Fig. 6.8. Structures of meprobamate, methaqualone, chloral hydrate and clomethiazole.

For reasons of space, we have renounced common comments to each subsection. Therefore, the reader should read these universal comments to subsections such as e.g. sample preparation of certain matrices, GC, LC or CE in the barbiturates or benzodiazepines sections.

6.6.1 Meprobamate

Meprobamate (2-methyl-2-propyl-trimethylene dicarbamate; Fig. 6.8) is a mild tranquilizer, which has been used since the early 1950s. It is also a metabolite of the centrally acting muscle relaxant carisoprodol. Today, it has lost its significance, since more modern medicaments such as benzodiazepines or benzodiazepine BZ1 (omega 1) receptor agonists show less side effects. Nevertheless, it is still in use and intoxications still occur [436,437]. Therefore, there still is need for analytical procedures. Meprobamate lacks any appreciable UV absorbance or fluorescence. Thus, it is not suitable for HPLC with UV or fluorescence detection without derivatization. Therefore, GC(-MS) methods are mainly used. However, problems encountered with GC are due to the heat instability of meprobamate at the injection port leading to thermal decomposition or the lack of derivatization which results in poor chromatography.

SPE and LLE procedures are described for extraction of meprobamate from biological matrices. Screening and identification of meprobamate is performed best using GC-MS in the EI full-scan mode.

Quantification in plasma is often performed using GC-FID [439–441]. GC-MS is also suitable for this purpose [278]. However, instead of full-scan mode SIM mode should be used, which provides higher sensitivity and precision. LC-MS has also been used [442]. Further prerequisite for precise quantification is the use of suitable internal standards. Etidocaine, lidocaine or vinylbital were used [438,439,443]. Daval *et al.* [436] used chloroform for LLE of meprobamate from plasma and carisoprodol as internal standard. Again, use of toxic chloroform should be avoided nowadays. Carisoprodol is not the best choice for internal standard, because it is a meprobamate prodrug.

6.6.2 Methaqualone

Methaqualone (2-methyl-3-*o*-tolyl-4(3*H*)-quinazolinone; Fig. 6.8) is a powerful sedative, which was widely used. It has a high potential for addiction and was therefore scheduled. Since modern sedative–hypnotics such as benzodiazepines or benzodiazepine BZ1 (omega 1) receptor agonists such as zopiclone, zaleplon or zolpidem show much less toxicity, its use in pharmacotherapy has decreased. Since methaqualone is relatively stable and since it shows favorable LC and GC properties, it has become a very widely used internal standard for GC and LC determinations of many different classes of analytes [350,444–448].

Relatively few papers on detection of methaqualone were published in the 1990s. In two papers, the use of immunoassays for methaqualone detection was described. Klinger *et al.* [449] proposed the addition of two volumes of *N,N*-dimethylformamide to serum, plasma, and postmortem blood with subsequent centrifugation. The resulting supernatant could be directly analyzed by EMIT® d.a.u.® urine reagents [449]. Brenner *et al.* [450] compared immunoassay and GC-MS results prior to and after cleavage of conjugates. They concluded, that the immunoassays cross-reacted with the conjugated hydroxy metabolites. Since the biggest part of the hydroxy metabolites is excreted as conjugates, cleavage of conjugates was necessary before GC-MS analysis [450].

Methaqualone makes no great demands on the analytical techniques. Different liquid–liquid and SPEs are described and they work quite well. Usual RP-LC and GC systems are suitable for separation. UV-DAD or, even better, MS detection is recommended. Several screening procedures using LC-DAD [384,389,451,452] or using GC-MS [17,48] cover besides other analytes the detection of methaqualone.

Since methaqualone reaches high concentrations in biomatrices, detection and quantification of methaqualone in urine, blood and gastric content using CE was possible [453]. Assessment of the stereoselective metabolism of methaqualone was successfully done by CE [454,455]. For determination in hair GC-MS was preferred [453].

6.6.3 Chloral hydrate

Chloral hydrate is the hydrate of trichloroacetaldehyde (Fig. 6.8). It was introduced into therapeutics more than 100 years ago. Today, it is still used as a sedative–hypnotic, especially in pediatrics. Trichloroethanol, its glucuronide, dichloroacetic acid and trichloroacetic acid have been identified as the metabolites of chloral hydrate. Trichloroethanol is the main pharmacologically active principle of chloral hydrate therapy and should therefore be included in the analysis.

The usual screening procedures used for systematic toxicological analysis do not cover chloral hydrate or its active metabolite trichloroethanol. Therefore, the Fujiwara reaction for halogenated hydrocarbons is recommended as qualitative test [456]. However, only chloral hydrate itself leads to the red reaction product, whereas the main metabolite, trichloroethanol, produces only a yellow color, which cannot be differentiated from typical urine color. Chloral hydrate and its phase I metabolites lack any appreciable UV absorbance or fluorescence. Thus, they are not suitable for HPLC with UV or fluorescence detection without derivatization. MS detection after LC is also possible [121]. GC methods with electron-capture detection are most often used [457–461]. Mass spectrometric detection is also successful [462,463]. A simple and effective method for determination of chloral hydrate and trichloroethanol in biosamples is the headspace GC(-MS) [464].

6.6.4 Clomethiazole

Clomethiazole (5-(2-chloroethyl)-4-methylthiazole; Fig. 6.8) has sedative, hypnotic and anticonvulsive effects. It is used in the treatment of the alcohol-withdrawal syndrome including delirium tremens. The main urinary metabolites are 4-methyl-5-thiazole acetic acid, 5-(1-hydroxy-2-chloroethyl)-4-methylthiazole and 5-(2-hydroxyethyl)-4-thiazole carboxylic acid lactone. GC procedures are suitable [17,465–467]. Clomethiazole can cause problems during sample preparation because of its high volatility. Evaporation of extraction solvents must be done carefully. Of course, it is also possible to use LC-MS/MS for determination of clomethiazole in liver microsome assays [468]. Sporkert and Pragst used HS-SPME and GC-MS for detection of clomethiazole in hair [404].

6.7 CONCLUDING SUMMARY AND PERSPECTIVES

Sedatives and hypnotics are lipophilic as most of the centrally acting drugs are. Therefore, analytical properties are similar. Both methods, LLE at suitable pH and SPE work well for extraction of such lipophilic drugs from biomatrices. If the problem of batch-to-batch inconformity of the SPE columns is monitored (e.g. using an IS (internal standard)), both methods for extraction can be recommended. Separation of sedatives and hypnotics works well using GC and LC techniques. Derivatization before GC is not necessary if only lipophilic parent compounds are targeted. However, lipophilic drugs are extensively metabolized and excreted mainly in metabolized form. Metabolites, especially hydroxylated ones, can be conjugated with glucuronic acid or sulfuric acid. Therefore, cleavage of conjugates may be necessary before analysis to improve the detection window in urine samples. In forensic toxicology, enzymatic hydrolysis using glucuronidase (and arylsulfatase) is recommended. Unfortunately, enzymatic hydrolysis is time-consuming and expensive. If results are requested in a short time (e.g. in clinical toxicology) rapid acid hydrolysis using hydrochloric acid is more convenient. However, forming of artifacts (e.g. benzophenones from benzodiazepines or carbinols from alkanolamine antihistamines) must be considered. The deconjugated metabolites and the hydrolysis artifacts must be derivatized prior to GC separation, which can be performed on standard fused-silica capillary such as low polarity dimethylpolysiloxane types or 5% diphenyl methylpolysiloxane types. For LC separation standard reversed-phase columns are suitable.

Today, mass spectrometric detectors for GC and LC have prevailed. Mass spectrometric detection provides best specificity of all methods, at least in the electron-impact full scan mode for qualitative analysis. For quantification, GC-MS should be run in the selected-ion monitoring mode, while LC-MS/MS should be run in multiple reaction monitoring (MRM) mode.

For quantification, suitable internal standards should be chosen. They can compensate for variability due to sample preparation, chromatography, or even ion suppression/enhancement and thus improve accuracy and precision data. As in any MS-based analytical methods, stable-isotope labeled analogues of the analytes are ideal IS. At this point it must be stressed that if no stable-isotope-labeled IS of the analyte is available and an alternative IS must be chosen, one should always avoid choosing a therapeutic drug for this purpose, at least not in analytical toxicology. In this field, it can never be excluded that the patient or defendant to be monitored has taken this drug. In such cases, the peak area of the IS would be overestimated leading to underestimation of the analyte concentration.

CE procedures for detection of drugs of forensic interest are more and more published. However, it must clearly be stated, that CE is still not a universal method. A strategy for sample preparation in CE has not yet been fully developed. There still are limitations in terms of reproducibility and concentration sensitivity. Mass spectrometric detection may help to solve these problems.

In summary, GC-MS, especially in the electron ionization full-scan mode, is still the method of choice for comprehensive screening providing best separation power, specificity and universality, although requiring derivatization. LC-DAD is also often used for screening, but its separation power and its specificity are still inferior to those of GC-MS. Finally, LC-MS has shown to be an ideal supplement, especially for the detection of more polar, thermolabile and/or low-dose drugs, especially in blood plasma. It may become the gold standard in clinical and forensic toxicology and doping control if the costs of the apparatus will be markedly reduced, the current disadvantages like irreproducibility of fragmentation, reduction of ionization by matrix, etc. will be overcome, and finally if one of the increasing number of quite different techniques will become the apparatus standard.

6.8 REFERENCES

1   F.T. Peters, J. Jung, T. Kraemer and H.H. Maurer, Ther. Drug Monit., 27 (2005) 334.
2   W.J. Cubala and J. Landowski, Prog. Neuropsychopharmacol. Biol. Psychiatry, 2 (2007) 539.
3   A. Flynn and D. Cox, Addiction, 101 (2006) 898.
4   C. Haasen, T. Mueller-Thomsen, T. Fink, A. Bussopulos and J. Reimer, Int. J. Neuropsychopharmacol., 8 (2005) 309.
5   G. Hajak, W.E. Muller, H.U. Wittchen, D. Pittrow and W. Kirch, Addiction, 98 (2003) 1371.
6   C.L. Kao, S.C. Huang, Y.J. Yang and S.J. Tsai, J. Clin. Psychiatry, 65 (2004) 1287.
7   T.H. Krueger, S. Kropp and T.J. Huber, Ann. Pharmacother., 39 (2005) 773.
8   I.A. Liappas, P.N. Malitas, N.P. Dimopoulos, O.E. Gitsa, A.I. Liappas, C.K. Nikolaou and G.N. Christodoulou, World J. Biol. Psychiatry, 4 (2003) 93.
9   I.A. Liappas, P.N. Malitas, N.P. Dimopoulos, O.E. Gitsa, A.I. Liappas, C. Nikolaou and G.N. Christodoulou, J. Psychopharmacol., 17 (2003) 131.
10 G. Quaglio, F. Lugoboni, A. Fornasiero, A. Lechi, G. Gerra and P. Mezzelani, Int. Clin. Psychopharmacol., 20 (2005) 285.
11 O. Quintela, F.L. Sauvage, F. Charvier, J.M. Gaulier, G. Lachatre and P. Marquet, Clin. Chem., 52 (2006) 1346.
12 L.R. Rappa, M. Larose-Pierre, D.R. Payne, N.E. Eraikhuemen, D.M. Lanes and M.L. Kearson, Ann. Pharmacother., 38 (2004) 590.
13 P.K. Sethi and D.C. Khandelwal, J. Assoc. Physicians India, 53 (2005) 139.
14 A. Meinitzer, W. Marz, H. Mangge and G. Halwachs-Baumann, J. Anal. Toxicol., 30 (2006) 196.
15 G. Frison, D. Favretto, L. Tedeschi and S.D. Ferrara, Forensic Sci. Int., 133 (2003) 171.
16 M. Juhascik, N.L. Le, K. Tomlinson, C. Moore, R.E. Gaensslen and A. Negrusz, J. Anal. Toxicol., 28 (2004) 400.

17 H.H. Maurer, J. Chromatogr., 530 (1990) 307.
18 H.H. Maurer, in: M. Bogusz (Ed.), Handbook of analytical separation sciences: forensic sciences, 2nd Ed., Elsevier Science, Amsterdam, 2007, p. 429.
19 H.H. Maurer, Clin. Chem. Lab. Med., 42 (2004) 1310.
20 M. Bogusz, R. Aderjan, G. Schmitt, E. Nadler and B. Neureither, Forensic Sci. Int., 48 (1990) 27.
21 P. Lillsunde, L. Michelson, T. Forsstrom, T. Korte, E. Schultz, K. Ariniemi, M. Portman, M.L. Sihvonen and T. Seppala, Forensic Sci. Int., 77 (1996) 191.
22 N. Matsumoto, Y. Komatsubara and H. Machishima, Jpn. J. Toxicol., 4 (1991) 57.
23 F. Moriya and Y. Hashimoto, Nippon. Hoigaku. Zasshi., 50 (1996) 50.
24 R.W. Schmid and C. Wolf, J. Pharm. Biomed. Anal., 7 (1989) 1749.
25 H. Levert, P. Odou and H. Robert, Biomed. Chromatogr., 16 (2002) 19.
26 K.M. Patil and S.L. Bodhankar, J. Pharm. Biomed. Anal., 39 (2005) 181.
27 G. Coppa, R. Testa, A.M. Gambini, I. Testa, M. Tocchini and A.R. Bonfigli, Clin. Chim. Acta, 305 (2001) 41.
28 H.H. Maurer, A. Weber and K. Pfleger, Z. Anal. Chem., 311 (1982) 414.
29 H.H. Maurer, . In: K. Pfleger, H.H. Maurer, A. Weber (Eds.), Mass spectral and GC data of drugs, poisons, pesticides, pollutants and their metabolites, VCH publisher, Weinheim, 1992, p. 3.
30 J.C. Spell, K. Srinivasan, J.T. Stewart and M.G. Bartlett, Rapid. Commun. Mass Spectrom., 12 (1998) 890.
31 J.N. Valenta, R.P. Dixon, A.D. Hamilton and S.G. Weber, Anal. Chem., 66 (1994) 2397.
32 F. Coudore, J.M. Alazard, M. Paire, G. Andraud and J. Lavarenne, J. Anal. Toxicol., 17 (1993) 109.
33 W.F. Ebling, W.L. Mills, S.R. Harapat and D.R. Stanski, J. Chromatogr., 490 (1989) 339.
34 R. Meatherall, J. Forensic Sci., 42 (1997) 1160.
35 S.J. Mule and G.A. Casella, J. Anal. Toxicol., 13 (1989) 13.
36 G.E. Blakey, P. Ballard, D.E. Leahy and M. Rowland, J. Pharm. Biomed. Anal., 18 (1999) 927.
37 U.M. Laakkonen, A. Leinonen and L. Savonen, Analyst, 119 (1994) 2695.
38 R. Meatherall and D. Ford, Ther. Drug Monit., 10 (1988) 101.
39 H. Wu, F. Guan and Y. Luo, Yaowu Fenxi Zazhi, 16 (1996) 316.
40 H.A. Adams, B. Weber, M.B. Bachmann, M. Guerin and G. Hempelmann, Anaesthesist, 41 (1992) 619.
41 T.C. Doran, Ther. Drug Monit., 10 (1988) 474.
42 M.J. Bogusz, R.D. Maier, B.K. Schiwy and U. Kohls, J. Chromatogr. B, 683 (1996) 177.
43 X.H. Chen, J. Wijsbeek, V.J. van, J.P. Franke and Z.R. de, J. Chromatogr., 529 (1990) 161.
44 S.D. Ferrara, L. Tedeschi, G. Frison and F. Castagna, J. Anal. Toxicol., 16 (1992) 217.
45 A. Polettini, A. Groppi, C. Vignali and M. Montagna, J. Chromatogr. B, 713 (1998) 265.
46 R.H. Liu, A.M. McKeehan, C. Edwards, G. Foster, W.D. Bensley, J.G. Langner and A.S. Walia, J. Forensic Sci., 39 (1994) 1504.
47 A. Namera, M. Yashiki, K. Okada, Y. Iwasaki, M. Ohtani and T. Kojima, J. Chromatogr. B, 706 (1998) 253.
48 R. Pocci, V. Dixit and V.M. Dixit, J. Anal. Toxicol., 16 (1992) 45.
49 J. Stephenson, Lab. 2000, 6 (1992) 76.
50 J. Stephenson and U. Tillmanns, GIT Fachz. Lab. 36 (1992) 130, 133.
51 X. Zhang, C. Fang and J. Jia, Sepu, 13 (1995) 182.
52 V. Ferranti, C. Chabenat, S. Menager and O. Lafont, J. Chromatogr. B, Biomed. Sci. Appl., 718 (1998) 199.
53 C. Pistos and J.T. Stewart, J. Pharm. Biomed. Anal., 36 (2004) 737.
54 D.J. Speed, S.J. Dickson, E.R. Cairns and N.D. Kim, J. Anal. Toxicol., 24 (2000) 685.
55 H. Hattori, N. Hoshino, H. Seno, O. Suzuki and T. Yamada, Hochudoku, 10 (1992) 16.
56 O. Suzuki, T. Kumazawa, H. Seno and H. Hattori, Med. Sci. Law, 29 (1989) 242.
57 M.D. Cantu, D.R. Toso, C.A. Lacerda, F.M. Lancas, E. Carrilho and M.E. Queiroz, Anal. Bioanal. Chem., 386 (2006) 256.
58 K. Ensing, J.P. Franke, A. Temmink, X.H. Chen and Z.R. de, J. Forensic Sci., 37 (1992) 460.

59   R.A. de Zeeuw, J. Wijsbeek and J.P. Franke, J. Anal. Toxicol., 24 (2000) 97.
60   J.P. Franke and R.A. de Zeeuw, J. Chromatogr. B, 713 (1998) 51.
61   C. Dietz, J. Sanz and C. Camara, J. Chromatogr. A, 1103 (2006) 183.
62   H. Kataoka, Anal. Bioanal. Chem., 373 (2002) 31.
63   R.M. Smith, J. Chromatogr. A, 1000 (2003) 3.
64   G. Vas and K. Vekey, J. Mass Spectrom., 39 (2004) 233.
65   V. Walker and G.A. Mills, Ann. Clin. Biochem., 39 (2002) 464.
66   H. Sachs and P. Kintz, J. Chromatogr. B, 713 (1998) 147.
67   P. Kintz, M. Villain and V. Cirimele, Ther. Drug Monit., 28 (2006) 442.
68   P. Kintz, Forensic Sci. Int., 142 (2004) 127.
69   F. Musshoff and B. Madea, Ther. Drug Monit., 28 (2006) 155.
70   F. Pragst and M.A. Balikova, Clin. Chim. Acta, 370 (2006) 17.
71   M. Villain, V. Cirimele and P. Kintz, Clin. Chem. Lab. Med., 42 (2004) 1265.
72   O. Dierich and M. Soyka, Fortschr. Neurol. Psychiatr., 73 (2005) 401.
73   M.L. Lo, S. Falaschini, G. Rappelli, F. Bambini, A. Baldoni, M. Procaccini and M. Cingolani, Int. J. Immunopathol. Pharmacol., 18 (2005) 567.
74   D.J. Crouch, Forensic Sci. Int., 150 (2005) 165.
75   D.A. Kidwell, J.C. Holland and S. Athanaselis, J. Chromatogr. B, 713 (1998) 111.
76   A.G. Verstraete, Forensic Sci. Int., 150 (2005) 143.
77   P. Kintz, A. Tracqui and P. Mangin, Int. J. Legal Med., 105 (1992) 1.
78   P. Kintz, B. Ludes and P. Mangin, J. Forensic Sci., 37 (1992) 328.
79   F. Tagliaro, B.Z. De, G. Lubli, C. Neri, G. Manetto and M. Marigo, Forensic Sci. Int., 84 (1997) 129.
80   A. Tracqui, . In: P. Kintz (Ed.), Drug testing in hair, CRC Press, FL, Boca Raton, 1996, p. 191.
81   K. Saisho, E. Tanaka and Y. Nakahara, Biol. Pharm. Bull., 24 (2001) 59.
82   V. Shankar, C. Damodaran and P.C. Sekharan, Forensic Sci. Int., 40 (1989) 45.
83   C.P. Campobasso, M. Gherardi, M. Caligara, L. Sironi and F. Introna, Int. J. Legal Med., 118 (2004) 210.
84   M. Cingolani, M. Cippitelli, R. Froldi, G. Tassoni and D. Mirtella, J. Anal. Toxicol., 29 (2005) 205.
85   P.M. Gannett, J.R. Daft, D. James, B. Rybeck, J.B. Knopp and T.S. Tracy, J. Anal. Toxicol., 25 (2001) 443.
86   J.T. Franeta, D. Agbaba, S. Eric, S. Pavkov, M. Aleksic and S. Vladimirov, Farmaco, 57 (2002) 709.
87   B. Dimitrova, I. Doytchinova and M. Zlatkova, J. Pharm. Biomed. Anal., 23 (2000) 955.
88   A. Haque, X. Xu and J.T. Stewart, J. Pharm. Biomed. Anal., 21 (1999) 1063.
89   M.A. Raggi, G. Casamenti, R. Mandrioli, C. Sabbioni and V. Volterra, J. Pharm. Biomed. Anal., 23 (2000) 161.
90   M. Terada, T. Shinozuka, H. Bai, M.N. Islam, Z. Tun, K. Honda, J. Yanagida and C. Wakasugi, Jpn. J. Forensic Toxicol., 13 (1995) 223.
91   E. Interschick, H. Wuest and H. Patscheke, Laboratoriumsmedizin, 18 (1994) 533.
92   A.D. Barbour, J. Anal. Toxicol., 15 (1991) 214.
93   M.M. Kushnir and F.M. Urry, J. Chromatogr. Sci., 39 (2001) 129.
94   J.M. Halket and V.G. Zaikin, Eur. J. Mass Spectrom. (Chichester, England), 9 (2003) 1.
95   J.M. Halket and V.V. Zaikin, Eur. J. Mass Spectrom. (Chichester, England), 10 (2004) 1.
96   J.M. Halket and V.G. Zaikin, Eur. J. Mass Spectrom. (Chichester, England), 11 (2005) 127.
97   J.M. Halket, D. Waterman, A.M. Przyborowska, R.K. Patel, P.D. Fraser and P.M. Bramley, J. Exp. Bot., 56 (2005) 219.
98   V.G. Zaikin and J.M. Halket, Eur. J. Mass Spectrom. (Chichester, England), 9 (2003) 421.
99   V.G. Zaikin and J.M. Halket, Eur. J. Mass Spectrom. (Chichester, England), 11 (2005) 611.
100   M. Iwai, H. Hattori, T. Arinobu, A. Ishii, T. Kumazawa, H. Noguchi, H. Noguchi, O. Suzuki and H. Seno, J. Chromatogr. B, Analyt. Technol. Biomed. Life Sci., 806 (2004) 65.
101   U. Staerk and W.R. Kulpmann, J. Chromatogr. B, Biomed. Sci. Appl., 745 (2000) 399.

102 H.H. Maurer, J. Chromatogr., 580 (1992) 3.
103 N.T. Lu and B.G. Taylor, Forensic Sci. Int., 157 (2006) 106.
104 K.S. Schwenzer, R. Pearlman, M. Tsilimidos, S.J. Salamone, R.C. Cannon, S.H. Wong, S.B. Gock and J.J. Jentzen, J. Anal. Toxicol., 24 (2000) 726.
105 H.H. Maurer, Spectroscopy Europe, 6 (1994) 21.
106 H.H. Maurer, K. Pfleger and A.A. Weber, Mass Spectral and GC Data of Drugs, Poisons, Pesticides, Pollutants and their Metabolites, Wiley-VCH, Weinheim, 2007.
107 H.H. Maurer, K. Pfleger and A.A. Weber, Mass Spectral Library of Drugs, Poisons, Pesticides, Pollutants and their Metabolites, Agilent Technologies, Palo Alto, CA, 2007.
108 M.I. Drost and L. Walter, J. Anal. Toxicol., 12 (1988) 322.
109 O.H. Drummer, A. Kotsos and I.M. McIntyre, J. Anal. Toxicol., 17 (1993) 225.
110 E.I. Minder, R. Schaubhut and D.J. Vonderschmitt, J. Chromatogr., 428 (1988) 369.
111 G. Quatrehomme, F. Bourret, Z. Liao and A. Ollier, J. Forensic Sci., 39 (1994) 1300.
112 P.P. Rop, J. Spinazzola, A. Zahra, M. Bresson, J. Quicke and A. Viala, J. Chromatogr., 427 (1988) 172.
113 S.H. Steiner, M.J. Moor and M.H. Bickel, Drug Metab. Dispos., 19 (1991) 8.
114 B. Brzakovic, M. Pokrajac, E. Dzoljic, Z. Levic and V.M. Varagic, Eur. J. Drug Metab. Pharmacokinet., 24 (1999) 233.
115 P. Kishore, K. Rajnarayana, M.S. Reddy, J.V. Sagar and D.R. Krishna, Arzneimittelforschung, 53 (2003) 763.
116 A. Martinavarro-Dominguez, M.E. Capella-Peiro, M. Gil-Agusti, J.V. Marcos-Tomas and J. Esteve-Romero, Clin. Chem., 48 (2002) 1696.
117 K.M. Matar, P.J. Nicholls, A. Tekle, S.A. Bawazir and M.I. Al-Hassan, Ther. Drug Monit., 21 (1999) 559.
118 B.Z. Budvar, G. Radeczky, A. Shalaby and G. Szasz, Acta Pharm. Hung., 59 (1989) 49.
119 H. Kanazawa, Y. Konishi, Y. Matsushima and T. Takahashi, J. Chromatogr. A, 797 (1998) 227.
120 Y. Hori, M. Fujisawa, K. Shimada, Y. Hirose and T. Yoshioka, Biol. Pharm. Bull., 29 (2006) 7.
121 H. Miyaguchi, K. Kuwayama, K. Tsujikawa, T. Kanamori, Y.T. Iwata, H. Inoue and T. Kishi, Forensic Sci. Int., 157 (2006) 57.
122 J.J. Jones, H. Kidwell and D.E. Games, Rapid Commun. Mass Spectrom., 17 (2003) 1565.
123 A. Ceccato, B. Boulanger, P. Chiap, P. Hubert and J. Crommen, J. Chromatogr. A, 819 (1998) 143.
124 C. Wolf and R.W. Schmid, J. Liq. Chromatogr., 13 (1990) 2207.
125 A. Alila, J.E. Heavner and P.H. Rosenberg, Am. J. Vet. Res., 49 (1988) 671.
126 G.A. Valaskovic, L. Utley, M.S. Lee and J.T. Wu, Rapid Commun. Mass Spectrom., 20 (2006) 1087.
127 W. Thormann, P. Meier, C. Marcolli and F. Binder, J. Chromatogr., 545 (1991) 445.
128 P. Wernly and W. Thormann, Anal. Chem., 64 (1992) 2155.
129 K.E. Ferslew, A.N. Hagardorn and W.F. McCormick, J. Forensic Sci., 40 (1995) 245.
130 M. Ivanova, A. Piunti, E. Marziali, N. Komarova, M.A. Raggi and E. Kenndler, Electrophoresis, 24 (2003) 992.
131 F.M. Lancas, M.A. Sozza and M.E. Queiroz, J. Anal. Toxicol., 27 (2003) 304.
132 K. Ohyama, M. Wada, G.A. Lord, Y. Ohba, O. Fujishita, K. Nakashima, C.K. Lim and N. Kuroda, Electrophoresis, 25 (2004) 594.
133 Y. Kataoka, K. Makino and R. Oishi, Electrophoresis, 19 (1998) 2856.
134 L. Yang, A.K. Harrata and C.S. Lee, Anal. Chem., 69 (1997) 1820.
135 U. Schmitt, J. Bojarski and U. Holzgrabe, Electrophoresis, 22 (2001) 3237.
136 K. Srinivasan, W. Zhang and M.G. Bartlett, J. Chromatogr. Sci., 36 (1998) 85.
137 F. Tagliaro and F. Bortolotti, Electrophoresis, 27 (2006) 231.
138 V. Pucci and M.A. Raggi, Electrophoresis, 26 (2005) 767.
139 J. Braun, S. Seyfert, J. Bernarding, A. Schilling, P. Marx and T. Tolxdorff, Neuroradiology, 43 (2001) 211.
140 M.A. ElSohly, W. Gul, K.M. Elsohly, B. Avula and I.A. Khan, J. Anal. Toxicol., 30 (2006) 524.
141 M. Cheze, G. Duffort, M. Deveaux and G. Pepin, Forensic Sci. Int., 153 (2005) 3.

142 M. Concheiro, M. Villain, S. Bouchet, B. Ludes, M. Lopez-Rivadulla and P. Kintz, Ther. Drug Monit., 27 (2005) 565.

143 P. Kintz, M. Villain, M. Cheze and G. Pepin, Forensic Sci. Int., 153 (2005) 222.

144 P. Kintz, M. Villain and B. Ludes, Ther. Drug Monit., 26 (2004) 211.

145 M. Villain, M. Cheze, V. Dumestre, B. Ludes and P. Kintz, J. Anal. Toxicol., 28 (2004) 516.

146 M. Villain, M. Cheze, A. Tracqui, B. Ludes and P. Kintz, Forensic Sci. Int., 145 (2004) 117.

147 A. Negrusz and R.E. Gaensslen, Anal. Bioanal. Chem., 376 (2003) 1192.

148 A. Negrusz, C.M. Moore, K.B. Hinkel, T.L. Stockham, M. Verma, M.J. Strong and P.G. Janicak, J. Forensic Sci., 46 (2001) 1143.

149 M.A. LeBeau, M.A. Montgomery, J.R. Wagner and M.L. Miller, J. Forensic Sci., 45 (2000) 1133.

150 S. Hegstad, E.L. Oiestad, U. Johansen and A.S. Christophersen, J. Anal. Toxicol., 30 (2006) 31.

151 K.L. Klette, R.F. Wiegand, C.K. Horn, P.R. Stout and J. Magluilo Jr., J. Anal. Toxicol., 29 (2005) 193.

152 D. Borrey, E. Meyer, L. Duchateau, W. Lambert, P.C. Van and L.A. De, Clin. Chem., 48 (2002) 2047.

153 D. Borrey, E. Meyer, W. Lambert, P.C. Van and A.P. De Leenheer, J. Chromatogr. B, Biomed. Sci. Appl., 765 (2001) 187.

154 M. Segura, J. Barbosa, M. Torrens, M. Farre, C. Castillo, J. Segura and T.R. de la, J. Anal. Toxicol., 25 (2001) 130.

155 O. Beck, Z. Lin, K. Brodin, S. Borg and P. Hjemdahl, J. Anal. Toxicol., 21 (1997) 554.

156 D.A. Black, G.D. Clark, V.M. Haver, J.A. Garbin and A.J. Saxon, J. Anal. Toxicol., 18 (1994) 185.

157 P.H. Dickson, W. Markus, J. McKernan and H.C. Nipper, J. Anal. Toxicol., 16 (1992) 67.

158 O.H. Drummer, J. Chromatogr. B, 713 (1998) 201.

159 J.G. Langner, B.K. Gan, R.H. Liu, L.D. Baugh, P. Chand, J.L. Weng, C. Edwards and A.S. Walia, Clin. Chem. (Winston-Salem), 37 (1991) 1595.

160 H.H. Maurer, GIT-Suppl., (1990) 3.

161 C. Moore, G. Long and M. Marr, J. Chromatogr. B, 655 (1994) 132.

162 S.B. Needleman and M. Porvaznik, Forensic Sci. Int., 73 (1995) 49.

163 H. Schutz, Z. Rechtsmed., 100 (1988) 19.

164 J. Baeumler and S. Rippstein, Helv. Chim. Acta, 44 (1961) 2208.

165 M. Katagi, M. Tatsuno and H. Tsuchihashi, Hochudoku, 9 (1991) 116.

166 H. Maurer and K. Pfleger, J. Chromatogr., 422 (1987) 85.

167 S.W.H. Toennes, H.H. Maurer, In: S.D. Ferrara (Ed.), Proceedings of the 35th International TIAFT Meeting in Padova, Centre of Behavioural and Forensic Toxicology, Padova, 1997, p. 227.

168 S.W.H. Toennes and H.H. Maurer, Clin. Chem., 45 (1999) 2173.

169 B.R. Simmons, Order No. DA9624079 From: Diss. Abstr. Int. (1995) 1814.

170 K. Takaichi, T. Shinohara and T. Nagano, Hochudoku, 13 (1995) 132.

171 H.R. Herraez, A.J. Louter, M.N. van-de and U.A. Brinkman, J. Pharm. Biomed. Anal., 14 (1996) 1077.

172 A. Baldacci and W. Thormann, J. Sep. Sci., 29 (2006) 153.

173 O. Papini, C. Bertucci, C.S. da, S.N. Dos and V.L. Lanchote, J. Pharm. Biomed. Anal., 40 (2006) 389.

174 O. Papini, S.P. da Cunha, A.C. da Silva Mathes, C. Bertucci, E.C. Moises, D.L. de Barros, C.R. de Carvalho and V.L. Lanchote, J. Pharm. Biomed. Anal., 40 (2006) 397.

175 A. Bugey, S. Rudaz and C. Staub, J. Chromatogr. B, Analyt. Technol. Biomed. Life Sci., 832 (2006) 249.

176 T. Gunnar, K. Ariniemi and P. Lillsunde, J. Mass Spectrom., 41 (2006) 741.

177 S. Pirnay, F. Herve, S. Bouchonnet, B. Perrin, F.J. Baud and I. Ricordel, J. Pharm. Biomed. Anal., 41 (2006) 1135.

178 V.A. Jabor, E.B. Coelho, N.A. Dos Santos, P.S. Bonato and V.L. Lanchote, J. Chromatogr. B, Analyt. Technol. Biomed. Life Sci., 822 (2005) 27.

179   M. Laloup, M.M. Ramirez Fernandez, B.G. De, M. Wood, V. Maes and N. Samyn, J. Anal. Toxicol., 29 (2005) 616.
180   S.N. Muchohi, K. Obiero, G.O. Kokwaro, B.R. Ogutu, I.M. Githiga, G. Edwards and C.R. Newton, J. Chromatogr. B, Analyt. Technol. Biomed. Life Sci., 824 (2005) 333.
181   S.N. Muchohi, S.A. Ward, L. Preston, C.R. Newton, G. Edwards and G.O. Kokwaro, J. Chromatogr. B, Analyt. Technol. Biomed. Life Sci., 821 (2005) 1.
182   O. Quintela, A. Cruz, A. Castro, M. Concheiro and M. Lopez-Rivadulla, J. Chromatogr. B, Analyt. Technol. Biomed. Life Sci., 825 (2005) 63.
183   M. Rouini, Y.H. Ardakani, L. Hakemi, M. Mokhberi and G. Badri, J. Chromatogr. B, Analyt. Technol. Biomed. Life Sci., 823 (2005) 167.
184   M. Villain, M. Concheiro, V. Cirimele and P. Kintz, J. Chromatogr. B, Analyt. Technol. Biomed. Life Sci., 825 (2005) 72.
185   M.H. Andraus, A. Wong, O.A. Silva, C.Y. Wada, O. Toffleto, C.P. Azevedo and M.C. Salvadori, J. Mass Spectrom., 39 (2004) 1348.
186   A. Bugey and C. Staub, J. Pharm. Biomed. Anal., 35 (2004) 555.
187   M. Cheze, M. Villain and G. Pepin, Forensic Sci. Int., 145 (2004) 123.
188   C. Kratzsch, O. Tenberken, F.T. Peters, A.A. Weber, T. Kraemer and H.H. Maurer, J. Mass Spectrom., 39 (2004) 856.
189   T.L. Laurito, G.D. Mendes, V. Santagada, G. Caliendo, M.E. de Moraes and N.G. De, J. Mass Spectrom., 39 (2004) 168.
190   S. Paterson, R. Cordero and S. Burlinson, J. Chromatogr. B, Analyt. Technol. Biomed. Life Sci., 813 (2004) 323.
191   B.E. Smink, J.E. Brandsma, A. Dijkhuizen, K.J. Lusthof, J.J. de Gier, A.C. Egberts and D.R. Uges, J. Chromatogr. B, Analyt. Technol. Biomed. Life Sci., 811 (2004) 13.
192   P.K. Kunicki, J. Chromatogr. B, Biomed. Sci. Appl., 750 (2001) 41.
193   J. Darius and P. Banditt, J. Chromatogr. B, Biomed. Sci. Appl., 738 (2000) 437.
194   I. Rasanen, I. Ojanpera and E. Vuori, J. Anal. Toxicol., 24 (2000) 46.
195   P. Marquet, O. Baudin, J.M. Gaulier, E. Lacassie, J.L. Dupuy, B. Francois and G. Lachatre, J. Chromatogr. B, Biomed. Sci. Appl., 734 (1999) 137.
196   P.G. Agbuya, L. Li, M.V. Miles, A.L. Zaritsky and A.D. Morris, Ther. Drug Monit., 18 (1996) 194.
197   R.L. Fitzgerald, D.A. Rexin and D.A. Herold, J. Anal. Toxicol., 17 (1993) 342.
198   W.E. Lambert, E. Meyer, P.Y. Xue and L.A. De, J. Anal. Toxicol., 19 (1995) 35.
199   R. Meatherall, J. Anal. Toxicol., 18 (1994) 369.
200   M. Shimizu, T. Uno, H.O. Tamura, H. Kanazawa, I. Murakami, K. Sugawara and T. Tateishi, J. Chromatogr. B, Analyt. Technol. Biomed. Life Sci.,.
201   N. Yasui-Furukori, Y. Inoue and T. Tateishi, J. Chromatogr. B, Analyt. Technol. Biomed. Life Sci., 811 (2004) 153.
202   V. Cirimele, P. Kintz and B. Ludes, J. Chromatogr. B, 700 (1997) 119.
203   A. Tracqui, P. Kintz and P. Mangin, J. Chromatogr., 616 (1993) 95.
204   I. Deinl, C. Franzelius, L. Angermaier, G. Mahr and G. Machbert, J. Anal. Toxicol., 23 (1999) 598.
205   H.L. Lord, M. Rajabi, S. Safari and J. Pawliszyn, J. Pharm. Biomed. Anal., 40 (2006) 769.
206   M.L. Nedved, G.S. Habibi, B. Ganem and J.D. Henion, Anal. Chem., 68 (1996) 4228.
207   S.H. Ahn, H.J. Maeng, T.S. Koo, D.D. Kim, C.K. Shim and S.J. Chung, J. Chromatogr. B, Analyt. Technol. Biomed. Life Sci., 834 (2006) 128.
208   L. Wang, H. Zhao, Y. Qiu and Z. Zhou, J. Chromatogr. A, 1136 (2006) 99.
209   J.C. Goncalves, T.M. Monteiro, C.S. Neves, K.R. Gram, N.M. Volpato, V.A. Silva, R. Caminha, M.R. Goncalves, F.M. Santos, G.E. Silveira and F. Noel, Ther. Drug Monit., 27 (2005) 601.
210   T. Gunnar, K. Ariniemi and P. Lillsunde, J. Mass Spectrom., 40 (2005) 739.
211   P. Valavani, J. tta-Politou and I. Panderi, J. Mass Spectrom., 40 (2005) 516.
212   N. Jourdil, J. Bessard, F. Vincent, H. Eysseric and G. Bessard, J. Chromatogr. B, Analyt. Technol. Biomed. Life Sci., 788 (2003) 207.

213 Y. Deng, J.T. Wu, T.L. Lloyd, C.L. Chi, T.V. Olah and S.E. Unger, Rapid Commun. Mass Spectrom., 16 (2002) 1116.

214 A. Miki, M. Tatsuno, M. Katagi, M. Nishikawa and H. Tsuchihashi, J. Anal. Toxicol., 26 (2002) 87.

215 T. Sano, K. Sato, R. Kurihara, Y. Mizuno, T. Kojima, Y. Yamakawa, T. Yamada, A. Ishii and Y. Katsumata, Leg. Med. (Tokyo), 3 (2001) 149.

216 H. Inoue, Y. Maeno, M. Iwasa, R. Matoba and M. Nagao, Forensic Sci. Int., 113 (2000) 367.

217 H. Nguyen and D.R. Nau, J. Anal. Toxicol., 24 (2000) 37.

218 A.A. Elian, Forensic Sci. Int., 101 (1999) 107.

219 J. Knapp, P. Boknik, H.G. Gumbinger, B. Linck, H. Luss, F.U. Muller, W. Schmitz, U. Vahlensieck and J. Neumann, J. Chromatogr. Sci., 37 (1999) 145.

220 K.K. Akerman, J. Jolkkonen, M. Parviainen and I. Penttila, Clin. Chem., 42 (1996) 1412.

221 M. Casas, L.A. Berrueta, B. Gallo and F. Vicente, J. Pharm. Biomed. Anal., 11 (1993) 277.

222 X.H. Chen, J.P. Franke, J. Wijsbeek and Z.R. de, J. Anal. Toxicol., 16 (1992) 351.

223 L.E. Edinboro and A. Poklis, J. Anal. Toxicol., 18 (1994) 312.

224 Y. Gaillard, M.J. Gay and M. Ollagnier, J. Chromatogr., 622 (1993) 197.

225 M. Kleinschnitz, M. Herderich and P. Schreier, J. Chromatogr. B, 676 (1996) 61.

226 C.K. Lai, T. Lee, K.M. Au and A.Y. Chan, Clin. Chem., 43 (1997) 312.

227 C.M. Moore and J.S. Oliver, in: J.S.Oliver (Ed.), Forensic toxicol., Proceedings of international meeting of the international association of the forensic toxicology, 26th, Scott. Acad. Press, Edinburgh, UK, 1992, p. 295.

228 F. Musshoff and T. Daldrup, Int. J. Legal Med., 105 (1992) 105.

229 B.C. Sallustio, C. Kassapidis and R.G. Morris, Ther. Drug Monit., 16 (1994) 174.

230 T.A. Biemer, J. Chromatogr., 410 (1987) 206.

231 P.G.M. Zweipfenning, K.S. Kruseman and C.J. Vermaase, in: J.S. Oliver (Ed.), Forensic toxicol., Proceedings of international meeteeting of the international association of the forensic toxicology, 26th, Scott. Acad. Press, Edinburgh, UK, 1992, p. 327.

232 M.A. El and C. Staub, J. Anal. Toxicol., 25 (2001) 209.

233 C.W. Huck and G.K. Bonn, J. Chromatogr. A, 885 (2000) 51.

234 B.E. Smink, M.P. Mathijssen, K.J. Lusthof, J.J. de Gier, A.C. Egberts and D.R. Uges, J. Anal. Toxicol., 30 (2006) 478.

235 J. Clarke and J.F. Wilson, Forensic Sci. Int., 150 (2005) 161.

236 H.H. Maurer, Anal. Bioanal. Chem., 381 (2005) 110.

237 P. Kintz, A. Tracqui, C. Jamey and P. Mangin, J. Anal. Toxicol., 20 (1996) 197.

238 V. Cirimele, P. Kintz and P. Mangin, J. Anal. Toxicol., 20 (1996) 596.

239 V. Cirimele, P. Kintz, C. Staub and P. Mangin, Forensic Sci. Int., 84 (1997) 189.

240 P. Kintz, V. Cirimele, F. Vayssette and P. Mangin, J. Chromatogr. B, 677 (1996) 241.

241 Y. Gaillard and G. Pepin, J. Chromatogr., 762 (1997) 251.

242 R.C. Irving and S.J. Dickson, Forensic Sci. Int., 166 (2007) 58.

243 E.I. Miller, F.M. Wylie and J.S. Oliver, J. Anal. Toxicol., 30 (2006) 441.

244 R. Kronstrand, I. Nystrom, J. Strandberg and H. Druid, Forensic Sci. Int., 145 (2004) 183.

245 R. Kronstrand, I. Nystrom, M. Josefsson and S. Hodgins, J. Anal. Toxicol., 26 (2002) 479.

246 M.D. Robertson and O.H. Drummer, J. Chromatogr. B, 667 (1995) 179.

247 K. Pien, M. Laloup, M. Pipeleers-Marichal, P. Grootaert, B.G. De, N. Samyn, T. Boonen, K. Vits and M. Wood, Int. J. Legal Med., 118 (2004) 190.

248 M. Wood, M. Laloup, K. Pien, N. Samyn, M. Morris, R.A. Maes, E.A. de Bruijn, V. Maes and B.G. De, J. Anal. Toxicol., 27 (2003) 505.

249 S.H. Hansen and Z.A. Sheribah, J. Pharm. Biomed. Anal., 39 (2005) 322.

250 M.S. Aurora Prado, M. Steppe, M.F. Tavares, E.R. Kedor-Hackmann and M.I. Santoro, J. Pharm. Biomed. Anal., 37 (2005) 273.

251 S.C. Bishop, M. Lerch and B.R. McCord, Forensic Sci. Int., 141 (2004) 7.

252 P. Perez-Lozano, E. Garcia-Montoya, A. Orriols, M. Minarro, J.R. Tico and J.M. Sune-Negre, J. Pharm. Biomed. Anal., 34 (2004) 979.

253 R.N. Rao, P. Parimala, S. Khalid and S.N. Alvi, Anal. Sci., 20 (2004) 383.

254 M. Bakavoli and M. Kaykhaii, J. Pharm. Biomed. Anal., 31 (2003) 1185.

255 C.F. Ferreyra and C.S. Ortiz, J. Pharm. Biomed. Anal., 29 (2002) 811.

256 I. Cepanec, H. Mikuldas, M. Litvic and I. Vukusic, Pharmazie, 56 (2001) 857.

257 C. Ferreyra and C. Ortiz, J. Pharm. Biomed. Anal., 25 (2001) 493.

258 S. Furlanetto, S. Orlandini, G. Massolini, M.T. Faucci, P.E. La and S. Pinzauti, Analyst, 126 (2001) 1700.

259 S. Stahlmann, T. Herkert, C. Roseler, I. Rager and K.A. Kovar, Eur. J. Pharm. Sci., 12 (2001) 461.

260 J. Segura, R. Ventura and C. Jurado, J. Chromatogr., B, 713 (1998) 61.

261 C.B. Eap, G. Bouchoux, G.K. Powell and P. Baumann, J. Chromatogr. B, Analyt. Technol. Biomed. Life Sci., 802 (2004) 339.

262 H.H. Maurer, Ther. Drug Monit., 24 (2002) 247.

263 H.H. Maurer, T. Kraemer, C. Kratzsch, F.T. Peters and A.A. Weber, Ther. Drug Monit., 24 (2002) 117.

264 A. Negrusz, C.M. Moore, T.L. Stockham, K.R. Poiser, J.L. Kern, R. Palaparthy, N.L. Le, P.G. Janicak and N.A. Levy, J. Forensic Sci., 45 (2000) 1031.

265 K. Kudo, N. Ikeda and Y. Hino, Leg. Med. (Tokyo), 1 (1999) 159.

266 S. Pirnay, S. Bouchonnet, F. Herve, D. Libong, N. Milan, P. D'athis, F. Baud and I. Ricordel, J. Chromatogr. B, Analyt. Technol. Biomed. Life Sci., 807 (2004) 335.

267 H. Gjerde, E. Dahlin and A.S. Christophersen, J. Pharm. Biomed. Anal., 10 (1992) 317.

268 A. Negrusz, C. Moore, D. Deitermann, D. Lewis, K. Kaleciak, R. Kronstrand, B. Feeley and R.S. Niedbala, J. Anal. Toxicol., 23 (1999) 429.

269 A. Negrusz, C.M. Moore, J.L. Kern, P.G. Janicak, M.J. Strong and N.A. Levy, J. Anal. Toxicol., 24 (2000) 614.

270 P. Kintz, A. Tracqui, P. Mangin and Y. Edel, J. Anal. Toxicol., 20 (1996) 393.

271 D.E. Moody, W.B. Fang, D.M. Andrenyak, K.M. Monti and C. Jones, J. Anal. Toxicol., 30 (2006) 50.

272 P. Kemp, G. Sneed, T. Kupiec and V. Spiehler, J. Anal. Toxicol., 26 (2002) 504.

273 S. Kerrigan and J.W. Phillipsk Jr., Clin. Chem., 47 (2001) 540.

274 A. Leino, J. Saarimies, M. Gronholm and P. Lillsunde, Scand. J. Clin. Lab. Invest., 61 (2001) 325.

275 B.A. Way, K.G. Walton, J.W. Koenig, B.J. Eveland and M.G. Scott, Clin. Chim. Acta, 271 (1998) 1.

276 R.E. West and D.P. Ritz, J. Anal. Toxicol., 17 (1993) 114.

277 C.E. Jones, J. Wians-FH, L.A. Martinez and G.J. Merritt, Clin. Chem., 35 (1989) 1394.

278 T. Gunnar, S. Mykkanen, K. Ariniemi and P. Lillsunde, J. Chromatogr. B, Analyt. Technol. Biomed. Life Sci., 806 (2004) 205.

279 A.J. Louter, E. Bosma, J.C. Schipperen, J.J. Vreuls and U.A. Brinkman, J. Chromatogr. B, 689 (1997) 35.

280 J. Hackett, A.A. Elian, Forensic Sci. Int. 2006.

281 N. Samyn, B.G. De, V. Cirimele, A. Verstraete and P. Kintz, J. Anal. Toxicol., 26 (2002) 211.

282 S. Pichini, R. Pacifici, I. Altieri, A. Palmeri, M. Pellegrini and P. Zuccaro, J. Chromatogr. B, Biomed. Sci. Appl., 732 (1999) 509.

283 B. Aebi, R. Sturny-Jungo, W. Bernhard, R. Blanke and R. Hirsch, Forensic Sci. Int., 128 (2002) 84.

284 G. Frison, L. Tedeschi, S. Maietti and S.D. Ferrara, Rapid Commun. Mass Spectrom., 15 (2001) 2497.

285 W.C. Cheng, T.S. Yau, M.K. Wong, L.P. Chan and V.K. Mok, Forensic Sci. Int., 162 (2006) 95.

286 I.F. Bares, F. Pehourcq and C. Jarry, J. Pharm. Biomed. Anal., 36 (2004) 865.

287 P. Ghosh, M.M. Reddy, B.S. Rao and R.K. Sarin, J. AOAC Int., 87 (2004) 569.

288 P. Proenca, H. Teixeira, J. Pinheiro, E.P. Marques and D.N. Vieira, Forensic Sci. Int., 143 (2004) 205.

289 B. Borggaard and I. Joergensen, J. Anal. Toxicol., 18 (1994) 243.

290 J.M. Duthel, H. Constant, J.J. Vallon, T. Rochet and S. Miachon, J. Chromatogr., 579 (1992) 85.

368   B. Levine, S.C. Wu and J.E. Smialek, J. Forensic Sci., 44 (1999) 369.
369   D.J. Pounder and J.I. Davies, Forensic Sci. Int., 65 (1994) 177.
370   B.A. El Zeany, A.A. Moustafa and N.F. Farid, J. Pharm. Biomed. Anal., 33 (2003) 393.
371   B. Paw and G. Misztal, J. Pharm. Biomed. Anal., 23 (2000) 819.
372   J.P. Bounine, B. Tardif, P. Beltran and D.J. Mazzo, J. Chromatogr., 677 (1994) 87.
373   K.M. Kelani, J. AOAC Int., 87 (2004) 1309.
374   T. Gunnar, K. Ariniemi and P. Lillsunde, J. Chromatogr. B, Analyt. Technol. Biomed. Life Sci., 818 (2005) 175.
375   E. Mannaert, J. Tytgat and P. Daenens, Forensic Sci. Int., 83 (1996) 67.
376   E. Mannaert, J. Tytgat and P. Daenens, Clin. Chim. Acta, 253 (1996) 103.
377   E. Mannaert and P. Daenens, Analyst, 121 (1996) 857.
378   B. Zhang, Z. Zhang, Y. Tian, F. Xu and Y. Chen, J. Pharm. Biomed. Anal., 40 (2006) 707.
379   F. Feng, J. Jiang, H. Dai and J. Wu, J. Chromatogr. Sci., 41 (2003) 17.
380   M.G. Gebauer and C.P. Alderman, Biomed. Chromatogr., 16 (2002) 241.
381   P.R. Ring and J.M. Bostick, J. Pharm. Biomed. Anal., 22 (2000) 495.
382   Q. Wang, L. Sun and C.E. Lau, J. Chromatogr. B, Biomed. Sci. Appl., 734 (1999) 299.
383   R.N. Gupta, J. Liq. Chromatogr. Relat. Technol., 19 (1996) 699.
384   W.E. Lambert, E. Meyer and A.P. De Leenheer, J. Anal. Toxicol., 19 (1995) 73.
385   P. Ptacek, J. Macek and J. Klima, J. Chromatogr. B, 694 (1997) 409.
386   F. Stanke, N. Jourdil and V.L. Bessard, J. Liq. Chromatogr. Relat. Technol., 19 (1996) 2623.
387   V. Ascalone, L. Flaminio, P. Guinebault, J.P. Thenot and P.L. Morselli, J. Chromatogr., 581 (1992) 237.
388   Y. Gaillard and G. Pepin, J. Chromatogr., 763 (1997) 149.
389   A. Tracqui, P. Kintz and P. Mangin, J. Forensic Sci., 40 (1995) 254.
390   G. Hempel and G. Blaschke, J. Chromatogr. B, 675 (1996) 131.
391   H. Maurer and K. Pfleger, J. Chromatogr., 428 (1988) 43.
392   F. Pragst, S. Herre and A. Bakdash, Forensic Sci. Int., 161 (2006) 189.
393   X. Chen, Y. Zhang and D. Zhong, Biomed. Chromatogr., 18 (2004) 248.
394   S. Dawling, N. Ward, E.G. Essex and B. Widdop, Ann. Clin. Biochem., 27 (1990) 473.
395   C.C. Pijnenburg, C.G. Barella, R.C. Philipse, J.M. Conemans and A.M. Duchateau, Ziekenhuisfarmacie, 10 (1994) 17.
396   G.R. Tonn, A. Mutlib, F.S. Abbott, D.W. Rurak and J.E. Axelson, Biol. Mass. Spectrom., 22 (1993) 633.
397   G.R. Tonn, F.S. Abbott, D.W. Rurak and J.E. Axelson, J. Chromatogr. B, 663 (1995) 67.
398   T.K. Walters and W.D. Mason, Pharm. Res., 9 (1992) 929.
399   C. Hasegawa, T. Kumazawa, X.P. Lee, M. Fujishiro, A. Kuriki, A. Marumo, H. Seno and K. Sato, Rapid Commun. Mass Spectrom., 20 (2006) 537.
400   H. Hattori, S. Yamamoto, M. Iwata, E. Takashima, T. Yamada and O. Suzuki, J. Chromatogr., 581 (1992) 213.
401   H. Seno, H. Hattori, T. Kumazawa and O. Suzuki, Forensic Sci. Int., 62 (1993) 187.
402   X.H. Chen, J.P. Franke, J. Wijsbeek and R.A. De Zeeuw, J. Anal. Toxicol., 18 (1994) 150.
403   C.M. Moore, J. Forensic Sci. Soc., 30 (1990) 123.
404   F. Sporkert and F. Pragst, Forensic Sci. Int., 107 (2000) 129.
405   M. Nishikawa, H. Seno, A. Ishii, O. Suzuki, T. Kumazawa, K. Watanabe and H. Hattori, J. Chromatogr. Sci., 35 (1997) 275.
406   O.W. Lau and Y.M. Cheung, Analyst, 115 (1990) 1349.
407   O.W. Lau and C.S. Mok, J. Chromatogr., 693 (1995) 45.
408   M.S. Ali, M. Ghori, S. Rafiuddin and A.R. Khatri, J. Pharm. Biomed. Anal., 43 (2007) 158.
409   C. Martinez-Algaba, J.M. Bermudez-Saldana, R.M. Villanueva-Camanas, S. Sagrado and M.J. Medina-Hernandez, J. Pharm. Biomed. Anal., 40 (2006) 312.
410   M.G. Orkoula, C.G. Kontoyannis, C.K. Markopoulou and J.E. Koundourellis, J. Pharm. Biomed. Anal., 41 (2006) 1406.
411   Y. Dong, X. Chen, Y. Chen, X. Chen and Z. Hu, J. Pharm. Biomed. Anal., 39 (2005) 285.

412 M.R. Gomez, L. Sombra, R.A. Olsina, L.D. Martinez and M.F. Silva, Farmaco, 60 (2005) 85.
413 R. Pascual I, R.M. Guardia and D.A. Molina, Anal. Sci., 20 (2004) 799.
414 A.F. Marchesini, M.R. Williner, V.E. Mantovani, J.C. Robles and H.C. Goicoechea, J. Pharm. Biomed. Anal., 31 (2003) 39.
415 M.R. Gomez, R.A. Olsina, L.D. Martinez and M.F. Silva, J. Pharm. Biomed. Anal., 30 (2002) 791.
416 P. Tipparat, S. Lapanantnoppakhun, J. Jakmunee and K. Grudpan, J. Pharm. Biomed. Anal., 30 (2002) 105.
417 H. Okamoto, A. Uetake, R. Tamaya, T. Nakajima, K. Sagara and Y. Ito, J. Chromatogr. A, 929 (2001) 133.
418 C. Barbas, A. Garcia, L. Saavedra and M. Castro, J. Chromatogr. A, 870 (2000) 97.
419 W.A. Korfmacher, T.A. Getek, E.B. Hansen Jr. and J. Bloom, LC-GC, 8 (1990) 538.
420 O.H. Drummer, S. Horomidis, S. Kourtis, M.L. Syrjanen and P. Tippett, J. Anal. Toxicol., 18 (1994) 134.
421 M. Valoti, M. Frosini, S. Dragoni, F. Fusi and G. Sgaragli, Methods Find. Exp. Clin. Pharmacol., 25 (2003) 377.
422 M. Bogusz, M. Erkens, R.D. Maier and I. Schroder, J. Liq. Chromatogr., 15 (1992) 127.
423 M. Bogusz, F.G. Neidl and R. Aderjan, J. Anal. Toxicol., 12 (1988) 325.
424 M. Bogusz, J. Anal. Toxicol., 15 (1991) 174.
425 M. Bogusz and M. Erkens, J. Anal. Toxicol., 19 (1995) 49.
426 M. Bogusz and M. Erkens, J. Chromatogr., 674 (1994) 97.
427 S.P. Elliott and K.A. Hale, J. Chromatogr. B, 694 (1997) 99.
428 R.D. Maier and M. Bogusz, J. Anal. Toxicol., 19 (1995) 79.
429 K.S. Kalasinsky, T. Schaefer and S.R. Binder, J. Anal. Toxicol., 19 (1995) 412.
430 M. Ohtsuji, J.S. Lai, S.R. Binder, T. Kondo, T. Takayasu and T. Ohshima, J. Forensic Sci., 41 (1996) 881.
431 N. Sadeg, G. Francois, B. Petit, H. Dutertre-Catella and M. Dumontet, Clin. Chem. (Washington), 43 (1997) 498.
432 K. Selinger, J. Prevost and H.M. Hill, J. Chromatogr., 526 (1990) 597.
433 C.L. Webb and M.A. Eldon, Pharm. Res., 8 (1991) 1448.
434 J.C. Hudson, M. Golin and M. Malcolm, J. Can. Soc. Forensic Sci., 28 (1995) 137.
435 L. Steinmann and W. Thormann, J. Capillary Electrophor., 2 (1995) 81.
436 S. Daval, D. Richard, B. Souweine, A. Eschalier and F. Coudore, J. Anal. Toxicol., 30 (2006) 302.
437 C. Charron, A. Mekontso-Dessap, K. Chergui, A. Rabiller, F. Jardin and A. Vieillard-Baron, Intensive Care Med., 31 (2005) 1582.
438 Y. Gaillard, F. Billault and G. Pepin, Forensic Sci. Int., 86 (1997) 173.
439 Y. Gaillard, M.J. Gay and M. Ollagnier, J. Chromatogr., 577 (1992) 171.
440 P. Kintz, P. Mangin, A.A. Lugnier and A.J. Chaumont, J. Anal. Toxicol., 12 (1988) 73.
441 T. Trenque, D. Lamiable, H. Millart, R. Vistelle and H. Choisy, J. Chromatogr., 615 (1993) 343.
442 T. Matsumoto, T. Sano, T. Matsuoka, M. Aoki, Y. Maeno and M. Nagao, J. Anal. Toxicol., 27 (2003) 118.
443 P. Kintz and P. Mangin, J. Anal. Toxicol., 17 (1993) 408.
444 P.S. Bonato, D De Carvalho, V.L. Lanchote, R.H.C. Queiroz and A. Cardozo dos Santos, Rev. Farm. Bioquim. Univ. Sao Paulo, 25 (1989) 95.
445 P. Fernandez, N. Lafuente, A.M. Bermejo, R.M. Lopez and A. Cruz, J. Anal. Toxicol., 20 (1996) 224.
446 G. Kauert, I. Herrle and M. Wermeille, J. Chromatogr., 617 (1993) 318.
447 J.M. Lamant, P. Kintz, A. Tracqui, P. Mangin, A.A. Lugnier and A.J. Chaumont, Ann. Biol. Clin (Paris), 46 (1988) 722.
448 M. Villain, V. Cirimele, A. Tracqui, F.X. Ricaut, B. Ludes and P. Kintz, J. Chromatogr. B, Analyt. Technol. Biomed. Life Sci., 798 (2003) 351.
449 R.A. Klinger, L.M. Blum and F. Rieders, J. Anal. Toxicol., 14 (1990) 288.
450 C. Brenner, R. Hui, J. Passarelli, R. Wu, R. Brenneisen, K. Bracher, M.A. ElSohly, V.D. Ghodoussi and S.J. Salamone, Forensic Sci. Int., 79 (1996) 31.

451 M. Balikova, in: B. Jacob and W. Bonte (Eds.), Advances in forensic science, Proceedings of meeting of the international association of the forensic science, 13th, Verlag Dr. Koester, Berlin, Germany, 1995, p. 72.

452 M. Bogusz and M. Wu, J. Anal. Toxicol., 15 (1991) 188.

453 O. Plaut, C. Girod and C. Staub, Forensic Sci. Int., 92 (1998) 219.

454 F. Prost and W. Thormann, Electrophoresis, 24 (2003) 2598.

455 F. Prost and W. Thormann, Electrophoresis, 22 (2001) 3270.

456 M. Geldmacher-von Malinckrodt, A.v. Heijst, C. Koeppel, . In: H.J. Gibitz, H. Schütz (Eds.), Einfache toxikologische Laboratoriumsuntersuchungen bei akuten Vergiftungen, VCH, Weinheim, 1995, p. 168.

457 T.C. Schmitt, J. Chromatogr. B, Analyt. Technol. Biomed. Life Sci., 780 (2002) 217.

458 D.J. Berry, J. Chromatogr., 107 (1975) 107.

459 D.K. Gorecki, K.W. Hindmarsh, C.A. Hall, D.J. Mayers and K. Sankaran, J. Chromatogr., 528 (1990) 333.

460 L. Humbert, M.C. Jacquemont, E. Leroy, F. Leclerc, N. Houdret and M. Lhermitte, Biomed. Chromatogr., 8 (1994) 273.

461 B. Levine, J. Park, T.D. Smith and Y.H. Caplan, J. Anal. Toxicol., 9 (1985) 232.

462 Z. Yan, G.N. Henderson, M.O. James and P.W. Stacpoole, J. Pharm. Biomed. Anal., 19 (1999) 309.

463 P.F. Heller, B.A. Goldberger and Y.H. Caplan, Forensic Sci. Int., 52 (1992) 231.

464 B. Koppen, L. Dalgaard and J.M. Christensen, J. Chromatogr., 442 (1988) 325.

465 M. Ende, G. Spiteller, G. Remberg and R. Heipertz, Arzneimittelforschung, 29 (1979) 1655.

466 E. Klug and V. Schneider, Z. Rechtsmed., 93 (1984) 89.

467 A.W. Jones, Forensic Sci. Int., 153 (2005) 213.

468 A. Lindqvist, S. Hilke and E. Skoglund, J. Chromatogr. A, 1058 (2004) 121.

M.J. Bogusz (Ed.). Forensic Science
Handbook of Analytical Separations, Vol. 6
© 2008 Elsevier B.V. All rights reserved.

287

CHAPTER 7

Antidepressants and antipsychotics

Donald R.A. Uges[1] and Jean M.H. Conemans[2]

[1]Laboratory for Clinical and Forensic Toxicology, University Medical Center, Groningen, P.O. Box 30.001,
9700 RB, The Netherlands.
[2]Central Department for Pharmacy, 's-Hertogenbosch Hospital, 5232 JL 's-Hertogenbosch,
The Netherlands

7.1 INTRODUCTION

Neuropsychiatric conditions have a point prevalence of 10% for adults. About 450 million people worldwide are estimated to be suffering from neuropsychiatric conditions (depression, schizophrenia, bipolar disorders, anxiety and related disorders, dementia, epilepsy, substance abuse, etc.). Many of these conditions require medication with psychotropic drugs. The distribution is equal between men and women, with the exception of depression (more women) and substance misuse (more men). The prevalence and severity, and the urgent need for treatment have recently been surveyed. The figures differ strongly from country to country.

Many psychiatric conditions require preventive medication for a long period.

So it was and will remain very interesting to the pharmaceutical industry to develop new chemical entities for the treatment of these frequently prevalent and chronic diseases. Progression in the knowledge of the pathophysiology of neuropsychiatric disorders enabled the development, testing and introduction of new (classes of) drugs for the mentally ill patient.

The antipsychotic lithium salts are not mentioned in this chapter.

7.1.1 Epidemiology

Psychiatric diseases are widespread. Depression is the most common form of mental disease. Another major psychiatric illness is schizophrenia. Anxiety is a third class. A fourth class is the bipolar mood disorder. Depression is intermittent, as is the psychotic form of schizophrenia; mood disorders are cycling. Psychiatric diseases have a chronic and recurrent pattern. More and more psychiatrists try to prevent relapses by the chronic (long-term) prescription of drugs. Worldwide 20 million

References pp. 314–317

people are suffering from depression (male/female ratio 1:2). Only 20% with receipt episodes are in treatment, 40% during lifetime. Half of all depressed individuals are undiagnosed, and 20% appear in general medical practices. So it became very interesting to the pharmaceutical industry to develop new chemical entities for the treatment of these frequently prevalent and chronic diseases [1]. Progression in the knowledge of the pathophysiology of psychiatric disorders and in neuropsychopharmacology enabled the synthesis, testing and introduction of new (classes of) drugs for the mentally ill patient.

7.1.2 Neuropsychopharmacology

Medicines used in psychiatry are chemically inhomogeneous. They are generally heterocyclic compounds, more or less lipophilic, extensively metabolized, with a high volume of distribution, largely differing in dosage and dosage frequency, in intended serum concentration, therapeutic index, and toxicity; they are intensively hepatically metabolized (by cytochromes); some have presystemic metabolism.

Several groups can be defined on the basis of chemical affinity: classic antidepressants: the tricyclics (TCA), antipsychotics: phenothiazines, thioxanthenes, butyrofenones, diphenylbutylpiperidines and benzamides. Other substances, mainly the currently used ones, are divided into pharmacological groups (e.g. selective serotonin reuptake inhibitor (SSRIs), MAO-inhibitors, atypical antipsychotics) rather than in chemical groups; they are chemically diverse.

Hence, it is not surprising that there is no single analytical technique to detect and measure all psychopharmacological active substances at once. In the mean time the analytical toxicologist should be aware of the fact that psychiatric medications can be prescribed in virtually all-thinkable combinations. Indeed, such combinations are seen in acute and chronic intoxications in the hospital and at the crime scene.

On the other side, these compounds show so much chemical and analytical similarity that they often can be determined in one run or can be screened for in one system.

A list of all available substances registered as drugs changes continuously. The older drugs fade away very slowly and the newer ones are introduced. The available antidepressants and antipsychotics differ from country to country. In Europe most borders have disappeared and people are traveling all over the world. So patients might have used unexpectedly combinations of old or very new drugs. In practice we see many patients with simultaneous prescriptions of several classes of psychotropic drugs, and even of concomitant prescription of drugs from one class. A list of antidepressants and antipsychotics is given in Table 7.1.

Intoxications with the newer antidepressants and antipsychotics immediately follow the introduction of the products by the pharmaceutical industry. The number of suicidal attempts with the now commonly prescribed SSRIs and atypical antipsychotics does not seem to have changed significantly. The modern drugs are generally less toxic. However, with a large overdose and an incorrect diagnosis, without analytical support or without fast treatment (anticonvulsants, cardiac care and antiserotonergic

TABLE 7.1

THE MOST USED ANTIDEPRESSANTS AND NEUROLEPTICS IN 2006 (A): T = TRICYCLIC ANTIDEPRESSANT; S = SSRI; O = ANTIDEPRESSANT, OTHER; P = PHENOTHIAZINE; TH = THIOXANTHENE; B = BUTYROPHENONE; BE = BENZAMIDE; D = DIPHENYLBUTYLPIPERIDINE; A = ANTIPSYCHOTIC, OTHERS; # = HARDLY USED ANYMORE; M = MAOI, (B): SUM = CONCENTRATION OF DRUG + METABOLITE TOGETHER, (C): S = STIP; HN = HPLC NORMAL PHASE; HR = HPLC REVERSED PHASE; G = GLC; GM = GCMS; LM = LCMSMS

Compound and Metabolite	Class (A)	Molecular Weight	CAS Number	T 1/2(h)	Concentration mg/l (B)			Analytical Method (C)		
					Therapeutic	Toxic	Lethal	1e	2e	3e
Alimemazine	P#	298.4	84-96-8	4–8	0.05–0.4	0.5		Hn		S
Amitriptyline	T	277.4	50-48-6	10–20	0.05–0.2	–		LM	G	S
Nortriptyline		263.4	72-69-5	15–50(90)	Sum 0.1–0.25	Sum 0.5	Sum 2	LM	G	
Aripiprazol	A	448.4	129722-12-9	75(146)	(0.075)0.1–0.5	–		LM		S
Bromperidol	B	420.3	10457-90-6	–34	0.002–0.02	–		LM	G	S
Butriptyline	T#	293.5	35941-65-2	–	0.07–0.15	0.4–0.5		Hn	G	S
Chlorpromazine	P	318.9	50-53-3	16–35	0.05–0.3; Child 0.04–0.08	(0.5) 1 (-2)	3	LM	G	S
Chlorprothixene	Th	315.9	113-59-7	8–12	0.03–0.3	0.7 (0.4–0.8)	1	LM	G	S
Citalopram	S	324.4	59729-33-8	30	(0,01)0.05–0.2	0,4–0,6	5	LM		
Desmethylcitalopram		288.4		48	0,01–0,1					
Didesmethylcitalopram		251.6		96	0,002–0,01					
Citalopram-N-oxide										
Clomipramine	T	314.9	303-49-1	12–36	0.1–0.25			LM	G	S
Desmethylclomipramine		300.8			Sum 0.15–0.55	Sum 0.6–0.8				
Clopentixol (zu)	Th	401.0	982-24-1	20	0.002–0.01(0,–15)	0.05–0.1		LM	G	
Clozapine	A	326.8	5786-21-0	20–30	0.2–0.8(1)	0.8–1.3	3	LM	G	S
Desmethylclozapine		312.1			<0.6	>0.6				
Desipramine	T	266.4	50-47-5	15–20(75)	0.075–0.25	0.5	3	LM	G	S
Dibenzepine	T#	295.4	4498-32-2	4	T 0.025–0.15; P 0.1–0.25;	–	18	G	Hn	
Desmethyldibenzepine		281.4			Sum 0.2–0.4					
Dosulepin	T	295.4	113-53-1	13–35	0.05–0.15	Sum 3; Sum 0.75	1	LM	G	S
Desmethyldosulepin		281.2		20–60	Sum 0.1–0.2	0.65–2				
Dosulepin-S-oxide		–			0.04–0.16					

TABLE 7.1
CONTINUED

Compound and Metabolite	Class (A)	Molecular Weight	CAS Number	T 1/2(h)	Concentration mg/l (B)			Analytical Method (C)		
					Therapeutic	Toxic	Lethal	1[e]	2[e]	3[e]
Doxepin	T#	279.4	1668-19-5	10–20	0.1–0.25	0.5–2	2	Hn	G	S
Nordoxepin		265.4	–	33–80	Sum 0.2–0.35					
Fluoxetine	S	309.3	54910-89-3	48–72	0.1–0,0,45(0,9)	Sum1,5–2		LM	G	S
Norfluoxetine		295.3	–	168	0.05–0,35; Sum 0,5					
Flupentixol	Th	434.5	2709-56-0	20–40	0.001–0.015	–		LM	G	S
Fluphenazine	P	437.5	69-23-8	15	–0.001–0.015	0,05–0,1				
Fluvoxamine	S	318.4	54739-18-3	7–60	0.05–0.25(0.5)	–		LM	G	S
Haloperidol	B	375.9	52-86-8	14–24	0.005–0.04	0.05–0.1		LM	G	S
Imipramine	T	280.4	50-49-7	6–20(30)	0.045–0.15	0.4–0.5		LM	G	S
Desipramine		266.4	50-47-5	15–20(75)	0.075–0.25; Sum 0.15–0.3	0.5				
Levomepromazine	P	328.5	60-99-1	15–30(78)	0.03–0.15	Sum 0.4–0.6	sum 2	LM	G	S
Desmethyllevopromazine					Sum 0.15–0.3	0.5	0.4			
Maprotiline	T	277.4	1062-69-8	20–60	0.075–0.25	0.3–0.8	2	LM	G	S
Desmethylmaprotiline		263.4			Sum 0.1–0.4	Sum 0.75–1				
Mianserin	O	264.4	24219-97-4	10–16	0.02–0.09	–		LM	S	
Desmethylmianserin		250.3	–	20–60	Sum 0.04–0.125	Sum 0.3–0.5	2			
Mirtazepine	O	265.4	61337-67-5	20–40	0.02–0.1	Sum 1		LM		
Desmethylmirtazepine		229.0		20–40	Sum 0.05–0.3					
Moclobemide	M	268.7	71320-77-9	1–2	P 1.5–5; T 0.4–1	5–8	16	LM	S	
Nefazodone	O#	470.0	83366-66-9	2–4				S		
Nortriptyline	T	263.4	72-69-5	15–50	0.075–0.25	0,5		LM	G	S
Olanzapine	A	312.4	132539-06-1	30–55	0.02–0.08	0,08–0,1		LM	G	S
Paroxetine	S	329.3	61869-08-7	12–40	0.01–0.075; P 0.015–0.15 (0.25) T0.01–0,035	0.3		LM	G	S
Perazine	P	339.5	84-97-9	8–16	0.025–0.1	0.5		LM	G	S
Perphenazine	P	404.0	58-39-9	8–12	0.0004–0.03	0.05		LM	G	S

Periciazine	P	365.5	2622-26-6	-12	0.005–0.03	0.1		LM	G	S
Pimozide	D	461.6	2062-78-4	48–150	0.001–0.02	–		LM	S	
Pipamperone	B	375.5	1893-33-0	11–35	0.1–0.4	0.5–0.6		LM	S	
Pipotiazine	P	475.7	39860-99-6	8–12	0.001–0.06	0.1		LM	G	S
Promazine	P	284.4	58-40-2	–	0.1–0.4	2–3	5	LM	G	S
Promethazine	P	284.4	60-87-7	7–14	0.1–0.4	1–2	2	Hn	G	
Protriptyline	T#	263.4	438-60-8	74	0.07–0.17	0.5–1	1–2	LM	G	
Quetiapine	A	767.0	111974-69-7	7	(0.04)0.075–0,17	1,8	7	LM	S	
Risperidone	A	410.5	106266-06-2	3	0,003–0,02	0,08		LM		
9-hydroxyrisperidone		426		24	Sum 0,02–0,06					
Sertraline	S	306.2	79617-96-2	26	0.05–0.3(0.5)	–		LM		
Desmethylsertraline										
Sulpiride	Be	341.4	15676-16-1	3–4 (6–8)	0.04–0.4 (0.6) P0,15–0,75	–		LM	Hr	S
Thioridazine	P#	370.6	50-52-2	24–36	0.2–1	2		Hn	S	
Sulforidazine		402.6	14759-06-9		0.3					
Mesoridazine		386.6	5588-33-0		Sum 0.75–1.5	Sum 3				
Tranylcypromine	M	133.2	155-09-9	1.5–3.2	0.05–0.3	0.3–0.5	5	LM	G	Hr
Trazodone	O	371.9	19794-93-5	6–13	T 0.3–1.5 P 1.5–2.5	4	15	LM	G	S
Trifluoperazine	P#	407.5	117-89-5	10	0.005–0.05	0.1–0.2		Hn	G	S
Triflupromazine	P#	352.4	146-54-3	–	0.03–0.1	0.3–0.5		Hn	G	S
Trimipramine	T#	294.4	739-71-9	7–9(30)	0.07–0.17	0,4–0.5		Hn	G	S
Venlafaxine	O	277.4	93413-69-5	5	Sum 0.25–0.75			LM	S	
o-desmethylvenlafaxine		263	–	11		Sum 1–1.5				

treatment) intoxicated patients still die, especially with antipsychotic drugs (e.g. olanzapine). The analytical toxicologist should be prepared by taking note of the literature, by applying for reference substances and by developing analytical methods before the first case presents.

Psychiatric diseases are commonly classified according to DSM-IV criteria. The differential diagnosis between the different brain disorders has developed substantially [2,3], but is scarcely applied to clinical practice.

7.1.2.1 Depression and antidepressants

In the medical management of depression pharmacotherapy plays an important role [4].

TCA, traditional antidepressants, have been available since the 1960s. Pharmacotherapy of depression was traditionally focused on norepinephrine. A newer generation of antidepressants was developed in the 1980s (fluoxetine 1986). These substances are designed based on molecule targeting, focused on serotonin; they are also designed to have a better tolerability and better safety indices.

In the Western world it is assumed that 50% of all suicide attempts are related to depression, and of all severe depressed patients 25% attempt suicide at least once [5]. Henry *et al.* [5] found a rate of 30 deaths by overdose per one million antidepressant prescriptions. Suicide is the 13th leading cause of death. Women attempt more suicides but men are doing this more successfully. Fourteen percent of those suicides have been related to antidepressants. An evidence based consensus guideline for TCA poisoning management is given in 2007 by Woolf c.s. [94].

Beside the mechanical way of suicide (hanging, shooting, drowning) women tend to commit more suicides with drugs, especially benzodiazepines, paracetamol and antidepressants. Tricyclic antidepressants are dangerous in acute intoxications, they are notorious for many drug-related deaths. Although the toxicity of the newer drugs is relatively low, there have been several case reports of deaths in which fluoxetine [5,6] and citalopram [7,8] and the not selective SSRIs (e.g. venlafaxine) [9,10] were strongly suspected as causative agents. There are many case reports on the Internet on people killing or dying under influence of modern antidepressants. In 2005 there was a lawsuit against Wyeth Pharmaceuticals on behalf of 10 families claiming wrongful death and personal injuries as a result of taking venlafaxine (Effexor®). Also several cases of SSRI-induced suicidal attempts and teenage deaths have been reported.

7.1.2.2 Schizophrenia, psychosis and antipsychotics

"Classical" or "typical" antipsychotics have been available since the 1950s. They are all dopamine antagonists. Nowadays the atypical antipsychotics are most commonly prescribed; this category of antipsychotics improves positive and negative symptoms of schizophrenia and causes the least side effects (extra pyramidal side effects). The first atypical antipsychotic drug was clozapine, which has been on the market internationally since 1969. However, it had been withdrawn, because of the too high risk of agranulocytosis. It has been available again for resistant psychosis in the Netherlands since 1989. Clozapine and its toxic metabolite desmethylclozapine concentration levels are checked by therapeutic drug monitoring. The classic neuroleptics are sometimes subdivided into high (specific) and low potency substances. Antipsychotics are also known as neuroleptics and psycholeptics. Compounds related to classical antipsychotics are found in preparations for nausea and vomiting (metoclopramide, prochlorperazine, thiethylperazine) and in cough remedies preparations (promethazine, oxomemazine).

Antipsychotics are generally less toxic than antidepressants. The older ones, such as the phenothiazines (promethazine, chlorpromazine, thioridazine) and butyrophenones (haloperidol) can cause a great number of side effects, but their intoxications are seldom life threatening. The current generations are mostly more effective and some are less and some more toxic. The toxicity of these drugs depends on the acute or chronic cause of overdose. The very effective antischizophrenic drugs, clozapine, olanzapine, quetiapine, risperidone and sertindole, have a narrow therapeutic window: higher concentrations often provoke seizures or an unacceptable rate of QT-prolongation. Wong [9] reported six cases of (the antipsychotic drug) olanzapine associated deaths. Anderson *et al.* [11] showed data of 35 case studies involving postmortem tissue distributors of olanzapine. Levine *et al.* [10] described three venlafaxine-related deaths.

Therefore the newer generations of these drugs might not always be safer. For the forensic toxicologist these drugs are of great interest. A great number of people died accidentally or intentionally due to these drugs. It is also important to find out whether people can be held responsible for criminal acts, caused by an improper use of prescribed psychopharmaceutic drugs. This could also be the case for driving under influence of this kind of medication.

7.1.2.3 Mood stabilizers and bipolar disorder

Besides antidepressants and antipsychotic drugs mood stabilizers are often required: lithium or anticonvulsant drugs such as carbamazepine, valproate and lamotrigine. In these cases the analytical toxicologist has to measure these substances as well. Therapeutic and toxic reference values of anticonvulsant levels are adapted from neurologic indications. As chromatographic methods cannot measure lithium, it is extremely important to carry out a special determination of lithium screening of psychotropic drugs is required. Lithium and anticonvulsants are outside the scope of this chapter [12].

7.1.2.4 Anxiolytics and axiety

Benzodiazepines and anxiolytics are commonly co-prescribed with antidepressants and antipsychotics [13,14].

7.1.3 Pharmacokinetics

In neuropsychopharmacology two subjects demand special attention: the role of cytochromes and active metabolites.

References pp. 314–317

7.1.3.1 Cytochromes

Antidepressants and neuroleptics are metabolized by the cytochromes CYP2D6, CYP2C19, CYP3A4 and/or CYP1A2. CYP2D6 and CYP2C19 are genetically polymorph. CYP2D6 can also be present as a duplication, with increased metabolism. The influence of genetically determined homo- or heterozygotic absence of enzymes and of a CYP2D6 multiplication on dosage and pharmacokinetics is well known [15–18].

TCAs are metabolized by CYP2D6 and some substances by CYP2C19, SSRIs by CYP2D6 and/or CYP2C19, neuroleptics by CYP2D6.

There is a special relation between metabolism, genetic influence on metabolism, and drug–drug interactions, e.g. enzyme inhibition will lower genetically inhibited metabolic capacity further and will enhance the risk on unwanted effects by CYP2D6. Determination of genotypes CYP2D6 and CYP2C19 was initially a research tool, but has increasingly become available for patient care. The analyses are easily performed by a molecular biological laboratory; for practical reasons a hospital laboratory can often detect only the most relevant variant genes. Phenotyping is a research tool.

It is uncommon to calculate dosages based on blood levels or to calculate levels at an earlier time or to forecast future levels without knowledge of a persons geno- or phenotype. Published pharmacokinetic data often neglect the differences in the cytochrome subgroups.

7.1.3.2 Metabolites

Antidepressants and neuroleptics are extensively metabolized. They can be active or inactive compared with the primary substance, or can be more or less toxic. They are sometimes incorporated in the traditional "therapeutic levels". The presence of active metabolites needs special attention from analytical point of view. Metabolites are also of interest in the interpretation of analytical data.

7.1.4 Why and when the analysis should be performed?

Case report: A schizophrenic girl (20 y) was admitted to a psychiatric clinic. There she received clozapine, some benzodiazepines and antibiotics. Her differential blood counts were checked on a regular basis, as clozapine can cause agranulocytosis. A new junior doctor sent a serum sample to our laboratory for a routine TDM analysis of clozapine We surprisingly found a lethal level of 4080 g/l clozapine. When we rang the doctor she told us her patient just died of sepsis. As 4080 g/l clozapine is not a "therapeutic" level, a forensic postmortem analysis was carried out. The pathologist concluded that this patient had advanced cancer with liver metastases. It seems that her liver failure caused this extremely high clozapine level and her psychiatric behavior. Probably the benzodiazepines had masked the seizures as first symptoms of this intoxication.

It is certain that every laboratory for TDM and toxicological analysis needs to have disposal of several reliable, flexible, selective and sensitive methods for the qualitative and quantitative determination of most old and new antidepressants and antipsychotic and their relevant metabolites in biological fluids.

TDM is a surrogate parameter for efficacy and toxicity; it is a better measure than "the dosage". TDM cannot replace other clinical markers to measure effects of drugs [19].

It is clear that a patient also can have his own personal ideal levels of psychotropic substances, regardless of reference "therapeutic levels" [20,21].

7.1.4.1 Analysis, issues and goals

Psychiatric drugs differ widely in chemical structure, in physical constants (lipophilicity, stability, spectra), molecular activity, toxicity, and pharmacokinetics (half life, protein binding, volume of distribution, metabolism) [13,14,22]. This composes a special challenge to the analytical toxicologist. It is obvious that it is hardly possible to describe a method for the determination of all these drugs in plasma. However, there is need for the determination of these drug levels in therapeutic drug monitoring (TDM) and in clinical and forensic toxicology.

How are therapeutic concentrations determined in practice? Are therapeutic levels evidence based? In the ideal situation a clinical study is designed to determine the optimal blood level for an optimal effect with minimal side effects. Some of these "therapeutic ranges" are very old; they date from the beginning of TDM, even from before the introduction of high-performance liquid chromatography (HPLC) [23]. Sometimes the "therapeutic range" is derived from experiments with volunteers. Sometimes therapeutic levels are established by other specialities as for the antiepileptics used in psychiatry. Newer methods are coming soon: measuring the receptor occupation in the brain with a positron emission tomography (PET) scan and relating the results to blood levels will give more precise and less broad "therapeutic range". This has already been done for some SSRIs and some newer neuroleptics.

As mentioned above TDM is required as substances have a relatively small therapeutic window and variable kinetics [24,25]. Beside genetic enzyme status, psychiatric blood levels also depend on drug interactions. During the chronic use patient compliance can often strongly decrease [26,27].

There are many studies showing a positive impact of TDM on patient response to drug, on efficacy and on safety of these drugs [28,29].

Many depressed patients commit suicide during the first weeks of pharmacotherapy. Most of these patients could survive these intoxications, if proper analysis and treatment is undertaken. The overdose with a SSRI may also cause violence against other people [30–32].

It will always be a debate between patients, doctors, forensic toxicologists, lawyers and pharmaceutical industries whether the assaults or murders under influence of an SSRI are caused in spite of or because of treatment. In forensic practice there can be another reason to determine psychiatric drugs, i.e. to identify substances responsible for side effects or toxic symptoms, e.g. hyperthermia: is it malignant hyperthermia, malignant neuroleptic syndrome, anticholinergic syndrome, sympathicomimetic intoxication or serotonergic syndrome? The difference does matter.

Publications on adverse reactions or intoxications without qualitative and quantitative analysis of the substances thought to be responsible for the effects and without a screening for other relevant substances should be treated with caution, as anecdotal evidence.

7.1.4.2 Metabolites

Traditionally, analysts regarded the presence of metabolites as inconvenient. Now metabolites are too important to neglect.

At the analytical stage, metabolites can be helpful in identifying a substance. They give additional spots or peaks in the chromatogram. If the metabolites are recognized they refer to the main substance and vice versa. This is less important when using the selective liquid chromatography-tandem mass spectrometry (LC-MS-MS) method.

Metabolites can be held responsible for the difference between immunoassays and chromatographic determinations, which is the case with the hydroxy metabolites of TCAs.

In addition, metabolites give additional information about the moment of ingestion, about metabolic capacity and about compliance. We think that immunoassays on antidepressants are hardly useful anymore. The only indication may be a fast exclusion of a TCA-overdose.

Finally, the metabolite can help in the explanation of findings and symptoms. Sometimes it is possible to use composite quantitative parameters. The effect and/or adverse effect and/or toxic manifestations of a metabolized compound correlate with sums of concentrations or AUCs, corrected for the activity factor(s) of each individual substance, if known from studies regarding these metabolites at their own pharmacological behavior.

For TDM it is very important to know which drugs and metabolites have to be quantitated and which are not active or available. Some metabolites are only quantitatively important in patients suffering from kidney failure.

7.1.4.3 Sampling

For clinical purposes the type of sample is probably not of great concern. Most laboratories just accept serum or plasma. Collection tubes for heparin plasma, EDTA plasma, normal serum tubes or gel separation tubes, glass or plastic tubes can all be used. There is no need to discard a sample, which has not been taken according to the laboratories' normal collecting instruction. However, during the validation procedure, different matrices and tubes have to be tested before use.

Some authors found differences, but these differences are clinically not important. There are many types (brands) of gel separation tubes; older publications describe tubes, which are no longer available. For specific purposes, gel separation tubes should be evaluated for the substances measured and the goal of the analysis (pharmacokinetic studies?). Not separated blood must never be stored too long, specially when collected in tubes with separation gel for substances can bind the separation gel [33].

From an analytical point of view, serum is the preferred material for chromatographic techniques, especially for LC-MS-MS (less ion suppression).

For lithium a T12-sampling time (12 hrs after intake) is generally accepted. For other psychiatric and neurological drugs this could also be a good habit; often the elimination half life is long enough to permit some variation in sampling time.

TDM must be performed in steady state situations.

It is important to mention that several of these drugs e.g. tricyclic antidepressants, accumulate in the heart tissue. Post mortem, these drugs are released from this tissue causing extremely high, seemingly lethal, heart blood levels, even with low therapeutic serum levels before death. Therefore, quantitative postmortem levels must always be determined in femoral blood [34–36].

In special cases, special samples such as urine, meconium, hair, organs, blood cells can be measured.

7.2 ANALYTICAL TECHNIQUES

In this chapter we will describe most of the analytical techniques, which are still being used or have been used for the bio-analysis of antidepressants and antipsychotics. These analyses will normally be done in serum or plasma for therapeutic drug monitoring (TDM, serum level determination for dose control) or clinical toxicology and in whole blood for postmortem determinations.

7.2.1 Considerations in choosing a method

As there are nowadays many psychotropic drugs used simultaneously with often several metabolites, we decided to change over to GC-MS for qualitative analysis of unknown substances and LC-MS-MS for the routine quantitative measurement of these drugs [37–40]. Using HPLC-UV (DAD) we noticed too many analytical interactions, e.g. for haloperidol and desclomipramine.

LC-MS-MS in TDM and toxicology has some advantages, like high selectivity and sensitivity, which does not require comprehensive extraction and purification of the sample, long linear range and high speed. On the other hand, some disadvantages were observed, like high cost of the equipment, matrix effects (ion suppression or enhancement) and need for highly qualified personnel.

Beside LC-MS-MS several other methods are available for the analysis of antidepressants and neuroleptics in plasma which are suitable for clinical and forensic laboratories [41–43].

There are immunoassays like the radioimmunoassays (hardly used anymore), the fluorescence polarization immunoassay (FPIA, TDx®, Abbott Diagnostics) and enzyme immunoassay (EMIT®, Syva, Dade Behring) available for the general detections of tricyclic antidepressants [28,29,44–46]. These assays are semiquantitative and selective for the whole group of tricyclic antidepressants. Banger *et al.* [47] found that the sum of clomipramine and desclomipramine concentrations measured by HPLC and by FPIA correlated significantly $r = 0.780$ and $p < 0.01$. However, 40% of individual FPIA determinations yielded results that differed by more than

50% from the HPLC concentrations. Also the Syva Rapid Test on tricyclic anti-depressants (Dade Behring) has a cross reactivity strongly depending on the drug or metabolic in the sample. The immunoassays could measure either higher or just lower concentrations than chromatography can. Immunoassays might be useful in clinical toxicology to give a first impression [43]. For TDM, forensic toxicology and for the follow up of a clinical toxicological case a chromatographic method is required.

Screening of these drugs is possible by thin-layer chromatography (TLC), but this technique has become less popular, because of the lack of sensitivity, time-consuming procedures and the lack of reliable quantitation. The HPLC with diode array and fluorescence detection is used for clinical toxicological analysis [42]. The GC-MS is most popular in forensic toxicological analysis [41,48].

7.2.2 Extraction procedures

It is important to choose the optimal extraction method for your laboratory. So far we know there is (still) no general extraction procedure to solve all your analytical dreams and nightmares! Our LC-MS-MS-method (see under) does not use an extraction procedure of the classic way. Therefore, it may be the solution for unreliable and time-consuming extraction procedures in the near future [39].

Most of the antidepressants and neuroleptics are administered as salts of a basic compound. At an alkaline pH they are soluble in an organic solvent. Their metabolites are more hydrophilic, and the recoveries in organic solvents are often lower than those of the mother compounds. We found that at pH 9.5 or higher nortriptyline was easily soluble in dichloromethane of analytical grade from Merck (Darmstadt, Germany) [49]. But after some time, when we used ultra pure dichloromethane from Radburn, we noticed just an insufficient recovery of nortriptyline, but an acceptable recovery of the mother compound amitriptyline. We assumed that pollutants in the dichloromethane from Merck formed a counter ion with the rather hydrophilic nortriptyline. This finding is clinically very important, because some patients, with an enzyme deficiency, can hardly metabolize amitriptyline to nortriptyline. Therefore, a low level caused by insufficient extraction might be interpreted totally incorrectly.

It is good to know that any of these drugs are not stable in the basic form, especially in daylight, and several ones tend to adsorb to the test tube. Therefore, we advice to use new or siliconized brown glass tubes. Tserng *et al.* [50] found that as much as 50% loss by adsorption can occur during the solvent evaporation step. Because of differential adsorption loss among parent drugs, metabolites and internal standards erroneous results could be obtained. The addition of as little as 0.05% diethylamine to the extract before evaporation may completely eliminate the adsorption loss of at least the tricyclic antidepressants and their metabolites.

As we are dealing with drugs, differing widely in their chemical structure and physical properties it is quite possible that some antidepressants or neuroleptics will require a special extraction procedure.

Depending on the quality of the specimen (e.g. old hemolytic blood from a postmortem) the chromatographic method, the urgency, the number of samples and the experiences of the laboratory one can choose between liquid–liquid extraction or solid-phase extraction (SPE). SPE is preferable if particular drugs have to be selectively isolated in serum and when hemolytic "red fluid" has to be analyzed. Liquid–liquid extraction procedures are preferable for screening procedures in emergencies, or in small amounts (TDM) [48,49].

7.2.2.1 Liquid–liquid extraction

In the toxicology laboratory we prefer the following liquid–liquid extraction for GC (gas chromatography) and HPLC, as it does not use a concentration step by evaporation at high temperature, and it provides very clean extracts, without the lipophilic, neutral benzodiazepines [41,49]:

Pipette in a clean brown test tube: 1.0 ml of plasma, 0.3 ml of sodium bicarbonate buffer (1 M, pH 9.6), 0.10 ml of a suitable internal standard solution and 5 ml of a mixture of heptane: isoamyl alcohol (98.5:1.5 v/v). Vortex for 2 min, or mix for 20 min in a shaking machine. Centrifuge for 5 min at 3000 rpm. Place the test tube for 5 min in a freeze-bath at approximately −50°C. Remove the water layer. Add to the organic phase 0.5 ml of hydrochloric acid 0.1 M mixture of heptane:isoamyl alcohol (85:15 v/v). Vortex for 1 min; centrifuge and freeze again at –50°C. Discard the organic layer. Add to the water phase 0.5 ml borax buffer (pH 9.0) and 150 µl of a mixture of hexane: amyl alcohol (85:15 v/v). Vortex, centrifuge and freeze again at –50°C.

Inject 60 µl of the organic layer into an HPLC and/or 1–2 µl into a GLC column. If no freeze bath is available it is also possible to separate the organic and the aqueous layers by quantitative pipetting or with vacuum. The use of a freeze bath (−45 to −60°C) is much faster and performs better and cleaner separation.

The volume of specimen (plasma, serum), the choice of the internal standard (e.g. promethazine HCl 0.5 mg/l or promazine HCl 0.6 mg/l in water) and the injection volume depend on the drug and the expected concentration range.

In clinical toxicology a fast extraction would be sufficient: 1 ml of plasma, with internal standard and 0.2 ml sodium hydroxide 1 M, is extracted with 10 ml hexane: isoamyl alcohol (99:1 v/v). If required, the organic layer might be back extracted with 0.2 ml of hydrochloric acid 0.05 M. The acid–water layer is injected in the HPLC [51]. We are used to extracting simply by shaking mechanically for 5 min with 0.1 ml of sodium hydroxide solution 2 M and 5 ml of dichloromethane (avoid emulsifying.) After centrifugation the intermediate and water phases are removed by suction. The organic layer in a clear test tube is evaporated to dryness with a stream of nitrogen above a water bath of about 60°C or under vacuum at 40°C [52].

Another simple liquid–liquid extraction of the antidepressants suitable for HPLC screening with diode array is as follows [22,53]: Add to 2.0 ml (blood, plasma or urine) in a 15 ml pyrex centrifuge tube 1.5 ml of ammonium chloride buffer (pH 9.5) and 5 ml of the extraction solvent (chloroform: 2-propanol: n-heptane = 10:14:26 v/v). Shake the mixture on a horizontal agitator for 10 min; centrifuge at 2800 g for 10 min.

Remove the aqueous layer, and evaporate the organic layer under vacuum to dryness. Dissolve the residue in 0.1 ml of the mobile phase.

7.2.2.2 Fast extraction for clinical toxicology (STIP extractions)

Extraction procedure: 1.00 ml sample (serum or plasma) + 1 ml acetonitrile (mix) + 7 ml dichloromethane (mix) + 100 µl 2 M HCl or NaOH in this sequence; mix gently 30 s. Centrifuge. Remove water and protein layer. Transfer the organic layer into a new tube. Evaporate the organic layer at 40°C under a stream of air. Dissolve the residue in 100 µl mobile phase. Inject 20 µl in the HPLC-system. Do one injection with the acidic and do one injection with the alkaline extract.

7.2.2.3 Solid-phase extraction

SPE is based on the principle of liquid chromatography, a physical extraction process that involves a solid phase and a liquid phase [54].

A SPE procedure normally involves five critical steps: column preconditioning, sample application, column wash, column drying (if required) and drug elution. In general this procedure requires much experience and a well-defined standard operation procedure. Differences between batches of columns, the non-standardized use of vacuum and the variable intensity of the drying step makes SPE a less robust procedure. Nevertheless, this technique can be preferred to liquid–liquid extraction in several circumstances. The advantages of SPE over liquid–liquid extraction include high selectivity, cleaner extracts, no emulsions, reduced solvent usage and higher throughput by automation. We do not agree with Lai *et al.* [55] that SPE provides generally more reproducible results. The higher throughput by SPE is mostly a fact, but the time required for development and troubleshooting might be longer for SPE than for liquid–liquid extraction. The enormous variation in columns, extraction fluids, extraction steps, adsorption characteristics of the different antidepressant and neuroleptic drugs and their metabolites makes it clear that the inter- and intra-day standard deviations and the recoveries for each analyte on the various SPE system have to be determined and validated.

There are many SPE procedures for antidepressants and neuroleptics in blood or plasma. As an example, the method of Chen is given [54]. The final extract is suitable for gas chromatographic analysis on a wide-bore capillary GC with nitrogen–phosphorus flame ionization detector (NPFID) or GC/MS, and for HPLC:

This extraction is performed on a Vac-Elut vacuum system (Varian) assembled with Bond Elut Certify columns (130 mg of sorbent mass, 10 ml of column volume, from Varian Sample Preparation Products, Harbor City, CA). These columns are preconditioned with 2 ml of methanol (all reagents used are of analytical grade e.g. Merck Darmstadt, Germany), followed by 2 ml of 0.1 M phosphate buffer (pH 6.0) under light vacuum (approx. 2 mm Hg). Then the samples are pretreated as follows:

After adding a suitable internal standard solution in water, 1 ml of plasma or urine is diluted with 4 ml of phosphate buffer (pH 6.0) and vortexed for 30 s; or 1 ml of whole blood (e.g. postmortem fluid) is sonicated for 15 min at room temperature. This sample is then diluted with 6 ml of phosphate buffer (pH 6.0) and vortexed for 30 s. The buffered matrix is centrifuged at 5000g for 10 min. The supernatant of the whole blood, or the diluted plasma, serum or urine is completely applied onto the SPE columns and drawn through at a flow rate of approximately 1.5 ml/min. The columns are washed with 1 ml of water and 1 ml of a 20% solution of acetonitrile in water (only for urine). The columns are acidified by passing through 0.5 ml of 0.001 M acetic acid, then dried under full vacuum (15 in. Hg) for 4 min. After adding 50 µl of methanol, drying under full vacuum is continued for 1 min.

After wiping the column outlets and manifold basin with tissue, 4 ml of acetone: chloroform (1:1 v/v) is added to each column, and this eluent was pulled through completely at a flow rate of 0.8 ml/min (Fraction A, only for acid drugs, therefore not for antidepressants or neuroleptics). After installing another set of brown evaporation tubes, 2 ml of 2% ammoniated ethyl acetate is added to each column and elutes completely at a flow rate of 0.5 ml/min (Fraction B, with the antidepressants and neuroleptics). The eluates are evaporated under a nitrogen stream (or under vacuum) until about 100 µl of solvent remains in each tube.

About 50 µl of the extract is injected into the HPLC or 1 µl into the GC.

Some authors prefer to add the internal standard to the eluate after SPE and before evaporation. The acid step (fraction A) is not strictly required, but provides cleaner extracts in fraction B. Specially the interfering high concentration of benzodiazepines (e.g. oxazepam) will be dramatically decreased in fraction B. The extra washing step with acetonitrile is required for urine. When this step is not performed too much interfering peaks will appear in the chromatogram. The recovery from plasma or whole blood is generally between 85 and 100% at a concentration of about 0.2 mg/l. Lai [55] uses a single-step elution of the same SPE column with ammonia–methanol to disrupt both ionic and non-polar interaction. As a result, most drugs and metabolites (at least in toxic concentrations) are quickly and efficiently recovered in a minimum volume of eluate suitable for direct injection into a HPLC-diode array system.

Some antidepressants, like the SSRIs fluoxetine, paroxetine and citalopram are highly protein bound. To obtain the desired interaction between analyte and sorbent, the analytes must be in an unbound form. Low flow rates and dilution of the sample have been reported to increase the concentration of the unbound drug available for sorbent interaction, and will also ease their passage through the sorbent bed [56].

Citalopram has several metabolites; some are not extractable under the standard alkaline conditions. Sometimes a high concentration of acetonitrile is required to increase the recovery (e.g. citalopram propionic acid) by increasing hydrophobic interactions and possibly also enhancement of polar interaction between the protonated analyte and the protonated silanol groups of the non-end capped C8 SPE columns [56].

It is possible to use SPE for the removal of the omnipresent benzodiazepines.

7.2.2.4 Supercritical fluid extraction

Supercritical fluids as an extraction medium provides a powerful, but relatively expensive alternative to traditional methods [57,58].

The advantage of the SPE/SFE approach is that you can improve the selectivity by changing the pressure and the temperature conditions of the extraction, and thereby change the solvating power of the supercritical fluid. As a result, chromatograms with fewer impurities are obtained [54]. A probably useful alternative to HPLC could be the packed column supercritical fluid chromatography. The techniques have been tried for antipsychotics, antidepressants and stimulants. Detection can be done with a diode array detector.

Carbon dioxide, the most widely used supercritical fluid is non-polar and has therefore limited solvation power for the more polar metabolites of the psychopharmaceutic agents. We can hardly believe that there is a place for supercritical fluid extraction (SFE) in TDM, clinical and forensic toxicology. We assume LC-MS-MS had made SFE superfluous.

7.2.2.5 Solid-phase micro extraction

Solid-phase micro extraction (SPME) employs a stationary phase of polymethylsiloxane coated on a fused-silica fiber to extract compounds from aqueous or volatile samples in a sealed vial. In the head space SPME method, the fiber can be directly injected into the port of a gas chromatograph unit for analysis after equilibration between the head space and the coated fiber [59].

Lee *et al.* [59] published an SPME method for the extraction of four tricyclic antidepressants from human whole blood. The recoveries of the four mother compounds were only 5.3–12.9% with a remarkable low coefficient of variation (CV) of 3–7%. This method might be useful for the analysis of very small sample volumes from forensic cases or animal studies. At this moment it is not clear to us whether these techniques, and/or capillary electrophoresis (CE), and/or micellar electro kinetic capillary chromatography (MECC) will be used on large scale in the bio-analysis of antidepressants.

7.2.2.6 Extraction of brain and liver

Tanaka *et al.* [60] described an easy extraction method of brain and liver suitable for the HPLC determination of tri- and tetracyclic antidepressants:

The tissue (1 g) is homogenized in a mixture of 9 ml hydrochloric acid of 0.1 M and 0.1 ml internal standard solution (diazepam 20 g/ml in water), and then centrifuged at 15000*g* for 10 min. The supernatant (1 ml) and 0.5 ml of 20% sodium carbonate and 4 ml of *n*-hexane:isoamyl alcohol (98.5:1.5 v/v) are mixed for 5 min; centrifuged at 1200*g* for 5 min. The organic phase is evaporated in a clean conical tube under nitrogen at 40°C. The residue is dissolved in 100 µl of mobile phase.

7.2.2.7 Extraction of antidepressants from hair

An increasing number of papers has been published on the detection of antidepressants in hair. The antidepressants amitriptyline, doxepin, clothiepin, imipramine, mianserin, moclobemide, fluoxetine, paroxetine and sertraline were extracted using butyl chloride and utilizing a clear-up back extraction. As we are dealing with very low concentration, LC-MS-MS or GC-MS(MS) are the methods of choice. Several cleaning and extraction steps have been published. Hair samples are often washed with 0.1% sodium dodecyl sulfate and water. After drying, the hair is weighed (about 50 mg). The extraction itself is mostly done at alkaline pH by sonification or just by liquid–liquid extraction [61].

Couper *et al.* [62] found a wide range of doses and hair concentrations, but little obvious correlation between total daily doses of amitriptyline and the hair concentrations. Total daily amitriptyline doses of 25; 50 and 75 mg/day correspond with hair concentrations of 3.5; 17 and 14 ng/mg, respectively.

7.2.2.8 Deglucuronidation

Several antidepressants and neuroleptics form glucuronides as metabolites, and are excreted in the urine as such.

For pharmacokinetic studies it would be of interest to determine the total amount of excreted metabolites. As glucuronides are normally not extracted by SPE or liquid–liquid extraction a deglucuronidation step before extraction is required. Urine specimens are diluted with the same volume of SPE diluent (phosphate buffer 0.4 M pH 6.0). To 6 ml of this mixture 50 μl of β-glucuronidase (from H-Helix-pomatia, about 90.000 kU/l) is added, mixed and incubated at 56°C for 2 h. The samples are cooled down to room temperature and then the normal SPE extraction is carried out [55].

7.2.3 LC-MS-MS Procedure

The Laboratory for clinical and forensic TDM and Toxicology, Dept. of Pharmacy, University Medical Center Groningen, The Netherlands (JP Thie, AMA Wessels, B Greijdanus and DRA Uges, 2005–2006) developed a general LC-MS-MS method for the bio-analysis of a large number of drugs [95]. The pharmacotherapeutic groups which can be qualitatively measured in serum, including their active metabolites are: antipsychotics, antiviral, included anti-HIV, cardiac drugs, including beta-blockers and antiagglutination drugs, antibiotics and fungicides, benzodiazepines, anticonvulsants and several other single-group drugs, such as tramadol, fentanyl, atropine, bupivacaine, idebenone and many others.

Principle: A mixture of solvents with internal standard and buffer precipitates the proteins in serum or plasma. After centrifugation the clear upper layer is injected into a LC-column. Ionization will take place by proton coupling during electron spray. The identity and concentration is calculated by means of the mass response at the specific mass +1 of the internal standard and of the drug and metabolite(s) involved.

Reagents:	all of HPLC quality
Eluent buffer:	10 g ammonium acetate + 10 ml acetic acid + 2 ml trifluoracetic acid anhydride + water till 1000 ml

Precipitation – internal standard mixture (P-IS):	0.05 mg cyanoimipramine (or other TCA) + 167 ml methanol + 833 ml of acetonitrile. Add 1 g ammonium acetate. Add sufficient acetic acid to the clear solution till pH = 6.

Apparatus:

HPLC pump:	standard gradient HPLC pump; flow: 0.2 ml/min
Injector:	automatic injector with cooling ($\pm 10°C$) and 5.0 µl injection
Column:	5 cm * 2.1 mm internal diameter (ID) C_{18}, 5 µm Hypurity Aquastar Javelin Express, Thermo Electron San Jose, CA (sometimes we use two of the same column in series)
Gradient:	a. Tricyclic AD: $t = 0$ to 1 min 5% buffer + 75% water + 20% acetonitrile b. SSRI and others: $t = 0$ to 1 min 5% buffer + 95% water Then for a and b: $t = 1$ to 5 min 5% buffer to 95% acetonitrile $t = 5$ to 6 min 5% buffer to 95% acetonitrile (constant) after 7 min new injection $t = 0$ again
Detector:	suitable and sufficient sensitive quadropole MSMS apparatus with Electron spray, e.g. Finnigan TSQ Quantum Discovery Mass selective see Table 7.2: name; LOQ = low level of quantitation; Mass = m/z M.ms is precursor ion; MSMS is product ion
Processor:	Xcalibur^{TM} Software Revision 1.4 SR1 Therma Electron San Jose, CA
Sample:	Serum (in most cases EDTA-plasma but whole blood is acceptable too). Peak: 2–3 h after oral administration; trough: just before new administration
Assay:	Take the optimal gradient, settings and masses (see Table 7.2). 50 µl serum + 375 µl P-IS mixture; vortex; Keep at –20°C for 30 min.; Centrifuge at 11,000 rpm for 5 min. Clear layer in injector vial. Inject 5.0 µl into the LC-column Every 7 min a new injection. (Do not inject faster: the chance of ion suppression or enhancement might be too high). Make calibrators and control samples in serum in the therapeutic and toxic ranges.
Clinical toxicology:	With this method, a first impression is provide within 30 min. including sample preparation.

TABLE 7.2
LIST OF ANTIDEPRESSANTS AND ANTIPSYCHOTICS, ANALYZED BY THE LCMSMS METHOD OF UGES *ET AL.*, WITH THEIR ANALYTICAL VALIDATED RANGE, LOW LEVEL OF DETERMINATION, MASS OF THE LCMS PROTON-DERIVATIVE AND MASS OF THE MSMS-ION AND THE COLLISION ENERGY OF THE MSMS

Name	Range	LOQ	Mass	m.MS	MSMS	Collision Energy
Amitriptyline	50–200	5	277.4	278.2	233.1	18
Bromperidol	2–20	2	420.3	420.0	165.0	27
Chlorprothixene	30–300	5	315.9	316.0	271.0	21
Citalopram	10–200		324.4	325.1	262.0	20
Clomipramine	25–150	5	314.9	315.2	227.0	47
Clopentixol	5–50	10	401.0	401.1	356.2	21
Clozapine	200–600	5	326.8	327.2	269.9	23
Cyanoimipramine			305.4	306.2	218.0	39
Desipramine	75–250	10		267.2	72.2	14
Dosulepin	50–150	5	295.5	296.2	223.0	24
Dosulepin, nor-	15–150	10		282.2	223.0	22
Dosulepin-*S*-oxide	40–150	5		312.2	211.0	17
Doxepine		5	279.4	280.0	107.1	23
Fluoxetine	100–450	25	309.3	309.9	148.2	10
Fluoxetine, desmethyl	100–450	25	295.3	296.0	134.2	11
Flupentixol	1–15	1	434.5	435.0	264.9	35
Fluvoxamine	50–200	5	318.3	318.9	200.1	21
Haloperidol	5–40	3	375.9	378.1	165.0	22
Imipramine	45–150	5		281.2	86.2	16
Levomepromazine		5	328.5	329.2	100.1	21
Levomepromazine, desmethyl-		10	314.5	315.2	186.0	43
Maprotiline	75–250	10	277.4	278.2	250.1	18
Maprotiline, desmethyl	75–250	25	263.4	264.2	169.1	18
Mianserin	20–90	5	264.4	265.1	208.1	22
Mianserin, desmethyl-	20–90	10	250.4	251.0	208.1	18
Mirtazepine	20–200	2	265.4	266.2	195.0	30
Mirtazepine, desmethyl-	20–200		251.4	252.1	195.0	22
Moclobemide	1500–3000		268.7	268.9	139.0	33
Nortriptyline	75–250	10	263.4	264.2	233.1	15
Olanzapine	20–100	3	312.4	313.2	256.1	24
Paroxetine	10–75	25	329.4	330.2	192.1	21
Perphenazine		3	404.0	404.1	171.1	23
Periciazine		3	365.5	365.9	142.1	24
Pimozide	1–20		461.6	461.9	328.2	28
Pipamperon	100–400	5	375.5	376.2	291.0	16
Promazine			284.4	285.1	86.1	19
Promethazine	100–400		284.4	285.1	198.0	30
Quetiapine	40–170		383.1	384.1	253.0	23
Risperidone	10–60	2	410.5	411.2	191.1	32
Risperidone,9-hydroxy	10–60	1	416.5	427.2	207.0	28
Sertraline	50–500	100	306.2	305.9	158.9	32
Sertraline,desmethyl	50–500	100	292.2	292.0	158.9	27
Sulpiride	40–750		341.4	342.1	112.1	25
Tranylcypromine	50–300		133.2	134.1	115.1	26
Trazodone	300–2500	50	371.9	372.0	176.0	24
Venlafaxine	100–1000	50	277.4	278.2	121.1	31
Venlafaxine, desmethyl	100–1000	100	263.4	264.2	107.1	40

References pp. 314–317

7.2.4 HPLC Procedure

Although normal phase HPLC provides high resolution, very stable and robust results, suitable for the quantification of the psychopharmaceuticals with most of their lipophilic and hydrophilic metabolites, most laboratories use only reversed-phase chromatography. The advantages of reversed-phase chromatography are the use of less toxic and cheaper mobile phases, the possibility to inject aqueous extracts and the possibility to use diode array detectors.

The organic mobile phase used for normalphase chromatography (with dichloromethane) has a UV cut off of around 230 nm, so an important part of the UV range of the diode array detector is not usable.

In our routine TDM laboratory we have used normal phase chromatography for determination of nearly all neuroleptics and antidepressants in serum since 1976. However, in 2005 we changed to LC-MS-MS for 118 substances in serum and since 2006 we do not use normal phase HPLC anymore.

Normal phase HPLC method for most tricyclic antidepressants, MAO-inhibitors, phenothiazines, thioxantenes, butyrophenones and their active metabolites in serum, suitable for TDM and clinical and forensic toxicology is as follows [52,54]:

Apparatus: Normal isocratic HPLC pump, with UV detector (generally 254 nm, 0.01 AUFS) and for some drugs a fluorescence detector in series.

Column: Microspher Si 100 × 4.6 mm, ID, (Chrompack, Middelburg, The Netherlands, no. 28400) or Lichrosorb 60 Si5 μm 150 × 3 mm, ID, (Merck, Darmstadt, Germany).

Mobile phase: Methanol:dichloromethane:buffer pH 3.2 = 10:90:0.15 v/v, with a flowrate of 1.0 ml/min(buffer pH 3.2 = 30% acetic diethylamide = 20:1 v/v; corrected to pH 3.2).

Detection: UV at 254 nm except: butriptyline at 265 nm; clopentixol at 240 nm; flupentixol at 240 nm; dibenzepin at 240 nm; fluoxetine at 240 nm; fluvoxamine at 245 nm; haloperidol at 245 nm; maprotiline at 265 nm; nomifensine at 293 nm; norfluoxetine at 240 nm; paroxetine at 293 nm; pipamperone at 244 nm; pipothiazine at 267 nm; protriptigline at 244 nm; thioproperazine at 265 nm and tiotixene at 240 nm.

Flowrate of the mobile phase is 1.0 ml/min., except: 0.8 ml/min: for hydroxyzine and mianserine; 1.3 ml/min: for chlorphenamine, fluoxetine, fluvoxamine, mequitazine, maprotiline, oxomemazine, pipamperone, protriptigline, tiapride and their main metabolites.

Fluorescence detection in series with UV: fluoxetine Ex 280 nm; Em 310 nm; maprotiline Ex 280 nm; Em 310 nm; orphenadrine Ex 265 nm; Em 310 nm; protriptigline Ex 280 nm; Em 310 nm.

Assay: Carry out one of the extraction procedures described above. Inject 60 μl of the organic phase into the HPLC.

Results: Fig. 7.1 shows a typical chromatogram of the tricyclic antidepressants and their metabolites. Although the mobile phase is water free the chromatogram is "pH-depending".

The lowest level of quantitation (LOQ) for most drugs is 5–10 μg/l serum or less and for the metabolites 25 μg/l or less. Haloperidol has a LOQ of 3 μg/l. The CV in the therapeutic range is mostly 1–5%. Also the new SSRIs (fluoxetine, fluvoxamine, paroxetine and sertraline) can be determined easily and reliably with this method.

Antidepressants and antipsychotics which are not measurable by this method are trazodone, droperidol, penfluridol, trifluperidol, fluspirilene and biperidene.

Reversed phase (RP) HPLC methods for antidepressants and neuroleptics.

An enormous amount of RP-HPLC methods has been published for toxicological screening or determination of the group of antidepressants and neuroleptics, or for TDM of a special drug and its metabolites [53,55,60,63–70].

For screening purposes a diode array detection is very useful, even necessary. It is important to note that a quite extensive number of neuroleptic drugs are pharmacologically active at very low plasma levels. Those drugs might easily be missed by a nonspecific, less sensitive, general assay by reversed phase-HPLC.

A simple and rapid HPLC screening procedure for 27 antipsychotics has been published by Tracqui *et al.* [53]: After an extraction [e.g. with a mixture of chloroform: 2-propanol:*n*-heptane (60:14:26 v/v] (see above) the residue is redissolved in 100 μl of the mobile phase. Of this solution 50 μl is then injected on a Nova Pack C18 (Waters) 4 μm (300 × 3.9 mm ID) at a constant temperature of 30°C. The mobile phase consisted of methanol:tetrahydrofuran:KH_2PO_4 0.01 M buffer (pH 2.6) = 65:5:30 v/v. Elution was isocratic with a flow rate of 0.8 ml/min (196.5 bar). The mobile phase was degassed and filtered through a 0.45 μm filter. The equilibration time of the system was 30 min before analysis. They used a UV/VIS diode array spectrophotometer with a wavelength ranging from 190 to 400 nm. The chromatography was monitored at 220 nm. The detection limit for haloperidol was 9 μg/l of blood (using the old definition of 3 times the background noise). In this article [53] the tricyclic antidepressants are not mentioned.

Balikova *et al.* [62] published a method for the determination of tricyclic antidepressants and the phenothiazines. They also used a diode array. After a SPE extraction the residue was redissolved in 0.2 ml of mobile phase, of which 0.15 ml was injected into the HPLC column. (We cannot recommend such a high injection volume!)

A guard (30 × 3 mm ID) and an analytical (150 × 3 mm ID) reversed-phase column with octadecylsilica Separon SG X G8, particle size 7 m (Tessek, Prague) was used. The mobile phase was acetonitrile:buffer pH 3 = 1:3 v/v, at 0.7 ml/min. (Mobile phase could be recycled for about one week.) The buffer was made from 1 l of NaH_2PO_4 0.01 M and 1.2 ml of nonylamine and phosphoric acid 1 M to pH 3.0. Tanaka *et al.* [60] used a new reversed-phase column with 2 μm silica gel (TSK gel Super-Octyl, from Toshoh, Tokyo). The very small particles give a higher column efficiency, therefore faster separation and better resolution than on the 5–10 μm silica gel. However, the retention capacity is lower than that of other conventional ODS columns, so that the content of organic modify in the mobile phase should be lowered, and the void volumes in the operating system must be reduced to a minimum. The retention times of most antidepressants are about 50–75% of those on a Hypersil C8-5 μm column (Both 100 mm × 4.6 mm ID). The mobile phase for both columns (2 and 5 μm) is methanol:KH_2PO_4 20 mM (pH 7) = 60:40 v/v, at a flow rate of 0.6 ml/min.

References pp. 314–317

There are many HPLC methods for special substances in serum [71].

HPLC with standardized retention times for toxicological bio-analysis. Nearly all Dutch hospital pharmacists use the same isocratic reversed-phase HPLC system with diode array (systematic toxicological identification procedure (STIP)) of 's-Hertogenbosch [72]. The advantage of this system is that it uses a very rapid and easy extraction. All these 60 clinical laboratories in Holland have the same retention times and diode array UV-spectra [42]. This STIP system is developed for rapid qualitative and quantitative determination of the 400 most commonly used drugs in overdose. But for many drugs with plasma concentrations above 50–75 µg/l this system can also be used for TDM. Sometimes an extraction procedure with a higher recovery is required [41,49]. If STIP is used for (semi-)quantitative screening the standard extraction procedure and no internal standard is used. But this flexible system is also used in routine TDM analysis. Then a more selective extraction procedure and an internal standard might sometimes be prereferable.

0.15 ml of internal standard + 1.0 ml acetonitrile; mix; add 6.0 ml of dichloromethane. Add 0.1 ml sodium hydroxide 2 M and mix 2 min. Dissolve the residue of the STIP extract (see under extractions) in 100 µl of the mobile phase and inject 40 µl into the HPLC.

The STIP column: Lichrospher RP – 18e 5 µm, 125 × 4.0 mm ID (Merck, Darmstadt, Germany no. 21568).

Mobile phase: 530 ml of ultra pure water, 146 µl triethylamine and 750 µl phosphoric acid 85%, mix; add 10% potassium hydroxide to pH 3.3. Add 470 ml acetonitrile (pH = ±4.0), degass by sonication. Flow 0.6 ml/min. The retention times in the library and on the chromatogram must be the same (window <10%); if not, the retention times must be corrected for by changing the phosphate concentration ('s-Hertogenbosch) or flowrate (Groningen) of the mobile phase. The most reliable method for the detection of drugs in HPLC is UV-detection; highest response for most substances is at 205 nm. The most sophisticated technique is a DAD; this uses the complete spectrum of the substance for discrimination between all substances present in the database. The spectrum is matched with the reference spectra present in the library of the system. The DAD-detection and interpretation system is named STIPSearch and is available at www.zanob.nl Conversion of software is available from E. Merck Amsterdam for non-Hitachi DAD-detectors. Included with this system is a toxicological database for most of the incorporated substances, referring to analytical data, half-lives and metabolites.

The MMD software uses an algorithm based on linear regression, so it is possible for this system to obtain (semi-)quantitative results: one can quickly get an indication of the quantity of the toxicological substances involved in the sample.

The STIP system is one of the most flexible and useful HPLC systems for quantitative screening for clinical toxicology and TDM. In the meantime other methods have been published as well. All these systems use a diode array detection with a retention index and spectra library [73–75].

A totally automated analytical system is REMEDI, which is commercially available [76–79]. We do not know whether this system is sensitive enough for TDM of antipsychotic drugs.

7.2.4.1 Stereospecific Procedure

Some antidepressants are only available as a racemic drug. Up till now, in clinical and forensic practice nearly only the racemic drug and metabolite concentrations are measured. However, in research, for very special cases and maybe for TDM, stereospecific determinations are required. The S-(+)-citalopram and to some extent the S-(+)-desmethylcitalopram mainly have SSRI-properties [80,81].

Rochat *et al.* [82] described a chiral LC determination of citalopram enantiomer in human plasma. They use an acetylated β-cyclobond, 5 μm, 0.46 × 25 cm ID chiral analytical column (cyclobond 2,000, ICT, Basel) with a fluorimetric detector at Ex 240 nm and Em 296 nm. As mobile phase is used 0.8 ml/min methanol:diethylamine buffer pH 6.1 = 65:35 v/v. With this method they found that the plasma concentrations of the distomer (*R*-(−)-citalopram) were higher than those of the active eutomer (*S*-(+)-citalopram). This may explain why almost all published studies did not show a relationship between clinical response and plasma levels [83,84]. The tricyclic antidepressant doxepin is marketed as a mixture of *cis*- and *trans*-geometric isomers in the ratio of 15:85. Yan *et al.* [85] published a normal phase HPLC system with a silica column and a mobile phase consisting of hexane:methanol:nonylamine = 95:5:0.3 v/v. They use a liquid–liquid extraction with a mixture of *n*-pentane:isopropanol = 95:5 v/v with a flowrate of 1.0 ml/min. They use a normal UV detector at 254 nm.

They found higher transisomer concentrations in plasma, and higher *cis-N*-desmethyldoxepine than that of the other isomers in healthy volunteers. A clear example of isomerization is hydroxynortriptyline. This metabolite itself is a mixture of *cis*- and *trans*, or *E* and *Z* enantiomers, which both exist as a (+) and a (-) optical isomer. All these four identities have their own pharmacokinetic and pharmacodynamic profile.

7.2.5 Gas chromatography

Nearly all GC systems use a capillary column with NPFID or mass selective (MS) detection. The older packed column is hardly used anymore as the modern capillary column yields to better results [86]. Generally speaking, GC provides a lower LOQ, especially of drugs with a low molecular UV extinction. This sensitivity is very important for neuroleptics which are administered by repository preparations (e.g. fluphenazine decanoate). In forensic toxicology, identification by means of GC/MS will be preferable to HPLC. It seems to us that in the next decades qualitative bioanalysis will be done with GC/MS and quantitative determinations by LC-MS-MS.

Probably, the GC-MS system most used in forensic toxicology is the one described by Maurer [48]. This method for systematic toxicological analysis of drugs and their metabolites is given elsewhere in this issue. When using urine samples, cleaning of the conjugates by rapid acid hydrolysis, or gently, but time consuming, enzymatic hydrolysis is required. Most antidepressants and their metabolites require derivatization (acetylation) before chromatography. The acetylation mixture can be evaporated before analysis so that the resolution power of the capillary columns does not decrease in contrast to other derivatization reagents. Molecular mass does not increase very much after acetylation, so compounds with a relatively high molecular mass and several derivatizable groups can be measured with inexpensive mass-selective detectors with a mass range of only up to 650 Da [87,88].

In our laboratory [41] for TDM and clinical toxicology we use Cp-SiL 5 CB, 10 m; 0.12 m film thickness 0.32 mm ID or Cp-SiL 19 CB, 10 m; 0.19 m film thickness 0.32 mm ID (Chrompack, Holland). Of the extract in organic solvent 3 μl is injected, splitless with a delay of 30 s at 250°C. The temperature program is 1 min 100°C/min till 230°C. Gas flows are: bypass 30 ml/min helium, septum purge 5 ml/min helium, overall 150 ml/min helium and column flow of 1 ml/min helium. The detection is NP/FID 300°C att. 8 × 1, column pressure is 10 psi. Most modern GL-apparatus are suitable.

As internal standard 0.2 ml (0.6 mg promazine HCl/l water) is mixed with 1 ml of patient plasma. The LOQ for amitriptyline is 6 μg/l and for nortriptyline 12 μg/l plasma. The CV at 100 μg/l is about 2%. Ulrich *et al.* [84] described a similar method. They used an injector in the split-splitless mode. The split (30 ml/min) was opened 0.05 min after injection. They sometimes also used a cold on-column program. This technique requires a special Hewlett Packard on-column equipment with a fused-silica needle and a duck-bill septum in the injector. They used a NP/FID at 300°C. They were also able to measure the different E-10-hydroxy-metabolites of amitriptyline and of nortriptyline. For the assay of these hydroxy-metabolites Ulrich *et al.* used a new insert in the injection port every day. The inserts are cleaned by standing in chromic-sulfuric acid for 30 min, carefully soaked with water and treated with a mixture of H_2O_2 30% and sulfuric acid (1:1 v/v) for 30 min. After again carefully washing with water and methanol the dry inserts are silinized by leaving to stand for 24 h in 5% dichlorodimethylsilane in toluene, then washing with methanol, soaking for 24 h in methanol and drying in at 80°C. They also pretreat all the glassware the same way. Ulrich [85] also published a GC method with an on column injection for the measurement of *cis*(*Z*)flupentixol in plasma. This method is able to separate the *cis*- and *trans*(*E*)isomer. By irradiation with UV light (365 nm) for 10 min the cisflupenthixol is partly converted into *trans*-flupenthixol by a photochemical reaction. The separation of *cis*- and *trans*-flupenthixol is only possible with a low injection volume. Fluphenazine is used as internal standard. This internal standard is not useful within 14 days after the last administration of fluphenazine–decanoate to the patient.

Trazodone is not easy to determine by HPLC, so especially for this drug and its metabolite we use a GC system. On the CP-SiL 5 CB column (with 15°C/min) the retention time of the metabolite (1-3-chlorofenyl-piperazine) is ± 3.84 min, promazine (IS) 7.61 min and trazodone 12.20 min, and on the CpSiL 19 CB 4.98; 8.68 and 14.40 min, respectively. The LOQ for trazodone is about 50 μg/l and for the metabolite 5 μg/l serum.

7.2.5.1 GC/MS Procedure

Elsewhere in this book the analytical procedure for GC/MS analysis is described. As internal standard the deuterium-labeled analogue of the antidepressant is used. It is mostly very difficult and expensive to obtain suitable deuterated internal standards. Sometimes the manufacturer of the drugs is able to provide the required amount. Both, SPE or liquid–liquid extractions are used [89,90]. Ackermann *et al.* [89] use derivatization with perfluoropropionic anhydride before the samples are analyzed by capillary GC with electron impact mass spectrometry with selected ion monitoring. The derivatization procedure is as follows: The absolutely dry residue from the extraction is dissolved in 0.2 ml of toluene and 20 μl of perfluoropropionic anhydride is added. After vortexing for 5 s the mixture is heated at 100°C for 30 min. Then, 1 ml of a mixture of methanol:water = 1:2 v/v and 1.4 ml of hexane is added, and then vortexed for 1 min, followed by a short centrifugation at 1250*g*. The aqueous phase is frozen at about −50°C and the organic phase transferred into a 1.5 ml conical vial. The residue is concentrated successively with 400, 200 and 80 μl of hexane, reconstituted in 10 or 20 μl of toluene and 1–2 μl is injected into the GC/MS. The GC/MS interface is maintained at about 260°C; the ion source temperature is set to 200°C. The MS is operated under EI ionization conditions.

7.2.6 Thin-layer chromatography

As far as we know TLC is hardly used in laboratories outside poor or underdeveloped countries. TLC is beginning to lose its position to HPLC. The three main disadvantages are the lack of sensitivity, of selectivity and of possibility of reliable quantitation. The concentrations of these kinds of drugs are mostly too low in plasma for using TLC. Also for urine, TLC would often be too insensitive to identify these drugs, particularly the neuroleptics. TLC is useful to identify unknown tablets and powders, or to control the purity of the reference substances. Most pharmacopoeias still use TLC for purity control. It is recommended to dissolve a small amount of the reference substance in methanol and to perform a TLC screen on impurities always before a standard is made, because many reference substances of the neuroleptics are rather unstable.

There are many systems for the determination of drugs in urine by TLC. The DFG Commission for Clinical Toxicological Analysis together with the TIAFT have published several suitable systems [91]. They set up 10 general TLC systems. Probably the most useful mobile phase is chloroform:methanol = 90:10 v/v on a silica F254 plate, impregnated with potassium hydroxide 0.1 M and dried. The committee advises to use several TLC systems together. The problem of this very comprehensive work of the DFG committee is that the new SSRIs are not mentioned in her lists.

Toxi-Lab® is a commercially available TLC system, including all the reagents and an up to date interpretation scheme. The list provides the drugs most used and their metabolites.

We have used a very fast TLC system, suitable for the identification of most basic drugs in urine. Mix 5 ml of urine, 0.5 ml of sodium hydroxide 4 M and 2.5 ml of dichloromethane and centrifuge. Filter through a Whatmann paper filter with 1 g of water-free sodium sulfate. Evaporate the filtrate to dryness. Dissolve the residue in 50 μl of dichloromethane or methanol. Put twice 5 μl of the extract, and of the

TABLE 7.3
LIST OF ANTIDEPRESSANTS AND ANTIPSYCHOTICS ANALYZED BY TLC AND THEIR QUALITIES

Drug	HR_f	254 nm	A.IPt		Mandelin spray	
	8 min	254 nm	Daylight	Daylight	5 min, 100°C, Daylight	5 min., 100°C, 366 nm
Alimemazine	79	+	+	Red	Red	Red/purple
Amitriptyline	69	Purple	Brown	Gray	Blue/gray	Yellow/red
Chlorpromazine	70	Purple	Brown	Red	Red	Red/purple
Chlorprothixene	73	Purple	Brown			Pink fluorescence
Citalopram	55	Deep blue	Brown	White	White	
Clomipramine	70	Deep blue	Brown	Blue	Blue	Green
Clozapine	55	Deep blue	Brown	Red/brown	Gray	Rose
Desipramine	37	Purple	Brown	Blue	Brown/blue	Yellow
Nordoxepin	39	Purple	Brown	Blue		
Doxepin	64	Purple	Brown	Yellow/gray	Brown	Brown/red
Fluphenazine	50	Purple	Brown	Orange	Red	Red/purple
Fluoxetine	42	Deep blue	Yellow/brown	White	White	Yellow
Fluvoxamine	44	Deep blue	Yellow/brown	-	White	Purple
Imipramine	65	Purple	Brown	Blue	Brown/blue	Yellow
Levomepromazine	73	Purple	Brown	Purple/brown	Blue	Red/purple
Maprotiline	35	+	Brown	Not visible (gray)	Beige/brown	Red/brown
Nefazadon	63	Deep blue	Yellow	Pink/red	Gray/pink	Pink
Norfluoxetine	40	Deep blue	Red/ brown	White	White	Yellow
Nortriptyline	42	Purple	Brown	Gray	Brown/blue	Yellow fluorescence
Olanzapine	53	Deep blue	Brown	Gray	Gray	Blue
Paroxetine	35	Deep blue	Brown	Blue	Green	Blue
Perazine	46	Purple	Blue	Orange/gray	Red	Red/purple
Periciazine	45	Blue	Brown	Brown	Brown	+
Perphenazine	36	Purple	Blue	Pink	Red	Red/purple
Promazine	60	Purple	Blue	Orange	Red	Red/purple
Protriptyline	35	Blue	Brown	+	+	Blue
Thioridazine	65	Purple	Brown	Blue	Blue	+
Trifluoperazine	52	Purple	Brown	Brown	Brown	+
Trimipramine	78	Purple	Brown	Blue	Blue	Yellow fluorescence
Venlafaxine	69	Deep blue	Yellow	Gray/pink	Gray/pink	Pink

standard solution on a plastic silicagel F254 nm TLC plate (Merck 5748). The mobile phase is ethyl acetate:cyclohexane:methanol:ammonia 25% = 70:15:10:5 v/v (mix shortly). Use 100 ml for one or two plates. Elution time 8.0 min (about 8 cm). Dry the plate on air, or use a hand warm blow-drier. The plate must be free of ammonia, which is achieved by placing it 2–5 min at 100°C before spraying. Examine the plate at 254 nm; cut the plate in two pieces. Then spray one part with acidified iodoplatinate (A.IPt). The other part is used to spray with a fresh solution of 50 mg ammonium vanadate in 20 ml of sulfuric acid 96% (Mandelin solution); examine at daylight; then place 5 min at 100°C; examine the plate again at daylight and under 366 nm respectively. The retention times (hR_f) are recalculated on nicotine $R_f = 0.55$ ($hR_f = 55$) Table 7.3.

7.2.7 Micellar electrokinetic capillary chromatography and capillary electrophoresis [92,93]

The CE and MECC methods have no significant advantages to HPLC or GC in TDM or clinical toxicology. The small sample volume required for this analysis might be an advantage in animal studies, in neonates, or in forensic cases with very small specimens. However, its relative advantage is overtaken by LC-MS-MS. The expensive apparatus and the lack of reliable or suitable methods for TDM and toxicology makes this technique not the method of choice for the determination of antidepressants and antipsychotics in human plasma or urine.

7.3 CONCLUSION

There are several good methods for the determination of antidepressants and antipsychotics in body fluids. Every laboratory with its own experiences, apparatus, skills and objective and subjective preferences has to make its own choice for its optimal methods. If there is a sufficient budget available LC-MS-MS and/or GC-MS are the methods of choice. Other techniques will lose their position in the bio-analysis of antidepressants and neuroleptics, except atomic absorption spectrophotometry (AAS), ICP or immunoassay for lithium.

7.4 ABBREVIATIONS

5-HT	5-hydroxytryptamine = serotonin
5-HT 1A; 2; 3	Serotonin receptoragonists subtypes
AAS	Atomic absorption spectrophotometry
A.Ipt	Acidified iodoplatinate spray
CZE	Capillary zone electrophoresis
CE	Capillary electrophoresis
CT	Computed tomography

CV	Coefficient of variation (relative standard deviation)
CYP	Cytochrome P450
FPIA	Fluorescent polarization immunoassay (e.g. TDx® or AxSym®)
GC or GLC	Gas chromatograph(y)
(HP)LC	(High performance) liquid chromatograph(y)
IC50	Intrinsic effective concentration, with 50% of the receptor activity
ID	Internal diameter
LC-MS-MS	Liquid chromatography-tandem mass spectrometry
LOQ	Low level of quantitation
hRf	Retention time × 100
Mandelin spray	Ammonia vanadate in 96% sulfuric acid
MECC = MEKC	Micellar electrokinetic capillary chromatography
MRI	Magnetic resonance imaging
MS	Mass selective detector (for GC or LC)
MSMS	Quadropole mass selective detection
NE	Norepinephrine (= noradrenaline)
(NP)FID	(Nitrogen–Phosphorus) flame ionization detector
PET	Positron emission tomography
SFE	Supercritical fluid extraction
SPE	Solid-phase extraction
SPME	Solid -phase Micro Extraction
SSRI	Selective serotonin reuptake inhibitor
STIP	Systematic toxicological identification procedure (reversed-phase liquid chromatography with diode array detection, drugs library and provided with constant retention times).
TDM	Therapeutic drug monitoring (combination of bio-analysis and pharmacokinetics for optimal dose regimen).
TLC	Thin-layer chromatography

7.5 REFERENCES

1   P. Mourilhe and P.E. Stokes, Drug Saf., 18 (1998) 57.
2   G. van den Brink, Pharm. Weekbl., 131 (1996) 337.
3   C. Nemeroff, Sci. Am., 278 (1998) 42.
4   J.J. Mann, N. Engl. J. Med., 353 (2005) 1819.
5   J.A. Henry, C.A. Alexander and E.K. Sener, BMJ, 310 (1995) 221.
6   D.R.A. Uges, J. Forensic Sci., 43 (1998) 1260.
7   K. Worm, C. Dragsholt, K. Simonsen and B. Kringsholm, Int. J. Legal Med., 111 (1998) 188.
8   M. Öström, A. Eriksson, J. Thorson and O. Spigset, Lancet, 348 (1996) 339.
9   S. Wong, Personal communication. Cases will be published.
10 B. Levine, A.J. Jenkins, M. Queen, R. Jufer and J.E. Smialek, J. Anal. Toxicol., 20 (1996) 502.
11 D.T. Anderson and T. Kuwahara. In: V. Spiehler (Ed.), Proceedings of the 1998 Joint SOFT/TIAFT International Meeting, Newport Beach, CA, 1998.

12 H.M. Neels, A.C. Sierens, K. Naelaerts, S.L. Scharpé, G.M. Hatfield and W.E. Lambert, Clin. Chem. Lab. Med., 42 (2004) 1228.

13 Martindale, The extra pharmacopoeia (Ed. XXXIV), Royal Pharmaceutical Society, London, 2005.

14 R. Lane, D. Baldwin and S. Preskorn, J. Psychopharmacol., 9 (1995) 163.

15 M.L. Dahl, L. Bertilsson and C. Nordin, Psychopharmacology, 123 (1996) 315.

16 K. Brøsen, Ther. Drug Monit., 18 (1996) 393.

17 M.H. Ensom, T.K. Chang and P. Patel, Clin. Pharmacokinet., 40 (2001) 783.

18 J. Kirchheiner, K. Nickchen, M. Bauer, M.L. Wong, J. Licinio, I. Roots and J. Brockmoller, Mol. Psychiatry, 9 (2004) 442.

19 M. Danhof, G. Alvan, S.G. Dahl, J. Kuhlmann and G. Paintaud, Pharm. Res., 22 (2005) 1432.

20 F. Bengtsson, Ther. Drug Monit., 26 (2004) 145.

21 P. Baumann, C. Hiemke, S. Ulrich, G. Eckermann, H.L. Kuss, G. Laux, B. Muller-Oerlingenhausen, M.L. Rao, P. Riederer and G. Zernig, Rev. Med. Suisse., 2.

22 A. Tracqui, P. Kintz, P. Kreissig and P. Mangin, J. Liq. Chromatogr., 15 (1992) 1381.

23 A.J.M. Loonen, J.M.H. Conemans and D.R.A. Uges, In: F. López-Muñoz, C. Alamo, (Eds.), Historia de la neuropsicofarmacología, Panamericana, Madrid, 2006, p. 405.

24 M.V. Rudorfer, H.K. Manji and W.Z. Potter, Drug Saf., 10 (1994) 18.

25 B.M. Power, L.P. Hackett, L.J. Dusci and K.F. Ilett, Clin. Pharmacokinet., 29 (1995) 154.

26 J.M. Perel, Clin. Chem., 34 (1988) 881.

27 A.C. Altamura and M. Percudani, J. Clin. Psychiatry, 54 (1993) 29.

28 D.R.A. Uges; M. von Clarmann; M.G. von Mallinckrodt; A.N.P. van Heijst; K. Ibe; M. Oellerich; H. Schültz; D. Stamm; F. Wunsch, Mitteilung XV der Senatskommission der Deutschen Forschungsgemeinschaft für Klinisch-toxikologische Analytik, VCH-Verlagsgesellschaft, Weinheim, 1990.

29 L.P. Hackett, L.J. Dusci and K.F. Ilett, Ther. Drug Monit., 20 (1998) 30.

30 J.G. Edwards, W.H.W. Inman, L. Wilton, G.L. Pearce and K. Kubota, Hum. Psychopharmacol., 12 (1997) 127.

31 J.N. Constantino, M. Liberman and M. Kincaid, J. Child Adoles. Psychopharmacol., 7 (1997) 31.

32 D.L. Frankenfield, S.P. Baker, W.R. Lange, Y.H. Caplan and J.E. Smialek, Forensic Sci. Int., 64 (1994) 107.

33 J. Karppi, K.K. Akerman and M. Parviainen, Clin. Chem. Lab. Med., 38 (2000) 313.

34 R. Jones, in: Proceedings of the 24th International Meeting of the 24th International Association of Forensic Toxicologists, Edmonton, Alberta, 1988.

35 J.M. Mayer, J. Kofoed and D.W. Robinson. In: G.R. Jones (Ed.), Proceedings of the 24th International Association of Forensic Toxicologists, Edmonton, Alberta, 1988.

36 O.H. Drummer and J. Gerostamoulos, Ther. Drug Monit., 24 (2002) 199.

37 J. Bhatt, A. Jangid, G. Venkatesh, G. Subbaiah and S. Singh, J. Chromatogr. B, Analyt. Technol. Biomed. Life Sci., 829 (2005) 75–81.

38 T. Shinozuka, M. Terada and E. Tanaka, Forensic Sci. Int., 162 (2006) 108.

39 P.A.M.M. Boermans, H.S. Go, A.M.A. Wessels and D.R.A. Uges, Ther. Drug Monit., 28 (2006) 295.

40 H.H. Maurer, Clin. Biochem., 38 (2005) 310.

41 P. Bouma, B. Greijdanus, H. Bloemhof, J. IJmker, D.R.A. Uges, D.R.A. Uges, in: R.A. de Zeeuw, (Eds.), Proceedings of the 25th International Association of Forensic Toxicologists, University Press, Groningen, The Netherlands, 1988, p. 411.

42 D.R.A. Uges and A. Messori, Europ. J. Hosp. Pharm., 2 (1996) 120.

43 D.R.A. Uges, A.C. Moffat and M.D. Osselton. In: B. Widdop, (Eds.), Clarke's analysis of drug and poisons in pharmaceuticals, body fluids and postmortem material, Pharmaceutical Press, London, 2004, p. 344.

44 M.L. Rao, U. Staberock, P. Baumann, C. Hiemke, A. Deister, C. Cuendet, M. Amey Härtters and M. Kraemer, Clin. Chem., 40 (1994) 929.

45 P. Nebinger and M. Koel, J. Anal. Toxicol., 14 (1990) 219.

46   W. Asselin and J. Leslie, J. Anal. Toxicol., 14 (1990) 168.
47   M. Banger, B. Hermes, S. Härtter and C. Hiemke, Pharmacopsychiatry, 30 (1997) 128.
48   H.H. Maurer, J. Chromatogr., 580 (1992) 580.
49   J.B.G.M. Noten and D.R.A. Uges. In: F.W.H.M. Merkus (Ed.), The serum concentrations of drugs, Excerpta Medica, Amsterdam, 1980, p. 125.
50   K.-Y. Tserng, R.J. McPeak, I. Dejak and K. Tserng, Ther. Drug Monit., 20 (1998) 646.
51   H.G.M. Westenberg, B.F.H. Drenth, R.A. de Zeeuw, H. de Cuyper, H.M. van Praag and J. Korf, J. Chromatogr., 142 (1977) 725.
52   D.R.A. Uges and J.B.G.M. Noten. In: F.W.H.M. Merkus (Ed.), The serum concentrations of drugs, Excerpta Medica, Amsterdam, 1980, p. 114.
53   A. Tracqui, P. Kintz, P. Kreissig and P. Mangin, Ann. Biol. Clin., 50 (1992) 639.
54   X.-H. Chen, Forensic Sci. Review, 4 (1992) 147.
55   C.-K. Lai, T. Lee, K.-M. Au and A. Yan.-Wo Chan, Clin. Chem., 43 (1997) 312.
56   L. Kristoffersen, A. Brugge, E. Lundanes and L. Slørdal, J. Chromatogr. B, Biomed. Sci. Appl., 734 (1999) 229.
57   S.B. Hawthorne, Anal. Chem., 62 (1990) 633.
58   J.W. King, J. Chromatogr. Sci., 27 (1989) 355.
59   X.-P. Lee, T. Kumazawa and K. Sato, J. Chromatogr. Sci., 35 (1997) 302.
60   E. Tanaka, M. Terada, T. Nakamura, S. Misawa and C. Wakasugi, J. Chromatogr. B, Biomed. Sci. Appl., 692 (1997) 405.
61   R.A. de Zeeuw, R.A. de Zeeuw, I.A. Hosani, S.A. Munthiri, in: A. Maqbool (Eds.), Proceedings of the 1995 International Conference and Workshop for Hair Analysis in Forensic Toxicology, Abu Dhabi, 1995.
62   M. Balikova, J Chromatogr. Biomed. Appl., 119 (1992) 75.
63   S. Joron and H. Robert, Biomed. Chromatogr., 8 (1994) 158.
64   G. Aymard, P. Livi, Y.T. Pham and B. Diquet, J. Chromatogr. B, 700 (1997) 183.
65   O.V. Olesen and K. Linnet, J. Chromatogr. B, Biomed. Sci. Appl., 698 (1997) 209.
66   C.B. Eap and P. Baumann, J. Chromatogr. B, Biomed. Sci. Appl., 686 (1996) 51.
67   Z.-L. Qin, J. Pharm. Biomed. Anal., 14 (1996) 1395.
68   K.K. Åkerman, J. Jolkkonen, H. Huttunen and I. Penttilä, Ther. Drug Monit., 20 (1998) 25.
69   J.-C. Alcarez, D. Bothau, I. Collignon, C. Advenier and O. Spreux-Varoquaux, J. Chromatogr. B, 707 (1998) 175.
70   J.A. Crifasi, N.X. Le and C. Long, J. Anal. Toxicol., 21 (1997) 415.
71   K.M. Kirschbaum, M.J. Müller, G. Zernig, A. Saria, A. Mobascher, J. Malevani and C. Hiemke, Clin. Chem., 51 (2005) 1718; 10 (2005) 1373.
72   R.A. Snoeren, C.G.J. Barella, R.C.A. Philipse, C.C. Pijnenburg, J.M.H. Conemans, F.D.A. Gerkens, A.M.J.A. Duchateau and D.R.A. Uges. In: R.A. de Zeeuw (Eds.), Proceedings of the 25th International Association of Forensic Toxicologists, University Press, Groningen, The Netherlands, 1988, p. 402.
73   M. Bogusz, M. Erkens, J.P. Franke, J. Wijsbeek and R.A. de Zeeuw, J. Liq. Chromatogr., 16 (1993) 1341.
74   M. Bogusz and M. Erkens, J. Chromatogr. A, 674 (1994) 97.
75   D.W. Hill and K.J. Langer, J. Liq. Chromatogr., 10 (1987) 377.
76   S.R. Binder, M. Regalia, M. Mazhar and J. Chromatogr, Biaggi-McEachern, 473 (1989) 325.
77   A. Turcant, A. Premel-Cabic, A. Cailleux and P. Allain, Clin. Chem., 37 (1991) 1210.
78   B.K. Logan, D.T. Stafford, I.R. Tebbett and C.M. Moore, J. Anal. Toxicol., 14 (1990) 154.
79   P.G.M. Zweipfenning and C. Verhulst. In: B. Kaempe, (Ed.), Proceedings of the 29th International Association of Forensic Toxicologists, Printing Mackeenzie, Denmark, 1991, p. 341.
80   T. Naitoh, M. Kakiki, S. Kawaguchi, Y. Kagei and T. Horie, J. Chromatogr. B, Biomed. Sci. Appl., 694 (1997) 153.
81   C.B. Eap, K. Powell, D. Campus-Souche, C. Monney, D. Baettig, W. Taeschner and P. Baumann, Chirality, 6 (1994) 555.
82   B. Rochat, M. Amey and P. Baumann, Ther. Drug Monit., 17 (1995) 273.

83 H. Dufour, M. Bouchacourt, P. Thermoz, A. Viala, P. Phak Rop, F. Gouezo, A. Durand and H.E. Hopfner Petersen, Int. Clin. Psychopharmacol., 2 (1987) 225.

84 S. Ulrich, T. Innenseen and U. Pester, J. Chromatogr. B, 655 (2006) 81.

85 S. Ulrich, J. Chromatogr. B, 668 (1995) 31.

86 H.H. Maurer, Drug Res., 39 (1989) 101.

87 H.H. Maurer and K. Pfleger. In: H.H. Maurer, A. Weber (Eds.), Mass spectral and GC data of drugs, poisons, pesticides, pollutants and their metabolites (2nd ed.)VCH Weinheim, Germany, 1992.

88 B.D. Paul, L.D. Mell Jr., J.M. Mitchell, I. Irving and A.J. Novak, J. Anal. Toxicol., 9 (1985) 222.

89 R. Ackermann, G. Kaiser, F. Schueller and W. Dieterle, Biol. Mass Spectrom., 20 (1991) 709.

90 D. Rogowsky, M. Marr, G. Long and C. Moore, J. Chromatogr. B, Biomed. Sci. Appl., 655 (1994) 138.

91 DFG Commission, Report XVII, VCH, Weinheim, Germany, 1992.

92 A. Aumatell and R.J. Wells, J. Chromatogr. B, Biomed. Sci. Appl., 669 (1995) 331.

93 K. Salomon, D.S. Burgi and J.C. Helmer, J. Chromatogr., 549 (1991) 375.

94 A.D. Woolf, A.R. Erdman, S. Nelson, E. Caravati, D.J. Cobaugh, L.L. Booze and P.M. Wax cs, Clinical Toxicology, 45 (2007) 203.

95 D.R.A. Uges, J.P. Thie, M.S. Bolhuis, A.M.A. Wessels, E.C.F. Dijkers and B. Greijdanus. Ther. Drug Monit. In press (2007).

References pp. 314–317

M.J. Bogusz (Ed.). Forensic Science
Handbook of Analytical Separations, Vol. 6
© 2008 Published by Elsevier B.V.

CHAPTER 8

Non-opioid analgesics

Thomas Kraemer[1] and Hans H. Maurer[2]

[1]*Institute of Legal Medicine, Forensic Toxicology, Saarland University, D-66421 Homburg, Germany*
[2]*Institute of Experimental and Clinical Pharmacology and Toxicology, Department of Experimental and Clinical Toxicology, Saarland University, D-66421 Homburg, Germany*

ABSTRACT

In this chapter, procedures for the detection of non-opioid analgesics in blood and urine as well as in alternative biomatrices such as oral fluid, sweat and hair are reported. Other matrices of forensic interest, such as autopsy samples or non-biological samples, are also considered. The non-opioid analgesics are divided here in 4 classes: paracetamol, acetylsalicylic acid, the so-called non-steroidal anti-inflammatory drugs (NSAIDs; including coxibs) and pyrazole derivatives. For each class, some chemical and pharmacological information is given. Sample preparation from different biomatrices, autopsy samples and non-biological matrices is discussed, as well as derivatization procedures for GC. In the analysis subchapter, GC, HPLC and CE procedures are reported with several methods for detection. Some figures are provided, showing the structures of important analytes and typical chromatograms and/or UV or mass spectra. The chapter ends with a conclusion and some comments on future perspectives.

8.1 INTRODUCTION

Non-opioid analgesics are among the most commonly consumed over-the-counter preparations all over the world. In this chapter, they are classified into paracetamol, acetylsalicylic acid (ASA), the so-called non-steroidal anti-inflammatory drugs (NSAIDs, including the coxibs) and pyrazole derivatives. They are used in the treatment of acute and chronic pain syndromes. Some of them are used also as antipyretic or antiphlogistic drugs. For more information on pharmacology of these

References pp. 348–356

drugs, their rational use in therapy and adverse drug reactions including cardio-vascular side effects of the coxibs, the review of Hinz and Brune [1] is recommended. Although these drugs are perceived to be safe drugs, they may lead to severe toxic effects in case of acute overdosage or in case of chronic abuse. They are also misused in doping of humans and horses. Therefore, non-opioid analgesics may be encountered in clinical and forensic toxicological analysis, as well as in doping control. A review on standards of laboratory practice in analgesic drug monitoring, including colorimetric and immunochemical tests as well as chromatographic procedures has been published by White and Wong [2].

8.2 PARACETAMOL

8.2.1 Introduction

Paracetamol (acetaminophen) is one of the most widely used antipyretic and an-algesic medicaments. It is safe at therapeutic doses, but large doses of paracetamol can result in severe liver damage (cf. Section 8.2.2). Since toxicity of paracetamol is correlated with its plasma levels, its monitoring is of great importance for toxico-logical assessment and for decision of *N*-acetylcysteine antidote therapy.

8.2.2 Structural features of paracetamol

Paracetamol (4′-hydroxyacetanilide, acetaminophen, Fig. 8.1) is metabolized prima-rily by conjugation with sulfate or glucuronic acid such as many other phenolic compounds. These conjugates are readily excreted into urine. The above-mentioned liver toxicity is mediated by *N*-acetyl-*p*-benzoquinone imine (NAPQI), a product of the cytochrome P-450 catalyzed *N*-hydroxylation. This metabolite is further bio-transformed to a glutathione conjugate, which is excreted into urine. In case of over-dose, the glutathione pool is drained and NAPQI both arylates and oxidizes cysteinyl thiol groups in proteins. Binding to proteins in the liver modifies their normal function and initiates processes that culminate in cell death. Nephrotoxicity is also reported. The antidote *N*-acetylcysteine can intercept NAPQI and avoid toxic effects.

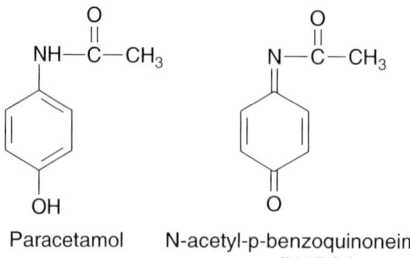

Fig. 8.1. Structure of paracetamol and *N*-acetyl-*p*-benzoquinone imine (NAPQI).

8.2.3 Sample preparation

8.2.3.1 Sample pretreatment and extraction of blood (serum, plasma) or urine

Suitable sample preparation is an important prerequisite for chromatography in biosamples. It involves isolation and, if necessary, cleavage of conjugates and/or derivatization. Prior to blood, serum or plasma extraction, precipitation of proteins may be useful, which can be achieved using solvents such as acetone or acetonitrile [3–8] or using strong acids such as perchloric acid [9–12]. Deproteinization by adding saturated sodium sulfate solution has further advantages: the organic phase is kept free from water and salting-out effects may improve the extraction rates of liquid–liquid extraction (LLE) [13–15]. Liquid–liquid and solid-phase procedures are used for extraction of paracetamol from biomatrices.

8.2.3.1.1 Liquid–liquid extraction procedures

Many different solvents were used for extraction of paracetamol from biological matrices at slightly acidic pH. They include diethyl ether, dichloromethane, isopropanol, ethyl acetate, acetonitrile or others. Often mixtures of these solvents were used [4,7,16–23]. In some procedures salting-out effects help improve extraction [15,22,24]. Since all these procedures work quite well with sufficient recoveries, no special recommendation is given here. However, use of toxic chloroform [23] should be avoided. For extraction of paracetamol at alkaline pH, addition of ion-pair reagents to the solvent were recommended [25,26]. Extractive alkylation merges ion-pair extraction and alkylation in one step. Maurer used this elegant technique for a screening procedure for detection of NSAIDs and their metabolites in urine as part of a systematic toxicological analysis procedure for acidic drugs and poisons by gas chromatography–mass spectrometry [17].

8.2.3.1.2 Solid-phase extraction procedures

Sample pretreatment for SPE depends on the sample type: whole blood and tissue (homogenates) need deproteinization and filtration/centrifugation steps before application to the SPE columns, whereas for urine usually a simple dilution step and/or centrifugation is satisfactory. Whatever SPE column is used, the analyst should keep in mind, that there are large differences from batch-to-batch, and that the same sorbents from different manufacturers also lead to different results [27]. Therefore, use of a suitable internal standard (e.g. deuterated analytes) is recommended.

Many solid-phase extraction (SPE) procedures are reported in the literature. Procedures most often include use of C8 or C18 phase columns [5,28–37]. Most of these procedures work quite well, so that the decision, what kind of column should be used for SPE, is often influenced by practical considerations (e.g. column type used in the laboratory for other determinations). A good overview over applications of solid-phase (micro-) extraction procedures can be found in several review articles [38–43].

8.2.3.2 Sample pretreatment and extraction of alternative matrices (oral fluid, sweat or hair)

Besides classical biomatrices such as blood and urine, alternative matrices such as oral fluid, sweat and hair have been tested for their usefulness in analytical toxicology. Development of more sensitive analytical equipment was a prerequisite for analysis in these matrices.

The value of drug testing or drug monitoring in alternative matrices should not be overestimated, especially for the following reasons. The general amount of sample is limited. The concentrations of drugs in oral fluid, sweat or hair usually are lower than in urine. The window of detection in sweat and oral fluid is shorter than that in urine. Nevertheless, sweat and oral fluid testing offers advantages over urine in the ease of collection. Hair samples also are easy to collect, and allow detection of chronic or past use of drugs and/or medicaments. However, still there is controversy on how to interpret the results, particularly concerning external contamination, cosmetic treatment or ethnical bias. Principles and kinetics of drug incorporation into hair are still under discussion as well as correlation between blood and hair concentrations [44]. Several reviews on the topic of hair analysis have been published in the last few years [45–49].

8.2.3.2.1 Sample preparation of oral fluid or sweat
For therapeutic drug monitoring, increased usage of oral fluid testing can be observed, whereas sweat was only minimally explored. Both oral fluid and sweat require extraction steps from the collection devices before analysis. Concentrations are lower than e.g., in urine. Therefore, the window of detection is shorter.

Oral fluid is collected by spitting or by using cotton swabs. Production of oral fluid can be stimulated by chewing (glass marbles, parafilm, teflone pieces) and/or by giving citric acid in the mouth. Oral fluid samples can be extracted in the same way as plasma samples. For separation and detection of paracetamol, GC and GC-MS or HPLC-UV and LC-MS are suitable [50–53]. Fujino *et al.* [54] fluorometrically determined paracetamol in small amounts of oral fluid after derivatization with a new fluorescence derivatization reagent for phenolic compounds (12–(3,5-dichloro-2,4,6-triazinyl)benzo[d]benzo[1′,2′–6,5]isoindolo[1,2-b][1,3]thiazolidine) after isocratic elution on a reversed-phase column.

Analysis of paracetamol in sweat is hardly mentioned in the literature. Therefore, recommendations cannot be given. Common considerations on oral fluid and sweat testing have been reviewed by Kidwell *et al.* [55] and more recently by Verstraete [56].

8.2.3.2.2 Sample preparation of hair
Before extraction, decontamination using solvents (e.g. acetone, dichloromethane, petrolether, water) and homogenization (e.g. in a ball mill) are necessary. Cleavage of conjugates during the extraction procedure is usually performed enzymatically by addition of glucuronidase/arylsulfatase or by simple acid hydrolysis using hydrochloric

acid. Determination of paracetamol in hair is seldom described [57], so that special recommendations cannot be given.

8.2.3.3 Sample pretreatment and extraction of body tissues and other autopsy material

Extraction of paracetamol from body tissues can be achieved by homogenizing the tissues in phosphate buffer followed by LLE or SPE as described above [8,58,59]. To improve the extraction rate, digestion steps using, e.g., papain, neutrase or collagenase can be useful [60].

8.2.3.4 Sample pretreatment and extraction of non-biological samples

Pure substances can usually be solved in solvents (e.g. methanol, ethanol, acetone, ethyl acetate, diethyl ether) and directly be analyzed by GC, LC or CE procedures. The same extraction methods, which are suitable for biomatrices can also be applied to extraction of paracetamol from non-biological matrices such as beverages, food or pharmaceutical formulations.

The use of supercritical CO_2 for the indirect isolation of paracetamol from the non-polar matrix of suppositories was demonstrated by Almodovar *et al.* [61]. Since paracetamol was not soluble in pure CO_2 at low pressure, but the waxy matrix was, the latter could be extracted, leaving the paracetamol behind. The remaining paracetamol was then removed from the extraction cell using ultrasound in warm water.

8.2.3.5 Derivatization for GC and GC-MS

Concentrations of paracetamol in blood and urine are relatively high. Therefore, derivatization of paracetamol prior to GC-(MS) analysis is often renounced. Nevertheless, derivatization can further improve determination of paracetamol. Acetylation or trifluoroacetylation are used most often, butylation has also been tested [15,31,62–67]. For more information also on more exotic derivatization procedures such as combined propionylation/propylation, the reviews of Segura *et al.* or Halket and Zaikin, Halket *et al.* and Zaikin and Halket on derivatization procedures are recommended [68–74]. Details on extractive alkylation for extraction and derivatization of acidic drugs in one step are given in Section 8.4.3.5.

8.2.4 Analysis of paracetamol

Concentrations of paracetamol in plasma and urine are very high. In case of intoxications, plasma concentrations of up to 1000 mg/l can be observed. Therefore, colorimetric assays are described for fast detection, when other methods are not available [75–80]. Some comparative studies were published, showing that even quantitative results are in accordance with those of chromatographic procedures [76–78]. Even if interferences are seldom or can partly be avoided by modifications of the procedure [75,80], more specific chromatographic procedures, which are described in the following, should be preferred [81–83].

Commercially available immunoassays provide sufficient quantitative information for decision of antidote therapy. In forensic toxicology, positive immunoassay results must be confirmed by a second independent method that is at least as sensitive as the screening test and that provides the highest level of confidence in the result.

8.2.4.1 GC and GC-MS procedures

Fused-silica capillary columns are suitable for separation of paracetamol. Standard low-polarity dimethylpolysiloxane types of column were used for systematic toxicological analysis [83]. The 5–10% diphenyl methylpolysiloxane types are more suitable for analysis of (underivatized) paracetamol [22,66,67,84–90]. A combination of an HP-5MS (cross-linked 5% phenyl-methylsiloxane) phase serially coupled to a second column of BPX50 (50% phenyl equivalent) phase was used for a comprehensive two-dimensional gas chromatography (GC-GC) time-of-flight mass spectrometry (TOFMS) screening procedure [91].

Simple flame ionization detectors (FID) provide sufficient sensitivity for detection of paracetamol [22,64] but not sufficient selectivity or specificity. The detector of choice is MS, providing good sensitivity and best specificity [17,31,63,65,83,91].

8.2.4.1.1 Screening and confirmation of paracetamol

Screening procedures using GC-FID [22] suffer from the lack of specificity, and should not be used. Screening and confirmation of non-opioid analgesics, including paracetamol, can be performed in one step using GC-MS in the electron-impact (EI), full-scan mode [63]. Such a procedure, including universal sample preparation for many different drug classes (acid hydrolysis, LLE and acetylation) as part of a general screening procedure, is described by Maurer [13,63]. For screening and confirmation mass chromatography based on full-scan recording was used. The identity of the peaks in the mass chromatograms was confirmed by computerized comparison of the peaks underlying mass spectra with reference spectra [14,92]. Besides paracetamol, the following groups of medicaments can simultaneously be covered by this procedure: amphetamine derivatives, barbiturates, benzodiazepines, opioids, other non-opioid analgesics, antidepressants, neuroleptics, antiparkinsonians, anticonvulsants, antihistamines, β-blockers, antiarrhythmics and laxatives. Maurer *et al.* [17] used extractive methylation for their screening procedure for detection of NSAIDs and their metabolites in urine as part of a systematic toxicological analysis procedure for acidic drugs and poisons by GC-MS. This method allowed the detection of therapeutic concentrations of acemetacin, acetaminophen (paracetamol), ASA, diclofenac, diflunisal, etodolac, fenbufen, fenoprofen, flufenamic acid, flurbiprofen, ibuprofen, indometacin, kebuzone, ketoprofen, lonazolac, meclofenamic acid, mefenamic acid, mofebutazone, naproxen, niflumic acid, phenylbutazone, suxibuzone, tiaprofenic acid, tolfenamic acid and tolmetin in urine samples. As already described above paracetamol can also be screened for by two-dimensional gas chromatography with TOFMS [91].

8.2.4.1.2 Quantification of paracetamol in blood, plasma or serum
GC-FID procedures after acetylation were used for quantification of paracetamol in blood [64]. GC-MS is also suitable for quantification of paracetamol. However, instead of full-scan mode single-ion monitoring (SIM) mode should be used, which provides higher sensitivity and precision. Further, prerequisite for precise quantification is the use of suitable internal standards [65]. Best results can be achieved with deuterated internal standards, which are commercially available. Today, mostly liquid chromatographic procedures are used for determination of paracetamol in blood, plasma or serum. Capillary electrophoresis (CE) is also suitable for that purpose because of the high plasma concentrations of paracetamol.

8.2.4.2 HPLC and LC-MS procedures

HPLC procedures are most often used for the analysis of paracetamol. Common C18 or C8 packing materials for LC columns have been most widely used and can be recommended for separation of paracetamol [4,10,11,19,23,36,59,93–108]. Isocratic or gradient elution procedures using acetonitrile or methanol/buffered water mixtures or mixtures of solvents (acetonitrile, hexane, isopropanol) with phosphate or acetate buffers resulted in sufficient separation. Slightly acidic pH is necessary to avoid deprotonation of the phenolic group of paracetamol. Separation of paracetamol could also be achieved using ion-pair reagents such as tetrabutylammonium salts [25,109–114]. For LC-MS volatile solvents and buffers such as methanol, acetonitrile, formate-buffers have to be used [4,29,93].

8.2.4.2.1 Screening and confirmation of paracetamol
Concentrations of paracetamol in plasma and urine are very high. Therefore, screening does not require sensitive techniques. Simple UV detectors provide enough sensitivity for detection, with the disadvantage of low specificity. Often the conjugates of paracetamol were included in these procedures [25,103,104,110,115–118]. Screening for drugs of abuse in plasma or serum was described using HPLC- UV-DAD [108]. This assay was capable of detecting and identifying therapeutic and toxic amounts of NSAIDs, barbiturates, anti-convulsants, diuretics, sulfonylurea anti-diabetic drugs, theophylline and analgesic drugs. Nevertheless, full-scan electron impact MS detection provides best specificity, which is important, especially in forensic toxicology. Hori [29] described simultaneous screening, identification and quantitative determination of drugs that frequently cause acute poisoning in Japan (salicylic acid, acetaminophen, theophylline, barbiturates and bromvalerylurea) using LC-MS.

8.2.4.2.2 Quantification of paracetamol in blood, plasma or serum
For separation and quantification of paracetamol in plasma, numerous LC procedures were described with different detection modes [8,11,19,21,23,33,102,108,119]. Since therapeutic plasma concentrations of paracetamol are relatively high, requirements on the analytical technique are not very high. Therefore, simple UV detection may be sufficient for determination of paracetamol in biosamples in terms

of sensitivity. Requirements on specificity are however not met. Determinations of paracetamol using LC-MS were also described [4,16,29,29,120,121]. In Fig. 8.2 representative SRM chromatograms are given for paracetamol (I), guaifenesin (II) and I.S. (osalmide, III) in human plasma samples: (A) blank plasma sample, (B) blank plasma sample spiked with paracetamol (100.0 ng/ml), guaifenesin (10.0 ng/ml) and I.S. (1.0 µg/ml) and (C) a volunteer plasma sample 2 h after an oral dose of 650 mg paracetamol, 200 mg guaifenesin, 60 mg pseudoephedrine and 20 mg dextrorphan (taken from Ref. [16]). The advantage of LC-MS over GC-MS is that the conjugates could be detected at the same time [122]. However, determination of the conjugates is not necessary for toxicological assessment.

8.2.4.3 CE procedures

Paracetamol is one of the ideal candidates for determination by CE due to its high plasma and urine levels. CE is a growing analytical technology. The great interest raised by CE is due to its high efficiency, mass sensitivity, minimum needs of solvent and sample volumes and in particular to the high versatility in terms of separation modes. However, sample matrix components and salts can have deleterious effects on CE separations. For urine, dilution of the sample can help, but this procedure works only when analytes such as paracetamol are present in high concentrations.

Direct injection of urine [123–125] is described as well as freeze drying of the samples before analysis [126,127]. Plasma or serum can be analyzed by direct injection techniques [102,128–130] or after deproteinization and/or extraction [10]. Deproteinization should not be performed using acids or salts, which would have deleterious effects on CE separations. Solvents such as acetonitrile are suitable. A rinsing step using, e.g., acetonitrile after direct plasma injection can be useful [102,128,130,131]. Sample preparation techniques used in CE are often transferred from LC or GC methods. However, mainly the small volumes to handle cause problems, when using devices not specifically tailored for such purpose. Some CE techniques allow sample enrichment during injection, known as sample stacking. In fact, a specific strategy for sample preparation in CE has not yet been fully developed.

For detection of paracetamol after CE separation, UV [3,102,130–132] and MS [123,124,126] have been used. In Fig. 8.3, an example for CE-ESI-MS determination of paracetamol is displayed (taken from Ref. [124]). More exotic detection modes such as dual electrode coulometric quantification in the redox mode [11] or the coupling of CE with NMR techniques [133] or with on-line Fourier transform infrared detection [134] were described, but they are not yet very suitable for routine analysis.

Fig. 8.2. Representative SRM chromatograms for paracetamol (I), guaifenesin (II) and I.S. (osalmide, III) in human plasma samples: (A) blank plasma sample, (B) blank plasma sample spiked with paracetamol (100.0 ng/mL), guaifenesin (10.0 ng/ml) and I.S. (1.0 µg/ml) and (C) a volunteer plasma sample 2 h after an oral dose of 650 mg paracetamol, 200 mg guaifenesin, 60 mg pseudoephedrine and 20 mg dextrorphan (taken from Ref. [16]).

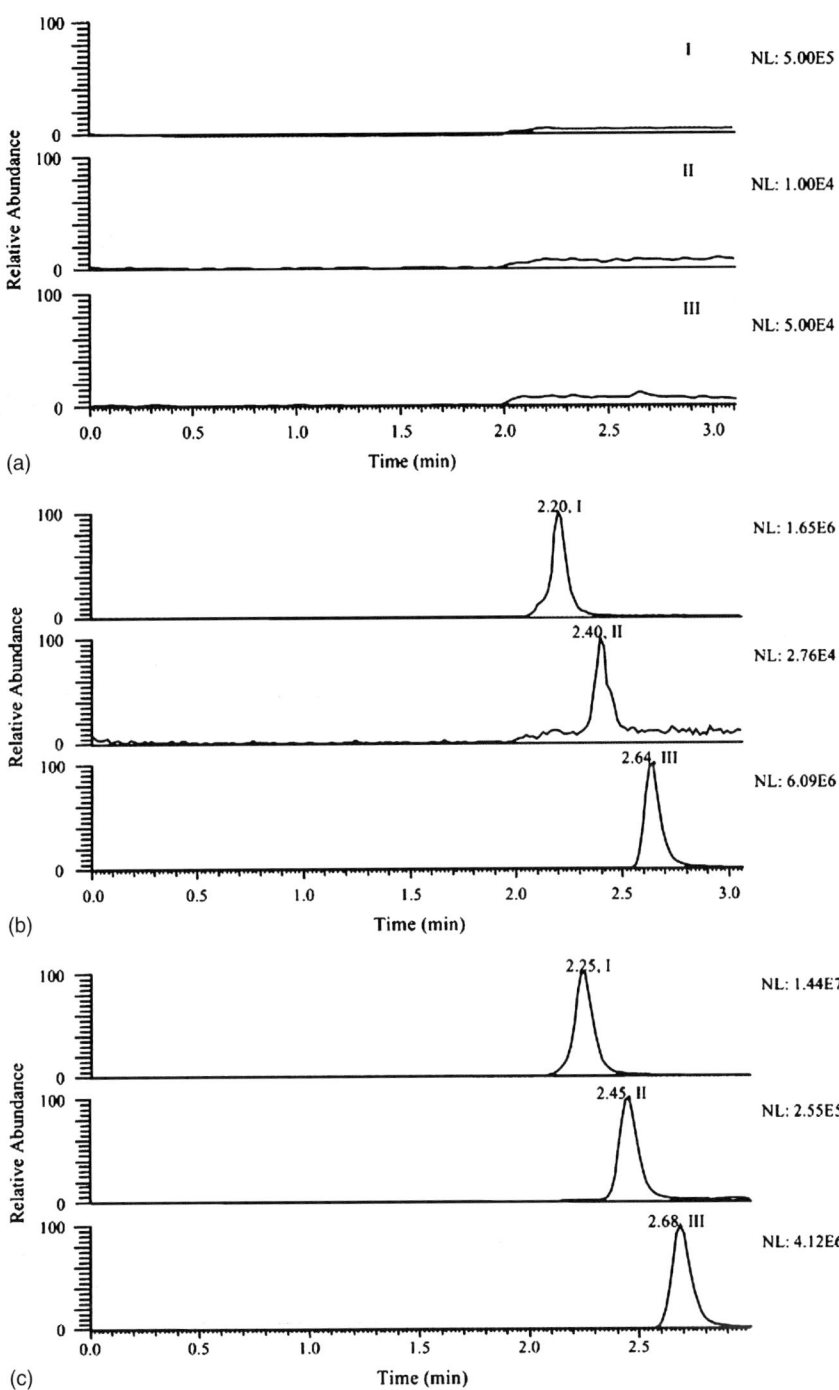

(a)

(b)

(c)

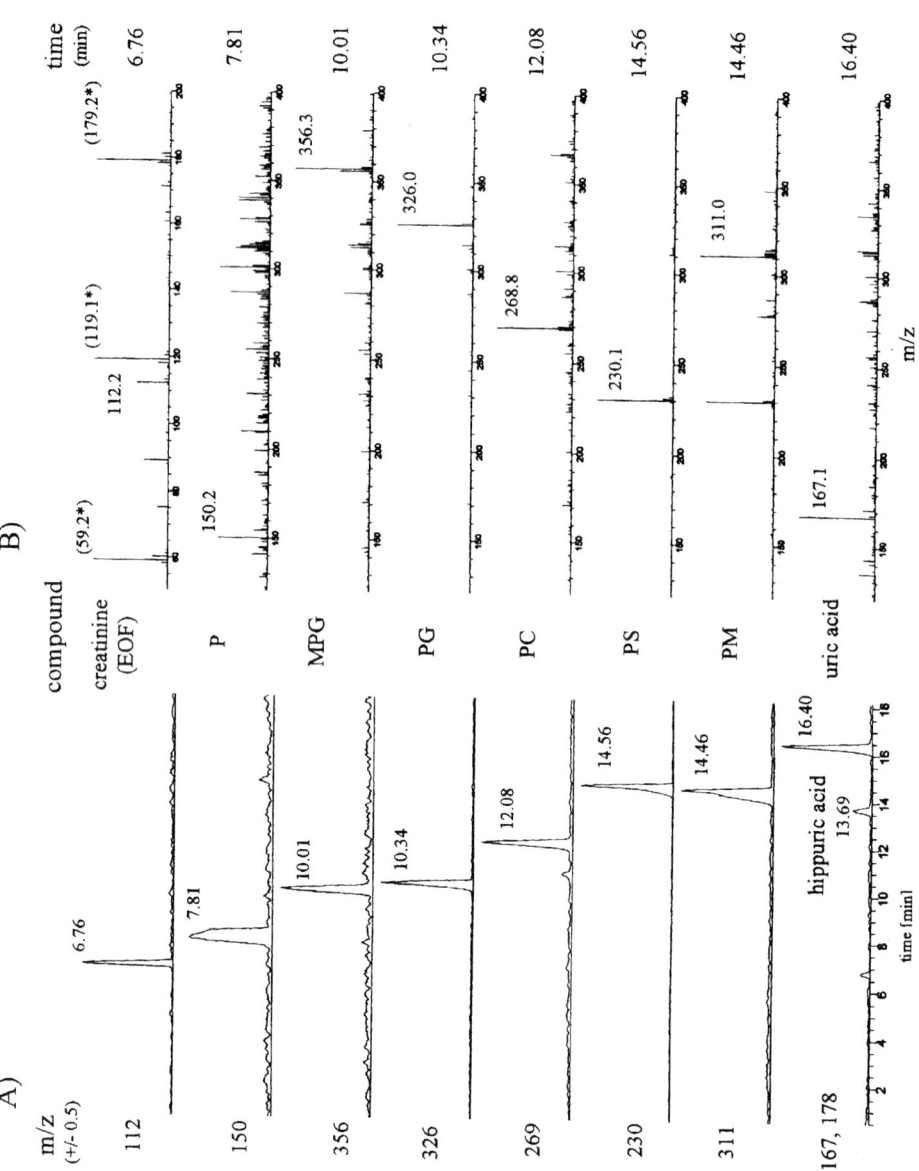

Fig. 8.3. Direct analysis of urine samples after the application of paracetamol by CE-ESI-MS. In (A), reconstructed ion electropherograms (RIEs) of the deprotonated species of paracetamol and its metabolites are presented. (B) shows the mass spectra at the apex of the peaks in the RIEs (taken from Ref. [124]).

The status of CE techniques in forensic toxicology has been displayed in detail in the review of Tagliaro and Bortolotti [135]. Today, CE techniques are still not very widely used in forensic toxicology. The hopes that all problems of the method will be solved and CE may be a useful tool in forensic toxicology did not come true. In a more recent review, Pucci and Raggi [136] concluded again that electrokinetic chromatographic methods might become a robust and widely accepted method and that their use might continue to grow.

8.3 ACETYLSALICYLIC ACID

8.3.1 Introduction

Acetylsalicylic acid (ASA) is very widely used as analgesic, anti-inflammatory and antipyretic drug. In addition, low-dosed ASA is used as antithrombotic agent. It is safe at therapeutic doses, but large doses of ASA can result in severe intoxication with fatal disturbances of the acid–base equilibrium of the body due to heavy metabolic acidosis. Monitoring of ASA is necessary for toxicological assessment. Chronic misuse can result in nephrotoxicity with total loss of both the kidneys. Detection of abuse is therefore necessary.

8.3.2 Structural features of acetylsalicylic acid

ASA (2-acetoxy benzoic acid, Fig. 8.4) has a pK_a value of 3.7. The phenolic ester bond is susceptible to hydrolysis, especially at alkaline conditions. In vivo, ASA is rapidly hydrolyzed by unspecific esterases of the plasma to the also pharmacologically active main metabolite salicylic acid (SA, Fig. 8.4). Half-life of ASA in plasma is about 1 h. SA is further metabolized by hydroxylation to gentisic acid, by conjugation with glycine to salicyluric acid and by other conjugation reactions (cf. Fig. 8.4).

Therefore, the target compound for analysis of ASA in biomatrices is SA, which is known to give deep purple complexes with iron-(III) ions, which can be used for salicylate spot test or colorimetric quantification ("Trinder test;" cf. Section 8.3.4).

8.3.3 Sample preparation

8.3.3.1 Sample pretreatment and extraction of blood (serum, plasma) or urine

Suitable sample preparation is an important prerequisite for chromatography in biosamples. It involves isolation and, if necessary, cleavage of conjugates and/or derivatization. After collection, plasma can first be treated with fluoride or physostigmine sulfate to inhibit enzymatic hydrolysis of ASA to SA [137]. Prior to blood, serum or plasma extraction, precipitation of proteins may be useful, which can be achieved using solvents such as acetone or acetonitrile [138] or using strong acids such as trichloroacetic acid [12,139].

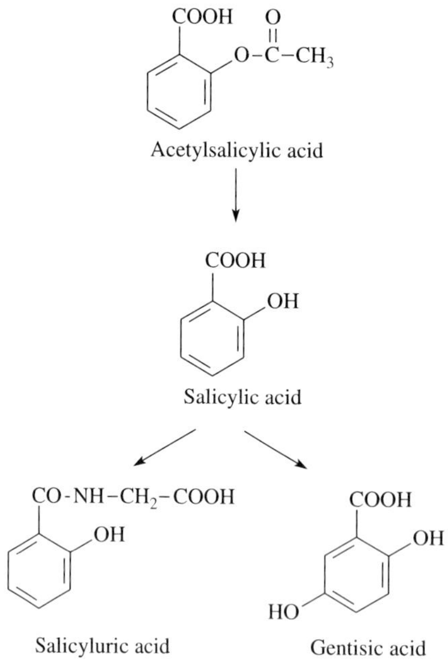

COOH   O
        ‖
         O–C–CH₃

Acetylsalicylic acid

COOH
        OH

Salicylic acid

CO-NH–CH₂–COOH
        OH

Salicyluric acid

COOH
        OH

HO

Gentisic acid

Fig. 8.4. Structure of acetylsalicylic acid and some of its metabolites.

Deproteinization by adding saturated sodium sulfate solution has further advantages: the organic phase is kept free from water and salting-out effects may improve the extraction rates of LLE [13–15]. Liquid–liquid and solid-phase procedures are used for extraction of ASA from biomatrices.

8.3.3.1.1 Liquid–liquid extraction procedures

ASA and SA are most often extracted from biological matrices at acidic pH by LLE. Many different solvents were used for this purpose. They include diethyl ether, dichloromethane, isopropanol, ethyl acetate, acetonitrile or others. Often mixtures of these solvents were used [58,138,140–152]. In some procedures salting out effects help improve extraction [15,24]. Since all these procedures work quite well with sufficient recoveries, no special recommendation is given here. Again, use of toxic chloroform [139,153–155] should be avoided. Extractive alkylation merges ion-pair extraction and alkylation in one step. Maurer used this elegant technique for a screening procedure for detection of NSAIDs and their metabolites in urine as part of a systematic toxicological analysis procedure for acidic drugs and poisons by gas chromatography-mass spectrometry [17].

8.3.3.1.2 Solid-phase extraction procedures

Sample pretreatment for SPE depends on the sample type: whole blood and tissue (homogenates) need deproteinization and filtration/centrifugation steps before application to the SPE columns, whereas for urine usually a simple dilution step and/or centrifugation is satisfactory. Whatever SPE column is used, the analyst should keep in mind, that there are large differences from batch-to-batch, and that the same sorbents from different manufacturers also lead to different results [27]. Therefore, use of a suitable internal standard (e.g. deuterated analytes) is recommended.

Some SPE procedures for ASA and SA are reported in the literature. Procedures most often include use of C8, C18 or mixed-mode phase columns [29,36,37,100, 156–161]. Most of these procedures work quite well, so that the decision, what kind of column should be used for SPE, is often influenced by practical considerations (e.g. column type used in the laboratory for other determinations). Using HPLC, SPE can also be performed on-line [162,163]. A good overview over applications of solid-phase (micro-) extraction procedures can be found in several review articles [38–43].

8.3.3.2 Sample pretreatment and extraction of alternative matrices (oral fluid, sweat or hair)

Common aspects of sample pretreatment of alternative matrices are already discussed (cf. Section 8.2.3.2).

8.3.3.2.1 Sample preparation of oral fluid or sweat
Oral fluid and sweat require common extraction steps from the collection devices before analysis. Concentrations are lower than those, e.g., in urine. Therefore, the window of detection is shorter.

Oral fluid is collected by spitting or by using cotton swabs. Production of oral fluid can be stimulated by chewing (glass marbles, parafilm, teflone pieces) and/or by giving citric acid in the mouth. Oral fluid samples can be extracted in the same way as plasma samples. For separation and detection of ASA and SA, mainly HPLC-UV and HPLC with fluorometric detection were used [51,148,157].

Analysis of ASA or SA in sweat is hardly mentioned in the literature. Therefore, recommendations cannot be given. Common considerations on oral fluid and sweat testing have been reviewed by Kidwell *et al.* [55], and more recently by Verstraete [56].

8.3.3.2.2 Sample preparation of hair
Before extraction, decontamination using solvents (e.g. acetone, dichloromethane, petrolether, water) and homogenization (e.g. in a ball mill) are necessary. Cleavage of conjugates during the extraction procedure is usually performed enzymatically by addition of glucuronidase/arylsulfatase or by simple acid hydrolysis using hydrochloric acid. Determination of ASA or SA in hair is not described, so that special recommendations cannot be given.

8.3.3.3 Sample pretreatment and extraction of body tissues and other autopsy material

Extraction of ASA or SA from body tissues can be achieved by homogenizing the tissues in phosphate buffer followed by LLE or SPE as described above [150,164,165]. To improve the extraction rate, digestion steps using, e.g., papain, neutrase or collagenase can be useful [60]. Van Hoof *et al.* [164] added a clean-up step using hydrophilic–lipophilic balanced SPE phase (Oasis HLB) to a simple acetonitrile extraction of muscle tissue.

8.3.3.4 Sample pretreatment and extraction of non-biological samples

Pure substances can usually be solved in solvents (e.g. methanol, ethanol, acetone, ethyl acetate, diethyl ether) and directly be analyzed by GC or LC procedures. The same extraction methods, which are suitable for biomatrices, can also be applied to extraction of paracetamol from non-biological matrices such as beverages, food or pharmaceutical formulations [166–173]. Special problems, such as determination of ASA from pressurized aerosol solutions in propellants for inhalation, are also described. For this purpose, it was necessary to separate ASA and surfactant, a necessary adjuvant in such preparations [174]. Di Pietra *et al.* [175] proposed diode-array detection (DAD) and on-line post-column photochemical derivatization after HPLC separation for determination of analgesics from pharmaceutical dosage forms. The characteristic photo-induced spectral modifications were useful for unambiguous identification of the various analgesic compounds. In Fig. 8.5, corresponding UV spectra are displayed. Mass spectrometric detection delivers the highest specificity. With LC-MS becoming a lab standard technique, detection of ASA and metabolites in non-biological samples has also been described [141,176–178]. Cartwright *et al.* [177] even employed a derivatization step (Tris(2,4,6-trimethoxyphenyl)phosphonium propylamine bromide) for the acidic compounds before LC-MS. However, reaction efficiencies for salicylic acid were poor.

8.3.3.5 Derivatization for GC and GC-MS

GC procedures for detection of ASA and SA are seldom described. Because of the free phenolic group and/or the carboxylic acid group, derivatization is necessary for determination of ASA and SA. Acetylation of the phenol and/or methylation of the carboxylic function as well as trimethylsilylation of the phenol and/or the acid function are used [17,67,83,155,176,179]. Acetylation allows no further differentiation of ASA and SA, since SA is acetylated to ASA. Pentafluorobenzylation is seldom used [180]. For more information on more exotic derivatization procedures such as combined propionylation/propylation, the review of Segura *et al.* or Halket and Zaikin, Halket *et al.* and Zaikin and Halket on derivatization procedures are recommended [68–74]. Details on extractive alkylation for extraction and derivatization of acidic drugs in one step are given in Section 8.4.3.5.

8.3.4 Analysis of acetylsalicylic acid

Concentrations of ASA in plasma and urine are very high. In case of intoxications plasma concentrations of up to 1000 mg/l can be observed. Therefore, colorimetric assays are described for fast detection, when other methods are not available. The known Trinder test and some modifications of this test were published [143, 181–183]. Comparative studies were published, showing that even quantitative results

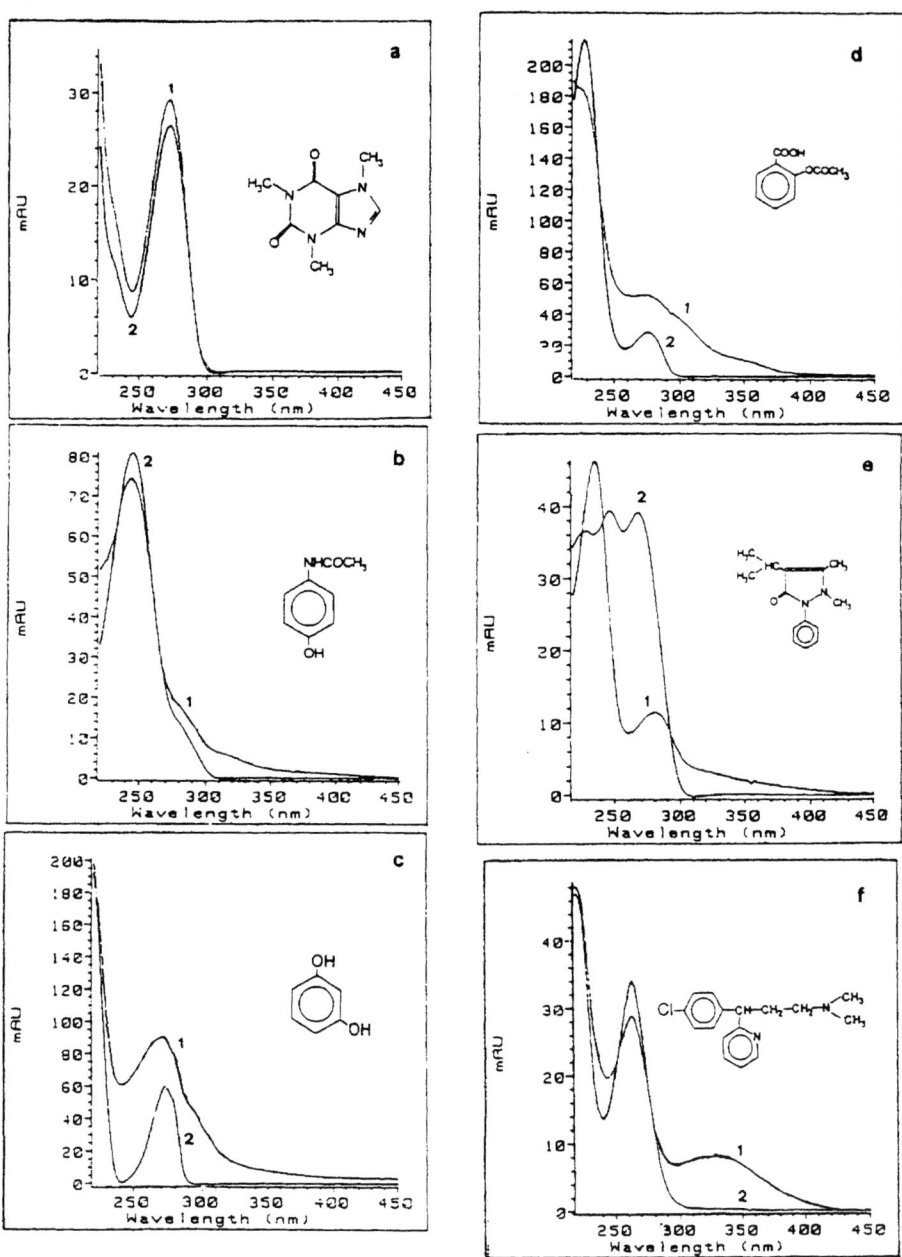

Fig. 8.5. UV spectra of analgesics and related compounds with on-line photoreactor switched (1) on and (2) off. (a) Caffeine; (b) paracetamol; (c) resorcinol; (d) acetylsalicylic acid; (e) propyphenazone; (f) chlorpheniramine (taken from Ref. [175]).

are in accordance with those of chromatographic procedures [184,185]. Commercially available immunoassays provide sufficient quantitative information for diagnosis of a salicylate intoxication. In forensic toxicology, positive immunoassay results must be confirmed by a second independent method that is at least as sensitive as the screening test and that provides the highest level of confidence in the result.

8.3.4.1 GC and GC-MS procedures

ASA and salicylic acid are seldom determined using GC or GC-MS techniques. Fused-silica capillary columns are suitable for separation of (derivatized) ASA and SA. Standard low-polarity dimethylpolysiloxane types of column were used, if ASA and SA should be detected during systematic toxicological analysis [83]. Polar columns of porous polymers (Porapak Q, S) were employed for quantitation of SA and analogues [186,187]. The detector of choice is MS, providing good sensitivity and best specificity [17,83,155,176,180,188–190].

8.3.4.1.1 Screening and confirmation of acetylsalicylic acid

Screening procedures for ASA and SA should employ GC-MS because of its superior sensitivity and specificity. Laakkonen *et al.* [191] used GC-MS for screening for NSAIDs, barbiturates and methyl xanthines in equine urine. Screening and confirmation of non-opioid analgesics, including ASA and SA, can be performed in one step using GC-MS in the EI, full-scan mode. Such a procedure, including universal sample preparation for many different drug classes (acid hydrolysis, LLE and acetylation) is part of a general screening procedure described by Maurer [13]. For screening and confirmation mass chromatography based on full-scan recording was used. The identity of the peaks in the mass chromatograms was confirmed by computerized comparison of the peaks underlying mass spectra with reference spectra [14,92]. Besides non-opioid analgesics such as ASA and SA, many other groups of medicaments can simultaneously be covered by this procedure. Indeed, this universal method was not specially designed for acidic drugs such as ASA or SA. However, because of the relatively high concentrations of these drugs, sensitivity is still sufficient for screening, even at low extraction rates. In addition, methylation occurs in the injection port, since methanol was used as solvent for the evaporated extracts. Maurer used extractive methylation for a screening procedure for detection of NSAIDs and their metabolites in urine as part of a systematic toxicological analysis procedure for acidic drugs and poisons by gas chromatography–mass spectrometry [17]. This technique proved to be a versatile method for STA of various acidic drugs, poisons and their metabolites in urine and it has also successfully been used for plasma analysis.

8.3.4.1.2 Quantification of acetylsalicylic acid in blood, plasma or serum

GC-MS should also be suitable for quantification of ASA or SA. However, instead of full-scan mode SIM mode should be used, which provides higher sensitivity and precision [155,180,189]. Further prerequisite for precise quantification is the use of suitable internal standards [180,189]. Best results can be achieved with deuterated internal standards [180].

8.3.4.2 HPLC and LC-MS procedures

HPLC procedures are most often used for the analysis of ASA or SA. Common C18 or C8 packing materials for LC columns have been most widely used and can be recommended for separation of ASA or SA. Isocratic or gradient elution procedures using acetonitrile or methanol/buffered water mixtures or mixtures of solvents (acetonitrile, methanol, isopropanol) with phosphate or acetate buffers resulted in sufficient separation. Acidic pH (down to pH 2.5) is necessary to avoid deprotonation of the carboxylic acid group of ASA and SA [100,137,138,140,147,148,152,163, 192–197]. Separation of ASA or SA could also be achieved using an ion-pair reagent (2-amino-2-(hydroxy-methyl)-1,3-propanediol), which forms a water-soluble ion-pair with SA. Separation is achieved on a usual C18 column [119]. For LC-MS volatile solvents and buffers such as methanol, acetonitrile, formate-buffers have to be used [29,164,177,198].

8.3.4.2.1 Screening and confirmation of acetylsalicylic acid

Concentrations of ASA and SA in plasma and urine are very high. Therefore, screening does not require sensitive techniques. Simple UV detectors provide enough sensitivity for detection with the disadvantage of low specificity.

Screening for drugs of abuse in plasma or serum was described using HPLC-UV-DAD [108]. This assay was capable of detecting and identifying therapeutic and toxic amounts of NSAIDs (including SA), barbiturates, anti-convulsants, diuretics, sulfonylurea anti-diabetic drugs, theophylline and analgesic drugs. Nevertheless, MS detection provides best specificity, which is important, especially in forensic toxicology. LC-MS procedures are available for screening and quantitation of non-opioid analgesics in biomatrices [29,140,141,198].

8.3.4.2.2 Quantification of acetylsalicylic acid in blood, plasma or serum

For separation and quantification of ASA and SA in plasma, numerous LC procedures were described with different detection modes [100,102,108,119,138, 140,141,147,152,163,195–197,199–202]. Since therapeutic plasma concentrations of SA are relatively high, requirements on the analytical technique are not very high. Therefore, simple UV detection may be sufficient for determination of SA in biosamples in terms of sensitivity. Requirements on specificity are however not met. With LC-MS becoming a lab standard technique, detection of ASA and metabolites in non-biological samples has also been described [141,176–178].

8.3.4.3 CE procedures

ASA and SA are also ideal candidates for determination by CE due to their very high plasma and urine levels. CE is a growing analytical technology. The great interest raised by CE is due to its high efficiency, mass sensitivity, minimum needs of solvent and sample volumes and in particular to the high versatility in terms of separation modes. However, sample matrix components and salts can have deleterious effects on CE separations. For urine, dilution of the sample can help, but this procedure works only when analytes such as paracetamol are present in high concentrations [127].

Plasma or serum can be analyzed by direct injection techniques [130] or after deproteinization and/or extraction [3,132,203]. Deproteinization should not be performed using acids or salts, which would have deleterious effects on CE separations. Solvents such as acetonitrile are suitable. Sample preparation techniques used in CE are often transferred from LC or GC methods. However, mainly the small volumes to handle cause problems, when using devices not specifically tailored for such purpose. Some CE techniques allow sample enrichment during injection, known as sample stacking. In fact, a specific strategy for sample preparation in CE has not yet been fully developed.

The separation of ASA and SA was demonstrated in a non-aqueous CE system with reversed electroosmotic flow. The flow was reversed by the addition of the polycation hexadimethrine bromide and thus a negative voltage was used. This separation method was applied to the assay of ASA and its major metabolites in plasma and urine [204].

For detection of ASA or SA after CE separation, UV detectors [130,203,204] or a combined fluorescence/absorbance detector [205], UV-DAD detectors [3] or laser-induced fluorescence detectors [206,207] have been used. CE-MS was also employed for direct assay of non-opioid analgesics and their metabolites including ASA and SA in human urine [123].

The status of CE techniques in forensic toxicology has been displayed in detail in the review of Tagliaro and Bortolotti [135]. Today, CE techniques are still not very widely used in forensic toxicology. The hopes that all problems of the method will be solved and CE may be a useful tool in forensic toxicology did not come true. In a more recent review, Pucci and Raggi [136] concluded again that electrokinetic chromatographic methods might become a robust and widely accepted method and that their use might continue to grow.

8.4 NON-STEROIDAL ANTI-INFLAMMATORY DRUGS

8.4.1 Introduction

In this chapter, the so-called non-steroidal anti-inflammatory drugs (NSAIDs) of different types are summarized. From a pharmacological point of view, they act as (non-selective) inhibitors of the cyclooxygenase (COX) system. Since the discovery of a second subtype of cyclooxygenase (COX-2), selective inhibitors (the so-called "coxibs") were developed. The COX-2 isoform was claimed to be inducible at the site of inflammation whereas COX-1 should be expressed constitutively in several tissues including gastric epithelium, thus being responsible for the gastrointestinal side effects of NSAIDs. Indeed, coxibs have caused less ulcerations of the gastro-intestinal mucosa compared to non-selective COX inhibitors. Increase of cardio-vascular events lead to withdrawal of rofecoxib (Vioxx) from the market. Discussion on safety of coxibs is still going on.

In Figs. 8.6 and 8.7, the structures of arylacetic acid derivatives such as indo-methacin or diclofenac, arylpropionic acid derivatives such as ibuprofen, naproxen or ketoprofen, oxicames such as piroxicam and some coxibs are given. For reasons of space, common comments to each subchapter have been renounced. Therefore,

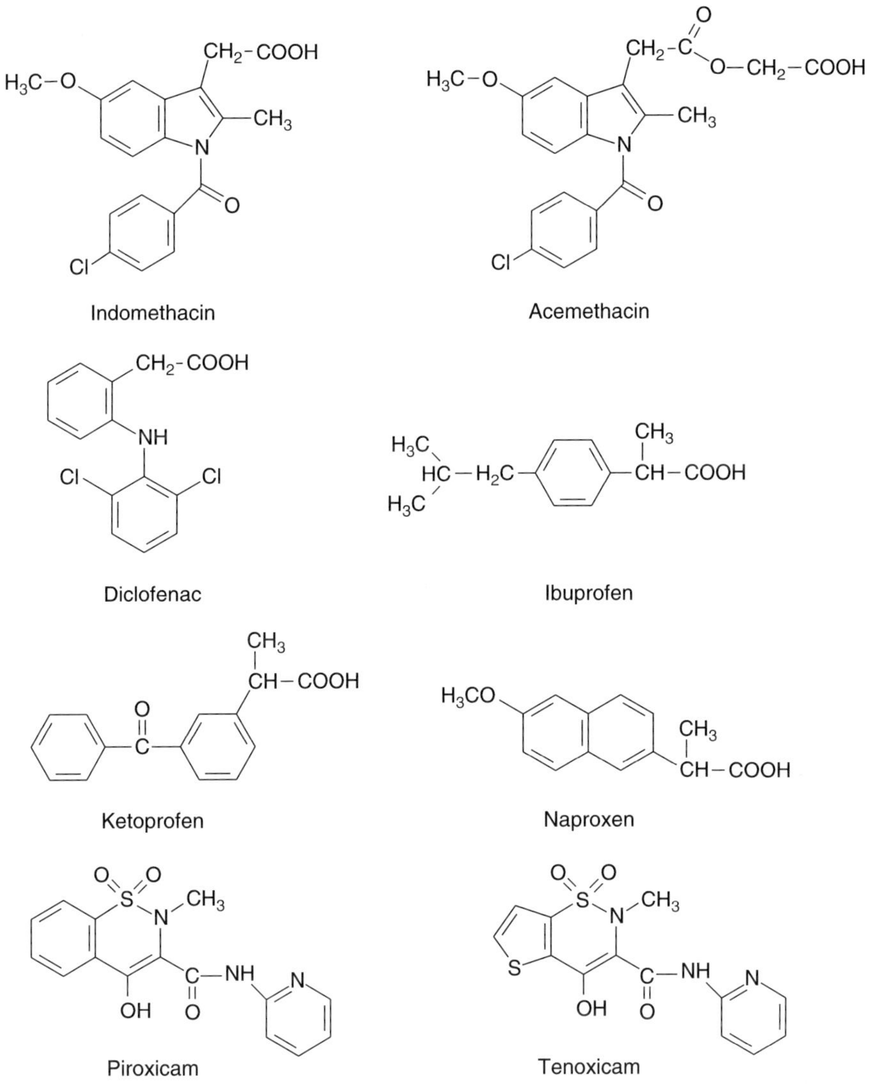

Fig. 8.6. Structures of some non-steroidal anti-inflammatory drugs (NSAIDs).

References pp. 348–356

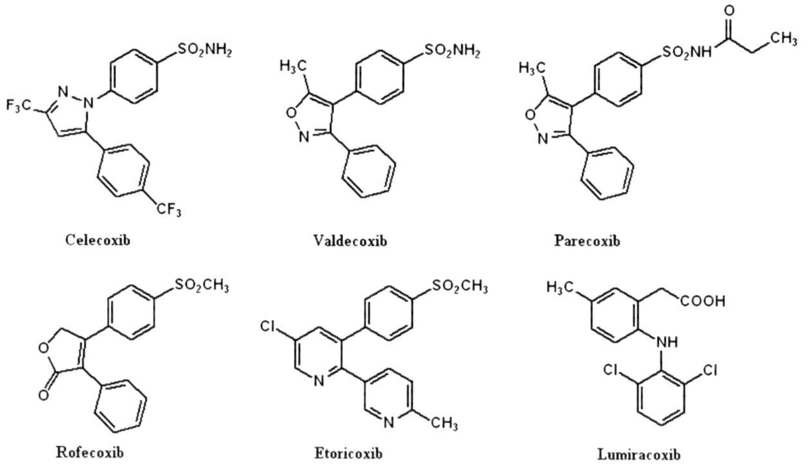

Fig. 8.7. Structures of some coxibs.

the reader should read these universal comments to subchapters, such as sample preparation of certain matrices, GC, LC or CE in the paracetamol or ASA sections.

8.4.2 Structural features of some NSAIDs

As shown in Fig. 8.6, the arylacetic and arylpropionic acid derivatives have free carboxylic acid groups, which determine the analytical properties of these substances. The oxicames also are acidic drugs, due to keto-enol tautomery. The mean daily doses are from 100 mg (diclofenac) to about 1800 mg (ibuprofen) for the arylcarboxylic acids and about 20 mg for the oxicames, so that plasma and urine levels are relatively high. Plasma half-life values are short for the arylcarboxylic acids (1.5–12 h) and much longer for the oxicames (70 h for tenoxicam).

As shown in Fig. 8.7, the coxibs belong to two distinct chemical classes: the diaryl-substituted cycles class for celecoxib, rofecoxib, valdecoxib (and parecoxib sodium, as a valdecoxib prodrug) and etoricoxib and the phenylacetic acid class for lumiracoxib (similar to diclofenac). Rofecoxib and etoricoxib contain a methylsulfone moiety while celecoxib and valdecoxib possess a sulfonamide group. The mean daily doses for the coxibs are from 40 mg (parecoxib) to about 400 mg (celecoxib) with terminal half-life values ranging from 6 h (lumiracoxib) to 22 h (etoricoxib). Many of the NSAIDs are chiral drugs. Most of them are marketed as racemates. It is known that the enantiomers have different pharmacodynamic and pharmacokinetic properties. The anti-inflammatory activity of NSAIDs has been shown to be largely stereospecific for the S-enantiomers [208]. However, this stereoselectivity of action is not manifest in vivo, due to the thus-far-unique unidirectional metabolic inversion of the chiral centre from the inactive R(−)-isomers to the *S*(+)-antipodes [208].

8.4.3 Sample preparation

8.4.3.1 Sample pretreatment and extraction of blood (serum, plasma) or urine

Common aspects of sample pretreatment of blood (serum, plasma) or urine are discussed in Sections 8.2.3.1 and 8.3.3.1. As described below, liquid–liquid and solid-phase procedures are used for extraction of NSAIDs from biomatrices.

Many different solvents were used for extraction of NSAIDs from biological matrices [7,22,140,145,191,209–219]. Since all these procedures work quite well with sufficient recoveries, no special recommendation is given here. The use of the toxic chloroform [212,220,221] should be avoided. Extraction and derivatization in one step can be achieved by extractive alkylation (cf. Section 8.4.3.5).

SPE procedures for NSAIDs are also reported in the literature [222–232]. Procedures include use of C8, C18 or mixed-mode Bond-Elut Certify® columns. Again, a good overview over applications of solid-phase (micro-) extraction procedures can be found in several review articles [38–43].

8.4.3.2 Sample pretreatment and extraction of alternative matrices (oral fluid, sweat or hair)

Common aspects of sample pretreatment of alternative matrices are already discussed in Sections 8.2.3.2 and 8.3.3.2. Detection of NSAIDs in oral fluid and sweat seems not to be of great interest, since it is hardly mentioned in the literature. Special recommendations cannot be given. Detection of some NSAIDs in hair integrated in a screening procedure using LC-DAD and GC-MS procedures is discussed in only one paper [57].

8.4.3.3 Sample pretreatment and extraction of body tissues and other autopsy material

Extraction of NSAIDs from body tissues can be achieved by homogenizing the tissues in phosphate buffer [233] followed by LLE or SPE as described above. To improve the extraction rate, digestion steps using, e.g., papain, neutrase or colla-genase can be useful [60]. Detection of NSAIDs in postmortem blood samples as part of screening procedures for different groups of drugs using HPLC-DAD is described in several papers [22,108,234].

8.4.3.4 Sample pretreatment and extraction of non-biological samples

Pure substances can usually be solved in solvents (e.g. methanol, ethanol, acetone, ethyl acetate, diethyl ether) and directly be analyzed by GC or LC procedures. The same extraction methods, which are suitable for biomatrices, can also be applied to extraction of NSAIDs from non-biological matrices such as beverages, food or pharmaceutical formulations [89,93,235–247]. Simple spectrophotometric detection of diclofenac, retained on a sephadex column [248] is described as well as normal-phase HPLC-UV [249] or even exotic methods such as packed column supercritical fluid chromatography with UV detection [250]. An extensive survey of the literature on the instrumental analytical methods for determination of COX-2 inhibitors in bulk drugs, formulations and biological fluids has been given by Nageswara [251]. NSAIDs are very widely used. Thus, they may become a problem for the environment via natural excretion with urine or feces. A lot of papers have been published dealing with detection of NSAIDs (and other therapeutically used substances) in river water, wastewater (especially from hospitals) or river sediment samples [28,176,178,252–267]. As only trace levels of drugs can be found in these matrices sensitive detection is necessary, which can be achieved by LC-MS/MS techniques [28,178,253–255,263,264], by GC-MS/MS techniques [176] or by simple GC-MS after special sample preparation (e.g. large volume online derivatization) [252,259,261,265].

8.4.3.5 Derivatization for GC and GC-MS

Derivatization of NSAIDs before GC is recommended to improve chromatographic properties and to avoid thermal decarboxylation of the drugs in the injection port of the GC. Most often, methylation is used [83,145,191,219,268]. The quantitative conversion of 26 NSAIDs to their corresponding *tert*-butyldimethylsilyl derivatives was examined by Kim *et al.* [269].

An effective extraction and derivatization method for acidic drugs is the so-called extractive alkylation, which was described for different groups of drugs such as anticoagulants, antidiabetics, calcium antagonists, ACE inhibitors, diuretics, laxatives and NSAIDs [270–276]. Mass spectra of such derivatives can be found in Refs. [14,92]. Maurer used this elegant technique for a screening procedure for detection of NSAIDs and their metabolites in urine as part of a systematic toxicological analysis procedure for acidic drugs and poisons by gas chromatography–mass spectrometry [17]. Again, for more information also on more exotic derivatization procedures such as combined propionylation/propylation, the reviews of Segura *et al.* or Halket and Zaikin, Halket *et al.* and Zaikin and Halket on derivatization procedures are recommended [68–74].

8.4.4 Analysis of NSAIDs

8.4.4.1 GC and GC-MS procedures

Fused-silica capillary columns are suitable for separation of (derivatized) NSAIDs. Standard dimethylpolysiloxane types of column were used [83,145,191,219]. The detector of choice is MS, providing good sensitivity and best specificity [17,83,145,176,191,219,252,259,268,277].

Screening procedures for NSAIDs should employ GC-MS because of its superior specificity. Laakkonen *et al.* [191] used GC-MS for screening for NSAIDs, barbiturates and methyl xanthines in equine urine, while Gonzalez *et al.* [145] tested equine plasma for NSAIDs using GC-MS. Maurer *et al.* [219] employed GC-MS for detection of ibuprofen and its metabolites in human urine. In Fig. 8.8, detection of ibuprofen and its metabolites by this GC-MS method in the full-scan mode after methylation is shown. Mass chromatograms indicating the presence of the analytes

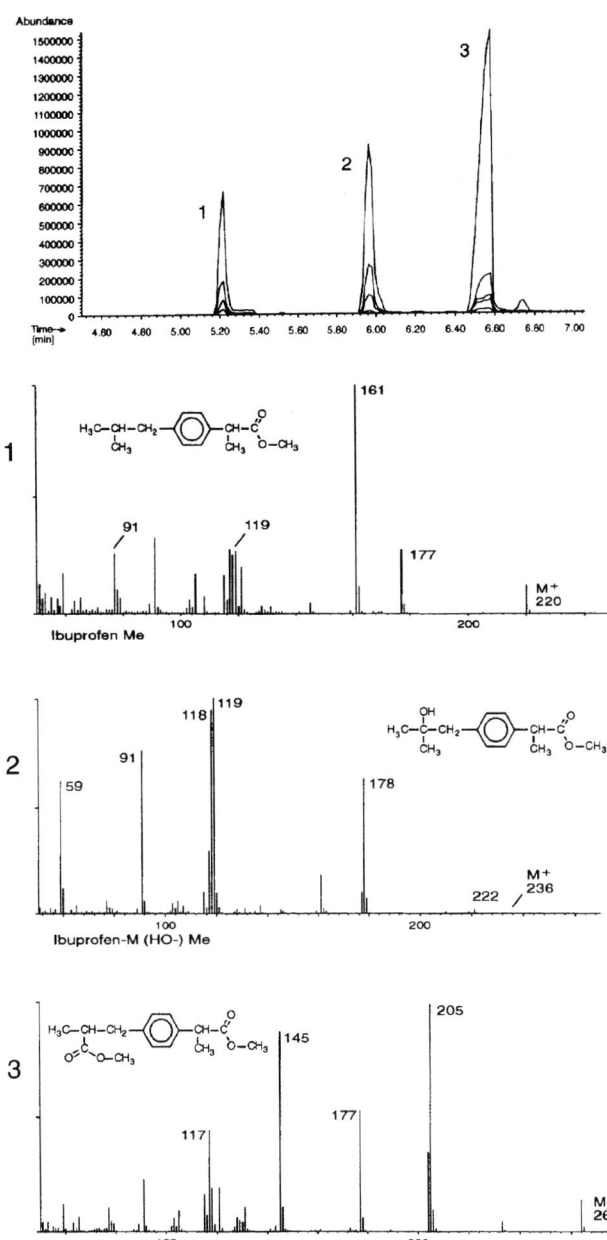

Fig. 8.8. Typical mass chromatograms with the masses m/z 264, 220, 178, 161, 145 and 119 indicating methylated ibuprofen (peak 1), methylated hydroxy ibuprofen (peak 2) and methylated carboxy ibuprofen (peak 3) in urine. The merged mass traces can be differentiated by their different colors on a color screen. In the bottom part reference full mass spectra of methylated ibuprofen (1), methylated hydroxy ibuprofen (2) and methylated carboxy ibuprofen (3) are displayed (taken from Ref. [219]).

References pp. 348–356

and the peaks underlying mass spectra are given. Screening and confirmation of non-opioid analgesics, including some NSAIDs, can be performed in one step using GC-MS in the EI, full-scan mode after extractive methylation [268]. The general screening procedure of Maurer [13,13,81,82,278], which is described above (cf. Section 8.2.4.1 and 8.3.4.1), also covers some NSAIDs. This universal method was not specially designed for this purpose. However, because of the relatively high concentrations of these drugs, sensitivity is still sufficient for screening, even at low extraction rates. In addition, methylation occurs in the injection port, since methanol is used as solvent for the evaporated extracts. A systematic toxicological analysis procedure for acidic drugs and poisons by GC-MS was established by Maurer *et al.* [17] using extractive methylation for their screening procedure for detection of NSAIDs and their metabolites in urine. This method allowed the detection of therapeutic concentrations of acemetacin, acetaminophen (paracetamol), ASA, diclofenac, diflunisal, etodolac, fenbufen, fenoprofen, flufenamic acid, flurbiprofen, ibuprofen, indometacin, kebuzone, ketoprofen, lonazolac, meclofenamic acid, mefenamic acid, mofebutazone, naproxen, niflumic acid, phenylbutazone, suxibuzone, tiaprofenic acid, tolfenamic acid and tolmetin in urine samples. GC-MS should also suitable for quantification of NSAIDs. However, instead of full-scan mode SIM mode should be used, which provides higher sensitivity and precision. Further prerequisite for precise quantification is the use of suitable internal standards. Best results can be achieved with deuterated internal standards.

Some NSAIDs are chiral drugs. The enantiomers may show different pharmacodynamic and pharmacokinetic properties. Therefore, enantioselective analysis may be necessary. Chiral derivatization to diastereomers (e.g. diastereomeric (R)-$(+)$-1-phenylethylamides) and separation on usual non-chiral columns can be used [279–281] as well as separation of the enantiomers on chiral stationary phases [277].

8.4.4.2 HPLC and LC-MS procedures

Many HPLC procedures for screening and quantification purposes have been described in the last years. Common C18 or C8 packing materials for LC columns have been most widely used and can be recommended for separation of NSAIDs. Isocratic or gradient elution procedures using standard solvents mixtures result in sufficient separation. Mobile phase systems containing ion-pair reagents were seldom used for determination of NSAIDs. Detection is preferably done by DAD and of course by mass spectrometry.

Simple UV detection is not suitable for screening purposes. Because of its higher specificity DAD is preferred [22,108,231,234,282–286]. However, the universality and specificity of GC-MS for STA [13,81–83] is not reached.

For separation and quantification of NSAIDs, several LC procedures were described with UV or UV-DAD detection [225,231,287,288]. Fig. 8.9 shows corresponding HPLC-UV chromatograms of extracts of (a) a blank plasma sample, (b) a quality control sample and (c) of a plasma sample after administration of 100 mg of celecoxib (taken from Ref. [225]).

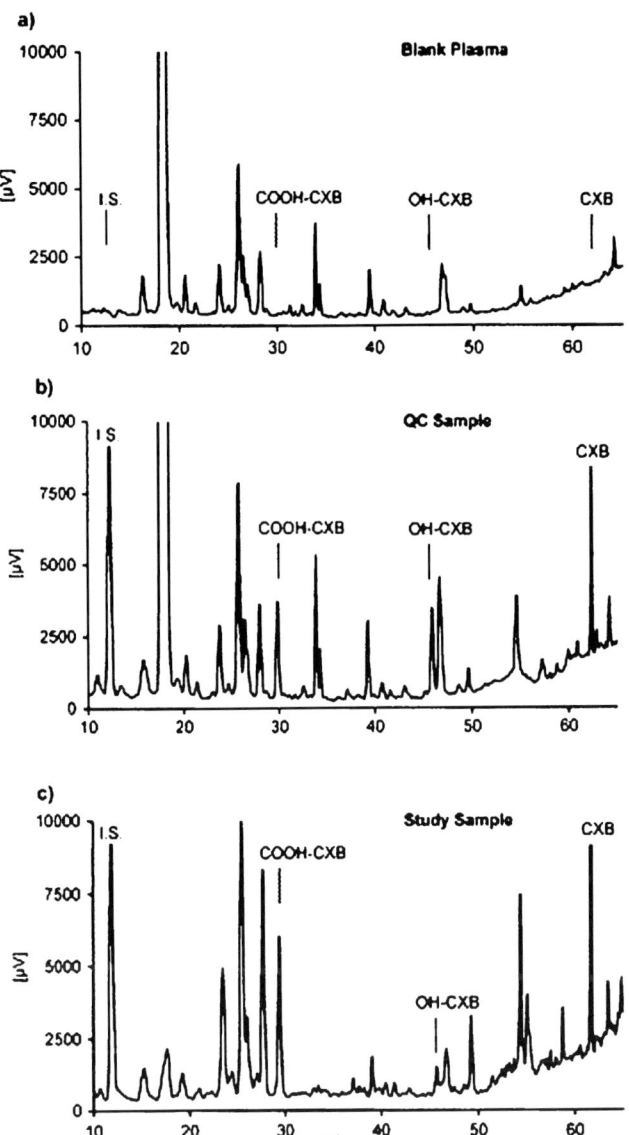

Fig. 8.9. HPLC traces for celecoxib (CXB), hydroxycelecoxib (OH-CXB), carboxycelecoxib (COOH-CXB) and phenacetin (I.S.). (a) Blank plasma sample, (b) quality control sample (80 ng/ml), (c) plasma sample 1.5 h post-administration of 100 mg celecoxib (152 ng/ml COOH-CXB, 18.9 ng/ml OH-CXB, 269 ng/ml CXB) (taken from Ref. [225]).

Especially for pharmacokinetic determinations, LC-MS/MS has proved to be a powerful tool. Already in 1998, Beaudry *et al.* [289] described a completely automated procedure using the Prospekt-LC-APCI/MS/MS system for quantifying NSAIDs (and many other pharmaceuticals) in biofluids. Today, a lot of LC-MS procedures for determination of NSAIDs in body fluids have been published [214,217,224,226,290–296].

As described above, some NSAIDs are chiral drugs. The enantiomers may show different pharmacodynamic and pharmacokinetic properties. Therefore, enantioselective analysis is performed. Chiral derivatization to diastereomers or addition of chiral mobile phase additives (e.g. norvancomycin [297]) and separation on usual non-chiral columns can be used as well as separation of the enantiomers on chiral stationary phases [214,298–302]. Detailed information on methods of analysis of chiral NSAIDs were given in the reviews of Davies [303] or Mullangi [304].

8.4.4.3 CE procedures

Common aspects of CE are already discussed above (Sections 8.2.4.3 and 8.3.4.3). Therefore, only special procedures are displayed here. For determination of NSAIDs in plasma or serum direct injection techniques were used [305,306]. Acetonitrile deproteinization prior to CE is also described [307,308]. Caslavska *et al.* [205] adapted a conventional, tunable UV–Vis absorbance detector for simultaneous fluorescence and absorbance detection. An electrokinetic immunoassay performed in a chip-based CE system for determination of naproxen in plasma was described by Phillips [309]. In this system, the analyte is captured by a fluorescently labeled antibody. Separation of NSAID enantiomers by CE methods using chiral selectors has successfully been done [216,310–312]. For detection of NSAIDs after CE separation, modern methods such as laser-induced fluorescence [233,305] or mass spectrometry [123,156,313] seem to prevail since they may overcome the disadvantages of CE in terms of sensitivity. Albrecht and Thormann [233] determined naproxen in liver and kidney tissues after homogenization using laser-induced fluorescence detection. Again, the status of CE techniques in forensic toxicology has been displayed in detail in the review of Tagliaro and Bortolotti [135].

8.5 PYRAZOLE DERIVATIVES

In this chapter, different pyrazole derivatives are summarized. In Fig. 8.10, the structures of (a) phenazone and derivatives such as propyphenazone and metamizol are given as well as (b) pyrazolidindione derivatives such as phenylbutazone and oxyphenbutazone, which is also metabolite of phenylbutazone. For reasons of space, common comments to each subchapter have been renounced. Therefore, the reader should read these universal comments to subchapters such as, e.g., sample preparation of certain matrices, GC, LC or CE in the paracetamol or ASA sections. As can be seen in Fig. 8.10, the phenazones do not have acidic properties. They are used as analgesics and antipyretics. The pyrazolidindione show additional antiphlogistic

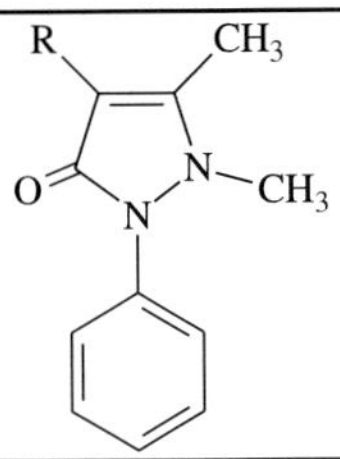

(a)

Name	R
Phenazone	-H
Propyphenazone	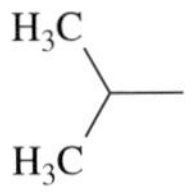 H_3C / H_3C (isopropyl)
Aminophenazone	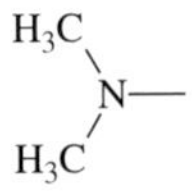 H_3C / H_3C N— (dimethylamino)
Metamizol	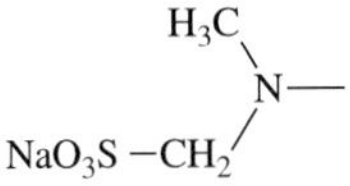 H_3C / NaO_3S-CH_2 N—

(b)

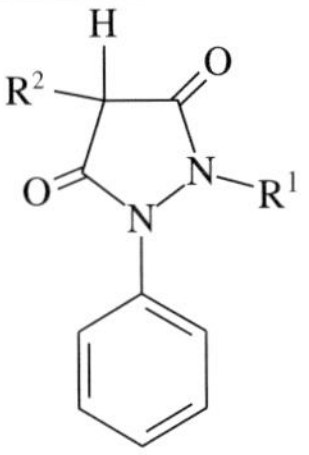

Name	R^1	R^2
Phenylbutazone	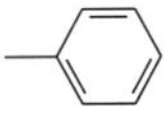 —phenyl	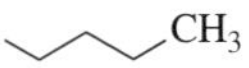 $\sim\sim CH_3$
Oxyphenbuta-zone	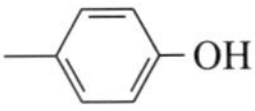 —phenyl—OH	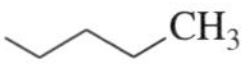 $\sim\sim CH_3$

Fig. 8.10. Structures of some pyrazole analgesics. (a) phenazone derivatives and (b) pyrazolidinedione derivatives.

References pp. 348–356

effects because of their acidic properties. The mean daily doses are from at least 100 mg (oxyphenbutazone) to about 1000 mg (phenazone and metamizol), resulting in relatively high plasma concentrations. Plasma half-lifes are short for the arylcarboxylic acids (1.5–12 h) and much longer for the pyrazolidindiones (70 h for phenylbutazone).

SPE and LLE procedures are described for extraction of pyrazoles from biological matrices. The phenazone derivatives are not acidic, so that they appear in usual screening methods for basic drugs. The pyrazolidindiones are best extracted at acidic pH. Because of their high concentrations in biomatrices, they are covered by usual screening procedures. Thus, screening of pyrazoles in biomatrices (partly of horses) is described using GC-MS in the EI full-scan mode [13,17,83,145], HPLC-DAD methods [234,284–286,314,315] and LC-MS [58,164,198].

Quantification in plasma is often performed using HPLC-UV or HPLC-DAD [316–322]. GC-MS should also be suitable for this purpose. However, instead of full-scan mode SIM mode should be used, which provides higher sensitivity and precision. Further, prerequisite for precise quantification is the use of suitable internal standards. For precise determination of trace amounts of phenazone and metabolites GC tandem MS and deuterated internal standards were also employed [323]. CE-MS was employed for direct assay of non-opioid analgesics and their metabolites including phenazone and propyphenazone in human urine [123]. In addition to their use for therapeutic purposes, pyrazoles and NSAIDs are used in animal doping (race horses, fighting bulls) or in animal production to improve quality characteristics of meat. Consequently, procedures for determination of such drugs in animal matrices have been published. They should also be suitable for human matrices. LC-MS is often used for these purposes [58,164,198,317]. Propyphenazone was determined even in beard hair of migraine patients, taking the drug [324].

Famprofazone (Fig. 8.11) is a pyrazole derivative of special interest in forensic toxicology, because the main metabolite is methamphetamine. Several GC(-MS) papers on the metabolism and the detection of famprofazone were published in the last years [325–331]. Detection of unchanged parent compound allowed the differentiation of famprofazone ingestion from illicit methamphetamine abuse [325]. More on this topic can be found in the reviews of Cody *et al.* [332–334].

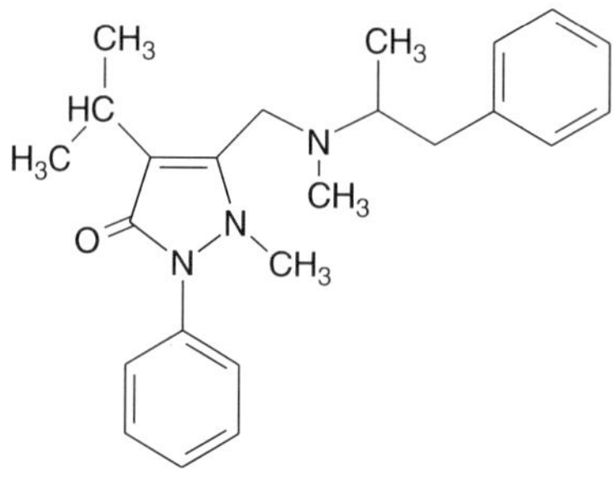

Fig. 8.11. Structure of famprofazone.

8.6 CONCLUDING SUMMARY AND PERSPECTIVES

Non-opioid analgesics are among the most commonly consumed over-the-counter preparations all over the world. Although they are perceived to be safe drugs, they may lead to severe intoxications in case of acute overdosage or in case of chronic abuse. Most of the non-opioid analgesics are acidic drugs, so that their analytical properties are similar. Both methods, LLE at suitable pH and SPE work well for extraction of such drugs from biomatrices. If the problem of batch-to-batch inconformity of the SPE columns is monitored (e.g. using an internal standard), both methods for extraction can be recommended. Separation of non-opioid analgesics works well using GC and LC techniques. Derivatization before GC is necessary. Parent compounds and/or phase I metabolites can be conjugated with glycine, glucuronic acid or sulfuric acid. Therefore, cleavage of glucuronide or sulfate conjugates may be necessary before analysis to improve the detection window in urine samples. In forensic toxicology, enzymatic hydrolysis using glucuronidase (and arylsulfatase) is recommended. Unfortunately, enzymatic hydrolysis is time-consuming and expensive. If results are requested in a short time (e.g. in clinical toxicology) rapid acid hydrolysis using hydrochloric acid is more convenient. However, forming of artifacts (e.g. ester hydrolysis, decarboxylation) must be considered. The analgesics, their deconjugated metabolites and/or their hydrolysis artifacts must be derivatized prior to GC separation, which can be performed on standard fused-silica capillary such as low polarity dimethylpolysiloxane types for systematic toxicological analysis or 5–10% diphenyl methylpolysiloxane types for special methods. For LC separation standard reversed-phase columns are suitable.

Today, mass spectrometric detectors for GC and LC have prevailed. Mass spectrometric detection provides best specificity of all methods, at least in the EI full-scan mode for qualitative analysis. For quantification, GC-MS should be run in the selected-ion monitoring mode while LC-MS/MS should be run in multiple reaction monitoring (MRM) mode.

For quantification, suitable internal standards should be chosen. They can compensate for variability due to sample preparation, chromatography, or even ion suppression/enhancement and thus improve accuracy and precision data. As in any MS-based analytical methods, stable-isotope labeled analogues of the analytes are ideal IS. At this point, it must be stressed that if no stable-isotope-labeled IS of the analyte is available and an alternative IS must be chosen, one should always avoid choosing a therapeutic drug for this purpose, at least not in analytical toxicology. In this field, it can never be excluded that the patient or defendant to be monitored has taken this drug. In such cases, the peak area of the IS would be overestimated leading to underestimation of the analyte concentration.

CE procedures for detection of drugs of forensic interest are more and more published. However, it must clearly be stated, that CE is still not a universal method. A strategy for sample preparation in CE has not yet been fully developed. There still are limitations in terms of reproducibility and concentration sensitivity. Mass spectrometric detection may help to solve these problems.

References pp. 348–356

In summary, GC-MS, especially in the electron ionization full-scan mode, is still the method of choice for comprehensive screening providing best separation power, specificity and universality, although requiring derivatization. LC-DAD is also often used for screening, but its separation power and its specificity are still inferior to those of GC-MS. Finally, LC-MS has shown to be an ideal supplement, especially for the detection of more polar, thermolabile and/or low-dose drugs, especially in blood plasma. It may become the gold standard in clinical and forensic toxicology and doping control if the costs of the apparatus will be markedly reduced, the current disadvantages like irreproducibility of fragmentation, reduction of ionization by matrix, etc. will be overcome, and finally if one of the increasing number of quite different techniques will become the apparatus standard.

8.7 REFERENCES

1   B. Hinz and K. Brune, Handb. Exp. Pharmacol., 177 (2007) 65.
2   S. White and S.H. Wong, Clin. Chem., 44 (1998) 1110.
3   K. Makino, T. Yano, T. Maiguma, D. Teshima, T. Sendo, Y. Itoh and R. Oishi, Ther. Drug Monit., 25 (2003) 574.
4   T. Matsumoto, T. Sano, T. Matsuoka, M. Aoki, Y. Maeno and M. Nagao, J. Anal. Toxicol., 27 (2003) 118.
5   E. Pufal, M. Sykutera, G. Rochholz, H.W. Schutz, K. Sliwka and H.J. Kaatsch, Fresenius. J. Anal. Chem., 367 (2000) 596.
6   P.M. Kabra, B.E. Stafford and L.J. Marton, J. Anal. Toxicol., 5 (1981) 177.
7   F. Nielsen-Kudsk, Acta Pharmacol. Toxicol., 47 (1980) 267 (Copenh).
8   J.C. West, J. Anal. Toxicol., 5 (1981) 118.
9   L.J. Brunner and S. Bai, J. Chromatogr. B, Biomed. Sci. Appl., 732 (1999) 323.
10 V. Bari, U.J. Dhorda and M. Sundaresan, Indian Drugs, 35 (1998) 222.
11 R. Whelpton, K. Fernandes, K.A. Wilkinson and D.R. Goldhill, Biomed. Chromatogr., 7 (1993) 90.
12 P. Lillsunde, L. Michelson, T. Forsstrom, T. Korte, E. Schultz, K. Ariniemi, M. Portman, M.L. Sihvonen and T. Seppala, Forensic Sci. Int., 77 (1996) 191.
13 H.H. Maurer, Spectrosc. Europe, 6 (1994) 21.
14 H.H. Maurer, K. Pfleger and A.A. Weber, Mass Spectral and GC Data of Drugs, Poisons, Pesticides, Pollutants and their Metabolites, Wiley-VCH, Weinheim, 2007.
15 H.H. Maurer, In: K. Pfleger, H.H. Maurer, A. Weber (Eds.), Mass spectral and GC data of drugs, poisons, pesticides, pollutants and their metabolites, VCH publisher, Weinheim, 1992, p. 3.
16 X. Chen, J. Huang, Z. Kong and D. Zhong, J. Chromatogr. B, Analyt. Technol. Biomed. Life Sci., 817 (2005) 263.
17 H.H. Maurer, F.X. Tauvel and T. Kraemer, J. Anal. Toxicol., 25 (2001) 237.
18 M.A. Campanero, B. Calahorra, E. Garcia-Quetglas, A. Lopez-Ocariz and J. Honorato, J. Pharm. Biomed. Anal., 20 (1999) 327.
19 S.M. Douidar and A.E. Ahmed, J. Clin. Chem. Clin. Biochem., 20 (1982) 791.
20 M.H. Hannothiaux, N. Houdret, M. Lhermitte, J. Izydorczak and P. Roussel, Ann. Biol. Clin., 44 (1986) 139 (Paris).
21 C.A. Korduba and R.F. Petruzzi, J. Pharm. Sci., 73 (1984) 117.
22 D.S. Lo, T.C. Chao, S.E. Ng-Ong, Y.J. Yao and T.H. Koh, Forensic Sci. Int., 90 (1997) 205.
23 R. Meatherall and D. Ford, Ther. Drug Monit., 10 (1988) 101.
24 H.H. Maurer, A. Weber and K. Pfleger, Z. Anal. Chem., 311 (1982) 414.
25 F. Kamali and B. Herd, J. Chromatogr., 530 (1990) 222.
26 F. Onur and N. Acar, J. Fac. Pharm. Gazi Univ., 7 (1990) 25.
27 M.J. Bogusz, R.D. Maier, B.K. Schiwy and U. Kohls, J. Chromatogr. B, 683 (1996) 177.

28 M.J. Gomez, M. Petrovic, A.R. Fernandez-Alba and D. Barcelo, J. Chromatogr. A, 1114 (2006) 224.
29 Y. Hori, M. Fujisawa, K. Shimada, Y. Hirose and T. Yoshioka, Biol. Pharm. Bull., 29 (2006) 7.
30 O.Q. Yin, S.S. Lam and M.S. Chow, Rapid Commun. Mass Spectrom., 19 (2005) 767.
31 D.J. Speed, S.J. Dickson, E.R. Cairns and N.D. Kim, J. Anal. Toxicol., 25 (2001) 198.
32 M. El Mouelhi and B. Buszewski, J. Pharm. Biomed. Anal., 8 (1990) 651.
33 B.R. Manno, J.E. Manno, C.A. Dempsey and M.A. Wood, J. Anal. Toxicol., 5 (1981) 24.
34 C.M. Moore and I.R. Tebbett, Forensic Sci. Int., 34 (1987) 155.
35 A.S. Wong, J. Anal. Toxicol., 7 (1983) 33.
36 S.D. Ferrara, L. Tedeschi, G. Frison and F. Castagna, J. Anal. Toxicol., 16 (1992) 217.
37 A. Polettini, A. Groppi, C. Vignali and M. Montagna, J. Chromatogr. B, 713 (1998) 265.
38 J.P. Franke and R.A. de Zeeuw, J. Chromatogr. B, 713 (1998) 51.
39 C. Dietz, J. Sanz and C. Camara, J. Chromatogr. A, 1103 (2006) 183.
40 H. Kataoka, Anal. Bioanal. Chem., 373 (2002) 31.
41 R.M. Smith, J. Chromatogr. A, 1000 (2003) 3.
42 G. Vas and K. Vekey, J. Mass Spectrom., 39 (2004) 233.
43 V. Walker and G.A. Mills, Ann. Clin. Biochem., 39 (2002) 464.
44 H. Sachs and P. Kintz, J. Chromatogr. B, 713 (1998) 147.
45 P. Kintz, M. Villain and V. Cirimele, Ther. Drug Monit., 28 (2006) 442.
46 P. Kintz, Forensic Sci. Int., 142 (2004) 127.
47 F. Musshoff and B. Madea, Ther. Drug Monit., 28 (2006) 155.
48 F. Pragst and M.A. Balikova, Clin. Chim. Acta, 370 (2006) 17.
49 M. Villain, V. Cirimele and P. Kintz, Clin. Chem. Lab Med., 42 (2004) 1265.
50 C.M.J. Berlin, S.J. Yaffe and M. Ragni, Pediatr. Pharmacol., 1 (1980) 135 (New York).
51 G. Drehsen and P. Rohdewald, J. Chromatogr., 223 (1981) 479.
52 F. Kamali, J.R. Fry and G.D. Bell, J. Pharm. Pharmacol., 39 (1987) 150.
53 P. Retaco, M. Gonzalez, M.T. Pizzorno and M.G. Volonte, Eur. J. Drug Metab. Pharmacokinet., 21 (1996) 295.
54 H. Fujino, H. Yoshida, H. Nohta and M. Yamaguchi, Anal. Sci., 21 (2005) 1121.
55 D.A. Kidwell, J.C. Holland and S. Athanaselis, J. Chromatogr. B, 713 (1998) 111.
56 A.G. Verstraete, Forensic Sci. Int., 150 (2005) 143.
57 Y. Gaillard and G. Pepin, J. Chromatogr., 762 (1997) 251.
58 I.R. Miksa, M.R. Cummings and R.H. Poppenga, J. Anal. Toxicol., 29 (2005) 95.
59 P. Colin, G. Sirois and S. Chakrabarti, J. Chromatogr., 413 (1987) 151.
60 V. Shankar, C. Damodaran and P.C. Sekharan, Forensic Sci. Int., 40 (1989) 45.
61 R.A. Almodovar, R.A. Rodriguez and O. Rosario, J. Pharm. Biomed. Anal., 17 (1998) 89.
62 B. Tienpont, F. David, T. Benijts and P. Sandra, J. Pharm. Biomed. Anal., 32 (2003) 569.
63 D.U. Ahn, C. Jo and D.G. Olson, J. Agric. Food Chem., 47 (1999) 2776.
64 A. Huggett, P. Andrews and R.J. Flanagan, J. Chromatogr., 209 (1981) 67.
65 S. Murray and A.R. Boobis, J. Chromatogr., 568 (1991) 341.
66 S.K. Pant and C.L. Jain, Indian Drugs, 28 (1991) 262.
67 E. Interschick, H. Wuest and H. Patscheke, Laboratoriumsmedizin, 18 (1994) 533.
68 J. Segura, R. Ventura and C. Jurado, J. Chromatogr. B, 713 (1998) 61.
69 J.M. Halket and V.G. Zaikin, Eur. J. Mass Spectrom., 9 (2003) 1 (Chichester).
70 J.M. Halket and V.V. Zaikin, Eur. J. Mass Spectrom., 10 (2004) 1 (Chichester).
71 J.M. Halket and V.G. Zaikin, Eur. J. Mass Spectrom., 11 (2005) 127 (Chichester).
72 J.M. Halket, D. Waterman, A.M. Przyborowska, R.K. Patel, P.D. Fraser and P.M. Bramley, J. Exp. Bot., 56 (2005) 219.
73 V.G. Zaikin and J.M. Halket, Eur. J. Mass Spectrom., 9 (2003) 421 (Chichester).
74 V.G. Zaikin and J.M. Halket, Eur. J. Mass Spectrom., 11 (2005) 611 (Chichester).
75 C.T. Archer and R.A. Richardson, Ann. Clin. Biochem., 17 (1980) 45.
76 R.R. Bridges, D.W. Kinniburgh, B.J. Keehn and T.A. Jennison, J. Toxicol. Clin. Toxicol., 20 (1983) 1.

77   P.A. Edwardson, J.D. Nichols and K. Sugden, J. Pharm. Biomed. Anal., 7 (1989) 287.
78   P.W.J. Hale and A. Poklis, J. Anal. Toxicol., 7 (1983) 249.
79   Z.K. Shihabi and R.M. David, Ther. Drug Monit., 6 (1984) 449.
80   M.B. Swanson and M.I. Walters, Clin. Chem., 28 (1982) 1171.
81   H. H. Maurer, in: M.Bogusz (Ed.), Handbook of analytical separation sciences: forensic Sciences (2nd ed.). Elsevier Science, Amsterdam, 2007, p. 429.
82   H.H. Maurer, Clin. Chem. Lab. Med., 42 (2004) 1310.
83   H.H. Maurer, J. Chromatogr., 580 (1992) 3.
84   A.B. Avadhanulu, A.R. Pantulu, R. Giridhar and Y. Anjaneyulu, East. Pharm., 36 (1993) 123.
85   A.B. Avadhanulu, A.R. Pantulu and Y. Anjaneyulu, Indian Drugs, 31 (1994) 201.
86   X. Guo, W. Qian, C. Yang and X. Zhu, Sepu, 16 (1998) 164.
87   W.H. Huang, A.R. Lee and K.S. Sung, Chung-hua Yao Hsueh Tsa Chih, 41 (1989) 325.
88   G.R. Rao, A.B. Avadhanulu, D.K. Vatsa and A.R.R. Pantulu, Indian Drugs, 27 (1990) 576.
89   G.R. Rao, A.B. Avadhanulu and A.R.R. Pantulu, East. Pharm., 34 (1991) 119.
90   R.T. Sane, S.R. Surve, M.G. Gangrade, V.V. Bapat and N.L. Chonkar, Indian Drugs, 30 (1993) 66.
91   S.M. Song, P. Marriott, A. Kotsos, O.H. Drummer and P. Wynne, Forensic Sci. Int., 143 (2004) 87.
92   H.H. Maurer, K. Pfleger and A.A. Weber, Mass Spectral Library of Drugs, Poisons, Pesticides, Pollutants and their Metabolites, Pollutants and their Metabolites, Wiley-VCH, Weinheim, CA, 2007.
93   A. Panusa, G. Multari, G. Incarnato and L. Gagliardi, J. Pharm. Biomed. Anal., 43 (2007) 1221.
94   D. Bose, A. Durgbanshi, A. Martinavarro-Dominguez, M.E. Capella-Peiro, S. Carda-Broch, J.S. Esteve-Romero and M.T. Gil-Agusti, J. Chromatogr. Sci., 43 (2005) 313.
95   L.S. Jensen, J. Valentine, R.W. Milne and A.M. Evans, J. Pharm. Biomed. Anal., 34 (2004) 585.
96   C. Pistos and J.T. Stewart, J. Pharm. Biomed. Anal., 36 (2004) 737.
97   D. Ivanovic, M. Medenica, A. Malenovic, B. Jancic and D. Misljenovic, Boll. Chim. Farm., 142 (2003) 386.
98   B.S. Nagaralli, J. Seetharamappa, B.G. Gowda and M.B. Melwanki, J. Chromatogr. B, Analyt. Technol. Biomed. Life Sci., 798 (2003) 49.
99   M.V. Vertzoni, H.A. Archontaki and P. Galanopoulou, J. Pharm. Biomed. Anal., 32 (2003) 487.
100   A.W. Abu-Qare and M.B. Abou-Donia, J. Pharm. Biomed. Anal., 26 (2001) 939.
101   L.A. Shervington and N. Sakhnini, J. Pharm. Biomed. Anal., 24 (2000) 43.
102   C.M. Dawson, T.W. Wang, S.J. Rainbow and T.R. Tickner, Ann. Clin. Biochem., 25 (1988) 661.
103   G.A. Di, W.M. O'Neill and I.W. Wainer, J. Pharm. Biomed. Anal., 17 (1998) 1191.
104   A.G. Goicoechea, M.J.L. De Alda and J.L. Villa-Jato, J. Liq. Chromatogr., 18 (1995) 325.
105   N. Iqbal, B. Ahmad, K.H. Janbaz, A.S. Ijaz and N.M. Raniha, Sci. Int., 3 (1991) 221 (Lahore).
106   K. Rona, K. Foldes and B. Gachalyi, Acta Pharm. Hung., 60 (1990) 156.
107   H.M. Stevens and R. Gill, J. Chromatogr., 370 (1986) 39.
108   O.H. Drummer, A. Kotsos and I.M. McIntyre, J. Anal. Toxicol., 17 (1993) 225.
109   M.I. Aguilar, S.J. Hart and I.C. Calder, J. Chromatogr., 426 (1988) 315.
110   P. Colin, G. Sirois and S. Chakrabarti, J. Chromatogr., 377 (1986) 243.
111   A. Esteban, M. Graells, J. Satorre and M. Perez-Mateo, J. Chromatogr., 573 (1992) 121.
112   S.J. Hart, R. Tontodonati and I.C. Calder, J. Chromatogr., 225 (1981) 387.
113   A.J. Quattrone and R.S. Putnam, Clin. Chem., 27 (1981) 129.
114   A.M. Rustum, J. Chromatogr. Sci., 27 (1989) 18.
115   S.J. Hart, M.I. Aguilar, K. Healey, M.C. Smail and I.C. Calder, J. Chromatogr., 306 (1984) 215.
116   D. Jung and N.U. Zafar, J. Chromatogr., 339 (1985) 198.
117   G. Ladds, K. Wilson and D. Burnett, J. Chromatogr., 414 (1987) 355.
118   J.M. Wilson, J.T. Slattery, A.J. Forte and S.D. Nelson, J. Chromatogr., 227 (1982) 453.
119   J. Osterloh and S. Yu, Clin. Chim. Acta, 175 (1988) 239.
120   L.D. Betowski, W.A. Korfmacher, J.O.J. Lay, D.W. Potter and J.A. Hinson, Biomed. Environ. Mass Spectrom., 14 (1987) 705.
121   Y. Teffera and F. Abramson, Biol. Mass Spectrom., 23 (1994) 776.

122   M. Ohta, N. Kawakami, S. Yamato and K. Shimada, J. Pharm. Biomed. Anal., 30 (2003) 1759.
123   S. Heitmeier and G. Blaschke, J. Chromatogr. B, Biomed. Sci. Appl., 721 (1999) 109.
124   S. Heitmeier and G. Blaschke, J. Chromatogr. B, Biomed. Sci. Appl., 721 (1999) 93.
125   W. Peng, T. Li, H. Li and E. Wang, Anal. Chim. Acta, 298 (1994) 415.
126   A.E. Ashcroft, H.J. Major, I.D. Wilson, A. Nicholls and J.K. Nicholson, Anal. Commun., 34 (1997) 41.
127   P. Wernly and W. Thormann, Anal. Chem., 64 (1992) 2155.
128   A. Kunkel and H. Watzig, Electrophoresis, 20 (1999) 2379.
129   A. Kunkel and H. Waetzig, Pharm. Unserer Zeit., 25 (1996) 275.
130   A. Kunkel, S. Gunter and H. Watzig, J. Chromatogr. A, 768 (1997) 125.
131   A. Kunkel, S. Guenter and H. Waetzig, Electrophoresis, 18 (1997) 1882.
132   K. Makino, Y. Itoh, D. Teshima and R. Oishi, Electrophoresis, 25 (2004) 1488.
133   K. Pusecker, J. Schewitz, P. Gfrorer, L.H. Tseng, K. Albert, E. Bayer, I.D. Wilson, N.J. Bailey, G.B. Scarfe, J.K. Nicholson and J.C. Lindon, Anal. Commun., 35 (1998) 213.
134   M. Kolhed, P. Hinsmann, B. Lendl and B. Karlberg, Electrophoresis, 24 (2003) 687.
135   F. Tagliaro and F. Bortolotti, Electrophoresis, 27 (2006) 231.
136   V. Pucci and M.A. Raggi, Electrophoresis, 26 (2005) 767.
137   R.A. Brandon, M.J. Eadie and M.T. Smith, Ther. Drug Monit., 7 (1985) 216.
138   F. Kees, D. Jehnich and H. Grobecker, J. Chromatogr. B, 677 (1996) 172.
139   D.G. Konstantianos and P.C. Ioannou, Analyst, 117 (1992) 877.
140   V. Pavan Kumar, M.C. Vinu, A.V. Ramani, R. Mullangi and N.R. Srinivas, Biomed. Chromatogr., 20 (2006) 125.
141   M. Sultan, G. Stecher, W.M. Stoggl, R. Bakry, P. Zaborski, C.W. Huck, N.M. El Kousy and G.K. Bonn, Curr. Med. Chem., 12 (2005) 573.
142   B. Alpertunga and E. Sariahmetoglu, Marmara Univ. Eczacilik Derg., 5 (1989) 131.
143   W.M. Asselin and J.D. Caughlin, J. Anal. Toxicol., 14 (1990) 254.
144   F. Gaspari and M. Locatelli, Ther. Drug Monit., 9 (1987) 243.
145   G. Gonzalez, R. Ventura, A.K. Smith, R. de la Torre and J. Segura, J. Chromatogr., 719 (1996) 251.
146   R.L. Kincaid, M.M. McMullin, D. Sanders and F. Rieders, J. Anal. Toxicol., 15 (1991) 270.
147   J. Klimes, J. Sochor, M. Zahradnicek and J. Sedlacek, J. Chromatogr., 584 (1992) 221.
148   M.E. Legaz, E. Acitores and F. Valverde, Tokai J. Exp. Clin. Med., 17 (1992) 229.
149   Y.H. Li, Z.H. Ren, X.L. Li, Y.F. Cai and Z.X. Zhai, Biomed. Chromatogr., 9 (1995) 155.
150   A. Marzo, G. Quadro, E. Treffner, M. Ripamonti, G. Meroni and C. Lucarelli, Arzneim. -Forsch., 40 (1990) 813.
151   F. Salinas, M. de la Pena, I. Duran-Meras and D.M. Soledad, Analyst, 115 (1990) 1007.
152   J. Sochor, J. Klimes and J. Sedlacek, Ceska Slov. Farm., 45 (1996) 8.
153   D.G. Konstantianos and P.C. Ioannou, Analyst, 121 (1996) 909.
154   S. Sriewoelan, T. Sri, S. Siam, T. Hermini, T. Armayati, A. Toeti and K. Sri, Acta Pharm. Indones., 14 (1989) 27.
155   T. Kakkar and M. Mayersohn, J. Chromatogr. B, 718 (1998) 69.
156   A.E. Ashcroft, H.J. Major, S. Lowes and I.D. Wilson, Anal. Proc., 32 (1995) 459.
157   V. Cavrini, R. Gatti and P.A. Di, Farmaco, 45 (1990) 683.
158   R. Karlicek, M. Gargos and P. Solich, J. Flow Inject. Anal., 13 (1996) 45.
159   J.A. Mongillo and J. Paul, Microchem. J., 55 (1997) 296.
160   X.H. Chen, J. Wijsbeek, V.J. van, J.P. Franke and Z.R. de, J. Chromatogr., 529 (1990) 161.
161   X.H. Chen, J.P. Franke, J. Wijsbeek and Z.R. de, J. Anal. Toxicol., 16 (1992) 351.
162   J.B. Quintana, J. Miro, J.M. Estela and V. Cerda, Anal. Chem., 78 (2006) 2832.
163   E. Yamamoto, S. Takakuwa, T. Kato and N. Asakawa, J. Chromatogr. B, Analyt. Technol. Biomed. Life Sci., 846 (2007) 132.
164   H.N. Van, W.K. De, S. Poelmans, H. Noppe and B.H. De, Rapid Commun. Mass Spectrom., 18 (2004) 2823.
165   W.N. Sloot and J.B. Gramsbergen, J. Neurosci. Methods, 60 (1995) 2.

166   J. Fogel, P. Epstein and P. Chen, J. Chromatogr., 317 (1984) 507.
167   R.N. Galante, A.J. Visalli and W.M. Grim, J. Pharm. Sci., 73 (1984) 195.
168   B.S. Kersten, T. Catalano and Y. Rozenman, J. Chromatogr., 588 (1991) 187.
169   K. Kitamura, M. Takagi and K. Hozumi, Chem. Pharm. Bull., 32 (1984) 1484 (Tokyo).
170   D.J. Krieger, J. Assoc. Off. Anal. Chem., 67 (1984) 339.
171   S. Torrado and R. Cadorniga, J. Pharm. Biomed. Anal., 12 (1994) 383.
172   A. Verstraeten, E. Roets and J. Hoogmartens, J. Chromatogr., 388 (1987) 201.
173   A. Villari, N. Micali, M. Fresta and G. Puglisi, J. Pharm. Sci., 81 (1992) 895.
174   F.E. Blondino and P.R. Byron, J. Pharm. Biomed. Anal., 13 (1995) 111.
175   P.A. Di, R. Gatti, V. Andrisano and V. Cavrini, J. Chromatogr. A, 729 (1996) 355.
176   S.S. Verenitch, C.J. Lowe and A. Mazumder, J. Chromatogr. A, 1116 (2006) 193.
177   A.J. Cartwright, P. Jones, J.C. Wolff and E.H. Evans, Rapid Commun. Mass Spectrom., 19 (2005) 1058.
178   A.A. Stolker, W. Niesing, E.A. Hogendoorn, J.F. Versteegh, R. Fuchs and U.A. Brinkman, Anal. Bioanal. Chem., 378 (2004) 955.
179   R.A.B. Muljono, A.M.G. Looman, R. Verpoorte and J.J.C. Scheffer, Phytochem. Anal., 9 (1998) 35.
180   D. Tsikas, K.S. Tewes, F.M. Gutzki, E. Schwedhelm, J. Greipel and J.C. Frolich, J. Chromatogr. B, 709 (1998) 79.
181   J.D. Charette, S. Zager and A.B. Storrow, Am. J. Emerg. Med., 16 (1998) 546.
182   S.A. Chubb, R.S. Campbell, J.R. Ramsay, P.M. Hammond, T. Atkinson and C.P. Price, Clin. Chim. Acta, 155 (1986) 209.
183   P.M. Hammond, J.R. Ramsay, C.P. Price, R.S. Campbell and S.A. Chubb, Ann. N. Y. Acad. Sci., 501 (1987) 288.
184   J.H. Clark, K. Nagamori and J.F. Fitzgerald, Clin. Chim. Acta, 145 (1985) 243.
185   D.R. Jarvie, R. Heyworth and D. Simpson, Ann. Clin. Biochem., 24 (1987) 364.
186   J. Plata, J.C. Orte, F. Martinez and J. Thomas, Analysis, 18 (1990) 146.
187   J. Plata, J.C. Orte, F. Martinez and J.M. Alvarez, An. Quim., 88 (1992) 374.
188   H.H. Maurer, Ther. Drug Monit., 24 (2002) 247.
189   X. Luo and D.C. Lehotay, Clin. Biochem., 30 (1997) 41.
190   A.K. Singh, Y. Jang, U. Mishra and K. Granley, J. Chromatogr., 568 (1991) 351.
191   U.M. Laakkonen, A. Leinonen and L. Savonen, Analyst, 119 (1994) 2695.
192   S.A. Chubb, R.S. Campbell and C.P. Price, J. Chromatogr., 380 (1986) 163.
193   D. Dadgar, J. Climax, R. Lambe and A. Darragh, J. Chromatogr., 342 (1985) 315.
194   Z. Krivosikova, V. Spustova and R. Dzurik, Methods Find. Exp. Clin. Pharmacol., 18 (1996) 527.
195   J.H. Liu and P.C. Smith, J. Chromatogr. B, 675 (1996) 61.
196   G.P. McMahon and M.T. Kelly, Anal. Chem., 70 (1998) 409.
197   R. Pirola, S.R. Bareggi and G. De Benedittis, J. Chromatogr. B, 705 (1998) 309.
198   F. Vinci, S. Fabbrocino, M. Fiori, L. Serpe and P. Gallo, Rapid Commun. Mass Spectrom., 20 (2006) 3412.
199   H.M. Adelman, P.M. Wallach and M.T. Flannery, J. Rheumatol., 18 (1991) 522.
200   C. Coudray, C. Mangournet, S. Bouhadjeb, H. Faure and A. Favier, J. Chromatogr. Sci., 34 (1996) 166.
201   D.M. Siebert and F. Bochner, J. Chromatogr., 420 (1987) 425.
202   T.B. Vree, v.E.-B. Kolmer, C.P.W.G. Verwey-van Wissen and Y.A. Hekster, J. Chromatogr. B, 652 (1994) 161.
203   Y. Goto, K. Makino, Y. Kataoka, H. Shuto and R. Oishi, J. Chromatogr. B, 706 (1998) 329.
204   S.H. Hansen, M.E. Jensen and I. Bjornsdottir, J. Pharm. Biomed. Anal., 17 (1998) 1155.
205   J. Caslavska, E. Gassmann and W. Thormann, J. Chromatogr. A, 709 (1995) 147.
206   S. Zaugg, X. Zhang, J. Sweedler and W. Thormann, J. Chromatogr. B, Biomed. Sci. Appl., 752 (2001) 17.
207   J. Caslavska and W. Thormann, Electrophoresis, 25 (2004) 1623.
208   J. Caldwell, A.J. Hutt and S. Fournel-Gigleux, Biochem. Pharmacol., 37 (1988) 105.

209 X. Bi, Z. Meng and G. Dou, J. Chromatogr. B, Analyt. Technol. Biomed. Life Sci., 850 (2007) 199.

210 O. Cheremina, K. Brune and B. Hinz, Biomed. Chromatogr., 20 (2006) 1033.

211 N.V. Ramakrishna, K.N. Vishwottam, S. Wishu and M. Koteshwara, J. Chromatogr. B, Analyt. Technol. Biomed. Life Sci., 816 (2005) 215.

212 H. Jalalizadeh, M. Amini, V. Ziaee, A. Safa, H. Farsam and A. Shafiee, J. Pharm. Biomed. Anal., 35 (2004) 665.

213 N.V. Ramakrishna, K.N. Vishwottam, S. Wishu and M. Koteshwara, J. Chromatogr. B, Analyt. Technol. Biomed. Life Sci., 802 (2004) 271.

214 P.S. Bonato, M.P. Del Lama and C.R. de, J. Chromatogr. B, Analyt. Technol. Biomed. Life Sci., 796 (2003) 413.

215 B. Hinz, D. Auge, T. Rau, S. Rietbrock, K. Brune and U. Werner, Biomed. Chromatogr., 17 (2003) 268.

216 V.A. Jabor, V.L. Lanchote and P.S. Bonato, Electrophoresis, 23 (2002) 3041.

217 C.M. Chavez-Eng, M.L. Constanzer and B.K. Matuszewski, J. Chromatogr. B, Biomed. Sci. Appl., 748 (2000) 31.

218 R.A. Carr, G. Caille, A.H. Ngoc and R.T. Foster, J. Chromatogr. B, 668 (1995) 175.

219 H.H. Maurer, T. Kraemer and A. Weber, Pharmazie, 49 (1994) 148.

220 F. Schonberger, G. Heinkele, T.E. Murdter, S. Brenner, U. Klotz and U. Hofmann, J. Chromatogr. B, Analyt. Technol. Biomed. Life Sci., 768 (2002) 255.

221 T. Velpandian, J. Jaiswal, R.K. Bhardwaj and S.K. Gupta, J. Chromatogr. B, Biomed. Sci. Appl., 738 (2000) 431.

222 H.H. Chow, N. Anavy, D. Salazar, D.H. Frank and D.S. Alberts, J. Pharm. Biomed. Anal., 34 (2004) 167.

223 M. Zhang, G.A. Moore, S.J. Gardiner and E.J. Begg, J. Chromatogr. B, Analyt. Technol. Biomed. Life Sci., 807 (2004) 217.

224 L. Brautigam, J.U. Nefflen and G. Geisslinger, J. Chromatogr. B, Analyt. Technol. Biomed. Life Sci., 788 (2003) 309.

225 E. Stormer, S. Bauer, J. Kirchheiner, J. Brockmoller and I. Roots, J. Chromatogr. B, Analyt. Technol. Biomed. Life Sci., 783 (2003) 207.

226 J.Y. Zhang, D.M. Fast and A.P. Breau, J. Chromatogr. B, Analyt. Technol. Biomed. Life Sci., 785 (2003) 123.

227 S. Liu, M. Kamijo, T. Takayasu and S. Takayama, J. Chromatogr. B, Analyt. Technol. Biomed. Life Sci., 767 (2002) 53.

228 C. Arcelloni, R. Lanzi, S. Pedercini, G. Molteni, I. Fermo, A. Pontiroli and R. Paroni, J. Chromatogr. B, Biomed. Sci. Appl., 763 (2001) 195.

229 J. Hermansson, A. Grahn and I. Hermansson, J. Chromatogr. A, 797 (1998) 251.

230 T. Shinozuka, M. Terada, S. Takei, N. Kuroda, K. Kurihara, C. Wakasugi and J. Yanagida, Hochudoku, 10 (1992) 126.

231 T. Shinozuka, S. Takei, N. Kuroda, A. Ogamo, M. Terada, C. Wakasugi and J. Yanagida, Jpn. J. Forensic Toxicol., 14 (1996) 43.

232 S.M.R. Stanley, N.A. Owens and J.P. Rodgers, J. Chromatogr. B, 667 (1995) 95.

233 C. Albrecht and W. Thormann, J. Chromatogr. A, 802 (1998) 115.

234 E.M. Koves, J. Chromatogr., 692 (1995) 103.

235 S. Azhagvuel and R. Sekar, J. Pharm. Biomed. Anal., 43 (2007) 873.

236 L. Brum Jr., M. Fronza, D.C. Ceni, T. Barth and S.L. Dalmora, J. AOAC Int., 89 (2006) 1268.

237 R. Hamoudova and M. Pospisilova, J. Pharm. Biomed. Anal., 41 (2006) 1463.

238 Y.H. Hsieh, S.J. Lin and S.H. Chen, J. Sep. Sci., 29 (2006) 1009.

239 M.S. Arayne, N. Sultana and F.A. Siddiqui, Pak. J. Pharm. Sci., 18 (2005) 58.

240 Y.L. Chen and S.M. Wu, Anal. Bioanal. Chem., 381 (2005) 907.

241 R.R. de Pablos, C. Garcia-Ruiz, A.L. Crego and M.L. Marina, Electrophoresis, 26 (2005) 1106.

242 M. Gandhimathi, T.K. Ravi and S.J. Varghese, J. Pharm. Biomed. Anal., 37 (2005) 183.

243 N. Kaul, S.R. Dhaneshwar, H. Agrawal, A. Kakad and B. Patil, J. Pharm. Biomed. Anal., 37 (2005) 27.

244   V.G. Nayak, V.R. Bhate, S.M. Purandare, P.M. Dikshit, S.N. Dhumal and C.D. Gaitonde, Drug Dev. Ind. Pharm., 18 (1992) 369.
245   H.L. Rau, A.R. Aroor and P.G. Rao, Indian Drugs, 29 (1991) 48.
246   A. Wainright, J. Microcolumn Sep., 2 (1990) 166.
247   A.E. Walily, S.F. Belal, M.H. Yakout and S. Zamel, Bull. Fac. Pharm., 31 (1993) 331 (Cairo Univ.).
248   d.C. Fernandez, B. Ortega and D. Molina, Anal. Chim. Acta, 369 (1998) 263.
249   B.M. Lampert and J.T. Stewart, J. Chromatogr., 504 (1990) 381.
250   V.R. Bari, U.J. Dhorda and M. Sundaresan, Talanta, 45 (1997) 297.
251   R.R. Nageswara, S. Meena and R.A. Raghuram, J. Pharm. Biomed. Anal., 39 (2005) 349.
252   J. Antonic and E. Heath, Anal. Bioanal. Chem., 387 (2007) 1337.
253   E. Botitsi, C. Frosyni and D. Tsipi, Anal. Bioanal. Chem., 387 (2007) 1317.
254   M.D. Hernando, E. Heath, M. Petrovic and D. Barcelo, Anal. Bioanal. Chem., 385 (2006) 985.
255   W. Seitz, W. Schulz and W.H. Weber, Rapid Commun. Mass Spectrom., 20 (2006) 2281.
256   E. Caro, R.M. Marce, P.A. Cormack, D.C. Sherrington and F. Borrull, J. Sep. Sci., 28 (2005) 2080.
257   J. Debska, A. Kot-Wasik and J. Namiesnik, J. Sep. Sci., 28 (2005) 2419.
258   T. Kosjek, E. Heath and A. Krbavcic, Environ. Int., 31 (2005) 679.
259   W.C. Lin, H.C. Chen and W.H. Ding, J. Chromatogr. A, 1065 (2005) 279.
260   A. Macia, F. Borrull, M. Calull and C. Aguilar, Electrophoresis, 25 (2004) 3441.
261   P.M. Thomas and G.D. Foster, J. Environ. Sci. Health A Tox. Hazard. Subst. Environ. Eng, 39 (2004) 1969.
262   X. Wen, C. Tu and H.K. Lee, Anal. Chem., 76 (2004) 228.
263   M.J. Hilton and K.V. Thomas, J. Chromatogr. A, 1015 (2003) 129.
264   S. Marchese, A. Gentili, D. Perret, G.D. Ascenzo and F. Pastori, Rapid Commun. Mass Spectrom., 17 (2003) 879.
265   I. Rodriguez, J.B. Quintana, J. Carpinteiro, A.M. Carro, R.A. Lorenzo and R. Cela, J. Chromatogr. A, 985 (2003) 265.
266   W. Ahrer, E. Scherwenk and W. Buchberger, J. Chromatogr. A, 910 (2001) 69.
267   S. Ollers, H.P. Singer, P. Fassler and S.R. Muller, J. Chromatogr. A, 911 (2001) 225.
268   D.C. Borrey, K.O. Godderis, V.I. Engelrelst, D.R. Bernard and M.R. Langlois, Clin. Chim. Acta, 354 (2005) 147.
269   K.R. Kim, W.H. Shim, Y.J. Shin, J. Park, S. Myung and J. Hong, J. Chromatogr., 641 (1993) 319.
270   J. Beyer, A. Bierl, F.T. Peters and H.H. Maurer, Ther. Drug Monit., 27 (2005) 509.
271   J. Beyer, F.T. Peters and H.H. Maurer, Ther. Drug Monit., 27 (2005) 151.
272   K.A. Alkhamis and D.E. Wurster, Pharm. Dev. Technol., 8 (2003) 127.
273   Y. Bamou, S. Bouhsain, S. Tellal, A. Dami, S. Yaakoubi, S. Mechtani and M. Derouiche, Ann. Biol. Clin., 59 (2001) 67 (Paris).
274   A.M. Lisi, R. Kazlauskas and G.J. Trout, J. Chromatogr. B, 692 (1997) 67.
275   H.H. Maurer, T. Kraemer and J.W. Arlt, Ther. Drug Monit., 20 (1998) 706.
276   H.H. Maurer and J.W. Arlt, J. Chromatogr. B, Biomed. Sci. Appl., 714 (1998) 181.
277   M. Petrovic, Z. Debeljak and N. Blazevic, J. Pharm. Biomed. Anal., 39 (2005) 531.
278   H.H. Maurer, J. Chromatogr., 530 (1990) 307.
279   M.J. Paik and K.R. Kim, Arch. Pharm. Res., 27 (2004) 820.
280   M.J. Paik, D.T. Nguyen and K.R. Kim, Arch. Pharm. Res., 27 (2004) 1295.
281   M.J. Paik, Y. Lee, J. Goto and K.R. Kim, J. Chromatogr. B, Analyt. Technol. Biomed. Life Sci., 803 (2004) 257.
282   S.P. Elliott and K.A. Hale, J. Chromatogr. B, 694 (1997) 99.
283   Y. Gaillard and G. Pepin, J. Chromatogr., 763 (1997) 149.
284   R.D. Maier and M. Bogusz, J. Anal. Toxicol., 19 (1995) 79.
285   A. Tracqui, P. Kintz and P. Mangin, J. Forensic Sci., 40 (1995) 254.
286   A. Turcant, A. Premel-Cabic, A. Cailleux and P. Allain, Clin. Chem., 37 (1991) 1210 (Winston-Salem).
287   M.A. Macia, J. Frias, A.J. Carcas, P. Guerra, R. Valiente and M.L. Lucero, Int. J. Clin. Pharmacol. Ther., 33 (1995) 333.

288   T. Shinozuka, M. Terada, A. Ogamo, S. Takei, R. Nakajima, M. Kato, M. Baba, A. Tomita, N. Kuroda, T. Murai, C. Wakasugi and J. Yanagida, in: T. Takatori and A. Takasu (Eds.), Current topics on forensic sciences, proceedings of the meeting of international associations of forensic sciences, 14th, Shunderson Communications, Ottawa, ON, 1996, p. 227.

289   F. Beaudry, J.C.Y. Le Blanc, M. Coutu and N.K. Brown, Rapid Commun. Mass Spectrom., 12 (1998) 1216.

290   H.M. Rigato, G.D. Mendes, N.C. Borges and R.A. Moreno, Int. J. Clin. Pharmacol. Ther., 44 (2006) 489.

291   Y. Alnouti, K. Srinivasan, D. Waddell, H. Bi, O. Kavetskaia and A.I. Gusev, J. Chromatogr. A, 1080 (2005) 99.

292   H.Y. Ji, H.W. Lee, Y.H. Kim, D.W. Jeong and H.S. Lee, J. Chromatogr. B, Analyt. Technol. Biomed. Life Sci., 826 (2005) 214.

293   W.Z. Shou, L. Magis, A.C. Li, W. Naidong and M.S. Bryant, J. Mass Spectrom., 40 (2005) 1347.

294   J.Y. Zhang, D.M. Fast and A.P. Breau, J. Pharm. Biomed. Anal., 33 (2003) 61.

295   M.E. Abdel-Hamid, L. Novotny and H. Hamza, J. Chromatogr. B, Biomed. Sci. Appl., 753 (2001) 401.

296   M.E. Abdel-Hamid, L. Novotny and H. Hamza, J. Pharm. Biomed. Anal., 24 (2001) 587.

297   Z. Guo, H. Wang and Y. Zhang, J. Pharm. Biomed. Anal., 41 (2006) 310.

298   A.R. de Oliveira, E.J. Cesarino and P.S. Bonato, J. Chromatogr. B, Analyt. Technol. Biomed. Life Sci., 818 (2005) 285.

299   F. Pehourcq, M. Matoga, C. Jarry and B. Bannwarth, Biomed. Chromatogr., 18 (2004) 330.

300   X.W. Teng, S.W. Wang and N.M. Davies, J. Pharm. Biomed. Anal., 33 (2003) 95.

301   L.O. Healy, J.P. Murrihy, A. Tan, D. Cocker, M. McEnery and J.D. Glennon, J. Chromatogr. A, 924 (2001) 459.

302   T.H. Eichhold, R.E. Bailey, S.L. Tanguay and S.H. Hoke, J. Mass Spectrom., 35 (2000) 504.

303   N.M. Davies, J. Chromatogr. B, Biomed. Sci. Appl., 691 (1997) 229.

304   R. Mullangi, M. Yao and N.R. Srinivas, Biomed. Chromatogr., 17 (2003) 423.

305   C. Albrecht, J. Reichen, J. Visser, D.K.F. Meijer and W. Thormann, Clin. Chem., 43 (1997) 2083 (Washington).

306   A. Schmutz and W. Thormann, Electrophoresis, 15 (1994) 1295.

307   S.J. Lin, Y.R. Chen, Y.H. Su, H.I. Tseng and S.H. Chen, J. Chromatogr. B, Analyt. Technol. Biomed. Life Sci., 830 (2006) 306.

308   Z.K. Shihabi and M.E. Hinsdale, J. Chromatogr. B, 683 (1996) 115.

309   T.M. Phillips and E.F. Wellner, Biomed. Chromatogr., 20 (2006) 662.

310   F.K. Glowka, J. Pharm. Biomed. Anal., 30 (2002) 1035.

311   C. Calvet, R. Cuberes, C. Perez-Maseda and J. Frigola, Electrophoresis, 23 (2002) 1702.

312   F.K. Glowka and M. Karazniewicz, J. Pharm. Biomed. Anal., 35 (2004) 807.

313   D.B. Strickmann, B. Chankvetadz, G. Blaschke, C. Desiderio and S. Fanali, J. Chromatogr. A, 887 (2000) 393.

314   M. Balikova, . In: B. Jacob, W. Bonte (Eds.), Advance forensic sciences, proceedings of the meeting of international associations of forensic sciences, 13th, Verlag Dr. Koester, Berlin, Germany, 1995, p. 72.

315   M. Bogusz and M. Wu, J. Anal. Toxicol., 15 (1991) 188.

316   E. Grippa, L. Santini, G. Castellano, M.T. Gatto, M.G. Leone and L. Saso, J. Chromatogr. B, Biomed. Sci. Appl., 738 (2000) 17.

317   S.B. Clark, S.B. Turnipseed, G.J. Nandrea, M.R. Madson, J.A. Hurlbut and J.N. Sofos, J. AOAC Int., 85 (2002) 1009.

318   M.C. Caturla and E. Cusido, J. Chromatogr., 581 (1992) 101.

319   A. Haque and J.T. Stewart, J. Pharm. Biomed. Anal., 16 (1997) 287.

320   L.M. Neto, M.H. Andraus and M.C. Salvadori, J. Chromatogr. B, Biomed. Appl., 678 (1996) 211.

321   K.E. Peck, A.C. Ray, G. Manuel, M.M. Rao and J. Foos, Am. J. Vet. Res., 57 (1996) 1522.

322   M.C. Rouan, J. Campestrini, J.B. Lecaillon and J. Godbillon, J. Chromatogr., 577 (1992) 387.

323   G. Engel, U. Hofmann and M. Eichelbaum, J. Chromatogr. B, Biomed. Appl., 666 (1995) 111.

324   M. Yegles and R. Wenning, Forensic Sci. Int., 107 (2000) 233.
325   F. Musshoff and T. Kraemer, Int. J. Legal. Med., 111 (1998) 305.
326   M. Neugebauer, A. Khedr, N. el-Rabbat, M. el-Kommos and G. Saleh, Biomed. Chromatogr., 11 (1997) 356.
327   E.S. Oh, S.K. Hong and G.I. Kang, Xenobiotica, 22 (1992) 377.
328   H.S. Shin, J.S. Park, P.B. Park and S.J. Yun, J. Chromatogr. B, Biomed. Appl., 661 (1994) 255.
329   H.S. Shin, Chirality, 9 (1997) 52.
330   H.S. Shin, B.B. Park, S.N. Choi, J.J. Oh, C.P. Hong and H. Ryu, J. Anal. Toxicol., 22 (1998) 55.
331   Y. Yoo, H. Chung and H. Choi, J. Anal. Toxicol., 18 (1994) 265.
332   J.T. Cody, J. Occup. Environ. Med., 44 (2002) 435.
333   T. Kraemer and H.H. Maurer, Ther. Drug Monit., 24 (2002) 277.
334   F. Musshoff, Drug Metab. Rev., 32 (2000) 15.

M.J. Bogusz (Ed.). Forensic Science
Handbook of Analytical Separations, Vol. 6
© 2008 Elsevier B.V. All rights reserved.

CHAPTER 9

Other therapeutic drugs of forensic relevance

Jet C. Van De Steene and Willy E. Lambert

Laboratorium voor Toxicologie, Universiteit Gent, Harelbekestraat 72 9000 Gent, Belgium

9.1 INTRODUCTION

Only a number of selected drugs or classes of drugs, which are toxicologically relevant are discussed in this chapter. It is impossible to mention all other classes. But after thoughtful considerations the following compounds were selected: cardiac glycosides, antiarrhythmics, oral antidiabetics and muscle relaxants.

Only recent applications of separation techniques, published from 2000 till now, will be discussed. In all cases not only sample preparation and chromatography will be considered, but also immunoassays.

9.2 CARDIAC GLYCOSIDES

The focus in this paragraph will be only on digoxin and digitoxin, the two major representatives within the group of the cardiac glycosides. They are obtained from *Digitalis lanata* Ehrhart and from *Digitalis purpurea* Linne (Fam. *Scrophulariaceae*).

9.2.1 Preparative applications

9.2.1.1 Crude materials, pure drugs and pharmaceutical preparations

Concerning pharmaceutical preparations, one article was found where dissolution tests for digoxin tablets were discussed [1]. A 40-mg tablet, containing 0.250 mg digoxin, was dissolved in 500 ml 0.01 M hydrochloric acid at 37°C for 30 min. After dissolution a solid-phase extraction (SPE) was performed on Sep-Pak C18 cartridges. After application of 30 ml of the dissolution medium, columns were washed with 5 ml of water, and digoxin was eluted with 2×1.0 ml of MeOH into a 5-ml volumetric flask. Holstege *et al.* [2] examined plant material. They extracted card-iotoxins (oleandrine, gitoxin, digitoxin, gitoxigenin and grayanotoxins I and II) with dichloromethane. Extracts were cleaned up with charcoal and reversed-phase SPE columns.

9.2.1.2 Blood, serum, urine and other materials taken from living subjects

Only one procedure was found where liquid–liquid extraction was used [3]. Extraction was performed by adding 50 μl ammonium chloride buffer (pH 8.8) to 0.1 ml of plasma. Then, 300 μl of acetonitrile and 150 μl of methylene chloride were added. In other publications, SPE was used for sample preparation [4–7].

Hafner *et al.* used, after protein precipitation with zinc sulfate, Oasis HLB columns (10 mg stationary phase): conditioning with 1 ml of acetonitrile, followed by 3 × 1 ml of water, sample (1 ml urine) application, washing with 2 × 1 ml of a 1 vol % solution of acetic acid in acetonitrile/water (10:90, v/v), and 2 × 1 ml of a 1 vol % solution of ammonium hydroxide in acetonitrile/water (10:90, v/v), 3 × 1 ml water. Elution occurred with 2 × 75 μl of acetonitrile [5].

Holstege *et al.* reported a multi-residue screening for cardiotoxins in gastro-intestinal contents and feces. Cardiotoxins were extracted with dichloromethane. Sample clean-up was done with charcoal and reversed-phase SPE columns [2].

9.2.2 Analytical applications

9.2.2.1 Crude materials, pure drugs and pharmaceutical preparations

9.2.2.1.1 Thin layer chromatography
Only one article was published since 2000. Pascual *et al.* described the screening for cardiac glycosides in plants based on thin layer chromatography (TLC)[8].

9.2.2.1.2 High-performance liquid chromatography
One article described the analysis of digoxin in tablets with HPLC-UV-detection [1]. Digoxin and derivatives were eluted from the Lichrosper RP-18e column (5 μm, 125 × 4.0 mm) with water and acetonitrile (72:28, v/v), with a flow rate of 1.1 ml/min. Detection wavelength was 218 nm. There was linearity in the range 60–3600 ng/ml.

Micellar electrokinetic chromatography was used to determine digoxin and digitoxin in drug formulations [9]. Detection limit was 4 μg/ml for digoxin and 6 μg/ml for digitoxin.

A highly efficient capillary electrochromatographic separation of cardiac glycosides and other steroids was presented by Mayer *et al.* [10]. Butyl-derivatized silica particles as stationary phase gave nearly three times faster electroosmotic flow compared to capillary electrochromatography. Detection limits of 10–80 nM were reached.

The following papers report newly developed methods based on high-performance liquid chromatography (HPLC) coupled to immunodetection or biochemical detection with digoxin as a test analyte. Tang and Karnes coupled immunoassay with

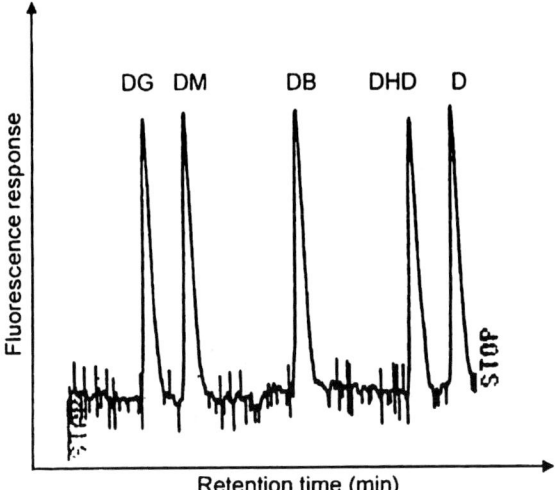

Fig. 9.1. Chromatogram showing inhibition of energy transfer by digoxigenin (DG), digoxigenin monodigitoxoside (DM), digoxigenin bisdigitoxoside (DB), dihydrodigoxin (DHD) and digoxin (D). R-Phycoerythrin labeled digoxin (digoxigenin-PE) concentration was 6.4×10^{-10} M and Cy5-labeled anti-digoxin antibody (Mab-Cy5) was $1.28 \times 10^{-9-}$ M. Fluorescence energy transfer (sensitized emission) was measured at 670 nm (from Graefe *et al.* [12], with permission).

liquid chromatography (LC) post-column [11]. They used an affinity column coupled to an immunoreaction detection system using magnetized beads and a laboratory-constructed electromagnetic separator. The dynamic range of the calibration curves in digoxin-spiked phosphate buffer was 0.25–12 ng/ml. Another approach was LC coupled to on-line post-column immunoreaction [12] (Fig. 9.1). The immunoreaction was monitored with fluorescence detection. The HPLC column used was a Zorbax ODS column (150 × 4.6 mm i.d., 5 μm particle size). Mobile phase consisted of tetra-hydrofuran/water (20:80, v/v). Limits of detection for digoxin and metabolites in phosphate buffer were in the picogram per ml range. At last, Schenk *et al.* [13] reported a liquid chromatographic method coupled on-line to flow cytometry for post-column homogeneous biochemical detection. Separation was performed isocratically on a Vydac C4 column (250 × 1.0 mm i.d., 5 μm particles) with methanol/water (40:60, v/v) at a flow rate of 100 μl/min. Detection limits of 0.1 nmol/l were achieved for digoxin.

9.2.2.2 Blood, serum, urine and other materials taken from living subjects

9.2.2.2.1 High-performance liquid chromatography
Since 2000, only HPLC methods with mass spectrometry (MS) as detection technique are published. Therapeutic levels of digoxin and digitoxin are in the low nanogram per ml range. In this way, sensitive detection by MS is necessary. Lacassie *et al.* studied a non-fatal case of intoxication with foxglove [14]. Seventeen cardiac glycosides, including digoxin, digitoxin, metabolites and others, were detected using

Nucleosil C18 (150 × 1 mm i.d., 5 μm particle size) coupled to a single quadrupole instrument. Gradient elution occurred with acetonitrile in 2 mM ammonium formate at a flow rate of 40 μl/min. Analytes were detected in positive ionization mode. Limits of quantification for digoxin and digitoxin were 2 and 10 ng/ml, respectively. However, the method is not suitable for therapeutic drug monitoring due to the high limit of quantification (LOQ) for digoxin and digitoxin. Other methods used tandem mass spectrometric detection. Because of the enhanced sensitivity obtained (and the low nanogram per ml therapeutic range for digoxin and digitoxin), these are the methods of choice. Malone *et al.* worked on an API 3000 and an API 365 LC-MS/MS system [6]. Digoxin quantification was linear in the range 0.1–10 ng/ml in human plasma and 1–100 ng/ml in human urine. Mitamura *et al.* determined digitoxin in human serum using stable isotope dilution liquid chromatography-electrospray-tandem mass spectrometry [7]. Analyses were done in the selected ion monitoring (SIM) mode. Digitoxin could be measured quantitatively in the range 5–100 ng/ml.

A rapid and sensitive LC/MS/MS assay for quantitative determination of digoxin in rat plasma was described with oleandrin as an internal standard [3]. Analytes were isocratically eluted from a 3-μm YMC ODS AQ analytical column with acetonitrile and 5 mM ammonium formate (pH 3.4) (50:50, v/v) at a flow rate of 0.2 ml/min. The API 3000 LC/MS/MS with turbo ion spray was used in the positive multiple reaction monitoring mode. The lower LOQ was 0.1 ng/ml.

In addition, also one capillary electrophoresis method was published [5]. Hafner *et al.* described a method, combining immunoassay and affinity probe capillary electrophoresis. Digoxin was used as a model analyte. Detection of 400 fM in 1 ml of serum was achieved.

9.2.2.2.2 Gas chromatography

Only one recent publication was found. Kiousi *et al.* described a gas chromatographic-mass spectrometric analysis for detection of cardiotonic glycosides in equine urine [4]. After silylation of the aglycon moieties of the analytes, they were detected with gas chromatography-mass spectrometry (GC-MS) and gas chromatography-high-resolution mass spectrometry (GC-HRMS). Detection limits of cardiotonic glycosides are in the range 55–200 ng/ml for GC-MS. For GC-HRMS the limits of detection were approximately five times more lower.

9.2.3 Immunoassays

After 2000, the following immunoassay techniques for serum digoxin have been reported: turbidimetric immunoassay [15,16], potentiometric immunoassay [17,18], radioimmunoassay [19], magnetic separation immunoassay [20,21], fluorescent immunoassay [22,23], enzyme-linked immunosorbent assay [24], chemiluminescence [25,26] and bioluminescence immunoassay [27,28].

In the case of digitoxin, library search gave us only one new immunoassay. Ikeda *et al.* [19] developed a radioimmunoassay for serum digitoxin. Interferences are of major concern when dealing with immunoassays. Many reports have been published concerning interferences by other drugs and endogenous digoxin-like immunoreactive factors. From the publications we focused on only some examined the possibility of cross-reactivity with other drugs [15,22], metabolites of digoxin and of other drugs [19], endogenous digoxin-like immunoreactive factors [24,29] and digoxin analogues [27,28]. Interferences are sometimes reduced, however, it is not clearly demonstrated that they are totally eliminated.

Approaches to minimize interference by cross-reacting molecules in immunoassays are described by Miller and Valdes [30]. Protein precipitation [15,24,29] and ultrafiltration can be used to decrease the concentration of endogenous digoxin-like immunoreactive factors which are mainly bound to high molecular weight proteins. Also SPE [27] and HPLC can be used to diminish the concentration of interferences in the sample, but this is – certainly in the case of chromatography – rather a time- and money-consuming approach. A more effective way to lower cross-reactivity is the increase of the incubation time [19,23] and/or temperature [25,26]: dissociation half-times of, e.g. digoxin-like immunoreactive factors are much less than those for digoxin. Immunoassay for cardiac glycosides is routinely used for therapeutic drug monitoring. It has been suggested that immunoassays for digoxin or digitoxin, which show cross-reactivity with active metabolites, are more related to the total biological activity of these drugs [31–33]. Alternatively, immunoassays or antibodies which are very specific to digoxin or digitoxin [19] may be useful for pharmacokinetic studies or for therapeutic immunoextraction of digoxin in patients with digoxin overdoses.

In general, the immunoassay technique is not regarded as the method of choice for determination of serum digoxin and digitoxin, due to interferences of other structural related drugs or endogenous digoxin-like immunoreactive factors which can lead to erroneous conclusions. Chromatographic analysis is regarded as a better option.

9.3 ANTIARRHYTHMICS

For the antiarrhythmic drugs we adhere the Vaughan Williams classification with the following representative agents: class I (Na^+-channel blockade antiarrhythmics) further divided into three subclasses, IA (quinidine, disopyramide and procainamide), IB (lidocaine and mexiletine), IC (flecainide and propafenone); class II, betaadrenergic blockade (propranolol and acebutolol); class III prolonged repolarization (amiodarone, sotalol and bretylium); class IV, Ca^{2+}-channel blockade (verapamil and diltiazem) [34].

9.3.1 Preparative applications

9.3.1.1 Crude materials, pure drugs and pharmaceutical preparations

Concerning the class I antiarrhythmics, quinidine was extracted in a protein-binding study from a buffer by solid-phase microextraction (SPME) [35]. Ravishankara *et al.* performed an extraction of quinidine from *Cinchona officinalis* stem bark, as described in the European Pharmacopoeia with certain modifications, i.e. the 25% sodium hydroxide solution was replaced by a 25% ammonia solution and tragacanth was replaced with carboxy methyl cellulose [36]. Pietras *et al.* reported the extraction of mexiletine from a formulation with methanol [37]. For the extraction of lidocaine from aqueous solutions, SPME with polydimethylsiloxane fibers was used [38]. For the class IV antiarrhythmic diltiazem, one article was found reporting the extraction from a formulation. Diltiazem was extracted from a gel with SPE on Oasis HLB cartridges [39]. For the extraction of verapamil from capsules methanol was used [40,41].

9.3.1.2 Blood, serum, urine and other materials taken from living subjects

For extraction of the class I antiarrhythmic disopyramide from human urine liquid–liquid extraction was performed with dichloromethane after alkalinization with sodium hydroxide [42] and dichloromethane [43]. From plasma the extraction was performed with toluene [44] or dichloromethane after protein precipitation with trichloroacetic acid [43]. One method was described for mexiletine [45]. Liquid–liquid extraction with ethyl acetate was performed on urine, after addition of a 1-M carbonate buffer (pH 9.0). Flecainide, a class IC antiarrhythmic, was frequently extracted with SPE on C18 phases from serum [46,47] or blood, urine and other toxicologically relevant matrices [48]. Doki *et al.* used liquid–liquid extraction with ethylacetate for the extraction of flecainide from serum [49]. To extract flecainide from hair, Takiguchi *et al.* used liquid–liquid extraction with ethyl acetate [50].

As sample preparation in the analysis of propafenone in serum, Afshar and Rouini used only protein precipitation with a mixture of zinc sulfate and methanol [51]. Hofmann *et al.* performed a more thorough sample preparation. They extracted propafenone from plasma and urine with SPE on a C18 sorbent [52]. For extracting lidocaine from plasma and urine, SPME was frequently applied. Different kind of fibers were used: carbowaxdivinylbenzene [53], polydimethylsiloxane [54–56] and polydimethylsiloxanedivinylbenzene [57]. An extraction from hair was performed by Sporkert and Pragst with headspace SPME. They used carbowax-divinylbenzene fibers [58]. Also de Jong and Koster published a headspace SPE; they used polydimethylsiloxane fibers for plasma and urine [59]. C18 [60] and C8 [61] SPE columns were also used. Van Hout *et al.* combined SPE (inside the GC-liner) with thermal desorption prior to injection of the sample onto the gas chromatography (GC); TENAX was used as sorbent [62]. Ultrafiltration of plasma was described for the determination of the free concentration in plasma [60,63,64].

Another technique applied for extraction of lidocaine from urine was liquid-phase microextraction [65–67]. On-line microextraction in packed syringes (MEPS) was another approach of microextraction of lidocaine, now used for extraction from plasma [68]. SPME was also performed in solid-phase extraction-pipette tips (SPE-PTs) [69]. Liquid–liquid extraction of lidocaine from serum was performed with dichloromethane [70]. A general extraction procedure for different kind of drugs (including lidocaine) from biological samples was described by Ahrens *et al.* [71]. For basic compounds, they used ethyl acetate as an extraction solvent, after alkalinization. In one publication, a reversed-phase pre-treatment column (MC-ODS) was used on-line [72].

For the determination of the class II antiarrhythmic propranolol in plasma and urine, different kind of approaches were reported: SPE (Oasis HLB [73,74], C18 [75]), SPME [76–78] and liquid–liquid extraction [79–82]. On-line extraction was done with SPE [74] and restricted access media (RP-8 ADS [83], MC-ODS [84], RP-18 ADS [85]). Propranolol, acebutolol and sotalol were extracted together with SPE on Isolute Confirm HCX cartridges [86]. Also Gergov *et al.* extracted these β-blockers with these SPE-cartridges [87]. Delamoye used for these compounds a liquid-liquid extraction with chloroform/pentanol/diethylether (6:2:1, v/v) [88]. For the analysis of acebutolol in hair SPE with Isolute C18 cartridges was used [89]. Josefsson and Sabanovic compared the extraction of a group of β-blockers from whole blood on different SPE cartridges: Oasis HLB, MCX and MAX [90]. The MCX-sorbent was the best. In another application on whole blood they applied protein precipitation with aceto-nitrile for the same group of β-blockers [90]. A popular extraction technique for sotalol from plasma was the on-line application of a restricted access material [91–93] or a precolumn [94]. Two methods described offline extractions, one method used protein precipitation as sample clean-up [95], the other one SPE on C8 cartridges [96].

For the extraction of the class III antiarrhythmic bretylium from equine urine, a SPE on Isolute CBA cartridges was described [97]. Amiodarone was extracted from plasma with protein precipitation with acetonitrile [98], liquid–liquid extraction ([99], using methyl-t-butylether [100], or isooctane/2-propanol (85:15, v/v) [101]) or SPE on Oasis MCX [102] or Isolute SCX cartridges [98].

Diltiazem, a class IV antiarrhythmic, was extracted with liquid–liquid extraction [103,104] and SPE on C18 cartridges [105]. Verapamil was extracted from plasma or urine by different approaches: SPE on cyano cartridges [106], SPME [107], liquid–liquid extraction [108–112], a pre-column [113], a restricted access material [114,115] or a molecularly imprinted polymer coupled on-line to a restricted access material precolumn [115]. One article described protein precipitation for sample clean-up [116].

9.3.1.3 Autopsy materials

For mexiletine, one article was published on human forensic samples [45]. Liver homogenate was extracted with ethylacetate, after adding a 1-M carbonate buffer (pH 9.0). For microsomal mixtures, extraction occurred with chloroform, after adjusting the pH to 12–13 with 40% NaOH.

One publication was found on the analysis of flecainide in liver, kidney, bile, stomach contents and vitreous humor [48]. After dilution and homogenization of the tissue samples, deproteinization occurred with ethanol. Extraction was done on SPE-C18 cartridges.

Analysis of propranolol in post-mortem human body fluids and tissue specimens was done with liquid–liquid extraction with acetonitrile, followed by SPE on Bond Elute Certify columns [117]. Angier *et al.* extracted propranolol from liver and kidney with liquid–liquid extraction with chloroform, followed by SPE on Bond Elut Certify material [79].

For the extraction of the class III antiarrhythmic amiodarone from liver micro-somes SPE on Oasis HLB cartridges was performed [118]. Verapamil, a class IV antiarrhythmic, was extracted from hepatocytes with a molecularly imprinted material coupled on-line to a restricted access material precolumn [115].

9.3.2 Analytical applications

9.3.2.1 Crude materials, pure drugs and pharmaceutical preparations

9.3.2.1.1 Thin layer chromatography

One high-performance thin layer chromatography (HPTLC) method was described for quinidine [36]. Pre-coated HPTLC silica gel G 60 plates were developed with choloroform/diethylamine (9.6:1.4, v/v) and scanned at 366 nm in fluorescence/reflectance mode (with cutoff filter K 400).

Five antiarrhythmics, i.e. disopyramide, flecainide, mexiletine, tocainide and verapamil, were separated by TLC on aluminum oxide 60 F_{254} with tetrahydrofuranhexane/25% ammonia (5:4.8:0.2, v/v) and on silicagel 60 F_{254} plates with chloroform/tetrahydrofuran/ethanol/25% ammonia (8.1:1.9:2:0.1, v/v). Substances were identified with UV irradiation at 254 nm [37]. A TLC method for local anesthetics and for the related drug procainamide was reported [119]. Separation occurred on Kieselgel 60 F_{254} plates with ethylacetate/methanol/32% ammonia solution (48:1:1.5, v/v) as eluent. Detection was first at 254 nm, then by spraying with a Co(II) thiocyanate solution, and subsequent spraying with Ehrlich's reagent.

Four TLC methods were published for the enantiomeric separation of propranolol [120,121]. Bhushan and Arora used silica gel plates impregnated with L-aspartic acid as chiral selector [120]. Aboul-Enein *et al.* [122] and Suedee *et al.* [123] used molecularly imprinted polymers (MIPs) as chiral stationary phase. Analysis of acebutolol was achieved with TLC by El-Gindy *et al.* on silica gel 60 F_{254} using ethanol/glacial acetic acid (4:1, v/v) as mobile phase [124] and UV-detection at 230 nm. In another publication on acebutolol and propranolol, TLC-separation was performed on silica gel 60 F_{254} using chloroform/methanol/ammonia (15:7:0.2, v/v). Limits of detection ranged from 30 to 400 ng [125]. The class IV antiarrhythmic verapamil was analyzed from capsules with HPTLC coupled with densitometric analysis [41]. Silicagel 60 F_{254} was used with ethylacetate/ethanol/acetic acid (8:2:0.5, v/v) as mobile phase. Detection and quantification limits were 0.15 and 0.45 μg/spot, respectively. Two TLC methods for the separation of the enantiomers of verapamil were published. In one article, molecular imprinted polymers of S-timolol were prepared as chiral stationary phase [122]. Another used vancomycine as chiral selector [126]. The limit of detection (LOD) for both enantiomers was 74 ng/spot.

9.3.2.1.2 High-pressure liquid chromatography

In the analysis of quinidine, reversed-phase columns were commonly used. An LC method on an Inertsil C18 column was used with 0.05 M aqueous ammonium acetate/acetonitrile (40:60, v/v) as mobile phase [35]. Fluorescence detection was performed: excitation at 280 nm, emission at 380 nm. Two methods using a cyanopropyl column for the analysis of procainamide were reported. A Lichrospher 100 CN packed into a Lichrocart was used with 0.01 M phosphate buffer (adjusted to pH 2.7 with 10% phosphoric acid/acetonitrile (95:5)) as mobile phase at a flow rate of 1 ml/min [127]. Detection occurred at 206 nm.

Needham and Brown analyzed procainamide on cyanopropyl- and pentafluorophenylpropyl stationary phases [128]. Mobile phases consisted of acetonitrile/5 mM ammonium formate, (pH 3.0) (90:10, v/v) and the flow rate was 0.4 ml/min. Detection occurred with MS with electrospray in single ion monitoring mode. For mexiletine, a reversed-phase HPLC method was developed on a C18 Microsorb column, using methanol/0.053 M sodium acetate buffer (50:50) at a flow rate of 1 ml/min [129]. Absorption was monitored at 254 nm. One article described the use of microparticulate strong cation-exchange material as stationary phase for the analysis of flecainide [130]. Only one HPLC method for the enantioselective HPLC analysis of propafenone and its main metabolites was published [131]. The analysis was performed on a Chiralpak AD column with hexane/ethanol (85:15) containing 0.1% diethylamine as mobile phase; detection occurred with a circular dichroism detection system. The analysis of lidocaine is mainly performed by HPLC coupled to either UV-detection [38,132–136] or diode array detection [137,138], while one method has been published using amperometric detection [139]. Reversed-phase C18 columns were used, with exception of C8 columns [38,133] and one stationary phase of liposome coated zirconia–magnesia [140]. Liawruangrath *et al.* described a HPLC method for lidocaine and another drug. Elution occurred on a Sperisorb ODS with 5.5% triethylamine in acetonitrile/water (70:30, v/v) as mobile phase and UV-detection at 254 nm [136]. Limits of detection and quantification were 5 and 12.5 µg/ml, respectively. For the determination of propranolol, a class II antiarrhythmic, HPLC coupled to UV-detection is a popular technique. While one publication used a C18 column [141], all others performed a chiral separation of the enantiomers of propranolol, using a chiralcel OD column [142,143], an amide-based chiral stationary phase XAD-4 [144], or a silicagel covalently derivatized with perphenylcarbamate β-cyclodextrine [145]. Ion-pair chromatography of propranolol was also reported [146,147]. Micellar LC with UV-detection at 225 nm was used to separate 16 β-blockers, including acebutolol, propranolol and sotalol [148]. Ruiz-Angel *et al.* published an HPLC-UV method for the determination of several β-blockers, acebutolol, propranolol and sotalol included, using a Spherisorb C18 column [149] and UV-detection at 225 nm. Bakhtiar *et al.* published an LC-MS/MS method for propranolol on a chirobiotic column (a chiral stationary phase with a bonded macrocyclic glycopeptide, like vancomycin and teicoplanin) [150]. In another publication, LC-MS and high-pressure liquid chromatography-diode array detector (HPLC-DAD) were used to determine acebutolol and propranolol [151]. Direct injection into the MS-detector was also performed. For the HPLC-DAD analysis, a Hypersil BDS C18 column was used, with methanol and a 1% acetic acid solution in water (4:1, v/v) as mobile phase. The absolute detection limits (amounts on column) for LC-MS (single ion monitoring) and HPLC were 0.2–0.5 and 10–25 ng, respectively. Wren and Tchelitcheff published an ultra performance liquid chromatography (UPLC)/MS method for the identification of seven β-blockers [152] (Fig. 9.2).

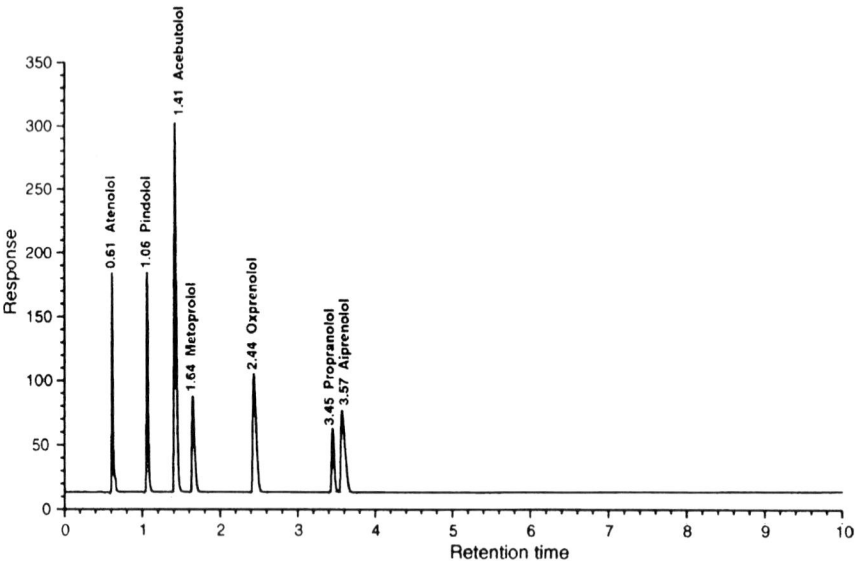

Fig. 9.2. UPLC chromatogram of the β-blockers (from Wren and Tchelitcheff [152], with permission).

All β-blockers were separated within four minutes on a column packed with Acquity C18 BEH particles. Solvents used were 0.1% trifluoroacetic acid in water (A) and 0.1% trifluoroacetic acid in acetonitrile (B). For the MS-measurements a single quadrupole instrument was used. Only one paper has been published for the determination of bretylium: an HPLC method with UV-detection at 220 nm [153]. The authors examined a C18 column and a pentafluorophenylpropylcolumn.

What concerns amiodarone, several HPLC methods have been published. Reversed-phase columns were commonly used [154–157]. One article described the use of microparticulate strong cation-exchange material as stationary phase [130]. Christopherson *et al.* used a Phenomenex Luna C8 (2) column with a 15 mM monobasic potassium phosphate and 30 mM triethylamine buffer (solvent A) and acetonitrile/methanol (1:1, v/v) (solvent B) as mobile phases and UV-detection at 240 nm [155]. Limits of detection and quantification were 0.025 and 0.050 μg/ml, respectively.

Diltiazem, a class IV antiarrhythmic, was commonly analyzed with HPLC coupled to UV-detection [39,158–160] or mass spectrometric detection [158,161]. Reversed-phase columns were used, with exception of one article describing a separation of the enantiomers of diltiazem on a chiral stationary phase [161]. Penmetsa *et al.* described an HPLC-MS method for the analysis of verapamil [162]. A chiral stationary phase, Chiralcel OD-R, was used. For the analysis of verapamil in capsules, an HPLC method coupled to UV-detection was published [40]. A Lichrosorb RP18 column was used with a mobile phase composed of acetonitrile/methanol/phosphate buffer (pH 2.7) (40:40:20, v/v). UV-detection was performed at 220 nm.

9.3.2.1.3 Other techniques
A continuous-flow method was published for the determination of quinidine and quinine based on the enhancement of their native fluorescence by on-line transitory retention on a solid support placed in a flow cell [163]. Detection limits for three injection volumes were determined: 3.9 μg/l for 40 μl, 0.4 μg/l for 600 μl and 0.2 μg/l for 1000 μl.

For the determination of propranolol, one supercritical fluid chromatographic method was published [164]. A nano-liquid chromatographic method for the enantiomeric resolution of propranol was also published, using a vancomycin modified silica stationary phase [165].

Diltiazem was also analyzed by supercritical fluid chromatography [166].

9.3.2.2 Blood, serum, urine and other materials taken from living subjects

9.3.2.2.1 Thin layer chromatography
For the determination of drugs, lidocaine included, in biological samples, Ahrens *et al.* used HPTLC in combination with fibre optical scanning densitometry [71].

9.3.2.2.2 High-pressure liquid chromatography
HPLC was the most commonly used technique to analyze antiarrhythmics in biological samples. To obtain enantioselective analysis of disopyramide and mono-*N*-dealkyldisopyramide in plasma and urine, Bortocan *et al.* used a chiralpak AD column with hexane/ethanol (91:9, v/v) + 0.1% diethylamine as mobile phase [167]. The quantification limit for disopyramide in plasma was 12.5 ng/ml. For the analysis of flecainide, HPLC was combined with UV-detection [47,48], fluorescence detection [49,50] and MS [46]. Doki *et al.* and Katori *et al.* used an ODS TSK gel column, with a mobile phase consisting of 0.1 M sodium 1-pentanesulfonate, acetonitrile and acetic acid (in different concentrations). UV-detection of flecainide [47] was at 298 nm, while for fluorescence detection of flecainide and its metabolites the excitation and emission wavelengths were 300 and 370 nm, respectively [49]. Takiguchi *et al.* used HPLC with fluorescence detection for hair analysis of flecainide [168]. A YMC-Pack Ph column was used, with a mobile phase of acetonitrile and 0.06% phosphoric acid (40:60, v/v). The excitation and emission wavelengths were 300 and 370 nm, respectively. Benijts *et al.* published an HPLC-DAD method for flecainide and its metabolites, not only in blood but also in urine, liver, kidney, bile, stomach contents and vitreous humor [48]. Separation was established on a Hypersil BDS phenyl column using water, methanol and 1.5 M ammonium acetate in a gradient system. The LOD for flecainide and its metabolites ranged from 2 to 8 μg/l blood and from 4 to 6 μg/l urine. LOQ ranged from 6.5 to 25 μg/l blood and from 13 to 20 μg/l urine. Breindahl used mass spectrometric detection and obtained a LOD and quantification in serum of 0.025 and 0.05 μg/ml, respectively [46]. A Supelcosil LC-CN was used for isocratic separation using 25 mM formic acid in water (pH 5.2)/acetonitrile (50:50), at a flow rate of 0.5 ml/min. Analyses of propafenone in plasma and urine were commonly done on C18 reversed-phase columns. Flores-Peres *et al.* used fluorescence detection, obtaining a LOD and quantification of 15 and 50 ng/ml,

respectively [169]. An HPLC assay on a Tracer Excel C18 column was developed using gradient elution with a 0.01 M phosphate buffer and acetonitrile at a flow rate of 1.7 ml/min [51]. The monitoring wavelength was 210 nm resulting in a lower LOQ of 10 ng/ml. Hofmann *et al.* used a mass spectrometer as detector [52]. Separation was performed on a Spherisorb ODS-2 column using 5 mM ammonium acetate in water and 5 mM ammonium acetate in methanol–tetrahydrofuran as mobile phase, at a flow rate of 0.3 ml/min. The mass spectrometer was used in the selective ion monitoring mode. The LOQ here was 10 pmol/ml corresponding to 3.4 ng/ml). The most popular technique for lidocaine analysis was HPLC, either with UV-detection [63–66,70,72,170] or with mass spectrometric detection [60,68,72,171,172]. In the case of UV-detection, C8 [70,170] or C18 [63–65,72] columns were used. Ma *et al.* achieved an LOD in urine of 0.05 μg/ml [65]. Gradient separation on the Johnsson spherigel C18 column occurred with a mixture of acetonitrile and triethylamine in water (11 mM)/0.1% phosphoric acid aqueous solution (10:90, v/v), and a mixture of acetonitrile and triethylamine in water (20 mM)/0.1% phosphoric acid aqueous solution (50:50, v/v). The detection wavelength was 210 nm. Kakiuchi *et al.* established an LOD of 20 ng/ml for free lidocaine in plasma [64]. An ODS column was used for isocratic elution with acetonitrile, methanol and a 0.05 M phosphate buffer (pH 4.0) (10:30:60, v/v) while the detection wavelength was 210 nm. Concerning the HPLC methods with mass detection, different kind of columns were used: C8 [60,68], C18 [72] and diol [172] columns. Kawano *et al.* used a Luna C18 (2) column for detection of lidocaine in plasma; gradient elution occurred with 10 mM ammonium acetate and acetonitrile [72]. SIM was used for detection. A YMC basic column was gradient eluted with 0.1% formic acid in acetonitrile/water (0.5:99.5) and 0.1% formic acid in acetonitrile/water (80:20) [60]. A triple quadrupole was used in multiple reaction monitoring mode. Limits of quantification in plasma and urine were 1.6 and 20 nmol/l, respectively. Arinobu *et al.* performed a gradient elution on a Lichrospher 100 DIOL column using water containing 0.09% formic acid and 20 mM ammonium acetate (solvent A) en acetonitrile (solvent B) [172]. An ion trap mass spectrometer was used in the positive mode and the LOD was 0.38 μg/ml, both for plasma and urine.

Several HPLC methods coupled to UV-detection for the analysis of propranolol in plasma or urine were published. Different columns were used: Inertsil ODS-2 for plasma [84]), Capcell Pack SG120 for urine with LOD 8 μg/ml [77], Separon SGX C18 for plasma with LOD 20 ng/ml [83] and Supelcosil C18 for serum: LOD 0.32 μg/ml [78]. Fluorescence detection is also used, with Nucleosil RP-18 column (plasma: LOQ 1.56 ng/ml [80]), Ultron ES-OVM (microdialysate: LOD for (R)- and (S)-propranolol 10 and 15 ng/ml, respectively [85]), Capcell Pak cyano (CN) UG120 column (rat plasma: LOD 1.34 ng/ml [81]), Spherisorb not endcapped ODS-2 (urine: LOQ 24 ng/ml [173]) and Capcell Pack SG120 (urine: LOD 0.4 μg/ml [77]). Also several HPLC methods coupled to MS were published [73–75,87]. Badaloni *et al.* analyzed propranolol and sotalol in plasma using HPLC-ESI-MS in SIM mode on a laboratory-made chiral stationary phase containing covalently bonded teicoplanin as chiral selector [73]. The LOQ achieved for both sotalol enantiomers was 4 ng/ml. Xia *et al.* published a method for propranolol in rat plasma, with a triple quadrupole as detector

[74]. A Chirobiotic T CSP column was used. The lower LOD was 0.5 ng/ml and the lower LOQ was 2 ng/ml for each enantiomer. For the analysis of acebutolol, propranolol and sotalol, three HPLC methods were published. Delamoye *et al.* used a Hypurity C18 column with UV-detection [88]. Limits of detection and quantification in plasma were 6 and 25 ng/ml for acebutolol, 8 and 25 ng/ml for sotalol and 6 and 25 ng/ml for propranolol. LC-MS/MS was also used [86,87]. Maurer *et al.* achieved the determination from plasma with Superspher 60 RP Select B in a Lichrocart column [86]. The LOD for acebutolol and propranolol was 10 ng/ml while for sotalol the LOD was below 100 ng/ml, respectively (Fig. 9.3). A group of β-blockers was analyzed in whole blood by Josefsson and Sabanovic by LC-MS/MS [90]. The separation was performed on a Hypersil Polar-RP column with acetonitrile and ammonium formate buffer (pH 3) (A – 10:90 and B – 80:20 v/v) as mobile phases. A triple quadrupole was used in positive multiple reaction monitoring.

Analysis of sotalol was mainly performed by HPLC, either combined with UV [92–94] or fluorescence detection [91,92,95,96]. One method used a chiral stationary phase (LOD 18 ng/ml; LOQ 37 ng/ml [92]); all others used a reversed-phase column.

Rbeida *et al.* achieved a LOQ of 5 ng/ml with HPLC-fluorescence detection [91]. Laër *et al.* achieved, also with HPLC-fluorescence detection, an LOQ of 90 ng/ml [96], while Chiap *et al.* had an LOQ of 25 ng/ml for their HPLC-UV method [94]. For bretylium, a class III antiarrhythmic, an LC-MS method in equine urine was published [97]. A reversed-phase column was used, with a mixture of aqueous ammonium formate (pH 3; 10 mM) and acetonitrile as mobile phase. For the analysis of amiodarone in plasma, HPLC coupled to UV-detection [99,100] or tandem mass spectrometric detection [101,102] was used. Juenke *et al.* described the analysis of amiodarone on a Plantinum Cyano column with acetonitrile/methanol/0.05 M ammonium acetate (40:56:3, v/v) as mobile phase at a flow rate of 3.0 ml/min and with UV-detection at 242 nm [100]. The run time was 2.2 min and the LOD was 0.3 mg/l. With the use of tandem mass spectrometry, lower limits of detection can be

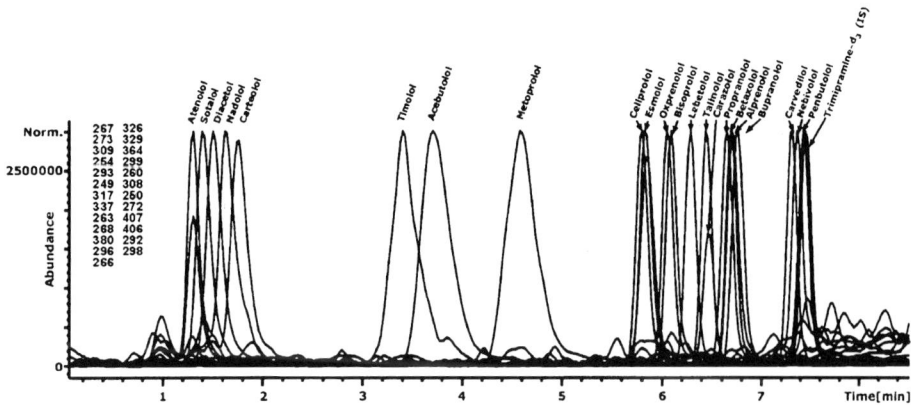

Fig. 9.3. Smoothed, normalized and merged mass chromatograms (scan mode) of a MEDIUM QC sample extract (from Maurer *et al.* [86], with permission).

achieved. Kollroser *et al.* published an LC-MS/MS method with a Symmetry C18 column and acetonitrile/0.1% acetic acid (46:54, v/v) as mobile phase [102]. An ion trap mass spectrometer was used and the limits of detection and quantification were 1 and 50 ng/ml, respectively. Maes *et al.* published an LC-UV and an LC-MS/MS method for amiodarone and desethylamiodarone in horse plasma [98].

An ODS Hypersil column was used for separation using isocratic elution with 0.01% diethylamine and acetonitrile (80:20, v/v) and UV-detection at 245 nm and ion-trap mass detection. The limits of detection for the LC-UV method and the LC-MS/MS method were 15 and 0.10 ng/ml, respectively, while the limits of quantification were 50 and 5 ng/ml, respectively. The class IV antiarrhythmic diltiazem was analyzed in plasma with HPLC coupled to UV-detection [103,104] or tandem mass spectrometry [105]. Reversed-phase columns were used in all methods. Verapamil was mainly analyzed in urine or plasma by HPLC coupled to UV-detection [107,116], fluorescence detection [109,111,113,174] and MS [106–108,110,112, 114,115]. Some articles used a chiral stationary phase for separation of the enantiomers [108,111,174], while others used C8 or C18 reversed-phase columns. Hedeland *et al.* performed a quantification of the enantiomers of verapamil and norverapamil in human plasma on a Chiral-AGP column with acetonitrile and aqueous ammonium acetate buffer (pH 7.4) (15:85, v/v) as mobile phase [108] (Figs. 9.4 and 9.5). Detection was in the SIM mode. Here tri-deuterated verapamil enantiomers were used as internal standards and the LOQ was 100 pg/ml. Borges *et al.* [112] and Von Richter *et al.* [106] both used a C8 column for analysis of verapamil. Mass spectrometric detection was performed in SIM. Borges *et al.* obtained a LOQ of 1 ng/ml while Von Richter *et al.* achieved an LOQ of 0.5 ng/ml.

9.3.2.2.3 Gas chromatography

LC is far more used than GC. The very few articles on gas chromatographic methods are presented below. Mexiletine was analyzed by capillary GC coupled to a mass spectrometric and a flame-ionization detector in urine, liver and microsomes [45]. Separation of the enantiomers was achieved by pre-column derivatization with *S*-(-)-*N*-(fluoroacyl)-prolyl chloride. For the analysis of lidocaine, several gas chromatographic methods were published. GC was combined with flame ionization detection (FID) [56,57], nitrogen–phosphorus detection (NPD) [53] and mass-selective detection [58,62,69]. Determination in plasma by GC-FID led to a detection limit of 5 ng/ml [57]; by GC-NPD the detection limit was 8 ng/ml [53]. Determination in urine with GC-MS led to detection limit of 0.5 ng/ml [62]; in plasma, detection limits of 0.75 ng/ml were achieved by van Hout *et al.* [69]. Sporkert and Pragst described the determination of lidocaine in hair with MS: detection and quantification limits were 0.1 and 0.4 ng/mg, respectively [58].

Two gas chromatographic methods for the determination of propranolol were published. Angier *et al.* used GC-MS to determine propranolol in blood and urine [79]. Li *et al.* used GC coupled to a flame ionization detector to analyze propranolol in urine [76]. The detection limit was 0.275 µg/l with headspace SPME and 0.193 µg/l with SPME. Acebutolol and propranolol were determined by GC/MS in hair [89].

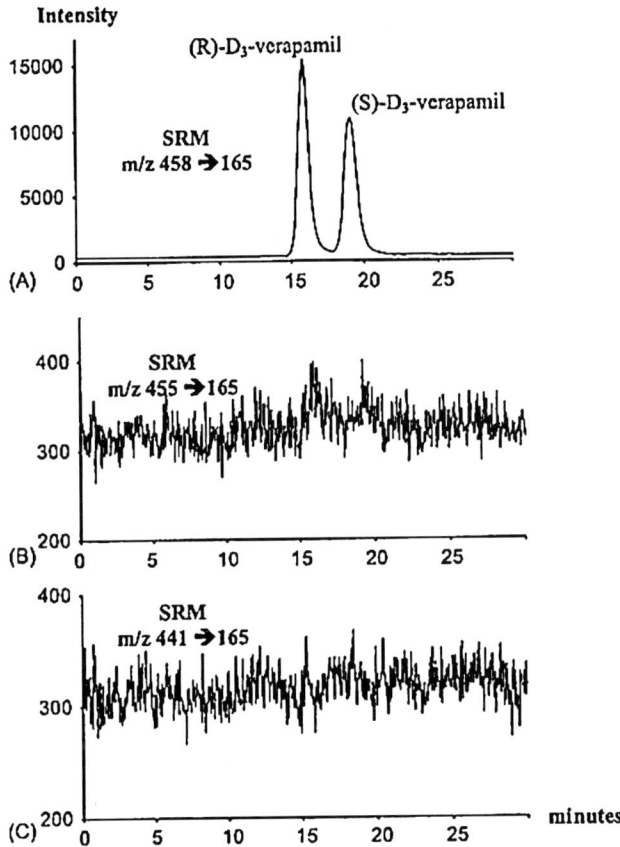

Fig. 9.4. LC-MS/MS chromatogram of an extract from blank plasma spiked with internal standard only. SRM channels: (A) D3-verapamil; (B) verapamil; (C) norverapamil (from Hedeland *et al.* [108], with permission).

Limits of detection were in the range 2–10 pg/mg. Fucci and Offidani published a GC/ MS method for the investigation of an unusual death by propranolol ingestion [175].

9.3.2.2.4 Other techniques

For the analysis of propranolol in a metabolization study, a chiral packed-column supercritical fluid chromatographic system coupled to a tandem mass spectrometer was used [176]. Another packed-column supercritical fluid chromatographic method with tandem mass spectrometry was published for the quantification of propranolol in mouse blood [177].

9.3.2.3 Autopsy materials

One paper described the determination of quinidine in rat liver slices. Elution occurred from a C8 column at a flow rate of 0.4 ml/min with 0.05% trifluoroacetic acid

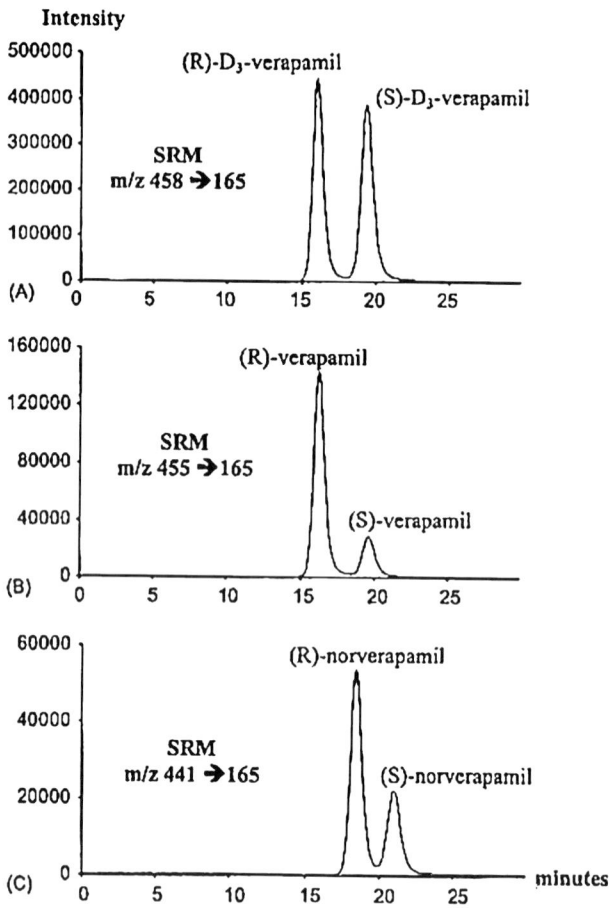

Fig. 9.5. LC-MS/MS chromatogram of a typical patient sample. SRM channels: (A) D3-verapamil; (B) verapamil; (C) norverapamil (from Hedeland *et al.* [108], with permission).

in double-distilled water (A) and 0.05% trifluoroacetic acid in acetonitrile (B) [178]. In this case mass spectrometric detection was used with electrospray ionization (ESI) and in single ion monitoring mode. Mexiletine was analyzed in liver and microsomes by capillary GC coupled to mass spectrometric and FID [45]. For the determination of propranolol in post-mortem human bodyfluids and tissue specimens, LC-atmospheric pressure chemical ionization (APCI)/MS was used [117]. The detection and quantification limits were 0.39 and 0.78 ng/ml, respectively. Angier *et al.* analyzed propranolol in liver and kidney with GC-MS [79]. Analysis of propranolol in microsomes was done with LC-MS/ MS by Shou *et al.*, using selected reaction monitoring on a triple quadrupole [179]. Two articles on the determination of amiodarone in liver microsomes were published. Both described the use of a C18 column; detection was either UV-detection [118] or tandem mass spectrometry [180].

For the determination of verapamil in hepatocytes, LC-UV and LC-MS were performed by Walles *et al.* [107]. The LOD for UV-detection was 52 ng/ml, while for MS-detection 5 ng/ml was obtained.

9.3.3 Immunoassays

Only three articles were published on the development of immunoassays for antiarrhythmics. Saita *et al.* described enzyme-linked immunoassays for amiodarone, sotalol and mexiletine [181–183]. Limits of detection were 16, 32 and 80 ng/ml, respectively.

9.4 ORAL ANTIDIABETICS

The two most important types of oral antidiabetics are sulfonylurea drugs and biguanides. Sulfonylurea drugs discussed here are glibenclamide, gliclazide, glimepiride, glipizide, gliquidone, chlorpropamide and tolbutamide. For the biguanides only metformin is discussed. Besides these two classes, also the new oral antidiabetics rosiglitazone, pioglitazone, nateglinide, acarbose and repaglinide are discussed.

9.4.1 Preparative applications

9.4.1.1 Crude materials, pure drugs and pharmaceutical preparations

Two procedures were published to extract sulfonylurea drugs from health food [184,185]. Both procedures used liquid–liquid extraction with acetone. For the extraction of metformin, a molecularly imprinted SPE was used [186]. This method was used for rapid screening for metformin in pure drugs.

9.4.1.2 Blood, serum, urine and other materials taken from living subjects

The most popular technique to extract sulfonylurea drugs from blood, plasma or urine was liquid–liquid extraction. Different kinds of solvents were used: diethylether/ethylacetate (1:1, v/v) [187,188], diethylether [189,190], 1,2-dichloroethane [191], diethylether/dichloromethane (70:30, v/v) [192], methanol [193], dichloromethane/hexane (1:1, v/v) [194], chloroform [195], dichloromethane [196], 1-chlorobutane/isopropanol/ethylacetate (88:2:10, v/v) [197] and methyl tertiary butylether/*n*-butylchloride (1:1, v/v) [198]. Protein precipitation was also applied, with ethanol [199] or acetonitrile [200]. Three articles were found on SPE. Hsieh and Selinger [201] and Magni *et al.* [202] used C18 SPE columns, while AbuRuz *et al.* [203] used Oasis HLB-SPE columns combined with ion-pairing agents. For the extraction of tolbutamide from plasma, Medvedovici *et al.* used two different approaches [204]. They compared liquid–liquid extraction with ethyl acetate and protein precipitation with methanol. Chen *et al.* also used protein precipitation [205].

For the extraction of tolbutamide from rat plasma, samples were extracted either on Empore C8 or C18 SPE plates [206]. The new oral antidiabetic rosiglitazone was extracted from plasma in three ways: by SPE C18 [207] columns, by liquid–liquid extraction with ethyl acetate [208,209] and by simple protein precipitation with acetonitrile [210] or perchloric acid [211]. Extraction from urine was done by SPE on C8 columns [212]. Pioglitazone was extracted from serum on Oasis HLB-SPE columns [213]. Extraction from plasma was performed with liquid–liquid extraction with methyl tert-butyl ether/*n*-butylchloride (1:1, v/v) [214]. Sample preparation for the analysis of nateglinide in plasma was limited to protein precipitation with acetonitrile [215,216]. Extraction of acarbose from plasma was performed on Oasis HLB-SPE columns [217]. Liquid–liquid extraction was the most popular technique to extract the biguanide metformin from plasma. As extraction solvent dichloro-methane [218], acetonitrile [219] or 1-butanol/n-hexane (50:50, v/v) (with back extraction into diluted acetic acid) [220] was used. Liquid–liquid extraction from dried blood spots was perfomed with 60% methanol in water [221]. Simple protein precipitation was achieved with acetonitrile [222–225]. SPE was also performed on cation exchange SPE-columns [226], ion-pair SPE on Oasis HLB columns [227] and by molecularly imprinted SPE [228]. The latter method was applied for rapid screening for metformin in plasma samples. Heinig and Bucheli used a cyanopre-column for on-line sample preparation [229].

9.4.1.3 Autopsy materials

One article described the sample preparation for the analysis of metformin in post-mortem tissues like blood, serum, bile, kidney and liver [230]. Protein precipitation with acetonitrile was used after homogenization of the tissues in water.

9.4.2 Analytical applications

9.4.2.1 Crude materials, pure drugs and pharmaceutical preparations

9.4.2.1.1 Thin layer chromatography

For several oral antidiabetics TLC methods have been published: glimepiride and pioglitazone [231], gliclazide [232] and chlorpropamide, tolbutamide, glibencl-amide, metformin, pioglitazone, rosiglitazone and repaglinide [233], glibenclamide [184], tolbutamide, chlorpropamide, gliclazide, glibenclamide and glimepiride [185]. Rosiglitazone was analyzed in pharmaceutical formulations with HPTLC [234]. Quantitative analysis of repaglinide in tablets was performed by TLC on RP-8 TLC plates with acetonitrile/(pH 6) phosphate buffer (60:40, v/v) and UV-detection at 225 nm [235]. Limits of quantification and detection were 0.27 and 0.08 µg per 10 µl, respectively. For the analysis of pioglitazone, rosiglitazone and rep-aglinide in pharmaceutical formulations, TLC was also used [236]. The deter-mination was performed on cyanopropyl plates with mobile phases comprising 1, 4-dioxane with phosphate buffers. Berecka *et al.* also analyzed these three new oral antidiabetic drugs with TLC, but on RP-8 adsorbent with UV-detection at 254 nm [237].

9.4.2.1.2 High-performance liquid chromatography

For the major part, the sulfonylurea drugs glibenclamide, gliclazide, glimepiride, glipizide and gliquidone were analyzed with HPLC. C18 columns were the most frequently used in combination with UV-detection [184,185,238–243], while only once mass spectrometric detection was used [184]. Besides C18 columns, also C8 columns were used [244,245]. In addition, one publication used a beta-cell membrane stationary phase [246] and another used a chiral column [247]. For tolbutamide, a nanospray-LC-MS/MS method was reported by Valaskovic *et al.* [248]. Allen *et al.* published an HPLC-UV-MS method, using a trapping column between the UV-detector and the mass spectrometric detector [249].

For the determination of rosiglitazone in bulk and pharmaceutical formulations, Radhakrishna *et al.* used HPLC with a C18 column and UV-detection at 245 nm [250]. Pioglitazone was analyzed in pharmaceutical formulations with HPLC coupled to UV-detection at 266 nm [251] or 228 nm [243]. Jedlicka *et al.* achieved hereby a LOD of 42 ng/ml. Analysis of nateglinide with HPLC was performed on molecularly imprinted polymeric stationary phases [252–254]. One method described the determination of the L-enantiomer of nateglinide (the D-enantiomer being the active form) in a purity control of the bulk drug substance [255]. Analysis was done on a Chiralcel OD-R column with a mobile phase consisting of 0.6 M sodium perchlorate and acetonitrile (48:52, v/v, pH 2), followed by UV-detection at 220 nm. The LOD and quantification were 0.3 and 0.8 µg/ml, respectively.

Repaglinide was determined in pharmaceutical formulations by HPLC with UV-detection [256]. Separation was performed on a C18 column with methanol and 0.1% triethylamine (50:50, v/v, pH adjusted to 7 with % ortho-phosphoric acid (v/v)) while UV-detection was at 235 nm. LP on a porous graphitic carbon stationary phase was performed to detect acarbose and its metabolite [257].

Fluorescence detection occurred after derivatization with 2-aminobenzamide. Metformin, a biguanide, was analyzed in pharmaceutical preparations with HPLC coupled to UV-detection, together with glipizide and gliclazide [258–260]. C18 columns were used, either with UV-detection at 225 [259] or 226 nm [258]. Metformin was also analyzed together with pioglitazone in pharmaceutical preparations, by HPLC and UV-detection [261,262]. Shankar *et al.* used a C18 column with UV-detection at 230 nm [261].

9.4.2.1.3 Gas chromatography

Only one method has been reported on the analysis of gliclazide in a pharmaceutical preparation by capillary GC [263].

9.4.2.2 Blood, serum, urine and other materials taken from living subjects

9.4.2.2.1 High-performance liquid chromatography

The sulfonylurea drugs glibenclamide, gliclazide, glimepiride, glipizide and gliquidone were analyzed with HPLC, either combined with UV [194–196,199,203], electrochemical [200], or mass spectrometric detection [187–193,197,198,201,202,264,265]. Reversed-phase columns were used, mostly C18 columns, with exception of phenyl

[198] and C8-stationary phases [191,194]. AbuRuz *et al.* analyzed simultaneously metformin, glipizide, gliclazide, glibenclamide or glimperide in plasma on a Discovery C18 column with 2 mM sodium dodecyl sulfate, acetonitrile (37.5%) and potassium dihydrogeniumphosphate (62.5%) as a mobile phase followed by UV-detection at 225 nm [203]. They obtained limits of quantification between 4 and 22.5 ng/ml. For the analysis of sulfonylurea drugs in plasma, Maurer *et al.* used a Superspher 60 RP Select B as stationary phase and ammonium formate (pH 3) and acetonitrile as mobile phase [187]. Detection occurred with APCI-MS in SIM mode. Limits of quantification varied between 0.01 and 30 mg/l. Thevis *et al.* described the identification of oral antidiabetics in urine by HPLC-MS/MS [264]. Limits of detection were between 10 and 30 ng/ml. For the identification and quantification of 8 sulfonylureas in plasma, Lc-ion-trap MS was used [189]. Limits of quantification varied between 7.8 and 78.1 µg/l. Ho *et al.* analyzed different kind of antidiabetics in equine plasma and urine: glipizide, glibenclamide, glimepiride, gliclazide, tolazamide, tolbutamide, nateglinide, repaglinide, rosiglitazone and pioglitazone [191]. Separation was achieved on a Supelcosil LC-8-DB column with ammonium formate in water (10 mM, pH 3) and methanol as mobile phase. Screening and confirmation of the drugs were performed in MS-MS mode, with full-scan acquisition. Limits of detection were below 1 ng/ml. For the analysis of glibenclamide in plasma Niopas and Daftsios obtained a LOQ of 10 ng/ml [194], while Ionescu *et al.* achieved a LOD of 25 ng/ml [266], both using reversed-phase columns and UV-detection. Chen *et al.* analyzed glibenclamide in plasma by LC-ESI-MS achieving a LOQ of 1 ng/ml [193]. Hsieh *et al.* established a LOQ of 10 ng/ml in serum, using LC-ESI-MS [201]. Gliclazide was analyzed on an Apollo C18 column with 70 mM disodium tetraborate (pH 7.5) containing 26.5% acetonitrile as mobile phase [200]. Electrochemical detection was performed with a glassy carbon electrode cell and an Ag/AgCl reference cell. This procedure resulted in a LOQ of 50 mM. Park *et al.* used also a C18 column, with 40 mM KH2PO4/acetonitrile/isopropyl alcohol (5:4:1, v/v) as mobile phase, and UV-detection at 229 nm [195]. The LOQ achieved here was 0.1 µg/ml. For the analysis of glimepiride, Rabbaa-Khabbaz *et al.* [196] (serum) and Song *et al.* [199] (plasma) both used a C18 column for separation, a UV-detector at 228 nm for quantification and glibenclamide as internal standard. They achieved a LOQ of 8.2 and 10 ng/ml, respectively. With the use of MS lower limits of quantification could be achieved in plasma. Salem *et al.* used ion-trap MS and achieved a LOQ of 5 ng/ml [190] while Pistos *et al.* achieved a LOQ of 0.5 ng/ml with a triple quadrupole [197]. For the simultaneous analysis of chlorpropamide, glibenclamide and glipizide in serum, Magni *et al.* used a C18 column and water and acetonitrile, both containing 0.05% acetic acid, as mobile phase [202]. For detection, electrospray ionization-mass spectrometry (ESI-MS) was used in SIM mode, resulting in a LOQ (with a signal to noise ratio of 3) of 10 ng/ml for all compounds.

Determination of free glipizide and rosiglitazone in plasma was done by Lin *et al.* [198]. Separation occurred isocratically on an Agilent Zorbax SB-phenyl column with acetontrile/water (50:50, v/v) with 10 mM ammonium acetate and 0.02% trifluoroacetic acid as mobile phase. Detection was with MS in the multiple reaction mode and a LOQ of 1 ng/ml was achieved. Three methods have been published concerning the analysis of tolbutamide. Medvedovici *et al.* analyzed tolbutamide in plasma with HPLC and diode array detection [204]. Balani *et al.* used LC-MS/MS for determination in rat plasma [206]. Supercritical fluid chromatography coupled to tandem mass spectrometry was done to determine tolbutamide in metabolism studies in different matrices [267]. A cyano column was used for separation and a triple quadrupole as detector. Chen *et al.* used high-speed gradient liquid chromatography/tandem mass spectrometry for the analysis of tolbutamide in plasma [205]. A Chromolith SpeedROD RP-18e was used for separation with acetonitrile, water and formic acid as mobile phase at a flow rate of 2.5 ml/min. Detection was with a triple quadrupole. Rosiglitazone, a new oral antidiabetic, was commonly analyzed with HPLC on a reversed-phase C18 column [207–209,211,212]. One method reported the analysis of rosiglitazone on a phenyl stationary phase [210]. Fluorescence detection was mostly used with an excitation wavelength at 247 nm and the emission wavelength at 367 nm [207–211]. Mamidi *et al.* achieved on a C18 column with fluorescence detection a lower LOQ of 5 ng/ml in plasma [209]. With the same configuration Muxlow *et al.* had a lower LOQ of 3 ng/ml in plasma [207]. Chou *et al.* used for the trace analysis of rosiglitazone in urine LC coupled to tandem mass spectrometry [212]. The LOD and lower LOQ achieved here were 0.03 and 0.1 ng/ml, respectively. Pioglitazone was analyzed with HPLC coupled to tandem mass spectrometry, using a reversed-phase C18 column. Xue *et al.* achieved a lower LOQ of 9 ng/ml in serum [213], while Lin *et al.* achieved a lower LOQ of 0.5 ng/ml in plasma [214]. Bauer *et al.* analyzed nateglinide in plasma by HPLC on a C18 column with UV-detection at 210 nm [215]. The LOQ here was 0.1 µg/ml. Nateglinide was also analyzed in animal plasma by micellar electrokinetic chromatography [216]. The LOD was 0.05 µg/ml. Acarbose was analyzed in plasma by LC coupled to tandem mass spectrometry [217]. Separation occurred on a C18 column while detection was in the SIM mode. Determination of the biguanide metformin in blood or serum was commonly done with either HPLC coupled to UV-detection [218–223,227,228,230,268], or HPLC coupled to MS [224–226,229,269]. C18 stationary phases were the most popular [218,219,221,224,227,230,268,269]. Besides C18, also C8 [225], silica [220,222], cyano [229] and cation exchange [223,226] were used as stationary phases. In the case of HPLC-UV, lowest quantification limits in plasma were in the low ng/ml range. Zhang *et al.* achieved an LOQ of 20 ng/ml [223], Amini *et al.* 15.6 ng/ml [220], Cheng *et al.* 10 ng/ml [222] and at last, AbuRuz *et al.* 5 ng/ml [227]. MS did not provide better limits of quantification. Heinig *et al.* [229] and Koseki *et al.* [226] both achieved a lower LOQ of 10 ng/ml. Koseki *et al.* was the only one who used a deuterated internal standard. Chen *et al.* obtained a lower LOQ of 2 ng/ml [225] (Fig. 9.6). Whang *et al.* achieved very low LOD of 250 pg/ml in plasma [269].

9.4.2.3 Autopsy materials

For the analysis of metformin in post-mortem specimens, one article was pubished [230]. HPLC analysis on a C18 column with UV-detection at 236 nm was done on serum, blood, bile, liver and kidney. The LOQ in blood was 0.5 µg/ml.

9.4.3 Immunoassays

No articles on immunoassays of oral antidiabetics were found.

9.5 MUSCLE RELAXANTS

Two major groups of muscle relaxants can be distinguished: depolarizing and non-depolarizing agents. For the first group only two compounds, succinylcholine and suxamethonium, are discussed. The second group can be divided in subgroups by two classifications. The first is based on the duration of the action. d-Tubocurarine and pancuronium are long-acting agents, vecuronium, atracurium, gallamine and cisatracurium are intermediate-acting agents while mivacurium and rocuronium belong to the group of short-acting agents. The non-depolarizing agents can also be classified on their chemical nature and include natural alkaloids (d-tubocurarine and alcuronium), the ammonio steroids (pancuronium, vecuronium and rocuronium) and the benzylisoquinolines (atracurium, mivacurium and gallamine). Besides these compounds, metaxalone and doxacurium are also discussed.

9.5.1 Preparative applications

9.5.1.1 Crude materials, pure drugs and pharmaceutical preparations

No sample preparation was found for muscle relaxants in crude materials, pure drugs or pharmaceutical preparations.

9.5.1.2 Blood, serum, urine and other materials taken from living subjects

Three techniques were used for sample preparation of muscle relaxants: protein precipitation with acetonitrile [270–273], liquid–liquid extraction [274–276] and SPE [97,270,277–280]. Liquid–liquid extraction of rocuronium from plasma was performed with dichloromethane, after ion pairing with iodide [274,275]. Nirogi *et al.* applied diethylether/dichloromethane (7:3, v/v) for liquid–liquid extraction of metaxalone from plasma [276]. Different solid-phase tubes were used for extraction of non-depolarizing muscle relaxants: Maxi-Clean IC-RP [270,280], Isolute CBA [97], Bond Elut C18 HF [278] and styrene-divinylbenzene polymer [279]. Bond Elut C1-SPE columns were used to extract succinylcholine from plasma [277], while Maxi-Clean IC-RP columns were applied for extraction of doxacurium from biological fluids [280].

Fig. 9.6. Representative SRM chromatograms of metformin (I) and IS (II) in human plasma: (A) A blank plasma sample; (B) a blank plasma spiked with metformin at the LOQ of 2 ng/ml and IS (200 ng/ml); (C) plasma sample from a volunteer 2 h after an oral administration of 500 mg metformin hydrochloride in combination with 2.5 mg glyburide (metformin, 36.8 ng/ml) (from Chen *et al.* [225], with permission).

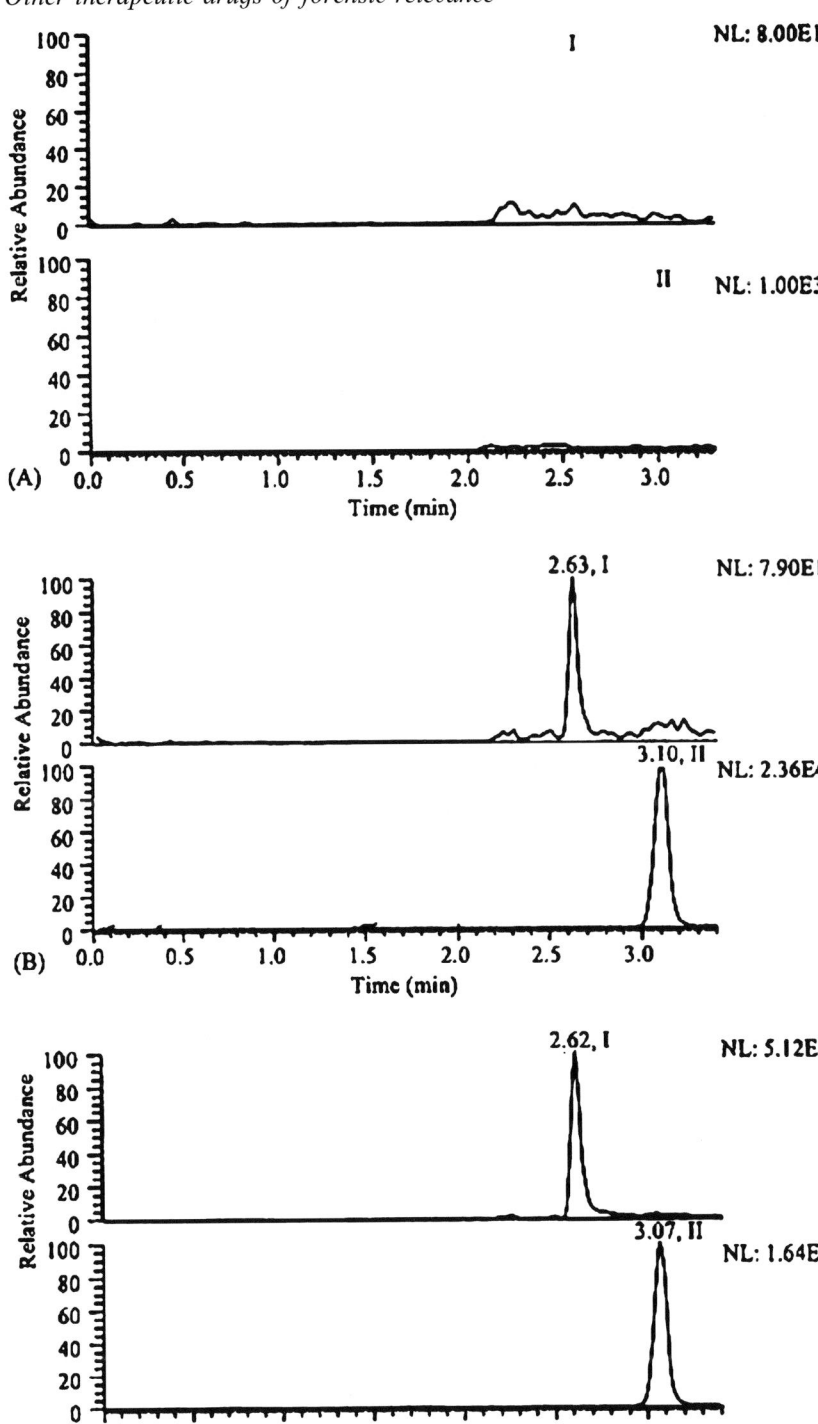

9.5.1.3 Autopsy materials

Extraction of groups of muscle relaxants from autopsy materials was done with SPE on a styrene-divinylbenzene polymer [279], Maxi-Clean IC-RP [270,280] or Bond Elut C18 HF [278].

9.5.2 Analytical applications

9.5.2.1 Crude materials, pure drugs and pharmaceutical preparations

Only one method was published on the determination of pancuronium bromide in a formulation. Zecevic *et al.* analyzed pancuronium bromide in Pavulon® injections with HPLC on a Supelcogel ODP-50 column with acetonitrile/methanol/water/trifluoroacetic acid (20:5:74.9:0.1, v/v) as mobile phase (adjusted to pH 2.0 with trifluoroacetic acid) and detection at 210 nm [281].

9.5.2.2 Blood, serum, urine and other materials taken from living subjects

9.5.2.2.1 High-performance liquid chromatography

Analysis of muscle relaxants was mostly done with HPLC on reversed-phase C18 columns coupled to mass spectrometric detection [97,270,272–274,278]. Kerskes *et al.* determined pancuronium, rocuronium, vecuronium, gallamine, suxamethonium, mivacurium and atracurium and its metabolites in whole blood, urine, bile, liver, brain and muscle [278]. Separation occurred on an Inertsil ODS-2 column using an ammonium acetate buffer (pH 5, 50 mM) and acetonitrile as mobile phase. As a detector an ion-trap mass spectrometer was used, either in MS, MS/MS or MS/MS/MS mode. Atracurium, rocuronium, pancuronium, vecuronium and mivacurium were analyzed in serum, blood, plasma, urine and gastric content by LC-ESI-MS on an X-terra MS C18 column with acetonitrile in 2 mM ammonium-formate (pH 3) as mobile phase [273]. Detection was in positive SIM mode, targeting one quantitation ion and one confirmation ion per compound. The LOQ in serum was 2.5 ng/ml for mivacurium, 5 ng/ml for rocuronium and pancuronium and 10 ng/ml for atracurium and vecuronium. Yiu *et al.* analyzed tubocurarine, alcuronium, vecuronium, mivacurium and pancuronium in equine urine with LC-MS/MS in the positive ESI mode [97]. Montgomery *et al.* analyzed mivacurium and metabolites in blood, urine and autopsy samples with HPLC coupled to fluorescence detection with excitation at 231 nm and emission at 315 nm, and coupled to mass spectrometric detection in multiple reaction monitoring mode [270]. The LOD for both methods was 50 ng/ml in whole blood and the LOQ for the LC-fluorescence method was 100 ng/ml. Rocuronium was analyzed in plasma by LC-ESI-MS [274] (Fig. 9.7). A Symmetry RP-18 column was applied for the separation with acetonitrile in water containing 0.1% trifluoroacetic acid as mobile phase, while single ion monitoring mode was used for mass spectrometric detection.

The LOQ here was 25 ng/ml. Epemolu *et al.* also analyzed rocuronium with LC-MS, but in pig plasma [272]. The lower LOQ here was also 25 ng/ml. Metaxolone

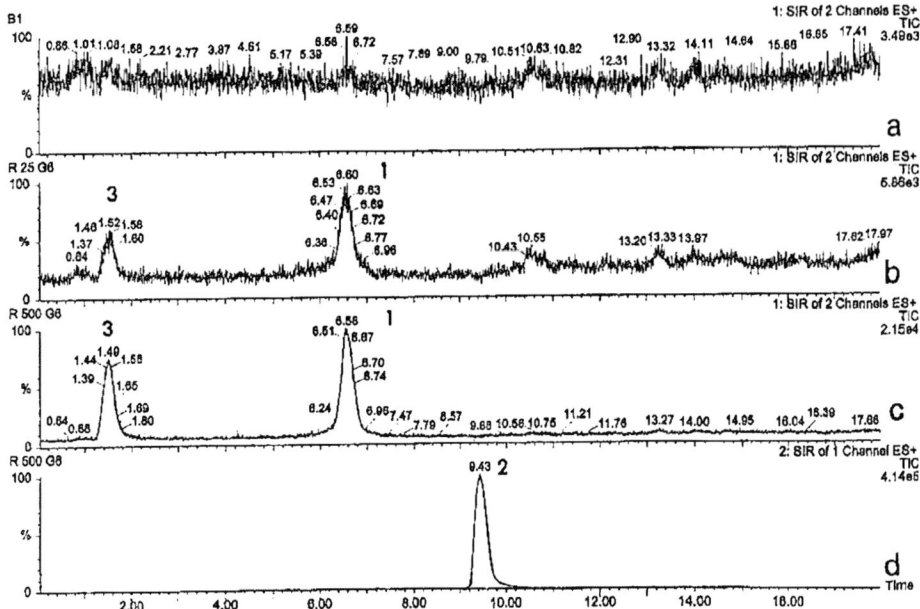

Fig. 9.7. Typical chromatograms of (a) blank human plasma, (b) blank plasma spiked with rocuronium at 25 ng/ml, (c) blank plasma spiked with rocuronium at 500 ng/ml, (d) blank plasma spiked with verapamil (internal standard) at 5 µg/ml. Peak 1 is rocuronium; peak 2 is verapamil; peak 3 is and additional product present in rocuronium (from Farenc *et al.* [274], with permission).

was analyzed in plasma by Nirogi *et al.* [276]. Separation occurred isocratically on a Waters Symmetry C18 column with 0.03% formic acid/acetonitrile (20:80, v/v) as mobile phase and detection was in the multiple reaction monitoring mode. The lower LOQ achieved was 50 ng/ml. Only one HPLC-UV method was described. Sasongko *et al.* determined gallamine in rat plasma, muscle and microdialysate samples on a C18 column with detection at 229 nm [271]. The LOQ in plasma was 1 µg/ml. Another method used only tandem mass spectrometric detection, without HPLC. Succinylcholine concentrations were determined by a stable isotope dilution assay using hexadeuterosuccinylcholine as internal standard with a triple quadrupole [277]. The lower LOQ in plasma was 25 ng/ml.

9.5.2.2.2 Gas chromatography
One GC-MS method was published for the determination of rocuronium in plasma [275]. The LOQ was 26 ng/ml.

9.5.2.3 Autopsy materials

Kerskes *et al.* analyzed pancuronium, vecuronium, rocuronium, gallamine, suxamethonium, mivacurium and atracurium in liver, bile, brain, muscle, urine and blood with LC-ESI-MS [278]. Separation occurred on a Inertsil ODS-2 column using an ammonium acetate buffer (pH 5, 50 mM) and acetonitrile as mobile phase. As detector an ion-trap mass spectrometer was used, in MS, MS/MS and MS/MS/MS mode. Determination of mivacurium in blood, urine, kidney, spleen, liver, muscle, diaphragm and wrist tissue was performed with HPLC coupled to fluorescence detection with excitation at 231 nm and emission at 315 nm, and to mass spectrometric detection in the multiple reaction monitoring mode [270]. Atracurium, rocuronium, pancuronium, mivacurium and vecuronium were analyzed in serum, plasma, blood, urine and gastric content by HPLC-MS on an X-terra MS C18 column with acetonitrile in 2 mM ammonium formate (pH 3) as mobile phase [273]. Detection was in positive SIM mode. Pancuronium was analyzed in aged autopsy samples by Andresen *et al.* with microbore HPLC coupled to ESI-MS with a triple quadrupole [279]. Montgomery *et al.* determined doxacurium in post-mortem fluids by LC-MS/MS [280]. A Licrosher 60 RP-select B column was used for separation with acetonitrile/water/methanesulfonic acid (40:60:0.025, v/v) and detection was in MS/MS mode.

9.5.3 Immunoassays

No articles on immunoassays were found.

9.6 CONCLUDING SUMMARY

Although a diversity of drugs was discussed here, a few general conclusions can be drawn as closure. First, what concerns the sample preparation, liquid–liquid extraction is less used in favor of SPE. SPE is applied more and more in on-line applications, because of benefits like precision, automation, solvent consumption and overnight applicability. New trends in sample preparation are the MIPs and the limited access materials. MIPs will be more important in the future when they become more commercially available.

For the cardiac glycosides, immunoassays remain valuable, however, chromatographic techniques are necessary to avoid false positives arising from interferences. For antiarrhythmics, oral antidiabetics and muscle relaxants no recent data were found on immunoassay analysis.

Overall, the trend of chromatographic analysis evolves to the application of chromatography combined with MS due to the major advantages of sensitivity and selectivity in MS-detection.

9.7 REFERENCES

1   A. Jedlicka, T. Grafnetterova and V. Miller, J. Pharm. Biomed. Anal., 33 (2003) 109.
2   D.M. Holstege, T. Francis, B. Puschner, M.C. Booth and F.D. Galey, J. Agric. Food Chem., 48 (2000) 60.
3   M. Yao, H.J. Zhang, S.H. Chong, M.S. Zhu and R.A. Morrison, J. Pharm. Biomed. Anal., 32 (2003) 1189.

4   P. Kiousi, Y.S. Angelis, M. Koupparis, D. Kouretas, N. Diakakis, A. Desiris and C.G. Georgakopoulos, Chromatographia, 59 (2004) S105.
5   F.T. Hafner, R.A. Kautz, B.L. Iverson, R.C. Tim and B.L. Karger, Anal. Chem., 72 (2000) 5779.
6   J. Malone, P. Dillon, D. Clements, L. McWhirter, J. Pollock, B. Bradley and P. Velagaleti, Drug Metab. Rev., 34 (2002) 163.
7   K. Mitamura, A. Horikawa, A. Nagahama, K. Shimada and Y. Fujii, J. Liq. Chromatogr. Relat. Technol., 28 (2005) 2839.
8   M.E. Pascual, M.E. Carretero, K.V. Slowing and A. Villar, Pharm. Biol., 40 (2002) 139.
9   H.H. Tseng, H.L. Wu, S.J. Lin and S.H. Chen, J. Sep. Sci., 26 (2003) 1693.
10 M. Mayer, A. Muscate-Magnussen, H. Vogel, M. Ehrat and G.J.M. Bruin, Electrophoresis, 23 (2002) 1255.
11 Z. Tang and H.T. Karnes, Biomed. Chromatogr., 17 (2003) 118.
12 K.A. Graefe, Z. Tang and H.T. Karnes, J. Chromatogr. B, 745 (2000) 305.
13 T. Schenk, A. Molendijk, H. Irth, U.R. Tjaden and J. van der Greef, Anal. Chem., 75 (2003) 4272.
14 E. Lacassie, P. Marquet, S. Martin-Dupont, J.M. Gaulier and G. Lachatre, J. Forensic Sci., 45 (2000) 1154.
15 P. Datta and A. Dasgupta, Ther. Drug Monit., 25 (2003) 478.
16 P. Datta and A. Dasgupta, Clin. Chem., 49 (2003) A82.
17 S. Grant, F. Davis, J.A. Pritchard, K.A. Law, S.P.J. Higson and T.D. Gibson, Anal. Chim. Acta, 495 (2003) 21.
18 D. Purvis, O. Leonardova, D. Farmakovsky and V. Cherkasov, Biosens. Bioelectron., 18 (2003) 1385.
19 K. Kudo, H. Tsuchihashi and N. Ikeda, Anal. Chim. Acta, 492 (2003) 83.
20 Z. Tang, K. Graefe, C. March and H.T. Karnes, Microchim. Acta, 144 (2004) 1.
21 Z. Tang and H.T. Karnes, Instrum. Sci. Technol., 30 (2002) 295.
22 P. Fernandez, J.S. Durand, C. Perez-Conde and G. Paniagua, Anal. Bioanal. Chem., 375 (2003) 1020.
23 F. Szurdoki, K.L. Michael and D.R. Walt, Anal. Biochem., 291 (2001) 219.
24 A. Paul, A. Wells and A. Dasgupta, Ther. Drug Monit., 22 (2000) 174.
25 C.X. Zhang, H.H. Zhang and M.L. Feng, Anal. Lett., 36 (2003) 1103.
26 H.L. Qi and C.X. Zhang, Anal. Chim. Acta, 501 (2004) 31.
27 S. Shrestha, I.R. Paeng, S.K. Deo and S. Daunert, Bioconjug. Chem., 13 (2002) 269.
28 Y.N. Shim and I.R. Paeng, Bull. Korean Chem. Soc., 24 (2003) 70.
29 P. Datta and A. Dasgupta, Ther. Drug Monit., 26 (2004) 85.
30 J.J. Miller and R. Valdes, Clin. Chem., 37 (1991) 144.
31 J.J. Miller, R.W. Straub and R. Valdes, Clin. Chem., 40 (1994) 1898.
32 P. Datta and F. Larsen, Clin. Chem., 40 (1994) 1348.
33 B. Bednarczyk, S.J. Soldin, I. Gasinska, M. Dcosta and L. Perrot, Clin. Chem., 34 (1988) 393.
34 E.M. Vaughan-Williams, Symposium on Cardiac Arrhythmias, 1970, p. 449.
35 G. Theodoridis, J. Chromatogr. B, 830 (2006) 238.
36 M.N. Ravishankara, N. Shrivastava, H. Padh and M. Rajani, Planta Med., 67 (2001) 294.
37 R. Pietras, H. Hopkala, D. Kowalczuk and A. Malysza, JPC-J. Planar Chromatogr, 17 (2004) 213.
38 E.H.M. Koster and G.J. de Jong, J. Chromatogr. A, 878 (2000) 27.
39 J.L. Buur, R.E. Baynes, J.L. Yeatts, G. Davidson and T.C. DeFrancesco, J. Pharm. Biomed. Anal., 38 (2005) 60.
40 A. Gumieniczek and H. Hopkala, J. Liq. Chromatogr. Relat. Technol., 24 (2001) 393.
41 D. Kowalczuk, J. AOAC Int., 88 (2005) 1525.
42 L. Fang, X. Yin, X. Sun and E. Wang, Anal. Chim. Acta, 537 (2005) 25.
43 R. Bortocan, V.L. Lanchote, E.J. Cesarino and P.S. Bonato, J. Chromatogr. B, 744 (2000) 299.
44 V.A.P. Jabor, V.L. Lanchote and P.L. Bonato, Electrophoresis, 22 (2001) 1406.
45 Q.F. Tao and S. Zeng, J. Biochem. Biophys. Methods, 54 (2002) 103.
46 T. Breindahl, J. Chromatogr. B, 746 (2000) 249.
47 K. Katori, M. Homma, K. Kuga, I. Yamaguchi, K. Sugibayashi and Y. Kohda, J. Pharm. Biomed. Anal., 32 (2003) 375.

48   T. Benijts, D. Borrey, W.E. Lambert, E.A. De Letter, M.H.A. Piette, C. Van Peteghem and A.P. De Leenheer, J. Anal. Toxicol., 27 (2003) 47.

49   K. Doki, M. Homma, K. Kuga, S. Watanabe, I. Yamaguchi and Y. Kohda, J. Pharm. Biomed. Anal., 35 (2004) 1307.

50   Y. Takiguchi, R. Ishihara, R. Kato, S. Kamihara, M. Yokota and T. Uematsu, J. Pharm. Sci., 90 (2001) 1891.

51   M. Afshar and M. Rouini, Japan Soc. Anal. Chem., 20 (2004) 1307.

52   U. Hofmann, M. Pecia, G. Heinkele, K. Dilger, H.K. Kroemer and M. Eichelbaum, J. Chromatogr. B, 748 (2000) 113–130..

53   M. Abdel-Rehim, M. Bielenstein and T. Arvidsson, J. Microcolumn Sep., 12 (2000) 308.

54   M.W.J. van Hout, V. Jas, H.A.G. Niederlander, R.A. de Zeeuw and G.J. de Jong, J. Sep. Sci., 26 (2003) 1563.

55   M.W.J. van Hout, V. Jas, H.A.G. Niederlander, R.A. de Zeeuw and G.J. de Jong, R. Soc. Chem., 127 (2002) 355.

56   E.H.M. Koster, I.S. Niemeijer and G.J. de Jong, Chromatographia, 55 (2002) 69.

57   E.H.M. Koster, C. Wemes, J.B. Morsink and G.J. de Jong, J. Chromatogr. B, 739 (2000) 175.

58   F. Sporkert and F. Pragst, J. Anal. Toxicol., 24 (2000) 316.

59   G.J. de Jong and E.H.M. Koster, Chromatographia, 52 (2000) S12.

60   M. Abdel-Rehim, M. Bielenstein, Y. Askemark, N. Tyrefors and T. Arvidsson, J. Chromatogr. B, 741 (2000) 175.

61   M.S. Anderson, B. Lu, M. Abdel-Rehim, S. Blomberg and L.G. Blomberg, Rapid Commun. Mass Spectrom., 18 (2004) 2612.

62   M.W.J. van Hout, R.A. de Zeeuw, J.P. Franke and G.J. de Jong, Chromatographia, 57 (2003) 221.

63   T. Fukuda, Y. Kakiuchi, M. Miyabe, S. Kihara, Y. Kohda and H. Toyooka, Reg. Anesth. Pain Med., 28 (2003) 215.

64   Y. Kakiuchi, T. Fukuda, M. Miyabe, M. Homma, H. Toyooka and Y. Kohda, Int. J. Clin. Pharmacol., 40 (2002) 493.

65   M. Ma, S. Kang, Q. Zhao, B. Chen and S. Yao, J. Pharm. Biomed. Anal., 40 (2006) 128.

66   S.Y. Kang, H.B. Wang, M. Ma, B. Chen and S.Z. Yao, Chin. J. Anal. Chem., 32 (2004) 1467.

67   W.D. Cao, J.F. Liu, H.B. Qiu, X.R. Yang and E.K. Wang, Electroanalysis, 14 (2002) 1571.

68   Z. Altun, M. Abdel-Rehim and L.G. Blomberg, J. Chromatogr. B, 813 (2004) 129.

69   M.W.J. van Hout, W.M.A. van Egmond, J.P. Franke, R.A. de Zeeuw and G.J. de Jong, J. Chromatogr. B, 766 (2001) 37.

70   J. Piwowarska, J. Kuczyñska and J. Pachecka, J. Chromatogr. B, 805 (2004) 1.

71   B. Ahrens, D. Blankenhorn and B. Spangenberg, J. Chromatogr. B, 772 (2002) 11.

72   S. Kawano, H. Murakita, E. Yamamoto and N. Asakawa, J. Chromatogr. B, 792 (2003) 49.

73   E. Badaloni, I. D'Acquarica, F. Gasparrini, S. Lalli, D. Misiti, F. Pazzucconi and C.R. Sirtori, J. Chromatogr. B, 796 (2003) 45.

74   Y.Q. Xia, R. Bakhtiar and R.B. Franklin, J. Chromatogr. B, 788 (2003) 317.

75   Y. Alnouti, K. Srinivasan, D. Waddell, H.G. Bi, O. Kavetskaia and A.I. Gusev, J. Chromatogr. A, 1080 (2005) 99.

76   X.J. Li, Z.R. Zeng, M.B. Hu and M. Mao, J. Sep. Sci., 28 (2005) 2489.

77   M. Katayama, Y. Matsuda, K. Shimokawa, S. Tanabe, I. Hara, T. Sato, S. Kaneko and H. Daimon, Anal. Lett., 34 (2001) 91.

78   W.M. Mullett, P. Martin and J. Pawliszyn, Anal. Chem., 73 (2001) 2383.

79   M.K. Angier, R.J. Lewis, A.K. Chaturvedi and D.V. Canfield, J. Anal. Toxicol., 29 (2005) 517.

80   A.J. Braza, P. Modamio and E.L. Marino, J. Chromatogr. B, 738 (2000) 225.

81   H.K. Kim, J.H. Hong, M.S. Park, J.S. Kang and M.H. Lee, Biomed. Chromatogr., 15 (2001) 539.

82   C. Misl'anova and A. Stefancova, J. Trace Microprobe Tech., 19 (2001) 163.

83   C. Misl'anova and M. Hutta, J. Chromatogr. B, 765 (2001) 167.

84   E. Yamamoto, K. Murata, Y. Ishihama and N. Asakawa, Anal. Sci., 17 (2001) 1155.

85   C. Misl'anova, A. Stefancova, J. Oravcova, J. Horecky, T. Trnovec and W. Lindner, J. Chromatogr. B, 739 (2000) 151.

86   H.H. Maurer, O. Tenberken, C. Kratzsch, A.A. Weber and F.T. Peters, J. Chromatogr. A, 1058 (2004) 169.
87   M. Gergov, J.N. Robson, E. Duchoslav and I. Ojanpera, J. Mass Spectrom., 35 (2000) 912.
88   M. Delamoye, C. Duverneuil, F. Paraire, P. de Mazancourt and J.C. Alvarez, Forensic Sci. Int., 141 (2004) 23.
89   P. Kintz, V. Dumestre-Toulet, C. Jamey, V. Cirimele and B. Ludes, J. Forensic Sci., 45 (2000) 170.
90   M. Josefsson and A. Sabanovic, J. Chromatogr. A, 1120 (2006) 1.
91   O. Rbeida, B. Christiaens, P. Chiap, P. Hubert, D. Lubda, K.S. Boos and J. Crommen, J. Pharm. Biomed. Anal., 32 (2003) 829.
92   M. Schlauch, K. Fulde and A.W. Frahm, J. Chromatogr. B, 775 (2002) 197.
93   P. Chiap, O. Rbeida, B. Christiaens, P. Hubert, D. Lubda, K.S. Boos and J. Crommen, J. Chromatogr. A, 975 (2002) 145.
94   P. Chiap, A. Ceccato, B.M. Buraglia, B. Boulanger, P. Hubert and J. Crommen, J. Pharm. Biomed. Anal., 24 (2001) 801.
95   S.R. Santos, O. Papini, C.E. Omosako, M.D. Pereira, T.B.G. Quintavalle, M.F. Riccio, A. Kurata, V.A. Pereira, G. Di-Pietro, O.E. Della-Paschoa, M. Danhof and P.L. Da-Luz, Brazilian J. Med. Biol. Res., 33 (2000) 199.
96   S. Laër, I. Wauer and H. Scholz, J. Chromatogr. B, 753 (2001) 421.
97   K.C.H. Yiu, E.N.M. Ho and T.S.M. Wan, Chromatographia, 59 (2004) S45.
98   A. Maes, K. Baert, S. Croubels, D. De Clercq, G. van Loon, P. Deprez and P. De Backer, J. Chromatogr. B, 836 (2006) 47.
99   A.S. Jun and D.R. Brocks, J. Pharm. Pharm. Sci., 4 (2001) 263.
100   J.M. Juenke, P.I. Brown, G.A. McMillin and F.M. Urry, J. Anal. Toxicol., 28 (2004) 63.
101   H.R. Ha, L. Bigler, B. Wendt, M. Maggiorini and F. Follath, Eur. J. Pharm. Sci., 24 (2005) 271.
102   M. Kollroser and C. Schober, J. Chromatogr. B, 766 (2002) 219.
103   K. Li, X. Zhang and F.L. Zhao, Biomed. Chromatogr., 17 (2003) 522.
104   D. Zendelovska, T. Stafilov and M. Stefova, Anal. Bioanal. Chem., 376 (2003) 848.
105   E. Molden, G.H. Boe, H. Christensen and L. Reubsaet, J. Pharm. Biomed. Anal., 33 (2003) 275.
106   O. von Richter, M. Eichelbaum, F. Schonberger and U. Hofmann, J. Chromatogr. B, 738 (2000) 137.
107   M. Walles, W.M. Mullett, K. Levsen, J. Borlak, G. Wünsch and J. Pawliszyn, J. Pharm. Biomed. Anal., 30 (2002) 307.
108   M. Hedeland, E. Fredriksson, H. Lennernas and U. Bondesson, J. Chromatogr. B, 804 (2004) 303.
109   W. Sawicki, J. Pharm. Biomed. Anal., 25 (2001) 689.
110   A. Tracqui, C. Tournoud, P. Kintz, M. Villain, C. Kummerlen, P. Sauder and B. Ludes, Hum. Exp. Toxicol., 22 (2003) 515.
111   P.C. Ho, D.J. Saville and S. Wanwimolruk, J. Liq. Chromatogr. Relat. Technol., 23 (2000) 1711.
112   N.C.D. Borges, G.D. Mendes, R.E. Barrientos-Astigarraga, P. Galvinas, C.H. Oliveira and G. De Nucci, J. Chromatogr. B, 827 (2005) 165.
113   O.H. Jhee, J.W. Hong, A.S. Om, M.H. Lee, W.S. Lee, L.M. Shaw, J.W. Lee and J.S. Kang, J. Pharm. Biomed. Anal., 37 (2005) 405.
114   M. Walles, J. Borlak and K. Levsen, Anal. Bioanal. Chem., 374 (2002) 1179.
115   W.M. Mullett, M. Walles, K. Levsen, J. Borlak and J. Pawliszyn, J. Chromatogr. B, 801 (2004) 297.
116   S.M. Shahriyar and C.A. Lau-Cam, J. Liq. Chromatogr. Relat. Technol., 23 (2000) 1253.
117   R.D. Johnson and R.J. Lewis, Forensic Sci. Int., 156 (2006) 106.
118   N. Hanioka, Y. Saito, A. Soyama, M. Ando, S. Ozawa and J. Sawada, J. Chromatogr. B, 774 (2002) 105.
119   M. Schmidt and F. Bracher, Pharmazie, 61 (2006) 15.
120   R. Bhushan and M. Arora, Biomed. Chromatogr., 17 (2003) 226.
121   M. Sajewicz, R. Pietka and T. Kowalska, J. Liq. Chromatogr. Relat. Technol., 28 (2005) 2499.
122   H.Y. Aboul-Enein, M.I. El-Awady and C.M. Heard, Pharmazie, 57 (2002) 169.

123 R. Suedee, T. Srichana, J. Saelim and T. Thavonpibulbut, JPC-J. Planar Chromatogr., 14 (2001) 194.

124 A. El-Gindy, A. Ashour, L. Abdel-Fattah and M.M. Shabana, J. Pharm. Biomed. Anal., 24 (2001) 527.

125 J. Krzek and A. Kwiecien, JPC-J. Planar Chromatogr., 18 (2005) 308.

126 R. Bhushan and D. Gupta, Biomed. Chromatogr., 19 (2005) 474.

127 M. Viale, M.O. Vannozzi, I. Pastrone, M.A. Mariggiò, A. Zicca, A. Cadoni, S. Cafaggi, G. Tolino, G. Lunardi, D. Civalleri, W.E. Lindup and M. Esposito, J. Pharmacol. Exp. Ther., 293 (2000) 829.

128 S.R. Needham and P.R. Brown, J. Pharm. Biomed. Anal., 23 (2000) 597.

129 S. Kaushik and K.S. Alexander, J. Liq. Chromatogr. Relat. Technol., 26 (2003) 1287.

130 R.J. Flanagan, E.J. Harvey and E.P. Spencer, Forensic Sci. Int., 121 (2001) 97.

131 P.S. Bonato, L.R. Pires de Abreu, C. Massetto de Gaitani, V.L. Lanchote and C. Bertucci, Biomed. Chromatogr., 14 (2000) 227.

132 A. Malenovic, D. Ivanovic, M. Medenica and B. Jancic, Acta Chim. Slov., 51 (2004) 559.

133 M.G. Gebauer, A.F. McClure and T.L. Vlahakis, Int. J. Pharm., 223 (2001) 49.

134 J.J. Martinez-Pla, Y. Martin-Biosca, S. Sagrado, R.M. Villanueva-Camanas and M.J. Medina-Hernandez, J. Chromatogr. A, 1047 (2004) 255.

135 L. Zivanovic, M. Zecevic, S. Markovic, S. Petrovic and I. Ivanovic, J. Chromatogr. A, 1088 (2005) 182.

136 S. Liawruangrath, B. Liawruangrath and P. Pibool, J. Pharm. Biomed. Anal., 26 (2001) 865.

137 H. Tsuchiya, M. Mizogami and K. Takakura, J. Chromatogr. A, 1073 (2005) 303.

138 J.M.L. Gallego and J.P. Arroyo, Anal. Bioanal. Chem., 374 (2002) 282.

139 Z. Fijalek, E. Baczynski, A. Piwonska and M. Warowna-Grzeskiewicz, J. Pharm. Biomed. Anal., 37 (2005) 913.

140 W.N. Zhang, Z.X. Hu, Y. Liu, Y.Q. Feng and S.L. Da, Talanta, 67 (2005) 1023.

141 Y.S. El-Saharty, J. Pharm. Biomed. Anal., 33 (2003) 699.

142 M.G. Schmid, O. Gecse, Z. Szabo, F. Kilar, G. Gubitz, I. Ali and H.Y. Aboul-Enein, J. Liq. Chromatogr. Relat. Technol., 24 (2001) 2493.

143 M. Santoro, H.S. Cho and E.R.M. Kedor-Hackmann, Drug Dev. Ind. Pharm., 27 (2001) 693.

144 Y.K. Agrawal and R.N. Patel, J. Chromatogr. B, 820 (2005) 23.

145 C.B. Ching, P. Fu, S.C. Ng and Y.K. Xu, J. Chromatogr. A, 898 (2000) 53.

146 X.Z. Xu, L. Zou, M.H. Hu and B.H. Shao, Chin. J. Anal. Chem., 29 (2001) 1295.

147 F. Gritti and G. Guiochon, Anal. Chem., 76 (2004) 7310.

148 A. Detroyer, Y.V. Heyden, S. Carda-Broch, M.C. Garcia-Alvarez-Coque and D.L. Massart, J. Chromatogr. A, 912 (2001) 211.

149 M.J. Ruiz-Angel, S. Carda-Broch, J.R. Torres-Lapasio, E.F. Simo-Alfonso and M.C. Garcia-Alvarez-Coque, Anal. Chim. Acta, 454 (2002) 109.

150 R. Bakhtiar and F.L.S. Tse, Rapid Commun. Mass Spectrom., 14 (2000) 1128.

151 M.E. Abdel-Hamid, Farmaco, 55 (2000) 136.

152 S.A.C. Wren and P. Tchelitcheff, J. Pharm. Biomed. Anal., 40 (2006) 571.

153 D.S. Bell and A.D. Jones, J. Chromatogr. A, 1073 (2005) 99.

154 N. Thyagarajapuram and K.S. Alexander, J. Liq. Chromatogr. Relat. Technol., 26 (2003) 1315.

155 M.J. Christopherson, K.J. Yoder and R.B. Miller, J. Liq. Chromatogr. Relat. Technol., 27 (2004) 95.

156 M.A. Khan, S. Kumar, J. Jayachandran, S.V. Vartak, A. Bhartiya and S. Sinha, Chromatographia, 61 (2005) 599.

157 A. Medvedovici, V. David, F. Albu and A. Farca, Rev. Roum. Chim., 49 (2004) 783.

158 M.G. Quaglia, E. Donati, S. Fanali, E. Bossu, A. Montinaro and F. Buiarelli, J. Pharm. Biomed. Anal., 37 (2005) 695.

159 M. Gil-Agusti, S. Carda-Broch, M.C. Garcia-Alvarez-Coque and J. Esteve-Romero, J. Chromatogr. Sci., 38 (2000) 521.

160 I. Quinones, A. Cavazzini and G. Guiochon, J. Chromatogr. A, 877 (2000) 1.

161 B. Chankvetadze, I. Kartozia and G. Blaschke, J. Pharm. Biomed. Anal., 27 (2002) 161.

162   K.V. Penmetsa, C.D. Reddick, S.W. Fink, B.L. Kleintop, G.C. DiDonato, K.J. Volk and S.E. Klohr, J. Liq. Chromatogr. Relat. Technol., 23 (2000) 831.

163   S. Ortega-Algar, N. Ramos-Martos and A. Molina-Diaz, Microchim. Acta, 147 (2004) 211.

164   A. Ellwanger, P.K. Owens, L. Karlsson, S. Bayoudh, P. Cormack, D. Sherrington and B. Sellergren, J. Chromatogr. A, 897 (2000) 317.

165   G. D'Orazio, Z. Aturki, M. Cristalli, M.G. Quaglia and S. Fanali, J. Chromatogr. A, 1081 (2005) 105.

166   K. Yaku and F. Morishita, J. Biochem. Biophys. Methods, 43 (2000) 59.

167   R. Bortocan, V.L. Lanchote, E.J. Cesarino and P.S. Bonato, J. Chromatogr. B, 744 (2000) 299.

168   Y. Takiguchi, R. Ishihara, M. Torii, R. Kato, S. Kamihara and T. Uematsu, Eur. J. Clin. Pharmacol., 58 (2002) 99.

169   C. Flores-Peres, H. Juarez-Olguin, B. Ramirez-Mendiola and J.B. Chavez, Chromatographia, 62 (2005) 373.

170   L. Manna, P. Bertocchi, L. Valvo and A. Bardocci, J. Pharm. Biomed. Anal., 29 (2002) 1121.

171   J.D. Feary, K.R. Mama, S.M. Thomasy, A.E. Wagner and R.M. Enns, Am. J. Vet. Res., 67 (2006) 317.

172   T. Arinobu, H. Hattori, A. Ishii, T. Kumazawa, X.P. Lee, O. Suzuki and H. Seno, Chromatographia, 57 (2003) 301.

173   M.J. Ruiz-Angel, P. Fernandez-Lopez, J.A. Murillo-Pulgarin and M.C. Garcia-Alvarez-Coque, J. Chromatogr. B, 767 (2002) 277.

174   E. Brandstetrova, G. Endersz and G. Blaschke, Pharmazie, 56 (2001) 536.

175   N. Fucci and C. Offidani, Am. J. Forensic Med. Pathol., 21 (2000) 56.

176   Y.S. Hsieh, L. Favreau, K.C. Cheng and J.W. Chen, Rapid Commun. Mass Spectrom., 19 (2005) 3037.

177   J.W. Chen, Y.S. Hsieh, J. Cook, R. Morrison and W.A. Korfmacher, Anal. Chem., 78 (2006) 1212.

178   H. Axelsson, C. Granhall, E. Floby, Y. Jaksch, M. Svedling and A.K. Sohlenius-Sternbeck, Toxicol. Vitro, 17 (2003) 481.

179   W.Z. Shou, L. Magis, A.C. Li and M.S. Bryant, J. Mass Spectrom., 40 (2005) 1347.

180   S.W. Myung, Y.J. Chang, H.K. Min, D.H. Kim, M. Kim, T. Kang, E.A. Yoo, Y.T. Sohn and Y.H. Yim, Rapid Commun. Mass Spectrom., 14 (2000) 2046.

181   T. Saita, H. Fujito and M. Mori, Biol. Pharm. Bull., 25 (2002) 954.

182   T. Saita, H. Fujito and M. Mori, Biol. Pharm. Bull., 26 (2003) 761.

183   T. Saita, H. Fujito, Y. Nakano and M. Mori, Biol. Pharm. Bull., 27 (2004) 94.

184   K. Kumasaka, T. Kojima, K. Doi and S. Satoh, Yakugaku Zasshi – J. Pharm. Soc. Jpn., 123 (2003) 1049.

185   K. Kumasaka, T. Kojima, H. Honda and K. Doi, J. Health Sci., 51 (2005) 453.

186   E.P.C. Lai and S.Y. Feng, Microchem. J., 75 (2003) 159.

187   H.H. Maurer, C. Kratzsch, T. Kraemer, F.T. Peters and A.A. Weber, J. Chromatogr. B, 773 (2002) 63.

188   Y. Dotsikas, C. Kousoulos, G. Tsatsou and Y.L. Loukas, Rapid Commun. Mass Spectrom., 19 (2005) 2055.

189   G. Hoizey, D. Lamiable, T. Trenque, A. Robinet, L. Binet, M.L. Kaltenbach, S. Havet and H. Millart, Clin. Chem., 51 (2005) 1666.

190   Salem II, J. Idrees and J.I. Al Tamimi, J. Chromatogr. B, 799 (2004) 103.

191   E.N.M. Ho, K.C.H. Yiu, T.S.M. Wan, B.D. Stewart and K.L. Watkins, J. Chromatogr. B, 811 (2004) 65.

192   M.R.L. Moura, G. de Nucci, S. Rath and F.G.R. Reyes, Anal. Bioanal. Chem., 378 (2004) 499.

193   B.M. Chen, Y.Z. Liang, F.Q. Guo, L.F. Huang, F.L. Deng, X. Chen and Y.L. Wang, Anal. Chim. Acta, 514 (2004) 185.

194   I. Niopas and A.C. Daftsios, J. Pharm. Biomed. Anal., 28 (2002) 653.

195   J.Y. Park, K.A. Kim, S.L. Kim and P.W. Park, J. Pharm. Biomed. Anal., 35 (2004) 943.

196   L. Rabbaa-Khabbaz, R.A. Daoud, D. Karam-Sarkis, C. Atallah and A. Zoghbi, J. Liq. Chromatogr. Relat. Technol., 28 (2005) 3255.

197 C. Pistos, M. Koutsopoulou and I. Panderi, Biomed. Chromatogr., 19 (2005) 394.
198 Z.P.J. Lin, D. Desai-Krieger and L. Shum, J. Chromatogr. B, 801 (2004) 265.
199 Y.K. Song, J.E. Maeng, H.R. Hwang, J.S. Park, B.C. Kim, J.K. Kim and C.K. Kim, J. Chromatogr. B, 810 (2004) 143.
200 C.Y. Kuo and S.M. Wu, J. Chromatogr. A, 1088 (2005) 131.
201 S. Hsieh and K. Selinger, J. Chromatogr. B, 772 (2002) 347.
202 F. Magni, L. Marazzini, S. Pereira, L. Monti and M.G. Kienle, Anal. Biochem., 282 (2000) 136.
203 S. AbuRuz, J. Millership and J. McElnay, J. Chromatogr. B, 817 (2005) 277.
204 A. Medvedovici, V. David, D. Miron and C. Mircioiu, Anal. Lett., 33 (2000) 2219.
205 Y.L. Chen, H. Junga, X.Y. Jiang and W. Naidong, J. Sep. Sci., 26 (2003) 1509.
206 S.K. Balani, N.V. Nagaraja, M.G. Qian, A.O. Costa, J.S. Daniels, H. Yang, P.R. Shimoga, J.T. Wu, L.S. Gan, F.W. Lee and G.T. Miwa, Drug Metab. Dispos., 34 (2006) 384.
207 A.M. Muxlow, S. Fowles and P. Russell, J. Chromatogr. B, 752 (2001) 77.
208 R. Mamidi, M.R. Chaluvadi, B. Benjamin, M. Ramesh, K. Katneni, A.P. Babu, J. Bhanduri, N.M.U. Rao and R. Rajagopalan, Arzneimittelforschung, 52 (2002) 560.
209 R. Mamidi, B. Benjamin, M. Ramesh and N.R. Srinivas, Biomed. Chromatogr., 17 (2003) 417.
210 M.W. Hruska and R.F. Frye, J. Chromatogr. B, 803 (2004) 317.
211 K.A. Kim and J.Y. Park, Biomed. Chromatogr., 18 (2004) 613.
212 C.C. Chou, M.R. Lee, F.C. Cheng and D.Y. Yang, J. Chromatogr. A, 1097 (2005) 74.
213 Y.J. Xue, K.C. Turner, J.B. Meeker, J. Pursley, M. Arnold and S. Unger, J. Chromatogr. B, 795 (2003) 215.
214 Z.P.J. Lin, W.H. Ji, D. Desai-Krieger and L.Y. Shum, J. Pharm. Biomed. Anal., 33 (2003) 101.
215 S. Bauer, E. Stormer, J. Kirchheiner, C. Michael, J. Brockmoller and I. Roots, J. Pharm. Biomed. Anal., 31 (2003) 551.
216 H.Y. Yan, G.L. Yang, F.X. Qiao and C. Yi, J. Pharm. Biomed. Anal., 36 (2004) 169.
217 B.B. Raut, B.L. Kolte, A.A. Deo, M.A. Bagool and D.B. Shinde, J. Liq. Chromatogr. Relat. Technol., 27 (2004) 1759.
218 F. Tache, V. David, A. Farca and A. Medvedovici, Microchem. J., 68 (2001) 13.
219 A. Zarghi, S.M. Foroutan, A. Shafaati and A. Khoddam, J. Pharm. Biomed. Anal., 31 (2003) 197.
220 H. Amini, A. Ahmadiani and P. Gazerani, J. Chromatogr. B, 824 (2005) 319.
221 S. AbuRuz, J. Millership and J. McElnay, J. Chromatogr. B, 832 (2006) 202.
222 C.L. Cheng and C.H. Chou, J. Chromatogr. B, 762 (2001) 51.
223 M. Zhang, G.A. Moore, M. Lever, S.J. Gardiner, C.M.J. Kirkpatrick and E.J. Begg, J. Chromatogr. B, 766 (2002) 175.
224 G.P. Zhong, H.C. Bi, S.F. Zhou, X. Chen and M. Huang, J. Mass Spectrom., 40 (2005) 1462.
225 X.Y. Chen, Q. Gu, F. Qiu and D.F. Zhong, J. Chromatogr. B, 802 (2004) 377.
226 N. Koseki, H. Kawashita, M. Niina, Y. Nagae and N. Masuda, J. Pharm. Biomed. Anal., 36 (2005) 1063.
227 S. AbuRuz, J. Millership and J. McElnay, J. Chromatogr. B, 798 (2003) 203.
228 S.Y. Feng, E.P.C. Lai, E. Dabek-Zlotorzynska and S. Sadeghi, J. Chromatogr. A, 1027 (2004) 155.
229 K. Heinig and F. Bucheli, J. Pharm. Biomed. Anal., 34 (2004) 1005.
230 K.A. Moore, B. Levine, J.M. Titus and D.R. Fowler, J. Anal. Toxicol., 27 (2003) 592.
231 R.S.T. Menon, S. Inamdar, M. Mote and A. Menezes, JPC-J. Planar Chromatogr., 17 (2004) 154.
232 J. Krzek, M. Dabrowska and U. Hubicka, JPC-J. Planar Chromatogr., 14 (2001) 183.
233 A. Gumieniczek, H. Hopkala, A. Berecka and D. Kowalczuk, JPC-J. Planar Chromatogr., 16 (2003) 271.
234 A. Gumieniczek, A. Berecka, H. Hopkala and T. Mroczek, J. Liq. Chromatogr. Relat. Technol., 26 (2003) 3307.
235 A. Gumieniczek, A. Berecka and H. Hopkala, JPC-J. Planar Chromatogr., 18 (2005) 155.
236 A. Gumieniczek, H. Hopkala and A. Berecka, J. Liq. Chromatogr. Relat. Technol., 27 (2004) 2057.
237 A. Berecka, A. Gumieniczek and H. Hopkala, JPC-J. Planar Chromatogr., 18 (2005) 61.
238 Y. Song, L.P. Niu, D.F. Wang, Y.P. Hu and D.Y. Hou, J. Sep. Sci., 26 (2003) 1595.
239 A. Berecka, A. Gumieniczek and H. Hopkala, J. AOAC Int., 89 (2006) 319.

240   S. Dhawan and A.K. Singla, J. Chromatogr. Sci., 41 (2003) 295.

241   P. Kovarikova, J. Klimes, J. Dohnal and L. Tisovska, J. Pharm. Biomed. Anal., 36 (2004) 205.

242   A. Dubey and I.C. Shukla, J. Indian Chem. Soc., 81 (2004) 84.

243   R.T. Sane, S.N. Menon, S. Inamdar, M. Mote and G. Gundi, Chromatographia, 59 (2004) 451.

244   M.A. Khan, S. Sinha, S. Vartak, A. Bhartiya and S. Kumar, J. Pharm. Biomed. Anal., 39 (2005) 928.

245   S.V. Saradhi, V.S. Kiran, V.H. Bindu and G.D. Rao, Asian J. Chem., 18 (2006) 1309.

246   G.D. Yang, L.C. He, X.L. Bian and L. Zhao, Chin. Sci. Bull., 50 (2005) 2427.

247   Y.R. Song, D.F. Wang, L.P. Niu, Y.P. Hu, Y.P. Yang and D.Y. Hou, Chin. J. Anal. Chem., 32 (2004) 435.

248   G.A. Valaskovic, L. Utley, M.S. Lee and J.T. Wu, Rapid Commun. Mass Spectrom., 20 (2006) 1087.

249   J.R. Allen, J.D. Williams, D.J. Burinsky and S.R. Cole, J. Chromatogr. A, 913 (2001) 209.

250   T. Radhakrishna, J. Satyanarayana and A. Satyanarayana, J. Pharm. Biomed. Anal., 29 (2002) 873.

251   A. Jedlicka, J. Klimes and T. Grafnetterova, Pharmazie, 59 (2004) 178.

252   J.F. Yin, G.L. Yang and Y. Chen, J. Chromatogr. A, 1090 (2005) 68.

253   J.F. Yin, G.L. Yang, Y.H. Zhang, H.Y. Liu and Y. Chen, Acta Chim. Sin., 62 (2004) 1922.

254   G.L. Yang, J.F. Yin, Z.W. Li, H.Y. Liu, L.P. Cai, D.X. Wang and Y. Chen, Chromatographia, 59 (2004) 705.

255   M.L. Qi, P. Wang, Y.J. Sun and Y. Li, J. Liq. Chromatogr. Relat. Technol., 26 (2003) 1839.

256   M. Gandhimathi, T.K. Ravi and S.K. Renu, Anal. Sci., 19 (2003) 1675.

257   Y. Daali, S. Cherkaoui, X. Cahours, E. Varesio and J.L. Veuthey, J. Sep. Sci., 25 (2002) 280.

258   B.L. Kolte, B.B. Raut, A.A. Deo, M.A. Bagool and D.B. Shinde, J. Liq. Chromatogr. Relat. Technol., 26 (2003) 1117.

259   M. Vasudevan, J. Ravi, S. Ravisankar and B. Suresh, J. Pharm. Biomed. Anal., 25 (2001) 77.

260   B.L. Kolte, B.B. Raut, A.A. Deo, M.A. Bagool and D.B. Shinde, J. Sep. Sci., 28 (2005) 2076.

261   M. Shankar, V.D. Modi, D.A. Shah, K.K. Bhatt, R.S. Mehta, M. Geetha and B.J. Patel, J. AOAC Int., 88 (2005) 1167.

262   B.L. Kolte, B.B. Raut, A.A. Deo, M.A. Bagool and D.B. Shinde, J. Chromatogr. Sci., 42 (2004) 27.

263   J. Krzek, J. Czekaj, M. Moniczewska and W. Rzeszutko, J. AOAC Int., 84 (2001) 1695.

264   M. Thevis, H. Geyer and W. Schanzer, Rapid Commun. Mass Spectrom., 19 (2005) 928.

265   H. Kim, K.Y. Chang, C.H. Park, M.S. Jang, J.A. Lee, H.J. Lee and K.R. Lee, Chromatographia, 60 (2004) 93.

266   M. Ionescu, T. Galaon, V. David and A. Medvedovici, Rev. Roum. Chim., 49 (2004) 351.

267   Y. Hsieh, L. Favreau, J. Schwerdt and K.C. Cheng, J. Pharm. Biomed. Anal., 40 (2006) 799.

268   V. David, C. Barcutean, I. Sora and A. Medvedovici, Rev. Roum. Chim., 50 (2005) 269.

269   Y.W. Wang, Y.B. Tang, J.K. Gu, J.P. Fawcett and X. Bai, J. Chromatogr. B, 808 (2004) 215.

270   M.A. Montgomery, M.A. LeBeau, M.L. Miller and R.A. Jufer, J. Anal. Toxicol., 29 (2005) 637.

271   L. Sasongko, I. Ramzan, K.M. Williams and A.J. McLachlan, J. Chromatogr. B, 754 (2001) 467.

272   O. Epemolu, I. Mayer, F. Hope, P. Scullion and P. Desmond, Rapid Commun. Mass Spectrom., 16 (2002) 1946.

273   H. Sayer, O. Quintela, P. Marquet, J.L. Dupuy, J.M. Gaulier and U. Lachatre, J. Anal. Toxicol., 28 (2004) 105.

274   C. Farenc, C. Enjalbal, P. Sanchez, F. Bressolle, M. Audran, J. Martinez and J.L. Aubagnac, J. Chromatogr. A, 910 (2001) 61.

275   L. Gao, I. Ramzan and B. Baker, J. Chromatogr. B, 757 (2001) 207.

276   R.V.S. Nirogi, V.N. Kandikere, M. Shukla, K. Mudigonda, W. Shrivastava and P.V. Datla, J. Anal.Toxicol., 30 (2006) 245.

277   J.J. Roy, D. Boismenu, H. Gao, O.A. Mamer and F. Varin, Anal. Biochem., 290 (2001) 238.

278   C.H.M. Kerskes, K.J. Lusthof, P.G.M. Zweipfenning and J.P. Franke, J. Anal. Toxicol., 26 (2002) 29.

279   B.D. Andresen, A. Alcaraz and P.M. Grant, J. Forensic Sci., 50 (2005) 196.

280   M.A. Montgomery, M.A. LeBeau and A.J. Jenkins, J. Anal. Toxicol., 30 (2006) 57.

281   M. Zecevic, L. Zivanovic and A. Stojkovic, J. Chromatogr. A, 949 (2002) 61.8.

M.J. Bogusz (Ed.). Forensic Science
Handbook of Analytical Separations, Vol. 6

CHAPTER 10

Mushroom toxins

Ilkka Ojanperä

Department of Forensic Medicine, University of Helsinki, P.O. Box 40, FI-00014, Finland

10.1 INTRODUCTION

Mushroom poisonings can be divided into non-life-threatening, early-onset poisonings, where symptoms appear within 6 h of eating a mushroom, and life-threatening, late-onset poisonings, where symptoms appear over 6 h after ingestion [1,2]. The toxins in the early-onset group include muscarine from *Inocybe* spp. (fiber caps) and *Clitocybe dealbata* (the sweater), muscimol and ibotenic acid from *Amanita muscaria* (fly agaric) and *A. pantherina* (the panther), hallucinogens – especially psilocybin – from *Psilocybe* spp. (magic mushrooms), coprine from *Coprinus atramentarius* (alcohol inky cap) and a possible allergen from *Paxillus involutus* (poison pax).

The toxins involved in the late-onset group include amatoxins from *A. phalloides* (death cap), *A. verna* (destroying angel), *A. virosa* and some *Galerina* and *Lepiota* species, orellanine from *Cortinarius orellanus* (deadly *Cortinarius*) and gyromitrin from *Gyromitra esculenta* (false morel). An extensive review of the occurrence and treatment of amatoxin poisonings, including the amatoxin-containing mushroom species, has recently been published by Enjalbert *et al.* [3].

In addition to clinical symptoms, the diagnosis of mushroom poisoning may be based on the identification of spores present in materials originating from mushrooms. The analysis of spores in stomach contents has been a frequently used procedure, especially in post-mortem forensic toxicology. However, immunoassay procedures and an increasing number of methods involving chromatography and mass spectrometry are also accessible today. More extensive literature is available on the toxicological analysis of two types of toxins: amanitins, a major factor in mushroom poisonings and psilocybin and psilocin, controlled substances found in magic mushrooms abuse.

10.2 AMATOXINS

Over 90% of lethal mushroom poisonings are attributed to one of the hepatotoxic species of amanita [4]. The various amanita species resemble edible mushrooms, so the poisonings are usually accidental. Toxic amanita contain amatoxins (amanitins, amanin, amanullin), phallotoxins (phalloidin, phalloin, phallisin, phallacidin) and virotoxins (viroidin, viroisin, desoxyviroidin desoxyviroisin). The chemical structures of α- and β-amanitin appear in Fig. 10.1.

A standard amanitin assay in body fluids has served as a radioimmunoassay (RIA) procedure sensitive to α-, β- and γ-amanitins with a limit of detection (LOD) of 3 ng/ml [5]. More recently, an enzyme-linked immunosorbent assay (ELISA) has been made commercially available for α- and γ-amanitins. This assay produced no false-positive findings with cases unexposed to mushrooms below a functional sensitivity value of 1.5 ng/ml in urine [6] and, with a urinary cut-off concentration of 5 ng/ml, researchers obtained the highest diagnostic efficacy in mushroom poisonings [7]. In 2004, Abuknesha and Maragkou [8] developed another ELISA sensitive to β-amanitin and moderately cross-reacting with α-amanitin.

For forensic purposes, however, immunoassay can be applied only as a preliminary test. Table 10.1 shows the chromatography methods published for amanitins that are feasible in both the clinical and forensic contexts. Thin layer chromatography (TLC) has been applied to amanita mushrooms, but not to intoxication samples as the detection limits of published methods seem too high for the analysis of amanitins in body fluids [9]. High performance liquid chromatography (HPLC) with UV or electrochemical (EC) detection is the most common technique reported, but researchers have also used capillary zone electrophoresis (CZE) and HPLC coupled with mass spectrometry (LC-MS). The sample preparation involves either a simple protein precipitation step using a water-soluble solvent or a mixture of solvents, or a solid-phase extraction (SPE) with reversed-phase or mixed-mode phases. The LOD in

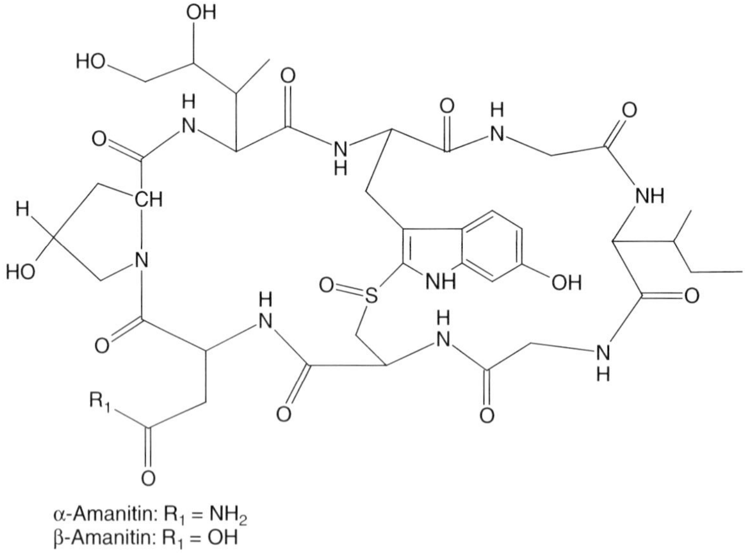

α-Amanitin: R_1 = NH_2
β-Amanitin: R_1 = OH

Fig. 10.1. Structures of α- and β-amanitin.

TABLE 10.1
ANALYTICAL METHODS FOR AMANITINS

Analyte	Specimen	Extraction	Analysis Method	Limit of Detection	Ref.
α-Amanitin	Serum	Protein precip.	HPLC RP-18, UV 302 nm	500 ng/ml	[10]
α-Amanitin	Serum	Protein precip.	HPLC RP-18, UV 303 nm	25 ng/ml	[11]
α-Amanitin β-Amanitin phalloidin	Serum, urine	Protein precip.	HPLC RP-18, UV 302 nm	10 ng injected	[12]
α-Amanitin β-Amanitin	Serum, urine, Stomach washings	protein precip.and SPE RP-18	HPLC RP-18, UV 280 nm	10 ng/ml	[13]
α-Amanitin β-Amanitin γ-Amanitin	Serum, urine	SPE RP-18	HPLC RP-4, ECD + 0.60 V	10 ng/ml	[14]
α-Amanitin phalloidin	Plasma	Protein precip.	HPLC 2 × RP-8, UV 303 nm column switching	10 ng/ml	[15]
α-Amanitin	Plasma	SPE RP-18	HPLC poly(styrene-divinylbenzene), ECD + 0.35 V	2 ng/ml	[16]
α-Amanitin β-Amanitin	Urine (mushrooms)	Dilution	CZE, UV 214 nm, DAD	1 µg/ml	[4]
α-Amanitin β-Amanitin	Urine	SPE RP-18	LC-MS electrospray α- m/z 919 β- m/z 920	10 ng/ml	[17]
α-Amanitin	Urine	SPE mixed mode	HPLC RP-18, ECD + 0.50 V	2 ng/ml LOD 10 ng/ml LOQ	[18]
α-Amanitin β-Amanitin	Urine (plasma)	SPE immunoaffinity	LC-MS RP-18, electrospray α- m/z 919, 920, 921 β- m/z 920, 921, 922	2.5 ng/ml LOD 5 ng/ml LOQ	[19]

Abbreviations: SPE, solid-phase extraction; RP, reversed phase; HPLC, high-performance liquid chromatography; ECD, electrochemical detection; CZE, capillary zone electrophoresis; DAD, diode array detection; LC-MS, liquid chromatography–mass spectrometry; LOD, limit of detection; LOQ, limit of quantitation; precip., precipitation.

body fluids is typically 10 ng/ml, which is higher than that obtained by immunoassay. The CZE LOD is too high for body fluids. A review of Dorizzi *et al.* [20] discusses the various methods in detail.

A sensitive HPLC method by Defendenti *et al.* [18], with a limit of quantitation (LOQ) of 10 ng/ml in urine, utilises mixed-mode (cation exchange-reversed phase) SPE, analytical separation with an RP-18 column and EC detection. In the SPE, the analyte is trapped at pH 6, interferences are washed with water, dilute HCl and chloroform–methanol, and the analyte is eluted with methanol. The analytical separation is carried out with an isocratic mobile phase consisting of phosphate buffer (pH 7.2) and acetonitrile. The linearity is good in the range of 10–200 ng/ml, and the average extraction recovery is 78%. The intra-assay quantitative precision (CV) at 10 ng/ml level is 3.6%. Chromatograms obtained from two intoxication cases appear in Fig. 10.2.

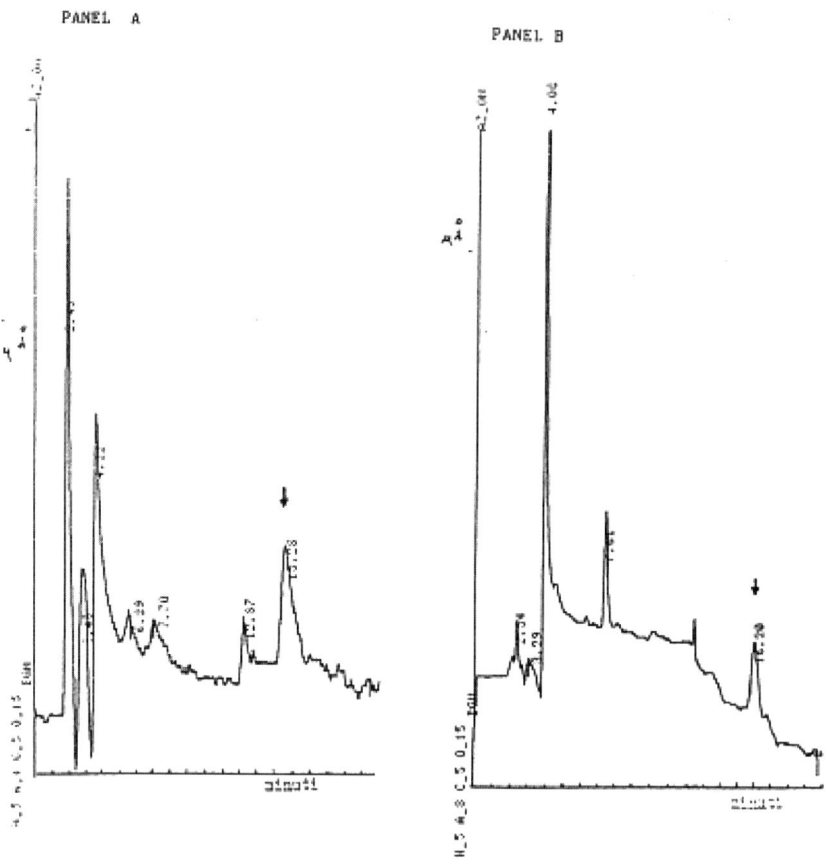

Fig. 10.2. Chromatograms of urine samples from two intoxicated patients showing (A) 28 ng/ml and (B) 12 ng/ml of α-amanitin [18]. In the latter case (B), the sum of α- and β-amanitins was 45 ng/ml by RIA.

Maurer *et al.* [19] applied immunoaffinity extraction and electrospray LC-MS to urine samples, using a 2.1 mm narrowbore RP-18 column, gradient elution and selected ion monitoring, and obtained an exceptional LOQ of 5 ng/ml and an LOD of 2.5 ng/ml. These low values resulted from the high selectivity of immunoaffinity extraction, which reduced the influence of the sample matrix by a factor of 100 (Fig. 10.3). The linear range was 5–75 ng/ml. At the level of LOQ, the absolute recoveries were 63% and 58% and the intra-day CV was 2.3% and 3.6% for α- and β-amanitin, respectively. The method compared favourably to RIA with authentic urine samples.

A. phalloides has a high content of amatoxins (5–8 mg in 25 g of fresh tissue, or 2.5–4 mg/g dry weight), hence the LD_{50} (humans) of 0.1 mg/kg: a full-grown mushroom may be sufficient to kill a human. In severe poisonings, acute hepatic failure will develop 3–4 days after ingestion, and patients die after 6–16 days [21]. Amanitins are bicyclic octapeptides and potent inhibitors of RNA polymerase II [22]. Jaeger *et al.* [23] have studied the kinetics of amatoxins in human poisonings. In 28 of the 45 patients intoxicated with *A. phalloides*, amatoxins were detected in at least one of the samples analysed. Amatoxins were more often detected in urine and faeces than in plasma and gastroduodenal fluid. The peak plasma concentrations ranged between 8–190 and 16–162 ng/ml for α- and β-amanitin, respectively, and in most patients amanitins could only be detected within 36 h after ingestion. Urine concentrations

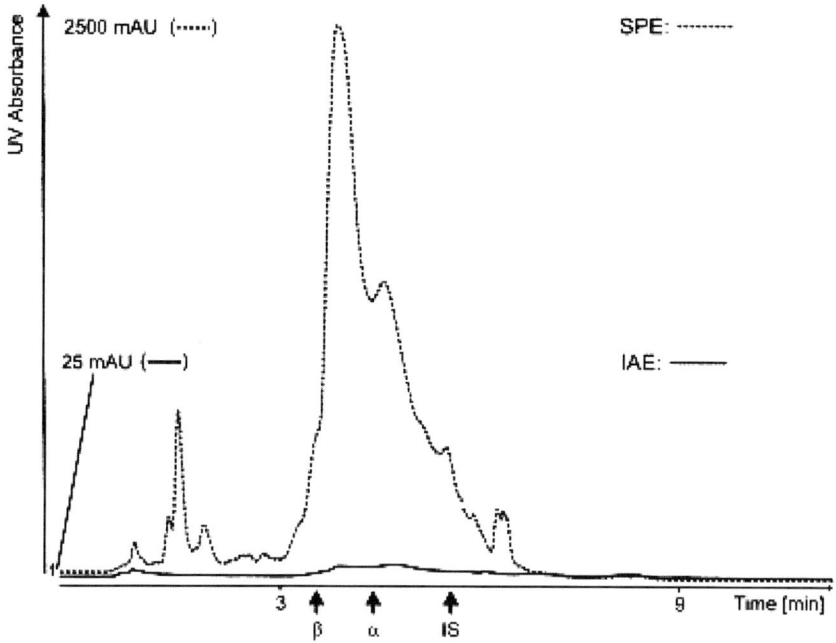

Fig. 10.3. LC-UV chromatograms at 302 nm of a blank urine sample after solid-phase extraction (dotted line) or immunoaffinity extraction (solid line). The arrows mark the retention times of α-amanitin, β-amanitin and internal standard (IS) [19].

were high during the 48–72 h immediately after ingestion, and then decreased. The mean ratios of urine/plasma concentrations were 60 and 19 for α- and β-amanitin, respectively. No correlation has been found between the plasma concentrations and the clinical severity or outcome. Some researchers have stated that only amatoxins, not phallotoxins and virotoxins, contribute to human mushroom poisonings [21,24].

10.3 ORELLANINE

Severe poisonings frequently occur due to the ingestion of cortinarius [25,26]. The main toxin, orellanine, has a bipyridyl skeleton containing two N-oxide groups (Fig. 10.4) and an LD_{50} (mouse) of 20 mg/kg [27]. It decomposes under UV light to non-toxic orelline. Orellanine poisonings are characterised by a long latent period (3–14 days) before symptoms of acute renal failure resulting from damage to the tubular epithelium [28]. Oubrahim *et al.* [29] gave a review on analytical methods for orellanine in mushroom samples and present their own results using TLC, gel electrophoresis and direct electron-spin resonance spectroscopy. TLC has also proven feasible in the analysis of human intoxication samples. One method by Ruedl *et al.* [30,31] involves the separation of orellanine with *n*-butanol-HCl-acetic acid–water on cellulose layers. Orellanine and its decomposition products, orellinine and orelline, are detected by UV irradiation at 366 nm and after spraying with $FeCl_3$/HCl. The absolute LOD is 10 ng. Orellanine and its decomposition products are seldom detected in urine, plasma and dialysis fluids after the onset of symptoms, by which time most of the toxin has concentrated in the kidneys. Orellanine is, however, detectable in small quantities of renal biopsy samples used for histological diagnosis: 160 μg/ml in a patient on day 9 after ingestion, and 35 μg/ml in another patient on day 60 [31]. In one study, unbelievably high amounts of orellanine have been reported in plasma and renal biopsies [32]. In general, little is known about the toxicokinetics of orellanine in humans.

10.4 GYROMITRIN

Gyromitrin (acetaldehyde methylformylhydrazone) and its homologues are toxic compounds that are converted in vivo into *N*-methyl-*N*-formylhydrazine, and then

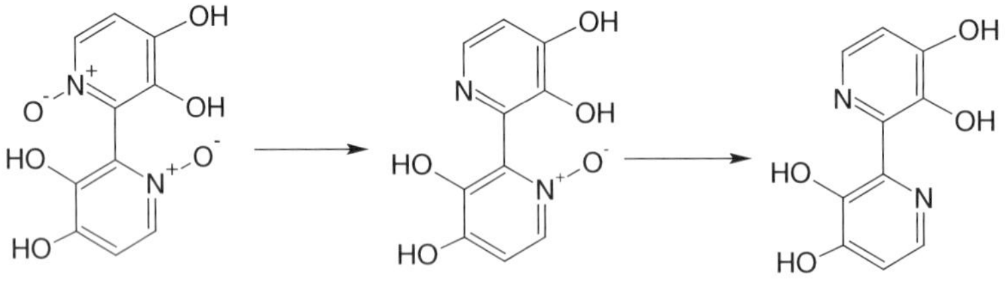

Fig. 10.4. Reduction of orellanine (1) to orellinine (2) and orelline (3) [26].

into *N*-monomethylhydrazine (MMH). These are mainly hepatotoxic and even carcinogenic compounds [33]. There are gas chromatographic (GC), HPLC and TLC methods available for measuring the toxin content in mushrooms. Dried *G. esculenta* mushrooms are commonly used as food even though they may still contain 3 mg/kg of gyromitrin [34]. Between 30 and 71% of MMH still remains in mushrooms after drying, and 10% after boiling twice in large amounts of water [35]. No established methods are available for determining the concentration of gyromitrin or the hydrazines in body fluids or tissues, but there is a report of post-mortem examinations of viscera by IR and UV spectroscopy and TLC [36].

10.5 PSILOCYBIN AND PSILOCIN

The ingestion of mushrooms containing psilocybin (Fig. 10.5) produces hallucinogenic effects and has become a popular form of substance abuse among some adolescents and young adults [37]. Fatalities from these mushrooms are rare. *P. semilanceata* (liberty cap) and *P. cubensis* (golden tops) contain psilocybin in amounts of 2–16 mg/g dry weight and psilocin, up to 10 mg/g dry weight [21]. Musshoff *et al.* [38] have studied in detail the psilocybin and psilocin content and morphology of hallucinogenic mushrooms on the German market.

One HPLC method of Lindenblatt *et al.* [39] allows the determination of psilocin, the active metabolite of psilocybin, in plasma with an LOQ below 1 ng/ml. The method, using bufotenine as an internal standard, involves cation-exchange SPE, analytical separation with an inert reversed-phase column with phosphate buffer pH 2.3 – acetonitrile – EDTA as the mobile phase, and EC detection at +675 mV. The method has an excellent linearity and precision, and the recovery for psilocin from plasma is nearly 100%. Maximum plasma concentrations in volunteers given 0.2 mg/kg psilocybin ranged from 6 to 21 ng/ml of psilocin. Hasler *et al.* [40,41] studied the pharmacokinetics of psilocin in plasma and urine by another HPLC-ECD method using a more complicated sample preparation step, which included protection of the unstable psilocin with ascorbic acid, freeze-drying and extraction with methanol. In plasma, psilocin and its metabolite, 4-hydroxyindole-3-acetic acid (4HIAA), could be measured with an LOQ of 0.8 ng/ml and 5 ng/ml, respectively.

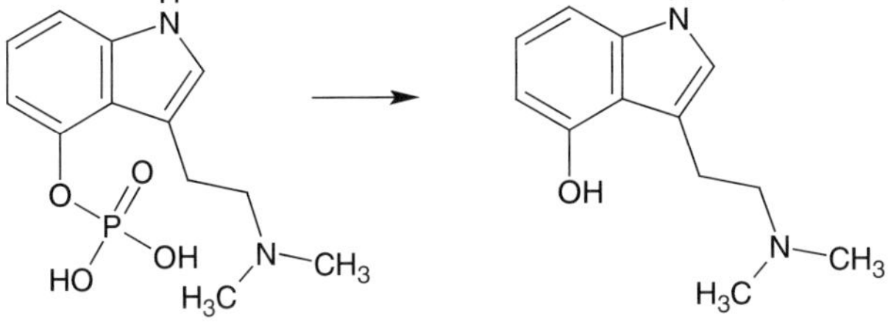

Psilocybin   Psilocin

Fig. 10.5. Transformation of psilocybin into psilocin in the gut [37].

References pp. 398–399

Single oral doses of 10–20 mg psilocybin produced an average peak plasma psilocin concentration of 8.2 ng/ml between 1 and 2 h after administration. For 4HIAA, the level was 20-fold higher. In urine, peak psilocin concentrations of up to 870 ng/ml were measured. Given the LOQ of 10 ng/ml, the detection time-window for unconjugated psilocin in urine was around 24 h. Incubation with β-glucuronidase led to 2-fold higher psilocin concentrations and subsequently extended detectability.

Psilocin can also be analysed by gas chromatography–mass spectrometry (GC-MS) after silylation of the phenolic hydroxyl and arylamino groups. Grieshaber *et al.* [42] developed a method with an LOQ of 10 ng/ml in urine. In suspected drug abusers, they found urine psilocin concentrations of 10–200 ng/ml after hydrolysis with β-glucuronidase, but no samples registered positive without hydrolysis and derivatisation. In an intoxicated car driver, Sticht and Käferstein [43] obtained total psilocin concentrations of 52 and 1760 ng/ml in serum and urine, respectively.

10.6 REFERENCES

1   M.J. Ellenhorn, Ellenhorn's medical toxicology: diagnosis and treatment of human poisoning, (2nd ed.). Williams & Wilkins, Baltimore, 2002, p. 1880.
2   C. Köppel, Toxicon, 31 (1993) 1513.
3   F. Enjalbert, S. Rapior, J. Nouguier-Soulé, S. Guillon, N. Amouroux and C. Cabot, J. Toxicol. Clin. Toxicol., 40 (2002) 715.
4   O. Brüggemann, M. Meder and R. Freitag, J. Chromatogr., 744 (1996) 167.
5   R.Y. Andres and W. Frei, Toxicon, 25 (1987) 915.
6   R.F. Staack and H.H. Maurer, Toxichem + Krimtech, 68 (2001) 68. Available at http://www. gtfch.org/tk/tk68_2/Staack.pdf (accessed July 2006).
7   R. Butera, C. Locatelli, T. Coccini and L. Manzo, J. Toxicol. Clin. Toxicol., 42 (2004) 901.
8   R.A. Abuknesha and A. Maragkou, Anal. Bioanal. Chem., 379 (2004) 853.
9   T. Stivje and R. Seeger, Z. Naturforsch. C, 34 (1979) 1133.
10 L. Pastorello, D. Tolentino, M. D'Alterio, R. Paladino, A. Frigerio, N. Bergamo and A. Valli, J. Chromatogr., 233 (1982) 398.
11 F. Belliardo and G. Massano, J. Liq. Chromatogr., 6 (1983) 551.
12 G. Caccialanza, C. Gandini and R. Ponci, J. Pharm. Biomed. Anal., 3 (1985) 179.
13 F. Jehl, C. Gallion, P. Birckel, A. Jaeger, F. Flesch and R. Minck, Anal. Biochem., 149 (1985) 35.
14 F. Tagliaro, S. Chiminazzo, S. Maschio, F. Alberton and M. Marigo, Chromatographia, 24 (1987) 482.
15 W. Rieck and D. Platt, J. Chromatogr., 425 (1988) 121.
16 F. Tagliaro, G. Schiavon, G. Bontempelli, G. Carli and M. Marigo, J. Chromatogr., 563 (1991) 299.
17 H.H. Maurer, T. Kraemer, O. Ledvinka, C.J. Schmitt and A.A. Weber, J. Chromatogr., 689 (1997) 81.
18 C. Defendenti, E. Bonacina, M. Mauroni and L. Gelosa, Forensic Sci. Int., 92 (1998) 59.
19 H.H. Maurer, C.J. Schmitt, A.A. Weber and T. Kraemer, J. Chromatogr. B., 748 (2000) 125.
20 R. Dorizzi, D. Michelot, F. Tagliaro and S. Ghielmi, J. Chromatogr., 580 (1992) 279.
21 H. Faulstich. In: W.M. Dabrowski, Z.E. Sikorski (Eds.), Toxins in food, CRC Press LLC, Boca Raton, 2005, p. 65.
22 T. Wieland, Int. J. Peptide Protein Res., 22 (1983) 257.
23 A. Jaeger, F. Jehl, F. Flesch, P. Sauder and J. Kopferschmitt, Clin. Toxicol., 31 (1993) 63.
24 T. Wieland and H. Faulstich, Crit. Rev. Biochem., 5 (1978) 185.
25 D. Michelot and I. Tebbett, Mycol. Res., 94 (1990) 289.
26 V.C. Danel, P.F. Saviuc and D. Garon, Toxicon, 39 (2001) 1053.
27 J. Holmdahl, J. Ahlmén, S. Bergek, S. Lundberg and S.-Å. Persson, Toxicon, 25 (1987) 195.

28   H. Prast and W. Pfaller, Arch. Toxicol., 62 (1988) 89.
29   H. Oubrahim, J.-M. Richard, D. Cantin-Esnault, F. Seigle-Murandi and F. Trécourt, J. Chromatogr., 758 (1997) 145.
30   C. Ruedl, M. Moser and G. Gstraunthaler, Mycol. Helv., 4 (1990) 99.
31   M. Rohrmoser, M. Kirchmair, E. Feifel, A. Valli, R. Corradini, E. Pohanka, A. Rosenkranz and R. Pöder, J. Toxicol. Clin. Toxicol., 35 (1997) 63.
32   S. Rapior, N. Delpech, C. Andary and G. Huchard, Mycopathologia, 108 (1989) 155.
33   D. Michelot and B. Toth, J. Appl. Toxicol., 11 (1991) 235.
34   H. Pyysalo and A. Niskanen, J. Agric. Food Chem., 25 (1977) 644.
35   B.K. Larsson and A.T. Eriksson, Z. Lebensm. Unters. Forsch., 189 (1989) 438.
36   G.V. Giusti and A. Carnevale, Arch. Toxicol., 33 (1974) 49.
37   R.H. Schwartz and D.E. Smith, Clin. Pediatr., 27 (1988) 70.
38   F. Musshoff, B. Madea and J. Beike, Forensic Sci. Int., 113 (2000) 389.
39   H. Lindenblatt, E. Krämer, P. Holzmann-Erens, E. Gouzoulis-Mayfrank and K.-A. Kovar, J. Chromatogr., 709 (1998) 255.
40   F. Hasler, D. Bourquin, R. Brenneisen, T. Bär and F.X. Vollenweider, Pharm. Acta Helv., 72 (1997) 175.
41   F. Hasler, D. Bourquin, R. Brenneisen and F.X. Vollenweider, J. Pharm. Biomed. Anal., 30 (2002) 331.
42   A.F. Grieshaber, K.A. Moore and B. Levine, J. Forensic Sci., 46 (2001) 627.
43   G. Sticht and H. Käferstein, Forensic Sci. Int., 113 (2000) 403.

PART 2:
SCREENING PROCEDURES
USED IN FORENSIC
TOXICOLOGY

M.J. Bogusz (Ed.). Forensic Science
Handbook of Analytical Separations, Vol. 6

403

CHAPTER 11

Forensic screening by gas chromatography

Ilkka Ojanperä and Ilpo Rasanen

Department of Forensic Medicine, P.O. Box 40, FI-00014 University of Helsinki, Finland

11.1 INTRODUCTION

Discovered already in 1952 by A.T. James and A.J.P. Martin [1], gas chromatography (GC) was introduced to forensic toxicological drug analysis during the 1960s [2], but it only became popular in the 1970s. The technique offered a long-awaited solution for quantitative analysis of drugs in blood, replacing earlier techniques in this area, such as UV spectrometry and thin-layer chromatography (TLC). In 1967, M.W. Anders and G.J. Mannering wrote in *Progress in Chemical Toxicology*: "The use of gas chromatography in chemical toxicology is in its infancy. Whereas its greatest usefulness will ultimately be realized in the screening for specimens in the 'general unknown' toxicological analysis, it is currently used largely in special procedures designed for the analysis of a single agent or class of agents ..." [3]. Indeed, as packed columns offered only limited separation efficiency, their use was first limited to quantitative target analysis of substances previously identified by TLC in the urine or liver. Along with the development of column oven temperature programming and better and more reproducible packed columns, comprehensive drug screening methods emerged [4]. This was followed by the establishment of comprehensive GC retention index (RI) libraries for the most common stationary phases, allowing cross-referencing with the retention data generated by other laboratories [5].

However, it was only the capillary column, and especially the fused silica column technology [6], that started the era of modern GC screening in the early 1980s. The most active years of developing and publishing screening methods by capillary GC were in the 1980s and early 1990s. That GC gradually became a mature technology and was more often connected with mass spectrometry (GC-MS) lowered the frequency of published non-MS methods but did not change the importance of GC as a workhorse in the laboratory. Instead, progress in electronics and pneumatics resulted

References pp. 423–424

in improved precision of retention parameters through retention time locking (RTL) [7], and this benefit has further strengthened the role of GC in toxicological screening.

This chapter concentrates on the screening of toxicological samples for drugs by capillary GC using techniques other than MS detection. The key published papers, listed in Tables 11.1 and 11.2, are divided according to the use of a single column or two parallel columns, respectively. These studies describe general methods for comprehensive drug screening or demonstrate a potential for such an application; narrow target analyses are not included. Screenings in pesticide and doping control are discussed in their respective chapters. Other screening techniques based on GC-MS, high-performance liquid chromatography (LC) and LC-MS are treated elsewhere in the book.

11.2 SAMPLE PREPARATION

Most GC screening methods have been developed mainly for blood (plasma, serum) samples. This is logical because GC by nature is at its best with non-polar analytes, such as the parent compounds of psychopharmaceuticals, which are present in blood at appropriate concentration levels. Urine, on the other hand, contains more polar metabolites that need derivatization to be detected by GC.

In comprehensive screening procedures aiming for high throughput, the sample preparation should be kept effortless. Although derivatization improves the limit of detection (LOD) and limit of quantification (LOQ) of compounds possessing an – OH or –NH group, particularly acids, phenols, imides and secondary and primary amines and amides, it has only occasionally been used in comprehensive screening procedures. Instead, simple liquid–liquid solvent extraction (LLE) with ethyl or butyl acetate, butyl chloride or methyl *tert*-butyl ether, is commonly carried out at pH 3–7 and pH 9–11 to fractionate acidic/neutral and basic drugs, respectively. Extraction with a fairly non-polar high-boiling solvent, such as butyl acetate, produces clean extracts, with the upper layer easily removable for direct injection into GC. Notably, LLE on a small scale, starting from ≤1 ml of blood and ≤0.5 ml of organic solvent, is both practical and economical. However, putrefied blood samples require a clean-up stage to prevent analytical columns from deteriorating.

Solid-phase extraction (SPE) has been used more rarely in connection with GC screening, as it is a more expensive, labour-intensive and vulnerable technique. However, an SPE method even for whole blood has been developed by Chen *et al.* [20]. Flanagan *et al.* [42] have recently discussed in detail various micro-extraction techniques in analytical toxicology, including micro-LLE, salting-out, extractive derivatization, protein precipitation, solid-phase micro-extraction and liquid-phase micro-extraction.

11.3 SEPARATION

11.3.1 General

Fused silica capillary columns have been used since the early 1980s for the screening of drugs and poisons in GC. Bonded polysiloxanes are the most popular stationary

TABLE 11.1
SINGLE-COLUMN CAPILLARY GAS CHROMATOGRAPHIC METHODS FOR DRUG SCREENING

Reference	Number of Drugs	Sample	Extraction	Limit of Detection (LOD) or Quantification (LOQ)	Column[a]	Detector	Identification Method	Retention Index Standards	Confirmation
Eklund et al.[b] (Sweden 1983) [8]	80 basic	Liver	LLE, butyl acetate		SE-52	NPD	RRT		GC-MS or TLC if necessary
Dunphy and Pandya (USA 1983) [9]	18 acidic and neutral	Serum	LLE, dichloromethane		DB-5	NPD	RRT		TLC or immunoassay
Anderson and Stafford (USA 1983) [10]	175 basic	Blood or urine	LLE, butyl chloride		SE-30	FID	RI	alkanes (external)	GC-MS
Ehresman et al. (USA 1985) [11]	56 acidic and basic	Body fluids	LLE, dichloro-methane	LOD generally ≤1 mg/l	SE-54	FID	RRT		GC-MS
Taylor et al. (USA 1986) [12]	25 basic	Urine	LLE, hexane: isoamyl alcohol 95:5	LOD ≥0.05 mg/l	DB-1701	NPD	RRT		confirmation method for TLC and immunoassay
Soo et al. (USA 1986) [13]	11 hypnotic-sedative	Serum	LLE, dichloro-methane (Toxi-Tube B)	LOD 0.5 mg/l	HP-5	NPD	RRT		GC-MS
Sharp (Canada 1986) [14]	170 basic	Blood	LLE, butyl chloride	LOD<0.2 mg/l	HP-1, 0.53 mm	NPD	RT		
Anderson and Fuller (USA 1987) [15]	52 acidic and neutral	Blood	LLE (Chem Elut)	LOD 1–20 mg/l	HP-1, 0.53 mm	FID	RI	alkanes (external)	GC-MS
Sharp (Canada 1987) [16]	60 acidic and neutral	Blood	LLE, ethyl acetate	LOD<10 mg/l	DB-1	FID	RI	alkanes (external)	GC-MS
Caldwell and Challenger (UK 1989) [17]	300 basic	Urine	LLE, butyl acetate	LOD<0.25 mg/l	HP-5	NPD	RRT		used with TLC and immunoassay
	119 basic	Blood	LLE, butyl acetate	LOD 0.1–0.2 mg/l	HP-17	NPD	RRT		GC-MS

TABLE 11.1
CONTINUED

Reference	Number of Drugs	Sample	Extraction	Limit of Detection (LOD) or Quantification (LOQ)	Column[a]	Detector	Identification Method	Retention Index Standards	Confirmation
Cox et al. (USA 1989) [18]									
Drummer et al. (Australia 1994) [19]	114 neutral and basic	Blood or plasma	LLE, butyl chloride	LOD 0.02–0.5 mg/l	BP-5	NPD	RRT		GC-MS or HPLC
Chen et al. (The Netherlands 1994) [20]	18 basic	Plasma, urine or blood	SPE (Bond Elut Certify)	LOD 0.1–0.2 mg/l	HP-1	NPD	RRT		
Lillsunde et al. Finland 1996 [21]	10 acidic	Blood	LLE, toluene:ethyl acetate 8:2. methylation (PHMAH)	LOD 0.25–1 mg/l	HP-5	NPD	RRT		GC-MS
Tokunaga et al. (Japan 1996) [22]	12 basic	Plasma	LLE, methyl t-butyl ether SPE (Sep-Pak C18)	LOD 0.01–0.2 mg/l LOD 0.005–0.05 mg/l	CBP-1, 0.53 mm	NPD	RT		
Soriano et al. (Spain 1996) [23]	34 basic and neutral	Urine	LLE, methyl t-butyl ether, (PrepStation)	LOD 0.1–0.8 mg/l	HP-5	NPD	RT		
Sanchez de la Torre et al. [b] (Spain 2005) [24]	8 psychiatric	Blood	SPE (Bond Elut Certify)	LOD 0.037–0.156 mg/l LOQ 0.12–0.52 mg/l	HP-1	NPD	RRT		

Abbrev.: LLE: liquid-liquid extraction; SPE: solid-phase extraction; PHMAH, phenyltrimethylammonium hydroxide;
[a]Column internal diameter is 0.2–0.32 mm, if not otherwise stated. NPD, nitrogen-phosphorus specific detector; FID: flame-ionization detector; RRT, relative retention time; RI, retention index; RT: absolute retention time; GC-MS, gas chromatography – mass spectrometry; TLC, thin-layer chromatography, HPLC, high-performance liquid chromatography.
[b]Includes validation data for quantification.

TABLE 11.2
DUAL-COLUMN CAPILLARY GAS CHROMATOGRAPHIC METHODS FOR DRUG SCREENING

Reference	Number of Drugs	Sample	Extraction	Limit of Detection (LOD) or Quantification (LOQ)	Columns[a]	Detectors	Identification method	Retention index standards	Reporting	Confirmation
Hime and Bednarczyk (USA 1982) [25]	48 basic	Urine	LLE, butyl chloride:isoamyl alcohol 99:1	LOD generally 0.05 mg/l	SE-54 OV-101	2 x NPD	RRT			other methods
Alm et al. (Sweden 1983) [26]	45	Illicit drug samples	LLE, ethyl acetate		SE-54 OV-215	NPD, FID	RRT		software	TLC and UV
Newton and Foery (USA 1984) [27]	33 acidic and neutral	Plasma	LLE, dichloro methane:isopropanol 95:5 (ClinElut)		Ultra 1 Ultra 2	2 x FID	RI	alkanes (external)	software	
Koves and Wells[b] (Canada 1985) [28]	102 basic	Blood	LLE, toluene	LOQ generally 0.2 mg/l	DB-1 DB-1701 non-simultaneous	2 x NPD	RRT			GC-MS
Fretthold et al. (USA 1986) [29]	47 basic	Serum, body fluids	LLE, hexane:isoamyl alcohol 98:2	LOD 0.01–0.2 mg/l	BP-1 DB-1701	2 x NPD	RRT			
Watts and Simonick (USA 1986) [30]	110 basic	Blood	LLE, butyl chloride	LOD 0.1 mg/l	Ultra 1 HP-17	2 x NPD	RRT			GC-MS
Cordonnier et al.[b] (Belgium 1987) [31]		Tissues	2 × SPE (Chem Elut)		CP Sil 8 CP Sil 19 CB	2 x NPD	RRT			
Turcant et al. (France 1988) [32]	200 neutral and basic	Plasma	LLE, diethyl ether	LOD generally 0.1 mg/l	Ultra 1 CP Sil 19 CB	2 x NPD	RRT		software	
Manca et al. (Canada 1989) [33]	172 neutral and basic	Plasma, urine or gastric	LLE, hexane:diethyl ether 1:1	Therapeutic concentrations	DB-1 DB-17	2 x NPD	RI	drugs (external)		
Lillsunde and Seppälä[b] (Finland 1990) [34]	21 benzodiazepines	Plasma or blood	LLE, hexane: dichloromethane 7:3	LOD 0.01–0.05 mg/l	2 × SE-54	ECD, NPD	RT		APL *Plus 5.5 Software	

TABLE 11.2
CONTINUED

Reference	Number of Drugs	Sample	Extraction	Limit of Detection (LOD) or Quantification (LOQ)	Columns[a]	Detectors	Identification method	Retention index standards	Reporting	Confirmation
Phillips et al. (USA 1990) [35]	200 neutral and basic	Blood or tissues	LLE, butyl chloride	LOD 0.01–0.3 mg/l	DB-1 DB-5 non-simultaneous	FID	RI	alkanes (external)		GC-MS
Ojanperä et al.[b] (Finland 1991) [36]	31 acidic and neutral	Blood	LLE, ethyl acetate; alkylbis(trifluoromethyl) phosphine sulfides (internal)	Therapeutic concentrations; Micman 4.03 software	NB-54 NB-1701	2 × NPD	RI			
Kim et al. (South Korea 1991,1993) [37,38]	26 acidic	Serum	LLE, diethyl ether, silylation (MTBSTFA)		DB-5 DB-17	2 × FID	RI	alkanes (internal)	software	GC-MS
Coudore et al.[b] (France 1993) [39]	9 barbiturates	plasma	LLE, chloroform	LOD 0.5 mg/l	SPB-1 SPB-20	NPD, FID	RRT			
Lillsunde et al.[b] (Finland 1996) [21]	40 basic	Blood	LLE, dichloromethane: toluene 1:9, acylation (HFBA)	LOD 0.01–0.1 mg/l	2 × HP-5	NPD, ECD	RI	drugs (internal)	SC-ChromBooster software	GC-MS
Rasanen et al.[b] (Finland 2000) [40] WorkStation 3.0 software	26 benzodiazepines	Blood	LLE, ethyl acetate, silylation (MTBSTFA)	LOD 0.003–0.05 mg/l LOQ 0.006–0.075 mg/l	DB-5 DB-17	2 × ECD	RI	benzodiazepine RI standards		SC-
Rasanen et al.[b] (Finland 2003) [41]	124 basic	Blood	LLE, butyl acetate	LOQ 0.02–5 mg/l	HP-5 DB-17	2 × NPD	RRT with RTL		SC-ChromBooster software	

LLE, liquid-liquid extraction; SPE, solid-phase extraction; MTBSTFA, *N*-methyl-*N*-(*tert*-butyldimethylsilyl)trifluoroacetamide; HFBA, heptafluorobutyric anhydride;

[a]column internal diameter is 0.2–0.32 mm, if not otherwise stated. NPD, nitrogen-phosphorus specific detector; FID, flame ionization detector; ECD, electron capture detector. RRT, relative retention time; RI, retention index; RT, absolute retention time; RTL, retention time locking. TLC, thin-layer chromatography; GC-MS, gas chromatography; GC-MS, gas chromatography – mass spectrometry.

[b]Includes validation data for quantification.

phases due to such favourable properties as high thermal stability, good diffusivity and suitable solvation characteristics (solvent strength or polarity and selectivity). The use of more polar polyethyleneglycol phases is limited by their maximum operating temperature of 250–280°C.

According to Table 11.1, the most popular stationary phases used in single column methods were non-polar 100% dimethylpolysiloxane (SE-30, HP-1, DB-1, etc.) and slightly polar 5% diphenylmethyl-polysiloxane (SE-52, HP-5, DB-5, etc.) or 1% vinyl 5% phenylmethyl-polysiloxane (SE-54), which possess the best thermal stability and a low column bleed. In only two methods were the intermediately polar columns 14% cyanopropylphenylmethylpolysiloxane (DB-1701) [12] and 50% phenylmethylpolysiloxane (HP-17) [18] used.

11.3.2 Dual-column approach

The use of dual capillary column GC was first described by Phillips *et al.* [43] for analysis of essential oils, and the concept was soon applied to drug screening [25]. According to Table 11.2, in dual-column methods, the stationary phase combinations were non-polar/slightly polar, non-polar/intermediately polar or slightly polar/ intermediately polar. Two methods had two similar slightly polar columns and two different detectors [21,34].

The use of two parallel capillary columns of different polarity makes GC even more powerful than using a single column, judging from the relatively high number of reports that utilize this approach in drug screening. However, most papers do not present well-founded arguments to support the dual-column concept. It has also been stated that a second phase in addition to dimethylpolysiloxane would be of limited use because the high degree of correlation would give little or no extra information in general screening procedures [44]. Manca *et al.* [33] evaluated their dual-column method using the concept of Discriminating Power by Moffat *et al.* [45,46]. They found that the probability of discriminating between two compounds chosen at random in the data base was 99.67%, using a tolerance of ± 5 Kováts RI units.

Rasanen *et al.* [47] evaluated their dual-column method consisting of separation on DB-5 and DB-1701. Although a high correlation (0.990) was found between the retention data, suggesting an apparent similarity between these columns, the correlation coefficient is not a sufficiently sensitive measure to judge the feasibility of using two columns in parallel. Instead, statistical calculations, based on the Mean List Length method [48], show that a 30–57% improvement in the identification power can be obtained using the dual-column approach at analytically relevant standard deviation (SD) levels. Interestingly, the highest gain is in the SD range of 4–8, and this decreases as the SD increases or decreases. An average intralaboratory SD of 1 was obtained for the present RI method during the 12-week period. In interlaboratory use, an SD value of 5–10 can be expected for a RI method [49].

References pp. 423–424

11.4 IDENTIFICATION

11.4.1 Retention parameters

High precision of the appropriate retention parameter is crucial in GC screening, in contrast to GC-MS, where broader tolerance may be allowed. Recently, the status of retention-based identification has been restored by advances in GC technology, leading to approaches like fast chromatography and method translation [7,50]. Three types of retention parameters have been used for identification in GC. The absolute retention time (RT), or the adjusted retention time (RT'), is the most imprecise identification method. Variations in carrier gas flow rate, oven temperature and film thickness influence the RT. However, the RT is simple to measure and may be useful in solving trivial identification problems.

Retention related to a single standard, the relative retention time (RRT), compensates for the effects mentioned above and allows more precise identification. However, the RRT is a retention parameter that is at its best in intralaboratory use under exactly the same instrumental conditions [51].

Obviously, the most precise identification parameter in GC is the retention index (RI). The RI system was originally based on the theory of Kováts [52]. The retention indices are calculated using at least two reference compounds that bracket the analytes. In GC drug screening methods, the aim is usually either to obtain Kováts indices using n-alkanes as RI standards, which can be compared with reference values in large databases using a search window of ± 25–60 units [44], or to obtain more precise retention parameters for daily intralaboratory use. The latter approach necessarily involves the use of non-alkane standards that are detectable by selective detectors and are structurally similar to analytes [49,53]. The RI methods reported in the literature predominantly use the external standard method; the standard series is injected daily before or between a sequence of analysis samples. The internal standard method, on the other hand, involves the injection of the standard series with each analysis sample.

11.4.2 Retention index systems

According to Kováts [52], the plot of the logarithm of retention volume versus the carbon number of a homologous series is linear under isothermal gas chromatographic conditions. Later, it was suggested that retention volumes can be replaced by adjusted retention times [54]. Kováts' equation for the calculation of retention index (RI_x) is

$$RI_x = 100_n + 100 \frac{\log RT'(x) - \log RT'(n)}{\log RT'(n+1) - \log RT'(n)} \tag{1}$$

where $RT'(x)$ is the adjusted retention time of the unknown substance, $RT'(n)$ and $RT'(n+1)$ are the adjusted retention times of the n-alkanes, with carbon number n used as a standard as follows: $RT'(n) < RT'(x) < RT'(n+1)$.

In linear temperature programming, the logarithms of adjusted retention times can be replaced by absolute retention times (RT) [55] and equation (1) can be rewritten as

$$RI_{lx} = 100_n + 100 \frac{RT(x) - RT(n)}{RT(n+1) - RT(n)} \tag{2}$$

where RI_{lx} is the linear retention index of x. However, for any homologous series of aliphatic organic compounds, there is a strong deviation from linearity in both cases for the first members of a series [54].

The temperature programs of instruments may deviate considerably from linearity, especially at the beginning and the end of the program, and the dependence of carrier gas flow upon temperature further enhances the non-linearity of the retention times of a homologous series. To compensate for these errors, cubic spline interpolation can be used. Cubic splines are functions composed of third-order polynomials. Linear retention indices may be calculated when using a non-linear temperature program, but more correct Kováts retention indices compared with linear behaviour can be calculated by using cubic splines [56].

Kováts indices can also be predicted from chemical structures by using chemometric methods. Garkani-Nejad [57] reported a standard error of prediction of about 80 RI units for a set of 846 organic compounds relevant in forensic analysis. This approach may be useful in confirming GC-MS results that are based on MS library search data only, when the reference standard is not available.

11.4.3 Retention index standards

Identification in drug screening is often based on RIs, with *n*-alkanes as standards, and comprehensive drug retention index libraries are available. Although the libraries can be used on an interlaboratory basis for tentative identification of unknown drugs, the daily precision obtained by this method is low.

In drug screening, medium polar stationary phases are preferred, and the retention of non-polar *n*-alkanes is reduced, which may result in the comparison of low-boiling polar solutes with *n*-alkanes of much higher boiling points. The high-boiling *n*-alkanes may require a significant increase in injector temperature, which can result in the formation of artefacts from temperature-sensitive sample components. Another problem is the weak response to *n*-alkanes of the selective detectors, mainly the nitrogen/phosphorous detector (NPD) and electron capture detector (ECD), required in drug analysis [51,54].

In temperature-programmed runs, the distribution constants and thereby the retention indices of different types of compounds are differently affected by changes in temperature. Small irregularities in a temperature program may thus lead to erroneous data. By using reference compounds with structures more similar to those of the samples, problems associated with differences in distribution constants can be minimized. Index standards should be stable both in solution and under the chromatographic conditions used. They should also bracket the solutes so that extrapolation is not required [33,54].

For drug screening, RI standards more similar to drugs in chemical nature have been suggested, including diisopropylamines [49], trialkylamines [58], nitroalkanes [59], alkylbis(trifluoromethyl)phosphine sulphides [36] and drug substances [33,60]. Franke *et al.* [49] found that drug substances, if selected carefully, were superior to the three first mentioned homologous series in terms of interlaboratory reproducibility. Rasanen *et al.* [61] proved that the internal standard RI method was regularly more precise than the external standard RI method. To avoid co-injection of commercially available drugs, they synthesized several dedicated homologue RI standard series for benzodiazepines [62], basic drugs [63] and acidic/neutral drugs [64] and demonstrated that the precision based on these internal standards was better than that of the external drug series.

11.4.4 Retention time locking (RTL)

Recently, the concepts of method translation and RTL in GC were introduced by Blumberg and Klee [7]. These are based on void time being a universal time unit in GC, and method translation being the scaling of the time axis of the temperature program relative to the void time. Method translation can be used for RTL, which allows chromatograms to be reproduced accurately from one GC to another or over a long period of time [65]. RTL has been successfully applied to multiresidue screening of pesticides in fruit and vegetable extracts by matching GC and GC-MS retention times to a common database [66].

Fig. 11.1 compares the long-term precision obtained by using three different retention parameters on HP-5 and DB-17 columns [41]. The retention parameters studied were RT, RRT related to dibenzepin and the internal RI based on the alkylfluoroaniline series [63]. The carrier gas program was set in the constant flow mode. The drug substances represented various secondary and tertiary aliphatic amine

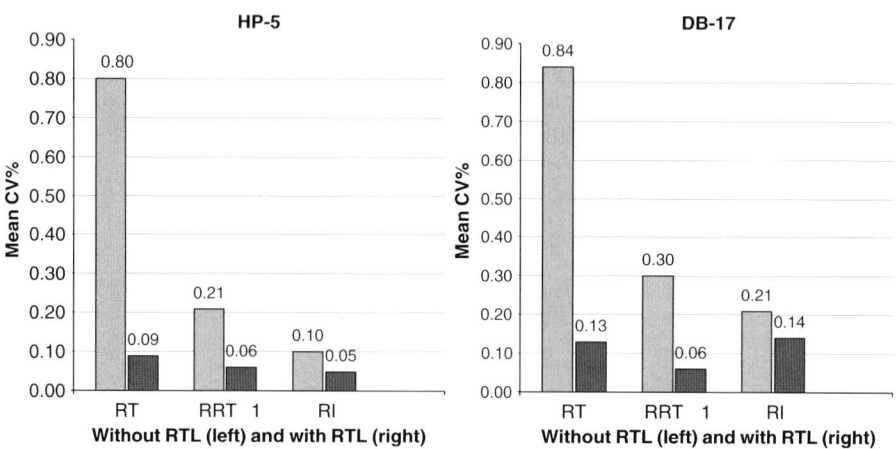

Fig. 11.1. Precision (CV%) of the retention parameters on HP-5 and DB-17 columns without and with retention time locking (RTL).

structures. All results were based on 128 repetitive runs of spiked bovine blood extracts during an 18-week period. A clear improvement was observed in the precision (CV%) of all three retention parameters on both columns using the RTL function, the benefit being largest with RT and smallest with RI. While all RTL-based retention parameters showed very high precision without large differences, RRT with an average CV below 0.1% on each column was in general the most precise approach for drug screening. The positive results obtained by RTL suggest that no column selectivity change caused chromatographic variation due to loading with biological extracts, as this could not be efficiently compensated by RTL.

11.5 DETECTION AND QUANTIFICATION

11.5.1 Sample introduction

Sample introduction into GC has a decisive influence on RT, LOD and quantitative results of analysis. For high precision, use of an autosampler is recommended. To obtain sufficiently low LODs with tolerable contamination of the analytical column, the splitless injection is generally used. Both the injection liner geometry and the position of glass wool to facilitate evaporation inside the liner may have a significant effect on the results. In addition, proper deactivation and cleaning of the liner and replacement of the glass wool at regular intervals are necessary. Choice of injection solvent, injection volume and initial column oven temperature should be carefully considered. The theory and practice of split and splitless injection have been thoroughly investigated by Grob [67].

11.5.2 Detection

Flame-ionization detector (FID) has a nearly universal response to organic compounds, a low LOD and a wide linear response range (10^7). The FID response results from the combustion of organic compounds in a small hydrogen-air diffusion flame. It is the most popular GC detector in current use. In drug screening, however, there may be interference from biological background, and this may be pronounced with post-mortem samples. In the absence of reference standards, FID can be used to predict relative response factors of known structures with reasonable accuracy using the effective carbon number concept, as shown with amphetamine-type compounds by Huizer *et al.* [68].

NPD is identical to FID, except that an alkali metal salt source is placed between the burner tip and the collector. Using low hydrogen flow rates, NPD is selective to nitrogen, having 10^4–10^5 times the response relative to carbon, and has a moderate linear response range (10^5). Tables 11.1 and 11.2 show that NPD is the most popular detector in toxicological drug screening. Most nitrogen-containing drugs give a satisfactory response, but nitro-compounds, amides and carbamates, such as meprobamate, are less favourable. Both the selectivity and the sensitivity of the detector are dependent on experimental variables, including the source heating current, source location, jet potential, air and hydrogen flow rates and choice of carrier gas. Especially the alkali bead lifetime may vary markedly depending on the individual bead and on the samples. With a workload of 30 runs per day in the authors' laboratory, the beads from a major manufacturer lasted from two weeks to four months. NPD is not compatible with silylation reagents and halogenated injection solvents such as dichloromethane.

Surface ionization detector (SID) is based on a similar principle to NPD, but it is substance-selective rather than element-selective. According to Kageura *et al.* [69], SID exhibited a high response for tertiary amines, a rather low response for secondary amines, no response for amides and little or no response for xanthines and benzodiazepines. Despite its potential in forensic toxicology, SID has not been used in comprehensive GC screening for drugs.

Electron capture detection (ECD) can be used for the sensitive analysis of compounds that have high electron affinities. Electrons produced from a radioactive ^{63}Ni source are selectively captured by, e.g., pesticides and drugs with certain structures, such as a halo group or the nitro group, or with lesser sensitivity, the carbonyl group. ECD is commonly applied to toxicological screening for benzodiazepines [34,40], but it produces no response with the 7-amino derivative of nitrazepam. The detector's linear response range is somewhat limited (10^4), which may result in a need to dilute the sample.

MS detection has, to a certain extent, replaced the detectors described above due to its unsurpassed properties in structural analysis. However, routine target screening of 50–200 compounds is often more feasible using selective detectors such as NPD and ECD. These detectors become even more attractive when comprehensive screening and simultaneous quantification are required at low concentrations in blood. In MS, the number of compounds that can be analysed in one run using selected ion monitoring (SIM) has been restricted, and the full scan mode may in turn lack sufficient sensitivity. This dilemma may be solved by splitting the GC-MS column effluent to a second selective detector like NPD.

11.5.3 Quantification

GC screening is amenable to simultaneous quantification. To maintain a comprehensive quantitative screening of over 100 compounds, one-point calibration once a month has been deemed practical. Compounds that do not appear frequently can be quantified as required. The calibration concentration in one-point calibration can be set at the borderline between therapeutic and toxic levels to produce the best performance at the range of interest. Target screenings aiming for low concentrations usually require multi-point calibration.

As shown in Tables 11.1 and 11.2, LOD of basic drugs is generally around 0.1–0.2 mg/l without derivatization. More precisely, as presented in Table 11.3, LOQ for aliphatic tertiary amines is 0.05–0.1 mg/l, and for aliphatic secondary and primary amines it is 0.2–0.5 mg/l. To detect and quantify the latter substances at therapeutic levels, derivatization is required. Lillsunde *et al.* [21] obtained an LOD of 0.01–0.1 mg/l for basic drugs in blood by GC-NPD/ECD after acylation with

TABLE 11.3
RETENTION AND QUANTIFICATION DATA FOR 124 BASIC DRUGS [41]

Compound	RRT HP-5	RRT DB-17	LOQ (mg/l)[a]
Amitriptyline	0.800	0.797	0.1
Amphetamine[b]	0.216	0.211	0.5
Biperiden	0.859	0.836	0.1
Bisoprolol	0.904	0.860	1.0
Brompheniramine	0.734	0.737	0.05
Bupivacaine	0.865	0.841	0.2
Buspirone	1.464	1.615	0.05
Caffeine	0.562	0.629	0.5
Carbamazepine[c]	0.894	0.985	1.0
Chlordiazepoxide[c]	1.045	1.054	0.2
Chloroquine	1.094	0.990	0.2
Chlorpheniramine	0.667	0.660	0.05
Chlorpromazine	1.028	0.981	0.05
Chlorprothixene	1.030	0.943	0.1
Cinchocaine	1.145	1.040	0.1
Cinnarizine	1.330	1.312	0.05
Citalopram	0.961	0.922	0.1
Clobutinol	0.530	0.490	0.2
Clomipramine	0.966	0.926	0.1
Clozapine	1.221	1.235	0.1
Cocaine	0.806	0.848	0.1
Codeine	0.943	0.966	0.1
Cyclizine	0.680	0.679	0.1
Dextrometorphan	0.757	0.757	0.1
Dextropropoxyphene	0.792	0.763	0.1
Diacetylmorphine	1.114	1.026	0.1
Diazepam	0.990	0.997	0.1
Dibenzepin[d]	1.000	1.000	
Diltiazem	1.261	1.314	0.1
Diphenhydramine	0.584	0.573	0.1
Disopyramide	1.035	1.000	0.2
Doxapram	1.230	1.219	0.1
Doxepin	0.825	0.837	0.05
Ethylmorphine	0.971	0.973	0.1
Fencamfamin	0.484	0.450	0.2
Fenfluramine	0.259	0.218	0.05
Fentanyl	1.155	1.061	0.05
Flecainide[c]	0.829	0.782	0.2
Fluconazole	0.736	0.791	0.1
Flumazenil	1.057	1.059	0.05
Fluoxetine	0.577	0.526	0.2
Fluvoxamine	0.589	0.529	1.0
Haloperidol	1.265	1.235	0.1
Hydrocone	0.994	1.011	0.1
Hydroxychloroquine	1.237	1.176	1.0
Hydroxyzine	1.221	1.177	0.2
Imipramine	0.821	0.823	0.05
Ketamine	0.577	0.604	0.1
Ketobemidone	0.704	0.729	0.2

TABLE 11.3
CONTINUED

Compound	RRT HP-5	RRT DB-17	LOQ (mg/l)[a]
Levomepromazine	1.050	0.995	0.1
Lidocaine	0.590	0.577	0.1
Maprotiline	0.908	0.907	0.2
MDMA	0.406	0.398	0.2
Meclozine	1.295	1.252	0.05
Melperone	0.597	0.557	0.05
Mepivacaine	0.712	0.728	0.2
Mesoridazine	1.547	1.779	0.2
Metamphetamine	0.237	0.225	0.2
Methadone	0.762	0.743	0.05
Methyl phenidate	0.513	0.513	0.05
Metoclopramide	1.107	1.087	0.05
Metoprolol[b,c]	0.684	0.666	0.5
Mexiletine	0.337	0.321	0.5
Mianserine	0.807	0.844	0.05
Milnasipram	0.931	0.862	0.1
Mirtazapin	0.840	0.885	0.05
Moclobemide	0.834	0.877	0.1
Molindone	1.020	1.023	0.1
Moperone	1.211	1.138	0.1
Nefazone	2.045	[e]	0.5
Nicotine	0.313	0.309	0.1
Nomifensine	0.769	0.850	0.2
Norcitalopram	0.982	0.952	0.5
Norclomipramine	0.987	0.958	0.5
Nordazepam	1.037	1.044	0.1
Nordextropropoxyphene amide	1.065	1.020	0.2
Nordoxepin	0.837	0.870	0.2
Norlevomepromazine	1.069	1.025	Qualitative
Normethadone	0.732	0.723	0.1
Normianserine	0.841	0.900	0.1
Norpromazine	0.922	0.944	Qualitative
Nortramadol	0.647	0.662	0.1
Nortrimipramine	0.840	0.850	0.2
Nortriptyline	0.813	0.834	0.1
Norverapamil[b]	1.421	1.541	1.0
Noscapine	1.381	1.569	0.2
Olanzapine	1.159	1.140	0.05
Orphenadrine	0.626	0.612	0.1
Oxycone	1.047	1.046	0.1
Pentazocine	0.859	0.849	0.1
Pentoxyverine	0.838	0.800	0.1
Pethidine	0.520	0.500	0.1
Phenazone	0.582	0.664	0.5
Phencyclidine	0.599	0.565	0.05
Pheniramine	0.554	0.551	0.05
Phentermine	0.229	0.219	0.05
Phenytoin	0.916	0.982	5.0
Pholcodine	1.380	1.532	0.5

TABLE 11.3
CONTINUED

Compound	RRT HP-5	RRT DB-17	LOQ (mg/l)[a]
Prilocaine	0.567	0.559	0.1
Procainamide	0.830	0.893	5.0
Promazine	0.900	0.909	0.1
Promethazine	0.862	0.879	0.1
Propranolol[b,c]	0.768	0.775	0.5
Quetiapine	1.425	1.604	0.2
Quinine	1.201	1.187	0.2
Reboxetine	0.913	0.942	0.5
Ropivacaine	0.802	0.793	0.1
Selegiline	0.353	0.335	0.1
Sertraline	0.929	0.918	0.1
Strychnine	1.376	1.593	0.1
Temazepam[c]	1.229	1.340	0.2
Thioridazine	1.357	1.438	0.1
Thioridazine, 5-sulfoxide	1.723	e	0.2
Tizanidine	1.011	1.081	0.5
Tramadol	0.630	0.620	0.1
Tramadol, O-desmethyl	0.675	0.688	0.2
Trazodone	1.501	1.732	0.2
Trimeprazine	0.880	0.878	0.1
Trimetoprim	1.082	1.096	1.0
Trimipramine	0.817	0.803	0.1
Venlafaxine	0.719	0.712	0.1
Verapamil	1.381	1.445	0.1
Zaleplon	1.297	1.416	0.1
Zolpidem	1.202	1.228	0.1
Zopiclone[b,c]	1.343	1.555	0.02

[a]Criteria for LOQ: 20% precision and 15% accuracy in the quantitative result of four parallel samples using single-point calibration.
[b]Quantitation preferably by another dedicated method.
[c]Several peaks produced; the main peak is indicated here.
[d]Internal standard.
[e]Not analysed on this column.

heptafluorobutyric anhydride. Benzodiazepines can be quantified at 0.006–0.075 mg/l by GC-ECD after silylation [40].

11.6 ESTABLISHED GC SCREENING FOR BASIC DRUGS

11.6.1 General

In the following, an established GC method is described that has been in routine use in the authors' laboratory for several years [41]. The method serves as a backbone of basic drug screening and quantification for both post-mortem and clinical samples. Target analyses by other techniques, such as GC-MS and LC-MS, are used to

References pp. 423–424

supplement the GC screening. Although developed for blood (whole blood, plasma, serum), the present GC screening is also useful in the analysis of urine. LC-based techniques are, however, preferable in urine analysis, as they are better suited to deal with the polar substances excreted into urine.

11.6.2 Sample preparation

Whole blood (1 ml) was transferred to a narrow centrifuge tube, Tris-buffer (1 M, pH 11, 0.3 ml) and the internal standard (dibenzepin 20 µg/ml in MeOH, 50 µl) were added and the mixture was shaken. The sample was extracted with butyl acetate (0.3 ml) in a multi-tube vortex mixer, centrifuged and an aliquot of the organic phase (150 µl) was transferred to an autosampler vial.

11.6.3 Gas chromatography

The GC was equipped with two parallel fused silica capillary columns, HP-5 and DB-17 (15 m × 0.32 mm, inner diameter, 0.25 µm film thickness), and two NP detectors (330°C). Uncoated deactivated fused silica precolumns of 10 m × 0.32 mm were connected to the analytical columns. The precolumns entered a single injector (270°C) through a dual column injector adapter. A deactivated straight liner with silanized glass wool was used in the injector. Automated injections were performed using a 2-µl apparent injection volume.

The carrier gas was helium, operated in the constant flow mode. The oven temperature was initially held at 100°C for 0.4 min, then increased by 25°C/min to 200°C, increased by 10°C/min to 240°C and increased by 25°C/min to 290°C, where it was held for 10 min. The carrier gas flow was 2 ml/min for 15 min and then increased by 2 ml/min^2 to 4 ml/min, which was held for 6.4 min. Under these conditions with new analytical columns and new 10-m precolumns, the RT of dibenzepin was 9.2 min on HP-5. This setting was locked by using five-point calibration data obtained with nominal initial pressure and with pressures of -20%, -10%, $+10\%$ and $+20\%$ from the nominal initial pressure. Relocking based on one scouting run was performed daily.

11.6.4 Data processing

The GC was operated and data were collected, integrated and saved using ChemStation software equipped with Retention Time Locking software (Agilent Technologies, Palo Alto, CA, USA). Data processing was performed by SC Chrombooster software (Sunicom, Helsinki, Finland).

11.6.5 Performance

Table 11.3 shows the RRT and LOQ values obtained by single-point calibration for 124 basic drugs and metabolites on HP-5 and DB-17. This list includes most

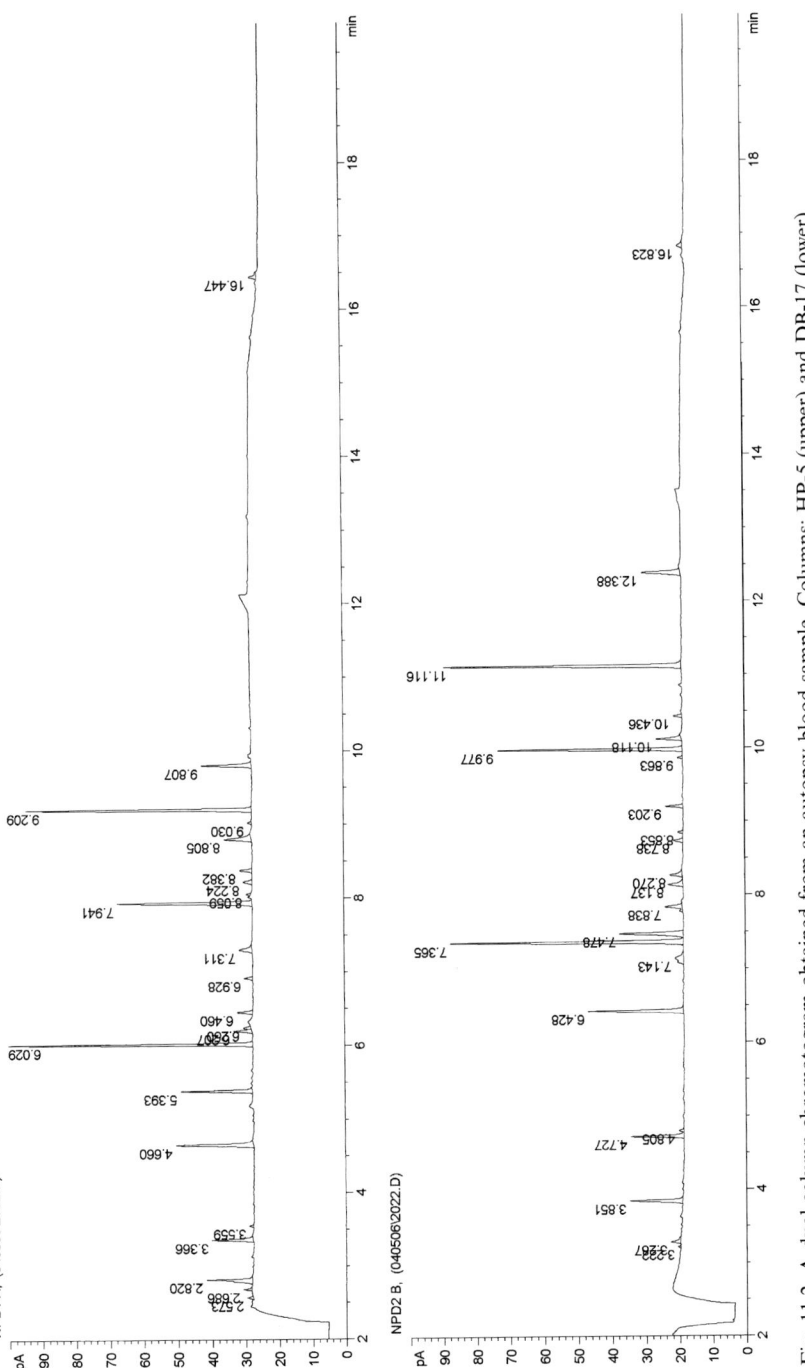

Fig. 11.2. A dual-column chromatogram obtained from an autopsy blood sample. Columns: HP-5 (upper) and DB-17 (lower).

```
Data  File     : 2022025.dta
Method         : basic.MTD
Date Created   : Thursday, May 4, 2006 at 23:58:43
Date Analyzed  : Friday, May 5, 2006 at 16:22:52
Description    : Sample Name   :
```

```
Channel    : 1
Library    :Basic1.LBY
```

Peak Compound	AbsRT	IdPara/Mtd	Diff	Area	mg/l
1 methamphetamine	2.573	0.279 R	0.00164	2.92	0.068
4 mexiletine	3.366	0.366 R	0.00420	20.95	0.423
5 selegiline	3.559	0.386 R	-0.00283	1.68	0.009
6 fencamfamin	4.660	0.506 R	-0.00216	48.28	0.785
7 caffeine	5.393	0.586 R	-0.00114	34.43	1.334
8 tramadol	6.029	0.655 R	-0.00111	112.86	1.528
9 nortramadol	6.207	0.674 R	-0.00376	8.88	0.375
11 cyclizine	6.460	0.701 R	0.00257	8.14	0.055
12 normethadone	6.928	0.752 R	0.00150	4.24	0.062
13 nomifensine	7.311	0.794 R	-0.00299	10.94	0.160
14 normianserine	7.941	0.862 R	-0.00063	77.81	1.536
15 pentazocine	8.059	0.875 R	0.00322	1.74	0.043
16 normirtazapin	8.224	0.893 R	-0.00204	5.98	0.000
17 carbamazepine	8.382	0.910 R	0.00156	6.90	0.438
18 codeine	8.805	0.956 R	-0.00009	21.31	0.973
19 ethylmorphine	9.030	0.981 R	-0.00156	1.78	0.059
20 dibenzepin/INT.STD	9.209	9.209 A	0.000	110.45	1.000
21 trimethoprim	9.807	1.065 R	0.00068	28.87	1.834
Total				**508.16**	**10.680**

```
Channel    : 2
Library    :Basic2.LBY
```

Peak Compound	AbsRT	IdPara/Mtd	Diff	Area	mg/l
	3.851	0.346 R	0.00262	30.12	0.232
5 MDMA	4.805	0.432 R	-0.00132	2.49	0.069
7 ketamine	7.143	0.643 R	0.00343	13.86	0.234
8 tramadol	7.365	0.663 R	-0.00199	132.98	1.527
9 caffeine	7.478	0.673 R	-0.00191	45.13	1.500
10 phenazone	7.838	0.705 R	0.00012	13.00	0.945
11 o-desmethyltramadol	8.137	0.732 R	-0.00275	10.78	0.221
14 dextromethorphan	8.853	0.796 R	-0.00017	2.69	0.046
15 amitriptyline	9.203	0.828 R	-0.00125	8.49	0.111
16 moclobemide	9.863	0.887 R	0.00116	2.36	0.030
17 mirtazapin	9.977	0.898 R	-0.00061	91.67	0.587
18 normianserine	10.118	0.910 R	-0.00032	14.23	0.253
19 reboxetine	10.436	0.939 R	0.00057	4.07	0.117
20 dibenzepin/INT.STD	11.116	11.116 A	1.000	128.84	1.000
21 trimethoprim	12.388	1.114 R	0.00029	29.02	1.789
22 pholcodine	16.823	1.513 R	-0.00191	4.54	0.520
Total				**534.28**	**9.181**

Fig. 11.3. Single-column reports from HP-5 and DB-17 for the case in Fig. 11.2, providing the best hit for each peak.

*** SC-Compare Report [Version 1.50] ***

Data File   : 2022025.dta
Method   : basic.MTD
Date Created : Thu May 4 2006 at 23:58:43
Date Analyzed : Fri May 5 2006 at 16:03:43

Compound	Ch.	Peak	AbsRT	IdPara	Diff	Area	Amount
tramadol	1	8	6.029	0.655	0.001	112.86	1.528
	2	8	7.365	0.663	0.002	132.98	1.527
caffeine	1	7	5.393	0.586	0.001	34.43	1.334
	2	9	7.478	0.673	0.001	45.13	1.500
nortramadol	1	9	6.207	0.674	0.003	8.88	0.375
	2	10	7.838	0.705	0.003	13.00	0.462
o-desmethyltramadol	1	11	6.460	0.701	0.002	8.14	0.190
	2	11	8.137	0.732	0.002	10.78	0.221
mirtazapin	1	14#	7.941	0.862	0.001	77.81	0.579
	2	17	9.977	0.898	0.000	91.67	0.587
normianserine	1	14#	7.941	0.862	0.000	77.81	1.536
	2	18	10.118	0.910	0.000	14.23	0.253
normirtazapin	1	16	8.224	0.893	0.002	5.98	0.000
	2	19	10.436	0.939	0.001	4.07	0.000
dibenzepin/INT.STD	1	20	9.209	9.209	0.000	110.45	1.000
	2	20	11.116	11.116	0.000	128.84	1.000
trimethoprim	1	21	9.807	1.065	0.000	28.87	1.834
	2	21	12.388	1.114	0.000	29.02	1.789

Fig. 11.4. Advanced dual-column comparison report for the case in Fig. 11.3.

psychotropic and other prescription drugs relevant in forensic toxicology that can be analysed without prior derivatization by GC even at the therapeutic concentration level. Maintenance comprises changing the injector liner glass wool daily and the liner weekly, shortening the precolumns weekly by 50 cm and quantitative calibration at one-month intervals.

References pp. 423–424

Fig. 11.2 displays a typical pair of chromatograms obtained from an autopsy case, representing elevated blood concentrations of the opioid tramadol and the antidepressant mirtazapine, together with their metabolites. In addition, the antibacterial agent trimethoprim and caffeine are present at therapeutic levels. The single-column reports, providing the best hit for each peak, and the advanced dual-column comparison report for the above case are shown in Figs 11.3 and 11.4, respectively. These figures demonstrate that the dual-column comparison report makes the results legible by indicating only those substances for which preselected detection windows fit the detected peaks on both columns [47]. The only false- positive finding in this report, normianserin, can be ruled out by the large concentration difference between the columns and by recognizing that the peak on column one is reserved for mirtazapine (marked with #).

Qualitative performance of the method is shown in Fig. 11.1, indicating excellent day-to-day precision of RRT (CV < 0.1%). Quantitative performance of the method is sufficient for most applications of forensic toxicology. The LOQ, typically ranging from 0.05 to 0.2 mg/l, is adequate for analysing most basic drugs at therapeutic levels (Table 11.3). The expanded uncertainty of measurement generally lies between 15% and 40%. Interlaboratory proficiency testing has been carried out with amitriptyline, citalopram, dextropropoxyphene, levomepromazine (methotrimeprazine) and methadone, all of which show good z-score values, always below 1.4 units (deviation from distribution's mean, expressed in units of SD) over a four-year period. The method has been accredited by the Finnish Centre for Metrology and Accreditation (FINAS).

11.7 CONCLUSIONS

Broad-scale screening for drugs and poisons is an integral part of forensic and hospital toxicology. The aim of drug screening is to identify and preferably also to quantify a maximal number of potentially toxic compounds present in a biological sample. Despite the increased attention directed at GC-MS and LC- MS techniques, GC screening with selective detection has maintained its position as a rapid and cost-effective tool for analysing the majority of common drugs involved in poisonings. Especially the dual-column approach, consisting of two parallel columns of different selectivity and two similar detectors, provides a high identification power close to GC-MS. Comparison of the quantitative response factors on the two columns provides an additional identification parameter. Applying RTL function significantly improves the precision of identification, not only between systems or laboratories, but also within an individual GC instrument and column. In addition, GC usually allows more substances to be included in quantitative analysis than GC-MS in the SIM mode. As 90% of drugs contain nitrogen, NPD is indispensable for detecting these substances, with minimal interference from the matrix. Particularly for tertiary amine drugs, GC-NPD can still be considered the optimal choice. Finally, the success of GC screening is very much dependent on the performance of the reporting software, especially in dual-column operation.

11.8 REFERENCES

1 A.T. James and A.J.P. Martin, Biochem. J., 50 (1952) 679.
2 K.D. Parker and P.L. Kirk, Anal. Chem., 33 (1961) 1378.
3 M.W. Anders, G.J. Mannering, in: A. Stolman (Ed.), Progress in Chemical Toxicology, Vol. 3. Academic Press, New York, 1967, 121..
4 B.S. Finkle, E.J. Cherry and D.M. Taylor, J. Chromatogr. Sci., 9 (1971) 393.
5 R.E. Ardrey and A.C. Moffat, J. Chromatogr., 220 (1981) 195.
6 R.D. Dandeneau and E.H. Zerenner, J. High Resolut. Chromatogr. Chromatogr. Commun., 2 (1979) 351.
7 L.M. Blumberg and M.S. Klee, Anal. Chem., 70 (1998) 3828.
8 A. Eklund, J. Jonsson and J. Schuberth, J. Anal. Toxicol., 7 (1983) 24.
9 M.J. Dunphy and M.K. Pandya, HRC & CC, 6 (1983) 317.
10 W.H. Anderson and D.T. Stafford, HRC & CC, 6 (1983) 247.
11 D.J. Ehresman, S.M. Price and D.J. Lakatua, J. Anal. Toxicol., 9 (1985) 55.
12 R.W. Taylor, C. Greutink and N.C. Jain, J. Anal. Toxicol., 10 (1986) 205.
13 V.A. Soo, R.J. Bergert and D.G. Deutsch, Clin. Chem., 32 (1986) 325.
14 M.E. Sharp, Can. Soc. Forens. Sci. J., 19 (1986) 83.
15 W.H. Anderson and D.C. Fuller, J. Anal. Toxicol., 11 (1987) 198.
16 M.E. Sharp, J. Anal. Toxicol., 11 (1987) 8.
17 R. Caldwell and H. Challenger, Ann. Clin. Biochem., 26 (1989) 430.
18 R.A. Cox, J.A. Crifasi, R.E. Dickey, S.C. Ketzler and G.L. Pshak, J. Anal. Toxicol., 13 (1989) 224.
19 O.H. Drummer, S. Horomidis, S. Kourtis, M.L. Syrjanen and P. Tippett, J. Anal. Toxicol., 18 (1994) 134.
20 X.-H. Chen, J.-P. Franke, J. Wijsbeck and R.A. de Zeeuw, J. Anal. Toxicol., 18 (1994) 150.
21 P. Lillsunde, L. Michelson, T. Forsström, T. Korte, E. Schultz, K. Ariniemi, M. Portman, M.-L. Sihvonen and T. Seppälä, Forensic Sci. Int., 77 (1996) 191.
22 H. Togunaga, K. Kudo and T. Imamura, Jpn J. Legal Med., 50 (1996) 196.
23 C. Soriano, J. Munoz-Guerra, D. Carreras, C. Rodriguez, A.F. Rodriguez and R. Cortes, J. Chromatogr. B, 687 (1996) 183.
24 C. Sanches de la Torre, M.A. Martinez and E. Almarza, Forensic Sci. Int., 155 (2005) 193.
25 G.W. Hime and L.R. Bednarczyk, J. Anal. Toxicol., 6 (1982) 247.
26 S. Alm, J. Jonson, H. Karlsson and E.G. Sundholm, J. Chromatogr., 254 (1983) 179.
27 B. Newton and R.F. Foery, J. Anal. Toxicol., 8 (1984) 129.
28 E.M. Koves and J. Wells, J. Forensic Sci., 30 (1985) 692.
29 D. Fretthold, P. Jones, G. Sebrosky and I. Sunshine, J. Anal. Toxicol., 10 (1986) 10.
30 V.W. Watts and T.F. Simonick, J. Anal. Toxicol., 10 (1986) 198.
31 J. Cordonnier, M. Van den Heede and A. Heyndrickx, Int. Analyst, May(3) (1987) 28.
32 A. Turcant, A. Premel-Cabic, A. Cailleux and P. Allain, Clin. Chem., 34 (1988) 1492.
33 D. Manca, L. Ferron and J.-P. Weber, Clin. Chem., 35 (1989) 601.
34 P. Lillsunde and T. Seppälä, J. Chromatogr., 533 (1990) 97.
35 A.M. Phillips, B.K. Logan and D.T. Stafford, J. High Resol. Chromatogr., 13 (1990) 754.
36 I. Ojanperä, I. Rasanen and E. Vuori, J. Anal. Toxicol., 15 (1991) 204.
37 K.-R. Kim, J.-H. Kim, H.-K. Park and C.-H. Oh, Bull. Korean Chem. Soc., 12 (1991) 87.
38 K.-R. Kim, W.-H. Shim, Y.-J. Shin, J. Park, S. Myung and J. Hong, J. Chromatogr., 641 (1993) 319.
39 F. Coudore, J.-M. Alazard, M. Paire, G. Andraud and J. Lavarenne, J. Anal. Toxicol., 17 (1993) 109.
40 I. Rasanen, I. Ojanperä and E. Vuori, J. Anal. Toxicol., 24 (2000) 46.
41 I. Rasanen, I. Kontinen, J. Nokua, I. Ojanperä and E. Vuori, J. Chromatogr. B, 788 (2003) 243.

42   R.J. Flanagan, P.E. Morgan, E.P. Spencer and R. Whelpton, Biomed. Chromatogr., 20 (2006) 530.

43   R.J. Phillips, R.J. Wolstromer, and R.R. Freeman, Hewlett-Packard Application Note AN 228–16 (1981).

44   R.A. de Zeeuw; J.P. Franke; H.H. Maurer; K. Pfleger, Gas-chromatographic retention indices of toxicologically relevant substances on packed or capillary columns with dimethylsilicone stationary phases, (3rd Ed.). DFG/TIAFT, VCH, Weinheim, 1992..

45   A.C. Moffat, W.S. Smalldon and C. Brown, J. Chromatogr., 90 (1974) 1.

46   A.C. Moffat, W.S. Smalldon and C. Brown, J. Chromatogr., 90 (1974) 19.

47   I. Rasanen, I. Ojanperä, J. Vartiovaara, E. Vuori and P. Sunila, J. High Resol. Chromatogr., 19 (1996) 313.

48   P.G.A.M. Schepers, J.P. Franke and R.A. de Zeeuw, J. Anal. Toxicol., 7 (1983) 272.

49   J.P. Franke, J. Wijsbeek and R.A. de Zeeuw, J. Forensic Sci., 35 (1990) 813.

50   P. Sandra and F. David, J. Chromatogr. Sci., 40 (2002) 248.

51   J.P. Franke, M. Bogusz and R.A. de Zeuw, Fresenius J. Anal. Chem., 347 (1993) 67.

52   E. Kováts, Helv. Chim. Acta, 41 (1958) 1915.

53   G. Castello, J. Chromatogr. A, 842 (1999) 51.

54   L. Blomberg, Adv. Chromatogr., 26 (1987) 229.

55   H. Van Den Dool and P.D. Kratz, J. Chromatogr., 11 (1963) 463.

56   W.A. Halang, R. Langlais and E. Kugler, Anal. Chem., 50 (1978) 1829.

57   Z. Garkani-Nejad, M. Karlovits, W. Demuth, T. Stimpfl, W. Vycudilic, M. Jalali-Heravi and K. Varmuza, J. Chromatogr. A, 1028 (2004) 287.

58   V.W. Watts and T.F. Simonick, J. Anal. Toxicol., 11 (1987) 210.

59   R. Aderjan and M. Bogusz, J. Chromatogr., 454 (1988) 345.

60   D.W. Christ, P. Noomano, M. Rosas and D. Rhone, J. Anal. Toxicol., 12 (1988) 84.

61   I. Rasanen, I. Ojanperä and E. Vuori, J. High Resolut. Chromatogr., 18 (1995) 66.

62   I. Rasanen, I. Ojanperä, E. Vuori and T.A. Hase, J. High Resolut. Chromatogr., 16 (1993) 495.

63   I. Rasanen, I. Ojanperä and E. Vuori, J. Chromatogr., 693 (1995) 69.

64   I. Rasanen, I. Ojanperä, E. Vuori and T.A. Hase, J. Chromatogr., 738 (1996) 233.

65   T. Sullivan and M. Klee, Am. Lab., 30 (1998) 20C.

66   J. Cook, M. Engel, P. Wylie and B. Quimby, J. AOAC Int., 82 (1999) 313.

67   K. Grob, Split and splitless injection for quantitative gas chromatography: Concepts, processes, practical guidelines, sources of error, (4th Ed.). Wiley-VCH, Weinheim, 2001..

68   H. Huizer, A.J. Poortman van der Meer and H.E. van Egmond, Sci. Justice, 41 (2001) 185.

69   M. Kageura, Y. Fujiwara, K. Hara, Y. Hieda and S. Kashimura, Japan. J. Forensic Toxicol., 10 (1992) 144.

M.J. Bogusz (Ed.). Forensic Science
Handbook of Analytical Separations, Vol. 6
© 2008 Published by Elsevier B.V.

CHAPTER 12

Forensic screening with GC-MS

Hans H. Maurer

Department of Experimental and Clinical Toxicology, Institute of Experimental and Clinical Pharmacology and Toxicology, Saarland University, D-66421 Homburg (Saar), Germany

12.1 INTRODUCTION

In this chapter, multi-analyte screening procedures are described for the simultaneous detection of several drug classes in blood or urine using gas chromatography-mass spectrometry (GC-MS). They are indispensable tools in forensic and clinical toxicology, because the compounds, which have to be analyzed, are often unknown. Therefore, the first step is screening for and identification of the compounds of interest followed by quantification.

12.2 SCREENING STRATEGIES

Immunoassays can be used for preliminary screening in order to differentiate between negative and presumptively positive samples, if only a single drug or drug class has to be monitored with so-called target screening procedures. Positive results must be confirmed by a second independent method that is at least as sensitive as the screening test and that provides the highest level of confidence in the result. This is ideally done by GC-MS or liquid chromatography (LC)-MS using the selected-ion monitoring (SIM) or the multiple-reaction monitoring (MRM) mode [1–5]. In order to increase the selectivity, at least three ions or two MRM transitions should be chosen representing different parts of the molecule preferably including the molecular ion. Identification criteria are the chromatographic retention time and the presence of the characteristic fragment ions with defined abundance ratios. This prescreening/confirmation strategy is employed, if only those drugs or poisons have to be determined, which are scheduled e.g. by law or by international sport organizations, and for which immunoassays are commercially available. Corresponding target screening and confirmation procedures are described for the particular drugs or drug classes in chapters 1–11.

References pp. 442–445

The strategy for non-target screenings must be more extensive, because several thousands of drugs or pesticides are on the market worldwide. For these reasons, so-called general unknown or systematic toxicological analysis (STA) procedures [1,3,6–9] are necessary that allow the simultaneous detection of as many toxicants as possible in biosamples. Without doubt, GC-MS, especially in the full-scan electron ionization (EI) mode, is still the reference method for comprehensive screening and reliable library-assisted identification, because huge libraries of reference EI mass spectra are available. They allow identification of unknown compounds even in absence of reference substances, if certain prerequisites are fulfilled [10]. LC-MS(-MS) applications for comprehensive screening are still rather limited, but some are looking promising [1–3,11,12] and have helped to extend the spectrum of sensitive and specific MS-based methods to analytes, which are not amenable to GC-MS analysis because of hydrophilic or thermolabile properties. Corresponding procedures are described in chapter 14.

12.3 SAMPLES AND THEIR WORK-UP

In most clinical toxicology cases, blood plasma and urine is available for screening, but in some forensic toxicology cases, only so-called alternative matrices are available. A number of GC-MS methods for analysis of drugs in alternative matrices like hair [13,14], sweat and saliva [15–17], meconium [18,19], or nails [20,21] have also been published and are described in chapter 21. A comprehensive screening for series of various drugs has not yet been described in alternative matrices, probably because the concentrations are too low for full-scan GC-MS detection. Blood (plasma, serum) is the sample of choice for quantification. However, if the blood concentration is high enough, screening can also be performed herein. This is especially advantageous, if only blood samples are available and/or the procedures allow simultaneous screening and quantification [5,22–24]. In driving under the influence of drugs (DUID) cases, blood analysis is even mandatory. Urine is still the sample of choice for comprehensive screening for and identification of unknown drugs or poisons, mainly because concentrations of drugs are relatively high in urine and the samples can be taken non-invasively [1,3,6]. However, the metabolites of these unknowns must be identified in addition or even exclusively.

Suitable sample preparation is an important prerequisite for GC-MS analysis in biosamples. It may involve cleavage of conjugates, isolation and derivatization preceded or followed by clean-up steps. Cleavage of conjugates can be performed by fast acid hydrolysis [25,26] or by gentle but time-consuming enzymatic hydrolysis [26,27]. However, the enzymatic hydrolysis of acyl glucuronides (ester glucuronides of carboxy derivatives such as non-steroidal anti-inflammatory drugs, NSAIDs) may be impossible due to acyl migration [28], an intramolecular transesterification at the hydroxy groups of the glucuronic acid, which leads to β-glucuronidase-resistant conjugates. If the analysis must be finished within a rather short time, like in emergency toxicology, it is preferable to cleave the conjugates by rapid acid hydrolysis [25,26,29]. Alkaline hydrolysis is only suitable for cleavage of acylalic (ester) conjugates such as 11-nor-delta-9-tetrahydrocannabinol-9-carboxylic acid glucuronide

[30]. However, the formation of artifacts during chemical hydrolysis must be considered [31]. Acyl glucuronides e.g. of acidic drugs were readily cleaved under the conditions of extractive alkylation (alkaline pH, elevated temperature) and needed no additional cleavage step [9,32].

Isolation can be performed by liquid–liquid extraction (LLE) at a pH at which the analyte is non-ionized or by solid-phase extraction (SPE) preceded or followed by clean-up steps. Sample pretreatment for SPE depends on the sample type: whole blood and tissue (homogenates) need deproteinization and filtration/centrifugation steps before application to the SPE columns, whereas for urine usually a simple dilution step and/or centrifugation is sufficient [33–38]. Whatever SPE column is used, the analyst should keep in mind, that there may be considerable differences from batch-to-batch, and that comparable sorbents from different manufacturers may also lead to different results [39]. Therefore, use of a suitable internal standard (e.g. deuterated analytes) is recommended. Solid-phase microextraction (SPME) is becoming a modern alternative to SPE and LLE. It is a solvent-free and concentrating extraction technique especially for rather volatile analytes. It is based on the adsorption of the analyte on a stationary phase coating a fine rod of fused silica. The analytes can be desorbed directly in the GC injector. Fast target screening procedures by GC-MS after SPME have been published in recent years [40–44].

Derivatization steps are essential, if relatively polar compounds containing e.g. carboxylic, hydroxy, primary or secondary amino groups are to be determined by GC-MS, and/or if electronegative moieties (e.g. halogen atoms) have to be introduced into the molecule for sensitive negative-ion chemical ionization (NICI) detection. The following procedures are typically used for basic compounds: acetylation (AC), trifluoroacetylation (TFA), pentafluoropropionylation (PFP), heptafluorobutyration (HFB), trimethylsilylation (TMS), or for acidic compounds: methylation (ME), extractive ME, pentafluoropropylation, TMS or *tert*-butyldimethylsilylation. Further details on derivatization methods can be found in refs. [26,30] and the pros and cons of derivatization procedures were discussed in a review of Segura *et al.* [45].

In the following, non-target GC-MS screening procedures are described covering basic and neutral as well as acidic drugs and/or their metabolites in blood or urine.

12.4 SCREENING IN BLOOD; SERUM; OR PLASMA

GC-MS full-scan screening procedures in blood principally allow to detect a wide range of analytes although their concentrations are generally lower than those in urine [1]. Target screening methods using the SIM mode have been published mainly for drugs of abuse, e.g. for new designer drugs including their validated quantification [33,37] or for application in the context of DUID [5]. GC-MS procedures relevant to clinical and forensic toxicology were published for screening and determination of organophosphorus pesticides in serum [46–48]. A GC-MS screening procedure has been described for about 100 acid, neutral and basic drugs in horse plasma [49]. Methods for postmortem drug analysis have been reviewed recently [50]. NICI allows to markedly lower the detection limits [38,51,52], but this technique is not suitable for comprehensive screening because the analytes must contain an electronegative moiety and the NICI mass spectra are less informative and reproducible than EI spectra.

Polettini has developed an automated screening procedure for barbiturates, benzodiazepines, antidepressants, morphine and cocaine in blood after SPE and TMS [24,53]. The sample preparation consisted of SPE and TMS derivatization both automated using an HP PrepStation. The samples were directly injected by the PrepStation and analyzed by full-scan GC-MS. Using macros, peak identification and reporting of results were also automated. This fully automated procedure takes about 2 h, which is acceptable for forensic drug testing or doping control, but not for emergency toxicology. Automation of the data evaluation is a compromise between selectivity and universality. If the exclusion criteria are chosen too narrow, peaks may be overlooked. If the window is too large, a series of proposals is given by the computer, which has to be revised by the toxicologist.

Maurer has described a rather comprehensive plasma screening procedure based on a standard LLE with diethyl ether-ethyl acetate at pH 7 and 12 after addition of the universal internal standard (IS) trimipramine-d3 [26,30,54–56]. This universal extract can be used for GC-MS as well as for LC-MS screening, identification and quantification [51,55,56]. The full-scan GC-MS screening is based on reconstructed mass chromatography using macros for selection of suspected drugs followed by identification of the unknown spectra by library search [57]. The selected ions for screening in plasma (and gastric content) have recently been updated by the author's coworkers using experiences from their daily routine work with this procedure and are summarized in ref. [1]. Table 12.1 lists 202 compounds detected so far by this procedure [57]. A rather universal SPE procedure has proved to be a good alternative for the LLE procedure leading to cleaner extracts [33–38]. With few exceptions, this SPE can also be used for the described plasma screening [26].

Of course, further compounds can be detected, if they are extractable under the conditions applied, volatile in GC and if their mass spectra are in the reference libraries [57–59]. In order to widen the screening window, comprehensive urine screening by full-scan GC-MS allowing detection of several thousand compounds is strictly recommended.

12.5 SCREENING IN URINE

Urine still remains the standard biosample for comprehensive screening, especially in general unknown cases, in forensic and clinical toxicology as well as in doping control. Corresponding procedures have recently been reviewed elsewhere [1]. Analytical quality criteria of the parent compound are of minor value, if the concentrations of the metabolites are much higher in urine than those of the parent drug and if the metabolites are primarily detected by the procedure. The procedure should be sufficiently sensitive to detect therapeutic concentrations at least over a 12–24 h period after ingestion. A rather comprehensive screening procedure was published for the detection of doping relevant stimulants, β-blockers, β-agonists and narcotics after enzymatic hydrolysis, SPE and combined TMS and TFA derivatization [60]. The

TABLE 12.1
202 COMPOUNDS DETECTED IN BLOOD PLASMA AFTER LLE AND FULL-SCAN GC-MS
USING MASS CHROMATOGRAPHY FOR SCREENING AND LIBRARY SEARCH FOR IDEN-
TIFICATION. DATA TAKEN FROM REF. [57]

Monitored Drugs	Monitoring Ions (m/z)
Acetylmethadol	353, 338, 225, 91, 72
Alimemazine	298, 198, 100, 84, 58
Allobarbital	208, 193, 167, 124, 80
Ambroxol	376, 279, 264, 262, 114
Amitriptyline	277, 215, 202, 91, 58
Amitriptylinoxide -$(CH_3)_2NOH$	232, 217, 202,
Amobarbital	211, 198, 197, 156, 141
Aprobarbital	210, 195, 167, 124
Atrazine	215, 200, 173, 68, 58
Bamipine	280, 182, 97, 91, 70
Barbital	156, 141, 112, 98, 83
Benoxaprofen	301, 256, 119, 91, 65
Bezafibrate-CO_2	317, 275, 139, 120, 107
Brallobarbital	245, 207, 165, 124, 91
Bromazepam	315, 286, 236, 208, 179
Buflomedil	307, 210, 195, 97, 84
Bupivacaine	288, 245, 140, 98, 84
Butabarbital	183, 156, 141
Butalbital	209, 181, 168, 167, 141
Butallylonal	223, 167, 124
Butinoline	291, 290, 115, 105, 70
Butobarbital	197, 184, 156, 141, 98
Cafedrine–H_2O	339, 277, 250, 207, 70
Caffeine	194, 109, 82, 67, 55
Canrenone	340, 325, 267, 227
Carbamazepine	236, 193, 165
Carbromal	208, 191, 165, 114, 69
Carisoprodol	260, 245, 158, 97, 55
Chlordiazepoxide	299, 282, 241, 124, 77
Chlormezanone	209, 152, 98
Chloroquine	319, 290, 245, 112, 86
8-Chlorotheophylline	214, 157, 129, 68
Chlorphenamine	274, 203, 167, 72, 58
Chlorpromazine	318, 272, 232, 86, 58
Chlorprothixene	315, 255, 221, 58
Citalopram	324, 238, 208, 190, 58
Clindamycin	388, 341, 126, 82
Clobutinol	255, 240, 130, 125, 58
Clomethiazole	161, 112, 85
Clomipramine	314, 269, 227, 85, 58
Clorazepate-M/artifact (nordazepam)	270, 269, 242, 241, 77
Clozapine	326, 256, 243, 192, 70
Codeine	299, 229, 162, 124
Cotinine	176, 118, 98
Crotamiton (*cis* and *trans*)	203, 188, 135, 120, 69
Crotylbarbital	210, 181, 156, 141, 55
Cyclobarbital	236, 207, 157, 141, 79

TABLE 12.1
CONTINUED

Monitored Drugs	Monitoring Ions (m/z)
Cyclopentobarbital	193, 169, 67
Cyclophosphamide-HCl	224, 175, 147, 69
o,p'-DDD	318, 235, 199, 165
o,p'-DDE	316, 281, 246, 210, 176
Dextropropoxyphene	250, 193, 178, 91, 58
Diazepam	284, 283, 256, 221, 77
Diazinon	304, 199, 179, 152, 137
Dichloroquinolinol	213, 185, 150
Diclofenac	295, 242, 214, 179, 108
Diethylallylacetamide	155, 140, 126, 69, 55
Dihydrocodeine	301, 244, 164, 115, 70
Diltiazem	414, 150, 121, 71, 58
Dimethoate	229, 125, 93, 87
Dimpylate	304, 199, 179, 152, 137
Diphenhydramine	227, 165, 152, 73, 58
Diphenhydramine-M (nor-)	167, 165, 152
Dipropylbarbital	170, 141, 98
Disopyramide	239, 212, 195, 167, 114
Dixyrazine-M (ring)	199, 167
Dosulepin	295, 234, 221, 202, 58
Doxepin	279, 234, 221, 202, 58
Ethaverine	395, 366, 352, 252, 236
Ethenzamide	165, 150, 120, 105, 92
Ethinamate	167, 124, 95, 91, 81
Ethosuximide	141, 113, 70, 55
Etomidate	244, 199, 105, 77
Fenfluramine	230, 216, 159, 72
Flecainide formyl artifact	426, 301, 218, 125, 97
Fluconazole	224, 155, 141, 127, 82
Furosemide-SO$_2$NH	251, 233, 96, 81, 53
Glutethimide	217, 189, 160, 132, 117
Granisetron	312, 159, 136, 110, 96
Guaifenesin	198, 124, 109
Halazepam	352, 324, 289, 241
Heptabarbital	221, 141
Hexethal	211, 156, 141, 55
Hexobarbital	236, 221, 157, 155, 81
Ibuprofen	206, 163, 161, 119, 91
Idobutal	181, 167, 124
Isoniazid formyl artifact	149, 122, 106, 78, 51
Kavain	230, 202, 104, 98, 68
Ketamine	237, 209, 180, 152, 102
Ketazolam-M	268, 239, 233, 205, 77
Lamotrigine	255, 185, 157, 123, 114
Laudanosine	357, 206, 190, 162, 151
Levacetylmethadol	353, 338, 225, 91, 72
Levetiracetam	170, 126, 112, 98, 69
Levomepromazine	328, 228, 185, 100, 58
Lidocaine	234, 120, 86, 72, 58

TABLE 12.1
CONTINUED

Monitored Drugs	Monitoring Ions (m/z)
Lisofylline	280, 236, 193, 180, 109
MCPA	200, 155, 141, 125, 77
Medazepam-M	270, 269, 242, 241, 77
Mephenytoin	218, 189, 104
Mepivacaine	246, 176, 120, 98, 70
Meprobamate	144, 114, 96, 83, 55
Mesuximide	203, 118, 103, 91, 77
Metamizol	215, 123, 91, 56
Methadone	309, 294, 223, 165, 72
Methaqualone	250, 235, 132, 91, 65
Metharbital	170, 155, 126, 112
Methohexital	261, 247, 221, 178, 79
Methylphenobarbital	246, 218, 146, 117
Methylprednisolone-$C_2H_4O_2$	314, 136, 121, 91, 77
Methylsalicylate	152, 120, 92, 65
Methyprylone	183, 155, 140, 98, 83
Metoprolol formyl artifact	279, 264, 127, 112, 56
Metronidazole	171, 124, 81, 54
Midazolam	325, 310
Mirtazapine	265, 208, 195, 180, 167
Mitotane	318, 235, 199, 165
Moclobemide	268, 139, 113, 100, 70
Moxaverine	307, 292, 248, 91
Naftidrofuryl	383, 368, 141, 99, 86
Naproxen–CO_2	184, 169, 141, 115
Narcobarbital	223, 181, 138, 124
Narconumal	209, 181, 167, 124, 97
Nealbarbital	223, 181, 167, 141, 57
Nefopam	253, 225, 179, 165, 58
Nicotine	162, 133, 84
Nifedipine	346, 329, 284, 268, 224
Nifenazone-M (desacyl-)	203, 93, 84, 56
Nisoldipine-M	328, 297, 282, 267, 250
Nitrendipine	360, 331, 238, 210, 150
Nordazepam	270, 269, 242, 241, 77
Oxazepam	268, 239, 233, 205, 77
Oxeladin	335, 320, 219, 144, 86
Papaverine	339, 338, 324, 308, 293
Paracetamol	151, 109, 81, 80
Parathion-ethyl-M (amino-)	261, 125, 109, 80
Pentifylline-M (HO-)	280, 236, 193, 180, 109
Pentobarbital	197, 156, 141, 98, 69
Pentoxifylline	278, 221, 193, 180, 109
Perazine	339, 238, 141, 113, 70
Pethidine	247, 218, 172, 71
Phenazone	188, 96, 77
Phencyclidine	243, 242, 200, 91, 84
Pheniramine	240, 196, 169, 72, 58
Phenobarbital	232, 204, 161, 146, 117

TABLE 12.1
CONTINUED

Monitored Drugs	Monitoring Ions (m/z)
Phenothiazine	199, 167
Phenprocoumon	280, 251, 189, 121, 91
Phenylbutazone	308, 252, 183, 77
Phenylmethylbarbital	218, 132, 104, 78
Phenytoin	252, 223, 180, 104, 77
Pholcodine	398, 114, 100, 70
Phosphamidon	264, 227, 193, 127, 72
Phoxim-M/artifact	196, 171, 143, 111, 97
Pinazepam-M (nordazepam)	270, 269, 242, 241, 77
Piperonyl butoxide	338, 193, 176, 149, 57
Prazepam-M (nordazepam)	270, 269, 242, 241, 77
Prednisolone	300, 122, 91
Prilocaine	220, 107, 86, 65
Primidone	218, 190, 161, 146, 117
Probarbital	169, 156, 141, 98
Procainamide	235, 120, 99, 86
Promazine	284, 199, 86, 58
Promethazine	284, 213, 198, 180, 72
Propofol	178, 163, 121, 117, 91
Propoxur	209, 152, 110, 81
Propoxyphene	250, 193, 178, 91, 58
Propyphenazone	230, 215, 56
Prothipendyl	285, 227, 200, 86, 58
Quinidine	324, 189, 173, 136
Roxatidine	306, 190, 116, 98, 84
Salicylamide	137, 120, 92, 65
Salicylic acid	138, 120, 92, 64
Secobarbital	209, 195, 168, 167, 141
Spironolactone-CH$_3$COSH	340, 325, 267, 227
Sulfanilamide	172, 156, 108, 92, 65
Sulfapyridine	184, 156, 108, 92, 65
Sultiame	290, 225, 184, 168, 104
Suxibuzone artifact	308, 252, 183, 77
Talbutal	167, 153, 124, 97
Temazepam	300, 271, 256, 228, 77
Tertatolol formyl artifact	307, 292, 141, 96, 57
Tetrazepam	288, 259, 253, 225
Theobromine	180, 137, 109, 82
Theophylline	180, 95, 68
Thiobutabarbital	228, 172, 157, 97, 57
Thiopental	242, 173, 172, 157, 69
Thioridazine	370, 126, 98, 70
Tiapride	328, 311, 213, 134, 86
Tiaprofenic acid–CO$_2$	216, 201, 139, 105, 77
Tilidine-M (nor-)	259, 83, 68
Tramadol	263, 218, 188, 135, 58
Tramadol–H$_2$O	245, 200, 141, 128, 58
Triflupromazine	352, 267, 86, 58
Trimethoprim	290, 259, 123

TABLE 12.1
CONTINUED

Monitored Drugs	Monitoring Ions (*m/z*)
Trimipramine	294, 249, 193, 99, 58
Valproic acid	144, 115, 102, 73
Verapamil	303, 260, 151, 58
Vinbarbital	195, 152, 141, 79, 67
Vinylbital	209, 195, 154, 83, 71
Xipamide–SO$_2$NH	275, 155, 121
Zolpidem	307, 235, 219, 92, 65
Zotepine	331, 299, 199, 72, 58

time-consuming enzymatic cleavage of conjugates is acceptable for doping analysis, since resul‎ ‎ot available as fast as in emergency toxicology. The authors did not focus the‎ ‎on the detection of the metabolites, even if they could be detected ‎ ‎ longer time and more sensitively than their parent compou‎ ‎analysis are given in chapter 22. The most comprehensive ST‎ ‎alysis after acid hydrolysis, LLE, AC and full-scan GC-‎ ‎ 2,000 different drugs, poisons and/or their metabolites ‎ ‎,22,25,61–67], thus fulfilling most demands of high-throughpu‎ ‎ in the meaning of analytical toxicology. As compliment to this STA, ‎ ‎prehensive GC-MS screening procedure was developed for the detection‎ ‎cidic drugs, poisons and/or their metabolites in urine after extractive ME [9,27,32,68–72] and finally one for amphophilic designer drugs of the pyrrolidinophenone-type and/or their metabolites [73–78].

12.5.1 General screening procedures for basic and neutral drugs in urine

As already mentioned, the most comprehensive GC-MS screening procedure allows detection of over 2,000 different drugs, poisons and/or their metabolites [1,6,25,61–67]. One prerequisite for this procedure is the mass spectral elucidation of the metabolism of the covered drugs and poisons. Another prerequisite for full-scan screening procedures is the availability of a suitable reference mass spectra collection as discussed below. Use of extensive mass spectral reference libraries from the fields of toxicology [57] and general chemistry [59] often allow to detect even unexpected compounds amenable to GC and EI.

This screening method was developed and has been improved and extended during the past years. Cleavage of conjugates is necessary before extraction since part of the drugs and/or their metabolites are excreted into urine as conjugates. For studies on the toxicological detection rapid acid hydrolysis was performed to save time, which is relevant e.g. in emergency toxicology. However, some compounds were destroyed or altered during acid hydrolysis [29,30,79]. Therefore, the standard procedure [31] had to be modified. Before extraction, half of the native urine volume is added to the previously hydrolyzed part. The extraction solvent used has proved to be very efficient in extracting compounds with very different chemical properties from biomatrices, so that it has been used for a STA procedure for basic and neutral analytes [1,6,22,25,61–67]. AC has proved to be very suitable for robust derivatization in order to improve the GC properties and thereby the detection limits of thousands of drugs and their metabolites [22,57]. The use of microwave irradiation reduced the incubation time from 30 min to 5 min [29,80], so that derivatization should no longer be renounced due to expense of time.

As listed in Table 12.2, this comprehensive full-scan GC-MS screening procedure allows within one run the simultaneous screening and library-assisted identification of the following categories of drugs: tricyclic antidepressants [81,82], selective serotonin reuptake inhibitors (SSRIs) [29], butyrophenone neuroleptics [83], phenothiazine neuroleptics [84], benzodiazepines [1,85], barbiturates [86]and other sedative-hypnotics [1,86], anticonvulsants [87], antiparkinsonian drugs [88], phenothiazine antihistamines [89], alkanolamine antihistamines [90], ethylenediamine antihistamines [91], alkylamine antihistamines [92], opiates and opioids [93], non-opioid analgesics [94], stimulants and hallucinogens [1,95–101], designer drugs of the amphetamine-type [25,61,62,95,102], the piperazine-type [102–109], the phenethylamine-type [64–67,110,111], the phencyclidine-type [63], *Eschscholtzia californica* ingredients [112], nutmeg ingredients [113], β-blockers [114], antiarrhythmics [114] and diphenol laxatives [115]. In addition, series of further compounds can be detected [22,57,116], if they are present in the extract and their mass spectra are contained in the used reference libraries [57–59]. As shown in Table 12.2, several ions per category were individually selected from the mass spectra of the corresponding drugs and their metabolites identified in authentic urine samples. Generation of the mass chromatograms can be started by clicking the corresponding pull down menu which executes the user defined macros [117].

The procedure is illustrated in Fig. 12.1. Typical reconstructed mass chromatograms with the given ions of an acetylated extract of a rat urine sample collected over 24 h after intake of 0.5 mg/kg BM of the designer drug 2C-E. They indicate the presence of acetylated 2C-E (peak no. 1), isomer 1 (peak no. 2) and 2 (peak no. 3) of the acetylated *O*-demethyl metabolite of 2C-E, isomer 1 (peak no. 9) and 2 (peak no. 10) of the acetylated *O*-demethyl oxo metabolite of 2C-E [65]. Fig. 12.2 shows the mass spectrum underlying the marked peak 2 in Fig. 12.1., the reference spectrum, the structure and the hit list found by computer library search [65]. This example was selected as it demonstrates that rat models can be used to develop such screening procedure, when human urine samples are not available. According to comparison studies [62,81,103,113,118], good agreement between rats and humans has been reported for the metabolic pathways and detectability of drugs in urine. Screening for further drugs is performed simply by monitoring other selected ions and identifying appearing peaks by library search.

It is evident, that compounds with acidic or zwitterionic properties cannot be analyzed with the general screening procedure described in chapter 13.3.1, because at the used extraction pH such compounds are ionized and unable to move to the organic phase. Therefore, alternative screening procedures have been developed and are presented below.

TABLE 12.2
CLASSES OF BASIC AND NEUTRAL DRUGS SIMULTANEOUSLY SCREENED FOR AND IDENTIFIED BY THE GC-MS STA WITH ACID HYDROLYSIS, LLE AND ACETYLATION OF URINE, THE MONITORING IONS, AND THE CORRESPONDING REFERENCES

Monitored Drug Classes	Monitoring Ions (*m/z*)	Refs.
Antidepressants, tricyclic	58, 84, 86, 100, 191, 193, 194, 205, and 120, 182, 195, 235, 261, 276, 284, 293	[81,82]
Antidepressants, SSRIs	58, 72, 86, 173, 176, 234, 238, 290	[29]
Neuroleptics, butyrophenones	112, 123, 134, 148, 169, 257, 321 and 189, 191, 223, 233, 235, 245, 287, 297	[83]
Neuroleptics, phenothiazine	58, 72, 86, 98, 100, 113, 114, 141 and 132, 148, 154, 191, 198, 199, 243, 267	[84]
Benzodiazepines	111, 205, 211, 230, 241, 245, 249, 257, 308, 312, 333, 340, 357	[1,85]
Barbiturates	83, 117, 141, 157, 167, 207, 221, 235	[86]
Other sedative-hypnotics	83, 105, 156, 163, 167, 172, 216, 235, 248, 261	[86]
Anticonvulsants	102, 113, 146, 185, 193, 204, 208, 241	[87]
Antiparkinsonian drugs	86, 98, 136, 150, 165, 196, 197, 208	[88]
Phenothiazine antihistamines	58, 72, 100, 114, 124, 128, 141, 199	[89]
Alkanolamine antihistamines	58, 139, 165, 167, 179, 182, 218, 260	[90]
Ethylenediamine antihistamines	58, 72, 85, 125, 165, 183, 198, 201	[91]
Alkylamine antihistamines	58, 169, 203, 205, 230, 233, 262, 337	[92]
Opiates and opioids	111, 138, 187, 245, 259, 327, 341, 343, 359, 420	[93]
Non-opioid analgesics	120, 139, 151, 161, 188, 217, 230, 231, 258, 308	[94]
Stimulants/hallucinogens	58, 72, 86, 82, 94, 124, 140, 192, 250	[1,95–101]
Designer drugs, amphetamine-type	58, 72, 86, 150, 162, 164, 176, 178	[25,61,62,95,102]

TABLE 12.2
CONTINUED

Monitored Drug Classes	Monitoring Ions (m/z)	Refs.
Designer drugs, piperazine-type	BZP: 91, 107, 137, 146, 191, 204 MDBP: 135, 137, 170, 262, 306 mCPP: 143, 145, 166, 182, 238, 254 TFMPP: 157, 161, 174, 200, 216, 330 MeOPP: 109, 148, 151, 162, 234, 262 Drugs with metabolites common with such designer drugs: Dropropizine: 132, 148, 175, 233, 320, 378 Fipexide: 135, 137, 141, 170, 262, 306	[102,104–109]
Designer drugs, phenethylamine-type	2C-B: 228, 287, 288 2C-D: 78, 238, 164, 223, 236, 295 2C-E: 192, 251, 178, 237 2C-I: 290, 349, 276, 335 2C-T-2: 224, 283, 256, 315 2C-T-7: 238, 297, 296, 355	[64–67,110,111]
Designer drugs, phencyclidine-type	232, 274, 273, 290	[63]
Eschscholtzia californica ingredients	136, 148, 165, 174, 188, 190	[112]
Nutmeg ingredients	150, 164, 165, 180, 194, 252, 266	[113]
β-Blockers	72, 86, 98, 140, 151, 159, 200, 335	[114]
Antiarrhythmics	72, 86, 98, 140, 151, 159, 200, 335	[114]
Laxatives	349, 360, 361, 379, 390, 391, 402, 432	[115]

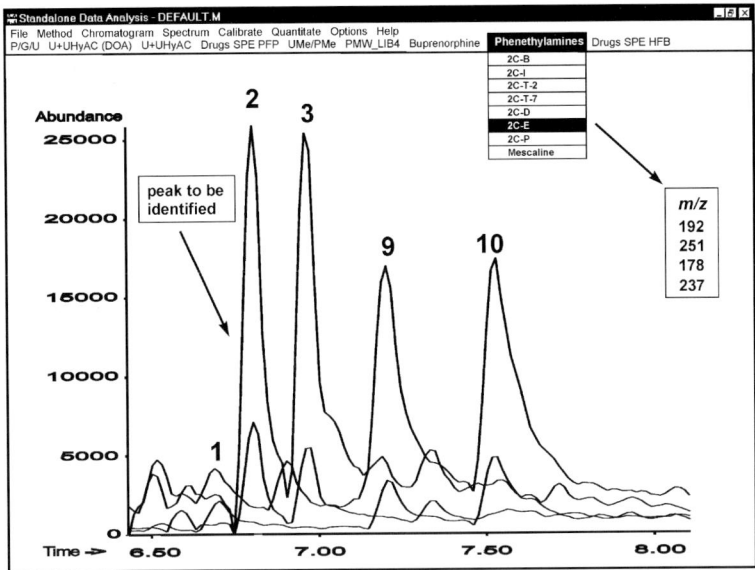

Fig. 12.1. Typical reconstructed mass chromatograms with the given ions of an acetylated extract of a rat urine sample collected over 24 h after intake of 0.5 mg/kg BM of 2C-E. They indicate the presence of acetylated 2C-E (peak no. 1), isomer 1 (peak no. 2) and 2 (peak no. 3) of the acetylated *O*-demethyl metabolite of 2C-E, isomer 1 (peak no. 9) and 2 (peak no. 10) of the acetylated *O*-demethyl oxo metabolite of 2C-E. Taken from ref. [65], with permission from Elsevier.

12.5.2 General screening procedure for acidic compounds after extractive methylation

Extractive alkylation has proved to be a powerful procedure for simultaneous extraction and derivatization of acidic compounds. The acidic compounds are extracted into the organic phase (toluene) at pH 12 as ion pairs with a phase-transfer catalyst (tetrahexyl ammonium). In the organic phase, the phase-transfer catalyst can easily be solvated due to its lipophilic hexyl groups, whereas poor solvation of the anionic analytes leads to a high reactivity with the alkylation reagent, most often methyl iodide. Part of the phase-transfer catalyst can also reach the organic phase as an ion pair with the iodide anion formed during the alkylation reaction or with anions of the urine matrix. Therefore, the remaining part had to be removed to prevent a loss of the GC column's separation power and to exclude interactions with analytes in the GC injection port. Several SPE sorbents and different eluents had been tested for efficient separation of the vestige of the phase-transfer catalyst from the analytes. A diol sorbent yielded best reproducibility and recovery under the described conditions [71,72].

The extracted and derivatized analytes were separated by GC and identified by GC-MS in the full-scan mode. As already described in chapter 13.3.1, the possible presence of acidic drugs and/or their metabolites could be indicated using mass chromatography with selective ions followed by peak identification using library search [57].

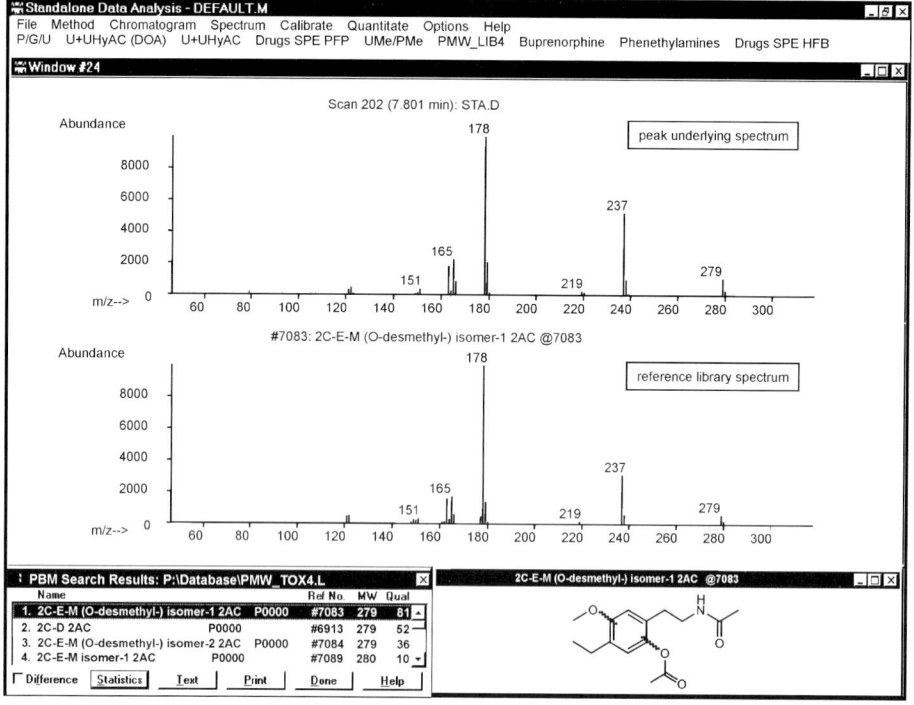

Fig. 12.2. Mass spectrum underlying the marked peak 2 in Fig. 12.1, the reference spectrum, the structure, and the hit list found by computer library search [57]. Taken from ref. [65], with permission from Elsevier.

As listed in Table 12.3, this full-scan GC-MS screening procedure allows within one run the simultaneous screening and library-assisted identification of the following categories of drugs: ACE inhibitors and AT_1 blockers [71], coumarin anticoagulants of the first generation [72], dihydropyridine calcium channel blockers [70], NSAIDs [32], barbiturates [69], diuretics [68], antidiabetics of the sulfonylurea type (sulfonamide part) [22,57], and finally after enzymatic cleavage of the acetalic glucuronides, anthraquinone and diphenol laxatives [27] or buprenorphine [119]. In addition, various other acidic compounds could also be detected [22,57].

12.5.3 General screening procedure for zwitterionic compounds after SPE and silylation

Pyrrolidinophenone derivatives such as R,S-α-pyrrolidinopropiophenone (PPP), R,S-4′-methyl-α-pyrrolidinopropiophenone (MPPP), 4′-methyl-α-pyrrolidinobutyr ophenone (MPBP), R,S-4′-methyl-α-pyrrolidinohexanophenone (MPHP), R,S-3′,4′-methylenedioxy-α-pyrrolidinopropiophenone(MDPPP) and R,S-4′-methoxy-α-pyrrolidinopropiophenone (MOPPP) are new designer drugs which have appeared on the illicit drug market [73,75–78,120]. Unfortunately, these drugs cannot be detected by common screening procedures due to the zwitterionic structure of their

TABLE 12.3

CLASSES OF ACIDIC DRUG SIMULTANEOUSLY SCREENED FOR AND IDENTIFIED BY GC-MS WITH EXTRACTIVE METHYLATION OF URINE, THE MONITORING IONS, AND THE CORRESPONDING REFERENCES

Monitored Drug Classes	Monitoring Ions (m/z)	Refs.
ACE inhibitors and AT_1 blockers	157, 160, 172, 192, 204, 220, 234, 248, 249, 262	[71]
Anticoagulants	291, 294, 295, 309, 313, 322, 324, 336, 343, 354	[72]
Calcium channel blockers (dihydropyridines)	139, 284, 297, 298, 310, 312, 313, 318, 324, 332	[70]
NSAIDs	119, 135, 139, 152, 165, 229, 244, 266, 272, 326	[32]
Barbiturates	117, 169, 183, 185, 195, 221, 223, 232, 235, 249	[69]
Diuretics	Thiazides: 267, 352, 353, 355, 386, 392 Loop diuretics: 77, 81, 181, 261, 270, 295, 406, 438 Others: 84, 85, 111, 112, 135, 161, 249, 253, 289, 363	[68]
Laxatives (after enzymatic cleavage of the acetalic glucuronides)	305, 290, 335, 320, 365, 350, 311, 326, 271, 346	[27]
Buprenorphine (after enzymatic cleavage of the acetalic glucuronides)	352, 384, 392, 424, 441, 481	According to ref. [119]

metabolites. Mixed-mode SPE has proven to be suitable for the extraction even of their zwitterionic metabolites [75]. Common TMS led to good GC properties.

As listed in Table 12.4, this comprehensive full-scan GC-MS screening procedure allows within one run the simultaneous screening and library-assisted identification of the following drugs and/or their metabolites: PPP [78], MPPP [78], MPHP [75], MOPPP [76], MDPPP [77] and MPBP [73].

12.6 REFERENCE MASS SPECTRAL LIBRARIES AS BASIS FOR FULL-SCAN GC-MS SCREENING PROCEDURES

As already mentioned, for general unknown screening procedures, full-scan mode is the method of choice. Reconstructed mass chromatography with selected ions is used for screening and library search for identification of peak underlying spectra after subtraction of the background. At least four selective ions including the molecular ion, if possible, should be present within acceptable tolerances. The match factor in case of computer aided library searching should be as high as possible. The final responsibility, however, lies with the toxicologist to decide, depending on the case, how and when the minimum of requirement for identity confirmation is reached [121].

TABLE 12.4

DRUGS AND/OR THEIR METABOLITES WITH ZWITTERIONIC PROPERTIES SIMULTANE-
OUSLY SCREENED FOR AND IDENTIFIED BY GC-MS AFTER SPE AND SILYLATION, THE
MONITORING IONS, AND THE CORRESPONDING REFERENCES

Monitored Drugs	Monitoring Ions (m/z)	Refs.
PPP	98, 105, 112, 193	[78]
MPPP	98, 112, 119, 221	[78]
MPHP	140, 178, 221, 228	[75]
MOPPP	98, 112, 135, 193	[76]
MDPPP	98, 112, 121, 306	[77]
MPBP	112, 126, 178, 318	[73]

Huge libraries of reference EI mass spectra are available [10]. They allow identifying unknown compounds even in absence of reference substances, if certain prerequisites are fulfilled. Suitable mass spectral reference libraries should be built under standardized conditions and not only be collected from various sources. In case of GC-MS, the spectra of derivatizable compounds should be recorded after common derivatization, because otherwise such compounds cannot sensitively be detected. Spectra of artifacts formed from the analyte during sample preparation or GC must be included as well as those of matrix compounds and of typical impurities/contaminants such as softeners. For use in clinical and forensic toxicology and doping control, the reference library should not only include spectra of reference substances of drugs, poisons, pesticides and pollutants, but also of their metabolites [22,57]. Finally, adequate procedures should be developed for isolation, (derivatization) and chromatographic separation to enable a sensitive and reliable detection of these compounds in biosamples. Different search algorithms are in use for electronic comparison of the spectra [10]. Nevertheless, visual comparison is still an indispensable part of compound identification *via* mass spectral library search. In principle, the mass spectral search may consist of a direct, one-to-one comparison of the unknown mass spectrum to each of the entries in the mass spectral library (forward searching) and/or of a reverse process in which each of the entries in the library is compared with the unknown mass spectrum (reverse searching) or a combination of both. Today, the most often used algorithms are that of the National Institute of Standards (NIST MS Search Program) and Probability-Based Matching (PBM search). These and others have been systematically compared [122–124]. Each of them has certain advantages and disadvantages, but there are no absolute criteria to prefer one particular algorithm. It is important to note, that the so-called match qualities of library searches always have to be used with utmost caution, especially in the case of samples with a high matrix load [125]. In such samples, overlaying mass spectra and thus comparatively low match qualities are the rule rather than the exception.

Several mass spectral reference libraries for GC/EI/MS have been published/or commercialized. The Wiley Registry of Mass Spectral Data containing over 270,000 entries is the foundation spectral library for all MS laboratories [59]. Most spectra are accompanied by the structure and trivial name, molecular formula, molecular weight, nominal mass and base peak. The NIST05 mass spectral database is a collection of over 190,000 EI mass spectra of over 160,000 unique compounds [58]. In addition to an expanded EI library and updated software, NIST05 includes a Kovats retention index library and a MS/MS spectral library. The Wiley Registry of Mass Spectral Data with NIST 2005 Spectral Data [126] is the biggest library and a combination of both, the Wiley Registry of Mass Spectral Data and the NIST05 library, containing over 460,000 spectra.

For use particularly in analytical toxicology, the Maurer/Pfleger/Weber Tox Library 2007 contains more than 7,800 entries [22,57]. The data have been recorded under standard conditions and collected in the author's laboratory. Relevant additional data are provided such as retention indices (Kovats Indexes) [127], chemical structures and empirical formulae, molecular mass, registry numbers of the Chemical Abstracts Services (CAS), major pharmacological categories as well as information in which biosample and after which sample preparation compounds could be detected. The most important data of this library are those of the metabolites, which have been recorded during over 20 years of research in the field of drug metabolism making this collection unique. The above-mentioned screening procedures are based on library-assisted identification. Using macros, the data evaluation can be facilitated [1], but the final decision on the identity of a compound will always rest in the hands of the experienced toxicologist or analyst who undertakes a visual comparison between the full mass spectrum of the measured compound and the reference spectra, by considering further chemical, analytical and toxicological aspects for plausibility [121]. Roesner [128] published a library with 1,700 mass spectra of parent compounds of designer drugs and chemical warfare agents taken from literature. The mass spectra collected by the American Association of Forensic Sciences (AAFS) and the International Association of Forensic Toxicologists (TIAFT) were included in the combined Wiley Registry of Mass Spectral Data with NIST 2005 Spectral Data.

12.7 CONCLUDING SUMMARY AND PERSPECTIVES

High-throughput drug screening procedures in analytical toxicology mean, that thousands of relevant toxicants can simultaneously be screened for, so-called systematic toxicological analysis (STA). GC-MS, especially in the EI full-scan mode, is still the reference method for comprehensive screening and reliable library-assisted identification, because universal sample work-up and GC separation as well as huge libraries of reference EI mass spectra are available. They allow identifying unknown compounds even in absence of reference substances, if certain prerequisites are fulfilled. However, LC-MS has shown to be an ideal supplement, especially for detection of more polar, unstable or low-dosed drugs, especially in blood plasma.

12.8 ACKOWLEDGEMENTS

The author likes to thank Dr. Frank T. Peters and Armin Weber for their discussion and help.

12.9 REFERENCES

1   H.H. Maurer, Clin. Chem. Lab. Med., 42 (2004) 1310.
2   H.H. Maurer, Clin. Biochem., 38 (2005) 310.
3   H.H. Maurer, J. Mass Spectrom., 41 (2006) 1399.
4   P. Marquet, Ther. Drug Monit., 24 (2002) 255.
5   M.R. Moeller and T. Kraemer, Ther. Drug Monit., 24 (2002) 210.
6   H.H. Maurer and F.T. Peters, Ther. Drug Monit., 27 (2005) 686.
7   H.H. Maurer, T. Kraemer, C. Kratzsch, L.D. Paul, F.T. Peters, D. Springer, R.F. Staack, A.A. Weber. In: W.J. Kleemann, J. Teske (Eds.), Toxicological analysis and certainty of results, Schmidt-Roemhild Verlag, Leipzig, 2003, p. 33.
8   H.H. Maurer. In: J. Yinon (Ed.), Advances in forensic applications of mass spectrometry, CRC Press LLC, Boca Raton, FL, 2003, p. 1.
9   H.H. Maurer, J. Chromatogr. B, Biomed. Sci. Appl., 733 (1999) 3.
10   H.H. Maurer and F.T. Peters. In: M. Gross and R.M. Caprioli (Eds.), Encyclopedia of mass spectrometry, Elsevier Science, Oxford, 2006, pp. 115–121.
11   F.L. Sauvage, F. Saint-Marcous, B. Duretz, D. Deporte, G. Lachatre and P. Marquet, Clin. Chem. (2006) [Epub ahead of print].
12   M. Gergov, I. Ojanpera and E. Vuori, J. Chromatogr. B, Analyt. Technol. Biomed. Life Sci., 795 (2003) 41.
13   F. Pragst and M.A. Balikova, Clin. Chim. Acta (2006) [Epub ahead of print].
14   P. Kintz, M. Villain and V. Cirimele, Ther. Drug Monit., 28 (2006) 442.
15   H.H. Maurer, Anal. Bioanal. Chem., 381 (2005) 110.
16   O.H. Drummer, Forensic Sci. Int., 150 (2005) 133.
17   P. Kintz and N. Samyn, Ther. Drug Monit., 24 (2002) 239.
18   E.M. Ostrea Jr., D.K. Knapp, L. Tannenbaum, A.R. Ostrea, A. Romero, V. Salari and J. Ager, J. Pediatr., 138 (2001) 344.
19   M.A. Huestis and R.E. Choo, Forensic Sci. Int., 128 (2002) 20.
20   Y.H. Caplan and B.A. Goldberger, J. Anal. Toxicol., 25 (2001) 396.
21   A. Palmeri, S. Pichini, R. Pacifici, P. Zuccaro and A. Lopez, Clin. Pharmacokinet., 38 (2000) 95.
22   H.H. Maurer, K. Pfleger, and A.A. Weber, Mass Spectral and GC Data of Drugs, Poisons, Pesticides, Pollutants and their Metabolites, Wiley-VCH, Weinheim, 2007.
23   A. Kankaanpaa, T. Gunnar, K. Ariniemi, P. Lillsunde, S. Mykkanen and T. Seppala, J. Chromatogr. B, Analyt. Technol. Biomed. Life Sci., 810 (2004) 57.
24   A. Polettini, A. Groppi, C. Vignali and M. Montagna, J. Chromatogr. B, 713 (1998) 265.
25   A.H. Ewald, F.T. Peters, M. Weise and H.H. Maurer, J. Chromatogr. B, Analyt. Technol. Biomed. Life Sci., 824 (2005) 123.
26   H.H. Maurer. In: H.H. Maurer, K. Pfleger and A.A. Weber (Eds.), Mass spectral and GC data of drugs, poisons, pesticides, pollutants and their metabolites, part 1, Wiley-VCH, Weinheim, 2007, pp. 4–32.
27   J. Beyer, F.T. Peters and H.H. Maurer, Ther. Drug Monit., 27 (2005) 151.
28   L.H. Spahn and L.Z. Benet, Drug Metab. Rev., 24 (1992) 5.
29   H.H. Maurer and J. Bickeboeller-Friedrich, J. Anal. Toxicol., 24 (2000) 340.
30   H.H. Maurer. In: K. Pfleger, H.H. Maurer, A. Weber (Eds.), Mass spectral and GC data of drugs, poisons, pesticides, pollutants and their metabolites, Wiley-VCH, Weinheim, 2000 part 4, p. 3.
31   H.H. Maurer. In: K. Pfleger, H.H. Maurer, A. Weber (Eds.), Mass spectral and GC data of drugs, poisons, pesticides, pollutants and their metabolites, VCH publisher, Weinheim, 1992, p. 3.
32   H.H. Maurer, F.X. Tauvel and T. Kraemer, J. Anal. Toxicol., 25 (2001) 237.
33   V. Habrdova, F.T. Peters, D.S. Theobald and H.H. Maurer, J. Mass Spectrom., 40 (2005) 785.
34   F.T. Peters, N. Samyn, C. Lamers, W. Riedel, T. Kraemer, G. de Boeck and H.H. Maurer, Clin. Chem., 51 (2005) 1811.
35   H.H. Maurer, O. Tenberken, C. Kratzsch, A.A. Weber and F.T. Peters, J. Chromatogr. A, 1058 (2004) 169.

36   C. Kratzsch, A.A. Weber, F.T. Peters, T. Kraemer and H.H. Maurer, J. Mass Spectrom., 38 (2003) 283.
37   F.T. Peters, S. Schaefer, R.F. Staack, T. Kraemer and H.H. Maurer, J. Mass Spectrom., 38 (2003) 659.
38   F.T. Peters, T. Kraemer and H.H. Maurer, Clin. Chem., 48 (2002) 1472.
39   M.J. Bogusz, R.D. Maier, B.K. Schiwy and U. Kohls, J. Chromatogr. B, 683 (1996) 177.
40   M. Walles, W.M. Mullett and J. Pawliszyn, J. Chromatogr. A, 1025 (2004) 85.
41   N. Fucci, N. De Giovanni and M. Chiarotti, Forensic Sci. Int., 134 (2003) 40.
42   S. Gentili, A. Torresi, R. Marsili, M. Chiarotti and T. Macchia, J. Chromatogr. B, Analyt. Technol. Biomed. Life Sci., 780 (2002) 183.
43   F. Musshoff, H.P. Junker, D.W. Lachenmeier, L. Kroener and B. Madea, J. Chromatogr. Sci., 40 (2002) 359.
44   A. Namera, M. Yashiki, T. Kojima and M. Ueki, J. Chromatogr. Sci., 40 (2002) 19.
45   J. Segura, R. Ventura and C. Jurado, J. Chromatogr. B, 713 (1998) 61.
46   E. Lacassie, M.F. Dreyfuss, J.M. Gaulier, P. Marquet, J.L. Daguet and G. Lachatre, J. Chromatogr. B, Biomed. Sci. Appl., 759 (2001) 109.
47   E. Lacassie, P. Marquet, J.M. Gaulier, M.F. Dreyfuss and G. Lachatre, Forensic Sci. Int., 121 (2001) 116.
48   F.A. Tarbah, H. Mahler, O. Temme and T. Daldrup, Forensic Sci. Int., 121 (2001) 126.
49   A. Takeda, H. Tanaka, T. Shinohara and I. Ohtake, J. Chromatogr. B, Biomed. Sci. Appl., 758 (2001) 235.
50   O.H. Drummer and J. Gerostamoulos, Ther. Drug Monit., 24 (2002) 199.
51   H.H. Maurer, T. Kraemer, C. Kratzsch, F.T. Peters and A.A. Weber, Ther. Drug Monit., 24 (2002) 117.
52   H.H. Maurer, Ther. Drug Monit., 24 (2002) 247.
53   A. Valli, A. Polettini, P. Papa and M. Montagna, Ther. Drug Monit., 23 (2001) 287.
54   H.H. Maurer, C. Kratzsch, A.A. Weber, F.T. Peters and T. Kraemer, J. Mass Spectrom., 37 (2002) 687.
55   H.H. Maurer, C. Kratzsch, T. Kraemer, F.T. Peters and A.A. Weber, J. Chromatogr. B, Analyt. Technol. Biomed. Life Sci., 773 (2002) 63.
56   C. Kratzsch, O. Tenberken, F.T. Peters, A.A. Weber, T. Kraemer and H.H. Maurer, J. Mass Spectrom., 39 (2004) 856.
57   H.H. Maurer, K. Pfleger and A.A. Weber, Mass Spectral Library of Drugs, Poisons, Pesticides, Pollutants and their Metabolites, Wiley-VCH, Weinheim, 2007.
58   U.S. Department of Commerce, NIST/EPA/NIH Mass Spectral Library 2005, Wiley, New York, NY, 2005.
59   F.W. McLafferty, Registry of Mass Spectral Data, (7th Ed). Wiley, New York, NY, 2001.
60   A. Solans, M. Carnicero, R. de-la-Torre and J. Segura, J. Anal. Toxicol., 19 (1995) 104.
61   A.H. Ewald, G. Fritschi and H.H. Maurer, J. Mass Spectrom., 41 [Epub ahead of print].
62   A.H. Ewald, G. Fritschi, W.R. Bork and H.H. Maurer, J. Mass Spectrom., 41 (2006) 487.
63   C. Sauer, F.T. Peters, R.F. Staack, G. Fritschi and H.H. Maurer, J. Mass Spectrom., 41 [Epub ahead of print].
64   D.S. Theobald, M. Putz, E. Schneider and H.H. Maurer, J. Mass Spectrom., 41 (2006) 872.
65   D.S. Theobald and H.H. Maurer, J. Chromatogr. B, Analyt. Technol. Biomed. Life Sci. (2006) [Epub ahead of print].
66   D.S. Theobald, S. Fehn and H.H. Maurer, J. Mass Spectrom., 40 (2005) 105.
67   D.S. Theobald, R.F. Staack, M. Puetz and H.H. Maurer, J. Mass Spectrom., 40 (2005) 1157.
68   J. Beyer, A. Bierl, F.T. Peters and H.H. Maurer, Ther. Drug Monit., 27 (2005) 509.
69   H.H. Maurer, F.X. Tauvel and T. Kraemer. In: I. Rasanen (Ed.), Proceedings of the 38th International TIAFT Meeting in Helsinki, TIAFT, Helsinki, 2001, pp. 316–323.
70   H.H. Maurer and J.W. Arlt, J. Anal. Toxicol., 23 (1999) 73.
71   H.H. Maurer, T. Kraemer and J.W. Arlt, Ther. Drug Monit., 20 (1998) 706.
72   H.H. Maurer and J.W. Arlt, J. Chromatogr. B, Biomed. Sci. Appl., 714 (1998) 181.

73   F.T. Peters, M.R. Meyer, G. Fritschi and H.H. Maurer, J. Chromatogr. B, Analyt. Technol. Biomed. Life Sci., 824 (2005) 81.

74   D. Springer, F.T. Peters, G. Fritschi and H.H. Maurer, J. Chromatogr. B, Analyt. Technol. Biomed. Life Sci., 773 (2002) 25.

75   D. Springer, F.T. Peters, G. Fritschi and H.H. Maurer, J. Chromatogr. B, Analyt. Technol. Biomed. Life Sci., 789 (2003) 79.

76   D. Springer, G. Fritschi and H.H. Maurer, J. Chromatogr. B, Analyt. Technol. Biomed. Life Sci., 793 (2003) 331.

77   D. Springer, G. Fritschi and H.H. Maurer, J. Chromatogr. B, Analyt. Technol. Biomed. Life Sci., 793 (2003) 377.

78   D. Springer, G. Fritschi and H.H. Maurer, J. Chromatogr. B, Analyt. Technol. Biomed. Life Sci., 796 (2003) 253.

79   H.H. Maurer. In: K. Pfleger, H.H. Maurer, A. Weber (Eds.), Mass spectral and GC data of drugs, poisons, pesticides, pollutants and their metabolites, VCH-Verlagsgesellschaft, Weinheim, 1992, p. 21.

80   T. Kraemer, A.A. Weber, H.H. Maurer, . In: F. Pragst (Ed.), Proceedings of the Xth GTFCh Symposium in Mosbach, Helm-Verlag, Heppenheim, 1997, p. 200.

81   J. Bickeboeller-Friedrich and H.H. Maurer, Ther. Drug Monit., 23 (2001) 61.

82   H. Maurer and K. Pfleger, J. Chromatogr., 305 (1984) 309.

83   H. Maurer and K. Pfleger, J. Chromatogr., 272 (1983) 75.

84   H. Maurer and K. Pfleger, J. Chromatogr., 306 (1984) 125.

85   H. Maurer and K. Pfleger, J. Chromatogr., 422 (1987) 85.

86   H.H. Maurer, J. Chromatogr., 530 (1990) 307.

87   H.H. Maurer, Arch. Toxicol., 64 (1990) 554.

88   H. Maurer and K. Pfleger, Fresenius' Z. Anal. Chem., 321 (1985) 363.

89   H. Maurer and K. Pfleger, Arch. Toxicol., 62 (1988) 185.

90   H. Maurer and K. Pfleger, J. Chromatogr., 428 (1988) 43.

91   H. Maurer and K. Pfleger, Fresenius' Z. Anal. Chem., 331 (1988) 744.

92   H. Maurer and K. Pfleger, J. Chromatogr., 430 (1988) 31.

93   H. Maurer and K. Pfleger, Fresenius' Z. Anal. Chem., 317 (1984) 42.

94   H. Maurer and K. Pfleger, Fresenius' Z. Anal. Chem., 314 (1983) 586.

95   H.H. Maurer, Ther. Drug Monit., 18 (1996) 465.

96   T. Kraemer, I. Vernaleken and H.H. Maurer, J. Chromatogr. B, Biomed. Sci. Appl., 702 (1997) 93.

97   T. Kraemer, G.A. Theis, A.A. Weber and H.H. Maurer, J. Chromatogr. B, Biomed. Sci. Appl., 738 (2000) 107.

98   T. Kraemer, R. Wennig and H.H. Maurer, J. Anal. Toxicol., 25 (2001) 1.

99   T. Kraemer, S.K. Roditis, F.T. Peters and H.H. Maurer, J. Anal. Toxicol., 27 (2003) 68.

100   H.H. Maurer and T. Kraemer, Arch. Toxicol., 66 (1992) 675.

101   H.H. Maurer, T. Kraemer, O. Ledvinka, C.J. Schmitt and A.A. Weber, J. Chromatogr. B, Biomed. Sci. Appl., 689 (1997) 81.

102   R.F. Staack, J. Fehn and H.H. Maurer, J. Chromatogr. B, Analyt. Technol. Biomed. Life Sci., 789 (2003) 27.

103   R.F. Staack, G. Fritschi and H.H. Maurer, J. Chromatogr. B, Analyt. Technol. Biomed. Life Sci., 773 (2002) 35.

104   R.F. Staack, G. Fritschi and H.H. Maurer, J. Mass Spectrom., 38 (2003) 971.

105   R.F. Staack and H.H. Maurer, J. Anal. Toxicol., 27 (2003) 560.

106   R.F. Staack and H.H. Maurer, J. Chromatogr. B, Analyt. Technol. Biomed. Life Sci., 798 (2003) 333.

107   R.F. Staack and H.H. Maurer, J. Mass Spectrom., 39 (2004) 255.

108   R.F. Staack and H.H. Maurer, J. Chromatogr. B, Analyt. Technol. Biomed. Life Sci., 804 (2004) 337.

109   R.F. Staack, D.S. Theobald and H.H. Maurer, Ther. Drug Monit., 26 (2004) 441.

110   D.S. Theobald, G. Fritschi and H.H. Maurer, J. Chromatogr. B, Analyt. Technol. Biomed. Life Sci. (2006), submitted.

111   D.S. Theobald and H.H. Maurer, J. Mass Spectrom. (2006), submitted.
112   L.D. Paul and H.H. Maurer, J. Chromatogr. B, Analyt. Technol. Biomed. Life Sci., 789 (2003) 43.
113   J. Beyer, D. Ehlers and H.H. Maurer, Ther. Drug Monit., 28 (2006) 568.
114   H. Maurer and K. Pfleger, J. Chromatogr., 382 (1986) 147.
115   H.H. Maurer, Fresenius' J. Anal. Chem., 337 (1990) 144.
116   H.H. Maurer, T. Kraemer, C. Kratzsch, L.D. Paul, F.T. Peters, D. Springer, R.F. Staack and A.A. Weber, in: M. Balikova and E. Navakova (Eds.), Proceedings of the 39th international TIAFT meeting in Prague, 2001, Charles University, Prague, 2002, p. 61.
117   H.H. Maurer, Spectroscopy Europe, 6 (1994) 21.
118   M. Balikova, Forensic Sci. Int., 153 (2005) 85.
119   A.M. Lisi, R. Kazlauskas and G.J. Trout, J. Chromatogr. B, 692 (1997) 67.
120   P. Roesner, T. Junge, G. Fritschi, B. Klein, K. Thielert and M. Kozlowski, Toxichem. Krimtech., 66 (1999) 81.
121   L. Rivier, Anal. Chim. Acta, 492 (2003) 69.
122   S.E. Stein and D.R. Scott, J. Am. Soc. Mass. Spec., 5 (1994) 859.
123   F.W. McLafferty, M.Y. Zhang, D.B. Stauffer and S.Y. Loh, J. Am. Soc. Mass Spectrom., 9 (1998) 92.
124   B. Aebi and W. Bernhard, J. Anal. Toxicol., 26 (2002) 149.
125   O.D. Sparkman, J. Am. Soc. Mass Spectrom., 7 (1996) 313.
126   F.W. McLafferty, Wiley registry of mass spectral data 7th edition with NIST 2005 spectral data, Wiley, New York, NY, 2005.
127   E. Kovats, Helv. Chim. Acta, 41 (1958) 1915.
128   P. Roesner, Mass spectra of designer drugs (SpecInfo), Wiley, New York, NY, 2003.

M.J. Bogusz (Ed.). Forensic Science
Handbook of Analytical Separations, Vol. 6

CHAPTER 13

High performance liquid chromatography in forensic toxicological analysis

Fritz Pragst

Institute of Legal Medicine, University Hospital Charité, Hittorfstr. 18, 14195 Berlin, Germany

13.1 INTRODUCTION

High performance liquid chromatography (HPLC) is one of the separation techniques most frequently used in forensic toxicology. In comparison to gas chromatography, it has particular advantages in case of non-volatile, thermally sensitive
and high molecular weight substances which can be analyzed under mild conditions
and without derivatization. The fundamentals and widespread applications of
HPLC are standard in analytical textbooks and were described in many mono-
graphs. In forensic toxicology, the identification and quantification of illegal and
therapeutic drugs, pesticides and other organic poisons from human body fluids and
tissue samples is the most dominant task. The parts of an HPLC device used for this
purpose are schematically shown in Fig. 13.1.

In this chapter, the progress of this method in the last years in general and with
respect to its forensic application in particular shall be reviewed. Mass spectrometric
detection in liquid chromatography (LC-MS) in its different variants will be excluded from this chapter and will be described in detail in Chapter 14.

13.2 PROGRESS IN HPLC TECHNIQUES

Since its invention in the mid-1970s, a steady development of HPLC techniques
occurred with respect to general performance parameters such as separation efficiency (increasing plate number), sensitivity, reproducibility, robustness, automation
and applicability to a wide range of analytes and matrices as well as to adaptation
for special analytical problems.

References pp. 485–489

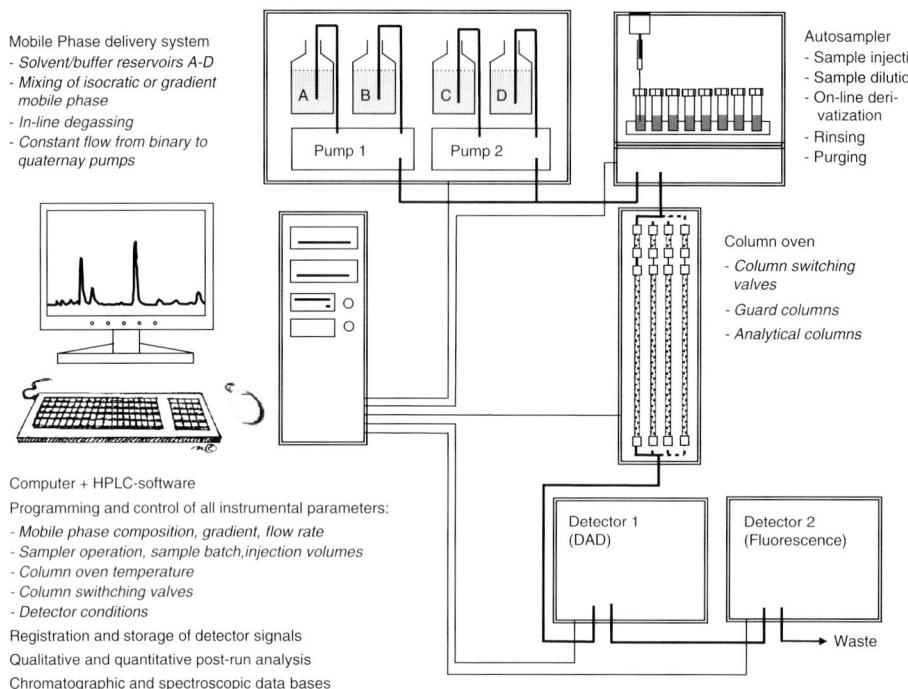

Fig. 13.1. Components of an up-to-date HPLC device.

Up-to-date HPLC devices are fully computer-operated and fulfill high analytical standards. This progress includes all kinds and steps of the HPLC analysis from sample injection to post-run analysis and data processing.

13.2.1 Mobile phase delivery system

The HPLC solvent delivery system consists of the mobile phase reservoirs, a degasser and one or more HPLC pumps. The reservoirs contain either the complete mobile phases for isocratic separations or the components from which the mobile phases are mixed in the programed ratio during the chromatographic run for isocratic or gradient elution. The mixing may occur in a mixing chamber after the pumps (high-pressure mixing) or prior to pumping by built-in high precision proportioning valves (low-pressure mixing). In the latter case, convenient access of up to four solvents for isocratic or gradient analysis is offered by quaternary pumps. The composition precision is described below 0.2%. The vacuum degasser is frequently incorporated into the pumps. Modern pumps have a variable flow rate between 1 µl/min and 5 ml/min with a precision <0.1% and an accuracy of ±0.5%. Dual floating piston mechanism and low dead volume pulse dampener as well as electronic residual pulsation absorption ensure a virtually pulse-free and stable solvent flow. The pressure limit of usual pumps is as a rule 400 bar, but pumps for more than 1000 bar were also constructed.

Highly pure solvents for HPLC are offered by many manufacturers, from which acetonitrile is most frequently used as a modifier in mobile phases in toxicological analysis because of its suitable physical properties and its excellent UV transmittance (> 65% at 195 nm).

13.2.2 Sampling

As a rule autosamplers are used. Up-to-date samplers are characterized by sample temperature control (e.g. 4–40°C), fast injection cycle (< 30 sec), ignorable carry over (< 0.01%), high precision of variable injection volume (0.1–100 μl ± 0.3%) and robustness, e.g., by needle-in-needle injection. The principle is still based on the Rheodyne valve. In special samplers, internal standard addition, sample dilution or derivatization steps are easily programed to reduce sample preparation time and errors associated with manual procedures.

13.2.3 HPLC columns and separation materials

The column is the most important part in HPLC separation. Depending on the diameter, analytical columns are classified into standard columns (typical internal diameter 4.6 or 4.0 mm), semi-micro columns (typical internal diameter 1–2 mm) and micro or capillary columns (internal diameter 0.1–1.0 mm). The micro columns are usually specified for MS detection. The enormous progress in the development of column-packing materials was yearly demonstrated on the Pittcon conferences and was reviewed by Stevenson [1,2]. Thousands of special stationary phases were developed and are offered and the silica particles were modified by several organic groups. Some frequently used examples, their particular properties and special application fields are given in Table 13.1.

Altogether, many new column-packing materials are offered by the manufacturers with improved properties for separation of complex substance mixtures of polar and non-polar as well as hydrophilic and hydrophobic substances in one sample, as they may occur in toxicological screening. However, systematic studies to evaluate these materials for this purpose are rather rare. Reasons for this could be the large effort necessary to develop new methods with hundreds of analytes, if the established method still sufficiently works and the illusion that the chromatographic separation becomes less important in context of the sophisticated and highly selective MS-MS and MS-TOF detections.

13.2.3.1 Reversed phases

The most frequently used column-packing materials are reversed phases, which consist of silica gel particles the surface of which is modified by a lipophilic layer of alkyl groups bound to Si-OH groups. In principle, materials with various alkyl

References pp. 485–489

TABLE 1
STATIONARY PHASES USED FOR HPLC

Material	Surface Bonded Group	Interaction	Remarks
RP 8	$O–Si(CH_3)_2–C_8H_{17}$	Hydrophobic	General use, superior for non-basic analytes
RP 18	$O–Si(CH_3)_2–C_{18}H_{37}$	Hydrophobic	General use, superior for non-basic analytes
Phenyl	$O–Si(CH_3)_2–(CH_2)_3–C_6H_5$	Hydrophobic, π–π	Enhanced selectivity for aromatics, moderately polar compounds, e.g., opioid alkaloids
Cyano	$O–Si(CH_3)_2–(CH_2)_3–C\equiv N$	Dipole, π–π, hydrophobic	General purpose under normal and reversed phase conditions
Nitrophenyl	$O–Si(CH_3)_2–(CH_2)_3–p–C_6H_4–NO_2$	π-charge tranfer	Chlorinated hydrocarbons
Diol	$O–Si(CH_3)_2–(CH_2)_3–O–CH_2–CHOH–CH_2OH$	Hydrogen bond	Steroids, phenols, substances with different polarity
Amino	$O–Si(CH_3)_2–(CH_2)_3–NH_2$	Weak anion exchange	Organic acids, carbohydrates
Dimethyl-amino	$O–Si(CH_3)_2–(CH_2)_3–N(CH_3)_2$	Weak anion exchange	Organic acids
Nucleosil SB	$O–Si(CH_3)_2–(CH_2)_3–C_6H_4–N(CH_3)_3^+$	Strong anion exchange	Inorganic anions, carbohydrates, oxalate
Nucleosil SA	$O–Si(CH_3)_2–(CH_2)_3–C_6H_4–SO_3$	Strong cation exchange	Metal ions, basic drugs
Phenyl	$O–Si(CH_3)_2–C_6H_5$	π–π, lipophilic	General purpose, neutral analytes
Biphenyl	$O–Si(CH_3)_2–C_6H_4–C_6H_5$	π–π, lipophilic	Highly retentive and selective for aromatics
HS-F5	$O–Si(CH_3)_2–(CH_2)_3–C_6F_5$	π–π, lipophilic	Substances with different polarity
RP-AmideC16	$O–Si(CH_3)_2–(CH_2)_3–NH–CO–(CH_2)_{14}–CH_3$	Polar imbedded, lipophilic, H-bond, dipole–dipole	Substances with different polarity and hydrophilicity
PEG	$O–Si(CH_3)_2–O–(CH_2–CH_2–O–)_n–CH_3$	Dipole, lipophilic	Substances with different polarity
HILIC	$O–Si(CH_3)_2–C_nH_{2n}–^+N(CH_3)_2–(CH_2)_3–SO_3^-$	Hydrogen bonds, ionic, dipole–dipole	Polar and hydrophilic compounds

groups from C1 to C18 are available but C8 and C18 materials are by far the most frequently used. Problems in batch-to-batch reproducibility and chromatographic performance of the initial materials were caused by non-uniform particle size, non-spherical particle shape and residual non-bound Si-OH groups as well as impurities by heavy metal ions which disturbed the separation by specific interaction with the analytes. These obstacles are completely overcome today. Highly pure and spherical silica of homogeneous size are synthetically produced from tetraethylsiloxane. The remaining free Si-OH groups are "endcapped" by trimethylsilyl groups. High coverage phases ($> 4 \, \mu mol/m^2$) lead to improved retention, pH stability and less tailing. In this way, retention data and chromatographic resolution can be well reproduced between different laboratories and over a long time. The separation power of a column increases with decreasing particle size but, unfortunately, this is bound to an increase of the pressure to enable a sufficient flow rate. A size of $5 \, \mu m$ is the most common compromise, but the trend goes to smaller particle diameters such as 1.7 μm in ultra performance liquid chromatography (UPLC). Alkyl C30 phases are the longest chain of monomeric reversed phase HPLC phases currently available (YMC, Kyoto, Japan). It is particularly suitable for carotenoids and tocopherols.

Besides the long-known phenylpropyl phases also phenylhexyl phases (Gemini[TM], Phenomenex) [3] and phenyl phases direct bonded to the silica are available which combine the advantages of an alkyl and a phenyl ligand. The problem of phase collapse of conventional RP 8 or RP 18 columns in mobile phases with more than 95% water is successfully overcome by hydrophilic endcapping (e.g. Nucleodur[®] C18 Pyramid, Machery Nagel; ODS-AQ column, YMC). The actual groups used for this purpose are still unknown.

13.2.3.2 Improved separation of polar and basic compounds

A problem of the RP 8 and the RP 18 columns in isocratic multi-component separation of analytes with different polarity is the low retention and resolution of polar substances and the high retention of non-polar compounds. Pentafluorophenylpropyl phase (PFP) and polar imbedded phases such as the polyethylene glycol bonded phase (PEG) and the RP amide phase enable a more uniform distribution of the retention times by increasing the retention of polar substances and avoiding excessive retention and wasted resolution of non-polar substances. In polar imbedded alkyl phases, the polar groups such as carbamide, carbamate, urea or ether are generally incorporated in the alkyl ligand close to the surface silica [3]. Also perfluorinated and fluorinated stationary phases (straight chain or branched aliphatic or phenyl) have shown novel selectivity for several compound classes as an alternative to traditional C8 and C18 phases [3,4].

Special selectivity features are also obtained by combined covalently bonding of two groups, e.g. octadecyl and phenylpropyl (Nucleodur[®] Sphinx, Macherey–Nagel) or octadecyl and ion exchange (SAX or SCX) in Hypersil Duet columns. The C18/SCX column was tested for toxicological screening procedures in acidic mobile phase and displayed a much better separation of many toxicologically relevant basic substances [5]. However, it was not superior to C8 columns because of extremely long retention times for lipophilic basic drugs such as amitriptyline because of the combined action of ionic and hydrophobic forces and since diprotonated coumpounds such as histamine or pirbuterol were not eluated at all.

However, pure SCX columns proved to be very useful for separation of basic drugs at optimized pH and ionic strength of the mobile phase [6].

13.2.3.3 Hydrophilic interaction chromatography

Many solutes, especially polar and hydrophilic compounds, are not retainable at RP materials. Normal phase liquid chromatography with non-aqueous mobile phases was used over a long time for this purpose. Instead of that, hydrophilic interaction liquid chromatography (HILIC) can advantageously be used with stationary silica phases, the SiOH groups of which are functionalized by hydrophilic groups [7–9]. Between them, materials with covalently bonded sulfobetaine type zwitterionic groups show unique separation capability and selectivity. Corresponding products are offered by SeQuant GmbH (Haltern, Germany). A ZIC®-HILIC phase carries a covalently to silica particles bonded, permanently zwitterionic group of the sulfobetaine type. In ZIC®-pHILIC phases, the group is polymer bonded with an increased pH stability between 2 and 10. Different from normal phase HPLC, semi-aqueous mobile phases consisting of an aqueous buffer and a polar hydrophilic modifier are applied in the same way as in reversed phase HPLC. The water content (typically between 3% and 60%) establishes a hydrophilic environment of the stationary phase. Typical applications include opiates and their glucuronides, flavanoids, quaternary ammonium compounds, ascorbic acid.

13.2.3.4 Monolithic columns

As an alternative to the packed particle-based columns, highly porous monolithic rod columns were developed [10]. The columns developed by Merck (Darmstadt, Germany) consist of a single piece of high purity polymeric silica gel with bimodal pore structure (Fig. 13.2) and are available as RP 8 and RP 18 endcapped as well as a normal phase in different lengths and diameters. Macropores with an average size of 2 µm allow rapid flow of the mobile phase, whereas mesopores with an average pore size of 13 nm create the large uniform area and the high performance chromatographic separation. These Chromolith® columns need a much lower operating pressure. Therefore, compared with a 5 µm particulate column, the speed of analysis can be typical four times faster.

13.2.3.5 Polymer-based separation materials

Fully polymeric reversed phases separation material on the basis of porous styrene/divinylbenzene with a high chemical and thermal stability and an unsurpassed pH stability between 1 and 14, a pore size of 100 Å and sufficient pressure capability

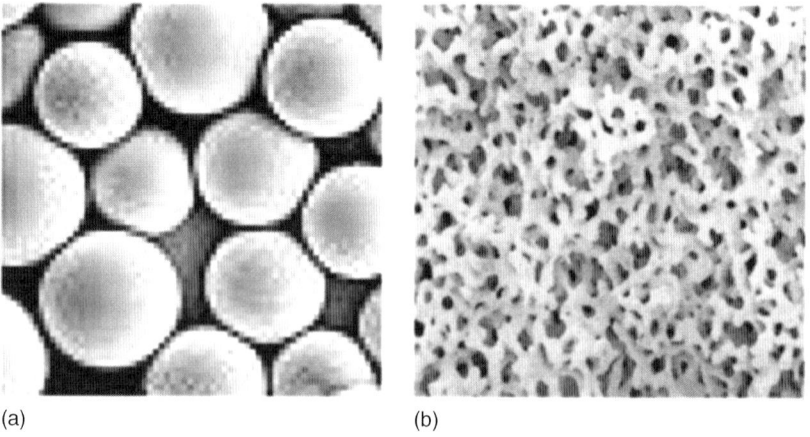

(a)   (b)

Fig. 13.2. HPLC separation materials. (a) Particle based and (b) monolithic.

(>300 bar) are offered by Polymer Laboratories, Inc. (Amhersts, MA, USA). These inherently hydrophobic materials have no bonded ligands and are available in particle sizes of 3, 5 and 8 µm and do not suffer from the typical silica problems of non-bonded acidic silanol groups, ionic impurities and incomplete endcapping. Similarly, a new line of columns (Gemini Twin) with C18 selectivity and pH stability from 1 to 12 was developed by Phenomenex (Torrance, CA) by grafting a unique silica-organic layer above the silica core.

13.2.3.6 Stationary phases for high-temperature HPLC separations

For extreme conditions (high temperature $>100°C$, pH 1 or 13) zirconia-based phases proved to be very resistant [11]. They are available, e.g., from Phenomenex as polybutadiene, polystyrene, C18 and carbon-coated phases. An overview about high-temperature LC separations was given by Yang and Lynch [12].

13.2.3.7 Chiral stationary phases

Over one-third of the marketed drugs have one or more asymmetric carbon atoms and, therefore, exist as a pair of enantiomers (optical antipodes, R/S-isomers or D/L-isomers) which cannot be separated under the usual non-chiral HPLC conditions. Since the enantiomers often have drastically different biological activities, their separated analytical determination becomes increasingly important. Besides the formation of diastereomeric derivatives with chiral reagents or the use of chiral additives to the mobile phase, chiral stationary phases (CSPs) are the most efficient solution to this problem.

An interaction at three groups between the analyte molecule and the CSP (ionic, dipole–dipole, π–π or hydrophobic interaction or hydrogen bond) is necessary for discrimination between both enantiomers (Fig. 13.3a). The enantioseparation on CSPs results from energy differences between transient diastereomeric complexes formed by these solute–CSP interactions. Since these interactions are always directly influenced by the mobile phase environments, an optimization of the HPLC conditions is always necessary.

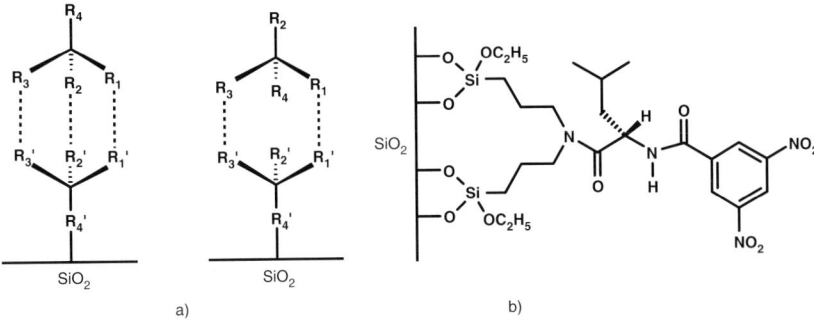

Fig. 13.3. Enantioseparation by HPLC. (a) Chiral three-point interaction between stationary phase and analyte in transient diastereomeric complexes. (b) Example of a Pirkle brush-type ligand described by Forjan *et al.* [17].

There are different types of CSP known [13–24]. In the Pirkle brush-type phase, the surface of the silica particles is modified by a chiral substance with a relatively small molecular weight [17]. An example is shown in Fig. 13.3b. It is a drawback of the brush-type CSPs that they separate only aromatic racemates. Further, frequently used types are derivatized polysaccharides (amylose, cellulose [18]), macrocyclic-type CSPs (cyclodextrines [19], glycopeptides [20] and chiral crown ethers [21] bonded to silica gel) and protein-based CSPs [22] (declining application because of lower stability). It is estimated that 1300 CSPs have been prepared and over 200 have been commercialized.

A further specific way for enantioseparation are molecularly imprinted polymers [23,24]. Imprinting of molecules occurs by the polymerization of functional and cross-linking monomers in the presence of the template/target ligand (e.g. one of the enantiomers). In this process, the template is initially allowed to establish bond formation with polymerizable functionality and the resulting complexes are subsequently copolymerized with cross-linkers into a rigid polymer. This is ground and the template molecules are removed by solvent extraction. The resultant imprints possess a steric and chemical memory for the template. Molecular imprinted separation materials can also be used for specific separation of non-chiral substance classes.

13.2.3.8 Bi- and multi-dimensional liquid chromatographic separation

In multi-dimensional liquid chromatography, the sample is subjected to more than one separation mechanism and each mechanism is considered an independent separation dimension. The theoretical basics and practical possibilities were reviewed by Dixon *et al.* [25]. The aim is to increase the peak capacity of the chromatogram and to improve the separation of multi-component samples. The coupled separations should behave orthogonal, that means, they are based on independent mechanisms without retention correlation. For example, separation in first dimension may be based on RP retention whereas an ion exchange or hydrophilic interaction occurs in the second dimension. In practice, common comprehensive multi-dimensional liquid chromatographic separation systems are configured around electronically activated 6–10 port valves (Fig. 13.4a). Comprehensive means that all fractions of the first column effluent are also submitted to the second separation. At these valves, two loops are connected in such a way that one is loaded with the eluate from the first dimension while the other is injecting the previously loaded eluate onto the second dimension. As the valve switches, the function of the two loops alternates in this within-loop automated fraction collection/reinjection concept. A condition for a comprehensive performance is that the second-dimension run-time is much faster than that of the first dimension. A three-dimensional plot of a corresponding two-dimensional chromatogram is shown in Fig. 13.4b.

13.2.4 HPLC detectors

HPLC detectors used in forensic investigations should on the one hand be suitable to detect a broad variety of different compounds with a high sensitivity and on the

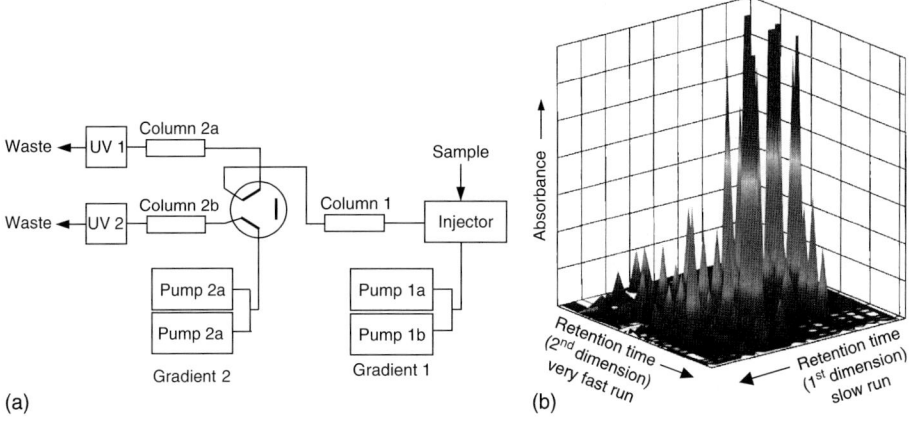

Fig. 13.4. Two-dimensional HPLC separation. (a) System configuration of chromatographic flows. (b) 3D plot of the two-dimensional separation of a human urine sample (from [25]). In the presented position of the switching valve, a fraction of the eluent from column 1 is separated in the second dimension on column 2b while the next fraction is loaded on the head of column 2a. After valve switching, the function of 2a and 2b is exchanged. Owing to the large run time of the gradient 1 and the very short-run time of the gradient 2, a large number of fractions can be separated in the second dimension.

other deliver substance specific signals as a contribution to the substance identification. These requirements are fulfilled by the available detectors to a different degree. Besides the MS detection light absorption in the ultraviolet and visible range, fluorescence and electrochemical oxidation or reduction are the most common molecular properties used for this purpose. Other principles are not sufficiently sensitive such as the refractive index detector and the electrical conductivity detector, or respond only to a very small number of substances such as the chemiluminescence detector. Optical rotation detectors or circular dichroism detectors for chiral detection of enantiomers are available but not have a sufficient sensitivity for analysis of traces.

13.2.4.1 Variable wavelength UV/VIS and photodiode array detectors (DAD)

Fixed and variable wavelength UV/VIS detectors have the advantage of a very high sensitivity (noise $<10^{-5}$ AU). However, a substance can be characterized only on the basis of the retention time. Therefore, these detectors are not suitable for screening procedures and are preferentially used if an identified analyte has to be determined in large series with always the same matrix. Some instruments enable to measure the UV spectrum of a peak by "on-the-fly spectral scanning" within about 1 sec.

The most universal detector beside the MSD is the UV/VIS diode array detector (DAD). This detector measures continuously the complete UV spectrum of the effluent from the column leading to a three-dimensional file of absorbance vs. wavelength and time. Up-to-date DADs contain a 512 or 1024-element diode array and are characterized by a wavelength accuracy of $<\pm 1$ nm, a resolution of

0.7–1.2 nm/diode, a noise $< 1.0 \times 10^{-5}$ absorbance units and a drift between 10^{-3} and 10^{-4} absorbance units/hour. In some detectors (e.g. those of manufacturers Thermoquest, Dionex or Knauer), fiber optic bundles are used, which provide for a more flexible arrangement and variability of the flow cell (e.g. changeable 10 and 50 mm path lengths). Furthermore, the optical fiber is used as a beam shaper: the light beam emerging from the flow cell in a circular pattern is almost completely collected by the fiber optic bundle and is reshaped into a narrow vertical beam by linear arrangement of the fibers at the other end of the bundle. In this way, a high-wavelength resolution with minimal light loss is achieved by avoidance of the usual optical slit. Sensitivity was increased at least five-fold by use of a 5 cm light-pipe flow cell made of special low refractive index material and working by the principle of total internal reflection (TIR, manufacturer Thermoquest, cell volume 10 μl).

Because of their high-wavelength resolution, wavelength accuracy and sensitivity, spectra measured with these modern detectors are highly reproducible. If the same substance is measured with detectors of different manufacturers, the spectra are in excellent agreement. Since HPLC-DAD is one of the most efficient techniques in systematic toxicological analysis (STA), its application will be described in more detail in Section 13.3.1.

13.2.4.2 Fluorescence detectors

An essential improvement in sensitivity and performance was also achieved for fluorescence detectors (FLD). In these detectors, the analyte molecules are transformed into the excited state by absorption of short wavelength light and return to the electronic ground state by emission of light at longer wavelengths which is used for the detection. Excitation and emission wavelengths are separately chosen by holographic grating monochromators.

In the detectors, stray light is as far as possible avoided and the fluorescence is measured by a photomultiplier. After optimum choice of the excitation and emission wavelengths, a sensitivity 1–3 order of magnitude higher than by UV absorption was achieved. For instance, the detection limits in urine for polycyclic aromatic hydrocarbons were of 0.1–0.5 pg/ml [26] and for opiates were 1.1–7 ng/ml [27].

The sensitivity is further increased by using a xenon flash lamp for excitation (e.g. Agilent) with a pulse frequency of 74 or 296 Hz. Excitation and fluorescence spectra of a peak can be measured within 0.6 sec with 10 nm wavelength steps. The chromatogram can simultaneously be recorded at four different fluorescence wavelengths. This enables to detect different analytes at optimum sensitivity and to contribute to substance identification on the basis of the characteristic intensity ratios between these wavelengths. Laser induced fluorescence detectors (LIF detector, Unimicro Technologies, Inc) are even more sensitive and can detect FITC (fluorescein isothiocyanate) derivatized substances at 10^{-12} mol/l with a signal-to-noise ratio above 10. The total fluorescence spectra could be recorded over the chromatogram by using an imaging spectrograph and a back illuminated CCD camera [28]. The identification of components in substance mixtures by LIF without complete chromatographic separation was achieved by using a dual fiber optic array in which the excitation and emission light is guided to and from the flow cell by optical fibers, and the fluorescence spectra from seven different excitation wavelengths were separately and simultaneously recorded using a spectrograph/CCD combination [29]. Non-fluorescing analytes can be derivatized by fluorescent agents (Section 13.2.6 [30]).

13.2.4.3 Chemiluminescence detectors, nitrogen specific detector

Chemiluminescence detectors are based on the measurement of light originating from a chemical reaction [31]. Different from fluorescence, it needs no exciting light source, and thus is not disturbed by any scattering light from it. This permits a larger signal-to-noise ratio and a high sensitivity. In practical performance, a fluorophor is pre- or post-column tagged to the analyte or the analyte is transformed into a fluorescent species and the chemiluminescence reactants are added in the detector. As an example, catecholamines were post-column converted to the strongly fluorescent heterocyclic products, which are sensitized by the bis-(3,6,9-trioxadecyloxy-carbonyl-4-nitrophenyl) oxalate/H_2O_2 chemiluminescence system [32]. The flow scheme of the fully automated HPLC device is shown in Fig 13.5a and the chemical reactions for derivatization and chemical excitation are given in Fig. 13.5b. 0.3 to 2 fmol (0.05–0.35 pg) of the catecholamines could be detected in this way. Other chemiluminescence processes exploited in such detectors are the luminol-H_2O_2-Co^{2+} system or the electron-transfer luminescence of the Ru(bpy)$_3^{3+}$ complex. Some substances such as antioxidants were also detected based on their property to inhibit the luminol emission (negative peaks) [33]. Chemiluminescence detectors have a wide dynamic range and a relatively simple instrumentation.

A particular advantage has the chemiluminescent nitrogen detector (CLND) as an elemental selective detector [34–36]. This commercially available detector (Antek Instruments Inc.) is based on combustion of the HPLC effluent in an oxygen-rich furnace at 1050°C to convert all organic species to oxides of carbon, nitrogen, sulfur, etc. and water (Fig. 13.6). The NO produced from nitrogen-containing compounds is then reacted with ozone to produce NO_2^* in excited state which emits light upon return to the ground state. This chemiluminescent response is proportional to the NO concentration and correspondingly to the amount of nitrogen originally present in the analyte. With the exception of compounds containing a $-N = N-$ group, the signal is independent of the structure of the substance and depends only on its nitrogen content. Therefore, quantification requires only one nitrogen-containing standard which needs not to be structurally related to the analyte. The detection limit is about 0.1 ng nitrogen. Of course, the mobile phase must be completely free of nitrogen-containing components.

13.2.4.4 Evaporative light scattering detector

For substances without light absorption, an evaporative light scattering detector can be an alternative [37,38]. The principle is shown in Fig. 13.7. The eluent is nebulized by a gas stream in a temperature-controlled evaporator tube where the solvent is removed leaving the less evaporative solute particles. These are detected by

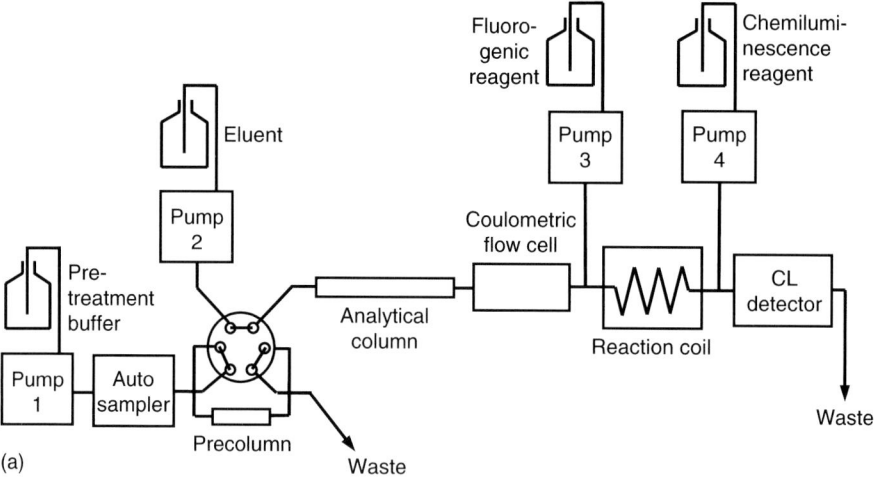

Coulometric oxidation and derivatization

Chemiluminescence detection of the derivative

(b)

Fig. 13.5. Fully automated determination of catecholamines with post-column electrochemical oxidation, cyclization with ethylenediamine as the fluorogenic reagent in a reaction coil and chemiluminescence detection. (a) Flow scheme of the HPLC device. (b) Formation of the luminescent derivative "Deriv" by coulometric oxidation and cyclization and chemical excitation by the diaryl oxalate/H_2O_2 chemiluminescent system.

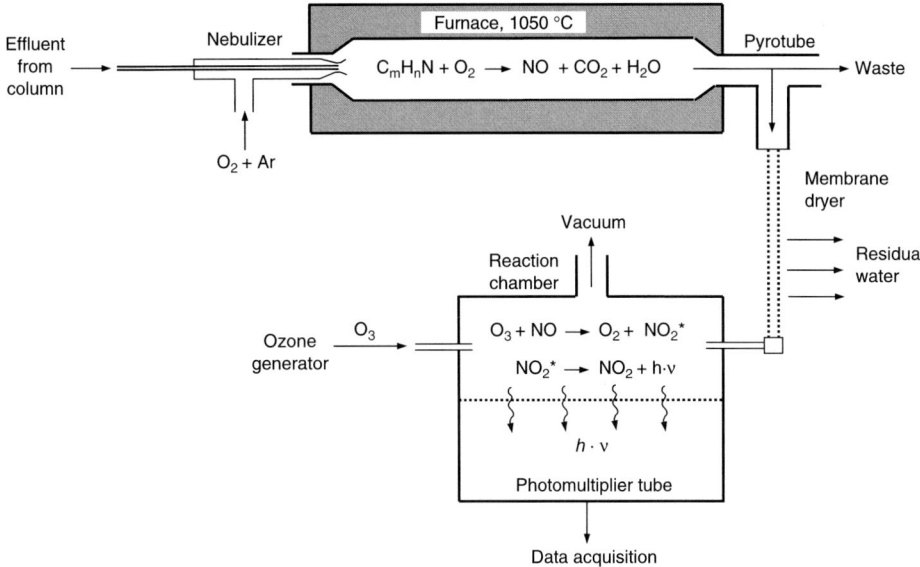

Fig. 13.6. Principle of the chemiluminescent nitrogen-specific detector [34].

measuring the scattered light as they pass the optical chamber. A disadvantage is still the relatively low sensitivity with 10 ng on column. Limits of detection around 0.3 µg/ml were determined for the analysis of several drugs from serum [39,40]. The mobile phase must be fully evaporable, e.g., water/acetonitrile or DMSO with formic acid/ammonium formiate buffer. The detector has a mass dependent response. However, a simple general calibration that is valid independent of the chemical structure as possible for the CLND could not be confirmed [36].

13.2.4.5 Electrochemical detectors

Electrochemical detectors (ECD) can be used for highly sensitive detection of oxidizable or reducible compounds down to femtogram levels. The fundamentals and applications in toxicology of this detector were described in a book of Flanagan, Perrett and Whelpton [41]. Beside the different retention times, a selectivity is also attained by the different oxidation or reduction potentials of the analytes in the mobile phase. In the flow cell or attached to the flow cell of the detector, three electrodes are arranged in a potentiostatic mode (Fig. 13.8). The analytes are detected by the current used for their oxidation or reduction at the working (or test) electrode. The potential of the working electrode is controlled by the reference electrode (an Ag/AgCl electrode or a Pd wire), but the current flows between test and counter electrode. Depending on the electrochemical reaction, the working electrode consists of glassy carbon, gold or platinum.

There are two principal types, the coulometric and the amperometric detector. In the *coulometric detector* (Fig. 13.8a), the working electrode has a very large surface,

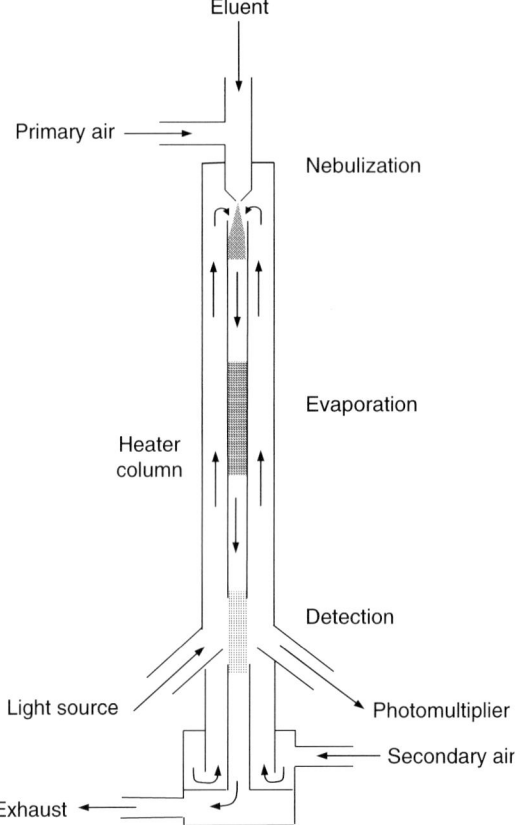

Fig. 13.7. Principle of the evaporative light-scattering detector.

e.g., a flow-through porous graphite electrode, leading to a complete reaction of the electroactive substance. In dual-electrode coulometric cells, two such working electrodes are arranged in series with independent control of the potentials and measurement of the currents. The potential of the first electrode can be chosen below the potential of the analyte leading to oxidation or reduction of undesirable electroactive sample constituents and in this way to an increase in selectivity. However, in case of reversible electrochemical reactions as for catecholamines, the analyte can be oxidized at the first electrode and re-reduced at the second. This reversal can be employed for even greater selectivity and sensitivity. Finally, up to eight coulometric detector cells were arranged in an electrode array detector with the potentials subsequently increasing, e.g., from $+450$ to $+1000\,\mathrm{mV}$ in case of illicit drugs [42]. In this way, a separate electrochemical detection of substances with different oxidation or reduction potential is possible. Because of the complete conversion of the analyte, coulometric detectors can be used for pre- or post-column derivatization.

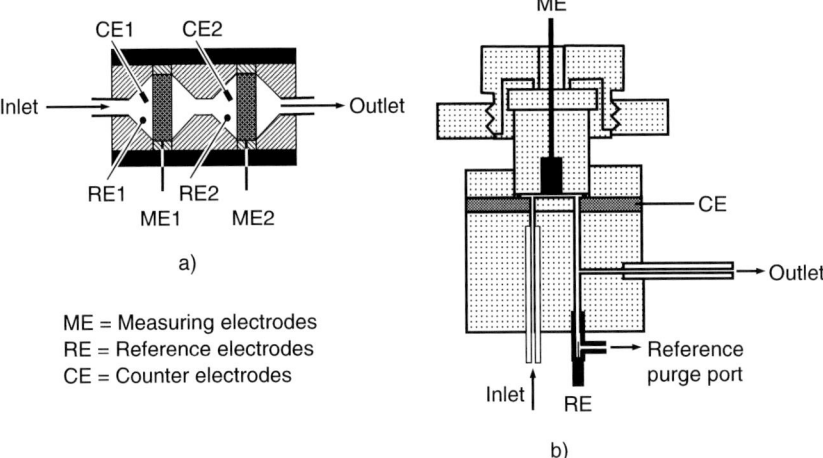

Fig. 13.8. Electrochemical HPLC detector cells. (a) Dual coulometric detector cell with porous carbon measuring electrodes. For instance, to improve the sensitivity, the potential of ME1 is set below the detection potential of the analyte and can eliminate potentially interfering compounds from detection at ME2. (b) Amperometric thin film cell. Reference electrodes: palladium or silver wire; amperometric measuring electrodes: glassy carbon, gold or platinum.

In *amperometric detectors*, the area of the working electrode is small and only a part of the analyte (up to 10%) is oxidized or reduced. The current is limited by the diffusion to the electrode surface and proportional to the concentration. An advantage as compared to coulometric detectors is a higher signal-to-noise ratio. Besides working at constant potential, the electrode can also be operated in a pulse mode in which different potentials are applied over adjustable time spans and the resulting currents are measured. This results, similar to pulse polarography, in an enhanced sensitivity. Furthermore, electrode deactivation by adsorption of sample constituents or their electrochemical products can be removed by short positive or negative potential pulses.

In principle, a voltammogram (current vs. potential curve) or cyclic voltammogram could be measured for each chromatographic peak and could be used for substance identification similar to the UV spectra in the DAD. Unfortunately, the change of the potential is bound to a high capacity current for charging or discharging the electrochemical double layer of the working electrode which is superimposed as a disturbing signal to the much lower characteristic Faraday current. Therefore, characteristic voltammograms from chromatographic peaks can be obtained only for high concentrations.

Beside the optimization of the cell design, the performance of ECDs has also been improved by the newer pulse-free HPLC pumps and by new possibilities of computerized potential control and data evaluation. The applicability of these detectors can be extended to non-electroactive analytes by suitable derivatization.

13.2.5 Software

For up-to-date HPLC instruments of all suppliers, the whole chromatography process including all operation parameters of the components is controlled by a comfortable chromatography software. This also contains method development, calibration, real time and post-run chromatogram evaluation and context-sensitive help. Data acquisition occurs with very high collection rates.

A detailed audit is possible for retrospective control of every command and instrumental response. Peak separation, integration and qualitative and quantitative evaluation are performed according to well-founded analytical standards.

For optimization of the HPLC conditions for specific separation problems, the software "ChromSword" (VWR International) was developed [43]. It calculates the RP retention times from the analyte structural formulas and solvent properties on the basis of an increment system of partial volumes of the involved atoms and groups and the dipole moments of the chemical bonds for usual mixtures of water and modifiers. The theoretically calculated retention times are adapted to practice by test runs and further optimization [44].

13.2.6 Fast liquid chromatography and ultra performance liquid chromatography

With the intention to increase chromatographic resolution and to shorten the measurement time, the fast liquid chromatography (Fast LC) or UPLC were developed based on particle sizes smaller than 2 μm [45]. These smaller particles dramatically increase column efficiency (mass sensitivity, analytical resolution and speed). It follows from the van Deemter equation that the column plate height decreases with decreasing particle size but becomes almost independent of the linear flow rate of the mobile phase at particle diameters below 2 μm (Fig. 13.9a). This means that in contrast to larger particle size, the resolution is not worsened at higher flow rate. As an example, using UPLC technique (column 1.7 μm RP 18, flow rate 1.4 ml/min) the same or even better resolution can be provided in a measurement time of 1 min as with a conventional HPLC (column 3 μm RP 18, flow rate 0.7 ml/min) in 10 min. However, both the decreased particle size and the increased flow rate require much higher pressure (up to 1200 bar) and improved instrumental and material capabilities. The conventional inversed phase material had either poor mechanical strength or only a poor loading capacity. This problem was met by the 1.7 μm XTerra® hybrid materials [46]. The particles are prepared from tetraethoxysilane and 1,2-bis(triethoxysilyl)ethane in sol/gel technique (Fig. 13.9b). The stabilization by ethylenedisiloxane bridges provides improved mechanical strengths even for fully porous particles and a broader pH operating range (pH 2–12). They are available as four bonded phases: C18, C8, Shield C18 (with embedded carbamate group for preferential retention of hydrogen bond donors) and ω-phenylhexyl with higher selectivity for aromatic substances. Typical UPLC columns have a diameter of, e.g., 2.1 mm and a length of 30–100 mm. Standard HPLC technology does not have the capability to take full advantage of sub-2 μm particles and advanced

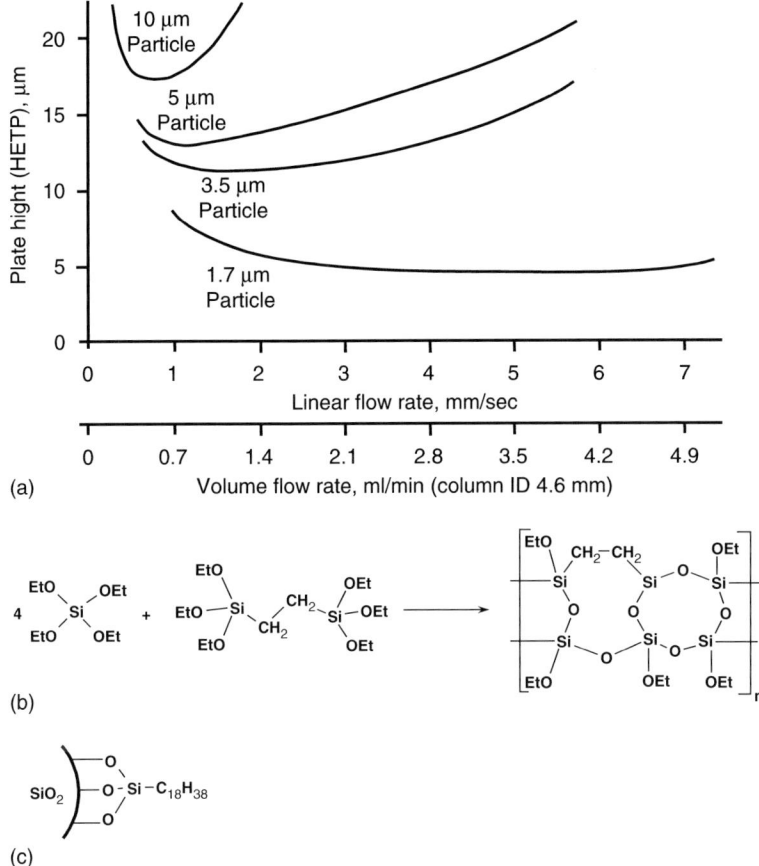

(a)

(b)

(c)

Fig. 13.9. Separation materials for Fast LC or UPLC. (a) Van Deemter plot for different particle sizes (HETP = height equivalent to a theoretical plate). Different from higher particle size, the resolution is not decreased with rising flow rate below 2 μm. (b) Bridged ethylsiloxane/silica hybride particles provide the mechanical stability to withstand the high pressure of UPLC. (c) Trifunctionally bonded alky groups ensure a very high hydrolytic stability.

technology in the solvent and sample manager, auto sampler, detector and data system had to be developed [45]. The pump was developed which is capable to deliver the mobile phase smoothly and reproducibly above 1000 bar and to compensate for solvent compressibility. With a run time of 1 or 2 min, the injection volume must be low (e.g. 10 nl) with a fast injection cycle. This is afforded by a pressure assist sample introduction. Needle-in-needle sampling improves ruggedness and a needle calibration sensor increases accuracy. With UPLC, half-height peak width of <1 sec can be obtained. Therefore, the detector sampling rate must be high enough to capture enough data points across the peak for a reproducible integration. Besides MS, only UV/VIS and DAD detectors meet these challenges using new electronics and Ethernet communication at the high data rates. The 10 mm detector light guided flow cell contains only 500 nL and is equivalent to an optical fiber. The light is efficiently transferred in an internal reflectance mode.

13.2.7 Derivatization in HPLC

Because of the mild separation conditions in dissolved state, derivatization is not a general prerequisite for HPLC. It is frequently used to enable the detection or to enhance the detection sensitivity (e.g. of fluorescence, chemiluminescence or ECD), to improve the chromatographic separation (e.g. increasing the RP retention times by lipophilic derivatization) or to enable enantioselective separation by formation of diastereomers.

An overview about HPLC methods with derivatization and fluorescence or chemiluminescence detection was given by Fukushima *et al.* [30]. A large number of derivatization reagents are commercially available. Highly fluorescent fluorenyl, anthryl, carbazolyl, dimethylaminonapthyl, 7-nitrobenzoxadiazolyl or acridonyl moieties are bound to amino groups of the analyte in pre-column derivatization by reactive acyl chloride, acyl azides, chloroformate, isothiocyanate, sulfonylchloride or activated fluoride groups. Conversely, most reagents for carboxyl groups have a primary or secondary amino group as nuclophiles to form an amide bond with the carboxylic group under the action of a carbodiimide activation reagent. For post-column derivatization, the reagent should not display an own fluorescence but form a strongly fluorescent product with the analyte in a fast and quantitative reaction. An example is *o*-phthaldialdehyde (OPA) which reacts with amino compounds in presence of 2-mercaptoethanol as a co-reactant under alkaline conditions at 80°C to the fluorescing mercaptoisoindols [47].

An example of post-column derivatization for chemiluminescent detection of catecholamines was already shown in Fig 13.5. Luminol-type reagents allow the sensitive detection of carboxylic acids [48] as well as 5-hydroxyindolylacetic acid [49] in the femtomol range. Derivatization for electrochemical detection has not gained much attention in forensic analysis. An overview was given by Buchberger [50]. Besides OPA, ferrocene and hydroquinone moieties are introduced for oxidative detection and dinitrobenzene and quinone moieties for reductive detection.

The pre-column derivatization with chiral reagents for HPLC separation of enantiomers was reviewed by Toyo'oka [51]. The advantages in comparison to the use of enantioselective stationary phases (Section 13.2.3.7) are that, in addition, the chiral reagent can also improve the detection by its high molar absorbance (UV labels) or fluorescence quantum yield (fluorescent labels) and that the usual non-chiral HPLC conditions can be applied. The elution order and the degree of separation of the two diastereomers are influenced by the distance between the two asymmetric centers in the substrate and the reagent; the distance should be as small as possible for best separation. The conformational rigidity of the diastereomers is another important factor. Besides the general standards for derivatization agents (specificity and quantitative reaction with the target functional group, stability of the

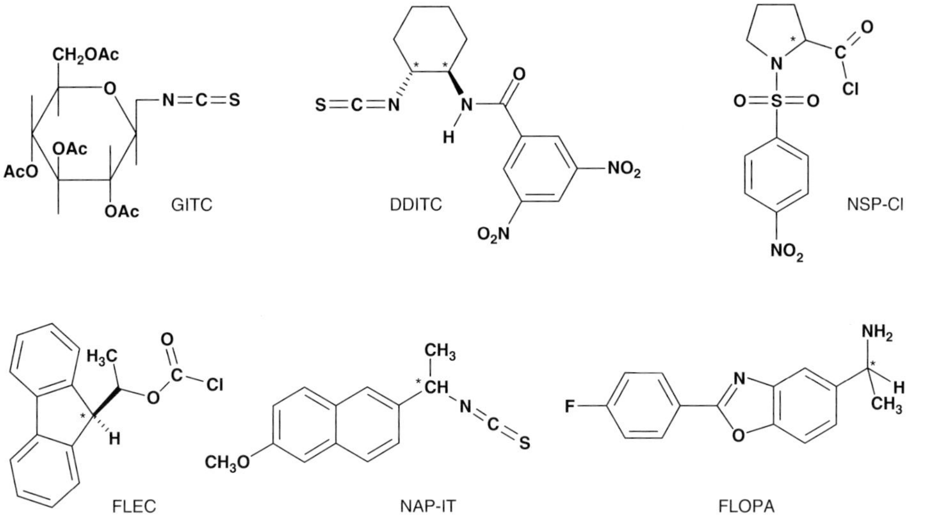

Fig. 13.10. Examples of frequently used chiral derivatization reagents for enantioselective HPLC. UV labels: GITC = 2,3,4,6-tera-*O*-acetyl-β-ᴅ-glucopyranosyl isothiocyanate, DDITC = (*S*,*S*)- or (*R*,*R*)-*N*-3,5-dinitrobenzoyl-*trans*-1,2-diaminocyclohexane isothiocyanate, NSP-Cl = 1-[(4-nitrophenyl)sulfonyl]prolyl chloride. Fluorescent labels: FLEC = 1-(9-fluorenyl)ethylchloroformate, NAP-IT = 1-(6-methoxy-2-naphthyl)ethyl isothiocyanate, FLOPA = 1-(*p*-fluorophenyl)-α-methyl-5-benzoxazoleethylamine. The chiral atoms are indicated by *.

derivative) also optical purity and resistance to racemization during the reaction are required. A large number of chiral UV as well as fluorescence labels are described for amino, carboxyl, hydroxy, mercapto or carbonyl groups [51]. Some examples of commercial chiral reagents are shown in Fig. 13.10. From forensic point of view, amphetamines are of particular interest [16,52] whereas cardiovascular drugs are most frequently investigated from pharmacological point of view [53].

13.3 APPLICATIONS OF HPLC IN FORENSIC ANALYSIS

13.3.1 Systematic toxicological analysis based on HPLC with photodiode array detector

The most important and also the most difficult task in the toxicological investigation of death or emergency cases is the unambiguous identification of the poison or the poisons, if there are no other indications from the case history. The corresponding search procedure is called "Systematic Toxicological Analysis" (STA) or "General Unknown Analysis." Since, despite a large progress, the resolution of the HPLC separation is limited, only a detector with a response specific to chemical structure is useful for this purpose. Besides MS detection, at present only the photodiode array detector (DAD) provides both sufficient specificity and sensitivity.

The use of HPLC-DAD in STA was first described by Demorest *et al.* [54] and later reported in several papers [55–67]. For peak identification both the UV spectra and the retention parameters are compared with a database of toxicologically relevant substances. Such databases with 100–600 substances were described in these papers [55–66]. A UV spectra library with 2682 compounds was produced by the author of this chapter and his colleagues [68] with an update of 600 further compounds in preparation.

HPLC-DAD is a basic technique used in forensic and clinical toxicological laboratories. It is known for its high robustness, easy handling, low capital and running costs, exact spectral reproducibility of concentration–absorbance relationships, low sensitivity to matrix interference by components such as fatty acids, cholesterol and carbohydrates, high detector sensitivity and high UV spectral specificity. Therefore, this method shall be presented more in detail.

13.3.1.1 *Specificity of substance identification based on UV spectra and retention times*

In HPLC-DAD, the substance identification is based on both UV spectrum and a retention parameter. From both, the UV spectrum is much more specific for this purpose (see below). UV spectra can be measured by up-to-date DADs with a high reproducibility, if the experimental conditions (composition of the mobile phase, wavelength resolution of the DAD) are not altered. Since the spectra of acidic or basic compounds can be completely different in their respective protonated and non-protonated state, the pH of the mobile phases used for building the spectra library and for the analysis should be in exact agreement. This is particularly important if the pK_a of the drugs are close to the pH of the buffer (e.g. diazepam with pK_a 3.5 in a mobile phase of pH 3.0). The UV spectrum depends also on the polarity of the solvent. However, since water is the most polar constituent, the type and concentration of the modifier is not very essential. That means that, as a rule, the spectrum is not changed to a measurable extent by increasing the proportion of the modifier in a gradient elution.

From the instrumental parameters, the wavelength accuracy is no problem in up-to-date DADs which have always a reliable technique for wavelength calibration based on holmium filters and deuterium or mercury emission lines. However, the spectrum can also be deteriorated by insufficient wavelength resolution. Therefore, all detector adjustments must be chosen in favor of maximum wavelength resolution to avoid wrong library search results.

The UV spectrum of a compound measured by DAD is a digital list of the absorbance as a function of the wavelength with a resolution of, as a rule, 1 nm. The identification of a spectrum is carried out by comparison with spectra from the library using all data pairs in the wavelength range measured (e.g. 181 pairs in the case of a spectrum measured from 200 to 380 nm). For assessment of spectral similarity, the description of the spectrum as a vector in *n*-dimensional space is preferred where *n* is the number of absorbance–wavelength pairs [69]. The direction of the vector in this space is determined by the specific shape of the spectrum and is independent of the concentration. In comparison between two spectra, the so-called "similarity index"

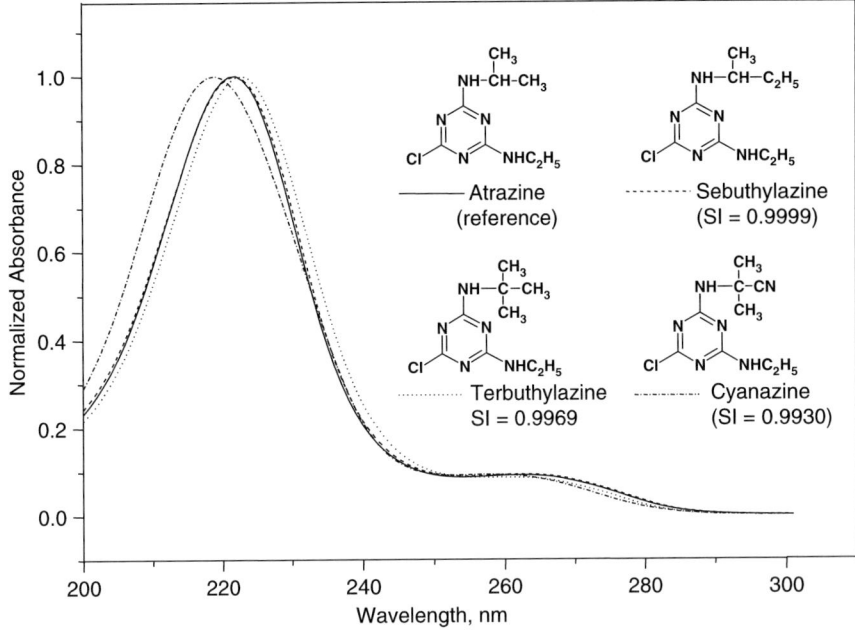

Fig. 13.11. Comparison of the UV spectrum of atrazine with those of sebuthylazine (SI = 0.9999), ter-buthylazine (SI = 0.9969) and cyanazine (SI = 0.9930). The alteration from a secondary to a tertiary alkyl substituent at the 4-amino group leads to a change of the UV spectrum, which is sufficient for discrimination between the compounds (SI = similarity index).

(SI) is defined as the cosine of the angel θ between the corresponding two vectors. It is numerically calculated by the DAD software from all data pairs of the two spectra.

In case of two identical spectra, both vectors point in exactly the same direction and SI is 1.0000 as the maximum possible value. In practical analysis, noise, matrix effects or fluctuations of experimental conditions may alter the spectrum to a certain extent. As the cosine function is relatively insensitive to small change of θ near 1, this leads only to a small decrease of SI. From a systematic evaluation followed that SI >0.9990 can be regarded as a threshold value for spectrum identity in case of an undisturbed chromatographic peak originating from an amount >1 ng [70]. The spectra can be clearly distinguished but are very similar between SI = 0.9900 and 0.9990. This is shown in Fig. 13.11 at the example of four triazine herbicides which differ only in the alkyl substituent at the 4-amino group. Atrazine cannot be distinguished by UV spectrum from sebuthylazine but differs clearly from ter-buthylazine and cyanazine.

First evaluations of HPLC-DAD with respect to the selectivity parameters "discrimination power" (DP) and "mean list length" (MLL) by Maier and Bogusz [63] and de Zeeuw *et al.* [71] led to rather poor results since the spectra were only compared by the wavelength of the maxima and not on the basis of the whole spectrum. A comprehensive evaluation of the selectivity of substance identification by HPLC-DAD was performed by Herzler *et al.* [70] using a UV spectra library of 2682 compounds. Besides statistical considerations about the diversity of the absorbing electron systems, the distinguishability of each spectrum from all other spectra stored in the library was evaluated by calculation of SI for all possible substance pairs (7.2×10^6 pairs).

The results are shown in Table 13.2. From altogether 2888 toxicologically relevant compounds involved in the study, 206 were found to have no UV absorption. Using the threshold SI of 0.9990 for identity, 1619 (60.4%) of the remaining 2682 substances showed a unique spectrum. For 399 compounds (14.9%) a second substance with an indistinguishable spectrum was found, and for 175 compounds (6.5%) there were two further substances with an identical spectrum. However, there were also less specific spectra with identity for 28 substances in one extreme case. On average, every substance was indistinguishable from 1.7 others of the 2682 compounds.

The long-term standard deviation of the relative retention time (RRT) (measured under isocratic conditions, corrected for death time and related to one reference substance) was found between 3 and 5% by the same authors [70]. By inclusion of RRT as a second parameter for identification and setting the error window to $\pm 2 s_{RT}$ (95% confidence interval), the selectivity of the method was further improved (Table 13.2). In this way, 84.2% of the substances were unambiguously identified, only 10.6% had a second and 2.6% had two further undistinguishable compounds. It was concluded by comparison with literature data of other techniques that HPLC-DAD is one of the most reliable methods for substance identification in toxicological analysis.

TABLE 2

EVALUATION OF THE SELECTIVITY OF HPLC-DAD FOR SUBSTANCE IDENTIFICATION [70]. EACH SUBSTANCE OF THE DATABASE WAS SEARCHED BY UV SPECTRUM OR BY UV SPECTRUM + RETENTION TIME IN THE DATABASE. CRITERIA FOR POSITIVE RESULTS WERE SI >0.9990 AND $\Delta RT \leqslant 2\, s_{RT}$, $s_{RT} = 0.05\, RT$. THE NUMBERS OF SUBSTANCES WITH 1–10 POSITIVE RESULTS ARE GIVEN

Number of Positive Results	By UV Spectrum		By UV Spectrum + RRT	
	Substance number (total 2682)	%	Substance number (total 1993)	%
1	1619	60.4	1678	84.2
2	399	14.9	212	10.6
3	175	6.5	52	2.6
4	127	4.7	24	1.2
5	72	2.7	19	0.9
6	50	1.9	8	0.4
7	31	1.2	0	0.0
8	26	1.0	0	0.0
9	27	1.0	0	0.0
10	23	0.8	0	0.0

13.3.1.2 Database of UV spectra and retention parameters

Although the UV spectrum of a compound is clearly determined by the geometry of the system of conjugated π-electrons and free-electron pairs (chromophor) and despite the large diversity of the spectra, it is not possible to deduce the molecular structure or the structure of molecular groups directly from the UV spectrum. Therefore, a database of UV spectra and retention parameters is an indispensable prerequisite for substance identification by HPLC-DAD. Since, in principle, each peak in a chromatogram should be identified, the database should contain as many toxicological relevant compounds as possible. This includes medical and illicit drugs, pesticides, alkaloids and all substances frequently occurring daily life.

The preparation of such a comprehensive database is beyond the possibilities of most laboratories. Commercial libraries were or are offered by some manufacturers together with the DAD (e.g. Merck-Hitachi, Waters and Dionex). A spectra library is also used in the Remedy instrument of BioRad. The HPLC-DAD database produced by the author and his co-workers [68] (2682 substances, update with further 600 substances in preparation) is the largest commercially available. The database contains also the RRT related to suitable reference substances and corrected for the dead time. The spectra and RRT were measured in acetonitrile/phosphate buffer pH 2.3 mobile phase by Shimadzu as well as Agilent instruments and are available in the formats EzChrome, ASCII and Agilent (ChemStation). They can now directly or after importation be used by the DADs of almost all DAD manufacturers. To include RRT into the search operation, the library was subdivided into 15 sublibraries according to RRT regions. In this way, the search can be restricted to one sublibrary that covers the retention time of the unknown peak.

Furthermore, the library was divided into 12 sublibraries according to pharmacological/toxicological effect or practical use (Table 13.3). This enables the investigator in a general unknown case to include into the search process indications about the kind of poisons suspected from case history (e.g. derived from symptoms, known previous diseases or profession of the individual) by choosing the appropriate sublibrary.

13.3.1.3 Metabolites in HPLC-DAD

As a rule, peaks of metabolites are present in the chromatograms of blood or urine samples in a toxicological screening. Only a small number of them are available as reference substances and are contained in spectra libraries. It is possible by HPLC-DAD to attribute a peak to a metabolite of a drug present in the same sample due to the similarity of the UV spectra and a characteristic shift of the retention time [72,73]. Metabolites have spectra very similar to the parent drug if the metabolization occurs at a group not in conjugation with the absorbing electron system. This is the case for hydroxylations and N- or O-dealkylations in the aliphatic part of the molecules. However, also dealkylations at N- or O-atoms of the conjugated molecule (formation of nor-metabolites) does not change the spectrum very much. An example is shown in Fig. 13.12a.

TABLE 3
SUBLIBRARIES AND NUMBER OF SUBSTANCES OF AN HPLC-DAD DATABASE OF TOXIC
COMPOUNDS ARRANGED WITH RESPECT TO EFFECT OR USE [68]

Sublibrary	Effect/Use	Substances
TOX01	Substances with addiction potential, analgesics (opioids), illegal drugs, psychoactive compounds, hypnotics (see also TOX02)	218
TOX02	Psychopharmaceuticals, neuroleptics, antiepileptics and drugs of similar effects (see also TOX01)	299
TOX03	Analgesics (without opioids), antirheumatics, antitussiva and drugs with similar effect (see also TOX 01)	220
TOX04	Antihistaminics, anti-allergic agents, further CNS active substances of various effect	150
TOX05	Cardiovascular agents (see also TOX04 and TOX06)	240
TOX06	Antidiabetics, diuretics, drugs with effect on blood coagulation and on the digestive system, further drugs of various effects	189
TOX07	Drugs with steroid structure, hormones, endogenous substances	262
TOX08	Cytostatics, antibiotics, other compounds with an activity against microorganisms, drugs with effect on the immune system	287
TOX09	Fungicides, disinfectants, adjuvants, diagnostics	240
TOX10	Insecticides, acaricides, nematicides and similar pesticides	264
TOX11	Herbicides	196
TOX12	Ecotoxic substances, PCBs, PAHs, fragrants, odorants, solvents, chemical reagents (see also TOX 10 and TOX 11)	117

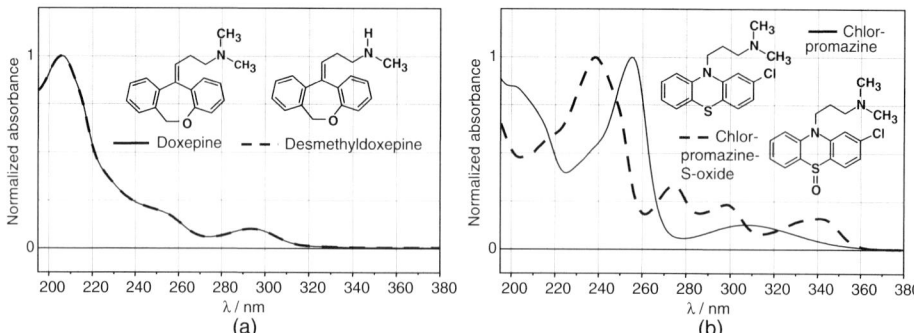

Fig. 13.12. Comparison of the UV spectra of drugs and their metabolites. (a) Desmethylation of doxepine at the isolated amino group causes no change in the UV spectrum (similarity index SI = 1.0000). (b) In chloropromazine, the S-atom is part of the light absorbing conjugated electron system. Oxidation to the S-oxide gives rise to a dramatic change of the spectrum (SI = 0.8838).

A drastic change of the spectrum can be observed if the metabolization leads to an alteration of the π-electron system of the chromophor as for instance in case of aromatic hydroxylation, formation of sulfoxides from thioethers, reduction of nitro to amino groups or oxidation of –CH$_3$ to –COOH, if these groups are attached to the chromophor (example in Fig. 13.12b). The attribution of a peak to a certain metabolite can also be supported by a characteristic change in retention time [73].

Most metabolization reactions (hydroxylations, formation of nor-metabolites) increase the hydrophilicity and, therefore, decrease the retention time on RP columns. An increase of the retention time occurs rather seldom, e.g., in the metabolic desamination of drugs. In this case, the hydrophilic ammonium group in acidic mobile phase is replaced by the less hydrophilic –OH or –COOH groups. An example is the formation of diphenylmethoxyacetic acid from diphenhydramine [74]. The identification of metabolites contributes to the drug identification and the quantitative determination can give additional evidence about the circumstances of drug use such as the time between drug application and collection of the blood sample (time of death in post-mortem cases) or distinguishing between oral intake and injection.

13.3.1.4 Sample preparation

The STA is performed for blood, urine and tissue samples as well as tablets, powders or syringe contents. The most important sample material in clinical as well as in forensic cases is venous blood (full blood, serum or plasma) because it is available from living individuals as well as from post-mortem cases and since the results allow an immediate interpretation concerning the severity of the poisoning. Therefore, this section is limited mainly to blood. The methods used for sample preparation in STA have been reviewed by Franke and de Zeeuw [75], Polettini [76] and Drummer [77]. They should, as far as possible, include all toxicologically relevant substances and exclude all interfering matrix constituents. It is an advantage of the UV detection that cholesterol, fatty acids, lipids and carbohydrates do not disturb the analysis, as they have no essential UV absorption above 195 nm. Because of the large variation in acidity (basicity) and hydrophilicity (lipophilicity), there is no universal technique for extraction of all possible organic poisons from the biological fluid. As a rule, more than one extraction has to be performed for a comprehensive analysis.

Liquid–liquid extraction is still most frequently used in HPLC-DAD analysis. A simple and very fast procedure for blood (serum, plasma) which was successfully applied by the author for many years is shown in Fig. 13.13 and involves acidic and basic extraction with CH_2Cl_2 and protein precipitation with acetonitrile [67,70,72,78]. CH_2Cl_2 proved to be a favorable compromise between extract purity and sufficient extraction yields between 60% and 100% of many non- and moderately polar substances [72]. Probably caused by formation of ion pairs with matrix constituents, basic drugs are also detected in the acidic extract and the other way round. Limits of detection between 10 and 50 ng/ml were achieved in routine measurements with CH_2Cl_2 extraction, depending on absorbance coefficients and retention time of the drug. More hydrophilic drugs such as paracetamol or salicylic acid are better determined from the supernatant of the protein precipitation with near quantitative yields. Since, in contrast to the extraction, the drugs are not enriched in the protein precipitation procedure the limits of detection are higher (0.1–0.3 µg/ml).

1-Chlorobutane was systematically studied for extraction by Demme *et al.* [79]. Owing to its lower polarity, this solvent provides purer extracts but it is not suitable for polar drugs. It proved to be particularly suitable for HPLC-DAD screening of benzodiazepines with detection limits below 5 ng/mg [80]. A mixture of hexane/

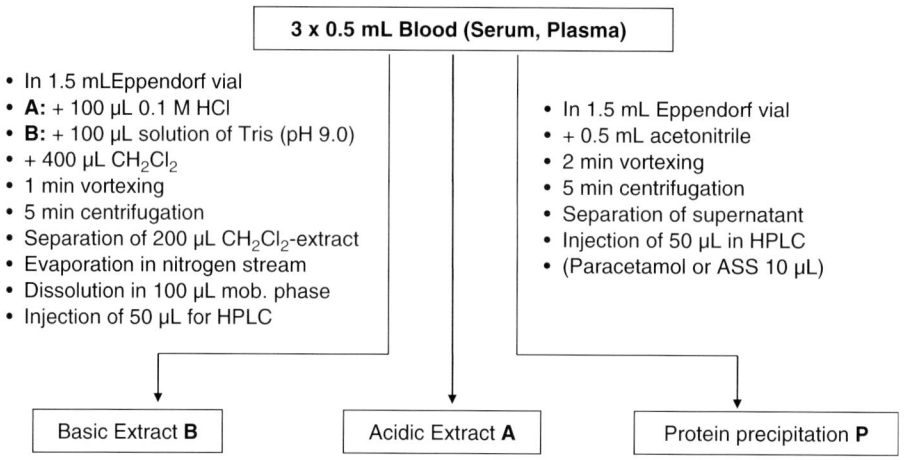

3 x 0.5 mL Blood (Serum, Plasma)

- In 1.5 mLEppendorf vial
- **A:** + 100 µL 0.1 M HCl
- **B:** + 100 µL solution of Tris (pH 9.0)
- + 400 µL CH$_2$Cl$_2$
- 1 min vortexing
- 5 min centrifugation
- Separation of 200 µL CH$_2$Cl$_2$-extract
- Evaporation in nitrogen stream
- Dissolution in 100 µL mob. phase
- Injection of 50 µL for HPLC

- In 1.5 mL Eppendorf vial
- + 0.5 mL acetonitrile
- 2 min vortexing
- 5 min centrifugation
- Separation of supernatant
- Injection of 50 µL in HPLC
- (Paracetamol or ASS 10 µL)

| Basic Extract **B** | Acidic Extract **A** | Protein precipitation **P** |

Different conditions in special cases:
- Carbonate buffer (pH 9.7); CHCl$_3$/Isopropanol (95:5 v/v) for Opiates
- 1 M NaOH; CH$_2$Cl$_2$for Amphetamines
- 0.1 M HCl; CHCl$_3$/Isopropanol (95:5 v/v) for Theophyllin
- 1-Chlorbutane for Benzodiazepines and lipophilic Pesticides
- At very low concentrations two times extraction and uniting the extracts

Fig. 13.13. Example of an efficient and fast sample preparation for systematic toxicological analysis by HPLC-DAD [67]. All three extracts are simultaneously prepared within 15 min.

isoamyl alcohol (99:1, v/v) was applied for basic drugs [81]. In a rapid screening procedure, Politi *et al.* [64] diluted simply gastric content with 0.01 N HCl (ratio 1:3 and 1:30) and, after centrifugation, injected 60 µl of both dilutions into the HPLC.

Solid-phase extraction (SPE) methods were developed for screening procedures by several authors [60,75,82–87]. Although these methods were partly developed for other detection methods, they should also be valid for HPLC-DAD. A mixed-phase bonded silica gel cartridge (C$_{18}$ + cation exchange) was used by Lai *et al.* [60] at pH 6 for extraction of basic, neutral and acidic drugs from blood or urine. The drugs were eluted by methanol/ammonia (10%) (5:1, v/v). The evaporated eluate was reconstituted with the mobile phase to 5% of the initial sample volume. Degel compared liquid/liqid extraction with ToxiTubes (a commercial organic solvent mixture), SPE with SPEC®C18 disc (hydrophobic interaction) and SPE with three mixed-mode columns (SPEC® Plus™ C18 AR/MP3 multi-modal disc, Bond-Elut CertifyTM LCR and Bond-Elut Isolute™ Confirm, always hydrophobic interaction + strong cation exchange) for 26 basic, neutral and acidic drugs from urine [84]. The mixed-mode columns delivered promising results with respect to broad-spectrum general screening, whereas for the C18 column no satisfying conditions for a universal extraction method could be found. Alabdalla [86] described a method based on two parallel extractions on C18 cartridges of the acidified blood (0.5 ml sample + 0.5 ml of 10 mM H$_3$PO$_4$ + 1 ml water) and of the alkalized blood (0.5 ml sample + 0.1 M

NaOH, diluted with water to a final volume of 2 ml). After eluation with methanol, both extracts were separately evaporated to a residual volume of 0.1 ml and analyzed by HPLC-DAD. The method proved to be useful for acidic, neutral and basic drugs.

The efficiency of two polystyrene resins Isolute 101 (copolymer with divinylbenzene) and OASIS HLB (copolymer with *N*-vinylpyrrolidone) for automated extraction of drugs from post-mortem blood and tissue samples was tested by Stimpfl [83]. The method delivered high extraction yields, a sufficient clean-up potential also for putrefied materials and avoided drug losses due to protein precipitation. The automated SPE of 17 basic and neutral drugs with seven non-polar (Isolute^{TM} C_2, C_4, C_8, C_{18}, $C_{18}MF$, PH and CN), three mixed-mode (Isolute HCX, HCX3 and HCX5) and two polymer-based (OASIS^{TM} HLB and MCX) column packings was systematically tested by Decaestecker *et al.* [82]. Best results were found for the C_8 column and the mixed-mode polymer sorbent OASIS^{TM} SCX. Computational techniques (Plackett–Burman screening design, rotatable central composite design, analysis of variance) were applied by the same authors to the same substance mixture, to optimize the extraction on the C_8 column with respect to 11 experimental parameters for providing clean extracts while maintaining high analyte recovery [87].

Column-switching systems in which a pre-column is coupled for on-line extraction to the analytical HPLC have been used successfully for analysis of drugs and metabolites in biological media and especially in plasma [88–96]. An example of a flow diagram is shown in Fig. 13.14. Restricted access materials (RAM) proved to be particularly suitable for this purpose and were reviewed by Souverain *et al.* [95]. These materials consist of silica particles with, for instance, a diameter of 25 µm and a pore diameter of 6 nm. The particles are modified outside by alkyldiol silica and within the

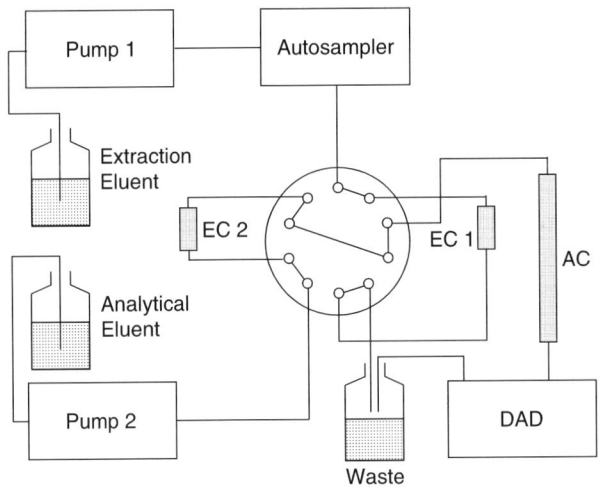

Fig. 13.14. Flow diagram of online SPE for HPLC-DAD with a 10-port valve and two equal extraction columns EC1 and EC2 [94]. In the presented position extraction and clean-up occurs on EC1 whereas the previous sample is eluted from EC2 in back-flush direction and separated on the analytical column AC. After switching of the valve, EC2 extracts the next sample and EC1 is eluted for separation on AC.

pores by usual RP materials. The outer layer prevents adsorption and denaturation of proteins and excludes macromolecules (e.g. $M > 15,000$) from the adsorption on the internal RP surface, whereas the target compounds of lower molecular weight can be extracted. RP materials for cation exchange [88] and anion exchange [89] were developed. Excellent recoveries were found for a series of drugs [88], but a systematic evaluation for the applicability in the STA was not yet described.

An automated qualitative screening method for identification of basic compounds from urine was described by Schoenberg *et al.* [96]. The urine was on-line extracted at pH 6.0 using an extraction column with polymer-based carboxylic acid functions (StrataX-CW, 20 mm × 2.1 mm, 35 μm, Phenomenex). The basic drugs are protonated at this pH and retained *via* ion–ion interactions by the carboxylate anions. The elution of the analytes occurs after switching to the acidic analytical mobile phase (pH 3.0) by protonation of the carboxylate anions. After chromatographic separation, the analytes were identified using the DAD spectra library [68]. The recoveries determined for seven typical compounds were between 73% and 97%. The major disadvantages of online SPE-HPLC couplings are the progressive deterioration of the reusable pre-column material and the risk of sample cross-contamination when biological fluids are analyzed. One solution to this problem is the use of single-use sorbent columns, which are changed before each new run in robotic sample processors [102]. Another new approach is an automated on-line renewable SPE based on multi-syringe flow injection-bead injection lab-on-valve analysis [103]. In this system, the on-line microcolumn is automatically packed before each sample from a suspension reservoir of uniformly sized spherical beads, e.g., *N*-vinylpyrrolinone-divinylbenzene copolymer (Oasis HLB, Waters, diameter 30 μm). After single use for extraction and elution, the beads are withdrawn to waste and replaced by a new packing. Recoveries of $>88\%$ and limits of detection between 0.05 and 0.15 μg/ml were described for six acidic drugs in combination with HPLC-DAD.

13.3.1.5 HPLC conditions

The chromatographic conditions used in STA by HPLC-DAD are shown in Table 13.4. As a rule, RP columns and mixtures of acetonitrile with an acidic aqueous buffer were preferred. Gradient elution with the proportion of acetonitrile increasing during the run has the advantage to elute compounds with very different lipophilicity in the same run, and the retention times of all toxicologically relevant compounds are more consistently distributed over the chromatogram [55,56,58,60–62,66,86,98–101]. Using up-to-date equipment (Section 13.2.1), the gradient can be exactly reproduced from run-to-run. As a parameter for reproducible characterization of the analyte retention, a system of retention indices (RI) was introduced which is based on the series of alkylphenylketones [104] by Smith or more advantageously of the nitroalkanes by Bogusz *et al.* [55,56] and Hill *et al.* [98] and is calculated according to equ. (1). Nitromethane has the RI 100, nitroethane 200 and so on. In practical analysis, a set of eight acidic drugs for the acidic extract and a set of 10 basic drugs (secondary standards) was proposed for calibration of the RI scale, the RI of which related to nitroalkanes were exactly determined before [56]. RI

TABLE 4
HPLC-CONDITIONS USED FOR SYSTEMATIC TOXICOLOGICAL ANALYSIS BY HPLC-DAD

Column	Eluent A/Eluent B	Isocratic/Gradient Retention Parameter Used[a]	Remarks	Ref.
250×4.0 mm LiChrospher RP 8, 5 µm	AN/0.025% H_3PO_4/TEA phosphate buffer[b]	Isocratic	272 drugs	Koves [59]
RP18	AN/0.05 M phosphate buffer pH 2.3 (37:63, v/v)	Isocratic MPPH	560 substances (only RRT)	Daldrup [97]
250×4.6 mm Lichrosorb RP8 5 µm	AN/0.05 M phosphate buffer pH 2.3 I 37: 63, v/v II 63:37, v/v III 20:80 v/v	Isocratic I: MPPH II: 4-phenylbenzophenone III: Salicylamide	2682 drugs, library is commercially available[c]	Pragst et al. [67,68, 70,72]
125×4 mm Supersphere RP 18, 4 µm	A: TEA-phosphate buffer (0.025 M, pH 3.0)[d] B: AN	Linear from 0% to 70% B in 30 min, 10 min conditioning Retention index system	880 drugs, library was commercially available	Bogusz et al. [55, 56]
150×4.6 mm Hypersil C_{18} 5 µm	A: 5% AN and B: 50% AN both in 0.05 M PP mit 375 mg/l sodium octylsulfate and 3 ml/l triethylamine, pH 3.0	Linear from 15% B to 90% B in 20 min, 5 min 90% B, in 3 min back to 15% B 6 min conditioning	300 drugs or metabolites	Lai et al. [60]
250×4.6 mm LiChrospher CH-8/II	A: 10% AN in phosphate buffer pH 3.2 B: AN	Linear from 100% A to 50% AN in 15 min, 5 min const. 50% AN back within 5 min	100 basic drugs	Logan et al. [62]
250×4.6 mm Waters C8, 5 µm	A: Phosphate buffer pH 3.8 B: AN	Step gradient: 15% B for 6.5 min; 1 ml/min 35% B for 18.5 min; linear flow rate increase to 1.5 ml/min 80% B for 3 min; 1.5 ml/min 7 min conditioning	600 drugd	Gaillard et al. [58]

TABLE 4
CONTINUED

Column	Eluent A/Eluent B	Isocratic/Gradient Retention Parameter Used[a]	Remarks	Ref.
125 × 4.0 mm Aluspher 5 μm	A: 0.0125 M NaOH in methanol B: 0.0125 M NaOH in H_2O	0–5 min 10% A, 5–20 min linear to 90% A, 20–25 min 90% A	130 basic drugs	Lambert [61]
250 × 4.6 mm Zorbax AX	A: 0.15 M H_3PO_4 + 0.05 M TEA in H_2O pH B: 0.15 M H_3PO_4 + 0.05 M TEA in H_2O/AN 20:80, v/v	0–100% B in 30 min 5 min at 100% B Retention index system	469 drugs	Hill et al. [98]
100 × 2.1 mm Hypersil ODS	A: 500 μl TEA in 1 l 0.02 M phosphate buffer	0–10 min from 15 to 40% B, 10–13.5 min to 75% B, 13.5–16 min to 80% B 2 min at 80% B	376 drugs	Turcant et al. [66]
150 × 4.6 mm Zorbax CN 5 μm	A: 0.025 M TEA buffer B: 99.9% AN	0–70% B in 20 min 5 min 70% B Retention index system	24 drugs	Alabdalla [86]
50 × 2.1 mm SB-C18 5 μm	A: 0.02 M $HClO_4$ in H_2O B: 0.02 M $HClO_4$ in AN/H_2O (80:20, v/v)	12.5–56.3% B in 2.5 min Retention index system	47 drugs, fast HPLC	Stoll et al., Porter et al. [99,100]

[a]Retention parameter: Standard substance for relative retention time RRT or system of retention indices based on nitroalkanes. No data are given, if absolute retention times were used.

[b]TEA-buffer: 9 ml conc. H_3PO_4 + 10 ml triethylamine in 900 ml H_2O.

[c]Library commercially available for DAD of Agilent, Beckmann, Dionex, Knauer, Shimadzu, Thermoquest and others.

[d]TEA-buffer: 25 ml 1 M triethylammonium phosphate pH 3.0 (Merck, Fluka) completed to 1000 ml.

could be reproduced in inter-laboratory comparison within ± 10 units (total range 1100 units). A disadvantage is the high daily effort for calibration.

$$RI_x = (RT_x - RT_A)\,(RI_B - RI_A)/(RT_B - RT_A) \qquad (1)$$

where RT, absolute retention time, corrected for the dead time t_0; RI, retention indices; A, B, standard substances with RI_A and RI_B neighbored to X; X, unknown substance.

A fast gradient HPLC-DAD system on a short narrow bore column (50×2.1 mm, RP18, $5\,\mu m$) for a high throughput screening was recently described by Stoll, Porter *et al.* [99,100]. The authors achieved a highly reproducible and very fast aqueous $HClO_4$/acetonitrile gradient by two binary pumps which delivered alternately exactly the same gradient to the analytical column [101]. Whereas the separation is performed by one pump, the second flushed the system with the initial eluent. This arrangement reduced the re-equilibration time of the column to less than 30 sec and allowed a total analysis time of only 2.80 min per gradient analysis with a reproducibility of the retention time better than 0.002 min. In this way, more than 20 analyses could be performed per hour. In analogy to [56], the authors used a retention index system based on 8 secondary standards and tested their system with a set of 47 target analytes (RI range from 140 to 571) with a medium standard deviation of RI between 3.3 and 5.9 over a time of 13 months. The evaluation of the HPLC-DAD method (spectrum + retention index) with these 47 substances leads to a MLL of 1.255 and a discriminating power of 0.997 (cf. Section 13.3.1.1).

Despite these advantages of a gradient, isocratic elution is frequently preferred [59,67,68] because of reproducible retention times, a constant spectroscopic background during the run, advantages in quantification and the possibility of recirculation of the mobile phase. A system of three mobile phases with different acetonitrile/phosphate buffer (pH 2.3) and a RP8 column was used by Pragst *et al.* [68] to cover the broad range of drugs varying widely in lipophilicity. The mobile phase can be chosen by switching valves to the DAD and the sampler. For reproducible characterization of the retention, the RRT is calculated after correction for the dead time according to equ. (2) related to only one reference substance for each of the three mobile phases (Table 13.4). For daily accuracy control, a solution containing $1\,\mu g$ histamine hydrochloride (determination of t_0), $1\,\mu g$ caffeine (control of the peak area constancy), $1\,\mu g$ of the RRT standard and $10\,\mu g$ benzene (control of spectral resolution and accuracy) is measured.

$$RRT_x = (RT_x - t_0)/(RT_{St} - t_0) \qquad (2)$$

where RRT, relative retention time; RT, absolute retention time; t_0, time of an unretained peak (dead time); St, standard substance; X, unknown substance.

13.3.1.6 Semi-quantitative determination of concentrations

With respect to quantification, UV detection in HPLC provides large advantages that are the long-term constancy of the peak area of a certain concentration and the general validity of the Lambert–Beer Law (proportionality between UV absorbance and concentration). This can be used for obtaining semi-quantitative results from HPLC peak areas in a relative short time. One possibility is the method of standard

cytostatics, immunosuppressants and a diversity of other drugs in many laboratories. Corresponding procedures are described in thousands of papers. This section shall be confined to illicit drugs and some other poisons which are of particular forensic interest and were advantageously determined by HPLC. Often, these methods are a reasonable and cost-efficient alternative to the more expensive GC-MS and LC-MS techniques.

13.3.2.1 Cannabinoids

HPLC methods for determination of cannabinoids were reviewed by Raharjo and Verpoorte [105]. Using UV detection, lower limits of quantification of THC from serum of 5 ng/ml were described [106]. However, electrochemical detection at +0.9 V vs. Ag/AgCl is preferred due to the oxidable phenolic structure of the compounds [102,107–109]. Limits of detection of 1 ng/ml in serum and 5 ng/ml in urine for THC and THC-COOH were determined and the methods proved to be suitable for routine confirmation of immunoassay results.

13.3.2.2 Opioids

The analysis of opiates from serum or urine was performed using DAD, variable wavelength UV, electrochemical and fluorescence detection. The determination of morphine, 6-acetylmorphine, codeine, methadone, its metabolite EDDP as well as cocaine, benzoylecgonine and cocaethylene from urine and plasma by SPE on Bond-Elut Certify cartridges and HPLC-DAD using an RP8 column and a phosphate buffer pH 6.53/acetonitrile gradient mobile phase was described by Fernández *et al.* [110,111]. The method with a limit of detection of 100 ng/ml for all substances in urine and between 10 and 50 ng/ml in plasma was applied in 23 drug fatalities. Similarly, morphine, morphine-3-glucuronide and morphine-6-glucuronide were determined from serum with limits of detection of 10, 60 and 90 ng/ml by UV detection [112]. A clearly higher sensitivity for these three substances was achieved by fluorescence detection (LOD 5 and 3 ng/mg, respectively) [113,114], and electrochemical detection (LOD 0.102, 0.135 and 0.135 µM) [115]. In addition to that, the combined use of fluorescence and ECD enabled an improved specificity [116].

HPLC with UV detection was regularly used for analysis of methadone and its metabolite EDDP. Since L-methadone is about 50 times more effective than D-methadone and both the L-enantiomere and the racemate are prescribed, the enantioselective analysis is frequently performed. A good separation was achieved on cyclodextrin bonded phases [117,118]. The low therapeutic range of buprenorphine (0.5–5 ng/ml) could sufficiently sensitively be measured by HPLC with ECD [119,120].

13.3.2.3 Cocaine and metabolites

Cocaine and its metabolites benzoylecgonine, and cocaethylene were determined in plasma and urine by SPE and HPLC-DAD [110,111,121,122]. The methods were used for confirmation analysis of immunoassay results. For inclusion of ecgonine methyl ester, the sample was derivatized with *p*-fluorobenzoyl chloride [123]. In another, more sensitive method, cocaine, benzoylecgonine, norcocaine and benzoylnorecgonine were determined from only 50 µl serum using a FLD with LODs of 0.5 ng/ml [124].

13.3.2.4 Amphetamines and designer drugs

HPLC methods with UV or diode array detection for determination of amphetamine, methamphetamine, methylenedioxymethamphetamine (MDMA), methylenedioxyethamphetamine (MDE), 4-methylthioamphetamine (MTA), mescaline or 4-bromo-3,5-dimethoxy-phenethylamine and some of their metabolites in urine were described [125–127]. The detection limits after usual SPE were between 20 and 80 ng/ml. However, a tremendous increase in sensitivity to 0.5 ng/ml was possible by specific sample preparation using single-drop liquid–liquid–liquid extraction (Fig. 13.16) [126]. The basic drugs were extracted from the alkaline sample into the hexane layer and from there simultaneously or successively re-extracted into a 5 µl drop of 0.02 H_3PO_4 with an enrichment factor of about 500.

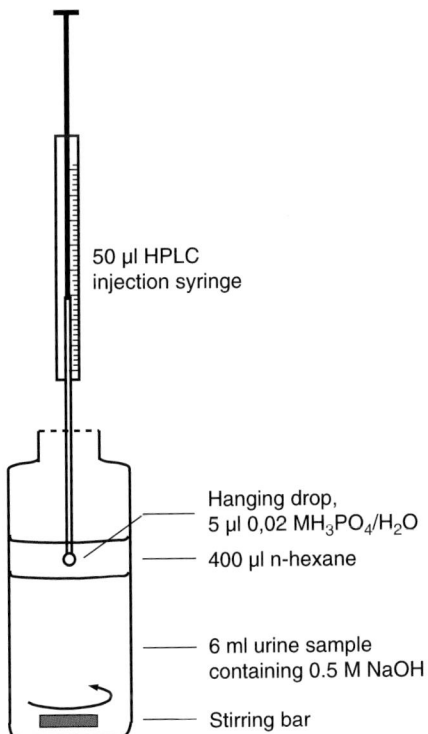

50 µl HPLC
injection syringe

Hanging drop,
5 µl 0,02 MH_3PO_4/H_2O

400 µl n-hexane

6 ml urine sample
containing 0.5 M NaOH

Stirring bar

Fig. 13.16. Single-drop liquid–liquid–liquid extraction of amphetamines from urine [126]. The analytes were 500-fold enriched in the drop leading to a detection limit of 0.5 ng/mg by HPLC-UV. Extraction and re-extraction time both 40 min.

A high sensitivity could also be obtained in detection of these compounds from plasma, urine oral fluid or hair using HPLC with fluorescence detection. Whereas the methylenedioxy-compounds MDMA, MDE and MDA display a sufficiently intense native fluorescence [128–131], amphetamine and metamphetamine were derivatized, e.g. with 4-[4,5-diphenyl-imidazolyl-(1)]-benzoylchloride [132–134]. Electrochemical detection was enabled by pre-column derivatization with 3,4-dihydroxybenzaldehyde [135]. Enantioselective analysis of amphetamines by HPLC were reviewed by Herraez-Hernandez *et al.* [16] and Liu *et al.* [136]. Fluorimetric detection was preferred also for this purpose [137,138].

13.3.2.5 Lysergic acid diethylamide

HPLC with FLD was used by several authors for determination of lysergic acid diethylamide (LSD) from blood, urine or hair [139–142]. Extraction by 1-chloro-butane from the alkalized sample and re-extraction into 0.1 M HCl lead to a detection limit of 0.5 ng/ml [139]. The sensitivity was improved to an LOD of 20 pg/ml for both LSD and nor-LSD by use of an LSD ImmunElute extraction kit [141,142]. A concentration of 1 pg/mg LSD could be detected in hair in this way [141].

13.3.2.6 Benzodiazepines

The analysis of benzodiazepines from biological fluids using HPLC were reviewed by Berrueta *et al.* [143]. Because of their characteristic and intense UV absorption, benzodiazepines including the low-dose drugs such as alprazolam or lorazepam can be analyzed with high sensitivity and selectivity using a diode array detector [80,144–148]. Liquid–liquid extraction with 1-chlorobutane [144,145] or *n*-hexane/ethyl acetate [146] as well as SPE with C18 extraction columns [147] or online SPE [148] proved to be suitable for this task. The detection limits were generally between 2 and 10 ng/ml and were sufficient for confirmation of positive immunoassay results in daily routine [149,150]. A sensitive determination of benzodiazepines with detection limits of 2–14 ng/ml was also possible using reductive electrochemical detection at a hanging mercury drop in combination with UV detection [151].

13.3.2.7 Pesticides

Many insecticides, acaricides, fungicides, herbicides and other pesticides are well extracted, separated and detected by HPLC-DAD and are involved in the corresponding screening procedures [65,67,68]. Specific problems arise with superwarfarin rodenticides because of their extreme lipophilicity and very low toxic concentrations. An HPLC method for bromadiolone, difenacoum, brodifacoum and difethialone based on combined UV and fluorescence detection after extraction with a chloroform/acetone mixture and quantification limits of 3–12 ng/ml was described by Kuijpers *et al.* [152]. In a similar way, HPLC with fluorescence by Chalermchaikit *et al.* [153] enabled detection limits of 1 ng/ml or 1 ng/g from serum and liver. However, with HPLC-DAD after diethylether/ethylacetate extraction from human serum, the detection limits were only between 25 and 100 ng/ml [154]. Practical application of these methods were described in several poisoning cases [155,156].

Also the detection of paraquat and analogous herbicides by HPLC is still a problem because of low extractability of the ionic compound and very short retention time under the usual RP-HPLC conditions. Ito *et al.* [157] solved this problem in an HPLC-DAD method by direct injection of the serum sample after deproteination with acetone and using a water/acetonitrile mobile phase containing the ion pair reagent with cation-releasing properties IPPC-MS3 (GL Sciences Inc., Tokyo, Japan, exact structure not described) and a usual ODS column for separation. In this way, the retention times were shifted to a well measurable region and paraquat and diquat were separated. In previous papers, sodium heptane sulfonate [158] or sodium octane sulfonate [159,160] had been used as ion pair reagents. The sample preparation was simply a deproteination with 6% perchloric acid [159] or an online SPE with a LiChroprep RP-8 [160]. The detection limits were 0.1 µg/ml. Another approach to the problem was used in the HPLC-DAD method by Arys *et al.* [161]. Blood and tissue homogenates were deproteinized with trichloroacetic acid and reduced with $NaBH_4$ at alkaline pH. The basic dihydropyridine products were amenable to extraction with diethyl ether and were analyzed after solvent evaporation and dissolution in the mobile phase. HPLC was performed at an Aluspher® column with an alkaline mobile phase with a detection limit of 0.06 µg/ml in blood.

13.3.2.8 Cyanide, thiocyanate and azide

The determination of cyanide in body fluids by HPLC with FLD after derivatization with 2,3-napthalenedialdehyde and taurine was described by several authors [162–165]. The reaction is shown in Fig. 13.17 and leads to a highly fluorescent benzoisoindole derivative. Fluorescence detection has the advantage in comparison to UV/VIS absorption that the reagents which are added in large excess do not fluoresce. Felscher and Wulfmeyer [163] isolated HCN from blood by microdistillation before derivatization and described a detection limit of 2 ng/ml. In a simpler and faster procedure, Mateus *et al.* [165] added 0.5 ml water and 2 ml methanol to 100 µl blood, vortexed and centrifugated the mixture and performed the derivatization directly in the supernatent. For analysis, 10 µl of the obtained solution were injected and detected at excitation and emission wavelengths of 418 and 460 nm, respectively. The detection limit was below 25 ng/ml.

In a similar way, thiocyanate was determined with high sensitivity (detection limit 0.16 nmol/l) from saliva and plasma by HPLC with FLD after derivatization with

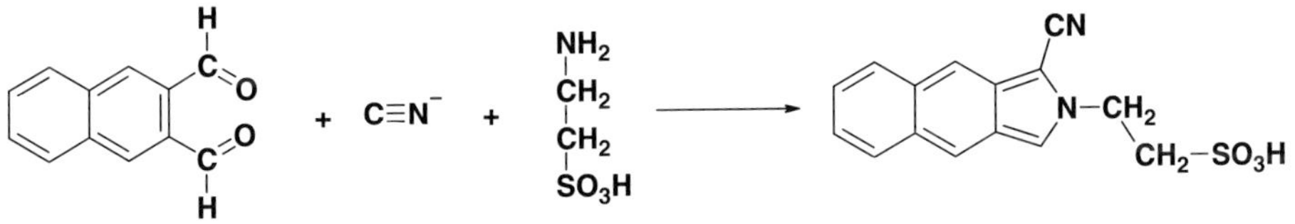

Fig. 13.17. Derivatization of cyanide for determination by HPLC with fluorescence detector.

3-bromomethyl-7-methoxy-1,4-benzoxazin-2-one [166]. However, in case of the plasma samples, SCN$^-$ had to be separated before derivatization using anion-exchange cartridges.

Azide was determined in body fluids and tissues of fatalities by HPLC-DAD after derivatization with 3,5-dinitrobenzoyl chloride [167] or benzoyl chloride [168]. In the first case, the samples were treated with acetonitrile and 0.02 N K$_2$CO$_3$ for protein precipitation and the supernatent adjusted to pH 5 with diluted HCl [167]. After 10 min reaction with 3,5-dinitrobenzoyl chloride, 50 μl of the solution were directly injected into HPLC. The detection limit was 0.08 μg/ml. The procedure with benzoyl chloride was very similar with a detection limit of 0.2 μg/ml [168]. In the chromatograms, the derivatives were well separated from the excessive reagents and no disturbing peaks from other sample constituents were seen.

13.4 CONCLUDING REMARKS

HPLC is a key technique in forensic toxicological analysis. An immense progress occurred in the last decade for all parts of this technique from sample preparation to data evaluation by sophisticated software programs. Most important, a wide variety of new or improved separation materials appeared with increased resolution, high reproducibility of retention behavior, better applicability for complicated mixtures of substance with very different polarity, lower flow resistance or increased stability to pressure, hydrolysis or high temperature. Measurement time was essentially decreased in fast LC and enantioselective separation has become an easily available tool. Sophisticated systems of column switching and online sample preparations can be realized based on exact reproducible computer control.

A series of different detectors is available for specific and sensitive detection of almost all conceivable substances. For toxicological analysis, particularly for general unknown cases and in laboratories with frequently changing analytes, a universal detector with applicability to as many as possible toxicologically relevant substances, with a high sensitivity and with the capability for unambiguous substance identification is preferred. The new generation LC-MS-MS combinations which were not dealt with in this chapter seem increasingly to fulfill this task. Nevertheless, other detectors were also much improved and are indispensable in a toxicological laboratory. The photodiode array detector proved to be a reliable and robust instrument for relative simple and fast identification and quantification of toxic compounds in biological samples and can manage a profound part of the daily toxicological screening cases of a laboratory also in the future, not only as a cheaper alternative but also as mutual completion to GC-MS or LC-MS-MS. The CLND seems to become a very efficient way of quantification in absence of the reference compound. On the other hand, the fluorescence and chemiluminescence detectors as well as the ECD are not suitable for general screening procedures. They need to be optimized to the specific analytes and should be considered if a large number of samples must be analyzed routinely for such analytes with extremely high sensitivity.

Altogether, HPLC offers a wide variety of interesting options for toxicological screening as well as for highly sensitive determination of specific substances. Future will show whether and to which extent these techniques have a chance to survive beside the increasingly dominating liquid chromatography–mass spectrometric methods in forensic laboratories, which are more and more controlled by economic restrictions.

13.5 REFERENCES

1   R. Stevenson, Am. Lab., 36 (2004) 21.
2   R. Stevenson, Am. Lab., 37 (2005) 14.
3   M. Przybyciel, LCGC North America. Column Technol. Suppl., 24 (June 2004) 26.
4   F.M. Yamamoto and S. Rokushika, J. Chromatogr. A, 898 (2000) 141.
5   M. Herzler, I. Fechner and F. Pragst, Toxichem + Krimtech, 65 (1998) 53.
6   P.E. Morgan, M. Hanna-Brown and R.J. Flanagan, Biomed. Chromatogr., 20 (2006) 765.
7   M.A. Strege, S. Stevenson and S.M. Lawrence, Anal. Chem., 72 (2000) 4629.
8   H. Tanaka, X. Zhou and O. Masayoshi, J. Chromatogr. A, 987 (2003) 119.
9   W. Jiang, G. Fischer, Y. Girmay and K. Irgum, J. Chromatogr. A, 1127 (2006) 82.
10 F. Svec, LCGC North America. Column Technol. Suppl., 24 (June 2004) 18.
11 H.A. Claessens and M.A. van Straten, J. Chromatogr. A, 1060 (2004) 23.
12 Y. Yang and and D.R. Lynch Jr., Stationary phases for high-temperature LC separations. LCGC North America. Column Technol. Suppl., 24 (June 2004) 34–38.
13 G. Cancelliere, I. D'Acquarica, F. Gasparrini, M. Maggini, D. Misitiand and C. Villani, J. Sep. Sci., 29 (2006) 770.
14 T.E. Beesley and J.T. Lee, LCGC North America. Column Technol. Suppl., 24 (June 2004) 30.
15 J. Bojarski, H.Y. Aboul-Enein and A. Ghanem, Curr. Anal. Chem., 1 (2005) 59.
16 R. Herraez-Hernandez, P. Campins-Falco and J. Verdu-Andres, J. Biochem. Biophys. Methods, 54 (2002) 147.
17 D.M. Forjan, D. Kontrek and V. Vinkovic, Chirality, 18 (2006) 857.
18 C. Yamamoto, S. Inagaki and Y. Okamoto, J. Sep. Sci., 29 (2006) 915.
19 I.W. Muderawan, T.T. Ong and S.C. Ng, J. Sep. Sci., 29 (2006) 1849.
20 T.J. Ward and A.B. Farris 3rd, J. Chromatogr. A, 906 (2001) 73.
21 R.J. Steffeck, Y. Zelechonok and K.H. Gahm, J. Chromatogr. A, 947 (2002) 301.
22 J. Haginaka, J. Chromatogr. A, 906 (2001) 253.
23 R.J. Ansell and K.L. Kuah, Analyst, 130 (2005) 179.
24 B. Sellergren, J. Chromatogr. A, 906 (2001) 227.
25 S.P. Dixon, I.D. Pitfield and D. Perret, Biomed. Chromatogr., 20 (2006) 508.
26 M. Buratti, O. Pellegrino, C. Valla, F.M. Rubino, C. Verducia and A. Colombi, Biomed. Chromatogr., 20 (2006) 971.
27 R. Dams, T. Benijts, W.E. Lambert and A.P. De Leenheer, J. Chromatogr. B, Analyt. Technol. Biomed. Life Sci., 773 (2002) 53.
28 Y. Hupka, J. Beike, J. Roegener, B. Brinkmann, G. Blaschke and H. Kohler, Int. J. Legal Med., 119 (2005) 121.
29 S.J. Hart, G.J. Hall and J.E. Kenny, Anal. Bioanal. Chem., 372 (2002) 205.
30 T. Fukushima, N. Usui, T. Santa and K. Imai, J. Pharm. Biomed. Anal., 30 (2003) 1655.
31 F. Li, C. Zhang, X. Guo and W. Feng, Biomed. Chromatogr., 17 (2003) 96.
32 M. Tsunoda, M. Nagayama, T. Funatsu, S. Hosoda and K. Imai, Clin. Chim. Acta, 366 (2006) 168.
33 T. Toyo'oka, A. Ogawa, H. Arai and H. Tanizawa, Biomed. Chromatogr., 13 (1999) 101.
34 M.A. Nussbaum, S.W. Baertschi and P.J. Jansen, J. Pharm. Biomed. Anal., 27 (2002) 983.
35 Y. Deng, J.T. Wu, H. Zhang and T.V. Olah, Rapid Commun. Mass Spectrom., 18 (2004) 1681.
36 S. Lane, B. Boughtflower, I. Mutton, C. Paterson, D. Farrant, N. Taylor, Z. Blaxill, C. Carmody and P. Borman, Anal. Chem., 77 (2005) 4354.

37 B.T. Mathews, P.D. Higginson, R. Lyons, J.C. Mitchell, N.W. Sach, M.J. Snowden, M.R. Taylor and A.G. Wright, Chromatographia, 60 (2004) 625.

38 H.E. Fries, C.A. Evans and K.W. Ward, J. Chromatogr. B, Analyt. Technol. Biomed. Life Sci., 819 (2005) 339.

39 A. Lemke and O. Kayser, Pharmazie, 61 (2006) 406.

40 N.C. Megoulas and M.A. Koupparis, Anal. Bioanal. Chem., 382 (2005) 290.

41 R.J. Flanagan; D. Perrett; R. Whelpton, Electrochemical detection in HPLC, analysis of drugs and poisons, The Royal Society of Chemistry, Cambridge, 2005.

42 G. Achilli, G.P. Cellerino, G.V. Melzi d'Eril and F. Tagliaro, J. Chromatogr. A, 729 (1996) 273.

43 W.-D. Beinert, V. Eckert, S. Galushko, V. Tanchuk and I. Shishkina, LCGC Europe on-line supplement, pp. 34–38, www.lcgceurope.com, 09.10.2006.

44 E.F. Hewitt, P. Lukulay and S. Galushko, J. Chromatogr. A, 1107 (2006) 79.

45 M.E. Swartz, LCGC Europe, 18 (May 2005) 5.

46 E.S. Grumbach, T.E. Wheat, M. Kele and J.R. Mazzeo, LCGC Europe, 18 (May 2005) 37.

47 P. Proenca, H. Teixeira, M.C. de Mendonca, F. Castanheira, E.P. Marques, F. Corte-Real and D. Nuno Vieira, Forensic Sci. Int., 146(Suppl.) (2004) 79.

48 M. Yamaguchi, H. Yoshida and H. Nohta, J. Chromatogr. A, 950 (2002) 1.

49 H. Yoshida, R. Nakao, T. Matsuo, H. Nohta and M. Yamaguchi, J. Chromatogr. A, 907 (2001) 39.

50 W. Buchberger, Elektrochemische Analysenverfahren, Grundlagen, instrumentation, Anwendungen, Spektrum Akademischer Verlag, Heidelberg, 1998, p. 225.

51 T. Toyo'oka, J. Biochem. Biophys. Methods, 54 (2002) 25.

52 D. Jirovsky, K. Lemr, J. Sevcik, B. Smysl and Z. Stransky, Forensic Sci. Int., 96 (1998) 61.

53 J. Bojarski, J. Biochem. Biophys. Methods, 54 (2002) 197.

54 D.M. Demorest, J.C. Fetzer, I.S. Lurie, S.M. Carr and K.B. Chatson, LCGC, 5 (1987) 128.

55 M. Bogusz and M. Erkens, J. Chromatogr. A, 674 (1994) 97.

56 M. Bogusz and M. Wu, J. Anal. Toxicol., 15 (1991) 188.

57 S.P. Elliott and K.A. Hale, J. Anal. Toxicol., 22 (1998) 279.

58 Y. Gaillard and G. Pépin, J. Chromatogr. A, 763 (1997) 49.

59 E.M. Koves, J. Chromatogr. A, 692 (1995) 103.

60 C.-K. Lai, T. Lee, K.-M. Au and A.Y.W. Chan, Clin. Chem., 43 (1997) 312.

61 W.E. Lambert, E. Meyer and A.P. De Leenheer, J. Anal. Toxicol., 19 (1995) 73.

62 B.K. Logan, D.T. Stafford, I.R. Tebbett and C.M. Moore, J. Anal. Toxicol., 14 (1990) 154.

63 R.D. Maier and M. Bogusz, J. Anal. Toxicol., 19 (1995) 79.

64 L. Politi, A. Groppi, A. Polettini and M. Montagna, Forensic Sci. Int., 141 (2004) 115.

65 A. Tracqui, P. Kintz and P. Mangin, J. Forensic Sci., 40 (1995) 254.

66 A. Turcant, A. Premel-Cabic, A. Cailleux and P. Allain, Clin. Chem., 37 (1991) 1210.

67 F. Pragst, M. Herzler and B.T. Erxleben, Clin. Chem. Lab. Med., 42 (2004) 1325–1340.

68 F. Pragst, M. Herzler, S. Herre, B-T. Erxleben and M. Rothe, UV-spectra of toxic compounds. Database of photodiode array UV spectra of illegal and therapeutic drugs, pesticides, ecotoxic substances and other poisons. Book and CD. Dieter Helm, Heppenheim, 2001.

69 H.-J.P. Sievert and A.C.J.H. Drouen. In: L. Huber, S.A. George (Eds.), Diode array detection in HPLC, Marcel Dekker, New York, 1993, p. 51.

70 M. Herzler, S. Herre and F. Pragst, J. Anal. Toxicol., 27 (2003) 233.

71 R.A. de Zeeuw, H. Hartstra and J.P. Franke, J. Chromatogr. A, 674 (1994) 3.

72 F. Pragst, H.H. Maurer, J. Hallbach, U. Staerk, W.R. Külpmann, F. Degel and H.J. Gibitz. In: W.R. Külpmann (Ed.), Klinisch-toxikologische Analytik–Verfahren, Befunde, interpretation, Wiley-VCH, Weinheim, 2002, p. 49.

73 S. Herre and F. Pragst, J. Chromatogr. B, 692 (1997) 111.

74 F. Pragst, S. Herre and A. Bakdash, Forensic Sci. Int., 161 (2006) 189.

75 J.P. Franke and R.A. de Zeeuw, J. Chromatogr. B, Biomed. Sci. Appl., 713 (1998) 51.

76 A. Polettini, J. Chromatogr. B, Biomed. Sci. Appl., 733 (1999) 47.

77 O.H. Drummer, J. Chromatogr. B, Biomed. Sci. Appl., 733 (1999) 27.

78 F. Pragst, M. Herzler and S. Herre, Klin. Biochem. Metab., 8 (2000) 13.
79 U. Demme, J. Becker, H. Bussemas, T. Daldrup, F. Erdmann, M. Erkens, P.X. Iten, H. Magerl, L. von Meyer, J. Teske, W. Weinmann, J.P. Weller. In: F. Pragst and R. Aderjan (Eds.), Proceedings of the GTFCh symposium 1999. Dr. Dieter Helm, Heppenheim, 1999, p. 213.
80 L. von Meyer, A. Schmoldt and W.R. Külpmann. In: W.R. Külpmann (Ed.), Klinisch-toxikologische Analytik – Verfahren, Befunde, interpretation, Wiley-VCH, Weinheim, 2002, p. 287.
81 K. Titier, S. Bouchet, F. Pehourcq, N. Moore and M. Molimard, J. Chromatogr. B, Analyt. Technol. Biomed. Life Sci., 788 (2003) 179.
82 T.N. Decaestecker, E.M. Coopman, C.H. Van Peteghem and J.F. Van Bocxlaer, J. Chromatogr. B, Analyt. Technol. Biomed. Life Sci., 789 (2003) 19.
83 T. Stimpfl, J. Jurenitsch and W. Vycudili, J. Anal. Toxicol., 25 (2001) 125.
84 F. Degel, Clin. Biochem., 29 (1996) 529.
85 R.A. de Zeeuw, J. Wijsbeek and J.P. Franke, J. Anal. Toxicol., 24 (2000) 97.
86 M.A. Alabdalla, J. Clin. Forensic Med., 12 (2005) 310.
87 T.N. Decaestecker, W.E. Lambert, C.H. Van Peteghem, D. Deforce and J.F. Van Bocxlaer, J. Chromatogr. A, 1056 (2004) 57.
88 P. Chiap, O. Rbeida, B. Christiaens, P. Hubert, D. Lubda, K.S. Boos and J. Crommen, J. Chromatogr. A, 975 (2002) 145.
89 O. Rbeida, B. Christiaens, P. Hubert, D. Lubda, K.S. Boos, J. Crommen and P. Chiap, J. Chromatogr. A, 1030 (2004) 95.
90 E. Yamamoto, K. Murata, Y. Ishihama and N. Asakawa, Anal. Sci., 17 (2001) 1155.
91 E. Yamamoto, H. Igarashi, Y. Sato, I. Kushida, T. Kato, T. Kajima and N. Asakawa, J. Pharm. Biomed. Anal., 42 (2006) 587.
92 M. Walles, J. Borlak and K. Levsen, Anal. Bioanal. Chem., 374 (2002) 1179.
93 Z. Yu, D. Westerlund and K.S. Boos, J. Chromatogr. B, Biomed. Sci. Appl., 704 (1997) 53.
94 R. Oertel, K. Richter, T. Gramatte and W. Kirch, J. Chromatogr. A, 797 (1998) 203.
95 S. Souverain, S. Rudaz and J.L. Veuthey, J. Chromatogr. B, Analyt. Technol. Biomed. Life Sci., 801 (2004) 141.
96 L. Schoenberg, T. Grobosch, D. Lampe and C. Kloft, J. Chromatogr. A, 1134 (2006) 177.
97 T. Daldrup, P. Michalke and W. Böhme, Angewandte Chromatographie Heft 37, Bodenseewerk Perkin Elmer & Co. GmbH, 1981.
98 D.W. Hill and A.J. Kind, J. Anal. Toxicol., 18 (1994) 233.
99 D.R. Stoll, C. Paek and P.W. Carr, J. Chromatogr. A, 1137 (2006) 153.
100 S.E. Porter, D.R. Stoll, C. Paek, S.C. Rutan and P.W. Carr, J. Chromatogr. A, 1137 (2006) 163.
101 A. P Schellinger, D. R Stoll and P.W. Carr, J. Chromatogr. A, 1064 (2005) 143.
102 E. Kramer and K.A. Kovar, J. Chromatogr. B, Biomed. Sci. Appl., 731 (1999) 167.
103 J.B. Quintana, M. Miro, J.M. Estela and V. Cerda, Anal. Chem., 78 (2006) 2832.
104 R.M. Smith, J. Chromatogr., 236 (1982) 313.
105 T.J. Raharjo and R. Verpoorte, Phytochem. Anal., 15 (2004) 79.
106 C. Abbara, R. Galy, A. Benyamina, M. Reynaud and L. Bonhomme-Faivre, J. Pharm. Biomed Anal., 41 (2006) 1011.
107 L.K. Thompson and E.J. Cone, J. Chromatogr., 421 (1987) 91.
108 Y. Nakahara and C.E. Cook, J. Chromatogr., 434 (1988) 247.
109 D.H. Fisher, M.I. Broudy and L.M. Fisher, Biomed. Chromatogr., 10 (1996) 161.
110 P. Fernandez, C. Vazquez, L. Moralesand and A.M. Bermejo, J. Appl. Toxicol., 25 (2005) 200.
111 P. Fernandez, L. Morales, C. Vazquez, A.M. Bermejo and M.J. Tabernero, Forensic Sci. Int., 161 (2006) 31.
112 J. Netriova, E. Blahova, Z. Johanesova, E. Brandsteterova, J. Lehotay, K. Serdt and J. Mocak, Pharmazie, 61 (2006) 528.
113 R. Aderjan, S. Hofmann, G. Schmitt and G. Skopp, J. Anal. Toxicol., 19 (1995) 163.
114 J. Beike, H. Kohler, B. Brinkmann and G. Blaschke, J. Chromatogr. B, Biomed. Sci. Appl., 726 (1999) 111.
115 A.W. Wright and M.T. Smith, Ther. Drug Monit., 20 (1998) 215.

116   Y. Rotshteyn and B. Weingarten, Ther. Drug Monit., 18 (1996) 179.
117   C. Pham-Huy, N. Chikhi-Chorfi, H. Galons, N. Sadeg, X. Laqueille, N. Aymard, F. Massicot, J.M. Warnet and J.R. Claude, J. Chromatogr. B, Biomed. Sci. Appl., 700 (1997) 155.
118   D.W. Boulton and C.L. Devane, Chirality, 12 (2000) 681.
119   L. Debrabandere, M. Van Boven and P. Daenens, J. Forensic Sci., 37 (1992) 82.
120   E. Schleyer, R. Lohmann, C. Rolf, A. Gralow, C.C. Kaufmann, M. Unterhalt and W. Hiddemann, J. Chromatogr., 614 (1993) 275.
121   K.M. Clauwaert, J.F. Van Bocxlaer, W.E. Lambert and A.P. De Leenheer, Anal. Chem., 68 (1996) 3021.
122   M. Balikova and J. Vecerkova, J. Chromatogr. B, Biomed. Appl., 656 (1994) 267.
123   S.C. Jamdar, C.B. Pantuck, J. Diaz and B. Mets, J. Anal. Toxicol., 24 (2000) 438.
124   L. Sun, G. Hall and C.E. Lau, J. Chromatogr. B, Biomed. Sci. Appl., 745 (2000) 315.
125   M.E. Soares, M. Carvalho, H. Carmo, F. Remiao, F. Carvalho and M.L. Bastos, Biomed. Chromatogr., 18 (2004) 125.
126   Y. He and Y.J. Kang, J. Chromatogr. A, 1133 (2006) 35.
127   H.J. Helmlin and R. Brenneisen, J. Chromatogr., 593 (1992) 87.
128   M. Concheiro, A. de Castro, O. Quintela, M. Lopez-Rivadulla and A. Cruz, Forensic Sci. Int., 150 (2005) 221.
129   R. Mancinelli, S. Gentili, M.S. Guiducci and T. Macchia, J. Chromatogr. B, Biomed. Sci. Appl., 735 (1999) 243.
130   K.M. Clauwaert, J.F. Van Bocxlaer, E.A. De Letter, S. Van Calenbergh, W.E. Lambert and A.P. De Leenheer, Clin. Chem., 46 (2000) 1968.
131   M. Brunnenberg, H. Lindenblatt, E. Gouzoulis-Mayfrank and K.A. Kovar, J. Chromatogr. B, Biomed. Sci. Appl., 719 (1998) 79.
132   O. al-Dirbashi, N. Kuroda, S. Inuduka, F. Menichini and K. Nakashima, Analyst, 124 (1999) 493.
133   S. Nakamura, M. Tomita, M. Wada, H. Chung, N. Kuroda and K. Nakashima, Biomed. Chromatogr., 20 (2006) 622.
134   K. Nakashima, A. Kaddoumi, Y. Ishida, T. Itoh and K. Taki, Biomed. Chromatogr., 17 (2003) 471.
135   N.A. Santagati, G. Ferrara, A. Marrazzo and G. Ronsisvalle, J. Pharm. Biomed. Anal., 30 (2002) 247.
136   J.T. Liu and R.H. Liu, J. Biochem. Biophys. Methods, 54 (2002) 115.
137   O. al-Dirbashi, N. Kuroda, F. Menichini, S. Noda, M. Minemoto and K. Nakashima, Analyst, 123 (1998) 2333.
138   J. Buechler, M. Schwab, G. Mikus, B. Fischer, L. Hermle, C. Marx, G. Gron, M. Spitzer and K.A. Kovar, J. Chromatogr. B, Analyt. Technol. Biomed. Life Sci., 793 (2003) 207.
139   M.M. McCarron, C.B. Walberg and R.C. Baselt, J. Anal. Toxicol., 14 (1990) 165.
140   D. Bergemann, A. Geier and L. von Meyer, J. Forensic Sci., 44 (1999) 372.
141   J. Rohrich, S. Zorntlein and J. Becker, Forensic Sci. Int., 107 (2000) 181.
142   T. Grobosch and U. Lemm-Ahlers, J. Anal. Toxicol., 26 (2002) 181.
143   L.A. Berrueta, B. Gallo and F. Vicente, J. Pharm. Biomed. Anal., 10 (1992) 109.
144   A. El Mahjoub and C. Staub, J. Pharm. Biomed. Anal., 23 (2000) 447.
145   A. Bugey and C. Staub, J. Pharm. Biomed. Anal., 35 (2004) 555.
146   W. He, N. Parissis and T. Kiratzidis, J. Forensic Sci., 43 (1998) 1061.
147   F. Musshoff and T. Daldrup, Int. J. Legal Med., 105 (1992) 105.
148   K.K. Akerman, J. Jolkkonen, M. Parviainen and I. Penttila, Clin. Chem., 42 (1996) 1412.
149   S.D. Ferrara, L. Tedeschi, G. Frison and F. Castagna, J. Anal. Toxicol., 16 (1992) 217.
150   L. Kroener, F. Musshoff and B. Madea, J. Anal. Toxicol., 27 (2003) 205.
151   M. Wilhelm, H.J. Battista and D. Obendorf, J. Chromatogr. A, 897 (2000) 215.
152   E.A. Kuijpers, J. den Hartigh, T.J. Savelkoul and F.A. de Wolff, J. Anal. Toxicol., 19 (1995) 557.
153   T. Chalermchaikit, L.J. Felice and M.J. Murphy, J. Anal. Toxicol., 17 (1993) 56.
154   H. Lotfi, M.F. Dreyfuss, P. Marquet, J. Debord, L. Merle and G. Lachatre, J. Anal. Toxicol., 20 (1996) 93.

155   R.B. Palmer, P. Alakija, J.E. de Baca and K.B. Nolte, J. Forensic Sci., 44 (1999) 851.
156   P.T. McCarthy, A.D. Cox, D.J. Harrington, R.S. Evely, E. Hampton, A.I. al-Sabah, E. Massey, H. Jackson and T. Ferguson, Hum. Exp. Toxicol., 16 (1997) 166.
157   M. Ito, Y. Hori, M. Fujisawa, A. Oda, S. Katsuyama, Y. Hirose and T. Yoshioka, Biol. Pharm. Bull., 28 (2005) 725.
158   K. Croes, F. Martens and K. Desmet, J. Anal. Toxicol., 17 (1993) 310.
159   P. Paixao, P. Costa, T. Bugalho, C. Fidalgo and L.M. Pereira, J. Chromatogr. B, Analyt. Technol. Biomed. Life Sci., 775 (2002) 109.
160   H.S. Lee, K. Kim, J.H. Kim, K.S. Do and S.K. Lee, J. Chromatogr. B, Biomed. Sci. Appl., 716 (1998) 371.
161   K. Arys, J. Van Bocxlaer, K. Clauwaert, W. Lambert, M. Piette, C. Van Peteghem and A. De Leenheer, J. Anal. Toxicol., 24 (2000) 116.
162   A. Sano, M. Takezawa and S. Takitani, Biomed. Chromatogr., 3 (1989) 209.
163   D. Felscher and M. Wulfmeyer, J. Anal. Toxicol., 22 (1998) 363.
164   Y. Hasuike, T. Nakanishi, R. Moriguchi, Y. Otaki, M. Nanami, Y. Hama, M. Naka, K. Miyagawa, M. Izumi and Y. Takamitsu, Nephrol. Dial. Transplant., 19 (2004) 1474.
165   F.H. Mateus, J.S. Lepera and V.L. Lanchote, J. Anal. Toxicol., 29 (2005) 105.
166   S.H. Chen, Z.Y. Yang, H.L. Wu, H.S. Kou and S.J. Lin, J. Anal. Toxicol., 20 (1996) 38.
167   W.E. Lambert, M. Piette, C. Van Peteghem and A.P. De Leenheer, J. Anal. Toxicol., 19 (1995) 261.
168   P. Marquet, S. Clement, H. Lotfi, M.F. Dreyfuss, J. Debord, D. Dumont and G. Lachatre, J. Anal. Toxicol., 20 (1996) 134.

M.J. Bogusz (Ed.). Forensic Science
Handbook of Analytical Separations, Vol. 6

CHAPTER 14

Forensic screening with liquid chromatography-mass spectrometry

Merja Gergov

Department of Forensic Medicine, PO Box 40, FI-00014, University of Helsinki, Helsinki, Finland

14.1 INTRODUCTION

Since the early 1990s, the coupling of liquid chromatography (LC) with mass spectrometry (MS) has become common in forensic toxicology. During the past ten years, the number of LC-MS applications in forensic toxicology has increased considerably, and several authors have reviewed the development and applications of this technique. In these reviews, the main features of LC-MS procedures reported for selected categories of drugs are summarised and critically evaluated [1–9]. Therefore, papers describing screening methods with limited coverage of drugs, are not cited here.

No single technique alone, such as gas chromatography-mass spectrometry (GC-MS) or LC-MS, covers all substances relevant in forensic studies, and therefore LC-MS, adequate as it is, cannot replace all other techniques, but will be used together with those recognised methods. The strength of using LC instead of gas chromatography (GC) in combination with MS is the feasibility of analysing thermolabile, polar and volatile compounds without time-consuming extraction and derivatization procedures. As in all kinds of screening procedures covering a wide range of chemically different compounds, the analysis design is a compromise of sample preparation and instrumental conditions for expected and unexpected relevant compounds. Comprehensive screening procedures cannot be optimised for every compound, and optimum conditions are sacrificed for coverage of as many kinds of compounds as possible.

The first applications of LC-MS have been conventional target analyses of one or two compounds and their metabolites, but the scope of applications has expanded to comprehensive target screenings of several compounds or groups of compounds simultaneously. Liquid chromatography-tandem mass spectrometry (LC-MS/MS) techniques also allow methods of combining screening with simultaneous quantitation or confirmation based on an automatic spectrum library search.

Comprehensive screening of toxicologically relevant compounds from a diverse matrix is a demanding task. LC-MS has proved amenable in forensic toxicology, in which a high reliability of results is categorical. LC-MS provides the specificity and sensitivity demanded for the screening of a wide range of compounds, even for substances that are not chromatographically separated. Three different approaches have been suggested: target screening, combined screening and confirmation by spectra, and screening based on accurate masses. In target-screening methods, only previously selected compounds are monitored using selected ion monitoring (SIM) or multiple reaction monitoring (MRM). In combined screening and confirmation methods, and in accurate mass-screening procedures, the survey scan is performed in full scan mode in order to detect any compound at relevant intensity. These types of screening methods are sometimes referred to as "general unknown" procedures. However, one of the major limiting factors remains the ion source, because none of the available sources is universal.

In forensic toxicology, however, extraction of drugs from biological material is a critical step related to the comprehensiveness of the screening. Other issues that may limit the coverage of a screening method include chromatography, ionisation, the pre-screening strategy, and the sensitivity and reliability of identification. No single technique or instrument is yet available that would cover all chemically different drugs with required selectivity, sensitivity, reliability and, preferably, possibility of automation.

14.2 SAMPLE PREPARATION FOR COMPREHENSIVE LC-MS SCREENING

A successful analytical method entails a seamless combination of extraction, separation, detection and reporting. The conventional approach has been to extract small groups of compounds separately and then to analyse the extracts by target methods dedicated to each compound group. This is a time-consuming and tedious procedure. Consequently, the trend has often leaned towards using universal multi-step extraction methods that would cover different categories of drugs. After combining these extracts which contain chemically different compounds, the next challenge is to separate and detect the drugs with a single method, or with as few methods as possible.

Drugs are present in blood plasma and inside the haemic cells and should be released before extraction. Hydrophilic drugs are usually free in solution, whereas lipophilic drugs are noncovalently bound to proteins or particles. Noncovalent bonds can be broken by dilution, pH change or organic solvents. Analysis of whole blood, plasma or serum is very important in forensic toxicology, because drug concentrations in blood represent an acute drug effect and can be used to estimate the level of intoxication. Conversely, many drugs and metabolites are present in urine as conjugates covalently bound to glucuronic acid, sulphate or glycine, and must be released prior to extraction by, for example, enzymatic hydrolysis.

Drummer [10] and Polettini [11] reviewed the literature available since 1990 concerning the extraction techniques suitable for systematic toxicological analysis. The screening of "general unknowns" requires appropriate extraction. Three different procedures are commonly used to isolate drugs from biological material: separate liquid–liquid extractions (LLEs) for basic and acidic compounds, solid-phase extraction (SPE) with mixed mode columns, or protein precipitation/dilution.

The most simple and rapid pre-treatment method for whole blood is precipitation with, for instance, acetonitrile or methanol. This is suitable for very polar compounds like metformin and atenolol, but more lipophilic analytes may stick to the surface of particles during precipitation, which lowers the recovery. Urine samples can be injected into the LC column directly after dilution, but urine is usually first hydrolysed to cleave conjugates, and then centrifuged before injection. Precipitation of blood and dilution of urine are rapid and simple methods, and reduce sample preparation time and the use of consumables, but on the other hand, large amounts of matrix material are co-injected and may interfere with the LC-MS analysis.

SPE is a commonly used extraction method that enables sample preparation in batches and automation of the extraction procedure, and generally provides good selectivity and clean extracts [12]. Mixed-mode columns (reversed phase (RP) + ion exchange) are especially feasible for extraction of a broad range of chemically different compounds. SPE is very suitable for urine samples, which are homogeneous, and thereby do not block the extraction columns. It can also be used for autopsy blood [13,14], but the material may be lumpy and sometimes even decayed, causing blockage of the SPE column, which can be decreased by sonication before SPE [15–17]. Plasma is a less complicated matrix, but in post-mortem samples it cannot be separated from whole blood due to haemolysis. Therefore, LLE is more generally suitable for autopsy blood samples [18]. Examples of choice of extraction method for comprehensive screening procedures appear in Table 14.1.

In screening methods, a wide variety of compounds with different polarities is compared with the analysed sample, and reasonable coverage with LLE can only be obtained through the separate extraction of acidic and basic compounds. When the sample amount is very limited, serial extraction steps are used, but multiple manipulations tend to reduce the recovery at every stage, and can cause unacceptable sensitivity for some compounds. To avoid this and to shorten the total analysis time, the acidic and basic extracts can be combined and subjected simultaneously to MS detection.

14.3 LIQUID CHROMATOGRAPHY PRIOR TO MASS SPECTROMETRY

Due to its universality, LC is a very suitable separation technique for multicomponent analysis, because the compounds need only to be dissolvable in a suitable solvent. On the other hand, the separation efficiency is not as good as in GC, although this disadvantage is not critical because MS provides good specificity and selectivity. The only special demands are the use of volatile buffers (e.g. ammonium

TABLE 14.1
COMPREHENSIVE SCREENING METHODS USING LC-MS AND LC-MS/MS ACCORDING TO INSTRUMENT TYPE. LATEST VERSIONS OF THE ESTABLISHED METHODS ARE LISTED

Compounds	Matrix	Extract. Method	Instrument type	Screening→ Confirmation	[Ref.] Year
39 sedative-hypnotics	Serum	SPE	Single quadrupole	SIM→database of target ions for 39 drugs	[30] 2006
>600 compounds	Blood	SPE	Single quadrupole	Single MS full scan pos/neg→in-source CID spectrum library (>600)	[26] 2003
70 psychoactive drugs	Serum	SPE	Single quadrupole	Single MS full scan→ in-source CID, two fragments	[28] 2001
40 anabolic steroids and 52 acidic drugs	Urine	SPE	Triple quadrupole	MRM→database of 90 MRMs	[33] 2006
238 drugs	Blood	LLE	Triple quadrupole	MRM→database of 238 MRMs	[31] 2003
70 drugs and metabolites	Body fluids	SPE	Triple quadrupole	MRM→database of >70 MRMs	[32] 2003
>400 compounds	Urine	SPE	Triple quadrupole	single MS full scan→ MS/MS spectrum library (>400)	[27] 1999
72 central nervous system stimulants	Plasma	Precipitation	TRAP	Single MS full scan→ in-source CID, two fragments	[40] 2004

Analytes	Matrix	Sample preparation	Instrument	Method	Reference/Year
>250 basic drugs	Urine	Direct injection	QTRAP	MRM→MS/MS spectrum library	[41] 2006
100 compounds	Blood	Precipitation	QTRAP	MRM→database of 100 MRMs and MS/MS library	[42] 2005
301 compounds	Blood, urine	LLE and SPE	QTRAP	MRM→MS/MS spectrum library	[38] 2005
"General unknown"	Serum	SPE	QTRAP	Single MS full scan→MS/MS spectrum library	[39] 2003
637 compounds	Urine	SPE	TOF	Single MS full scan→database of accurate masses (637)	[44] 2003
433 compounds	Urine	SPE	TOF	Single MS full scan→database of accurate masses (433)	[43] 2001
Unknown designer steroids	Urine	SPE	QTOF	Single MS full scan→database of *Merck Index*	[47] 2006
>300 compounds	Blood	SPE	QTOF	Single MS full scan→MS/MS spectrum library (300)	[45] 2004

acetate) and low flow rates ($< 300\,\mu l/min$) with an ion spray source, which allows the liquid to be evaporated in the ion source.

The LC-system, prior to MS, usually consists of at least a binary pump, with one channel used for buffer and one for organic solvent, a column oven and a vacuum degasser to keep the separation conditions stable. Also, the separation column should be universally applicable and inert in screening procedures, and typically RP C-18 columns with a short guard column of the same phase are commonly used.

In contrast to the early suggestions, the chromatographic separation step is important even when a mass spectrometer is used as a detector. Chromatographic separation is usually performed using a gradient run, and the gradient is adjusted to resolve most of the expected compounds from each other. However, separation of all components is impossible in screening methods in which hundreds of different compounds are analysed, and in which the main objectives are to separate analytes from the LC-front peak, from each other, and from the peaks originating from the biological material. Another aim is to obtain reasonable retention and as good a peak shape as possible for all compounds. Therefore, selection of the LC conditions for a screening method is always a compromise between a reasonable total analysis time, peak shape and separation.

More important than the actual retention times is the reproducibility of absolute and relative retention times, because in most of the screening applications, retention times serve as one criterion for having a "hit" in the preliminary survey scan. In a typical LC system, the retention time window is 0.3 min or less. Also, in the confirmatory analysis, retention times serve as one identification criterion. Detailed optimisation of LC conditions and eluent composition is unnecessary for the development of spectral libraries, since several studies showed that LC conditions do not influence the appearance of MS/MS spectra [19–22].

Keeping these practical aspects in mind, very simple and general LC procedures can be used in the LC-MS screening methods. The mobile phase typically consists of acetonitrile as the organic solvent and an aqueous solution of ammonium acetate or ammonium formate (e.g. 2–10 mmol, 0.1% formic acid) as a buffer, and the total eluent flow varying from 200 to $400\,\mu l/min$ for applications with an ESI ion source and up to 2 ml/min with an APCI ion source. Typically, a slow linear gradient (e.g. acetonitrile from 15% to 100% in 10 to 30 min) and RP column (10–15 cm) have been used to obtain an acceptable separation of various types of compounds in a reasonable total analysis time.

14.4 IONISATION TECHNIQUES

ESI (electrospray ionisation) and APCI (atmospheric pressure chemical ionisation) have been predicted to be of major future interest [9,23] and, indeed, the majority of recent papers concerning established applications of LC-MS in forensic toxicology use ESI or APCI. Recently, instruments with both ion sources mounted in the same source housing have come to the market, which permits periodical switching between ionisation modes during LC elution. Selecting the more suitable ion source

individually for each compound enhances sensitivity for many compounds, but this option can only be used in target screenings with previously selected compounds for which ionisation preferences have been studied.

The ESI ion source consists of a capillary needle, through which the sample solution from LC is introduced and to which a voltage is applied. The needle is inside another capillary, which is used to introduce the nebulising gas. The eluent flow rate is usually 0.2–0.4 ml/min in a normal column (i.d. 1–5 mm). A drying gas (e.g. nitrogen) is applied to assist droplet formation and evaporation, and splitting of the eluent flow is unnecessary. In an ESI ion source, molecules are ionised in the liquid phase and evaporated to the gas phase under atmospheric pressure, then introduced into a mass spectrometer.

In APCI, eluent from LC flows through a heated probe (100–500°C) in which the solvent and sample are volatilised. Solvent molecules are ionised in the gas phase by a "corona discharge electrode" (metallic needle), and after several ion–molecule reactions, ionised solvent molecules transfer a charge to the analytes. The high APCI operating temperatures permit the use of high flow rates (up to 2 ml/min).

ESI and APCI are both soft ionisation techniques, and therefore molecules are only slightly, or not at all, fragmented. Protonated $[M + H]^+$ or deprotonated $[M–H]^-$ molecules are mainly observed, depending on the polarity of the electric field applied. Adduct ions, formed within the ion source by interaction of the molecule with other atoms or molecules, can also occur intensive and characteristic enough to be used for identification. Typical adduct ions are ammonium $[M + NH_4]^+$, sodium $[M + Na]^+$ and potassium $[M + K]^+$ adducts in positive mode, and acetate $[M + HCOO]^-$ and chlorine $[M + Cl]^-$ adducts in negative mode. Generally, the positive mode is used for basic compounds and the negative mode for acidic compounds. Structural information can also be obtained even with single-mass spectrometers, for example, by using the so-called in-source collision-induced dissociation (in-source CID) techniques in which molecules are fragmented inside the ion source by raising the orifice or fragmentor voltage.

ESI and APCI have both been used in forensic applications. ESI is suitable for molecules of a large range of polarity, from moderately non-polar to highly polar molecules, and also for thermally labile and high molecular weight compounds. Therefore, it is more commonly used in screening methods that involve a wide range of chemically different compounds. The polarity range of APCI is more limited, but has the benefit of suitability for non-polar compounds as well. Another advantage of APCI over ESI is that it is less susceptible to matrix effects in samples containing a high concentration of salts, or an excess of other compounds that can ionise in the operating conditions ("ion suppression"). The possibility of using non-polar solvents with APCI is convenient, because after screening with LC-MS, the same extract can be used in confirmation analysis with another technique, such as GC-MS. The combination of these two different techniques would fulfil the prerequisite of using two independent techniques in forensic analysis. Examples of the choices between ESI and APCI in forensic applications appear in a review by Maurer [4].

14.5 COMPREHENSIVE SCREENING PROCEDURES

Several strategies for the comprehensive screening of relevant drugs in forensic samples with LC-MS have been proposed in the literature. Screening procedures have been based on different mass analysers: single quadrupole, triple quadrupole, time-of-flight (TOF), a hybrid of quadrupole and time-of-flight (QTOF), ion trap (TRAP), and a hybrid of quadrupole and ion trap (QTRAP). Each of the techniques has advantages and limitations. Quadrupoles have a long history of being used in various applications, and have shown high reliability in quantitation. QTOF and QTRAP have also become of great interest and many of the most recent screening methods in forensic toxicology have been based on these techniques. Recent prices of these instruments made them more affordable for routine laboratories.

14.5.1 Screening with LC-quadrupole mass spectrometers

Quadrupole mass spectrometers consist of one or three quadrupoles (a set of four coaxial rods) in sequence. In a triple quadrupole instrument, the centre quadrupole is used for fragmentation and is referred to as the collision cell. A selected mass range can be scanned, or only selected ions can be monitored with the first and third of the identical mass filter quadrupoles. The practical resolution of a quadrupole instrument is one mass unit.

The first comprehensive screening procedure using a single quadrupole mass spectrometer was presented by Marquet *et al.* 1998 [24], and was later developed further by the same authors [6,25,26]. The method included LC-RP chromatographic separation and detection by a single-mass spectrometer with an ionspray interface and using the in-source CID fragmentation mode. Four separate spectra were generated simultaneously: two positive and two negative spectra at low (weak fragmentation) and at high (extensive fragmentation) orifice voltage. After acquisition, the low- and high-energy spectra were summed at both polarities, and these two reconstructed spectra were searched against positive and negative mass spectral libraries. Other authors also presented a similar concept [27–29]. In one of the applications, even though a triple quadrupole mass spectrometer was used, the survey was based on a single-MS full scan with orifice voltage switching between low and high in each scan, but the final confirmation step was based on MS/MS product ion spectra from the molecular ions found. The MS/MS spectra were acquired at three collision energies and searched against the laboratory's own spectrum library [27].

The benefit of using single-MS for the survey scan and in-source CID for obtaining spectra is that while acquisition is performed in full scan mode, it is not restricted to previously selected compounds. The disadvantage, however, is that co-eluting compounds result in interfering spectra, making them unidentifiable from the libraries of pure compounds. It has also been impossible to reproduce these spectra without extensive tuning of the instrument, and researchers have reported differences in fragmentation degree between separate instruments [19]. A critical step in using the single-MS full scan mode for screening is the detection of small peaks in the background; however, a data-handling procedure was developed to overcome this problem [26]. Even though multi-step methods exist for single-MS instruments, a traditional and simple procedure is to use SIM to screen selected compounds, as was reported recently [30].

The most common way to use triple quadrupole instruments is to use SIM or MRM for target screening and quantitation of a limited number of drugs. SIM mode is the only option with single quadrupole instruments, but is not very selective, since all compounds with the same parent ion (m/z value) will pass the survey scan, which may result in several false positives. Therefore, positive findings must be confirmed in a second step, based on product ion spectra searched against a spectral library or by at least two fragment ions. MRM mode in triple quadrupoles is more selective, and in many applications the positive findings are verified by monitoring a second MRM transition during the same run. Recently, several authors employed MRM to screen a wide range of drugs (Table 14.1). The most comprehensive multicomponent screening method using MRM includes 238 compounds [31]. All transitions were monitored during a single chromatographic time period, but in three consecutive experiments with different collision energies (20, 35 and 50 eV). Positive identification was based on three criteria: the correct precursor ion, fragment ion and retention time. In another analogous method, the sensitivity of the screening was enhanced by splitting the chromatographic run into time windows [32]. In this way, fewer transitions are monitored at the same time, and therefore higher dwell times can be used to improve sensitivity, or alternatively, to increase the number of compounds to be searched. Another way to improve sensitivity is to optimise individually the collision energy for every compound [32,33].

There are some benefits of using MRM with several fragments instead of a single-MS full scan and in-source CID spectra in screening methods: higher sensitivity, better specificity because the fragments originate only from the selected precursor ion, and the option of including quantification in the automatic procedure. On the other hand, one weak point is that no matter how many compounds the MRM procedure covers, it never constitutes "general unknown" screening because preselection of the precursor and fragment ions limits the coverage of the survey scan while the monitored ions are selected in advance.

Even though MRM is known to be sensitive and reliable for a large number of compounds, some important factors demand special attention: cross-talk, which may occur if two compounds co-elute and have the same fragment ions; the instrument scanning speed, which limits the number of compounds that can be detected during a single chromatographic run; and the dwell time, which affects sensitivity and also limits the number of compounds. Cross-talk is a known feature of older triple quadrupole models, but can be avoided by careful planning of the order of MRM transitions [31,34]. Scanning speed and the applied dwell time together contribute to the sensitivity. With modern instruments, dwell times as low as 5 ms can be used. Increasing the dwell time promotes sensitivity but leads to fewer data points across the chromatographic peak and thereby too long dwell times make detection insecure. Also the instrument scanning speed, especially in the older instruments, can be decisive and determine the number of transitions that can be monitored during a single run.

The fourth option with triple quadrupoles is to use either of the two special scan modes: the precursor ion scan or the neutral loss scan. The precursor ion scan searches for homologous compounds (precursor ions), which have common product ions. The instrument looks for a predefined m/z product ion and associates it back to the precursor ion it originated from. This mode is useful for compounds originating from certain drug groups, such as steroids [35]. The neutral loss mode searches masses with a predefined mass difference (loss of a neutral molecule), reducing the need for knowledge of the parent molecule's structure.

14.5.2 Screening with LC-ion trap mass spectrometers

During the last few years, the suitability of LC-MS ion trap instruments for forensic applications has undergone active investigation. The capabilities of MS^2 and MS^3 together with comparable prices for triple quadrupoles, have likely fostered a growing interest in these techniques. Ion trap instruments have two configurations: the conventional three dimensional (3D) form and the linear (2D) form constituting four parallel electrodes (a quadrupole). A conventional 3D ion-trap mass spectrometer (single-MS TRAP) uses three electrodes to trap ions in a small volume. A mass spectrum is obtained by changing the electrode voltages to eject the ions from the trap. In a 2D linear ion trap, however, ions are confined in the axial dimension by means of an electric field at the ends of the quadrupole. A quadrupole ion trap instrument (QTRAP) is a hybrid quadrupole mass spectrometer which combines the scan speed (up to 4000 amu/s) and the sensitivity of the ion trap MS while retaining the selective scanning modes of the triple quadrupole. Consequently, it provides the high selectivity and sensitivity required in screening and quantitation based on MRM. LC coupled with ion trap MS was first used in forensic toxicology to identify a wide range of basic drugs from urine samples by Fitzgerald *et al.* in 1999 [36], and since then, several authors have reported screening methods for small drug groups using ion trap technology. Only recently have applications been reported for a larger selection of toxicologically relevant drugs (Table 14.1).

The single-MS TRAP in the full scan product ion mode enables screening and quantitation with MRM, and identification by product ion spectra simultaneously in one run without sacrificing sensitivity [37]. This is possible because ion traps measure all the ions retained during the trapping. This differs from quadrupoles, which show considerable differences in sensitivities between MRM and full scan modes. However, the MRM trace of a single-MS TRAP is not "real", as in the triple quadrupole, but "extracted" by the post-acquisitional selection of the fragment ion from the full scan spectra, and consequently, signals tend to be more noisy. Also, with these "extracted" MRM transitions, the total acquisition time (duty cycle) is slower than in quadrupoles, which is an important disadvantage in the context of large screenings, which require duty cycle short enough to ensure proper sensitivity [38].

Therefore, in comprehensive target screenings based on MRM, quadrupole instruments would be more feasible.

On the other hand, the enhanced mass accuracy and better sensitivity in scan mode make ion trap instruments an interesting alternative to the conventional triple quadrupoles in forensic screening. The mass accuracy of a QTRAP instrument ($\sim$0.01 Da for pure compounds at m/z $\sim$500 Da) is much better than that of conventional triple quadrupoles ($\sim$0.7 Da), but not as good as that of a single-MS TOF and QTOF ($\sim$0.0025 Da). Other advantages of the ion trap mass spectrometers include the ability to trap and accumulate ions to increase the signal-to-noise ratio, and the capability of producing MS^3 fragment ions. Both features are beneficial in the identification of drugs in forensic samples, for which the biological matrix sets high selectivity and sensitivity requirements.

Marquet *et al.* presented the first forensic application of comprehensive target screening using QTRAP [39]. The first step is the "survey scan", run in single-MS full scan mode. The ions are accumulated and filtered, and all ions exceeding the previously set criteria for peak area are detected. Once an ion is detected, the instrument instantly switches to MS/MS mode and acquires product ion spectra both in positive and negative mode and with high and low collision energy. The high- and low-collision energy product ion spectra are combined and the summary spectra are searched against positive and negative spectrum libraries. All this is done automatically using information dependent acquisition (IDA). A difficult issue here, as in all methods based on a single-MS full scan, is the detection of small signals from the background and the difficulty of setting the threshold for peak area, because background varies between samples. In another application, the same survey strategy was used, but because the instrument was a single-MS TRAP, the confirmation was based on fragments obtained from in-source CID fragments in a second injection [40]. Other recent comprehensive forensic screening applications with ion trap instruments use MRM for the survey scan [38,41,42], followed by IDA-triggered MS/MS product ion spectra [38,41] or by comparison of the MRM transitions to an MRM database [42].

14.5.3 Screening based on accurate mass measurement

TOF-MS is the third common alternative for the multicomponent drug screening of biological samples. Accurate mass measurements are used to determine the elemental composition of all ions of interest in the sample. The calculation of possible molecular structures facilitates the screening of genuine unknowns by comparing the calculated structures to those in electronic databases, which essentially are lists of accurate masses and names of the corresponding chemical compounds. TOF instruments provide relatively high mass accuracy ($\sim$5 ppm) and reasonable resolution (5000–10000 fwhm) and have been used in combinatorial chemistry and in the analysis of drugs and their metabolites.

In single-MS TOF instruments (single-MS TOF), quadrupoles focus the ion beam on the TOF analyser, which is positioned perpendicularly, and ions are pulsed into the flight tube by applying voltage. The ions separate according to their m/z values, as high-mass ions have longer flight times. A quadrupole instrument (QTOF) is a hybrid mass spectrometer with MS/MS capability. The quadrupole operates as an ion guide in MS mode and is used for selecting parent ions in MS/MS mode. A collision cell is located between the quadrupole and the TOF analyser to induce fragmentation in MS/MS experiments.

Comprehensive target screening methods have been reported both for single-MS TOF [43,44] and QTOF instruments [45–47]. A typical concept is to screen over the desired mass range (e.g. m/z 100–700) with a single-MS in full scan mode. Based on these acquired accurate masses, the elemental composition of detected peaks can be calculated and searched against an electronic target database, such as the *Merck Index* [47] or the laboratory's own list of relevant compounds and their accurate masses [43,44]. Thus, the compounds can be tentatively identified without reference compounds. Using single-MS TOF instruments, a mass accuracy of 5–10 ppm has been reported for the majority of drugs in authentic samples [44]. If a metabolic pattern for detected drugs is used as an additional criterion, a mass window of 20 ppm in real urine samples was considered sufficient to detect the correct findings, and, with the use of dedicated software, to rule out apparent false–positive findings [43,44]. In a further development of the method, mass spectral identification was based not only on matching measured accurate masses, but also on the isotopic pattern (SigmaFit) match between the sample component and the theoretical values in the database. A new generation TOF instrument allowed the use of a mass window as narrow as ± 10 ppm in screening [48]. By using accurate masses, even single-MS TOF instruments enable a large number of toxicants to be screened during a single injection without sacrificing sensitivity or selectivity. One limitation, set by the user, is the selection of compounds in the database. Some complexity in interpreting the results stems from the fact that two or even more drugs are often obtained for the same molecular formula. The number of candidates differing by formula can be further decreased by using an instrument with higher mass accuracy (e.g. FTMS). In any case, retention time is a necessary criterion for identification.

An alternative concept for identification is to use QTOF instruments with tandem mass capability. After the first screening step, the confirmation of tentatively found compounds is based on product ion spectra, obtained using IDA-triggering, in which all masses exceeding the preset threshold of intensity are automatically se-lected for precursor ions [46]. The product ion spectra should be acquired with more than one collision energy, if genuine unknowns are screened, because the most favourable fragmentation conditions are unknown in advance. The product ion spectra obtained are usually searched against an in-house spectral library, but this again restricts identification to those compounds included in the library. To include automatic quantitation in these procedures, based on the accurate mass of the pre-cursor, is also possible [46].

An obvious benefit of TOF mass spectrometers is that they acquire the entire mass spectrum simultaneously instead of scanning at preset steps, as with quadrupoles and ion traps. Any compound of interest can be extracted after acquisition from the original run without the need for re-analysing the sample. This is beneficial when the sample amount is limited or when insufficient sample is left for further studies. Acquiring the entire mass spectrum is an important feature when very large numbers of compounds are screened. Even though TOF mass spectrometers provide accurate molecular weight, screening methods require good chromatographic separation. Co-eluting peaks, as well as unintentional in-source CID fragmentation, lead to a mixture of mass peaks of unknown origin (drugs, endogenous compounds or their fragments).

An even higher mass accuracy can be obtained with Fourier-transform mass spectrometry (LC-FTMS). It takes advantage of ion-cyclotron resonance to select and detect ions. Ions in a magnetic field will move in a circular path and the signal produced in the process is Fourier-transformed to produce the emitted frequencies, and thereby the masses of the ions present. FTMS can provide very high resolution (10^6), which is its main advantage over other mass analysers. However, apparently due to the high price of the instrumentation and maintenance, this technique is seldom used in forensic studies. In one recent method for the confirmation of drugs in urine samples the mass accuracy obtained in urine samples was reported to be as good as 3 ppm [49]. With such ultimate accuracy, the number of possible candidate drugs with a corresponding structure was only one, despite the use of a very large database of 7640 compounds.

The possibility of detecting toxicants without reference compounds is a major advantage of high resolution mass spectrometers. In forensic toxicology, maintaining a large selection of reference compounds is very expensive and practically impossible, because most of the metabolites are not commercially available. Therefore, LC-QTOF and LC-FTMS appear to be very suitable for this type of application, and when used for comprehensive target analysis, these techniques can be considered the closest approach to what is meant by "general unknown screening".

14.6 IDENTIFICATION USING LC-MS AND LC-MS/MS SPECTRAL LIBRARIES

In forensic studies, positive findings should be confirmed using another independent technique. For example, the results of immunological screenings for illegal drugs are traditionally verified with GC-MS. Alternatively, another sample material should be studied when possible. In practice, this typically means a qualitative screening for relevant compounds and their metabolites from urine followed by a quantitative measurement from the blood. When LC-MS/MS is used for these tasks, according to the generally accepted rules and guidelines for reliable identification [50,51], the number of MRM transitions should be at least two, thus providing a total of three diagnostic ions. With the additional criterion for retention time, these four identification points provide a high confidence level in both screening and quantitation. Alternatively, identification can be based on the comparison of MS/MS product ion spectra with spectra in a reference spectrum library, or it can be based on accurate mass measurement. One should keep in mind, however, that using the same type of technique (GC-MS or LC-MS) for both screening and confirmation excludes all those drugs that are not amenable to the selected technique. If a second technique is not used, analysing another sample material with the same technique, does not verify the absence of relevant compounds.

Comparison of an unknown spectrum to spectra in large spectrum libraries has been used for years with GC-MS to identify unknown compounds. This has been possible because electron impact (EI) ionisation spectra are very reproducible, and therefore huge commercial spectral libraries exist that include as many as several hundred thousand compounds. For example, the *NIST 05* (National Institute of Standards and Technology, http://sisweb.com) contains EI spectra for approximately 163,000 different chemical compounds. The situation is different with LC-MS, because both ESI and APCI are soft ionisation techniques, and therefore molecules are only slightly, or not at all, fragmented. Protonated $[M + H]^+$ or deprotonated $[M–H]^-$ molecules are mainly observed, depending on the polarity of the electric field applied. Product ion spectra, obtained with instruments having MS/MS capability, contain as much information as GC-MS/EI spectra, and can serve in identification. The main challenge is that common and universal LC-MS/MS spectral libraries still do not exist. The obvious reason for this is that there is still no agreement on the fragmentation conditions to be employed, and on the way in which these conditions should be standardised with different instruments. Another important issue is the reproducibility of spectra within and between instruments.

Several laboratories have developed LC-MS/in-source CID and LC-MS/MS spectral libraries with quadrupole, ion trap and TOF instruments for in-house use, but only a few of these have been published in scientific papers or on the Internet. MS/MS spectral libraries with QTOF are seldom reported because in most of the screening methods, accurate masses are preferred for identification of compounds. A summary of the largest published spectral libraries appear in Table 14.2.

14.6.1 In-source CID spectral libraries

Several in-source CID libraries have been created based on summarising spectra obtained with and without fragmentation (Table 14.2). The number of compounds mentioned in Table 14.2 are those originally reported, and their number has probably increased because these libraries are updated continuously. Because in-source CID spectra are created inside the ion source, several parameters affect fragmentation. Declustering potential (the difference between voltages at the orifice plate and skimmer), which mainly affects in-source CID spectra, can be very effectively used to obtain fragmentation. Various combinations of low and high voltages (e.g. $+25/+90$ V [52], $+20/50/+80$ V [27], $\pm20/\pm80$ V [25], $\pm10/\pm50/\pm100$ V [20] or even linear ramping of the voltage) have been proposed [21], but in general, only two different levels are needed for relevant information. Combined low and high voltage spectra contain both fragment ions and an intensive protonated molecule, which is usually weak or entirely absent from the LC-MS/MS product ion spectra at medium or high collision energy levels. The presence of the protonated molecule in the spectrum is an undeniable advantage of in-source CID spectra over LC-MS/MS

TABLE 14.2

COMPREHENSIVE LC-MS SPECTRAL LIBRARIES CONTAINING TOXICOLOGICALLY RELEVANT COMPOUNDS. THE APPROXIMATE NUMBER OF COMPOUNDS MAY HAVE BEEN UPDATED SINCE ESTABLISHMENT OF THE LIBRARY

Instrument Type	Fragmentation Type	Compound Type	Approximately Number of Compounds	[Ref.] year
Single quadrupole	In-source CID	Drugs	~400	[52] 2000
Single quadrupole	In-source CID	Pesticides, explosives	~800	[20] 2000
Single quadrupole	In-source CID	Drugs	~400	[27] 1999
Single quadrupole	In-source CID	Drugs, toxicants	~600	[24] 1998
Triple quadrupole	MS/MS	Drugs	~400	[52] 2000
Triple quadrupole	MS/MS	Drugs	~400	[55] 2000
Ion trap	MS/MS	Miscellaneous	~1000	[65] 2001
Ion trap	MS/MS	Miscellaneous	~600	[64] 2000
Ion trap	MS/MS	Drugs	Not available	[36] 1999
QTRAP	MS/MS	Drugs	~2000	[37] 2005
QTRAP	MS/MS	Drugs	~300	[38] 2005
QTOF	MS/MS	Drugs, toxicants	~300	[45] 2004

spectra. On the other hand, MS/MS spectra are more specific because the only origin of the fragments is the precursor ion chosen. In practical work with real samples, a problem arises from co-eluting compounds, which leads to a summary of in-source spectra of all these compounds. Such a mixture of unknown spectra cannot be resolved from any library.

Several groups [20–22,25,27,53] have investigated the effect of different variables on in-source CID spectra, such as mobile phase composition, pH, flow rate, analyte concentration, heater gas temperature, ion source dirtiness and needle position. In addition to the most important variable (declustering potential), the needle position and heater gas temperature also significantly affect the appearance of the spectrum, while the other variables have a greater effect on the intensity of mass signals. The settings of these parameters are not always reproducible, and comparison of in-source CID spectra between instruments demands effective tuning of the instruments. Glafenine and haloperidol were suggested as tuning compounds [19,20,25,27,54]. The reproducibility of spectra has also been questionable, and indeed poor reproducibility was reported [19,50], even within a single instrument [53].

14.6.2 MS/MS product ion spectral libraries

The appearance of an MS/MS spectrum is independent of the design of the ion source because fragmentation occurs after it, inside the collision cell or ion trap. The efficiency of ionisation and the transfer of ions into the gas phase, however, influence the intensity of the entire spectrum. Once a precursor ion is selected in the first quadrupole and introduced into the collision cell, applied voltages and collision gas pressure affect the fragmentation. Both these parameters can be precisely and reproducibly adjusted. Usually, two to four collision energies have been applied (e.g. 20/30/40/50 eV [55] and 20/35/50 eV [56]), yet higher or lower collision energies add no new information to the product ion spectra. Detailed optimisation of LC conditions and eluent composition is unnecessary in developing spectral libraries, as several studies have shown that LC composition does not affect the appearance of MS/MS spectra [19–22].

The earliest standard operation protocols for acquiring MS/MS spectra were established twenty years ago [57], and later it was shown that with fixed settings, it is possible to create instrument-independent spectral libraries [22,59]. Recent studies have shown this to be possible with modern LC-MS/MS triple quadrupole instruments [55,56]. The question of using fragment peak intensity ratios as an identification criterion has been under discussion among research groups developing spectral libraries. Because ion ratios vary depending on the ionisation and fragmentation techniques used, a different approach was developed to ignore their effect on library search results: a library was created by putting the m/z values of the fragments in the library at 100% or 50% intensity only, chosen based on the peak area counts. This simplified procedure allowed the comparison of MS/MS spectra obtained with different triple quadrupole instruments to those obtained with an ion trap mass spectrometer [58,59].

In contrast to previous opinions [19,25], judgment on similarity should include criteria for peak intensity ratios, at least when conclusions are drawn on the performance of instruments. Mass spectra are generally used for identification of unknown compounds, and the criteria for positive identification have been discussed in many contexts. However, no common agreement yet exists on the acceptance criteria for a reliable match, although approximately 70–80% would presumably represent a fair or good match [21,22]. According to studies on the subject, that would result in a satisfactory confidence level, presuming that the peak intensity would allow detection of the entire spectrum.

The long-term reproducibility of spectra is a critical issue because the identification of unknowns in forensic cases demands continuity and high confidence level. With the exception of one study [53], good within-instrument reproducibility of quadrupole mass spectrometers over three to eight months was reported for LC-MS/ in-source CID spectra [21,22], and over thirty months for LC-MS/MS spectra [55,56]. The MS/MS spectra are assumed to be more reproducible, because only selected ions are fragmented in controlled conditions in the collision cell; nevertheless, one recent study has reported poor reproducibility of spectra with a QTRAP instrument [60]. Taking into account the huge amount of labour required to create a large spectral library, reproducibility should be ensured for years, as has been established for GC-MS spectral libraries.

Comparisons of spectra between different quadrupole mass spectrometers have been reported both with in-source CID spectra and with MS/MS product ion spectra. In-source CID spectra were shown to contain the same fragments, but significant differences in their relative ion ratios were noted even after tuning of the instruments [19,21,22,53,54,61]. MS/MS spectra created with triple quadrupoles from the same and from different manufacturers, appeared compatible after harmonising the fragmenting conditions [56]. Ion trap libraries have also been created (Table 14.2), and a recent study has shown that a library, set up with a triple quadrupole instrument, can be transferred to a linear QTRAP mass spectrometer [62]. Also, in another study, spectra obtained with ion trap were successfully searched against a triple quadrupole library [63]. Different results have also been reported recently: spectra from linear QTRAP instrument were compared with spectra from the same and from two other triple quadrupole instruments [60], and the authors concluded that inter-instrument differences in relative intensity of fragment ions were too substantial to be used as criteria for identification. However, in this study the initial standardising of the fragmentations conditions with a tuning compound was reported unsatisfactory.

Comparison between spectra is difficult when instruments use incompatible software programs. Also, algorithms for data processing prior to the library search are different and alter the data in different ways, leading to unacceptable library search results. For instance, the software may use centroiding parameters that delete important small peaks, such as those for the protonated molecules.

As can be concluded from the variety of libraries, co-operation in combining existing libraries or creating new and comprehensive ones will be difficult, but would benefit all laboratories.

14.7 DISCUSSION AND CONCLUSIONS

In routine forensic analysis, all positive findings from screening procedures must be confirmed. This has been based on product ion spectra, which presume the existence of large spectral libraries. However, commercial libraries are currently unavailable for LC-MS/MS spectra, mainly due to disagreement over the conditions in which to obtain reference LC-MS spectra, and because of doubts about the reproducibility of spectra. Therefore, the use of accurate masses for identification seems the most promising alternative for the future.

An important issue with all LC-MS methods is the phenomenon called "ion suppression", which may occur but pass unseen in the chromatograms, and therefore can very insidiously ruin the reliability of an LC-MS method. Ion suppression takes place in the ion source and is caused by less volatile compounds in the sample matrix that prevent analytes from reaching the surface of the droplets, and thus from evaporating and reaching the mass analyser. Also, part of the analytes may precipitate during solvent evaporation or remain unevaporated, and thereby never reach the mass spectrometer. In addition, if the protons in the eluent are more attracted by the matrix, then some of the analytes will remain unionised and will be drawn out of the interface by the vacuum [66]. In such situations, suppression of the analytes may lead to such a low peak in intensity that the compounds will pass completely undetected during the survey scan, or the measured concentration will be too low.

Target ion monitoring by SIM or MRM very effectively "cleans" the chromatogram from co-eluting, interfering substances. However, the compounds originating from the matrix do not exit the system, but become "invisible". Not only endogenous compounds from the matrix, but all other co-eluting compounds, including other drugs, metabolites, salts and the isotope-labelled internal standards, may cause ion suppression [67–69]. Ion suppression is often connected only to quantitation, but it affects the screening procedures in the same way: suppressed compounds will pass undetected during the survey scan and, consequently, are excluded from the confirmation steps. Both ESI and APCI ion sources experience ion suppression, but research has shown ESI to be more susceptible than APCI [70]. Because errors caused by ion suppression are sample-dependent and incidental, thorough validation in applications for biological matrices should contain spike tests for a wide variety of authentic samples, including high and low spike levels, to estimate the extent of suppression [71]. In several applications for combinatorial chemistry, fast chromatography ("high throughput") has been suggested, but this includes the obvious risk of ion suppression, which has been demonstrated to appear mostly during the LC-front peak where the analytes are not separated from the polar and unretained components of the matrix [71,72]. Therefore, fast chromatography cannot be recommended in the forensic context. Instead, using an efficient LC gradient reduces suppression by separating the analytes from the LC-front peak and from other matrix-related peaks, which may cause suppression also later during the run [72].

Automation of the screening and confirmation procedures, from starting a sample batch to printing an explicit summary report of the results, is necessary for a routine analysis when large numbers of samples are to be analysed daily. Automation also decreases the possibility of human errors. The use of IDA-triggering has proven to be a very practical tool for automatic acquisition, but special programs are often needed for automatic processing of the obtained data, because the original software may not be sufficiently flexible.

A wide diversity of screening methods for single drug classes has been developed for routine use in forensic studies. All of the reported techniques (quadrupole, TOF or ion trapping) have proven to be suitable for this purpose. Because TOF and ion trap mass spectrometers offer higher resolution, their selectivity is better than that of traditional quadrupole mass spectrometers. From the practical point of view, however, maintaining mass accuracy at a high level demands continuous calibration and stable external temperature conditions. Also, the complexity of some of the reported methods with hybrid instruments may prevent them from becoming routine applications in forensic laboratories.

The screening of real unknowns by accurate mass measurement can be considered the closest approach to tentative structure evaluation because neither the availability of reference compounds nor spectral libraries restrict the number of compounds to be screened. On the other hand, the selection of relevant xenobiotic compounds routinely found in post-mortem forensic toxicological cases may not be very wide, consisting of only a few hundred compounds, which in turn favours target screening. Therefore, the screening method in forensic laboratories can be chosen according to the instrumentation available and the preferred screening strategy.

14.8 ABBREVIATIONS

APCI	Atmospheric pressure chemical ionisation
CID	Collision-induced dissociation
EI	Electron impact
ESI	Electrospray ionisation
FTMS	Fourier-transform mass spectrometry
GC	Gas chromatography
IDA	Information dependent acquisition
LC	Liquid chromatography (high performance)
LLE	Liquid–liquid extraction
MRM	Multiple reaction monitoring
MS	Mass spectrometry
MS/MS	Tandem mass spectrometry
QTOF	Quadrupole time-of-flight instrument
QTRAP	Quadrupole ion-trap instrument
RP	Reversed phase
SPE	Solid-phase extraction
SIM	Selected ion monitoring
TOF	Time-of-flight
TRAP	Ion-trap instrument

References pp. 509–511

14.9 REFERENCES

1   H.H. Maurer, Anal. Bioanal. Chem., 38 (2005) 110–118.
2   M. Thevis and W. Schaenzer, J. Chromatogr. Sci., 43 (2005) 22–31.
3   L. Politi, A. Groppi and A. Polettini, J. Anal. Toxicol., 29 (2005) 1–14.
4   H.H. Maurer, Clin. Chem. Lab. Med., 42 (2004) 1310–1324.
5   W.M.A. Niessen, J. Chromatogr. A, 1000 (2003) 413–436.
6   P. Marquet, Ther. Drug Monit., 24 (2002) 255–276.
7   P. Marquet, Ther. Drug Monit., 24 (2002) 125–133.
8   M.R. Moeller and T. Kraemer, Ther. Drug Monit., 24 (2002) 210–221.
9   J.F. van Bocxlaer, K.M. Clauwaert, W.E. Lambert, D.L. Deforce, E.G. Van den Eeckhout and A.P. De Leenheer, Mass Spectrom. Rev., 19 (2000) 165–214.
10 O.H. Drummer, J. Chromatogr. B, 733 (1999) 27–45.
11 A. Polettini, J. Chrom. B, 733 (1999) 47–63.
12 R.A. de Zeeuw and J.P. Franke, General unknown analysis, in: M.J. Bogusz (Vol. Ed.), R.M. Smith (Series Ed.), Forensic science – Handbook of analytical separations (Vol. 2), Elsevier, Amsterdam, 2000, pp. 567–599.
13 S.H. Cosbey, I. Craig and R. Gill, J. Chromatogr. B, Biomed. Appl., 669 (1995) 229–235.
14 J. Yawney, S. Treacy, K.W. Hindmarsh and F.J. Burczynski, J. Anal. Toxicol., 26 (2002) 325–332.
15 X.H. Chen, J.P. Franke, J. Wijspeek and R.A. de Zeeuw, J. Anal. Toxicol., 16 (1992) 351–355.
16 X.H. Chen, J.P. Franke, K. Ensing, J. Wijsbeek and R.A. de Zeeuw, J. Anal. Toxicol., 17 (1993) 421–426.
17 X.H. Chen, J.P. Franke, K. Ensing and R.A. de Zeeuw, J. Chromatogr. B, Biomed. Appl., 617 (1993) 147–151.
18 R.A. Cox, J.A. Crifasi, R.E. Dickey, S.C. Ketzler and G.L. Pshak, J. Anal. Toxicol., 13 (1989) 224–228.
19 W. Weinmann, M. Stoertzel, S. Vogt and J. Wendt, J. Chromatogr. A, 926 (2001) 199–209.
20 A. Schreiber, J. Efer and W. Engewald, J. Chromatogr. A, 869 (2000) 411–425.
21 A.G.A.M. Lips, W. Lameier, R.H. Fokkens and N.M.M. Nibbering, J. Chromatogr. B, 759 (2001) 191–207.
22 J.M. Hough, C.A. Haney and R.D. Voyksner, Anal. Chem., 72 (2000) 2265–2270.
23 M.J. Bogusz, J. Chromatogr. B, 733 (1999) 65–91.
24 P. Marquet, J.L. Dupyi, G. Lachâtre, B. Shushan, E. Duchoslav, C. Monasterios, P. Ilisieu and J. Anacleto, Proceedings of the 46th ASMS Conference on Mass Spectrometry and Allied Topics, Orlando, FL, 1998.
25 P. Marquet, N. Venisse, N. Lacassie and G. Lachâtre, Analusis, 28 (2000) 925–937.
26 F. Saint-Marcoux, G. Lachâtre and P. Marquet, J. Am. Soc. Mass Spectrom., 14 (2003) 14–22.
27 W. Weinmann, A. Wiedemann, B. Eppinger, M. Renz and M. Svoboda, J. Am. Soc. Mass Spectrom., 10 (1999) 1028–1037.
28 M. Rittner, F. Pragst, W.-R. Borg and J. Neumann, J. Anal. Toxicol., 25 (2001) 115–124.
29 N. Venisse, P. Marquet, E. Duchoslav, J.L. Dupuy and G. Lachâtre, J. Anal. Toxicol., 27 (2003) 7–14.
30 H. Miyaguchi, K. Kuwayama, K. Tsujikawa, T. Kanamori, Y.T. Iwata, H. Inoue and T. Tohru, Forensic Sci. Int., 157 (2006) 57–70.
31 M. Gergov, I. Ojanperä and E. Vuori, J. Chromatogr. B, 795 (2003) 41–53.
32 M. Josefsson, R. Kronstrand, J. Andersson and M. Romas, J. Chromatogr. B, 789 (2003) 151–167.
33 E.N.M. Ho, D.K.K. Leung, T.S.M. Wan and N.H. Yu, J. Chromatogr. A, 1120 (2006) 38–53.
34 X. Tong, I.E. Ita, J. Wang and J.V. Pivnichny, J. Pharm. Biomed. Anal., 20 (1999) 773–784.
35 M. Thevis, H. Geyer, D. Bahr and W. Schänzer, Eur. J. Mass Spectrom., 11 (2005) 419–427.
36 R.L. Fitzgerald, J.D. Rivera and D.A. Herold, Clin. Chem., 45 (1999) 1224–1234.
37 G. Hoizey, D. Lamiable, T. Trenque, A. Robinet, L. Binet, M.L. Kaltenbach, S. Havet and H. Millart, Clin. Chem., 51 (2005) 1666–1672.
38 C.A. Mueller, W. Weinmann, S. Dresen, A. Schreiber and M. Gergov, Rapid Commun. Mass Spectrom., 19 (2005) 1332–1338.

39   P. Marquet, F. Saint-Marcoux, T.N. Gamble and J.C.Y. Leblanc, J. Chromatogr. B, Anal. Technol. Biomed. Life Sci., 789 (2003) 9–18.

40   J.L.E. Reubsaet and S. Pedersen-Bjergaard, J. Chromatogr. A, 1031 (2004) 203–211.

41   S.M.R. Stanley and H.C. Foo, J. Chromatogr. B, Anal. Technol. Biomed. Life Sci., 836 (2006) 1–14.

42   G.L. Herrin, H.H. McCurdy and W.H. Wall, J. Anal. Toxicol., 29 (2005) 599–606.

43   M. Gergov, B. Boucher, I. Ojanperä and E. Vuori, Rapid Commun. Mass Spectrom., 15 (2001) 521–526.

44   A. Pelander, I. Ojanperä, S. Laks, I. Rasanen and E. Vuori, Anal. Chem., 75 (2003) 5710–5718.

45   T.N. Decaestecker, S.R. Vande Casteele, P.E. Wallemacq, C.H. Van Peteghem, D.L. Defore and J.F. Van Bocxlaer, Anal. Chem., 76 (2004) 6365–6373.

46   T.N. Decaestecker, K.M. Clauwaert, J.F. Van Bocxlaer, W.E. Lambert, E.G. Van den Eeckhout, C.H. Van Peteghem and A.P. De Leenheer, Rapid Commun. Mass Spectrom., 14 (2000) 1787–1792.

47   M.W.F. Nielen, T.F.H. Bovee, M.C. van Engelen, P. Rutgers, A.R.M. Hamers, J.A. van Rhijn and L.A.P. Hoogenboom, Anal. Chem., 78 (2006) 424–431.

48   S. Ojanpera, A. Pelander, M. Pelzing, I. Krebs, E. Vuori and I. Ojanpera, Rapid Commun. Mass Spectrom., 20 (2006) 1161–1167.

49   I. Ojanperä, A. Pelander, S. Laks, M. Gergov, E. Vuori and M. Witt, J. Anal. Toxicol., 29 (2005) 34–40.

50   L. Rivier, Anal. Chim. Acta, 492 (2003) 69–82.

51   B. Maralikova and W. Weinmann, J. Chromatogr. B, Anal. Technol. Biomed. Life Sci., 811 (2004) 21–30.

52   M. Gergov, J.N. Robson, E. Duchoslav and I. Ojanperä, J. Mass Spectrom., 35 (2000) 912–918.

53   M.J. Bogusz, R.-D. Maier, K.D. Krüger, K.S. Webb, J. Romeril and M.L. Miller, J. Chromatogr. A, 844 (1999) 409–418.

54   W. Weinmann, M. Stoertzel, S. Vogt, M. Svoboda and A. Schreiber, J. Mass Spectrom., 36 (2001) 1013–1023.

55   W. Weinmann, M. Gergov and M. Goerner, Analusis, 28 (2000) 934–941.

56   M. Gergov, W. Weinmann, J. Meriluoto, J. Uusitalo and I. Ojanperä, Rapid Commun. Mass Spectrom., 18 (2004) 1039–1046.

57   P.H. Dawson and W.-F. Sun, Int. J. Mass Spectrom. Ion Process., 55 (1983/1984) 155–170.

58   P.G.M. Kienhuis and R.B. Geerdink, Trends Anal. Chem., 19 (2000) 460–474.

59   P.G.M. Kienhuis and R.B. Geerdink, J. Chromatogr. A, 974 (2002) 161–168.

60   R. Jansen, G. Lachatre and P. Marquet, Clin. Biochem., 38 (2005) 362–372.

61   A.W.T. Bristow, W.F. Nichols, K.S. Webb and B. Conway, Rapid Commun. Mass Spectrom., 16 (2002) 2374–2386.

62   A. Schreiber, M. Gergov and W. Weinmann, Poster 52th ASMS Conference on Mass Spectrometry and Allied Topics, Nashville, TN, 2004.

63   J.L. Josephs and M. Sanders, Rapid Commun. Mass Spectrom., 18 (2004) 743–759.

64   C. Baumann, M.A. Cintora, M. Eichler, E. Lifante, M. Cooke, A. Przyborowska and J.M. Halket, Rapid Commun. Mass Spectrom., 14 (2000) 349–356.

65   M. Cintora, E. Lifante, M. Eichler, S. Rodriguez, M. Cooke, A. Przyborowska, S. Down, R. Patel and J. Halket, Adv. Mass Spectrom., 15 (2001) 609–610.

66   R. King, R. Bonfiglio, C. Fernandez-Mezler, C. Miller-Stein and T. Olah, J. Am. Soc. Spectrom., 11 (2000) 947–950.

67   C.R. Mallet, Z. Lu and J.R. Mazzeo, Rapid Commun. Mass Spectrom., 18 (2004) 49–58.

68   T.M. Annesley, Clin. Chem., 49 (2003) 1041–1044.

69   H.R. Liang, R.L. Foltz, M. Meng and P. Bennett, Rapid Commun. Mass Spectrom., 17 (2003) 2815–2821.

70   R. Dams, M.A. Huestis, W.E. Lambert and C.M. Murphy, J. Am. Soc. Mass Spectrom., 14 (2003) 1290–1294.

71   B.K. Matuszewski, M.L. Constanzer and C.M. Chavez-Eng, Anal. Chem., 70 (1998) 882–889.

72   C. Müller, P. Schäfer, M. Störtzel, S. Vogt and W. Weinmann, J. Chromatogr. B, 733 (2002) 47–52.

M.J. Bogusz (Ed.). Forensic Science
Handbook of Analytical Separations, Vol. 6
© 2008 Elsevier B.V. All rights reserved.

CHAPTER 15

Forensic toxicological screening with capillary electrophoresis and related techniques

Ameriga Fanigliulo, Federica Bortolotti, Jennifer Pascali and
Franco Tagliaro

*Department of Medicine and Public Health, Section of Forensic Medicine, University of Verona, 37134
Verona, Italy*

15.1 INTRODUCTION

Capillary electrophoresis (CE) is gaining popularity as a proficient and effective technique for analytical separations, alternative and often complementary to chromatography.

Speed of analysis, separation efficiency, versatility and simplicity of application, and cost effectiveness are the peculiar advantages generally recognised to this still relatively novel technique, which allowed CE to come on the scene of various application fields, ranging from pharmaceutical and pharmacological analysis to clinical chemistry, molecular biology and analytical toxicology.

Although the use of CE in forensic science is still in its infancy, this technique has already received major attention in the forensic science literature. Also, CE has been recognised as potentially admissible in the US courts as a method of evidence, in application of the federal rules of evidence following the Daubert Standard (1996) [1].

The peculiar features of CE indeed make it attractive when simplicity and rapidity of method development, along with minimal sample requirements are to be fulfilled. Moreover, and probably most importantly, CE has proved an exceptional versatility, both in separation/detection modes and in variety of applications. Indeed, no other technique can provide, with the same hardware, the analysis of a range of analytes spanning from inorganic ions to large biomolecules (proteins, DNA, etc.) as CE does.

Being CE still novel in most forensic toxicology laboratories, this chapter is intended to the illustration of the basic principles of this technique, to the description of the key features of its different modes of separation, and, far from any expectation of

References pp. 532–534

completeness, to the review of a selected number of applications in the field of forensic toxicology, as examples of the great potential of this technique as a practical and productive investigation tool in the hands of the forensic scientist (see Table 15.1).

15.2 TECHNICAL FEATURES AND INSTRUMENTATION

The peculiarity of CE with respect to other separation techniques relies on the electrophoretic/electrokinetic processes upon which separations take place. Charged species tend to migrate in an electric field driven by the electrostatic force towards the oppositely charged electrode. Their velocity at a given applied electric field depends on the individual *electrophoretic mobility*, which differs from one ionic species to another, on the basis of the mass-to-charge ratios. Besides, interactions taking place between analytes and other molecules from the medium (typically a buffer solution, a micellar system and/or a polymer gel) in which separation occurs, contribute to differentiate the migration velocities of different chemical species. It follows that the running buffer properties, such as pH, ionic strength and temperature, are critical parameters in any electrophoretic separation.

Carrying out electrophoresis inside a capillary-shaped compartment offers neat advantages over traditional electrophoresis in slab gels. In particular, better control of Joule heating, minimal lateral diffusion and minimal dead volumes allow the achievement of excellent separation efficiencies, reaching up to 10^5–10^6 theoretical plates. CE capillaries are usually circular cross-sectional tubes made up of fused silica, having diameters between 25 and 100 μm. These geometrical characteristics and the chemical nature of the inner wall give rise to an important phenomenon, affecting analyte migration, known as *electroosmosis*. Whenever a solid surface is in contact with an aqueous solution, it exhibits an excess of surface charge due to ionisation processes or adsorption of species from the medium. This surface charge in fact produces the formation of a double electrical layer near the solid wall, defined by the ζ potential. In the specific case of fused silica capillaries, the inner wall exposes to the buffer a great number of silanol groups, which are ionised (as SiO^-) at pH higher than 2, thus showing a net negative charge. Under these conditions, because of tendency to charge balance, cations from the solution will be attracted at the interface with the capillary wall and will build up a double electrical layer. When a potential difference is established between the ends of the capillary, they will migrate towards the cathode (negative electrode), drawing water molecules with them by osmosis. This generates a flow of liquid inside the capillary, named *electroosmotic flow* (EOF), showing a flat (piston-like) profile, which results particularly beneficial for molecular separations. In fact, it limits to a minimum band broadening, which, on the contrary, is a typical drawback of capillary liquid chromatography (LC), where the pressure driven flow of the mobile phase, hindered by shear forces at the wall, yields parabolic flow profiles, generating a relevant band broadening. Electrophoretic migration and EOF may have either the opposite or the same direction. Thus, they sum as vectors, inducing both charged and neutral species to move, with velocities depending on the prevailing force in each case. This latter

TABLE 15.1
SELECTED APPLICATIONS OF CE IN FORENSIC TOXICOLOGICAL ANALYSIS

Drug/Analyte	Matrix	Separation	Detection	LOD	Ref.
Codeine, dihydrocodeine and glucuronides	Urine	MEKC / CZE	LIF / MS	10–40 ng/mL (LIF) / 100–200 ng/mL (MS)	[28]
Methamphetamine	Urine	CZE, chiral	MS	30–50 ng/mL	[30]
BZP	—	CZE	MS	5.0×10^{-7}–4.0×10^{-6}	[31]
Morphine and metabolites	Urine	CZE	MS	100–200 ng/mL	[32]
Nicotine and metabolites	Urine	CZE	MS	0.55–11.25 ng/mL	[33]
Amphetamines	Plasma	CZE	MS	1 ppb	[41]
Amphetamine, trazodone, salbutamol, ephedrine, codeine	Urine (spiked)	CZE	DAD / MS	50 ppb (MS)	[42]
Basic drugs	—	NACE	MS	—	[43]
Barbiturates	—	CZE / MEKC	UV	0.2–5 µg/mL	[54]
Basic drugs	—	CZE / MEKC	UV	0.3–3 µg/mL (CZE) / 1–9 µg/mL (MEKC)	[55]
Acidic drugs	—	CZE / MEKC	UV / DAD	—	[56]
Acidic drugs	—	CZE / MEKC	UV	—	[57]
(73) Pharmaceutical (basic) drugs	—	CZE	UV	—	[58]
Amphetamines	—	CZE	UV	—	[59]
Acidic and basic drugs	Urine, serum (spiked)	CZE / MEKC / NACE	DAD	—	[60]
(326) Basic and neutral drugs	Blood	CZE	UV	10 ng/mL	[61]
Opiates	Urine (spiked)	CZE	LIF	50–100 pg/mL	[62]
Amphetamines	Blood (spiked)	CZE	DAD	0.5–3.1 µg/mL	[63]
Amphetamines	Urine	CZE, chiral	UV	50 ng/mL	[64]
Amphetamines, codeine, morphine	Urine	CZE, chiral	DAD	20–40 ng/mL	[65]
Amphetamines	Urine	CZE	DAD / MS	20–70 ng/mL (DAD) / 50–200 ng/mL (MS)	[66]

concentration is important as well, since increasing buffer molarity increases its ionic strength, thus reducing the EOF. A high ionic strength increases heat generation and reduces adsorption of analytes to the wall. These two phenomena affect separations in opposite directions, the former favouring peak broadening, the latter hindering it.

Commonly used CE buffers are borate, phosphate, citrate, acetate and Tris (trishydroxymethylamino methane), which can be adjusted to different pH values. Care must be taken to use volatile buffers when mass spectrometric detectors have to be employed.

It is not uncommon to use buffer additives, which typically include detergents, viscosity modifiers, organic solvents and denaturants. Their presence is exploited to affect the EOF, to improve solubility of analytes, to retard some molecules with respect to others in the electromigration, to alter analyte–wall interactions and to modify selectivity. In some particular applications, the addition of additives (surfactants, cyclodextrins, etc.) to the running buffer introduces additional mechanisms of separation and new criteria of selectivity.

Fused silica is undoubtedly the most widely used material for the production of CE capillaries. Fused silica capillaries, in fact, fulfil the requirements of chemical and physical resistance, high thermal conductivity, precision of diameter dimension along all their length and UV transparency (allowing optical detection even in the low UV). To accomplish the needed mechanical resistance, capillaries are externally coated with a protective polyimide layer, which is optically opaque.

Fused silica capillaries can be internally "uncoated" (naked) or "coated" with a thin layer of proper polymers. Usually, coating is used to eliminate the EOF or to change its direction and to reduce solute–wall interactions with the wall sylanol groups. The most common coatings are polyacrylamide, cellulose, PVA, amino acids, amines, surfactants, aryl pentafluoro compounds, polyvinylpyrrolidinone, polyethyleneimine, etc. Also, polymers well known as liquid chromatographic (C_2, C_8, C_{18}) or gas chromatographic (PEG, phenylmethyl silicone) stationary phases can be used for this purpose.

Once the external polyimide coating has been removed, the UV transparency of the silica wall can be exploited to perform "in-capillary" optical detection, thus avoiding post-separation added volumes and consequent band broadening. The main problem with detection for CE is certainly sensitivity, because of the minimal quantities of injected analytes and minimal volumes (nL) involved. In the case of absorbance-based detection modes, reduced optical path length ($32\,\mu m$ effective for a 50-μm diameter capillary, because of the curved surface) is a further heavy detection drawback. In real cases, 10^{-6} M is the maximum sensitivity gained with UV detection, corresponding to picogram amounts of injected analytes.

However, much higher sensitivity can be achieved with other detection methods, and, particularly, with laser-induced fluorescence or mass spectrometry.

15.2.2 Modes of separation

A number of different separation modes can be performed with the same CE hardware, by simple tuning of a few analytical parameters. This unique characteristic makes CE the ideal tool to face the large and varied spectrum of problems which typically are encountered in forensic science, ranging from small ion (CN^-, K^+, etc.) to protein and DNA fragment analysis.

15.2.2.1 Capillary zone electrophoresis (CZE)

CZE is the simplest way of operation that can be realised in CE. It gains molecule separation by exploiting the different mobility of ions under an electric field. In this way, analytes migrate in discrete zones along the capillary, with different velocities, depending on individual electrophoretic mobility. Electrophoretic mobility of analytes (μ_e) is described by the following equation:

$$\mu_e = \frac{q}{6\pi\eta r}$$

where q is the ion charge, r the ion radius and η the solution viscosity.

Besides electrophoretic migration, the presence of the EOF produces an electrically driven flow of liquid, which, in fused silica capillaries (exposing SiO^- groups), is directed from the anode to the cathode. The electroosmotic mobility is described by the following equation:

$$\mu_{eof} = \frac{\varepsilon\zeta}{4\pi\eta}$$

where ε is the dielectric constant, ζ the zeta potential and η the solution viscosity.

Because of the presence of EOF, as already mentioned, both anionic and cationic solutes move towards the detector, which is typically located at the cathodic end of the capillary. Neutral species move, without separation, at the same velocity as the EOF. EOF velocity can be modified or reduced almost to zero or direction reversed by modifying buffer pH, composition or ionic strength, and by using additives or capillary wall modifiers. In this case, a permanent modification of the inner capillary surface can be accomplished by chemical alteration or, preferably, a "dynamic coating" can be obtained by simply adding proper modifiers to the running buffer. A major advantage of dynamic coating is reproducibility, since it is continuously regenerated, being the modifier in the running buffer.

CZE can also be adjusted to perform chiral separations, by adding special modifiers to the running buffer, acting as chiral selectors. This way of operation is largely less expensive and more efficient than in high-performance liquid chromatography (HPLC) or gas chromatography (GC), where the use of chiral stationary phases is needed. The most commonly used chiral selectors are cyclodextrins (CDs), i.e. neutral cyclic oligosaccharides consisting of six, seven or eight glucose units, named α-, β- and γ-CDs, respectively. They realise chiral selectivity by formation of reversible inclusion complexes with each enantiomer with different affinity.

Although aqueous buffers are usually employed as running electrolytes in CZE, the alternative use of organic solvents, added with small amounts of organic acids or buffer salts, has proved to be effective in a wide range of applications [2]. In *non-aqueous capillary electrophoresis* (NACE) organic solvents not only improve molecule solubility, but, in many cases, also affect advantageously the EOF and ion mobility. What is dramatically changed in an organic medium, with respect to water environment, is molecules' pK_a. In this perspective, molecules hard to be separated in aqueous buffers because of similar pK_a, may show completely different values in non-aqueous environments [3].

15.2.2.2 Micellar electrokinetic chromatography (MEKC)

MEKC is known as an electrokinetic technique able to separate neutral molecules, by exploiting their partitioning between BGE and micelles of a surfactant, added to the BGE. Micelles have a charged outer surface and transport uncharged species through the capillary in their hydrophobic core, towards the electrode with opposite charge. In MEKC, the EOF is oriented towards the detector whereas micelle migration direction is opposite. Thus, micelles will exert a selective "retarding" effect on neutral molecules, depending on their partitioning with their hydrophobic core. The mostly used surfactant for this purpose is sodium dodecyl sulphate (SDS), but cationic surfactants, such as cetyl trimethyl ammonium bromide (CTAB) and dodecyl trimethyl ammonium bromide (DTAB), are also used. MEKC can be seen as a very flexible mode of CE, capable of separating both neutral and charged species at the same time. Thus it proves to be a suitable tool for a wide range of applications.

15.2.2.3 Capillary electrochromatography (CEC)

CEC is a rather new technique, which has recently attracted much attention. It is in fact a kind of "hybrid" between capillary LC and CE. In simple words, it is capillary zone electrophoresis carried out inside a capillary containing a stationary phase, similarly to an HPLC column. Thus, separation occurs through different mechanisms, both chromatographic (partitioning of molecules between stationary and mobile phase) and electrophoretic (differential mobility of charged species in an electric field). The driving force of the mobile phase is the EOF generated at the interface between the mobile phase and the stationary phase. This minimises band spreading typically occurring in capillary LC in which the mobile phase is forced through the column by an external pump.

Although in theory very efficient and selective, this technique has not yet found spread application, most probably because of problems in availability and reproducibility of CEC columns.

15.2.2.4 Capillary isoelectric focusing (cIEF)

cIEF is mostly used to separate proteins (or other ampholites) on the basis of their isoelectric points (pI), that is the pH value at which their charged moieties compensate each other and thus the protein does not migrate under an electrical field. The separation in this case is not attained by migration differences, but through the focusing of analytes in discrete zones of the capillary. The separation compartment (the capillary) is filled in with an amphoteric species mixture (known as Ampholine®). When the electric field is applied in such a system, a gradient of pH is formed in the capillary, trapped between the acidic end at the anode and the basic end at the cathode. The migration of analytes stops as each of them settles (*focusing*) where the solution pH equals its pI. After focusing, an external force is needed to drive all analyte zones out of the capillary. This is typically accomplished by the application of an external pressure or by electroosmosis.

This mode of operation is particularly used for the separation of isoforms of the same protein, which often differs too slightly in net charge or in mass-to-charge ratio, but exhibit different pIs.

15.2.2.5 Capillary isotachophoresis (cITP)

Isotachophoresis means "moving at the same velocity". Indeed, this technique attains a steady state in which analytes form discrete zones moving in series at the same speed towards the detector. This process takes place between a *leading electrolyte* solution, containing ions with the highest mobility among all the other species in the system, and a *terminating electrolyte*, containing ions with the lowest mobility than all. It is interesting to note that in this case, differently from electrophoresis, the electric field is different in each point of the capillary. cITP has proved to be not only a valid separation mode, but also an excellent method for analyte preconcentration to be coupled with other CE separation modes.

15.2.2.6 Capillary gel electrophoresis (CGE)

Directly derived from the slab gel electrophoresis, CGE performs separation of species in a viscous medium, typically a linear or cross-linked polymer. In this situation molecules migrate at different velocities on the basis of their size, being not affected by their mass-to-charge ratio. This characteristic is particularly useful to discriminate molecules which, in spite of a different size, show identical mass-to-charge ratios (typically biopolymers). Fragments of DNA of different length are an example. CGE has rapidly become extremely popular in molecular biology, being today the standard approach for fragment length DNA analysis and for DNA sequencing.

15.2.3 Modes of detection

A major drawback generally attributed to CE is the lack of sensitivity, in terms of minimal detectable concentration. In fact, as any other capillary separation system, CE deals with minute amounts of sample (nanolitre), which inherently limit the mass of analyte reaching the detector and consequently the sensitivity. A fundamental requirement to maintain an acceptable sensitivity is to avoid zone broadening, by limiting any "dead volumes" in the system.

Different detection techniques, both "in-capillary" and "off-capillary", have been applied, with different advantages and disadvantages, with the aim of maximising sensitivity and selectivity.

References pp. 532–534

15.2.3.1 Optical methods

Measuring absorption of UV-visible light is the most frequently used detection method employed in CE instruments. Absorbance detection is usually performed "in-capillary", that is performing detection directly on a section of the capillary, thus avoiding the presence of unwanted dead volumes. Since UV-visible light absorbance linearly depends on the path length (Beer-Lambert's law), and since in CE the optical cell is the capillary itself, sensitivity will depend on the capillary diameter. Nevertheless, increasing capillary diameter is not a viable solution to improve sensitivity, because analyte diffusion increases dramatically.

Special technical solutions, based on peculiar capillary design, have been implemented to overcome this problem. Z-cells [4] and bubble cells [5] are examples. In the former case, an increased optical path length is gained through a double right angle in the capillary region where detection occurs. Such a design realises a segment of the capillary in which the light beam can run axially for a length much higher than the capillary section. In bubble cells, on the other hand, an expanded region ("bubble") is realised directly inside the capillary. When the analyte region enters the bubble, it expands radially and, consequently, contracts longitudinally, realising an optical window with the same concentration but increased path length.

Diode array detectors (DADs) have also been implemented in commercially available CE instruments. The availability of a multichannel, dispersive optical detector can greatly enhance the information content of each analytical run.

Alternatively, analyte fluorescence can be exploited for detection in CE. Fluorescence detectors measure the intensity of the light emitted by molecules when they relax from an excited state induced by an external energy source. A laser beam is often used as the source of excitation (laser-induced fluorescence, LIF). In this case, sensitivity for fluorescent molecules, or for molecules that can be made fluorescent after a chemical reaction, rises up to 10^{-12} M or higher.

In the case of non-UV-absorbing molecules, indirect detection can be performed, adding a detectable additive (e.g. UV-absorbing) to the running buffer, with the same charge and similar mobility as the analyte of interest. What actually occurs in the capillary is a displacement of the additive by the non-absorbing analyte, from the zone occupied by it. This displacement may be recorded at the detector as a negative peak (i.e. increase in optical transparency) [6,7]. Drawbacks of this detection mode are sensitivity, which is almost constantly lower than in the corresponding "direct" mode, and linearity, which displays a narrower range.

15.2.3.2 Electrochemical methods

Among the possible choices, electrochemical (EC) techniques, including amperometric, potentiometric and conductimetric detection, are one of the most sensitive detection tools that can theoretically be coupled with CE.

In particular, amperometric detectors, based on an electrochemical reaction occurring in analyte molecules, gain high sensitivity and selectivity towards electroactive analytes [8]. Besides, this technique is mass sensitive and therefore independent on cross-sectional path length in the capillary.

EC detection can be performed on-capillary, when capillaries with internal diameter smaller than 25 μm are employed, and off-capillary, when larger capillaries are used. In the latter case, in fact, the faradic current would be too low to be detected, with respect to the separation current.

The crucial problem in EC detection is actually the isolation of the separation circuit from the detection circuit. Several technical solutions have been proposed to address this problem, including porous glass junctions, porous graphite joints, cellulose acetate and palladium. In this case, the transferring of analyte from the capillary terminal end to the detection cell is accomplished by the EOF.

A further drawback of EC detection mode is the necessity of using miniaturised electrodes, compatible with capillary dimensions, and so-built devices are not commercially available so far.

15.2.3.3 Mass Spectrometry

The hyphenation of CE with mass spectrometry (CE-MS) has opened new perspectives of application to capillary electrophoretic and electrokinetic techniques. Since it first appeared in the literature in 1987, with a pioneering work of Olivares [9], the technological development of CE-MS has grown exponentially, in particular in the latest few years, and now stands as one of the most relevant fields of innovation in the applied analytical sciences.

One of the first reviews on this technique, mainly focused on the technological issues and counting a few noticeable applications, appeared in 1995 [10]. The number of reviews published in the latest years has incredibly risen, especially from 2003 onward [11–17]. This success is certainly a consequence of latest developments of the Nobel-prized electrospray ionisation (ESI) [18], which offers a powerful ionisation source for molecules in liquid phase. Interfacing a high-voltage-driven separation technique with a voltage-based ionisation system and transferring the so-obtained ions to a mass spectrometer is not a trivial issue. The key points to be successful in such an operation are the following:

a. to ensure the electrical continuity that closes the CE circuit,
b. to build up a stable flow of ionised molecules that enter the spectrometer,
c. to use a fast-scanning detector, suitable for the high efficiency (i.e. narrow peaks) of CE,
d. to choose a proper running buffer, meeting both CE separation and ion source requirements.

Most of these points deal with a proper design of the interface. So far, several ion sources have been coupled to CE. Experiments have been performed with atmospheric pressure chemical ionisation (APCI) [19], atmospheric pressure photoionisation (APPI) [20], matrix-assisted laser desorption (MALDI) [21] and electrospray ionisation (ESI) interfaces. As a matter of fact, the last one has proved to be the most suitable interface for CE and it is thereof the most widely used in CE-MS applications.

References pp. 532–534

The coupling between the outlet of the CE capillary and the ESI interface has been historically realised in different ways, basically with or without the use of an additional liquid to support the flow and hence the electric contact.

In a sheath liquid interface [22,23] the flow coming out of the separation capillary (in the order of magnitude of nL/min) is diluted in a coaxial flow (µL/min) of a suitable solution, running in a stainless steel capillary that surrounds the separation capillary. At the tip of this metal tube a Taylor cone is induced by an external voltage, which realises a spray, as in typical LC-ESI-MS. A gas flow, external and concentric with respect to the liquid ones, can be used to further support the spray. In this arrangement, two basic results are obtained: (i) the electrical continuity between the electropherograph and the mass detector and (ii) the build-up of a suitable flow for the onset of a stable electrospray. Besides, the dilution effect allows for the use, in small amounts, of non-volatile running buffers or detergents, since their presence becomes negligible in the main stream of the sheath liquid. Because of its versatility, instrumental simplicity and capability to sustain a stable spray, sheath liquid interface is the most commonly employed in CE-MS applications.

Sheathless interfaces [24] realise the electrical contact through a metal junction, made up by metal coating of the capillary end or via stainless steel connections or microelectrodes inserted in the capillary. This setting often takes an advantage of nanoelectrospray tips to spray the sample directly in the mass spectrometer inlet. New instrumental applications in this sense have recently been reported [25,26]. Sheathless interfaces have the evident advantage of improving concentration sensitivity, since no dilution occurs after CE separation, and gaining a better ionisation efficiency. The drawbacks of such a coupling are merely practical and can be summarised in a lack of reproducibility and robustness, since delicate manipulation and fabrication of miniaturised components is needed. Capillary junctions often suffer from misalignment and metal coatings or metal tips have limited durability.

From the first trial works in the 1980s until now, CE has been coupled to magnetic sector [27], quadrupole [41], ion trap [28–32], triple quadrupole [33], Fourier transform ion cyclotron resonance [34,35] and time-of-flight (TOF) mass spectrometers [36–38].

Actually, at present, ion traps (ITs) are the favourite mass detectors used in combination with CE, for their relative speed of scan, reasonable costs and, moreover, for the possibility to perform MS^n experiments, which give access to the structural information of analytes, when they are softly ionised as in ESI.

The hyphenation of CE with TOF, although very recent, is seen as particularly promising, especially after the advent of bench-top TOF analysers. In fact, the high-scan rate (10–200 µs per spectrum), resulting in high-sampling frequency, is advantageously suitable for the fast CE separations and allows for good peak definition [39,40]. Besides, TOF, having resolution of 10,000 and mass accuracy lower than 5 ppm, is so far the cheapest instrument allowing accurate mass determination, with consequent improvement in identification power, even in the absence of reproducible fragmentation. This characteristic can be advantageously exploited in the forensic and toxicological fields, where rapid and reliable molecule identification is crucial for effective diagnosis.

Under specific operating conditions, also chiral separations [30,41], dynamic coating [42] and NACE [43,44] can be performed without impairing the mass spectrometer performance by source contamination and/or ion suppression.

15.3 CAPILLARY ELECTROPHORESIS IN TOXICOLOGICAL SCREENING

Toxicological analysis can be aimed at searching specific substances or classes of substance (*direct search*) which may or may not be comprised in a scheduled list of toxicants, or at detecting and identifying "any possible" harmful compounds "whose presence is uncertain and whose identity is unknown" (*general unknown analysis*) [45].

In the case of direct search, where most often the presence of definite drugs must be ascertained or excluded in order to fulfil a law or regulatory constrain, a two-step strategy is generally employed. In the first step, a screening test, generally based on immunoassays (when available), has the purpose of maximising the *diagnostic sensitivity* that means detecting all possible "positive" specimens (even if some "negative" samples can erroneously be classified as "positives"). The presumptively "positive" samples from the screening are then confirmed by a second, independent analytical method, at least as sensitive as the first one but far more selective, thus maximising *diagnostic specificity*. The purpose of this second test is to identify all the "false positives" among the presumed positives from the screening test.

The general unknown search, being aimed at either establishing or excluding the presence of as many toxicants as possible in biosamples, needs a definitely more extensive and comprehensive screening strategy, since the number of possible molecules to be detected or discarded is necessarily huge.

Such an objective may look too demanding and, for this reason, it has most frequently been translated into a systematic logical-analytical approach, which makes use of standardised procedures, well-characterised analytical methods, databases of analytical information and, finally, statistical evaluation of both techniques and results. This approach is called *systematic toxicological analysis* (STA) and has recently been reviewed with respect to the most commonly employed separation techniques [46–48] including chromatographic and electrophoretic/electrokinetic methods.

A comprehensive critical review of such techniques, applied to the clinical and forensic toxicological screening, was published by Drummer [46]. The author in particular comments on the inadequacy of a single separation/detection technique to identify a full range of drugs/toxicants, thus at least two, possibly orthogonal, techniques are recommended. In this perspective, CE is seen as a promising analytical technique for toxicological screening, because of its peculiar separation and selectivity mechanisms, which make it complementary to gas and LC.

In a rigorous definition, STA can be regarded as a methodology offering the capability to identify as many compounds as possible through the comparison of their physico-chemical characteristics with reference substances of toxicological relevance. Mathematical approaches have been applied to evaluate the suitability of a generic analytical method to the requirements of STA. One of them, fervently sustained by de Zeeuw, implies the calculation of the *identification power* (IP) of a technique. Among the parameters contributing to the IP, the *discriminating power* (DP) is defined as the probability that two randomly selected substances from the test set are discriminated. On the other hand, the *mean list length* (MLL) is the average length of the list of feasible candidates for a given analytical figure. In other words, the shorter is the list, the higher the probability of unambiguously identifying a substance. It is interesting to note that ideally both DP and MLL tend to 1, but they approach this value from opposite directions [45,49].

Several separation techniques, mostly chromatographic, have been evaluated with regard to their suitability for STA, among which TLC, GC and HPLC [50–53]. More recently, the same evaluation criteria have been applied to CE [54]. A test set of 25 barbiturates was analysed using both CZE and MEKC and the performance of the two techniques was compared. The identification parameter considered for the calculation of DP, MLL and IP was the "corrected effective mobility" ($\mu_{\text{eff}}^{\text{c}}$), that is the effective mobility corrected by the interpolation between reference and measured values of μ_{eff} of standards. The inter-day reproducibility of $\mu_{\text{eff}}^{\text{c}}$ proved to be the most critical parameter affecting the IP, being $<0.6\%$ for CZE and $<0.5\%$ for MEKC, respectively. In the present conditions, the two considered CE techniques were found to have good IPs, especially when used in combination. Besides, they showed no or little correlation with chromatographic techniques, such as GC and HPLC, thus indicating that combinations of chromatographic and electrophoretic or electrokinetic techniques may be particularly effective for STA.

Mutual independency of CZE and MEKC was further evaluated by Tagliaro *et al.* [55] in the separation of a panel of drugs including caffeine, morphine, barbital, pentobarbital, codeine, nalorphine, lidocaine, procaine, heroin, flunitrazepam, acetylcodeine, papaverine, amphetamine, narcotine, cocaine, diazepam, tetracaine, narceine, 6-monoacetylmorphine (6-MAM) and thebaine. Three different CE mehods were tested, namely CZE with an acidic buffer (CZE1), CZE with a basic buffer (CZE2) and MEKC with SDS in a basic medium. The separation patterns of the three methods were compared with Spearman's test and with principal component analysis, showing that CZE1 and CZE2 were significantly and directly correlated ($r = 0.749$), whereas MEKC and CZE2 were also significantly, but inversely correlated ($r = -0.865$). MEKC and CZE1 (limitedly to the basic drugs) appeared non-correlated ($r = -0.131$) and therefore the two techniques proved suitable for combined use in STA.

Intra- and inter-instrument reproducibility of migration parameters for identification purposes in STA were further evaluated in [56] and [57].

Moreover, IP proved enhanced for CE in the toxicological screening of basic drugs by the use of a dynamic coating system, creating a double layer on the inner capillary surface [58]. Such a system, already used by Lurie *et al.* [59] for the separation of methamphetamine, amphetamine, MDA (methylenedioxyamphetamine), MDMA (methylenedioxymethamphetamine), MDEA (methylenedioxyethylamphetamine) and cocaine in seized drugs, was reported to stabilise the EOF, to reduce analysis times and, most importantly, to dramatically increase the reproducibility of $\mu_{\text{eff}}^{\text{c}}$. The improved reproducibility, its effects on MLL, and thus on IP, were demonstrated by applying the same CZE method, with and without the addition of dynamic coating agents, to the screening of a pool of 73 basic drugs [58].

A further step in the evaluation of CE for STA was made by the same group in the attempt to develop analytical procedures capable of discriminating both acidic and basic compounds [60]. Five different methods were tested, based on CZE, MEKC and NACE, on a mixture consisting of six acidic and six basic drugs. Only the MEKC method with positively charged micelles (CTAB) resulted capable of discriminating both acidic and basic drugs in the same run. The single most interesting conclusion of the work was that the different selectivities of the tested CE methods could be advantageously exploited for screening purposes in analytical toxicology, since the above-mentioned methods could be easily applied in sequence on the same CE instrument, with little effort for set-up change, which is accomplished by simply changing buffer and capillary.

As we go beyond the formal/mathematical approach to STA, extensively studied by the de Zeeuw's group, we can consider as "general unknown analysis", more in general, also any screening procedure capable of detecting and identifying as many toxicants as possible, with a validated, rapid and effective analytical method. From this point of view, a large number of CE applications and methods can be accounted in the recent literature.

One of the first attempts to show the suitability of CE, and specifically CZE, for comprehensive forensic drug screening was reported by Hudson *et al.* [61], showing the capability of this technique of complementing or even replacing GC, HPLC and immunoassay methods. A panel of 326 basic and neutral drugs were separated and detected with CZE using phosphate-running buffers at pH 2.5 and pH 9.5, UV detection and electrokinetic injection. A procedure for preparing blood extracts was also proposed. Limit of detection was estimated to be lower than 10 ng/mL for most of the tested drugs.

Alnajjar *et al.* [62] described a CE method with fluorescence detection (either native or after derivatisation) for the analysis of multiple drugs of abuse in biological fluids. Morphine, normorphine, 6-acetyl-morphine and codeine were analysed without any derivatisation step and their detectability was found better than that obtained by GC and HPLC.

The development of a rapid CE-DAD method has been proposed by Nieddu *et al.* [63] for the screening of a class of amphetamine designer drugs, comprising 10 methylenedioxy derivatives of amphetamines and phenethylamine. The detection in blood samples and separation was performed within 15 min. No interference from the matrix was observed.

Chiral electrokinetic methods have widely been used for the determination of amphetamine and methamphetamine enantiomers [64], alone or in combination with other illicit drugs [65]. In some cases, mass spectrometry was used for detection [66,67].

A factorial design to screen for important design variable and a simplex calculation were used by Dahlén *et al.* [68] to optimise the CZE separation of amphetamine and 13 amphetamine analogues.

Similarly, an experimental design based on a multivariate approach [69] was applied to study the effects of a chiral selector (HDMS-β-CD) on enantioresolution of several basic drugs (β-blockers, local anaesthetics, sympathomimetics and one tricyclic antidepressant) in the presence of an ion-pairing agent (camphorsulphonate).

To overcome the well-known limitations in sensitivity of CE methods, efforts have been spent by several authors to concentrate the analytes before CE separation. A preconcentration and on-line extraction system coupled to CZE was proposed by Wei *et al.* [70] for the determination of opiates in human urine. It comprises a monolithic capillary tube where sample is loaded, microextracted and finally concentrated to an appropriate aliquot for injection in CZE. A CZE method for opiate search in hair was developed by Manetto *et al.* [71], using a binary system (0.1 M sodium phosphate, pH 2.5, with 40% ethylene glycol) as running buffer. A significant enhancement in sensitivity was gained by performing head-column field amplified sample stacking (FASS) injection. The validated method was applied to the detection of opiates in real user samples.

The detection and quantitation of D-lysergic acid diethylamide (LSD), one of the most potent hallucinogenic drugs, represent an analytical challenge for forensic toxicologists, because of the extremely low doses typically used and the extensive metabolism of the drug. LIF detection, employing a He–Cd laser at 325 nm wavelength, was used to gain enough sensitivity for the detection of LSD in specimens of human blood [72]. The CE method used a citrate–acetate system as running buffer. Further improvements in sensitivity were achieved by sweep-MEKC, also using fluorescence detection [73]. Sensitivity tests showed detection limits of 60 pg/mL and 0.6 ng/mL in model water–methanol solutions and in real blood samples, respectively.

Another class of abused drugs difficult to be detected in biological fluids is represented by benzodiazepines (BZP), which have been approached either with CZE or MEKC. The separation of 10 different BZP was afforded by a MEKC method, making use of a mixture of SDS and dextran sulphate as running buffer [74]. A double-dynamic coating CZE method with IT-MS detector was developed for the separation of six BZP in spiked urine, and MS^2 experiments were performed for confirmation [75].

From a fairly technological point of view, a considerable step forward was permitted to STA in the last 10 years, by the development of the so-called hyphenated chromatographic-spectroscopic techniques, as remarked by Polettini [47]. In fact, the availability of multichannel detectors, able to record the whole spectrum of the eluting analytes, adds a third dimension to chromatography. Among them, the mass selective detectors have rapidly become a "golden standard" in toxicological analysis, for unequalled selectivity, sensitivity and identification power. So far, the MS detector has most often been combined with GC and LC techniques. The combination of the fast and efficient CE separation techniques with the inherently specific and sensitive MS detection looks as a powerful tool in toxicological analysis, particularly promising for screening purposes, since it provides an incomparable possibility unravelling complex biological samples.

McClean *et al.* [31] optimised a CZE-ESI-IT-MS method for the determination of selected 1,4-benzodiazepines, identified by sequential product ion fragmentation. An analogous approach, exploiting MS^3 fragmentation experiments, was used to unambiguously identify codeine, morphine, dihydrocodeine and their glucuronides in urine samples, previously submitted to immunoassays for opiates [28]. The method was further optimised in 2001 [32], by using 25 mM ammonium acetate adjusted at pH 9, achieving detection limits of 100–200 ng/mL for free opioids (starting from 2 mL urine samples).

A simple and rapid CZE-ESI-IT-MS has recently been proposed by Gottardo *et al.* [76] for the simultaneous detection and quantitation of many of the most abused drugs in human hair, including 6-monoacetylmorphine, morphine, amphetamine, MA, MDA, MDMA, benzoylecgonine, ephedrine and cocaine. With an analytical run of 20 min, the method exhibits limits of detection well below the most severe cut-off concentrations adopted in hair analysis (0.1 ng/mg) and good linearity in the concentration range from 0.025 to 5 ng of each analyte/mg of hair.

Nicotine and eight of its metabolites were detected and identified by CE-MS in smokers' urine using triple quadrupole as detector. Detection limits and reproducibility obtained with hydrodynamic and sample stacking electrokinetic injections were compared [33].

The enantioseparation and discrimination of amphetamines and other pharmaceutical drugs was obtained through the on-line coupling of partial filling CE and ESI-MS by Cherkaoui *et al.* [77], testing different chiral selectors. The method was validated by the same group and a fast and efficient sample preparation step for plasma specimens was reported [41].

The complete filling technique was instead exploited for the enantioseparation of methamphetamine and its metabolites in urine [30]. The CE-MS method used 1 M formic acid adjusted at pH 2.5 as BGE, added with a diluted mixture of 3 mM β-cyclodextrin and 10 mM heptakis-(2,6-di-*O*-methyl)-β-cyclodextrin. Detection limits ranging from 0.03 to 0.05 µg/mL were thus obtained. The complete filling technique was also applied for the enantioselective determination of anaesthetics [78] and barbiturates [79].

The double dynamic coating, already used in CE separation with UV detectors [59], was experimented on a CE-MS system [42,75]. The double coating was realised by subsequent flushing through the capillary of a polycation buffer solution ("initiator") first and of a polyanion solution afterwards ("accelerator"), which also acted as the running buffer. The initiator adsorbed on the inner wall of bare silica capillary and the polyanion created a double layer adsorbing onto it. The net effect of this tough coating was a high and stable EOF, and a drastic reduction of analysis time. This setting, with proper choice of buffers for MS suitability, was tested for the separation of five basic drugs, including amphetamine, ephedrine, trazodone, codeine and salbutamol. It is worth noting that in such a procedure the inter-run rinsing step becomes critical, because the frequent introduction of NaOH and polycation solution in the MS source would be detrimental for MS detection and continuous opening of the ESI source between runs is not feasible. For this reason, electrolyte composition and rinsing procedures were optimised for both DAD and MS detectors and reproducibility of migration times and peak areas were evaluated accordingly [42].

The coupling of NACE with ESI-MS was applied to several problems relevant to toxicological analysis. Peri-Okonny *et al.* [43] evaluated this technique in the separation of nine basic drugs, including tricyclic antidepressant and bronchodilator drugs. Steiner *et al.* [44] reported a comparison among several organic solvents in the preparation of ammonium acetate buffer for the separation of four model drugs, including quinine, 2-aminobenzimidazole, procaine and propanolol. Anderson *et al.* [80] developed a NACE-MS method for the determination of lidocaine and two of its metabolites in human plasma, and examined the influence of sheath liquid composition, drying gas temperature and nebulising gas pressure on the separation efficiency. Repeatability and limits of detection were compared with those obtained in the same conditions with UV detection. Optimisation of an NACE-MS method for the analysis of fluoxetine and related compounds was carried out by Cherkaoui *et al.* [81], who observed a neat increase in sensitivity with respect to UV detection.

MS detection has also been proficiently exploited for the direct detection of analytes exhibiting poor UV absorbance. Gottardo *et al.* [82] developed a CE-IT-MS method for the determination in human urine of γ-hydroxybutyric acid (GHB), a psychotropic substance recently introduced in the market of illicit drugs. Separation buffer pH and composition were optimised for MS detection, modifying a previous method with indirect detection [83]. The selectivity of the MS detector was such that diluted urine without any other clean-up could be directly injected and analysed.

The growing list of specific CE applications in forensic toxicology reported in a noticeable number of recent reviews [84–88] is a symptom of the relevance that such technique is acquiring in this field.

Some applications are worth to be cited, although not strictly "toxicological", since they do not deal with biological specimens, but with seized drugs or powder street samples. Nevertheless, the separation efficiency and the potential for drug discrimination exhibited in such methods, make them good candidates for systematic toxicological screening. As an example, the double dynamic coating procedure reported by Lurie *et al.* [89] was used for the detection and quantitation of a great variety of seized drugs, using the same capillary and only varying the composition of the running buffer, in order to tune the method to the chemical nature of the analytes. In another CZE application, CDs were used, added to the phosphate-running buffer to achieve rapid determination of heroin, heroin impurities and additives in clandestine heroin samples [90].

A complete review of the use of CE profiling methods applied for both pharmaceutical and forensic purposes is reported by Hilhorst *et al.* [91]. CE separation techniques are presented here as a valid alternative, capable to overcome limitations of traditional GC (thermolability of compounds) and HPLC (lack of efficiency) methods. As already stated, many of the reported methods show potential for application in toxicological analysis as well. Table 15.1 presents selected applications of CE in toxicological analysis.

15.4 CONCLUSIONS

CE is a separation technique unequivocally characterised by versatility of application, ease of use, speed of analysis and high separation efficiency. These features show distinct advantages in forensic toxicological analysis, even with respect to well-established GC and HPLC methods. In addition, the minimal requirements of CE in terms of both injected amounts and sample pre-treatment offer the possibility of preserving most of samples for future cross-testing with alternative methods, a need which is particularly strict in the forensic environment.

A special interest has recently been devoted to the hyphenation of CE with MS, which, notwithstanding excellent applications in other areas, in forensic sciences, including forensic toxicology, still looks to be in its infancy. Once the technological problems related to such an interfacing and the difficulties originating from the poor "electrophoretic background" of the forensic toxicology analysts had been overcome, the application of this technique based on standardized commercial instrumentation, will certainly offer new important developments, exploiting both the separation capability of CE and the excellent specificity and sensitivity of mass spectrometry.

The present overview shows that CE still represents an interesting *niche* of promising analytical developments in forensic science, as it is proved by a constantly growing number of applications. From this point of view, the still unresolved problem of the toxicological screening (particularly as general unknown analysis) looks to offer to CE-MS (particularly in the TOF configuration) a crucial, difficult challenge, which however, in our opinion, with the existing up-to-date instrumentation, can be won.

15.5 ABBREVIATIONS

APCI	Atmospheric pressure chemical ionisation
APPI	Atmospheric pressure photoionisation
BGE	Background electrolyte
BZP	Benzodiazepines
CD	Cyclodextrin
CE	Capillary electrophoresis
CEC	Capillary electrochromatography
CGE	Capillary gel electrophoresis
CIEF	Capillary isoelectric focusing
CITP	Capillary isotachophoresis
CTAB	Cetyl trimethyl ammonium bromide
CZE	Capillary zone electrophoresis
DAD	Diode array detector
DP	Discrimination power
DTAB	Dodecyl trimethyl ammonium bromide
EC	Electrochemical
EOF	Electroosmotic flow
ESI	Electrospray ionisation
FASS	Field amplified sample stacking
GC	Gas chromatography
GHB	γ-Hydroxybutyric acid

References pp. 532–534

HPLC	High-performance liquid chromatography
IP	Identification power
IT	Ion trap
LC	Liquid chromatography
LIF	Laser-induced fluorescence
LOD	Limit of detection
MA	Methamphetamine
MALDI	Matrix-assisted laser desorption
6-MAM	6-Monoacetylmorphine
MDA	Methylenedioxyamphetamine
MDEA	Methylenedioxyethylamphetamine
MDMA	Methylenedioxymethamphetamine
MEKC	Micellar electrokinetic chromatography
MLL	Mean list length
MS	Mass spectrometry
NACE	Non-aqueous capillary electrophoresis
PEG	Polyethyleneglycol
PVA	Polyvinyl alcohol
SDS	Sodium dodecyl sulphate
STA	Systematic toxicological analysis
TLC	Thin layer chromatography
TOF	Time-of-flight
UV	Ultraviolet

15.6 REFERENCES

1   C.A. Kuffner Jr., E. Marchi, J. Morgado and C.R. Rubio, Anal. Chem., 68 (1996) 241A–246A.
2   M.-L. Riekkola, Electrophoresis, 23 (2002) 3865–3883.
3   S.P. Porrai, M.-L. Riekkola and E. Kenndler, Chromatographia, 53 (2001) 290–294.
4   J.P. Chervet, R.E.J. Van Soest and M. Ursem, J. Chromatogr., 543 (1991) 439–449.
5   G.B. Gordon, US Patent 5,061,361 (1991).
6   C. Johns, M. Macka and P.R. Haddad, Electrophoresis, 24 (2003) 2150–2167.
7   J.L. Beckers and P. Boček, Electrophoresis, 24 (2003) 518–535.
8   T. You, X. Yang and E. Wang, Electroanalysis, 11 (1999) 459–464.
9   J.A. Olivares, N.T. Nguyen, C.R. Yonker and R.D. Smith, Anal. Chem., 59 (1987) 1230–1232.
10 J. Cai and J. Henion, J. Chromatogr. A, 703 (1995) 667–692.
11 P. Schmitt-Kopplin and M. Frommberger, Electrophoresis, 24 (2003) 3837–3867.
12 W.F. Smyth and P. Brooks, Electrophoresis, 25 (2004) 1413–1446.
13 W.F. Smyth, Electrophoresis, 26 (2005) 1334–1357.
14 W.F. Smyth, Electrophoresis, 27 (2006) 2051–2062.
15 S.A. Shamsi and B.E. Miller, Electrophoresis, 25 (2004) 3927–3961.
16 C.W. Huck, G. Stecher, H. Scherz and G. Bonn, Electrophoresis, 26 (2005) 1319–1333.
17 A.C. Servasi, J. Crommen and M. Fillet, Electrophoresis, 27 (2006) 2616–2629.
18 M. Yamashita and J.B. Fenn, J. Phys. Chem., 88 (1984) 4451–4459.
19 Y. Tanaka, K. Otsuka and S. Terabe, J. Pharm. Biomed. Anal., 30 (2003) 1889–1895.
20 R. Mol, G.J. de Jong and G.W. Somsen, Electrophoresis, 26 (2005) 146–154.
21 C.W. Huck, R. Backry, L.A. Hubr and G.K. Bonn, Electrophoresis, 27 (2006) 2067–2074.

22   R.D. Smith, C.J. Baringa and H.R. Udseth, Anal. Chem., 60 (1988) 1948–1952.
23   J.H. Wahl, D.R. Goodlett, H.R. Udseth and R.D. Smith, Anal. Chem., 64 (1992) 3194–3196.
24   J.H. Wahl and R.D. Smith, J. Capillary Electrop., 1 (1994) 62–71.
25   Z. Kele, G. Ferenc, E. Klement, G.K. Tòth and T. Janàky, Rapid Commun. Mass Spectrom., 19 (2005) 881–885.
26   G.M. Janini, T.P. Conrads, K.L. Wilkens, H.J. Issaq and T.D. Veenstra, Anal. Chem., 75 (2003) 1615–1619.
27   J.R. Perkins and K.B. Tomer, Anal. Chem., 66 (1994) 2835–2840.
28   A.B. Wey, J. Caslavska and W. Thormann, J. Chromatogr. A, 895 (2000) 133–146.
29   S.L. Nelsson, C. Andersson, P.J.R. Sjoeberg, D. Bylund, P. Petersson, M. Joerntén-Karlsson and K.E. Markides, Rapid Commun. Mass Spectrom., 17 (2003) 2267–2272.
30   R. Iio, S. Chinaka, S. Tanaka, N. Takayama and K. Hayakawa, Analyst, 128 (2003) 646–650.
31   S. McClean, E.J. O'Kane and W.F. Smyth, Electrophoresis, 21 (2000) 1381–1389.
32   A.B. Wey and W. Thormann, J. Chromatogr. A, 916 (2001) 225–238.
33   E.E.K. Baidoo, M.R. Clench, R.F. Smith and L.W. Tetler, J. Chromatogr. B, 796 (2003) 303–313.
34   S.A. Hofstadler, J.H. Wahl, R. Bakhtiar, G.A. Anderson, J.E. Bruce and R.D. Smith, J. Am. Soc. Mass Spectrom., 5 (1994) 894–899.
35   S.A. Hofstadler, J.C. Severs, R.D. Smith, F.D. Swanek and A.G. Ewing, Rapid Commun. Mass Spectrom., 10 (1996) 919–922.
36   S. Ullsten, A. Zuberovic, M. Wetterhall, E. Hardenborg, K.E. Markides and J. Bergquist, Electrophoresis, 25 (2004) 2090–2099.
37   I.M. Lazar, G. Naisbitt and M.L. Lee, Analyst, 123 (1998) 1449–1454.
38   I.M. Lazar, G. Naisbitt and M.L. Lee, Chomatographia, 50 (1999) 188–194.
39   I.M. Lazar, A.L. Rockwood, E.D. Lee and M.L. Lee, J. Chromatogr. A, 829 (1998) 279–288.
40   I.M. Lazar, A.L. Rockwood, E.D. Lee, J.C.H. Sin and M.L. Lee, Anal. Chem., 71 (1999) 2578–2581.
41   J. Schappler, G. Guillarme, J. Prat, J.-L. Veuthey and S. Rudaz, Electrophoresis, 27 (2006) 1537–1546.
42   G. Vanhoenacker, F. de l'Escaille, D. De Keukeleire and P. Sandra, J. Chromatogr. B, 799 (2004) 323–330.
43   U.L. Peri-Okonny, E. Kenndler, R.J. Stubbs and N.A. Guzman, Electrophoresis, 24 (2003) 139–150.
44   F. Steiner and M. Hassel, J. Chromatogr. A, 1068 (2005) 131–142.
45   R.A. de Zeeuw and J.P. Franke, 'General unknown' analysis, in: M.J. Bogusz (Ed.), Handbook of analytical separations, Vol. 2 – Forensic Science. Elsevier, Amsterdam, 2000, pp. 567–599.
46   O.H. Drummer, J. Chromatogr. B, 733 (1999) 27–45.
47   A. Polettini, J. Chromatogr. B, 733 (1999) 47–63.
48   H.H. Maurer, J. Chromatogr. B, 733 (1999) 3–25.
49   P.G.A.M. Schepers, J.P. Franke and R.A. de Zeeuw, J. Anal. Toxicol., 7 (1983) 272–278.
50   R.A. de Zeeuw, J. Chromatogr. B, 689 (1997) 71.
51   H.F.J. Hegge, J.P. Franke and R.A. de Zeeuw, J. Forensic Sci., 36 (1991) 1094.
52   J.P. Franke, M. Bogusz and R.A. de Zeeuw, Fresenius J. Anal. Chem., 347 (1993) 67.
53   J.P. Franke, M. Bogusz and R.A. de Zeeuw, Fresenius J. Anal. Chem., 347 (1993) 73.
54   C.M. Boone, J.P. Franke, R.A. de Zeeuw and K. Ensing, J. Chromatogr. A, 838 (1999) 259–272.
55   F. Tagliaro, F.P. Smith, S. Turrina, V. Equisetto and M. Marigo, J. Chromatogr. A, 735 (1996) 227–235.
56   C.M. Boone, J.P. Franke, R.A. de Zeeuw and K. Ensing, Electrophoresis, 21 (2000) 1545–1551.
57   C.M. Boone, G. Manetto, F. Tagliaro, J.C.M. Waterval, W.J.M. Underberg, J.-P. Franke, R.A. de Zeeuw and K. Ensing, Electrophoresis, 23 (2002) 67–73.
58   C.M. Boone, E.Z. Jonkers, J.P. Franke, R.A. de Zeeuw and K. Ensing, J. Chromatogr. A, 927 (2001) 203–210.
59   I.S. Lurie, M.J. Bethea, T.D. McKibben, P.A. Hays, P. Pellegrini, R. Sahai, A.D. Garcia and R. Weinberger, J. Forensic Sci., 46 (2001) 1025–1032.
60   C.M. Boone, J.W. Douma, J.P. Franke, R.A. de Zeeuw and K. Ensing, Forensic Sci. Int., 121 (2001) 89–96.
61   J.C. Hudson, M. Golin and M. Malcolm, Can. Soc. Forensic Sci. J., 28 (1995) 137–152.

62   A. Alnajjar, J.A. Butcher and B. McCord, Electrophoresis, 25 (2004) 1592–1600.
63   M. Nieddu, G. Boatto, A. Carta, A. Sanna and M. Pisano, Biomed. Chromatogr., 19 (2005)
      737–742.
64   Y.J. Heo, Y.S. Wang, M.K. In and K.-J. Lee, J. Chromatogr. B, 741 (2000) 221–230.
65   A. Ramseier, J. Caslavska and W. Thormann, Electrophoresis, 20 (1999) 2726–2738.
66   A. Ramseier, C. Siethoff, J. Caslavska and W. Thormann, Electrophoresis, 21 (2000) 380–387.
67   G. Boatto, M. Nieddu, A. Carta, A. Pau, M. Palomba, B. Asproni and R. Cerri, J. Chromatogr. B,
      814 (2005) 93–98.
68   J. Dahlén and S. von Eckardstein, Forensic Sci. Int., 157 (2006) 93–105.
69   A.-C. Servais, M. Fillet, P. Chiap, W. Dewé, P. Hubert and J. Crommen, Electrophoresis, 25
      (2004) 2701–2710.
70   F. Wei, M. Zhang and Y.-Q. Feng, Electrophoresis, 27 (2006) 1939–1948.
71   G. Manetto, F. Tagliaro, F. Crivellente, V.L. Pascali and M. Marigo, Electrophoresis, 21 (2000)
      2891–2898.
72   M. Frost and H. Koehler, Forensic Sci. Int., 92 (1998) 213–218.
73   C. Fang, J.-T. Liu, S.-H. Chou and C.-H. Lin, Electrophoresis, 24 (2003) 1031–1037.
74   Y. Suzuki, H. Arakawa and M. Maeda, Biochem. Chromatogr., 18 (2004) 150–154.
75   G. Vanhoenacker, F. de l'Escaille, D. De Keukeleire and P. Sandra, J. Pharm. Biochem. Anal., 34
      (2004) 595–606.
76   R. Gottardo, F. Bortolotti, G. De Paoli, J.P. Pascali, I. Mikšík and F. Tagliaro, J. Chromatogr. A
      (2007), in press, available online 10 January 2007.
77   S. Cherkaoui, S. Rudaz, E. Varesio and J.-L. Veuthey, Electrophoresis, 22 (2001) 3308–3315.
78   Y. Al-nouti and M.G. Bartlett, J. Am. Soc. Mass Spectrom., 13 (2002) 928–935.
79   K. Srinivasan and M.G. Bartlett, Rapid Commun. Mass Spectrom., 14 (2000) 624–632.
80   M.S. Anderson, B. Lu, M. Abdel-Rehim, S. Blomberg and L.G. Blomberg, Rapid Commun. Mass
      Spectrom., 18 (2004) 2612–2618.
81   S. Cherkaoui and J.-L. Veuthey, Electrophoresis, 23 (2002) 442–448.
82   R. Gottardo, F. Bortolotti, M. Trettene, G. De Paoli and F. Tagliaro, J. Chromatogr. A, 1051
      (2004) 207–211.
83   F. Bortolotti, G. De Paoli, R. Gottardo, M. Trattene and F. Tagliaro, J. Chromatogr. B, 800
      (2004) 239–244.
84   F. Tagliaro, S. Turrina, P. Pisi, F.P. Smith and M. Marigo, J. Chromatogr. B, 713 (1998) 27–49.
85   W. Thormann, I.S. Lurie, B. McCord, U. Marti, B. Cenni and N. Malik, Electrophoresis, 22 (2001)
      4216–4243.
86   F. Tagliaro and F. Bortolotti, Electrophoresis, 27 (2006) 231–243.
87   N. Anastos, N.W. Barnett and S.W. Lewis, Talanta, 67 (2005) 269–279.
88   J.F. van Bocxlaer, Ther. Drug Monit., 27 (2005) 752–755.
89   I.S. Lurie, P.A. Hays and K. Parker, Electrophoresis, 25 (2004) 1580–1591.
90   M. Macchia, G. Manetto, C. Mori, C. Papi, N. Di Pietro, V. Salotti, F. Bortolotti and F. Tagliaro,
      J. Chromatogr. A, 924 (2001) 499–506.
91   M.J. Hilhorst, G.W. Somsen and G.J. de Jong, Electrophoresis, 22 (2001) 2542–2564.

M.J. Bogusz (Ed.). Forensic Science
Handbook of Analytical Separations, Vol. 6

CHAPTER 16

HPLC-ICP-MS screening for forensic applications

Kevin M. Kubachka, Douglas D. Richardson and Joseph A. Caruso

Department of Chemistry, University of Cincinnati, Cincinnati, OH 45221-0172, USA

16.1 INTRODUCTION

There is a wide range of substances that pose a threat to human safety. These toxins exist in the environment both naturally and created by human activities. Some man-made toxins are designed with destructive applications in mind (anthrax, chemical warfare agents), others with good intentions, are toxic to humans (pesticides), while more are byproducts from industry/manufacturing (lead from fossil fuels). Regardless of their route to the environment and consequentially humans, scientific methods for the detection of these compounds are of great importance. In regards to forensics, analytical methods are needed with the following expectations in mind:

1. Establishing environmental levels of contaminants to ensure safety by:
   a. Assessing risk
   b. Minimizing exposure
   c. Developing countermeasures
2. Determining the identification of unknown substances at a crime scene.
3. Early detection of toxins in humans with the hopes of preventing chronic exposure.
4. Establishing the cause of death, as in the case of a poisoning.

These methods need to be applicable in a variety of sample matrices including environmental (water, soil, air, etc.), foods and beverages, and biological samples (hair, blood, urine, etc.). Due to the large variety of sample types, methods should be selective and able to resolve analytes from possible interferences. Qualitative analysis is important when identifying unknowns, while quantitation is necessary to establish toxic levels and enforce regulatory limits of various compounds. Also analysis at low levels is needed as many compounds are considered a hazard at even

References pp. 559–563

part per billion levels. For example, in the US, as mandated by the Environmental Protection Agency (EPA) in 2001, arsenic (As) in drinking water is limited to 10 parts per billion [1].

There are several elements of high concern as they are occupational or residential exposure hazards. Often times, to monitor these elements they are analyzed with respect to their total element concentration. Many EPA restrictions are currently based on this information, as in the case of uranium, in which the limit in water is at $30\,\mu g\ L^{-1}$ [1]. However, there are several elements in which more information is useful based on the fact that an element may have varying toxicities in different "forms." One specific example is the toxicity of chromium (Cr), as Cr^{VI} is considered carcinogenic and mutagenic and Cr^{III} is an essential nutrient [2]. Research devoted to the investigation of various chemical forms or species of the same element has been termed "elemental speciation." In this chapter, the elements antimony (Sb), arsenic (As), chromium (Cr), lead (Pb), mercury (Hg), selenium (Se), tellurium (Te), tin (Sn) and vanadium (Vn) will be discussed as they are of most interest and relate most heavily to forensic applications.

Typical analysis of the previously mentioned elements is performed through the use of instruments with element-specific detection capabilities, where these elements are used as "elemental tags." One of the more common elemental analysis techniques includes atomic absorption spectroscopy (AAS) with flame ionization (FAAS) or electrothermal ionization (ETAAS). Optical emission spectroscopy used other ionization sources including as microwave-induced plasma (MIP) and inductively coupled plasma (ICP). The inability of these spectroscopic methods to reach low detection limits and the presence of spectral interferences enhanced the need for more sensitive and selective detectors [3]. With the use of ICP with mass spectrometry (ICP-MS) for element-specific detection experiments, sensitivity, and selectivity are significantly improved. Other methods such as molecular mass spectrometry (MS) with softer ionization methods, such as electrospray ionization (ESI), chemical ionization (CI), or atmospheric pressure ionization (API) can also be used, however, a decrease in both sensitivity and selectivity are typically observed compared to ICP-MS element-specific detection [3].

ICP-MS is very sensitive with detection limits for some analytes at low part per trillion (ppt) levels. As previously mentioned, it is an element-specific detection method, with the ability to detect almost all elements in the periodic table virtually simultaneously depending upon the mass analyzer. The use of this harsh ionization technique in conjunction with a mass selective detector provides a tool for isotope-specific experiments. Quantitation is easily achieved using an external calibration curve, standard addition, or isotope dilution methods, with a dynamic range of up to 9 orders of magnitude. Also, when speciation analysis is required, it is easily interfaced with chromatographic techniques such as high performance liquid chromatography (HPLC), gas chromatography (GC), and capillary electrophoresis (CE). This chapter will focus upon forensic applications of HPLC with ICP-MS elemental-specific detection.

16.2 INSTRUMENTATION

16.2.1 Inductively coupled plasma mass spectrometer

What follows is a brief summary of ICP-MS from sample introduction to analyte detection as explained from left to right in accordance with Fig. 16.1. The initial part is the sample introduction, typically involving nebulization of the liquid sample, in which gas forces the solvent out of the end of a narrow inner diameter tube, thus forming a "mist" of sample. The sample then is transported into the ionization source, which in most cases is an argon plasma, by the carrier gas, typically argon. Here, the sample is desolvated, atomized, and ionized to positive charge by the plasma which is at temperatures from 6000 to 10,000 K. The ions then pass through the interface region consisting of both a sampler and skimmer cone (composed of either nickel or platinum depending upon the application). The interface region is the transition point from atmospheric pressure to the vacuum region of the instrument. This vacuum region consists of ion optics for focusing, collision/reaction cell for interference removal (described in the next section), mass analyzer (quadrupole, time of flight, sector field), and detector (electron multiplier).

16.2.1.1 Resolving interferences

As previously mentioned, ICP-MS is an element-specific technique with the mass analyzer detecting an analyte based on its mass to charge ratio (m/z). However, there are interferences present including isobaric interferences (^{54}Cr and ^{54}Fe) and polyatomic interferences ($^{40}Ar_2^+$ and $^{80}Se^+$). The $^{40}Ar_2^+$ polyatomic interference at 80 m/z is always present at high levels in an argon plasma and interferes with the most abundant isotope of Se at 80 m/z (49.61%). There are several approaches to solving this problem. One involves the use of introducing a collision or reaction gas in conjunction with an energy barrier (He, H_2, Xe, NH_3), to collide/react with the

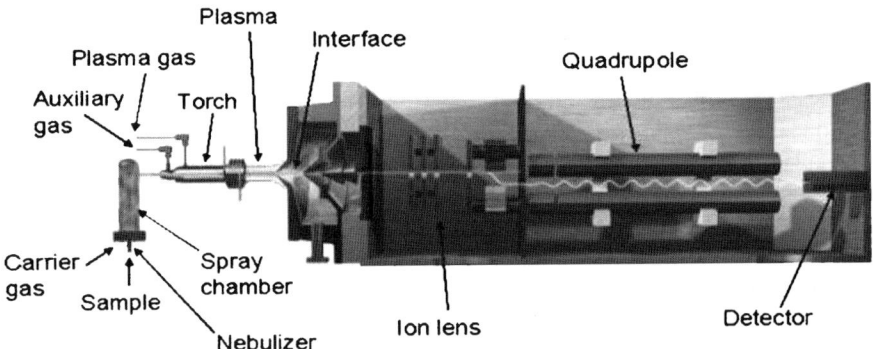

Fig. 16.1. Agilent ICP-MS block diagram. Reaction/collision cell is located after the ion lens and before the mass analyzer (in this case, a quadrupole). From Agilent Technologies with permission.

TABLE 16.1

ANALYTES OF INTEREST WITH THEIR MAJOR ISOTOPES. THE RELATIVE ISOTOPIC ABUNDANCE OF EACH ISOTOPE AND MAJOR CORRESPONDING POLYATOMIC INTERFERENCES ARE ALSO LISTED. THERE ARE NO POLYATOMIC INTERFERENCES FOR MERCURY

Isotopes	Percent Abundance	Polyatomic Interferences
^{75}As	100	^{40}Ar^{35}Cl^{+}
^{52}Cr	83.8	^{35}Cl^{16}O^{1}H^{+}
^{206}Pb, ^{207}Pb, ^{208}Pb	24.1, 22.1, 52.4	Pt oxides
^{121}Sb, ^{123}Sb	57.2, 47.8	Pd oxides
^{78}Se, ^{80}Se	23.5, 49.8	^{40}Ar$_2^{+}$
^{116}Sn, ^{118}Sn, ^{120}Sn	14.5, 24.2, 32.6	Ru and Pd oxides
^{126}Te, ^{128}Te, ^{130}Te	19.0, 31.7, 33.8	Ru and Pd oxides
^{51}Vn	99.8	^{35}Cl^{16}O^{+}, ^{34}S^{16}OH^{+}

Adapted from reference [9].

interference or analyte. Further specifics on this topic can be found in the literature [4–7]. For example, the use of H_2 can reduce the background from 10,000,000 cps to 10 cps, producing a background equivalent concentration of approximately 1 ppt for ^{80}Se [8].

There are other strategies of working around these interferences such as monitoring a less abundant isotope (^{82}Se) or calculating the contribution of possible interferences based isotopic information (^{40}Ar^{35}Cl^{+} and ^{75}As^{+} by monitoring ^{40}Ar^{37}Cl^{+}). Another option is the use of a high-resolution sector field detector that can resolve, for example, the ^{80}Se^{+} at (79.9165 Da) from ^{40}Ar^{40}Ar^{+} at (79.9248 Da) [5,6]. For elements discussed in this chapter, the common interferences are shown in Table 16.1.

16.2.1.2 Hydride generation

As previously mentioned above, the most common sample introduction technique for HPLC-ICP-MS is pneumatic nebulization. A major drawback of this sample introduction method is the transport efficiency of the sample aerosol to the plasma ionization source. Efforts to overcome this limitation have led to the development of alternative sample introduction methods such as laser ablation [10], electrothermal evaporation [11], and hydride generation [12,13]. These techniques allow for improved analyte transport efficiency, separation from matrix interferences, and improved detection limits by a factor of 10 or more compared to conventional pneumatic nebulization. Story and Caruso, in their critical review of hydride generation techniques reported a decrease in detection limits for hydride forming elements of up to three orders of magnitude compared to pneumatic nebulization techniques [14].

Hydride generation (HG), the most popular alternative sample introduction technique, has been utilized for nearly 40 years as a derivatization method for the analysis of trace elements capable of forming hydrides (As, Bi, Ge, Hg, Pb, Sb, Se, Sn, Te, In, and Tl). Generation of volatile metal hydrides can be accomplished in a variety of ways including: electrochemical generation, photoinduced generation, thermochemical generation, metal-acid reduction, and sodium borohydride reduction [15–17]. The suspected mechanism for hydride generation is discussed elsewhere [15,16].

Arsenic is the most common element subjected to analysis by HPLC-HG-ICP-MS [12,18–23]. Arsenic speciation experiments utilizing this technique have been performed in a wide range of sample matrices including; urine [12], river water [18], seawater [22], pepper plants [19], biological tissue [20,23], and soil [21].

Typically, hydride generation for arsenic was thought to be limited to the two inorganic forms (As^{III} and As^V) as well as three methylated forms (methylarsonate (MA), dimethylarsinate (DMA), and trimethylarsine oxide). This work demonstrated quantitative analysis of four arsenosugars without a decomposition step by HPLC-HG-ICP-MS for the first time [13].

Selenium speciation by HPLC-HG-ICP-MS has previously been applied to multiple inorganic and organic species [24–27]. These species include selenite, selenate, selenocyanate, trimethylselenonium, selenomethionine, and multiple selenosugars.

Analysis of mercury species with element-specific detection through the generation of a volatile species is typically referred to as cold vapor generation [15].

Developed in the 1960s this process results in the reduction of mercury species to the elemental state Hg^0 [15]. The resulting Hg^0 possesses an extremely high vapor pressure allowing simple separation from an aqueous matrix and efficient transport to the detector. Typical reducing agents for this reaction include tin chloride and sodium tetrahydroborate [15]. Analysis of mercury species by HPLC-CV-ICP-MS has previously been applied to seawater, soil, and fish tissue matrices [28–31].

Hydride generation was used to help eliminate NaCl from the matrix prior to ICP-MS detection when analyzing antimony, thus leading to better detection limits [32].

16.2.2 Interfacing HPLC with ICP-MS

With all the advantages of ICP-MS, one main disadvantage is that only elemental information is gained. Therefore, speciation was developed by which the theory is to separate various compounds, based on its species-specific properties using some type of chromatography, then pass the eluent into the ICP-MS for mass-specific detection.

The most common separation technique coupled to ICP-MS is HPLC. The interfacing is very simple and well documented in the literature [33], as just a PEEK tubing capillary from the end of the column to the nebulizer of the ICP-MS is typically all that is needed. The liquid eluent is ideal for ICP-MS nebulizers, which are chosen based on their compatibility with the method specific chromatographic flow rate.

Compositions of the mobile phases in HPLC are the most crucial of factors when interfacing with ICP-MS. Problems occur with high concentrations of organic modifiers such as methanol (MeOH) and acetonitrile (ACN). Typically these should be kept ≤20% v/v as higher levels can result in plasma instability, thus leading to a noisy baseline, and at high enough levels, the plasma will extinguish. Some research has cited that the addition of MeOH can actually increase the signal to noise of the analyte as it can increase ionization efficiency of the plasma source. Due to this reason and less instability of the plasma, as caused by MeOH compared to ACN, MeOH is typically the organic solvent of choice for liquid chromatographic methods applied to ICP-MS. This problem can be overcome by desolvation or dilution of the mobile phase prior to introduction into the ICP-MS [34] among other methods [35,36]. This premise also applies to solutions with high salt concentrations, as high enough levels will cause the similar problems and introduction of easily ionized elements such as sodium and potassium can lead to decreased ionization for analytes of interest. It is also important to note that exceeding any of these solution limitations could cause instrument components to dirty faster thus decreasing ion transmission and lowering sensitivity.

In HPLC-ICP-MS, species identification is accomplished through comparison to standards both by matching retention times and/or spiking a sample with a known standard and monitoring an increase in response. If standards are not commercially available or capable of being synthesized, the eluting unknown compound may be collected from the HPLC and analyzed with molecular mass spectrometry. This may require special sample preparation prior to analysis due to the difference in ionization source and mass analyzer, typically involving the removal of mobile phase constituents.

HPLC-ICP-MS has many advantages that make it very powerful for speciation analysis. HPLC is less time consuming and simple in regards to sample preparation in that no derivatization is needed either prior to analysis or post-column. The interfacing of the two techniques is quite user-friendly. ICP-MS is the ultimate in element-specific detection as far as analyzing a wide range of elements with high sensitivity and selectivity. A plethora of HPLC separation schemes can be interfaced with ICP-MS, thus leading to a wide variety of applications. Applications are constantly being improved upon and new samples/matrices are being explored all the time. Most of the applications discussed in this chapter involve analytical scale chromatography (flow rates $0.1 - 2.0 \, \text{mL} \, \text{min}^{-1}$). With the growing use of capillary scale ($0.002–0.1 \, \text{mL} \, \text{min}^{-1}$) and nano scale ($2–0.05 \, \mu\text{L} \, \text{min}^{-1}$) chromatography and compatible low flow nebulizers, these applications are being published more often, but the current lack of developed methods precludes them from this chapter.

16.3 APPLICATIONS

16.3.1 Arsenic

Arsenic has been an element of much concern throughout history. One source of infamy arises from the speculation that Napoleon met his demise at the hands of arsenics' toxicity. Although arsenic poisonings are not frequent, it will always carry high interest in the publics' eye. Arsenic is also under heavy surveillance as it is a common environmental pollutant. The main source of arsenic contamination for the aquatic environment is from geological sources, either surface weathering or underground deposits [37]. Humans play a part in the problem with the use of pesticides and herbicides, pressure-treated lumber, industrial manufacturing, and roxarsone (a growth promoter in chicken feed). The problem is especially prevalent in less developed countries, in which arsenic contamination of village water supplies is all too common. Because of these reasons, scientific methods are needed to analyze body fluids for diagnosing exposure as a result of a criminal poisoning (i.e. determining the cause of death) or from environmental sources. Methods should also be applicable for environmental samples to assess risk of contamination.

As mentioned in the introduction, many regulatory limits of arsenic (arsenic in drinking water) are based on total arsenic levels. However, different species of arsenic have varying levels of toxicity. Arsenic is most commonly found in these forms: arsenite (As^{III}), arsenate (As^{V}), monomethylarsonic acid (MMA^{V}), dimethylarsinic acid (DMA^{V}). As^{III} and As^{V} are viewed as the acutely toxic species. MMA^{V} and DMA^{V} compounds are carcinogenic. However, arsenobetaine (AsB) and arsenocholine (AsC) are viewed as virtually non-toxic [37]. There are several other arsenic-containing compounds that are of interest in various areas of research; they include, but are not limited to the species in Table 16.2.

The most common route of exposure for arsenic is oral ingestion. The commonly accepted detoxification pathway of arsenic is methylation. Fig. 16.2 shows a proposed methylation pathway in the body [38].

When a normal person is exposed to trace levels of inorganic arsenic it is partly methylated into MMA and DMA, which are excreted largely in the urine [39]. Arsenic also frequently enters the blood. One study explored the distribution of various arsenic species in human organs following fatal acute intoxication by arsenic trioxide. The majority of the arsenic found in the organs was As^{III} [39]. Arsenic is commonly analyzed in food and water samples, with a wide distribution of species

TABLE 16.2
COMMON ARSENIC SPECIES FOUND IN SPECIATION ANALYSIS

Name	Abbreviation	Chemical Formula
Inorganic compounds		
Arsenite (arsenous acid)	As^{III}	$As(OH)_3$
Arsenate (arsenic acid)	As^{V}	$AsO(OH)_3$
Organic compounds		
Monomethylarsonous acid	MMA^{III}	$CH_3As(OH)_2$
Monomethylarsonic acid	MMA^{V}	$CH_3AsO(OH)_2$
Dimethylarsinous acid	DMA^{III}	$(CH_3)_2AsOH$
Dimethylarsinic acid	DMA^{V}	$(CH_3)_2AsO(OH)$
Arsenobetaine	AsB	$(CH_3)_3As^+CH_2COO^-$
Arsenocholine	AsC	$(CH_3)_3As^+CH_2COO^-$
Trimethylarsine oxide	TMAO	$(CH_3)_4As^+$
Tetramethylarsonium ion	Me_4As^+	$(CH_3)_4As^+$
Arsenic-containing ribosides	Arseno sugars	Various sugar structures

Reprinted from [37] with permission from Elsevier.

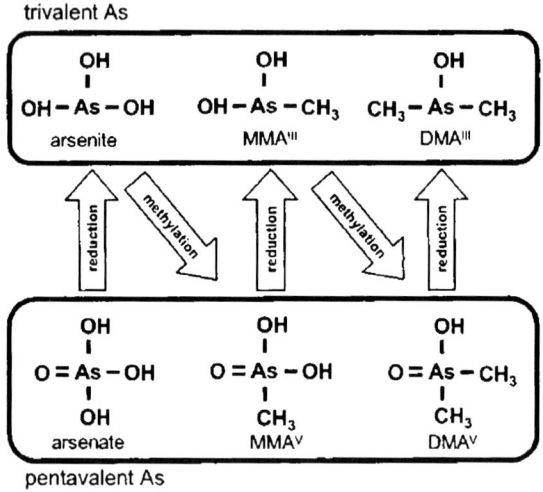

Fig. 16.2. Reduction and methylation pathways for arsenic. From reference [38] with permission.

present throughout various samples. For example, arsenic is found in high levels in seafood, but the majority of the arsenic is in AsB [40,41].

16.3.1.1 Sample preparation

One main problem in any speciation analysis is the preservation of the natural species at time prior and right up to analysis. This is a particular problem in the case of oxidation of As^V to As^{III}. In a study of urine samples, it was concluded that samples could be kept at low temperatures (4–20°C) with no additives or acidification for up to 2 months without inter-species conversion [42]. For liquid samples such as drinking water and urine, analyte extraction is viewed as unnecessary. Typically, the samples are diluted with deionized water (DIW), filtered through 0.45 μm filters to remove any particulate[1] and then injected into the HPLC-ICP-MS system. This is typically referred to as the "dilute and shoot" method and has been used by Shraim *et al.* [43] (5X dilution) among others [44]. He *et al.* [45] analyzed blood samples by partitioning the blood into serum and erythrocyte portions, followed by de-proteinization and filtration through a 0.45 μm membrane. No other pretreatment procedures were needed before injection. These methods can only be used for samples in which the matrix is liquid based and are compatible with the chromatographic method.

When dealing with a more complicated matrix, extraction of the analytes is necessary. The key to extraction for arsenic speciation is not to alter the natural

---

[1] It goes without saying that all liquid samples/extracts should be filtered through a 0.45 μm or 0.2 μm filter prior to injection into the HPLC system, even if not mentioned. This also applies to mobile phases and is especially crucial with capillary and nano HPLC applications.

species. Again, the problem of the conversion of As^V into As^{III} is the main concern and sometimes cannot be avoided. Some reports will only state the sum of the inorganic arsenic compounds rather than the individual species due to the uncontrollability of the As^V to As^{III} conversion.

Solid samples incur more challenges, in which extractions are necessary. Typically samples are lyophilized, made as homogeneous as possible, and then extracted with some type of solvent, typically DIW and/or organic solvents such as acetonitrile, methanol, and/or chloroform. Supercritical fluid extractions (SFE) and microwave-assisted extractions (MAE) are also among those used [46,47]. Once the analytes are in the liquid portion, it can be injected into the HPLC.

When analyzing human organs, extractions were performed using a 50:50 MeOH/DIW (v/v) solution. [39] There are a variety of extraction procedures involving food and environmental samples. A multi-step extraction for arsenic from algae uses methanol, acetone, and diethyl ether as used by Madsen *et al.* [48] Poultry waste was extracted with DIW, and solid phase extraction was used to clean up the samples by removing the hydrophobic organic compounds that would complicate the separation. [49] For freeze-dried apples, several extraction solvents were evaluated including 50:50 MeOH/DIW (v/v), α-amylase + 50:50 MeOH/DIW (v/v), and 40:60 ACN/DIW (v/v). Each was in combination with sonication and all three produced high percent recoveries. [50] In work done by Ackerman *et al.* [51] an extraction using trifluoroacetic acid (TFA) is compared with an enzymatic extraction using pepsin and pancreatin. These methods were applied to rice samples with the TFA extraction proving slightly more efficient; however, the use of TFA will convert As^V into As^{III}.

16.3.1.2 Chromatography

As arsenic is of high concern, there are a great number of speciation techniques; only several of the more common methods are discussed here. For a more complete listing refer to the review by B'Hymer *et al.* [37] Ion exchange is the most commonly used HPLC technique for arsenic speciation. Anion-exchange is often used to separate As^{III}, As^V, MMA^V, and DMA^V, while cation-exchange is frequently used to separate AsB, AsC, trimethylarsine oxide (TMAO) and tetramethylarsonium ion (Me_4As^+). In ion exchange, the main factors that influence the separation are pH of the buffer, ionic strength and concentration of the buffer solutions, and temperature. As previously mentioned, buffers that are compatible with ICP-MS must be chosen. The most commonly used buffers for ion exchange with ICP-MS are phosphate, [44,50] carbonate, [52,53] phthalic acid, [54] tetramethylammonium hydroxide, [55] formate, [38] and nitrate [56] buffers. Ammonium is the cation of choice as it is the most volatile and leaves less residual material on the detector's components.

A few specific examples of ion exchange methods are present in the following paragraphs. The Dionex IonPak AS7 is commonly used and Kohlmeyer *et al.* [57] were able to separate 17 arsenic compounds using a nitric acid, 0.05 mM benzene-1, 2-disulfonic acid dipotassium salt, 0.5% MeOH v/v mobile phase with gradient altering the nitric acid concentration. This type of separation is used in several applications with nitric acid levels typically from 0.1 to 50 mM. Other additives to the

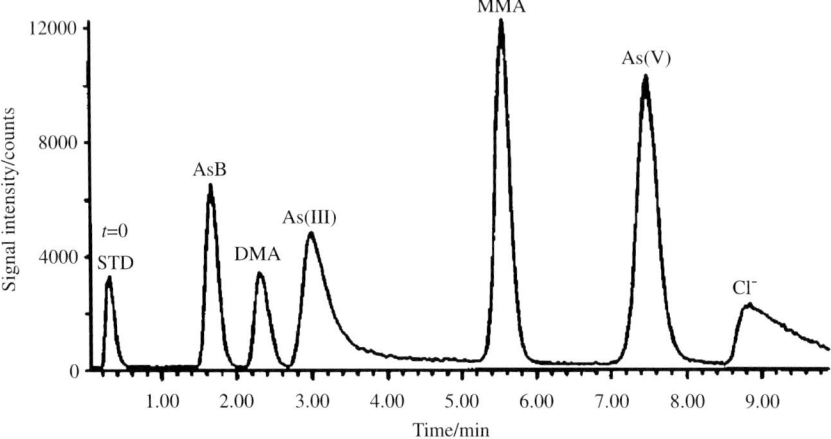

Fig. 16.3. Anion exchange chromatogram of five arsenic species in a 0.1% Cl-matrix resulting from gradient elution with ICP-MS detection. AsB, 2 ppb; DMA, 1 ppb; As^{III}, DMA and As^V, 5 ppb. From reference [59], reproduced with permission of The Royal Society of Chemistry.

mobile phases have been used such as acetate buffers including ion pairs like benzene-1,2-disulfonic acid. Small levels of MeOH have been added to increase sensitivity (0.5% v/v in DIW). Creed and coworkers reported $(NH_4)_2CO_3$ mobile-phase-based separations using a Hamilton PRP-X100 column and InterAction Ion-120 column interfaced with ICP-MS (shown in Fig. 16.3) and ESI-MS for identification of previously uncharacterized arsenosugars [58,59]. Madsen *et al.* [48] developed a cationic exchange methods for algal extract using a Hewlett Packard Zorbax 300 SCX column using 20 mM pyridine, pH 2.2 with ICP-MS and ESI-MS for detection.

Reversed-phase ion-pairing chromatography is also commonly used for arsenic speciation and is explained in depth elsewhere [60–65]. Le and Ma [66] used a C_{18} column with 10 mM propanesulfonate, 4 mM malonic acid, and 0.1% MeOH v/v to separate 7 arsenic standards (shown in Fig. 16.4).

16.3.1.3 Detection limits

Detection limits for arsenic species are very low at part per trillion levels with absolution detection limits reported around 50–200 pg of arsenic [37].

16.3.2 Selenium

High interest involving selenium centers on its dual personality as it can be both toxic or beneficial. Most recently it was discovered that Se has anti-cancer properties as shown in research by Clark *et al.* [67]. Selenium has also been investigated in slowing progression from AIDS and HIV along with proper immune functioning [68]. The problem with selenium is that its toxic and the beneficial range is quite

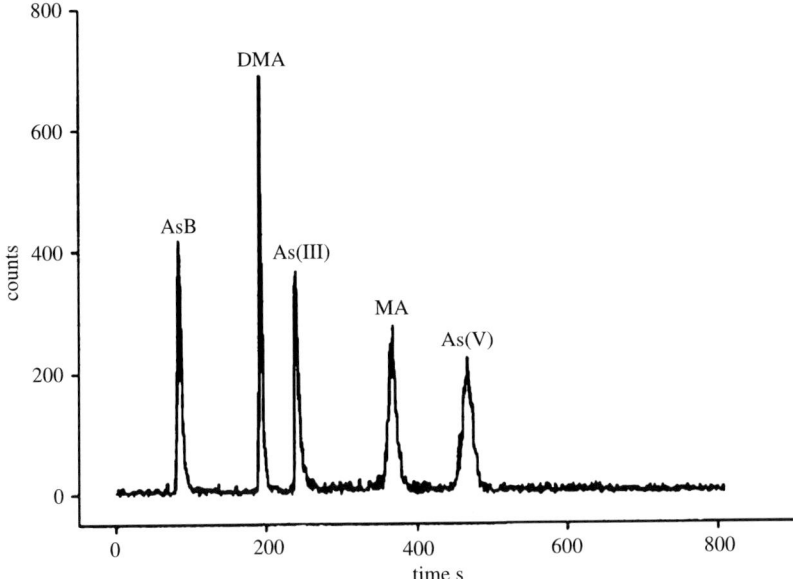

Fig. 16.4. Anion exchange chromatogram of an aqueous standard solution containing As, DMA, As[III], MMA (MA), and As[V] (1 μg As L^{-1} each). From reference [55] with kind permission of Springer Science and Business Media.

small, in general it is considered to be between 40 and 100 μg kg^{-1} of body weight per day [69]. Due to this narrow toxicity range, accurate and precise methods are needed to correctly assess selenium levels.

Selenium reaches the environment mainly from coal mining and irrigation water that extracts the selenium from underground shale. It is generally accepted that the inorganic forms, selenate and selenite, are considered toxic while the organic forms are non-toxic or even beneficial. Many supplements are now sold containing the most common beneficial Se compound, selenomethionine (SeMet). Other compounds, such as selenomethylcysteine (SeMeCys), are also thought to have anti-cancer properties. Recently, there have been several selenosugars that have been reported. [70,71] A more complete list of common selenium compounds is shown in Table 16.3. Selenium also has the capability of entering proteins by both specific coding via the UGA codon as selenocysteine, and nonspecifically in the form of SeMet as selenium replaces sulfur in methionine. There are approximately 35 selenoproteins that have been found in mammals [72], while others have been found in yeast [73].

Cases of selenium poisoning are very rare [75]. Such cases involving selenic acid or sodium selenite have been reported [76]. The bulk of the research involving humans and selenium focuses on the bioavailability of various selenium species and metabolism.

Environmental samples and food stuffs receive the most attention related to selenium speciation. Important applications of selenium speciation include insuring

TABLE 16.3
SOME INORGANIC AND ORGANIC COMPOUNDS OF INTEREST IN SPECIATION ANALYSIS

Chemical Name	H_2Se (volatile)
Hydrogenselenide	SeO_3H_2 (SeO_3^{2-})
Selenous acid (selenite)	SeO_4H_2 (SeO_4^{2-})
Selenocyanate	HSeCN
Trimethylselenonium cation	$(CH_3)_3Se^+$
Dimethylselenide	$(CH_3)_2Se$ (volatile)
Dimethyldiselenide	$(CH_3)Se-Se(CH_3)$ (volatile)
Dimethylseleniumsulfide	$(CH_3)Se-S(CH_3)$ (volatile)
Dimethylseleniumdioxide	$(CH_3)_2SeO_2$ (volatile)
Dimethylselenopropionate	$(CH_3)_2Se^+CH_2CH_2COOH$
Methylselenol	CH_3SeH
Methylseleninic acid	$CH_3Se(O)OH$
Methylseleninic acid	CH_3SeOH
Selenocysteine	$HOOCCH(NH_2)CH_2-Se-H$
Selenomethylcysteine	$HOOCCH(NH_2)CH_2-Se-CH_3$
Selenocystine	$HOOCCH(NH_2)CH_2-Se-Se-CH_2CH(NH_2)COOH$
Selenomethionine	$HOOCCH(NH_2)CH_2CH_2-Se-CH_3$
Selenoethionine	$HOOCCH(NH_2)CH_2CH_2-Se-CH_2CH_3$
γ-Glutamyl-Se-methylselenocysteine	$H_2NCH_2CH_2-CO-NHCH(COOH)CH_2-Se-CH_3$
Selenocystathionine	$HOOCCH(NH_2)CH_2CH_2-Se-CH_2CH(NH_3)COOH$
Selenohomocysteine	$HOOCCH(NH_2)CH_2CH_2-Se-H$
Se-adenoxylselenohomocysteine	$HOOCCH(NH_2)CH_2CH_2-Se-CH_2C_4H_5C_5N_4NH_2$
Selenosugars	Various sugar structures
Selenoproteins	Various proteins and enzymes (i.e., GPX, Selenoprotein P, TR)

that foods contain the beneficial forms of selenium rather than toxic species and confirming the selenium content of commercial selenium supplements.

16.3.2.1 Sample preparation

Liquid samples that have been analyzed include urine, blood, and natural waters. Urine samples can simply be diluted [71]. Blood samples can be centrifuged down, and their supernatant collected, then injected [77]. Water samples can just be injected directly.

Extraction of selenium must be performed to preserve the original species. Typically for food samples, hot water extracts are performed to mimic food preparation techniques [78]. This is typically deemed adequate for freeing selenium species not associated with larger molecules [79]. There is a wide range of sample preparation techniques used to extract selenoamino acids (SeMet, SeMeCys, etc.) from proteins and other larger molecules. These include the use of various combinations sodium dodecyl sulfate, driselase, proteinase K [80], protease [81], trypsin [82], and lipase [83]. Yang *et al.* [84] evaluated 14 extraction techniques applied to yeast samples and determined that the use of 4 M methanesulfonic acid was the most efficient at extracting SeMet. Concern should be taken when looking at previous research in regards to SeMet content, as when enzymes are used, the SeMet content could be from proteins rather than native SeMet [79].

16.3.2.2 Chromatography

Several speciation methods have been designed. A few of the more commonly used methods are summarized here, for a more complete listing refer reviews by B'Hymer and Caruso [74] and Polatajko *et al.* [79]. Anionic exchange is not as commonly used, due to the relatively high pKa's of the majority of selenium compounds of interest, but it is able to resolve selenite and selenate. Cationic exchange is more frequently used. Larsen *et al.* [85] have achieved the separation of a mixture of 12 selenium species comprising of selenoamino acids, selenonium ions, and inorganic selenium. Separation of anionic species was carried out with an anion-exchange column with isocratic elution with an aqueous salicylate-TRIS mobile phase at pH = 8.5. The cationic species were separated using cation-exchange column with gradient elution with aqueous pyridinium formate, pH = 3 [79].

Perhaps, the most commonly used separation method of selenium species is ion-pairing reversed phase chromatography. Although selenate and selenite typically cannot be resolved [79], a wide variety of selenium compounds can be separated. An assortment of ion-pairs have been used including formic acid, trifluoroacetic acid (TFA), heptafluorobutyric acid (HFBA), hexane sulfonic acid, citric acid, and malonic acid. TFA and HFBA concentrations are typically $\leq 1\%$ (0.1%) but up to 5% has been used. Isocratic runs have been used with the addition of an organic modifier. Kotrebai *et al.* [86] compared different ion pairs, with the eventual separation of 22 selenium compounds using ion-pairing RP chromatography, shown in Fig. 16.5. Montes-Bayon *et al.* [87] and Uden and coworkers [86,88] use the combination of ICP-MS and ESI-MS which can help to identify selenium species which could not be identified by ICP-MS due to the lack of available standards. For this, mobile phases must be compatible with both methods including at or below 0.1% (v/v) TFA or HFBA (for electrospray) and low organic (for ICP). Only the relatively high abundant selenium species can be identified by ESI-MS due to sensitivity differences or preconcentration is necessary.

16.3.2.3 Detection limits

Common detection limits for selenium species are typically low part per billion levels. Encinar *et al.* [89] reported detection limits below 0.5 ppb for human serum samples using a collision cell. B'Hymer and Caruso reported detection limits of 0.5–2 ppb.

16.3.3 Chromium

Chromium is a perfect example of this variation in species toxicity in that the chromic ion (Cr^{III}) is an essential nutrient involved in the regulation of glucose, cholesterol, and fatty acid metabolism, while hexavalent chromium, typically chromate, is mutanogenic and carcinogenic [90]. Alternatively, chromium particles in air

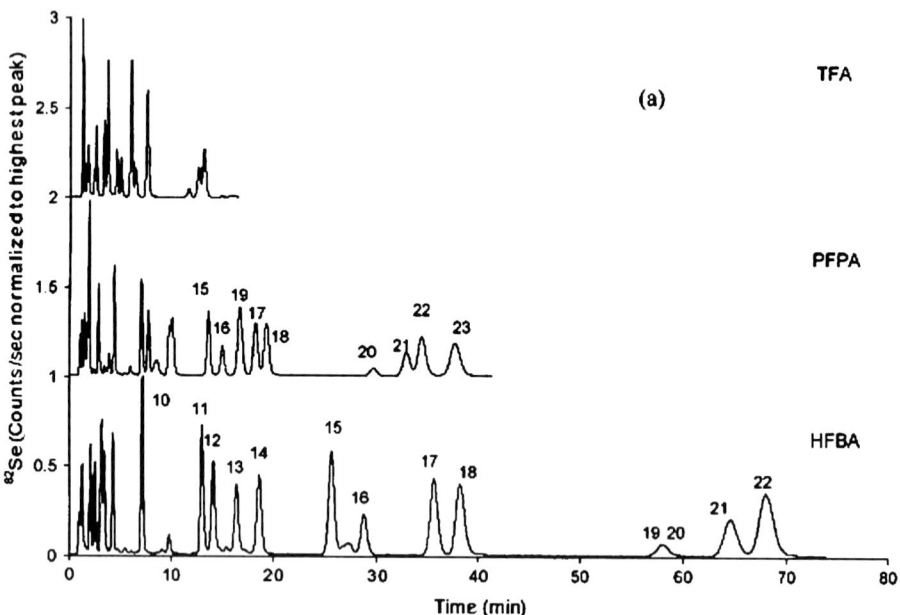

Fig. 16.5. HPLC-ICP-MS chromatograms of selenium standards using 0.1% TFA, PFPA or HFBA as ion-pairing agents. Se^{IV}, Se^{VI}, and selenocystine elute before 7 min. Se-methylselenocysteine (9) and selenomethionine (11). Reprinted from [86] with permission from Elsevier.

aid in the oxidation of sulfur dioxide (SO$_2$) resulting in the production of acidic gases which produce acid rain [90]. Hexavalent chromium is widely utilized in a variety of industries including: plastics, metal working, and paints/inks/dyes [91]. Due to the widespread use and health risk associated with this element, environmental and occupational regulatory agencies have mandated strict guidelines for the storage, use, and disposal of anything containing hexavalent chromium [91]. Moreover, the toxicity differences between Cr^{III} and Cr^{VI} requires an analytical method capable of species differentiation and low level detection.

16.3.3.1 Sample preparation

Separation of the two chromium species is a challenging task in that the chromic ion (Cr^{III}) exists as a cationic aqua-hydroxo complex, while Cr^{VI} exists as anionic chromate. The two redox forms of chromium pose limitations in sample collection, storage, and pretreatment procedures due to possible interconversion between the two oxidation states. Pantsar-Kallio and Manninen [93] stress the importance of sample analysis immediately following collection.

16.3.3.2 Chromatography

Efforts to separate both redox species of chromium have led to the investigation of reverse-phase chromatography or ion chromatography coupled with element-specific detection [90–101]. Byrdy *et al.* [95] were the first to demonstrate HPLC-ICP-MS for chromium speciation. This work consisted of a reverse-phase separation on a Dionex AS7 column with an NG1 guard [95]. The mobile phase was comprised of 35 mM ammonium sulfate with ammonium hydroxide at pH 9.2 [95]. Prior to analysis, a chelation procedure was utilized to stabilize the Cr^{3+} in the standard solutions with EDTA. The authors cited previous work of successful application of the developed method to dietary supplements, urine standard reference material, and chromium dyes [95,96].

Pantsar-Kallio and Manninen performed anion exchange for the separation of chromium species from multiple interfering ions such as chloride, sulfate, carbonate, cyanide, and some organic species was discussed. They used a Waters IC-Pak A column with a 4–40 mM HNO_3 gradient [93].

Seby *et al.* [94] investigated the influence of interfering ions for chromium speciation following work by Barnowski *et al.* [97] in which the use of a IonPac CS5A with a IonPac CG5A guard column with both anion and cation exchange capabilities using a nitric acid eluent. Separation of the two chromium species was achieved in less than 7 min [94]. This is shown in Fig. 16.6. Due to the increase in pH accompanied with the addition of hydrogen carbonate resulted in hydrolysis of the Cr^{III} into various hydroxycomplexes (Fig. 16.7). The newly formed peaks were suspected to be $Cr(OH)(H_2O)_5^{2+}$ and $Cr(OH)_3^0$ due to the instability of $Cr(OH)_2(H_2O)^{4+}$.

16.3.3.3 Detection limits

Byrdy *et al.* [95] concluded that the single ion monitoring capability of ICP-MS provided ultra-trace detection levels (ng) previously never achieved. Seby *et al.* [94] achieved detection limits of 0.38 and 0.20 $\mu g\,L^{-1}$ for Cr^{III} and Cr^{VI} respectively. Several other HPLC-ICP-MS techniques were reported with varying detection limits [98–101], with the lowest being 0.063 and 0.061 $\mu g\,mL^{-1}$ for Cr^{III} and Cr^{VI}, respectively [100].

16.3.4 Lead

In the past, lead was used in paints and as an anti-knocking agent in fuels. However, its use is now heavily regulated, due to its many health concerns. Of the utmost importance are its implications to the health of children as they are at the highest risk for lead contamination according to the EPA. The main routes of exposure include deteriorating lead-based paint, lead-contaminated dust, and lead-contaminated residential soil [1]. Lead species of high interest include inorganic Pb (Pb^{2+}), trialkyl lead and tetraalkyl lead compounds. Tetraethyl lead (TTEL) is a central nervous system toxin that produces an acute toxic psychosis [102]. Inorganic lead is much less toxic than its organic forms with varying toxicities within organic lead compounds, for example tetraethyl lead is approximately 10 fold more toxic than tetramethyl lead in rats [103]. When coming from

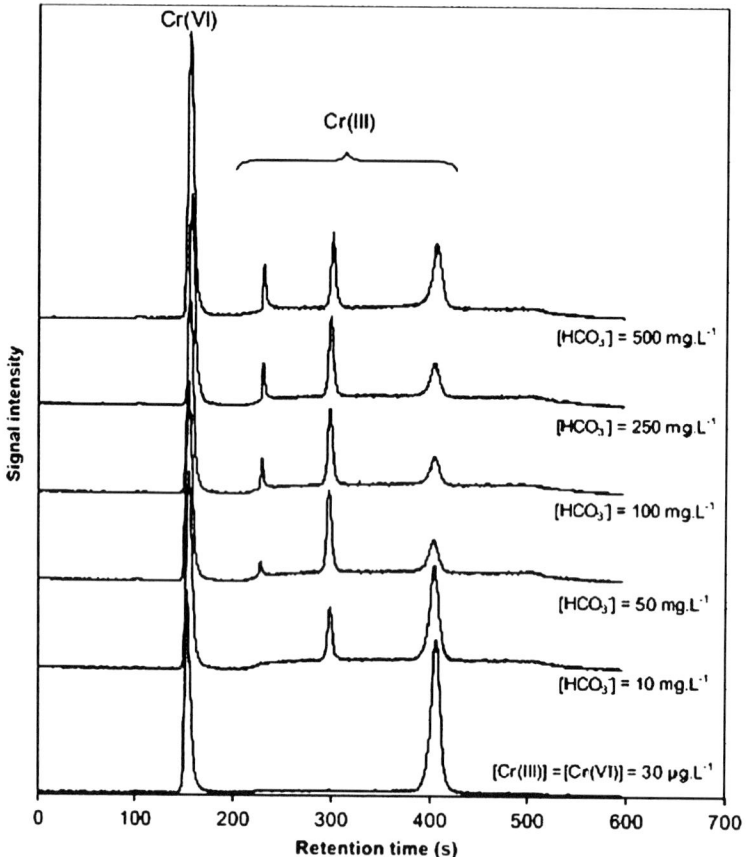

Fig. 16.6. Description of the effect of hydrogen carbonate ions on the developed chromium speciation method. From reference [94], reproduced with permission of The Royal Society of Chemistry.

$$Cr(H_2O)_6^{3+} + H_2O \rightleftharpoons Cr(OH)(H_2O)_5^{2+} + H_3O^+$$

$$Cr(OH)(H_2O)_5^{2+} + H_2O \rightleftharpoons Cr(OH)_2(H_2O)_4^+ + H_3O^+$$

$$Cr(OH)_2(H_2O)_4^+ + H_2O \rightleftharpoons Cr(OH)_2^0 + H_3O^+$$

Fig. 16.7. Hydrolysis of Cr^{III} in solution over the pH range of 4–10. From reference [94], reproduced with permission of The Royal Society of Chemistry.

auto emissions lead enters the environment as tetraalkyl lead compounds and is easily broken down by the environment into trialkyl lead compounds. Eventually, biological systems will metabolize the lead into the least toxic form, inorganic lead.

16.3.4.1 Sample preparation

Ideally, samples for lead speciation should be analyzed as soon as possible. If not, they should be stored in Teflon or polyethylene containers. When analyzing biological samples concern should be taken because acidification changes the physico-chemical distribution of lead species, and therefore must not be used prior to speciation [104]. Ebdon and coworkers [105] found that using properly cleaned containers, unacidified natural water samples stored at 4°C in absence of light for up to 3 months exhibited no measurable inter-species conversion.

16.3.4.2 Chromatography

In recent years, the speciation of lead has been analyzed more using GC rather than HPLC. The majority of HPLC-ICP-MS methods for lead speciation methods were developed prior to 1999. HPLC methods for the separation of lead compounds have dealt almost exclusively with the separation of the trialkyl forms. These separations exploit the difference in hydrophobicity of the species, typically utilizing an ion-pair in combination with high percent organic with a RP column. Ebdon *et al.* [105] used an acetate buffer system with sodium 1-pentanesulfonic acid (SPSA) and 60% MeOH (v/v) (shown in Fig. 16.8). Pan *et al.* [106] used similar conditions without SPSA. In two studies by Al-Rashdan *et al.* [107,108] Pb (Pb^{2+}), trimethyllead chloride (TML), triethyllead chloride (TEL), triphenyllead chloride (TPhL), and (TTEL) were separated with isocratic or gradient elution involving the use of 30% v/v MeOH. Brown *et al.* [109] used only a DIW/MeOH mobile phase system to separate Pb^{II}, TML and TEL using isotope dilution for quantification. Shum *et al.* [110] separated the same using a microbore RP column with 5 mM ammonium pentanesulfonate in 20% ACN (v/v) (shown in Fig. 16.9).

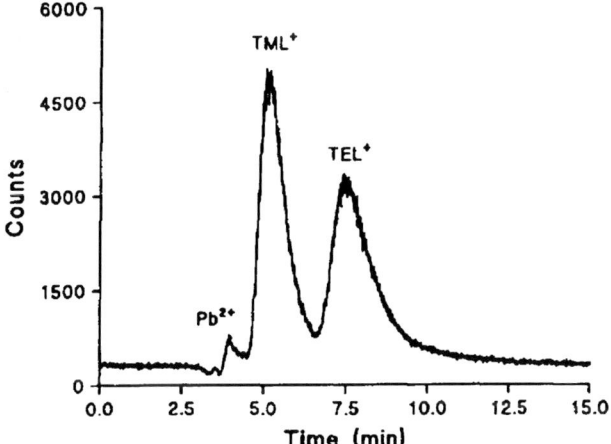

Fig. 16.8. Isocratic ion-pairing chromatogram of inorganic lead, TML and TEL. Reprinted from [105] with permission from Elsevier.

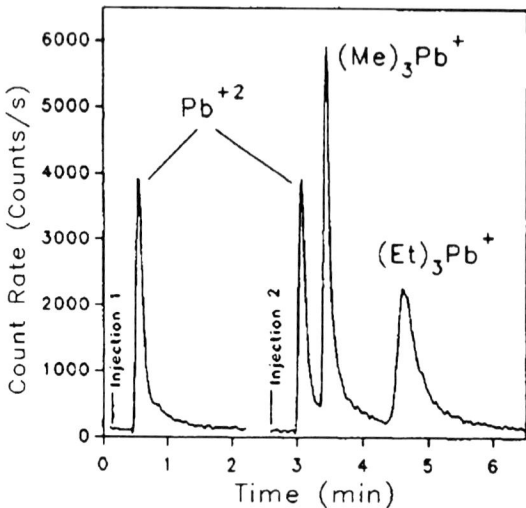

Fig. 16.9. Isocratic ion-pairing chromatogram of separation of Pb species in NIST SRM 2670 freeze-dried urine (normal level). EDTA was added to the sample at 10 mg L-1 before injection. *Injection* 1: NIST SRM 2670 freeze-dried urine (normal level). *Injection* 2: NIST SRM 2670 freeze-dried urine spiked with 40 pg (as Pb) of each of the trialkyllead species. From reference [110] with permission.

16.3.4.3 Detection limits

Although baseline resolution is somewhat difficult to achieve for each of the above mentioned methods, detection limits were reported at 0.5–2 ppb.

16.3.5 Mercury

Mercury is one of the more infamous metals of concern. Mercury mainly enters the environment from industrial applications. Coal-fired power plants are the largest remaining source of human-caused mercury emissions [1]. Criminal mercury poisonings are extremely rare and in one reported case only total Hg in blood was examined rather than mercury speciation [111]. Organomercury compounds are more toxic than inorganic mercury, Hg^{II} [112]. Mercury compounds of high interest include methyl mercury (CH_3Hg^+), ethyl mercury ($C_2H_5Hg^+$), dimethyl mercury, diethyl mercury and phenyl mercury ($C_6H_5Hg^+$).

16.3.5.1 Sample preparation

The containers most suitable for Hg sample storage are Pyrex or Teflon containers and must be cleaned rigourously (e.g., with aqua regia, chromic acid, nitric acid, and BrCl). A final soaking of Teflon in 1% HCl at 70 °C removes all traces of oxidizing compounds that may interfere with CH_3Hg^+ [113].

Liquid samples are quite simplistic in analysis by HPLC-ICP-MS as samples can be directly injected or diluted, then injected. Extracting mercury species from non-liquid samples is the major source of difficulty in mercury speciation. One problem is extracting both organic and inorganic species in the sample extraction. Also it has been shown that conversion of Hg^{II} into CH_3Hg^+ is possible in many extraction techniques, thus leading to overestimates of CH_3Hg^+ levels [113].

Most extraction procedures focus on removing CH_3Hg^+ and Hg^{II} from the matrix. Mercury species were extracted from hair samples using 2 mL of HNO_3 (concentrated) and 1 mL of H_2O_2 (30% v/v) overnight at room temperature. The solution was diluted to 10 mL with DIW and injected directly onto the column [114]. Most extraction techniques derive from one presented by Westoo in which the sample is heated in acid (HCl) at high temperatures (microwave-assisted extraction), followed by non-polar (toluene) extraction. Some authors have recommended back extraction using some type of aqueous phase to further clean up the sample [112,113,115]. Two of the more successful procedures in terms of extraction efficiency of CH_3Hg^+ are by Horvat *et al.* [116] with $95 \pm 4\%$ extracted from soil via a distillation method; and by Tseng *et al.* [117] with 95–105% extracted from fish tissue by alkaline digestion using tetramethyl ammonium hydroxide (TMAH) with focused microwave power.

16.3.5.2 Chromatography

As with lead, the majority of Hg speciation is currently performed using GC. However, there are still several HPLC methods, most of which involve the use of some type of ion pair with RP separation (IP-RP) and an organic modifier to help elute hydrophobic analytes from the column.

Morton *et al.* [114] separated Hg^{II} from CH_3Hg^+ using a C_{18} RP column with 0.06 M ammonium acetate, 5% v/v methanol, 0.1% v/v 2-mercaptoethanol as a mobile phase. Shum *et al.* [110] separated Hg^{II}, CH_3Hg^+, and methyl ethyl mercury using a microbore RP column with 5 mM ammonium pentanesulfonate in 20% ACN (v/v). Separation of 5 mercury compounds was presented by Falter and Ilgen using a C_{18} column with 65% ACN (v/v) with ammonium acetate to adjust pH to 5.5 [118] (shown in Fig. 16.10). Tu *et al.* [31] used a Dionex PCX-500 guard column with 65% MeOH v/v with 0.45 M HCl.

16.3.5.3 Detection limits

Detection limits of species are quite low at >1 ppb per compound. Tu *et al.* [31] reported detection limits of 35 pg Hg mL^{-1} and 73 pg Hg mL^{-1} for Hg^{II} and CH_3Hg^+, respectively. Preconcentration was used by Falter and Ilgen [118] to get absolute detection limits of 10–20 pg of Hg for various mercury species. Shum *et al.* [110] reported absolute detection limits for Hg^{II}, CH_3Hg^+ and MEM of 6–18 pg of Hg or 3–9 ppb in regards to Hg.

16.3.6 Tin

Tin is also a heavily regulated element. The toxicity of tin is different from species to species as the organic form is far more toxic than the inorganic forms, which are

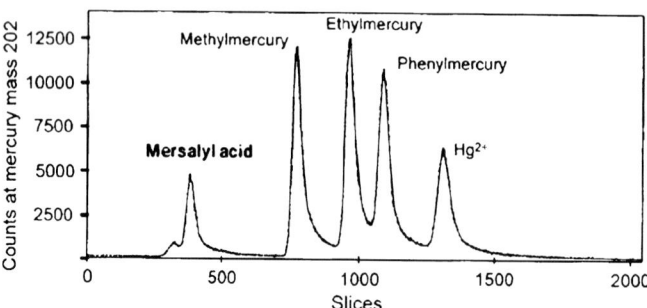

Fig. 16.10. Separation of 5 mercury standards. The concentrations of each standard are: mersalyl acid at 50 pg Hg mL^{-1}; CH$_3$Hg$^+$, C$_2$H$_5$Hg$^+$, and C$_6$H$_5$Hg$^+$ are at 200 pg Hg mL^{-1}; and Hg^{II} at 220 pg Hg mL^{-1}. From reference [118] with kind permission of Springer Science and Business Media.

regarded as virtually non-toxic. The toxicity of the organic forms increase as the alkyl chains increase. Maximum toxicity is achieved at tributyltin (TBT) [119]. The plastics industry is the source of most organotin usage [120]. They are also found in anti-fouling paints and pesticides [121]. These compounds tend to accumulate in the soil, thus making them a constant hazard [122]. Species of interest include the butyltins (mono (MBT), di (DMT), and (TBT)), phenyltins (mono (MPhT), di (DPhT), and tri (TPhT)), and triorganotins (methyl (TMT), ethyl (TET), and propyl (TPT)).

16.3.6.1 Sample preparation

Sample preparation remains a challenge for extracting organotins from solid samples. Solvent extractions have some success at approximately 70–100%. These usually involve combinations of sonication, stirring, hydrochloric acid, and/or acetic acid with sequential non-polar extraction [122,123]. A method used by Arnold *et al.* [124] involving pressurized liquid extraction (PLE) offers high recovery with little degradation of original organotin species. Microwave-assisted extraction was shown to cause extensive degradation of phenyltin compounds [125]. Optimum solvents for PLE were shown to be methanol with 0.5 M acetic acid and 0.2% (w/v) tropolone [122] and extraction efficiencies were above 90%, except for DPhT.

16.3.6.2 Chromatography

For HPLC-ICP-MS, a few genres of separations have been applied to the separation of organotin compounds. Again, as with lead and mercury, the majority of current speciation analysis is carried out by GC. The main species of interest are TBT and the phenyltin compounds. The use of ion exchange is popular but currently, none have separated all the triorganotins, butyl tin species, and phenyl tin species. This is most likely due to the incompatibility of the mobile phases to elute every-thing off the column with ICP-MS [121]. Reversed phase separations have also been used. Chiron *et al.* [122] used a TSK gel ODS TM column and a combination of tropolone and triethylamine at an optimum ratio of 0.075% (w/v):0.1% (v/v) in a

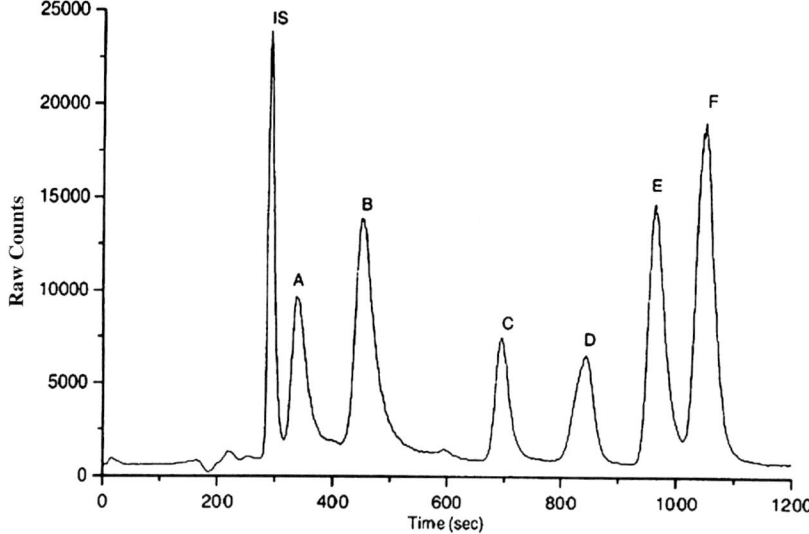

Fig. 16.11. Chromatogram obtained after extraction of a spiked sediment by PLE. Spiking level: BTs—1 µg g^{-1}; PhTs—0.5 µg g^{-1}. (IS) internal standard; (A) Monophenyltin; (B) monobutyltin; (C) triphenyltin; (D) diphenyltin; (E) tributyltin; (F) dibutyltin. Reprinted from [122] with permission from Elsevier.

methanol–DIW–acetic acid (72.5:21.5:6, v/v/v) mobile phase (shown in Fig. 16.11). Sodium pentane sulfonate has been employed as an ion pair in IP-RP [126,127]. The majority of these methods for speciating tin on RP columns use high organic levels, which require special considerations, previously mentioned, when interfacing with ICP-MS. A method by White *et al.* [36] used chromatography that was compatible with both API-MS and ICP-MS thus achieving molecular confirmation in conjunction with low detection limits. They used a C_{18} column with ACN, acetic acid, DIW, and triethylamine (at 65:10:25:0.5 v/v) to resolve DPhT, TPhT, DBT, and TBT.

16.3.6.3 Detection limits

Detection limits for species of tin in terms of absolute detection limits vary from 1.5 to 55 pg of Sn [121].

16.3.7 Vanadium

The beneficial and harmful effects of vanadium are not well understood. At trace levels, vanadium is an essential element for normal cell growth, but at higher concentrations, it can be toxic [128]. Vanadium is mainly used in steel manufacturing and is also released into the environment from the burning of fossil fuels. The possibility of vanadium's insulin type behavior has also been explored [129]. The daily estimated recommended daily amount for humans is 10–60 µg [130]. One big concern is the presence of vanadium in natural waters. In natural waters, it normally exists as V^V or V^{IV}, with V^V being more toxic than V^{IV} [131]. These compounds usually are present at part per billion levels, therefore selective and sensitive methods of analysis are necessary to evaluate vanadium contamination.

16.3.7.1 Sample preparation

Vanadium^{IV}, as the vanadyl cation VO_2^+, may be present in reducing environment. It is most stable in acidic solution below pH 2, but is oxidized to the pentavalent state by atmospheric oxygen at higher pH values [128]. This conversion between oxidation states becomes one of the main problems facing vanadium speciation analysis. Since the majority of the applications in the literature involve water, the main sample preparation techniques are to acidify the sample, keep at low temperatures, and in an oxygen-free environment. Other reports believe it is necessary to analyze the samples immediately after collection without acidification [128]. Other techniques involve the complexation of the vanadium species with ethylenediaminetetraacetic acid (EDTA) [132] or (1,2-cyclohexylene-dinitrilo)tetraacetic acid (CDTA) [133] to stabilize each species.

16.3.7.2 Chromatography

There are only a limited number of HPLC-ICP-MS methods in the literature. Wann and Jiang [132] used EDTA to complex and stabilize V^{IV} and V^V to form $[VOY]^{2-}$ and $[VO_2Y]^{3-}$ (Y represents deprotonated EDTA). Separation was then carried out using RP C_8 column with 3 mM EDTA, 0.5 mM tetrabutyl ammonium phosphate (TBAP), pH 6.5 and 12% MeOH (v/v) as a buffer. The separation is shown in Fig. 16.12. Other HPLC methods have been presented and could be adapted to use ICP-MS for Vn detection. Work by de Beer and Coetzee [133] separated V^{IV} and V^V using a Dionex Ionpac AG5 guard column. The buffer was 5.6 mM NaHCO$_3$,

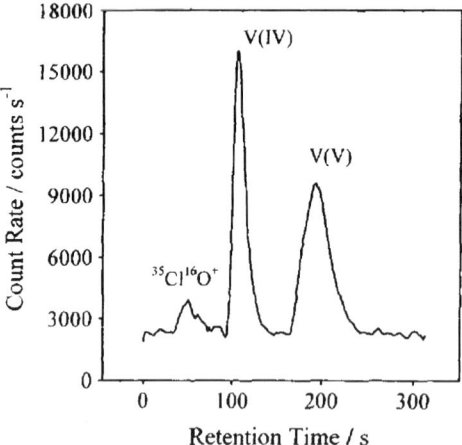

Fig. 16.12. HPLC-ICP-MS chromatogram for: ClO$^+$, V^{IV}, and V^V. Vanadium species at 5 ng mL^{-1} and Cl$^-$ at 0.2 M. Reprinted from [132] with permission from Elsevier.

4.6 mM Na_2CO_3, and 20 mM CDTA. In this method, UV detection was used, but using ammonium salts instead of sodium salts would make this method more compatible with ICP-MS. Gaspar and Posta [134] used 0.1 M KH-phthalate in DIW with a RP C_{18} column using FAAS for detection.

16.3.7.3 Detection limits

Gaspar and Posta [134] reported detection limits for V^V and V^{IV} as 0.18 and 0.15 mg/mL, respectively, with FAAS detection. With the substitution of ICP-MS, detection limits could certainly be lowered. Of the methods surveyed here, Wann and Jiang [132] had the best detection limits for online HPLC-ICP-MS determination for V^{IV} and V^V at 0.025 and 0.041 ppb, respectively.

16.3.8 Antimony

Natural sources of antimony (Sb) in the environment come via soil runoff and rock weathering. Mining, fossil fuel combustion, and other industrial processes also contribute to antimony entering the environment [135]. The toxicity of antimony is heavily dependent on the species with the inorganic Sb^{III} oxyanion being the most toxic [135], Sb^{III} is approximately 10 fold more toxic than Sb^V, with inorganic species typically being more toxic than organic forms [136]. Sb^{III}, Sb^V, and trimethylantimony dichloride are the most commonly analyzed species.

16.3.8.1 Sample preparation

For urine samples, dilutions ranging from 1:3 to 1:50 with DIW and filtration were the only necessary preparation steps prior to injection [32]. Extraction of Sb from solid samples is problematic due to the low extraction efficiency of usually only a few percent [32]. Amerieh *et al.* [137] extracted Sb from soil using 100 mM citric acid, pH 2.1 at room temperature for 45 min in which ~40% total Sb was extracted. Extraction of sewage samples were carried out by Lintschinger *et al.* [138] using combinations of DIW, MeOH and KOH with adequate extraction efficiencies. When extracting Sb from fly coal ash, 1 M citric acid was used with efficiencies of 22–36% [139].

Krachler and Emons [32] reported that the high NaCl content of urine complicates anionic separation, but they were eventually able to overcome it using a PRP-X100 column (20 mM EDTA, pH 4.7) to separate Sb^V and Sb^{III}, and an ION-120 column (2 mM NH_4HCO_3 and 1 mM tartaric acid, pH 8.5) to separate Sb^V and $TMSbCl_2$. Additionally, hydride generation was used to help eliminate NaCl from the matrix prior to ICP-MS detection thus leading to better detection limits [32].

Some methods involve the complexation of Sb^{III} and Sb^V with citric acid [137] prior to separation. The PRP-X100 [138,139] was the most commonly used column, with others such as Synchropak Q300 [136] and Dionex AS4A-SC4 [138]. Mobile phases typically consisted of combinations of EDTA [136,137], phthalic acid [138], and tetraethylammonium hydroxide [138], among others [139].

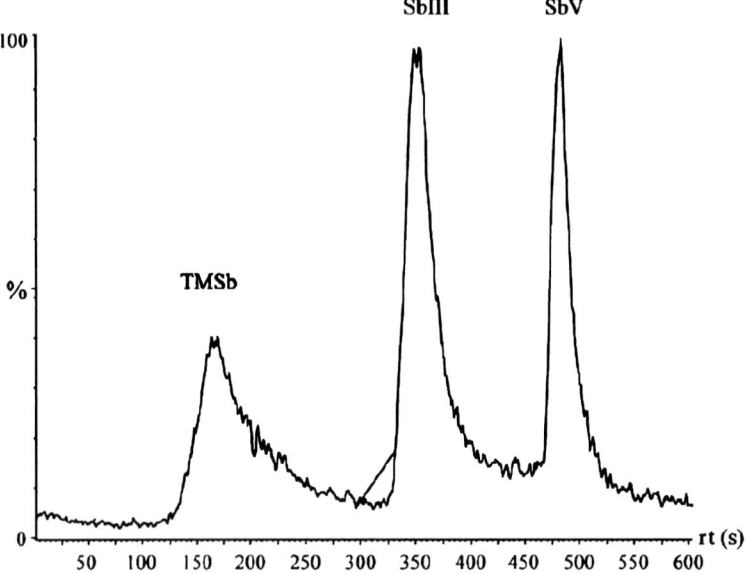

Fig. 16.13. The separation of Sb^V, Sb^{III} and TMSb all at $20\,ng\,mL^{-1}$. From reference [140], reproduced with permission of The Royal Society of Chemistry.

Nash and coworkers were able to separate all three compounds of interest in less than 10 min. A gradient elution was used at 60°C using Alltec HAAX column with 100 mM ammonium tartrate mobile phase with pH gradient from pH 2.3 to 1.5 [140] (shown in Fig. 16.13).

16.3.8.2 Detection limits

Detection limits for Sb^V, Sb^{III}, TMSb ranged between 0.005 and 0.3 ppb with the lowest detection limits being reported by Lintschinger [138].

16.3.9 Tellurium

Tellurium is similar to selenium in terms of its toxicity, as Te^{IV} is 10 times more toxic than Te^{VI}. [141] Tellurium is commonly used in electronics, metallurgy, and pharmaceuticals [141]. Little research has been presented using HPLC-ICP-MS to separate both species simultaneously. Two such methods have been used to detect only Te^{IV} using HPLC-ICP-MS. [142,143] One method by Viñas *et al.* [144] involves complexing the tellurium species with 50 mM citric acid. Speciation is then carried out using a PRP-X100 column with 8 mM EDTA and 2 mM potassium hydrogenphthalate as a mobile phase. The separation is less than 8 min, with atomic fluorescence spectrometry used as the detection method; it is feasible that ICP-MS could easily be substituted. Detection limits were $0.69\,\mu g\,Te\,L^{-1}$ and $0.76\,\mu g\,Te\,L^{-1}$ for Te^{VI} and Te^{IV}, respectively.

16.4 CONCLUDING REMARKS

In summation, HPLC-ICP-MS can be applied to large number of sample types while analyzing for numerous elements. There are several important factors to consider when choosing the speciation method. Sample preparation is crucial, as preserving the native form of the species must be balanced with extraction efficiency. A wide range of chromatographic methods can be coupled with ICP-MS, with ion exchange and reverse phase ion pairing being the most popular. Mobile phase constituents must be properly selected to ensure compatibility with ICP-MS. ICP-MS offers low detection limits, which are critical when doing trace analysis, and with the use of hydride generation, further reduction of detection limits for many elements can be achieved. Due to the versatility of HPLC-ICP-MS new applications are constantly being developed and can easily be tailored to fit individual cases.

16.5 REFERENCES

1   EPA; URL http://www.epa.gov/. Environmental Protection Agency, 2006.
2   D.T. Gjerde and H.C. Mehra, J. Chromatogr. Library, 47 (1991) 213–237.
3   M. Montes-Bayon, K. DeNicola and J.A. Caruso, J. Chromatogr., A, 1000 (2003) 457–476.
4   Thomas, R. Spectroscopy (Duluth, MN, United States) 2001, p. 16.
5   S.D. Tanner, V.I. Baranov and D.R. Bandura, Spectrochim. Acta, Part B, 57B (2002) 1361–1452.
6   I. Feldmann, N. Jakubowski and D. Stuewer, Fresenius J. Anal. Chem., 365 (1999) 415–421.
7   T. Yamada and N. Yamada. The ICP-MS Journal 2002, Agilent publication number 5988–7502EN.
8   E. McCurdy and D. Potter, Agilent Technologies ICP-MS Journal 2001, 10.
9   T.W. May and R.H. Wiedmeyer, At. Spectrosc., 19 (1998) 150–155.
10 L. Yang, R.E. Sturgeon and Z. Mester, J. Anal. At. Spectrom., 20 (2005) 431–435.
11 L.F. Dias, G.R. Miranda, T.D. Saint'Pierre, S.M. Maia, V.L.A. Frescura and A.J. Curtius, Spectrochim. Acta, Part B: At. Spectrosc., 60B (2005) 117–124.
12 T. Nakazato and H. Tao, Anal. Chem., 78 (2006) 1665–1672.
13 E. Schmeisser, W. Goessler, N. Kienzl and K.A. Francesconi, Anal. Chem., 76 (2004) 418–423.
14 W.C. Story and J.A. Caruso, In: Z.B. Alfassi and C.M. Wai (Eds), Preconcentration techniques for trace elements, CRC Press, Boca Raton, FL, 1992.
15 R.E. Sturgeon, Mester. Z. Appl. Spectrosc., 56 (2002) 202A–213A.
16 A.G. Howard, J. Anal. At. Spectrom., 12 (1997) 267–272.
17 L.K. Olson, N.P. Vela and J.A. Caruso, Spectrochim. Acta. Part B: At. Spectrosc., 50B (1995) 355–368.
18 D. Sanchez-Rodas, J. Luis Gomez-Ariza, I. Giraldez, A. Velasco and E. Morales, Sci. Total Environ., 345 (2005) 207–217.
19 J. Szakova, P. Tlustos, W. Goessler, D. Pavlikova and J. Balik, Appl. Organometallic Chem., 19 (2005) 308–314.
20 J. Kirby, W. Maher, M. Ellwood and F. Krikowa, Austr. J. Chem., 57 (2004) 957–966.
21 S. Garcia-Manyes, G. Jimenez, A. Padro, R. Rubio and G. Rauret, Talanta, 58 (2002) 97–109.
22 T. Nakazato, H. Tao, T. Taniguchi and K. Isshiki, Talanta, 58 (2002) 121–132.
23 T. Dagnac, A. Padro, R. Rubio and G. Rauret, Talanta, 48 (1999) 763–772.
24 D. Kuehnelt, N. Kienzl, D. Juresa and K.A. Francesconi, J. Anal. At. Spectrom., 21 (2006) 1264–1270.
25 D. Kuehnelt, D. Juresa, N. Kienzl and A. Francesconi Kevin, Anal. Bioanal. Chem., 386 (2006) 2207–2212.
26 D. Wallschlaeger and J. London, J. Anal. At. Spectrom., 19 (2004) 1119–1127.

27 J.M. Gonzalez LaFuente, J.M. Marchante-Gayo, M.L. Fernandez Sanchez and A. Sanz-Medel, Talanta, 50 (1999) 207–217.

28 R. Clough, S.T. Belt, B. Fairman, T. Catterick and E.H. Evans, J. Anal. At. Spectrom., 20 (2005) 1072–1075.

29 C.S. Chiou, S.J. Jiang and K.S. Kumar Danadurai, Spectrochim. Acta, Part B: At. Spectrosc., 56B (2001) 1133–1142.

30 C.-C. Wan, C.-S. Chen and S.-J. Jiang, J. Anal. At. Spectrom., 12 (1997) 683–687.

31 Q. Tu, W. Johnson Jr. and B. Buckley, J. Anal. At. Spectrom., 18 (2003) 696–701.

32 M. Krachler and H. Emons, J. Anal. At. Spectrom., 16 (2001) 20–25.

33 W.R. LaCourse, Anal. Chem., 74 (2002) 2813–2831.

34 L.C. Alves, M.G. Minnich, D.R. Wiederin and R. Houk, S. J. Anal. At. Spectrom., 9 (1994) 399–403.

35 B. Bouyssiere, Y.N. Ordonez, C.-P. Lienemann, D. Schaumloeffel and R. Lobinski, Spectrochim. Acta, Part B, 61 (2006) 1063–1068.

36 S. White, T. Catterick, B. Fairman and K. Webb, J. Chromatogr., A, 794 (1998) 211–218.

37 C. B'Hymer and J. Caruso, A. J. Chromatogr., A, 1045 (2004) 1–13.

38 Y. Shiobara, Y. Ogra and K.T. Suzuki, Chem. Res. Toxicol., 14 (2001) 1446–1452.

39 L. Benramdane, M. Accominotti, L. Fanton, D. Malicier and J.-J. Vallon, Clin. Chem. (Washington, DC, U. S.), 45 (1999) 301–306.

40 K.S. Park, J.S. Kim, H. Lee, H. Pyo, S.-T. Kim and K.B. Lee, Key Eng. Mater., 277-279 (2005) 431–437.

41 R. Schaeffer, C. Soeroes, I. Ipolyi, P. Fodor and N.S. Thomaidis, Anal. Chim. Acta, 547 (2005) 109–118.

42 J. Feldmann, V.W. Lai, W.R. Cullen, M. Ma, X. Lu and X.C. Le, Clin. Chem. (Washington, DC, U. S.), 45 (1999) 1988–1997.

43 A. Shraim, X. Cui, S. Li, C. Ng Jack, J. Wang, Y. Jin, Y. Liu, L. Guo, D. Li, S. Wang, R. Zhang and S. Hirano, Toxicol. Lett., 137 (2003) 35–48.

44 D. Heitkemper, J. Creed, J. Caruso and F.L. Fricke, J. Anal. At. Spectrom., 4 (1989) 279–284.

45 B. He, G.-b. Jiang and X.-b. Xu, Fresenius J. Anal. Chem., 368 (2000) 803–808.

46 S. Garcia Salgado, M.A. Quijano Nieto and M.M. Bonilla Simon, J. Chromatogr., A, 1129 (2006) 54–60.

47 D. Sanchez-Rodas, J. Luis Gomez-Ariza and V. Oliveira, Anal. Bioanal. Chem., 385 (2006) 1172–1177.

48 A.D. Madsen, W. Goessler, S.N. Pedersen and K.A. Francesconi, J. Anal. At. Spectrom., 15 (2000) 657–662.

49 B.P. Jackson and P.M. Bertsch, Environ. Sci. Technol., 35 (2001) 4868–4873.

50 J.A. Caruso, C. B'Hymer and D.T. Heitkemper, Analyst (Cambridge, UK), 126 (2001) 136–140.

51 A.H. Ackerman, P.A. Creed, A.N. Parks, M.W. Fricke, C.A. Schwegel, J.T. Creed, D.T. Heitkemper and N.P. Vela, Environ. Sci. Technol., 39 (2005) 5241–5246.

52 B.S. Sheppard, J.A. Caruso, D.T. Heitkemper and K.A. Wolnik, Analyst (Cambridge, UK), 117 (1992) 971–975.

53 C. B'Hymer and J.A. Caruso, J. Liquid Chromatogr. Relat. Technol., 25 (2002) 639–653.

54 B.S. Sheppard, W.L. Shen, J.A. Caruso, D.T. Heitkemper and F.L. Fricke, J. Anal. At. Spectrom., 5 (1990) 431–435.

55 J. Lintschinger, P. Schramel, A. Hatalak-Rauscher, I. Wendler and B. Michalke, Fresenius' J. Anal. Chem., 362 (1998) 313–318.

56 D.T. Heitkemper, N.P. Vela, K.R. Stewart and C.S. Westphal, J. Anal. At. Spectrom., 16 (2001) 299–306.

57 U. Kohlmeyer, J. Kuballa and E. Jantzen, Rapid Commun. Mass Spectrom., 16 (2002) 965–974.

58 P.A. Gallagher, X. Wei, J.A. Shoemaker, C.A. Brockhoff and J.T. Creed, J. Anal. At. Spectrom., 14 (1999) 1829–1834.

59 X. Wei, C.A. Brockhoff-Schwegel and J.T. Creed, J. Anal. At. Spectrom., 16 (2001) 12–19.

60 J.S. Fritz, J. Chromatogr., A, 1085 (2005) 8–17.

61   J. Stahlberg, J. Chromatogr., A, 855 (1999) 3–55.
62   G. Schill, J. Biochem. Biophys. Methods, 18 (1989) 249–270.
63   W.R. Melander and C. Horvath, Chromatogr. Sci. Ser., 31 (1985) 27–75.
64   M.T.W. Hearn, Chromatogr. Sci. Ser., 31 (1985) 1–26.
65   J. Stahlberg, J. Chromatogr., A, 855 (1999) 3–55.
66   X.C. Le and M. Ma, J. Chromatogr., A, 764 (1997) 55–64.
67   L.C. Clark, G.F. Combs Jr., B.W. Turnbull, E.H. Slate, D.K. Chalker, J. Chow, L.S. Davis, R.A. Glover, G.F. Graham, E.G. Gross, A. Krongrad, J.L. Lesher Jr., H.K. Park, B.B. Sanders Jr., C.L. Smith and J.R. Taylor, JAMA, 276 (1996) 1957–1963.
68   M.P. Rayman, Lancet, 356 (2000) 233–241.
69   Office of Dietary Supplements, N. C. C., National Institutes of Health, 2003.
70   D. Juresa, J. Darrouzes, N. Kienzl, M. Bueno, F. Pannier, M. Potin-Gautier, K.A. Francesconi and D. Kuehnelt, J. Anal. At. Spectrom., 21 (2006) 684–690.
71   D. Kuehnelt, N. Kienzl, P. Traar, N.H. Le, K.A. Francesconi and T. Ochi, Anal. Bioanal. Chem., 383 (2005) 235–246.
72   K. DeNicola Cafferky, D. D. Richardson and J. A. Caruso, Spectroscopy (Duluth, MN, United States) 21 (2006) 18,20,22–24.
73   P. Giusti, D. Schaumloeffel, H. Preud'homme, J. Szpunar and R. Lobinski, J. Anal. At. Spectrom., 21 (2006) 26–32.
74   C. B'Hymer and J.A. Caruso, J. Chromatogr., A, 1114 (2006) 1–20.
75   D.M. Hunsaker, H.A. Spiller and D. Williams, J. Forensic Sci., 50 (2005) 942–946.
76   A. Gasmi, R. Garnier, M. Galliot-Guilley, C. Gaudillat, B. Quartenoud, A. Buisine and D. Djebbar, Veterinary Hum. Toxicol., 39 (1997) 304–308.
77   Y. Kobayashi, Y. Ogra and K.T. Suzuki, J. Chromatogr., B. Biomed. Sci. Appl., 760 (2001) 73–81.
78   Y. Ogra, K. Ishiwata, J.R. Encinar, R. Lobinski and K.T. Suzuki, Anal. Bioanal. Chem., 379 (2004) 861–866.
79   A. Polatajko, N. Jakubowski and J. Szpunar, J. Anal. At. Spectrom., 21 (2006) 639–654.
80   C. B'Hymer and J.A. Caruso, J. Anal. At. Spectrom., 15 (2000) 1531–1539.
81   K. Wrobel, S.S. Kannamkumarath, K. Wrobel and J.A. Caruso, Anal. Bioanal. Chem., 375 (2003) 133–138.
82   V. Gergely, K.M. Kubachka, S. Mounicou, P. Fodor and J.A. Caruso, J. Chromatogr., A, 1101 (2006) 94–102.
83   A. Polatajko, B. Banas, J.R. Encinar and J. Szpunar, Anal. Bioanal. Chem., 381 (2005) 844–849.
84   L. Yang, R.E. Sturgeon, S. McSheehy and Z. Mester, J. Chromatogr., A, 1055 (2004) 177–184.
85   E.H. Larsen, M. Hansen, T. Fan and M. Vahl, J. Anal. At. Spectrom., 16 (2001) 1403–1408.
86   M. Kotrebai, J.F. Tyson, E. Block and P.C. Uden, J. Chromatogr., A, 866 (2000) 51–63.
87   M. Montes-Bayon, E.G. Yanes, C. Ponce de Leon, K. Jayasimhulu, A. Stalcup, J. Shann and J.A. Caruso, Anal. Chem., 74 (2002) 107–113.
88   M. Kotrebai, J.F. Tyson, P.C. Uden, M. Birringer and E. Block, Analyst (Cambridge, UK), 125 (2000) 71–78.
89   J.R. Encinar, D. Schaumloeffel, Y. Ogra and R. Lobinski, Anal. Chem., 76 (2004) 6635–6642.
90   M.V. Balarama Krishna, K. Chandrasekaran, S.V. Rao, D. Karunasagar and J. Arunachalam, Talanta, 65 (2005) 135–143.
91   M. Leist, R. Leiser and Toms, A., *Spectroscopy (Duluth, MN, United States)* 2006, pp. 29–31.
92   H. Guerleyuek and D. Wallschlaeger, J. Anal. At. Spectrom., 16 (2001) 926–930.
93   M. Pantsar-Kallio and P.K.G. Manninen, J. Chromatogr., A, 750 (1996) 89–95.
94   F. Seby, S. Charles, M. Gagean, H. Garraud and O.F.X. Donard, J. Anal. At. Spectrom., 18 (2003) 1386–1390.
95   F.A. Byrdy, L.K. Olson, N.P. Vela and J.A. Caruso, J. Chromatogr., A, 712 (1995) 311–320.
96   G.K. Zoorob and J.A. Caruso, J. Chromatogr., A, 773 (1997) 157–162.
97   C. Barnowski, N. Jakubowski, D. Stuewer and J.A.C. Broekaert, J. Anal. At. Spectrom., 12 (1997) 1155–1161.

98 S. Saverwyns, K. Van Hecke, F. Vanhaecke, L. Moens and R. Dams, Fresenius' J. Anal. Chem., 363 (1999) 490–494.

99 A.G. Coedo, T. Dorado, I. Padilla and F.J. Alguacil, J. Anal. At. Spectrom., 15 (2000) 1564–1568.

100 Y.-L. Chang and S.-J. Jiang, J. Anal. At. Spectrom., 16 (2001) 858–862.

101 B. Chardin, F. Chaspoul, P. Gallice and M. Bruschi, J. Liquid Chromatogr. Relat. Technol., 25 (2002) 877–887.

102 D.A. Gidlow, Occup. Med. (Oxford, England), 54 (2004) 76–81.

103 H.P. Schipulle, Compendium of environmental standards, in: Environmental handbook, Vol. 3. German Federal Ministry for Economic Cooperation and Development, Germany, 2006.

104 D.C. Baxter and W. Frech, Pure Appl. Chem., 67 (1995) 615–648.

105 L. Ebdon, S.J. Hill and C. Rivas, Spectrochim. Acta, Part B, 53B (1998) 289–297.

106 Y. Pan, X. Liu, X. He and C. Wang, Fenxi Huaxue, 33 (2005) 1560–1564.

107 A. Al-Rashdan, D. Heitkemper and J.A. Caruso, J. Chromatogr. Sci., 29 (1991) 98–102.

108 A. Al-Rashdan, N.P. Vela, J.A. Caruso and D.T. Heitkemper, J. Anal. At. Spectrom., 7 (1992) 551–555.

109 A.A. Brown, L. Ebdon and S.J. Hill, Anal. Chim. Acta, 286 (1994) 391–399.

110 S.C.K. Shum, H.M. Pang and R.S. Houk, Anal. Chem., 64 (1992) 2444–2450.

111 L. Labat, V. Dumestre-Toulet, J.P. Goulle and M. Lhermitte, A fatal case of mercuric cyanide poisoning. Laboratoire de Biochimie et de Biologie Moleculaire, CHRU de Lille, Avenue du Pr Leclercq, 59307 Lille Cedex, France, 2004.

112 C.F. Harrington, TrAC, Trends Anal. Chem., 19 (2000) 167–179.

113 M. Leermakers, W. Baeyens, P. Quevauviller and M. Horvat, TrAC, Trends Anal. Chem., 24 (2005) 383–393.

114 J. Morton, V.A. Carolan and P.H.E. Gardiner, J. Anal. At. Spectrom., 17 (2002) 377–381.

115 E. Bramanti, C. Lomonte, M. Onor, R. Zamboni, A. D'Ulivo and G. Raspi, Talanta, 66 (2005) 762–768.

116 M. Horvat, N.S. Bloom and L. Liang, Anal. Chim. Acta, 281 (1993) 135–152.

117 C.M. Tseng, A. De Diego, F.M. Martin, D. Amouroux and O.F.X. Donard, J. Anal. At. Spectrom., 12 (1997) 743–750.

118 R. Falter and G. Ilgen, Fresenius J. Anal. Chem., 358 (1997) 401–406.

119 G.-b. Jiang, Q.-f. Zhou and B. He, Environ. Sci. Technol., 34 (2000) 2697–2702.

120 K.M. Attar, Appl. Organomet. Chem., 10 (1996) 317–337.

121 E. Gonzalez-Toledo, R. Compano, M. Granados and M. Dolors Prat, TrAC, Trends Anal. Chem., 22 (2003) 26–33.

122 S. Chiron, S. Roy, R. Cottier and R. Jeannot, J. Chromatogr., A, 879 (2000) 137–145.

123 S.J. Hill, L.J. Pitts and A.S. Fisher, TrAC, Trends Anal. Chem., 19 (2000) 120–126.

124 C.G. Arnold, M. Berg, S.R. Mueller, U. Dommann and R.P. Schwarzenbach, Anal. Chem., 70 (1998) 3094–3101.

125 O. Donard, B. Lalere, F. Martin and R. Lobinski, Anal. Chem., 67 (1995) 4250–4254.

126 U.T. Kumar, N.P. Vela, J.G. Dorsey and J.A. Caruso, J. Chromatogr., A, 655 (1993) 340–345.

127 W.-S. Chao and S.-J. Jiang, J. Anal. At. Spectrom., 13 (1998) 1337–1341.

128 K. Pyrzynska and T. Wierzbicki, Talanta, 64 (2004) 823–829.

129 H. Seiler, A. Sigel and H. Sigel, Handbook on Metals in Clinical and Analytical Chemistry. New York, 1994.

130 D. Barceloux, J. Toxicol., Clin. Toxicol., 37 (1999) 265–278.

131 R.G. Wuilloud, J.C. Wuilloud, R.A. Olsina and L.D. Martinez, Analyst (Cambridge, UK), 126 (2001) 715–719.

132 C.-C. Wann and S.-J. Jiang, Anal. Chim. Acta, 357 (1997) 211–218.

133 H. de Beer and P.P. Coetzee, Fresenius J. Anal. Chem., 348 (1994) 806–809.

134 A. Gaspar and J. Posta, Fresenius J. Anal. Chem., 360 (1998) 179–183.

135 M.J. Nash, J.E. Maskall and S.J. Hill, J. Environ. Monit., 2 (2000) 97–109.

136 J. Zheng, M. Ohata and N. Furuta, Anal. Sci., 16 (2000) 5–80.

137 S. Amereih, T. Meisel, E. Kahr and W. Wegscheider, Anal. Bioanal. Chem., 383 (2005) 1052–1059.

138 J. Lintschinger, O. Schramel and A. Kettrup, Fresenius J. Anal. Chem., 361 (1998) 96–102.
139 R. Miravet, J.F. Lopez-Sanchez and R. Rubio, Anal. Chim. Acta, 576 (2006) 200–206.
140 M.J. Nash, J.E. Maskall and S.J. Hill, Analyst (Cambridge, UK), 131 (2006) 724–730.
141 C. Yu, Q. Cai, Z.-X. Guo, Z. Yang and S.B. Khoo, Analyst (Cambridge, UK), 127 (2002) 1380–1385.
142 T. Guerin, M. Astruc, A. Batel and M. Borsier, Talanta, 44 (1997) 2201–2208.
143 T. Lindemann, A. Prange, W. Dannecker and B. Neidhart, Fresenius J. Anal. Chem., 368 (2000) 214–220.
144 P. Vinas, I. Lopez-Garcia, B. Merino-Merono and M. Hernandez-Cordoba, Appl. Organomet. Chem., 19 (2005) 930–934.

PART 3:
ACTUAL AND EMERGING PROBLEMS OF FORENSIC TOXICOLOGY

M.J. Bogusz (Ed.). Forensic Science
Handbook of Analytical Separations, Vol. 6
© 2008 Elsevier B.V. All rights reserved.

CHAPTER 17

Analytical markers of acute and chronic alcohol consumption

Anders Helander[1] and Olof Beck[2]

[1]*Department of Clinical Neuroscience, Karolinska Institute and Karolinska University Hospital, SE-171 76 Stockholm, Sweden*
[2]*Department of Medicine, Division of Clinical Pharmacology, Karolinska Institute and Karolinska University Hospital, SE-171 76 Stockholm, Sweden*

17.1 INTRODUCTION

17.1.1 Biological markers

By definition, biological markers, or biomarkers, are parameters measured in the blood, other body fluids, or tissues as indicators of a biologic state. Biomarkers are used as objective tools and diagnostic tests to monitor the presence of a disease, its severity, and progression over time and upon treatment. Biomarkers are also commonly used as screening tests for risk assessment and predictors of morbidity and mortality.

Many of the routine blood tests performed in hospital and central laboratories are clinically important biomarkers. A wide range of modern technologies and approaches are used for biomarker discovery and identification, and new candidate markers from readily accessible biological fluids are continuously evaluated for their value (e.g. improved sensitivity and/or specificity compared with the standard tests) in clinical studies.

17.1.2 Need for detecting alcohol use and abuse

Biomarkers have found uses in detection and follow-up of alcohol-related problems [1,2]. Alcohol programs focusing on early recognition and treatment of problem drinking by low-cost screening and brief interventions are important and may have greater health impact than expensive tertiary treatment of chronic heavy alcohol use and alcohol dependence [3,4]. To collect data of alcohol intake, self-report and

References pp. 583–588

standardized questionnaires (e.g. CAGE, MAST, and AUDIT) [5–7] are popular ways because of their relative ease and low cost. However, considering the well-known fact that many patients fail to provide accurate information of their alcohol consumption, due to denial and deliberate or unintentional underreport [8,9], relying solely on data collected by interview or questionnaires sometimes creates a validity problem. In this connection, the information obtained from various clinical laboratory tests, including measurement of alcohol (ethanol) in breath or body fluids and blood- or urine-based alcohol biomarkers, represent a useful objective complement to self-report measures and other clinical information.

17.1.3 Ethanol testing

A standard way to determine without doubt that a person has recently consumed alcoholic beverages is to verify the presence of ethanol in samples of body fluids (blood, urine, or saliva) or breath [10–11]. The ethanol distributes into all body fluids and tissues in proportion to their water content and, after absorption and distribution of ethanol is complete, there is normally a close correlation between the concentrations in saliva, blood, and urine. The time frame for a positive identification by ethanol testing is rather limited [12], because the concentration decreases rapidly over time, mainly owing to oxidative metabolism in the liver by the action of alcohol dehydrogenase (ADH) and excretion processes. Another problem is that ethanol may be produced naturally in the body because of abnormal yeast proliferation after ingesting carbohydrate-rich meals (the "auto-brewery syndrome") [13,14], and also synthesized after sampling because of microbial contamination and fermentation [15].

The urine ethanol concentration reflects the average concentration in blood prevailing during the time that urine was produced and stored in the bladder since the previous void. Accordingly, there is usually a time lag between the blood and urine concentration-time profiles [12]. The first morning urine void after an evening's drinking might be positive for ethanol, although the concentrations in blood or breath have already returned to zero [16]. The breath ethanol test result is routinely translated into the presumed coexisting blood–ethanol concentration, using a blood-to-breath conversion factor [17,18].

In the central laboratory, measurement of ethanol in blood and urine samples is usually carried out by gas chromatography (GC) or enzymatic (i.e. ADH) methods. For clinical purposes as a rapid screening test for recent drinking, a large number of methods and devices for on-site testing are available, including breath-alcohol analyzers (breathalyzers) and saliva tests [11,19].

17.1.4 Established alcohol biomarkers

Given that impairment of liver function and damage to liver cells are well-known effects of long-term heavy alcohol consumption, standard measures of liver dysfunction such as elevated levels of gamma-glutamyl transferase (GGT), and aspartate and alanine aminotransferase (AST and ALT) are widely employed as biomarkers of chronic drinking [20]. Together with another traditional alcohol test, the mean corpuscular volume of the erythrocytes (MCV), they are also included as part of routine blood-chemistry profiles taken during medical examinations and on admission to the hospital and therefore readily available at low cost from most clinical laboratories.

For use as alcohol biomarkers, important disadvantages are the limited specificity for alcohol. Abnormal values are also seen in liver disorders of non-alcohol origin and after taking common medications (e.g. barbiturates and antiepileptics) [21]. Furthermore, reference intervals need to be adjusted for gender, age and body weight [22], and for lifestyle factors such as smoking, which is very common in alcoholics. This reduces their diagnostic specificity for recognition of alcohol-related medical conditions. In addition, because of the considerable time delay before these tests turn positive [23], they are not sensitive enough for early detection of recent heavy drinking or relapse, but may be useful for follow-up of alcoholic patients with diagnosed liver disease, and in combination with other alcohol-related biochemical parameters. Usually it also takes long time for the liver function tests and MCV to recover to normal values after cessation of chronic heavy drinking. Accordingly, more sensitive and specific alcohol biomarkers are required and requested.

This chapter will provide more detailed information about a number of newer alcohol biomarkers that are already available for routine use, or currently in the process of being evaluated for routine application.

17.2 TESTS OF ACUTE ALCOHOL CONSUMPTION

17.2.1 Ethyl glucuronide (EtG) and ethyl sulfate (EtS)

17.2.1.1 Background

Both ethyl glucuronide (EtG) and ethyl sulfate (EtS) have been known as chemical entities since long. The potential use of EtG and EtS as alcohol biomarkers is, however, more recent and associated with the development of sensitive and specific mass spectrometric methods for quantitative determination and to the undisputable demonstration of their occurrence in man as ethanol metabolites [24,25]. The interest in EtG was initially focused on urine, but later also serum and hair have received attention [26–28]. The promising development of EtG as an alcohol marker triggered interest also in EtS [25,29,30], ethyl phosphate [31], and ethyl nitrite [32].

17.2.1.2 Biochemistry

EtG and EtS are produced as direct phase II metabolites of ethanol (Fig. 17.1), besides the more important oxidative pathway *via* ADH. The estimated proportion of each of these is about 0.05% of the ingested ethanol dose [25,33]. The enzymes responsible for forming EtG and EtS are UDP-glucuronosyltransferase (UGT) and

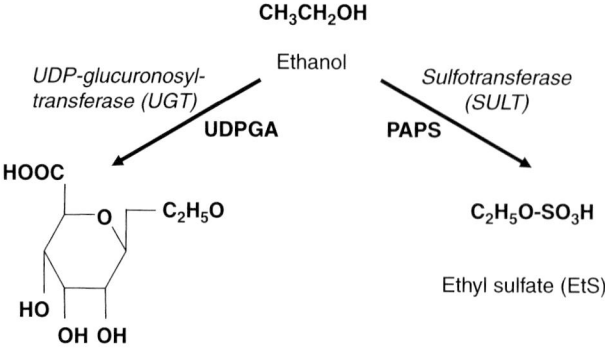

Fig. 17.1. Enzymatic pathways for the conjugation of ethanol with UDP-glucuronic acid (UDPGA) to form ethyl glucuronide (EtG) and with 3′-phosphoadenosine 5′-phosphosulfate (PAPS) to form ethyl sulfate (EtS).

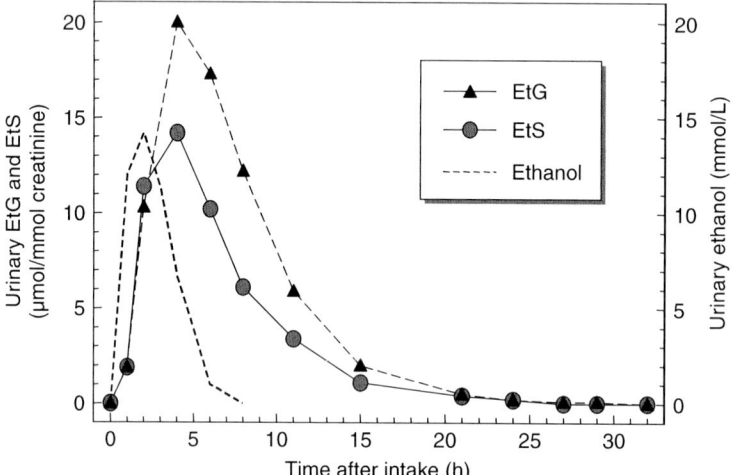

Fig. 17.2. Time course of urinary excretion of ethanol, ethyl glucuronide (EtG), and ethyl sulfate (EtS) after drinking alcohol (0.5 g ethanol/kg body weight) in a fasted state.

sulfotransferase (SULT), respectively [34,35]. Both theses enzymes are polymorphic in humans. Multiple forms of UGT are responsible for EtG formation with highest activity observed for the UGT1A1 and 2B7 isoforms [35]. Also multiple forms of SULT are responsible for EtS formation with SULT1A3 being the most important in vivo [34]. No endogenous formation of EtG or EtS has been detected.

The high-water solubility of EtG and EtS makes them ready for excretion in urine [36,37]. The excretion profile of EtG and EtS in blood and urine has been well documented in studies of healthy volunteers [29,33,37–39] (Fig. 17.2). In controlled experiments with oral administration of ethanol to healthy volunteers, a detection time for EtG of 24–48 h after ingestion has been observed. The detection time of EtG in urine after ethanol is undetected in blood was about 15–25 h [33]. Longer detection time (>48 h) was reported from another experiment using higher doses of ethanol [39]. A study of 23 alcoholic patients during detoxification suggested a typical detection time of about 48 h after ethanol is eliminated [38]. The detection time for EtS is about the same as for EtG [29,40]. In serum, a terminal elimination half-life of 2–3 h was estimated for EtG, which in agreement with the kinetics for urine [41]. No accumulation occurs of EtG following repeated ethanol ingestion [42]. EtG is excreted in urine in a process influenced by water-induced diuresis [33], making it possible to include correction of urine levels to creatinine concentrations for some applications [43]. The serum kinetics if EtG has been analyzed in detail [44], and a mathematical model was developed with potential use in problem cases as a complement to blood–alcohol data.

Although EtG initially was found to be a stable analyte in human urine [41,45], a study of patients suffering from urinary tract infection [46] demonstrated that EtG was unstable on storage in samples infected with three different bacterial species, one being the most prevalent human uropathogen *Escherichia coli*. Interestingly, EtS was completely stable under these conditions [46]. More recently, a preliminary observation has indicated that EtG but not EtS may also be produced in vitro under certain circumstances (Helander *et al.*, unpublished). Taken together, this indicates that EtS might be a more robust parameter than EtG.

The presence of EtG in hair was originally reported in 1994 [47] and this has later been confirmed by several research groups [26,48–51]. Recently, using methods with high sensitivity, EtG was found in all cases with reported high (>40 g) daily consumption of ethanol. Also hair collected from subjects with more moderate alcohol consumption contained detectable EtG. The EtG content in hair correlates with the daily consumption of ethanol, and a cutoff level of 4–5 pg/mg hair has been proposed [52]. The level of EtG in hair collected from alcoholic patients ranged from <5 to 13,100 pg/mg. No report exists regarding the possible presence of EtS in hair.

17.2.1.3 Analysis

GC-MS. The first mass spectrometric (MS) methods for EtG were based on GC-MS and applied for serum, urine, and hair samples [24,37,53,54]. These procedures were based on protein precipitation with methanol or acetone for both serum and urine, evaporation to dryness, and subsequent derivatization with acylation or silylation resulting in rather crude extracts for chromatographic analysis. The development of a solid-phase extraction (SPE) procedure, which was needed due to unwanted interference from matrix components, provided a desired increase in selectivity [54].

LC-MS. EtG and EtS are more ideal analytes for negative ion liquid chromatography-mass spectrometry (LC-MS) due to their polar nature. The first LC-MS application for EtG was demonstrated by Nishikawa *et al.* [55], who developed a procedure for its determination in serum. The method employed electrospray ionization and protein precipitation with methanol. Detection was performed in the selected ion monitoring (SIM) mode of the deprotonated molecule (negative mode).

At about the same time, a tandem LC-MS (LC-MS/MS) method for EtG in samples of urine and serum was developed independently [53]. This procedure, which was not fully validated but employed the pentadeuterated EtG as internal standard, used electrospray ionization and SIM and provided increased selectivity. A fully validated method combining LC-MS and LC-MS/MS and using direct injection of diluted urine was reported by Stephanson *et al.* [41]. This method employed a graphite column (Hypercarb), which enabled a more suitable and reproducible chromatographic solution for EtG, and used single MS for quantification and MS/MS to confirm the identity. This method is still in routine use in the authors' laboratory for analysis of clinical samples and is found to be very robust. For forensic applications, however, such a strategy cannot be used for conclusive results, as the recommended requirements are to monitor three ions in LC-MS or two product ions in LC-MS/MS [56]. Therefore, the first fully validated application for measuring EtG with LC-MS/MS, meeting forensic standards, was developed [57]. The method used pentadeuterated EtG as internal standard, a minimal sample preparation, and reversed phase chromatography (phenyl-propyl column material).

EtS testing was easily adopted to fit with the existing LC-MS procedures used for EtG [25]. The most recent method for simultaneous determination of EtG, EtS, and ethyl phosphate uses a simple 1:20 dilution of urine and combines the use of stable isotope-labeled internal standards, a mixed model stationary phase suitable for these polar compounds, with the use of LC-MS/MS and thereby also meeting requirement of forensic standards [31]. Analysis of EtG in hair has predominantly been performed by LC-MS/MS, as very low levels must be determined. For extraction of EtG, most investigators have used water as extraction medium after washing the hair with organic solvent [50,52,58,59].

Immunochemical assays. An initial effort to develop an ELISA method for EtG was not successful [60]. However, a promising DRI prototype for EtG in urine has been presented but is not yet commercially available [61].

17.2.1.4 Applications

EtG was early suggested to have potential value for both clinical and forensic applications [62]. The fact that EtG and EtS are direct ethanol metabolites makes these analytes attractive from a forensic perspective [63]. The possible use of other body fluids than urine and of tissues in post-mortem investigations has been demonstrated [64].

The potential of using EtG to reveal post-mortem formation of ethanol has been studied in more detail [27]. In 93 cases where alcohol ingestion was suggested by clinical information and ethanol was present in blood, EtG was always detectable. In another 53 cases where alcohol ingestion was unlikely, no EtG was present and this was used to conclude post-mortem formation of detected blood ethanol in 11 cases.

EtG testing has also been proposed to be of value in cases of suspected impaired driving [15,44]. The primary value was to verify in vivo ethanol metabolism in order to exclude contamination from other sources. The relation between urinary EtG levels normalized to creatinine with blood alcohol levels in 100 cases of suspected drunk driving has been reported [43]. Such information can potentially be of value for evaluation of problem cases.

A further demonstration of the value of EtG in problem cases was shown by Klys *et al.* [65]. The involvement of chronic alcohol consumption for the fatal outcome was suggested from the detection of EtG in hair collected from the victim. It was concluded that an interaction from alcohol with the clomipramine treatment had occurred.

As the application of EtG as an alcohol intake marker develops, it is important to document the reliability, for example to consider reporting limits as development of analytical methodology may allow detection of irrelevant levels. A usual reporting limit of EtG in urine has been 0.1 mg/L. Such a low-reporting limit is questioned by the occurrence of EtG in levels above this limit after mouthwash with ethanol-containing solution [66] and use of hand sanitizers. A more appropriate limit for urine would be 0.5 mg/L and, in addition, the creatinine concentration should be taken into consideration.

17.2.2 5-Hydroxytryptophol (5-HTOL)

17.2.2.1 Background

5-Hydroxytryptophol (5-HTOL) is a naturally occurring substance in mammals and arises as a metabolite of the hormone and neurotransmitter serotonin (5-hydroxy-tryptamine, 5-HT). The main metabolite of 5-HT is 5-hydroxyindoleacetic acid (5-HIAA). Prior to the discovery of 5-HTOL as a 5-HT metabolite, a significant reduction by ethanol on the formation of 5-HIAA had been observed [67]. Further studies [68] demonstrated 5-HTOL to be formed at the expense of 5-HIAA. This shift in 5-HT metabolism was due to an interaction of ethanol with the pathways involved in monoamine metabolism [69] (Fig. 17.3).

17.2.2.2 Biochemistry

The biochemistry behind the alcohol-induced increase in 5-HTOL formation is partly due to the competitive inhibition of acetaldehyde on 5-HIAA formation (Fig. 17.3). The increased NADH levels that result from ethanol metabolism is another factor for the increased 5-HTOL formation, possibly even more important than the inhibition by acetaldehyde [70].

Metabolism of 5-HT occurs by action of monoamine oxidase (MAO) to form the intermediate 5-hydroxyindole-3-acetaldehyde (5-HIAL) (Fig. 17.3). This inter-mediate aldehyde is either oxidized by aldehyde dehydrogenase (ALDH) into 5-HIAA or reduced by ADH into 5-HTOL. The most important enzyme for 5-HTOL formation is the class-I ADH, which is ~30 times more active in the reduction of 5-HIAL than for oxidation of 5-HTOL [70]. The enzyme activity is highly dependent on the concentration of the co-factor NADH and under normal conditions the redox state does not favor the pathway leading to formation of

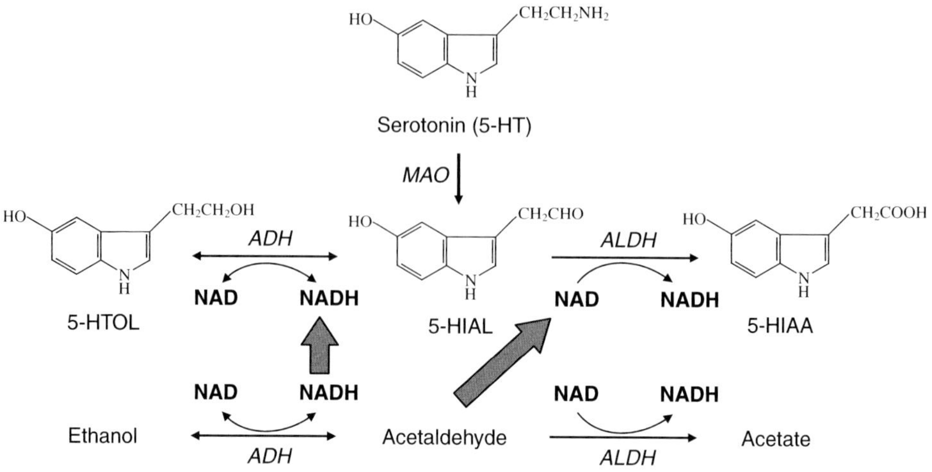

Fig. 17.3. Metabolic pathways for serotonin (5-hydroxytryptamine, 5-HT) to form 5-hydroxytryptophol (5-HTOL) and 5-hydroxyindoleacetic acid (5-HIAA), and the interaction with ethanol metabolism.

5-HTOL. However, at an elevated NADH/NAD$^+$ ratio that occurs during ethanol metabolism, 5-HTOL formation is dramatically favored.

The genetic variability of the ADH and ALDH enzymes has important influences on ethanol and acetaldehyde elimination [71]. However, it was found that this variability does not influence the urinary ratio between 5-HTOL and 5-HIAA in man [72]. The same cutoff level for an increased 5-HTOL level can therefore be used for different populations [73]. Furthermore, the accuracy of the 5-HTOL test is not influenced by factors such as the age, gender, ethnic origin, or common diseases or medications [73]. Apart from alcohol consumption, disulfiram (Antabuse) therapy, which is a potent inhibitor of ALDH, represents the only known cause of a raised 5-HTOL/5-HIAA ratio [74].

Unlike 5-HIAA, which is excreted to urine in a free form, 5-HTOL is excreted almost entirely in conjugated form [75,76]. 5-HTOL can become conjugated with either sulfuric or glucuronic acid by action of phenol SULT and UGT, respectively. The proportion of 5-HTOL being conjugated with glucuronic acid into 5-HTOL glucuronide (GTOL) is 2–4 times greater than the sulfate-conjugated fraction. About 80% of total 5-HTOL is excreted as GTOL [76,77], making GTOL an attractive target analyte for direct quantification.

17.2.2.3 Analysis

GC-MS. Use of the GC-MS technique has been essential for studying 5-HTOL [78–80]. Measurement of 5-HTOL in urine by GC-MS requires the liberation of conjugated forms by enzymatic hydrolysis. An updated summary of the experimental protocol for 5-HTOL determination in urine by GC-MS has been published [81]. In brief, the method involves hydrolysis with *E. coli* β-glucuronidase, solvent extraction with diethyl ether, formation of pentafluoropropionyl derivatives, and SIM using a tetradeuterated internal standard.

HPLC. Methods based on HPLC have also been developed, using fluorimetric and electrochemical detection [82]. However, in routine application, chromatographic interferences sometimes occurred. More recently, the successful application of a modified HPLC method with electrochemical detection has been reported [39,83].

LC-MS. The development of liquid LC-MS has opened a possibility for direct measurement of GTOL [77]. The use of an electrospray ionization interface and monitoring of negative ions enabled the direct measurement of GTOL in urine at elevated concentrations. Measurement of the low baseline levels of GTOL was, however, not possible without solid-phase preparation of urine prior to analysis. In another LC-MS application, 5-HTOL and 5-HIAA were measured simultaneously in a procedure employing enzymatic hydrolysis, solvent extraction, and derivatization [84]. The advantage over GC-MS was a much reduced analysis time. Further development of LC-MS will offer an analytical procedure for simultaneous determination of GTOL/5-HIAA ratio by direct injection of urine (Stephanson *et al.*, unpublished).

Immunochemical assay. A mouse monoclonal antibody directed towards GTOL has been developed and used to produce immunochemical methods (RIA and ELISA) for direct measurement of GTOL in urine [85–87]. A recent paper successfully validated this prototype assay [83] and found it to be a viable alternative for routine GTOL determination.

17.2.2.4 Applications

The first proposal of 5-HTOL as a sensitive biomarker for acute drinking was made 15 years ago [88]. In subsequent experimental studies, the increase in urinary 5-HTOL was demonstrated to correlate with ethanol kinetics, however, with a considerable time lag [89] (Fig. 17.4). Routine use of the ratio of 5-HTOL/5-HIAA was also introduced, to compensate for interferences by urine dilution and dietary intake of 5-HT [90], and a limit of >0.015 for the molar ratio was proposed to indicate any intake of alcohol [91].

In a series of validation studies involving both healthy volunteers and alcohol-dependent patients undergoing detoxification, the performance and accuracy of the urinary 5-HTOL/5-HIAA ratio as acute alcohol marker was investigated. Overall, the sensitivity to detect any recent drinking was at least 3-fold higher compared with conventional analysis of ethanol and the specificity was indicated to be close to 100% [12,16,92]. The response to a given dose of ethanol was also demonstrated to be highly reproducible [93].

A comparison between 5-HTOL and self-report for detection of recent alcohol intake was done within a European cancer and nutrition study [94]. In 102 subjects, a close linear relationship existed between the urinary 5-HTOL/5-HIAA ratio and the amount of alcohol ingested during the previous 24 h. The results verified that intake of alcohol above 0.1 g/kg is required to give an elevated 5-HTOL level in a

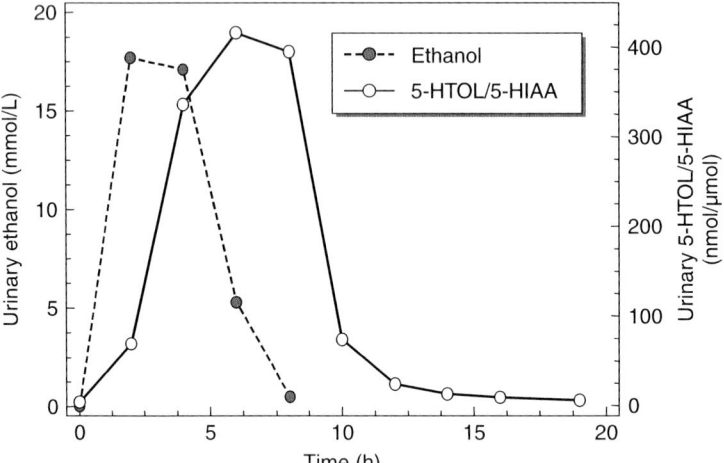

Fig. 17.4. Time course of urinary ethanol and the 5-hydroxytryptophol to 5-hydroxyindoleacetic acid ratio (5-HTOL/5-HIAA) after drinking alcohol (0.5 g ethanol/kg body weight) in a fasted state.

urine sample taken the following day. Accordingly, when used for monitoring of any drinking in the previous day, the 5-HTOL level in morning urine samples may provide a rough estimate of the total amount of alcohol ingested.

The 5-HTOL test has also gained interest in forensic toxicology as a way to settle whether the ethanol identified in a blood or urine specimen originates from alcohol ingestion prior to death, or sampling, or has been generated artifactually (post-mortem) due to microbial action of yeasts and bacteria [95,96]. The 5-HTOL test has been of value in a number of legal cases in Sweden [97]. A similar use of 5-HTOL was made in five problem cases where alcohol was suspected in fatal civil aviation accidents [98].

17.3 TESTS OF CHRONIC ALCOHOL CONSUMPTION

17.3.1 Carbohydrate-deficient transferrin (CDT)

17.3.1.1 Background

Carbohydrate-deficient transferrin (CDT) refers to changes in the microheterogeneity of the iron transport glycoprotein transferrin that are seen after prolonged heavy drinking [99]. Initial studies revealed structural changes in transferrin in both CSF and serum of alcoholic patients [100], and these abnormalities were also demonstrated to normalize during periods of abstinence. An important advantage of CDT compared with the established alcohol biomarkers, such as the liver function tests, was the higher specificity for alcohol with lower risk for false identifications of heavy alcohol consumption in patients suffering from non-alcohol-related liver disease [99,101–104]. Structural modifications in the glycosylation pattern of transferrin are also seen in patients with a group of rare hereditary disorders known as congenital disorders of glycosylation (CDG). Measurement of transferrin microheterogeneity has become used both for screening and diagnosis of CDG [105], and as a sensitive and specific routine alcohol biomarker for detection and follow-up of heavy drinking [99,106].

17.3.1.2 Biochemistry

Transferrin consists of a single polypeptide chain of 679 amino acids, 0–2 *N*-linked oligosaccharide units (*N*-glycans), and two iron-binding sites [107]. The *N*-glycans show a complex structure and different transferrin glycoforms are named based on the total number of terminal, negatively charged sialic acid residues. Tetrasialotransferrin, which contains two biantennary *N*-glycans each terminated by a sialic acid, normally accounts for ~75–80% of total transferrin in human serum [108,109]. Other common glycoforms are pentasialo- (~15%), trisialo- (~5%), hexasialo- (~2%), and disialotransferrin (~2%), whereas asialo-, monosialo-, heptasialo-, and octasialotransferrin usually are found in trace amounts or are not detectable at all [108,109]. There are also several genetic B, C and D homozygous, and heterozygous variants of transferrin, of which transferrin C is the most common phenotype [110,111].

After a period of sustained heavy drinking, corresponding to an average daily intake of at least ~50–80 g ethanol over 1–2 weeks, increased relative amounts of transferrin molecules that contain only one complete *N*-glycan (i.e. disialotransferrin), or lack both *N*-glycans (asialotransferrin) [112–114], are commonly detected [99]. When drinking is discontinued, the glycoform pattern slowly recovers to normal [99,115], and the time to reach a stable baseline level could take one month or longer [23]. The biological mechanism(s) behind the alcohol-mediated elevation of disialo- and asialotransferrin may be inhibition of enzymes responsible for glycosyl synthesis and transfer [116,117].

17.3.1.3 Analysis

Because the main CDT glycoforms (i.e. disialo- and asialotransferrin) miss two or four terminal sialic acid residues, respectively, compared with tetrasialotransferrin, and are therefore less negatively charged, they are readily separated by charge-based analytical techniques such as isoelectric focusing (IEF), ion-exchange chromatography (HPLC), and capillary (zone) electrophoresis (CE/CZE) [101,118]. Over the years, many different methods have been in use for determination of CDT and this has often created confusion in the clinical application. There have been method-dependent discrepancies in the definition of the analyte, the way test results are reported (absolute or relative amounts), and in their accuracy to distinguish a normal from an elevated value (i.e. the diagnostic sensitivity and specificity). Moreover, the lack of traceability of reference intervals has hampered the clinical implementation and acceptance of this laboratory test.

IEF. In the original CDT studies, IEF was the standard method used for separation, identification, and quantification of the transferrin glycoforms in samples of serum and cerebrospinal fluid (CSF) [100]. The glycoforms reported to be alcohol-related showed isoelectric points at pH 5.7 or above after complete iron saturation, corresponding to disialo-, monosialo-, and asialotransferrin, and were collectively named CDT [99]. For routine quantitative purposes, a limitation of the IEF technique is that there is no linear relationship between the concentration and the staining intensity of tetrasialotransferrin, the major band as determined by densitometric analysis. This makes accurate relative quantification of the minor glycoforms impossible. The method is also too laborious and time-consuming to be suitable for routine use [119]. However, IEF is useful for visualization and identification of genetic transferrin variants and for preliminary diagnosis of CDG.

HPLC. An HPLC method for quantification of transferrin glycoforms and CDT was introduced in 1993 [115]. The method was based on anion-exchange chromatographic separation of iron-saturated serum transferrin glycoforms by salt gradient elution followed by photometric detection. The relative amount of any single glycoform, or combination of glycoforms, in relation to total transferrin (all glycoforms) can be calculated from the peak areas in the chromatograms. Use of the relative CDT amount (%CDT) instead of an absolute amount is important to achieve a high diagnostic specificity, because having a very high total transferrin concentration might otherwise render falsely high and false-positive CDT results [120–122].

Several variants of the original HPLC method have been published [109,123,124] and commercial CDT kits introduced [125]. A common feature of all HPLC methods is that quantification relies on the selective absorbance of the iron–transferrin complex at ~460–470 nm [126]. This represents an analytical advantage over the CE methods, because they use detection at ~200 nm where all proteins absorb. Today, sensitive and specific HPLC measurement of CDT is routinely applied in many hospital and research laboratories (Fig. 17.5).

CE. As for HPLC, the CE/CZE methods for CDT allow for reproducible separation and quantification of single transferrin glycoforms. Both commercial standard kits and optimized versions have been used for CDT testing [127–129]. The relative amounts of disialotransferrin determined by HPLC and CE are highly correlated, but the analytical sensitivity of CE appears to be lower compared with HPLC [130], thereby yielding lower values for the CDT glycoforms [109,131]. A disadvantage is the unspecific measurement of the peptide bond at ~200 nm, whereby many other biomolecules with similar isoelectric points (e.g. C-reactive protein, light chains, and complement factors) could interfere when present at increased concentrations [131–133]. However, improved CE methods showing better resolution between disialo- and trisialotransferrin and producing values for the CDT glycoforms approaching those obtained by HPLC have recently been published [128,129].

Immunoassays. Immunochemical assays for CDT have been much used and are convenient for routine application. A disadvantage is that they cannot distinguish single glycoforms but measure a fraction of transferrin glycoforms as CDT. In the original CDTect method that was introduced in 1992 [134], but is no longer used, a

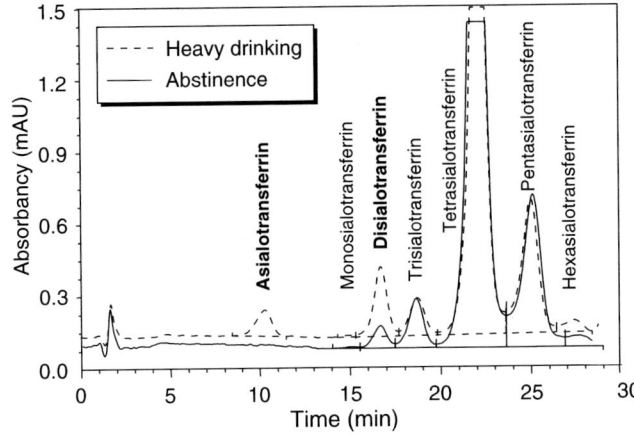

Fig. 17.5. HPLC chromatograms showing the transferrin glycoform pattern in serum from an alcoholic patient after a period of heavy drinking, and after several weeks of abstinence from alcohol. Heavy alcohol consumption increases the relative amounts of disialo- and asialotransferrin, which is used as an alcohol biomarker (known as carbohydrate-deficient transferrin, CDT)

CDT fraction was separated from the non-CDT glycoforms using a small anion-exchange column, followed by measurement of the absolute amount of CDT (in units/L) by a transferrin immunoassay. Subsequent immunoassays (e.g. %CDT TIA, %CDT) retained the original column separation-transferrin immunoassay test principle, but involved modifications in the separation or quantification procedures and expressed CDT in relative (%CDT) concentration [135]. A drawback with these methods is that they separate CDT from non-CDT glycoforms based on differences in the net charge of the molecule, without any control of the individual glycoforms eluted. Accordingly, these methods may give inaccurate results for any shift in charge, for example in genetic transferrin variants and CDG [99,101,111,136].

A major advancement was the recent introduction of the first direct immunoassay for determination of %CDT (N Latex CDT), which at the same time appears not to be influenced by genetic transferrin variants [137,138].

17.3.1.4 Applications

After CDT was first introduced in the 1970s as a more specific alcohol biomarker than the established tests, it has become widely used in different settings for routine detection of heavy alcohol consumption and follow-up of treatment, and as an objective outcome measure in a large number of alcohol research studies [139,140]. Besides the original applications of the CDT test in the treatment and long-term monitoring of alcohol-dependent outpatients [141–143], CDT has been successfully applied as a screening tool for alcohol use disorders in surgical and trauma patients [144,145], and for monitoring of relapse drinking in patients following liver transplantation for alcoholic cirrhosis [146,147]. CDT testing has also significant potential for enhancing the quality of medical treatment in primary and company health care settings [4,148], and to identify and intervene with patients being treated for Type 2 diabetes and hypertension and other chronic medical problems adversely affected by heavy alcohol use [149,150]. In this respect, the addition of routine CDT screening to patient self-report may even provide positive net economic benefits [151]. Another rapidly growing use of the CDT test is in confirmation of abstinence and detection of relapse in connection with rehabilitation of convicted drunk drivers and regranting of driving licenses [152–154].

Still a limitation of CDT testing is the lack of standardization regarding the analytical procedures, and also in the definition of the analyte. This prompted the International Federation of Clinical Chemistry and Laboratory Medicine (IFCC) to convene a working group for standardization of CDT measurement that addresses such issues as definition of the analyte, establishment of reference methodology, and production of standardized calibrators [155].

17.3.2 Phosphatidylethanols (PEth)

17.3.2.1 Background

Phosphatidylethanols (PEth) are abnormal phospholipids formed in the presence of ethanol and become accumulated in cell membranes [156]. PEth were early proposed as a promising new biomarker of chronic heavy alcohol consumption [157]. Subsequent studies of alcohol-dependent patients undergoing detoxification revealed detectable levels of PEth in the blood for up to 14 days after admission, while no PEth was detected in control subjects who had abstained from alcohol [158]. The average half-life for PEth disappearance was 4 days, but there was large variation in the time course between alcoholic individuals [159]. The quantity and frequency of drinking needed to generate a detectable PEth level in blood appears to be similar to that required for a positive CDT value, and the sensitivity of PEth in alcoholic patients has been reported to be at least equal to, or greater than, that of CDT [160,161].

17.3.2.2 Biochemistry

PEth are negatively charged diacyl phospholipids that are formed slowly from other naturally occurring phospholipids such as phosphatidylcholines *via* a transphosphatidylation reaction catalyzed by phospholipase D (Fig. 17.6) [156,162]. Presence of PEth is well known to affect many biochemical properties of the cell membranes in humans. High-density lipoprotein (HDL) particles have been suggested to act as carriers of PEth and possibly mediate some of the antiatherogenic effects of moderate alcohol consumption [163].

A serious limitation with this biomarker concerns its stability, as PEth may be both degraded and formed on storage of specimens [164]. An in vitro study using human erythrocytes demonstrated that PEth were formed at physiologically relevant ethanol concentrations [165]. Blood obtained from alcoholic subjects accumulated

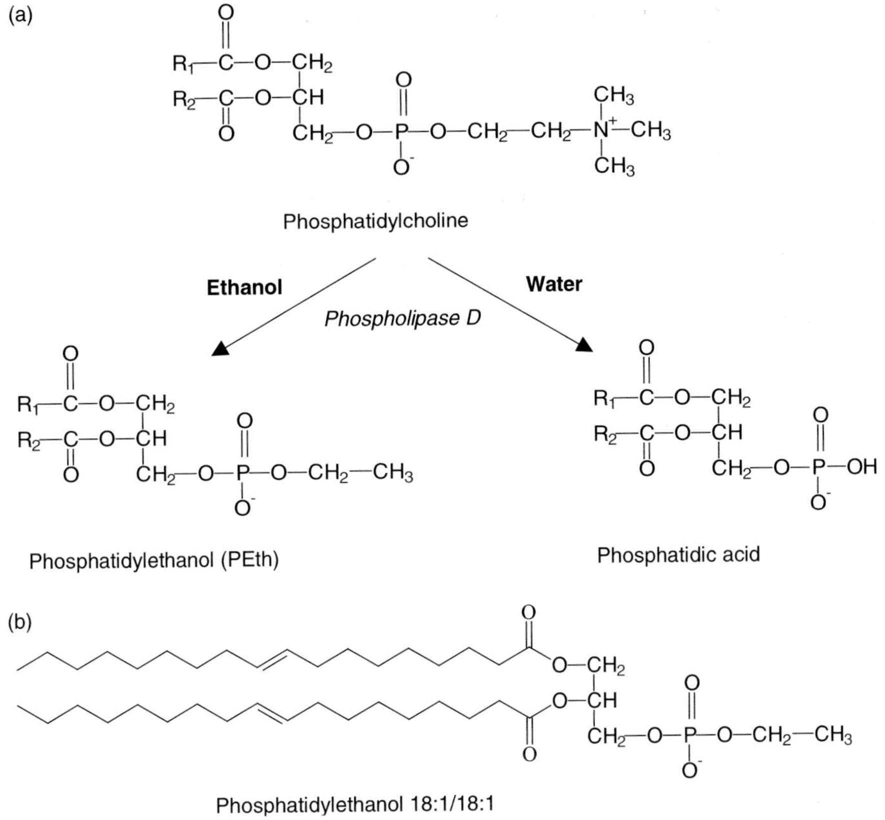

(a)

Phosphatidylcholine

Ethanol   Water

Phospholipase D

Phosphatidylethanol (PEth)   Phosphatidic acid

(b)

Phosphatidylethanol 18:1/18:1

Fig. 17.6. (a) Metabolic pathways for the phospholipase D-catalyzed formation of phosphatidylethanols (PEth) in the presence of ethanol and phosphatidic acid in the presence of water and (b) structural illustration of one commercially available PEth molecule (18:1/18:1).

about twice as much PEth than control blood, suggesting that the sensitivity of PEth as biomarker may be influenced by prior heavy drinking. The risk of artifactual formation of PEth during storage of blood samples containing ethanol influences the specificity of this test [166], and the samples must therefore be stored properly prior to analysis. For example, refrigerated blood samples that contain ethanol are reported to be stable for 72 h, whereas PEth may be formed artifactually if they are frozen at −20°C [166].

17.3.2.3 Analysis

In most studies published to date, HPLC and evaporative light-scattering detection has been used for quantification of PEth [167,168]. A quantitative LC-MS method using time-of-flight (TOF) detection has also been developed [163]. In the MS methods, the analysis focused on one single PEth molecule and not on a group of PEth as with the HPLC method. However, to allow for widespread clinical use of the PEth biomarker, more simple analytical procedures that are appropriate for routine use are needed.

17.3.2.4 Applications

Measurement of PEth in blood has long been proposed as a specific and sensitive alcohol biomarker, but only few studies have evaluated its usefulness in clinical settings. Studies of alcohol-dependent patients have shown promising results, and also that the PEth concentrations in blood are correlated to the amount of alcohol consumed [161,169]. PEth have also been evaluated as a biomarker of previous heavy drinking in post-mortem blood [170].

17.4 FUTURE ASPECTS ON ALCOHOL TESTING AND ALCOHOL BIOMARKERS

A number of sensitive and specific biomarkers of acute alcohol consumption exist for different applications, but analytical methods and also new markers with further improved specifications are likely to be developed in the future. The experiences obtained with the current alcohol biomarkers will then be very important, for understanding of the most important characteristics (e.g. regarding sensitivity, specificity, time window, methodology and biological specimen) to be focused on.

The acute markers should not be considered mainly as stand-alone biomarkers but, rather, as a complement to other tests with different dose and time-window characteristics, and also with self-report measures and questionnaires. For example, the results of the WHO/ISBRA study clearly demonstrated the value of having access to a wide variety of data of alcohol use to verify each case [73]. In this respect, better knowledge of individual drinking habits might contribute to improved treatment and, in the end, a better outcome.

While biomarkers of prolonged heavy alcohol intake such as CDT and GGT might be preferable in situations of a clinical suspicion of harmful drinking habits, and in general health screening for high-risk drinking, test sensitive enough to detect recent drinking are better suited for close monitoring of the diagnosed patient. By introducing regular and/or random testing for excessive and harmful alcohol consumption with the aid of a sensitive acute marker, early detection of relapse drinking can also be achieved. This strategy is of major value in outpatient treatment of patients that are able to work.

17.5 CONCLUSION

A perfect alcohol biomarker should be completely specific for alcohol and also exhibit high sensitivity to harmful drinking habits within a given time period prior to testing. Furthermore, the test should be generally available, inexpensive, and yield rapid results with an easily accessible biological specimen such as urine, saliva, or blood. Measuring ethanol in body fluids or breath is a highly specific and simple alcohol test, but because ethanol is eliminated fairly rapidly from the body, the sensitivity of this method is rather low. Even intake of a bottle of wine or corresponding amounts of other alcoholic beverages in the evening will usually not be detectable the following morning.

Most of the currently available alcohol biomarkers in routine or research use perform quite well, when selected high-risk populations are compared with low drinking or abstinent subjects, while they perform less satisfactory in the general population. However, having access to a panel of alcohol test with well-defined dose-response curves and detection times for use with different biological specimens, the best single marker or test combination in each case can be chosen. To control for acute alcohol intake or relapse, the currently recommended biomarkers besides ethanol testing are measurement of EtG, EtS, or 5-HTOL in a sample of urine. These sensitive and specific tests offer unique possibilities to disclose intake of even small to moderate amounts of ethanol for a considerable time period. They are also useful to distinguish between in vivo ethanol metabolism and artifactual ethanol formation for forensic and clinical applications. Caution must, however, be taken as the experience of these relatively new tests is still emerging. For example, recent data suggest that EtS might be a more robust and reliable parameter than EtG.

Among the biomarkers used for detection and follow-up of sustained excessive and harmful drinking levels, CDT is the currently most alcohol-specific routine test that has received a widespread use. In clinical settings, CDT is preferably combined with GGT to allow for detection of alcoholic liver disease. The recent introduction of simple analytical methods for measurement of CDT, together with the ongoing international standardization work, are two important factors to improve further the value of this biomarker. Development of more simple analytical methods is also important for PEth analysis to allow for further research studies and clinical implementation of this test.

17.6 REFERENCES

1   A. Helander, J. Neural. Transm., (2003) 15–32.
2   O. Niemelä, Clin. Chim. Acta, 377 (2007) 39.
3   T.F. Babor, M. Aguirre-Molina, G.A. Marlatt and R. Clayton, Am. J. Health Promot., 4 (1999) 98–103.
4   U. Hermansson, A. Helander, A. Huss, L. Brandt and S. Rönnberg, Alcohol Clin. Exp. Res., 24 (2000) 180–187.
5   D. Mayfield, G. McLeod and P. Hall, Am. J. Psychiat., 131 (1974) 1121–1123.
6   M.L. Selzer, Am. J. Psychiat., 127 (1981) 89–94.
7   J.B. Saunders, O.G. Aasland, T.F. Babor, J.R. de la Fuente and M. Grant, Addiction, 88 (1993) 791–804.
8   R.K. Fuller, K.K. Lee and E. Gordis, Alcohol Clin. Exp. Res., 12 (1988) 201–205.
9   A. Helander, J. von Wachenfeldt, A. Hiltunen, O. Beck, P. Liljeberg and S. Borg, Drug Alcohol Depen., 56 (1999) 33–38.
10   P. Bendtsen, J. Hultberg, M. Carlsson and A.W. Jones, Alcohol Clin. Exp. Res., 23 (1999) 1446–1451.
11   R. Swift, Addiction, 98(Suppl. 2) (2003) 73–80.

12   A. Helander, O. Beck and A.W. Jones, Clin. Chem., 42 (1996) 618–624.
13   B.K. Logan and A.W. Jones, Med. Sci. Law, 40 (2000) 206–215.
14   E. Jansson-Nettelbladt, S. Meurling, B. Petrini and J. Sjolin, Acta Paediatr., 95 (2006) 502–504.
15   F.C. Kugelberg and A.W. Jones, Forensic Sci. Int., 165 (2007) 10–29.
16   P. Bendtsen, A.W. Jones and A. Helander, Alcohol Alcoholism, 33 (1998) 431–438.
17   A. Berger, BMJ, 325 (2002) 1403.
18   A.W. Jones and L. Andersson, Forensic Sci. Int., 132 (2003) 18–25.
19   A.W. Jones, J. Anal. Toxicol., 19 (1995) 169–173.
20   K.M. Conigrave, P. Davies, P. Haber and J.B. Whitfield, Addiction, 98(Suppl. 2) (2003) 31–43.
21   J.B. Whitfield, Crit. Rev. Clin. Lab Sci., 38 (2001) 263–355.
22   K. Puukka, J. Hietala, H. Koivisto, P. Anttila, R. Bloigu and O. Niemela, Alcohol Alcoholism, 41 (2006) 522–527.
23   A. Helander and S. Carlsson, Alcohol Clin. Exp. Res., 20 (1996) 1202–1205.
24   G. Schmitt, R. Aderjan, T. Keller and M. Wu, J. Anal. Toxicol., 19 (1995) 91–94.
25   A. Helander and O. Beck, Clin. Chem., 50 (2004) 936–937.
26   G. Skopp, G. Schmitt, L. Potsch, P. Dronner, R. Aderjan and R. Mattern, Alcohol Alcoholism, 35 (2000) 283–285.
27   G. Hoiseth, R. Karinen, A.S. Christophersen, L. Olsen, P.T. Normann and J. Morland, Forensic Sci. Int., 165 (2007) 41–45.
28   L. Morini, L. Politi, A. Zucchella and A. Polettini, Clin. Chim. Acta, 376 (2007) 213–219.
29   A. Helander and O. Beck, J. Anal. Toxicol., 29 (2005) 270–274.
30   S. Dresen, W. Weinmann and F.M. Wurst, J. Am. Soc. Mass Spectrom., 15 (2004) 1644–1648.
31   W. Bicker, M. Lammerhofer, T. Keller, R. Schuhmacher, R. Krska and W. Lindner, Anal. Chem., 78 (2006) 5884–5892.
32   X.S. Deng, P. Bludeau and R.A. Deitrich, Alcohol, 34 (2004) 217–223.
33   H. Dahl, N. Stephanson, O. Beck and A. Helander, J. Anal. Toxicol., 26 (2002) 201–204.
34   H. Schneider and H. Glatt, Biochem. J., 383 (2004) 543–549.
35   R.S. Foti and M.B. Fisher, Forensic Sci. Int., 153 (2005) 109–116.
36   P.I. Jaakonmaki, K.L. Knox, E.C. Horning and M.G. Horning, Eur. J. Pharmacol., 1 (1967) 63–70.
37   G. Schmitt, P. Droenner, G. Skopp and R. Aderjan, J. Forensic Sci., 42 (1997) 1099–1102.
38   F.M. Wurst, S. Seidl, D. Ladewig, F. Muller-Spahn and A. Alt, Addict. Biol., 7 (2002) 427–434.
39   K. Borucki, R. Schreiner, J. Dierkes, K. Jachau, D. Krause, S. Westphal, F.M. Wurst, C. Luley and H. Schmidt-Gayk, Alcohol Clin. Exp. Res., 29 (2005) 781–787.
40   F.M. Wurst, S. Dresen, J.P. Allen, G. Wiesbeck, M. Graf and W. Weinmann, Addiction, 101 (2006) 204–211.
41   N. Stephanson, H. Dahl, A. Helander and O. Beck, Ther. Drug Monit., 24 (2002) 645–651.
42   T. Sarkola, H. Dahl, C.J. Eriksson and A. Helander, Alcohol Alcoholism, 38 (2003) 347–351.
43   J. Bergström, A. Helander and A.W. Jones, Forensic Sci. Int., 133 (2003) 86–94.
44   P. Droenner, G. Schmitt, R. Aderjan and H. Zimmer, Forensic Sci. Int., 126 (2002) 24–29.
45   H. Schloegl, S. Dresen, K. Spaczynski, M. Stoertzel, F.M. Wurst and W. Weinmann, Int. J. Legal Med., 120 (2006) 83–88.
46   A. Helander and H. Dahl, Clin. Chem., 51 (2005) 1728–1730.
47   R. Aderjan, H. Besserer, H. Sachs, G. Schmitt and G. Skopp, Proc TIAFT-SOFT Meeting in Tampa, 1994.
48   A. Alt, I. Janda, S. Seidl and F.M. Wurst, Alcohol Alcoholism, 35 (2000) 313–314.
49   F. Pragst, K. Spiegel, F. Sporkert and M. Bohnenkamp, Forensic Sci. Int., 107 (2000) 201–223.
50   C. Jurado, T. Soriano, M.P. Gimenez and M. Menendez, Forensic Sci. Int., 145 (2004) 161–166.
51   L. Morini, L. Politi, A. Groppi, C. Stramesi and A. Polettini, J. Mass Spectrom., 41 (2006) 34–42.
52   L. Politi, L. Morini, F. Leone and A. Polettini, Addiction, 101 (2006) 1408–1412.
53   F.M. Wurst, C. Kempter, S. Seidl and A. Alt, Alcohol Alcoholism, 34 (1999) 71–77.
54   I. Janda and A. Alt, J. Chromatogr. B, Biomed. Sci. Appl., 758 (2001) 229–234.

55   M. Nishikawa, H. Tsuchihashi, A. Miki, M. Katagi, G. Schmitt, H. Zimmer, T. Keller and R. Aderjan, J. Chromatogr. B, Biomed. Sci. Appl., 726 (1999) 105–110.

56   B. Maralikova and W. Weinmann, J. Chromatogr. B, Analyt. Technol. Biomed. Life Sci., 811 (2004) 21–30.

57   W. Weinmann, P. Schaefer, A. Thierauf, A. Schreiber and F.M. Wurst, J. Am. Soc. Mass Spectrom., 15 (2004) 188–193.

58   I. Janda, W. Weinmann, T. Kuehnle, M. Lahode and A. Alt, Forensic Sci. Int., 128 (2002) 59–65.

59   M. Yegles, A. Labarthe, V. Auwarter, S. Hartwig, H. Vater, R. Wennig and F. Pragst, Forensic Sci. Int., 145 (2004) 167–173.

60   H. Zimmer, G. Schmitt and R. Aderjan, J. Anal. Toxicol., 26 (2002) 11–16.

61   L. Anne, C. Bih, S. Mitra, V. Bodepudi, M. Datuin and R. Ruzicka, Development of a homogenous enzyme immunoassay for the detection of ethyl glucuronide in urine. Abstract at the TIAFT Meeting, Seoul, Korea, August 29–September 2, 2005, p. 60.

62   F.M. Wurst, C. Kempter, J. Metzger, S. Seidl and A. Alt, Alcohol, 20 (2000) 111–116.

63   S. Seidl, F.M. Wurst and A. Alt, Addict. Biol., 6 (2001) 205–212.

64   H. Schloegl, T. Rost, W. Schmidt, F.M. Wurst and W. Weinmann, Forensic Sci. Int., 156 (2006) 213–218.

65   M. Klys, M. Scislowski, S. Rojek and J. Kolodziej, Leg. Med. (Tokyo), 7 (2005) 319–325.

66   A. Costantino, E.J. Digregorio, W. Korn, S. Spayd and F. Rieders, J. Anal. Toxicol., 30 (2006) 659–662.

67   G. Rosenfeld, Proc. Soc. Exp. Biol. Med., 103 (1960) 144–149.

68   V.E. Davis, H. Brown, J.A. Huff and J.L. Cashaw, J. Lab. Clin. Med., 69 (1967) 132–140.

69   M.J. Walsh and N.Y. Ann, Acad. Sci., 215 (1973) 98–110.

70   S. Svensson, M. Some, A. Lundsjo, A. Helander, T. Cronholm and J.O. Hoog, Eur. J. Biochem., 262 (1999) 324–329.

71   D.P. Agarwal, Pathol. Biol. (Paris), 49 (2001) 703–709.

72   A. Helander, C. Walzer, O. Beck, L. Balant, S. Borg and J.P. von Wartburg, Life Sci., 55 (1994) 359–366.

73   A. Helander and C.J. Eriksson, Alcohol Clin. Exp. Res., 26 (2002) 1070–1077.

74   O. Beck, A. Helander, S. Carlsson and S. Borg, Pharmacol. Toxicol., 77 (1995) 323–326.

75   O. Beck, S. Borg, L. Eriksson and A. Lundman, Naunyn-Schmiedeberg's Arch. Pharmacol., 321 (1982) 293–297.

76   A. Helander, O. Beck and L. Boysen, Life Sci., 56 (1995) 1529–1534.

77   N. Stephanson, H. Dahl, A. Helander and O. Beck, J. Chromatogr. B, Analyt. Technol. Biomed. Life Sci., 816 (2005) 107–112.

78   H.C. Curtius, M. Wolfensberger, U. Redweik, W. Leimbacher, R.A. Maibach and W. Isler, J. Chromatogr., 112 (1975) 523–531.

79   S. Takahashi, D.D. Godse, A. Naqvi, J.J. Warsh and H.C. Stancer, Clin. Chim. Acta, 84 (1978) 55–62.

80   O. Beck, S. Borg, B. Holmstedt, H. Kvande and R. Shroder, Acta Pharmacol. Toxicol. (Copenh), 47 (1980) 305–307.

81   O. Beck and A. Helander, Addiction, 98(Suppl. 2) (2003) 63–72.

82   A. Helander, O. Beck and S. Borg, J. Chromatogr. Biomed. Appl., 579 (1992) 340–345.

83   J. Dierkes, M. Wolfersdorf, K. Borucki, W. Weinmann, G. Wiesbeck, O. Beck, S. Borg and F.M. Wurst, Clin. Biochem., 40 (2007) 128–131.

84   R.D. Johnson, R.J. Lewis, D.V. Canfield and C.L. Blank, J. Chromatogr. B, Analyt. Technol. Biomed. Life Sci., 805 (2004) 223–234.

85   E. Unger, E. Åkerblom, R.-M. Jönsson, A. Helander, O. Beck and R. Brandt, A radioimmunoassay for the 5-hydroxytryptophol-glucuronide. 6th ESBRA Meeting, Stockholm, Sweden, 1997.

86   A. Helander, H. Dahl, E. Unger, S. Borg and O. Beck, Evaluation of an immunoassay for 5-hydroxytryptophol glucuronide in urine. 7th ESBRA Congress, Barcelona, Spain, 1999.

87   A. Helander, C. Gäredal, H. Dahl and O. Beck, Development of an enzyme immunoassay for glucuronide-conjugated 5-hydroxytryptophol (GTOL) in human urine. RSA/ISBRA Congress, San Francisco, USA, 2002.

88   A. Voltaire, O. Beck and S. Borg, Alcohol Clin. Exp. Res., 16 (1992) 281–285.

89   A. Helander, O. Beck, G. Jacobsson, C. Löwenmo and T. Wikström, Life Sci., 53 (1993) 847–855.

90   A. Helander, T. Wikström, C. Löwenmo, G. Jacobsson and O. Beck, Life Sci., 50 (1992) 1207–1213.

91   A. Helander, O. Beck and Borg S, Alcohol Alcoholism (Suppl. 2) (1994) 497–502.

92   R.L. Hagan and A. Helander, Aviat. Space Environ. Med., 68 (1997) 30–34.

93   A.W. Jones and A. Helander, Alcohol Clin. Exp. Res., 23 (1999) 1921–1926.

94   A. Kroke, K. Klipstein-Grobusch, K. Hoffmann, I. Terbeck, H. Boeing and A. Helander, Br. J. Nutr., 85 (2001) 621–627.

95   A. Helander, O. Beck and A.W. Jones, Lancet, 340 (1992) 1159.

96   A. Helander, O. Beck and A.W. Jones, J. Forensic Sci., 40 (1995) 95–98.

97   A. Helander and A.W. Jones, Lakartidningen, 99 (2002) 3950–3954.

98   R.D. Johnson, R.J. Lewis, D.V. Canfield, K.M. Dubowski and C.L. Blank, J. Forensic Sci., 50 (2005) 670–675.

99   H. Stibler, Clin. Chem., 37 (1991) 2029–2037.

100 H. Stibler, C. Allgulander, S. Borg and K.G. Kjellin, Acta Med. Scand., 204 (1978) 49–56.

101 T. Arndt, Clin. Chem., 47 (2001) 13–27.

102 G.J. Meerkerk, K.H. Njoo, I.M. Bongers, P. Trienekens and J.A. van Oers, Alcohol Clin. Exp. Res., 22 (1998) 908–913.

103 G.J. Meerkerk, K.H. Njoo, I.M. Bongers, P. Trienekens and J.A. van Oers, Alcohol Clin. Exp. Res., 23 (1999) 1052–1059.

104 M. Salaspuro, Alcohol, 19 (1999) 261–271.

105 H.H. Freeze and M. Aebi, Curr. Opin. Struct. Biol., 15 (2005) 490–498.

106 K. Golka and A. Wiese, J. Toxicol. Environ. Health B, Crit. Rev., 7 (2004) 319–337.

107 G. de Jong, J.P. van Dijk and H.G. van Eijk, Clin. Chim. Acta., 190 (1990) 1–46.

108 O. Mårtensson, A. Härlin, R. Brandt, K. Seppä and P. Sillanaukee, Alcohol Clin. Exp. Res., 21 (1997) 1710–1715.

109 A. Helander, A. Husa and J.-O. Jeppsson, Clin. Chem., 49 (2003) 1881–1890.

110 M.I. Kamboh and R.E. Ferrell, Hum. Hered., 37 (1987) 65–81.

111 A. Helander, G. Eriksson, H. Stibler and J.-O. Jeppsson, Clin. Chem., 47 (2001) 1225–1233.

112 E. Landberg, P. Påhlsson, A. Lundblad, A. Arnetorp and J.-O. Jeppsson, Biochem. Biophys. Res. Commun., 210 (1995) 267–274.

113 J. Peter, C. Unverzagt, W.D. Engel, D. Renauer, C. Seidel and W. Hösel, Biochim. Biophys. Acta, 1380 (1998) 93–101.

114 C. Flahaut, J.C. Michalski, T. Danel, M.H. Humbert and A. Klein, Glycobiology, 13 (2003) 191–198.

115 J.-O. Jeppsson, H. Kristensson and C. Fimiani, Clin. Chem., 39 (1993) 2115–2120.

116 C.S. Lieber, Alcohol, 19 (1999) 249–254.

117 P. Sillanaukee, N. Strid, J.P. Allen and R.Z. Litten, Alcohol Clin. Exp. Res., 25 (2001) 34–40.

118 F. Bortolotti, G. De Paoli and F. Tagliaro, J. Chromatogr. B, Analyt. Technol. Biomed. Life Sci., 841 (2006) 96–109.

119 Y. Xin, A.S. Rosman, J.M. Lasker and C.S. Lieber, Alcohol Alcoholism, 27 (1992) 425–433.

120 K. Sorvajärvi, J.E. Blake, Y. Israel and O. Niemelä, Alcohol Clin. Exp. Res., 20 (1996) 449–454.

121 J. Keating, C. Cheung, T.J. Peters and R.A. Sherwood, Clin. Chim. Acta, 272 (1998) 159–169.

122 A. Helander, Clin. Chem., 45 (1999) 131–135.

123 E. Werle, G.E. Seitz, B. Kohl, W. Fiehn and H.K. Seitz, Alcohol Alcoholism, 32 (1997) 71–77.

124 U. Turpeinen, T. Methuen, H. Alfthan, K. Laitinen, M. Salaspuro and U.H. Stenman, Clin. Chem., 47 (2001) 1782–1787.

125 A. Helander and J.P. Bergström, Clin. Chim. Acta, 371 (2006) 187–190.

126 J.-O. Jeppsson, Biochim. Biophys. Acta, 140 (1967) 468–476.

127 F.J. Legros, V. Nuyens, M. Baudoux, K. Zouaoui Boudjeltia, J.L. Ruelle, J. Colicis, F. Cantraine and J.P. Henry, Clin. Chem., 49 (2003) 440–449.

128 C. Lanz, M. Kuhn, V. Deiss and W. Thormann, Electrophoresis, 25 (2004) 2309–2318.

129 J. Joneli, C. Lanz and W. Thormann, J. Chromatogr. A, 1130 (2006) 272–280.

130 A. Helander, J.P. Wielders, R. Te Stroet and J.P. Bergström, Clin. Chem., 51 (2005) 1528–1531.

131 F.J. Legros, V. Nuyens, E. Minet, P. Emonts, K.Z. Boudjeltia, A. Courbe, J.L. Ruelle, J. Colicis, F. de L'Escaille and J.P. Henry, Clin. Chem., 48 (2002) 2177–2186.

132 C. Lanz and W. Thormann, Electrophoresis, 24 (2003) 4272–4281.

133 B. Ramdani, V. Nuyens, T. Codden, G. Perpete, J. Colicis, A. Lenaerts, J.P. Henry and F.J. Legros, Clin. Chem., 49 (2003) 1854–1864.

134 H. Stibler, S. Borg and M. Joustra, Alcohol Alcoholism (Suppl. 1) (1991) 451–454.

135 A. Helander, M. Fors and B. Zakrisson, Alcohol Alcoholism, 36 (2001) 406–412.

136 A. Helander, J. Bergström and H.H. Freeze, Clin. Chem., 50 (2004) 954–958.

137 A. Helander, H. Dahl, I. Swanson and J. Bergström, Alcohol Clin. Exp. Res., 28 (2004) 33A.

138 J.R. Delanghe, A. Helander, J.P. Wielders, J.M. Pekelharing, H.J. Roth, F. Schellenberg, C. Born, E. Yagmur, W. Gentzer and H. Althaus, Clin. Chem., 53 (2007) 1115–1121.

139 N. Ait-Daoud, B.A. Johnson, M. Javors, J.D. Roache and N.A. Zanca, Alcohol Clin. Exp. Res., 25 (2001) 847–849.

140 R.F. Anton and M. Youngblood, Alcohol Clin. Exp. Res., 30 (2006) 1878–1883.

141 S. Borg, A. Helander, A. Voltaire Carlsson and A.M. Högstrom Brandt, Alcohol Clin. Exp. Res., 19 (1995) 961–963.

142 A. Helander, A.V. Carlsson and S. Borg, Alcohol Alcoholism, 31 (1996) 101–107.

143 B. Hock, M. Schwarz, I. Domke, V.P. Grunert, M. Wuertemberger, U. Schiemann, S. Horster, C. Limmer, G. Stecker and M. Soyka, Addiction, 100 (2005) 1477–1486.

144 M.J. Martin, C. Heymann, T. Neumann, L. Schmidt, F. Soost, B. Mazurek, B. Bohm, C. Marks, K. Helling, E. Lenzenhuber, C. Muller, W.J. Kox and C.D. Spies, Alcohol Clin. Exp. Res., 26 (2002) 836–840.

145 H. Tonnesen, M. Carstensen and P. Maina, Eur. J. Surg., 165 (1999) 522–527.

146 A. Heinemann, M. Sterneck, R. Kuhlencordt, X. Rogiers, K.H. Schulz, B. Queen, F. Wischhusen and K. Puschel, Alcohol Clin. Exp. Res., 22 (1998) 1806–1812.

147 G.A. Berlakovich, T. Soliman, E. Freundorfer, T. Windhager, M. Bodingbauer, P. Wamser, H. Hetz, M. Peck-Radosavljevic and F. Muehlbacher, Transplant Int., 17 (2004) 617–621.

148 P. Sillanaukee, M. Aalto and K. Seppa, Alcohol Clin. Exp. Res., 22 (1998) 892–896.

149 M. Fleming, R. Brown and D. Brown, J. Stud. Alcohol, 65 (2004) 631–637.

150 P.M. Miller and R.F. Anton, Addict. Behav., 29 (2004) 1427–1437.

151 K.S. Dillie, M. Mundt, M.T. French and M.F. Fleming, Alcohol Clin. Exp. Res., 29 (2005) 2008–2014.

152 H. Kristenson and J.-O. Jeppsson, Läkartidningen, 95 (1998) 1429–1430.

153 B. Bjerre, S. Borg, A. Helander, J.-O. Jeppsson, G. Johnson and G. Karlsson, Lakartidningen, 98 (2001) 677–683.

154 B.M. Appenzeller, S. Schneider, A. Maul and R. Wennig, Drug Alcohol Depen., 79 (2005) 261–265.

155 J.-O. Jeppsson, T. Arndt, F. Schellenberg, J.P.M. Wielders, R.F. Anton, J.B. Whitfield and A. Helander, Clin. Chem. Lab. Med., 45 (2007) 558–562.

156 C. Alling, L. Gustavsson and E. Änggård, FEBS Lett., 152 (1983) 24–28.

157 G.C. Mueller, M.F. Fleming, M.A. LeMahieu, G.S. Lybrand and K.J. Barry, Proc. Natl. Acad. Sci. USA, 85 (1988) 9778–9782.

158 P. Hansson, M. Caron, G. Johnson, L. Gustavsson and C. Alling, Alcohol Clin. Exp. Res., 21 (1997) 108–110.

159 A. Varga, P. Hansson, G. Johnson and C. Alling, Clin. Chim. Acta, 299 (2000) 141–150.

160 A. Varga, P. Hansson, C. Lundqvist and C. Alling, Alcohol Clin. Exp. Res., 22 (1998) 1832–1837.

161 S. Aradottir, G. Asanovska, S. Gjerss, P. Hansson and C. Alling, Alcohol Alcoholism, 41 (2006) 431–437.

162   S. Li, H. Lin, G. Wang and C.H. Huang, Arch. Biochem. Biophys., 385 (2001) 88–98.
163   A. Tolonen, T.M. Lehto, M.L. Hannuksela and M.J. Savolainen, Anal. Biochem., 341 (2005) 83–88.
164   A. Bruhl, A. Faldum and K. Loffelholz, Biochim. Biophys. Acta, 1633 (2003) 84–89.
165   A. Varga and C. Alling, J. Lab. Clin. Med., 140 (2002) 79–83.
166   S. Aradottir, K. Moller and C. Alling, Alcohol Alcoholism, 39 (2004) 8–13.
167   T. Gunnarsson, A. Karlsson, P. Hansson, G. Johnson, C. Alling and G. Odham, J. Chromatogr. B, Biomed. Sci. Appl., 705 (1998) 243–249.
168   S. Aradottir and B.L. Olsson, BMC Biochem., 6 (2005) 18.
169   F.M. Wurst, S. Alexson, M. Wolfersdorf, G. Bechtel, S. Forster, C. Alling, S. Aradottir, K. Jachau, P. Huber, J.P. Allen, V. Auwarter and F. Pragst, Alcohol Alcoholism, 39 (2004) 33–38.
170   P. Hansson, A. Varga, P. Krantz and C. Alling, Int. J. Legal Med., 115 (2001) 158–161.

M.J. Bogusz (Ed.). Forensic Science
Handbook of Analytical Separations, Vol. 6
© 2008 Elsevier B.V. All rights reserved.

CHAPTER 18

Toxicological aspects of herbal remedies

Maciej J. Bogusz and Mohammed Al-Tufail

Department of Pathology and Laboratory Medicine, King Faisal Specialist Hospital and Research Center, P.O. Box 3354, MBC10, 11211 Riyadh, Saudi Arabia

18.1 INTRODUCTION

Plants and herbs have been used since earliest times to cure various ailments. By the middle of the nineteenth century, about 80% of all medicines were derived from herbs. With the development of the modern pharmaceutical industry came the domination of chemically defined drugs, which in many cases were isolated or developed from natural sources, some examples being atropine, codeine, colchicine, digoxin, ephedrine, morphine, physostigmine, pilocarpine, quinine, quinidine, reserpine, strychnine, taxol, tubocurarine, vincristine, or vinblastine. Nevertheless, herbal remedies have never been off the scene, particularly in developing countries. According to the World Health Organization (WHO), about three-quarters of the world's population still rely on traditional remedies (mainly from herbs). In past decades, a renaissance in the use of herbal remedies has also been observed in industrialized societies. The use of herbal remedies has increased not only because of their perceived effectiveness and safety, but also because of negative feelings relating to rejection of science and technology, desperation and dissatisfaction with pharmacology. The use of self-prescribed herbal medicines within the general population of the U.S. increased from 2.5% in 1990 to 12.1% in 1997 and it amounted to $4–5 billion in the late 1990s [1]. Pierce [2] addressed the increasing use of herbal remedies in Western society from a forensic toxicological point of view. The pharmaceutical industry is willing to invest in research into herbal remedies in the hope that drugs isolated from natural materials may be brought to the market faster than synthesized ones [3]. Herbal preparations are often aggressively advertised as "soft medicines" or "pure natural remedies" as a viable alternative to pharmaceutical drugs called "chemical clubs." However, as Ernst wrote, "the notion that natural can be equated with harmless is as prevalent as it is misleading" [4]. In the past, several

References pp. 608–610

handbooks have been published summarizing the chemistry, pharmacology, applications, and legal status of various herbal remedies [5–8]. Several concerns have been raised regarding the efficacy of herbal preparations. Ernst divided herbal remedies into three groups: those with proven efficacy (St. John's wort, gingko, kava, and horse chestnut), those with doubtful efficacy (ginseng, valerian, echinacea, saw palmetto), and those showing no documented relevant efficacy (garlic, guar gum) [4,9].

In papers on the future of ethnopharmacology, several leading scientists in the field gave their views on the place of traditional medicine in therapy, showing the great possibilities as well as the dangers [10–13]. The need for an organized scientific approach was stressed as a key to future success. For example, studies on dart and arrow poisons used by primitive tribes led to the discovery of a multitude of various compounds, which are very useful in the therapy of malaria, chronic pain, and heart diseases, among others [14].

However, ethnoparmacological studies are very difficult; herbal medicines usually contain a range of pharmacologically active compounds originating from various plants and in some cases it is not known which of these constituents produces the therapeutic effect. It is also known that the same plant species, growing in different conditions, can have different phytochemical compositions. Hence, testing for efficacy is obviously more complex than with synthetic drugs [15,16].

Apart from a questionable efficacy, herbal remedies often put prospective consumers at serious risk associated with various potential dangers such as intrinsic toxicity, herb–drug interaction, adulteration with synthetic drugs, and contamination with toxic metals or pesticides. Several meta-analyses and books have been published on the interaction between herbal remedies and prescribed drugs and on the contamination of Asian herbal remedies with heavy metals [17–23]. However, as the authors of one review state, the evidence presented in their article may represent only the tip of a much bigger iceberg, due to under-reporting and a lack of primary, well-documented, systematic investigations [22]. The problems of safety and contamination of herbal remedies have been addressed in primary studies from Taiwan [24], Singapore [25], Hong Kong [26], and Saudi Arabia [27]. Each of these studies demonstrated the serious risk ensuing from the uncontrolled use of herbal preparations, mainly of Asiatic origin.

The ever-increasing use of herbal remedies of questionable quality became a concern for various health agencies at both national and international levels. The World Health Organization and the European Union have issued several guidelines and acts concerning the safe and appropriate use of herbal medicines [28–30]. Safety issues related to herbal medicines are complex and include the possible toxicity of natural herbal constituents, the presence of contaminants or adulterants, and the potential interactions between herbs and prescription drugs. The production of herbal remedies is not controlled or regulated. Studies performed in Israel and Jordan have demonstrated that persons involved in the production, distribution, and application of herbal remedies ("herbalists") very often do not have any basic education, knowledge, or ethics [31,32]. It must be added that in the age of worldwide information access via the Internet, information concerning toxic plant material is easily available to anyone with suicidal tendencies [33]. Special attention should be given to the exposure of children to herbal remedies since many parents, influenced by the"naturalist" lobby, will often uncritically consider the use of herbal preparations [34].

The role of analytical toxicology in the field of pharmacology and toxicology of herbal remedies is of primary importance. Chromatographic methods have made it possible to identify some known natural compounds in unrelated species and have shed new light on their functions [35]. Modern separation methods are frequently applied in the quality control of herbal remedies produced in industrial conditions. Thin layer chromatography (TLC) has been used for the analytical fingerprinting of herbal extracts [5]. Quality control systems based on chromatographic or electrophoretic fingerprinting have been recommended for the evaluation of herbal remedies [36]. Wen *et al.* [37] reviewed the use of separation methods for quality control of natural antibacterial and antirheumatic compounds occurring in traditional Chinese medicines. Among the methods reviewed, gas chromatography (GC) and high pressure liquid chromatography (HPLC) with various detectors were regarded as the most valuable techniques. TLC was of lesser importance, whereas CE and other electromigration methods are likely to replace chromatographic methods if the disadvantages in reproducibility and sensitivity of electromigration methods can be resolved. Solid phase extraction (SPE) and solid-phase microextraction (SPME) appeared as the most important pretreatment techniques. In the review of Gaillard and Pepin on poisoning by plant material, LC-MS-MS was mentioned as the most important analytical method [38].

18.2 INTRINSIC TOXICITY OF HERBAL REMEDIES

A multitude of toxic compounds can be found in plants and herbal remedies used in traditional treatments. Some selected plants are shown in Table 18.1.

Most of the information on the toxicological aspects of the use of traditional medicines comes from South Africa where the widespread use of plants and herbs by the African population has been combined with local medical vigilance. The health services of South Africa are aware of the dangers associated with the use of plant remedies and have provided a great deal of hard data in the last few decades. Fennell *et al.* [39] gave a systematic assessment of African plant remedies concerning their efficacy and safety. Chromatographic techniques were recognized as being the most valuable for the authentication of plants and plant parts used in traditional therapies. It was estimated that between 60 and 85% of the native population use traditional medicines, usually in combination. Cases of acute poisoning due to traditional medicines have frequently been reported [40] and mortality has been estimated to be as high as 10,000–20,000 per annum [41]. Venter and Joubert [42] analyzed cases of acute poisoning admitted to hospital in Pretoria between 1981and 1985. Poisoning with traditional medicines resulted in the highest mortality, accounting for 51.7% of all deaths. Patients were predominantly children between the ages of 1 and 5 years. In their study, Luyckx *et al.* [43] collected clinical data concerning the adverse effects associated with the use of South African traditional

TABLE 18.1
TOXIC COMPOUNDS IN SELECTED PLANTS

Plant	Toxic Compounds
Aconitum napellus – Aconite	Aconitine
Acorus calamus – Calamus	*cis*-β-Asarone
Aloe vera – aloe	Aloe-emodin, emodin, chrysophanic acid, rhein, salicylic acid
Areca catechu – betel	Arecoline, arecaidine
Aristolochia – snakeroot, wild ginger	Aristolochic acid
Atractylis glummifera—glue thistle, *Callilepis laureola* – impila	Atractyloside
Cascara sagrada – cascara	Chrysophanol, emodin, thein, cascarosides
Cassia senna – senna	Chrysophanol, emodin, rhein, sennosides
Catha edulis – khat	Norpseudoephedrine, cathinone, norephedrine
Colchicum autumnale – autumn crocus	Colchicines, demecolcine
Digitalis purpurea – digitalis	Acetyldigitoxin, digitoxin, digoxin, digitoxigenin, digoxigenin
Ephedra – Ma Huang	Ephedrine, norephedrine, norpseudoephedrine, methylephedrine
Myristica fragrans – nutmeg	Myristicin, safrole, eugenol, methyleugenol
Piper methysticum – kava	Kavain, methysticin
Sassafras officinale – sassafras	Safrole, isosafrole, dihydrosafrole
Strychnos species – dart and arrow poison	Strychnine, strychnogucine B
Symphytum officinale – comfrey, *Senecio jacobea* – ragwort	Pyrrolizidine alkaloids
Syzygium aromaticum – cloves	Eugenol

remedies. The study included 103 patients of academic hospitals in Johannesburg. The most common clinical features on presentation were dehydration, vomiting, jaundice, diarrhea, altered mental status, and oligoanuria. Renal dysfunction was present in 76% of patients and liver dysfunction in 48%. The overall mortality was 34%. It was indicated that in view of the large numbers of African individuals living in the United States and Europe, it is important for physicians elsewhere to be aware of the potential toxicity of African folk remedies, and to inquire about their use. Nephrotoxic action of African remedies was studied by Luyckx *et al.* [44]. Clinical data concerning acute renal failure were evaluated retrospectively in 78 patients who had recently used plant remedies. Overall mortality in patients was 41%. Mortality was higher in adults (45.5%) than in infants (36.6%), in patients with both renal and liver dysfunction (62.5%), and in HIV-positive (44.4%) patients. Steenkamp and Stewart [45] reviewed reports of herbal remedies causing nephrotoxicity. The indications for use of the remedies, signs and symptoms in poisoned patients, and the methods used to detect toxic compounds in plant specimens or in biological fluids were covered. Liquid chromatography (LC) with diode array detector (DAD) or Mass Spectrometry (MS) detection was most frequently used for detecting these compounds. The spectrum of substances included aloesin, aristolochic acid, cathinone, colchicine, amatoxins, and phallotoxins, among others.

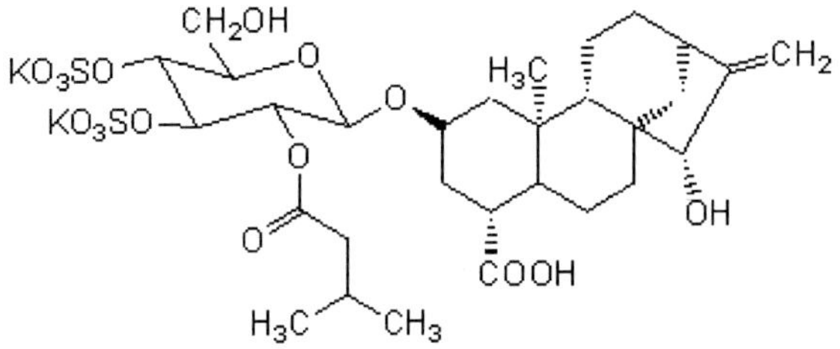

Fig. 18.1. Structure of atractyloside.

18.2.1 Atractyloside

Atractyloside, a diterpenoid glycoside (Fig. 18.1.), occurs in *Callilepis laureola*, a plant known by its traditional Zulu name of "impila" (meaning "good health"), which is used as a multipurpose remedy, protecting not only against various diseases but also against evil and bad ghosts. Impila is most often prepared using the tuberous root-stock of the plant. The decoctions may be prepared from the fresh or dried tuber. The administration of impila-based remedies has been associated with acute hepatic and renal failure, resulting in high mortality. Atractyloside occurs also in *Atractylis glummifera* (glue thistle). It was demonstrated that tissue slices prepared from male domestic pig kidney and liver showed concentration-dependent biochemical alterations after incubation with atractyloside [46]. Popat *et al.* [47] incubated human hepatoblastoma Hep G2 cells with increasing concentrations of an aqueous extract of *C. laureola* and observed that cytotoxicity was produced in a concentration-dependent manner. Stewart and Steenkamp [48] reviewed the biochemistry and toxicity of atractyloside and indicated that the inhibition of the mitochondrial ADP transporter is the primary mechanism of poisoning. The same research team [49] reported the case of poisoning by impila in a mother and child. Simple TLC screening was developed for detecting atractyloside in urine. Gaillard and Pepin [38] detected atractyloside using negative ionization LC-ESI-MS in gradient of acetonitrile and formate buffer (Fig. 18.2).

18.2.2 Pyrrolizidine alkaloids

Pyrrolizidine poisoning, manifesting mainly as hepatic veno-occlusive disease, is well-known in Africa, since plants containing these compounds are given to children to induce sleep. Stewart and Steenkamp [50] reviewed the occurrence, chemistry, and metabolism of pyrrolizidine alkaloids as well as methods used for detection of these compounds. Conradie *et al.* [51] presented two cases of acute intoxication of two sets of twins, treated with traditional remedies containing pyrrolizidines. In one family, both 1-month-old twins survived with severe hepatic damage. In the other family, one of the 1-month-old twins died due to liver necrosis. GC-MS analysis of the remedies applied revealed the presence of pyrrolizidine alkaloid retrorsine in both cases. Crews *et al.* [52] determined pyrrolizidine alkaloids in honey. The compounds

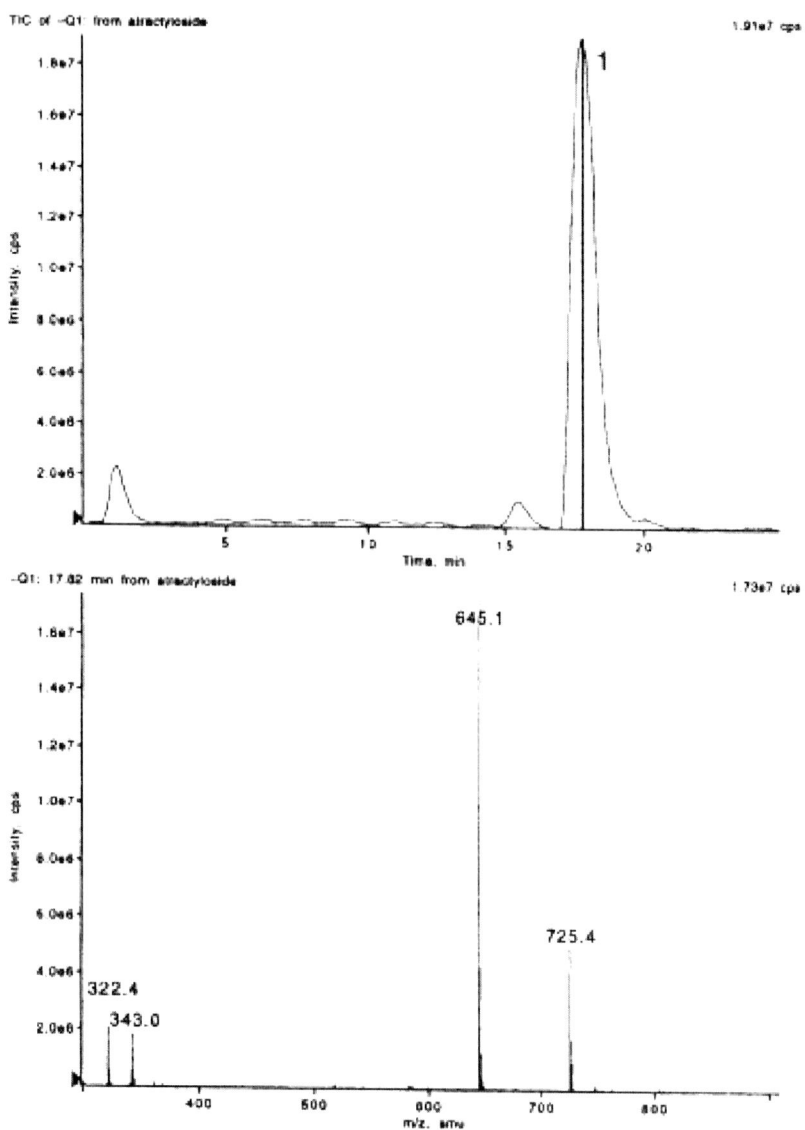

Fig. 18.2. Chromatogram and mass spectrum (negative ions) of atractyloside. (From ref. 38 with permission of Elsevier Science.)

originated from *Senecio jacobea* (ragwort) and were brought by bees. The procedure consisted of Extrelut extraction and LC-APCI-MS determination in selected ion monitoring (SIM) mode. Six alkaloids were identified. Lin *et al.* [53] presented an LC-ESI-MS and an LC-ESI-MS-MS method for determining two types of hepatotoxic pyrrolizidine alkaloids and also for distinguishing the hepatotoxic and nontoxic alkaloids. The methods were applied for the determination of alkaloids in blood samples obtained from rats treated with toxic plants and plant materials. Pyrrolizidine alkaloids in crude extracts from *Senecio* species were separated with reversed-phase HPLC and detected with thermospray MS and nuclear magnetic resonance (NMR). Retrorsine was identified as the main alkaloid [54].

18.2.3 Aristolochic acid

Aristolochic acids, which occur in *Aristolochia* and *Asarum* species, are widely used in traditional Chinese medicines as diuretics and analgesics. In particular remedies containing aristolochic acids were often used for reducing body weight. Their use may cause severe renal failure [55,56]. Preparations containing high contents of aristolochic acids were removed from the Chinese pharmacopeia in 2005. Nevertheless, these remedies are still widespread all over the world and have been used to such an extent that their toxic effect is known as "Chinese herbs nephropathy". Besides nephrotoxicity, aristolochic acids induced malignant tumors in experimental animals [57] and in humans [58,59]. For this reason a worldwide ban on all products containing aristolochic acids was proposed. The use of *Aristolochia* species is currently forbidden in Australia, Canada, and in most European and Asian countries. Nevertheless, herbal remedies containing aristolochic acids are still in use.

Chan *et al.* [60] applied LC-ESI-MS-IT for the determination of aristocholic acids in plant material and herbal remedies. A limit of detection (LOD) of 12 and 15 ng/ml was achieved. Aristolochic acids were analyzed in 37 herbal samples using capillary zone electrophoresis (CZE) by Li *et al.* [61]. The LOD was around 3 μg/ml using UV detection. Koh *et al.* [62] examined a herbal remedy named "Fangji," which was adulterated with aristolochic acid. HPLC-DAD was used for analytical fingerprinting of methanolic extracts of remedies and for the detection of aristolochic acid I. An LC-MS-MS procedure was applied for confirmation. Chan *et al.* [63] published the HPLC-DAD procedure for quantitative determination of aristolochic acids in Chinese herbal remedies. The substances were isolated by methanolic sonication and separated on an RP 18 column in a gradient of methanol and 0.5% acetic acid. The results varied from zero to almost 2 mg of aristolochic acid per gram of dry weight. The method was also used for chromatographic fingerprinting of herbs.

Zhou *et al.* [64] determined aristolochic acids in plant material using CE with electrochemical detection. The LOD for both compounds was 14 and 31 ng/ml. This method was found to be suitable for analytical fingerprinting of various plant materials (Figs. 18.3 and 18.4).

18.2.4 Safrole

Safrole (4-allyl-1,2-methylene dioxybenzene) occurs in essential oils originating in many plants used for seasonings, e.g., sassafras, camphor, nutmeg, and black pepper.

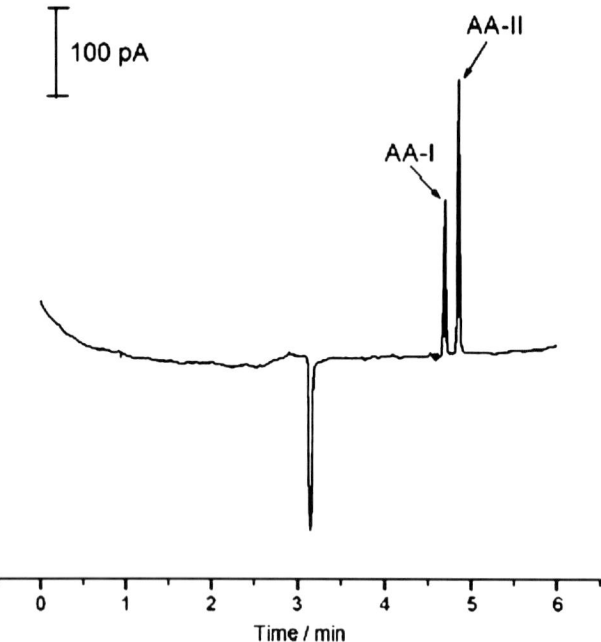

Fig. 18.3. Aristolochic acid standard solution. (From ref. 63 with permission of Elsevier Science.)

The major toxicity of safrole and isosafrole comes from their carcinogenic nature after oxidation. Safrole is oxidized into 1-hydroxysafrole, isosafrole and dihydrosafrole, which are all carcinogenic [65]. Safrole is used as seasoning in various beverages. Heikes [66] analyzed safrole in sassafras tea, using SFE combined with GC-MS. SFE with CO_2, with subsequent HPLC, was used by Ehlers *et al.* [67] for determination of safrole, myristicin, eugenol, and other compounds from nutmeg and mace oils. Carlson and Thompson [68] isolated safrole and isosafrole from sassafras herbal products with solvent extraction, followed by HPLC-UV. The LOD was 1 ng/ml. The method was applied for various commercial herbal samples. Safrole and isosafrole once used extensively as a seasoning in soft drinks (e.g., in Coke) have been prohibited in the USA since the 1970s, but are allowed in China in concentrations below 1 mg/l. Choong and Lin [69] developed a GC-FID method for the determination of safrole and isosafrole in soft drinks, using 1,4-dihydrobenzene as industrial standard (IS). In the Taiwanese market 20 out of 25 soft drinks contained safrole in concentrations 3–5-fold higher than the limit of 1 mg/l (Fig. 18.5).

18.2.5 Salvinorine A

Salvia divinorum, a member of the mint family (Lamiaceae) originates from Mexico. Local Indians used it as a psychedelic and hallucinogenic drug. The plant is known under the names: Magic Mint, Diviner's Sage, Spa Maria Pastora. In the 1990s, it

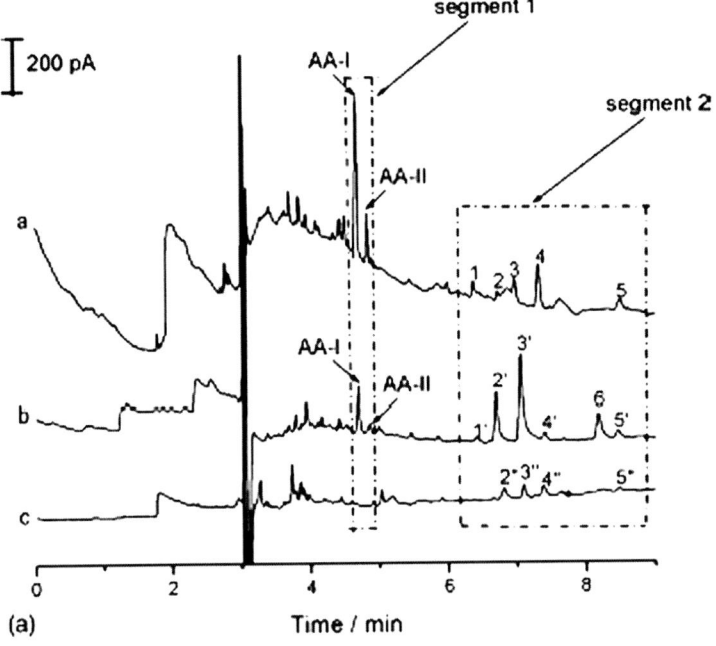

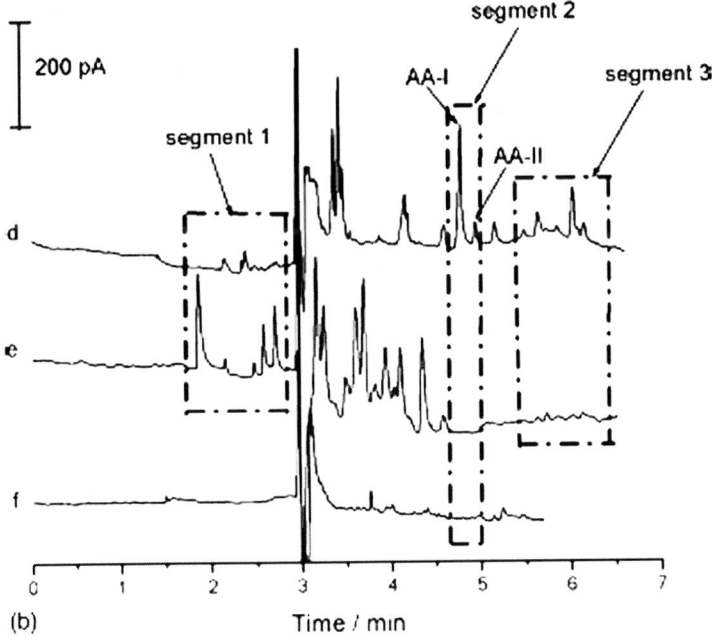

Fig. 18.4. Aristolochic acid in plant material. (From ref. 63 with permission of Elsevier Science.)

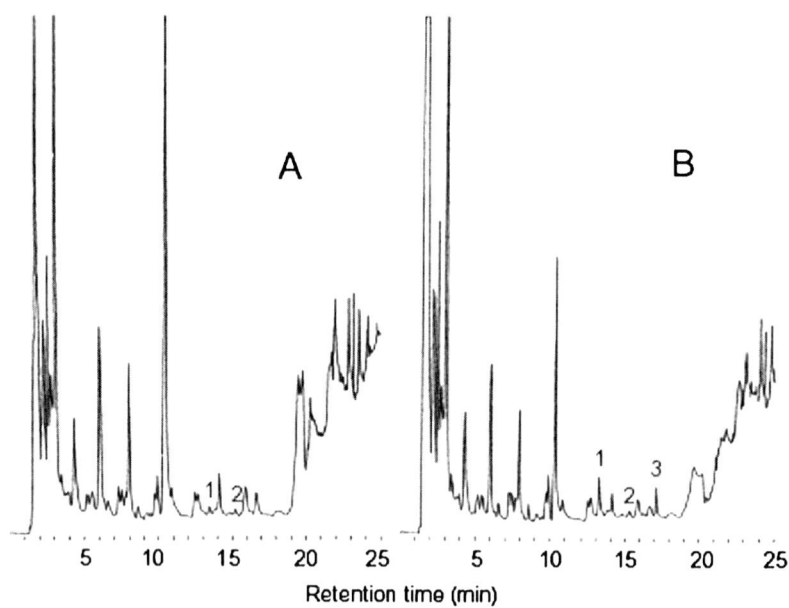

Fig. 18.5. GC of soft drink containing safrole. (From ref. 68 with permission of the publisher.)

became popular as a recreational better as drug. Strictly speaking, *Salvia divinorum* should be considered together with drugs of abuse of natural origin; its description in this chapter is because of the fact that various Internet sources and "alternative" herbal shops are offering this herb as a herbal dietary supplement [70,71]. Salvia is advertised as a legal hallucinogen; however, its legality differs in various countries (www.sagewisdom.org/legalstatus.html).

The main active compound present in *Salvia divinorum* is Salvinorin A, which is the only known non-nitrogenous κ-opioid receptor agonist [72]. Lee *et al.* [73] isolated several other neoclerodane diterpenoids from salvia, also showing agonistic activity on κ-opioid receptor. Salvinorine is deactivated in the gastrointestinal system; traditionally, Mexican Indians chewed fresh salvia leaves and at present an inhalatory route is preferred. Salvia leaves were smoked as a marijuana substitute and cultivated for this purpose in Switzerland [74]. Gonzalez *et al.* [75] collected demographic data and information on patterns of use among 32 recreational users of salvia in Spain. The herb was predominantly smoked, which causes psychedelic activity such as modified visual perception, changes in mood, or various somatic sensations. Imanshahidi and Hosseinzadeh [76] recently reviewed the pharmacological effects of salvia on the central nervous system. A plant root extract showed sedative, hallucinogenic, analgesic, and antiparkinsonian action and inhibited opiate withdrawal syndrome.

The analysis of salvinorins is based on mass spectrometric determination. Medana *et al.* [77] applied LC-ESI-MS(n) for the determination of six salvinorins and three divinatorins isolated from *S. divinorum* leaves. Pichini *et al.* [78] determined Salvinorin A in urine, saliva, and sweat taken from two consumers who smoked

75 mg of salvia leaves. The drug was quantitated by GC-EI-MS in SIM mode, after chloroform–isopropanol extraction. The method was fully validated; salvinorin A was found in urine and saliva, but not in sweat.

18.3 TOXIC METALS IN HERBAL REMEDIES

The origin of toxic metals in herbal remedies may be of various types. Some metals are deliberately added to the formulation; in other cases contamination may occur from grinding weights or brewing pots. The herbs might also have been grown in environmentally contaminated areas. Ernst *et al.* [79,80] presented a systematic review of contamination of Asiatic herbal remedies with heavy metals. Twenty-two reports were included, showing high incidences of contamination particularly in Chinese and Indian preparations. Espinoza *et al.* [81] examined the mercury and arsenic content in Chinese herbal balls used as traditional medicines. Mercury levels ranged from 7.8 to 421 mg and arsenic levels from 0.1 to 36.6 mg, analyzed by atomic absorption spectroscopy and X-ray fluorescence spectroscopy.

Traditional Indian remedies, known as Ayurveda medicines, create particular toxicological problems due to deliberate addition of toxic metals to the formulations. These preparations may contain very high concentrations of lead, arsenic, or mercury. Dunbabin *et al.* [82] presented the case of a 37-year-old man, who returned to Australia from a trip to India where he had been taking two different herbal tonics. He developed abdominal pain, anorexia, and malaise. Investigation revealed low-grade hepatitis and normocytic anemia with prominent basophilic stippling. The blood lead concentration was high, and analysis of the herbal tablets revealed a very high lead content. The patient was treated with calcium ethylenediaminetetraacetic acid (EDTA), which resulted in a high urinary excretion of lead and the resolution of his symptoms over a period of several days. Another case from Australia was reported by Tait *et al.* [83]and concerned a 24-year-old pregnant woman with anemia with normal iron level. Blood Pb was 5.2 μmol/l and the blood picture showed morphological changes typical of lead poisoning. She gave premature birth at 30 weeks of gestation. The infant showed a starting blood Pb concentration of 7.6 μmol/l, which rose during chelation therapy to 11.8 μmol/l and then decreased. The mother admitted that she had taken several tablets prescribed by an Indian Ayurvedic "doctor." These tablets contained 8.9% of lead (45 mg per tablet) and 0.003% mercury (15 μg per tablet). The Center for Disease Control (CDC) presented a report of 12 cases of lead poisoning associated with ayurvedic medications between 2000 and 2003 [84]. The details are presented in Table 18.2. All patients originated from India or Nepal. The need for culturally appropriate educational efforts was stressed in order to protect people using traditional or folk medications.

Saper *et al.* [85] analyzed the content of toxic metals in Ayurvedic remedies manufactured in South Asia and sold in Boston-area stores. The concentrations of lead, mercury, and arsenic in each remedy were measured by X-ray fluorescence spectroscopy, and daily metal ingestion for adults and children was estimated. Fourteen out of seventy remedies contained heavy metals: lead (5–37,000 μg/g),

TABLE 18.2
REPORTED CASES OF LEAD POISONING RELATED TO AYURVEDIC MEDICATIONS [74]

State	Year	Age (years)	Sex	Patient's Country of Origin	BLL[a] at Presentation (μg/dl)	Type of Ayurvedic Medications Ingested	Lead Concentration of Medications (ppm)	Received Chelation Therapy
New Hampshire	2001	37	Female	India	81	Two powders, three tablets	Powders: 12,000–17,000 Tablets: 60–100	Yes
California	2003	31	Female	India	112	Nine medications, including pill taken four times daily	Pill taken four times daily: 73,900; three others: 21, 65, and 285	Yes
California	2003	34	Male	India	80	10 powders, tablets, syrups	Tablet: 78,000; pill: 36	Yes
Massachusetts	2002	89	Male	India	62	Guglu tablets	14,000	Yes
Massachusetts	2002	60	Female	India	56	Guglu tablets	14,000	Yes
Massachusetts	2003	46	Female	Nepal	19	Sundari Kalp (pill and liquid)	Pill: 96,000; liquid: 0	Yes
New York	2000	25	Female	India	91	Pill	79,000	Yes
New York	2001	52	Male	India	49	Unknown form	Not known	Not known
New York	2000	57	Female	India	27	Unknown form	Not known	No
New York	2000	40	Female	India	92	Jambrulin	44,000	Yes
New York	2001	56	Male	India	100	Powder	Not known	Yes
Texas	2003	50	Male	Not stated	92	Jambrulin	22,700–26,700	Yes

[a]Blood lead level.

mercury (28–104,000 µg/g), and/or arsenic (37–8130 µg/g). If taken as recommended by the manufacturers, each of these 14 could result in heavy metal intakes above the published regulatory standards.

As well as Ayurvedic medicines, there are other traditional remedies that may contain high concentrations of toxic metals. In Malaysia [86], The Drug Control Authority has implemented a registration scheme for traditional medicines, with special emphasis on the quality, efficacy, and safety. As such, a total of 100 products in various pharmaceutical dosage forms of a herbal preparation, containing Tong-kat Ali (traditional aphrodisiac remedy), were analyzed for mercury content using a cold vapor atomic absorption spectrophotometer. The results showed that 36% of the analyzed products possessed 0.52–5.30 ppm of mercury and, therefore, do not comply with the quality requirement for traditional medicines in Malaysia. More-over, very distinct batch-to-batch inconsistency was observed. Caldas and Machado [87] analyzed 130 samples of various Brazilian herbal remedies for the content of cadmium, mercury, and lead, using atomic absorption spectrophotometry. The study showed high concentrations of lead in numerous remedies. Cadmium and mercury exposure through the herbal medicines did not appear to be a health concern. Steenkamp *et al.* [88] analyzed traditional South African herbal remedies for selenium, manganese, copper, lead, zinc, and mercury content. Only a few showed high levels of toxic metals. Urine samples obtained from 65 patients admitted to hospital following treatment with a traditional herbal remedy were also analyzed for metals. Only a small number of the patients had abnormally high levels of metal excretion. These data suggest that, in contrast to the experience with traditional Chinese and Indian preparations, metal contamination from plants seems not to be a problem in traditional South African remedies. South Africa contains some of the world's largest deposits of uranium, which is mined as a by-product of gold production. Thirty South African herbal remedies were analyzed for uranium content using the adsorptive stripping voltammetry method. In five samples, the levels were greatly elevated, showing concentrations above 40,000 ppb. The mean uranium concentration of the remainder of the specimens was of the order of 15,000 ppb [89].

18.4 ORGANIC CONTAMINANTS OF HERBAL REMEDIES

Pesticides are the most important organic contaminants of herbal remedies. Leung *et al.* [90] performed a systematic study of the contamination of organochlorine insecticides of four selected authentic Chinese materia medica, namely: Radix Angelicae Sinensis, Radix Notoginseng, Radix Salviae Miltiorrhizae, and Radix Ginseng.. Altogether 10 representative batches of samples were analyzed for each herb. Six batches were collected in the major cultivation areas of the Mainland whilst the remaining four batches were procured in the Hong Kong herbal market. All except Radix Angelicae Sinensis have been identified as containing quintozene and hexa-chlorocyclohexane at various levels. Hexachlorobenzene and lindane were also reported in samples of Radix Ginseng. The presence of DDT, the banned pesticide, and its derivatives, was also observed in one of the Radix Notoginseng samples. All the results were gathered for setting up regulatory permissible limits of organo-chlorine pesticide residues in Chinese materia medica used in Hong Kong. Ling *et al.* [91] developed a method involving the simultaneous extraction and clean-up of 13 organochlorine pesticides from Chinese herbal medicines using supercritical fluid extraction (SFE) followed by gas chromatography electron capture detector (GC-ECD) and GC-MS confirmation. The pesticides in the study consisted of α-, β-, γ-, and δ-benzene hexachloride, heptachlor, aldrin, heptachlor epoxide, endosulfan I, 4,4′-DDE (1,1-dichloro-2,2-bis(*p*-chlorophenyl)ethene), dieldrin, endrin, 4,4′-DDD (1,1-dichloro-2,2-bis(*p*-chlorophenyl)ethane), endosulfan II, 4,4′-DDT (2,2-bis (*p*-chlorophenyl)1,1,1-trichloroethane), endrin aldehyde, and endosulfan sulfate. Florisil sorbent was placed with the sample in the SFE vessel to provide a simple and effective clean-up approach. Mean recoveries that ranged from 78 to 121% were obtained for the pesticides except for endosulfan II, endosulfan sulfate, and endrin aldehyde. The method was applied to analyze herbal remedies sold in Taiwan. Wu and Li [92] described a method for determining 11 organophosphorus insecticides (dichlorvos, methamidophos, acephate, diazinon, dimethoate, chlorpyrifos, malath-ion, parathion, quinalphos, methidathion, and ethion) in the Chinese herbal med-icine "Job's-tears." The organophosphorus insecticides were extracted with dichloromethane and cleaned up with a mixture of Celite 545-activated carbon (4 + 1). The extracts were analyzed by GC using a nitrogen–phosphorus detector. Analysis of fortified Job's-tears showed average recoveries ranging from 73.90 to 104.23%. The minimum detectable amount ranged from 0.1 to 0.5 ng, and the limit of quantitation for the method was 0.05 mg/kg. Zuin *et al.* [93] applied headspace SPME (HS-SPME) using fiber coated with polydimethylsiloxane-poly(vinyl alcohol) (PDMS/PVA) for the trace determination of organochlorine and organophosphorus pesticides in herbal infusions of *Passiflora* L. by GC-ECD. The capacity of the PDMS/PVA coating for the pesticides was compared with that of commercial PDMS fibers, with advantageous results. The effects of parameters such as the sample ionic strength, dilution of the infusion, extraction temperature, and time were investigated. The analytical curves for the range between 0.04 and 6 ng/ml of each compound were presented. The detection limits in these matrices varied from 0.01 ng/ml (β-endosulfan) to 1.5 ng/ml (malathion).

18.5 ADULTERATION OF HERBAL REMEDIES WITH PHARMACEUTICALS

Adulteration of herbal remedies with undeclared synthetic drugs is a common problem, which may potentially cause serious adverse effects. Huang *et al.* [24] found that around 24% of 2609 samples of traditional Chinese medicines analyzed in Taiwan were adulterated with synthetic drugs of various pharmacological activ-ities. The most frequently used are nonsteroid anti-inflammatory drugs (NSAID), steroids, and analgesics. Koh *et al.* [25] and Liu *et al.* [94] reported a similar adul-teration profile in Chinese "herbal" remedies analyzed in Singapore. They used the HPLC-DAD library comprising 266 common synthetic adulterants and GC-MS screening procedure for routine examination of herbal remedies. Ernst [95] published a systematic review of 22 studies carried out on adulteration of Chinese herbal remedies with synthetic drugs from 1990 to 2000. The review summarized the type of herbal remedy and identified its adulterants, patients' data, symptoms, and outcome. It was concluded that adulteration is potentially a serious problem, putting consumers of herbal remedies at risk. In several reports, the presence of codeine [96], dexamethasone and indomethacin [97] phenylbutazone [98,99], chloropamide [100], or fenfluramine [101] was described in "herbal" remedies. Bogusz *et al.* [27] found synthetic drugs in 8 out of 247 herbal remedies present in the Saudi Arabian market. Herbal remedies used to enhance sexual activity were frequently adulterated with sildenafil or its analogues. Blok-Tip *et al.* [102] applied LC-ESI-MS-MS and NMR for structure elucidation of sildenafil analogues in herbal remedies. Shin *et al.* [103] applied NMR for the identification of sildenafil analogue present in a commercial herbal drink in Korea. Gratz *et al.* [104] applied LC-ESI-MS in full-scan mode for the detection of sildenafil, tadalafil, and vardenafil in dietary supplements and herbal remedies. Reepmeyer and Woodruff [105] identified a novel synthetic vardenafil designer drug ("piperidenafil") in a herbal "natural" remedy, which is marketed via the Internet as a sexual enhancer. The structure was elucidated by LC-MS and GC-MS analysis. Liang *et al.* [106] published a study on the detection of synthetic adulterants in herbal remedies, using LC-ESI-MS-MS in multiple reaction monitoring (MRM) mode (Fig. 18.6). The nine most common adulterants present in Chinese herbal remedies are sildenafil, famotidine, ibuprofen, promethazine, diazepam, nifedipine, capropril, amoxicillin, and dextromethorphan. The pharmacological action of the drugs identified was in agreement with the declared promises of "herbal" preparations. Snyman *et al.* [107] described two cases from South Africa in which herbal remedies consumed by patients admitted to hospital were adulterated with trimethadione, propofol, and diclofenac. The adulterants were identified using HPLC-DAD and GC-MS. Bogusz *et al.* [108] developed an LC-ESI-MS-MS procedure for detection of most common synthetic adulterants in herbal remedies. Eighty drugs belonging to various pharmacological classes were included in the study and analyzed in positive or negative ionization mode. For most drugs two transitions were monitored, using protonated or deprotonated molecules as precursor ions. The drugs were isolated from herbal remedies using simple methanol extraction. Chromatographic separation was done in a gradient of acetonitrile – 10 mM ammonium formate buffer (pH 3.0). Drugs tested were grouped in suites, comprising analgesic drugs, antibiotics, antidiabetic drugs, antiepileptic drugs, aphrodisiacs, hormones and anabolic drugs, psychotropic drugs, and weight-reducing compounds. These suites were used according to the declared benefits of examined preparations. Limits of detection ranged from 5 pg to 1 ng per injected sample. A drug-free herbal remedy spiked with eight various pharmaceuticals occurring in adulterated herbal preparations was used for internal proficiency testing. The recoveries of spiked drugs ranged from 63 to 100%. The procedure was applied in everyday casework, and several undeclared drugs, such as sildenafil, tadalafil, testosterone (Fig. 18.7), or glibenclamide, were identified in herbal remedies. The pharmacological properties of detected drugs always corresponded with the claims

Q. Liang et al. / Journal of Pharmaceutical and Biomedical Analysis xxx (2005) xxx–xxx

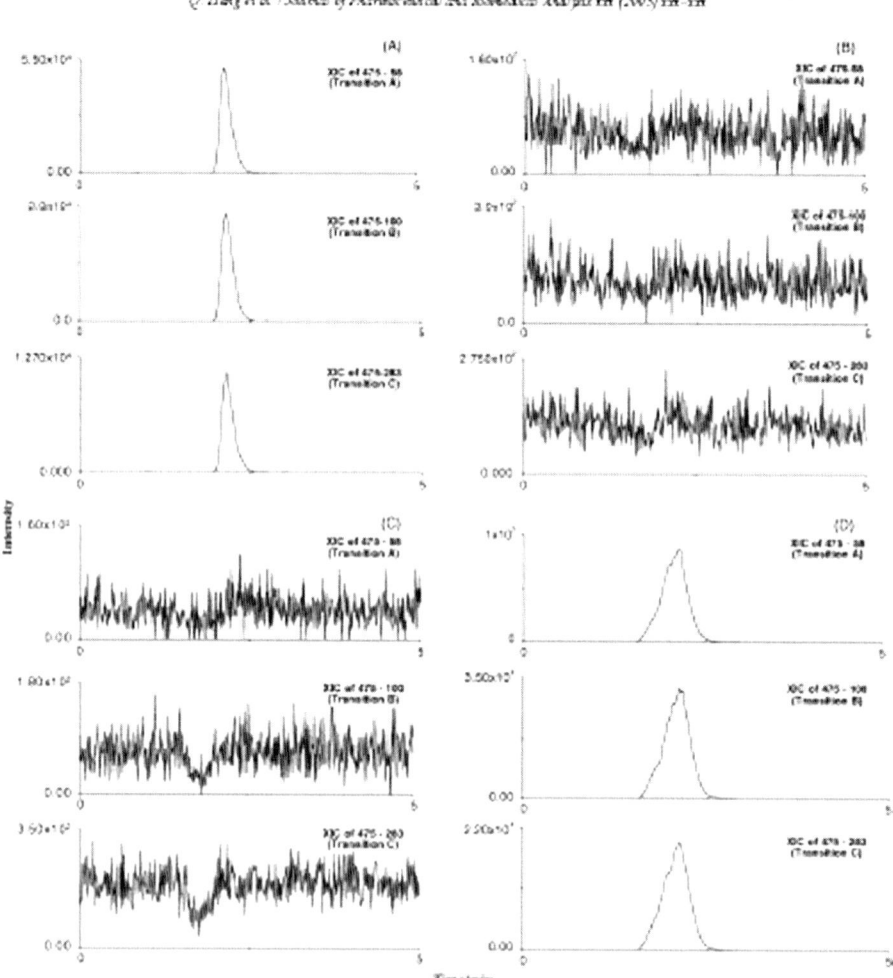

Fig. 18.6. LC-MS-MS chromatogram of sildenafil (A), blank sample (B), negative herbal sample (C), and sample adulterated with sildenafil (D). (From ref. 106 with permission of Elsevier Science.)

made for the "natural" remedies. For example, recently distributed "natural slim-ming herbal remedies" contained the synthetic drug sibutramine – an active ingre-dient in commercial pharmaceutical preparations of Reductil® or Meridia® that are used in the management of obesity (Fig. 18.8). The presence of sibutramine was also reported by Jung et al. [109] in Chinese herbal medicine obtained in Germany via the Internet. The content of sibutramine in one capsule, analyzed by GC-MS and HPLC-DAD, was 27.4 mg, and exceeded twice the highest recommended dose. The patient taking this remedy developed symptoms of sibutramine overdose: vertigo, headache, and numbness (Table 18.3).

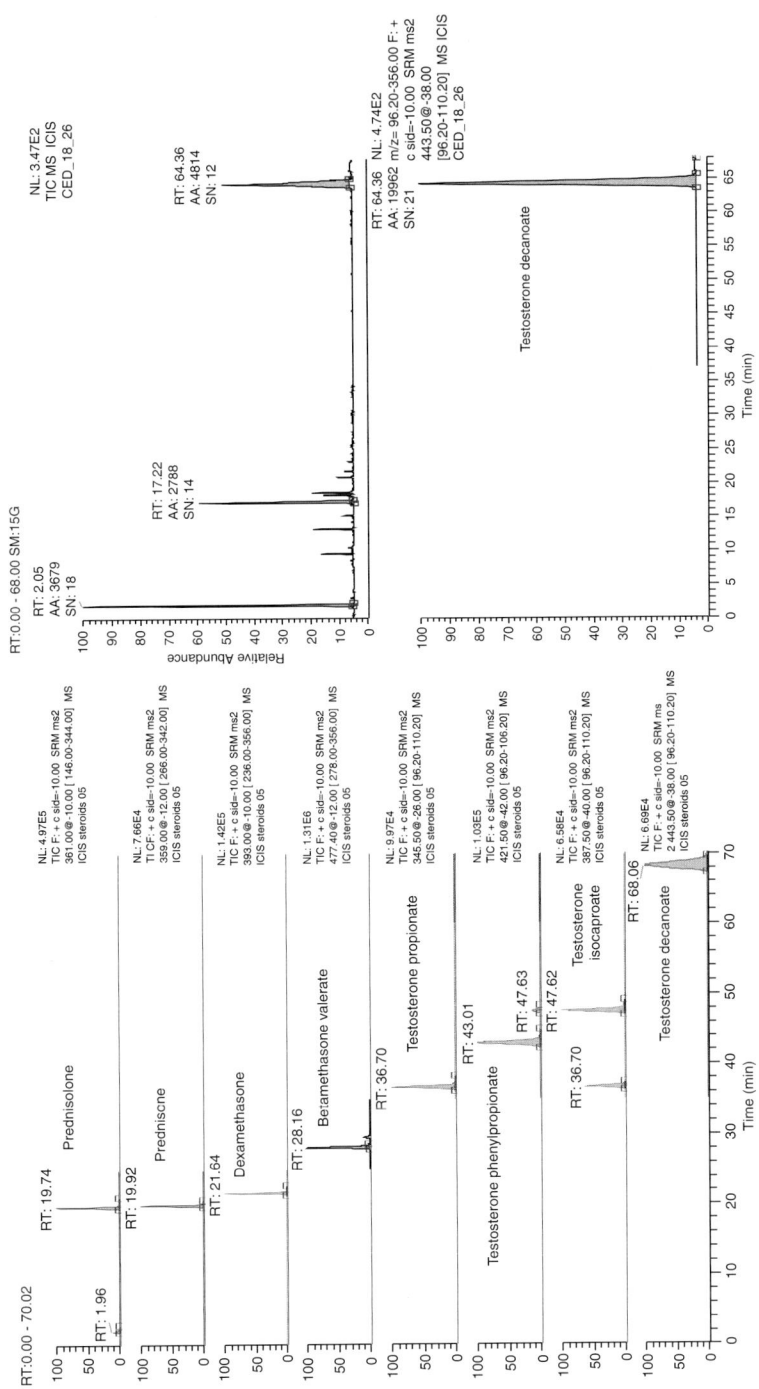

Fig. 18.7. LC-MS-MS chromatogram of the mixture of selected steroids (left), and mass chromatogram of the extract from a Japanese remedy "for female sexuality," containing testosterone decanoate (right). For each peak, a total ion chromatogram (TIC) of two transitions is shown. (From ref. 97 with permission of Elsevier Science.)

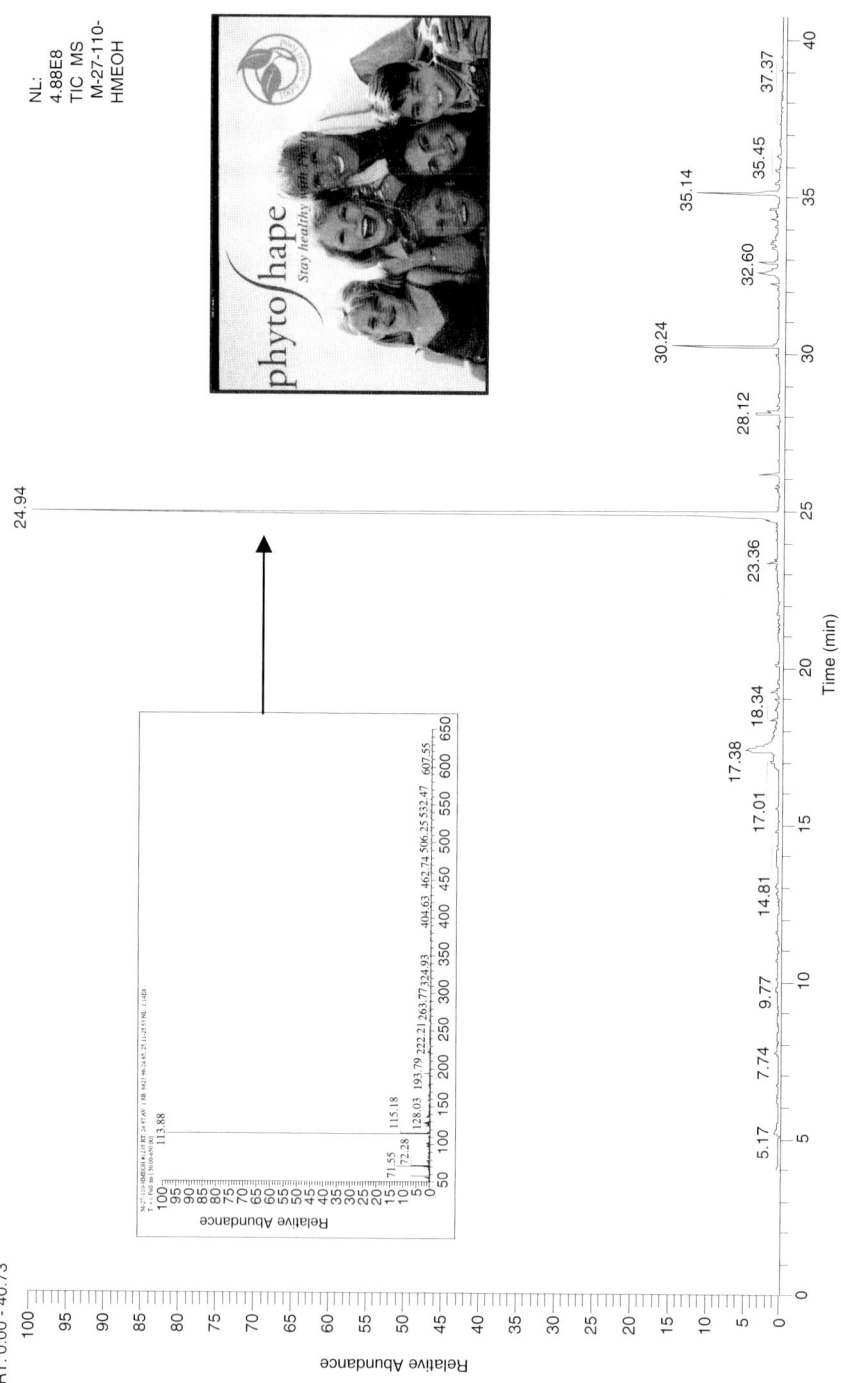

Fig. 18.8. GC-MS full scan mass chromatogram of the extract of "phyto shape" capsule. Peak at Rt 24.94 was identified as sibutramine.

TABLE 18.3
SELECTED ADULTERANTS IDENTIFIED IN HERBAL REMEDIES IN OWN CASEWORK MATERIAL

Remedy	Claim	Analytical Finding
Chinese herbal capsules "*Yong Gang*"	Good health and stamina food supplement	Tadalafil, sildenafil
Chinese herbal capsules "*Vigorous*"	For natural general strength	Sildenafil 49 mg/capsule
Herbal capsules "*Phyto Andro*" from Etumax, Malaysia	Tongkat Ali and other Asian herbs to nourish the body and fortify the male sexual function	Sildenafil 18 mg/capsule
Instant herbal powder "*XKL*" Japanese fluid	Drink "to enhance male strength" For women sexuality	Sildenafil 91 mg/g Testosterone decanoate
Unlabelled herbal tablets from Lebanon	Against diabetes. Recommended daily dose 15 tablets!	Glibenclamide 7.5 mg/tablet
Unlabelled herbal powder from Jordan	Against diabetes	Glibenclamide 4.5 mg/g
Unlabelled herbal capsules	Herbal slimming remedy	Fenfluramine, phentermine, caffeine, nordazepam
Herbal powder "*Jamu Ragel*" from Indonesia	Against rheumatism and pain	Phenylbutazone, aminopyrine, dipyrone, chlorpheniramine
Herbal capsules "*Phyto Shape*" from Etumax, Malaysia	Food supplement used to enhance energy and slim without rebound	Sibutramine
"*Royal honey*" from Etumax, Malaysia	Etumax royal honey for health and vitality	Vardenafil

References pp. 608–610

18.6 CONCLUSIONS

Herbal remedies certainly cannot be regarded as "soft" medicines. In many cases, they contain high concentrations of toxic metals, or are contaminated with pesticides. Dangerous and toxic natural compounds are often present. The frequent presence of undeclared synthetic drugs may jeopardize unwary consumers.

Generally, there is an urgent need to regulate and control the production and distribution of herbal remedies as has been the case with pharmaceutical compounds. Since the majority of herbal remedies are distributed through the Internet or by some uncontrolled shops offering alternative healing, the present situation poses a serious and growing danger to society.

18.7 REFERENCES

1   D.M. Eisenberg, R.B. Davis, S.L. Ettner, S. Appel, S. Wilkey, M. Van Rompay and R.C. Kessler, J. Am. Med. Ass., 280 (1998) 1569.
2   A. Pierce, Bull. TIAFT, 35 (2005) 75.
3   B.M. Mantz, Venture capitalists warm to herbal. Wall Street J., 10 Jan (2002).
4   E. Ernst and M.H. Pittler, Complement. Alternat. Med., 86 (2002) 149.
5   F. Gaedcke and B. Steinhoff, Herbal medicinal products, Medpharm Scientific Publishers, CRC Press, Stuttgart, 2003.
6   J. Barnes; L.A. Anderson; J.D. Philipson, Herbal medicines, (2nd ed.). Pharmaceutical Press, London, 2002.
7   M. Wichtl (Ed.), Herbal drugs and phytopharmaceuticals. (3rd ed.)Medpharm Scientific Publishers, CRC Press, Stuttgart, 2004.
8   X. Yan, J. Zhou and G. Xie, Traditional chinese medicines. Molecular structures, natural sources, and applications. Ashgate Publ. Co., Aldershot, UK, 1999.
9   E. Ernst, Ann. Intern. Med., 136 (2002) 42.
10 H.S. Kim, J. Ethnopharmacol., 100 (2005) 37.
11 A.H. Gilani and A.ur. Rahman, J. Ethnopharmacol., 100 (2005) 43.
12 G.A. Cordell and M.D. Colvart, J. Ethnopharmacol., 100 (2005) 5.
13 A.K. Jäger, J. Ethnopharmacol., 100 (2005) 3.
14 G. Philippe and L. Angenot, J. Ethnopharmacol., 100 (2005) 85.
15 E. Ernst and M.H. Pittler, Complemen. Alternat. Med., 86 (2002) 149.
16 E. Ernst. In: M.N.G. Dukes, J.K. Aronson (Eds.), Meyler's side effects of drugs (14th ed.)Elsevier Science, Amsterdam, 2000, pp. 1649–1681.
17 M.J. Cupp (Ed.), Toxicology and clinical pharmacology of herbal products. Humana Press, Totowa, NJ, 2000.
18 E. Ernst, Herbal medicine, Heinemann Publ., Oxford, 1999.
19 B. Tomlinson, T.Y.K. Chan, J.C.N. Chan, J.A. Critchley and P.H.P. But, J. Clin. Pharmacol., 40 (2000) 451.
20 A.A. Izzo and E. Ernst, Drugs, 61 (2001) 2163.
21 A. Fugh-Berman and E. Ernst, J. Clin. Pharmacol., 52 (2001) 587.
22 E. Ernst and L. Thompson, Coon Clin. Pharmacol. Ther., 70 (2001) 497.
23 E. Ernst, Trends Pharmacol. Sci., 23 (2002) 136.
24 W.F. Huang, K.C. Wen and M.L. Hsiao, J. Clin. Pharmacol., 37 (1997) 344.
25 H.L. Koh and S.O. Woo, Drug Safety, 23 (2000) 351.
26 T.Y. Chan, J.C. Chan and B. Tomlinson, Lancet, 342 (1993) 1532.
27 M.J. Bogusz, M. Al-Tufail and H. Hassan, Adverse Drug React. Tox. Rev., 21 (2002) 219.

28   Guidelines for the appropriate use of herbal medicines. WHO Regional Publications, WHO Regional Office for The Western Pacific, Manila, 1998.

29   WHO guidelines on safety monitoring of herbal medicines in pharmacovigilance systems. WHO, Geneva, 2004.

30   Mandate for the EMEA working party on herbal medicinal products. www.emea.eu.int

31   H. Azaizeh, S. Fulder, K. Khalil and O. Said, Fitoterapia, 74 (2003) 98.

32   B.E. Abu-Irmaileh and F.U. Afifi, J. Ethnopharmacol., 89 (2003) 193.

33   W. Kawohl and E. Habermeyer, J. Ethnopharmacol., 100 (2005) 138.

34   A.D. Woolf, Pediatrics, 112 (2003) 240.

35   M. Tulp and L. Bohlin, Trends Pharm. Sci., 26 (2005) 176.

36   Y.Z. Liang, P. Xie and K. Chan, J. Chromatogr. B, 812 (2004) 53.

37   D. Wen, Y. Liu, W. Li and H. Liu, J. Chromatogr. B, 812 (2004) 101.

38   Y. Gaillard and G. Pepin, J. Chromatogr. B, 733 (1999) 181.

39   C.W. Fennell, K.L. Lindsey, L.J. McGaw, S.G. Sparg, G.I. Stafford, E.E. Elgorashi, O.M. Grace and J. van Staden, J. Ethnopharmacol., 94 (2004) 205.

40   V. Steenkamp, M.J. Stewart and M. Zuckerman, Ther. Drug Monit., 20 (1998) 510.

41   S. Thomson, Traditional African medicine: Genocide and ethnopiracy against the African people. Report to the South African Medicines Control Council. Gaia Research Institute, 2000.

42   C.P. Venter and P.H. Joubert, Biomed. Environ. Sci., 1 (1988) 388.

43   V.A. Luyckx, V. Steenkamp, J.R. Rubel and M.J. Stewart, Cent. Afr. J. Med., 50 (2004) 46.

44   V.A. Luyckx, V. Steenkamp and M.J. Stewart, Ren. Fail., 27 (2005) 35.

45   V. Steenkamp and M.J. Stewart, Ther. Drug Monit., 27 (2005) 270.

46   D.K. Obatomi and P.H. Bach, Food Chem. Toxicol., 36 (1998) 335.

47   A. Popat, N.H. Shear, I. Malkiewicz, M.J. Stewart, V. Steenkamp, S. Thomson and M.G. Neuman, Clin. Biochem., 34 (2001) 229.

48   M.J. Stewart and V. Steenkamp, Ther. Drug Monit., 22 (2000) 641.

49   V. Steenkamp, M.J. Stewart and M. Zuckerman, Hum. Exp. Toxicol., 18 (1999) 594.

50   M.J. Stewart and V. Steenkamp, Ther. Drug Monit., 23 (2001) 698.

51   J. Conradie, M.J. Stewart and V. Steenkamp, Ann. Clin. Biochem., 42 (2005) 141.

52   C. Crews, J. Startin and P.A. Clarke, Food Addit. Contam., 14 (1997) 419.

53   G. Lin, K.-Y. Zhou, X.-G. Zhao, Z.-T. Wang and P. But, Rapid Comm. Mass Spectrom., 12 (1998) 1445.

54   K. Ndjoko, J.L. Wolfender, E. Röder and K. Hostettmann, Planta Med., 65 (1999) 562.

55   G.M. Lord, R. Tagore, T. Cook, P. Gower and C.D. Pusey, Lancet, 2354 (1999) 481.

56   R.J. Ko and R. Ko, Clin. Toxicol., 37 (1999) 697.

57   J.-P. Cosyns, R.-M. Goebbels, V. Liberton, H.H. Schmeiser, C.A. Bieler and A.M. Bernard, Arch. Toxicol., 72 (1998) 738.

58   A. Pfohl-Leszkowicz, T. Petkova-Bocharova, I.N. Chernozemsky and M. Castegnaro, Food Addit. Contam., 19 (2002) 282.

59   V.M. Arlt, M. Stiborova and H.H. Schmeiser, Mutagenesis, 17 (2002) 265.

60   S.A. Chan, M.J. Chen, T.Y. Liu, M.R. Fuh, J.F. Deng, M.L. Wu and S. Hsieh S.-J., Talanta, 60 (2003) 679.

61   W. Li, S. Gong, D. Wen, B. Che, Y. Liao, H. Liu, X. Feng and S. Hu, J. Chromatogr. A, 1049 (2004) 211.

62   H.L. Koh, H. Wang, S. Zhou, E. Chan and C.O. Woo, J. Pharm. Biomed. Anal., 40 (2006) 653.

63   W. Chan, K.M. Hui, W.T. Poon, K.C. Lee and Z. Cai, Anal. Chim. Acta, 576 (2006) 112.

64   X. Zhou, C. Zheng, J. Sun and T. You, J. Chromatogr. A, 1109 (2006) 152.

65   M.S. Benedetti, A. Malnoe and A.L. Broillet, Toxicology, 7 (1977) 69.

66   D.L. Heikes, J. Chromatogr. Sci., 32 (1994) 253.

67   D. Ehlers, J. Kirchhoff, D. Gerard and K.W. Quirin, Int. J. Food Sci. Technol., 33 (1998) 215.

68   M. Carlson and R.D. Thompson, J. AOAC Int., 80 (1997) 1023.

69   Y.M. Choong and H.J. Lin, J. Food Drug Anal., 9 (2001) 27.

70   C.E. Dennehy, C. Tsourounis and A.E. Miller, Ann. Pharmacother., 39 (2005) 1634.

71 T.E. Prisinzano, Life Sci., 78 (2005) 527.

72 D.J. Siebert, J. Ethnopharmacol., 43 (1994) 53.

73 D.Y. Lee, Z. Ma, L.Y. Liu-Chen, Y. Wang, Y. Chen, W.A. Carlezon and B. Cohen, Bioorg. Med. Chem., 13 (2005) 5635.

74 C. Giroud, F. Felber, M. Augsburger, B. Horisberger, L. Rivier and P. Mangin, Forensic Sci. Int., 112 (2000) 143.

75 D. Gonzalez, J. Riba, J.C. Bouso, G. Gomez-Jarabo and M.J. Barbanoj, Drug Alcohol Depend. May (2006) (Epub ahead of print).

76 M. Imanshahidi and H. Hosseinzadeh, Phytother. Res., 20 (2006) 427.

77 C. Medana, C. Massolino, M. Pazzi and C. Baiocchi, Rapid Comm. Mass Spectrom., 20 (2006) 131.

78 S. Pichini, S. Abanades, M. Farre, M. Pellegrini, E. Marchei, R. Pacifici, R. de la Torre and P. Zuccaro, Rapid Comm. Mass Spectrom., 19 (2005) 1649.

79 E. Ernst and L. Thompson Coon, Heavy metals in traditional Chinese medicines: A systematic review. Clin. Pharmacol. Ther., 70 (2001) 497.

80 E. Ernst, Toxic heavy metals and undeclared drugs in Asian herbal medicines. Trends Pharmacol. Sci., 23 (2002) 136.

81 E.A. Espinoza, M.J. Mann, B. Bleasdell, J. DeKorte and M. Cox, Toxic metals in selected traditional Chinese medicinals. J. Forensic Sci., 41 (1996) 453.

82 D.W. Dunbabin, G.A. Tallis, P.Y. Popplewell and R.A. Lee, Med. J. Aust., 157 (1992) 835.

83 P.A. Tait, A. Vora, S. James, D.J. Fitzgerald and B.A. Pester, Med. J. Aust., 177 (2002) 193.

84 Anonymous, MMWR Weekly Rep., 53 (2004) 582.

85 R.B. Saper, S.N. Kales, J. Paquin, M.J. Burns, D.M. Eisenberg, R.B. Davis and R.S. Phillips, JAMA, 292 (2004) 2868.

86 H.H. Anf, E.L. Lee and H.S. Cheang, Int. J. Toxicol., 23 (2004) 65.

87 E.D. Caldas and L.L. Machado, Food Chem. Toxicol., 42 (2004) 599.

88 V. Steenkamp, M. von Arb and M.J. Stewart, Forensic Sci. Int., 114 (2000) 89.

89 V. Steenkamp, M.J. Stewart, L. Chimuka and E. Cukrowska, Health Phys., 89 (2005) 679.

90 K.S. Leung, K. Chan, C.L. Chan and G.H. Lu, Phytother. Res., 19 (2005) 514.

91 Y.C. Ling, H.C. Teng and C. Cartwright, J. Chromatogr. A, 835 (1999) 145.

92 J. Wu and L. Li, J. AOAC Int., 87 (2004) 1260.

93 V.G. Zuin, A.L. Lopez, J.H. Yariwake and F. Augusto, J. Chromatogr. A, 1056 (2004) 21.

94 S.Y. Liu, S.O. Woo and H.L. Koh, J. Pharm. Biomed. Anal., 24 (2001) 983–992.

95 E. Ernst, J. Int. Med., 252 (2002) 107–113.

96 S.Y. Liu, S.O. Woo, M. Holmes and H.L. Koh, J. Pharm. Biomed. Anal., 22 (2000) 481.

97 Health Sciences Authority Press Release, Singapore, 5 April 2004 (www.hsa.gov.sg).

98 Health Sciences Authority Press Release, Singapore, 23 January 2002 (www.hsa.gov.sg).

99 A.J. Lau, M.J. Holmes, S.O. Woo and J.H.L. Koh, J. Pharm. Biomed. Anal., 26 (2003) 401.

100 D.M. Wood, S. Athwal and A. Panahloo, Diabet. Med., 21 (2004) 625.

101 C. Corns and K. Metcalfe, J. Roy. Soc. Health, 122 (2002) 213.

102 L. Blok-Tip, B. Zomer, F. Bakker, K.D. Hartog, M. Mamzink, J. ten Hove, M. Vredenbregt and and D. de Kaste, Food Addit. Contam., 21 (2004) 737.

103 C. Shin, M. Hong, D. Kim and Y. Lim, Magn. Res. Chem., 42 (2004) 1060.

104 S.R. Gratz, C.L. Flurer and K.A. Wolnik, J. Pharm. Biomed. Anal., 36 (2004) 525.

105 C.Z. Reepmeyer and J.T. Woodruff, J. Chromatogr. A, 1125 (2006) 67.

106 Q. Liang, J. Qu, G. Luo and Y. Wang, J. Pharm. Biomed. Anal., 40 (2005) 305.

107 T. Snyman, M.J. Stewart, A. Grove and V. Steenkamp, Ther. Drug Monit., 27 (2005) 86.

108 M.J. Bogusz, H. Hassan, E. Al-Enazi, Z. Ibrahim and M. Al-Tufail, J. Pharm. Biomed. Anal., 41 (2006) 554.

109 J. Jung, For. Sci. Int., 161 (2006) 221.

M.J. Bogusz (Ed.). Forensic Science
Handbook of Analytical Separations, Vol. 6
611

CHAPTER 19

Drugs and driving

Elke Raes[1], Alain Verstraete[1,2] and Robert Wennig[3]

[1]*Department of Clinical Biology, Microbiology and Immunology, 185 De Pintelaan, Ghent University,*
Ghent, Belgium
[2]*Laboratory of Clinical Biology – Toxicology, Ghent University Hospital, 185 De Pintelaan, Ghent, Belgium*
[3]*Toxicology Laboratory, Laboratoire National de Santé, Université du Luxembourg, Luxembourg*

19.1 INTRODUCTION

Every year more than a million people in the world are killed and many millions
more are injured in traffic crashes [1]. Traffic crashes are the consequence of many
factors, which can be classified into three categories: the road, the vehicle and the
driver. A crash is rarely attributable to only one factor, so it is very difficult to
determine in what percentage of crashes alcohol or drugs have contributed. Toxicological analysis can be performed on samples of drivers suspected of driving under
the influence of drugs (DUID) by the police, drivers injured or killed as a consequence of a traffic accident and on samples of drivers that were collected during
roadside surveys. By comparing the proportion of positive samples of injured or
killed drivers with those that were collected during roadside surveys, some studies
were even able to assess the risk of being involved in or responsible for a traffic
accident when driving under the influence of a particular drug. Analysis of drugs of
abuse in biological matrices is thus very important, not only to assess the size of the
problem, but also to support law enforcement on DUID.

19.1.1 Epidemiology

Recent epidemiological studies show that drugs are present in the blood, urine or
oral fluid of approximately 1–12% [2–4] of the general driving population, 8–51% of
injured drivers [3,5–7] and 6–35% of killed drivers [3,8–11]. In the USA in 2004, 10.6
million persons reported driving under the influence of an illicit drug during the past
year [12]. Recent studies also revealed that DUID is associated to significantly
elevated risks of being involved in or responsible for a traffic accident. For example,
the IMMORTAL-study in the Netherlands revealed that drivers under the influence

References pp. 646–651

of benzodiazepines alone have a relative risk for an accident that is three times greater than the risk of a drug-free driver [3]. The SAM-study in France found that driving under the influence of cannabis doubles the risk of being responsible for an accident and that every year 180 fatal accidents are attributable to DUID of cannabis [13]. Many other studies have been published on the effects of drug use on driving (general psychoactive agents [14–16], antihistamines [17], methadone [18], antiepileptics [19], methylphenidate [20], ecstasy [21] or chlormethiazole [22]), on epidemiology risk factors [19], driving conditions [23], chronic medical conditions [24] and severity of traffic accidents [25]. These data as well as our own data [15,26] clearly show that DUID is not a rare phenomenon and that it represents a substantial risk for traffic safety.

19.1.2 Legal issues

Generally there are two possible types of DUID legislation, namely "impairment" legislation and "*per se*" legislation. With "impairment" legislation, the prosecution needs to demonstrate the state of impairment of the driver, and the analysis of drugs in body fluids only provides corroborating evidence as to the cause of the impairment. Unfortunately, this remains subjective and requires the assessment by a medical doctor or a specially trained police officer. Many countries experience difficulties in obtaining convictions. For this reason, and in analogy to alcohol, many countries have introduced "*per se*" laws. A "*per se*" law prohibits driving if drugs are present in blood, serum, plasma or oral fluid above a certain threshold concentration. This facilitates the enforcement of legislation on DUID, since the prosecution does not have to prove that the driver was impaired. The cut-offs used are analytical cut-offs. This means that any detectable concentration of a drug is sufficient to be sanctioned: a "zero-tolerance" law. At the moment, the following countries have "*per se*" legislation in addition to the existing "impairment" legislation:

- *Germany* was the first country to introduce a "*per se*" law: §24a of the Road Traffic Act was amended in March 1998, so that any person driving a vehicle in road traffic under the influence of cannabis, heroin, morphine, cocaine, amphetamine or designer amphetamines commits an offence. This does not apply if the substance originates from having taken prescribed medication as intended for a specific illness. The Federal Constitutional Court decided in 2004 that the cut-off for δ-9-tetrahydrocannabinol (THC) could be lowered to 1 ng/ml.
- In *Belgium*, a similar law was adopted in March 1999. A driver can be stopped by the police and asked to perform a standardised test battery to detect recent drug consumption. If this is positive, an immunoassay is performed on a urine sample and if this is positive, blood is taken and sent to the lab for gas chromatography–mass spectrometry (GC-MS) analysis.
- *Sweden* also introduced a "*per se*" law in 1999. There exists zero-tolerance for narcotics (including benzodiazepines), except if the drugs are taken according to a medical prescription, the dose is not too high and no impairment is present. The

detection of driving under the influence is performed by an eye examination, followed by further examination if there is reasonable suspicion. If drugs are found in blood, the driver is also sanctioned for drug use.

- In *France*, a *"per se"* legislation was implemented in 2001 for drivers involved in a fatal car accident. In 2003, this was expanded to drivers involved in an accident, suspected of impaired driving or when the driver has committed an offence. A driver is sanctioned if blood analysis shows prior exposure to an illicit drug.
- *Finland* also introduced a *"per se"* legislation in 2003. The drugs covered are those listed in the UN conventions on narcotics, but the law is not applicable if drug are used according to a physician's prescription.
- In *Luxembourg* an "impairment law" has been introduced in 1955 (together with ethanol) but it is not implemented. However, a *"per se"* law with "zero tolerance" covering all psychotropic drugs of the UN-Conventions is still under discussion by the competent authorities.
- In *Switzerland*, a driver is considered unfit to drive since 2004 if it is proven that his blood contains THC, free morphine, cocaine, amphetamine, methamphetamine, N-ethyl-3,4-methylenedioxy-N-ethylamphetamine (MDEA) or 3,4-methylenedioxy methamphetamine (MDMA). The cut-offs are based on the results of proficiency testing, taking into account a measurement uncertainty of 30%. In special cases (e.g. consumption of several drugs, symptoms of withdrawal), an expert advice based on the "three pillars" (police observations, medical examination and toxicology results) will be sought.
- Until Act n° 111/2003, which was implemented in 2004, the state of *Victoria* had as the other states of Australia, an impairment law. Since then a *"per se"* law became effective.
- In the *USA* 14 states have implemented a *"per se"* legislation. In 2005, the president has signed a proposal that allows the use of federal transportation money to persuade states to adopt zero-tolerance laws. Legal changes in the USA are discussed by Kadehjian [27].

The analytical cut-offs of Germany, Belgium, France, Sweden and Switzerland are given in Table 19.1. There is no consensus on the analytical cut-offs between the different countries. This lack of consensus can be partially attributed to the use of different biological matrices (serum in Germany; plasma in Belgium and whole blood in France, Sweden and Switzerland) and the different consequences that follow a positive result: for example, in Belgium there is a penal sanction that follows a positive result, while in Germany there is an administrative sanction [28].

The effectiveness of *"per se"* laws in increasing the number of prosecutions has already been demonstrated in some countries:

- In Germany there was a five-fold increase in the number of persons suspected of DUID from 1997 to 1999 after a constant evolution during several years. Moeller [29] attributes this to two facts:
  - a training programme, which was introduced for police officers to detect subjects who are DUID and
  - a *"per se"* law, which came into effect in Germany in August 1998.

TABLE 19.1
ANALYTICAL CUT-OFF LIMITS IN BLOOD, PLASMA OR SERUM AS AGREED UPON OR
PROPOSED IN DIFFERENT COUNTRIES

	Germany[a]	Belgium	France	Sweden	Switzerland[b]
Amphetamine	50	50	50	30	15 (22)[c]
MDMA	50	50	50	20	15 (22)
MDEA	50	50	50	20	15 (22)
MDA				20	
MBDB		50		20	
Cocaine		50	50	20	15 (22)
Benzoylecgonine	150	50	50	20	–
Morphine (free)	20	20	20	5	15 (22)
THC	1[d]	2	1	0.3	1.5 (2.2)

Note: All concentrations in ng/ml, except Sweden: ng/g.
[a]Lower cut-offs have been proposed, but they are not yet used everywhere.
[b]The numbers between parenthesis are the cut-offs that take the measurement uncertainty into account.
[c]For Switzerland also 15 ng/ml for methamphetamine.
[d]Decision by the Federal Constitutional Court on the 21st of December 2004 (1 BvR 2652/03).

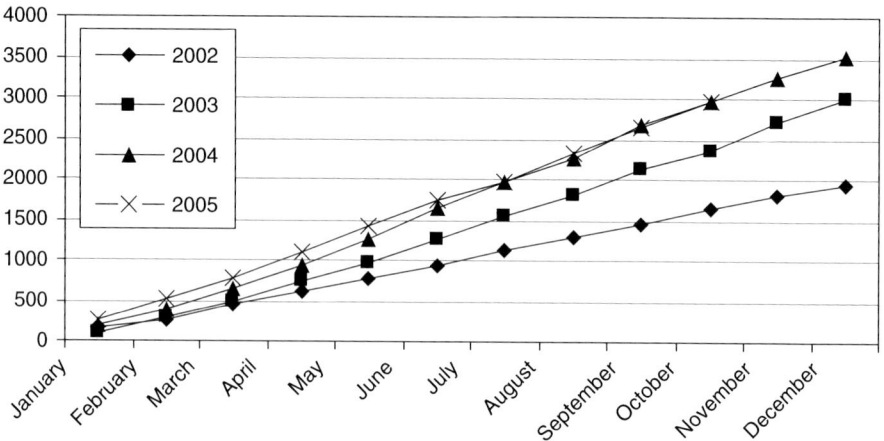

Fig. 19.1. The cumulative monthly increase in the number of samples after introducing the zero-tolerance law in 2003 in Finland. (From ref. [188] with permission of the authors.)

- In Finland, until 2002 there was a slow increase in the number of samples that were investigated. Since the introduction of a *"per se"* legislation in 2003 there was a cumulative monthly increase in the number of samples (Fig. 19.1).
- In Sweden, immediately after the zero-limit law came into force, the number of cases of DUID submitted by the police for toxicological analysis increased sharply and was ten-fold higher in 2005 than before the new legislation (Fig. 19.2). Nevertheless, Jones [30] concludes that Sweden's zero-concentration limit has done nothing to reduce DUID or deter the typical offender because recidivism is high in

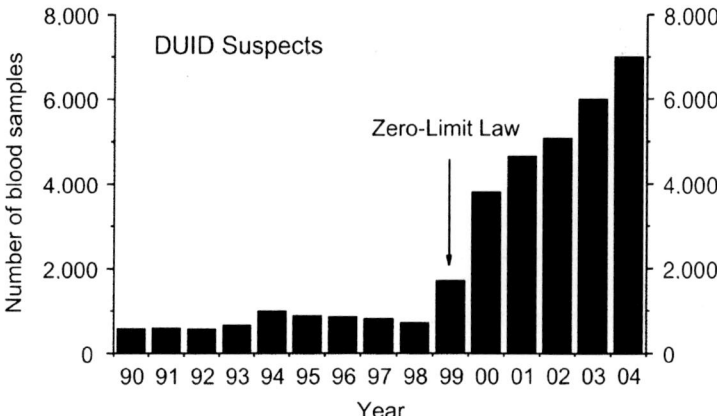

Fig. 19.2. Development in the number of blood samples from DUID suspects in Sweden submitted for toxicological analysis before and after the zero-limit law was introduced on 1 July 1999. (From ref. [30] by permission of Taylor & Francis Group.)

this population of individuals (40–50%). Many traffic delinquents in Sweden are criminal elements in society with previous convictions for drunk and/or drugged driving as well as other offences. The spectrum of drugs identified in blood samples from DUID suspects has not changed much since the zero-limit law was introduced [30].

In Germany and Italy, hair analysis is already a part of the procedure for granting or re-granting the driver's license. In contrast to other illicit drugs, recreational cannabis use in Germany is not generally incompatible with possessing a driving license. Therefore as a prerequisite, every case with a positive cannabis hair result must be checked in a medical psychological examination whether the applicant is able to separate cannabis use and driving [31]. In Italy, hair analysis is included in a panel of clinical and laboratory tests to investigate the toxicological behaviour of the driver license applicant. According to the Italian law, former users must provide evidence of drug cessation and demonstrate no risk of relapse [31].

19.2 ANALYTICAL METHODS

19.2.1 Introduction

A typical procedure of toxicological analysis involves two steps: screening and confirmation. Usually an immunoassay followed by a chromatographic method is performed, the first allows for a preliminary monitoring of a large number of samples in a reduced period of time, whilst the second step provides the required specificity. There has recently been an increase in the repertoire of immunoassays now available with the techniques becoming increasingly more specific for the drugs and/or metabolites being monitored [32]. Drug screening through urinalysis is a widely accepted tool for rapid detection of potential drug use at a relatively low cost [33]. Nevertheless, for the confirmation of drug use more sensitive and more specific methods are required. For the confirmation of drugs in blood of drivers, the French Society for Analytical Toxicology (SFTA) has recommended various methods by GC-MS [34–36] because of its sensitivity and specificity. However, the procedure is very labour-intense and time-consuming, particularly as sample preparation, i.e. solid phase extraction (SPE) and derivatisation are unavoidable [37]. Recently, liquid chromatography–tandem mass spectrometry (LC-MS/MS) has become a powerful tool for quantitative confirmatory analysis of drugs of abuse and has begun to spread in the field of forensic toxicology [38]. Guidelines for confirmatory analysis by GC-MS and LC-MS/MS have been published recently by several organisations (world anti-doping agency (WADA) [39], the International Olympic Committee (IOC) [40], the Society of Forensic Toxicologists (SOFT) [41], the German Society of Toxicological and Forensic Chemistry (GTFCh) [42] and the European Union (EU) [43,44]). The prerequisites for forensic confirmatory analysis by LC-MS/MS with respect to EU guidelines are chromatographic separation, a minimum number of two MS/MS transitions to obtain the required identification points and predefined thresholds for the variability of the relative intensities of the MS/MS transitions (multiple reaction monitoring (MRM) transitions) in samples and reference standards [38].

19.2.2 Relevant reviews

In 2002, Marquet published a review about the progress of LC-MS in clinical and forensic toxicology [45]. The review gives an overview of the analytic methods for xenobiotics that could be set up with these instruments for clinical or forensic toxicology and it also discusses new applications of LC-MS in the field of toxicology, such as general unknown screening (GUS) procedures and mass spectral libraries using LC-APCI-MS(/MS). The author concluded that almost all drugs of abuse can be analysed in biological fluids or tissues with LC-MS or LC-MS/MS instruments, with equal or better specificity and generally better sensitivity than GC-MS, using simpler sample preparation.

In 2005, Jansen *et al.* [46] published a review on the comparison of MS/MS spectra obtained by either a linear-ion-trap (LIT) tandem MS (QTRAP^{TM}) operated in enhanced product ion scan (EPI) mode with those obtained in the triple-quadrupole product ion scan mode run on the same as well as on two other instruments. The obtained MS/MS spectra did not offer the expected inter-instrument reproducibility despite an attempt at standardising the fragmentation conditions using a reference compound. However, although the inter-instrument differences in ion relative intensity were significant, the obtained spectra looked almost similar. The authors suggested that in library searching algorithms, higher weight would be assigned to the m/z ratios than to their relative intensity in the spectra.

In 2002, Maurer [47] reviewed the role of GC-MS-negative ion chemical ionisation (NICI) in clinical and forensic toxicology, doping control and biomonitoring. The studies that are included for forensic toxicology describe GC-MS-NICI methods for the analysis of benzodiazepines, THC and opioids in whole blood, plasma or urine, and in alternative matrices such as sweat, hair, bone or muscle samples of humans or rats. The author concluded that NICI can improve the sensitivity by a factor of several thousands compared with electron ionisation (EI) for the determination of compounds that contain halogen atoms or that can be transformed to halogenated derivatives.

Butler and Guilbault [48] made a review in 2004 on the background chemistry, metabolism, pharmacology, toxicology, effects on humans, diagnosis and treatment, followed by a listing of the then current chemical and bioanalytical methods for ecstasy, including GC, high pressure liquid chromatography (HPLC), capillary electrophoresis (CE) and immunochemical techniques. They concluded that immunoassay techniques serve only for screening purposes, and that confirmation is largely based on GC.

Tagliaro and Bortolotti [49] recently published a review about the advances in the applications of CE to forensic sciences from 2001 until the first part of 2005. The overview includes the most relevant examples of analytical applications of CE and electro kinetic techniques in several fields, including forensic drugs and poisons. The author concluded that CE with ultraviolet (UV) detection typically lacks sufficient sensitivity and selectivity in order to avoid interferences from matrix components in biological fluids, but that a major improvement of CE is represented by the coupling with MS, which fulfils the strictest requirement of forensic toxicology in terms of analytical accuracy.

Van Bocxlaer [50] also states that CE is a re-emerging approach and that it makes a worthwhile addition to, e.g. LC because of the increasingly polar and ionic molecules encountered, as illustrated for the analysis of γ-hydroxybutyric acid (GHB). The author also indicates LC-time-of-flight (TOF)-MS and matrix-assisted laser desorption/ionisation (MALDI)-MS as promising approaches.

19.2.3 Detection of drugs in urine

Possible biological matrices for the detection of drugs of abuse are urine, blood (or serum or plasma), oral fluid, sweat or hair, and they all have their specific advantages and limitations. At first, urine was the sample of choice because of the high concentrations of drug metabolites it contains. Disadvantages of the use of urine are the delayed appearance of drugs, the need of supervision during sampling, the risk of adulteration and the absence of correlation with impairment. For example, for some substances such as cannabis, the metabolites can be detected for a long time, especially after chronic use. The presence of a drug in urine is thus only an indication of previous drug use, and this does not necessarily mean that the person is under the influence. As there exist several reliable on-site tests for the detection of drugs in urine, this matrix is primarily used as biological matrix for roadside drug screening [16,28]. The project ROSITA revealed that when a urine test is performed after selection of drivers based on a test battery of several physical external signs and tests, the analysis of urine by GC-MS correlates well to the presence or absence of drugs in blood in 94% of the cases for amphetamines, in 89% for benzodiazepines, in 86% for cannabis, in 97% for cocaine and in 86% for opiates [51]. Moreover, a study by Raes and Verstraete [26] showed that roadside urine screening significantly decreases the number of unnecessary blood analyses in DUID cases.

The recent progress since the reviews mentioned in the relevant reviews above will be discussed in what follows, and will be divided in methods for one group of drugs of abuse and methods for various groups of drugs of abuse.

19.2.3.1 Methods for one group of drugs of abuse

A method using SPE and LC-MS/MS with negative APCI for the confirmation of 11-nor-9-carboxy-δ-9-THC (THC-COOH) in urine samples was developed by Weinmann *et al.* [52]. The conjugates of THC-COOH were hydrolysed prior to SPE. The LC-MS runtime was 6.5 min by gradient elution with a retention time of 2.4 min. Linearity of calibration was obtained in the range between 0 and 500 ng/ml ($r^2 = 0.998$). The limits of detection (LOD) and limits of quantitation (LOQ) were 2.0 ng/ml and 5.1 ng/ml, respectively. Day-to-day precision was tested at 15 and 250 ng/ml and were 3.3% and 4.5%, respectively.

A review published in 2005 by Nakashima [53] focuses on HPLC methods used for the practical analysis of drugs of abuse, mainly for amphetamine derivatives and MDMA in biological samples such as urine, blood and hair. For blood there were 13 studies reported, using methods such as fluorescence detection (FL), HPLC-FL, HPLC-UV, fluorescence polarisation immunoassay (FPIA), rapid emerging drug identification (REMEDi) or LC-ESI-MS in the selected ion-monitoring (SIM) mode. For FL the LOD equalled 10 ng/ml in two different studies, for HPLC-UV 25 ng/ml and for LC-ESI-MS 5–50 ng/ml.

An application note by Applied Biosystems reports the quantification of 27 benzodiazepines, zolpidem and zopiclone in serum and urine using LC-MS/MS. Sample preparation of urine was performed by hydrolysis, dilution, vortexing and centrifugation. The linear regression coefficients were between 0.9927 and 0.9995 for all the analytes. The LOD for each analyte was between 0.01 and 0.5 ng/ml [54].

19.2.3.2 Methods for various groups of drugs of abuse

Lu and Taylor [33] evaluated the sensitivity and specificity of two immunoassays, namely an enzyme multiplied immunoassay technique (EMIT® II) and a kinetic interaction of microparticle in solution assay (KIMS) for the detection of a 10-drug panel (amphetamine, barbiturates, benzodiazepines, marijuana, cocaine, methadone, methaqualone, opiate, phencyclidine (PCP) and propoxyphene) in urine samples. GC-MS was used as a confirmation method for all positives. Both immunoassays generated fairly consistent results, and the concordance rate against each of the 10 drugs was relatively high (97.4–100%). The discrepancies occurred mostly at concentrations near the cut-off.

A *direct injection LC-MS/MS* method for the simultaneous quantification of opioids, cocaine and metabolites in urine was developed and validated in 2003 by

Dams *et al.* [55]. A 10 µl aliquot of urine was directly injected onto the LC-MS/MS system. The absence of sample preparation substantially reduced total analysis time. All analytes were separated in 26 min. Identification and quantification was based on selected reaction monitoring (SRM). LOQs were established between 10–100 ng/ml and linearity was obtained up to a maximum of 10,000 ng/ml with an average $r^2 > 0.99$.

In 2004, Nordgren and Beck [56] developed a direct injection LC-APCI-MS/MS method for the direct screening for 23 different substances (phenylethylamines, hypnotics and *N*-benzylpiperazine) in urine samples. The achieved LODs were lower compared with commercial immunochemical methods. There was a linear response for all analytes with an intraday coefficient of variation (CV) of about 16% and the gradient elution gave variability in relative retention time of about 1%. Positive results were confirmed by reanalysis including sample preparation by SPE. The method was applied to 529 urine samples from suspected drug users, of which 35 samples were screened positive for phenylethylamines, and 20 for hypnotics. Of these, 23 samples were confirmed positive for phenylethylamines and 11 for hypnotics. The authors concluded that LC-MS/MS is a valuable complement to immunochemical screening analysis, especially for substances for which immunochemical methods are not yet available or when an increased sensitivity is needed.

In 2005, Mueller *et al.* [57] published a method for multi-target screening in blood and urine for 301 drugs using a QTRAP™ LC-MS/MS system and automated library searching. The extraction was performed either by liquid–liquid extraction (LLE) or SPE. An MRM as survey scan and an EPI scan as dependent scan were performed in an information-dependent acquisition (IDA) experiment. The EPI scan was carried out at the following three different collision energies: 20, 35 and 50 eV. This method was successfully applied to samples of forensic cases. All compounds were first identified using established methods. The multi-target screening method yielded identical results in less than half the time required by conventional techniques.

In Table 19.2 an overview is given of the methods used by different types of recent epidemiological studies to detect, confirm and quantify various groups of drugs of abuse in urine. Most of these studies used an immunoassay technique as screening method. As confirmation GC was mostly used followed by various detection techniques (MS, DAD, UV, nitrogen–phosphorus detection (NPD)). Some studies used thin layer chromatography (TLC) or HPLC for confirmation.

19.2.4 Detection of drugs in blood, serum or plasma

The advantages of blood are that identification and quantification can be performed in one matrix, that in most cases the unchanged drug is detectable and that the blood matrix is relatively homogeneous. Blood is considered the best matrix for confirmation analysis, because the presence of drugs in blood corresponds best with recent use and impairment. Difficulties may arise when only aged or haemolysed blood is available. Other disadvantages are the invasive way of sampling [58] and the

TABLE 19.2
DIFFERENT TYPES OF EPIDEMIOLOGICAL STUDIES THAT ANALYSED FOR SEVERAL
GROUPS OF DRUGS IN URINE

Author	Year	N	Method
Roadside surveys			
Mathijssen [166]	1997–1998	893	(a) EMIT [R]
			(b) GC/MS (GC/DAD for cannabis)
Dussault *et al.* [4]	1999–2000	5931	(a) Immunoassay
			(b) HPLC/MS and GC/MS
Assum *et al.* [3]	2000–2004	2873	(a) EMIT [R] (benzodiazepines: special high sensitivity protocol with on-line deglucuronidation)
			(b) GC-MS
(Fatally) injured drivers			
Lowenstein and Koziol-McLain [167]	1995–1996	414	Various methods: EIA, TLC, GC and GC-MS
Cheng *et al.* [9]	1996–2000	197	(a) Immunoassay
			(b) GC-MS or LC-MS
Vignali *et al.* [168]	1997–1999	53	(a) EMIT [R]
			(b) GC-MS
Kintz *et al.* [90]	1999	162	(a) FPIA
			(b) GC-MS
Assum *et al.* [3]	2000–2004	63	(a) EMIT [R] (benzodiazepines: special high sensitivity protocol with on-line deglucuronidation)
			(b) GC-MS
Walsh *et al.* [5]	2003	108	EIA and Roche TesTcup[TM] Pro5 rapid POC immunoassay
Drivers suspected of driving under the influence of drugs			
Plaut and Staub [169]	1995–1999	311	(a) Immunoassay
			(b) GC-MS
Ceder [170]	1999	Unknown	(a) Immunoassay
			(b) GC-MS
Thorsdottir *et al.* [171]	2000–2002	162	(a) FPIA or GC-MS
			(b) GC-MS, GC-NPD or HPLC-UV

Note: a: screening, b: confirmation/quantification.

difficulties encountered in some countries from the legal point of view in obtaining abusers' blood samples [53].

A review on drugs of abuse monitoring in blood for control of DUID was published in 2002 by Moeller and Kraemer [58]. This review describes procedures for detection of the following drugs of abuse in whole blood, plasma and serum: (meth)amphetamine, MDMA, MDEA, 3,4-methylenedioxyamphetamine (MDA), THC, 11-hydroxy-δ-9-THC (11-OH-THC), THC-COOH, cocaine, benzoylecgonine (BE), ecgonine methyl ester (EME), cocaethylene, the opiates (heroin, 6-monoacetylmorphine (6-MAM), morphine, or codeine) and methadone as well as GHB, lysergic acid diethylamide (LSD), PCP and psilocybin/psilocin. The authors concluded that for many of the analytes, sensitive immunologic screening methods are available and that GC-MS is the state-of-the-art method for confirmatory analysis of positive immunoassay results or for screening and confirmation in one step. Exceptions included confirmation of the glucuronides of morphine, which are more suitable for LC-MS analysis. More LC-MS procedures had been developed for screening and confirmation of the drugs of abuse in blood, but there was still a need for validated procedures to increase the role of LC-MS in this field. The recent progress since this review will be discussed in what follows, and will be divided in methods for one group of drugs of abuse and methods for various groups of drugs of abuse.

19.2.4.1 Methods for one group of drugs of abuse

A method for the determination of THC in whole blood was published by Chu and Drummer [59] in 2002. The sample preparation consisted of LLE, followed by confirmation using GC-MS in electron impact mode. The substance was derivatised with pentafluoropropionic anhydride in pentafluropropanol. The LOD was 0.5 ng/ml for a 1 ml specimen, with recovery greater than 70%. The intra-assay CV was 3.1–5.2%, and the inter-assay CV was 6.4–9.5%, at THC concentrations of 1, 5 and 25 ng/ml. The accuracy was between 95% and 97%.

Samyn *et al.* [60] published a study in 2002 to assess ecstasy concentrations in controlled and in real life conditions. In a double-blind placebo controlled study a 75 mg dose MDMA was administered orally to 12 healthy volunteers who were known to be recreational MDMA users. Screening was performed in urine by FPIA. Quantitative analyses were performed in plasma using the deuterated analogues of the analytes of interest and a GC-MS in SIM mode. The LOQs were 20 ng/ml for MDA and 10 ng/ml for the other amphetamines. The resulting plasma concentrations varied from 21 to 295 ng/ml, with an average peak concentration of 178 ng/ml observed between 2 and 4 h after administration.

Laloup *et al.* [61] validated an enzyme-linked immunosorbent assay (ELISA^{TM})-based screening assay for the detection of amphetamine, MDMA and MDA in blood and oral fluid. Authentic plasma samples from 360 drivers were screened, using a 1:5-fold dilution. True positive, true negative, false positive and false negative were determined using GC-MS as a reference technique. Analysis showed that the ELISA^{TM} technique has a sensitivity of 98.3% and a specificity of 100% at a cut-off value of 66.5 ng/ml.

In 2003, Wood *et al.* [62] developed an LC-MS/MS method for the quantification of amphetamines in human plasma and oral fluid. The amphetamines were isolated from human plasma using a simple methanol precipitation step. The developed method, which requires only 50 μl of biological sample, has a total analysis time of less than 20 min (including sample preparation) and LODs of 2 ng/ml or better.

Maralikova and Weinmann [38] published in 2004 a fully validated HPLC-MS/MS method for the simultaneous confirmation of THC, 11-OH-THC and THC-COOH in human plasma. Sample clean-up was performed by automated silica-based SPE. Data were acquired with an LC-MS/MS system equipped with a

TurboIonSpray® interface and triple-quadrupole mass analyser using positive elect-rospray ionisation and MRM. The LOD and LOQ of THC and 11-OH-THC were 0.2 and 0.8 ng/ml, respectively, and for THC-COOH 1.6 and 4.3 ng/ml. This method was developed and validated following the EU guidelines on LC-MS/MS valid in 2004 and was sensitive enough to fulfil the recommendations of the GTFCh. Finally the method was successfully applied to the identification of these drugs in blood, plasma and urine samples.

A review published in 2005 by Nakashima [53] focuses on HPLC methods used for the practical analysis of drugs of abuse, mainly for amphetamine derivatives and MDMA in biological samples such as urine, blood and hair. For blood nine studies were reported, using methods such as FL, HPLC-FL, HPLC-ESI, LC with sonic spray ionisation MS and LC-APCI-MS/MS. For FL the LOD was less than 0.87 ng/ml plasma. For LC-APCI-MS/MS the first study reported an LOD of 0.8 ng/ml serum and a lower LOQ of 3 ng/ml serum. The LC-APCI-MS/MS method in the second study had a lower LOQ of 0.2 ng/ml plasma.

An application note by Applied Biosystems reports the quantification of 27 ben-zodiazepines, zolpidem and zopiclone in serum and urine using LC-MS/MS. Sample preparation of serum was performed by dilution, vortexing and centrifugation. The linear regression coefficients were between 0.9927 and 0.9995 for all the analytes. The LOD for each analyte was between 0.01 and 0.5 ng/ml [54].

19.2.4.2 Methods for various groups of drugs of abuse

Gunnar *et al.* [63] developed a validated procedure for the simultaneous semiquan-titative/quantitative screening of benzodiazepines, cannabinoids, opioids, cocaine, antidepressants, antipsychotics and antiepileptics. The analytes were isolated by high-yield LLE. The dried extracts were derivatised by two-step silylation and an-alysed by the combination of two different GC separations with both EC and MS operating in a SIM mode. Intra- and inter-day precisions were within 2.5–21.8 and 6.0–22.5%, and the determination coefficients of linearity ranged from 0.9896 to 0.9999. The LOQ varied from 2 to 2000 ng/ml due to a variety of the relevant concentrations of the analysed substances in blood. The method was reliable and rapid requiring only a small sample volume, even in the presence of a high matrix content of whole blood. The method has been routinely used for 1500 samples of drivers suspected of DUID or clinical samples during 2002–2003 and has been accredited by the Finnish Centre for Metrology and Accreditation (FINAS).

Maralikova and Weinmann [38] developed and validated an HPLC-MS-MS method for the simultaneous confirmation of (nor)morphine, morphine-3-glucur-onide, codeine, 6-MAM, BE, methadone and cocaine in plasma samples. Sample preparation consisted of automated silica-based SPE. The LC-MS/MS system con-sisted of a triple-quadrupole mass spectrometer equipped with TurboIonSpray® interface using positive electrospray ionisation and MRM. The LOD and LOQ of morphine-3-glucuronide were 2.4 and 6.6 ng/ml, respectively, for normorphine 3.1 and 8.0 ng/ml, for morphine 4.0 and 6.3 ng/ml, for codeine 2.3 and 3.1, for MAM 1.2 and 5.3 ng/ml, for BE 0.3 and 0.9 ng/ml, for cocaine 1.5 and 4.7 ng/ml and for methadone 0.7 and 1.8 ng/ml. This method was developed and validated following the EU guidelines on LC-MS/MS valid in 2004 and was sensitive enough to fulfil the recommendations of the GTFCh. Finally, the presented method was successfully applied to the identification of these drugs in blood, plasma and urine samples.

A review on LC-MS or LC-MS/MS methods for the screening and quantification of various illicit and therapeutic drugs in blood, plasma or serum was published in 2005 by Maurer [64]. The author reports 15 studies on LC-MS screening and detection and 13 studies on LC-MS quantification. He concluded that LC-MS is an ideal supplement to GC-MS, especially for the detection and quantification of more polar, thermolabile or low-dosed drugs in blood plasma.

Mueller *et al.* [57] published in the same year a method for multi-target screening in blood and urine for 301 drugs using a QTRAP™ LC-MS/MS system and automated library searching. The extraction was performed either by LLE or SPE. An MRM as survey scan and an EPI scan as dependent scan were performed in an IDA experiment. The EPI scan was carried out at the following three different collision energies: 20, 35 and 50 eV. This method was successfully applied to samples of forensic cases. All compounds were fist identified using established methods. The multi-target screening method yielded the identical results in less than half the time required by conventional techniques.

Recently, an LC-MS method in the presence of deuterated analogues was developed by Concheiro *et al.* [65] for the simultaneous determination of morphine, 6-MAM, amphetamine, methamphetamine, MDA, MDMA, MDEA, MBDB, BE and cocaine. The sample preparation consisted of SPE. The method was fully validated, including linearity (2–250 ng/ml, $r^2 > 0.99$), recovery ($> 50\%$), within- and between-day precision and accuracy (CV and bias $< 15\%$), LOD (0.5 ng/ml for methamphetamine, MDMA, BE and cocaine and 1 ng/ml for morphine, 6-MAM, MDA, MDEA and MBDB) and LOQ (2 ng/ml). Only 0.2 ml of plasma was needed to do the analysis and no matrix effect was observed. This procedure was applied to 156 road fatalities, of which 11 cases tested positive to cocaine, 1 case to amphetamine derivatives and 57 cases to alcohol. Appenzeller *et al.* [66,67] recently published a national survey on alcohol, drugs and driving. For this study they used HPLC, GC-MS and CE methods.

Table 19.3 gives an overview of the methods used by different types of recent epidemiological studies to detect, confirm and quantify various groups of drugs of abuse in blood, serum or plasma.

19.2.5 Detection of drugs in oral fluid

In recent years, the interest in the use of oral fluid as biological matrix has increased significantly, as this matrix displays some particularly interesting properties. First of all, oral fluid can be obtained easily by non-medical personnel in a non-invasive and observable way. There is also some correlation with impairment [16]. A study that compared the quantitative results by GC-MS in oral fluid of drivers suspected of driving under the influence of drugs to the corresponding results in blood revealed

TABLE 19.3

DIFFERENT TYPES OF EPIDEMIOLOGICAL STUDIES THAT ANALYSED FOR SEVERAL
GROUPS OF DRUGS IN BLOOD, SERUM OR PLASMA

Author	Year	N	Method
Roadside surveys			
Assum *et al.* [3]	2000–2004	501	Cannabis and opiates: (a) EIA (Cozart[R]) (b) GC-MS Benzodiazepines, cocaine, TCA, amphetamines and methadone: HPLC
(Fatally) injured drivers			
del Rio *et al.* [10]	1991–2000	5,745	(a) Immunological or chromatographic methods (b) GC-MS or GC-NPFID
Longo *et al.* [172]	1995–1996	2,500	(a) RIA (b) GC(-MS) or HPLC
Cheng *et al.* [9]	1996–2000	197	(a) GC-NPD, GC-MS or HPLC-DAD (b) Appropriate validated methods
Vignali *et al.* [168]	1997–1999	119	(a) EMIT[R] (b) GC-MS
Kintz *et al.* [90]	1999	198	(a) In urine, if available (FPIA) (b) GC-MS, LC-DAD or LC-MS
Brault *et al.* [11]	1999–2002	855	(a) Immunoassay (b) HPLC-MS and GC-MS
Mura *et al.* [134]	2000–2001	900	(a) Immunoassay (b) GC-MS (drugs of abuse) or HPLC-DAD (psychoactive therapeutic drugs)
Holmgren *et al.* [173]	2000–2002	855	(a) In urine (immunoassays) (b) GC-MS
Assum *et al.* [3]	2000–2004	121	Cannabis and opiates: (a) EIA (Cozart[R]), (b) GC-MS Benzodiazepines, cocaine, TCA, amphetamines and methadone: HPLC
Logan and Schwilke [8]	2001–2002	370	(a) Immunoassay or EMIT[R] (b) GC-MS or GC-FID
Bernhoft *et al.* [6]	2002–2004	45	(a) EIA (Cozart[R] Drugs of Abuse Microplate EIA) (b) GC-MS or LC-MS/MS
Assum *et al.* [3]	2003–2004	87	Opiates and cannabis: (a) EMIT[R], (b) GC-MS Cocaine, amphetamines and benzodiazepines: (a–b) LC-MS
Mura *et al.* [174]	2003–2004	2,003	(a) Immunoassay (b) GC-MS
Drivers suspected of driving under the influence of drugs			
Smink *et al.* [175]	1995–1998	1,665	(a) In urine (FPIA) (b) HPLC-electrochemical detection, GC-MS, HPLC-UV or GC-NPD

TABLE 19.3
CONTINUED

Author	Year	N	Method
Seymour and Oliver [176]	1999	156	(a) EIA (buprenorphine and methadone) GC and HPLC (acidic, basic and neutral drugs) (b) GC-MS
Maes *et al.* [177]	2000–2001	896	(a) In urine (immunoassay) (b) GC-MS using deuterated internal standards
Thorsdottir *et al.* [171]	2000–2002	162	(a) FPIA or GC-MS (b) GC-MS, GC-NPD or HPLC-UV
Kotsos *et al.* [178]	2000–2003	311	(a) ELISA^{TM} (b) GC-MS
Toennes *et al.* [179]	2001	131	(a) EIA (Bio-Rad CODA® system and serum drug screening tests) (b) GC-MS
Appenzeller *et al.* [67]	2001–2002	210	Illicit drugs and methadone: (a–b) GC-MS Medicinal drugs: (a–b) HPLC
Laumon *et al.* [13]	2001–2003	10,748	(a) In urine (b) GC-MS
Augsburger *et al.* [180]	2002–2003	440	(a) In urine (immunological tests: RIA, EMIT®, etc.) (b) GC-MS, GC-ECD, HPLC-DAD or HPLC-MS

Note: a: screening, b: confirmation/qualification.

that the positive predictive value of oral fluid was 98, 92 and 90% for amphetamines, cocaine and cannabis, respectively [68]. Furthermore, the results of the European project ROSITA indicated that for most drugs of abuse the correlation with blood is better for oral fluid than for urine [51]. Nevertheless, the results of ROSITA and the follow-up project ROSITA 2 indicated that none of the currently available onsite oral fluid drug testing devices are reliable enough to be recommended for roadside screening for drivers. A study by Dierich and Soyka [69] found that drug testing in oral fluid by cloned enzyme donor immunoassay (CEDIA^{TM}) and EMIT® still is not as accurate as urine analytics. However, the experience in the state of Victoria in Australia shows that random roadside oral fluid testing of drivers for methamphetamine and cannabis has a deterrent effect: the level of awareness of drivers of random oral fluid testing increased from 78% to 92%, 33% of illicit drug users stated that the drug tests had influenced them, primarily to avoid taking drugs when they are going to drive and the proportion of drug-using respondents who drove while under the influence of drugs dropped in the after period from 45% to 35% [70].

References pp. 646–651

Other advantages of oral fluid analysis are that there is less interference caused by endogenous compounds than in blood or urine [71] and that the parent drug is present [16]. On the other hand the oral cavity can be contaminated by intranasal and smoked drug use, leading to extremely high concentrations in oral fluid. It is also difficult to obtain sufficient sample volume for the analysis [72], and the concentrations of THC and benzodiazepines in this matrix are low [16].

In 2004, Choo and Huestis [73] published a review on the use of oral fluid as a diagnostic tool. This review describes advances over the previous 5 years in the area of oral fluid as a diagnostic tool for several applications, including the monitoring of drugs of abuse. All the studies reported used GC-MS(/MS) for the analysis of drugs of abuse in oral fluid. Recent relevant studies that were not included in this review, or were published since this review will be discussed in what follows, and will be divided in methods for one group of drugs of abuse and methods for various groups of drugs of abuse.

19.2.5.1 Methods for one group of drugs of abuse

Cooper *et al.* [74] validated the Cozart® Amphetamine Microplate enzyme immuno-assay (EIA) for the analysis of amphetamines in oral fluid. Oral fluid samples were collected by Cozart® RapiScan from 135 drugs users and 35 healthy volunteers. The samples were analysed by EIA (cut-off = 45 ng/ml amphetamine equivalents in neat oral fluid) in the laboratory and confirmed by GC-MS (cut-off = 30 ng/ml). The intra-assay precision of the EIA for forty assays was 2.74−7.1% CV (within assay) and 3.4−7.0% CV (within day). A total of 78 samples were positive for various amphetamines and related designer drugs. The EIA had a sensitivity of $91.7 \pm 3.3\%$ and a specificity of $95.9 \pm 1.9\%$.

In 2003, Wood *et al.* [62] developed an LC-MS/MS method for the quantification of amphetamines in human plasma and oral fluid. The amphetamines were isolated from oral fluid using a simple methanol precipitation step. The developed method, which requires only 50 µl of biological sample, has a total analysis time of less than 20 min (including sample preparation) and LODs of 2 ng/ml or better.

A sensitive enantioselective LC-MS method was developed by Rosas *et al.* [75] for the simultaneous determination of oral fluid concentrations of (*R*)- and (*S*)-methadone and (*R*)- and (*S*)-2-ethylidene-1,5-dimethyl-3,3-diphenyl-pyrrolidine (a primary metabolite of methadone). The samples were collected by means of Salivette® devices and centrifuged. Collected oral fluid was spiked with deuterated internal standards and directly injected into the LC-MS. Enantioselective separations were achieved on a liquid chromatographic chiral stationary phase based upon immobilised α_1-acid glycoprotein using a mobile phase composed of acetonitrile: ammonium acetate buffer (10 mM, pH 7.0) in a ratio of 18:82 (v/v), a flow rate of 0.9 ml/min and a temperature of 25°C. Under these conditions, enantioselective separations were achieved within 15 min. The linear relationship between peak height ratio and drug-enantiomer concentrations were obtained for methadone in the range of 5.0–600.0 ng/ml and for its metabolite from 0.5–15.0 ng/ml per enantiomer with CVs > 0.9994. The lower LOQ for methadone was 5 ng/ml and for its metabolite

0.5 ng/ml. Acceptable intra- and inter-day precision of the method (CVs < 4%) and accuracy (CVs < 4%) were obtained.

An analytical method using SPE and LC-MS was developed and validated for the confirmation of THC in oral fluid samples [76]. Selected ion-recording was used for the quantification. Intra- and inter-day analyses had CVs less than 10%, and the calculated extraction efficiencies for THC ranged from 76% to 83%. Calibration standards spiked with THC between 2 and 100 ng/ml showed a linear relationship ($r^2 = 0.999$). The method was applied to oral fluid samples of 40 cannabis users either collected by expectoration or by Salivette®. In 55% of the samples collected by expectoration, THC was detected with concentration ranges from 1033 to 6552 ng/ml and in 45% of the cases THC was found at concentrations between 51 and 937 ng/ml. The results obtained after Salivette® collection differed from those after expectoration: THC concentration was undetectable in 26 of 40 volunteers.

Laloup *et al.* [61] validated an ELISA™-based screening assay for the detection of amphetamine, MDMA and MDA in blood and oral fluid. Oral fluid samples, collected with the Intercept® collection device were obtained from a controlled study performed on 18 volunteers. Each subject received either placebo or a high (100 mg) or a low (75 mg) dose of MDMA in combination with or without alcohol, according to a double-blind, 6-way cross-over design. The resulting 216 oral fluid samples were screened, using a 1:5 dilution. True positive, true negative, false positive and false negative were determined using LC-MS/MS as a reference technique. Combined results of the analysis of the high- and low-dose oral fluid samples indicated a screening cut-off of 51 ng/ml D-amphetamine equivalents with both a sensitivity and specificity of 98.6%.

In the same year, Laloup *et al.* [77] developed and fully validated a method for the quantitative analysis of THC in preserved oral fluid by LC-MS/MS. Oral fluid was collected with Intercept® and sample preparation consisted of a simple LLE, and this led to significant decreases in the interferences present in the matrix. Validation was performed using both 100 and 500 µl of oral fluid, with resulting LOQs of 0.5 and 0.1 ng/ml, respectively. This method was applied to the analysis of 102 oral fluid samples of volunteers who had received either a placebo cigarette or a marijuana cigarette. Only the presence of THC had to be confirmed, thus 500 µl of oral fluid was used. For samples where the response exceeded the upper limit of the standard curve, reanalysis of only 100 µl was performed. At 30 min before drug administration all specimens were negative, except for three subjects in which low concentrations were found. However, in both the placebo and marijuana condition, THC could be detected, probably due to incomplete removal of THC for the preparation of the placebo cigarette. The method was also applied to 48 oral fluid samples collected during roadside controls. The measured THC concentrations varied considerably and some samples had to be reanalysed after dilution.

A procedure for the screening by LC-MS/MS of 17 benzodiazepines and hypnotics in oral fluid after collection by the Intercept® device was presented by Kintz *et al.* [78]. The method involves extraction of 0.5 ml of oral fluid treated with 0.5 ml of phosphate buffer (pH 8.4) in the presence of 5 ng diazepam-d$_5$ used as internal standard, with 3 ml of diethyl ether/methylene chloride (50/50) and separation using LC-MS/MS. The LOQ for all benzodiazepines and hypnotics ranged from 0.1 to 0.2 ng/ml. Linearity was observed from the LOQ of each component to 20 ng/ml ($r^2 > 0.99$). Ion suppression was lower than 10% for all drugs except zopiclone (93%).

Yegles and Wennig [79] described concentration ranges of 15 benzodiazepines using GC-MS-NCI in 103 patients from a substitution programme.

19.2.5.2 Methods for various groups of drugs of abuse

Mortier *et al.* [80] developed a method for the simultaneous and quantitative determination of opiates, amphetamines, cocaine and BE in oral fluid by LC-MS/MS. The samples were prepared by SPE and MS detection was performed on a quadrupole TOF mass spectrometer by electrospray ionisation, without the use of a splitter. The total runtime was 34 min. A validation study was carried out. The LOQ was 2 ng/ml for all compounds.

A quantitative LC-APCI-MS/MS method for the simultaneous determination of multiple illicit drugs, methadone and their metabolites in oral fluid was developed and validated by Dams *et al.* [81]. Sample pre-treatment was limited to acetonitrile protein precipitation. LC separation was performed in 25.5 min, with a total analysis time of 35 min. Identification and quantification were based on MRM. LOD and LOQ were established between 0.25 and 5 ng/ml and 0.5 and 10 ng/ml, respectively. Linearity was obtained with an average correlation coefficient > 0.99, over a dynamic range from the LOQ up to maximum 500 ng/ml.

Wood *et al.* [82] developed a method for the quantitative analysis of multiple illicit drugs in preserved oral fluid by SPE and LC-MS/MS. The applied HPLC gradient ensured the elution of all the drugs examined within 14 min and produced chromatographic peaks of acceptable symmetry. After SPE the processed samples were demonstrated to be stable for 48 h, except for cocaine and BE, where a slight negative trend was observed, but this did not compromise the quantification.

An application note released by Applied Biosystems describes an LC-MS/MS method for the simultaneous screening of 23 drugs of abuse in oral fluid. The sample preparation consisted of LLE [83].

A qualitative and quantitative analytical method was developed and validated for the determination of 49 licit and illicit drugs in oral fluid by Wylie *et al.* [84]. Small oral fluid samples, volume 1 ml, were collected from volunteers using a modified Omni-Sal® device and the analytes were extracted by SPE. LC-MS/MS and GC-MS were used in parallel to analyse the extracts for the targeted drugs. Extracts were analysed by GC-MS in their underivatised form and as their pentafluoropropionyl derivatives. Deuterated internal standards were used for quantification by LC-MS/MS to minimise matrix effects. Methadone-d9 and tumoxetine were used as the internal standards for the quantification of non-derivatised and derivatised analytes, respectively, by GC-MS. Linearity was demonstrated over the range 5–200 ng/ml and LODs were less than 4 ng/ml for each drug analysed.

TABLE 19.4
DIFFERENT TYPES OF EPIDEMIOLOGICAL STUDIES THAT ANALYSED FOR SEVERAL
GROUPS OF DRUGS IN ORAL FLUID

Author	Year	*N*	Method
Roadside surveys			
Behrensdorff and Steentoft [2]	2000	896	(a) EIA (Cozart® Drugs of Abuse Microplate EIA) (b) Benzodiazepines: GC-ECD or LC-DAD other drugs of abuse: GC-MS
Wylie *et al.* [84]	2003–2004	1312	LC-MS/MS and GC-MS
Swann [70]	2004–2005	4200	(a) Immunoassay (Drugwipe® and Cozart® RapiScan) (b) GC-MS
(Fatally) injured drivers			
Kintz *et al.* [90]	1999	197	(a) In urine, if available (FPIA) (b) GC-MS
Bernhoft *et al.* [6]	2002–2004	300	(a) EIA (Cozart® Drugs of Abuse Microplate EIA) (b) GC-MS or LC-MS/MS
Drivers suspected of driving under the influence of drugs			
Toennes *et al.* [179]	2001	168	(a) Immunoassay (Dräger DrugTest®) (b) GC-MS

Note: a: screening, b: confirmation/qualification.

In Table 19.4 there is an overview given of the methods used by different types of recent epidemiological studies to detect, confirm and quantify various groups of drugs of abuse in oral fluid.

19.2.6 Detection of drugs in sweat

Sweat offers as oral fluid a non-invasive way of screening at the roadside with the possibility of direct supervision. This is a major advantage in comparison to urine testing. Another advantage compared to urine is the presence of the parent drug. Sweat testing is very easy but false positives are possible due to external contamination of the skin (e.g. by being in a room where a lot of cannabis is smoked). As for urine, sweat has a long detection window. This can be an advantage, as for the investigation of prior exposure to psycho-active drugs after late sampling, but this also means that there is less correlation with the presence of a pharmacological effect at the time of sampling. Other disadvantages of sweat are the delayed appearance of drugs in sweat, the low concentrations of THC and benzodiazepines and the limited number of on-site tests that are available and research that has been performed [16,51,85].

References pp. 646–651

19.2.6.1 Methods for one group of drugs of abuse

The usefulness of sweat testing for the detection of MDMA was also investigated by Pichini *et al.* [86]. Nine healthy male subjects and recreational users of MDMA were given a single 100 mg dose. Sweat was collected up to 24 h with sweat patches from which drugs were eluted and then analysed by immunoassay and GC-MS using deuterated internal standards. MDMA was detected as early as 1.5 h after consumption and peaked at 24 h. Intersubject variability was large. Peak MDMA concentrations for the same dose varied 30-fold in magnitude, and the concentrations ranged between 3.2 and 1326.1 ng/patch.

The use of fastpatches for the detection of crack and other cocaine use was studied by Liberty *et al.* [87]. Through the use of mild heating and a slightly larger collection pad than standard sweat patches, fastpatches may significantly decrease necessary wearing times while increasing the time window in which drugs can be detected. One hundred and eighty active cocaine users wore fastpatches. For detection and quantification of cocaine and BE, the eluate from the patch was extracted and analysed by GC-MS analysis in SIM mode using deuterated internal standards. The cut-off for the GC-MS assay was 10 ng/ml. The use of crack cocaine was detected by means of its unique pyrolysis products, which were also analysed from the patch eluate by GC-MS. Cocaine use was detected in sweat of 92% of subjects through GC-MS, comparing favourably with 91% with EMIT® urinalysis. Crack metabolites were detected in 54% of subjects. There were no significant differences in detection rates between 15-, 20- and 30-min wear periods.

Kacinko *et al.* [88] studied the disposition of cocaine and its metabolites in human sweat after controlled administration. Nine participants received three low doses (75 mg/70 kg) of cocaine HCl subcutaneously within 1 week and 3 weeks later three high doses (150 mg/70 kg). Six additional participants completed portions of the study. The patches underwent subsequent elution, SPE, derivatisation and finally aliquots of 2 μl were analysed by GC-MS. Cocaine was the primary analyte detected with 24% of patches positive at a LOQ of 2.5 ng/patch and 7% of patches positive at the proposed Substance Abuse and Mental Health Services Administration cut-off of 25 ng/patch. EME was detected more often and at generally higher concentrations than BE. During washout, two participants' weekly patches tested positive (≥25 ng/patch) during the first week, one remained positive during second week and none were positive during third week. Cocaine and EME were detectable within 2 h and BE was not detected until 4–8 h after low doses and slightly sooner after high doses. The majority of the drugs was excreted within 24 h. Over 70% of weekly patches worn during low doses were positive for cocaine (≥25 ng/patch), increasing to 100% during high doses.

A method by LC-MS/MS for the detection and quantification of amphetamines in sweat was developed by Samyn *et al.* [60]. LOQs were estimated by using moistened cotton fleeces wiped on the forehead of non-drug users and spiked with decreasing concentrations of amphetamines. For MDMA the LOQ was 5 ng/wipe. This method was applied to sweat wipes of 12 healthy volunteers who had received a single 75 mg dose of MDMA. Even 4–5 h after ingestion, the MDMA levels in sweat only averaged 25 ng/wipe.

Keller *et al.* [89] present the use of ion mobility spectrometry (IMS) technology in cases of forensic interest, e.g. the analysis of post-mortem sweat samples for drugs. Specimens of sweat were obtained from deceased cocaine addicts by means of a roll-shaped oral fluid collector that was slightly moistened with ethanol. The sweat pad was pressed on the Teflon membrane filter, the solvent was evaporated and the filter directly inserted into the IMS system. The authors compared the results obtained by IMS with those obtained by GC-MS. For IMS analysis, the threshold value for cocaine was set between 2 and 50 digital units, and the results were obtained in 8 s. The authors could conclude from the results that IMS is a simple and reliable tool for the detection of cocaine in sweat of deceased drug addicts.

19.2.6.2 Methods for various groups of drugs of abuse

Samyn *et al.* [68] evaluated the positive predictive value of sweat samples. Therefore, 180 sweat samples of drivers suspected of DUID were collected by wiping the forehead with a fleece moistened with isopropanol and quantitatively analysed by GC-MS. The probability that a positive sweat wipe result matched a positive blood result was over 90% for cocaine and amphetamines and 80% for cannabis.

In 2000, Kintz *et al.* [90] published an epidemiological study in which sweat was collected as biological matrix. Sweat was collected from 197 drivers that were injured by a traffic accident. The collected sweat was subsequently analysed by GC-MS. Sixteen drivers were positive for THC, one for ecstasy, one for cocaine, one for heroin and three for codeine. The results showed that the parent drug is predominantly present in sweat. The authors thus concluded that, when urine is absent, sweat is an excellent alternative to confirm the use of heroin, as 6-MAM is found in this matrix.

19.2.7 Detection of drugs in hair

Since 1979, hair has been used to document chronic drug exposure, with applications in forensic toxicology, clinical toxicology, occupational medicine and doping control. The major practical advantage of hair testing compared with urine and blood testing for drugs is its larger detection window, which is weeks to months, depending on the length of hair shaft analysed. Other advantages of hair are its stability and ease-of-transportation. A disadvantage of the use of hair as biological matrix is the possible contamination by exposure to drugs in the air and the absence of correlation with recent use because of the delayed appearance of drugs in hair. Recently, Kintz [91] edited a review on analytical and practical aspects of drug testing in hair. Hair can be applied as a biological sample in the context of DUID for driving license regranting [53,92].

19.2.7.1 Methods for one group of drugs of abuse

To investigate the *in vitro* contamination of hair by marijuana smoke, Thorspecken *et al.* [93] determined the cannabinoid concentration in hair by GC-MS. The calibration curves covered a concentration range of 0.5–50 µg/g hair. The results indicated that cannabinoids were deposited in detectable concentrations on the hair fibres from marijuana smoke. Cannabinoid concentrations were dependent on air concentration and hair pre-treatment. Uptake was less in untreated than in pre-treated hair. Concentrations were increased in damp hair, but were even higher in greased hair. External contaminants were completely removed by washing with methanol and dichloromethane in untreated hair only. Washing with dodecyl sulphate in water was insufficient in all cases. The authors concluded that environmental marijuana smoke exposure may produce false positive or falsely increased test results in hair.

Uhl and Sachs [94] examined 66 hair segments of forensic cases by ELISA^{TM}, GC-MS and GC-MS/MS. The authors concluded that the best strategy to prove cannabis exposure/consumption is to screen by ELISA^{TM}, to prove exposure by GC/MS and to prove consumption by determination of THC-COOH by GC-MS/MS.

A procedure was developed in 2004 by Cairns *et al.* [95] for the *removal* and identification of drug *contamination* in the analysis of human hair. The wash procedure consisted of the addition of 2 ml dry isopropanol to 12 mg hair, subsequent shaking for 15 min, removal of isopropanol, addition of 0.01 M phosphate buffer/ 0.01% bovine serum albumin (BSA) pH 6, shaking for 30 min and removal of buffer. This 30 min wash is repeated twice more, followed by two 60 min washes using the same conditions. After the final wash, the hair sample is enzymatically digested prior to radioimmunoassay (RIA) analysis or confirmation by GC-MS of LC-MS/MS. The last phosphate buffer-BSA wash is saved and analysed. The amount of drug per milligram hair in the last wash is multiplied by 5 (or 3.5 for methamphetamine) and this result is subtracted from the amount of drug per milligram hair in the hair digest. The result of subtracting the indicated multiple of the last wash drug value from the digest value is termed the Wash Criterion, and is an estimate of the amount of drug that would remain in the hair if further washing were to be applied (five additional 1 h washes in the cases of cocaine, morphine and PCP, and 3.5 additional hours of washing in the case of methamphetamine). If the result after the subtraction is less than the cut-off for the parent drug, the result is considered negative in indicating drug use. The parent-drug cut-off values for the drugs cocaine, opiates, PCP and methamphetamine were 500, 200, 300 and 500 pg/mg hair, respectively.

The same research group determined amphetamines in washed hair samples of demonstrated users and workplace subjects. A first aliquot of 7–9 mg of hair was screened by RIA, and confirmation was performed by LC-MS/MS on a second aliquot of 12 mg of hair. Before LC-MS/MS analysis, the washing procedure described in the study above was performed on the hair aliquots. The LOD and LOQ for the method were 25 pg/mg hair for amphetamine, 100 pg/mg hair for methamphetamine and 10 pg/mg hair for both MDMA and MDA [96].

A method for the determination of a single exposure to bromazepam and clonazepam in hair was developed by Chèze *et al.* [97]. Head hair was collected 1 month after the exposure and treated by incubation with Soerensen buffer pH 7.6

followed by LLE. Analysis was performed by LC-MS/MS. After single exposure, bromazepam was present in powdered hair at 28 pg/mg and 7-aminoclonazepam at 22 pg/mg in the first 1 cm segment, while no clonazepam was detectable. The method was successfully applied in two forensic cases.

A laboratory in France developed a method to detect zopiclone in hair by LC-MS/MS. The LOD was 0.3 pg/mg [98]. A year later the same research group developed an LC-MS/MS method for the detection of alprazolam in hair with an LOD of 0.4 pg/mg and an LOQ of 2.0 pg/mg [99], and for the detection of 23 benzodiazepines and benzodiazepine-like hypnotics in hair from victims of drug-facilitated crimes [100]. Clonazepam-d_4 was added as internal standard to 20 mg hair. LODs are lower than 2 pg/mg. Villain *et al.* [101] developed in the same year a screening method for 16 benzodiazepines and hypnotics in human hair by LC-MS/MS. The method involved decontamination of hair with methylene chloride, hair cut into small pieces, incubation of 20 mg in phosphate buffer in the presence of 1 ng diazepam-d_5 used as internal standard, LLE and separation by LC-MS/MS. The LOQ for all benzodiazepines and hypnotics ranged from 0.5 to 5 pg/mg using a 20 mg hair sample.

A similar method for the screening of benzodiazepines using LLE and LC-MS/MS was published in 2006 by Irving and Dickson [102]. Hair samples were obtained from patients who had recently discontinued using or were currently prescribed one or more of the targeted drugs. Prazepam was used as internal standard. Some components in the hair matrix gave the same transitions as some of the analytes but this did not compromise the analyses because their retention times differed from those for the target compounds. The analytical runtime was 8–10 min.

Miki *et al.* [103] reviewed in 2003 three types of screening methods and a column-switching LC-MS method for the determination of methamphetamine and designer drugs and their metabolites in hair samples that were developed by his team. The screening methods included IMS-based, EMIT®- and immunochromatography-based methods. The calibration curves of the LC-MS method for methamphetamine and its possible metabolites amphetamine and *p*-hydroxymethamphetamine were linear ($r^2 > 0.9995$) over the range of 0.02–30 ng/mg. The lower LODs of these analytes were 0.02 ng/mg and 0.1–0.2 ng/mg in the SIM and full-scan modes, respectively, using a 100 μl hair extract sample, which corresponds to a 2.5 mg hair sample.

A review published in 2005 by Nakashima [53] focuses on HPLC methods used for the practical analysis of drugs of abuse, mainly for amphetamine derivatives and MDMA in biological samples such as urine, blood and hair. For hair eight studies were reported, using methods such as FL, HPLC-FL, chemiluminescence detection and LC-APCI-MS. For HPLC-FL one study reported LODs ranging from 11 to 200 pg/mg hair, and for LC-APCI-MS another study reported LODs from 0.05 to 0.2 pg/mg hair.

Martins *et al.* [104,105] described a method for the enantioselective quantification of amphetamine-type stimulants in hair. The data showed in most cases higher concentrations of (*S*)-methamphetamine and (*S*)-amphetamine than those of the corresponding (*R*)-enantiomers.

19.2.7.2 Methods for various groups of drugs of abuse

In Italy, hair analysis is used for license (re-)granting to drug addicts. The hair samples are first screened for opiates, cocaine and ecstasy by RIA. Positives are then confirmed by HPLC, CE, GC-MS or MS-MS [106]. Montagna *et al.* [107] developed a sensitive GC-MS method for the simultaneous determination of opiates, cocaine and metabolites in hair at a cut-off level of 0.1 ng/mg to assess past exposure to these drugs in applicants for driving licenses. The sampling consisted of collection of one hair (5 cm length) and one urine sample analysed for opiates and cocaine. No cases with positive urine and negative hair were observed in the 214 examined cases within the period September 1995–February 1999. In the Italian province of Brescia, a programme including analysis of opiates and cocaine in two hair segments (0–3 and 3–6 cm) and in urine was adopted in order to reissue the driving license to former drug addicts or occasional users [108]. The hair samples were analysed by GC-MS in the SIM mode, according to Montagna *et al.* [107]. Testing for cannabinoids was not part of these programs due to its slow clearance [31].

Musshoff *et al.* [109] compared results of hair analyses for drugs of abuse with self-reports and urine tests. Analysis of urine was performed by means of immuno-assays. Hair samples were subsequently washed in 5 ml of deionised water, di-chloromethane and petroleum ether, respectively. After drying, the hair samples were cut into small pieces and analysed by GC-MS. The washing solutions were also analysed by GC-MS to exclude contamination. Except for opiates the comparison between self-reported drug use and urinalysis showed a low correlation. Hair tests revealed consumption in even more cases than urinalysis. Comparing self-reports and the results of hair analyses, drug use was dramatically under-reported, especially for cocaine. Cocaine hair tests appeared to be highly sensitive and specific in iden-tifying past cocaine use even when urine was negative. On the other hand, hair lacked sensitivity to detect cannabinoids.

A method for the determination of opiates and cocaine in hair using automated EIA screening methods (ELISA^{TM}) followed by GC-MS confirmation was devel-oped in 2006 by Lachenmeier *et al.* [110]. No interfering peaks from the hair matrix were observed. For heroin, acetylcodeine and morphine there was a linear behaviour throughout the whole range (0.5–25 ng/mg). For MAM, codeine, cocaine, BE and cocaethylene two linear calibration curves, one for a high and one for a low con-centration range, were established. The LODs and LOQs were lower than 0.06 and 0.27 ng/mg, respectively, for all analytes. For ELISA^{TM}, the intra-assay precision was 11% for cocaine and 3% for the opiates, while inter-assay precision was 12% for cocaine and 9% for the opiates.

Kronstrand *et al.* [111] published in 2004 an LC-MS/MS method for the simul-taneous analysis of several drugs of abuse in human hair. The method included nicotine, cotinine, morphine, codeine, 6-acetylmorphine, ethylmorphine, amphet-amine, methamphetamine, MDA, MDMA, BE, cocaine, 7-aminoflunitrazepam and diazepam with an LOD–LOQ of 24–80 pg/mg, 25–83 pg/mg, 9–30 pg/mg, 15–50 pg/mg, 25–83 pg/mg, 4–13 pg/mg, 33–110 pg/mg, 6–20 pg/mg, 12–40 pg/mg, 4–13 pg/mg, 3–10 pg/mg, 5–16 pg/mg, 70–232 pg/mg and 17–58 pg/mg, respectively.

19.2.8 Quality control

Quality control is very important, especially when the results of analyses are used as legal evidence of drug use. Therefore, schemes to assess external quality can be established. This was first done to monitor quantitative analysis of drugs in urine. Schemes were subsequently introduced for a range of other matrices of interest to toxicologists, but few are really dedicated to the field of DUID. A recent publication by Penders and Verstraete [112] gives an overview of the existing standards and guidelines for analytical toxicology. In France, the SFTA organises an annual external quality control for detecting and quantifying seven molecules from four drug families including their metabolites in blood samples [113]. The German GTFCh organises every year three controls on three samples. Commercial controls are available from Medichem (Steinenbronn, Germany) [28]. In Switzerland, the laboratories that are allowed to perform toxicological analyses are obliged to participate in two controls every year, which are organised by an independent and accredited organism: the Swiss Centre of Quality Control in Genève [114]. In Australia, the National Association of Testing Authorities (NATA) is the national laboratory accreditation authority. The laboratories from a study that was published by Drummer *et al.* [115] took part in proficiency trials and have been accredited by NATA in Forensic Science. In Finland FINAS accredits the competence of laboratories. Competence is established on the basis of international standards and consistent procedures [116].

The need for external quality control for drugs-of-abuse urinalysis has long been recognised. In 1987, a proficiency-testing programme was introduced in Spain. The first year's results verified a beneficial influence of the programme in regard to reducing errors when adequate techniques are used, the need for using confirmatory techniques and the importance of experience and laboratory resources for optimising results [117]. In 1995, an Italian proficiency-testing programme for the analysis of psychoactive substances in urine was created on an educational basis. Batches of six urine samples, validated by reference laboratories, are sent every 3 months to participating laboratories. In 1999, a study by Ferrara *et al.* [118] revealed that the mean percentage of correct results was 96.8%, with a yearly improvement of 0.4%.

Clarke and Wilson [119] report the results of proficiency testing of drug detection in oral fluid. Eighteen external quality assessment (proficiency testing) samples were prepared from client samples collected with the Intercept® oral fluid collection device and by spiking drug-free oral fluid. Samples were circulated in pairs at quarterly intervals to 13 USA and UK based laboratories for analysis by a panel of OraSure micro-plate Intercept® EIA kits and hyphenated MS techniques. The sensitivity was 50% for the amphetamine specific assay, 93% for methyl-amphetamines, 64% for barbiturates, 73% for cannabinoids, 100% for cocaine and metabolites, 69% for benzodiazepines, 95% for methadone, 79% for opiates and 93% for PCP. A small number of the sensitivity errors were attributable to chromatographic errors.

One of the aims of the Society of Hair Testing is the development of proficiency tests, so that all laboratories that perform hair analysis can produce comparable results, or at least detect the same compounds. Since the society started in 1995, several proficiency tests have been organised. In 2001, 18 laboratories participated in a proficiency test organised by the Society of Hair Testing. Samples included one drug-free hair sample and two samples from drug users previously checked for homogeneity by three reference laboratories. The compounds present in the samples included opiates, cocaine and BE, cannabinoids and amphetamines. All the laboratories used GC-MS, with the exception of two which used GC-MS/MS and LC-MS/MS. Six laboratories performed initial screening tests by RIA, ELISA^{TM} or EMIT$^{®}$. The laboratories performed well qualitatively with the exception of eight false positives. Unfortunately the scatter of quantitative results was high [120]. In 2004, the Society of Hair Testing published recommendations for hair testing in forensic cases, involving sampling, shipping, storage, decontamination, hair disintegration and extraction, screening, mass spectrometric analysis, and internal and external quality control. They also describe criteria such as recommended LOQs for the analysis of specific drug classes such as opiates, cocaine, amphetamines and cannabinoids [121].

The Istituto Superiore di Sanità of Rome, Italy, in cooperation with Institut Municipal d'Investigaciò Mèdica of Barcelona, Spain, set up an external quality control programme (HAIRVEQ) to evaluate reliability in hair testing for drug abuse by laboratories from the Italian National Health Service. Samples included in the programme were real hair samples from drug consumers. Prior to sending, the samples were reduced to powdered form, mixed to ensure homogeneity and tested by GC-MS by four reference laboratories. Up till then, four different exercises had been concluded and 23 laboratories participated. Samples containing high and low concentrations of opiates, cocaine and metabolites, low concentrations of MDMA and two blank samples were included in the intercomparison exercises performed in the first year of HAIRVEQ activities. The results showed an insufficient performance of the participating laboratories, with about 82% of laboratories reporting incorrect qualitative results and only between 35 and 55% of the reported quantitative results being considered as satisfying. The authors concluded that the Italian authorities need to provide guidelines for method validation and recommended cut-off concentrations to orientate laboratories in their quality objectives when developing analytical methodologies [122].

19.3 INTERPRETATION

19.3.1 Analytical limitations

It is important to be aware of and to take into consideration the analytical limitations when developing or applying a method. The limitations to GC-MS(/MS) are that this technique is very labour-intense and time-consuming, particularly as sample preparation, i.e. SPE and derivatisation are needed [37]. With LC-MS(/MS) on the other hand, the choice of chromatographic systems and thus the separation power is often limited because only certain volatile buffers and mobile-phase additives can be used [123].

Maurer [64] mentions some limitations to LC-MS screening procedures, namely the limited spectral information of ESI and/or APCI when compared to EI mass spectra, the varying collision-induced dissociation (CID) fragmentation between different instruments and the possibility of ion suppression in ESI. Criteria for compound identification by single or multiple stage LC-MS were reviewed by Rivier [124]. Maurer [64] agrees with the final conclusion of Rivier [124] that the responsibility lies with the toxicologist to decide, depending on the case, how and when the minimum requirement for identity confirmation has been reached, and not to rely exclusively on match quality parameters.

Problems in the interpretations of analytical results due to overlapping peaks of atropine and cocaine and similar MS spectra in LC-MS/MS have been discussed by Uges *et al.* [125].

19.3.2 Type of biological matrix

The type of biological matrix that is sampled and analysed can have an influence on the concentration of the analyte of interest. For example for some drugs the concentration in plasma is different from the concentration in whole blood (Table 19.5). Skopp *et al.* [126] discovered that the blood-to-plasma ratio of morphine glucuronides were strongly influenced by variations in haematocrit and water content, indicating that conclusions drawn from pharmacokinetic studies and transferred to parent drug to metabolite ratios resulting from forensic blood samples may be biased by the particular biological matrix under investigation. The detection times of drugs of abuse depend on the type, pH and concentration of the matrix that is analysed, but also on the dose, the sensitivity of the method, the preparation, the route of administration, the duration of use (acute or chronic), the molecule or metabolite that is looked for and the interindividual variation in metabolic and renal clearance. In general, the detection time is longest in hair, followed by urine, sweat, oral fluid and blood [127].

TABLE 19.5
BLOOD TO PLASMA RATIOS

Drug	Ratio
Methamphetamine [181]	0.7 (1 overdose case)
THC [182]	0.6
Cocaine [183]	1.0
Morphine [126]	1.0
Codeine [184]	0.9
Diazepam [185]	0.6
Oxazepam [186]	1.0
Flunitrazepam [185][a]	0.8

[a]In Clarke's analysis of drugs and poisons [187] the reciprocal value is given by mistake [185].

References pp. 646–651

Ion suppression effects can be caused by endogenous compounds, metabolites, formulation agents, etc. [141]. Mortier *et al.* [80] even discovered interference from a polymer originating from the sampling device used to collect oral fluid. The authors recommend further oral fluid samples being taken by the spitting method, or if an oral fluid collection device is considered mandatory, to leave its choice depending not only on ease-of-use but have analytical considerations participate in the selection process. Wood *et al.* [82] also observed that Intercept®, an FDA approved sampling device that is used on a large scale in the USA for workplace drug testing, contains some ingredients (stabilisers and preservatives) that can cause substantial interferences, e.g. ion suppression or enhancement during LC-MS/MS analysis, in the absence of suitable sample pre-treatment. The use of SPE was demonstrated to be highly effective in decreasing the interferences.

Neerman [143] described the possible causation of interference and/or cross-reactivity by ingested foods/drugs, thereby contributing to false positive/negative results. An example of interferences caused by ingested drugs in LC-MS/MS analysis is the possibility of false positive results for the opioid drug tramadol in cases where subjects were being treated with the antidepressant venlafaxine [144]. Problems relating to the confirmatory analysis of drugs of abuse in biological matrices using LC-MS/MS were already highlighted by Maralikova and Weinmann [38]. They suggested that at least four identification points are used including retention time, two MRM transitions and the co-elution of a stable isotopic analogue. Allen [144] only relied on the retention time and a single MRM transition that resulted in the misidentification of tramadol. This work demonstrates thus the problem with drug specificity in drug confirmation using LC-MS/MS, particularly when dealing with complex biological matrices. Increasing the time over which chromatographic separation occurs offers one possible solution, but this places an extra demand on the use of very expensive LC-MS/MS equipment. As tramadol undergoes similar metabolism to venlafaxine, resulting in both *N*- and *O*-desmethyl metabolites, the author concluded that identification of both the parent drug and its metabolite enables the confirmation of tramadol in urine samples using shorter chromatography conditions.

Little *et al.* [145] examined the matrix ionisation effects that are caused in LC-MS/MS analysis by glycerophosphocholines, the major phospholipids in plasma. They developed a method, in-source MRM, for detecting glycerophosphocholines during LC-MS/MS method development. The approach uses in-source CID to yield trimethylammonium-ethyl phosphate ions, which are formed from mono- and disubstituted glycerophosphocholines. The resulting ion is selected by the first quadrupole, passed through the collision cell in the presence of collision gas at low energy to minimise fragmentation, and selected by the third quadrupole. The approach can be combined with standard MRM transitions with little compromise in sensitivity during method development and sample analysis.

19.3.4 Storage of samples

Preanalytical stability of a drug and its major metabolites in biological matrices is an important consideration whenever the analyte pattern is used to estimate drug use.

The conditions and the length of storage of biological samples can influence the concentration of drugs it contains, and the influence can differ between different kinds of drugs. In 1995, Giorgi and Meeker [146] published a study on the stability of some drugs in stored *blood* samples over a 5-year period at room temperature. They found that cocaine and BE had poor stability. Methamphetamine and PCP were fairly stable, whereas unconjugated morphine showed wide variation throughout the study. Initially the morphine concentration decreased, then increased at the 3-year interval, and finally decreased at the 4- and 5-year intervals. Skopp *et al.* [147] found that the concentration of all benzodiazepines and some of their metabolites in blood and plasma had decreased to at least 60% of the original levels after 240 days stored at 4°C. A clear pattern of breakdown could not be established. A general problem for the interpretation of analytical results in *urine* is the fact that cannabinoids are relatively unstable in this matrix. Fraga *et al.* [148] investigated the stability of cannabinoids in urine samples using several storage temperatures. Losses higher than 22% were observed in some urine samples after storage at room temperature for 10 days. Lower losses (8%) were observed when the samples were refrigerated for 4 weeks. Important mean losses of 8, 16 and 20% were found when the urine samples were frozen during 40 days, 1 year and 3 years, respectively. Losses higher than 5% could be observed after 1 day, which the authors attribute to the decrease of the solubility of THC-COOH or adsorption processes of cannabinoid molecules to the plastic storage containers. Skopp and Pötsch [149] found that during storage THC-COOH is liberated from its glucuronide in *urine and plasma* in a time- and temperature-dependent manner. The glucuronide was stable at −20°C in both matrices, whereas its concentrations decreased at all other storage conditions (4°C, 20°C and 40°C up to 10 days). For a given storage time and temperature, the decrease in plasma was higher than that in urine. Degradation of the glucuronide followed an apparent first-order process and led to the formation of THC-COOH. In a subsequent study, Skopp and Pötsch [150] found that there is a loss in mean total THC-COOH concentration in *urine* samples after 15 days of storage at 20°C. THC-COOH was stable at pH 5, while its concentration slightly decreased with time at higher pH values. However, liberation of THC-COOH from its glucuronide could not be completely prevented by adjusting pH. In general, frozen storage was demanded to be able to consider free THC-COOH as a reliable marker to identify recent cannabis consumption. Investigation of the short-term stability of LSD and its metabolites 2-oxo-3-OH-LSD and nor-LSD in urine under different storage conditions (−20°C, 4°C, 22°C for 7 days protected from light; 40°C protected from light and 22°C exposed to natural light behind window glass for 3 days) showed a clear temperature-dependent degradation of these compounds. In samples exposed to natural sunlight the concentrations decreased very rapidly. Both metabolites were more stable than the parent drug, and 2-oxo-3-OH-LSD was far more stable than nor-LSD. The authors strongly recommend that 2-oxo-3-OH-LSD and nor-LSD would be included in future confirmation protocols, and that urine samples that need to be analysed for LSD need to be protected from light and transported rapidly [151].

Dickson *et al.* [152] recently studied the effects of storage systems on drug stability and recovery in oral fluid using three different collection devices (Cozart, Immunalysis and Microgenics). Substantial drug losses during collection and storage of samples were shown to occur in drug-spiked oral fluid. Of the three systems tested, the Cozart collection system was the only one acceptable for THC. Storage at either 5°C or room temperature had no significant effect on drug recoveries. The authors strongly recommend that researchers determine recoveries before further drug studies in oral fluid are conducted.

19.3.5 Correlation between concentration in biological fluid and effect on performance

For amphetamines and opiates, there exists a strong relationship between the blood concentrations and the effects. For cannabis and cocaine on the other hand, there is a shift in time between the evolution of the concentration in blood and the physical and psychological effects the user experiences. As a consequence it is difficult to establish a relationship between the blood concentrations and the effects [153]. Huestis *et al.* [154] developed two mathematical models to predict the time of cannabis use from the analysis of a single plasma sample for cannabinoids. The first model was based on THC concentrations and the second model on the ratio of THC-COOH to THC in plasma. The two models were validated with cannabinoid data of nine published and unpublished clinical studies. The accuracy of model prediction was evaluated by comparison of the predicted time of prior drug use to the actual time of exposure. Both models correctly predicted the time of exposure within the 95% confidence interval for about 90% of the samples. These findings were confirmed by a controlled cannabis smoking study from Manno *et al.* [155].

Daldrup [156] introduced in 1996 the Cannabis Influence Factor (CIF) to interpret the concentrations of THC and its metabolites in blood. It correlates better with the acute effect of cannabis than the blood THC concentration alone. Drivers with a CIF greater than 10 are considered to be unfit to drive. The CIF is calculated as follows: $CIF = (THC \ [ng/ml]/314.5 + 11\text{-}OH\text{-}THC \ [ng/ml]/330.5) \times 100/(THC\text{-}COOH \ [free, ng/ml]/344.5)$ [156,157].

Bramness *et al.* [158] studied in Norway the possible relation between benzodiazepine concentrations and impairment in apprehended drivers. The legislation in Norway concerning DUID is an "impairment" legislation. Physicians perform a clinical test for drunkenness and take blood samples from drivers suspected of DUID. In this study, 818 samples from drivers suspected of DUID, containing only one benzodiazepine and a reference group consisting of 10,759 samples containing only alcohol were included. Drivers that were considered impaired by the physician had significantly higher blood levels of diazepam, oxazepam and flunitrazepam than those not impaired and the risk of being assessed impaired did rise with increasing benzodiazepine blood level, with odds ratios for being assessed as impaired of 1.61, 3.65 and 4.11 for the three supratherapeutic drug levels (mildly elevated, moderately elevated and highly elevated). In a subsequent study, the same authors published results on the relationship between benzodiazepine concentration and the results of simple clinical tests for impairment in the same sample of drugged drivers. They found that 9 subtests and observations were significantly related to blood benzodiazepine concentrations, even after adjustment for a variety of background variables, namely the Romberg's test, one observation concerning alertness, four tests on motor and coordination, two observations on speech and one observation regarding appearance [159]. Gustavsen *et al.* [160] published in 2006 a study on the relation between impairment and blood amphetamine and/or methamphetamine concentrations in 878 Norwegian drivers suspected of DUID with amphetamine or methamphetamine as the only drug present in the blood samples. They found a modest, but significant relationship. A ceiling effect as observed above blood concentrations of 0.27–0.53 mg/l. Khiabani *et al.* [161] published in 2006 a similar study for cannabis. In this study 589 samples from drivers suspected of DUID containing THC as the only drug were included. Drivers who were considered being impaired by the physician had higher blood THC concentrations than the drivers who were judged as not impaired. Furthermore, drivers with blood THC concentrations above 3 ng/ml had an increased risk for being judged impaired compared to drivers with lower concentration ranges. These data suggest a relationship between blood concentrations of these drugs and the effects on performance.

For THC, Grotenhermen *et al.* [162] recommended a cut-off of 5–10 ng/ml serum to effectively separate unimpaired drivers from those driving under the influence of cannabis. However, several studies generated results that plead for a lower cut-off. Ramaekers *et al.* [163] concluded that 2 and 5 ng/ml serum should establish the lower and upper range of a legal THC limit. Laumon *et al.* [13] found a statistically significant increased risk of being responsible for a fatal accident from THC concentrations of 1 ng/ml blood or higher. Khiabani *et al.* [161] discovered that drivers with blood THC concentrations above 3 ng/ml blood have an increased risk for being judged impaired by a physician than drivers with lower concentration ranges. Recently, Mura *et al.* [164] even found that active constituents of cannabis may be present in the brain regions when they are no longer detectable in blood. Moreover, the brain regions assayed were those where cannabis CB1 receptors are present in high densities, consistent with postulated consequences of cannabis use on behavioural and cognitive functions.

19.4 CONCLUSION

DUID is an increasingly important issue in the field of traffic safety, it is thus essential to be able to detect drugs of abuse in samples of drivers. The analytical techniques used depend on the aim of the determination, but also on several other factors such as the biological matrix investigated, the type of legislation, the cut-offs applied, the kind of drugs of abuse. More and more countries add "*per se*" legislation to the existing "impairment" legislation. In countries with "impairment" legislation or for epidemiological studies, the analytical techniques will involve narrow or broad screening, while the analysis in countries with "*per* se" legislation will focus on specific compounds. The recent methods developed mostly concern

LC-MS(/MS) techniques, while the methods described in the law are mostly based on GC-MS(/MS). LC coupled to MS/MS allows the achievement of high-throughput analysis because it combines high specificity and increased signal-to-noise with short chromatographic run times and a potential to reduce sample preparation because there is no need for derivatisation [82]. It is a very powerful analytical tool for confirmatory analysis especially of polar, thermally labile and not volatile compounds and their metabolites [38]. The sample preparation that often remains the rate-limiting step can be simplified and speeded up by means of extraction supports such as restricted access media, large size particle and monolithic phases that allow the direct injection of complex matrices in column-switching configurations [165]. Guidelines for confirmatory analysis by LC-MS/MS have been published recently by several organisations, but these have not yet been included in procedures for drug analysis [38]. When interpreting the analytical results, several factors need to be taken into consideration, such as the analytical limitations, the characteristics of the type of biological matrix sampled, the possibility of matrix effects, the influence of the length and conditions of storage and the relationship between the concentration and the effects on performance.

In countries with "impairment" legislation, the demonstration of the presence of a drug of abuse is sufficient. However, in countries with a "*per se*" legislation there are established cut-offs. There is no consensus between the different countries on these cut-offs. The difficult correlation between the effects and the concentration in blood makes it difficult if not impossible to establish cut-offs for some drugs. Moreover these cut-offs would give drug users the impression that the usage of drugs is legal, which it is not! Therefore all governments allowing "*per se*" legislation opted for analytical cut-offs [28]. There is however still much research needed to be able to determine cut-offs based on the risk of being involved in or responsible for an accident, as is now the case for alcohol.

19.5 ABBREVIATIONS

6-MAM:	6-Monoacetylmorphine
11-OH-THC:	11-hydroxy-δ-9-THC
BE:	Benzoylecgonine
BSA:	Bovine serum albumin
CE:	Capillary electrophoresis
CEDIA^{TM}:	Cloned enzyme donor immunoassay
CID:	Collision-induced dissociation
CIF:	Cannabis Influence Factor
CV:	Coefficient of variation
ECD:	Electron capture detection
EI:	Electron ionisation
EIA:	Enzyme immunoassay
ELISA^{TM}:	Enzyme-linked immunosorbent assay
EME:	Ecgonine methyl ester
EMIT$^{\circledR}$:	Enzyme multiplied immuno technique

EPI:	Enhanced product ion scan
EU:	European Union
FINAS:	Finnish Centre for Metrology and Accreditation
FL:	Fluorescence detection
FPIA:	Fluorescence polarisation immunoassay
GC:	Gas chromatography
GHB:	γ-Hydroxybutyric acid
GTFCh:	German Society of Toxicological and Forensic Chemistry
GUS:	General unknown screening
HPLC:	High pressure liquid chromatography
IDA:	Information-dependent acquisition
IMS:	Ion mobility spectrometry
IOC:	International Olympic Committee
KIMS:	Kinetic interaction of microparticle in solution assay
LC:	Liquid chromatography
LLE:	Liquid–liquid extraction
LOD:	Limit of detection
LOQ:	Limit of quantitation
LSD:	Lysergic acid diethylamide
MALDI:	Matrix-Assisted Laser Desorption/Ionisation
MDA:	3,4-methylenedioxyamphetamine
MDEA:	N-ethyl-3,4-methylenedioxy-N-ethylamphetamine
MDMA:	3,4-methylenedioxy methamphetamine
MRM:	Multiple reaction monitoring
MS:	Mass spectrometry
MS/MS:	Tandem mass spectrometry
NATA:	National Association of Testing Authorities
NICI:	Negative ion chemical ionisation
NPD:	Nitrogen-phosphorus detection
NPFID:	Nitrogen-phosphorus flame-ionisation detection
PCP:	Phencyclidine
POC:	Point of collection
REMEDi:	Rapid emerging drug identification
RIA:	Radioimmunoassay
SIM:	Selected ion-monitoring
SOFT:	Society of Forensic Toxicologists
SPE:	Solid phase extraction
SRM:	Selected reaction monitoring
TCA:	Tricyclic antidepressants
THC:	Delta-9-tetrahydrocannabinol
THC-COOH:	11-nor-9-carboxy-δ-9-THC
TLC:	Thin layer chromatography
TOF:	Time-of-flight
UV:	Ultraviolet detection
WADA:	World anti-doping agency

References pp. 646–651

19.6 REFERENCES

1   World Health Organization and World Bank, 2004, Ch. 5.
2   I. Behrensdorff and A. Steentoft, Accid. Anal. Prev., 35 (2003) 851–860.
3   T. Assum, M. P. M. Mathijssen, S. Houwing, S. C. Buttress, B. Sexton, R. J. Tunbridge and J. Oliver, IMMORTAL Deliverable D-R4.2: The prevalence of drug driving and relative risk estimations. A study conducted in the Netherlands, Norway and United Kingdom, 2005.
4   C. Dussault, M. Brault, J. Bouchard and A.M. Lemire, The contribution of alcohol and other drugs among fatally injured drivers in Québec: some preliminary results. La Société de l'Assurance Automobile du Québec. The 16th International Conference on Alcohol, Drugs and Traffic Safety, Montréal, Canada, 2002.
5   J.M. Walsh, R. Flegel, R. Atkins, L.A. Cangianelli, C. Cooper, C. Welsh and T.J. Kerns, Accid. Anal. Prev., 37 (2005) 894–901.
6   I.M. Bernhoft, A. Steentoft, S.S. Johansen, N.A. Klitgaard, L.B. Larsen and L.B. Hansen, Forensic Sci. Int., 150 (2005) 181–189.
7   A. Sukhai, A preliminary investigation into the nexus between substance abuse and non-fatal road traffic injury outcome in South-Africa, in: J. Oliver, P. Williams and A. Clayton (Eds.), Proceedings of the 17th International Conference on Alcohol, Drugs and Traffic Safety, Glasgow, Scotland, UK, 2004.
8   B.K. Logan and E.W. Schwilke, Changing patterns of alcohol and drug use in fatally injured drivers in Washington State 1992–2002, in: J. Oliver, P. Williams and A. Clayton (Eds.), Proceedings of the 17th International Conference on Alcohol, Drugs and Traffic Safety, Glasgow, Scotland, UK, 2004.
9   J.Y.K. Cheng, D.T.W. Chan and V.K.K. Mok, Forensic Sci. Int., 153 (2005) 196–201.
10   M.C. del Rio, J. Gomez, M. Sancho and F.J. Alvarez, Forensic Sci. Int., 127 (2002) 63–70.
11   M. Brault, C. Dussault, J. Bouchard, and A.M. Lemire, The contribution of alcohol and other drugs among fatally injured drivers in Québec: final results. in: J. Oliver, P. Williams and A. Clayton (Eds.), The 17th International Conference on Alcohol, Drugs and Traffic Safety, Glasgow, United Kingdom, 2004.
12   http://www.drugabusestatistics.samhsa.gov/nsduh/2k4Results/2k4Results.htm#2.15.
13   B. Laumon, B. Gadegbeku, J.L. Martin and M.B. Biecheler, BMJ, 331 (2005) 1371–1376.
14   C. Chowaniec, K. Rygol, M. Kobek and M. Albert, Forensic Sci. Int., 147 (2005) S53–S55.
15   R. Wennig, Bull. Soc. Sci. Med. Grand Duche Lux., 1 (2005) 23–53.
16   J.M. Walsh, J.J. de Gier, A.S. Christopherson and A.G. Verstraete, Traffic Inj. Prev., 5 (2004) 241–253.
17   J.C. Verster and E.R. Volkerts, Ann. Allergy Asthma Immunol., 92 (2004) 675.
18   M. Bilban and C.B. Jakopin, Coll. Antropol., 26 (2002) 107–117.
19   G.L. Krauss, A. Krumholz, R.C. Carter, G. Li and P. Kaplan, Neurology, 52 (1999) 1324–1329.
20   R.A. Barkley, K.R. Murphy, T. O'connell and D.F. Connor, J. Safety Res., 36 (2005) 121–131.
21   K.A. Brookhuis, D. de Waard and N. Samyn, Psychopharmacology, 173 (2004) 440–445.
22   A.W. Jones, Forensic Sci. Int., 153 (2005) 213–217.
23   J. Vanakoski, M.J. Mattila and T. Seppala, Eur. J. Clin. Pharmacol., 56 (2000) 453–458.
24   G. McGwin, R.V. Sims, L. Pulley and J.M. Roseman, Am. J. Epidemiol., 152 (2000) 424–431.
25   B.E. Smink, B. Ruiter, K.J. Lusthof, J.J. Gier, D.R. Uges and A.C. Egberts, Accid. Anal. Prev., 37 (2005) 427–433.
26   E. Raes and A.G. Verstraete, J. Anal. Toxicol., 29 (2005) 632–636.
27   L. Kadehjian, Forensic Sci. Int., 150 (2005) 151–160.
28   A. Verstraete and N. Samyn, Ann. Toxicol. Anal., 15 (2003) 83–91.
29   M.R. Moeller, Ann. Toxicol. Anal., 15 (2003) 145–150.
30   A.W. Jones, Traffic Inj. Prev., 6 (2005) 317–322.
31   F. Pragst and M.A. Balikova, Clin. Chim. Acta,, 370 (2006) 17–49.
32   S. George, Clin. Chem. Lab. Med., 42 (2004) 1288–1309.
33   N.T. Lu and B.G. Taylor, Forensic Sci. Int., 157 (2006) 106–116.

34   Y. Gaillard, G. Pepin, P. Marquet, P. Kintz, M. Deveaux and P. Mura, Toxicorama, 8 (1996) 17–22.
35   P. Kintz, V. Ciriminele, G. Pepin, P. Marquet, M. Deveaux and P. Mura, Toxicorama, 8 (1996) 33.
36   P. Marquet, G. Lachâtre, P. Kintz, G. Pepin, M. Deveaux and P. Mura, Toxicorama, 8 (1996) 23–28.
37   M. Wood, M. Morris, D.P. Cooper, G. De Boeck, N. Samyn and J. Claereboudt, Measurement of amphetamines in biological samples by LC/MS-MS. TIAFT Young Scientists Workshop 2001: "Oral fluid in toxicology," Ghent, Belgium, 2001.
38   B. Maralikova and W. Weinmann, J. Chromatogr. B: Anal. Technol. Biomed. Life Sci., 811 (2004) 21–30.
39   http://www.wada-ama.org.
40   http://www.olympic.org.
41   http://www.soft-tox.org/docs/Guidelines.2002.final.pdf.
42   R. Aderjan, T. Briellman, T. Daldrup, U. Demme, K. Harzer, M. Herbold, H. Käferstein, G. Kauert, L. von Meyer, M. Möller, F. Musshoff, G. Schmitt and W. Weinmann, Toxichem. Krimtech., 65 (1998) 18–24.
43   EU-Amtsblatt der Europaïschen Gemainschaften, 2002/657/EG. 17-8-2002.
44   A.A.M. Stolker, R.W. Stephany and L.A. van Ginkel, Analusis, 28 (2000) 947–951.
45   P. Marquet, Ther. Drug Monit., 24 (2002) 255–276.
46   R. Jansen, G. Lachatre and P. Marquet, Clin. Biochem., 38 (2005) 362–372.
47   H.H. Maurer, Ther. Drug Monit., 24 (2002) 247–254.
48   D. Butler and G.G. Guilbault, Anal. Lett., 37 (2004) 2003–2030.
49   F. Tagliaro and F. Bortolotti, Electrophoresis, 27 (2006) 231–243.
50   J.F. Van Bocxlaer, Ther. Drug Monit., 27 (2005) 752–755.
51   A. Verstraete and G. Brusini, ROSITA, Roadside testing assessment, pp. 1–397, 2001.
52   W. Weinmann, M. Goerner, S. Vogt, R. Goerke and S. Pollak, Forensic Sci. Int., 121 (2001) 103–107.
53   K. Nakashima, J. Health Sci., 51 (2005) 272–277.
54   Applied Biosystems, Application note: Quantitation of 27 benzodiazepines, zolpidem and zopiclone in serum and urine using liquid chromatography/tandem mass spectrometry (LC/MS/MS), 2005.
55   R. Dams, C.M. Murphy, W.E. Lambert and M.A. Huestis, Rapid Commun. Mass Spectrom., 17 (2003) 1665–1670.
56   H.K. Nordgren and O. Beck, Ther. Drug Monit., 26 (2004) 90–97.
57   C.A. Mueller, W. Weinmann, S. Dresen, A. Schreiber and M. Gergov, Rapid Commun. Mass Spectrom., 19 (2005) 1332–1338.
58   M.R. Moeller and T. Kraemer, Ther. Drug Monit., 24 (2002) 210–221.
59   M.H. Chu and O.H. Drummer, J. Anal. Toxicol., 26 (2002) 575–581.
60   N. Samyn, G. De Boeck, M. Wood, C.T.J. Lamers, D. de Waard, K.A. Brookhuis, A.G. Verstraete and W.J. Riedel, Forensic Sci. Int., 128 (2002) 90–97.
61   M. Laloup, G. Tilman, V. Maes, G. De Boeck, P. Wallemacq, J. Ramaekers and N. Samyn, Forensic Sci. Int., 153 (2005) 29–37.
62   M. Wood, G. De Boeck, N. Samyn, M. Morris, D.P. Cooper, R.A.A. Maes and E.A. de Bruijn, J. Anal. Toxicol., 27 (2003) 78–87.
63   T. Gunnar, S. Mykkanen, K. Ariniemi and P. Lillsunde, J. Chromatogr. B: Anal. Technol. Biomed. Life Sci., 806 (2004) 205–219.
64   H.H. Maurer, Clin. Biochem., 38 (2005) 310–318.
65   M. Concheiro, A. de Castro, O. Quintela, M. Lopez-Rivadulla and A. Cruz, J. Chromatogr. B: Anal. Technol. Biomed. Life Sci., 832 (2006) 81–89.
66   B.M.R. Appenzeller, S. Schneider, A. Maul and R. Wennig, Drug Alcohol Depend., 79 (2005) 261–265.
67   B.M.R. Appenzeller, S. Schneider, M. Yegles, A. Maul and R. Wennig, Forensic Sci. Int., 155 (2005) 83–90.

68 N. Samyn, G. De Boeck and A.G. Verstraete, J. Forensic Sci., 47 (2002) 1380–1387.
69 O. Dierich and M. Soyka, Fortschritte der Neurologie Psychiatrie, 73 (2005) 401–408.
70 P.Swann, Rosita 2 meeting, Baltimore, Maryland, USA, 5-12-2005 (personal communication).
71 P. Kintz, N. Samyn, A. Verstraete, . In: P. Mura (Ed.), Alcool, médicaments, stupéfiants et conduite automobile, Elsevier, Paris, 1999, pp. 143–169.
72 A.G. Verstraete, Forensic Sci. Int., 150 (2005) 143–150.
73 R.E. Choo and M.A. Huestis, Clin. Chem. Lab. Med., 42 (2004) 1273–1287.
74 G. Cooper, L. Wilson, C. Reid, C. Hand and V. Spiehler, Forensic Sci. Int., 159 (2006) 104–112.
75 M.E.R. Rosas, K.L. Preston, D.H. Epstein, E.T. Moolchan and I.W. Wainer, J. Chromatogr. B: Anal. Technol. Biomed. Life Sci., 796 (2003) 355–370.
76 H. Teixeira, P. Proenca, A. Verstraete, F. Corte-Real and D.N. Vieira, Forensic Sci. Int., 150 (2005) 205–211.
77 M. Laloup, M.D.R. Fernandez, M. Wood, G. De Boeck, U. Henquet, V. Maes and N. Samyn, J. Chromatogr. A, 1082 (2005) 15–24.
78 P. Kintz, M. Villain, M. Concheiro and V. Cirimele, Forensic Sci. Int., 150 (2005) 213–220.
79 M.A. Yegles and R. Wennig, GC/MS/NCI Screening for benzodiazepines in oral fluid, in: F. Pragst (Ed.), GTFCh Mosbach Symposium Proceedings, pp. 107–111, 14–16 April 2005.
80 K.A. Mortier, K.E. Maudens, W.E. Lambert, K.M. Clauwaert, J.F. Van Bocxlaer, D.L. Deforce, C. Van Peteghem and A.P. De Leenheer, J. Chromatogr. B: Anal. Technol. Biomed. Life Sci., 779 (2002) 321–330.
81 R. Dams, C.M. Murphy, R.E. Choo, W.E. Lambert, A.P. De Leenheer and M.A. Huestis, Anal. Chem., 75 (2003) 798–804.
82 M. Wood, M. Laloup, M.D.R. Fernandez, K.M. Jenkins, M.S. Young, J.G. Ramaekers, G. De Boeck and N. Samyn, Forensic Sci. Int., 150 (2005) 227–238.
83 Applied Biosystems, Application note: simultaneous screening of 23 drugs of abuse in oral fluid using an LC-MS/MS method, 2005.
84 F.M. Wylie, H. Torrance, R.A. Anderson and J.S. Oliver, Forensic Sci. Int., 150 (2005) 191–198.
85 P. Kintz, M. Villain, V. Cirimele, J.P. Goulle and B. Ludes, Acta Clin. Belg., Suppl. (2002) 24–30.
86 S. Pichini, M. Navarro, R. Pacifici, P. Zuccaro, J. Ortuno, M. Farre, P.N. Roset, J. Segura and R. de la Torre, J. Anal. Toxicol., 27 (2003) 294–303.
87 H.J. Liberty, B.D. Johnson, N. Fortner and D. Randolph, Addict. Biol., 8 (2003) 191–200.
88 S.L. Kacinko, A.J. Barnes, E.W. Schwilke, E.J. Cone, E.T. Moolchan and M.A. Huestis, Clin. Chem., 51 (2005) 2085–2094.
89 T. Keller, A. Keller, E. Tutsch-Bauer and F. Monticelli, Forensic Sci. Int., 161 (2006) 130–140.
90 P. Kintz, V. Cirimele, F. Mairot, M. Muhlmann and B. Ludes, Presse Med., 29 (2000) 1275–1278.
91 P. Kintz, Analytical and practical aspects of drug testing in hair. CRC Taylor and Francis, Boca Raton, London, New York, 2006.
92 P. Kintz, M. Villain, V. Cirimele, C. Janey and B. Ludes, Ann. Toxicol. Anal., 15 (2003) 117–122.
93 J. Thorspecken, G. Skopp and L. Pötsch, Clin. Chem., 50 (2004) 596–602.
94 M. Uhl and H. Sachs, Forensic Sci. Int., 145 (2004) 143–147.
95 T. Cairns, V. Hill, M. Schaffer and W. Thistle, Forensic Sci. Int., 145 (2004) 97–108.
96 T. Cairns, V. Hill, M. Schaffer and W. Thistle, Forensic Sci. Int., 145 (2004) 137–142.
97 M. Chèze, M. Villain and G. Pepin, Forensic Sci. Int., 145 (2004) 123–130.
98 M. Villain, M. Chèze, A. Tracqui, B. Ludes and P. Kintz, Forensic Sci. Int., 145 (2004) 117–121.
99 P. Kintz, M. Villain, M. Chèze and G. Pepin, Forensic Sci. Int., 153 (2005) 222–226.
100 M. Chèze, G. Duffort, M. Deveaux and G. Pepin, Forensic Sci. Int., 153 (2005) 3–10.
101 M. Villain, M. Concheiro, V. Cirimele and P. Kintz, J. Chromatogr. B: Anal. Technol. Biomed. Life Sci., 825 (2005) 72–78.
102 R.C. Irving and S.J. Dickson, The detection of sedatives in hair and nail samples using tandem LC-MS-MS. Forensic Sci. Int., (15-5-2006).
103 A. Miki, M. Katagi and H. Tsuchihashi, J. Health Sci., 49 (2003) 325–332.
104 L.F. Martins, M. Yegles, H. Chung and R. Wennig, J. Chromatogr. B: Anal. Technol. Biomed. Life Sci., 842 (2006) 98–105.

105   L. Martins, M. Yegles, H.S. Chung and R. Wennig, J. Chromatogr. B: Anal. Technol. Biomed. Life Sci., 825 (2005) 57–62.

106   F. Tagliaro, R. Valentini, G. Manetto, F. Crivellente, G. Carli and M. Marigo, Forensic Sci. Int., 107 (2000) 121–128.

107   M. Montagna, C. Stramesi, C. Vignali, A. Groppi and A. Polettini, Forensic Sci. Int., 107 (2000) 157–167.

108   M.C. Ricossa, M. Bernini and F. De Ferrari, Forensic Sci. Int., 107 (2000) 301–308.

109   F. Musshoff, F. Driever, K. Lachenmeier, D.W. Lachenmeier, A. Banger and B. Madea, Forensic Sci. Int., 156 (2006) 118–123.

110   K. Lachenmeier, F. Musshoff and B. Madea, Forensic Sci. Int., 159 (2006) 189–199.

111   R. Kronstrand, I. Nystrom, J. Strandberg and H. Druid, Forensic Sci. Int., 145 (2004) 183–190.

112   J. Penders and A. Verstraete, Accred. Qual. Assur., 11 (2006) 284–290.

113   G. Pépin, G. Duffort, N. Rommel, P. Kintz, V. Dumestre-Toulet, M.-F. Kergueris, G. Lachatre, M. Moulsma, J.P. Goullé, C. Lacroix, I. Ricordel, P. Mura, F. Vincent, A. Gruson, M. Lhermitte, B. Capolaghi, A. Turcant, M.-H. Ghysel and P. Corteel, Ann. Toxicol. Anal., 15 (2003) 108–116.

114   M. Augsburger, Ann. Toxicol. Anal., 15 (2003) 138–144.

115   O.H. Drummer, J. Gerostamoulos, H. Batziris, M. Chu, J.R. Caplehorn, M.D. Robertson and P. Swann, Forensic Sci. Int., 134 (2003) 154–162.

116   http://www.mikes.fi.

117   J. Segura, T.R. de la, M. Congost and J. Cami, Clin. Chem., 35 (1989) 879–883.

118   S.D. Ferrara, G. Brusini, S. Maietti, G. Frison, F. Castagna, S. Allevi, A.M. Menegus and L. Tedeschi, Int. J. Legal Med., 113 (1999) 50–54.

119   J. Clarke and J.F. Wilson, Forensic Sci. Int., 150 (2005) 161–164.

120   C. Jurado and H. Sachs, Forensic Sci. Int., 133 (2003) 175–178.

121   Society of Hair Testing, Forensic Sci. Int., 145 (2004) 83–84.

122   S. Pichini, M. Ventura, M. Pujadas, R. Ventura, M. Pellegrini, P. Zuccaro, R. Pacifici and T.R. de la, Forensic Sci. Int., 145 (2004) 109–115.

123   L.J. Langman and B.M. Kapur, Clin. Biochem., 39 (2006) 498–510.

124   L. Rivier, Anal. Chimica Acta, 492 (2003) 69–82.

125   D.R.A. Uges, J.-P. Thie and B. Greijdanus, Unreliable results can always occur unexpectedly, even if an extensive validated method on LC/MS/MS is used, in: M.Z. Karlovsek (Ed.), The International Association of Forensic Toxicologists (TIAFT) 44th International Meeting, Ljubljana, Slovenia, 2006.

126   G. Skopp, L. Pötsch, B. Ganssmann, R. Aderjan and R. Mattern, J. Anal. Toxicol., 22 (1998) 261–264.

127   A.G. Verstraete, Ther. Drug Monit., 26 (2004) 200–205.

128   O.H. Drummer, Forensic Sci. Int., 150 (2005) 133–142.

129   A.W. Jones and L. Karlsson, Hum. Exp. Toxicol., 24 (2005) 615–622.

130   O.H. Drummer, Drugs in drivers killed in Australian road traffic accidents. Melbourne, 1994.

131   R.W. Elder, R.A. Shults, D.A. Sleet, J.L. Nichols, R.S. Thompson and W. Rajab, Am. J. Prev. Med., 27 (2004) 57–65.

132   C.A. Soderstrom, P.C. Dischinger, J.A. Kufera, S.M. Ho and A. Shepard, Annu. Proc. Assoc. Adv. Automot. Med., 49 (2005) 315–330.

133   O.H. Drummer, J. Gerostamoulos, H. Batziris, M. Chu, J. Caplehorn, M.D. Robertson and P. Swann, Accid. Anal. Prev., 36 (2004) 239–248.

134   P. Mura, P. Kintz, B. Ludes, J.M. Gaulier, P. Marquet, S. Martin-Dupont, F. Vincent, A. Kaddour, J.P. Goulle, J. Nouveau, M. Moulsma, S. Tilhet-Coartet and O. Pourrat, Forensic Sci. Int., 133 (2003) 79–85.

135   M.C. Longo, C.E. Hunter, R.J. Lokan, J.M. White and M.A. White, Accid. Anal. Prev., 32 (2000) 623–632.

136   C.L. O'Neal, D.J. Crouch, D.E. Rollins and A.A. Fatah, J. Anal. Toxicol., 24 (2000) 536–542.

137   D.J. Crouch, Forensic Sci. Int., 150 (2005) 165–173.

138   N. Takayama, S. Tanaka, R. Kizu and K. Hayakawa, Biomed. Chromatogr., 13 (1999) 257–261.

139 N. Takayama, S. Tanaka, R. Kizu and K. Hayakawa, Jpn. J. Toxicol. Environ. Health, 44 (1998) 116–121.

140 P.J. Taylor, Clin. Biochem., 38 (2005) 328–334.

141 P.J. Larger, M. Breda, D. Fraier, H. Hughes and C.A. James, J. Pharm. Biomed. Anal., 39 (2005) 206–216.

142 R. Dams, M.A. Huestis, W.E. Lambert and C.M. Murphy, J. Am. Soc. Mass Spectrom., 14 (2003) 1290–1294.

143 M.E. Neerman, Labmed, 37 (2006) 358–361.

144 K.R. Allen, Clin. Toxicol., 44 (2006) 147–153.

145 J.L. Little, M.F. Wempe and C.M. Buchanan, J. Chromatogr. B: Anal. Technol. Biomed. Life Sci., 833 (2006) 219–230.

146 S.N. Giorgi and J.E. Meeker, J. Anal. Toxicol., 19 (1995) 392–398.

147 G. Skopp, L. Pötsch, I. Konig and R. Mattern, Int. J. Legal Med., 111 (1998) 1–5.

148 S.G. Fraga, J.F.D.F. Estevez and C.D. Romero, Ann.Clin. Lab. Sci., 28 (1998) 160–162.

149 G. Skopp and L. Pötsch, Clin. Chem., 48 (2002) 301–306.

150 G. Skopp and L. Pötsch, J. Anal. Toxicol., 28 (2004) 35–40.

151 G. Skopp, L. Pötsch, R. Mattern and R. Aderjan, Clin. Chem., 48 (2002) 1615–1618.

152 S. Dickson, A. Park, S. Nolan, S. Kenworthy, C. Nicholson, J. Midgley, R. Pinfold and S. Hampton, Forensic Sci. Int. (2006).

153 M. Deveaux, J.P. Goullé and M. Lhermitte, Ann. Toxicol. Anal., 15 (2003) 98–107.

154 M.A. Huestis, J.E. Henningfield and E.J. Cone, J. Anal. Toxicol., 16 (1992) 283–290.

155 J.E. Manno, B.R. Manno, P.M. Kemp, D.D. Alford, I.K. Abukhalaf, M.E. McWilliams, F.N. Hagaman and M.J. Fitzgerald, J. Anal. Toxicol., 25 (2001) 538–549.

156 T. Daldrup, Cannabis im Strassenverkehr. 1996.

157 T. Daldrup, The Cannabis Influence Factor (CIF). The 41st International TIAFT Meeting, Melbourne, Australia, 2003.

158 J.G. Bramness, S. Skurtveit and J. Morland, Drug Alcohol Depend., 68 (2002) 131–141.

159 J.G. Bramness, S. Skurtveit and J. Morland, Eur. J. Clin. Pharmacol., 59 (2003) 593–601.

160 I. Gustavsen, J. Morland and J.G. Bramness, Accid. Anal. Prev., 38 (2006) 490–495.

161 H.Z. Khiabani, J.G. Bramness, A. Bjorneboe and J. Morland, Traffic Inj. Prev., 7 (2006) 111–116.

162 F. Grotenhermen, G. Leson, G. Berghaus, O.H. Drummer, H.-P. Krüger, M.C. Longo, H. Moskowitz, B. Perrine, J.G. Ramaekers, A. Smiley and R. Tunbridge, Developing per se laws for driving under the influence of cannabis (DUIC), in: J. Oliver, P. Williams and A. Clayton (Eds.), Proceedings of the 17th international conference on Alcohol, Drugs and Traffic Safety, Glasgow, Scotland, UK, 2004.

163 J.G. Ramaekers, M.R. Moeller, P. van Ruitenbeek, E.L. Theunissen, E. Schneider and G. Kauert, Drug Alcohol Depend., (2006).

164 P. Mura, P. Kintz, W. Dumestre, S. Raul and T. Hauet, J. Anal. Toxicol., 29 (2005) 842–843.

165 J.L. Veuthey, S. Souverain and S. Rudaz, Ther. Drug Monit., 26 (2004) 161–166.

166 M.P.M. Mathijssen, Drug-, medicijn- en alcoholgebruik van automobilisten in Nederland, 1997/1998. R-99-5. Leidschendam, Stichting Wetenschappelijk Onderzoek Verkeersveiligheid SWOV, 1999.

167 S.R. Lowenstein and J. Koziol-McLain, J. Trauma, 50 (2001) 313–320.

168 C. Vignali, A. Groppi, A. Polettini, A. Valli, A. Sali, B. Cancelli, and M. Montagna, Drugs and driving. Toxicological findings in 119 fatally injured drivers. 2001. Prague, Czech Republic. The International Association of Forensic Toxicologists (TIAFT) 39th International Meeting.

169 O. Plaut and C. Staub, Driving under the influence of drugs in the canton of Geneva, Switzerland. Results and roadside survey project, in: H. Laurell and F. Schlyter (Eds.), Proceedings of the 15th International Conference on Alcohol, Drugs and Traffic Safety, Stockholm, Sweden, 2000.

170 G. Ceder, Drugged driving in Sweden – Effects of new legislation concerning zero-tolerance for narcotic drugs, in: H. Laurell and F. Schlyter (Eds.), The International Council on Alcohol, Drugs and Traffic Safety (ICADTS). Proceeding of the 15th International Conference on Alcohol, Drugs and Traffic Safety, Stockholm, 2000.

171 G. Thorsdottir, K. Magnusdottir and J. Kristinsson, Alcohol, drugs and driving in Iceland during years 2000 to 2002, in: J. Oliver, P. Williams and A. Clayton (Eds.), Proceedings of the 17th International Conference on Alcohol, Drugs and Traffic Safety,Glasgow Scotland, UK, 2004.

172 M.C. Longo, C.E. Hunter, R.J. Lokan, J.M. White and M.A. White, Accid. Anal. Prev., 32 (2000) 613–622.

173 P. Holmgren, A. Holmgren and J. Ahlner, Forensic Sci. Int., 151 (2005) 11–17.

174 P. Mura, C. Chatelain, V. Dumestre, J.M. Gaulier, M.H. Ghysel, C. Lacroix, M.F. Kergueris, M. Lhermitte, M. Moulsma, G. Pepin, F. Vincent and P. Kintz, Forensic Sci. Int. (2005).

175 B.E. Smink, B. Ruiter, K.J. Lusthof and P.G. Zweipfenning, Forensic Sci. Int., 120 (2001) 195–203.

176 A. Seymour and J.S. Oliver, Drugs and driving on Strathclyde roads – An update, in: H. Laurell and F. Schlyter (Eds.), Proceedings of the 15th International Conference on Alcohol, Drugs and Traffic Safety, Stockholm, Sweden, 2000.

177 V. Maes, N. Samyn, M. Willekens, G. De Boeck and A. Verstraete, Ann. Toxicol. Anal., 15 (2003) 128–137.

178 A. Kotsos, J. Gerostamoulos, M. Boorman and O.H. Drummer, Prevalence of drugs in impaired Victorian drivers. The 41st International TIAFT Meeting, Melbourne, Australia, 2003.

179 S.W. Toennes, G.F. Kauert, S. Steinmeyer and M.R. Moeller, Forensic Sci. Int., 152 (2005) 149–155.

180 M. Augsburger, N. Donze, A. Menetrey, C. Brossard, F. Sporkert, C. Giroud and P. Mangin, Forensic Sci. Int., 153 (2005) 11–15.

181 S. Katsumata, K. Sato, H. Kashiwade, S. Yamanami, H. Zhou, I. Yonemura, H. Nakajima and H. Hasekura, Forensic Sci. Int., 62 (1993) 209–215.

182 M. Widman, S. Agurell, M. Ehrnebo and G. Jones, J. Pharm. Pharmacol., 26 (1974) 914–916.

183 A.R. Jeffcoat, M. Perezreyes, J.M. Hill, B.M. Sadler and C.E. Cook, Drug Met. Dispos., 17 (1989) 153–159.

184 W.D. Darwin, B.C. Parker and E.J. Cone, Codeine concentrations in plasma versus aged whole blood. Annual meeting of the International Association of Forensic Toxicologists, Tampa, FL, 1994.

185 P.X. Iten, Fahren unter drogen- und medikamenteneinfluss – forensische interpretation und begutachtung, Universität Zürich, Switzerland, 1994.

186 H.J. Shull, G.R. Wilkinson, R. Johnson and S. Schenker, Ann. Intern. Med., 84 (1976) 420–425.

187 A.C. Moffat, M.D. Osselton and B. Widdop, Clarke's analysis of drugs and poisons (3rd ed.). London, United Kingdom, 2004.

188 P. Lillsunde, T. Gunnar and H. Seppa, Rosita 2 in Finland. Rosita 2 meeting, Baltimore, MD, USA, 5-12-2005.

M.J. Bogusz (Ed.). Forensic Science
Handbook of Analytical Separations, Vol. 6
© 2008 Elsevier B.V. All rights reserved.

CHAPTER 20

Unconventional samples and alternative matrices

Marleen Laloup, Gert De Boeck and Nele Samyn

*Section of Toxicology, Federal Public Service Justice, National Institute of Criminalistics and Criminology
(N.I.C.C.), Vilvoordsesteenweg 100, 1120 Brussels, Belgium*

It has been known for several decades that trace metals in the body accumulate in the hair and can be detected weeks or even months after the original exposure. The search for drugs in human hair started in the late 1970s, resulting ultimately in the routine use of hair analysis as a tool for the detection of long-term drug use in forensic science, traffic medicine, occupational medicine and clinical toxicology. Due to the increasing knowledge of drug incorporation mechanisms in hair and the availability of more sensitive and more specific analytical techniques, the number of studies relating to hair analysis has increased exponentially during the last decade. This chapter focuses mainly on the available literature of the last 6 years; the reader is referred to a number of excellent reviews [1–7] on previously published work.

Oral fluid analysis for drugs was first used almost 30 years ago for the purpose of therapeutic drug monitoring. At that time, it was already known that it is possible to predict plasma concentrations of drugs by analysis of the corresponding oral fluid sample. Interest in the use of oral fluid for drug testing in the workplace – as an alternative to urine – and for testing of intoxicated drivers has been increasing since 1999. Recent developments in the field of collection devices, the use of screening and confirmation techniques and the application of cut-off values will be discussed.

Several reports have demonstrated that sweat is a suitable alternative biological matrix for monitoring recent drug use and provides an additional tool for monitoring drug use, e.g. in detoxification centers. An external continuous monitor, through the use of a sweat patch, offers substantial advantages over frequent urine testing, such as convenience, cost and longer detection windows. Recent methods dealing with the collection of this biological matrix and their analysis are detailed further in this chapter.

Drug abuse during pregnancy is a major health problem because the associated perinatal complications are high. Identification of maternal drug use should be

References pp. 690–697

established soon after birth for appropriate intervention and follow-up. During the past decades, urine has been the specimen of choice for drugs of abuse screening also at delivery. However, drugs present in the urine reflect consumption or exposure during the preceding 1 to 4 days. Therefore, the use of meconium as a test specimen in the screening of newborns for drugs of abuse will be shown at the end of this chapter.

After a short review on the physiology, mechanisms of drug incorporation and advantages and disadvantages of the matrix, this chapter will focus on sampling methods, preparative application and analytical procedures. The analytical part of recent papers will be highlighted, without dealing in detail with the authors' results.

20.1 HAIR

20.1.1 Introduction

The major practical advantage of this matrix compared to urine or blood testing is the larger surveillance window – weeks to months, depending on the length of the hair shaft – making the detection of long-term use of drugs feasible. Given a mean hair growth rate of 1 cm/month, it is actually possible to extrapolate a personal record of drug use through the analysis of hair segments [8,9]. In addition, drug incorporation in hair is very stable and the collection of a hair strand is fairly simple, non-invasive and replicable.

During the analysis, the solid hair is transferred into a liquid phase (the hair itself or the molecules of interest) and then analyzed like a urine or blood sample. For the quantification of drugs and drug metabolites in hair, the majority of laboratories currently use gas chromatography-mass spectrometry (GC-MS). However, many of the more challenging assays (e.g. benzodiazepines, cannabinoids) require MS-MS technology to achieve the required limits of detection. Accordingly, in the past few years, the use of liquid chromatography-mass spectrometry (LC-MS) and liquid chromatography-tandem mass spectrometry (LC-MS-MS) for hair analysis has been introduced.

20.1.1.1 Physiology of hair

Hair is an annexe of the skin, originating in a hair follicle, where actively dividing matrix cells form the germination centre of the hair. Hair follicles are embedded in the dermis of the skin and are highly vascularized in order to provide the growing hair root (bulb) metabolic fuel. The hair shaft consists of an outer cuticle, an inner medulla and a central cortex and is mostly composed of fibrous proteins (mostly α-keratins, 65–95%), lipids (1–9%) and small quantities of trace elements, polysaccharides and water. Matrix cells may also acquire pigment (melanin), which will eventually determine the colour of the hair shaft.

Human hair grows in circles, known as the anagen (active hair growth, 4–8 years), catagen (short transition stage, few weeks) and telogen (resting period, 10 weeks) phases. On the scalp of an adult, approximately 85% of the hairs are in the anagen growing phase. It is generally assumed that scalp hair grows 1 cm/month. However, the growth rate and the length of the telogen resting period are dependent on different factors such as race, disease, nutritional deficiencies, anatomical localization and age. Indeed, several authors have shown that pubic hair, arm hair, axillary hair and beard hair can be used as alternative sources for toxicological analysis in the absence of head hair [10,11]. Differences in drug concentrations in hair from different anatomical regions were noted.

20.1.1.2 Mechanisms of drug incorporation into hair

The mechanisms of drug incorporation into hair have not fully been clarified but require (i) entry of the drug into a hair cell and (ii) fixation of the drug into that cell. Different possible routes of drug entry into a hair cell have been suggested by different authors and can be summarized into the following processes: passive diffusion from blood (endogenous pathway), diffusion from sweat and sebum (endogenous–exogenous pathway) and contamination from the environment (exogenous pathway).

Permeation of a drug molecule into a cell requires cross-over of the plasma membrane. Preferentially, free, non-ionized molecules that are sufficiently lipid soluble diffuse across the cell membrane. Once inside the cell, the uncharged drug will dissociate depending upon its pKa value. Since hair is considered to be of an acidic nature (isoelectric point of keratinized hair fibres is close to pH 6), this explains the preferential incorporation of basic drugs into hair [12]. It has been postulated that the generally high parent drug to metabolite ratio found in hair is mainly due to the transport phenomena across the plasma membranes in the hair follicle [13].

An important factor for the fixation of a drug within a cell is the drug interaction with the components in the matrix and/or pigments. For instance, melanin is a polyanionic polymer consisting of eumelanin, which is responsible for black and brown hair colour, and pheomelanin, which is responsible for red hair colour. It is generally accepted that melanin binds a variety of drugs through both ionic and possibly van der Waals interactions [13–16]. However, the question is to what extent this melanin binding contributes to the overall sequestration of drugs in hair. This hair colour bias (defined as an increased likelihood of detecting drug use in one group of individuals over another based on their hair colour when both have comparable use and/or exposure) remains an area of continued debate in the hair testing field [17–19]. A number of studies [14,20–22] have shown that the degree of bias may differ from drug to drug. As such, several authors proposed to normalize the concentrations of drugs found in hair for the hair colour effect by expressing it as a function of the melanin concentration rather than by hair weight [22,23]. On the other hand, it has been shown that this melanin binding mechanism alone cannot account for the presence of drugs in this matrix, since drugs are also trapped into the hair of albino animals, which lack melanin [24].

In conclusion, it may be said that the main factors influencing drug incorporation in hair are basicity (pKa), lipophilicity and affinity for binding to melanin of drugs.

20.1.1.3 Distinguishing passive exposure from consumption

The history of hair testing for illicit drugs has always been accompanied by the crucial problem of differentiating passive exposure from consumption, especially in the case of smoked drugs. Among the proposed methods to discriminate between passive exposure and active drug consumption, there are two major criteria: the adoption of a cut-off value along with the use of metabolite-to-parent ratios, and the use of procedures for decontamination.

In the first case, efforts have been made for the definition of cut-off values for each drug and criteria for the presence of metabolites have been recommended (Table 20.1) [25–28].

Two metabolites that are definitive indicators of use even without washing because they are formed *in vivo*, are cocaethylene in the case of ingestion of ethanol along with cocaine and 11-nor-delta9-tetrahydrocannabinol-9-carboxylic acid (THC-COOH) from use of cannabinoids.

The washing of hair samples has been well documented particularly for the analysis of cocaine. However, there is no consensus or uniformity in the washing procedures. Care should be taken to avoid that washing methods extract important amounts of drugs from inside the hair matrix. It has been shown that this passive contamination is very persistent and some authors even suggested that none of the decontamination procedures described is capable to remove all passive contamination. This problem, though largely debated, is still not fully resolved and is the subject of fierce scientific discussion [29–31].

Finally, it appears that the use of extensive washing procedures and the application of certain wash criteria, in combination with cut-off and metabolite criteria, may be an appropriate policy to interpret a positive analytical result [32,33].

As such, it can be concluded that a positive result should be interpreted with extreme caution, and that, in absence of a general consensus, every laboratory must validate its own decontamination procedure.

TABLE 20.1
RECOMMENDED CUT-OFF VALUES AND METABOLITE-TO-PARENT DRUG RATIOS BY THE SOHT [25]

	Recommended LOQ	Metabolites
Opiates	⩽0.2 ng/mg for each compound	Distinction between the consumption of heroin from codeine or morphine use by the presence of 6-AM.
Cocaine	⩽0.5 ng/mg for cocaine ⩽0.05 ng/mg for other compounds	At least one of the following compounds should be present in addition to cocaine: BE, cocaethylene, norcocaine or EME.
Amphetamines	⩽0.2 ng/mg for each compound	Caution is warranted for the ingestion of legal drugs producing positive results for methamphetamine and amphetamine.
Cannabinoids	⩽0.1 ng/mg for THC ⩽0.2 pg/mg for THC-COOH	Confirmation of THC-COOH is required to definitively prove the use of cannabinoids.

20.1.1.4 Interpretation of dose and time relationships

Whereas some authors presented data indicating a linear relationship between drug dose and amount of parent drugs and/or metabolites found in hair [34,35], some others showed the lack of dose–concentration association [36,37]. Various reasons have been postulated for the lack of relationship such as inter-subject variability of drug incorporation in hair, drug moving along the hair shaft with time and finally incorporation of drug into hair by multiple mechanisms. Furthermore, an alternation in drug concentrations can be induced by cosmetic treatments of hair. Studies including bleaching, colouring, perming and UV-radiation have shown a reduction of drug levels or, alternatively, under favourable environmental contamination conditions, a higher incorporation ratio [32,38,39]. In addition, a certain decrease in concentration over time can occur resulting from normal hair hygiene and wear [40].

Nonetheless, it seems apparent that hair analysis can distinguish categorized patterns of drug consumption [41,42]. For some drugs, an increased incorporation was reported after chronic (ab)use, leading to different parent/metabolite ratios [40].

20.1.2 Collection and sample pre-treatment

Hair is best collected from the area at the back of the head, called the *vertex posterior*. This area has less variability in hair growth than other areas and the hair is less subject to age- and sex-related influences. Hair strands should be tightened together and cut as close as possible to the scalp. The sample size can vary, but samples of 30–50 mg are most current. Furthermore, in case of segmental analysis, proximal and distal portions must be identified.

Sample preparation and sample extraction are the most important steps of an analytical procedure. After decontamination (washing) and drying of the hair strands, hair segments of 1–2 cm can be cut into snippets or powdered before extraction with a ball mill. Several methods have been proposed for the extraction of drugs from the pulverized or cut hair matrix, including chemical hydrolysis, enzymatic hydrolysis, direct solvent extraction or supercritical fluid extraction. The method to be used depends on the chemical nature and stability of the drug. In this respect, caution should be taken with respect to the hydrolytic conversion of certain compounds.

20.1.3 Screening techniques

20.1.3.1 Immunological procedures

Several groups have analyzed hair samples using a radioimmunoassay (RIA) technique [43–45]. RIA is a common sensitive and reliable immunological technique, but the use of radioactively labelled material prevents its use out of safe areas. Enzyme immunoassays (EIAs) are a good alternative because of the high sensitivity, the safe and simple use and its relatively low cost for high-throughput analysis.

Conventional urinary immunoassays were shown to lack the required sensitivity and specificity for hair analysis [46]. In contrast, oral fluid tests appear to be more appropriate for hair analysis, because of the cross-reactivity with the parent drug and lipophilic metabolites found in saliva [47]. At the optimal cut-off concentration, a hair immunoassay should have a sensitivity and specificity of higher than 90% [46]. In general, after extraction of the hair matrix, no further sample clean-up is needed and immunoassays with a high sensitivity and an acceptable reproducibility have been published for opioids [48–50], cocaine [50], amphetamines [51], cannabinoids [52] and benzodiazepines [53,54] in hair.

20.1.3.2 Chromatographic procedures

The analytical method most frequently used for hair analysis is GC-MS, in electron impact (EI) mode and single ion monitoring (SIM), but positive and negative chemical ionization (PCI and NCI) mass detectors have also been used. Because of the different physical and chemical properties of the compounds investigated, most papers dealing with broad spectrum screening of drugs of abuse in hair involve two or more procedures for the extraction and detection [55,56].

Paterson *et al.* described the detection of several drugs of abuse using a simple methanolic extraction, without any further clean-up steps [57] using a HP-5MS GC column and EI in SIM mode. A relative high sensitivity was obtained, but, in order to cope with the dirty extracts, the maintenance of the mass spectrometer has to be vigorous. In order to reduce the time needed for sample extraction, Girod and Staub have developed an automated solid-phase extraction (SPE), after hydrolysis of the powdered hair samples with HCl [58]. Analysis was performed on a DB-5MS column, using full-scan GC-EI-MS and ion trap GC-PCI-MS, allowing lower limits of quantitation (LOQ). In an attempt to avoid the time-consuming derivatization step, Gentili *et al.* described the use of headspace solid-phase microextraction (HS-SPME) coupled to a GC-EI-MS [59]. This one-step procedure allows the simultaneous determination of cocaine, several amphetamines, ketamine and methadone in hair samples. Headspace solid-phase dynamic extraction (HS-SPDE), coupled to a GC-EI-MS-MS was applied to the analysis of methadone, cannabinoids and amphetamines [60]. Two methods were needed: methadone and 1,5-dimethyl-3,3-diphenylpyrrolidine (EDDP) were analyzed without derivatization in the same run as the silylated cannabinoids; the amphetamines were acylated with slightly changed SPDE conditions. However, as no time-consuming wet chemistry was involved in the SPDE step, a reduction in the total analysis time was obtained.

In recent years, LC-MS(-MS) has been used for hair analysis in order to increase the sensitivity (higher signal-to-noise) and specificity of the analysis and to be able to detect GC-unstable compounds. Indeed, Welch *et al.* [61] confirmed the usefulness of LC-MS for the analysis of hair. These authors used both GC-EI-MS and LC-electrospray ionization (ESI)-MS to determine the concentrations of drugs of abuse in two standard reference materials. Both methods used 0.1 M HCl for extraction of all the analytes from the hair, except for delta9-tetrahydrocannabinol (THC), which was extracted with 1 M NaOH. Although different clean-up procedures were used, the GC-MS and LC-MS results were in good agreement.

Recently, Kronstrand *et al.* [62] described a simple and sensitive method appropriate for the screening of several drugs of abuse in human hair with LC-ESI-MS-MS. The hair samples were, without any pre-treatment or decontamination, extracted by incubation with a mixture of acetonitrile, methanol and formate buffer, which conveniently was the mobile phase A used in their chromatographic method. No further clean-up was performed and a 10 μL aliquot was injected into the LC-MS-MS system, using a Zorbax phenyl analytical column with gradient elution. This simple method was shown to have a good sensitivity and a high extraction recovery.

20.1.4 Confirmation analysis

20.1.4.1 Cocaine and opioids

Table 20.2 summarizes the procedures for the analysis of cocaine and opiates in hair. Alkaline digestion has to be avoided since it hydrolyses acetylated compounds and cocaine [63,64]. Therefore, these drugs are usually extracted from the hair matrix using an acid digestion followed by either liquid–liquid extraction or SPE, or by direct methanolic extraction without further purification (Table 20.2). According to some authors, acid digestion also leads to a partial hydrolysis [63,65]. Minimal hydrolysis of acetylated opiates and cocaine can be achieved by incubation in a neutral medium such as methanol [63,66], Soerensen buffer pH 7.4 [64] or phosphate buffer pH 5 or 6 [65]. In these cases, however, a lower extraction recovery was noted [63]. Hydrophobic solvents like dioxane and acetonitrile lower the extraction rate [67,68]. When using supercritical CO_2 modified with 10% of methanol, Brewer *et al.* were able to simultaneously extract cocaine, benzoylecgonine (BE), morphine and codeine, resulting in higher recoveries for cocaine in comparison with acid hydrolysis or sonication with methanol [69].

Recently, LC-MS and LC-MS-MS, both in ESI and APCI mode, have been used for the determination of opiates and cocaine in human hair, allowing a reduction in analysis time with comparable sensitivities as with GC-MS(-MS) [33,70,71].

The presence of other opioids in hair has also been investigated (Table 20.3). Most recent publications concern methadone and its major metabolites (EDDP and 2-ethyl-5-methyl-3,3-diphenylpyrroline (EMDP)). Generally, lower concentrations of EDDP than of methadone are detected [72,73]. EMDP is detected only in trace-amounts. GC-PCI-MS had a significant advantage over GC-EI-MS for analysis of methadone and its metabolites, in avoiding fragmentation and providing a high intensity molecular ion at $m/z = 310$ for methadone [37].

20.1.4.2 Amphetamine-type stimulants

Most techniques dealing with the detection of amphetamine, methamphetamine and the designer amphetamines in hair use acid or alkaline hydrolysis, or a combination

TABLE 20.2
ANALYTICAL METHODS FOR THE DETERMINATION OF OPIATES, COCAINE AND METABOLITES IN HAIR

Reference	Drugs	Extraction	Sample Clean-up	Derivatization	Column	Chromatography	LOQ
[74]	Heroin, 6-AM, morphine, codeine, cocaine, BE, cocaethylene	Methanol (56 C, 18 h)	–	BSTFA (+1% TMCS)	Restek Rtx-5MS	GC-EI-MS-MS	10–20 pg/mg
[75]	Codeine, morphine, 6-AM, cocaine, BE	0.1 N HCl (45 C, overnight)	Mixed-mode SPE	MSTFA	HP-5MS	GC-EI-MS	NR
[76]	Cocaine, cocaethylene, BE, EME, norcocaine	0.1 N HCl (45 C, overnight)	Mixed-mode SPE	HFIP TFAA	DB-5MS	GC-PCI-MS-MS	50 pg/mg
[77,78]	6-AM, morphine, codeine, cocaine, BE, EME	0.1 N HCl (45 C, overnight)	Mixed-mode SPE	MSTFA	HPUltra-2	GC-EI-MS	20–90 pg/mg
[79]	Morphine, morphine-3-glucuronide, morphine-6-glucuronide	MeOH:H$_2$O (20:1), 1 h sonication followed by 24 h incubation at room temperature	–	–	Mightysil RP-18	LC-ESI-MS	NR
[69]	Morphine, codeine, cocaine, BE	Supercritical fluid extraction using CO$_2$ modified with methanol (10%)	–	PFPA	DB-1	GC-EI-MS	NR
[80]	6-AM, morphine, codeine, acetylcodeine	0.1 N HCl (60 C, 12 h)	Automated mixed-mode SPE	PA	DB-5MS	GC-EI-MS	80–100 pg/mg
[33]	Cocaine, BE	Proteinase-K (40 C, overnight)	Mixed-mode SPE	–	Keystone Scientific Betasil C8	LC-ionspray-MS-MS	NR
[65]	6-AM, morphine, codeine, cocaine, BE	0.1 N phosphate buffer (pH 5) (45 C, 18 h)	Mixed-mode SPE	MSTFA:TMCS (100:2)	HP-5MS	GC-EI-MS	NR, LOD: 50–200 pg/mg

Ref.	Analytes	Extraction	Clean-up	Derivatization	Column	Detection	LOD/Range
[81]	Cocaine, BE, cocaethylene	Methanol (50°C, 18 h)	SPME	butylchloroformate	HP-5MS	GC-EI-MS	100–500 pg/mg
[64]	6-AM, morphine, codeine, dihydrocodeine, hydrocodone, acetylcodeine	Soerensen buffer (pH 7.4) (50°C, 5 h)	Mixed-mode SPE	MSTFA	HP-5MS	GC-EI-MS	100–200 ng/mg, higher for acetylcodeine
[70]	6-AM, morphine, codeine	Acetonitrile (room temperature, overnight)	*n*-Butyl chloride/acetonitrile (4:1)	–	YMC ODS-AQ	LC-ESI-MS	50–80 pg/mg
[82]	6-AM, morphine, codeine	0.1 N HCl (45°C, overnight)	Mixed-mode SPE	HFBA	DB-5MS	GC-NCI-MS(-MS)	NR, LOD: 2–5 pg/mg
[66]	6-AM, morphine, codeine, norcodeine, cocaine, BE, EME, cocaethylene, norcocaine	Sonication in methanol (37°C, 3 h)	Mixed-mode SPE	–	Synergi Hydro RP	LC-APCI-MS-MS	17–80 pg/mg
[83]	Morphine	Sonication with methanol/TFA (9:1) (45°C, 18 h)	–	MSTFA (+1% TMIS)	DB-1MS	GC-EI-MS	30 pg/mg
[84]	Cocaine, AEME, EME, cocaethylene	0.1 N HCl (60°C, overnight)	Automated mixed-mode SPE	–	DB-5MS	GC-PCI-MS-MS	50–100 pg/mg
[50]	Heroin, 6-AM, morphine, codeine, acetylcodeine, cocaine, BE, cocaethylene	Sonication in methanol (50°C, 5 h)	Mixed-mode SPE	MSTFA	HP-5MS	GC-EI-MS	40–260 pg/mg
[85]	Cocaine, cocaethylene	Pronase (37°C, overnight)	SPME	–	HP-5	GC-EI-MS	400 pg/mg

Note: NR, not reported.

TABLE 20.3
ANALYTICAL METHODS FOR THE SEPARATION AND ANALYSIS OF OPIOIDS IN HAIR

Reference	Drugs	Extraction	Sample Clean-up	Derivatization	Column	Chromatography	LOQ
[73]	Methadone EDDP	Pronase E (37°C, 12 h)	SPME	–	HP-5	GC-EI-MS	3.46 ng/mg 0.36 ng/mg
[86]	Methadone EDDP EMDP	1 N NaOH (110°C, 20′)	HS-SPME	–	HP-5MS	GC-EI-MS	0.10 ng/mg 0.16 ng/mg 0.16 ng/mg
[37]	Methadone EDDP	0.01 N HCl (60°C, overnight)	Automated mixed-mode SPE	–	DB-5MS	GC-PCI-MS	0.05 ng/mg 0.20 ng/mg
[87,88]	Buprenorphine Norbuprenorphine	0.1 N HCl (56°C, overnight)	Chloroform/isopropanol/n-heptane (25:10:65)	–	Novapak C18	LC-ionspray-MS	NR
[89]	Codeine Morphine 6-AM Hydrocodone Hydromorphone Oxycodone	0.1 N HCl (55°C, overnight)	SPE mixed-mode	(1) Methoxyamine (2) BSTFA	DB-5	GC-EI-MS	0.32 ng/mg 0.15 ng/mg 1.10 ng/mg 0.65 ng/mg 0.15 ng/mg 0.14 ng/mg
[90]	Tramadol	3 N HCl (60°C, overnight)	SPE mixed-mode	–	HP-5MS	GC-EI-MS	NR, LOD: 0.2 ng/mg
[91]	Fentanyl	Saturated phosphate buffer (pH 8.4)	Dichloromethane/isopropanol/n-heptane (50:17:33)	–	HP-5MS	GC-EI-MS-MS	0.001 ng/mg
[72]	R- and S- Methadone EDDP EMDP	MeOH/TFA (9:1) (37°C, overnight)	–	–	Chiral AGP	LC-ESI-MS-MS	0.05 ng/mg 0.03 ng/mg 0.30 ng/mg

Note: NR, not reported.

of hydrochloric acid and methanol, followed by a purification step (liquid–liquid extraction or SPE) and derivatization before injection into the GC-MS system using different types of analytical columns [7,92]. Pujadas *et al.* compared four different extraction procedures [42]. These authors concluded that digestion with 1 M sodium sulfide, followed by a liquid–liquid extraction with *tert*-butyl methyl ether and mixed-mode SPE offered the best compromise between recovery of the analytes from the matrix, clean-up of extracts and absence of chromatographic interferences. Nishida *et al.* [93] suggested the use of propylchloroformate as a derivatization reagent to avoid some problems associated with trifluoroacetylation (e.g. loss of derivatives during derivatization or instability of the derivatives). More recently, Villamor *et al.* [94] developed a GC-MS method for the simultaneous determination of amphetamine, methamphetamine, 3,4-methylenedioxymethamphetamine (MDMA), 3,4- methylenedioxyamphetamine (MDA) and 3,4-methylenedioxyethyl-amphetamine (MDEA) in hair. Hair was hydrolyzed in 1 M NaOH at 40°C, subjected to extraction with dichloromethane/isopropanol (4:1) and derivatized with pentafluoropropionic anhydride (PFPA). The reported LOQs were between 0.023 ng/mg for MDEA and 0.151 ng/mg for amphetamine. The analysis of hair specimens from 24 users showed markedly higher concentrations of MDMA than MDA, with an average of MDMA/MDA ratio of 6.8. The average methamphetamine concentration was 11.5 times higher than that of its metabolite amphetamine. Frison *et al.* [95] presented an analytical approach for the quantitative analysis of 15 amphetamine-related drugs, based on SPE, subsequent derivatization with 2,2,2-trichloroethyl chloroformate and GC-MS. Sample preparation involved extraction of analytes with methanol/HCl (98:2, v/v) from pulverized hair before SPE. These authors argued that this derivatization procedure offers some important advantages such as an important increase in the molecular weight of the original analytes (+ 174 Da), leading to improved chromatographic selectivity (long retention times and non-tailing peaks). In addition, the introduction of three chlorine atoms gives rise to distinctive MS signatures and as such unambiguous analyte identification. Limit of detection (LODs) were in the range from 0.1 to 0.2 ng/mg hair.

A sensitive GC-NCI-MS assay was developed for the quantification and the determination of enantiomeric ratios of amphetamine, methamphetamine, MDMA, MDA or MDEA enantiomers in hair using (*S*)-heptafluorobutyrylprolyl chloride ((*S*)-HFBPCl) as derivatization agent [96]. Reported LOQs ranged from 0.002 to 0.046 ng/mg. In most cases higher concentrations of (*R*)-methamphetamine and (*R*)-amphetamine than those of the corresponding (*S*)-enantiomers were observed.

It appears that the sensitivity of high-pressure liquid chromatography (HPLC) with direct fluorescence detection is high enough to determine MDA, MDMA and MDEA in hair, obtaining a LOD lower than 1 ng/mg [97]. This technique was also successfully applied for the chiral analysis of methamphetamine and amphetamine [98,99].

The possibility of coupling LC with tandem mass spectrometers can offer increased sensitivity and specificity. However, until now, the use of LC-MS-MS for the determination of amphetamines has not been extensively studied in hair. Only recently, Cairns *et al.* proposed a LC-ESI-MS-MS method for the confirmation of methamphetamine, amphetamine, MDMA and MDA in enzymatically digested hair samples [100]. Separation was achieved using a Betasil C8 HPLC column, eluted with a mixture of water and acetonitrile containing 0.1% formic acid (FA) (80:20, v/v). The reported LOQ was 100 pg/mg hair for methamphetamine, 25 pg/mg hair for amphetamine and 10 pg/mg hair for both MDMA and MDA.

Different authors described the use of HS-SPME procedures [101–103] for the extraction of amphetamine, methamphetamine and designer amphetamines from hair. Incubation of the hair matrix with NaOH, followed by an adsorption on the extraction fibre in the needle of the SPME device and on-fibre derivatization with MBTFA, enhanced the detection limits to 0.01 ng/mg for amphetamine and methamphetamine [103]. These authors also presented a fully automated HS-SPDE-GC-MS method, using the same sample extraction method and on-coating derivatization [104]. The reported LODs were similar with the values obtained using the corresponding SPME method.

Only one report using supercritical fluid extraction for the detection of amphetamine and its analogues has been published [105]. The modifier was chloroform and isopropyl alcohol (90:10, v/v), pumped at 10% in 90% of CO_2. PFPA was used as derivatization reagent and analysis was by GC-MS. Sample extraction was performed overnight with methanol.

Kronstrand *et al.* used a validated GC-EI-MS method to investigate the incorporation of selegiline and its metabolites (desmethylselegiline, amphetamine and methamphetamine) into human hair [40,106]. After extraction with potassium hydroxide (80°C, 10 min), sample clean-up was performed using liquid–liquid extraction with isooctane and back-extraction in sulphuric acid before derivatization with TFA. This back-extraction step was needed to avoid a co-eluting interference with desmethylselegiline. This study showed that methamphetamine was the predominant metabolite in hair, amphetamine being the second most common. They concluded that the measurement of desmethylselegiline as well as interpretation of the ratio between amphetamine and methamphetamine in hair might be a way to distinguish abuse from therapeutic medication, in the case that an enantioselective analysis is not possible.

Finally, Sporkert *et al.* developed the only method dealing with the simultaneous determination of cathinone, norpseudoephedrine (cathine) and norephedrine in human hair [107]. The compounds were extracted for 4 h with phosphate buffer pH 2.0, followed by SPE on a mixed-mode column, derivatization with heptafluorobutyric anhydride (HFBA) and GC-EI-MS analysis, interfaced with a classical HP-5MS analytical column. Sample preparation was performed with an aqueous buffer, since an alkaline digestion of the hair samples proved not to be suitable for the extraction of cathinone, the latter being not stable under these conditions. As a rule, the highest concentrations were found for norpseudoephedrine, followed by norephedrine and much lower concentrations of cathinone.

20.1.4.3 Cannabis

In 1999, Staub published a summary on the determination of cannabinoids in biological samples, with a special attention to alternative matrices such as hair [5]. Since then, a number of reports have been published, all of them using GC-MS(-MS), mostly with NCI (Table 20.4). As THC, cannabinol (CBN) and cannabidiol (CBD) are present in marijuana smoke, the detection of the endogenous metabolite, THC-COOH, in hair, is indicative of drug ingestion as opposed to environmental contamination. Usually, single MS systems do not achieve the required sensitivity of 0.2 pg/mg for THC-COOH, as recommended by the Society of Hair Testing (SOHT) (Table 20.1) [25]. However, different approaches are offered to increase the sensitivity. Sachs and Dressler obtained an LOD of 0.3 pg/mg, using a costly and time consuming extraction and clean-up step with both a liquid–liquid extraction and a HPLC purification [108]. The same LOD could be achieved, using GC-NCI-MS equipped with a high-volume injector (25 µL-injection) [109]. Generally, average concentrations of CBN in hair are superior to these of THC [110–112]. Baptista *et al.* reported CBD concentrations inferior to those of THC and CBN [110]. This was in contrast with other findings in which CBD was the major analyte with the highest concentrations [111–114].

Usually, liberation of cannabinoids from the matrix is performed with NaOH, followed by liquid–liquid extraction or SPE (Table 20.4). It appears, however, that the severe digestion conditions (1 M NaOH at 95°C for 10 min) affect the stability of CBD [115]. No degradation of THC or CBN were noted under these conditions. More recently SPME and SPDE have been used for the rapid screening of cannabinoids in hair [111,112].

In order to approach the required sensitivity using a single quadropole mass spectrometric system, Moore *et al.* employed a Deans switch two-dimensional GC/GC instrument [116]. The second analytical column is passed through a cryogenic focussing device that allows 'cold-trapping' of the drug, followed by a rapid release after the background has been separated. In addition, by using electron capture chemical ionization (ECCI), less fragmentation occurs at low source and quadrupole temperatures, allowing a higher sensitivity in comparison with other sources. Indeed, this is the first reported procedure able to routinely detect the Substance Abuse and Mental Health Services Administration (SAMHSA) proposed concentrations for THC-COOH in hair (i.e. 0.05 pg/mg) [28].

Until now, no LC-MS-MS method describing the specific analysis of THC and/or THC-COOH in hair has been published.

20.1.4.4 Therapeutic drugs

20.1.4.4.1 Benzodiazepines

Until recently, the detection of benzodiazepines in human hair, the most abused pharmaceutical drugs in the world, was not well documented. A number of benzodiazepines are chemically unstable and degrade in acid and especially in alkaline conditions. Several earlier publications on the analysis of one or more compounds describe an incubation of cut or pulverized hair in Soerensen buffer pH 7.6, followed by a liquid–liquid extraction with a mixture of ether and chloroform [119]. A comparison of extraction procedures revealed that the highest recoveries from rat hair were obtained with an overnight methanol–TFA (50:1) extraction followed by a

TABLE 20.4
PROCEDURES FOR THE DETECTION OF CANNABINOIDS IN HAIR

Year	Compound	Extraction	Clean-up	Derivatization	Column	Chromatography	LOD (pg/mg)	Reference
2000	THC-COOH	2 N NaOH (95°C, 30')	n-Hexane/ethyl acetate + HPLC clean-up (RP C8 column)	PFPA/HFIP	HP-5MS	GC-NCI-MS	0.3	[108]
2000	THC-COOH	1 N NaOH (80°C, 1h)	n-Hexane ethyl acetate (9:1)	BSTFA	NR	GC-EI-MS-MS	NR	[117]
2001	THC-COOH	1 N NaOH (80°C, 30')	SPE	TFAA/HFIP	HP-5MS	GC-NCI-MS	0.3	[109]
2002	THC CBN CBD	β-glucuronidase/arylsulfatase (40°C, 2h)	Chloroform/isopropyl alcohol (97:3) (after alkalinization) + n-hexane/ethyl acetate (90:10) (after acidification)	PFPA/PFPOH	HP-Ultra 2	GC-EI-MS	20 50 20	[110]
	THC-COOH				HP-Ultra 2	GC-EI-MS/GC-NCI-MS	500/5	
2002	THC CBN CBD	1 N NaOH (90°C, 5')	HS-SPME	MSTFA (on-fibre)	HP-5MS	GC-EI-MS	50 140 80	[111]
2003	THC CBN CBD	1 N NaOH (90°C, 5')	HS-SPDE	MSTFA (on-coating)	DB-5MS	GC-EI-MS	140 120 90	[112]
2004	THC CBN	Methanol (ultrason, 5 h)	–	MTBSTFA	DB-5	GC-EI-MS	40 40	[52]
	THC-COOH	Methanol/10 N KOH (1:1) (70°C, 30')	SPE	PFPA/HFIP	DB-5	GC-NCI-MS-MS	0.1	
2005	THC-COOH	1 N NaOH (80°C, 1h)	Heptane ethyl acetate (9:1)	PFPA	Equity-1	GC-NCI-MS-MS	50	[118]
2005	THC CBN CBD	1 N NaOH (95°C, 10')	n-Hexane/ethyl acetate (75:25)	MSTFA/TMCS/TMSI	Ultra-1	GC-EI-MS	6 2 5	[115]
2006	THC-COOH	H₂O/2 N NaOH (1:1)	SPE	TFAA/HFIP	1st GC: DB35-MS 2nd GC: DB-1	GC-GC-ECCI-MS	0.05	[116]

Note: NR, not reported.

liquid–liquid extraction step with dichloromethane [120]. Until 2002, in most cases, GC-MS in NCI mode was used, with *N,O*-bis(trimethylsilyl)-trifluoroacetamide (BSTFA) and HFBA as derivatization reagents.

El Mahjoub *et al.* have demonstrated the use of on-line HPLC with photodiode-array (PDA) detection [121]. After extraction with methanol/ammonia (97.5:2.5, v/v), a sample clean-up was performed on-line using a restricted access extraction column. The benzodiazepines included in their method were retained on a pre-column and then eluted in back-flush mode and separated on a C8 column. The extracting mobile phase consisted of a mixture of 30 mM phosphate buffer (pH 7.2) and acetonitrile (94:6, v/v), whereas for the analytical mobile phase a gradient of 20 mM phosphate buffer (pH 2.1) and acetonitrile was used. The measured LOQ was between 0.3 and 0.45 ng/mg.

The use of LC-MS-MS for the detection of benzodiazepines in hair was first reported by McClean *et al.* in 1999 [122]. It was concluded that, while capillary electrophoresis (CE) remains superior to LC in terms of separation efficiency, LC-MS-MS has proved the most useful tool in sensitivity and selectivity terms for identification of ng/mg levels in hair. From 2002 onwards, LC-MS-MS is the instrument of choice for the detection of benzodiazepines in hair. Indeed, several reports involving this specific technique were published since then. Kronstrand *et al.* established an analytical procedure for the screening of seven benzodiazepines and metabolites, using LC-ESI-MS-MS in hair of psychiatric patients [123]. Chromatographic separation was achieved using a Zobax Phenyl analytical column, eluted with a gradient of acetonitrile/methanol/20 mM formate buffer. Typical limits of quantification were 25–125 pg/mg, after digestion with proteinase K and mixed-mode SPE.

However, it was the group of Kintz that demonstrated the utility of LC-ESI-MS-MS to test for benzodiazepines in hair at pg/mg level. Several methods were published involving one or more benzodiazepines, after overnight incubation with phosphate buffer (pH 8.4) and liquid–liquid extraction with ether/dichloromethane (10/90, v/v) using a 20-mg hair sample [9,124–129]. The analytical column was always a XTerra MS C18, eluted with a mixture of acetonitrile with either 0.1% FA in water or formate buffer. The reported LOQ ranged from 0.5 to 5 pg/mg hair.

To improve the recoveries and the repeatability for the 7-aminobenzodiazepines in hair, Chèze *et al.* performed a matrix extraction using 0.1 N NaOH (15' at 95°C), whereas for the other compounds, a classical extraction using Soerensen buffer at pH 7.6 (14 h at 56°C) was used [130,131]. Compounds were separated on an Uptisphere ODB C18 column, using a gradient of 2 mM formate buffer (pH 3) and acetonitrile. The LOQs ranged from 2 to 5 pg/mg.

Finally, our laboratory reported a similar LOQ for an extended range of 28 hypnotics using a triple-quad LC-MS-MS system with a XTerra MS C18 column and a gradient consisting of a mixture of methanol with 0.1% FA in water, applying a shorter incubation procedure (2 h with methanol at 45°C) followed by a liquid–liquid extraction with 1-chlorobutane [132].

Due to the sensitivity and the specificity of these methods, the applicability to monitor single use of benzodiazepines was demonstrated, enabling as such the possibility to detect a drug-facilitated assault (DFA) [133].

immunoglobulins and DNA. The major constituent of the oral fluid, however, is water. Under healthy conditions, adults will produce 500–1500 mL saliva per day. The electrolyte concentrations and the volume of saliva which is being produced are not only influenced by the moment of the day, but also by the type of the salivation stimulus [151]. This is the major difference between oral fluid and blood, in which the concentrations of the various components can only vary between narrow border values. In addition, the low protein concentration in oral fluid makes drug binding minimal compared to that observed in plasma.

Unstimulated oral fluid has a pH in the range of 5.6–7 but with stimulation the pH increases to approximate the pH of blood, i.e. 7.4, and even higher [152].

20.2.1.2 Mechanisms of drug transfer in oral fluid

It is generally accepted that there are three major possible routes for drugs to enter saliva from the blood stream: a passive diffusion process, an active process against a concentration gradient and ultrafiltration through pores in the membrane. The predominant mechanism appears to be the simple passive diffusion process, which is characterized by the transfer of drug molecules down a concentration gradient. Consequently, there is a known relationship between oral fluid and plasma concentrations depending on the pH of the two specimens, the degree of protein binding in plasma, lipophilicity of the drug and the pKa of the drug [153].

In addition, it has been demonstrated that shortly after drug use by smoking, oral ingestion or nasal insufflation (e.g. heroin, methamphetamine, marijuana and cocaine), contamination of the oral cavity can lead to dramatically elevated drug concentrations measured in oral fluid. In some cases, after a period of approximately 2 h, the oral fluid levels more readily reflect blood levels [154].

Oral fluid contains predominately the parent drug because of the higher lipid solubility and, therefore, higher potential for passive diffusion [154,155]. Oral fluid concentrations for weakly basic drugs with a pKa close to the pH of saliva (e.g. cocaine, opiates) can be significantly higher than the corresponding plasma concentration, especially in unstimulated saliva [143,156,157]. In contrast, there is very little partitioning of THC between plasma and oral fluid, yet similarity in time profiles occurs. This is possibly due to reserves of THC deposited in the oral mucosa that are leached out with time [158].

The saliva-to-plasma (S/P) ratios are also highly affected by the manner in which the oral fluid is collected [154]. In addition, a significant intra-individual variability of the S/P ratio of several drugs has to be taken into account. This factor limits the usefulness of single oral fluid measurements to predict concurrent plasma concentrations for therapeutic drug monitoring [157].

20.2.2 Collection of oral fluid

Common methods of oral fluid collection are spitting, draining, suction, and collection on various types of absorbent swabs. Spitting itself is usually a sufficient stimulus to elicit a flow, but certain drugs, e.g. some tricyclic antidepressants and amphetamines can pharmacologically reduce salivation. In these circumstances, the flow can be stimulated mechanically (Teflon™, chewing gum etc.) or chemically (lemon drops or citric acid crystals), resulting in an oral fluid sample, which differs in composition from oral fluid collected without stimulation.

There are a variety of devices available that have been marketed to facilitate the collection of oral fluid and to provide a cleaner specimen which is more suitable for analysis [154]. As a general rule, they consist of a sorbent material that becomes saturated in the mouth of the donor, and after removal, the oral fluid is recovered by centrifugation or by applying pressure. Commercial devices include Omni-Sal® (Cozart Biosciences Ltd., Abingdon, UK), Salivette® (Sarstedt AG, Rommelsdorf, Germany), Intercept® (OraSure Technologies, Bethlehem, PA, USA), Finger Collector® (Avitar Technologies, Inc, Canton, MA, USA) and ORALscreen™ (Avitar Technologies, Inc) and Quantisal™ (Immunalysis Corporation, Pomona, CA, USA). Several reviews report significant differences in % recovery from the sorbent material for various drugs and a large variability in the ultimately measured oral fluid concentrations [154,155,159]. A typical example is the observation that THC remains bound to the sampling device and that an organic solvent is needed to release it [142,160].

To increase the stability of the analytes during storage, a stabilizing buffer is often used to dilute the neat saliva specimen. Several authors have investigated the stability of specific analytes like THC [154,161], morphine and 6-acetylmorphine (6-AM) [162], and MDMA, MDA, MDEA and MBDB [163] in the collected oral fluid sample. An additional problem for the interpretation of quantitative results is the difficulty in estimating the actual volume of oral fluid collected when a dilution buffer is used in the device [146,154].

As a conclusion, any method for oral fluid needs a thorough understanding of the chosen collection method in order to interpret test results. In addition, the choice of a collection device should not only depend on the ease-of-use, but analytical considerations have to be taken into account as well. SAMHSA is presently recommending oral fluid sampling by spitting into a neat tube [28], but this process is less hygienic and more time-consuming both for the collector and the donor. Table 20.6 shows the proposed SAMHSA cut-off concentrations for screening and confirmation of oral fluid.

20.2.3 Screening techniques

20.2.3.1 Immunological procedures

A number of laboratory-based microtiter plate EIAs applied for the screening of blood and hair are also useful for the analysis of oral fluid specimens because of their high sensitivity and cross-reactivity with the parent drugs. Some commercially available EIA kits were designed to be used with specific collection devices. These include the OraSure micro-plate Intercept® EIA kits and the Cozart® micro-plate EIA. However, it was shown that their application was not restricted to the use of one single type of collection device (Table 20.7).

References pp. 690–697

TABLE 20.6

SAMHSA PROPOSED CUT-OFF CONCENTRATIONS FOR EACH DRUG IN ORAL FLUID (*DRAFT 4*) [28]

Drug class	Initial Test Cut-off Concentration (ng/mL)	Confirmatory Test Cut-off Concentration (ng/mL)
COCAINE	20	
Cocaine		8[a]
BE		8
OPIATES	40[b]	
Morphine		40
Codeine		40
6-AM		4
AMPHETAMINES	50	
Amphetamine		50
Methamphetamine		50[c]
MDMA		50
MDA		50
MDEA		50
CANNABIS	4	
THC		4

[a]Cocaine or BE.

[b]Labs are permitted to initial test all specimens for 6-AM using a 4 ng/mL cut-off.

[c]Specimen must also contain amphetamine at a concentration $\geqslant$ LOD.

Niedbala *et al.* [162] compared the results obtained by the OraSure opiate Intercept EIA with those obtained with a urine EIA screening. Using a cut-offs of 10 and 2000 ng/mL respectively, an agreement of 93.7% was obtained. After confirmation of the presumptive positive samples with GC-MS-MS (10 ng/mL cut-off), 92% were found positive for one or more opiates. However, no data on FN results were reported. When comparing the oral fluid data obtained with the STC Cocaine metabolite EIA with the urine EIA, a sensitivity and specificity of 73 and 85%, respectively was obtained [164].

When using the Cozart® EIA Cocaine Oral Fluid, Kim *et al.* [165] showed that the low pH produced in specimens collected with the Salivette® with citric acid-treated cotton swab (mean pH (SD) of specimens: 2.8 (0.3)) yielded a lower specificity compared with specimens collected after citric acid candy expectoration (mean pH (SD) of specimens: 4.3 (0.8)). However, Niedbala *et al.* [164] reported that there was no significant effect of oral fluid pH on the STC Cocaine Metabolite Micro-Plate EIA when the sample pH was between 5.0 and 9.0. In addition, no effect of common foodstuffs and household chemicals was noted when using these EIA kits. Following the evaluation of an external quality assessment scheme involving the Intercept® collection device and the corresponding EIA kits, Clarke and Wilson [166] noted that the major source of error observed was the sensitivity where the EIA failed to achieve the cut-off values specified by the manufacturer.

Because of the advantages of oral fluid as a testing matrix, on-site testing devices have been developed to screen for drugs of abuse in oral fluid and adapted for the

TABLE 20.7
IMMUNOLOGICAL SCREENING PROCEDURES FOR THE DETECTION OF DRUGS IN ORAL FLUID SPECIMENS

Study Detail	Collection Method	Screening Technique	Evaluated Compounds for Confirmation	Optimal Cut-offs for Screening/ Confirmation in Oral Fluid (ng/mL)	Sensitivity (%)	Specificity (%)	Reference
Controlled administration of codeine (n = 1406)	Citric candy stimulation and spitting or Salivette® (neutral or citric acid-treated)	Cozart® opiate EIA	Codeine, norcodeine, morphine, normorphine	20/20	82.9	98.7	[172]
Methadone-treatment program (n = 216)	Cozart® Rapiscan Saliva Collector	Cozart® opiate EIA	Heroin, 6-AM, codeine, morphine, dihydrocodeine	30/30	99.1	94.4	[173]
Methadone-treatment program (n = 198) and volunteer donors (n = 40)	Cozart® Rapiscan Saliva Collector	Cozart® d-methadone EIA	Methadone, EDDP	30/30	91.3	100	[174]
Chronic cocaine users (n = 149) and volunteers receiving single doses of cocaine (n = 163)	Intercept®	STC Cocaine Metabolite EIA	Cocaine, BE	10/10	95	82	[164]
Controlled administration of cocaine (n = 1100)	Citric candy stimulation and spitting or Salivette® (neutral or citric acid-treated)	Cozart® cocaine EIA	Cocaine, BE, EME, cocaethylene	20/8 30/15	91.8 89.8	85.0 89.6	[165]
Methadone-treatment program (n = 217)	Cozart® Rapiscan Saliva Collector	Cozart® cocaine EIA	Cocaine, BE, cocaethylene MDMA, MDA	30/30	95.7	100	[175]
Controlled study with MDMA (n = 216)	Intercept®	Cozart® amphetamine EIA		51/10	98.6	98.6	[176]
Treatment program (n = 135) and volunteer donors (n = 35)	Cozart® Rapiscan Saliva Collector	Cozart® amphetamine EIA	Amphetamine, methamphetamine, MDMA, MDA, MBDB	45/30	91.7	95.9	[177]

workplace, drug dependency clinics and roadside driving under the influence [147,167–171].

20.2.3.2 Chromatographic procedures

Although GC-MS(-MS) is the standard technique for drug testing in oral fluid, LC-MS(-MS) has also been used for analysis of this sample matrix [178]. On the one hand, it contains considerably less proteins and lipids in comparison with blood or plasma; on the other hand, it was also characterized as having a higher protein, amino acid, and especially mucin content, as compared to urine [152]. The presence of these compounds can be a hurdle in the development of fast and easy sample preparation methods [179], showing that to avoid matrix suppression, a more time-consuming SPE method using mixed-mode cartridges was required [180]. Samples (200 μL) were analyzed after separation on a Hypersil BDS phenyl column in MS-MS mode on a quadrupole-time-of-flight (QTOF) instrument. Due to the limited linear range of the TOF instrument used in this study, quadratic regression curves were shown to generate the best fit for all compounds. Unfortunately, results showed that the developed SPE method was not able to eliminate the concomitant substances when analyzing oral fluid collected with a specific device in comparison with neat oral fluid. Also Dams *et al.* [181] showed that none of their evaluated sample preparation procedures provided sufficient clean-up of oral fluid for LC-ESI-MS-MS to be used for quantification. However, this seemed not to be the case for LC-APCI-MS-MS. In general, oral fluid was shown to have more interferences than urine, mainly in ESI, and residual matrix components were both of a hydrophilic and hydrophobic nature. In a later report, these authors described the use of LC-APCI-MS-MS in combination with protein precipitation using acetonitrile for the simultaneous quantification of 27 compounds, including illicit drugs, methadone and their metabolites in 200 μL of spitted oral fluid [182]. The analytes were separated on a Synergi Polar RP column using a mixture of ammonium formate, FA and acetonitrile.

In addition to the endogenous substances present in oral fluid, specific collection devices mostly include a preservation buffer containing other compounds e.g. stabilizing salts, non-ionic surfactants and anti-bacterial agents. Their presence can cause severe ion suppression/enhancement effects during LC-ESI-MS-MS analyses with diminished precision and accuracy of subsequent measurements [183]. As such, sample clean-up is a very important factor in LC-ESI-MS(-MS) and the assessment of matrix effect should be performed in any case. Taking this into account, Wood *et al.* [183] developed a fully validated LC-ESI-MS-MS method for the quantification of multiple basic drugs in oral fluid samples collected with the Intercept® device. Chromatographic separation was achieved using a C18 analytical column and gradient elution with 10 mM ammonium bicarbonate and methanol. The optimized SPE (mixed-mode) procedure proved to be highly effective for the elimination of matrix effects. In addition, these authors showed that the pH of the processed samples was of major concern for the stability of cocaine and 6-AM and that attention should be paid to this factor when developing methods for the simultaneous determination of drugs with important differences in chemical characteristics.

Recently, two validated GC-EI-MS methods for the simultaneous analysis of multiple drugs and drugs of abuse were described [160,184]. In the first report, oral fluid samples were collected using a modified Omni-Sal® device and extracted using a single mixed-mode extraction cartridge [184]. Elution of the 49 licit and illicit drugs from the cartridge was performed using two different elution mixtures: acetone/chloroform (50:50, v/v), fraction 1; and ethyl acetate/ammonia (98:2, v/v), fraction 2. GC-MS of the underivatized and pentafluoropropionyl derivatives and LC-ESI-MS-MS analyses were used in parallel: the first fraction contained diazepam and its three metabolites, and was analyzed with LC-ESI-MS-MS; the second fraction (including the amphetamines, cocaine, opioids and their metabolites) was divided into two portions to be analyzed by LC-ESI-MS-MS and GC-MS. In order to acquire sufficient data points across the peaks, extracts had to be injected twice into the LC-MS-MS instrument. Gunnar *et al.* [160] described the simultaneous quantification of 30 drugs of abuse using 250 µL of spitted oral fluid. The drugs were also eluted in two phases, the first one with a mixture of toluene/ethyl acetate (80:20, v/v) eluting THC and most of the benzodiazepines, and the second one with acetonitrile/ammonia (100:4, v/v), containing amphetamines, opiates, cocaine, BE, midazolam, alprazolam and zolpidem. The lower vaporization volume of toluene in comparison with acetonitrile allowed the use of higher injection volumes without peak distortion or broadening. The second eluate fraction was divided into two parts, allowing the application of two distinct derivatization procedures: HFBA which was selected for the amphetamines and MSTFA for the other compounds in this fraction. *N*-methyl-*N*-*tert*-butyldimethylsilyl-trifluoroacetamide (MTBSTFA) was chosen for the derivatization of the benzodiazepines and THC. Furthermore, various GC-MS parameters were optimized, including fast temperature programming and carrier gas flow-rate, pulsed flow split/splitless injection, target tuning and SIM time windows. GC columns (30 m) with a larger internal diameter of 0.32 and 0.53 mm (DB-35MS and DB-5MS columns) were selected due to higher reproducibility and sensitivity for certain molecules (e.g. benzodiazepines). The same method was also applied to samples collected with the Intercept® device. Due to the presence of non-volatile macromolecules in the blue buffer of this device, the selectivity of the method was decreased in comparison with neat oral fluid. Although this did not deteriorated the accuracy of the method, long-term problems are obvious in the form of liner, column and ion source contamination.

In 2003, two different groups have investigated the use of SPME for oral fluid testing of various drugs [185,186]. The reported procedures had limitations related to the nature of the drugs that could be determined.

20.2.4 Confirmation analysis

20.2.4.1 Cocaine and opioids

Several publications have shown useful correlations among the concentration levels of cocaine and metabolites in oral fluid with those encountered in blood and urine samples. In the past 5 years, only three papers concerning the quantification of cocaine and its metabolites in oral fluid were published (Table 20.8). In the first one

[187], samples used for validation were collected by spitting, whereas authentic samples were obtained with Salivette®. The drug recoveries from the Salivette for ecgonine methyl ester (EME), cocaine and BE ranged from 23 to 84%, hampering as such accurate quantification of these compounds. The same problem was encountered when using the same procedure for the analysis of opiates [188].

Several methods for opiates were developed to investigate the pharmacokinetics in this matrix after controlled administration (Table 20.8) [189–191]. Jones *et al.* [89] described the simultaneous determination of several opiates in both hair and oral fluid samples using methoxime/BSTFA derivatives, as such enabling the chromatographic separation of the prescription opiates. This method was successfully applied to the analysis of oral fluid samples collected after the ingestion of poppy seeds [190].

Three reports dealing with the quantitative analysis of methadone were published in the past 5 years [192–194], two of them being able to separate chromatographically the *R*- and *S*-enantiomers of this compound. Both used a chiral AGP column, providing a good chromatographic separation between the two enantiomers. In addition, the report of Rosas *et al.* included the simultaneous enantioselective determination of EDDP. Application of this method for the analysis of samples from patients enrolled in a methadone maintenance program showed an *R/S* ratio for methadone ranging between 1.3 and 2.2 regardless of the dose; for EDDP it was noted that this ratio was < 1.0 for patients receiving 100 mg/day and > 1.0 for those receiving 70 mg/day.

20.2.4.2 Amphetamines

Until 2001, only a few reports dealing with the quantification of amphetamines in oral fluid were published. In 2001, Navarro *et al.* [143] applied a previously published method for plasma samples to the detection of MDMA and its metabolites MDA and 4-hydroxy-3-methoxymethamphetamine (HMMA) in spitted oral fluid samples obtained from a controlled study. Analysis of the samples was performed using GC-EI-MS after mixed-mode SPE and derivatization with MBTFA. Similarly, Schepers *et al.* [157] used an extended version of an earlier published method for the analysis of sweat samples for the quantification of methamphetamine and amphetamine. Oral fluid samples from a controlled study with methamphetamine were first worked-up using mixed-mode SPE and derivatized using MTBSTFA before GC-EI-MS analysis.

A rapid GC-EI-MS assay for the simultaneous determination of 15 psychoactive amines using only 100 μL of spitted sample was developed and validated by Kankaanpää *et al.* [197]. Extraction and derivatization occurred in a single step using a mixture of toluene, methylmexiletine (as IS) and HFBA with direct injection of an aliquot of the solvent phase into the GC-MS system. Different columns were tested, with the bonded, cross-linked non-polar DB-5MS column yielding satisfactory separation of all compounds. With an LOQ of 20 ng/mL for all compounds tested, this method offers the rapidity of an immunoassay combined with the versatility and accuracy of a confirmation technique.

Recently, Scheidweiler and Huestis [198] described a validated GC-EI-MS method for the quantification of amphetamine, methampethamine, MDMA, MDA, MDEA, HMMA and 3-hydroxy-4-methoxyamphetamine (HMA) in spitted oral fluid samples. Sample clean-up was achieved using mixed-mode SPE cartridges and the compounds were derivatized with heptafluorobutyric acid (HFAA). Using 400 µL of sample, a LOQ of 5 ng/mL was obtained for all analytes, except for HMA and HMMA, where an LOQ of 25 ng/mL was observed.

One report describes the use of HPLC with fluorescence detection for the detection of the designer amphetamines MDMA, MDA, MDEA and MBDB in spitted oral fluid samples after liquid/liquid extraction with Toxitube A® [163]. The analytical method allowed good separation of the compounds in an isocratic mode with phosphate buffer (pH 5) and acetonitrile (75:25, v/v) in only 10 min using a C8 analytical column. The reported LOQ was 10 ng/mL for all compounds. However, no internal standard was used due to the difficulty to find a compound with native fluorescence and good chromatographic properties under isocratic conditions. This could be the reason for the higher %RSD observed during the inter-assay precision experiments.

Finally, Wood *et al.* [199] developed a very sensitive LC-MS-MS method for the quantification of amphetamines in oral fluid samples. Following a simple protein precipitation step with methanol, a LOQ of 0.5–1.0 ng/mL could be achieved using only 50 µL of sample. All analyses were performed using a Hypersil BDS C18 column eluted isocratically using ammonium acetate and acetonitrile (75:25, v/v).

20.2.4.3 Cannabinoids

Very little is known about the composition of cannabinoid compounds in oral fluid samples. In the process of smoking cannabis, THC is deposited in the oral cavity, and it appears that this oral mucosal depot of active THC is the primary source of THC that is ultimately collected and measured during oral fluid testing [158]. In addition to THC, CBD and CBN are detected in oral fluid after smoking of cannabis. Recently, Day *et al.* [200] presented data on the detection of THC-COOH in oral fluid using gas chromatography-tandem mass spectrometry (GC/MS/MS) to achieve the sensitivity required for the detection of THC-COOH in oral fluid. They reported a quantitation limit of 10 pg/mL and concentrations up to 240 pg/mL present in the oral fluid specimens.

In 2000, Kintz *et al.* [201] described the use of a GC-MS procedure to test for THC in oral fluid samples collected with the Salivette®. Extraction of this compound from the cotton roll was performed using a mixture of hexane/ethyl acetate (90:10, v/v), followed by derivatization with TBAH/DMSO and methyliodide (LOQ = 1 ng/Salivette). This method was subsequently applied to specimens obtained from injured drivers. Later on, these authors reported the detection of THC in oral fluid samples collected with the Intercept® device using the same analytical procedure [202]. The use of GC-EI-MS-MS after liquid/liquid extraction with hexane/ethyl acetate (90:10, v/v) and derivatization with BSTFA, or SPE with C18 cartridges before HFPA derivatization, has also been reported [203–205].

TABLE 20.8
ANALYTICAL PROCEDURES FOR THE DETECTION OF COCAINE AND OPIOIDS IN ORAL FLUID SAMPLES

Reference	Drugs	Sample Collection	Sample Clean-up	Derivatization	Column	Chromatography	LOQ (ng/mL)
[187]	Cocaine EME BE	Spiting, Salivette®	ToxiTube® A	BSTFA	Capillary column with 5% phenyl-methylsiloxane	GC-PCI-MS	7.4 3.0 0.8
[195]	BE Cocaine Cocaethylene	Salivette®	Mixed-mode SPE	—	Hypersil BDS C18	LC-EI-MS-MS	10 10 10
[196]	Cocaine AEME EME Cocaethylene	Salivette®	Automated mixed-mode SPE	—	DB-5	GC-PCI-MS-MS	2 — 5 2
[191]	Dihydrocodeine Dihydrocodeine-6-glucuronide Dihydromorphine Dihydromorphine-6-glucuronide Dihydromorphine-3-glucuronide N-nordihydrocodeine	Salivette®	C8 SPE	—	Nucleosil 100 C18	HPLC with fluorescence detector	19.8 20.0 14.8 15.0 15.0 14.9

Ref.	Analytes	Sample	Extraction	Derivatization	Column	Method	Value
[189]	Codeine / Norcodeine / Morphine / Normorphine	Citric acid candy stimulation, citric acid-treated and neutral Salivette®	Mixed-mode SPE	(1) MTBSTFA (2) BSTFA	HP-1 or Phenomenex ZB1	GC-EI-MS	2.5 / 2.5 / 2.5 / 2.5
[89,190]	Codeine / Morphine / 6-AM / Hydrocodone / Hydromorphone / Oxycodone	Artificial oral fluid	Mixed-mode SPE	(1) Methoxime (2) BSTFA	DB-5	GC-EI-MS	2 / 2 / 3 / 10 / 3 / 3
[188]	Codeine / Morphine / 6-AM	Spitting, Salivette®	ToxiTube® A	BSTFA	Capillary column with 5% phenyl-methylsiloxane	GC-PCI-MS	2.3 / 6.7 / 2.0
[194]	R-/S-methadone / R-/S-EDDP	Salivette®	—	—	α1-acid glycoprotein (AGP)	LC-EI-MS	5.0 / 0.5
[193]	R-/S-methadone	Salivette®	Filtration	—	α1-acid glycoprotein (AGP)	LC-ESI-MS	5
[192]	Methadone / EDDP	Spitting?	SPME	—	HP-5	GC-EI-MS	45 / 18

Recently, several authors used LC-MS(-MS) for the confirmation of THC positive samples using different collection protocols [206–209]. The XTerra MS C18 column eluted with a mixture of acetonitrile/0.05% NH$_3$ (70:30, v/v) [208,209], acetonitrile/ 0.1% FA (85:15, v/v) [206] or methanol/1 mM ammonium formate (90:10) [207] was used in these methods. Sample clean-up was performed using either mixed-mode SPE and elution with hexane/ethyl acetate (80:20, v/v) [208,209] or liquid/liquid extraction with hexane [206,207]. The LOQ obtained using tandem MS technology was markedly lower in comparison with single MS methods. Using only 100 µL of sample, an LOQ of 0.5 ng/mL for THC could be obtained [207].

20.2.4.4 Benzodiazepines and neuroleptics

Because of their extensive protein binding (95–99%), benzodiazepines are only present in oral fluid in very low concentrations (low ng or pg range). Therefore, sensitive assays are needed for the analysis of these samples.

Only one paper deals with the determination of flunitrazepam and its metabolite, 7-aminoflunitrazepam in oral fluid samples using GC-NCI-MS [210]. After the controlled administration of 1 mg flunitrazepam, samples obtained by spitting were derivatized using HFBA following extraction using mixed-mode SPE cartridges. Due to the low pKa of flunitrazepam, cartridges had to be washed with 1 N phosphoric acid (pH 1) before washing with methanol to retain this compound on the column. Chromatography was achieved using a HP-1 capillary column and the reported LOQs were 0.1 and 0.15 ng/mL for flunitrazepam and 7-aminoflunitrazepam, respectively. These authors showed that the stability of flunitrazepam in spitted oral fluid samples was poor when no NaF was added to the samples as preservative.

The past 2 years, several LC-ESI-MS(-MS) methods dealing with the detection of one or more benzodiazepines in oral fluid samples have been published. The group of Kintz has described the detection of zolpidem, lorazepam and tetrazepam in this matrix after the controlled administration of these drugs [124,125,211]. These authors published a validated procedure for the screening of 17 benzodiazepines and hypnotics in oral fluid after collection with the Intercept® device by LC-MS-MS [212]. Extraction of 500 µL of this preserved oral fluid was performed with diethyl ether/dichloromethane (50:50, v/v). Compounds were separated on a XTerra MS C18 analytical column using a gradient of acetonitrile/0.1% FA in water. The LOQ ranged from 0.1 to 0.2 ng/mL.

Using a single MS instrument, Quintela et al. [213,214] obtained slightly higher LOQs for the detection of up to nine selected benzodiazepines in oral fluid samples collected with the Salivette®. In these methods, chromatography is achieved using a XTerra RP$_{18}$ column eluted with a gradient of acetonitrile in 0.1% FA.

Finally, Flarakos et al. [215] developed a robust LC-MS-MS quantification assay for the new neuroleptic drugs risperidone and 9-hydroxyrisperidone both in human plasma and oral fluid. The method used 25 µL of sample precipitated with acetonitrile. Analyses were performed using column switching for online clean-up of the extracts. Both the loading and analytical column were a Zorbax SB C18 and different mixtures of ammonium acetate/acetonitrile were used for loading and analytical purposes. The plasma assay was cross validated for oral fluid analyses. Both analytes were present in the oral fluid samples of patients treated with risperidone, providing evidence of recent adherence with therapy.

20.3 SWEAT

20.3.1 Introduction

Although the use of sweat for drug testing has been hampered by difficulties in sample recovery and sensitivity of analytical methods, significant advances facilitating sample collection and improving the accuracy of diagnostic techniques have permitted successful sweat testing for several drugs of abuse. Sweat testing is relatively non-invasive and identification of drug in sweat may serve as a means of monitoring recent drug use with a window of detection which can be somewhat wider than that provided by urine testing. Disadvantages, however, include the generally low volume of collected sample and the difficulty to provide a quantitative test when using traditional collection systems (cotton wipes, sweat patches).

Sweat testing may serve as a useful tool in the surveillance of individuals in treatment and probation programs due to the fact that monitoring illicit drugs in sweat on a weekly basis may provide sufficient detection sensitivity [216–218]. Another application of sweat testing has been the roadside on-site testing of potentially intoxicated drivers [142,219].

20.3.1.1 Physiology of sweat

Sweat is secreted from ecrine and apocrine glands originating deep within the skin dermis and terminating in secretory ducts that empty onto the skin surface and into hair follicles. Sweat is secreted onto the skin surface and evaporates causing convectional body heat loss. Water is the primary constituent of sweat, approximately 99%, and sodium chloride is the most concentrate solute. Sweat also contains albumin, gamma globulins, waste products, trace elements, drugs and many other substances found in blood. The amount of sweat secreted is highly variable and dependent upon daily activity, emotional state and environmental temperature. Similarly to oral fluid, sweat has a pH lower than that of plasma (mean 5.8). With the increased flow rate, sweat pH has been found to increase to between 6.1 and 6.7 [216].

20.3.1.2 Mechanisms of drug transfer in sweat

There are several potential mechanisms by which drugs may be secreted in sweat including passive diffusion from blood into sweat glands and transdermal migration of drugs across the skin [152]. Non-ionized basic drugs diffuse into sweat and become ionized as a result of the lower pH of sweat as compared to blood, which is the same mechanism as for oral fluid. It appears that molecular mass, pKa, degree of protein binding and lipophilicity primarily determine drug and metabolite disposition.

There is some evidence that after chronic exposure, lipophilic drugs may be stored in adipose tissue, which can serve as a drug depot [220]. As such, storage in skin and subcutaneous fat can occur and produce a positive sweat patch result several days after cessation of drug abuse. Although this phenomenon has been contradicted by others [221], controlled studies with cocaine and codeine indicate that a positive sweat patch can still be obtained during the third week after abstinence [222,223].

In general, the primary analyte found in sweat is the parent drug [216,218, 222,224–227]. The measured concentration of drug is likely to be dependent upon the collection method and the site of collection [216,222–224]. SAMHSA has proposed a confirmatory cut-off concentration of 25 ng/patch for cocaine (or BE), opiates and amphetamines [28]. In order to reduce the false positive rate during their study, Kidwell *et al.* [228] proposed to increase the cut-off of cocaine to 75 ng/patch. However, as indicated by the results of Kacinko *et al.* [222], this would significantly increase the number of false negative results. For THC the proposed cut-off is 1 ng/patch [28].

Several studies have assessed the sensitivity, specificity and efficiency of sweat patch versus urine testing [216–218,228–231]. The results obtained were variable within one drug class and a fierce debate is ongoing about the preferential use of the sweat patch in detecting new episodes of drug use in formerly chronic drug users. In general, no correlation with the consumed dose could be established, in part due to a large inter- and intrasubject variability noted in most of these studies.

20.3.2 Collection of sweat

Specialized collection devices have been developed to improve sweat collection and the recovery of drug analytes. PharmChek® Sweat Patch from PharmChem Laboratories Inc. (Fort Worth,TX, USA) has been introduced in 1990 and successfully applied to sweat testing for drugs of abuse. This patch can be worn for an extended time period and concentrates solutes on a collection pad while allowing water to evaporate. The recovery of several drugs from the sweat patch was examined and was higher than 70% [227,232]. Only for THC it appears that a substantial fraction of the drug remains bound to the pad, which was to be expected taking into account the lipophilic nature of the drug and its avidity to glass and plastic surfaces [233].

In theory, sweat patches are sealed to the skin and exclude environmental contamination. However, Kidwell and Smith showed in an *in vitro* study that several drugs of abuse applied directly to the skin of drug-free individuals may persist there for several days [232]. Neither normal hygiene nor recommended cleaning procedures (with 70% isopropanol) before application of the patch completely remove drugs deposited on the skin. Thus, it might be argued that environmental contamination before patch application is a possible occurrence. In addition, several studies suggest that there is a time-dependent loss of drugs during patch-wearing over time [224,227]. As characterized for cocaine and MDMA, results were consistent with a reabsorption back into the skin. The metabolization or degradation on the surface of the skin and the loss to the environment appear to be minor pathways [224]. These phenomena limit one of the goals of sweat patch testing, i.e. cumulative drug detection.

Sweat can also be collected using Hand-held or Torso Fast Patches (Sudormed). In these patches, a metallic activation is used to initiate sodium acetate crystallization, which is an exothermic process. Generally, the patches are worn for 30 min to increase the amount of sweat collected and to improve sensitivity [216].

For epidemiological surveys and for testing of potentially impaired drivers a punctual sweat collection can be provided using a cosmetic pad moistened with isopropanol [142,156,201].

20.3.3 Sweat testing procedures

Both RIA and EIA were successfully applied to perform a first screening of the sweat samples [217,227,228,234,235]. Moody and Cheever [234] evaluated two types of immunoassays, RIA and EIA, for their ability to detect and quantitate cocaine and metabolites or heroin and metabolites in extracts of sweat patches. RIA showed the highest sensitivity for cocaine and its metabolites with a LOD of 2.5–100 ng/patch. For the EIA, the highest sensitivity was obtained for 6-AM and heroin with LODs for opiates ranging from 1.7 to 24.7 ng/patch. Acceptable intra- and inter-run precisions were obtained for both assays. The usefulness of RIA for the semiquantitative screening of cocaine and metabolites was confirmed in a later report by these authors [235]. Kidwell *et al.* [228] evaluated various EIAs for screening cocaine in diverse matrices, with substantial modifications to the manufacturers' procedures. The Cozart (Cozart Biosciences Ltd.) and the OraSure (OraSure Technologies) EIAs showed a sensitivity and specificity of $\geqslant 93.8$ and 90%, respectively, using an EIA cut-off of 25 ng/mL (GC-MS cut-off: 10 ng/mL cocaine). In contrast, although the Microgenics cocaine assay (Microgenics Corporation, Fremont, CA, USA) showed the requisite sensitivity for the matrices examined, it had poor precision when run in a manual mode.

Finally, Pichini *et al.* [227] used the STC microplate EIA kit for the semiquantitative determination of methamphetamine in serum for the screening of MDMA in sweat. Following a single dose of MDMA, positive results were observed at 1.5 h and as long as 24 h after drug administration. No false positive or false negative results were observed by applying a cut-off of 4.4 ng MDMA/patch. In addition, a good correlation between values obtained by EIA and GC-MS confirmation was noted.

Until now, the only device successfully applied for the on-site detection of drugs of abuse in sweat is the Drugwipe® (Securetec, Ottobrunn, Germany) [142,219,236,237].

In general, GC-MS with either EI or PCI after SPE remains the preferred method for the confirmatory analysis of these samples (Table 20.9).

20.4 MECONIUM

It is well established that illicit and licit substances can cross the placenta, the primary physiological link between mother and foetus. Meconium is the first faecal matter passed by a neonate. Its formation starts between the 12th and 16th week of

TABLE 20.9
ANALYTICAL METHODS FOR THE DETECTION OF DRUGS IN SWEAT SAMPLES (NR, NOT REPORTED)

Reference	Drugs	Sample Collection + Elution	Sample Clean-up	Derivatization	Column	Chromatography	LOQ
[232]	Cocaine, BE, heroin, amphetamine, methamphetamine, MDMA	PharmChek® Sweat Patches and skin swabs, 3 × eluted with 0.1 M HCl	Mixed-mode SPE	Acetic acid anhydride/PFPOH	DB-5MS	GC-PCI-MS	Ca. 2 ng/patch or swab
[216]	Cocaine and 11 metabolites, codeine, norcodeine, morphine, normorphine, 6-AM	Sudormed Hand-held and Torso Fast Patches, 2 × elution with acetate buffer 0.5 M (pH 5)	Mixed-mode SPE + further clean-up with acetonitrile	MTBSTFA (1% TBDMCS) and BSTFA (1% TMCS)	HP-1MS	GC-EI-MS	Ca. 2. ng/patch
[217]	Heroin, morphine, codeine, 6-AM	PharmChek® Sweat Patches, 1 × eluted with acetate buffer 0.2 M (pH 5) with methanol (25:75, v/v)	NR	NR	NR	GC-EI-MS	Heroin: 12.5 ng/patch, morphine, codeine and 6-AM: 7.5 ng/patch
[231,225]	Cocaine, BE, AEME, ecgonidine	Hand-held Fast Patches, 1 × eluted with acetate buffer 0.2 M (pH 5) with methanol (25:75, v/v)	Mixed-mode SPE	Cocaine, BE and AEME: HFIP; PFPA; ecgonidine: BSTFA	NR	GC-EI-MS	NR
[228]	Cocaine, BE	PharmChek® Sweat Patches and skin swabs, 2 × eluted with 0.1 M HCl	Mixed-mode SPE	Acetic acid anhydride/PFPOH	DB-5MS	GC-PCI-MS	NR; LOD: cocaine: 4 ng/mL, BE: 2 ng/mL
[235]	Cocaine, BE, EME	PharmChek® Sweat Patches, 1 × eluted with acetate buffer 0.2 M (pH 5) with methanol (25:75, v/v)	Mixed-mode SPE	Hexafluoroisopropanol and PFPA	DB-5MS	GC-PCI-MS	4 ng/patch

Ref	Analytes	Sample preparation	Extraction	Derivatization	Column	Instrument	LOD/Result
[224]	Cocaine, BE	PharmChek® Sweat Patches 1 × eluted with acetate buffer 0.2 M (pH 5) with methanol (25:75, v/v)	Mixed-mode SPE	MTBSTFA	DB-5MS	GC-PCI-MS	2 ng/patch
[238]	Cocaine, cocaethylene	PharmChek® Sweat Patches, 1 × elution with acetate buffer 0.2 M (pH 5)	SPME	–	HP-5MS	GC-EI-MS	NR; LOD: 12.5 ng/patch
[222]	Cocaine, BE, EME, cocaethylene, egonine ethyl ester, norcocaine, norcocaethylene, m- and p-hydroxycocaine, m- and p-hydroxy-BE	PharmChek® Sweat Patches, 3 × elution with acetate buffer 0.2 M (pH 4)	Mixed-mode SPE	MTBSTFA (1% TBDMCS) and BSTFA (1% TMCS)	HP-1MS	GC-EI-MS	2.5–5.0 ng/patch
[223]	Codeine, norcodeine, morphine, normorphine, 6-AM	PharmChek® Sweat Patches, 3 × elution with acetate buffer 0.5 M (pH 4)	Mixed-mode SPE	MTBSTFA (1% TBDMCS) and BSTFA (1% TMCS)	HP-1MS	GC-EI-MS	2.5–5.0 ng/patch
[156]	MDMA	Sweat wipe, 1 × eluted with methanol	Filtration	–	Hypersil BDS C18	LC-ESI-MS-MS	5 ng/wipe
[227]	MDMA, MDA, HMMA	Pharm Chek® Sweat Patches, 1 × eluted with 0.1 M phosphate potassium buffer (pH 6)	Before EIA: C18 SPE; Before confirmation: Mixed-mode SPE	MBTFA	Ultra-2	GC-EI-MS	MDMA: 3.2 ng/patch, MDA: 2.4 ng/patch, HMMA: 3.2 ng/patch
[201]	THC	Cosmetic pads, direct extraction	hexane/ethyl acetate (90:10, v/v)	TBAH/DMSO + methyliodide	HP5-MS	GC-EI-MS	1 ng/wipe
[233]	THC	PharmChek® Sweat Patches 1 × eluted with acetate buffer 0.2 M (pH 5) with methanol (25:75, v/v)	mixed-mode SPE	TFAA	Rtx-1	GC-NCI-MS	0.4 ng/patch

TABLE 20.10
ANALYTICAL METHODS FOR THE DETECTION OF DRUGS IN MECONIUM

Reference	Drugs	Extraction From Matrix	Sample Clean-up	Derivatization	Column	Chromatography	LOQ (ng/mg)
[248]	Cocaine + 15 metabolites	0.5 g, sonication with methanol	Mixed-mode SPE	–	Zorbax Eclipse XDB-C8	LC-ESI-MS-MS	0.005
[240]	6-AM, morphine, codeine	1 g, (ultrasonic) disruption in methanol	Comparison of 4 mixed-mode SPE cartridges	BSTFA	DB-1/DB-5MS	GC-EI-MS	0.005–0.020
[246]	Methadone, EDDP, EMDP	1 g, ultrasonic disruption in methanol	Before EIA: clean-up with methanol; before GC-MS: mixed-mode SPE	–	DB-5MS	GC-EI-MS	0.025
[249]	Arecoline	1 g, addition of saturated NH$_4$Cl (pH 9.5) before 5′ extraction with chloroform/ isopropanol (95:5) in horizontal shaker	Back-extraction with 0.5 M HCl and re-extraction with chloroform/ isopropanol (95:5)	–	Luna C18	LC-ESI-MS	0.005
[241]	Morphine, 6-AM, codeine, cocaine, BE, cocaethylene (1); Morphine-3-glucuronide, morphine-6-glucuronide (2)	1 g, 20′ extraction with methanol in horizontal shaker	(1) Mixed-mode SPE, (2) Ethyl SPE	–	Zorbax Eclipse XDB-C8	LC-ESI-MS	0.001–0.004
[242]	Amphetamine, methamphetamine,		Mixed-mode SPE	–	XTerra RP 18	LC-ESI-MS	0.004–0.005

Ref	Analytes	Sample preparation	SPE/cleanup	Derivatization	Column	Detection	LOD
[250]	MDMA, MDA, HMMA, MDEA, MBDB; m- and p-hydroxy-BE	1 g, extraction with methanol-HCl in horizontal shaker; 1 g, 20' extraction with methanol in horizontal shaker	Mixed-mode SPE	–	Zorbax Eclipse XDB-C8	LC-ESI-MS	0.0013 and 0.0045
[247]	Oxycodone	0.5 g, mixed with methanol	Mixed-mode SPE	BSTFA + 1% TMCS after addition of 10% methoxyamine HCl	DB-5MS	GC-EI-MS	0.05
[251]	THC-COOH, 11-OH-THC	1 g, homogenized with methanol, followed by hydrolysis with β-glucuronidase	Cerex THC columns	MTBSTFA	DB-5MS	GC-EI-MS	0.01
[252]	Methadone, EDDP, EMDP, methadol	0.5 g, ultrasonic disruption in methanol, followed by incubation in ultrasonic bath for 30'	Mixed-mode SPE	–	Synergi Hydro-RP 80A	LC-APCI-MS-MS	0.005–0.025

gestation and usually accumulates in foetal bowel until birth and is passed by the neonate 1 to 5 days after birth. For this reason, meconium analysis allows the detection of maternal drug use to approximately the last 20 weeks of gestation and consequently provides information of foetal chronic exposure to drugs, being more informative than urine for the detection of drug exposure in pregnancy [239]. However, meconium samples are more difficult to analyze, and additional processing steps are required to disrupt the tissues and to extract the analytes from this non-uniform matrix [240].

In recent years, drug testing in meconium has been successfully applied to assess intra-uterine exposure to drugs to provide the basis for appropriate treatment and follow-up of newborns, which can present symptoms of drug withdrawal and impairment in physical and mental development [241–244]. Recently, two large scale projects using meconium analysis to estimate the prevalence of drug use by pregnant women and the effects of exposure to illicit drugs during pregnancy on the foetus and infant were published [243,245].

Recent screening methods have utilized EIA [246–248] and confirmation was obtained by GC-MS and LC-MS(-MS) (Table 20.10).

20.5 CONCLUSION

Recent advances in analytical techniques have enabled the detection of drugs and drug metabolites in alternative biological specimens for investigating current and past exposure to drugs by individuals. The detection of analytes in the low ng or pg range in these matrices offers additional information in some areas where the analysis of conventional samples such as urine and blood is oblivious to some major disadvantages (e.g. DFSA, DUID, drug use during gestation). Although GC-MS is still the method of choice in practice, GC-MS-MS or LC-MS-MS methods are today in use in several laboratories, even for routine cases, particularly to target low dosage compounds.

Some basic physiological issues and analytical problems still remain to be resolved. For hair samples, the passive contamination of the sample, the distribution along the hair strand, the dose-to-concentration correlation, the hair color bias, are problems that require a result to be interpreted with great attention and caution.

The usability of oral fluid and sweat depend largely on their collection method. The availability and ease-of-use of a (commercial) device is often an important factor, but other factors such as the actual volume of oral fluid collected, the recovery of the analytes from the device and drug stability should be taken into account. Moreover, the environmental contamination and the individual variability hamper a straightforward interpretation of the results. In this respect, the application of cut-off values is of utmost importance.

Meconium appears to be one of the most promising matrices for monitoring *in utero* drug exposure, but there are many issues that need to be resolved prior to optimal use of this biological matrix. It is necessary to establish if drug and metabolite concentrations correlate with maternal drug usage and, importantly, with neonatal outcomes.

20.6 ABBREVIATIONS

6-AM	6-Acetylmorphine
AEME	Anhydroecgonine methyl ester
APCI	Atmospheric pressure chemical ionization
BE	Benzoylecgonine
BSTFA	*N*,*O*-bis(trimethylsilyl)-trifluoroacetamide
CBD	Cannabidiol
CBN	Cannabinol
CE	Capillary electrophoresis
DFA	Drug-facilitated assault
DMSO	Dimethyl sulfoxide
ECCI	Electron capture chemical ionization
EDDP	1,5-Dimethyl-3,3-diphenylpyrrolidine
EDMP	2-Ethyl-5-methyl-3,3-diphenylpyrroline
EI	Electron impact
EIA	Enzyme immunoassay
EME	Ecgonine methyl ester
ESI	Electrospray ionization
FA	Formic acid
GC-MS	Gas chromatography-mass spectrometry
GC-MS-MS	Gas chromatography-tandem mass spectrometry
HFAA	Heptafluorobutyric acid
HFBA	Heptafluorobutyric anhydride
HFIP	1,1,1,3,3,3-Hexafluoro-2-propanol
HMA	3-Hydroxy-4-methoxyamphetamine
HMMA	4-Hydroxy-3-methoxymethamphetamine
HPLC	High-pressure liquid chromatography
HS-SPDE	Headspace solid-phase dynamic extraction
HS-SPME	Headspace solid-phase microextraction
LC-MS	Liquid chromatography-mass spectrometry
LC-MS-MS	Liquid chromatography-tandem mass spectrometry
LOD	Limit of detection
LOQ	Limits of quantitation
MBTFA	*N*-methyl-bis(trifluoramide)
MDA	3,4-Methylenedioxyamphetamine
MDEA	3,4-Methylenedioxyethyl-amphetamine
MDMA	3,4-Methylenedioxymethamphetamine
MSTFA	*N*-methyl-*N*-trimethylsilyl-trifluoroacetamide
MTBSTFA	*N*-methyl-*N*-*tert*-butyldimethylsilyl-trifluoroacetamide
NCI	Negative chemical ionization
PA	Propionic anhydride
PCI	Positive chemical ionization
PDA	Photodiode-array
PFPA	Pentafluoropropionic anhydride

PFPOH	Pentafluoropropanol
QTOF	Quadrupole-time-of-flight
RIA	Radioimmunoassay
SAMHSA	Substance Abuse and Mental Health Services Administration
SIM	Single ion monitoring
(S)-HFBPCl	(S)-heptafluorobutyrylprolyl chloride
SOHT	Society of Hair Testing
SPE	Solid-phase extraction
TBAH	Tetrabutyl-ammonium hydroxide
TFAA	Trifluoroacetic anhydride
THC	delta9-tetrahydrocannabinol
THC-COOH	11-nor-delta9-tetrahydrocannabinol-9-carboxylic acid
TMCS	Trimethylchlorosilane
TMIS	Trimethyliodosilane
TMSI	N-trimethylsilylimidazole

20.7 REFERENCES

1   V. Cirimele. In: P. Kintz (Ed.), Drug testing in hair, CRC Press, Boca Raton, FL, 1996, p. 181.
2   D. Garside and B.A. Goldberger. In: P. Kintz (Ed.), Drug testing in hair, CRC Press, Boca Raton, FL, 1996, p. 151.
3   A. Tracqui. In: P. Kintz (Ed.), Drug testing in hair, CRC Press, Boca Raton, FL, 1996, p. 191.
4   H. Sachs and P. Kintz, J. Chromatogr. B, Biomed. Sci. Appl., 713 (1998) 147.
5   C. Staub, J. Chromatogr. B, Biomed. Sci. Appl., 733 (1999) 119.
6   Y. Nakahara, J. Chromatogr. B, Biomed. Sci. Appl., 733 (1999) 161.
7   P. Kintz and N. Samyn. In: M.J. Bogusz (Ed.), Unconventional samples and alternative matrices, Elsevier Science B.V., Amsterdam, The Netherlands, 2000, p. 459.
8   S. Pichini, O. Garcia-Algar, L. Munoz, O. Vall, R. Pacifici, C. Figueroa, J.A. Pascual, D. Diaz and J. Sunyer, J. Expo. Anal. Environ. Epidemiol., 13 (2003) 144.
9   M. Villain, M. Cheze, V. Dumestre, B. Ludes and P. Kintz, J. Anal. Toxicol., 28 (2004) 516.
10   J.P. Goulle, M. Cheze and G. Pepin, J. Anal. Toxicol., 27 (2003) 574.
11   E. Han, W. Yang, J. Lee, Y. Park, E. Kim, M. Lim and H. Chung, Forensic Sci. Int., 147 (2005) 21.
12   C.R. Robbins, Chemical and physical behavior of human hair, Springer, New York, 1988.
13   L. Potsch, G. Skopp and M.R. Moeller, Forensic Sci. Int., 84 (1997) 25.
14   C.R. Borges, J.C. Roberts, D.G. Wilkins and D.E. Rollins, J. Anal. Toxicol., 27 (2003) 125.
15   D.J. Claffey and J.A. Ruth, Chem. Res. Toxicol., 14 (2001) 1339.
16   D.L. Dehn, D.J. Claffey, M.W. Duncan and J.A. Ruth, Chem. Res. Toxicol., 14 (2001) 275.
17   R.C. Kelly, T. Mieczkowski, S.A. Sweeney and J.A. Bourland, Forensic Sci. Int., 107 (2000) 63.
18   D.A. Kidwell, E.H. Lee and S.F. DeLauder, Forensic Sci. Int., 107 (2000) 39.
19   T. Mieczkowski and R. Newel, Forensic Sci. Int., 107 (2000) 13.
20   L. Gautam, K.S. Scott and M.D. Cole, J. Anal. Toxicol., 29 (2005) 339.
21   D.L. Hubbard, D.G. Wilkins and D.E. Rollins, Drug Metab. Dispos., 28 (2000) 1464.
22   D.E. Rollins, D.G. Wilkins, G.G. Krueger, M.P. Augsburger, A. Mizuno, C. O'Neal, C.R. Borges and M.H. Slawson, J. Anal. Toxicol., 27 (2003) 545.
23   R. Kronstrand, S. Forstberg-Peterson, B. Kagedal, J. Ahlner and G. Larson, Clin. Chem., 45 (1999) 1485.

24 M.H. Slawson, D.G. Wilkins and D.E. Rollins, J. Anal. Toxicol., 22 (1998) 406.
25 Society of hair testing, Forensic Sci. Int., 145 (2004) 83.
26 http://www.soht.org.
27 M. Uhl, Forensic Sci. Int., 107 (2000) 169.
28 Department of Health and Human Services. Substance Abuse and Mental Health Services Administration. Proposed Revisions to Mandatory Guidelines for Federal Workplace Drug Testing Programs, 69 FR 19673, 2004.
29 G. Romano, N. Barbera and I. Lombardo, Forensic Sci. Int., 123 (2001) 119.
30 G. Romano, N. Barbera, G. Spadaro and V. Valenti, Forensic Sci. Int., 131 (2003) 98.
31 T. Cairns, V. Hill, M. Schaffer and W. Thistle, Forensic Sci. Int., 145 (2004) 97.
32 M. Schaffer, V. Hill and T. Cairns, J. Anal. Toxicol., 29 (2005) 319.
33 M.I. Schaffer, W.L. Wang and J. Irving, J. Anal. Toxicol., 26 (2002) 485.
34 J.D. Ropero-Miller, B.A. Goldberger, E.J. Cone and R.E. Joseph Jr., J. Anal. Toxicol., 24 (2000) 496.
35 E.A. Welp, I. Bosman, M.W. Langendam, M. Totte, R.A. Maes and E.J. van Ameijden, Addiction, 98 (2003) 987.
36 S. Paterson, R. Cordero, M. McPhillips and S. Carman, J. Anal. Toxicol., 27 (2003) 20.
37 C. Girod and C. Staub, Forensic Sci. Int., 117 (2001) 175.
38 M. Yegles, Y. Marson and R. Wennig, Forensic Sci. Int., 107 (2000) 87.
39 G. Skopp, L. Potsch and M. Mauden, Clin. Chem., 46 (2000) 1846.
40 R. Kronstrand, M.C. Andersson, J. Ahlner and G. Larson, J. Anal. Toxicol., 25 (2001) 594.
41 G. Pepin and Y. Gaillard, Forensic Sci. Int., 84 (1997) 37.
42 M. Pujadas, S. Pichini, S. Poudevida, E. Menoyo, P. Zuccaro, M. Farre and T.R. de la, J. Chromatogr., B, Analyt. Technol. Biomed. Life Sci., 798 (2003) 249.
43 O. Quintela, A.M. Bermejo, M.J. Tabernero, S. Strano-Rossi, M. Chiarotti and A.C. Lucas, Forensic Sci. Int., 107 (2000) 273.
44 F. Tagliaro, R. Valentini, G. Manetto, F. Crivellente, G. Carli and M. Marigo, Forensic Sci. Int., 107 (2000) 121.
45 L.D. Katikaneni, F.R. Salle and T.C. Hulsey, Biol. Neonate, 81 (2002) 29.
46 V. Spiehler, Forensic Sci. Int., 107 (2000) 249.
47 A. Miki, M. Katagi, N. Shima and H. Tsuchihashi, J. Anal. Toxicol., 28 (2004) 132.
48 V. Cirimele, S. Etienne, M. Villain, B. Ludes and P. Kintz, Forensic Sci. Int., 143 (2004) 153.
49 G. Cooper, L. Wilson, C. Reid, D. Baldwin, C. Hand and V. Spiehler, J. Anal. Toxicol., 27 (2003) 581.
50 K. Lachenmeier, F. Musshoff and B. Madea, Forensic Sci. Int., 159 (2006) 189.
51 S.A. Sweeney, R.C. Kelly, J.A. Bourland, T. Johnson, W.C. Brown, H. Lee and E. Lewis, J. Anal. Toxicol., 22 (1998) 418.
52 M. Uhl and H. Sachs, Forensic Sci. Int., 145 (2004) 143.
53 A. Negrusz, C. Moore, D. Deitermann, D. Lewis, K. Kaleciak, R. Kronstrand, B. Feeley and R.S. Niedbala, J. Anal. Toxicol., 23 (1999) 429.
54 J. Segura, C. Stramesi, A. Redon, M. Ventura, C.J. Sanchez, G. Gonzalez, L. San and M. Montagna, J. Chromatogr., B, Biomed. Sci. Appl., 724 (1999) 9.
55 C. Gambelunghe, R. Rossi, C. Ferranti, R. Rossi and M. Bacci, J. Appl. Toxicol., 25 (2005) 205.
56 L. Skender, V. Karacic, I. Brcic and A. Bagaric, Forensic Sci. Int., 125 (2002) 120.
57 S. Paterson, N. McLachlan-Troup, R. Cordero, M. Dohnal and S. Carman, J. Anal. Toxicol., 25 (2001) 203.
58 C. Girod and C. Staub, Forensic Sci. Int., 107 (2000) 261.
59 S. Gentili, M. Cornetta and T. Macchia, J. Chromatogr., B, Analyt. Technol. Biomed. Life Sci., 801 (2004) 289.
60 D.W. Lachenmeier, L. Kroener, F. Musshoff and B. Madea, Rapid Commun. Mass Spectrom., 17 (2003) 472.
61 M.J. Welch, L.T. Sniegoski and S. Tai, Anal. Bioanal. Chem., 376 (2003) 1205.
62 R. Kronstrand, I. Nystrom, J. Strandberg and H. Druid, Forensic Sci. Int., 145 (2004) 183.

63   A. Polettini, C. Stramesi, C. Vignali and M. Montagna, Forensic Sci. Int., 84 (1997) 259.
64   M.A. Balikova and V. Habrdova, J. Chromatogr. B, Analyt. Technol. Biomed. Life Sci., 789 (2003) 93.
65   F.S. Romolo, M.C. Rotolo, I. Palmi, R. Pacifici and A. Lopez, Forensic Sci. Int., 138 (2003) 17.
66   K.B. Scheidweiler and M.A. Huestis, Anal. Chem., 76 (2004) 4358.
67   M. Rothe and F. Pragst, J. Anal. Toxicol., 19 (1995) 236.
68   M. Schaffer, V. Hill and T. Cairns, J. Anal. Toxicol., 29 (2005) 76.
69   W.E. Brewer, R.C. Galipo, K.W. Sellers and S.L. Morgan, Anal. Chem., 73 (2001) 2371.
70   B.K. Charles, J.E. Day, D.E. Rollins, D. Andrenyak, W. Ling and D.G. Wilkins, J. Anal. Toxicol., 27 (2003) 412.
71   E. Vinner, J. Vignau, D. Thibault, X. Codaccioni, C. Brassart, L. Humbert and M. Lhermitte, Forensic Sci. Int., 133 (2003) 57.
72   T. Kelly, P. Doble and M. Dawson, J. Chromatogr. B, Analyt. Technol. Biomed. Life Sci., 814 (2005) 315.
73   A.C. Lucas, A.M. Bermejo, M.J. Tabernero, P. Fernandez and S. Strano-Rossi, Forensic Sci. Int., 107 (2000) 225.
74   S. Pichini, R. Pacifici, I. Altieri, M. Pellegrini and P. Zuccaro, J. Anal. Toxicol., 23 (1999) 343.
75   M.C. Ricossa, M. Bernini and F. DE Ferrari, Forensic Sci. Int., 107 (2000) 301.
76   J.A. Bourland, E.F. Hayes, R.C. Kelly, S.A. Sweeney and M.M. Hatab, J. Anal. Toxicol., 24 (2000) 489.
77   M. Montagna, C. Stramesi, C. Vignali, A. Groppi and A. Polettini, Forensic Sci. Int., 107 (2000) 157.
78   M. Montagna, A. Polettini, C. Stramesi, A. Groppi and C. Vignali, Forensic Sci. Int., 128 (2002) 79.
79   T. Toyo'oka, M. Yano, M. Kato and Y. Nakahara, Analyst, 126 (2001) 1339.
80   C. Girod and C. Staub, J. Anal. Toxicol., 25 (2001) 106.
81   F.C. de Toledo, M. Yonamine, R.L. Moraes Moreau and O.A. Silva, J. Chromatogr. B, Analyt. Technol. Biomed. Life Sci., 798 (2003) 361.
82   A. Acampora, C.E. Della, G. Martone and N. Miraglia, J. Mass Spectrom., 38 (2003) 1007.
83   O. Sabzevari, K. Abdi, M. Amini and A. Shafiee, Anal. Bioanal. Chem., 379 (2004) 120.
84   E. Cognard, S. Rudaz, S. Bouchonnet and C. Staub, J. Chromatogr. B, Analyt. Technol. Biomed. Life Sci., 826 (2005) 17.
85   A.M. Bermejo, P. Lopez, I. Alvarez, M.J. Tabernero and P. Fernandez, Forensic Sci. Int., 156 (2006) 2.
86   F. Sporkert and F. Pragst, J. Chromatogr. B, Biomed. Sci. Appl., 746 (2000) 255.
87   P. Kintz, Forensic Sci. Int., 121 (2001) 65.
88   P. Kintz, M. Villain, A. Tracqui, V. Cirimele and B. Ludes, J. Anal. Toxicol., 27 (2003) 527.
89   J. Jones, K. Tomlinson and C. Moore, J. Anal. Toxicol., 26 (2002) 171.
90   K.A. Hadidi, J.K. Almasad, T. Al Nsour and S. Abu-Ragheib, Forensic Sci. Int., 135 (2003) 129.
91   P. Kintz, M. Villain, V. Dumestre and V. Cirimele, Forensic Sci. Int., 153 (2005) 81.
92   N. Takayama, R. Iio, S. Tanaka, S. Chinaka and K. Hayakawa, Biomed. Chromatogr., 17 (2003) 74.
93   M. Nishida, A. Namera, M. Yashiki and T. Kojima, J. Chromatogr. B, Analyt. Technol. Biomed. Life Sci., 789 (2003) 65.
94   J.L. Villamor, A.M. Bermejo, P. Fernandez and M.J. Tabernero, J. Anal. Toxicol., 29 (2005) 135.
95   G. Frison, L. Tedeschi, D. Favretto, A. Reheman and S.D. Ferrara, Rapid Commun. Mass Spectrom., 19 (2005) 919.
96   L. Martins, M. Yegles, H. Chung and R. Wennig, J. Chromatogr. B, Analyt. Technol. Biomed. Life Sci., 825 (2005) 57.
97   F. Tagliaro, Z. De Battisti, A. Groppi, Y. Nakahara, D. Scarcella, R. Valentini and M. Marigo, J. Chromatogr. B, Biomed. Sci. Appl., 723 (1999) 195.
98   O. Al Dirbashi, N. Kuroda, S. Inuduka, F. Menichini and K. Nakashima, Analyst, 124 (1999) 493.

99 O.Y. Al Dirbashi, N. Kuroda, M. Wada, M. Takahashi and K. Nakashima, Biomed. Chromatogr., 14 (2000) 293.
100 T. Cairns, V. Hill, M. Schaffer and W. Thistle, Forensic Sci. Int., 145 (2004) 137.
101 S. Gentili, A. Torresi, R. Marsili, M. Chiarotti and T. Macchia, J. Chromatogr. B, Analyt. Technol. Biomed. Life Sci., 780 (2002) 183.
102 J. Liu, K. Hara, S. Kashimura, M. Kashiwagi and M. Kageura, J. Chromatogr. B, Biomed. Sci. Appl., 758 (2001) 95.
103 F. Musshoff, H.P. Junker, D.W. Lachenmeier, L. Kroener and B. Madea, J. Chromatogr. Sci., 40 (2002) 359.
104 F. Musshoff, D.W. Lachenmeier, L. Kroener and B. Madea, J. Chromatogr. A, 958 (2002) 231.
105 D.L. Allen and J.S. Oliver, Forensic Sci. Int., 107 (2000) 191.
106 R. Kronstrand, J. Ahlner, N. Dizdar and G. Larson, J. Anal. Toxicol., 27 (2003) 135.
107 F. Sporkert, F. Pragst, R. Bachus, F. Masuhr and L. Harms, Forensic Sci. Int., 133 (2003) 39.
108 H. Sachs and U. Dressler, Forensic Sci. Int., 107 (2000) 239.
109 C. Moore, F. Guzaldo and T. Donahue, J. Anal. Toxicol., 25 (2001) 555.
110 M.J. Baptista, P.V. Monsanto, E.G. Pinho Marques, A. Bermejo, S. Avila, A.M. Castanheira, C. Margalho, M. Barroso and D.N. Vieira, Forensic Sci. Int., 128 (2002) 66.
111 F. Musshoff, H.P. Junker, D.W. Lachenmeier, L. Kroener and B. Madea, J. Anal. Toxicol., 26 (2002) 554.
112 F. Musshoff, D.W. Lachenmeier, L. Kroener and B. Madea, Forensic Sci. Int., 133 (2003) 32.
113 V. Cirimele, H. Sachs, P. Kintz and P. Mangin, J. Anal. Toxicol., 20 (1996) 13.
114 S. Strano-Rossi and M. Chiarotti, J. Anal. Toxicol., 23 (1999) 7.
115 J.Y. Kim, S.I. Suh, M.K. In, K.J. Paeng and B.C. Chung, Arch. Pharm. Res., 28 (2005) 1086.
116 C. Moore, S. Rana, C. Coulter, F. Feyerherm and H. Prest, J. Anal. Toxicol., 30 (2006) 171.
117 M. Chiarotti and L. Costamagna, Forensic Sci. Int., 114 (2000) 1.
118 R. Marsili, S. Martello, M. Felli, S. Fiorina and M. Chiarotti, Rapid Commun. Mass Spectrom., 19 (2005) 1566.
119 V. Cirimele, P. Kintz and B. Ludes, J. Chromatogr. B, Biomed. Sci. Appl., 700 (1997) 119.
120 K.S. Scott and Y. Nakahara, Forensic Sci. Int., 133 (2003) 47.
121 A. El Mahjoub and C. Staub, Forensic Sci. Int., 123 (2001) 17.
122 S. McClean, E. O'Kane, J. Hillis and W.F. Smyth, J. Chromatogr. A, 838 (1999) 273.
123 R. Kronstrand, I. Nystrom, M. Josefsson and S. Hodgins, J. Anal. Toxicol., 26 (2002) 479.
124 M. Concheiro, M. Villain, S. Bouchet, B. Ludes, M. Lopez-Rivadulla and P. Kintz, Ther. Drug Monit., 27 (2005) 565.
125 P. Kintz, M. Villain, V. Cirimele, G. Pepin and B. Ludes, Forensic Sci. Int., 145 (2004) 131.
126 P. Kintz, M. Villain, V. Dumestre-Toulet and B. Ludes, J. Clin. Forensic Med., 12 (2005) 36.
127 P. Kintz, M. Villain, M. Cheze and G. Pepin, Forensic Sci. Int., 153 (2005) 222.
128 M. Villain, M. Cheze, A. Tracqui, B. Ludes and P. Kintz, Forensic Sci. Int., 145 (2004) 117.
129 M. Villain, M. Concheiro, V. Cirimele and P. Kintz, J. Chromatogr. B, Analyt. Technol. Biomed. Life Sci., 825 (2005) 72.
130 M. Cheze, M. Villain and G. Pepin, Forensic Sci. Int., 145 (2004) 123.
131 M. Cheze, G. Duffort, M. Deveaux and G. Pepin, Forensic Sci. Int., 153 (2005) 3.
132 M. Laloup, M.M. Ramirez Fernandez, G. De Boeck, M. Wood, V. Maes and N. Samyn, J. Anal. Toxicol., 29 (2005) 616.
133 P. Kintz, M. Villain and B. Ludes, Ther. Drug Monit., 26 (2004) 211.
134 M. Shen, P. Xiang, H. Wu, B. Shen and Z. Huang, Forensic Sci. Int., 126 (2002) 153.
135 S. McClean, E.J. O'Kane and W.F. Smyth, J. Chromatogr. B, Biomed. Sci. Appl., 740 (2000) 141.
136 C. Muller, S. Vogt, R. Goerke, A. Kordon and W. Weinmann, Forensic Sci. Int., 113 (2000) 415.
137 V. Cirimele, P. Kintz, O. Gosselin and B. Ludes, Forensic Sci. Int., 107 (2000) 289.
138 W. Weinmann, C. Muller, S. Vogt and A. Frei, J. Anal. Toxicol., 26 (2002) 303.
139 M. Josefsson, R. Kronstrand, J. Andersson and M. Roman, J. Chromatogr. B, Analyt. Technol. Biomed. Life Sci., 789 (2003) 151.

140 S.W. Toennes, S. Steinmeyer, H.J. Maurer, M.R. Moeller and G.F. Kauert, J. Anal. Toxicol., 29 (2005) 22.

141 A.G. Verstraete and M. Puddu. In: A.G. Verstraete (Ed.), Rosita. Roadside testing assessment, Rosita Consortium, Gent, 2001, p. 167.

142 N. Samyn, G. De Boeck and A.G. Verstraete, J. Forensic Sci., 47 (2002) 1380.

143 M. Navarro, S. Pichini, M. Farre, J. Ortuno, P.N. Roset, J. Segura and T.R. de la, Clin. Chem., 47 (2001) 1788.

144 S.W. Toennes, G.F. Kauert, S. Steinmeyer and M.R. Moeller, Forensic Sci. Int., 152 (2005) 149.

145 Y.H. Caplan and B.A. Goldberger, J. Anal. Toxicol., 25 (2001) 396.

146 E.J. Cone, L. Presley, M. Lehrer, W. Seiter, M. Smith, K.W. Kardos, D. Fritch, S. Salamone and R.S. Niedbala, J. Anal. Toxicol., 26 (2002) 541.

147 A.G. Verstraete, Forensic Sci. Int., 150 (2005) 143.

148 G.A. Bennett, E. Davies and P. Thomas, Drug Alcohol Depend., 72 (2003) 265.

149 E.T. Moolchan, A. Umbricht and D. Epstein, J. Addict. Dis., 20 (2001) 55.

150 J. Neale and M. Robertson, Drug Alcohol Depend., 71 (2003) 57.

151 J.K. Aps and L.C. Martens, Forensic Sci. Int., 150 (2005) 119.

152 D.A. Kidwell, J.C. Holland and S. Athanaselis, J. Chromatogr. B, Biomed. Sci. Appl., 713 (1998) 111.

153 N. Samyn, A. Verstraete, C. van Haeren and P. Kintz, Forensic Sci. Rev., 11 (1999) 1.

154 D.J. Crouch, Forensic Sci. Int., 150 (2005) 165.

155 O.H. Drummer, Forensic Sci. Int., 150 (2005) 133.

156 N. Samyn, G. De Boeck, M. Wood, C.T. Lamers, D. de Waard, K.A. Brookhuis, A.G. Verstraete and W.J. Riedel, Forensic Sci. Int., 128 (2002) 90.

157 R.J. Schepers, J.M. Oyler, R.E. Joseph Jr., E.J. Cone, E.T. Moolchan and M.A. Huestis, Clin. Chem., 49 (2003) 121.

158 M.A. Huestis and E.J. Cone, J. Anal. Toxicol., 28 (2004) 394.

159 P. Kintz and N. Samyn, Ther. Drug Monit., 24 (2002) 239.

160 T. Gunnar, K. Ariniemi and P. Lillsunde, J. Mass Spectrom., 40 (2005) 739.

161 C. Moore, M. Vincent, S. Rana, C. Coulter, A. Agrawal and J. Soares, Forensic Sci. Int., 164 (2006) 126.

162 R.S. Niedbala, K. Kardos, J. Waga, D. Fritch, L. Yeager, S. Doddamane and E. Schoener, J. Anal. Toxicol., 25 (2001) 310.

163 M. Concheiro, A. de Castro, O. Quintela, M. Lopez-Rivadulla and A. Cruz, Forensic Sci. Int., 150 (2005) 221.

164 R.S. Niedbala, K. Kardos, T. Fries, A. Cannon and A. Davis, J. Anal. Toxicol., 25 (2001) 62.

165 I. Kim, A.J. Barnes, R. Schepers, E.T. Moolchan, L. Wilson, G. Cooper, C. Reid, C. Hand and M.A. Huestis, Clin. Chem., 49 (2003) 1498.

166 J. Clarke and J.F. Wilson, Forensic Sci. Int., 150 (2005) 161.

167 A. J. Jenkins and B. A. Goldberger (Eds), On-site drug testing, Humana Press, Totowa, NJ, 2002.

168 G. Cooper, L. Wilson, C. Reid, L. Main and C. Hand, Forensic Sci. Int., 150 (2005) 239.

169 M. Laloup, M.M. Ramirez Fernandez, M. Wood, G. De Boeck, V. Maes and N. Samyn, Forensic Sci. Int., 161 (2006) 175.

170 D.J. Crouch, J.M. Walsh, R. Flegel, L. Cangianelli, J. Baudys and R. Atkins, J. Anal. Toxicol., 29 (2005) 244.

171 J.M. Walsh, R. Flegel, D.J. Crouch, L. Cangianelli and J. Baudys, J. Anal. Toxicol., 27 (2003) 429.

172 A.J. Barnes, I. Kim, R. Schepers, E.T. Moolchan, L. Wilson, G. Cooper, C. Reid, C. Hand and M.A. Huestis, J. Anal. Toxicol., 27 (2003) 402.

173 G. Cooper, L. Wilson, C. Reid, D. Baldwin, C. Hand and V. Spiehler, Forensic Sci. Int., 154 (2005) 240.

174 G. Cooper, L. Wilson, C. Reid, D. Baldwin, C. Hand and V. Spiehler, J. Forensic Sci., 50 (2005) 928.

175 G. Cooper, L. Wilson, C. Reid, D. Baldwin, C. Hand and V. Spieher, J. Anal. Toxicol., 28 (2004) 498.

176 M. Laloup, G. Tilman, V. Maes, G. De Boeck, P. Wallemacq, J. Ramaekers and N. Samyn, Forensic Sci. Int., 153 (2005) 29.

177 G. Cooper, L. Wilson, C. Reid, C. Hand and V. Spiehler, Forensic Sci. Int., 159 (2006) 104.

178 H.H. Maurer, Anal. Bioanal. Chem., 381 (2005) 110.

179 K.A. Mortier, K.M. Clauwaert, W.E. Lambert, J.F. Van Bocxlaer, E.G. Van den Eeckhout, C.H. Van Peteghem and A.P. De Leenheer, Rapid Commun. Mass Spectrom., 15 (2001) 1773.

180 K.A. Mortier, K.E. Maudens, W.E. Lambert, K.M. Clauwaert, J.F. Van Bocxlaer, D.L. Deforce, C.H. Van Peteghem and A.P. De Leenheer, J. Chromatogr. B, Analyt. Technol. Biomed. Life Sci., 779 (2002) 321.

181 R. Dams, M.A. Huestis, W.E. Lambert and C.M. Murphy, J. Am. Soc. Mass Spectrom., 14 (2003) 1290.

182 R. Dams, C.M. Murphy, R.E. Choo, W.E. Lambert, A.P. De Leenheer and M.A. Huestis, Anal. Chem., 75 (2003) 798.

183 M. Wood, M. Laloup, M.M. Ramirez Fernandez, K.M. Jenkins, M.S. Young, J.G. Ramaekers, G. De Boeck and N. Samyn, Forensic Sci. Int., 150 (2005) 227.

184 F.M. Wylie, H. Torrance, R.A. Anderson and J.S. Oliver, Forensic Sci. Int., 150 (2005) 191.

185 N. Fucci, N. De Giovanni and M. Chiarotti, Forensic Sci. Int., 134 (2003) 40.

186 M. Yonamine, N. Tawil, R.L. Moreau and O.A. Silva, J. Chromatogr. B, Analyt. Technol. Biomed. Life Sci., 789 (2003) 73.

187 P. Campora, A.M. Bermejo, M.J. Tabernero and P. Fernandez, J. Anal. Toxicol., 27 (2003) 270.

188 P. Campora, A.M. Bermejo, M.J. Tabernero and P. Fernandez, Rapid Commun. Mass Spectrom., 20 (2006) 1288.

189 I. Kim, A.J. Barnes, J.M. Oyler, R. Schepers, R.E. Joseph Jr., E.J. Cone, D. Lafko, E.T. Moolchan and M.A. Huestis, Clin. Chem., 48 (2002) 1486.

190 T.P. Rohrig and C. Moore, J. Anal. Toxicol., 27 (2003) 449.

191 G. Skopp, L. Potsch, K. Klinder, B. Richter, R. Aderjan and R. Mattern, Int. J. Legal Med., 114 (2001) 133.

192 A.C. dos Santos Lucas, A. Bermejo, P. Fernandez and M.J. Tabernero, J. Anal. Toxicol., 24 (2000) 93.

193 D. Ortelli, S. Rudaz, A.F. Chevalley, A. Mino, J.J. Deglon, L. Balant and J.L. Veuthey, J. Chromatogr. A, 871 (2000) 163.

194 M.E. Rosas, K.L. Preston, D.H. Epstein, E.T. Moolchan and I.W. Wainer, J. Chromatogr. B, Analyt. Technol. Biomed. Life Sci., 796 (2003) 355.

195 K. Clauwaert, T. Decaestecker, K. Mortier, W. Lambert, D. Deforce, C. Van Peteghem and J. Van Bocxlaer, J. Anal. Toxicol., 28 (2004) 655.

196 E. Cognard, S. Bouchonnet and C. Staub, J. Pharm. Biomed. Anal., 41 (2006) 925.

197 A. Kankaanpaa, T. Gunnar, K. Ariniemi, P. Lillsunde, S. Mykkanen and T. Seppala, J. Chromatogr. B, Analyt. Technol. Biomed. Life Sci., 810 (2004) 57.

198 K.B. Scheidweiler and M.A. Huestis, J. Chromatogr. B, Analyt. Technol. Biomed. Life Sci., 835 (2006) 90.

199 M. Wood, G. De Boeck, N. Samyn, M. Morris, D.P. Cooper, R.A. Maes and E.A. de Bruijn, J. Anal. Toxicol., 27 (2003) 78.

200 D. Day, D. Kuntz and M. Feldman, THCA detection in oral fluid down to 10 pg/ml. Presented at Society of Forensic Toxicologists Annual Meeting, Nashville, TN, 2005.

201 P. Kintz, V. Cirimele and B. Ludes, J. Anal. Toxicol., 24 (2000) 557.

202 P. Kintz, W. Bernhard, M. Villain, M. Gasser, B. Aebi and V. Cirimele, J. Anal. Toxicol., 29 (2005) 724.

203 R.S. Niedbala, K.W. Kardos, D.F. Fritch, S. Kardos, T. Fries, J. Waga, J. Robb and E.J. Cone, J. Anal. Toxicol., 25 (2001) 289.

204 R.S. Niedbala, K.W. Kardos, D.F. Fritch, K.P. Kunsman, K.A. Blum, G.A. Newland, J. Waga, L. Kurtz, M. Bronsgeest and E.J. Cone, J. Anal. Toxicol., 29 (2005) 607.

205 S. Niedbala, K. Kardos, S. Salamone, D. Fritch, M. Bronsgeest and E.J. Cone, J. Anal. Toxicol., 28 (2004) 546.

206   M. Concheiro, A. de Castro, O. Quintela, A. Cruz and M. Lopez-Rivadulla, J. Chromatogr. B, Analyt. Technol. Biomed. Life Sci., 810 (2004) 319.

207   M. Laloup, M.M. Ramirez Fernandez, M. Wood, G. De Boeck, C. Henquet, V. Maes and N. Samyn, J. Chromatogr. A, 1082 (2005) 15.

208   H. Teixeira, P. Proenca, A. Castanheira, S. Santos, M. Lopez-Rivadulla, F. Corte-Real, E.P. Marques and D.N. Vieira, Forensic Sci. Int., 146(Suppl) (2004) S61.

209   H. Teixeira, P. Proenca, A. Verstraete, F. Corte-Real and D.N. Vieira, Forensic Sci. Int., 150 (2005) 205.

210   N. Samyn, G. De Boeck, V. Cirimele, A. Verstraete and P. Kintz, J. Anal. Toxicol., 26 (2002) 211.

211   P. Kintz, M. Villain and B. Ludes, J. Chromatogr. B, Analyt. Technol. Biomed. Life Sci., 811 (2004) 59.

212   P. Kintz, M. Villain, M. Concheiro and V. Cirimele, Forensic Sci. Int., 150 (2005) 213.

213   O. Quintela, A. Cruz, M. Concheiro, A. de Castro and M. Lopez-Rivadulla, Rapid Commun. Mass Spectrom., 18 (2004) 2976.

214   O. Quintela, A. Cruz, A. Castro, M. Concheiro and M. Lopez-Rivadulla, J. Chromatogr. B, Analyt. Technol. Biomed. Life Sci., 825 (2005) 63.

215   J. Flarakos, W. Luo, M. Aman, D. Svinarov, N. Gerber and P. Vouros, J. Chromatogr. A, 1026 (2004) 175.

216   M.A. Huestis, J.M. Oyler, E.J. Cone, A.T. Wstadik, D. Schoendorfer and R.E. Joseph Jr., J. Chromatogr. B, Biomed. Sci. Appl., 733 (1999) 247.

217   M.A. Huestis, E.J. Cone, C.J. Wong, A. Umbricht and K.L. Preston, J. Anal. Toxicol., 24 (2000) 509.

218   T.M. Winhusen, E.C. Somoza, B. Singal, S. Kim, P.S. Horn and J. Rotrosen, Addiction, 98 (2003) 317.

219   N. Samyn and C. van Haeren, Int. J. Legal Med., 113 (2000) 150.

220   J.A. Levisky, D.L. Bowerman, W.W. Jenkins and S.B. Karch, Forensic Sci. Int., 110 (2000) 35.

221   L. Lester, N. Uemura, J. Ademola, M.R. Harkey, R.P. Nath, S.J. Kim, E. Jerschow, G.L. Henderson, J. Mendelson and R.T. Jones, J. Anal. Toxicol., 26 (2002) 547.

222   S.L. Kacinko, A.J. Barnes, E.W. Schwilke, E.J. Cone, E.T. Moolchan and M.A. Huestis, Clin. Chem., 51 (2005) 2085.

223   E.W. Schwilke, A.J. Barnes, S.L. Kacinko, E.J. Cone, E.T. Moolchan and M.A. Huestis, Clin. Chem., 52 (2006) 1539.

224   N. Uemura, R.P. Nath, M.R. Harkey, G.L. Henderson, J. Mendelson and R.T. Jones, J. Anal. Toxicol., 28 (2004) 253.

225   H.J. Liberty, B.D. Johnson and N. Fortner, J. Anal. Toxicol., 28 (2004) 667.

226   E.J. Cone, M.J. Hillsgrove, A.J. Jenkins, R.M. Keenan and W.D. Darwin, J. Anal. Toxicol., 18 (1994) 298.

227   S. Pichini, M. Navarro, R. Pacifici, P. Zuccaro, J. Ortuno, M. Farre, P.N. Roset, J. Segura and T.R. de la, J. Anal. Toxicol., 27 (2003) 294.

228   D.A. Kidwell, J.D. Kidwell, F. Shinohara, C. Harper, K. Roarty, K. Bernadt, R.A. McCaulley and F.P. Smith, Forensic Sci. Int., 133 (2003) 63.

229   J.A. Levisky, D.L. Bowerman, W.W. Jenkins, D.G. Johnson, J.S. Levisky and S.B. Karch, Forensic Sci. Int., 122 (2001) 65.

230   D.J. Crouch, R.F. Cook, J.V. Trudeau, D.C. Dove, J.J. Robinson, H.L. Webster and A.A. Fatah, J. Anal. Toxicol., 25 (2001) 625.

231   H.J. Liberty, B.D. Johnson, N. Fortner and D. Randolph, Addict. Biol., 8 (2003) 191.

232   D.A. Kidwell and F.P. Smith, Forensic Sci. Int., 116 (2001) 89.

233   T. Saito, A. Wtsadik, K.B. Scheidweiler, N. Fortner, S. Takeichi and M.A. Huestis, Clin. Chem., 50 (2004) 2083.

234   D.E. Moody and M.L. Cheever, J. Anal. Toxicol., 25 (2001) 190.

235   D.E. Moody, A.C. Spanbauer, J.L. Taccogno and E.K. Smith, J. Anal. Toxicol., 28 (2004) 86.

236   R. Pacifici, M. Farre, S. Pichini, J. Ortuno, P.N. Roset, P. Zuccaro, J. Segura and T.R. de la, J. Anal. Toxicol., 25 (2001) 144.

237 S. Pichini, M. Navarro, M. Farre, J. Ortuno, P.N. Roset, R. Pacifici, P. Zuccaro, J. Segura and T.R. de la, Clin. Chem., 48 (2002) 174.

238 M.J. Follador, M. Yonamine, R.L. Moraes Moreau and O.A. Silva, J. Chromatogr. B, Analyt. Technol. Biomed. Life Sci., 811 (2004) 37.

239 M.A. Huestis and R.E. Choo, Forensic Sci. Int., 128 (2002) 20.

240 M.Y. Salem, S.A. Ross, T.P. Murphy and M.A. ElSohly, J. Anal. Toxicol., 25 (2001) 93.

241 S. Pichini, R. Pacifici, M. Pellegrini, E. Marchei, E. Perez-Alarcon, C. Puig, O. Vall and O. Garcia-Algar, J. Chromatogr. B, Analyt. Technol. Biomed. Life Sci., 794 (2003) 281.

242 S. Pichini, R. Pacifici, M. Pellegrini, E. Marchei, J. Lozano, J. Murillo, O. Vall and O. Garcia-Algar, Anal. Chem., 76 (2004) 2124.

243 A.M. Arria, C. Derauf, L.L. Lagasse, P. Grant, R. Shah, L. Smith, W. Haning, M. Huestis, A. Strauss, S.D. Grotta, J. Liu and B. Lester, Matern. Child Health J., 10 (2006) 293.

244 B.M. Lester, M. ElSohly, L.L. Wright, V.L. Smeriglio, J. Verter, C.R. Bauer, S. Shankaran, H.S. Bada, H.H. Walls, M.A. Huestis, L.P. Finnegan and P.L. Maza, Pediatrics, 107 (2001) 309.

245 S. Pichini, C. Puig, P. Zuccaro, E. Marchei, M. Pellegrini, J. Murillo, O. Vall, R. Pacifici and O. Garcia-Algar, Forensic Sci. Int., 153 (2005) 59.

246 M.A. ElSohly, S. Feng and T.P. Murphy, J. Anal. Toxicol., 25 (2001) 40.

247 N.L. Le, A. Reiter, K. Tomlinson, J. Jones and C. Moore, J. Anal. Toxicol., 29 (2005) 54.

248 Y. Xia, P. Wang, M.G. Bartlett, H.M. Solomon and K.L. Busch, Anal. Chem., 72 (2000) 764.

249 S. Pichini, M. Pellegrini, R. Pacifici, E. Marchei, J. Murillo, C. Puig, O. Vall and O. Garcia-Algar, Rapid Commun. Mass Spectrom., 17 (2003) 1958.

250 S. Pichini, E. Marchei, R. Pacifici, M. Pellegrini, J. Lozano and O. Garcia-Algar, J. Chromatogr. B, Analyt. Technol. Biomed. Life Sci., 820 (2005) 151.

251 R. Coles, T.T. Clements, G.J. Nelson, G.A. McMillin and F.M. Urry, J. Anal. Toxicol., 29 (2005) 522.

252 R.E. Choo, C.M. Murphy, H.E. Jones and M.A. Huestis, J. Chromatogr. B, Analyt. Technol. Biomed. Life Sci., 814 (2005) 369.

M.J. Bogusz (Ed.). Forensic Science
Handbook of Analytical Separations, Vol. 6
© 2008 Published by Elsevier B.V.

CHAPTER 21

Doping substances in human and animal sport

Jordi Segura[1,2], Rosa Ventura[1,2], José Marcos[1] and
Ricardo Gutiérrez Gallego[1,2]

[1]*Department of Experimental and Health Sciences, Universitat Pompeu Fabra, Dr. Aiguader 88, 08003
Barcelona, Spain*
[2]*Pharmacology Research Unit, Institut Municipal d'Investigació Mèdica IMIM-Hospital del Mar,
Dr. Aiguader 88, 08003 Barcelona, Spain*

21.1 INTRODUCTION

The misuse of drugs in human and animal sports in an attempt to enhance performance, usually known as doping, is unfortunately an extensive and old practice. Ethical and health aspects are of particular concern. Sports, fair play, medical ethics and potential risks for the health of the athlete leads to consider this practice unacceptable.

The definition of doping adopted in the World Anti-doping Code [1] is the occurrence of one or more of a set of doping rules violations. Among them, reference is made to: the presence of a prohibited substance in an athlete's body specimen, the use of a prohibited substance or method, refusing or evading sample collection, tampering, trafficking or even assisting or any other type of complicity with a doping violation.

The List of classes of prohibited substances and methods of doping according to the World Anti-doping Agency is updated at least once a year, usually more often. The effective version of 2007 is presented in Table 21.1 [2]. Substances prohibited in all situations are anabolic agents (mainly steroids), hormones, β2-agonists, agents with anti-estrogenic activity, diuretics and masking agents. Methods to increase oxygen transfer, manipulations of the sample and the practice of gene doping are also prohibited. In addition, stimulants, narcotics, cannabis derivatives and glucocorticosteroids are also prohibited in a competitive situation. Other groups are subjected to certain restrictions (see Table 21.1).

All these banned substances must not be present in tested urine samples, therefore, laboratories report the presence of such compounds in the samples on a qualitative basis. For some specific compounds, it is difficult to distinguish between the social or

TABLE 21.1
WADA LIST OF CLASSES OF PROHIBITED SUBSTANCES AND METHODS OF DOPING, 2007
[2]

SUBSTANCES AND METHODS PROHIBITED AT ALL TIMES (IN- AND OUT-OF-
  COMPETITION)
Prohibited Substances
  S1. Anabolic androgenic agents
    a. Exogenous steroids
    b. Endogenous steroids
    c. Other anabolic agents
  S2. Hormones and related substances
    a. Erythropoietin (EPO)
    b. Growth hormone (hGH), insuline-like growth factors (e.g. IGF-I), mechano growth factors
    (MGFs)
    c. Gonadotrophins (LH, hCG)
    d. Insulin
    e. Corticotrophins
  S3. Beta-2 agonists
  S4. Agents with anti-estrogenic activity
    a. Aromatase inhibitors
    b. Selective estrogen receptor modulators
    c. Other anti-estrogenic substances
  S5. Diuretics and other masking agents
Prohibited Methods
  M1. Enhancement of oxygen transfer
    a. Blood doping
    b. Artificially enhancing the uptake, transport or delivery of oxygen
  M2. Chemical and physical manipulation
    a. Tampering (catheterisation, urine substitution and/or alteration ...)
    b. Intravenous infusions
  M3. Gene doping
SUBSTANCES AND METHODS PROHIBITED IN-COMPETITION
In addition to the categories S1 to S5 and M1 to M3, the following categories:
Prohibited Substances
  S6. Stimulants
  S7. Narcotics
  S8. Cannabinoids
  S9 Glucocorticosteroids
SUBSTANCES AND METHODS PROHIBITED IN PARTICULAR SPORTS
  P1. Alcohol
  P2. Beta-blockers

---

therapeutic use and the misuse, and, therefore, threshold concentrations have been established. In other cases, the threshold is used to differentiate between physiological values and exogenous administration of the compound. Compounds with a threshold concentration are listed in Table 21.2 [3].

Among the prohibited methods are blood doping (use of autologous, homologous or heterologous blood or red blood cells products) and the artificial increase of oxygen uptake by different means. Also forbidden are pharmacological, chemical and physical manipulations that alter the integrity and validity of urine samples.

TABLE 21.2
SUMMARY OF URINARY CONCENTRATIONS ABOVE WHICH WADA-ACCREDITED LAB-
ORATORIES MUST REPORT FINDINGS FOR SPECIFIC SUBSTANCES [3]

Carboxy-THC (cannabinoids metabolite)	$> 15\,\text{ng/ml}$
Cathine	$> 5\,\mu\text{g/ml}$
Ephedrine	$> 10\,\mu\text{g/ml}$
Epitestosterone	$> 200\,\text{ng/ml}$
Methylephedrine	$> 10\,\mu\text{g/ml}$
Morphine	$> 1\,\mu\text{g/ml}$
19-Norandrosterone	$> 2\,\text{ng/ml}$
Salbutamol	$> 1\,\mu\text{g/ml}$
T/E (testosterone/epitestosterone)	> 4 (ratio)

Examples are catheterisation, urine substitution or tampering. A recent addition is gene doping, defined as the non-therapeutic use of cells, genes, genetic elements, or the modulation of gene expression, having the capacity to enhance athletic performance.

The list of prohibited substances internationally used for animal sports is comprehensive and, in general, the use of any drug during competition is forbidden. As an example, the list of forbidden substances of the International Equestrian Federation (FEI) is presented in Table 21.3 [4]

The real incidence of the misuse of drugs to improve performance is not known. The most reliable data can be obtained from the statistics of the doping controls performed. In 2006 [5] 33 IOC accredited laboratories analysed more than one hundred ninety eight thousand samples from athletes. The percentage of positive cases was 1.96%. Anabolic agents had the highest incidence, accounting for 45.4% of the positive cases followed by cannabinoids (12.8%), stimulants (11.5%), diuretics (6.7%), glucocorticosteroids (6.5%) and hormones (1.0%). The percentage of positive findings for β2-agonists was 14.6%, but most of them corresponded to declared salbutamol or terbutaline, allowed by inhalation. The percentage of other groups was lower than 1%. Drugs most often reported in horses are phenylbutazone, morphine, flunixin, furosemide, dexamethasone, theobromine, caffeine, procaine and clenbuterol [6].

Urine samples are routinely used to perform doping control in athletes. At present, blood samples are collected also by some sport federations or organisations for specific analyses such as synthetic haemoglobins, homologous blood transfusion, human growth hormone, or indirect markers of other hormone abuse. In animal sports, blood is collected as a second choice when urine is not obtainable. Urine is the preferred fluid because its collection is non-invasive (although witnessed collection may represent some intrusion of privacy), it is generally available in sufficient quantity, and the drugs and/or their metabolites are in general present in relatively high concentrations. The main disadvantage of urine is that the drug can be excreted either as free or conjugated metabolites, and the parent drug may be present in relatively low concentrations.

References pp. 739–744

TABLE 21.3
LIST OF PROHIBITED SUBSTANCES OF THE INTERNATIONAL EQUESTRIAN FEDERATION [4]

Substances capable at any time of acting on one or more of the following mammalian body systems:
   the nervous system
   the cardiovascular system
   the respiratory system
   the digestive system other than certain specified substances
   for the oral treatment of gastric ulceration
   the urinary system
   the reproductive system other than certain specified substances
   the musculoskeletal system
   the skin (e.g. hypersensitising agents).
   the blood system
   the immune system, other than those in licensed vaccines against infectious agents
   the endocrine system
Antipyretics, analgesics and anti-inflammatory substances
Cytotoxic substances
Endocrine secretions and their synthetic counterparts
Masking agents

For some substances, findings below the following thresholds are not actionable:

Plasma carbon dioxide	36 mmol
Urine boldenone (male horses)	0.015 µg/ml
Dimethyl sulphoxide	15 µg/ml in in urine
	1 µg/ml in plasma
Urine estranediol (male horses)	1
(for the following ratio in urine: 5a-estrane-3b,17a-diol/5(10)-estrene-3b, 17a-diol	
Urine hydrocortisone	1 µg/ml
Salicylic acid	625 µg/ml in urine
	5.4 µg/ml in plasma
Urine testosterone (geldings)	0.02 µg/ml
Urine testosterone (fillies and mares not in foal)	0.055 µg/ml
Urine theobromine	2 µg/ml

Owing to the high number of banned substances in human and animal sports, a series of screening methods addressed to the detection of groups of a wide number of compounds and/or metabolites with similar physicochemical properties have been developed. These screening procedures are applied in routine doping control to all samples and are designed to eliminate "true negative" specimens. If the presence of a compound or metabolite is suspected in some sample, a second confirmation method specific for the compound detected is applied. Confirmation methods must be performed, as far as possible, using mass spectrometry (MS), which allows a true identification of the compounds. Criteria to demonstrate the presence of prohibited substances have been established by the WADA [3].

The analytical methodology used by doping control laboratories has been extensively described, especially in major sport events [7–14]. The techniques used for screening and confirmation purposes depend on each specific group of compounds.

Physicochemical properties of the parent compounds or their metabolites and the sensitivity and specificity required are the most important factors to take into account. From a general point of view, capillary gas chromatography (GC) or high performance liquid chromatography (HPLC) coupled to mass spectrometry (MS) are the most used techniques for separation. For screening purposes, immunological methods such as enzyme-linked immunosorbent assay (ELISA) or fluorescence polarisation immunoassay (FPIA) are also employed, mainly in the field of doping control in animals. In recent years, due to the high requirements in sensitivity and specificity, chromatography coupled to high resolution MS or to tandem MS (MS-MS) has also been used for anabolic agents to achieve the detection limits required by international organisations. With the incorporation of peptide hormone doping, other methodologies usually associated with proteomic analyses are also being introduced in the anti-doping field. Genomic and transcriptomic approaches most probably will play a key role for the detection of gene doping.

In addition to the ability to detect dope agents or their metabolites, other important factors that should be taken into account when dealing with dope testing are the analysis capacity and the elapsed time between the receipt of the samples and the delivery of results. These factors are especially important at major sport events and also when the number of samples is high. Around three to four thousand samples are analysed in 14 days during Olympic Games, and around twelve hundred samples in the same period of time during major regional events.

The demand of doping controls after competition or out-of-competition (unannounced doping control in athletes during training periods) by National and International Sports Federations is increasing in the last decade and that is leading to an increasing number of laboratories that perform doping analyses. Owing to the high mobility of athletes and the high number of controls, athletes can be tested by different laboratories in short periods of time. Therefore, the need for harmonisation in the quality of analytical results is obvious. To ensure a high level of quality, laboratories must be accredited according to international quality standards (ISO 17025). Moreover, laboratories performing doping control in athletes should be additionally accredited by the WADA.

The objective of this chapter is to provide a guide to the separation and detection methods employed in human and animal doping control to detect the misuse of the main groups of prohibited substances: anabolic steroids, stimulants, diuretics, β-adrenergic drugs (agonists and antagonists), anti-inflammatory drugs, peptide hormones, corticosteroids, enhancers of oxygen transfer and other doping agents. Some of the drugs included in the list of prohibited substances in the stimulant group (cocaine, amphetamines) have been specifically treated in other chapters of this handbook, and only those not described before will be reviewed. Other groups also tested in doping control, such as narcotic analgesics, have also been specifically treated in other chapters of this handbook. Usually a brief description on the physicochemical properties and metabolic and excretion data for each kind of compounds is given; subsequently, a review of analytical methodologies is presented.

21.2 ANABOLIC ANDROGENIC STEROIDS

Synthetic anabolic steroids belong to the pharmacological group with major impact on drug abuse in sport. All compounds belonging to this class resemble testosterone in their chemical structure [15] but with a series of modifications that have major impact on the way metabolic enzymes of the human and animal body modify their structures during biotransformation process. The metabolism of anabolic steroids in humans has been documented [16,17]. Generally, it involves a series of phase I reactions which include the C-4,5 double bond reduction to 5α and 5β-androstane isomers and the rapid reduction of the 3-keto group by 3-hydroxy-dehydrogenase enzymes, and in some cases the introduction of hydroxyl groups in different positions of the molecule. In phase II, those metabolites are further conjugated by glucuronidation and, to a lesser extent, sulphation. Since only small amounts of unchanged anabolic steroids are excreted unchanged in urine, the hydrolysis of conjugates is a fundamental step in many of the analytical methods used.

The hydrolysis of conjugates of anabolic steroids is accomplished successfully by enzymatic processes. Acidic hydrolysis may produce dehydrations and rearrangements of unstable compounds [16]. Enzymes more often used are β-glucuronidase from *Escherichia coli* or *Helix pomatia*. In case of the latter a substantial sulphatase activity is also present. Owing to some side reactions described with preparations from *H. Pomatia* [18], β-glucuronidase from *E. coli* is the preferred enzyme. Conditions for hydrolysis involve usually 1 h at about 55°C, although some of the glucuronides may be hydrolysed with lower reaction times and temperatures. Currently the hydrolysis step is carried out on urine samples previously cleaned by solid-phase extraction, with polystyrene-divinylbenzene polymer (Amberlite XAD-2) or reverse phase material (octadecylsilane bonded silica). These previous cleaning processes are carried out as, for some urine samples, yield of hydrolysis has been shown to be higher when compared with direct hydrolysis on native urine. As sulphates are a relevant part of steroid conjugation in horses, methanolysis (e.g. 1 M anhydrous methanolic HCl added to urine and heated at 50–60°C for 10–15 min) is the preferred hydrolysis procedure in equine anti-doping testing [19].

When metabolites of anabolic steroids are free from their conjugates, or in those few cases where non-conjugated metabolites are directly excreted in urine, a simple extraction from the matrix interferences is needed. Liquid–liquid extraction is often used after adjusting pH to values of around 9–10 [11]. Different solvents have been used but diethyl ether or *tert*-butylmethyl ether is the preferred ones [15]. The latter has the advantage of no peroxides being formed. For specific confirmation in some cases, and in order to further reduce biological background interferences, *n*-pentane is also an adequate solvent. Alternatively, solid–liquid extraction (again Amberlite XAD-2 or C_{18} cartridges) can be used [18] after the hydrolysis step to obtain extracts for further analysis. Immunoaffinity chromatography (IAC) and HPLC fractionation have also been described to isolate some steroids from biological extracts [20,21].

Many of the isolated metabolites do not exhibit good gas chromatographic behaviour because of the presence of hydroxyl and keto groups in their structure.

Among the multiple reagents described for the derivatisation of hydroxyl or enolysed keto groups of anabolic steroids, trimethylsilylation has been particularly useful [22,23]. *N*-methyl-*N*-(trimethylsilyl)trifluoroacetamide (MSTFA) is the reagent of choice, especially when a catalyst (e.g. trimethylsilylimidazole (TMSIm)) is present, for the derivatisation of tertiary alcohols. Also, the possibility of forming silyl groups for enolic forms is catalysed by the use of ammonium iodide [24]. Addition of reducing agents such as dithioerythritol, ethanethiol, or 2-mercaptoethanol minimises the formation of iodine. When higher increases in masses are desirable, *tert*-butylsilyl derivatives can be of use [25]. Acyl derivatives are also interesting for some specific anabolic steroids and metabolites containing nitrogen, such as stanozolol [26].

When analysing trimethylsilyl (TMS) derivatives of anabolic steroids by GC, a careful derivatisation of glass-liner in the injection port can dramatically affect the chromatographic behaviour of some compounds. If MS is used as detection technique, the mass spectra corresponding to the TMS derivatives show the presence of high mass molecular ions, the presence of ions corresponding to losses of 90 mass units (TMS-OH). As in many methylsilyl derivatives, M^+-15 is a usual peak [27]. Main gas chromatographic and mass spectrometric characteristics of urinary detectable compounds after synthetic anabolic steroids administration are collected in Tables 21.4 and 21.5. The number of TMS substituents, including the enolisation of keto groups, is also indicated. Metabolites excreted in free form are collected in Table 21.4 while those excreted free and conjugated are listed in Table 21.5. Alternatively, the protection of the keto group with the formation of alkyloximes, and the subsequent analysis of the combination of trimethylsilyl ether-oxime derivative [28] can be used.

In some instances, the use of catalysts for enolisation of the keto groups needs to be avoided because of serious side reaction. This is the case of anabolic steroids containing double bonds conjugated to a keto group, e.g. trenbolone and tetrahydrogestrinone (THG). THG is a designer steroid undetectable until 2003 when its use by high-ranking athletes was discovered turning into one of the biggest drug scandals in sports history [29]. Since then sensitive and specific methods for rapid screening of urine samples have been developed and evaluated [30]. This sub-family includes steroids that have been used in animals for many years (trenbolone), and new chemically modified steroids (the so-called designer steroids) like THG, norbolethone, gestrinone, dihydrogestrinone, methyltrienolone, methyldienolone and propyltrenbolone. Fig. 21.1 shows the chemical structure of these steroids.

In addition to conventional GC-low resolution MS, the use of GC-high resolution MS [20,24] is applicable when lower concentrations are to be detected. Alternatively, GC-MS-MS [31] also allows high specificity down to the level of nanograms and subnanograms per millilitre of urine.

In recent years, the administration of metabolic precursors of testosterone is increasing among athletes who try to improve their performance with anabolic androgenic steroids and to escape from easy detection. Findings in urine of abnormal concentrations of these compounds, which are always normally present in the human or animal body, are indicative of a prohibited use, but sometimes several

TABLE 21.4
GAS CHROMATOGRAPHIC AND MASS SPECTOMETRIC DATA OF EXOGENOUS ANABOLIC STEROIDS AND/OR THEIR METABOLITES (FREE FRACTION)

Precursor (DCI and Systematic Name)	Excreted Substance (DCI or Systematic Name)	Derivative	RRT[a]	Monitored Ions (m/z)		
4-Chlorometandienone (4-chloro-17β-hydroxy-17-methyl-androsta-1,4-dien-3-one)	4-Chloro-17α-methylandrosta-1,4-dien- 6β,17β-diol-3-one	bis O-TMS	1.010	243	315	317
Danazol (17α-pregna-2,4-dien-20-yno[2,3-d]isoxazol-17β-ol)	Ethisterone (17α-ethynyl-4-androsten-17β-ol-3-one)	mono O-TMS	0.863	369	196	384
Fluoxymesterone (9α-fluoro-11β,17β-dihydroxy-17α-methylandrost-4-en-3-one)	9α-Fluoro-17α-methylandrost-4-en-3α,6β,11β,17β-tetraol	tetrakis O-TMS	0.898	462	552	642
Formebolone (2-formyl-11α,17β-dihidroxy-17α-methylandrost-1,4-dien-3-one)	2-Hydroxymethylen-17α-methylandrosta-1,4-dien-11α,17β-diol-3-one	tris O-TMS	1.102	457	562	547
Metandienone (17β-hydroxy-17α-methylandrosta-1,4-dien-3-one)	6β-Hydroxymetandienone	bis O-TMS	0.921	281	209	460
	17α-Epimetandienone	mono O-TMS	0.807	282	194	
Oxandrolone (17β-hydroxy-17α-methyl-2-oxa-5α-androstan-3-one)	Oxandrolone	mono O-TMS	0.899	308	321	363
	17α-Epioxandrolone	mono O-TMS	0.836	308	321	363
Stanozolol (17β-hydroxy-17α-methyl-5α-androst-2-eno[3,2-c]pyrazole)	3′-Hydroxystanozolol	bis O-TMS, N-TMS	1.129	669	684	594

[a]Relative retention times to Stanozolol (RT: 5.0 min) obtained with the following chromatographic conditions: a CG Hewlett Packard Ultra 1 cross-linked methyl silicone capillary column, length 16.5 m, I.D. 0.2 mm and film thickness 0.11 μm, helium carrier gas at a flow of 0.8 ml/min (measured at oven temperature 180°C) and an oven temperature programme from 200 to 300°C at 30°C/min.

TABLE 21.5
GAS CHROMATOGRAPHIC AND MASS SPECTROMETRIC DATA OF EXOGENOUS ANABOLIC STEROIDS AND/OR THEIR METABOLITES EXCRETED FREE AND CONJUGATED (TOTAL FRACTION)

Precursor (DCI and Systematic Name)	Excreted Substance (DCI or Systematic Name)	Excretion in Urine[a]	Derivative	RRT[b]	Monitored Ions (m/z)		
Bolasterone (17β-hydroxy-7α,17α-dimethylandrost-4-en-3-one)	7α,17α-Dimethyl-5β-androstane-3α,17β-diol	Conjugated (G)	bis O-TMS	0.920	374	284	143
Boldenone (17β-hydroxyandrosta-1,4-dien-3-one)	Boldenone	Conjugated (G)	bis O-TMS	0.860	430	415	206
	17β-Hydroxy-5β-androst-1-en-3-one	Conjugated (G,S)	bis O-TMS	0.615	432	417	194
Calusterone (17β-hydroxy-7β,17α-dimethylandrost-4-en-3-one)	7β,17α-Dimethyl-5α-androstane-3α,17β-diol	Conjugated (G)	bis O-TMS	0.877	374	284	143
Clostebol (4-chloro-17β-hydroxyandrost-4-en-3-one)	4-Chloro-3α-hydroxyandrost-4-en-17-one	Conjugated (G)	bis O-TMS	0.924	468	466	451
	4-Chloro-3α-hydroxy-5α(β)-androstan-17-one	Conjugated (G, S)	bis O-TMS	1.003	470	468	453
4-Chlorometandienone (4-chloro-17β-hydroxy-17-methyl-androsta-1,4-dien-3-one)	4-Chloro-17α-methylandrosta-1,4-dien-6β,17β-diol-3-one	Free	bis O-TMS	1.199	317	315	143
Danazol (17α-ethynyl-17β-hydroxy-4-androsteno[2,3-d]-isoxazole)	Ethisterone (17α-ethynyl-4-androsten-17β-ol-3-one)	Free	bis O-TMS	1.034	456	316	301
	2ξ-Hydroxymethylethisterone (17α-ethynyl-2ξ-hydroxymethyl-4-androsten-17β-ol-3-one)	Conjugated (G)	tris O-TMS	1.185		558	543
Drostanolone (17β-hydroxy-2α-methyl-5α-androstan-3-one)	3α-Hydroxy-2α-methyl-5α-androstan-17-one	Conjugated (G)	bis O-TMS	0.743	448	433	343

TABLE 21.5
CONTINUED

Precursor (DCI and Systematic Name)	Excreted Substance (DCI or Systematic Name)	Excretion in Urine[a]	Derivative	RRT[b]	Monitored Ions (m/z)		
Fluoxymesterone (9α-fluoro-11β,17β-dihydroxy-17α-methylandrost-4-en-3-one)	Fluoxymesterone		tris O-TMS	1.175	143	552	462
	9α-Fluoro-18-nor-17,17-dimethyl-11β-hydroxyandrosta-4.13-diene-3-one	Conjugated (G)	bis O-TMS	0.806	208	357	462
Furazabol (17β-hydroxy-17α-methyl-5-androstano[2,3-c]-[1,2,5]-oxadiazole)	Furazabol	Conjugated (G)	mono O-TMS	1.151	143	387	402
	16β-Hydroxyfurazabol	Conjugated (G)	bis O-TMS	1.236	218	231	490
Formebolone (2-formyl-11α,17β—dihydroxy-17α-methylandrost-1,4-dien-3-one)	2-Hydroxymethyl-11α,17β-dihydroxy-17α-methylandrost-1,4-dien-one	Free	tetrakis-O-TMS	1.245	143	351	634
4-Hydroxytestosterone	4-Hydroxytestosterone	Conjugated (G)	tris-O-TMS	1.141	520	505	415
	4-Hydroxyandrostenedione	Conjugated (G)	tris-O-TMS	1.125	518	503	296
	4-Hydroxynandrolone	Conjugated (G)	tris-O-TMS	1.066	506	491	416
Mesterolone (17β-hydroxy-1α-methyl-5α-androstan-3-one)	3α-Hydroxy-1α-methyl-5α-Androstan-17-one	Conjugated (G)	bis O-TMS	0.808	343	433	448
Mestanolone (17β-hydroxy-17α-methyl-5α-androstan-3-one)	17α-Methyl-5α-androstane-3α,17β-diol	Conjugated (G)	bis O-TMS	0.813	143	345	435
Methandriol (17α-methylandrost-5-en–3β,17β-diol)	17α-Methyl-5β-androstane-3α,17β-diol	Conjugated (G)	bis O-TMS	0.811	143	345	435
Metandienone (17β-hydroxy-17α-methylandrosta-1,4-dien-3-one)	6β-Hydroxymetandienone	Free, Conjugated (G)	tris O-TMS	1.151	317	517	532
	17β-Methyl-5β-androst-1-ene-3α,17α-diol	Conjugated (G)	bis O-TMS	0.617	143	358	448
	18-Nor-17,17-dimethyl-5β-androsta-1,13-dien-3α-ol	Conjugated (G)	mono O-TMS	0.390	216	253	358

Metenolone (17β-hydroxy-1-methyl-5α-androst-1-en-3-one)	Metenolone	Conjugated (G)	bis *O*-TMS	0.923	195	208	446
	3α-Hydroxy-1-methylen-5α-Androstan-17-one	Conjugated (G)	bis *O*-TMS	0.779	431	446	
Methyltestosterone (17β-hydroxy-17α-methylandrost-4-en-3-one)	17α-Methyl-5α-androstane-3α,17β-diol	Conjugated (G)	bis *O*-TMS	0.813	143	345	435
	17α-Methyl-5β-androstane-3α,17β-diol	Conjugated (G)	bis *O*-TMS	0.821	143	345	435
18-Methyl-19-nortestosterone	5-Estran-18-methyl-3α-ol-17-one	Conjugated (G)	bis *O*-TMS	0.749	434	405	315
7α-Methyl-19-nortestosterone	5-Estran-7α-methyl-3α-ol-17-one	Conjugated (G)	bis *O*-TMS	0.743	434	419	329
Mibolerone (17β-hydroxy-7α,17α-dimethylestr-4-en-3-one)	19-Nor-7α,17α-dimethyl-5α(β)-androstane-3α(β),17β-diol	Conjugated (G)	bis *O*-TMS	0.867	143	270	360
	Dihydromibolerone	Conjugated (G)	bis *O*-TMS	1.103	143	358	448
Nandrolone (17β-hydroxyestr-4-en-3-one)	Norandrosterone (3α-hydroxy-5α-estran-17-one)	Conjugated (G)	bis *O*-TMS	0.598	422	405	420
	Noretiocholanolone (3α-hydroxy-5β-estran-17-one)	Conjugated (G)	bis *O*-TMS	0.661	315	405	420
Norethandrolone (17β-hydroxy-17α-ethylestr-4-en-3-one)	17α-Ethyl-5β-estrane-3α,17β-diol	Conjugated (G)	bis *O*-TMS	0.927	157	331	421
	17α-Ethyl-5β-estrane-3α,17β,20-triol	Conjugated (G)	tris *O*-TMS	1.134	245	331	421
Norandrostenedione	Norandrosterone (3α-hydroxy-5α-estran-17-one)	Conjugated (G)	bis *O*-TMS	0.598	422	405	420
	Noretiocholanolone (3α-hydroxy-5β-estran-17-one)	Conjugated (G)	bis *O*-TMS	0.661	315	405	420
Norclostebol (4-estren-4-chloro-17β-ol-3-one)	4-Chloro-4-estren-3α-ol-17-one	Conjugated (G)	bis *O*-TMS	0.878	452	437	417
Norbolethone 13β-Ethyl-17β-hydroxy-18,19-dinor-17α-pregn-4-en-3-one	18α-homo-17α-ethyl-5α-estrane-3α,17β-diol	Conjugated (G)	bis *O*-TMS	1.002	464	435	255
	18α-homo-17α-ethyl-5β-estrane-3α,17β-diol	Conjugated (G)	bis *O*-TMS	1.049	464	345	255

TABLE 21.5
CONTINUED

Precursor (DCI and Systematic Name)	Excreted Substance (DCI or Systematic Name)	Excretion in Urine[a]	Derivative	RRT[b]	Monitored Ions (m/z)		
Oxandrolone (17β-hydroxy-17α-methyl-2-oxa-5α-androstan-3-one)	Oxandrolone	Free	mono O-TMS	1.040	143	308	363
Oxymesterone (4,17β-dihydroxy-17α-methylandrost-4-en-3-one)	Oxymesterone	Conjugated (G)	tris O-TMS	1.176	389	519	534
Oxymetholone (17β-hydroxy-2-hydroxymethylene-17α-methyl-5α-androstan-3-one)	2ζ-Hydroxymethylene-17α-methyl-5α-androstane-3ζ,(4/6)ζ,17β-triol	Conjugated (G)	tetrakis O-TMS	1.204	143	460	550
Stanozolol (17β-hydroxy-17α-methyl-5α-androst-2-eno[3,2-c]pyrazole)	3'-Hydroxystanozolol	Free, Conjugated (G)	bis O-TMS,N-TMS	1.261	254	545	560
	16β-Hydroxystanozolol	Conjugated (G)	tris O-TMS	1.279	218	231	560
Stenbolone (17β-hydroxy-2-methyl-5α-androst-1-en-3-one)	Stenbolone	Conjugated (G)	bis-O-TMS	0.862	446	431	208
	3α-Hydroxy-2-methyl-5α-androst-1-en-17-one	Conjugated (G)	bis-O-TMS	0.787	446	431	275
1-Testosterone (17β-hydroxy-androst-1-en-3-one)	1-Testosterone	Conjugated (G)	bis-O-TMS	0.794	432	417	327

[a]G: glucuronide conjugate, S: sulphate conjugate.
[b]Relative retention times to methyltestosterone (RT: 15.0 min) obtained with the following chromatographic conditions: a GC Hewlett Packard Ultra 1 cross-linked methyl silicone capillary column, length 16.5 m, I.D. 0.2 mm and film thickness 0.11 μm, helium carrier gas at a flow of 0.8 ml/min (measured at oven temperature 180°C) and an oven temperature programme from 181 to 230°C at 3 and 40°C/min to a final temperature of 310°C.

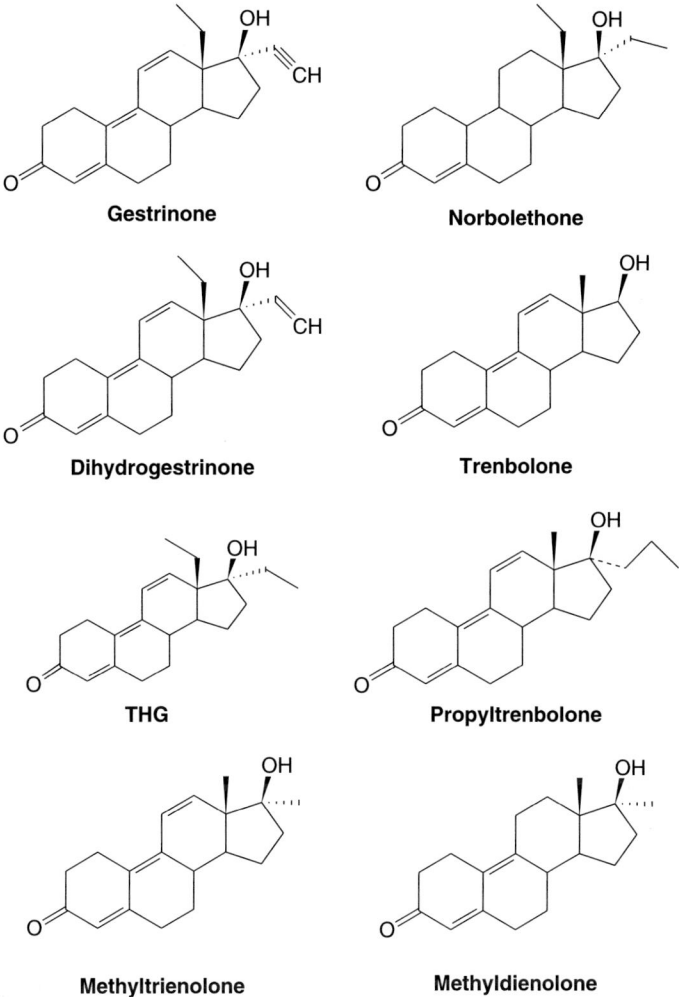

Fig 21.1. Chemical structure of some steroids.

physiological or pathological factors could be considered to explain such results. The need to distinguish the exogenous origin of such endogenous-like substances has made necessary the development of more elaborated analytical procedures than those explained above for synthetic anabolic steroids. One of the more powerful methods is the differentiation of a different ^{13}C content of the exogenous substance in urine from natural precursors (e.g. cholesterol, pregnanediol) by means of isotope ratio mass spectrometry [32–34]. For this purpose, the direct analysis of the administered substance or of a marker metabolite is carried out. The later is usually more successful due to the extensive metabolism of these compounds. For example, abuse of testosterone can be detected by measuring the abnormally low δ^{13}C values for its metabolites 5β-androstane-3α,17β-diol and 5α-androstane-3α,17β-diol [35], androsterone and etiocholanolone [36]. Similar approaches have been described for the detection of other endogenous steroids administration to humans [37,38], and to cattle [39,40].

The isolation of metabolites for isotope ratio measurement may be carried out by conventional liquid–liquid or solid–liquid extraction. Interestingly, the formation of hydrazones may be used to isolate anabolic steroids containing the keto group [41]. If the hydrazone is water soluble, its removal by liquid extraction may be used to concentrate only these hydroxyl-containing metabolites in the organic phase. Final acetylation of hydroxyls groups renders suitable compounds for $^{13}C/^{12}C$ isotope ratio measurement by means of GC/combustion/isotope ratio MS (GC/C/IRMS). Alternatively, semi-preparative HPLC may be used after the first extraction. HPLC fractions are then derivatised to form acetyl derivatives and analysed by GC/C/IRMS [40]. This sample preparation is needed to obtain high purity peaks by GC/C/IRMS in order to have accurate δ^{13}C‰ values.

As many of endogenous-like steroids are administered as esters of medium or long chain acids, their derivatisation either as TMS or perfluoroacyl derivatives allows the direct detection of the small amount of unchanged ester present in blood, which is a definitive demonstration of doping by this kind of substances. Enolysed mono-TMS derivatives of nine different testosterone esters show a base peak corresponding to their molecular ion, which offers low limits of detection either in GC-MS with selective ion monitoring or in GC-MS-MS [42]. Alternatively, the formation of enolysed trifluoroacetyl (TFA), pentafluoropropionyl (PFP) or heptafluorobutyryl (HFB) derivatives offers multiple possibilities for structural confirmation and for detection by negative chemical ionisation (NCI) [42].

Owing to their chemical properties, an adequate approach consists on analysing the underivatised steroids by liquid chromatography/tandem mass spectrometry (LC/MS/MS) [43]. As compared with GC, LC has found limited applicability in the detection of anabolic steroids in doping control. Its coupling to MS by a particle beam interface (PB) was proposed [44]. Coupling by electrospray (ES) or atmospheric pressure chemical ionisation (APCI) emerged as a suitable alternative to GC, especially for the detection of intact glucuronide or sulphate-conjugates of polar metabolites [45,46].

This unstoppable move towards LC/MS/MS for steroid analysis have already produced some methods for detecting a limited number of steroids, e.g. the designer steroid sub-family explained above [30,43,47,48]. However, a general LC/MS/MS method for all the forbidden steroids has not been developed yet. The elevated number of metabolites to be detected, the low limits of detection required and the high concentrations of naturally occurring steroids which have similar chemical structure have made, so far, impossible to obtain such a method. Thus, most likely GC/MS comprehensive analyses may always be important, particularly in evaluating the steroid profile of athletes. The database of steroid mass spectra in GC-MS is unsurpassed and steroid identification by this technique will remain as the gold standard.

21.3 STIMULANTS

Stimulants, such as amphetamines and other sympatomimetic drugs, was one of the firsts groups of drugs prohibited in the doping control field. Analytical methods used to screen for unconjugated nitrogen-containing compounds were described long ago [49–51]. The structure of these drugs is closely related to that of endogenous cathecolamines. They are phenylalkylamine derivatives with substituents in different positions: in the amine group (e.g. methamphetamine, dimethamphetamine), in the phenyl ring (e.g. fenfluramine, chlorphentermine, methoxyphenamine) and in α and β carbon atoms of the side chain (methylephedrine, ephedrine [α-hydroxy-phenylethylamine]).

Stimulants, in general, suffer extensive metabolism, and a small percentage of the drug is excreted unconjugated. They are metabolised in the liver by reactions of hydroxylation affecting the aromatic ring and the alkyl chain, deamination and N-dealkylation reactions, and further conjugation with glucuronic acid and, to a lesser extent, with sulphate. Recovery of the parent drug is urinary flow- and pH-dependent. In the case of amphetamine and methamphetamine, acidic urine increases both the rate of excretion and the percentage of unchanged drug excreted. As their pharmacological effects are produced by relatively high oral doses of the drug and the misuse is expected to be relatively close in time to the competition, this would result in urine concentrations of the drug and/or their metabolites high enough to be detected in biological fluids.

Immunological methods are useful as screening methods when the expected prevalence of positive cases for each individual substance is low, such in sports drug testing. FPIA or ELISA allow the direct analysis of the urine samples and give good results for screening purposes [52]. These methods can be applied to plasma samples but require a previous alkaline extraction with organic solvents of the samples before the application of the immunological tests.

Chromatographic methods described to detect stimulants in plasma and urine require a previous extraction from the matrix interferences. These compounds share some chemical properties which makes possible a unified analysis. They are basic compounds (pK_a 7–10), mainly because they are nitrogen-containing compounds, volatiles (molar mass < 350 g/mol and amenable to GC), and have functional groups that can be derivatised to enhance their gas chromatographic properties.

The screening procedure most often described is based on a liquid–liquid extraction from the biological fluid at strong alkaline pH (pH of the samples should be adjusted for optimum recovery around pH 14, where the amino group is non-ionised). The organic solvent used could be freshly distilled ether or *tert*-butylmethyl ether [11,53]. The addition of anhydrous sodium sulphate to increase the ionic strength of the aqueous phase increases the extraction into the ethereal phase of many of those compounds [11,53]. Care must be taken when concentrating extracts containing volatile compounds (e.g. amphetamine), as the free bases may be lost during evaporation of the solvent. These losses may be avoided by adding methanolic hydrochloric acid or a drop of trimethylchlorosilane (TMCS) to the extract to form the corresponding hydrochloride salts of the drugs before evaporating. If further purification of the extract is required, an additional clean-up step based on back-extraction into acid may be used. The organic extracts are analysed by GC-NPD and the suspected substances are identified by GC-MS. Both analyses are performed in the same extract or the GC-MS analysis is done after further concentration of the organic extract. Owing to the extraction at a high pH level and the use of an NPD, the chromatographic biological background is very low.

The screening procedure enables the detection and identification of many drugs (see Table 21.6) and their metabolites, including most of the stimulants, β-adrenergic agents and narcotics (methadone, pentazocine and pethidine). In addition it is possible to detect other nitrogen-containing drugs such as anti-histaminics, benzodiazepines, tricyclic antidepressants and local anaesthetics. Local anaesthetics are particularly important in the field of animal doping.

Diphenylamine and *N,N*-diisopropyl-1-amino-*n*-dodecane (DIPA-12) have been the most usual internal standards [11,53]. The GC separation, usually carried out using fused-silica capillary columns gives high reproducible retention behaviour, which is very useful for the identification of the compounds detected. The column most used is a cross-linked 5% phenylmethyl silicone capillary column (12.5 m × 0.2 mm internal diameter (I.D.), 0.33 μm film thickness).

When analysing the extracts obtained by GC-MS, the mass spectrum of many of the stimulants show one major low mass ion and other poor diagnostic ions because of its very low intensity. Characteristic fragment ions at m/z 44, 58, 72, 86 or 100 can be obtained for different phenylethylamine compounds depending on the substituents at the nitrogen atom and at the carbon atom in the α position (see Table 21.6).

The confirmation procedure for substances detected by the screening method is often focused on the detection of substances and their urinary metabolites [54]. These procedures allow the detection of more polar and less volatile drugs and metabolites. In these cases, a hydrolysis of the conjugates is the initial step previous to the extraction of the samples. An acidic hydrolysis could be performed and mercaptoacetic acid or cysteine is useful to minimise oxidation during the hydrolysis [55]. However, some side reactions have been described in these conditions for some β-blockers [56] and 6-monoacetilmorphine [57]. Alternatively, an enzymatic hydrolysis with *H. pomatia* juice with β-glucuronidase and arylsulphatase activity is the most usual procedure [11,54] . After hydrolysis, extraction of the samples, previously conditioned to basic pH, is done with polymeric bonded-phase silica columns with lipophylic and ion-exchange properties, such as Bond-Elut Certify® [53,54]. The detection by GC-MS of ritalinic acid (acidic metabolite of methylphenidate) has been described using specific extraction conditions [58].

Mass spectra of the underivatised molecule of many stimulants, as mentioned above, show poor diagnostic ions. The preparation of suitable derivatives, which give unique retention times, often provide additional information to help the identification of the compound present because the fragmentation pattern can be strongly altered [27]. The derivatisation technique commonly used is based on double derivatisation [23,59], where sequential addition of MSTFA and *N*-methyl-bis-trifluoroacetamide (MBTFA) gives the TMS derivatives of hydroxyl, acidic and phenolic groups and the TFA derivatives of primary and secondary amines

TABLE 21.6
GAS CHROMATOGRAPHIC, MASS SPECTROMETRIC DATA OF STIMULANTS, OTHER DRUGS AND/OR THEIR METABOLITES

Compound	Systematic	RRT[a]	m/z[b]
Heptaminol	6-Amino-2-methyl-2-heptanol	0.36	44
Amphetamine	α-Methylbenzeneethanamine	0.38	44
Norfenfluramine	α-Methyl-3-(trifluoromethyl)benzeneethanamine	0.39	44
Phentermine	α,α-Dimethylbenzeneethanamine	0.41	58
Menthol	(1α,2β,5α)-5-Methyl-2-(1-methylethyl)-cyclohexanol	0.42	138
Propylhexedrine	N,α-dimethylcyclohexaneethanamine	0.43	58
Methamphetamine	N,α-dimethylbenzeneethanamine	0.44	58
Etilamfetamine	N-ethyl-α-methylbenzeneethanamine	0.51	72
Fenfluramine	N-ethyl-α-methyl-3-(trifluoromethyl)benzeneethanamine	0.51	72
Mephentermine	N,α,α-trimethylbenzeneethanamine	0.51	72
Dimetamfetamine	N,N,α-trimethylbenzeneethanamine	0.52	72
Dinordiethylpropion	2-Amino-1-phenyl-1-propanone	0.60	44
Chloramphetamine	4-Chloro-α-methyl-benzeneethanamine	0.62	56
Norpseudoephedrine (Cathine)	[S-(R*,R*)]-α-(1-aminoethyl)-benzenemethanol	0.62	44
Norephedrine	[R-(R*,S*)]-α-(1-aminoethyl)-benzenemethanol	0.63	44
Normethoxyphenamine	2-Methoxy-α-methyl-benzeneethanamine	0.63	44
Nicotine	3-(1-Methyl-2-pyrrolidinyl)pyridine	0.65	84
Bemegride	4-Ethyl-4-methyl-2,6-piperidinedione	0.66	55
Eugenol	2-Methoxy-4-(2-propenyl)phenol	0.66	166
Chlorphentermine	4-Chloro-α,α-dimethylbenzeneethanamine	0.67	58
Ephedrine/Pseudoephedrine	[R-(R*,S*)]-α-[1-(methylamino)ethyl]benzenemethanol/ [S-(R*,R*)]-α-[1-(methylamino)ethyl]benzenemethanol	0.69	58
Methoxyphenamine	2-Methoxy-N,α-dimethyl-benzeneethanamine	0.70	58
Nicotinamide	3-Pyridinecarboxamide	0.70	106
Nordiethylpropion	2-(Ethylamino)-1-phenyl-1-propanone	0.71	72
Methylephedrine	[R-(R*,S*)]-α-[1-(dimethylamino)ethyl]benzenemethanol	0.74	72

TABLE 21.6
CONTINUED

Compound	Systematic	RRT[a]	m/z[b]
N-Ethylnorephedrine	[R-(R*,S*)]-α-[1-(ethylamino)ethyl]benzenemethanol	0.76	72
Phenmetrazine	3-Methyl-2-phenylmorpholine	0.78	71
N-desethylnikethamide	N-ethyl-3-pyridinecarboxamide	0.81	106
Phendimetrazine	3,4-Dimethyl-2-phenylmorpholine	0.81	85
Hordenine	4-[2-(Dimethylamino)ethyl]phenol	0.82	58
Etafedrine	α-[1-(Ethylmethylamino)ethyl]benzenemethanol	0.84	86
Pholedrine	4-[2-(Methylamino)propyl]phenol	0.84	58
Amfepramone (Diethylpropion)	2-(Diethylamino)-1-phenyl-1-propanone	0.85	100
Piracetam	2-Oxo-1-pyrrolidineacetamide	0.85	98
MDMA	N,α-dimethyl-1,3-benzodioxole-5-ethanamine	0.89	58
Nikethamide	N,N-diethyl-3-pyridinecarboxamide	0.90	106
Mephenesine	3-(2-methylphenoxy)-1,2-propanediol	0.93	108
N,N-diethylnorephedrine	[R-(R*,S*)]-α-[1-(diethylamino)ethyl]benzenemethanol	0.93	100
Benzocaine	4-Aminobenzoic acid ethyl ester	0.94	120
MDEA	N-ethyl-α-methyl-1,3-benzodioxole-5-ethanamine	0.95	72
Pentetrazole	6,7,8,9-Tetrahydro-5H-tetrazolo[1,5-a]azepine	0.95	55
Mefenorex	N-(3-chloropropyl)-α-methylphenethylamine	0.96	120
Fenproporex	3-[(1-Methyl-2-phenylethyl)amino]-propanenitrile	0.98	97
Bupropion	1-(3-Chlorophenyl)-2-[(1,1-dimethylethyl)amino]-1-propanone	0.98	100
Diphenylamine (ISTD)	N-phenylbenzeneamine	1.00	169
Prolintane	1-[1-(Phenylmethyl)butyl]pyrrolidine	1.01	126
Methocarbamol artifact	3-(2-Methoxyphenoxy)-1,2-propanediol	1.04	124
Fencamfamin	N-ethyl-3-phenylbicyclo[2.2.1]-heptan-2-amine	1.07	215
Crotetamide	N-[1-[(dimethylamino)carbonyl]-propyl]-N-ethyl-2-butenamide	1.08	86
Cotinine	1-methyl-5-(3-pyridinyl)-2-pyrrolidinone	1.10	98
Cropropamide	N-[1-[(dimethylamino)carbonyl]-propyl]-N-propyl-2-butenamide	1.13	100
Methylphenidate	α-Phenyl-2-piperidineacetic acid methyl ester	1.16	84
Pethidine	1-Methyl-4-phenyl-4-piperidinecarboxylic acid ethyl ester	1.17	71
Amfetaminil	α-[(1-Methyl-2-phenylethyl)amino]benzeneacetonitrile	1.18	132
Norpethidine	4-Phenyl-4-piperidinecarboxylic acid ethyl ester	1.20	57

Substance	Chemical name		
Lidocaine met	2-(Ethylamino)-N-(2,6-dimethylphenyl)acetamide	1.23	58
Alphaprodine	1,3-Dimethyl-4-phenyl-4-piperidinol propanoate	1.24	172
Caffeine	3,7-Dihydro-1,3,7-trimethyl-1H-purine-2,6-dione	1.26	194
Benzphetamine	N,α-dimethyl-N-(phenylmethyl)-benzeneethanamine	1.28	148
Fluoxetine	(±)-N-methyl-γ-[4-(trifluoromethyl)phenoxy]benzenepropanamine	1.28	44
Antipyrine	1,2-Dihydro-1,5-dimethyl-2-phenyl-3H-pyrazol-3-one	1.31	188
Etamivan	N,N-diethyl-4-hydroxy-3-methoxybenzamide	1.32	151
Lidocaine	2-(Diethylamino)-N-(2,6-dimethylphenyl)acetamide	1.32	86
Pyrovalerone	1-(4-Methylphenyl)-2-(1-pyrrolidinyl)-1-pentanone	1.32	126
Oxprenolol	1-[(1-Methylethyl)amino]-3-[2-(2-propenyloxy)phenoxy]-2-propanol	1.35	72
Tramadol	2-[(Dimethylamino)methyl]-1-(3-methoxyphenyl)cyclohexanol	1.39	58
Clobenzorex	N-[(2-chlorophenyl)methyl]-α-methylbenzeneethanamine	1.41	168
Tramadol met (N-desmethyl-)	2-[(Methylamino)methyl]-1-(3-methoxyphenyl)cyclohexanol	1.42	44
Dipyrone artifact	1,2-Dihydro-1,5-dimethyl-4-methylamino-2-phenyl-3H-pyrazol-3-one	1.44	217
Procaine	4-Aminobenzoic acid 2-(diethylamino) ethyl ester	1.47	86
Amiphenazole	5-Phenyl-2,4-thiazolediamine	1.48	191
Metoprolol	1-[4-(2-Methoxyethyl)phenoxy]-3-[(1-methylethyl)amino]-2-propanol	1.48	72
Tramadol met (O-desmethyl-)	2-[(Dimethylamino)methyl]-1-(3-hydroxyphenyl)cyclohexanol	1.48	58
EDDP (methadone met)	2-Ethylidene-1,5-dimethyl-3,3-diphenylpyrrolidine	1.49	277
Alprenolol	1-[(1-Methylethyl)amino]-3-[2-(2-propenyl)phenoxy]-2-propanol	1.54	72
Methadone	6-Dimethylamino-4,4-diphenyl-3-heptanone	1.60	72
Pipradrol	α,α-Diphenyl-2-piperidinemethanol	1.61	84
Diclofenac	2-[(2,6-Dichlorophenyl)amino]benzeneacetic acid	1.62	214
Propranolol	1-[(1-Methylethyl)amino]-3-(1-naphthalenyloxy)-2-propanol	1.62	72
Pyrimethamine	5-(4-Chlorophenyl)-6-ethyl-2,4-pyrimidinediamine	1.62	247
D-Propoxyphene	[S-(R*,S*)]-α-[2-(dimethylamino)-1-methylethyl]-α-phenylbenzeneethanol propanoate (ester)	1.64	58
Amitriptyline	3-(10,11-Dihydro-5H-dibenzo[a,d]cyclohepten-5-ylidene)-N,N-dimethyl-1-propanamine	1.65	58
Imipramine	10,11-Dihydro-N,N-dimethyl-5H-dibenz[b,f]azepine-5-propanamine	1.67	234
Tetracaine	4-(Butylamino)benzoic acid 2-(dimethylamino)ethyl ester	1.67	58
Desipramine	10,11-Dihydro-N-methyl-5H-dibenz[b,f]azepine-5-propanamine	1.70	234
Mepyramine	N-[(4-methoxyphenyl)methyl]-N',N'-dimethyl-N-2-pyridinyl-1,2-ethanediamine	1.71	121
Pindolol	1-(1H-indol-4-yloxy)-3-[(1-methylethyl)amino]-2-propanol	1.71	133
Aminoglutethimide	3-(4-Aminophenyl)-3-ethyl-2,6-dioxopiperidinedione	1.71	203
Pentazocine	(2α,6α,11R*)-1,2,3,4,5,6-hexahydro-6,11-dimethyl-3-(3-methyl-2-butenyl)-2,6-methano-3-benzazocin-8-ol	1.75	217
Betaxolol	1-[4-[2-(Cyclopropylmethoxy)ethyl]4,2-phenoxy]-3-[(1-methylethyl)amino]-2-propanol	1.79	72
Bisoprolol	1-[[2-(Methylethoxy)ethoxy]-methyl]phenoxy]-3-[(1-methylethyl)amino]-2-propanol	1.79	72

TABLE 21.6
CONTINUED

Compound	Systematic	RRT[a]	m/z[b]
Propoxycaine	4-Amino-2-propoxybenzoic acid 2-(diethylamino)ethyl ester	1.79	86
Codeine	(5α,6α)-7,8-Didehydro-4,5-epoxy-3-methoxy-17-methylmorphinan-6-ol	1.83	299
Clomipramine	3-Chloro-10,11-dihydro-N,N-dimethyl-5H-dibenz[b.f]azepine-5-propanamine	1.84	58
Mazindol	5-(4-Chlorophenyl)-2,5-dihydro-3H-imidazo[2,1-a]isoindol-5-ol	1.86	266
Pentoxifylline	3,7-Dihydro-3,7-dimethyl-1-(5-oxohexyl)-1H-purine-2,6-dione	1.86	221
Ethylmorphine	(5α,6α)-7,8-Didehydro-4,5-epoxy-3-ethoxy-17-methylmorphinan-6-ol	1.87	313
Hydrocodone	(5α)-4,5-Epoxy-3-methoxy-17-methylmorphinan-6-one	1.90	299
Chlorpromazine	2-Chloro-N,N-dimethyl-10H-phenothiazine-10-propanamine	1.91	58
Oxycodone	(5α)-4,5-Epoxy-14-hydroxy-3-methoxy-17-methylmorphinan-6-one	1.97	315
Amineptine met	Dihydro-10,11-dibenzo[a,d]cycloheptenyl-5-amino-5-pentanoic	2.01	192
Chloroquine	N^4-(7-chloro-4-quinolinyl)-N^1,N^1-diethyl-1,4-pentanediamine	2.01	86
7-Amino-flunitrazepam	5-(2-Fluorophenyl)-1,3-dihydro-1-methyl-7-amino-2H-1,4-benzodiazepin-2-one	2.06	283
Metoclopramide	4-Amino-5-chloro-N-[2-(diethylamino)ethyl]-2-methoxybenzamide	2.08	86
Ambroxol	4-[[(2-Amino-3,5-dibromophenyl)methyl]amino]cyclohexanol	2.10	264
Azaperone	1-(4-Fluorophenyl)-4-[4-(2-pyridinyl)-1-piperazinyl]-1-butanone	2.12	107
Propionylpromazine	1-[10-(3-(Dimethylamino)-propyl]-10H-phenothiazin-2-yl]-1-propanone	2.17	58
Quinine	6'-Methoxycinchonan-9-ol	2.21	136
Doxapram	1-Ethyl-4-[2-(4-morpholinyl)ethyl]-3,3-diphenyl-2-pyrrolidinone	2.26	100
Fenetylline	3,7-Dihydro-1,3-dimethyl-7-[2-[(1-methyl-2-phenylethyl)amino]ethyl]-1H-purine-2,6-dione	2.48	250
Pholcodine	(5α,6α)-7,8-Didehydro-4,5-epoxy-17-methyl-3-[2-(4-morpholinyl)ethoxy]morphinan-6-ol	2.60	114
Strychnine	Strychnidin-10-one	2.62	334
Morazone	1,2-Dihydro-1,5-dimethyl-4-[3-methyl-2-phenyl-4-morpholinyl)methyl]-2-phenyl-3H-pyrazol-3-one	2.74	201

[a]Relative retention times to diphenylamine (RT: 4.77 min) obtained with the following chromatographic conditions: a GC Hewlett Packard Ultra 2 cross-linked 5% phenylmethyl silicone capillary column, length 12.5 m, I.D. 0.2 mm and film thickness 0.11 μm, helium carrier gas at a flow of 0.8 ml/min (measured at 180°C) and an oven temperature programme from 90 to 300°C at 20°C/min (final time 4 min).

[b]Diagnostic ion analysis is performed in the SCAN acquisition mode (m/z 40–500).

[11,54,60]. These derivatives are very stable in solution and show excellent gas chromatographic properties.

Doping control is, sometimes, concerned with the interpretation of results associated with the quirality of the substances detected. For instance, *S*-(+)-methamphetamine is strongly active as stimulant while *R*-(–)-methamphetamine, present in some medications for cold, has little activity at the central nervous system level. Also, interpretation of analytical results on selegiline metabolism, an anti-Parkinsonian drug which can be converted by metabolism to *R*-(–)-amphetamine and *R*-(–)-methamphetamine and their conjugated *p*-hydroxy derivatives, can be confusing if they are not properly identified [61].

GC separation of the optical isomers could be based on two different strategies: conversion into diastereomers by reaction with an optically pure reagent and then separation on achiral chromatographic phases, or direct separation using chiral stationary phases [62]. Some chiral reagents used for analysing stimulants (fenfluramine, amphetamine derivatives and methylphenidate and its metabolite ritalinic acid) to produce diastereomers are *S*-(–)-trifluoroacetylprolylchloride [61], *S*-(–)-heptafluorobutyrylprolylchloride [63,64] or methoxytrifluoromethylphenylacetyl chloride (Mosher acid chloride reagent) [65].

Optical diastereomers of ephedrines (ephedrine vs. pseudoephedrine, phenylpropanolamine vs. cathine) are not well resolved by the usual GC-NPD screening conditions. A change of the oven program in the GC analysis gives good results in the separation between these diastereomers [66]. Also, N-TFA-O-TMS derivatives formed in the procedure described for the confirmation of stimulants allow the separation of both diastereomers [54]. Pentafluorobenzoyl chloride, often used as a derivatisation reagent for amines, could be used as a reagent for an extractive acylation reaction directly on the urine to give the pentafluorobenzoyl (PFB) derivatives of ephedrine and pseudoephedrine with a good separation of both diastereomers [65]. Alternative approaches to qualitatively and quantitatively determine these compounds in doping control samples were recently developed using LC-MS-MS instrumentation by direct analysis of the urine sample, without previous sample preparation [67,68].

Apart from the common used GC-NPD and GC-MS methods, a screening method for the detection of stimulants in human urine using LC-MS-MS has been recently described [69]. After a liquid–liquid extraction at strong alkaline pH, samples were analysed by LC-MS using a high efficiency reversed-phase column and detected with an APCI interface operated in positive ionisation mode.

21.4 DIURETICS

Large differences in molecular structures and, thus, in physicochemical properties can be found among the compounds belonging to the pharmacological group of diuretics: compounds with carboxylic acid functions (etacrynic acid), with carboxylic acid and sulphonamide functions (furosemide, bumetanide, piretanide), with sulphonamide groups (thiazide diuretics), steroidal structure (spironolactone, canrenone), or diuretics with amino functions (amiloride, triamterene), among others [70]. These differences make it difficult the development of single screening and confirmation methods. Regarding the metabolism and urinary excretion, most of these compounds are excreted unchanged in human urine to a variable extent (from 99% of the dose for acetazolamide to 4–12% for triamterene). Therefore, procedures to screen for diuretics can be, in general, addressed to the detection of the parent drug.

Extensive methodology has been described for the detection and quantitation of single diuretics, however, the number of screening methods for the whole group is more limited. Liquid–liquid extraction procedures have been the most widely described to isolate the parent drugs from the urine matrix. Two different strategies have been followed to extract all the compounds (acidic, basic and neutral compounds) from the urine. One of them is the application of two separate liquid–liquid extractions at acidic and basic pH and the extracts are mixed afterwards, but poor detection limits have been reported for some of the compounds [71–74]. The second strategy is the use of a single extraction at neutral or alkaline pH with salting-out effect. Extraction at pH 7 with diethyl ether and anhydrous sodium sulphate to promote salting-out effect allowed the recovery of most of the compounds although amiloride was not extracted [75]. However, the use of a more polar solvent (ethylacetate) with sodium chloride for salting-out effect and alkaline pH has proven to be useful for the detection of all the compounds [76].

Solid-phase extraction has also been evaluated. Sep-Pak C_{18} columns with a mixture of diethyl ether and methanol as elution solvent have been employed to screen for diuretics [75,77]. However, there is a difficulty to achieve single solid-phase extraction conditions to recover all the compounds of the group due to the differences in polarity as it was shown using different Bond-Elut columns (octadecyl, octyl, ethyl, cyclohexyl, phenyl, cyanopropyl) [78]. A column switching procedure using octadecyl cartridges directly coupled to LC detection has also been reported with the same limitations due to the differences in polarity [79]. However, these solid-phase extraction procedures can be successfully used to detect single compounds or a group of them with similar characteristics [80–82].

Isolation of diuretics from urine by adsorption on a polystyrene resin (XAD-2) and elution with methanol has been used before derivatisation and GC-MS analysis [83] or before LC-MS analysis [84]. In this case, the lack of specificity of the extraction procedure is compensated by the use of selective techniques for separation and detection as GC/MS or LC-MS/MS.

Both LC with DAD [76] and GC-MS after methylation [83,85–88] were used to screen for diuretics in urine. The use of LC avoids derivatisation, which is in most instances a time-consuming step, and offers a more comprehensive screening than GC-MS due to the inability of the latter to form suitable derivatives for some of the compounds. In the last years, the availability of robust and relatively inexpensive mass spectrometric detectors for LC has promoted the use of LC-MS systems for screening and confirmation of these compounds [71,84,89–91].

LC separation of diuretics has been usually performed with octadecylsilane columns with 5 µm particle size. A substantial reduction of the analysis time was achieved using columns with 3 μm particle size. The reduction of the particle size allows the reduction of the column length to obtain the same chromatographic efficacy and, thus, a reduction in the analysis time. The separation of 20 diuretics can be accomplished in approximately 8 min using an Ultrasphere ODS column, 3 μm particle size, 7.5 × 0.46 cm [76].

Mobile phases containing an acidic aqueous buffer and acetonitrile as organic modifier have been usually reported [70]. Propylamine [92] or an ammonium salt [76] have been added to the acidic aqueous phase to improve the chromatographic behaviour of diuretics with amino groups (amiloride, triamterene). Gradient elution is needed to obtain adequate run times, due to the differences in polarity of the different diuretics.

Capillary GC-MS offers a more selective approach than LC-DAD, and the high degree of standardisation of this technique made GC-MS one of the most widely used tools to screen for diuretics in urine. The need for derivatisation is the main limitation factor. Owing to the inability of silylating reagents to form stable derivatives with sulphonamide groups, methylation is the method of choice to derivatise most of the diuretic compounds [70]. Trimethylsilylation with MSTFA is used for specific compounds such as amiloride [88]. Three main methylation procedures have been described for screening purposes: extractive methylation, pyrolytic methylation and methylation with methyl iodide in acetone. Comparison of the three methylation procedures to analyse diuretics revealed that derivatisation with methyl iodide in acetone is the most comprehensive methylation method and the best choice for screening purposes [83]. Extractive and pyrolytic methylations were found to be faster and more effective procedures for some compounds and can be of interest for confirmation purposes.

Methylation with methyl iodide in acetone and dry potassium carbonate allows the derivatisation of amine functions, such as those of triamterene, in addition to carboxylic acids, sulphonamide groups and alcohols. The main drawback is the long incubation time needed to derivatise diuretics with sulphonamide or amino functions, in contrast to the derivatisation of compounds with only carboxylic acid functions, which can occur without incubation [88]. Fast methylation reactions and improvement in detection limits were achieved using microwave assisted derivatisation in comparison with thermal incubation [77].

Extractive methylation consists of the extraction of the organic acid as an ion pair with a quaternary ammonium salt from the alkaline aqueous phase into an aprotic organic solvent (containing the methylation reagent, methyl iodide) where the methylation reaction occurs. Extractive methylation can be directly applied to the urine sample [85,86,93]. The quaternary ammonium salt must be removed to avoid interferences during GC analysis and premature loss in column efficacy. An efficient clean-up procedure based on solid-phase extraction with a macroreticular acrylic copolymer of the organic extract obtained after extractive methylation has been described [86].

In pyrolytic methylation, the residue obtained after the extraction procedure is dissolved in the methylation reagent, normally a quaternary ammonium hydroxide solution, such as trimethylaniline hydroxide, tetramethylammonium hydroxide, or a mixture of both. The reaction occurs in the injector of the gas chromatograph, which is kept at high temperatures. For some diuretics a high degree of methylation has been described [87].

Electron impact is the preferred ionisation technique when using GC-MS. EI mass spectra with high diagnostic value are obtained, and three ions are monitored for each compound for screening and confirmation purposes [70]. GC-MS with NCI has been used in some cases to improve sensitivity. Mass spectra with less fragmentation have been obtained using NCI [94].

LC-MS analysis of diuretics was early studied using thermospray and particle beam interfaces [88,95]. Procedures with suitable sensitivity were developed, however, an application to routine analysis was impeded by technical limitations of these interfaces. In the last years, comprehensive screening methods based on LC-MS using an electrospray interface have been developed [71,84,89–91]. In electrospray ionisation, negative ion mode is generally preferred for compounds of acidic nature, however, for the basic diuretics positively charged ions are mainly formed. Co-elution of positively and negatively charged ions makes necessary scan-to-scan polarity switching. Regarding the sensitivity of detection, LC-MS offers better limits of detection than GC-MS. For some compounds, MS/MS detection is necessary to increase sensitivity. Mass spectrometric behaviour of thiazide-based diuretics after electrospray ionisation and collision-induced dissociation has been extensively studied [96]. Nowadays, LC-MS/MS is the technique of choice for comprehensive screening of diuretics.

21.5 β-ADRENERGIC DRUGS (β-BLOCKERS AND β-AGONISTS)

The structure of β-agonists is closely related to that of endogenous cathecholamines. They are (β-hydroxyphenylethyl)amines bearing different substituents at the amino group (mainly, *tert*-butyl or isopropyl groups) and in the phenyl ring. β-Blocker drugs (antagonists of β-adrenergic receptor) have a general structure of *N*-alkyl-β-hydroxy-aryloxypropylamine. Different aromatic rings with different substituents are linked to the oxygen atom, and *tert*-butyl and isopropyl groups are also the most common substituents of the amino function. Some β-blockers, such as labetalol and sotalol, have a (β-hydroxyphenylethyl)amino structure like β-agonists.

β-Adrenergic agonists are eliminated from the body by metabolism and by renal excretion of the unchanged drug. There are two important metabolic pathways depending on the structure of the compound: methylation by the enzyme cathecol-*O*-methyltransferase and phenolic conjugation with sulphate [97]. Metabolism of β-adrenergic antagonists is very extensive and involves reactions consisting of hydroxylations affecting the aromatic ring and N-dealkylations, and conjugations with glucuronic acid or sulphate [98,99].

In general terms, the pharmacological effects of β-blockers are produced by relatively high oral doses of drugs and concentrations of drug and/or metabolites found in urine are relatively high (μg/ml level). However, other compounds of the group (e.g. most of β₂-agonists) are active at lower doses and some of them can be administered by inhalation, leading to concentrations in urine at the ng/ml or sub-ng/ml level.

Immunological methods, such as ELISA, applied directly to the urine sample are widely used to detect β-agonists for screening purposes. An alkaline extraction with organic solvents is needed to extract β-agonist drugs from plasma samples previous to the application of the ELISA test. The use of the ELISA test designed to the detection of β-agonists drugs has proven to be also useful for the detection of β-blockers in urine due to the similarities in chemical structures of both groups of substances [100].

When dealing with chromatographic methods, a sample preparation is needed and hydrolysis of the conjugates is usually the first step in the analysis of β-adrenergic drugs. Hydrolysis in acidic conditions has been used to analyse β-blocker drugs [7,55] but some of them, such as atenolol, pindolol or timolol, are unstable in these conditions. The most usual hydrolysis procedure is enzymatic hydrolysis using *H. pomatia* juice containing β-glucuronidase and arylsulphatase activity. These conditions are milder than an acidic hydrolysis and, thus, avoid degradations of labile β-blockers [11,54,101]. Incubations at pH 5.2, at 55°C for 2 h have been described [11,54].

Originally, methods to extract β-adrenergic drugs have been based on liquid–liquid extraction procedures. The urine is adjusted at alkaline pH (pH 9 or higher) to achieve deprotonation of the amino group and then extracted with an organic solvent. Diethyl ether or mixtures of diethyl ether with *tert*-butanol [11,55], or *tert*-butylmethyl ether and *n*-butanol [102] have been used. The development of solid-phase extraction procedures, that allow a substantial reduction of the amount of organic solvent per sample and are more suitable for automation, has led doping control analysts to use them as an alternative to traditional liquid–liquid extraction methods. Extraction with co-polymeric bonded-phase silica columns with lipophylic and ion-exchange properties, such as Bond-Elut Certify or Clean Screen DAU have been successfully used to extract a few number of β-blockers and β-agonists drugs from human and animal urine [101,103,104]. A comprehensive solid-phase extraction procedure based on Bond-Elut Certify columns has been described to analyse β-adrenergic drugs in human urine together with stimulants and narcotic analgesics [54]. Solid-phase extractions with C_{18} cartridges based on lipophylic interactions [105] or cation-exchange mechanisms [106] have also been described for the recovery of β-agonists from urine. In the last case, the acidification of the urine to pH 3 was needed to get the amino function protonated. IAC has also been described to isolate β-agonists from biological extracts [107].

Another strategy used in most laboratories in routine doping control is the analysis of β-agonists and, in some instance, β-blockers using the same sample preparation as described for the analysis of synthetic anabolic steroids in previous sections.

GC-MS in EI conditions has been for many years the method of choice to analyse β-adrenergic drugs in doping control laboratories. Owing to the presence of polar functional groups (amino and hydroxyl groups), a derivatisation step is needed to make these compounds amenable for GC analysis. Formation of TMS-ether derivatives by reaction with MSTFA alone has been described for β-agonists and β-blockers [101]. Other silylating reagents such as *N,O*-bis(trimethylsilyl)trifluoroacetamide (BSTFA) [105] or the use of catalysts such as 1% TMCS [107,108] have been described for β-agonists. In these conditions, amino groups are not derivatised, and the EI mass spectra of the TMS-ether derivatives of compounds with *tert*-butylamino or isopropylamino substituents at the nitrogen atom show base peaks at *m/z* 86 or *m/z* 72, respectively, formed by α-cleavage.

It is common practice for the analysis of β-blockers to perform selective derivatisation with MSTFA, which enables the formation of TMS derivatives of hydroxyl and phenolic functions, followed by MBTFA, which forms TFA derivatives with primary and secondary amines [11,55]. In this case, when an isopropyl group is linked to the amino function, the *O*-TMS-*N*-TFA derivative is formed and the base peak is *m/z* 284. However, for compounds with *tert*-butylamino groups the amino function is not derivatized and the base peak in the EI mass spectrum of these compounds is *m/z* 86, resulting from α-cleavage. The monitorisation of these common and characteristic fragment ions is used to screen for these compounds by GC-MS [54]. Derivatisation with trifluoroacetic anhydride to form *N,O*-TFA derivatives has also been described [11].

To increase selectivity and sensitivity, the formation of cyclic derivatives of the bifunctional β-hydroxyamino group has been proposed. The formation of 2-(dimethyl)silamorpholine derivatives [101] or alkylboronate derivatives, mainly methylboronate derivatives [102,103,106,109,110] has been successfully applied. 2-(Dimethyl)silamorpholine derivatives can be formed by reaction with diethylamine and chloromethyldimethylchlorosilane. Alkylboronate derivatives can be formed by reaction with the corresponding boronic acids (methylboronic, butylboronic, and phenylboronic). The cyclic anhydride of methylboronic acid, trimethylboroxine, has also been employed to form cyclic methylboronates. Mass spectra of these cyclic derivatives show the most abundant ions in the high mass range, providing more structural information, less chemical noise and better sensitivity. Detection limits at sub-ng/ml level have been obtained for clenbuterol-methylboronate in human urine [103].

The main drawback of cyclic derivatisation is the unsuitability to be used as a comprehensive screening. Molecules containing additional hydroxyl groups can form side derivatives. Salbutamol and salmeterol are two of the exceptions, as the relative position of the two additional hydroxyl groups allows the formation of the bis-methylboronate derivatives.

In reference to data acquisition, analyses by GC-MS in EI conditions in scan mode offers adequate sensitivity for the detection of β-blocker drugs [54]. However, a more sensitive GC-MS-MS analysis have also been proposed for the detection of this group of substances in human urine [111].

On the other hand, for the detection of β-agonists the use of selective ion monitoring is mandatory to achieve the limits of detection required. As for some anabolic agents, for some specific β-agonists such as clenbuterol, the use of GC-medium resolution MS, GC-MS-MS or even GC-MS-MS is needed to fulfil the criteria of detection required for doping control in athletes [31,112]. Chemical ionisation with ammonia has also been described for some β-agonists to increase sensitivity in their detection [105].

Owing to the availability of several ionisation sources (ESI, APCI) and robust tandem MS (e.g. triple-stage quadrupole or ion trap instruments) many new LC/MS/MS screening and confirmation methods have been developed during the last years [67,113]. β-blocker drugs can be easily determined by such methods. They can be detected either as a single group using an APCI interface [114], or together with the diuretic drugs by means of an electrospray interface [89].

In the field of veterinary analysis, LC/MS/MS has been used to achieve adequate sensitivity for β-agonists detection [106,115]. Using a similar approach, an LC/ESI-MS/MS procedure that allows simultaneous detection of 18 β2-agonists has been developed [116].

Different authors have also studied the analysis of β-adrenergic antagonists by CE but its application in routine doping control is in this moment very limited [117].

The route of administration is another important aspect when dealing with β-agonists. The use of some β-agonists drugs such as salbutamol, terbutaline and salmeterol is allowed by inhalatory route in athletes for the treatment of asthma or asthma induced exercise and forbidden by the oral route. All these compounds have an asymmetrical carbon atom and are administered as a racemic mixture. In the case of salbutamol, which is excreted in urine as an unchanged drug and as a sulphate-conjugate, it has been demonstrated that the urinary excretion of the enantiomers in free form is different depending on the route of administration. Taking into account the concentration of total salbutamol, total free salbutamol and the ratio of $S(+)/R(-)$, a criteria can be established to distinguish between both administration routes [118]. The chiral separation of salbutamol can be accomplished by LC using an urea type silica-bonded chiral phase column (Chirex^{TM} 3022) and a mobile phase containing hexane:dichloromethane:methanol:trifluoroacetic acid (250:218:31:1, v/v), and fluorimetric detection [119]. Determination of enantiomers of salbutamol and its 4-O-sulphate metabolites in human plasma and urine by chiral LC-MS-MS has also been described [120].

21.6 PEPTIDE HORMONES

Peptide and protein hormones have gained much interest within the scientific anti-doping society with the increasing awareness of their potential appeal to the (semi)professional athletic community. A direct consequence has been the growing number of substances, or rather subclasses, in the paragraph dedicated to these types of molecules on the international list of prohibited substances [2]. One has to coin the word subclasses as, in contrast to low-molecular weight steroid-like drugs, these type of molecules usually comprise a more-or-less discrete number of structurally distinct isoforms due to the occurrence of co- and post-translational modifications (C/PTMs). The fast majority of the list-members are being produced through re-combinant technologies with an additional process for hCG that can be also isolated from urine of pregnant women. The production process gives rise to potential structural differences between the endogenous and exogenous variants as the occurring PTMs (so-called secondary gene events) that depend on environmental than genetic factors. Thus, in addition to the rather traditional approach of quantifying hormonal levels through immunological techniques much effort is being devoted to the development of separation and analysis protocols that emphasise, or ideally pinpoint, the differences between self and non-self. The difficulty of this enterprise is emphasised if the outcome of nearly a decade of research is considered with a single screening protocol firmly implemented in anti-doping analysis. This protocol concerns erythropoietin (EPO) and was introduced for the Olympic Games of 2000. Although two distinct approaches were presented simultaneously; one based on blood markers (concentrations of haemoglobin, erythropoietin and serum transferrin receptor, as well as percentage of reticulocytes) of altered erythropoiesis and using blood [121], and the second one using urine and the differences in p*I* values between rEPO and uEPO [122], only the latter has found a wider acceptance. The most probable reason is that it is a direct method resulting in an iso-electric focussing profile for each substance (Fig. 21.2) that can directly be related to its endogenous or exogenous origin. Nevertheless, the implementation of this screening protocol caused a small revolution in the anti-doping community. Not only did it concern a totally different type of molecule (high molecular weight, heterogeneous, and complex in structure), but it requires a large amount of starting material (20 ml of urine), is very laborious (more than 150 different steps in the standard operating procedure) and time-consuming (up to 3 days). As such it requires highly skilled persons and nearly exclusive dedication. Noteworthy of the protocol are the required concentration factor of 400–1000 times, in order to enable detection, that is achieved by three sequential molecular weight filtrations ((I) 0.22 μm to remove cellular debris, (II) 30 kDa filtration to reduce 20 to ≤2 ml, (III) 30 kDa filtration to reduce the volume to 20–50 μl). This concentration has to be combined with optimum sample recovery from the filters and total absence of sample degradation due to proteolytic enzymes. Likewise, the development of a patented double-blotting strategy [123], in order to minimise background noise and enhance detection, is ingenious. One of the main drawbacks of the method reside in the fact that a decision of potential abuse has to be made on the basis of the ratio between isoforms with particular p*I* values. This has prompted the study of specially designed algorithms and accompanying software for accurate interpretation of the results [124,125]. A second point of possible critique is inherent to the immunological approach. Even though a highly specific monoclonal antibody is employed, recently doubts have been raised about the possible cross reactivity with other proteins that could be secreted under particular circumstances [126]. Regardless the veracity of these observations, the slightest doubt accompanying any publications with respect to the outcome of a doping analysis results in re-enforced efforts to prove doubts are false or develop yet another unambiguous method. The most promising results in this direction are being obtained with research aimed at finding structural differences in post-translational modifications glycosylation [127–130].

A second hormone that has received much attention for the development of a reliable analytical protocol is somatotropin. Excellent reviews have appeared

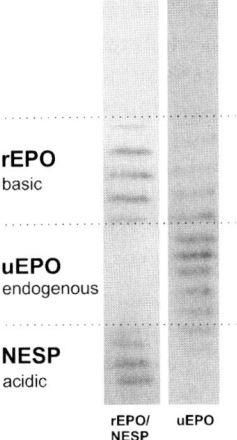

rEPO
basic

uEPO
endogenous

NESP
acidic

rEPO/   uEPO
NESP

Fig 21.2. Iso-electro-focusing profile of rEPO and uEPO.

recently covering the developments thus far [131,132]. Once more, two distinct strategies have been pursued but till this day all efforts have not resulted in a robust method. This reflects the reluctance from the governing bodies to accept a test for this kind of naturally occurring substances. The first approach aims at detection of middle-term effects through the quantification of indirect markers such as pro-collagen III peptide (P-III-P) or insulin-like growth factor I (IGF-I) [133–136]. The disadvantage of using the secondary-effect approach is largely counter-balanced by the larger window of opportunity to suspect abuse. While the half-life in circulation of hGH is around 15 min the markers remain altered for several days after the last administration (personal observation). The other approach focuses on the direct analysis of GH in circulation through immunological measurement. As endogenous hGH exists as a mixture of isoforms due to alternative splicing [137] and/or post-translational processing [138] the administration of a single, recombinant, isoform of 22 kDa results in an altered ratio of the circulating isoforms through the administration in itself and the subsequent inhibited endogenous production. Already in 1999 a paper described how this phenomenon could be exploited to detect GH abuse using a set of two monoclonal antibodies to evaluate the ratio between 22 kDa GH and all GH isoforms [139]. While the normal ratio appeared to oscillate between 0.196 and 0.788 using this assay, after rGH administration the ratio resulted higher than 1 for all tested individuals. This test has been further refined, with the generation of two different sets of monoclonal antibodies, and has entered the validation phase. In a parallel approach the ratio between the 22 and 20 kDa isoform has been used for the same purpose. In this case the standard ratios (22:20 kDa) in healthy individuals turned out to be 11.7 ± 7.2 and 10.4 ± 6.0, in male and female, respectively, and remained fairly constant. With administration of recombinant 22 kDa hGH it was observed that after 3 h the ratios increased and remained at higher levels in the following 72 h [140]. A ratio, however, is difficult to calculate due to decreased levels of 20 kDa hGH. When these fall below the limit-of-quantitation of 5.0 pg/ml no true ratio can be provided. Nevertheless, such finding should trigger a follow-up study where the ratio is expected to return to normal values after administration is discontinued. Finally, also this hormone has been subjected to a meticulous structural analysis, by means of state-of-the-art MS techniques, in order to find structural elements that may permit unambiguous discrimination between endogenous and exogenous GH. In this respect, at least for one of the recombinant preparations the approach appears promising with the discovery of a methionine-to-valine substitution at position 14 in ca. 1% of the material [141]. This observation has been attributed to the infidelity of translation for which the M_{14}–V_{14}-exchange could also be present in all other preparations that employ *E. coli* as expression system, and opens yet another avenue to address hGH abuse.

From all other substances present on the 2007 list, only the gonadotropins are addressed in a regular way employing readily available kits developed for clinical approaches. Despite being a well-accepted analytical tool in clinical settings the results obtained with different kits and different instruments show large variations for identical samples. Also replicate analyses show significant variations for singular samples for which development of alternative approaches is recommendable. In the case of chorionic gonadotropin (CG), liquid chromatography hyphenated to MS has been proposed for the quantification using particular tryptic peptide of the β-subunit (T5 and the peak area of three fragments) [142]. Despite the fact that this approach appears much more reliable that the strict immunological approach, antibodies are still required in the sample work-up transferring the potential problem to a different part of the analytical process. This is precisely the Achilles-heel of all structural–analytical approaches for the analysis of peptide hormones but as long as general purification strategies are not applicable because of down-stream sample handling or detection limits, immunoglobulins will still be included in the most appropriate protocols.

21.7 ANTIINFLAMMATORY DRUGS

21.7.1 Non-steroidal anti-inflamatory drugs

The use of substances acting on the musculo-skeletal system such as non-steroidal anti-inflammatory drugs (NSAIDs) is prohibited for different veterinary regulations to protect animals. The use of anti-inflammatory drugs is a common practice that might influence the horse's performance or mask an underlying health problem, and could falsely affect the outcome of a competition.

Among the NSAIDs, phenylbutazone remains the most popular and most widely used in equine medicine. At first, FEI (Fédération Equestre Internationale) established a threshold concentration for the detection of phenylbutazone, but at present its use is completely forbidden. For salicylic acid, as it is normally present in horse samples, a threshold concentration has been established in equine plasma (5.4 μg/ml) and urine (625 μg/ml) to distinguish between normal concentrations and those arising from medication.

Metabolic biotransformation had been described for some NSAIDs in horses however, most of them can be detected as unchanged compounds [143–147] Although the NSAIDs have high differences in their molecular structures (salicylates, phenylalkyl-, heteroaryl- and indolealkyl acids, oxicams, pyrazoles…), they share an acidic character (pK_a <5.0). Methods allowing the simultaneous detection of different NSAIDs have been described by GC-MS [148–152] or LC-MS-MS [153].

To detect NSAIDs in biological samples an extraction from the matrix interferences is needed. The pH of samples must be previously conditioned to a value of 2–3. Conventional methodology involves usually liquid–liquid extractions with solvents such as *tert*-buthylmethyl ether or ethyl ether as the most used [149]. After discarding the aqueous phase, a saturated solution of sodium hydrogencarbonate is added to the organic layer and mixed. The extracts obtained with *tert*-buthylmethyl ether or diethyl ether are cleaner than those obtained with other organic solvents such as ethylacetate or chloroform. Also, a partially automated methodology based in a flow system for liquid–solid extraction with XAD-2 and back-elution methylation has been described [150]. Flufenamic acid [149] or zomepirac [150] have been described as internal standards.

Although some of the NSAIDs could be analysed without derivatisation, such as phenylbutazone, the formation of derivatives improves their gas chromatographic properties. Unified derivatisation procedures had been described even considering the great variety of chemical structures of the different NSAIDs, the most common being methylation [149,151,152]. The methylation procedure allows the formation of methyl esters or ethers of the different functions: carboxylic acid functions (diclofenac, meclofenamic), menolic acid groups (phenylbutazone and oxyphenbutazone) and phenolic groups (oxyphenbutazone and salicylic acid). Methyl derivatives are stable during at least 1 week after its formation [149]. Methylation is usually performed with methyl iodide in acetone and dry potassium carbonate, which acts as a catalyst for providing the alkaline medium needed. Description of kinetics of derivatisation has been presented [149] and a more extensive description of the methodologies to form methyl derivatives is included elsewhere in this chapter (Section 21.4). In fact, the procedure described allows the simultaneous detection of NSAIDs and diuretics in equine plasma and urine. Extractive methylation and GC-MS screening to analyse NSAIDs has also been used [152].

A comprehensive procedure has been recently described to analyse NSAIDS and other acidic drugs, among other compounds such as anabolic steroids or corticosteroids, using LC-MS-MS in horse urine [153]. The sample preparation consist of an hydrolysis with β-glucuronidase, followed by a solid-phase extraction using Bond-Elut Certify columns with elution of different fractions to recover basic and acidic and neutral drugs. The fraction of acidic drugs was purified using a liquid–liquid extraction. The detection of NSAID was achieved using LC-MS-MS with electrospray ionisation in negative ion mode. The high efficiency reversed-phase column (with 3 µm particles) coupled to the fast scanning capability of the LC-MS-MS instrument used allowed the analysis of multiple analytes in a single time segment resulting in turnaround times as lower as 10 min per sample.

21.7.2 Corticosteroids

Synthetic corticosteroids are structural analogues of the endogenous hormone cortisol. They produce strong anti-inflammatory effect and, in consequence, small doses are required to achieve the desired effect. The analysis of synthetic corticosteroids in biological samples is difficult due to the low concentrations expected as a consequence of the low therapeutic doses and its extensive metabolism. Methods for detecting corticosteroids administration must be effective in the nanogram and subnanogram per millilitre of urine concentration range.

The screening of corticosteroids in biological samples can be performed using immunological methods such as ELISA [154–156]. These methods suit the requirements of sensitivity and turnaround time, but a confirmatory analysis based on MS is required to eliminate false positive results.

The first step of sample preparation is the hydrolysis of the corticosteroid metabolite-conjugates, which is usually performed enzimatically with preparations having β-glucuronidase and sulphatase activity such as *H. pomatia* juice [157–159].

Isolation of corticosteroids from biological matrixes has been achieved in different ways. Liquid–liquid extractions of plasma or urine with organic solvents such as dichloromethane, diethyl ether or ethylacetate [160–163], or solid-phase extractions using C_{18} cartridges or Bond-Elut Certify columns have demonstrated to give good recoveries [159,164–167]. IAC has been applied to eliminate the masking effect of the large amount of co-extracted material in order to achieve detection limits at the sub-ng/ml range [168,169]. In some cases, LC fractionation has been applied after liquid–liquid or solid-phase extractions. In an attempt of automation, IAC combined on-line with LC-MS was described to detect dexamethasone and flumethasone [170]. Extraction of 11 corticosteroids and 2 steroid-conjugates from urine with solid-phase microextraction using four types of fibers has also been evaluated [171].

GC-MS in EI or CI conditions has been used for the determination of corticosteroids in biological fluids. The direct analysis of corticosteroids by GC-MS is, in general, unsuitable due to the thermal instability of the dihydroxyacetone side chain at C17, which is lost to yield the 17-oxo steroid, and usually a derivatisation is performed to protect some or all of the functional groups. The most extensively employed derivatisation procedures for corticosteroids are trimethylsilylation [159,172–175] and methoxymation of the ketone functions followed by trimethylsilylation of the hydroxyl groups [160,176–180].

Different degrees of silylation have been obtained depending on the reagent and catalisers used: mixtures of *N,O*-bis(trimethylsilyl)acetamide (BSA), TMSIm and TMCS allow the derivatisation of hydroxyl and ketone groups [172], while reaction with TMSIm in pyridine and formamide leads to the derivatisation of only hydroxyl functions [159]. Methoxime-TMS derivatives have been formed by reaction with methoxyamine hydrochloride in pyridine followed by reaction with silylating reagents such as BSTFA [160] or TMSIm [179,180].

Chemical oxidation to the 1,4-androstandiene-3,11,17-trione analogues and analysis by GC-MS under NCI conditions has been described for dexamethasone

[157,181,182], and for several corticosteroids with previous purification on C_{18} cartridges and IAC [169].

Direct GC analysis with on-column injection and NCI has also demonstrated to valuable in the detection of dexamethasone and flumethasone in equine urine after IAC extraction [168]. Formation of bismethylenedioxy derivatives of the 17-dihydroxyacetone side chain followed by acylation with heptafluorobutyric anhydride (HFBA) of the ketone in C3 *via* enol formation has been described for some synthetic and natural corticosteroids [183].

During the last years, these GC-MS methods have been gradually replaced by LC-MS approaches, both in urine and plasma. The first efforts were made in order to optimise the separation of several natural and synthetic corticosteroids using reversed-phase conditions [183]. LC with UV detection was used to analyse corticosteroids in plasma or urine samples in the ng/ml level [155,184]. Fluorescence detection with previous derivatisation using 9-anthroyl nitrile allowed the detection of several corticosteroids in the sub-ng/ml level [167]. In addition to that, LC coupled to MS with different interfaces was evaluated: TSP [161,185], PB [170], ES [171] and APCI [158].

Nowadays the use of either APCI or ESI sources, combined with the very selective SRM mode of operation, has facilitated the development of screening procedures for large numbers of compounds in a single analysis. These procedures have been successfully applied to detect corticosteroids in equine [186,187], bovine [158] and human urine [188,189].

The fast scanning capabilities of the new LC/MS instrumentation has enabled the elaboration of more comprehensive LC-MS-MS methods. Thus, the same extract can be simultaneously tested for over 200 target analytes [153]. In this method, a total of 36 MRM transitions are monitored within a single time segment, proving the pre-eminence of LC-MS-MS over GC-MS approaches. The simpler sample preparation and the shorter analysis time allowed a substantial increase on the number of samples analysed per day.

21.8 ENHANCEMENT OF OXYGEN TRANSFER

21.8.1 Haemoglobin based oxygen carriers

Blood substituents based on haemoglobin (HBOCs) are developed for clinical practice in case of surgery or impaired blood donation. As such these therapeutic agents result highly attractive to the athletic community to increase the circulating oxygen concentration artificially [190]. In turn, the scientific community has responded to this interest with the development of several analytical tools for the detection of these compounds. A comprehensive review on this subject was published recently by Kazlauskas and coworkers [191]. The far most simple approach relies on the visual inspection of plasma sample in search for sample extra coloration. This screening protocol has an apparent limit of detection of 1% HBOC in plasma but the reading may be altered as a result of partial or total haemolysis. As such a confirmatory process is always required. Three different analytical approaches have appeared recently to fulfil this task. One of them is an electrophoretic approach based on immuno-elimination of interfering haptoglobin, electrophoretic separation and Western blot, followed by the elegant detection employing the peroxidase activity of the haem moiety of haemoglobin and analogues [192]. A different strategy was published by the same authors targeting polymerised HBOCs through size-exclusion HPLC [193]. Even though this technique is able to differentiate between endogenous and modified haemoglobin the procedure results appear rather lengthy (chromatograms of over 1 h) and lack sensitivity. Alternatively, the third approach is based on the proteolytic (either trypsin or endo-Glu C) breakdown of a plasma work-up followed by LC-MS analysis of the mixture. The sequences, derived from native, exogenous or haemoglobin adducts are sufficiently different to allow for the unambiguous detection of the HBOC's origin [194,195].

21.8.2 Blood doping

Blood doping is used in endurance sports to increase the concentration of haemoglobin in blood and thus enhance the amount of oxygen that can be transported to the muscle. Blood doping includes the use of transfusions with autologous or homologous blood or with red blood cells products of any origin.

The detection of the use of homologous blood transfusion is performed by flow cytometry analysis [196,197]. The method is based on the detection of red blood cells with different antigenic profiles. The antigenic profile of the cells is under genetic control and, thus, red blood cells from a single individual have an identical spectrum of antigens. The method measures a series of red blood cells antigens (between 8 and 10 different antigens). The detection of mixed populations of red blood cells (one population from the receptor and another from the donor) is a demonstration of a homologous transfusion. Results obtained by flow cytometry of a sample obtained after homologous transfusion is shown in Fig. 21.3. A mixture of populations of antigens c, Jk^b, Fy^a and Fy^b were detected in this sample, demonstrating a previous blood transfusion.

21.9 MISCELLANEA

21.9.1 Agents with anti-estrogenic activity

Drugs with anti-estrogenic activity were added to the list of banned substances in 2001. This group comprises aromatese inhibitors, selective oestrogen receptor modulators (SERMs) and other anti-estrogenic substances such as clomiphene and cyclofenil. They have different medical applications, including the treatment of breast cancer and the treatment of infertility [198,199]. Some athletes take them in order to counteract the effects of anabolic steroid abuse by: firstly stopping the minor metabolic pathway of the anabolic steroids leading to estrogens which give

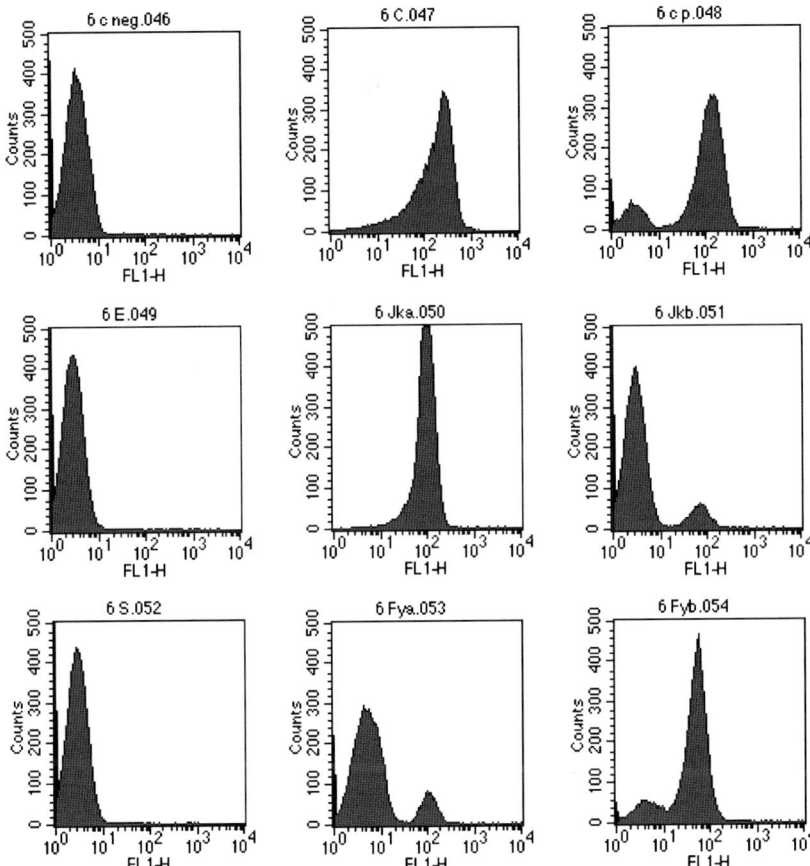

Fig. 21.3. Flow cytometry analysis of a blood sample collected after homologous blood transfusion. The following antigens are shown from right to left, from top to bottom: negative control, antigen C, antigen c, antigen E, antigen Jk[a], antigen JK[b], antigen S, antigen Fy[a], and antigen Fy[b].

rise to side effects such as breast development in males, and secondly stimulating the hypothalamus to release hormones which stimulate the testes to produce testosterone. Detection of some of these compounds can be accomplished adding them in the same screening method used for anabolic steroids. This is the case for clomiphene, tamoxiphen and cyclofenil, being hydroxy-clomiphene, hydroxy-methoxy-tamoxifen and hydroxy-bis-desacetyl-cyclofenil the monitored metabolites, respectively [200]. Aromatase inhibitors such as anastrazole, letrozole, exemestrane and aminogluthetimide can also be detected in the same screening method. Nevertheless, more specific methodologies have been established for letrozole [201,202] and aminoglu-thetimide [202]. Anastrazole and exemestane determination is also possible by LC/MS/MS [203].

References pp. 739–744

21.9.2 5α-Reductase inhibitors

5α-Reductase inhibitors are a class of drugs approved for the treatment of symptomatic benign prostatic hyperplasia [204]. They were included in the WADA 2005 list of prohibited substances as a *masking agent* and exemplified with finasteride and dutasteride. The possible use of finasteride as a masking agent has already been demonstrated with a combined administration of finasteride and nandrolone [205]. As a consequence, the concentrations of norandrosterone are markedly decreased compared with a single nandrolone administration.

The changes in the steroid profile that may have influence on the interpretation of longitudinal studies [206], the prevention on the detection of steroids mainly metabolised through 5α-reductase and lowering the time window of detection of some compounds (5α-DHT) are other observations that justified the prohibition of these compounds. Indirect evidences of their intake are easily achieved through the determination of the ratios between epimeric 5α- and 5β-steroids for example, androsterone/etiocholanolone or 5α-androstane-3α,17β/5β-androstane-3α,17β-diol ratios. Unquestionable prove for finasteride use can be obtained through detection of finasteride metabolite ω-carboxyfinasteride by LC/MS/MS [207].

On the other hand, due to the pharmacokinetic properties of dutasteride, neither dutasteride nor its metabolites are detected in urine. In order to find direct proofs other biological samples are needed (faeces or plasma). Consequently, its detection in urine is only possible by evaluating the steadiness of the 5α-/5β ratios in longitudinal studies.

21.9.3 Hydroxyethyl starch

Plasma volume expanders (PVE) have been developed as plasma substitutes in the treatment of haemorrhage [208] or hypovolaemic shock [209] as well as for cryoprotection of biological material [210]. The first generation PVE (albumin, gelatin and dextran) were rapidly substituted by chemically modified polysaccharides such as acetyl starch and hydroxyethyl starch. Hydroxyethyl starch (HES) has found widespread acceptance because of the limited adverse reactions [211] and the fact that the half-life time can be carefully modulated based on the average degree of molar substitution [212]. Athletes too have learned that this substance prevents dehydration and as such enhances endurance [213,214] and that the combined use of HES and recombinant erythropoietin can be advantageous, as an elevated blood volume is achieved while heamatocrit and haemoglobin levels may remain within the legal range. As HES is a polysaccharide the detection of this substance in urine was performed employing standard protocols in carbohydrate chemistry such as the conventional sugar analysis by means of GC-MS of acid released, tri-methyl silyl derivatised monosaccharides [215]. The confirmatory protocol was derived from the linkage analysis including per-methylation of the polysaccharide, acid hydrolysis, sodium borodeuteride reduction of the anomeric centre do distinguish symmetric monosaccharides and per-acetylation of the resulting alditol [216]. Again, analysis is effectuated by GC-MS. As elevated doses of HES must be infused to be effective [as much as 500 ml of a 6% (w/v) solution (personal communication)] as little as 20 μl of urine is sufficient for either analysis. The main drawback of these methods lie in the labour-intensity, time required and the use of one or more derivatisation steps. Furthermore, the fact that the identifier ions are identical to those present in other, glycoprotein derived, monosaccharides that are released with the same protocol renders rather complex chromatograms. An alternative approach was developed based on a carefully controlled partial acid hydrolysis and direct analysis by MALDI-TOF mass spectrometry [217]. This approach yields a characteristic profile of any carbohydrate-derived polysaccharide covering the presence of dextran, HES and acetyl starch simultaneously but the analysis is difficult to automate. In this respect the use of LC-MS for the same purpose has resulted in two additional protocols; one following the traditional chemical hydrolysis [218] and a second one including an enzymic (dextranase) degradation and per-acetylation of the generated isomaltose [219]. Finally, efforts to even further reduce the workload with respect to this compound have resulted in an ultra-rapid protocol for the quantification of glucose in urine. This approach permits simultaneous screening of up to 96 samples in less than 1 h uses the acid-catalysed hydrolysis of polysaccharides followed by a triple dehydration of the monosaccharides to yield [5-(hydroxymethyl)]-2-furalde-hydes that subsequently react *in situ* with anthrone to give an UV-absorbing complex at 620 nm [220]. As the amount of free glucose in urine should be negligible due to the re-absorbance from the renal tubuli, values surpassing a certain threshold indicate either a diabetic condition or the presence of a PVE and triggers a more specific analysis by one of the above mentioned methods.

21.10 LEGAL ASPECTS

Results of doping controls usually generate decisions that can affect the sport career of an athlete. The sanction generally imposed for a doping offence when a serious dope substance is found is two years of suspension [1]. A second offence may represent even suspension for life in the sport activity of the subject involved. Thus, doping controls have been since years very stringent in those aspects that represent warranty of quality and optimum operating methodologies. Nevertheless, it is note-worthy that other kind of analysis in areas which can give rise to even worse consequences for the person involved (e.g. imprisonment based on forensic or med-ico-legal analysis) sometimes fail to maintain a similar level of quality assurance.

Dope testing results are sometimes taken to court by the athlete affected in an attempt to find weak points in the whole process of testing, which might invalidate the results. Usually, the focus is addressed to administrative aspects related to sam-ple collection, sealing and transportation. However, concerns about the analytical task done in the laboratories sometimes arise and the analyst is faced then with legal challenges on his/her activities. WADA accredited laboratories are presumed to have conducted sample analysis and custodial procedures in accordance with the International Standards for laboratory analysis [3]. If an athlete rebuts the

LC	liquid chromatography
LIMS	Laboratory Information Management System
MALDI	matrix assisted laser desorption ionisation
MBTFA	N-methyl-bis-trifluoroacetamide
MRM	multiple reaction monitoring
MS	mass spectrometry
MSD	mass selective detectors
MS/MS	tandem mass spectrometry
MSTFA	N-methyl-N-(trimethylsilyl)trifluoroacetamide
MTP	methoxytrifluoromethylphenyl
NCI	negative chemical ionisation
NPD	nitrogen-phosphorus detector
NSAID	non-steroidal antiinflammatory drug
ODS	octadecil
PB	particle beam
PFB	pentafluorobenzoyl
PFP	pentafluoropropionyl
PTM	post translational modification
PVE	plasma volume expander
RhGH	recombinant human growth hormone
REPO	recombinant erythropoietin
RIA	radioimmunoassay
SERM	selective oestrogen receptor modulator
SRM	selected reaction monitoring
TFA	trifluoroacetyl
THG	tetrahydrogestrinone
TMCS	trimethylchlorosilane
TMS	trimethylsilyl
TMSIm	trimethylsilylimidazole
TOF	time of flight
TSP	thermospray
UEPO	urinary eryhtropoietin
UV	ultraviolet
WADA	World Anti-Doping Agency

21.12 ACKNOWLEDGEMENTS

The authors are grateful to financing received from the Spanish (Ministerio de Educación y Ciencia; Consejo Superior de Deportes) and Catalan (Departament d'Universitats, Recerca i Societat de l'informació; Consell Catalá de l'Esport) authorities for supporting the preparation of the present review. Special thanks are addressed also to the World Anti-doping Agency for supporting several research projects mentioned in this chapter. The scientific contribution from other staff members of the Drug Research Unit at IMIM is also deeply acknowledged.

21.13 REFERENCES

1   World Anti-doping Agency (WADA). World antidoping code. World Anti-Doping Agency, Montreal, 2003.
2   World Anti-doping Agency (WADA). The 2007 prohibited list, 2006.
3   World Anti-doping Agency (WADA). International standards for laboratories. Montreal, 2005.
4   International Equestrian Federation. Equine anti-doping and medication control rules. FEI web site. FEI, 1-6-2006.
5   World Anti-doping Agency (WADA). 2006 Adverse analytical findings reported by accredited laboratories. WADA web site. 2007.
6   R.A. Sams. Medication case reports. 2003. Association of Offcial Racing Chemists AORC.
7   S.C. Chan, G.A. Torok-Both, D.M. Billay, P.S. Przybylski, C.Y. Gradeen, K.M. Pap and J. Petruzelka, Clin. Chem., 37 (1991) 1289.
8   D.H. Catlin, R.C. Kammerer, C.K. Hatton, M.H. Sekera and J.L. Merdink, Clin. Chem., 33 (1987) 319.
9   J. Park, S. Park, D. Lho, H.P. Choo, B. Chung, C. Yoon, H. Min and M.J. Choi, J. Anal. Toxicol., 14 (1990) 66.
10 L.D. Bowers and J. Segura, Clin. Chem., 42 (1996) 999.
11 J. Segura, J.A. Pascual, R. Ventura, J.I. Ustaran, A. Cuevas and R. Gonzalez, Clin. Chem., 39 (1993) 836.
12 B. Nikolin, M. Lekic and M. Sober, Bosn. J. Basic Med. Sci., 3 (2003) 45.
13 B. Berglund and L. Wide, Scand. J. Med. Sci. Sports, 12 (2002) 354.
14 G.J. Trout and R. Kazlauskas, Chem. Soc. Rev., 33 (2004) 1.
15 W. Schänzer, . In: S.B. Karch (Ed.), Drug abuse handbook, CRC, Boca Raton, 1998 Chapter 9, pp. 671–689.
16 W. Schanzer, G. Opfermann and M. Donike, Steroids, 57 (1992) 537.
17 W. Schanzer, Clin. Chem., 42 (1996) 1001.
18 R. Masse, C. Ayotte and R. Dugal, J. Chromatogr., 489 (1989) 23.
19 Y. Bonnaire, L. Dehennin, P. Plou, P.L. Toutain, A. Ginn, L. Grainger, R. Armstrong, M.C. Dumasia, A. Neddermann, E. Houghton, P.W. Tang, W.C. Law and D.L. Crone, in: D.E. Aver and E. Houghton (Eds.), Proceedings of the 11th International Conference of Racing Analyzts and Veterinarians, Queensland, Australia, 1996, p. 56.
20 W. Schanzer, P. Delahaut, H. Geyer, M. Machnik and S. Horning, J. Chromatogr. B, Biomed. Appl., 687 (1996) 93.
21 J. Marcos, X. de la Torre, J.C. Gonzalez, J. Segura and J.A. Pascual, Anal. Chim. Acta, 522 (2004) 79.
22 C. Ayotte, D. Goudreault and A. Charlebois, J. Chromatogr. B, Biomed. Appl., 687 (1996) 3.
23 M. Donike, J. Chromatogr. A, 115 (1975) 591.
24 W. Schanzer, S. Horning, G. Opfermann and M. Donike, J. Steroid Biochem. Mol. Biol., 57 (1996) 363.
25 S.J. Gaskell, A.W. Pike and K. Griffiths, Steroids, 36 (1980) 219.
26 W. Schanzer, G. Opfermann and M. Donike, J. Steroid Biochem., 36 (1990) 153.
27 J. Segura, R. Ventura and C. Jurado, J. Chromatogr. B, Biomed. Sci. Appl., 713 (1998) 61.
28 C.H. Shackleton, J. Chromatogr., 379 (1986) 91.
29 T.C. Malvey and T.D. Armsey, Curr. Sports Med. Rep., 4 (2005) 227.
30 D.H. Catlin, M.H. Sekera, B.D. Ahrens, B. Starcevic, Y.C. Chang and C.K. Hatton, Rapid Commun. Mass Spectrom., 18 (2004) 1245.
31 J. Marcos, J.A. Pascual, X. de la Torre and J. Segura, J. Mass Spectrom., 37 (2002) 1059.
32 R. Aguilera, M. Becchi, L. Grenot, H. Casabianca and C.K. Hatton, J. Chromatogr. B, Biomed. Appl., 687 (1996) 43.
33 X. de la Torre, J.C. Gonzalez, S. Pichini, J.A. Pascual and J. Segura, J. Pharm. Biomed. Anal., 24 (2001) 645.
34 X. de la Torre, J.C. Gonzalez, S. Pichini, J.A. Pascual and J. Segura, J. Pharm. Biomed. Anal., 24 (2001) 645.

35   R. Aguilera, T.E. Chapman, B. Starcevic, C.K. Hatton and D.H. Catlin, Clin. Chem., 47 (2001) 292.
36   R. Aguilera, T.E. Chapman and D.H. Catlin, Rapid Commun. Mass Spectrom., 14 (2000) 2294.
37   A.T. Cawley, R. Kazlauskas, G.J. Trout, J.H. Rogerson and A.V. George, J. Chromatogr. Sci., 43 (2005) 32.
38   R. Aguilera, C.K. Hatton and D.H. Catlin, Clin. Chem., 48 (2002) 629.
39   M. Hebestreit, U. Flenker, C. Buisson, F. Andre, B. Le Bizec, H. Fry, M. Lang, A.P. Weigert, K. Heinrich, S. Hird and W. Schanzer, J. Agric. Food Chem., 54 (2006) 2850.
40   C. Buisson, M. Hebestreit, A.P. Weigert, K. Heinrich, H. Fry, U. Flenker, S. Banneke, S. Prevost, F. Andre, W. Schaenzer, E. Houghton and B. Le Bizec, J. Chromatogr. A, 1093 (2005) 69.
41   C.H. Shackleton, A. Phillips, T. Chang and Y. Li, Steroids, 62 (1997) 379.
42   X. de la Torre, J. Segura, A. Polettini and M. Montagna, J. Mass Spectrom., 30 (1995) 1393.
43   M. Thevis, H. Geyer, U. Mareck and W. Schanzer, J. Mass Spectrom., 40 (2005) 955.
44   D. Barron, J. Barbosa, J.A. Pascual and J. Segura, J. Mass Spectrom., 31 (1996) 309.
45   L.D. Bowers and Sanaullah, J. Chromatogr. B, Biomed. Appl., 687 (1996) 61.
46   T. Kuuranne, T. Kotiaho, S. Pedersen-Bjergaard, R.K. Einar, A. Leinonen, S. Westwood and R. Kostiainen, J. Mass Spectrom., 38 (2003) 16.
47   A. Leinonen, T. Kuuranne and R. Kostiainen, J. Mass Spectrom., 37 (2002) 693.
48   K. Deventer, P.V. Eenoo and F.T. Delbeke, Biomed. Chromatogr., 20 (2006) 429.
49   A.H. Beckett, G.T. Tucker and A.C. Moffat, J. Pharm. Pharmacol., 19 (1967) 273.
50   M. Donike, Chromatographia, 3 (1970) 422.
51   M. Donike and D. Stratmann, Chromatographia, 7 (1974) 182.
52   R. de la Torre, R. Badia, G. Gonzalez, M. Garcia, M.J. Pretel, M. Farre and J. Segura, J. Anal. Toxicol., 20 (1996) 165.
53   P. Hemmersbach and R. de la Torre, J. Chromatogr. B, Biomed. Appl., 687 (1996) 221.
54   A. Solans, M. Carnicero, R. de la Torre and J. Segura, J. Anal. Toxicol., 19 (1995) 104.
55   D.S. Lho, J.K. Hong, H.K. Paek, J.A. Lee and J. Park, J. Anal. Toxicol., 14 (1990) 77.
56   H.H. Maurer, J. Chromatogr., 580 (1992) 3.
57   A. Solans, R. de la Torre and J. Segura, J. Pharm. Biomed. Anal., 8 (1990) 905.
58   A. Solans, M. Carnicero, R. de la Torre and J. Segura, J. Chromatogr. B, Biomed. Appl., 658 (1994) 380.
59   M. Donike, J. Chromatogr. A, 103 (1975) 91.
60   D.S. Lho, H.S. Shin, B.K. Kang and J. Park, J. Anal. Toxicol., 14 (1990) 73.
61   H.H. Maurer and T. Kraemer, Arch. Toxicol., 66 (1992) 675.
62   H.L. Jin and T.E. Beesley, Chromatographia, 38 (1994) 595.
63   N.R. Srinivas, J.W. Hubbard and K.K. Midha, J. Chromatogr., 530 (1990) 327.
64   S.D. Roy and H.K. Lim, J. Chromatogr., 431 (1988) 210.
65   R. Kazlauskas, A.M. Lisi and G. Trout, in W. Schanzer, H. Geyer, A. Gotzmann and U. Mareck-Engelke (Eds.), Recent advances in doping analysis. Sport und Buch Strauß, 1999, p. 431.
66   M.J. Pretel, J.A. Pascual, M. Mestres, R. de la Torre and J. Segura, Second IOC World Congress on Sport Sciences. COOB 92 (1991) 287.
67   M. Thevis and W. Schanzer, J. Chromatogr. Sci., 43 (2005) 22.
68   M. Spyridaki, P. Kiousi, A. Vonaparti, P. Valavani, V. Zonaras, M. Zahariou, E. Sianos, G. Tsoupras and C. Georgakopoulos, Anal. Chim. Acta, (2006) online.
69   K. Deventer, P. Van Eenoo and F.T. Delbeke, Rapid Commun. Mass Spectrom., 20 (2006) 877.
70   R. Ventura and J. Segura, J. Chromatogr. B, Biomed. Appl., 687 (1996) 127.
71   K. Deventer, F.T. Delbeke, K. Roels and P. Van Eenoo, Biomed. Chromatogr., 16 (2002) 529.
72   S.F. Cooper, R. Masse and R. Dugal, J. Chromatogr., 489 (1989) 65.
73   F.Y. Tsai, L.F. Lui and B. Chang, J. Pharm. Biomed. Anal., 9 (1991) 1069.
74   W. Schänzer, in: P. Bellotti, G. Benzi and A. Ljungqvist (Eds.), Proceedings of the International Athletics Foundation World Symposium on Doping in Sport, 1988, p. 89.
75   S.J. Park, H.S. Pyo, Y.J. Kim, M.S. Kim and J. Park, J. Anal. Toxicol., 14 (1990) 84.
76   R. Ventura, T. Nadal, P. Alcalde, J.A. Pascual and J. Segura, J. Chromatogr. A, 655 (1993) 233.

77   L. Amendola, C. Colamonici, M. Mazzarino and F. Botre, Anal. Chim. Acta, 475 (2003) 125.
78   P. Campíns, R. Herráez and A. Sevillano, J. Liq. Chromatogr., 14 (1991) 3575.
79   P. Campins-Falco, R. Herraez-Hernandez and A. Sevillano-Cabeza, Anal. Chem., 66 (1994) 244.
80   M.B. Barroso, H.D. Meiring, A. de Jong, R.M. Alonso and R.M. Jimenez, J. Chromatogr. B, Biomed. Sci. Appl., 690 (1997) 105.
81   S. Salado and L.E. Vera-Avila, J. Chromatogr. B, Biomed. Sci. Appl., 690 (1997) 195.
82   P. Campíns, R. Herráez and A. Sevillano, J. Chromatogr. B, Biomed. Appl., 612 (1993) 245.
83   D. Carreras, C. Imaz, R. Navajas, M.A. Garcia, C. Rodriguez, A.F. Rodriguez and R. Cortes, J. Chromatogr. A, 683 (1994) 195.
84   D. Thieme, J. Grosse, R. Lang, R.K. Mueller and A. Wahl, J. Chromatogr. B, Biomed. Sci. Appl., 757 (2001) 49.
85   A.M. Lisi, G.J. Trout and R. Kazlauskas, J. Chromatogr., 563 (1991) 257.
86   A.M. Lisi, R. Kazlauskas and G.J. Trout, J. Chromatogr., 581 (1992) 57.
87   H.W. Hagedorn and R. Schulz, J. Anal. Toxicol., 16 (1992) 194.
88   R. Ventura, PhD Thesis, Universitat de Barcelona, Spain, 1994.
89   K. Deventer, P. Van Eenoo and F.T. Delbeke, Rapid Commun. Mass Spectrom., 19 (2005) 90.
90   V. Sanz-Nebot, I. Toro, R. Berges, R. Ventura, J. Segura and J. Barbosa, J. Mass Spectrom., 36 (2001) 652.
91   S.D. Garbis, L. Hanley and S. Kalita, J. AOAC Int., 81 (1998) 948.
92   F. De Croo, B.W. Van den and P. De Moerloose, J. Chromatogr., 325 (1985) 395.
93   J. Beyer, A. Bierl, F.T. Peters and H.H. Maurer, Ther. Drug Monit., 27 (2005) 509.
94   J.D. Ehrhardt, Rapid Commun. Mass Spectrom., 6 (1992) 349.
95   R. Ventura, D. Fraisse, M. Becchi, O. Paisse and J. Segura, J. Chromatogr., 562 (1991) 723.
96   M. Thevis, M.H. Schmickler and W. Schanzert, Anal. Chem., 74 (2002) 3802.
97   D.J. Morgan, Clin. Pharmacokinet., 18 (1990) 270.
98   B.B. Hoffman, R.J. Lefkowitz, . In: A. Goodman Gilman, T.W. Rall, A.S. Nies, P. Taylor (Eds.), Goodman and Gilman's The Pharmacological Basis of Therapeutics, Pergamon Press Inc, New York, 1990, p. 221.
99   R.C. Baselt; R.H. Cravey, Disposition of toxic drugs and chemicals in man, Chemical Toxicology Institute, Foster City, 1995.
100 R. Ventura, G. Gonzalez, M.T. Smeyers, R. de la Torre and J. Segura, J. Anal. Toxicol., 22 (1998) 127.
101 M.C. Dumasia and E. Houghton, J. Chromatogr., 564 (1991) 503.
102 A. Polettini, M.C. Ricossa, A. Groppi and M. Montagna, J. Chromatogr., 564 (1991) 529.
103 A. Polettini, A. Groppi, M.C. Ricossa and M. Montagna, Biol. Mass Spectrom., 22 (1993) 457.
104 R. Ventura, L. Damasceno, M. Farre, J. Cardoso and J. Segura, Anal. Chim. Acta, 418 (2000) 79.
105 J.A. van Rhijn, H.H. Heskamp, M.L. Essers, H.J. van de Wetering, H.C. Kleijnen and A.H. Roos, J. Chromatogr. B, Biomed. Appl., 665 (1995) 395.
106 G. Van Vyncht, S. Preece, P. Gaspar, G. Maghuin-Register and E. DePauw, J. Chromatogr. A, 750 (1996) 43.
107 H. Hooijerink, R. Schilt, E.O. van Bennekom and F.A. Huf, J. Chromatogr. B, Biomed. Appl., 660 (1994) 303.
108 F.J. Couper and O.H. Drummer, J. Chromatogr. B, Biomed. Appl., 685 (1996) 265.
109 J. Zamecnik, J. Anal. Toxicol., 14 (1990) 132.
110 G.D. Branum, S. Sweeney, A. Palmeri, L. Haines and C. Huber, J. Anal. Toxicol., 22 (1998) 135.
111 L. Amendola, F. Molaioni and F. Botre, J. Pharm. Biomed. Anal., 23 (2000) 211.
112 L. Amendola, C. Colamonici, F. Rossi and F. Botre, J. Chromatogr. B, Analyt. Technol. Biomed. Life Sci., 773 (2002) 7.
113 L. Politi, A. Groppi and A. Polettini, J. Anal. Toxicol., 29 (2005) 1.
114 M. Thevis, G. Opfermann and W. Schanzer, Biomed. Chromatogr., 15 (2001) 393.
115 A.F. Lehner, J.D. Harkins, W. Karpiesiuk, W.E. Woods, N.E. Robinson, L. Dirikolu, M. Fisher and T. Tobin, J. Anal. Toxicol., 25 (2001) 280.
116 M. Thevis, G. Opfermann and W. Schanzer, J. Mass Spectrom., 38 (2003) 1197.

117   N.T. Nguyen and R.W. Siegler, J. Chromatogr. A, 735 (1996) 123.

118   R. Berges, J. Segura, R. Ventura, K.D. Fitch, A.R. Morton, M. Farre, M. Mas and X. de la Torre, Clin. Chem., 46 (2000) 1365.

119   R. Berges, J. Segura, X. de la Torre and R. Ventura, J. Chromatogr. B, Biomed. Sci. Appl., 723 (1999) 173.

120   K.B. Joyce, A.E. Jones, R.J. Scott, R.A. Biddlecombe and S. Pleasance, Rapid Commun. Mass Spectrom., 12 (1998) 1899.

121   R. Parisotto, C.J. Gore, K.R. Emslie, M.J. Ashenden, C. Brugnara, C. Howe, D.T. Martin, G.J. Trout and A.G. Hahn, Haematologica, 85 (2000) 564.

122   F. Lasne and J. de Ceaurriz, Nature, 405 (2000) 635.

123   F. Lasne, J. Immunol. Methods, 253 (2001) 125.

124   I. Bajla, I. Hollander, M. Minichmayr, G. Gmeiner and C. Reichel, Comput. Methods Programs Biomed., 80 (2005) 246.

125   I. Bajla, I. Hollander, G. Gmeiner and C. Reichel, Med. Biol. Eng Comput., 43 (2005) 403.

126   M. Beullens, J.R. Delanghe and M. Bollen, Blood, 107 (2006) 4711.

127   M. Nagano, G. Stubiger, M. Marchetti, G. Gmeiner, G. Allmaier and C. Reichel, Electrophoresis, 26 (2005) 1633.

128   G. Stubiger, M. Marchetti, M. Nagano, R. Grimm, G. Gmeiner, C. Reichel and G. Allmaier, J. Sep. Sci., 28 (2005) 1764.

129   E. Llop, R. Gutiérrez Gallego, V. Belalcazar, G.W. Gerwig, J.P. Kamerling, J. Segura and J.A. Pascual, Submitted (2007).

130   V. Belalcazar, R. Gutiérrez Gallego, E. Llop, J. Segura and J.A. Pascual, Electrophoresis, 27 (2006) 4387.

131   C.M. McHugh, R.T. Park, P.H. Sonksen and R.I. Holt, Clin. Chem., 51 (2005) 1587.

132   A.E. Rigamonti, S.G. Cella, N. Marazzi, L. Di Luigi, A. Sartorio and E.E. Muller, Trends Endocrinol. Metab, 16 (2005) 160.

133   R. Abellán, R. Ventura, S. Pichini, R. Di Giovannandrea, M. Bellver, R. Olivé, R. Pacifici, J.A. Pascual and P. Zuccaro, J. Segura. Int. J. Sports Med, 27 (2006) 976.

134   R. Abellan, R. Ventura, S. Pichini, J.A. Pascual, R. Pacifici, S. Di Carlo, A. Bacosi, J. Segura and P. Zuccaro, Clin. Chem. Lab Med., 43 (2005) 75.

135   A. Sartorio, F. Agosti, N. Marazzi, N.A. Maffiuletti, S.G. Cella, A.E. Rigamonti, L. Guidetti, L. Di Luigi and E.E. Muller, Clin. Endocrinol. (Oxf), 61 (2004) 487.

136   P.H. Sonksen, J. Endocrinol., 170 (2001) 13.

137   F.M. DeNoto, D.D. Moore and H.M. Goodman, Nucleic Acids Res., 9 (1981) 3719.

138   G. Baumann, Horm. Res., 51 (Suppl. 1) (1999) 2.

139   Z. Wu, M. Bidlingmaier, R. Dall and C.J. Strasburger, Lancet, 353 (1999) 895.

140   M. Sato, N. Junko, T. Shimazaki, Y. Takao, Y. Otsuka and M. Ueki, in: W. Schänzer, R. Geyer, A. Gotzmann and U. Mareck (Eds.), Recent advances in doping analysis (13). Sport und Buch Strauß, 2005, p. 13.

141   F. Hepner, E. Csaszar, E. Roitinger, A. Pollak and G. Lubec, Proteomics, 6 (2006) 775.

142   L.H. Gam, S.Y. Tham and A. Latiff, J. Chromatogr. B, Analyt. Technol. Biomed. Life Sci., 792 (2003) 187.

143   T.M. Dyke, R.A. Sams and K.W. Hinchcliff, Am. J. Vet. Res., 59 (1998) 1481.

144   T. Tobin, Drugs and performance in the horse, Charles C Thomas, Springfield, Illinois, 1981.

145   E. Benoit, P. Jaussaud, S. Besse, B. Videmann, D. Courtot, P. Delatour and Y. Bonnaire, J. Chromatogr., 583 (1992) 167.

146   P. Jaussaud, D. Guieu, D. Courtot, B. Barbier and Y. Bonnaire, J. Chromatogr., 573 (1992) 136.

147   P. Jaussaud, D. Courtot, J.L. Guyot and J. Paris, J. Chromatogr., 423 (1987) 123.

148   E.G. de Jong, J. Kiffers and R.A. Maes, J. Pharm. Biomed. Anal., 7 (1989) 1617.

149   G. Gonzalez, R. Ventura, A.K. Smith, R. de la Torre and J. Segura, J. Chromatogr. A, 719 (1996) 251.

150   S. Cardenas, M. Gallego, M. Valcarcel, R. Ventura and J. Segura, Anal. Chem., 68 (1996) 118.

151   U.M. Laakkonen, A. Leinonen and L. Savonen, Analyst, 119 (1994) 2695.

152   H.H. Maurer, F.X. Tauvel and T. Kraemer, J. Anal. Toxicol., 25 (2001) 237.
153   E.N. Ho, D.K. Leung, T.S. Wan and N.H. Yu, J. Chromatogr. A, 1120 (2006) 38.
154   C.L. Chen, D. Zhu, K.D. Gillis and M. Meleka-Boules, Am. J. Vet. Res., 57 (1996) 182.
155   L.M. Ribeiro Neto, H.S. Spinosa and M.C. Salvadori, J. Anal. Toxicol., 21 (1997) 393.
156   M.L. Rodriguez, I. McConnell, J. Lamont, J. Campbell and S.P. FitzGerald, Analyst, 119 (1994) 2631.
157   D. Courtheyn, J. Vercammen, H. De Brabander, I. Vandenreyt, P. Batjoens, K. Vanoosthuyze and C. Van Peteghem, Analyst, 119 (1994) 2557.
158   S.R. Savu, L. Silvestro, A. Haag and F. Sorgel, J. Mass Spectrom., 31 (1996) 1351.
159   R. Bagnati, V. Ramazza, M. Zucchi, A. Simonella, F. Leone, A. Bellini and R. Fanelli, Anal. Biochem., 235 (1996) 119.
160   A.K. Singh, B. Gordon, D. Hewetson, K. Granley, M. Ashraf, U. Mishra and D. Dombrovskis, J. Chromatogr., 479 (1989) 233.
161   S.J. Park, Y.J. Kim, H.S. Pyo and J. Park, J. Anal. Toxicol., 14 (1990) 102.
162   A. Santos-Montes, A.I. Gasco-Lopez and R. Izquierdo-Hornillos, J. Chromatogr., 620 (1993) 15.
163   W.J. Jusko, N.A. Pyszczynski, M.S. Bushway, R. D'Ambrosio and S.M. Mis, J. Chromatogr. B, Biomed. Appl., 658 (1994) 47.
164   H. Shibasaki, T. Furuta and Y. Kasuya, J. Chromatogr. B, Biomed. Sci. Appl., 692 (1997) 7.
165   H. Hirata, T. Kasama, Y. Sawai and R.R. Fike, J. Chromatogr. B, Biomed. Appl., 658 (1994) 55.
166   B.C. McWhinney, G. Ward and P.E. Hickman, Clin. Chem., 42 (1996) 979.
167   N. Shibata, T. Hayakawa, K. Takada, N. Hoshino, T. Minouchi and A. Yamaji, J. Chromatogr. B, Biomed. Sci. Appl., 706 (1998) 191.
168   S.M. Stanley, B.S. Wilhelmi, J.P. Rodgers and H. Bertschinger, J. Chromatogr., 614 (1993) 77.
169   P. Delahaut, P. Jacquemin, Y. Colemonts, M. Dubois, J. De Graeve and H. Deluyker, J. Chromatogr. B, Biomed. Sci. Appl., 696 (1997) 203.
170   C.S. Creaser, S.J. Feely, E. Houghton and M. Seymour, J. Chromatogr. A, 794 (1998) 37.
171   D.A. Volmer and J.P. Hui, Rapid Commun. Mass Spectrom., 11 (1997) 1926.
172   P.M. Simpson, J. Chromatogr., 77 (1973) 161.
173   Y. Kasuya, J.R. Althaus, J.P. Freeman, R.K. Mitchum and J.P. Skelly, J. Pharm. Sci., 73 (1984) 446.
174   K. Minagawa, Y. Kasuya, S. Baba, G. Knapp and J.P. Skelly, J. Chromatogr., 343 (1985) 231.
175   J. Girault, B. Istin and J.B. Fourtillan, Biomed. Environ. Mass Spectrom., 19 (1990) 295.
176   G.M. Rodchenkov, V.P. Uralets and V.A. Semenov, J. Chromatogr., 423 (1987) 15.
177   G.M. Rodchenkov, V.P. Uralets, V.A. Semenov and V.A. Gurevich, J. Chromatogr., 432 (1988) 283.
178   G.M. Rodchenkov, V.P. Uralets and V.A. Semenov, J. Chromatogr., 426 (1988) 399.
179   G.M. Rodchenkov, A.N. Vedenin, V.P. Uralets and V.A. Semenov, J. Chromatogr., 565 (1991) 45.
180   J.M. Midgley, D.G. Watson, T. Healey and M. Noble, Biomed. Environ. Mass Spectrom., 15 (1988) 479.
181   G.R. Her and J.T. Watson, Anal. Biochem., 151 (1985) 292.
182   K. Kayganich, J.T. Watson, C. Kilts and J. Ritchie, Biomed. Environ. Mass Spectrom., 19 (1990) 341.
183   H. Shibasaki, T. Furuta and Y. Kasuya, J. Chromatogr., 579 (1992) 193.
184   S.A. Doppenschmitt, B. Scheidel, F. Harrison and J.P. Surmann, J. Chromatogr. B, Biomed. Appl., 674 (1995) 237.
185   S. Steffenrud and G. Maylin, J. Chromatogr., 577 (1992) 221.
186   P.W. Tang, W.C. Law and T.S. Wan, J. Chromatogr. B, Biomed. Sci. Appl., 754 (2001) 229.
187   G.N. Leung, E.W. Chung, E.N. Ho, W.H. Kwok, D.K. Leung, F.P. Tang, T.S. Wan and N.H. Yu, J. Chromatogr. B, Analyt. Technol. Biomed. Life Sci., 825 (2005) 47.
188   R.L. Taylor, S.K. Grebe and R.J. Singh, Clin. Chem., 50 (2004) 2345.
189   K. Fluri, L. Rivier, A. Dienes-Nagy, C. You, A. Maitre, C. Schweizer, M. Saugy and P. Mangin, J. Chromatogr. A, 926 (2001) 87.
190   D.R. Spahn and R. Kocian, Curr. Pharm. Des., 11 (2005) 4099.

191 C. Goebel, C. Alma, C. Howe, R. Kazlauskas and G. Trout, J. Chromatogr. Sci., 43 (2005) 39.

192 F. Lasne, N. Crepin, M. Ashenden, M. Audran and J. de Ceaurriz, Clin. Chem., 50 (2004) 410.

193 E. Varlet-Marie, M. Ashenden, F. Lasne, M.T. Sicart, B. Marion, J. de Ceaurriz and M. Audran, Clin. Chem., 50 (2004) 723.

194 M. Thevis, R.R. Ogorzalek Loo, J.A. Loo and W. Schanzer, Anal. Chem., 75 (2003) 3287.

195 M. Gasthuys, S. Alves and J.C. Tabet, Anal. Chem., 77 (2005) 3372.

196 M. Nelson, M. Ashenden, M. Langshaw and H. Popp, Haematologica, 87 (2002) 881.

197 M. Nelson, H. Popp, K. Sharpe and M. Ashenden, Haematologica, 88 (2003) 1284.

198 O.C. Freedman, S. Verma and M.J. Clemons, Breast Cancer Res. Treat., 99 (2006) 241.

199 O. Karaer, S. Oruc and F.M. Koyuncu, Acta Obstet. Gynecol. Scand., 83 (2004) 699.

200 U. Mareck-Engelke, G. Sigmund, G. Opfermann, H. Geyer and W. Schänzer, in: W. Schänzer, H. Geyer, A. Gotzmann and U. Mareck-Engelke (Eds.), Recent advances in doping analysis. Sport und Buch Strauß, 2001, pp. 53–61.

201 U. Mareck, G. Sigmund, G. Opfermann, H. Geyer, M. Thevis and W. Schanzer, Rapid Commun. Mass Spectrom., 19 (2005) 3689.

202 U. Mareck, G. Sigmund, G. Opfermann, H. Geyer and W. Schanzer, Rapid Commun. Mass Spectrom., 16 (2002) 2209.

203 U. Mareck, H. Geyer, S. Guddat, N. Haenelt, A. Koch, M. Kohler, G. Opfermann, M. Thevis and W. Schanzer, Rapid Commun. Mass Spectrom., 20 (2006) 1954.

204 T.H. Tarter and E.D. Vaughan Jr., Curr. Pharm. Des., 12 (2006) 775.

205 H. Geyer, E. Nolteernsting and W. Schänzer, in: W. Schänzer, H. Geyer, A. Gotzmann and U. Mareck (Eds.), Recent advances in doping analysis. Sport und Buch Strauß, 1999, p. 71.

206 M.A.S. Marques, C.H.B. Bizarri, J.N. Cardoso and F.R. De Aquino Neto, in: W. Schanzer, H. Geyer, A. Gotzmann and U. Mareck (Eds.), Recent advances in doping analysis. Sport und Buch Strauß, 1999, p. 317.

207 S. Simoes, B. Vitoriano, C. Manzoni and X. de la Torre, in: W. Schänzer, H. Geyer, A. Gotzmann and U. Mareck-Engelke (Eds.), Recent advances in doping analysis. Sport und Buch Strauß, 2005.

208 S.K. Nakasato, Clin. Pharm., 1 (1982) 509.

209 S. Lazrove, K. Waxman, C. Shippy and W.C. Shoemaker, Crit. Care Med., 8 (1980) 302.

210 A. Sputtek, C. Bacher, R. Langer, W. Kron, H.A. Henrich and G. Rau, Infusionsther. Transfusionsmed., 19 (1992) 276.

211 M.L. Cittanova, J. Mavre, B. Riou and P. Coriat, Intensive Care Med., 27 (2001) 1830.

212 H.P. Ferber, E. Nitsch and H. Forster, Arzneimittelforschung, 35 (1985) 615.

213 S.J. Montain and E.F. Coyle, J. Appl. Physiol., 73 (1992) 903.

214 International Olympic Committee. Prohibited classes of substances and prohibited methods, 2003.

215 M. Thevis, G. Opfermann and W. Schanzer, J. Chromatogr. B, Biomed. Sci. Appl., 744 (2000) 345.

216 M. Thevis, G. Opfermann and W. Schanzer, J. Mass Spectrom., 35 (2000) 77.

217 R. Gutiérrez Gallego and J. Segura, Rapid Commun. Mass Spectrom., 18 (2004) 1324.

218 K. Deventer, P. Van Eenoo and F.T. Delbeke, J. Chromatogr. B, Analyt. Technol. Biomed. Life Sci., 834 (2006) 217.

219 S. Guddat, M. Thevis and W. Schanzer, Biomed. Chromatogr., 19 (2005) 743.

220 A. Laurentin and C.A. Edwards, Anal. Biochem., 315 (2003) 143.

221 J. Segura, Ther. Drug Monit., 18 (1996) 471.

222 J.A. Pascual, R.R. Ewin and J. Segura, in: M. Donike, H. Geyer, A. Gotzmann, U. Mareck-Engelke and C. Reichel (Eds.), Proceedings of the 10th Cologne Workshop on Dope Analysis. Sport und Buch Strauß, 1993, p. 345.

M.J. Bogusz (Ed.). Forensic Science
Handbook of Analytical Separations, Vol. 6
© 2008 Elsevier B.V. All rights reserved.

CHAPTER 22

Pharmacogenomics for forensic toxicology in enabling personalized medicine ☆

Steven H.Y. Wong

Department of Pathology, Medical College of Wisconsin, USA
Toxicology Department, Milwaukee County Medical Examiner's Office Milwaukee, WI, USA

22.1 INTRODUCTION – INTER-RELATIONSHIP OF PHARMACOGENOMICS, FORENSIC TOXICOLOGY, AND PERSONALIZED MEDICINE

With the completion of the human genome project, one of the most tangible benefits is the emerging practice of pharmacogenomics as part of genomic medicine [1–10] and personalized medicine [11]. In a recently published commentary in the *Journal of the American Medical Association*, the benefit of genomic medicine was demonstrated by reviewing its use in preventive medicine [12]. By performing preimplantation genetic for cancer, the molecular diagnosis may be regarded as one of the assisted reproductive technology. Embryos without a familial variation are chosen for implantation. On the opposite end of the spectrum, the use of molecular/genetic testing in postmortem forensic science has included DNA fingerprinting for identity testing. Recently, pharmacogenomics as molecular autopsy has been used for the assessment of genetic contribution to drug toxicity in postmortem forensic toxicology. In common with other applications of clinical and scientific findings in forensic science, the findings might add to the understanding of disease mechanism, and optimization of treatment including drug therapy. Thus, the use of pharmacogenomics in forensic toxicology may add to the understanding of drug toxicity due to

---

☆ Modified and updated from a previous publication: S.H.Y. Wong, Pharmacogenomics enabling personalized medicine, in: A. Dasgupta (Ed.), Handbook of drug monitoring methods. Humana Press, Totowa, NJ. P. 211–223.

References pp. 758–760

genetically predisposed impaired drug metabolism, and may provide findings which may be "back-extrapolated" for the benefits of optimization of "antemortem" drug therapy. In doing so, pharmacogenomics in forensic toxicology would thus provide better interpretation, indirectly enabling the emerging personalized medicine.

While pharmacogenomics and pharmacogenetics are currently used interchangeably, pharmacogenetics is readily defined as the study of the genetic effect, e.g. single nucleotide polymorphism (SNP), on an individual's ability to metabolize a drug or compound, whereas pharmacogenomics is concerned with the whole genome effect on drug metabolism and efficacy. Pharmacogenomic biomarker, in combination with well-accepted biomarkers such as therapeutic drug monitoring and other functional testing, emerging proteomic biomarkers and possibly molecular imaging, enables personalized medicine. The combination identifies the right patient, with the right diagnosis/treatment, matching with the right drug, the right dose and at the right time, thus achieving clinical efficacy with no or minimized toxicity.

The emerging clinical applications of pharmacogenetics/pharmacogenomics may be directly verified by several US Food Drug Administration (FDA) approved genotyping methodologies/platforms, the frequent inclusion of this topic in scientific and clinical meetings, and the upcoming availability of a 2007 pharmacogenomic survey program offered by the College of American Pathologists. All these positive developments, however, should be interpreted with some probing questions of the available evidence based studies to support clinical pharmacogenomic applications. What is emerging is the reality that pharmacogenomics serves as an "adjunct" to other testings and practice. The term "convergence" is often mentioned in its use in combination with functional testing such as therapeutic drug management. Thus, this chapter would attempt to present pharmacogenomics, not only as an emerging, inter-dependent discipline, but also as a complementing field in enhancing drug therapy and as an adjunct to forensic pathology/toxicology. Following the introduction, a section is devoted to the pharmacogenomics space and the enabling drivers. Then, the principles of pharmacogenetics/pharmacogenomics are introduced, with update on the candidate pharmacogenomics testing. References are provided for the readers interested in the details of the clinical findings of the applications of pharmacogenomics and the technical details of pharmacogenomics protocol. Forensic application section would begin with an algorithm proposed for the use of pharmacogenomics in forensic pathology/toxicology, with emphasis on inter-relationship to case history with drug use/abuse history, scene investigation, and autopsy findings. Recent examples would include opioids and antidepressants.

22.2 PHARMACOGENOMICS "SPACE"

In assessing the status of pharmacogenomics for drug discovery/development and clinical adaptation, a recent market analysis was completed by O'Dell and Doyle in 2004 [13]. The market analysis was conducted by surveying 53 out of 200 person-contact database. These individuals were representatives in pharmaceutical and diagnostic companies, regulatory and clinical colleagues, and others. The market was small, $800 million in 2002 as compared to the pharmaceutical market of $433 billion in 2003. Some of the key findings included the encouraging and engaging roles of the FDA, and the need for physician education. It concluded that the pharmacogenomics approached a "tipping" point in 2003/2004. With the recent developments of several new FDA approvals of pharmacogenomics testing/devices, pharmacogenomics "tipped" forward in the clinical pharmacogenomics/personalized medicine space, in part due to the proactive roles of the FDA in collaboration with other professional organizations for the past several years.

Another recent study funded by the European Commission examined the status of pharmacogenetics and the challenges for applications [14]. It reviewed the science and industry base in the US, Europe, and Japan. Pharmacogenomics is regarded as interdisciplinary in 60 research institutions. The countries with more than 10 institutions are: US with 73, Germany with 35, UK with 27, Japan with 25, The Netherlands with 21, Sweden with 14, Italy and Switzerland with 13, and France with 12. The US institutions are usually better funded by National Institute of Health. European institutions received funding from governments but not from the European Union. The major areas of research study are: drug metabolism, disease mechanisms, and disease predisposition. These institutions collaborated more with other research groups than with companies, with the possible outcome of limiting the clinical applications. The commercial sector was comprised of about 47 companies, mostly small to medium sizes with a high turnover rate of about 40%! However, the influx of new companies seemed to have maintained the total number. The business model may be divided into 12 options under the areas: drug discovery, drug safety in development, drug efficacy in development, marketed drug safety, and stratification of diseases and infectious agents. Further, the top five areas are:

1. CNS
2. Drug metabolism/toxicity
3. Cardiovascular
4. Cancer
5. Infection.

Both FDA and the European Agency for the Evaluation of Medicinal Products (EMEA) are proactive but follow different paths. Whereas FDA provides guidance documents, EMEA conducts meetings with sponsors. The established clinical pharmacogenomics tests for the European countries would include HER2 testing for breast cancer, and thiopurine methyltransferase (TPMT) for acute lymphoblastic leukemia (ALL). It concluded that many interdependent variables would contribute to clinical applications of pharmacogenomics.

By applying the genomic medicine for drug discovery and development, pharmacogenomics has been regarded as a "new science". FDA has proactively outreached to the scientific colleagues in pharmaceutical companies, partially for the purpose of developing a rational approval process. A series of workshops were held in the Washington, DC area (Pharmacogenomics: 1st workshop, May, 2002, 2nd workshop, November, 2003, and 3rd workshop, April, 2005, Pharmacogenomic drug-diagnostic co-development workshop, July, 2004, and Application and Validation of Genomic Markers, October, 2005), following by publications of guidance documents (First draft, November 2003 and final draft, March, 2005), and white/ concept papers (Critical Path Initiative, March, 2004 and Drug-diagnostic co-development, April, 2005) [15–26].

The topics included co-development of drug and diagnostics, sometimes regarded as "theranostics", and the voluntary genomics data submission (VGDS) process. More recently, VGDS was modified to VXDS – voluntary "X" data submission, with "X" representing diagnostic proteomic and other "omics" biomarkers in the future. In September 2006, the Center for Clinical Device and Radiological Health (FDA-CDRH) issued a draft of a guidance document "In Vitro Diagnostic Multivariate Index Assays" which addresses test system and data process, with implications to genetic testing. For practitioners of clinical pharmacogenomics, it would be important to follow the outcome of the final draft of this and other documents. Two recent chapters by the FDA-CDRH [27] and FDA-Center for Drug Evaluation and Research [28] provided guidance of the various regulatory issues related to use of pharmacogenomic biomarkers. In summary, the FDA workshops, the guidance documents and publications of concept and white papers serve as enabling tools toward the practice of clinical pharmacogenomics.

As a result of the decreasing number of submissions as shown in Fig. 22.1, the critical path has been advocated by the FDA to facilitate the co-development of drugs along with genomic and proteomic diagnostic biomarkers [29].

As an example and extension of that concept/practice, the Critical Path Institute (C-Path), founded by the University of Arizona as part of the Arizona Biosciences Roadmap in July 2005, is an independent, neutral, community funded, non-profit/tax exempt organization. Other key members are SRI International and FDA. Funding sources included public sector, foundation, FDA, and Agency for Healthcare Research on Quality. It also has a partnership with universities such as George Washington University, and professional organizations such as the Drug Information Association, the American College of Clinical Pharmacology, and others. It has a

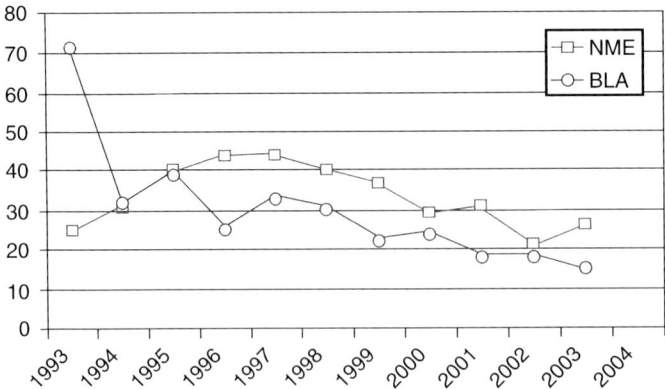

Fig. 22.1. Submissions to FDA by year for biologics and NME drugs [29].

Collaborations to Solve Common Roadblocks
in Medical Product Development

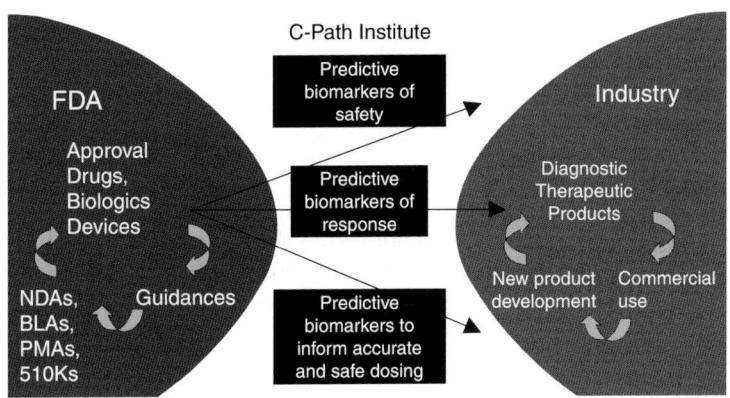

Fig. 22.2. FDA-C-Path-Industry [29].

consortium of 13 pharmaceutical companies – Merck, Johnson & Johnson, Pfizer, Novartis, GlaxoSmithKline, Schering, Roche, AstraZeneca, Boehringer-Ingelheim, Amge, Sanofi-Aventis, Bristol-Myers Squibb, and Abbott.

As shown by the model of collaboration in Fig. 22.2, the C-Path Institute would share methods, data, and strategies in order to ensure the safety of newly marketed drugs. Some of the projects enlisted consortium of diagnostic and pharmaceutical companies to develop better treatment of lung cancer, and to develop an approach for strokes treatment and to lower the incidence of death due to embolism *via* warfarin dosing therapy possibly by using pharmacogenomic biomarkers. According to the article [29], C-Path is currently undertaking 76 projects in 6 main categories

1. Better evaluation tools
2. Streamlining clinical trials
3. Harnessing bioinformatics
4. Modernizing manufacturing
5. Developing products to address urgent public health needs
6. Specific at-risk populations – pediatrics.

Another major factor in adapting clinical pharmacogenomics is the education outreach to both patient and clinician. To that end, the International Society of Pharmacogenomics recently published a position paper recommending the education effort for the deans of schools of medicine, pharmacy and allied healthcare [30]. The following 10 recommendations are proposed:

1. Encourage the deans to include the teaching of pharmacogenomics
2. Global outreach to policy makers and government leaders to educate physicians, pharmacists, and nurses
3. Basic medical teaching to include four to eight hours of lecture

4. Graduate school
5. Continuing medical education
6. Pharmacogenomics for oncology
7. Pharmacogenomics update
8. Pharmacogenomics dedicated issues in journals
9. Educational tools using web-based learning
10. Better general education to outreach to patient and general public.

If these recommendations are adopted by the deans, it would pave way to prepare graduating physicians for clinical pharmacogenomics. For example, a Personalized Medicine e-Symposium held on June 21, 2006 addressed the various issues [12].

Another major educational effort in enhancing the practice is the Laboratory Medicine Practice Guidelines for Clinical Pharmacogenetics, prepared by the National Academy of Clinical Biochemistry (NACB), the academy of the American Association for Clinical Chemistry (AACC) [31]. The document was drafted with input from NACB, AACC members as well as selected colleagues from other professional societies and regulatory agencies both here in the US and Europe. There are 10 sections: introduction, pharmacology and populations genetics, methodologies and quality assurance, services consideration, reporting, clinical considerations, TDM and PGx interface, ancillary considerations – dose and forensic, regulatory and glossary.

22.3 PRINCIPLES OF PHARMACOGENETICS/PHARMACOGENOMICS

The basic principles of pharmacogenetics and pharmacogenomics have been reviewed in depth by recent articles and chapters [4–11]. According to the central dogma of molecular biology, the genetic code of DNA is passed, through transcription, onto *mRNA*. The information in *mRNA* is passed, through translation, in protein synthesis. These proteins may be drug metabolizing enzymes, transporters, and receptors. As a result, DNA genetic variations would determine the enzyme activity, or transporters and receptor sensitivity. For drug metabolizing enzyme, the lack of and the presence of genetic variation would result in normal to deficient or higher enzyme activities. Genetic variations might include SNPs, deletion, duplications, and other variations. According to Evans and McLeod, the polygenic determinants of drug response are illustrated by Fig. 22.3 [10]. By comparing an individual with two wild type allele, an extensive metabolizer, on the left to an individual with two variant alleles, a poor metabolizer, on the right, the genetic variations would result in lower enzymes activity and elevated area under curve (AUC) with corresponding increased toxicity, and decreased receptor sensitivity and efficacy. The heterozygous individual in the middle with one variant allele, an intermediate metabolizer, with resultant AUC, toxicity and efficacy intermediate between those of the extensive and poor metabolizers. With a possible combination of 9 metabolism and receptor genotypes, the therapeutic index would range from 13 to 0.125. Further, individuals with multiple copies of the genes correspond to

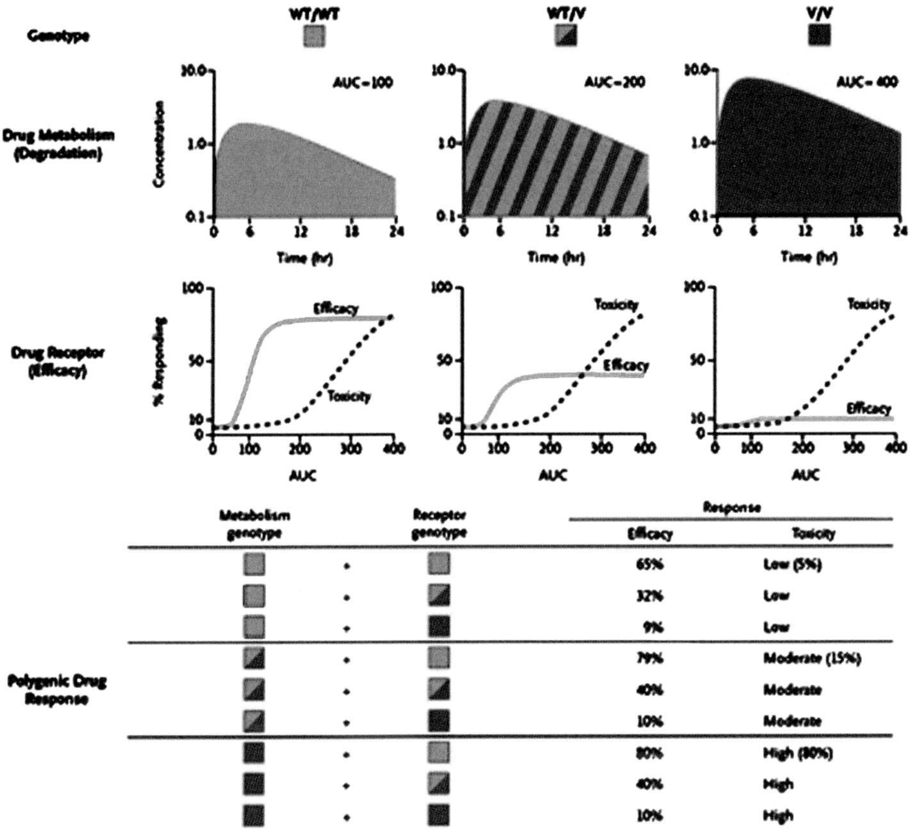

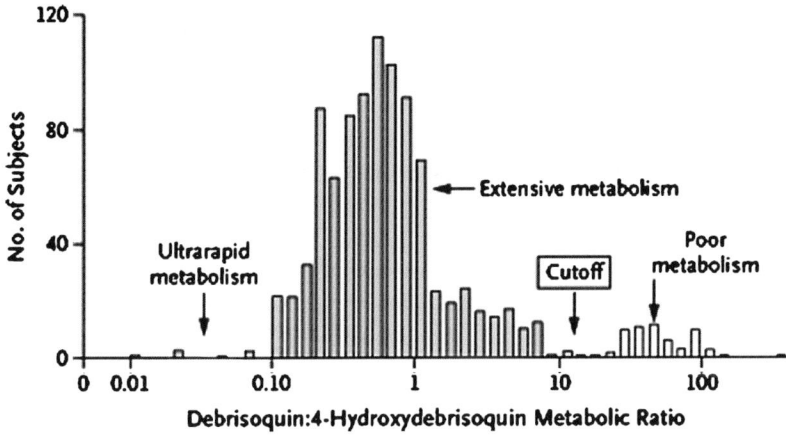

Fig. 22.3. Polygenic determinants of drug response (with permission from ref. [10] copyright © 2003 Massachusetts Medical Society).

Fig. 22.4. Pharmacogenetics of CYP 2D6 [32].

ultra-rapid metabolizers, and Fig. 22.4 show the debrisoquin metabolic ratios of these phenotypes [32].

This relationship might be readily further conceptualized by a pharmacology triangle, proposed by Linder and Valdes Jr. Pharmacogenetics provides the fundamental basis, the independent variable for the two interrelated, dependent variables – pharmacokinetics (drug metabolism) and pharmacodynamics (drug action) [33].

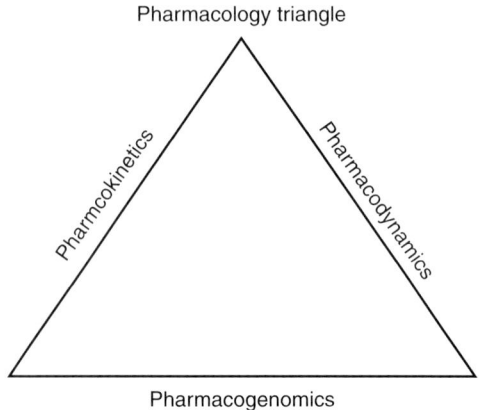

Weinshilboum reviewed the pharmacogenetics of phase I drug metabolism enzymes including CYP 2D6, 2C9, 2C19, dihydropyrimidine dehydrogenase and butyrylcholinesterase, and phase II enzymes: *N*-acetyltransferase 2, uridine diphosphate-glucuronosyltransferase 1A1, thiopurine *S*-methyltransferase, and catechol *O*-methyltransferase [9]. Evans and McLeod reviewed the effect of polymorphism on drug target gene on drug effect: angiotensin-converting enzyme, arachidonate 5-lipoxygenase, β2-adrenergic receptor, bradykinin B2 receptor, dopamine receptors, estrogen receptor-α, glycoprotein, and serotonin transporter, and the effect of polymorphism in disease – or treatment-modifying genes: adducin, apolipoprotein E, HLA, cholesterol ester transfer protein, ion channels, methylguanine methyl transferease, parkin, prothrombin and factor V, and stromelysin-1 [10]. Wong and Jannetto supplemented the information in a recent chapter [34].

22.4 PHARMACOGENOMICS TESTS AND METHODOLOGIES

The majority of the current testing might be classified as pharmacogenetics (PGx). In attempting to guide the possible planning of clinical pharmacogenomics testing, AACC conducted a survey of the top 10 tests in 2005, showing the following:

1. CYP 2D6
2. TPMT
3. CYP 2C9
4. CYP 2C19
5. NAT

6. CYP 3A5
7. UGT1A1
8. MDR1
9. CYP 2B6
10. MTHFR

With recent pharmacogenomics findings for warfarin therapy, it would be appropriate to add the *VKORC1* gene to this list [35]. Within the list, CYP and other phase II enzymes such as UGT1A1 accounted for the majority of drug/substrate metabolism for drugs approved in the US, about 75% involving CYP 3A4 and CYP 2D6 enzymes. According to the draft of LMPG by NACB, listed in the website:

http://www.nacb.org/lmpg/LMPG_Pharmacogenetics.pdf

the proposed alleles to be initially included for clinical pharmacogenetics: *CYP2D6—*1 to *12, *17* and **2A., 2C9—*1 to *6*, and *2C19—*1 to *8* and **17*. The final recommendations are pending. During the 3rd FDA-DIA workshop, O'Kane presented an assessment of possible routine pharmacogenomics testing for medical care [36]. He summarized some of the barriers including the current findings of more than 160 alterations for *CYP 2D6* genes. Assay problems would include allelic drop-out, intra-allelic recombination, the need for specific assays not affected by pseudogenes *CYP 2D7* and *CYP 2D8*, and to address gene conversion of *CYP 2D6* from *CYP 2D7*. Cautions were recommended that *CYP 2D6* genotyping might not be that routine!

Some of the AACC top 10 pharmacogenomic tests are readily performed either by home-brew assay or some commercially available, FDA approved test/platform. Payne recently reviewed how to choose a method [37], Jannetto et al. [38], Weber [39] and other recently published chapters [6] also summarized the currently available technologies. The approaches included non-amplification, e.g. fluorescent in situ hybridization (FISH), target and signal amplification methods including endpoint PCR detection, allele-specific primers, length analysis using RFLP and OLA, real-time PCR, signal amplifications, and new methods including solid-phase microarrays and fluorescent-based bead assay (liquid microarray). Other technologies reviewed include: pyrosequencing and in the future nanotechnology. The manufacturers and the status of FDA approval are listed in Table 22.1.

From personal communications, some labs performing genotyping has adapted the PCR liquid bead-based detection. The choice of the platform and assays seem to reflect, similar to the selection of clinical chemistry analyzers and tests, on the ease of "home-brew" assay development and the cost of the instrument and reagents.

22.5 CLINICAL APPLICATIONS

In the recently published book [6], the clinical applications of pharmacogenomics are classified according to drug group, specialties and diseases including: opioids, pain management, nicotine addiction, HIV treatment, immunosuppressants, TPMT

TABLE 22.1
METHODOLOGIES FOR PHARMACOGENETICS TESTING [37]

Method	Company	FDA Cleared or Approved
Sequencing[a]	Abbott (Abbott Park, IL)	Yes
Real-time PCR	Applied Biosystems (Foster City, CA)	–
PCR arrays	Autogenomics (Carlsbad, CA)	–
Sequencing[a]	Bayer Healthcare (Tarrytown, NY)	Yes
Pyrosequencing	Biotage AB (Uppsala, Sweden)	–
Real-time PCR	Celera Diagnostics (Alemeda, CA)	–
Real-time, allele-specific PCR	DxS Genotyping (Manchester, UK)	–
PCR	Gentris (Morrisville, NC)	–
User-developed PCR arrays	Nanogen (San Diego, CA)	–
Nanoparticles	Nanosphere (Northbrook, IL)	–
PCR arrays	Roche Diagnostics (Indianapolis, IN)	Yes
Invader assay	Thirdwave Technologies (Madison, WI)	Yes
PCR bead-based detection	Tm Biosciences Corp. (Toronto, Ontario)	–
FISH	Vysis[b] (Des Plaines, IL)	Yes

PCR: polymerase chain reaction; FISH: florescent *in situ* hybridization.
[a]Sequencing for HIV drug resistance.
[b]Vysis is now Abbott Molecular Diagnostics.

for ALL, psychiatry, clinical and forensic toxicology. Another previous publication by Jicinio and Wong [4] offered extensive basic and clinical information for pharmacogenomics. Readers are directed to these references for detail. Other important and emerging areas include cancer, cardiovascular disorders, and hematology. It would be important to recognize the role of TDM as a global phenotypic index including contributing pharmacokinetic, pharmacodynamic, drug–drug interaction, and other environmental factors. Thus, pharmacogenomic biomarkers might be readily characterized as an adjunct to enable the practice of personalized medicine. In order to update on these applications, a summary of recent examples would include pharmacogenomics for warfarin therapy, and the treatment of colorectal cancer by irinotecan.

Warfarin, an antithrombotic agent, has narrow index and large inter-individual variation. Recent publications proposed a new dosing regimen based on pharmacogenetics of genes of *CYP 2C9* and *vitamin K epoxide reductase complex protein 1 (VKORC1)* [35,41,42]. Warfarin is racemic, with the active enantiomer, *S*-warfarin metabolized by CYP 2C9. Variant *CYP 2C9*2* and *3* correspond to decreased enzyme activities. For Caucasians, the prevalence of extensive, intermediate, poor, and ultra-rapid metabolizers are 58, 38, 4, and 4–18%, respectively. *CYP 2C9* genotype accounts for 6–10% of warfarin dosing variability [40,41]. VKORC1 mediates the reduction of vitamin K, and its genetic variations account for 25% of warfarin dose variability. Mean dose for *VKCORC1 A/A, A/B*, and *B/B* genotypes are 2.7, 4.8, and 6.1 mg/day [35]. The additional contribution from CYP 2C9 and non-genetic factor account for up to 60% of warfarin variability. [42]. A dosage adjustment model is proposed along with INR measurement with dosage reduction to 33% for *CYP 2C9 *3/*3* genotype.

In the using pharmacogenomics for cancer, the latest example is the FDA approved test for *uridine diphosphate-glucuronosyltransferase 1A1* (Third Wave Technologies) for stratifying patients undergoing colorectal cancer treatment with irinotecan. UGT1A1 medicates the conjugation of irinotecan active metabolite, SN-38 to a glucuronide metabolite [43–46]. Individual homozygous for *UGT1A1*28* allele would have reduced enzyme activity, therefore requiring lower dose.

22.6 FORENSIC APPLICATIONS

Pharmacogenomics application in postmortem forensic toxicology was initially demonstrated by Druid, Holmgren, Carlsson, and Ahlner in a 1999 Swedish study [47], followed other studies from Finland and US. The experiences and observations were summarized in three chapters of a book on pharmacogenomics [48–50]. Collectively, these publications emphasized pharmacogenomics as an adjunct in forensic pathology, complementing information including autopsy findings, case history including medication and scene investigation. As an adjunct, molecular autopsy is proposed as a concept, similar to the use of molecular analysis of *SCN5A* variant as a contributing factor to sudden infant death syndrome [51]. Further, the selective application of pharmacogenomics as molecular autopsy is demonstrated in Fig. 22.5 (The proposed Milwaukee pharmacogenomic algorithm for forensic toxicology.). Case selection is initiated during the forensic pathology review [52]. The model assesses various co-variables such as acute or chronic toxicity, autopsy findings, sample collection sites, postmortem intervals, co-administered drugs, case/medical/medication histories, scene investigations, and possible intent. As the case review continues with developing toxicological findings, elevated drug concentrations and identification of known and unknown drug/metabolites, which might have interacted are prime criteria for case selection for pharmacogenomics. Further, the postmortem intervals are also considered in case selection and in interpretation. Once the case is selected, whole blood samples are then transferred, with chain of custody, for pharmacogenomics testing by the molecular and pharmacogenomics laboratory. Currently, the testing platform is based on Pyrosequencing™ and includes: *CYP 2D6*2-*8, CYP 2C9*2*3, CYP 2C19*2-*4, CYP 3A4*1B*, and *CYP 3A5*3*. The forensic applications are illustrated by three opioid cases; methadone and antidepressants, oxycodone, and fentanyl from recent publications.

22.6.1 Methadone and antidepressants

The decendent was a 41-year-old female, six months in her pregnancy [53]. She had heart murmur and rheumatoid arthritis treated with methadone. Further, amitriptyline was prescribed for her depression. She celebrated New Year' Eve with her husband. On the following morning, she was found dead in her living room. Scene investigation revealed her ingestion of nine 50-mg tablets of amitriptyline within 17 days, and two to three 95-mg dose of methadone. Several years prior, she did attempt suicide by drug ingestion. Toxicological analysis of iliac blood showed the

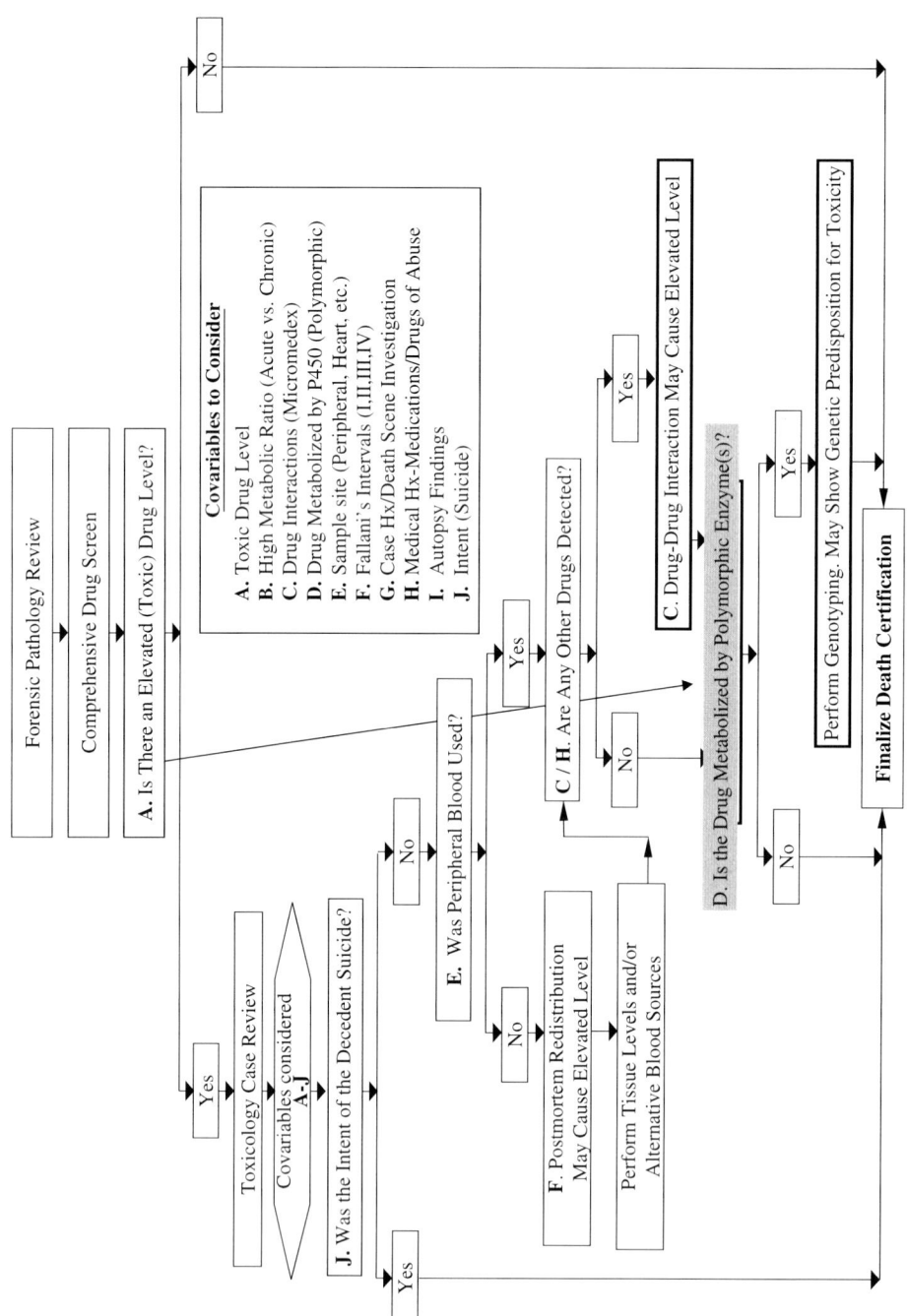

Fig. 22.5. The proposed Milwaukee pharmacogenomic algorithm for forensic toxicology.

following drugs and concentrations in mg/l: methadone, 0.7, amitriptyline, 1.5, nortriptyline, 2.2, diazepam, 0.19, *N*-desmethyl diazepam, 0.13, and alcohol, non-detected. The elevated antidepressant concentrations of iliac, peripheral blood, would not be due to postmortem drug redistribution and more attributable to acute drug ingestion. Molecular autopsy by pharmacogenomics showed that she was homozygous for *CYP 2D6*4*, corresponding to a poor metabolizer phenotype. This would result in the lack of hydroxylation of methadone, amitriptyline, and nortriptyline, thus resulting in elevated parent drug concentrations. Death certification was: cause, mixed drug toxicity, and manner, accident.

22.6.2 Oxycodone

A 49 year-old-male with a history of alcoholism and chronic lower back pain was prescribed OxyContin™ and Percocet™ [52]. He also had depression, post-traumatic stress disorder, and attempted suicide once. The scene investigation revealed only 12 of the 60 oxycodone pills which were obtained from the day before. Last seen by his roommate in the morning, the decedent was found unresponsive later that afternoon. Toxicological analysis of subclavian blood showed: methadone, 0.437 mg/l, and alcohol, non-detected. Molecular autopsy by pharmacogenomics showed that he was *CYP 2D6 *4* homozygous, corresponding to a poor metabolizer phenotype. Autopsy also showed that he had hepatic cirrhosis and atherosclerotic heart disease. Given the short postmortem interval and subclavian blood source, the elevated oxycodone was not due to postmortem drug re-distribution. It might be due to poor metabolizer phenotype and hepatic cirrhosis, both contributing to impaired drug metabolism. Death was certified as follows: cause of death, oxycodone overdose, and manner of death, accident.

22.6.3 Fentanyl

A 44-year old white female complained about her knee pain and was treated with Duragesic™ fentanyl patches [54]. She appeared to be "goofy" and went to bed. She was found dead 24 h later. The decedent was a drug abuser with psychiatric history. Previously, she cut her arm to obtain drugs and also expressed suicidal ideation. Toxicological analysis of subclavian blood showed the following drugs and concentrations in mg/l: fentanyl, 0.019, norfentanyl, 0.008, cyclobenzaprine, 0.16, tramadol, 0.06, diphenhydramine, 0.08, citalopram, 0.22, olanzapine, positive and alcohol, non-detected. The metabolic ratio (MR) of fentanyl/norfentanyl was 2.5. Molecular autopsy by pharmacogenomics showed that she was heterozygous for *CYP 3A4*1B* and *CYP 3A5*3*, different from the majority of Caucasians *CYP 3A4, WT* and *CYP 3A5*3 HM*. In this study with limited number of fentanyl cases, the MR of this case is lower than the MRs of majority of the cases with high fentanyl concentrations. Together, these findings suggested that for the first time, CYP 3A5 co-mediated with CYP 3A4 the metabolism of fentanyl to norfentanyl. Death certification for the above case was: cause of death, mixed drug toxicity, and manner of death, accident.

References pp. 758–760

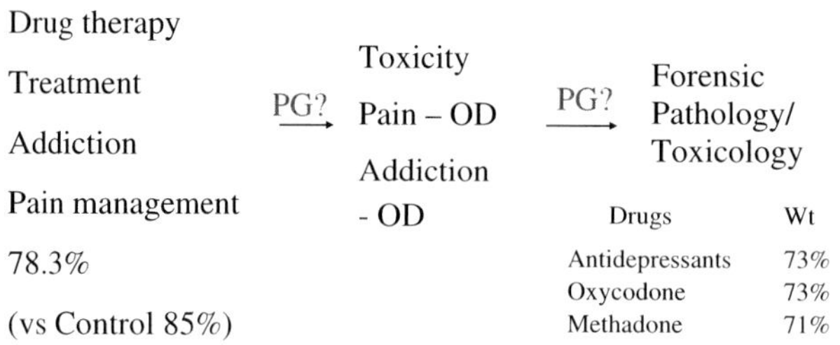

Drug therapy

Treatment	Toxicity		Forensic

Drug therapy

Treatment   Toxicity   Forensic
       PG? → Pain – OD PG? → Pathology/
Addiction   Toxicology
          Addiction

Pain management   - OD

	Drugs	Wt
78.3%	Antidepressants	73%
	Oxycodone	73%
(vs Control 85%)	Methadone	71%

Fig. 22.6. Pharmacogenomics converging continuum?

From these studies, the wild type prevalences, defined as non-*CYP 2D6 *2, *3, *4 and *5*, showed the following trend: antidepressants and oxycodone, 73%, and methadone, 71%, lower than the 85% in controlled group, and 78.3% in pain management study. Together, the genetic contribution to drug metabolism impairment may be interpreted as gene dose effect, resulting in more pronounced drug toxicity. Such an effect might be readily demonstrated as a pharmacogenomics converging continuum as shown in Fig. 22.6, and is helpful to understand the effect in "live" antemortem population.

22.7 CONCLUSIONS

The emerging practice of personalized medicine is dependent on pharmacogenomics and other biomarkers, and might take longer time to realize. While pharmacogenomic, proteomic, RNA interference, and other molecular and functional biomarkers are increasingly used in drug discovery and development as encouraged governmental agencies, and scientific and professional organizations, pharmacogenomics has been used to enhance patient classification for enhanced safety in therapies with approved drugs. Thus, the use of pharmacogenomics as an adjunct for molecular autopsy would add to the understanding of potential genetic contribution to metabolism of approved drug such as methadone and oxycodone, thus enabling and improving the practice of "antemortem" drug therapy, a tangible benefit for family members of decedents. With the upcoming availability of proficiency survey program by the College of American Pathologists in 2007, and quality assurance/control from commercial sources, the clinical adaptations will soon be readily achieved by clinical laboratories. NACB guidelines would certainly pave the way. Challenges remain for adequate reimbursement, clinical interpretation, and ethical guidelines.

22.8 REFERENCES

1   International Human Genome Sequencing consortium, Initial sequencing and analysis of the human genome. Nature, 409 (2001) 860.

2   J.C. Venter, M.D. Adams, E.W. Myers, P.W. Li, R.J. Mural, G.G. Sutton, H.O. Smith M. Yandell; *et al.*, Science, 291 (2001) 1304.

3   A.E. Guttmacher, F.S. Collins and J.M. Drazen (Eds.), Genomic medicine – articles form the *New England Journal of Medicine*. Johns Hopkins University Press, Baltimore, MD, and *New England Journal of Medicine*, Boston, MA, 2004, pp. 1–179.

4   J. Jicinio; M.L. Wong, Pharmacogenomics, Wiley-VCH, Weinheim, Germany, 2002, pp. 1–559.

5   M.W. Linder, R.A. Prough and R. Valdes Jr., Clin. Chem., 43 (1997) 254.

6   S.H.Y. Wong, M.W. Linder and R. ValdesJr. (Eds.), Pharmacogenomics and proteomics – enabling the practice of personalized medicine. AACC Press, Washington, DC, 2006 pp. 1–386.

7   R. White and S.H.Y. Wong, Pharmacogenomics and its clinical applications. MLO, 37 (2005) 20–27.

8   W.W. Weber, Pharmacogenetics, Oxford University Press, Oxford, UK, 1997, pp. 1–344.

9   R. Weinshilboum, N. Engl. J. Med., 348 (2003) 529.

10   W.E. Evans and H.L. McLeod, N. Engl. J. Med., 348 (2003) 538.

11   Personalized Medicine e-Symposium, June 21, 2006 (http://www.e-symposium.com/pm/archive.php).

12   K. Offit, M. Sagi and K. Hurley, JAMA, 296 (2006) 2727.

13   L. O'Dell and J. Doyle, Opportunities in pharmacogenomics – market research analysis, May 2004.

14   M.M. Hopkins, D. Ibarreta, S. Gaisser, M. Enzing CM, J. Ryan, P.A. Martin, G. Lewis, S. Detmar M.E. van den Akker-van Marle; *et al.*, Nat. Biotechnol., 24 (2006) 403.

15   R.A. Salerno and L.J. Lesko, Pharmacogenomics, 5 (2004) 25.

16   www.fda.gov/cder/genomics/ (1. Personalized medicine: What is it? How will it affect health care? Felix W. Frueh, presented April 26, 2005 to the FDA Science Forum, 2. Concepts and tools in pharmacogenomics, and 3. Pharmacogenomics from the ground up: Submissions and labels in regulatory pharmacogenomics).

17   L.J. Lesko and J. Woodcock, Nat. Rev. Drug Discovery, 3 (2004) 763.

18   R.A. Salerno, Developing the regulatory pathway for pharmacogenomics. Regulat. Affairs Focus, 8 (2004) 12.

19   L.J. Lesko, R.A. Salerno, B.B. Spear, D.C. Anderson, T. Anderson, C. Brazell, J. Collins, A. Dorner, D. Essayan; *et al.*, J. Clin. Pharm., 43 (2003) 342.

20   R.A. Salerno and L.J. Lesko, Pharmacogenom. J., 5 (2004) 503.

21   J.K. Leighton, J. DeGeorge, D. Jacobson-Kram, J. MacGregor, D. Mendrick and A. Worobec, Pharmacogenom. J., 5 (2004) 507.

22   R. Gualberto, J. Collins, A.J. Dorner, S.J. Wang, R. Guerciolini and S.M. Huang, Pharmacogenom. J., 5 (2004) 513.

23   W.L. Trepicchio, G.A. Williams, D. Essayan, S.T. Hall, L.C. Harty, P.M. Shaw, B.B. Spear, S.J. Wang and M.L. Watson, Pharmacogenom. J., 5 (2004) 519.

24   R.A. Salerno and L.J. Lesko, Pharmacogenom. J., 6 (2006) 1.

25   S.J. Wang, N. Cohen, D.A. Katz, G. Ruano, P.M. Shaw and B. Spear, Retrospective validation of genomic biomarkers – What are the questions, challenges and strategies for developing useful relationships to clinical outcomes – workshop summary. Pharmacogenom. J., 6 (2006) 82.

26   W.L. Trepicchio, D. Essayan, S.T. Hall, G. Schechter, Z. Tezak, S.J. Wang, D. Weinreich and R. Simon, Designing prospective clinical pharmacogenomic (PG) trials: meeting report on drug QJ;development strategies to enhance therapeutic decision making. Pharmacogenomics. J., 6 (2006) 689.

27   Z. Tezak, J. Hackett, Biomarker-based diagnostic devices in therapeutic applications (marketed therapeutics). In: S.H.Y. Wong, M. Linder, Valdes, R. Jr. (Eds.), Pharmacogenomics and proteomics: Enabling the practice of personalized medicine, AACC Press, Washington, DC, 2006, pp. 37–40.

28   F. Goodsaid, S.M. Huang, F. Frueh, R. Temple, L.J. Lesko, Regulatory guidance and application of genomic biomarkers in drug development. In: S.H.Y. Wong, M. Linder, Valdes, R. Jr. (Eds.), Pharmacogenomics and proteomics: Enabling the practice of personalized medicine, AACC Press, Washington, DC, 2006, pp. 41–52.

29   E.G. Feigal, J. Cossman and R.L. Woosley, Drug Discovery Develop., 9 (2006) 34.
30   D. Gurwitz, J.E. Lunshof, D. Dedoussis, C.S. Flordellis, U. Fuhr, J. Kirchheiner, J. Licinio, L. Llerena, V.G. Manolopoulos, L.J. Sheffield, G. Siest, F.T. Torricelli, V. Vasiliou and S. Wong, Pharmacogenom. J., 5 (2005) 221.
31   http://www.nacb.org/lmpg/LMPG_Pharmacogenetics.pdf.
32   L. Bertillsson, Y.W. Lou, Y.L. Du, Y. Liu, T.Y. Kuang, X.M. Liao, K.Y. Wang, J. Reviriego, L. Iselius and F. Sjöqvist, Clin. Pharm. Ther., 51 (1992) 388.
33   M.W. Linder and R. Valdes Jr., Ann. Clin. Lab. Sci., 29 (1999) 140.
34   S.H.Y. Wong, P.J. Jannetto, Pharmacogenomics. In: A. Wu (Ed.), Tietz's applied laboratory medicine (4th ed.)St. Louis, Saunders Elsevier, 2006, pp. 1713–1742.
35   M.J. Rieder, A.P. Reiner, B.F. Gage, D.A. Nickerson, C.S. Eby, H.L. McLeod, D.K. Blough, K.E. Thummel, D.L. Veenstra and A.E. Rettie, N. Engl. J. Med., 352 (2005) 2285.
36   D.J. O'Kane, Use of PG in routine medical care. 3rd FDA-DIA Workshop, Bethesda, MD, April 13, 2005.
37   D. Payne, Clin Lab. News, 7 (2006) 14.
38   P.J. Jannetto, E. Laleli-Sahin, B.C. Schur, S.H. Wong, Enabling pharmacogenomics: Methodologies for genotyping. In: S.H.Y. Wong, M.W. Linder, Valdes, R. Jr. (Eds.), Pharmacogenomics and proteomics – enabling the practice of personalized medicine, AACC Press, Washington, DC, 2006, pp. 1–386.
39   W.W. Weber, Techniques for analyzing pharmacogenetic variation. In: S.H.Y. Wong, M.W. Linder, Valdes, R. Jr. (Eds.), Pharmacogenomics and proteomics – enabling the practice of personalized medicine, AACC Press, Washington, DC, 2006, pp. 1–386.
40   M.K. Higashi, D.L. Veenstra, L.M. Kondo, A.K. Wittkowsky, S.L. Srinouanaprachanh, F.M. Farin and A.E. Rettie, JAMA, 287 (2002) 1690.
41   E.A. Sconce, T.I. Kahn, H.A. Wynne, P. Avery, L. Monkhouse, B.P. King, P. Wood, P. Kesteven A.K. Daly; *et al.*, Blood, 106 (2005) 2329.
42   S. Marsh and H.L. McLeod, Human Mol. Gen., 15 (2006) R89.
43   L.P. Rivory, M.R. Bowles, J. Robert and S.M. Pond, Biochem. Pharmacol., 52 (1996) 1103.
44   L. Iyer, D. Hall, S. Das, A. Melissa, B.A. Mortell, J. Ramírez, S. Kim, A.D. Rienzo and M.J. Ratain, Clin. Pharmacol. Ther., 65 (1999) 576.
45   E. Araki, M. Ishikawa, M. Iigo, T. Koide, M. Itabashi and A. Hoshi, Jap. J. Cancer Res., 84 (1993) 697.
46   L. Iyer L, S. Das, L. Janisch, J. Ramírez, T. Karrison, G.F. Fleming, E.E. Vokes, R.L. Schilsky and M.J. Ratain, Pharmacogenet. J., 2 (2002) 43.
47   H. Druid, P. Holmgren, B. Carlsson and J. Ahlner, Forensic Sci. Int., 99 (1999) 25.
48   P. Holmgren, J. Ahlner, Pharmacogenomics for forensic toxicology – Swedish experience. In: S.H.Y. Wong, M.W. Linder, Valdes, R. Jr. (Eds.), Pharmacogenomics and proteomics – enabling the practice of personalized medicine, AACC Press, Washington, DC, 2006, pp. 295–300.
49   A. Sanjantila, P. Lunetta, I. Ojanpera, Postmortem pharmacogenetics – towards molecular autopsies. In: S.H.Y. Wong, M.W. Linder, Valdes, R. Jr. (Eds.), Pharmacogenomics and proteomics – enabling the practice of personalized medicine, AACC Press, Washington, DC, 2006, pp. 301–310.
50   S.H.Y. Wong, S.B. Gock, R.Z. Shi, M. Jin, M.A. Wagner, B.C. Schur, P.J. Jannetto, E. Sahin, J. Bjerke, N. Nuwayhid, J.M. Jentzen, Pharmacogenomics as an aspect of molecular autopsy for forensic pathology/toxicology. In: S.H.Y. Wong, M.W. Linder, Valdes, R. Jr. (Eds.), Pharmacogenomics and proteomics – enabling the practice of personalized medicine, AACC Press, Washington, DC, 2006, pp. 311–320.
51   M.J. Ackerman, B.L. Siu, W.E. Sturner, D.J. Tester, C.R. Valdivia, J.C. Makielski and J.A. Towbin, JAMA, 286 (2001) 2264.
52   P.J. Jannetto, S.H. Wong, S.B. Gock, E. Laleli-Sahin, B.C. Schur and J.M. Jentzen, J. Anal. Tox., 26 (2002) 438.
53   S.H. Wong, M.A. Wagner, J.M. Jentzen, B.C. Schur, J. Bjerke, S.B. Gock and C.C. Chang, J. Forensic Sci., 48 (2003) 1406.
54   M. Jin, S.B. Gock, P.J. Jannetto, J.M. Jentzen and S.H. Wong, J. Anal. Tox., 29 (2005) 590.

WEBSITES

http://www.genome.gov/glossary.cfm.
http://www.geneclinics.org.
http://www.cdc.gov/genomics/hugenet/reviews.htm.
http://www.cancer.gov/cancer_information/pdq.
http://www.ncbi.nlm.nih.gov/omin/.
http://www4.od.nih.gov/oba/sacgt.htm.
http://www.nhlbi.nih.gov/resources/docs/cht-book.htm.
http://www.nhlbi.nih.gov/about/factpdf.htm.
http://www.cardiogenomics.med.harvard.edu.
http://www.nhgri.nih.gov/Policy_and _public_affairs/Legislation/insure.htm.
http://medicine.iupui.edu/flockhart/table.htm.
http://www.imm.ki.se/CYPalleles/.
http://www.aidsinfonyc.org/tag/science/pgp.html.

M.J. Bogusz (Ed.). Forensic Science
Handbook of Analytical Separations, Vol. 6
© 2008 Published by Elsevier B.V.

CHAPTER 23

Aspects of quality assurance in forensic toxicology

Rolf E. Aderjan

Institute of Legal Medicine and Traffic Medicine, University Hospital – Ruprecht-Karls University of Heidelberg 69115 Heidelberg, Voßstr. 2, Germany

23.1 INTRODUCTION

Aristotle (384–322 B.C.) distinguished 10 basic categories of human being and thinking, one of which is "quality." Its elements are taken to be either objective distinguishing marks or stipulated by human perception. John Locke (1632–1704) drew a distinction between (external) sensation and (internal) reflection. He assigned their quality to (objective) primary elements, such as dimensions in space or time, and to secondary subjective ones, such as color or olfaction (due to imagination). In other words, in addition to measurable ones, various subjective components of quality exist whose weighting may entail affirmative, negative, and limiting utterance, respectively. Correspondingly, any decision is subject to either intentional or unconscious considerations of the quality of a matter, an object, or an activity, in order to approve or refuse it. If several alternatives are available for choice, they need to be compared and evaluated by ranking their quality elements. Depending on the individual matter and the way of decision, more or less weighted subjective and objective factors may contribute to quality ranking. After all, in contrast to prevailing objective quality signs, any decision may be governed by subjective factors. Such consideration is relevant for the understanding of quality aspects.

In the view of customers and manufacturers, quality has to do primarily with the relation between cost and benefit. Goods may be preferred according to their being "fit for use." From his observation of nature, Darwin deduced the principle of evolution, "survival of the fittest." Based on environmentally supportive and selective conditions, both replication (related to replicability, repeatability, and reproducibility) and mutation (as a base of accommodation) are genetic quality-raising processes by which present information, abilities, and knowledge are passed into the future. In humans, as in other organizational structures in nature, quality generally is a moving

References pp. 814–819

target demanding a change. Therefore, an aphorism, attributed to Lichtenberg (1742–1799), is still true:

> I cannot say whether things will get better if we change; what I can say is they must change if they are to get better.

Clearly, all quality assurance aspects in forensic toxicology should refer to the primary "quality product" expected from a forensic toxicologist (FT). This product is to provide objective analytical chemical results, including the bare analytical values, and their interpretation as within definite limits obtained also by appropriately trained staff in other forensic analytical laboratories. To serve as a means of evidence, results must correspond to legal requirements and be capable of being conclusively testified to in court. Even if it should be admitted that courts have limited capacity to review the FT's work, that work is always open to cross-examination. FTs have always been concerned about the quality of their results, most likely more so than other laboratory organizations. Therefore, forensic science laboratories have probably developed quality assurance and quality control activities as long as they have existed. While accreditation and certification was requested in calibration and testing laboratories for several reasons, such as safety, improvement, or equal conditions, in forensic science laboratories quality assurance and control were always thought of as good laboratory practice (GLP).

The development towards worldwide quality terms and the standards of today, such as GLP and the ISO standards, were initiated in order to eliminate the bad reputation of national products using quality badges, such as "made in …," and also because of complaints about inadequate drug toxicology studies performed by contract laboratories for big manufacturing companies. New regulations, which formulated the expectations and requirements for an organization engaged in non-clinical laboratory studies aimed at the safety of regulated products, were established in GLP by the American Food and Drug Administration (FDA) in the U.S. Federal Register in 1978 [1–3]. The U.S. Environmental Protection Agency (EPA), too, had issued partly more specific GLP standards for toxic substances and pesticides in parallel with those of the FDA. Similar regulations entitled GLP in the Testing of Chemicals were adopted by the Organization of Economical Cooperation and Development (OECD) in 1982 and also by the European Community (CE) in 1987. Since 1967, U.S. Clinical Laboratories have been subjected to regulation by the U.S. Clinical Laboratories Improvement Act (CLIA). Any laboratory handling human specimens for analysis must hold a license issued by the U.S. Secretary of Health and Human Services.

Since 1989, the general requirements for the accreditation of analytical chemical laboratories in Europe have been established in the standard called the General Criteria for the Operation of Testing Laboratories (EN 45001), which has also become the national standard of individual European countries. In addition, at an international level, General Requirements for the Competence of Calibration and Testing Laboratories (ISO/IEC Guide 25) have been created [4]. These requirements are designed to be applied to the performance of all objective measurements. According to the requirements, in cases of dispute, the accreditation bodies should adjudicate unresolved matters. With reference to EN 45001 and ISO/IEC Guide 25, an interpretation of the accreditation requirements was produced by a joint working group of the European Cooperation of Accreditation of Laboratories (EAL/CEOC). It was aimed at specific guidance on accreditation regarding staff, equipment and calibration, test procedures, handling of items and components, measurement uncertainty, proficiency testing, as well as at the subcontracting of tests. Guidelines according to ISO 9000, ISO 17025, and ISO 15189 in their updated versions are now authoritative documents for laboratory accreditation [5], which in 2005 were subjected to amendments concerning the effectiveness of the management system and its correction or improvement. However, in order to fit with forensic demands, additional requirements were being introduced and developed in the ILAC – Guidelines for Forensic Laboratories and for dealing with management and technical requirements of objective testing, as practiced in laboratories involved in forensic analysis and examination [6,8]. The International Laboratory Accreditation Conference (ILAC) was initiated by the United States in 1979 and holds meetings biannually. ILAC's efforts are concerned with coordinating the use, competence, and reliability of national testing bodies in more than 40 countries, as follows:

1. legal aspects of international reciprocity of test data,
2. the format of an international directory of laboratory accreditation systems, and other schemes for assessment of laboratories, and
3. definitions and common criteria for national accreditation systems.

In order to give guidance in forensic quality terms, professional organizations such as The International Association of Forensic Toxicologists (TIAFT) [7],[1] the Gesellschaft für Toxikologische und Forensische Chemie (GTFCh) [8], or Society of Forensic Toxicologists (SOFT) [9] and Swiss experts [10] were formulating and updating national and international guidelines for (objective testing in) forensic or analytical toxicological laboratories, based on both scientific and quality aspects, even before the need for accreditation was disseminated. Necessary additional guidelines had to be developed, introduced, and complemented, taking national legal regulations and jurisdictions into account, for instance, the specific detection of substances referred to in legislation covering driving under the influence in several European countries.[2]

Due to the widely recognized need in many western industrial countries, private, professional, and governmental organizations promote the evaluation, upgrading, and periodic checking of the performance of laboratories and also examine their testing ability for conformance to standards. Quality assurance principles of testing laboratories are now generally accepted. Accreditation bodies now (at a cost) certify

---

[1] TIAFT-Guidelines: a still valid version was adopted in 1993 and republished in 2006 [7].

[2] For instance, Section 24 a.2 of the German Traffic Act and corresponding guidelines referring to the problems of detecting definite compounds in blood after the usage of heroin, cocaine, amphetamines, and cannabis.

References pp. 814–819

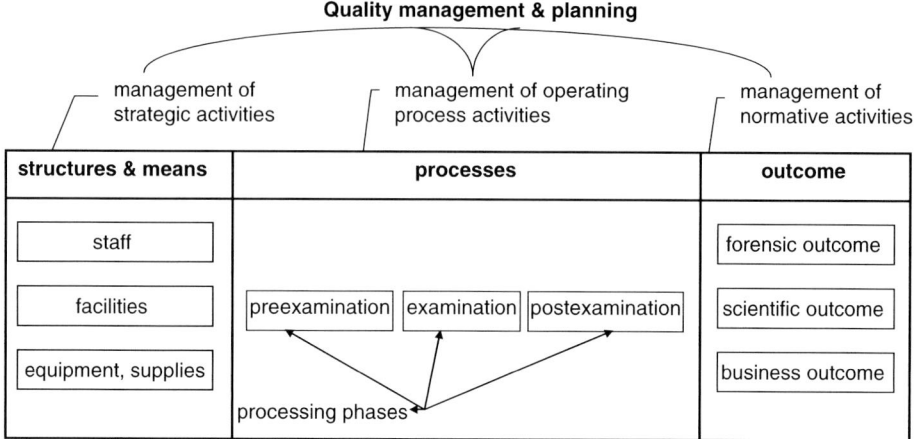

Fig. 23.1. Important quality areas that need to be managed in forensic-toxicological laboratories with respect to quality.

laboratories, providing assurance of their testing ability and adherence to procedures on inspection.

Besides formal safety regulations, quality concepts were developed from a more basic point of view. While earlier in the 1960s, quality control predominantly focused on the product being manufactured, it is now clear that quality skills themselves need to be produced. Quality experts have described the important essentials of producing quality. In 1980, Donabedian [11] analyzed the elements of quality for health care services. He proposed a concept in which three sections of quality were distinguished: (1) structural quality: the availability and organization of human, physical, and financial resources, such as facilities, space, staff, equipment, and consumables; (2) processing quality: the activities necessary to deliver a product or service including examination, quality assurance, reporting, interpretation, and consultation; and (3) the quality of the outcome with respect to patients' health.

As shown in Fig. 23.1, these elements can also be applied to create a comprehensive quality concept in analytical toxicology.

23.1.1 Organizing quality and quality concepts

To introduce such new principles to manage quality, it may be necessary to overcome a high activation energy in a laboratory. First, the generally accepted guidance documents need to be studied. For the realization of a quality management system (QMS), one member of the staff needs to be enabled and nominated as a quality supervisor. Step by step all the members of the staff need to be informed and involved in the coming activities. In addition to the creation and implementation of standard operating procedures (SOP), all regulations, including internal audits, have to be explicitly written in a quality management manual. The SOPs refer to all logistical activities, examinations, and equipment as well as quality control procedures. In order to comply with actual practical procedures, SOPs and quality documents should first and must preferably be written by the corresponding coworkers, supported by experienced external accreditation experts. Then all documents need to be examined and released by the previously trained superior "in-house experts." Implementation of a QMS can only be achieved by continuously performing all the required steps. Besides its necessity in a high-throughput/workload laboratory, a well-designed computerized electronic laboratory information and management system (LIMS), which models the in-house organization principles, is an indispensable quality management tool, after whose implementation a step-by-step introduction of the quality concept can be effectively transferred into action.

Finally, accreditation and certification may be required, by more or less voluntary applications for external audits organized by authorized national accreditation bodies. Usually, certification is possible after introduction of a QMS according to the ISO EN DIN 9000 series. In contrast, accreditation means a quality proof being based on the EN 45000 series and now on the ISO/IEC 17025 standard. These require suitable internal quality controls and successful participation in external quality controls as well as an approval of the laboratory's conformity with its QMS. In its 2005 Edition, ISO/IEC 17025 introduces some changes referring to QMS qualification and developing appropriate QMS monitoring and improvement strategies, including customers' satisfaction. For further developing quality and QMS issues and to eliminate any deficits, which are identified, Deming [12] (according to Malorny [13]) has recognized the PDCA cycle (a continuous sequence of planning, doing, checking, and adjusting).

In contrast to industry, where high quality should usually delight the customer and consumers, in forensic science the toxicological expert may have several "customers," some of whom may not be pleased by his product Prosecutors, judges, defenders, and last, but not least, the accused have different and even conflicting interests in the FT's reporting. Therefore, in case of doubt, analytical results may be questioned from diametrically opposed points of view. This fact has a major influence on quality management in forensic toxicology.

Forensic toxicologists must anticipate potentially relevant facts. Do they keep in mind all known difficulties or interference of the methods that are used when they start to work on a case? Reasonable doubts must accompany the toxicologist's daily casework. Is he or she always aware of possibly working on a case in which afflicted people seriously distrust results or conclusions? Have all of the substances been determined, which will be necessary for conclusive case-based reasoning? The multitude of such questions makes it understandable that educational ring tests as well as true proficiency testing are accepted as a highly important means to increase the degree of certainty of analyses aiming at the detection of previously unknown substances and their quantitative determination. However, along with progress in science and increasing knowledge about metabolism, human formation, and elimination kinetics of an increasing number of toxicologically relevant parent compounds and their metabolites, new concepts and methods will be developed, entailing a permanent involvement in continuing advanced education. That is why the GTFCh has introduced an acceptance procedure including the appointment as "Forensic Toxicologist GTFCh," attesting to a qualifying knowledge according to updated guidelines [14] and requiring a continuing advanced education in the future. According to German accreditation regulations, forensic accreditation requires the acquisition of this appointment for laboratories and their responsible toxicological directors [8].

Limited laboratory services may be selected because of large case numbers and economic pressures. Also, the pressure of the high responsibility required for forensically demanding and time-consuming cases may stimulate a wish to limit laboratory services more towards the analysis of definite analytical parameters. Sometimes in automated laboratory analysis indicators for questionable results are carried along with the analytical method, including sample consumption. In contrast to clinical samples, however, forensic samples must be carefully treated as unique specimens, and analytical pitfalls should be anticipated by experienced staff as much as possible. In laboratories performing a selected number of standardized clinical analytical methods, services, such as blood alcohol determination and drugs of abuse in urine, are being offered. The analyst may then obtain objective, precise, and accurate results. In contrast, in forensic toxicology laboratories the analyst often must first solve the fundamental task of poison detection in human organs and body fluids, at times in very complicated cases, before being able to apply or, if necessary, to work out and validate a new analytical method. As justice should remain affordable and because the economical selection of analytes entails a loss of analytical experience, it needs to be stressed that time-consuming and less cost-effective tasks must not be excluded from forensic-toxicological standard methods and moved to limited-service laboratories due to economics, including some risks of quality.

Forensic analysts are sometimes confronted with unexpected or difficult cases including decomposed bodies. According to the results of a general unknown analysis or to a question, new or problem substances need to be determined, requiring methods previously not tried and tested. Consequently, the repertoire of methods needs to be continuously increased by developing appropriate approaches including, at times, the costly validation of methods applied only a few times, or even once.

Running parallel to activities since the 1960s aimed at defining standards, scientific societies continuously work out their recommendations and guidelines for forensic analytical work and laboratories.[3] However, merely acting according to working guidelines or formally fulfilling standards does not guarantee that the top quality levels needed in forensic toxicology are permanently achieved and in specific cases.

Forensic toxicology is a scientific discipline in which ongoing efforts to improve methods of poison detection point out the close relationship between quality and scientific development. This becomes obvious regarding frequently occurring problems and deficits in interpretations experienced during casework.

Besides the formal and theoretical principles of quality assurance, there is one more facultative requirement regarding the FT's management activities. This requirement

---

[3] Newly structured Guidelines of the GTFCh are presently being prepared (2006) including appendices for specific analytes, method validation,, minimum limits of detection and quantification.

refers to goal finding in science and corresponding organizing activities in a forensic toxicological laboratory. At universities, for example, scientific work usually needs to be carried out in addition to casework. Approved methods are often developed by one's own efforts. However, analysts trying to run "solutions" offered in the literature often experience the need to vary or improve them. Therefore, the quality concept must include the FT's maintaining competence in qualification and expertise, and a need for continuous self-education by updating and improving his or her scientific knowledge, especially by doing his or her own research in analytical and human toxicology.

23.2 A QUALITY CONCEPT FOR ANALYTICAL TOXICOLOGICAL LABORATORIES

Donabedian understood quality care to aim at a successful change in patients' present and future health status. In order to reach such an outcome, the appropriate quality of both the structures and the processes of health care services are required. For a toxicology laboratory, similar needs can be seen (Fig. 23.1).

23.2.1 Quality of the structure

23.2.1.1 The staff-management and basic requirements

One key to maintaining quality and suitable performance of any chemical analytical laboratory is successful interaction of all staff members. This is achieved by the performance of the laboratory director as well as by the selection of qualified supervisors and trained non-supervisory staff. The laboratory director must have appropriate education and experience regarding proper planning, organizing, staffing, leading, and controlling. Most of these activities involve abilities in personnel management. In order to be able to apply management concepts and techniques, it is recommended to the FT that his postgraduate education should ideally include a sequence of training courses in current personnel management skills.

As all work needs to be organized into individual jobs, an optimal level of laboratory performance is only achieved if appropriate personnel are recruited and placed into the correct positions. In addition, realistic goals and fair pay policies in combination with a bi-directional flow of information within a laboratory are important quality requirements. The staff must be aware of the meaning and importance of their tasks and know how to contribute to the goals of the management system. Appropriate communication pathways need to be implemented and communication about the effectiveness of the management system must take place. Regular evaluation of the employees and a positive attitude to their work is crucial to improve motivation. The management needs to create an appropriate environment to encourage the staff members to participate in decisions including the perception of their vital interests. As mistakes inseparably belong to daily work, an effective strategy of improvement of the individual is probably more important than trusting in the usefulness of rewards or discipline as a situation appears to demand.

23.2.1.2 Core competencies of a forensic toxicologist as a laboratory director

Because they are the most important aspects for quality assurance, an organization's attention must mainly be directed towards personnel qualifications. Besides the primary laboratory management skills, the FT basically needs scientific skills, technical skills such as observation of, knowledge of, and the introduction of new technologies, continuous self-education, and participation in and sharing of knowledge in the scientific society.

Guidelines have been developed by the GTFCh for the fundamental skills required for certification as an FT [14]. To be certified, an applicant must demonstrate that he or she fulfills a series of minimum requirements. They include:

- an academic degree in a subject of natural science related to forensic toxicology, such as chemistry or pharmaceutical science;
- seven years of professional toxicological work including testifying in courts (presently updated to only five years of FT-coached in-house training including examination by GTFCh experts.);
- participation in scientific symposia and in workshops of the GTFCh as well as appropriate postgraduate education;
- the ability to suitably understand and discuss analytical procedures;
- the ability to correctly interpret forensic toxicological testing results; and
- sufficient forensic testifying experience by presentation of advanced written expert opinions taken from his or her professional work.

The size and organization of a laboratory influences management structures. Therefore, the qualification and experience of superiors and coworkers depend on the requirements and responsibilities of each position. As a minimum, the laboratory director must overlook all analytical activities undertaken and have a comprehensive scientific analytical and methodological knowledge and responsibility. According to the full analytical scope of forensic toxicology, he needs to be supported by at least two coworkers in supervisory positions. Except for the maintenance of complex analytical instruments, most of the basic analytical activities presuppose coworkers with a sound technical-analytical but not necessarily academic education. As a consequence of the scientific development of a laboratory, highly evolved analytical instrumentation purchased for scientific reasons has become or will possibly be commonly used in the future. For this reason, a laboratory's list of supervisory positions must include specialists who can take care of one or more of such complex instruments, analytical methods, and the tasks to be carried out by them. Unfortunately, forensic toxicology laboratories of some university institutes of legal medicine have become relatively small. In these, because of the lack of appropriate positions, the working scientist at the bench may belong to the supervisory staff without a clear-cut definition of his or her responsibility. This understanding of responsibility is critical with respect to science, quality, and accreditation requirements as, on the one hand, relevant scientific research interests may easily be obstructed by casework overload and, on the other hand, a non-analytically educated head of the department takes responsibility for reports, but transfers analytical responsibility. In order to revise such deficient structures, some universities have started to establish positions for a fully responsible FT. Developing more of such well-tailored positions will be a major challenge in these times of increasing knowledge requirements and decreasing financial resources.

The rapid scientific development of forensic toxicology and the use of more and more advanced technical analytical instrumentation is another reason why it seems advisable to exclusively employ a scientifically qualified FT for the position of a fully responsible laboratory director. Along with a rating according to scientific evaluation requirements, the level of performance and quality is becoming decisive for both the existence of a forensic toxicology laboratory and its structural conditions.

23.2.1.3 Quality aspects regarding the role of an expert witness

Independent of the national legal system, a scientist's unique role as an expert witness (EW) is to be unbiased in helping a judge (and the jury in the case of the United States and United Kingdom as well as other related adversary legal systems) to understand the scientific truth. This duty, however, may not always be simple. In specific cases, a thorough preparation and updating of current scientific knowledge is required. Interestingly, there is little information available about the professional training of an EW, which is particularly important for FTs testifying in an adversary system. Recently, a condensed overview has been presented in order to summarize the training requirements for an EW in the United States [15]. Eight "qualities" needed to be developed [16]. The expert must:

- perform a thorough investigation;
- be personable, genuine, and natural;
- have an ability to teach;
- be generally competent;
- be believable;
- persuade without advocacy;
- be prepared; and
- demonstrate enthusiasm.

In Ref. 17, the "expert witness (EW)" has been defined as follows: "... a witness having special knowledge of the subject about which he is to testify. That knowledge must generally be such as it is not normally possessed by the average person. This expertise may derive from either study and education or from experience and observation. An expert witness must be qualified by the court to testify as such." Compared to this definition, some of the qualities refer to behavioral and technical skills, which, however, need to be based on scientific knowledge. Among some advice regarding pre-trial procedures, such as discovery of evidence and deposition or trial procedures such as cross-examination, useful hints are given in order to keep current and competent, improve personal presentation or avoid taking sides, and help an attorney win the case.

As the opposing attorney in the adversary system may have several goals during cross-examination, the EW needs to focus on this part of the trial. According to Refs. 15 and 18, such goals may consist of:

- discreditation of the EW by attacking conduct and character;
- attacking qualifications by establishing gaps in the EW's professional resume;
- showing inconsistencies in the EW's statements;
- exposing the EW's bias; giving reasons why testimony is slanted;
- attacking the witness's fact basis; tests conducted were inadequate, inappropriate, or incorrect; and
- discrediting the EW's conclusions by showing that it is confused and therefore can be wrong (changing the hypothetical used on direct).

In contrast to the adversary system, e.g., in the German legal system, one or more expert witnesses may be accepted or appointed by the judge and need to be present in the courtroom in order to observe the entire taking of evidence and then give his or their expert opinions in order to support the judge in his or her assessment of evidence. Such a procedure has major advantages. First, the opposing attorney's experts can directly be examined by each other and also by the judge regarding the scientific quality of their statements. Second, the judge can appoint an expert of his or her own choice to obtain a superior scientific expertise including examination of written opinions and the excerpts of the procedure.

23.2.1.4 Non-supervisory staff

The selection of assistant and technical staff working under the supervision of the scientific staff is very important for a laboratory. Appropriate professional, chemical, and analytical education of technicians, selection according to an appropriate pre-employment interview, a knowledge of the principles of the methods used in a laboratory and the ability to apply them, handling of equipment, as well as under-standing the importance of quality assurance are major requirements to be expected at this level. The recognition and reporting of potential sources of error and of situations where quality standards may be missed are other expectations addressed by technical staff. As such qualifications are usually based on experience, repeated teaching and training of staff members is indispensable.

23.2.1.5 Facilities, equipment, supplies

All guidelines demand suitable facilities, equipment, and supplies belonging to the important structural presuppositions. All of them need to be managed skillfully. The availability of a sufficient number of clean vented laboratory rooms suitable for trace analysis and the prevention of contamination during all analytical processes is a fundamental requirement. Additionally, in forensic toxicology, all laboratory rooms need to be protected in order to exclude access by unauthorized persons.

The equipment and supplies depend on laboratory functions, which need to be described in a QMS. In high-throughput laboratories, running a suitable laboratory information system is obligatory for properly monitoring the state of progress of individual casework, which may consist of only one, several, or even multiple analytical tasks, including tasks given to subcontractors. However, even if the decision to purchase the right equipment and supplies suitable for chemical toxicological analysis clearly belongs to quality aspects, focusing on details or the criteria of the corresponding considerations would be beyond the scope of this chapter.

In a fitting and appropriately developed QMS, all procedures are clearly described and updated analytical method SOPs are used, which include their validation data. QMS introduction will be an investment whose benefit may not immediately be obvious in terms of the costs. However, clear and completely written information and easy access to contact persons,, manufacturers, providers, and supplies will help to optimize organizational procedures by clarified responsibilities, and a rapid understanding of the pathways will save the staff's time in favor of performing all quality assurance procedures necessary, including instrument checks, monitoring of control charts, giving helpful comments, and undertaking corrections when parameters have become out of range or whenever necessary, and by avoiding errors. Therefore, regarding visible and hidden costs, an effective quality management supported by all staff member may turn out to be an economical necessity.

23.2.2 Quality of the processes

23.2.2.1 Preanalytical phase procedures

Regardless of the objectives of a chemical analysis, suitable collection of the samples is the most important preanalytical activity. Only in very few cases, this action, regrettably, is carried out by FTs themselves. Therefore, it is recommended that detailed advisory preanalytical information is provided to all personnel involved in the collection, storing and transportation of samples [19,20].

In Germany, for example, taking of not only urine samples or hair strands but also venous blood samples regularly serves as a means of evidence in criminal cases. As decreed according to the code of criminal procedures, the taking of blood samples by physicians ordered by the police has proved a powerful means to clarify both the individual state of influence as well as the respective situation in which the afflicted person has been. However, the instability of drugs such as heroin and cocaine as well as irreversible binding of cannabinoids to matrix constituents and other factors, such as temperature, oxygen, or UV-radiation sensitivity of analytes, are commonly encountered pre-analytical problems. All physicians and police officers responsible for sampling supplies, taking of samples, and their transportation to laboratories must be informed about such details in order to understand and to comply with procedures as issued by the governmental agency. Besides this, the stability or degradation and decomposition of analytes by various influences is an important matter of interest to the FT, easily being neglected in routine analysis.

Post-mortem sampling procedures need to be strongly distinguished from collecting samples from living subjects. The proper documentation of both the origin and quality of a sample is particularly important to the analyst. Clear-cut sampling and its documentation are both obligatory with respect to the interpretation of analytical results and the conclusions drawn. Site dependence, post-mortem changes, and redistribution of drugs and their metabolites are important phenomena [21–23] with which medical examiners need to be familiar when they collect the specimens. The publication of literature reporting on post-mortem redistribution, including animal experiments, continues [24–34].

In particular, substances with high distribution volume and specific tissue binding, such as cardiac glycosides or cannabinoids [35], and many other drugs are subject to redistribution after death. The mechanisms are diffusion, breakdown of membranes or decomposition of tissue, changes of water content, and agglutination of blood constituents, etc. Therefore, blood concentration measured is often unpredictably site dependent. Finally, the compounds themselves will decay with time after death. At autopsy, the case history may not necessarily be known in detail. Therefore, well-documented sites, comprehensive specimen collection, and the taking of relevant samples must be carried out particularly in cases of possibly previously unknown homicidal poisoning. Hence, the early exchange of information at autopsy between the medical examiner and the laboratory needs to be stressed. There is still a great need for studies aimed at detailed knowledge about substance stability or the occurrences and causes of concentration changes in samples. As cannabinoid analysis in blood and urine has become an analytical task whose frequency nearly equals that of ethanol analysis in blood, relevant reporting on the stability of the major oxidative THC metabolite is taken as one example among many others [36].

In addition, more knowledge of specific non-metabolic post-mortem breakdown products needs to be gathered and analyzed.

In a forensic laboratory, clearly defined and well-documented procedures are major points of specific care such as:

- advising customers in proper sample storage and transport;
- treatment, starting at the arrival in the laboratory;
- specimen registration, sample portioning, including sample treatment, and how to avoid mistakes;
- sample storage, including the decision about the necessity of freezing or refrigerating only;
- registration and documentation of samples with respect to the chain of custody; and
- security checks and storage of samples for additional examination and further investigations.

Proper sampling of post-mortem specimens has recently being reviewed and, accordingly, GTFCh recommendations for post-mortem sampling and handling of samples have been published [37].[4]

---

[4] Versions in English are being prepared.

23.2.2.2 Examination and methods

Forensic toxicologists need to perform poison detection in human organs and body fluids. Therefore, a careful selection of appropriate methods and their suitable application is a major point regarding laboratory performance in forensic toxicology. Two kinds of validated methods and corresponding equipment need to be maintained regarding the tasks to be carried out. First, approved general unknown analysis and selective screening methods need to be deployed. Second, various quantitative methods need to be applied according to specific requests or in cases in which particular results were previously obtained by general unknown analysis.

For the introduction of a new method into regular casework, following a logical sequence of the PDCA cycle, several steps need to be taken:

- elaboration of a new analytical method;
- testing of the method using spiked samples;
- measures of internal quality control;
- participation in external quality control;
- development of a plan for corrective actions; and
- steps to define a problem, investigate and determine its causes, initiate corrective actions to eliminate the problem, and check if the problem has been eliminated.

23.2.2.3 Screening methods and general unknown toxic substances

Depending on the scope of substances that need to be detected or identified, so-called general unknown analysis and screening methods are used. As it is most important to the analyst to have indicators of substances present in a material, a screening procedure should but must not necessarily include substance identification. Based on the purpose of screening procedures, their diagnostic selectivity and sensitivity need to be high enough to avoid too many negative confirmation results.

A universal screening purpose and ideal general unknown analytical method would include all toxic substances. Such a claim has remained an ideal comparable to the medieval legend of the "Holy Grail." In contrast, usually a sequence of screening procedures is carried out using a combination of separation and detection methods to find the toxic substances present. The term "screening" refers to a limited number of substances related to each other by chemical, physical, and/or structural properties. Immunoassays for drugs or GC-head space analysis for screening of volatile compounds are examples of different areas of screening. Analytical screening is more or less extensive and its extent can be regarded as a quality. In general, complete information about the practical performance of screening methods for qualitative poison detection in practical cases is difficult to obtain. Some of the methods propagated for use in systematic toxicological analysis (STA) are based on the parent compound but not on their metabolites. In addition, the testing of practical detection limits in difficult matrices is lacking.

The identification of unknown substances and the amounts that need to be determined are more critical in cases under suspicion of homicidal poisoning than in suicidal or accidental poisonings because of higher amounts of substances usually present in such cases. At present, for the general unknown analysis of toxic low-molecular-weight substances, their possible detection is based on suitable sample preparation methods and the extraction of as many substances as possible. Upon chromatographic separation, any unknown substances need to be identified by consequent application of a spectroscopic method. Based on sufficient chromato-graphic separation prior to recording, the identification power of spectra mainly depends on their informational specificity.

GC and HPLC are now the most common separation methods which, depending on the chromatographic principle, compete with each other effectively regarding the palette of accessible substances. The additional use of GC [38,39] or HPLC and retention indices [40] of reference substance mixtures helps to characterize the chro-matographic column and improve the discrimination power of the analytical procedure needed to identify unknown substances. However, in contrast to sub-stance quantization, for screening purposes, the baseline resolution of chromato-graphic peaks is not required under any circumstances.

Gas chromatographic separation in combination with mass spectroscopy (GC/MS) using several libraries for comparison of standardized 70 eV mass spectra is generally available in forensic-toxicological laboratories [41]. A dual-column chromatography using MS and nitrogen-sensitive detection can be a useful extension of GC/MS screening. Matrix interference plays an important role in substance discovery and identification issues. However, knowing the limits of identification of individual sub-stances in screening procedures used for various materials is more a matter of experience [42] than a result of circumstantial and nearly impossible method valida-tion. Recently, in a consensus approach, the U.S. American Society of Forensic Toxicologists (SOFT) defined the maximum detection limit for drugs recognized to be used in sexual assault. (Table 23.1, [43]). Obviously, such minimum detection limits need to refer to specific search runs using single ion monitoring and also to general unknown screening procedures.

Similar recommendations for minimum detection limits have been published for analytes listed referring to diagnosis and confirmation of brain death (Table 23.2).

Clearly, the use of recovery-sensitive internal standards is recommended. In gen-eral unknown analyses, acidic and basic deuterium-labeled compounds can also be used (e.g. barbiturates or tricyclic antidepressants).

It is recommended, as well to check an instrument's performance and calibration with special attention to amphoteric substances with critical chromatographic peak shapes, such as benzoylecgonine or morphine [44]. The analyst's attention needs to be drawn to suitable recording of mass spectra or ion monitoring of xenobiotics as both may be disturbed in the presence of high amounts of substances co-eluting in GC (e.g., polyethylene glycols of different chain lengths liberated from pharmaceutical formulations).

Due to interference within the ion source, the extraction efficiency of charged particles may be reduced, as was observed for parent drug substances and their deuterated analogues [45,46] and for environmental compounds [47]. Other problems of mass spectra generation may arise by the uncritical use of an ion trap (and other types of) MS that needs careful evaluation of data in order to generally achieve

TABLE 23.1
INTERNET PRESENTATION (ONLY FIRST PAGE) OF RECOMMENDED URINARY MAXIMUM DETECTION LIMITS REFERRING TO COMMON DRUG FACILITATED SEXUAL ASSAULT (DFSA) DRUGS AND METABOLITES (HTTP://WWW.SOFT-TOX.ORG/DOCS/SOFT%20-DFSA%20LIST.PDF). SOCIETY OF FORENSIC TOXICOLOGISTS (SOFT). DRUG-FACILITATED SEXUAL ASSAULT COMMITTEE. RECOMMENDED MAXIMUM DETECTION LIMITS FOR COMMON DFSA DRUGS AND METABOLITES IN URINE SAMPLES

Target Analytes	Parent Drug	Trade Names/Street Names!	Recommended Maximum Detection Limit
Ethanol			
Ethanol	Ethanol	Alcohol, ethyl alcohol, "booze"	10 mg/dL
GHB and analogs			
Gamma-hydroxybutyrate	Gamma-hydroxybutyrate	Xyrem, "GHB," Easy Lay," "G," "Georgia Home Boy," "Grievous Bodily Harm," "Liquid Ecstasy," "Liquid E," "Liquid G," "Liquid X," "Salty Water," "Scoop," "Soap"	10 μg/mL
	1,4-Butanediol	"1,4-BD," "Enliven," "Inner G," "Revitalize Plus," "Serenity," "SomatoPro," "Sucol B," "Thunder Nectar," "Weight Belt Cleaner," "White Magic"	
	Gamma-butyrolactone	"GBL," "Blue Nitro," "G3," "Gamma G," "G.H. Revitalizer," "Insom-X," "Invigorate," "Remforce," "Renewtrient,""Verve"	
Benzodiazepines			

Many benzodiazepines are biotransformed into glucuronide-conjugated metabolites. To improve detection limits and times, it is recommended that laboratories use instrumental techniques that will detect the glucuronide metabolites or hydrolyze urine specimens to free the conjugate before extraction.

Alprazolam	Alprazolam	Xanax, Niravam	10 ng/mL
α-Hydroxy-alprazolam			
Chlordiazepoxide	Chlordiazepoxide	Librium, Libritabs	10 ng/mL

TABLE 23.1
CONTINUED

Target Analytes	Parent Drug	Trade Names/Street Names!	Recommended Maximum Detection Limit
Clonazepam	Clonazepam	Clonapin, Klonopin, Rivotril	5 ng/mL
7-Aminoclonazepam			
Diazepam	Diazepam	Valium, Diastat, Dizac	10 ng/mL
Flunitrazepam	Flunitrazepam	Rohypnol	5 ng/mL
7-Aminoflunitrazepam			
Lorazepam	Lorazepam	Ativan	10 ng/mL
Nordiazepam	Diazepam, Chlrodiazepoxide		10 ng/mL
Oxazepam	Oxazepam, Diazepam, Chlrodiazepoxide, Nordiazepam, Temazepam	Serax	10 ng/mL
Temazepam	Temazepam, Diazepam	Normison, Restoril	10 ng/mL
Triazolam	Triazolam	Halcion	5 ng/mL
4-Hydroxy-triazolam			

TABLE 23.2
RECOMMENDED MINIMUM DETECTION LIMITS TO SUPPORT BRAIN DEATH DIAGNO-
SIS ACCORDING TO HALLBACH ET AL. [2002] TOXICHEM & KRIMTECH 69 (3) 124–127

Wirkstoff	Untere Grenze Destherapeutischen Bereiches (mg/L)	Empfohlene Untere Messbereichsgrenze (mg/L)
Thiopental	1.0	0.5
Pentobarbital (Thiopental-Metabolite)	1.0	0.5
Phenobarbital	10.0	5.0
Methohexital	0.5	0.25
Midazolam	0.04	0.02
Diazepam	0.2	0.1
Nordazepam (Diazepam-Metabolite)	0.2	0.1

acceptance [48]. Recently introduced mass spectrometric techniques, such as time-of-flight (TOF) spectrometers, need to be evaluated as useful new tools for screening analysis operating with full mass spectra even in trace analysis (see below).

As a consequence of a laboratory's need to assess the performance of the screening methods used, and due to the lack of simple validation criteria for generally unknown analytical methods, the participation in external quality assessment schemes including case report-oriented analysis of general unknowns is strongly recommended [49]. Among substances often overlooked in screening strategies may likely be the highly water-soluble compounds, such as glycols [50], and low-chain aliphatic acids such as glycolic acid and beta- or gamma-hydroxybutyric acid.

There is increasing use in general unknown analysis of MS in combination with HPLC [51]. Depending on individual substances, suitable ionization modes, such as electrospray and atmospheric pressure chemical ionization systems [52–54], are now increasingly used and are affordable. Besides costs, however, the comparability of mass spectra recorded under specific conditions is still a limiting factor with respect to the widespread use of HPLC/MS and data collection for libraries. As spectra may vary between instruments of different manufacturers, more work needs to be carried out in this area. The use of such techniques for the detection of common drugs of abuse in serum or urine has been published [55,56]. Progress in the detection and quantitation of toxic plant [57] or animal peptides and proteins [58] is expected, including increasing forensic use of HPLC/MS. In spite of far less discrimination and identification power, HPLC and UV spectra are commonly used in combination with retention time or retention index concepts and library search techniques [59].

Usually using specific flame ionization, but increasingly the mass-selective detectors, head space GC is successfully applied to sensitively detect and quantitatively determine gases and volatiles in body fluids or organ tissues. For other groups of chemical substances, such as organic and inorganic toxic anions, practically no comprehensive systematic methods for poison detection have been proposed and approved. Separate testing of compounds under suspicion seems to be usual. Even if high-performance ion liquid chromatography has become an approved method for analysis of drinking water, up to now no satisfactory application of this method has been described for use in post-mortem samples. As molecular weight information is obtained from mass spectra, HPLC/MS/MS is, as expected, becoming a powerful tool for substance identification and confirmation of results of screening methods for toxic cations, quaternary ammonium compounds and various anions, and for toxic proteins.

Chemical diagnosis of metal poisoning is another area of toxicological analysis in which quality assurance is urgently needed. Instead of separate testing for the presence of an individual metal, inductively coupled plasma mass spectroscopy (ICP-MS) seems appropriate for use in screening procedures and the identification of metals. As acute metal poisoning is rare, such expensive instrumentation may be used in collaboration with specialized laboratories for economic reasons.

In conclusion, quality results of a general unknown analysis will strongly depend not only on a shrewd analytical strategy as conducted by an experienced toxicologist but also on the availability of appropriate equipment, the use of approved methods, and a regularly updated reference data collection. Progress in general unknown analysis continues. A laboratory's participation in external quality control (EQC) and an educational ring test is the best means to examine performance and to compare and improve methods of performance in general unknown screening.

New approaches in general unknown analysis include using GC 70 eV ionization and LC ionization techniques and time of flight mass spectrometry (GCMS-TOF, LC/MS-TOF) in which the drift velocity of ions in an electric field is measured. Precise molecular weight, peak recognition, and deconvolution software is applied. By such aids, all ions of mass scans can be recorded for substance identification in one run [60–64].

23.2.2.4 Quantitative determination of toxic substances and their metabolites

Quantification of parent compounds and, as often required, their metabolites presumes:

- qualified and trained staff, appropriate equipment, supplies, and reagents;
- the use of approved methods validated regarding their purpose including calibration; and
- corresponding stocks of certified pure reference substances including proper storage conditions (or at least easy access to substances possibly found in general unknown analysis).

GC/MS is routinely used for quantification of licit and illicit drugs, synthetic chemicals, and poisons. In order to achieve highly precise and accurate determinations, the analysis needs to be conducted using internal standards with chemical properties similar to the analyte and, if available, the deuterated analogues of the analytes. For an acceptable precision of a method, the recovery of spiked substances should be not less than 50%. Depending on the selectivity of the chromatographic separation, selected ion monitoring (SIM) based on four or at least three ions is sufficient for identification and quantification of trace amounts. The ion ratio is acceptable if varying within $\pm 20\%$ of the theoretical value. This criterion seems to be derived from a practical approach rather than from a systematic investigation serving as a decisive base. In qualitative trace analysis, however, little above the limit of detection and below the limit of quantitation (LOQ) small signal intensities will be detected. As statistical errors in the reading of less intensive ions become more pronounced, enlarging deviations must be considered. Some details regarding substance identification will be discussed below.

Tandem-MS-MS and high-resolution MS (HRMS) may help to diminish the time needed for quantitative batch analysis and chromatographic separation using detection techniques (e.g., selected reaction monitoring, neutral loss scan, and precursor ion scan after collision gas-induced decomposition in triple-stage quadrupole instruments). The use of such techniques, far-reaching elimination of the background, and high discrimination power are other advantages that allow substance identification by specific monitoring of a single ion transition instead of several correlated ion intensities in single low-resolution quadrupole MS instruments. With increasing use of LC/MS/MS instruments, however, pitfalls [65] have recently become obvious, which should lead to increased validation efforts in HPLC/MS including more individual real samples to check for matrix-dependent effects of cross-talk [66] and ion suppression [67].

23.2.2.5 Validation of methods

Regardless of the use of approved isolation, purification, or separation steps, any *new* analytical method needs to be validated before its use in order to ensure that it is capable of yielding acceptable results.

Validation is a procedure that determines whether an analytical instrument or method as applied in a laboratory provides the analyst and customer with results fitting their intended purpose. In general, validation activities refer to sampling, stability of substances or samples, performing analytical methods, calibration of instruments and methods, suitability of reference materials, calibrators, reagents, data acquisition, data documentation, and data statistics.

Usually, the term "validation" is not used for elements of quality assurance in a laboratory other than analytical elements.. Primarily, validation seems necessary where new elements are taken into a procedure, such as new products, new analytical methods, new analytical instruments, or new quality testing criteria. That is why chemical analysts generally need to know three major validation aspects:

1. the performance of the instruments used;
2. the performance characteristics of the method; and
3. by whom and for which purpose will the measurements be used?

Suitable forensic validation should include interlaboratory comparability and traceability as needed for objective testing in forensic toxicology. If correct procedures have been followed by different laboratories at different times, the same chemical analysis performed with the material under investigation should produce results agreeing with each other within definite limits. Traceability means that this comparability of results can be traced back to appropriate national or international measurement standards via an unbroken chain of comparisons. Typically, validation characteristics can be classified with respect to *identity, pureness,* and *content of a substance.*

Data obtained during method validation should cover the performance characteristics as described in the following sections. They are some of the key ingredients that contribute to the analyst being in a position to understand and control the uncertainties that will affect the measurement. Where generally accepted validation data do not exist, e.g., in the development of a new standard method for a regulatory authority or for legal decisions, the proper course of action is to validate a method through a collaborative study involving at least eight laboratories [68].

In order to support scientific development, the definition of analytical standards, which need to be met by a scientist's own method, has proved more useful in forensic toxicology than establishing standard methods. Regarding such standards, the introduction of other improved methods may be delayed and circumstantial. In addition, in forensic toxicology, many scientists are strongly convinced of their own method. It may become difficult to find a corresponding number of laboratories using identical methods for purposes of necessary collaborative studies.

Under urgent circumstances, quantitative analytical methods previously not applied and not appropriately tested may sometimes be applied for rapid information for customers, e.g., an intensive care unit or the police. If such results are later used for legal decisions, for obvious reasons the methods should be validated as soon as possible. As in complex cases, new or unusual questions may arise later in the course of criminal procedures, and it is recommended to update validation and internal quality control data frequently with respect to detailed examination.

Practical method validation requires investigation of performance characteristics that depend on the method's purpose. As shown in Table 23.3, several performance levels of analytical methods in relation to validation can be distinguished.

As has been demonstrated for polychlorinated biphenyls [69], the critical points of control in environmental quantitative trace analysis procedures were the following:

- It must not miss any substance present at or above the limit of detection, and possibly at or above a selected threshold limit concentration.
- It must not report false positives.
- The recovery must be 60–100%, with a coefficient of variation of > 20%.
- Values found, corrected for recovery, must agree within 20% with the recovery-corrected values obtained by the official method.
- There must be a documented validation that the extraction procedure removes compounds of interest from the sample matrix.

Both the progress and the practice of method validation were recently reviewed [70]. Based on this survey, a generally accepted approach to method validation was given as an appendix to the GTFCh guidelines [71].[5] Accordingly, a practical solution and scheme to calculate corresponding validation parameters by reporting

---

[5] English version given in: Bulletin of the TIAFT XXXII(1), 16–23 (2003).

TABLE 23.3
PERFORMANCE LEVELS OF ANALYTICAL METHODS IN RELATION TO VALIDATION

Action/ Level	Propositions Needed for Proper Working	Deviations Due to Instruments and Methods	Acquaintance with Resolution/ Separation Power	Testing Method Performance in Relation to Definite Requirements	Interlaboratory Comparison	Reaction to Adjusting, Necessary Changes
Measurement of physical parameters Identification	Calibration of the instrument	Measurement precision	Selectivity		Selectivity	
Measurement of physical-chemical parameters		Measurement precision and method precision	Robustness	Linearity, precision accuracy	Accuracy	Robustness
Qualitative (trace) analysis (one substance or more)		Measurement precision and method precision	Selectivity robustness	Limit of detection, limit of identification	Selectivity completeness of detection and identification	Robustness
Content, quantitative trace determination		Measurement precision and method precision	Selectivity robustness	Possible range of measurements, linearity, precision, accuracy, limit of detection, limit of identification, limit of quantification	Reproducibility, accuracy	Robustness

References pp. 814–819

the appropriate analytical data into an easily taught Excel®-based computer program with fill-in sheets has been developed [72].

Another tool useful in forensic toxicology is regular participation in external quality control trials as well as use of certified reference material (CRM), both of which are now available for illicit drugs in serum as a result of round robin testing and external quality control trials organized by the GTFCh.

23.2.2.6 Extraction recovery

The correct use of deuterated internal standards is an excellent method to check the recovery. In order to yield acceptable precision and detection certainty, the recovery should be as high as possible. Currently deuterated standards are used, and an overall recovery of better than 50% with variability of $\pm 10\%$ may be sufficient.

23.2.2.7 Selectivity and specificity

The investigation of the method's selectivity aims at low-interference frequency possibly resulting from normal and atypical tissue constituents such as xenobiotics and contamination, particularly at low concentration levels. No definite procedure exists for completely investigating the selectivity with respect to economical validation. First, it is recommended to analyze around 100 previously tested blank samples for typical metabolites in order to find co-eluting peaks. Second, in order to find the origin of and kind of co-eluting material in samples containing the analyte, the analyst may check the whole mass spectrometric background at the retention time of a poorly resolved peak and change the chromatographic conditions if necessary.

23.2.2.8 Calibration linearity and analytical sensitivity

In order to calibrate properly, a linear relationship is expected between signal intensity and amount measured. The slope of the calibration line is a measure of the power of discrimination between concentration levels (analytical sensitivity). High recovery and narrowing of the chromatographic peak shape will increase sensitivity and yield better measurement precision and fit of the calibration line. The calibrators used should cover the whole expected concentration range. In human and other biological samples ranges depend on substance toxicity, dosage, body weight, species, individual tissues, and pharmacokinetic or toxicokinetic properties.

As concerns calibration and analytical limits, the statistically based German DIN 32645 [73] states that the midpoint of the calibration should be near the statistical mean of the concentrations to be tested. Five ideally equidistant calibrator levels should be chosen. With respect to both measurement precision and detection limit, for trace analysis in forensic toxicology the calibrator concentration should be shifted towards the lowest levels. With use of a computer program and statistical functions as described in the DIN, the calculation of GC/MS-SIM calibration data gained by peak by integration during frequent trace determinations of tetrahydrocannabinol in human serum in actual cases has shown that four of these calibrator levels may be chosen at 15, 35, 50, and 75% of a maximum concentration of $10\,\mu g/L$ [74]. Additionally, higher concentrations in a control calibrator can be analyzed separately.

23.2.2.9 Limits of detection, quantitation, and identification limits

Interrelated critical limits of a quantitative analytical method need to be determined during validation. The limit of detection as well as the LOQ are fundamental data to document assay performance. In forensic toxicology, the detection of a substance often must include its identification, especially when the presence of a substance is in question.

Regarding the determination of the limit of detection, empirical and statistical approaches exist. Empirically, the limit of detection of a method was assessed by measuring decreasing concentrations of a single analyte in order to establish the lowest concentration that can be detected with an acceptable response. According to the definition of the International Federation of Clinical Chemistry [75], the detection limit is "the smallest single result which, with a stated probability of (commonly) 95%, can be distinguished from a suitable blank." A similar definition was given by the International Union of Pure and Applied Chemistry [76].

For practical reasons and in order to certify new GC/MS instruments for analysis of illicit drugs in urine, the limits of detection (LOD) were defined as "the lowest concentration of drug that a laboratory can detect in a specimen with forensic certainty at a minimum of 85% of the time." Overall batch acceptance criteria were as follows: acceptable quantitation of control material within 20% of the target; chromatography (peak shape, symmetry, integration, peak, and baseline resolution); retention time within 0.1% of the extracted reference compound; mass ion ratio within $\pm 20\%$ of the extracted item; and individual specimen. The acceptance criteria excluded quantitation requirements in the individual specimen.

According to the German DIN 32365 standard, the limit of detection of an analytical method can be calculated daily using calibration data. This approach assumes a normal ("Gaussian") frequency distribution of readings at a definite concentration. The terms α and β represent the error probability levels to (falsely) classify an analyte as present or not present, respectively. Each quantitative method has a critical value representing a concentration at which the probability of an α error reading in a sample not containing the analyte is 0.01. In contrast to the above approaches, DIN defines two LOD. The lower is defined as a concentration at which the analyte is classified as present or not present with equal probability. Consequently, this concentration is calculated from calibration data taking $\beta = 0.5$ and $\alpha = 0.01$. Depending on the analytical problem, other values of β and α can be chosen. That is why the statistical solution provides a second and higher LOD. As defined by the German DIN 32365, this limit refers to the smallest quantity that can be detected at a probability level of $1-\beta$ (detection certainty of 95% or 99% if $\beta = 0.05$ or 0.01, respectively). This concentration threshold has higher coverage. The certainty to detect an analyte is 99%, meaning that the analyte is regularly detected when present at this concentration.

23.2.2.10 GC/MS substance identification using selected ion monitoring

The LOQ refers to the minimum quantity being determined with both a defined probability level (e.g., $\beta \leqslant 0.01$) and an acceptable relative uncertainty.

In the DIN approach, both the lower LOD and the critical value are of theoretical interest and no values are suitable for the customer's use: Only the LOD has high detection certainty. However, it refers to a single signal as detected using MS/MS-techniques or high-resolution MS. For methods using low-quadrupole mass resolution, LOD is not a suitable parameter to guarantee substance identification. The appropriate detection of three or even four characteristic ions is required to identify an analyte in such quantitative GC/MS analysis. For GC/MS-SIM quantitation, usually, the most intensive signal is chosen as the target ion. Regarding substances with less than four intensive ions, SIM may not be sufficient for identification. The corresponding identification power depends on the availability of qualifiers with suitable abundance. In order to identify an analyte, the least intensive of three qualifiers must be recorded at least at a 90% detection certainty level, including suitable mass ion ratios (within $\pm 20\%$ deviation of theoretical ratio). Regarding the calculation of the LOD (i.e., certainty to detect and identify[6]), the calibration data (peak areas) of least intense ion being monitored need to be used. In contrast, the calculation of the LOQ can be based on the most intensive ion.

In pharmacokinetic studies, for instance, the detection and even quantitation of a known substance may be needed in terminal elimination at concentrations at which the substance can no longer be identified in a forensic sense, e.g., assuming monitoring of the 100% intensive target ion (and far less intensive qualifiers). Consequently, for forensic purposes, the LOQ must not be chosen lower than the limit at which identification is possible.

23.2.2.11 Cut-offs

In contrast to clear characterization of LOD or quantitation of qualitative and quantitative methods, the term "cut-off" is used more or less arbitrarily in screening procedures using immunoassays. The meaning of a cut-off is mainly a decision limit defined by a nominally given concentration level of the analyte used for calibration. The analyte, however, may not necessarily be present in real samples (e.g., cannabinoids and benzodiazepines in urine, where instead of parent compounds their glucuronides represent the target analyte). However, the analytical meaning of a cut-off is different from a diagnostic meaning.

It may be useful to avoid false-positive readings in screening tests at both a concentration level as low as possible (optimum diagnostic sensitivity) and at a high probability level of analytical trueness (commonly at least of 95%). An analytical cut-off often includes possible readings of cross-reactions within the substance group and should exclude responses of material not representing any analytical target. Any false sources of response in a test result usually cannot be recognized or

---

[6] There is semantic difficulty in translating the limit terms as used in the German DIN standard.

distinguished from correct ones. By using cut-offs, however, testing should exclude analytical false-positives as far as possible.

In contrast, diagnostic cut-offs may be needed to:

1. separate collectives, e.g., healthy and ill persons;
2. avoid misleading interpretation of results, e.g., regarding the origin of illicit drug metabolite excretions in urine;
3. allow a corresponding application of appropriate confirmation methods;
4. have comparable testing preconditions regarding regulatory, authority, and legal decisions; and
5. have comparable testing preconditions in proficiency testing.

Regarding the above topics for workplace testing, national and international proposals for cut-offs exist for immunoassay screening and for confirmation of illicit drugs and their metabolites in urine, as shown in Tables 23.2 and 23.3. NIDA cut-offs were the first ones used for abuse testing of drugs in urine [77]. The U.S. Department of Health and Human Services and its Substance Abuse and Mental Health Administration (SAMHSA) have defined federal standards for urine drug testing. On the Internet [78], SAMHSA's Workplace Resource Centre gives access to cut-off concentrations as specified in the Mandatory Guidelines for Federal Workplace Drug Testing Programs that include immunoassays and GC/MS confirmation. These cut-offs (2001 version) for immunoassays and GC/MS confirmation are presented in Table 23.4. In order to yield comparable results as well as unavoidable measurement uncertainty, no information is presented about minimum accuracy and precision performance needed at a level of 15 ng/mL, as these were defined for cannabinoids earlier.

In Germany, no mandatory regulations have been enacted or accepted to date. Nevertheless, for abstinence testing of drugs of abuse, the cut-off regulation is useful

TABLE 23.4
CONFIRMATORY TEST LEVELS FOR WORKPLACE TESTING OF SUBSTANCE ABUSE AND QUANTITATIVE CONFIRMATION AS PUBLISHED BY THE US DEPARTMENT OF HEALTH AND HUMAN SERVICES

	Initial IA Test Levels ng/mL (Equivalents of Reference Substance)	GC/MS Confirmatory Test	Test Levels (ng/mL)
Marijuana metabolites	50	Delta-9-tetrahydro-canna-binol-9-carboxylic acid	15
Opiate metabolites	300	Morphine	300
		Codeine	300
Cocaine metabolites	300	Benzoylecgonine	150
Phencyclidine	25	Phencyclidine	25
Amphetamines	1000	Amphetamine	500
		Methamphetamine, and coincidend with amphetamine	500
			200

TABLE 23.5
URINE THRESHOLD CONCENTRATIONS AS PROPOSED IN CLINICAL AND EU WORK-
PLACE TESTING ACCORDING TO [81] AND [105]

Analyte or Group	KKGT Dutch Health Care Cut-off	UKNEAS – Clinical Thresholds (µg/L)	EU – Workplace Testing Threshold (µg/L)
Screening tests			
	(Reference substance >)		
Amphetamine group	1000 (Methylamphetamine)	1000	300
Barbiturate group	300 (Secobarbital)	500	–
Cannabinoide group	100 (11-nor-THC delta 9 COOH)	300	50
Benzodiazepine group	300 (Morphine)	500	–
Opiate group	300 (Oxazapam)	500	300
Methadone	300 (Methadone)		
Single analytes			
Amphetamine		1000	200
Methamphetamine		1000	200
Methylendioxyamphetamine		1000	200
Methylendioxymethamphetamine		1000	200
Methylendioxyethylamphetamine		1000	200
Specific barbiturate		500	–
Delta-9-THC-carboxylic acid		15	15
Cocaine		3000	150
Benzoylecgonine		300	
Specific benzodiazepine		300	
Methadone		500	
Morphine		500	200 after splitting of conjugates
Dihydrocodeine		1000	
Buprenorphine		1	–
Phencyclidine		25	–
Lysergic acid diethylamide		5	–

and recommended; however, it should include monitoring of the performance of both test kits and testing Tables 23.4 and 23.5.

In 1998, a regulation in Germany within the Road Traffic Act has been promulgated. According to an appendix to §24a II of the Road Traffic Act, the presence of illicit drugs in the blood of drivers is regarded as a violation. Therefore, analytical threshold limit values for GC/MS analysis of illicit drugs (parent compounds) were proposed by an expert group in order to combine the analytical and minimum risk assessment aspects in regard to drivers and drug abuse [79]. These concentrations were halved in 2001 [80]. Accordingly, in 2004, the German Federal Constitutional Court judged a limit of 1 ng THC/mL serum to be the minimum concentration at which a driver can operate a vehicle in spite of being impaired. This limit is said to include a safety margin of 100% relative to a limit of detection of 0.5 ng THC/mL serum. Some questions are still open regarding the introduction, size, and role of measurement uncertainty at this concentration and about a single value measured with a correspondingly validated high-precision method.

As clinical testing aims at appropriate therapeutic treatment, performance criteria other than forensic testing may be suitable. In emergency toxicology and cases of drug overdose, rapid, low-cost, and possibly less sensitive methods may be chosen. In specific cases, the identification of substances and trace analysis may be less important than the exclusion of their presence. The usefulness of higher clinical cut-offs than those for forensic or EU workplace testing has been reported [81] (Table 23.5).

Clinical thresholds reflect the use of less sensitive methods for the confirmation of immunoassay results by clinical laboratories. For clinical testing, some single analyte thresholds for parent compounds were chosen that were higher than screening thresholds, indicating possibly nonrealistic testing conditions, e.g., for unchanged cocaine in high concentration in relation to its major metabolites. Immunoassays usually show cross-reaction to several known and possibly unknown metabolites. Some tests are not capable of detecting the parent compound when applied to human body fluids or tissues, which is why confirmatory testing in urine must usually detect concentrations lower than the cut-offs of the immunoassays used. As a consequence, forensic or workplace testing laboratories must not refer to clinical quality assessment. As some sites may provide both clinical and workplace testing, an optional selection of cut-offs would be in contrast to appropriate proficiency testing.

As a consequence of the lowering of detection limits by more and more powerful analytic instrumentation, future forensic work concerning cut-offs must concentrate on the definition of thresholds for substance traces or substance metabolites in order to discriminate exogenous substances that are taken or given from those that originate from natural or systemic sources or from small substance doses taken up unknowingly. One known example is the systemic excretion of γ-hydroxy butyric acid and its discrimination from traces excreted after criminal poisoning [82–85]. Another example illustrates the problem: While painting, a painter who is under alcohol abstinence control may inhale ethyl acetate. Does this compound produce ethyl glucuronide, which tests as an alcohol marker throughout the world? If excreted in urine, can ethyl glucuronide be present in measurable concentration and, if so, at which maximum level?

23.2.2.12 Immunochromatographic tests: lateral flow assay

23.2.2.12.1 Easy Use of LFA?
For many trained and untrained users immunochromatographic lateral flow assay (LFA) technology is a welcomed alternative to collection, transport of urine samples, and subsequent quality-controlled laboratory analysis. Lateral flow immunochromatographic testing and targeting of the more commonly abused substances became a multimillion-dollar business; however, as a designated qualitative test LFA may entail some losses of quality. LFA formats range from dipsticks or pipette strip tests to multiple test cards or plastic cassettes to cups. The amount of sample needed for testing ranges from a few drops of urine to a minimum amount of 20–30 mL as dictated by sample collection vessels or cup dimensions. To screen for the presence of

a single drug or drugs of an abuse class or to detect several groups of drugs such as amphetamines, benzodiazepines, cannabinoids, cocaine, and opiates, all known devices use the immunoassay principle. The devices utilize several antibody reaction principles such as agglutination reactions, fluorescence, or chromogenic (gold) labeling of antibodies and labeled drug conjugates. It is common to distinguish them according to their visible response, usually a test line, which is generated by a drug present at or above the designated nominal concentration threshold.

In laboratory testing, the chosen nominal cut-off (numerical reading) combines the exclusion of minimum (potentially unspecific) test reactions with diagnostic group differentiation needs (e.g., passive consumption of cannabis). In contrast, lateral flow test cut-offs are numbers greatly differing from laboratory test cut-offs with calibrated readings. In reality, in corresponding tests, LFA cut-off numbers mean a generally precisely reached threshold saturation of a labeled antibody by present drug molecules fully suppressing the antibody's binding at the test line. (Fig. 23.2) Similar to drug radioimmunoassay with low-molecular-weight compounds (haptens) in which the loss of competitive antibody binding of a radioactive tracer can be counted in relation to the concentration of the hapten present, the absence of a line or color means the test result is positive. At the lateral flow cut-off, the test reaction on the strip should have developed close to its maximum. Therefore, even the development of a less intensive line or color indicates that the drug is present (below the threshold).

According to saturation analysis principles related to an antibody's affinity, the lateral flow immune response consists of two steps: a primary (pre)incubation of the

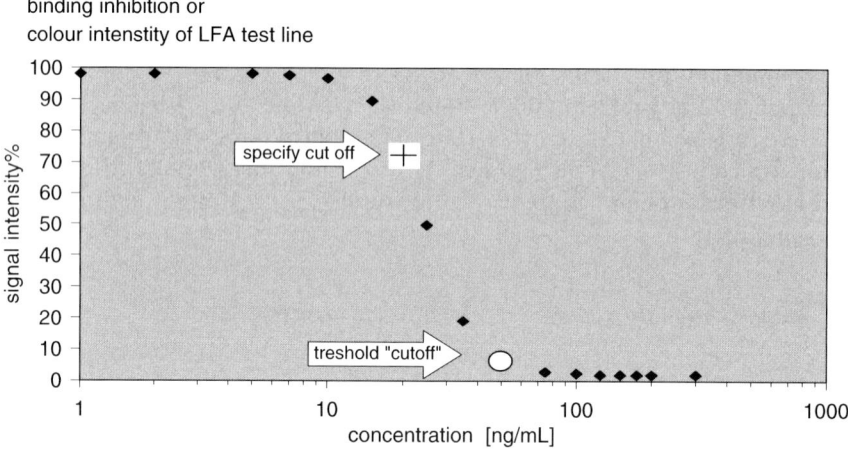

Fig. 23.2. Difference between cut-off numbers in machine-based liquid-phase immunoassays and lateral flow immunochromatography tests: While the conventional understanding of cut-off principles refers to cross-reactivity and specificity issues (shown by $+$), the lateral flow "cut-off" means test line visibility (disappearance, open circle) at a (technically adjustable) threshold. As can be seen from the flattening shape of the dose–response curve at high binding inhibition, lateral flow visual readings at thresholds ("cut-offs") must have relatively high imprecision due to the inborn variability of both binding kinetics of the antibodies and antibody desorption from the analyte contact zone into the chromatographic flow.

antibody with the target analyte, and a secondary competitive reaction between the unbound (and also the target analyte bound) antibody sites and the hapten-protein conjugate at the test line. Because of their in-line combination, these two steps may merge into each other. This disadvantage may also depend on the still incomplete release of antibody sites from the sampling pad, and the incomplete saturation of antibody binding sites when reaching the hapten-conjugate at the test line. Both the inborn variability of the antibody binding kinetics and the lateral flow kinetics at a relatively high target analyte concentration may give rise to a relatively imprecise calibration curve, hence, visual readings at the designated cut-off.

Unlike other qualitative applications, most LFA devices for drugs of abuse give a negative visual sign when the drug of interest is at or above the defined threshold. To avoid quantitative measures, such testing was called qualitative testing. At present, there are only a few devices that indicate the presence of the designated drug with the appearance of a line. The term "cut-off" needs to be considered further in selecting a device, as these devices will impact the number of samples requiring confirmation. Because of the above disadvantage, the statistical likelihood of obtaining a negative result for a sample containing the drug or its metabolites near the cut-off should be defined by the manufacturer. Validation studies during selection and implementation should include testing of the defined cut-off.

According to an antibody's specificity, as with laboratory-based methods, most tests detect drug metabolites instead of the parent drug and often detect other metabolites than those present in the sample. One example is, again, the calibration using 11-nor-delta-9-carboxytetrahydrocannnabinol (which in practice is not present) instead of its acylic glucuronide (the usual major cannabinoid metabolite present in urine).

Although urine samples may need to be transferred, most of the steps for lateral flow testing require an operator's intervention such as a sample application, possible transfer of the sample to another portion of the device, timing of the immune reaction, reading or interpreting the visual endpoint, and recording and documenting the result. Visual endpoint readings clearly depend on the chosen technological approach. Turnaround times seen from the initial sample application to reading of the result are up to 15 min, but they usually take much less than this time.

The operator is responsible for all potential manipulations of the sample, the reagent application, timing, interpretation, and recording of the results, including quality control indicators. At times there may be little control of the sample volume applied, depending on the principle: Some dipstick devices use dipping time; cassette tests use small plastic pipettes to introduce the sample. This technique greatly differs from automated immunoassay-based laboratory methods for serum/plasma or urine with primary tube sampling.

There may well be a potential need for regulations for the application of immunochromatographic lateral flow (LFA) testing in different forensic, clinical, and non-clinical situations (e.g., at the workplace or for abstinence testing, for testing of drug users in road traffic, for testing of former drug users whose driving licenses have been withdrawn, or for rapid drug testing in emergency situations).

References pp. 814–819

One problem may be that some commercial providers of LFA devices may have a limited background and limited information to correctly train purchasers or users on function, application, and especially careful interpretation of results. Such training should include quality issues as well as well-known device limitations. All users of LFA devices should understand the devices' limitations, including statistical and analytical sensitivity, specificity, and nomenclature used.

Users need to be aware of commonly experienced interferences resulting from the presence of substance-related or unrelated drugs or metabolites that could have an impact on readings of results as well as the interpretation of results. In practice, however, such necessary information is sometimes not available or is incomplete and therefore is rarely or inadequately given. Aimed at a relevant diagnosed population, a careful evaluation needs to be conducted by well-educated test users and needs to take place in those environments where the tests will be used. For instance, roadside testers may check their tests at various low and high outdoor temperatures and also after storage at various automobile interior temperatures. Controlled studies on population and the purposes of testing are rarely published, perhaps because test providers and distributors have limited information about the background knowledge of their customers, the test applicants, and the environments in which the purchasers want to use the tests.

Testers need to be aware of any known interferences from reagents, chemicals, and other methods of adulteration or manipulation that could influence the results and interpretation. Procedures need to be adopted within a protocol framework, which should ensure that specimens remain tamper-free. If required, the type of testing chosen should enable the tester to detect manipulation and adulteration of the sample by the donor.

According to accreditation principles, all analyses must be subject to quality control and quality assurance, which should encompass a quality system that includes effective training, record keeping, and review. As costs come into play, a legally defensible approach and the recording of data need to be considered, and insufficient evidence for or against specimen stability as a justification for the testing location must be taken into account.

Although generally reliable in comparison to automated screening methods for drugs of abuse, immunochromatographic devices do not have sufficient specificity to be used for legal or forensic applications. Results may be subject to legal challenge unless positive results are confirmed by a generally accepted additional method.

LFA for drugs of abuse may, under some circumstances, provide limited but adequate information for clinical purposes or intervention. In case of doubt, when a definitive penal or legal action is to be feared, laboratory confirmation is unavoidable. Screening using LFA can only be effective when issues of quality and data recording are adequately addressed.

When used by trained laboratory personnel, currently available immunochromatographic test strip devices for drug screening in urine may produce results comparable to liquid-phase and machine-based laboratory screening methods, according to some evidence.

When used by trained non-laboratory personnel, as is becoming more and more common, outcome and results are much poorer. Increases in testing by experienced laymen run into the risk of a serious loss of quality in drugs of abuse testing.

23.2.2.12.2 Lateral flow tests and alternative matrices

For LFA, urine is presently the best established matrix. Cut-off thresholds, interferences, and interactions have been established and studied more in urine than in other matrices. If alternate matrices are to be used for LFA, the antibodies and cut-offs must be optimized with respect to the most abundant parent drug or metabolite in that matrix. [One example is the possible presence of tetrahydrocannabinol (THC) in saliva or oral fluid resulting from contamination by cannabis smoke, but without the presence of original substance or metabolites, which antibodies for urine should target.]

Unsatisfactory results for certain highly serum protein bound drugs, especially for THC, benzodiazepines, or opiates detection, are reported using oral fluids for drug screening by LFA. There is a lack of evidence regarding the obvious quantitative and qualitative limitations of oral fluid testing. As routines for sampling at specific sites and different collection methods for oral fluids or sweat definitely exist, procedures need to be standardized.

23.2.2.13 Substance identification and confirmatory tests

In forensic toxicology, one important principle is substance identification including confirmation of qualitative and quantitative results by the application of at least two independent methods or specific detection techniques. Identification of a substance can be achieved applying a step-by-step isolation procedure to the material, including extraction and chromatographic or other separation, which allows safe discrimination of the substance of interest and possible interference. After chromatographic separation and, preferably, calculation of retention indices, substance identification is usually safely accomplished by recording appropriate spectroscopic information, such as mass spectral or infrared spectral data, and their comparison to data of a pure reference substance.

As the suitable combination of two or more quantitative determination methods and an agreement of results is regarded as a useful tool of objective testing, a valid result cannot be obtained by multiple applications of screening or re-screening tests. There is no gain in validity if specimens initially tested for drug screening using immunoassay techniques are forwarded for a drug immunoassay with different specificity in order to eliminate false-positive or false-negative measurements. A specimen found positive for a substance or class of substances in an initial screening test needs to be confirmed by using a method that ensures substance identification, such as chromatographic or other separation techniques in combination with MS, selected four-ion monitoring, or high-resolution MS and MS/MS techniques. The usefulness of UV-spectra for substance identification depends on the specificity of an individual spectrum. A UV spectrum shows less characteristic information of a chemical substance than its IR spectrum (e.g., in the fingerprint range). As the discrimination power in HPLC is less than that in GC [86,87], a careful use of HPLC and more or less nonspecific detection and further confirmation may be required.

23.2.2.14 Accuracy and precision

Both accuracy and precision are terms related to error. Strictly speaking, the extent of inaccuracy and imprecision needs to be determined. Assuming that a set of observations is made under the same testing conditions, the arithmetic mean and the standard deviation will provide the information required. While the standard deviation refers to the precision and the agreement of results of a set of replicate measurements among themselves, accuracy means the closeness of an individual result as well as of the arithmetic mean of several results to the true, expected, or accepted value.

Two classes of error – systematic error and random error – contribute to the total error of a result. While random error reflects fluctuations in the use of the method, which are unpredictable and unavoidable, systematic errors result from a multiplicity of causes due to poor analytical practice or failure in the application of methods and instruments.

Internal quality control measures using control charts as described in the literature [88] is one of the means to identify sources of error. Especially for analyses of large numbers of the same type of samples, and if they cover a sufficient period of time, control charts may provide useful information on accuracy and precision as well as on the occurrence of unexpected analytical trends and a lack of randomness.

23.2.2.15 Internal quality control

Internal quality control (IQC) in a forensic toxicology laboratory includes all activities by which conditions and procedures can be controlled in order to warrant reliable and suitable analytical results. IQC needs to concentrate on the quality of processing. Any successful analytical work is based on using approved or validated methods. IQC is to check whether in casework the individual analytical runs met the performance characteristics as determined by method validation. IQC is also to inform about changes.

In order to provide appropriate analytical preconditions the analyst must be sure that:

- analytical instruments are regularly checked, well maintained, properly calibrated, and adjusted;
- instruments for sample preparation or subsampling are clean and trouble-free;
- refrigerators and freezers for sample storage or reagents are in acceptable condition;
- suitable materials, reagents, and solutions are available and prepared for use;
- appropriate control samples, blank samples, and reference material are run in the same way as the case samples;

- procedures are implemented in order to exclude carryover of the analytes and contamination of an individual testing sample; and
- staffs comply with operation procedures such as subsampling, sample preparation, analytical methods, and documentation.

Regarding IQC, different areas of attention can be distinguished in qualitative and quantitative analysis.

23.2.2.16 Internal quality control of qualitative testing

First, the purpose of forensic chemical analysis is to give an answer to the question of whether there is one or more analytes of interest present in a specimen or a dead human body. Besides the sample being representative, in order to be found, the analyte must be present in a sample in a concentration above the limit of detection of the method applied. That is why any quantitative analytical method is basically a quantitative one.

Second, the concentration of the analytes needs to be determined in order to know at which level they have been present and to draw plausible conclusions regarding their origin, toxicological action, etc. Accordingly, the performance of a qualitative method can be checked by analyzing one or more samples that do not contain the analytes in order to compare the results to a control sample containing the analytes in a concentration at and little above the limit of detection. Additionally, the application of control samples with two different higher concentration levels is recommended.

Control samples containing a highly immunoreactive reference substance are also used for immunoassays (IA) and other related tests. However, IA need to be distinguished from other methods used as qualitative screening tests. IA are indirect methods using a competitive reaction principle which is usually sensitive to more than one substance present if their chemical structures are very closely related. Thus, in a way, IA are specific but not selective methods. IA readings may be subject to possible reaction failure or specific, unspecified cross-reactions and matrix effects. Therefore, especially near the theoretical limit of detection, IA readings usually show relevant imprecision. Consequently, as mentioned, several specific reasons exist as to why cut-offs need to be used (e.g., a 95% confidence limit to avoid a very high rate of false-positive results). In forensic terms and also in general, because of limited specificity, IA cannot be used to identify a single compound and cannot even safely indicate the presence of chemicals structurally related to the reference compound.

As for screening tests based on chromatography, like general unknown analysis, both the analytical instrument's performance of substance detection and the efficiency of the extraction method need to be checked. In GC/MS, usually the limit of detection should refer to the full scan mode and ten most intensive diagnostic ions, respectively, including a signal-to-noise ratio of 3. However, as stated in a review [42], most of the papers referring to chromatographic toxicological analysis and detection limit only contain data on pure substances. Detection limits are only useful if they are measured using spiked biosamples and if the drug itself is the predominant compound in the sample. In many instances, compared to the parent

or cannabinoids in serum or urine without comment may lead to a misunderstanding by analytical laymen, as this result is easily changed by heroin consumption and recent marijuana intoxication, respectively. Nevertheless, the reporting of immunoassay results without any forensic-toxicological or clinical explanation of their meaning is not a rare event.

Casework in forensic analysis more than in clinical toxicology includes appropriate documentation of analytical results. Clearly, documentation must provide the original analytical data. The data need to be stored in such a form, which allows duplication of qualitative and quantitative information and conclusions when an analytical expert is asked to examine the data.

23.2.3 External quality control (EQC)

External quality control has a bi-directional purpose. Laboratories participating in external quality assessment schemes and proficiency testing need to know and also to show whether their methods are sufficiently applicable to detect a wide range of analytes in toxicological analysis. The need for quality assurance and control was not just a quest of the last decades of the twentieth century. More than 150 years ago, the German chemist Remigius Fresenius complained about the lack of guidelines for quality control: "Nevertheless, it seems to me as if in this part of the science still much is left to be carried out, both in respect of the assessment of the most secure methods for poison detection," … (it would be helpful) "if the state authorities would – in form of a standard – stipulate well-tested methods which have been approved for poison detection, and each chemist performing toxicological analysis would be asked to follow these" [96].

In the United States, since the late 1940s, the College of American Pathologists (CAP) has conducted interlaboratory comparisons designed to assess the state of the art in clinical laboratory practice [97,98]. Some reports on proficiency testing in forensic toxicology were published in 1976 and 1977 [99–102]. However, in 1985, the need for adequate external quality control over laboratory performance was dramatized in an article referring to studies conducted in the 1970s on laboratories providing drug screen services to federally funded drug treatment programs [103]. Furthermore, the study had focused attention on the general lack of quality control programs. Consequently, several such testing programs for drugs of abuse in urine were initiated. Programs were conducted by the U.S. Department of Defense, Armed Forces Institute of Pathology (both open proficiency-testing samples), and, starting in 1984, by the College of American Pathologists. Since 1980, a urine toxicology program became available from the American Association of Bioanalysts. State programs as conducted by the states of California, Pennsylvania, and New York are providing blind sample testing and open proficiency testing. Other individual and commercial programs also exist [97]. In a feasibility study on proficiency testing in forensic toxicology [104], it was shown that the methods used by the participants satisfied requirements of accuracy, although they were tested in Europe [105].

23.2.3.1 The relation between clinical drug of abuse testing and forensic or clinical–toxicological proficiency testing

There may be a need to harmonize both the criteria of successful participation in ring tests and the corresponding substance panels tested. Harmonization, however, may easily be upset by the goal of testing and comparing as many European drug testing laboratories as possible. For comprehensive proficiency testing as needed in forensic toxicology, neither limitation to commercial IA-based drug testing with confirmation analysis nor a European centralized survey of the most frequent drug tests (including the corresponding laboratories) will help warrant an effective screening of abused substances. Regarding national legal requirements, different drug-taking behavior in individual countries, and the subsequent necessity for national quality control schemes to exist, it seems advisable not to mix the goals of clinical chemical, workplace, or forensic toxicological proficiency testing.

In comparing the purpose of proficiency tests, it must be stated that there are relevant differences between forensic and clinical testing. Diagnostically sensitive, rapid, and cost-effective drug testing in clinical chemistry differs from the need for substance identification in forensic testing and for legal issues. As the detection window is longer than in blood and the concentrations are higher, clinical testing by using commercially available IA test kits for frequently abused drugs is preferably performed in urine. Other clinical proficiency testing is focused on therapeutic drug monitoring (TDM) in serum. Clinical chemical proficiency testing therefore demands a selection of analytical parameters that need to be performed rapidly and often with respect to a definite concentration range. However, ring tests restricted to a set of drugs of abuse will focus more on the performance of commonly used specific tests than on general screening performance. Analytical results often need to be interpreted in order to make it possible for public authorities to make decisions and to judge if further expert knowledge is required. For this reason, forensic proficiency testing needs to focus on laboratories being able to provide reliable screening and identification as well as suitably interpreted results.

As an example, the toxic effects of benzodiazepines, their different potencies and dosages, their concentration in blood, the need for substance identification, and the qualitative and quantitative testing of various benzodiazepines, as well as a lack of suitable external quality control in both forensic and clinical toxicology [106] have been obvious for a long time. Only recently have further details of the problem been characterized [107,108].

Forensic testing has focused on the detection and identification of poisons and intoxicants whose presence was previously unknown. In addition to urinalysis, in forensic toxicological analysis quantitative substance determination in blood (and also other tissues) is frequently performed. Forensic toxicological PT schemes should conduct the corresponding ring tests. Since IA is used for screening of substance groups only, but not for substance identification, in forensic toxicological PT no IA results need to be reported.

In addition, forensic PT schemes should organize the collaborative studies necessary in order to:

- investigate performance criteria for various methods and standards fitting with the requirements as needed in qualitative and quantitative analysis [109] for objective testing including trace determination; and
- test the use of reference material.

Besides the question of which method used by laboratories is best for identifying a particular chemical compound, both forensic toxicological analysis and proficiency testing must refer to frequently and also rarely observed analytes. The panel of substance needs to be open to newly discovered illicit and licit drugs which must first be qualitatively detected and, if necessary, quantitatively determined to pass the ring test. Additionally, the identification of metabolites rather than of parent compounds may cause identification difficulties.

The presence of a new drug, such as *N*-methyl-3,4-methylendioxyphenyl-2-butanamine (MBDB,) in a urine ring test has caused disastrous analytical results including numerous false-positive "identifications" of MBDB when amphetamines were present [110].

In a clinical context, the question posed is generally not whether the subject has abused drugs at some unspecified time in the past, but more typically whether a non-prescribed compound was taken within the past 24–72 h? Therefore, diagnostic sensitivity and specificity of available analytical techniques are factors that need to be taken into account in a setting of the desirable standards of the performance of illicit and licit drug assays.

In regard to both these definitions, the techniques used for the detection of drug abuse in urine, external quality assessment in the UK has shown in 1991 [111] that as far as clinical understanding of sensitivity is concerned, the use of confirmation methods such as chromatography had inadequate diagnostic sensitivity. In a report on the same scheme, gas chromatography with mass spectroscopy as used in the participating laboratories, was reported to be significantly less sensitive than other techniques for the detection of 0.5 mg/L of benzoylecgonine (71%) and 1.5 mg/L of morphine (88%). Interestingly, in contrast, high-performance liquid chromatography turned out to be the most (100%) sensitive for amphetamine. Regarding diagnostic sensitivity, commercially available immunoassays for drugs of abuse performed well when operating above their specified cut-off concentrations. The authors state that threshold concentrations are therefore greater than those of the field of employee testing [112], though the consequences of analytical errors are no less serious in the damage they may cause to the patient. The data demonstrated that the laboratories did not always achieve the goals of easily confirming the presence of an analyte by the use of two techniques based on different physical chemical principles.

23.2.3.2 Choice of proficiency-testing schemes

As laboratories may change to the most convenient scheme, proficiency testing should cover the services offered. However, due to increasing competition and for economic reasons, various laboratories may offer their toxicological analytical services, primarily designed for clinical purposes, to non-professional customers with the belief that the results will be sufficient for forensic evidence. Laboratories without suitable (expensive) confirmation facilities probably frequently report drug abuse positives exclusively based on the use of IA techniques. Consequently, IA confirmation is performed by more qualified laboratories and on request only.

Thus, voluntary choice of the scheme and participation of laboratories with respect to favorable performance criteria and well-tailored testing schemes may create problems regarding quality and use of results and forensic expertise. As a distinction is drawn between routine clinical drug abuse testing and forensic-toxicological analytical work, the same discrimination should also be made for proficiency testing.

In a forensic as well as in a clinical toxicological context, a clear distinction needs to be made between educational ring tests helping laboratories to improve their methods and other tests aimed at laboratory certification. As mentioned above, approved case-based proficiency testing for clinical toxicology laboratories was successfully introduced in the Netherlands. In this scheme, reasonable toxicological testing was supported by a case history.

23.2.3.3 GTFCh's forensic toxicological proficiency-testing scheme

In the Federal Republic of Germany, method comparison and toxicological ring tests for various drugs were introduced in 1969 [113]. Consequently, ring tests were performed in 1970–1975 [114–118] and continued after a gap of several years [119]. Parallel to those, ring tests for the detection of both lead and thallium were organized [120,121]. Since 1985, ten ring tests were organized by the committee "Qualitätskontrolle" of the GTFCh, which referred to opiates, splitting of their glucuronides, benzodiazepines, and hypnotics in solution, urine, and serum [122].

Since 1982, ring tests for blood alcohol determination have been introduced to the German Society of Clinical Chemistry and Biochemistry (DGKC). As no other PT scheme was available, forensic laboratories also participated in these ring tests. Unfortunately, they did not conform to forensic requirements. According to (recently updated) guidelines for forensic blood alcohol determination [123], the mean of four independent values must be reported by forensic laboratories to which the blood samples are submitted. Both the values of two separate determinations using, e.g., head-space gas chromatography and two separate determinations using enzymatic ADH-method need to be within a precision range of 0.1 g/kg below a BAC of 1 g/kg and within ±5% of the mean (above 1 g/kg). As the scheme is designed for clinical alcohol determination, it refers to the accuracy of a single value but not to the precision of four values as required in forensic blood alcohol determination. As a consequence, based on internationally harmonized protocol [124], the GTFCh organized a forensic blood alcohol proficiency-testing program in 1995 and, at the same time, qualitative and quantitative proficiency testing of drugs of abuse in blood and serum.

The introduction of proficiency testing of drugs of abuse in blood and serum became necessary because of a new paragraph in the German Road Traffic Act (§24a II StVG), at present referring to illicit drugs and driving. The GTFCh proficiency-testing program is organized according to the above reflections and offers ring tests in the following areas:

- illicit drugs in blood (serum);
- benzodiazepines in blood (serum);
- ethanol in serum and whole blood;
- substance identification in urine;
- markers of alcoholism; and
- qualitative general unknown analysis (in combination with a brief case history).

Before the program was started, collaborative studies on THC trace determinations were performed [125,126] followed by the introduction of quantitative ring tests [127]. Supported by the German Ministry of Transportation and the Bundesanstalt für Straßenwesen (BASt), the program has given rise to a significant increase of performance with respect to interlaboratory comparability of quantitative results which are just above the detection limit. In Fig. 23.3, the results of the first round-robin testing and the 10th testing round are compared. Since then, a regularly used external analytical quality control scheme has been established, as scientifically conducted by the GTFCh and technically achieved by Arvecon GmbH [128]. Figure 23.4 shows the year 2005 performance of GCMS determination for cannabinoids in serum using deuterated standards of 64 participating laboratories seen in the ring tests of the GTFCh.

The data show that:

- the continued use in daily casework and proficiency testing led to a change of the methods used in the beginning;
- the variability of the concentration measurements has improved; and
- using GC/MS and deuterated internal standards, the coefficients of variation can nowadays be less than in 1982 [129].

This underlines both the necessity and the effectiveness of proficiency testing. Clearly such proficiency testing must refer to the threshold limit value concept as preferred by the German legislation and jurisdiction and to acceptable practical

---

Fig. 23.3. The comparison of (A) the 1st ring test (1995) and (B) the 11th ring test of GTFCh-EQA-scheme of Δ-9-tetrahydrocannabinol in serum (1998) shows impressive improvement of the participating laboratories due to the introduction of external quality control. 1st ring test: reference labs mean $= 3.77 \pm 1.04$ (SD), coefficient of variation $= 0.27$, participants mean $= 9.21 \mu g$ THC/L serum ± 9.13 (SD). 11th ring test: *consensus mean* $= 2.6 \mu g$ THC/L serum ± 0.5 (SD), coefficient of variation $= 0.19$. The upper bar charts show the values determined; the lower bar charts refer to the z-score ($=$ (measured value $-$ mean)/target standard deviation). A z-score of two or less means an acceptable or better result. The preliminary "educational" testing did show not only some false high IA results due to the metabolite concentration present, but also some questionable GC/MS results. (No use of a standard method requested in both ring tests.)

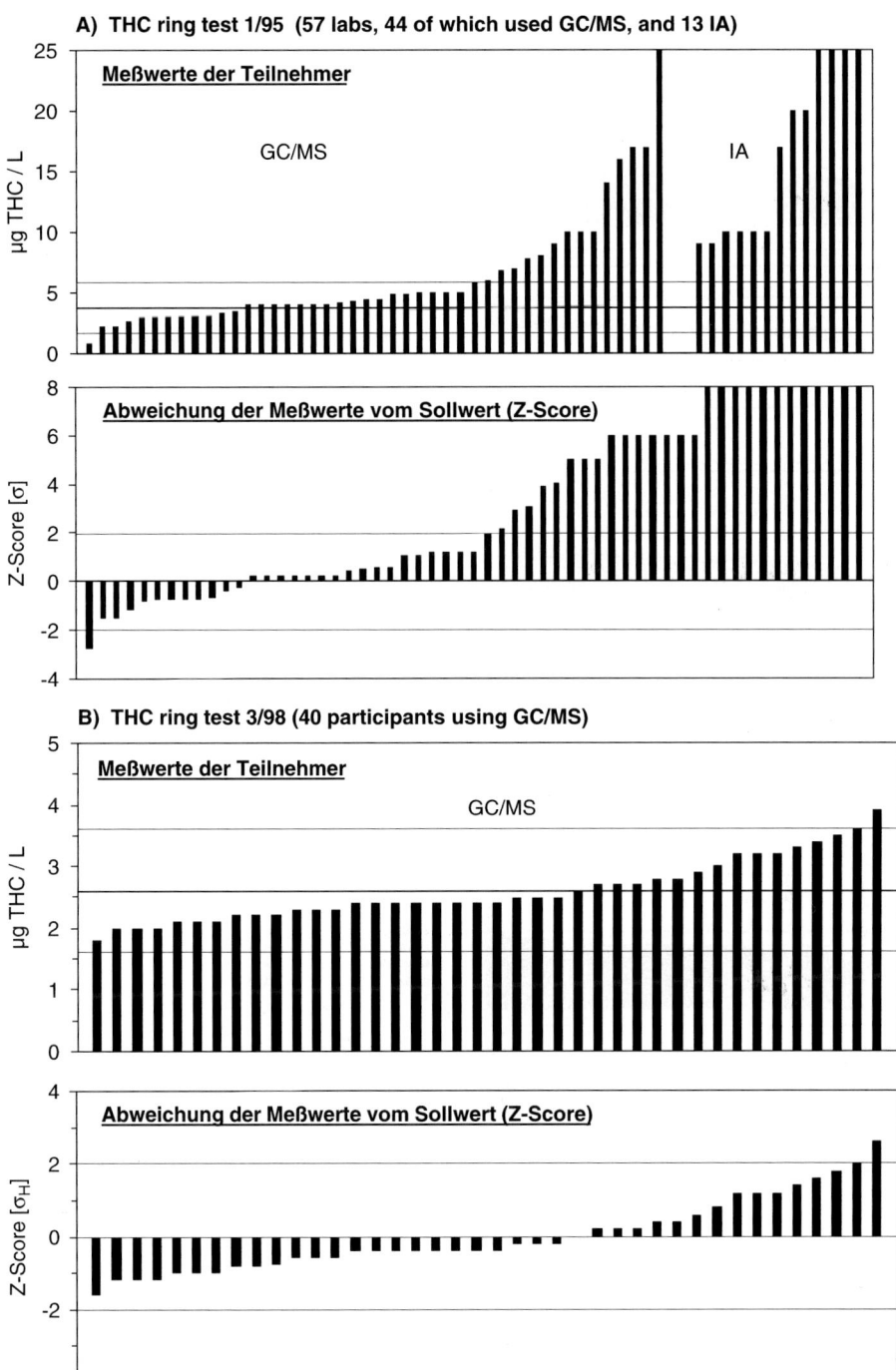

A) THC ring test 1/95 (57 labs, 44 of which used GC/MS, and 13 IA)

Meßwerte der Teilnehmer

GC/MS

IA

Abweichung der Meßwerte vom Sollwert (Z-Score)

B) THC ring test 3/98 (40 participants using GC/MS)

Meßwerte der Teilnehmer

GC/MS

Abweichung der Meßwerte vom Sollwert (Z-Score)

References pp. 814–819

GTFCh - BTMF 3/05   **Lab code: 1328**

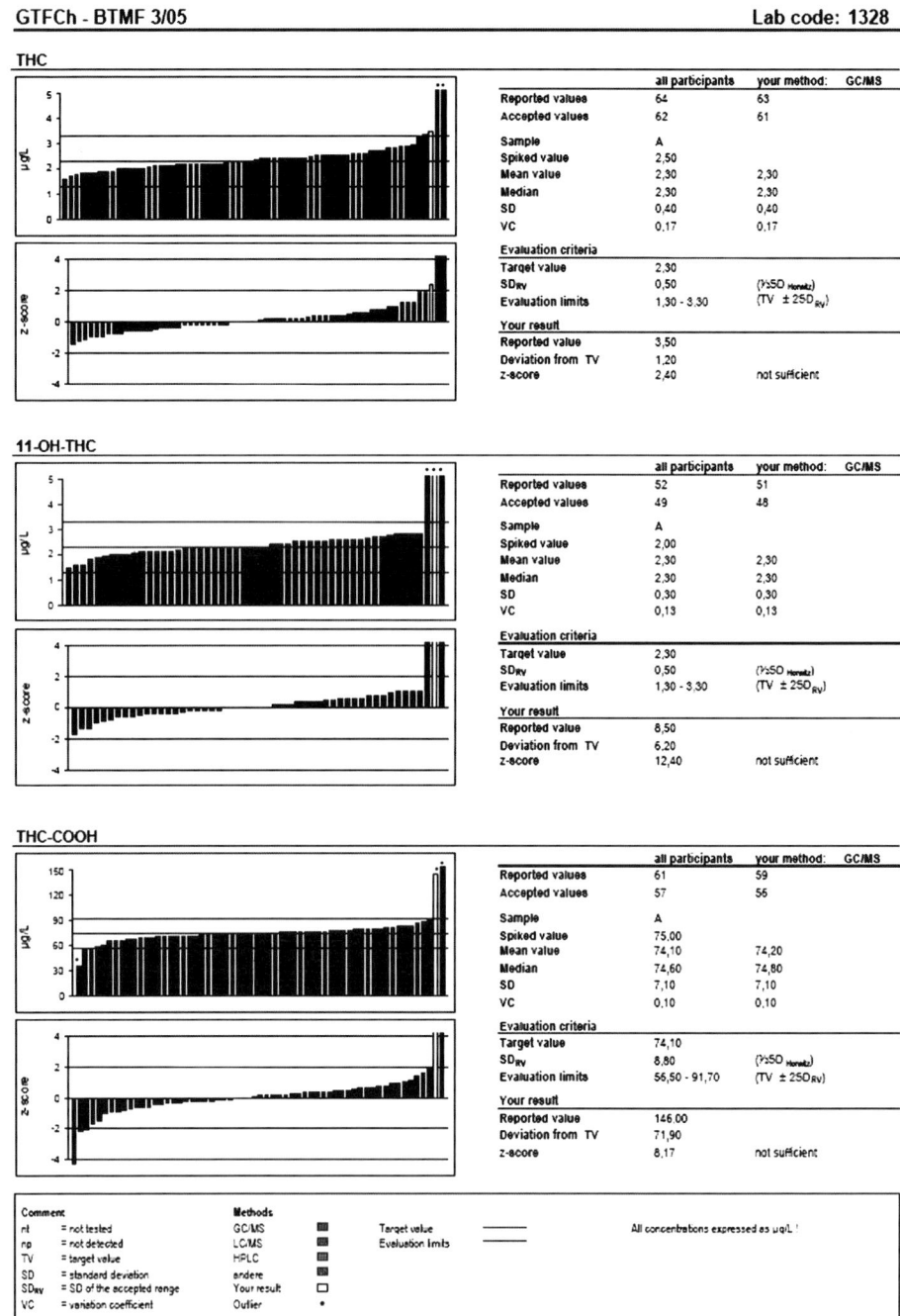

Fig. 23.4. Ring test results and performance of MS determination of THC and cannabinoids in serum in the year 2005.

identification criteria (Appendix to the guidelines of the GTFCh, in preparation [130]).

23.3 QUALITY OF THE OUTCOME

For a toxicology laboratory, outcome is measured not only by successful casework and testifying in lawsuits to discover the scientific truth, but also by progress in scientific research and development of the scientific discipline. Scientific outcome of an organization has substantial relevance with respect to the scientific reputation of an institution. Cooperation between institutions, their influence on legal procedures, or even legislation on a national level need to be considered. Besides a scientist's efforts, the working processes chosen, and research areas, scientific outcomes also depends on policies and on given internal and external structures of a laboratory's organization. The evaluation of the outcome may become difficult due to the elements of quality which can be a mixture of various objective but also some subjective elements. More simply, outcome can be regarded in terms of business results.

23.3.1 Toxicological analytical services and business results

Business results have an important influence on the satisfaction of the customer. The quality of a product is usually judged by the customer in terms of its performance and cost. In order to have proper business results, short-term and long-term business strategies need to be developed, which are not within the scope of this article. Nevertheless, business strategies need to involve cost-effective organizing of both analytical work and quality. Quality management brings about visible and invisible costs. Therefore, administrative instruments will be needed by which quality improvement can be observed and the costs of quality deficits and the cost-effectiveness of their prevention can be determined and lowered to a minimum.

If quality-oriented formal procedures are installed in small laboratories in order to apply for future accreditation, additional personnel and increased paperwork and possibly the help of a consulting company need to be considered. Accreditation costs are relatively high; however, working without being accredited may become a major disadvantage with respect to business competition with potentially equivalent providers. Therefore, economical accreditation and quality costs are important for an acceptable business result.

Besides internal efforts to achieve objective analytical quality, it must be remembered that customers compare the service of different providers. Regarding a forensic analytical laboratory in terms of a service company, not only the internal procedures must run satisfactorily. Additional factors play an important role about which individual staff members may be unaware. The nature of individual coworkers often determines whether they can be motivated, are interested in participating in the goals of their organization, and properly understand the decisions of the management. Some key factors are considered essential with respect to the satisfaction of the customer (Table 23.6 and 23.7).

TABLE 23.6

CUT-OFF CONCENTRATIONS AS SPECIFIED IN THE MANDATORY GUIDELINES FOR FEDERAL WORKPLACE DRUG TESTING PROGRAMS

Agency: Substance Abuse and Mental Health Services Administration, Date: November 1, 2001

The following cut-off concentrations are used by certified laboratories to test urine specimens collected by Federal agencies and by employers regulated by the Department of Transportation

Initial test cut-off concentration (ng/mL)

Marijuana metabolites 50
Cocaine metabolites 300
Opiate metabolites 2000

Phencyclidine 25
Amphetamines 1000

Confirmatory test cut-off concentration (ng/mL)

Marijuana metabolite[a] 15
Cocaine metabolite[b] 150
Opiates
  Morphine 2000
  Codeine 2000
  6-Acetylmorphine[d] 10
Phencyclidine 25
Amphetamines
  Amphetamine 500
  Methamphetamine[c] 500

[a]Delta-9-tetrahydrocannabinol-9-carboxylic acid.
[b]Benzoylecgonine.
[c]Specimen must also contain amphetamine at a concentration and gt; = 200 ng/mL.
[d]Test for 6-AM when morphine concentration exceeds 2000 ng/mL.

TABLE 23.7

IMPORTANT KEY FACTORS IN RESPECT OF THE CUSTOMER'S SATISFACTION

Key Factor	Example
Access	Easy contact by phone or fax is possible
Comprehension	Individual awareness
Communication	Ready for and capable of correct information
Competence	Technical ability shown during contact
Politeness	Respectfulness and kindness of the staff
Credibility	Reputation of a lab, personal attitude of coworkers
Responsiveness	Rapid reaction of the staff in time
Reliability	Accuracy and precision of measurements and values
Keeping of time-limits, meeting of deadlines	
Security	Physical and financial security
Material environment	Facilities, equipment, technical aids

23.3.2 Scientific outcome in forensic toxicology

23.3.2.1 Scientific work

Practicing forensic toxicology and performing toxicological analysis is in fact not possible without a strong scientific background. Sensitivity to new views and questions arising during casework must be developed. Forensic and clinical casework are both sources and destinations of new solutions tried for in scientific work. Bringing in new approaches and publishing specific casework is a must for the FTs, who need to share their knowledge. That is why a research vision and goal-finding as well as planning and performing relevant studies oriented towards the future needs of forensic toxicology are most important concerning qualified outcome. Like other researchers in natural sciences, FT must focus on important abilities such as:

- knowledge of the state of the art;
- inquiring and knowing scientifically relevant needs of research in his or her subject;
- developing a working hypothesis;
- assessing its feasibility;
- proper and timely applications for grants;
- developing new analytical methods or tools (of poison or drug detection) and their application in order to improve (forensic-toxicological) data acquisition; and
- applying validated and approved analytical methods in order to solve recognized (forensic-toxicological interpretation) problems.

Purchasing of expensive analytical tools usually depends on the scientific evaluation of previous work and on peer-reviewed application for grants. Especially when a scientist applies for grants, the fulfillment of both formal and intrinsic scientific quality aspects is indispensable. When independent experts of scientific commissions finally decide about research grants, planned projects will usually be rated by a peer group of leading scientists regarding:

- scientific relevance;
- innovation;
- originality;
- feasibility;
- scientific qualification of the applicant(s); and
- previous grants.
- Criteria for a clear-cut design the study are:
- the objective(s) of the study are precisely stated;
- actual state of research in relation to the study;
- state of the previous investigations carried out;
- all working steps necessary;
- details of the way to achieve them;
- available staff, material, and technical (analytical) devices;
- allowance to use the basic equipment of the institution;
- realistic requirement of additional staff materials and devices;
- realistic schedule; and
- acceptable relation between objectives, costs, and outcome.

Forensic toxicology units are often associated with institutes of forensic medicine. In concurrence with other scientific subjects, scientific activities in forensic toxicology are evaluated not only in regard to casework, but also in relation to their scientific impact. Researchers, suspicious that scientific work in forensic toxicology is not financially supported as much as other more currently appreciated scientific subjects, need to remember that the instruments for scientific evaluation are at least congruent with the criteria of good quality. To be supported with equal priority and because of the undisputed need for proper forensic-toxicological investigations, not only the scientific community but also society in general needs to be well informed. With respect to public health and jurisdiction, society should strongly be convinced about our scientific progress and the importance of the goals of forensic toxicology.

23.3.2.2 Impact of scientific work within the subject and in relation to other subjects

Along with an increasing reduction of national research funds and scientists competing harder for positions, nowadays finding the means to move forward becomes increasingly difficult. One of the criteria by which the work of scientists eventually can be measured may be found using the "Science Citation Index" (SCI). It has been brought up by one of the pioneers of bibliometric methods for evaluation research performance [131]. Scrutinizing the reference lists for all articles published in thousands of scientific journals led to a huge database. By storing and indexing the bibliographical information it became possible to calculate, document, and compare:

1. the number of times the work of a certain first author of an article is cited; and
2. the number of times a certain journal is cited.

By monitoring the productivity of individual scientists (articles per author) over a period of time it is also possible to determine the bibliographical impact of:

1. a particular article (citations per paper); and
2. a first author (citations per author).

As a measure of frequency with which the average journal article has been cited, the corresponding Impact Factor (IF) can be regarded as a reflection of a journal's importance and influence within the scientific community. As a consequence, it seems possible that important work may be overlooked if it is published in a less favored journal. As other modes of scientific impact may exist, this specific kind of interaction in the literature must not be identified with the true influence of the work of a scientist.

According to SCI journal citation reports, the IF of a journal is determined by dividing the number of all current citations of source items published in the journal during the previous two years by the number of articles that the journal publishes in these two years. When publishing articles that become highly cited, a journal will get a high IF. Publication of articles rarely or never cited will result in a low IF. In fact, however, most scientific articles are cited only once (possibly self-citations) or are not cited at all.

Some scientists complained that in spite of some obvious deficiency, the IF has been unintelligibly picked as a scientometric measure at German universities [132]. Others drew a negative picture regarding the IF and exposed problems of the IF including its role in possibly marking a gloomy future for young scientists in less broad subjects

[133]. As the IF is extracted from journal citation reports, the base of calculation was thought to be inappropriate.

The uncritical uses of citation analysis to compare the importance of scientific journals or to judge research significance have been frowned upon by those who prefer traditional peer assessment [134]. The use and misuse of journal impact factors for assessing the work of individual scientists and their professional standing and esteem among their peers has been much debated in popular journals and weekly magazines [135–137]. One point of criticism was that it is the editorial board of a private Institute of Scientific Information that, without entering into criteria, decides on journals to be taken as source journals, the citations of which are used for calculation purposes [132]. However, it seems that most peer-reviewed journals are included in the database. Other problems seen are outlined below:

- Possible frequent self-citation or alleged citation circles in source journals as well as shallow or limited literature search might have decisive influence on the data used for the index.
- Favouring of frequent negative citations.
- Non-English languages and non-Latin alphabets having relevant disadvantage, and consequently, possibly rapid scientific development in such countries may not be recognized in Anglo-American countries.
- Articles aiming at professional education are necessarily published in a national language; however, they will not be cited as those to whom they are addressed will not publish in their part.
- Specific subjects like forensic toxicology will not reach IF in their journals as high as those of the basic and substantial scientific subjects that lead the hit list.
- The time during which citations are registered is two years. Thus, short half-lives of information and predominantly the most favorite and broad publication organs are supported. (It was complained that in 1995 out of 10 of the mostly cited relevant medical journals, only two were among those with the highest IF. In contrast, the impact winner had a very low number of citations.)
- The quality of a reference list may be a problem of the IF. (How many authors do really read the original publications and how many of them read results of literature searches only or copy the reference lists of others to save time and efforts? How many of them bring in a complete reference list after a comprehensive studying of the literature?)

In the final analysis [138], impact was shown to reflect the ability of journals and editors to attract the best papers available. However, the IF gives limited information about the relevance of a journal in its own discipline. Forensic science and forensic toxicology are applied sciences. In general, interest in forensic toxicology is attracted only by a relatively small number of scientists. Compared to basic research subjects, there are only a few corresponding forensic toxicology departments at European universities with appropriate permanent research positions including undergraduate and postgraduate education. This results in fewer papers and lower frequency of citation. Furthermore, forensic scientists are usually employed at government laboratories that have different qualification rules and, to some extent, these organizations tend to discourage open publication research. This is another reason why forensic science journals have low impact factors [139]. The present climate of performance evaluation, quality assessment, and competitiveness will hopefully help to make scientists in government and university laboratories more inclined not only to publish their work but also to publish in peer-reviewed journals. However, prolific authorship obviously does not carry as much weight for recognition and esteem in all fields of forensic science as it does in academic medicine, where with similar effort leading academic positions can be reached.

In order to judge the relative importance of different scientific journals, misleading information will be obtained when simply comparing the size of their IF. As shown in [140], by information also gleaned from SCI, a scope-adjusted impact factor can be obtained by dividing the conventional IF for a particular journal by the number of citing journals and multiplying by 1000. A calculation including original and new ranking of the scope-adjusted impact factor of 11 journals of interest to forensic scientists is given in [139].

Some heads of departments encouraged young scientists to publish their work in impact journals above a threshold of 2.0 in order to virtually boost its impact. Interestingly, none of the forensic journals met this demand [141]. However, some authors may have experienced that the IF of forensic science journals may change unexpectedly according to publication and consequent citation fluctuations [139,141].

For assessing the track records and the quality of published work, instead of counting the sum and average of IF of journals in which an author has published, the number of citations to individual articles should be investigated. This exercise, however, would entail much more effort [142,143]. In addition to this, for adjusting IF and citation counts, there will be the problematical handling of names and numbers of coauthors contributing to articles. Perhaps the ordering of the names of multiauthor articles should be considered as well [144,145].

Whatever the pros and cons of citation counting may be, there is no need to discredit an objective method of evaluating the scientific literature that can be one important pillar of scientific quality assessment [139,141,146]. The use of the IF in this respect presumes the knowledge of its rules, the limit, and the limitations. Uncritically using the IF like scientific currency should not be accepted, as it would be a non-scientific onset leading to ignoring of the competence at universities in favour of a pseudo-objective measure. All in all, a standardization is needed, particularly within a scientific subject, considering peer-reviewed publications in national journals as well. Some research results may need a long time until their significance is fully understood. As the interest in a particular work may increase if its usefulness has been recognized from a changing point of view, the relevance of scientific work may be disregarded at first. Therefore, the IF should not be used exclusively as a unique parameter indicating the research activities of scientists, their institution, and their quality. In contrast, the SCI may by helpful as a contribution to any careful individual technical examination.

Besides being always justified, evaluation of forensic toxicology facilities needs to be rated from a distinguished point of view when compared to other medical subjects. Like legal medicine, forensic toxicology has many elements of an interdisciplinary subject. Forensic toxicology is related to a sequence of other subjects such as forensic pathology, clinical and emergency toxicology, analytical chemistry, pharmacology, clinical chemistry, clinical pharmacology, pharmaceutical science, environmental chemistry, and also criminal and behavioral science. Therefore, the impact of forensic toxicological scientific work within these subjects needs to be increased and weighted. The effort of scientists to strengthen their positions needs to focus on sound projects attracting the attention of the scientific community.

23.3.2.3 The impact of forensic toxicology on society

The improvement of analytical instruments and laboratory methods as a means of evidence in cases of violation or in criminal cases and procedures has important relevance in a society. In this context, forensic toxicology services contribute to public health and to risk prevention with respect to crime fighting, crime investigation, traffic accidents, drug addiction, etc. Some national regulations, specific problems of legislation, jurisdiction, and police activities may be based on the availability and proper use of valid chemical-analytical testing methods. Costs and benefit always need to be considered.

In Germany, for example, according to the code of criminal procedures, drawing and analyzing a blood sample in addition to urine analysis is generally accepted. Consequently, results of blood analyses serve as a regular means of evidence in criminal procedures, e.g., concerning the various time-dependent states of drunkenness and inebriation evoked by drugs, as are euphoria, loss of inhibition and self-control, intoxication, withdrawal, symptoms, dysphoria, and aggressiveness.

Assuming a definite blood alcohol concentration–effect relationship and road accident risk assessment, in 1990 the German Supreme Court proclaimed (lowered) the scientifically based threshold limit value of the BAC as 1.1 (g/kg). As absolute driving inability is assumed [147] at or above this concentration, drivers are penalized including a time-limited withdrawal of their driver's license. In contrast, driving and BAC values above 0.5 g/kg is a violation which entails fines, and – above 0.8 g/kg – a limited driving prohibition. Following requests from the police, in spite of obvious biological disagreement between blood alcohol and breath alcohol values [148], in 1998 the German government ignored these differences and introduced a regulation in which the measurements of blood alcohol or breath alcohol concentrations entails identical legal consequences [149]. In spite of this, the regulation was consequently adopted in the jurisdiction. In 2007, a "zero" alcohol tolerance regulation was introduced for young drivers. Accordingly, a zero BAC of 0.2 ‰ has been proposed including a basic level of 0.1‰ and a safety margin of 100%. As can be seen from preliminary ring test results (Fig. 23.5), this safety margin can be regarded sufficient: When a spike value of 0.1 g ethanol/liter serum was distributed, 94 participating laboratories clearly confirmed the spike value with both the GC-head space method and the enzymatic ADH-method (consensus value of 0.1 g/L), and did not report any level above 0.15 g ethanol/liter serum.

For illicit drugs, up to now, no concentration–effect relationship comparable to ethanol can be deduced. Satisfactory toxicological models aiming at such a

Probe A - GC

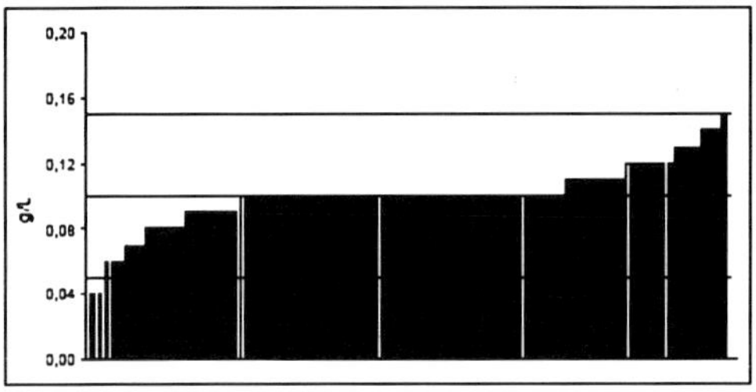

Anzahl Ergebnisse		
Gemeldete Werte	94	
Akzeptierte Werte	94	
Gesamtergebnisse	Teilnehmer	
Mittelwert	0,100	
Median	0,100	
SD	0,019	
VC	0,187	
Bewertungskriterien		
Zielwert	0,100	
SD $_{RV}$	0,025	
VC $_{RV}$	0,250	
Bewertungsgrenzen	0,050 - 0,150	
Ihre Ergebnisse	1	2
Messwerte	0,100	0,120
Abweichung vom MW	0,000	0,020
z-score	0,000	0,800
Präzision (VB ≤ 0,1)	ok	
Richtigkeit (|z-score| ≤ 2)	ok	

Probe A - ADH

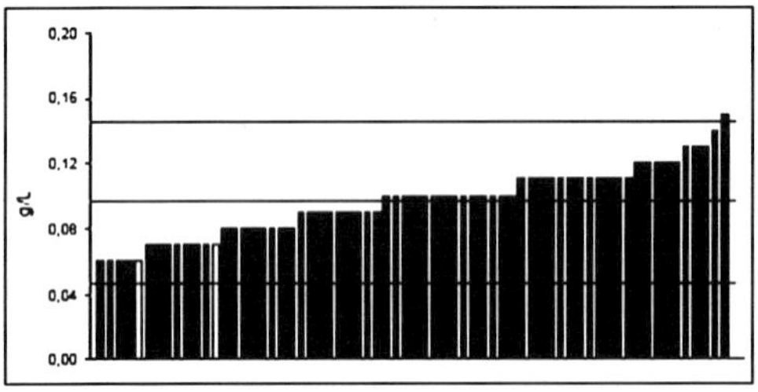

Anzahl Ergebnisse		
Gemeldete Werte	66	
Akzeptierte Werte	66	
Gesamtergebnisse	Teilnehmer	
Mittelwert	0,096	
Median	0,100	
SD	0,021	
VC	0,216	
Bewertungskriterien		
Zielwert	0,096	
SD $_{RV}$	0,025	
VC $_{RV}$	0,260	
Bewertungsgrenzen	0,046 - 0,146	
Ihre Ergebnisse	1	2
Abweichung vom MW	0,070	0,060
z-score	-0.026	-0.036
z-score	-1.040	-1.440
Präzision (VB ≤ 0,1)	ok	
Richtigkeit (|z-score| ≤ 2)	ok	

Fig. 23.5. Results of the 4th blood alcohol ring test in 2006 using serum spiked with 0.1 g ethanol/liter serum.

relationship could not be developed. Consequently, no similar threshold values could be determined. Alternatively, for blood ethanol concentration in drivers above 0.8 g/kg, the mere presence of illicit drugs in the blood of drivers is now a violation entailing fines and limited driving prohibition; however, no immediate withdrawal of the driving license takes place. Analytical Threshold Limit Values were proposed by relevant scientific societies, reflecting both the analytical identification of the parent compounds and acute risks caused by the consumption of illicit drugs [79]. An appropriate metabolite, e.g., benzoylecgonine (cocaine) or morphine (heroin), needs to be determined if the parent compound is not stable enough in an unpreserved blood sample. As the determination of an inactive metabolite like benzoylecgonine would not be consistent with intentions of the §24a StVG, an Analytical Threshold Limit Value considerably higher than an acceptable limit of detection was proposed. With respect to the biological plausibility of the analytical result, however, relevant metabolites generally need to be determined in addition to the parent or primary compound, respectively. As threshold limit values may have some cut-off function and are not identical with analytical detection limits, it remains to be seen if they will be accepted in jurisdiction. Therefore, analytical methods for the determination of alcohol, illicit drugs, as well as markers of alcoholism in blood (serum) need to be validated according to specific standards [150] to meet scientific and legal requirements.

Improving evidence and the interpretation of results presumes that the human metabolic, toxicokinetic, and toxicodynamic background must be investigated and sometimes reinvestigated for new and old substances. Forensic toxicological laboratories help to prepare both scientifically based legislation and jurisdiction by successful implementation of new substance identification techniques and investigation of the metabolic fate of toxic chemicals and intoxicants in humans under various consuming conditions, drugging, and poisoning. This has considerable relevance with respect to public health.

23.4 CONCLUSIONS

Forensic toxicology is a science in which – like other sciences – providing quality is more than developing or adopting procedures and analytical methods that merely need to be performed according to international standards and guidelines or technical requirements of objective testing. Accreditation is important. It requires a great deal of energy and expense but does not, however, guarantee all of the quality levels needed. The relation of costs to benefits of laboratory accreditation needs to be carefully considered. Nevertheless, the conformity of a laboratory with accepted quality and management structures will be required in future.

Forensic toxicologists take care of the science of poison detection in human organs and body fluids. Our scientific research means acquiring knowledge to develop and improve forensic toxicological tools and better evidence. What scientists think today will be the reality of the future. Therefore, a careful development and selection of new appropriate methods and instrumentation as well as their suitable application is one major crucial point regarding the claims and laboratory performance in forensic toxicology. Two kinds of approaches, methods, and corresponding equipment need to be further developed, validated, if possible, and maintained regarding open tasks to be carried out:

1. Approved general unknown analysis and selective screening methods.
2. Various quantitative and confirmation methods need to be applied according to specific requests and in cases in which specific qualitative results were previously obtained by general unknown analysis.

There are a practically unlimited number of poisons that may be present in individual cases and under particular circumstances. Therefore, forensic toxicology is a scientific discipline in which permanent efforts to complete and improve the methods of poison detection show its close relation to raising quality. How can the analyst effectively test the power of the methods used in his or her laboratory? The most important means to increase the degree of certainty of analyses aimed at the detection of previously unknown substances and their quantitative determination consist of a

References pp. 814–819

combination of educational ring tests and true proficiency testing. External quality control and ring tests in forensic toxicology need specific guidance because of the multiplicity of the possible analytes and matrices presenting considerable difficulty compared to other areas of testing. Clinical chemical testing and external quality control prefers a selection of analytical parameters which need to be performed rapidly and often with respect to a definite concentration range. Ring tests restricted to a set of drugs of abuse will focus more on the performance of commonly used specific tests than on general substance screening performance.

Differences between post-mortem sampling and sample taking from living persons need to be considered as well as the corresponding interpretation of results. Forensic toxicological analytical results therefore need to be carefully interpreted to make it possible for public authorities to make decisions and investigate whether further expert knowledge is needed. The difficult, often time-consuming scientific analytical tasks in forensic toxicology and casework should not be separated from drug of abuse testing with respect to large-scale testing of a few analytes and some kind of outsourcing for economic reasons. Improving evidence and the interpretation of results presumes that human metabolic, toxicokinetic, and toxicodynamic background must be investigated and sometimes reinvestigated for new and old substances in both case samples and scientific experiments.

Forensic toxicological laboratories and qualified research help to prepare both scientifically based legislation and jurisdiction by the successful implementation of new substance identification techniques and investigation of the metabolic fate of toxic chemicals and intoxicants in humans under various circumstances of drug abuse and poisoning. This has considerable social importance. By being financially supported with appropriate priority and because of the undisputed need of proper forensic toxicological investigations, not only the scientific community but all of society need to be permanently informed about the importance of scientific goals and progress in forensic toxicology.

23.5 REFERENCES

1   U.S. Federal Register (December 1978), "Nonclinical Laboratory Studies; Good Laboratory Practice Regulations," p. 60013 (21 CFR Part 58).
2   U.S. Federal Register (November 29, 1983), "Toxic Substances Control, Good Laboratory Practice Standards," p. 53922 (40 CFR Part 792); and "Pesticide Program, Good Laboratory Practice Standards," p. 53946 (40 CFR Part 160).
3   U.S. Federal Register (December 28, 1987), Federal Insecticide, Fungicide, and Rodenticide Act (FIFRA) and Toxic Substance Control Act (TSCA), Good Laboratory Practice Standards, p. 48920 (40 CFR Parts 160 and 792).
4   International Standards Organisation, Guide 25 (3rd ed. 1990), American National Standards Institute, New York, NY.
5   General requirements for the competence of testing and calibration laboratories ISO/IEC DIS 17025: 2005.
6   International Laboratory Accreditation Cooperation: Guidelines for forensic laboratories, Draft 1.1, August 1998.
7   First Version: A.C. Moffat and M.D. Osselton, Guidelines for the performance of quantitative methods for the analysis of drugs in blood in forensic toxicology, 26th Meeting of the TIAFT

Glasgow, UK, Aug 14–19, 1989; the 1993 version was republished in 2006: Bull. Int. Assoc. Forensic Toxicol., XXXI(4) (1993) 23–26.

8   GTFCh: Laborrichtlinien für chemisch-toxikologische Untersuchungen. Toxichem. Krimtech., 58, (1991) 19–22 (These guidelines are being updated at present).

9   Society of Forensic Toxicologists (SOFT)/American Academy of Forensic Sciences (AAFS), Toxicology section, Forensic toxicology laboratory guidelines, Draft copy, 1997.

10   Richtlinien für die Suchstoffanalytik. Arbeitsgruppe Schweizerischer Institutionen: Bundesamt für Gesundheit, Schweizer Apothekenverein, Gruppe der Leiter gerichtschemischer Laboraotorien, Schweizer Verband der Diagnositka-Industrie, Schweizerische Union für Labormedizin, Schweizerischer Verband für klinische Chemie, Schweizerischer Verband für medizinisch-analytische Laboratorien, Gesundheitsamt Basel, 1996.

11   A. Donabedian, The definition of quality and approaches to its assessment, Health Administration Press, Ann Arbor, MI, 1980.

12   W.E. Deming, Out of the crisis, Mass, Cambridge, 1986.

13   C. Marlorny; K. Kassebohm, Brennpunkt TQM, Schäffer-Poschel Verlag, Stuttgart, 1994, p. 64.

14   Richtlinien für die Erteilung der Anerkennung als Forensischer Toxikologe – GTFCh, www.gtfch.org

15   J.B. Buchan, The expert witness. In: Ken Habben (Ed.), Current approaches in forensic toxicology, Rev. 2. The forensic toxicologists certification board (FTCB) of the American Academy of Forensic Sciences, Inc., 1996.

16   H.A. Feder, Succeeding as an expert witness. Tageh Press, Glennwood Springs, Colorado, 1993, 12, (cited according to 15).

17   S.H. Gifis, Law dictionary, (3rd ed). Barrons Educational Series, Inc., Hauppauge, NY, 1991, (cited according to 15).

18   D. Poynter, The expert witness handbook, Para Publishing, Santa Barbara, CA, 1987.

19   G. Skopp and L. von Meyer, Toxichem. Krimtech., 71 (2004) 100–105.

20   G. Skopp, Forensic Sci. Int., 142 (2004) 75–100.

21   W.H. Anderson and R.W. Prouty. In: R.C. Baselt (Ed.), Advances in analytical toxicology, vol. 2. Year Book Medical, Chicago, IL, 1989, pp. 70–72.

22   D.J. Pounder. In: C.H. Wecht (Ed.), Legal medicine, Butterworth Legal Publishers, Salem, 1993, pp. 163–192.

23   R.W. Prouty and W.H. Anderson, J. Forensic Sci., 35 (1990) 243–270.

24   T. Hildberg, S. Rogde and J. Morland, J. Forensic Sci., 44 (1999) 3–9.

25   F. Moriya and Y. Hashimoto, J. Forensic Sci., 44 (1999) 10–16.

26   K.E. Rodda and O.H. Drummer, Forensic Sci. Int., 164 (2006) 235–239.

27   A.L. Pelissier-Alicot, N. Coste, C. Bartoli, M.D. Piercecchi-Marti, A. Sanvoisin, J. Gouvernet and G. Leonetti, Forensic Sci. Int., 156 (2006) 35–39.

28   A.L. Pelissier-Alicot, J.M. Gaulier, C. Dupuis, M. Feuerstein, G. Leonetti, G. Lachatre and P. Marquet, Int. J. Legal Med., 120 (2006) 226–232.

29   A.L. Pelissier-Alicot, M. Fornaris, C. Bartoli, M.D. Piercecchi-Marti, A. Sanvoisin and G. Leonetti, Forensic Sci. Int., 150 (2005) 81–83.

30   E.A. De Letter, M.P. Bouche, J.F. Van Bocxlaer, W.E. Lambert and M.H. Piette, Forensic Sci. Int., 141 (2004) 85–90.

31   J.R. Caplehorn and O.H. Drummer, Drug and Alcohol Rev., 21 (2002) 329–333.

32   E.A. De Letter, F.M. Belpaire, K.M. Clauwaert, W.E. Lambert, J.F. Van Bocxlaer and M.H. Piette, Int. J. Legal Med., 116 (2002) 225–232.

33   E.A. De Letter, K.M. Clauwaert, F.M. Belpaire, W.E. Lambert, J.F. Van Bocxlaer and M.H. Piette, Int. J. Legal Med., 116 (2002) 216–224.

34   B. Bailey, R. Daneman, N. Daneman, J.M. Mayer and G. Koren, Forensic Sci. Int., 110 (2000) 61–70.

35   R. Aderjan and R. Mattern, Z. Rechtmed., 86 (1980) 13–20.

36   G. Skopp and L. Pötsch, Clin. Chem., 48 (2002) 301–306.

37   G. Skopp and L. von Meyer, Toxichem. Krimtech., 71 (2004) 101–107.

38   Report XVIII of the DFG commission for clinical toxicological analysis; Special issue of the TIAFT Bulletin. Gaschromatographic retention indices of toxicologically relevant substances on packed or capillary columns with dmethysilicone stationary phases analysis, R.A. de Zeeuw, J.P. Franke, H.H. Maurer, K. Pfleger (Eds.), 3rd revised and enlarged edition, VCH Verlagsgesellschaft D6940, Weinheim, FRG (1992).

39   Report XIX of the DFG commission for clinical toxicological analysis. Retention indices of solvents and other volatile substances for use in toxicological analysis; Special issue of the TIAFT Bulletin, R.A. de Zeeuw, J.P. Franke, G. Machata, M.R. Möller, R.K. Müller, A. Graefe, D. Tiess, K. Pfleger, M. Geldmacher-v. Mallinckrodt,. VCH Verlagsgesellschaft D6940 Weinheim, FRG (1992).

40   M. Bogusz, G. Neidl-Fischer and R.E. Aderjan, J. Anal. Toxicol., 12 (1988) 325–329.

41   K. Pfleger; H.H. Maurer; A. Weber, Mass spectral data and GC data of drugs, poisons and their metabolites, Verlag Chemie, Weinheim, FRG, 1992.

42   H.H. Maurer, J. Chromatogr. Biomed. Appl., 580 (1992) 3–41.

43   http://www.soft-tox.org/docs/SOFT%20DFSA%20List.pdf.

44   H.H. Maurer, personal communication.

45   D. Ostheimer, M. Cremese, A.H.B. Wu and D.W. Hill, J. Anal. Toxicol., 21 (1997) 17–22.

46   A.H.B. Wu, D. Ostheimer, M. Cremese, E. Forte and D. Hill, Clin. Chem., 40 (1994) 216–220.

47   Y. Tondeur, P.W. Albro, J.R. Hass and D.J. Harvan, Anal. Chem., 56 (1984) 1344–1347.

48   D.J. Borts and L.D. Bowers, J. Anal. Toxicol., 22 (1998) 258–259.

49   A.K. Chaturvedi, J. Forensic Sci., 45 (2000) 422–428.

50   A.W. Jones and L. Hard, Scand. J. Clin. Lab. Invest., 64 (2004) 629–634.

51   H.H. Maurer, J. Chromatogr., B713 (1998) 3–25.

52   J.B. Fenn, M. Mann, C.K. Meng, S.F. Wong and C.M. Whitehouse, Science, 246 (1989) 64–71.

53   P. Kebarle and T. Liang, Anal. Chem., 65 (1993) A 972–A 986.

54   H. Hoja, P. Marquet, B. Verneuil, H. Lofti, B. Pernicault and G. Lachatre, J. Anal. Toxocol., 21 (1997) 116–126.

55   W. Weinmann and M. Svoboda, J. Anal. Toxicol., 22 (1998) 319–328.

56   M. Bogusz, R.D. Mayer, K.D. Krüger and U. Kohls, J. Anal. Toxicol., 22 (1998) 549–557.

57   H.H. Maurer, T. Kraemer, O. Ledvinka, C.J. Schmidt and A.A. Weber, J. Chromatogr. B, 689 (1997) 81–89.

58   M. Prybylski and M.O. Glocker, Angw. Chem. Int. Ed. Engl., 35 (1996) 806–822.

59   M. Bogusz and M. Erkens, J. Chromatogr., 674 (1994) 97–126.

60   B. Aebi, R. Sturny-Jungo, W. Bernhard, *et al.*, Forensic Sci. Int., 14;128 (1–2) (2002) 84–89.

61   K.D. Ballard, W.E. Vickery and L.T. Nguyen; *et al.*, J. Am. Soc. Mass Spectrom., 17(10) (2006) 1456–1468.

62   T.N. Decaestecker, K.M. Clauwaert and J.F. Van Bocxlaer; *et al.*, Rapid Commun. Mass Spectrom., 14(19) (2000) 1787–1792.

63   R. Kaneko, S. Hattori, S. Furuta; *et al.*, J. Mass Spectrom., 41(6) (2006) 810–814.

64   S.G. Roussis, Anal. Chem., 73(15) (2001) 3611–3623.

65   M. Jemal and Y.Q. Xia, Curr. Drug Metab., 7 (2006) 491–502.

66   L. Leclercq, C. Delatour, I. Hoes, F. Brunelle, X. Labrique and J. Castro-Perez, Rapid Commun. Mass Spectrom., 19 (2005) 1611–1618.

67   B.K. Matuszewski, M.L. Constanzer and C.M. Chavez-Eng, Anal. Chem., 70 (1998) 882–889.

68   IUPAC, Pure Appl. Chem., 60 (1988) 855–864.

69   Chemistry quality assurance handbook, Vol. 2, (1982), US Department of Agriculture, Food Safety and Inspection Service, Washington, DC, sec. 2.5.10.

70   F.T. Peters and H.H. Maurer, Toxichem. Krimtech., 68(3) (2001) 116–127.

71   F.T. Peters, G. Schmitt, M. Hartung, M. Herbold, T. Daldrup and F. Musshoff, Toxichem. Krimtech., 71 (2004) 146–154.

72   Valistat[R] mit Buch "Methodenvalidierung im forensisch-toxikologischen Labor." Programm auf Excel-Basis zur Validierung nach den Richtlinien der GTFCh unter Berücksichtigung der ISO 5725. To be ordered from: www.arvecon.de; info@arvecon.de

73 DIN 32645 (DIN = Deutsches Institut für Normung), Beuth-Verlag, Berlin, 1992.
74 G. Schmidt, M. Herbold, B.E.N. Version 2.03, Programm auf Excel-Basis zur Berechnung der Nachweis, Erfassungs-und Bestimmungsgrenze nach DIN 32645 und Auswertung von Messdaten über die Kalibrationsdaten, Arvecon GmbH.
75 J. Büttner, R. Both, J.H. Boutwell and P.M.G. Broughton, Clin. Chem., 22 (1976) 532–540.
76 IUPAC, Nomenclature, symbols units and their usage in spectrochemical analysis-II, Spectrochim. Acta Part B, 33 (1978) 242.
77 U.S. National Institute of Drug Abuse: NIDA-Guidelines, Fed. Reg. 53, Nr. 69, 11970–11989, 1988.
78 http://workplace.samhsa.gov/DrugTesting/RegGuidance/UrineConcen.htm.
79 Ergebnisbericht der gemeinsamen Arbeitsgruppe für Grenzwertfragen und Qualitätskontrolle. Blutakolhol, 35 (1998) 372–375.
80 Grenzwertkommision: Beschluss zu § 24a II StVG vom 21.11.2002. Toxichem. Krimtech., 69 (3) (2002) 127.
81 J. F. Wilson in: I. Dijkhuis and B. Widdop (Eds.), First DOA News Bulletin, 1998, p. 3.
82 A.A. Elian, Forensic Sci. Int., 128(3) (2002) 120–122.
83 D.T. Yeatman and K. Reid, J. Anal. Toxicol., 27(1) (2003) 40–42.
84 S. Elliott, P. Lowe and A. Symonds, Forensic Sci. Int., 139(2–3) (2004) 183–190.
85 P. Kintz, M. Villain V. Cirimele; *et al.*, Forensic Sci. Int., 143(2–3) (2004) 177–181.
86 R. Aderjan and M. Bogusz, J. Chromatogr., 454 (1988) 345–351.
87 M. Bogusz and R. Aderjan, J. Anal. Toxicol., 12 (1988) 67–72.
88 F.M. Garfield, Quality assurance principles for analytical laboratories, 3rd printing 1994, AOAC International.
89 Ringversuche der GTFCh, c/o Institute of Legal Medicine and Traffic Medicine, University of Heidelberg, D 69115 Heidelberg, FRG.
90 H.H. Maurer and J.W. Arlt, J. Anal. Toxicol., 22 (1999) 1.
91 S.M. Dysel, C.M. Erikson and L.V.S. Hood, J. Anal. Toxicol., 17 (1993) 190–191.
92 W.A. Shewhart, The economic control of the quality of manufactured products, D. van Nostrand Company Inc., New York, 1931.
93 S. Levey and F.R. Jenning, Am. J. Clin. Pathol., 20 (1950) 1059.
94 W. Funk; V. Dammann; G. Donnevert, Qualitätssicherung in der analytischen chemie, VCH-Verlag, Weinheim, FRG, 1992.
95 J.T. Cody, J.C. Garriott, R.L. Foltz, M.A. Peat and M.I. Schaffer, J. Forensic Sci., 35 (1990) 236–242.
96 R. Fresenius, Ann. Chem. Pharm., XLIX (1844) 275–286.
97 R.W. Willette, Proficiency testing and quality control programs, in: R.L. Hawks and C.N. Chang (Eds.), Urine testing of drugs of abuse, National Institute of Drug Abuse Research monograph 73, DHHS Publication No. (ADM) 87–1481, Washington, DC, Supt. of Doc. Govt. Print. Off., 1986.
98 D. Sohn, J. Anal. Toxicol., 1 (1977) 111–117.
99 E.C. Dinovo and L.A. Gottschalk, Clin. Chem., 20 (1976) 846.
100 R.C. Kelly and I. Sunshine, Clin. Chem., 22 (1976) 1413.
101 E.C. Dinovo and L.A. Gottschalk, Clin. Chem., 22 (1976) 2056.
102 K.L. McCloskey and B.S. Finkle, J. Forensic Sci., 22 (1977) 675.
103 S. Hansen, P. Cowdil and J. Boon, JAMA, 253 (1985) 2382–2387.
104 M.A. Peat, J.S. Finnigan and B.S. Finkle, J. Forensic Sci., 28 (1983) 139–157.
105 First DOA News Bulletin, I. Dijkhuis and B. Widdop (Eds.), 1998.
106 DRUID – European research project under the guidance of the German BASt Bundesanstalt für Straßenwesen, Contract n° TREN-05-FP6TR-S07.61320 (relating to project "Driving Under the Influence of Drugs, Alcohol and Medicines") signed between the Commission of the European Community and Bundesanstalt für Straßenwesen (BASt).
107 A. Verstrate, F.M. Belpaire and G.G. Leroux-Roels, J. Anal. Toxicol., 22 (1998) 27–31.
108 F. Divanon, D. Debruyne and A. Moulin, J. Anal. Toxicol., 22 (1998) 559–566.

109   IUPAC: Harmonized protocols for the adoption of standardized analytical methods and for the presentation of their performance characteristics, W.D. Pocklington, Pure Appl. Chem., 62 (1990) 149.

110   D.S. Ferrara, I.L. Tedeschi, G. Frison and G. Brusini, J. Chromatogr., B, 713 (1998) 227–243.

111   J.F. Wilson, J. Williams, G. Walker, P.A. Toseland, P.L. Smith, A. Richens and D. Burnett, Clin. Chem., 37 (1991) 442–447.

112   T.C. Quong, R.P. Chamberlain, D.L. Frederick, B. Kabour and I. Sunshine, Clin. Chem., 34 (1988) 605–632.

113   G. Machata, Beitr. Gerichtl. Med., 27 (1970) 192–198.

114   G. Machata, Beitr. Gerichtl. Med., 28 (1971) 367–368.

115   G. Machata, Beitr. Gerichtl. Med., 29 (1972) 427–428.

116   G. Machata, Beitr. Gerichtl. Med., 32 (1974) 191–192.

117   G. Machata, Beitr. Gerichtl. Med., 28 (1971) 367–368.

118   G. Machata, Beitr. Gerichtl. Med., 33 (1975) 228–229.

119   K. Harzer, Toxichem. Krimtech., 10 (1980) 9.

120   M. Geldmacher-v. Mallinckrodt, Toxichem. Krimtech., 12 (1980) 9.

121   M. Geldmacher-v. Mallinckrodt, Toxichem. Krimtech., 22 (1982) 2.

122   E. Schneider and K. Harzer, Toxichem. Krimtech., 22.

123   Blutalkoholbestimmung für forensische Zwecke, Empfehlungen zur Anpassung der Richtlinien des Bundesgesundheitsamtes von 1966 an Gesetze, Verordnun-gen und Rechtsprechung, edited by Deutsche Gesellschaft für Rechtsmedizin and Gesellschaft für Toxikologische und Forensische Chemie, 1999.

124   M. Thompson and R. Wood, J. AOAC Int., 76 (1993) 927.

125   R. Aderjan, K. Harzer, M.R. Möller, E. Schneider, M. Herbold, Proceedings of the 48th Annual Meeting of the American Academy of Forensic Sciences (AAFS), Vol. II., K31, Nashville, TN, Colorado Springs, CO, p. 219 (1996).

126   R. Aderjan, Th. Daldrup, M. Herbold, H. Käferstein, G. Kauert, M.R. Möller, in: L. Farell and J.B. Zettl (Eds.), Abstracts of the Meeting Society of Forensic Toxicologists (SOFT), Denver, CO (1996).

127   R. Aderjan, M. Herbold, Qualitätskontrolle für quantitative Analysen von Betäubungsmitteln im Blut, Bericht der Bundesanstalt für Straßenwesen – Mensch und Sicherheit M 87. Hrsg.: Bundesanstalt für Straßenwesen, Bergisch-Gladbach,Verlag für neue Wissenschaft GmbH, Bremerhaven (1998).

128   Annual programme: www.gtfch.org; www.arvecon.de (info.@arvecon.de).

129   W. Horwitz, Anal. Chem., 54(1) (1982) 67A.

130   www.gtfch.org.

131   E. Garfield, Science, 178 (1972) 471–479.

132   N.M. Meenen, Unfallchirurgie, 23 (1997) 128–133.

133   U.K. Lindner and V. Oehm, Rechtsmedizin, 7 (1997) 35–36.

134   M.H. MacRoberts and B.R. MacRoberts, Trends Biol. Sci., 14 (1989) 8–11.

135   G. Motta, Nature, 376 (1995) 720.

136   N.B. Metcalfe, Nature, 376 (1995) 720.

137   S. Hanson, Lancet, 346 (1995) 906.

138   E. Garfield, Br. Med. J., 313 (1996) 411–413.

139   A.W. Jones, Forensic Sci. Int., 62 (1993) 173–178.

140   E.J. Huth. In: S. Lock (Ed.), The future of medical journals, British Medical Journal, London, 1991, pp. 81–92.

141   A.W. Jones, J. Forensic Sci., 43 (1998) 439–444.

142   E.G. Cohn and D.P. Farrington, Br. J. Criminol., 34 (1994) 204–206.

143   H.B. Hansen and J.H. Hendriksen, Clin. Physiol., 17 (1995) 409–418.

144   P.O. Seglen, J. Int. Med., 229 (1991) 109–111.

145   A.W. Jones, Alcohol Alcohol, 31 (1996) 11–15.

146   M.J. Bogusz, Rechtsmedizin, 8 (1998) 195.

147   German Bundesgerichtshof (BGH), 4 StR 297/90.
148   N. Bilzer and K.D. Hatz, Blutalkohol, 35 (1998) 321–330.
149   H.J. Bode, Blutalkohol, 35 (1998) 220–238.
150   These method standards will be worked out at present by an expert group of the GTFCh. They will
      be published in an appendix to the laboratory guidelines.

PART 4:
FORENSIC CHEMISTRY

M.J. Bogusz (Ed.). Forensic Science
Handbook of Analytical Separations, Vol. 6
© 2008 Elsevier B.V. All rights reserved.

CHAPTER 24

Explosives

Jehuda Yinon

Department of Environmental Science, Weizmann Institute of Science, Rehovot 76100, Israel

24.1 INTRODUCTION

Analysis of explosives is an important part of forensic chemistry. Its major applications include the analysis of post-blast residues and identification of traces of explosives on suspects' hands, clothing, and other related items. The results of these analyses are not only necessary for the investigation of a bombing, but serve also as evidence in court.

Modern methodologies for forensic analysis of explosives are based on gas chromatography (GC) and high performance liquid chromatography (HPLC) with specific detectors, on ion mobility spectrometry and on mass spectrometry (MS), as GC/MS, LC/MS, and MS/MS. As most explosives are thermally labile compounds, HPLC and LC/MS have an obvious advantage over GC and GC/MS, as the chromatography is carried out at room temperature. However, when using GC methods, one has to take precautions to minimize thermal decomposition in the GC column and injector.

Thin-layer chromatography (TLC) is still used in some forensic laboratories because it is simple, fast, and inexpensive [1–3]. It is a good screening technique but has to be backed up by an identification technique for evidence in court.

The main explosives forensic laboratories have to deal with include standard explosives, plastic explosives, and improvised explosives. Table 24.1 shows a representative list of these explosives and their characteristics [4,5].

24.2 EXTRACTION AND CLEAN-UP PROCEDURES

General schemes for the analysis of post-explosion residues have been described in the literature [3,6].

The schemes include screening of the collected debris by visual and microscopic inspection, extraction, clean-up procedures, and analysis.

In all post-explosion schemes, the debris to be analyzed are subjected to two extractions: aqueous and organic. Aqueous extraction is used for inorganic compounds

References pp. 838–838

TABLE 24.1
LIST OF EXPLOSIVES AND THEIR CHARACTERISTICS

	Explosive	Name/Contents	Formula	Molecular Weight	Vapor Pressure [Torr] at 25°C	Relative Vapor Density
Standard explosives	TNT	2,4,6-Trinitrotoluene	$C_7H_5N_3O_6$	227.15	5.8×10^{-6}	7.7 ppb
	RDX	1,3,5-Trinitro-1,3,5-triazacyclohexane	$C_3H_6N_6O_6$	222.6	4.6×10^{-9}	6.0 ppt
	HMX	1,3,5,7-Tetranitro-1,3,5,7-tetrazacyclooctane	$C_4H_8N_8O_8$	296.16	3×10^{-9} (at 100°C)	3.95 ppt
	Tetryl	2,4,6,N-tetranitro-N-methylaniline	$C_7H_5N_5O_8$	287.15	5.7×10^{-9}	
	PETN	Pentaerythritol tetranitrate	$C_5H_8N_4O_{12}$	316.2	1.4×10^{-8}	18 ppt
	NG	Nitroglycerin glycerol trinitrate	$C_3H_5N_3O_9$	227.1	3.0×10^{-4}	0.39 ppm
Improvised explosives	EGDN	Ethylene glycol dinitrate	$C_2H_4N_2O_6$	152.1	0.07	92.6 ppm
	ANFO	Ammonium nitrate + fuel oil	NH_4NO_3	80.05	9.1×10^{-6}	12 ppb
	Urea nitrate	Urea nitrate	CH_4N_2O HNO_3			
	TATP	Triacetone triperoxide	$C_9H_{18}O_6$	222.23	0.052	68.5 ppm
Plastic explosives	C-4	RDX + plasticizer				
	Semtex	RDX + PETN + plasticizer				
	Detasheet	PETN + plasticizer				

and for some water-soluble organic ingredients, such as sugars. Organic extraction, using mainly acetone, is used for organic explosives and related compounds. Organic impurities, which are co-extracted with the explosive residues in the organic extraction, have to be eliminated by clean-up procedures. Most clean-up procedures are based on adsorption of the explosives from organic extracts of debris or handswabs onto a solid sorbent and the subsequent elution of the explosives from the sorbent [2].

Solid-phase microextraction (SPME) has been used for the extraction of explosives from aqueous solutions and from headspace [7]. SPME involves exposing a sorbent-coated silica fiber to an aqueous solution containing explosives or from the headspace of a heated sample. The fiber is then thermally desorbed into the injection port of a gas chromatograph or introduced into the desorption chamber of a high performance liquid chromatograph. SPME eliminates the need of large volumes of solvent and relatively large sample sizes as often required for liquid-liquid extraction and solid-phase extraction of explosives.

24.3 ANALYSIS OF EXPLOSIVES BY GAS CHROMATOGRAPHY (GC)

Gas chromatography (GC) with specific detectors has been found to be a good method for the separation and analysis of many organic explosives [3]. This can be achieved when using the GC under controlled experimental conditions, such as type and length of the capillary column, special injections techniques, temperature of column and injector, and the use of selective detectors.

Low polarity columns are commonly used for the separation and analysis of explosives, to minimize polar interactions of the nitro groups, which can produce irreversible adsorption on the stationary phase or decomposition of the explosives at high temperatures. Compounds which evaporate only at higher temperatures, such as RDX and PETN, should be eluted from the column as fast as possible, in order to minimize their decomposition. This can be done by either increasing the flow rate of the carrier gas or shortening the length of the column, which will have an adverse effect on the separation capability of the chromatography. Columns as short as 1.5 m have been used [8].

Temperature-programmable injectors are preferred in order to minimize thermal decomposition of the analyzed explosive.

The GC detectors, which are mainly used for the analysis of explosives, are the electron capture detector (ECD) and the thermal energy analyzer (TEA) detector.

24.3.1 GC with electron capture detector (ECD)

ECD have a fast response and are highly sensitive for most electron capturing compounds. However, their specificity for explosives is low. Two examples are presented, as follows:

SPME coupled with on-column GC-ECD was used for the analysis of post-blast explosive residues [9]. Investigated explosives included amongst other compounds

TNT, PETN, RDX, NG, and tetryl. GC column was a Restek $RT_X^{®}$–TNT, 6 m long, 0.53 mm I.D., 1.5 μm film thickness. The best fiber to trap the majority of organic explosives was found to be the PDMS/DVB.

In another example, a method for the detection of nitroaromatic and nitramine explosives from a PTFE wipe was developed using thermal desorption, a cooled injection system, and GC-ECD [10].

A dual column (Restek $RT_X^{®}$–TNT2 and Restek $RT_X^{®}$–TNT, both 6 m long, 0.53 mm I.D., 1.5 μm film thickness) and dual ECD were used to enable simultaneous confirmation analysis of the explosives desorbed.

24.3.2 GC with thermal energy analyzer (TEA) detector

The TEA detector, also known as the chemiluminescence detector, is a nitro and nitroso-specific detector. In the TEA detector, nitro compounds are pyrolyzed to form $NO^{•}$radicals, which pass into a reaction chamber where they are oxidized by ozone to form electronically excited nitrogen dioxide (NO_2^*). The excited nitrogen dioxide decays back into its ground state with emission of chemiluminescent light in the near-infrared region ($\lambda \cong 0.6$–2.8 μm). The intensity of the emitted ight is proportional to the NO concentration, hence to the nitrocompound concentration.

The TEA detector, although more specific for explosives than the ECD, is less sensitive.

A survey of high explosives traces in public places was carried out using GC-TEA with GC/MS for confirmation [11,12]. The detector was a Thermedics TEA Model 610 detector. Three different columns were used: SGE 12QC2/BP1, SGE 12QC2/BP5, and Chrompack CP-Sil-19CB. A positive explosive detection was recorded only if analyses on all three columns gave consistent results. A TEA standard solution containing 12 nitrocompounds was used for retention time comparisons. Fig. 24.1 shows a GC-TEA chromatogram of a TEA standard with a BP5 column. (FNT is 2-fluoro-5-nitrotoluene reference marker. MT is "Musk Tibetine" – 2,6-dinitro-3,4,5-trimethyl-*tert*-butylbenzene – reference marker). Detection limits for NG, TNT, PETN, and RDX were estimated to be in the low ng range.

24.4 ANALYSIS OF EXPLOSIVES BY HIGH-PERFORMANCE LIQUID CHROMATOGRAPHY (HPLC)

HPLC has an advantage over GC for the analysis of explosives, especially for the less volatile and thermally labile ones, because it operates at room temperature. The most popular detector for HPLC is the UV detector, at 254, 230, and 214 nm, as well as the photodiode-array UV detector. For specific applications other detectors, such as the electrochemical and TEA detectors have been used.

Columns mainly used for explosives are C8 and C18, but also CN columns. Mobile phases are methanol–water or acetonitrile–water, at various ratios, with or without a buffer, with various flow rates, depending on the column I.D.

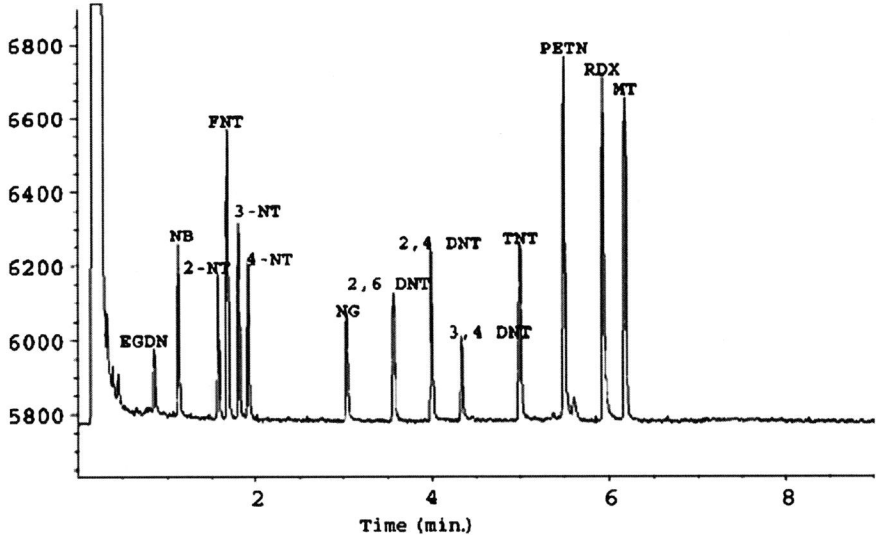

Fig. 24.1. GC-TEA chromatogram of a TEA standard with a BP5 column. (FNT is 2-fluoro-5-nitrotoluene reference marker; MT is "Musk Tibetine" – 2,6-dinitro-3,4,5-trimethyl-tert-butylbenzene – reference marker.) Reproduced from H.E. Cullum *et al.*, J. Forensic Sci., 49 (2004) 684. Copyright ASTM INTERNATIONAL. Reprinted with permission.

A Pinnacle II® C18 HPLC column (250 mm × 4.6 mm I.D., particle size 5 μm) was used for the analysis of 14 nitrocompounds, including TNT, RDX, HMX, and tetryl [13]. Mobile phase was water-methanol (50:50) at a flow rate of 1.5 ml/min. Detector was UV at 254 nm. Confirmation was carried out with a Pinnacle II® CN column.

HPLC with a Supelcosil octyl C-8 column (150 mm × 4.6 mm I.D., 5 μm particle size) and Supelcosil LC-8 guard column (20 mm × 4.6 mm I.D., 5 μm particle size) with a diode array UV detector was used to separate a series of explosives, including TNT and its metabolites [14]. The influence of temperature (35–55°C) and the use of an ion-pair reagent on the chromatographic resolution and retention were investigated. The organic mobile phase was methanol. The aqueous mobile phase consisted of 0.025 M sodium phosphate buffer (pH 7) with or without 0.1% or 0.5% w/w 1-octanesulfonic acid (ion-pair reagent). Mobile phase flow rate was 1 ml/min. Fig. 24.2 shows two HPLC chromatograms, with and without ion-pair reagent added, of 14 explosives and metabolites.

A study was done on rapid screening of explosives by HPLC with monolithic stationary phases and UV photodiode array detection [15]. The columns were silica-based octadecyl silane monolithic Chromolith SpeedROD RP-18e (50 mm × 4.6 mm I.D.) and Chromolith Performance RP-18e (100 mm × 4.6 mm) columns. Mobile phase consisted of methanol-water gradients at flow rates of 0.2–10 ml/min. Column temperature was investigated over the range 20–60°C. Increasing temperature caused a decrease of 15–30% in retention for all explosives. Fig. 24.3 shows the HPLC chromatograms of seven explosives on a 5 cm ODS monolithic column at a column temperature of 60°C. Separation was obtained in less than 2 min.

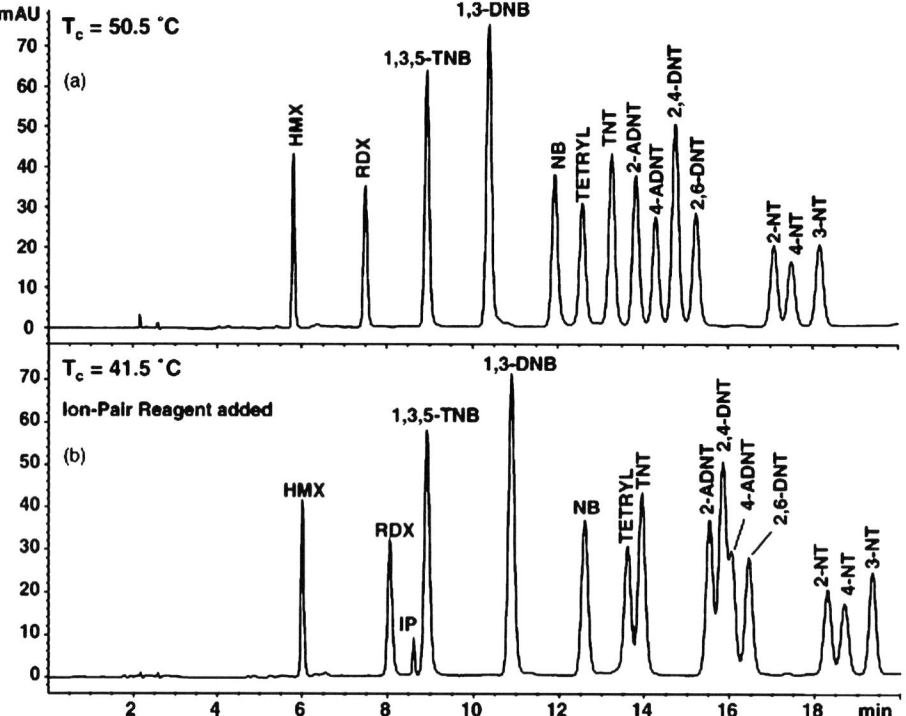

Fig. 24.2. Two HPLC chromatograms, with and without ion-pair reagent added, of 14 explosives and metabolites. Reprinted from T. Borch *et al.*, J. Chromatogr. A, 1022 (2004) 83, Copyright 2003, with permission from Elsevier.

Solid phase micro extraction (SPME) coupled with on-column HPLV-UV was evaluated for the analysis of post-blast explosive residues [7]. A modified SPME/HPLC interface using dual six-port valves and a 200 μl inner volume SPME desorption chamber was used. The HPLC columns consisted of a combination of a cyano Res-Elut CN (30 mm × 4.6 mm I.D., 5 μm particle size) column and an octadecyl Bondesil C-18 (250 mm × 4.6 mm I.D., 5 μm particle size) column. The UV detector was operated at a wavelength of 220 nm for EGDN, NG, and PETN and 254 nm for other investigated explosives, such as TNT, RDX, and tetryl. Limits of detection were at the ppb level for determination of explosives in water.

24.5 ANALYSIS OF EXPLOSIVES BY GC/MS

Although GC/MS is limited by the thermal decomposition of some of the explosives, it is still a widely used method in many forensic laboratories.

GC/MS has been used in three different ionization modes: electron ionization (EI), chemical ionization (CI), and negative-ion chemical ionization (NCI), depending on the type of explosives to be analyzed.

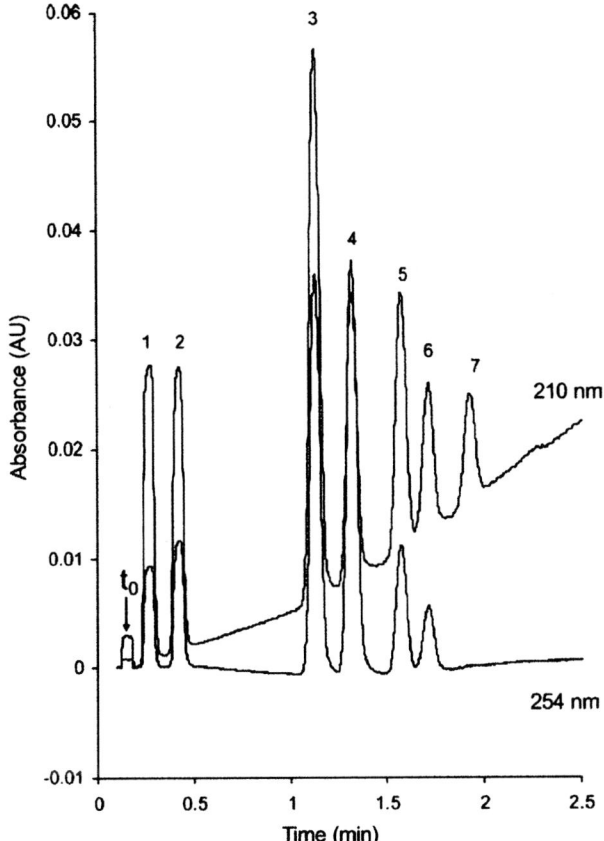

Fig. 24.3. HPLC chromatograms at two different UV wavelengths of 7 explosives on a 5 cm ODS monolithic column at a column temperature of 60°C. Reproduced from B. Paull *et al.*, J. Forensic Sci., 49 (2004) 1184. Copyright ASTM INTERNATIONAL. Reprinted with permission.

While EI is mainly used for organic peroxide explosives, CI and NCI is being used for nitroaromatic, nitrate ester, and nitramine explosives.

A study comparing detection limits for GC/MS analysis of 2,4-DNT, TNT, RDX, and PETN, using EI, CI, and NCI, showed that NCI gave the lowest detection limits, which were between 0.18 and 1.11 ng [16].

GC/MS-NCI with chlorinated additives, including chloromethane (CH_3Cl), dichloromethane (CH_2Cl_2), chloroform ($CHCl_3$), and carbon tetrachloride (CCl_4), was evaluated for the analysis of explosives [17].

Results were compared with those using only methane as reagent gas.

A J&W DB-5MS column (6 m × 0.25 mm I.D., 0.25 µm film thickness) was used with helium as carrier gas. The temperature program for the GC started at 40°C for 1 min, then 25°C/min to 250°C with 1 min hold. Mass spectra containing chloride adduct anions [M + Cl]⁻ were obtained for most investigated explosives. Abundant anions observed were at m/z 46, 62, and 187 [M + Cl]⁻ in EGDN, m/z 46, 62, and 262 [M + Cl]⁻ in NG, m/z 46, 62, 315 [M-H]⁻, and 351 [M + Cl]⁻ in PETN, m/z 46, 102, 129, 176, and 257 [M + Cl]⁻ in RDX, and m/z 46, 102, 176, and 331 [M + Cl]⁻ in HMX. The mass spectra of TNT and tetryl did not produce any chlorinated adduct anions and were similar to NCI mass spectra with methane.

Most noteworthy were the formation of molecular adduct anions in EGDN, NG, and PETN. In contrast, only the anions at m/z 46 [NO₂]⁻ and 62 [NO₃]⁻ were observed when methane was used as reagent gas, which makes their forensic identification inconclusive – being based only on differences in chromatographic retention times. Fig. 24.4 compares the mass spectra of EGDN and NG, using NCI with methane and dichloromethane as reagent gases.

Triacetone triperoxide (TATP) is a powerful explosive, manufactured in clandestine laboratories and used by terrorists. As TATP sublimes easily, analysis was performed by SPME trapping of its vapor, using polydimethylsiloxane/divinyl benzene (PDMS/DVB) fiber, followed by desorption into a GC/MS [18]. An Alltech BP-1 column (30 m × 0.25 mm I.D., 0.25 μm film thickness) was used with helium as carrier gas. Column temperature was held at 50°C for 3 min, then heated to 230°C at a rate of 5°C/min and held at 230°C for 5 min.

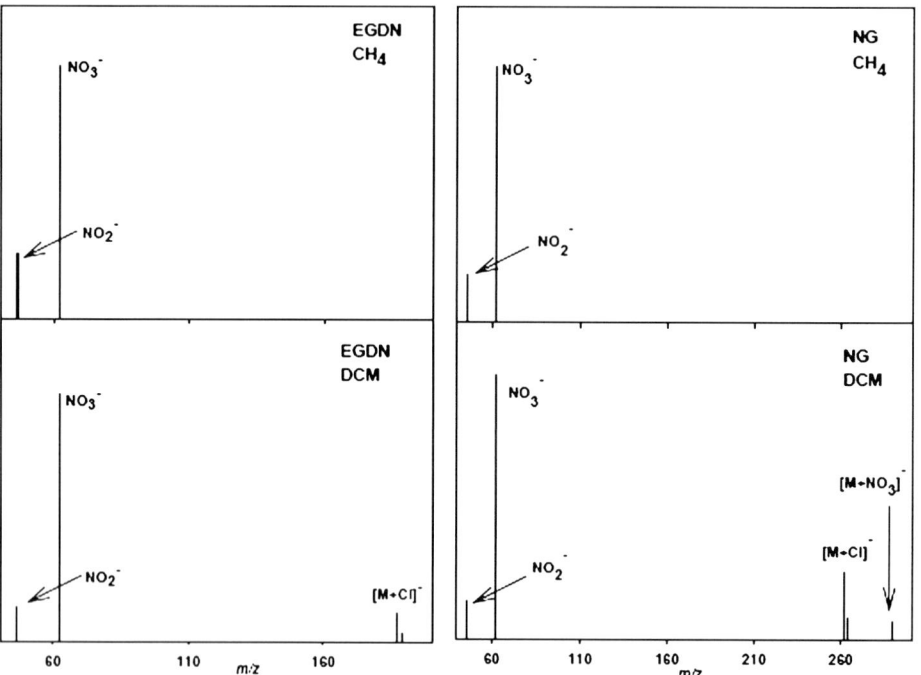

Fig. 24.4. Mass spectra of EGDN and NG, using NCI with methane and dichloromethane as reagent gases. Reproduced from J.G. McDonald *et al.*, Proceedings of the 8th International Symposium on Analysis and Detection of Explosives, Ottawa, Canada, June 2004, p. 85. With permission.

The EI mass spectrum of TATP contained a low-abundant (less than 1%) molecular ion at m/z 222 and several fragment ions, at m/z 75, 59, and 43. Results showed that the use of SPME for recovery of TATP was ~1500 times more efficient than headspace, and 50 times more efficient than adsorption on Amberlite XAD-7 [18].

GC/MS-CI analysis of TATP was done with methane and ammonia as reagents [19]. With methane, a low-abundant $[M + H]^+$, at m/z 223, and several fragment ions are observed, while with ammonia, an adduct ion at m/z 240 $[M + NH_4]^+$ (base peak) and a low-abundant $[M + H]^+$, at m/z 223, are observed.

24.6 ANALYSIS OF EXPLOSIVES BY LC/MS

The thermal lability of many explosives, along with the requirements of high sensitivity, especially in the analysis of post-explosion residues, makes LC/MS a method of choice for the analysis of explosives. Both electrospray ionization (ESI) and atmospheric pressure chemical ionization (APCI) are being used, depending on the type of explosives [20].

Several additives have been used in LC/MS-ESI of explosives in order to enhance the sensitivity. Nitramine and nitrate ester explosives showed enhanced response for ammonium nitrate additive, by forming $[M + NO_3]^-$ adduct ions in the negative-ion mode [21,22]. Fig. 24.5 shows the LC/MS-ESI mass chromatograms of a 25 pg/μl Semtex sample (a plastic explosive containing RDX and PETN) with post column introduction of ammonium nitrate [22]. HPLC separation was achieved with a C18 column (100×2.1 mm I.D., 5 μm particle size), using an isocratic mobile phase of methanol-water (70:30), at a flow rate of 150 μl/min.

LC/MS with electrospray ionization in the negative-ion mode was used to detect RDX and its degradation products in contaminated groundwater [23]. Chromatographic separation was achieved with a Nova-Pak C18 column (150×2.1 mm I.D., 4 μm particle size), using an isocratic mobile phase of methanol-water (50:50), at a flow rate of 200 μl/min. The $[M + 75]^-$ and $[M + 45]^-$ cluster ions were the most intense ions in the mass spectrum of RDX.

The formation of RDX cluster ions in LC/MS and the origin of the clustering agents have been studied in order to determine whether the clustering anions originate from self-decomposition of RDX in the source or from impurities in the mobile phase [24]. Isotopically labeled RDX ($^{13}C_3$-RDX and $^{15}N_6$-RDX) was used in order to establish the composition and formation route of RDX adduct ions produced in ESI and APCI sources. Results showed that in ESI, RDX clusters with formate, acetate, hydroxyacetate, and chloride anions, were present in the mobile phase as impurities at ppm levels. In APCI, part of the RDX molecules decompose, yielding NO_2^- species, which in turn cluster with a second RDX molecule, producing abundant $[M + NO_2]^-$ cluster ions. Fig. 24.6 shows the ESI mass spectrum of RDX. The ions $[M + 45]^-$, $[M + 59]^-$, $[M + 75]^-$, $[2M + 59]^-$, and $[2M + 75]^-$ are adduct ions formed as a result of the presence of these impurities, probably in the HPLC-grade methanol of the mobile phase.

However, as mentioned above, the addition of an additive, such as ammonium nitrate or ammonium chloride [22,25] will produce significant adduct ions at abundances higher by several orders of magnitude than the impurity cluster ions.

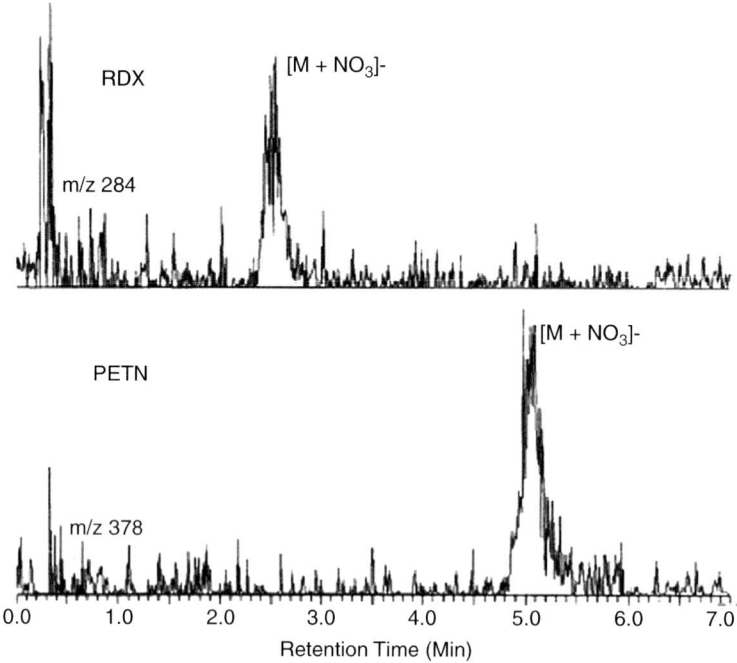

Fig. 24.5. LC/MS-ESI mass chromatograms of a 25 pg/μl Semtex sample (a plastic explosive containing RDX and PETN) with post column introduction of ammonium nitrate. Reprinted from X. Zhao *et al.*, J. Chromatogr. A, 977 (2002) 59, Copyright 2002, with permission from Elsevier.

Fig. 24.7 shows the negative ion APCI mass spectrum of RDX with addition of ammonium chloride [25]. The major ions are $[M + {}^{35}Cl]^-$ and $[M + {}^{37}Cl]^-$.

LC/MS-ESI in the negative ion mode with additives forming adduct ions of explosives with chloride, formate, acetate, and nitrate was investigated [26]. HPLC separation was obtained with a Hypersil ODS C18 column (100×2.1 mm I.D.), using an isocratic mobile phase of methanol-aqueous mixture (50:50), at a flow rate of 150 μl/min. Fig. 24.8 shows the negative ion ESI mass spectrum of a mixture of explosives containing EGDN, NG, TNT, PETN, RDX, and HMX. The aqueous mixture contained 0.3 mM ammonium chloride, ammonium formate, and ammonium nitrate.

LC/MS-ESI in the positive ion mode was used for trace analysis of TATP [27]. Chromatographic separation was carried out with a YMC ProC18 (150×2.0 mm I.D., 3 μm particle size) column, using methanol-water (70:30) with 5 mM ammonium acetate buffer as mobile phase, at flow rates between 100 and 200 μl/min. Samples were injected as acetonitrile solutions. The peaks observed in the LC/MS mass spectrum were at m/z 75, 89, 90, 91, 102, 107, 194, 240, and 252. The ion at m/z 240 was believed to be the $[M + NH_4]^+$ adduct ion, formed because of the use of nitrogen drying gas. This ion was enhanced when using ammonium acetate buffer. Detection limit of TATP was in the 100 pg/μl range.

Mass spectrometry with ESI and APCI ionization in the negative ion mode was used for the analysis of urea nitrate, a powerful improvised explosive [28]. Methanol

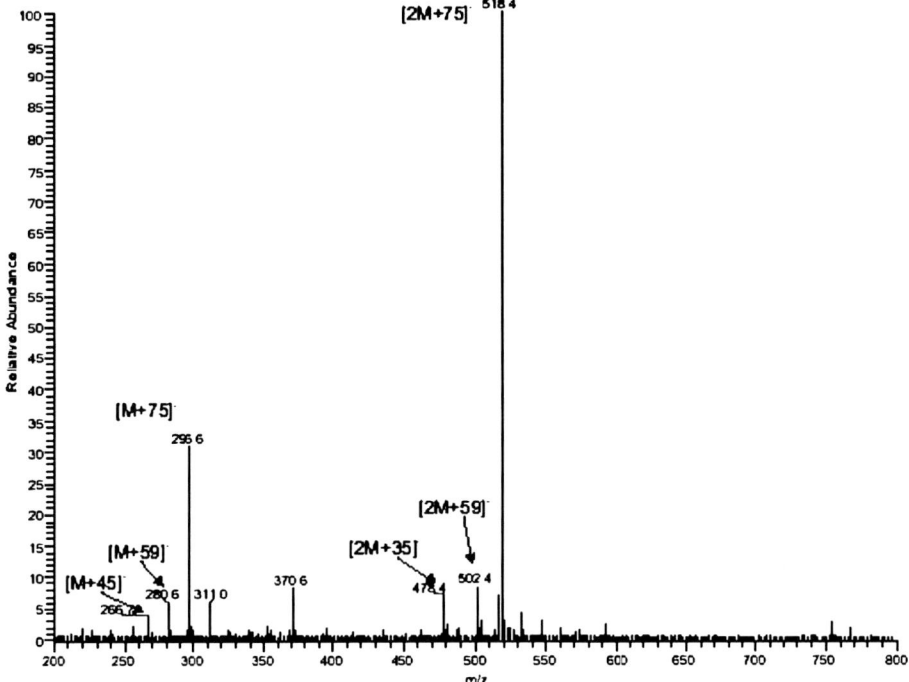

Fig. 24.6. ESI mass spectrum of RDX without additive. Reprinted from A. Gapeev *et al.*, Rapid Commun. Mass Spectrom., 17 (2003) 943, Copyright 2003, with permission from Wiley.

solutions of the explosive were introduced by loop injection and by HPLC, using a Synergi MAX-RP 80A column (150 × 2 mm I.D., 4 μm particle size). Optimal results were obtained in negative ion ESI at a heated capillary temperature of 50°C and in negative ion APCI at a vaporizer temperature of 180°C. Under these conditions, characteristic adduct ions were observed, in both ESI and APCI, at m/z 185, $[M + NO_3]^+$, and m/z 248, $[M + NO_3 + HNO_3]^+$.

Fig. 24.9 shows the negative ion APCI mass spectrum of urea nitrate, obtained by loop injection at a vaporizer temperature of 180°C and capillary temperature of 70°C.

Desorption electrospray ionization (DESI) (without LC) has been used for trace detection of TNT, RDX, HMX, PETN, and the plastic explosives C-4, Semtex-H, and Detasheet, directly from a wide variety of surfaces (metal, plastic, paper, polymer) without sample preparation or pretreatment [29]. The DESI ionization technique is based on directing a pneumatically assisted electrospray onto a surface and collecting the secondary ions generated by the interaction of charged microdroplets or gas-phase ions derived from the electrospray, with neutral molecules of analyte present on the surface. Limits of detection were in the subnanogram to subpicogram range. Addition of reagents in the spray solution formed characteristic adduct ions that improved selectivity and identification capability. Fig. 24.10a shows the negative ion DESI spectrum for a plastic explosive C-4 fingerprint on glass, using

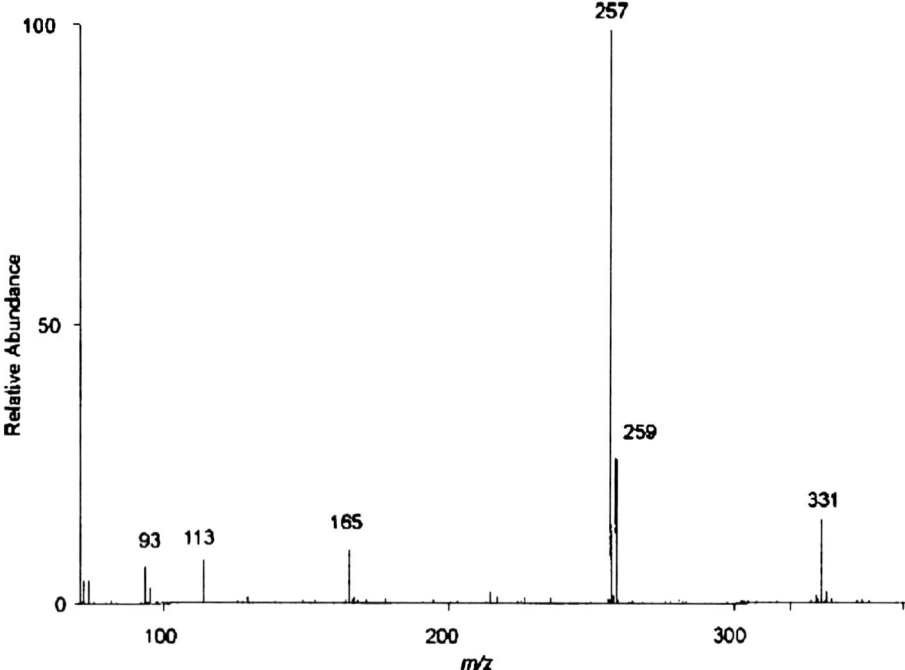

Fig. 24.7. Negative ion APCI mass spectrum of RDX with addition of ammonium chloride. Reprinted from G.S. Evans *et al.*, Rapid Commun. Mass Spectrom., 16 (2002) 1883, Copyright 2002, with permission from Wiley.

methanol–water–HCl (1:1:0.05%) as spray solvent. Identity of the ions was confirmed by tandem mass spectrometry (MS/MS). Fig. 24.10b shows the negative ion DESI spectrum of a C-4 fingerprint after five transfers onto a glass slide. The absolute abundance of the ion at m/z 257 $(RDX + ^{35}Cl)^-$ for the five transfers is more than one order of magnitude lower than the one of the original fingerprint.

24.7 ANALYSIS OF EXPLOSIVES BY ION MOBILITY SPECTROMETRY (IMS)

Although IMS of explosives is mainly used for the detection of hidden explosives, it can also be used in the forensic laboratory for the screening of suspected items for traces of explosives.

The IMS consists of a sample inlet system, an atmospheric pressure ion source, followed by an ion-molecule reactor, an ion drift spectrometer, and a detector.

Sample ions formed in the reactor are injected into the drift region by an applied electric field, where they are separated according to their mobility as they travel through a drift gas. The ion mobility spectrum, called plasmagram, consists of a plot of ion current as a function of drift time, which depends on the ionic mass [30].

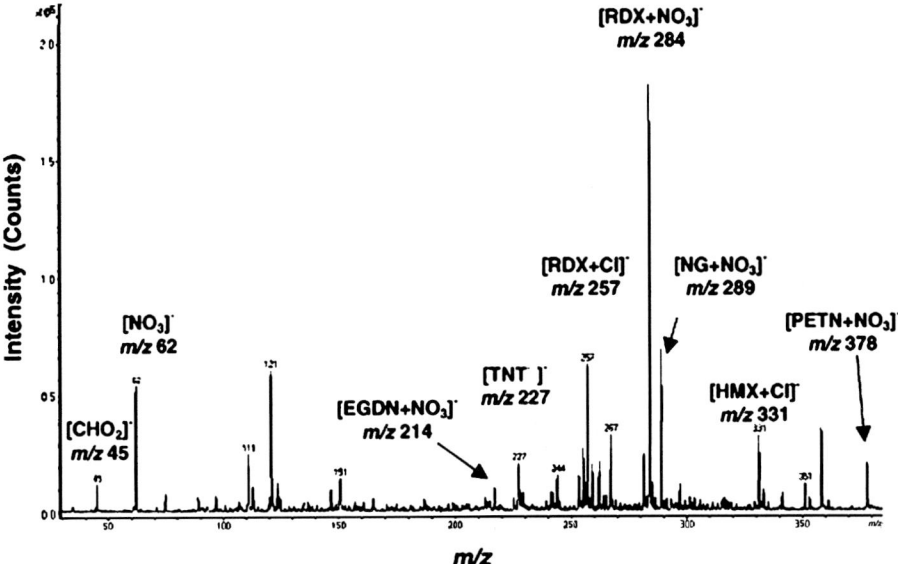

Fig. 24.8. Negative ion ESI mass spectrum of a mixture of explosives containing EGDN, NG, TNT, PETN, RDX and HMX. The aqueous mixture contained 0.3 mM ammonium chloride, ammonium formate, and ammonium nitrate. Reprinted from J.A. Mathis *et al.*, Rapid Commun. Mass Spectrom., 19 (2005) 99, Copyright 2004, with permission from Wiley.

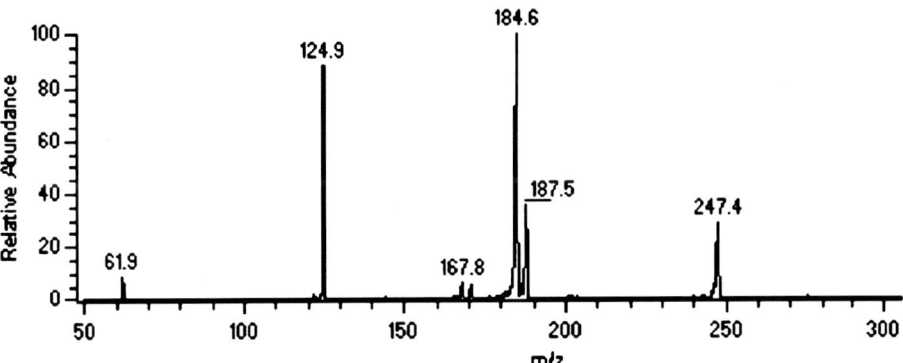

Fig. 24.9. Negative ion APCI mass spectrum of urea nitrate, obtained by loop injection at a vaporizer temperature of 180°C and capillary temperature of 70°C. Reprinted from T. Tamiri, Rapid Commun. Mass Spectrom., 19 (2005) 2094, Copyright 2005, with permission from Wiley.

Coupling of solid-phase extraction (SPE) with IMS in the negative ion mode has been found useful for the identification of trace amounts of explosives in liquid matrices, since the extraction provides a means of pre-concentrating the explosives and removing interferents [31]. Extraction was carried out with 3 M Empore SDB-RPS

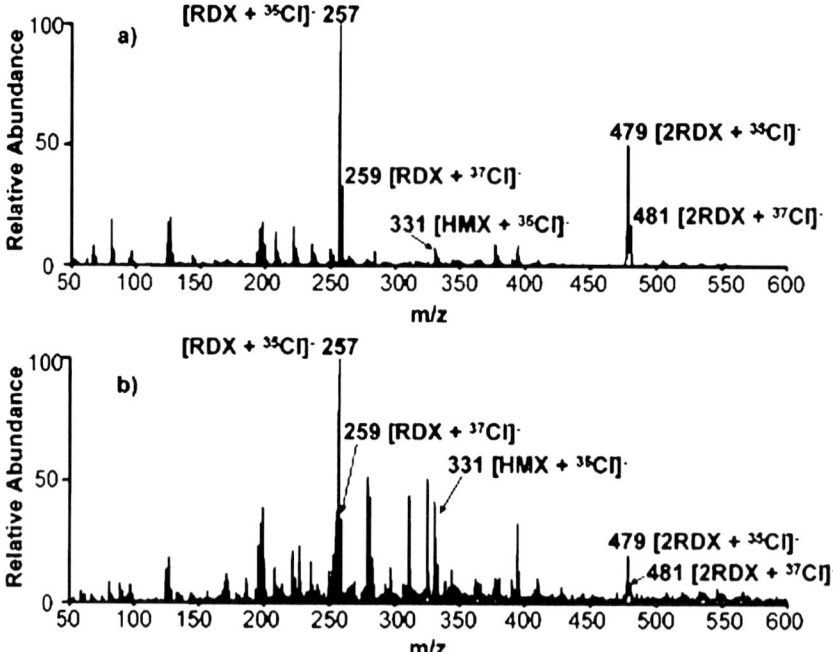

Fig. 24.10. (a) Negative ion DESI spectrum for a plastic explosive C-4 fingerprint on glass, using methanol–water–HCl (1:1:0.05%) as spray solvent. (b) Negative ion DESI spectrum of a C-4 fingerprint after five transfers onto a glass slide. Reprinted with permission from I. Cotte-Rodriguez *et al.*, Anal. Chem., 77 (2005) 6755, Copyright 2005, American Chemical Society.

SPE disks. The disk was then inserted into the IMS, where the explosives are thermally desorbed from the disk. In addition to the calibration gas, methyl salicylate, a reagent, and hexachloroethane was used, which produced chloride ions that selectively attach to the nitro containing molecules. Detection limits were in the low ppt range for TNT, RDX, and HMX.

The use of IMS for forensic explosive residue analysis was investigated [32]. Standard solutions of TNT, RDX, PETN, and Tetryl were analyzed by IMS, followed by qualitative analyses of explosive residues and a handswab sample taken from a suspect, which showed traces of TNT and Tetryl.

Fig. 24.11 shows the IMS plasmagrams of a standard mixture (a) and the handswab sample (b).

TATP was analyzed by IMS in the positive ion mode without the addition of a reactant ion [33]. TATP was detected as a cluster of three peaks. Highest sensitivity was obtained when TATP was dissolved in toluene. Detection limit was $187\,\mu g/ml$.

The IMS was then coupled to a mass spectrometer. The mass spectrum consisted mainly of one intense ion at m/z 223, MH^+.

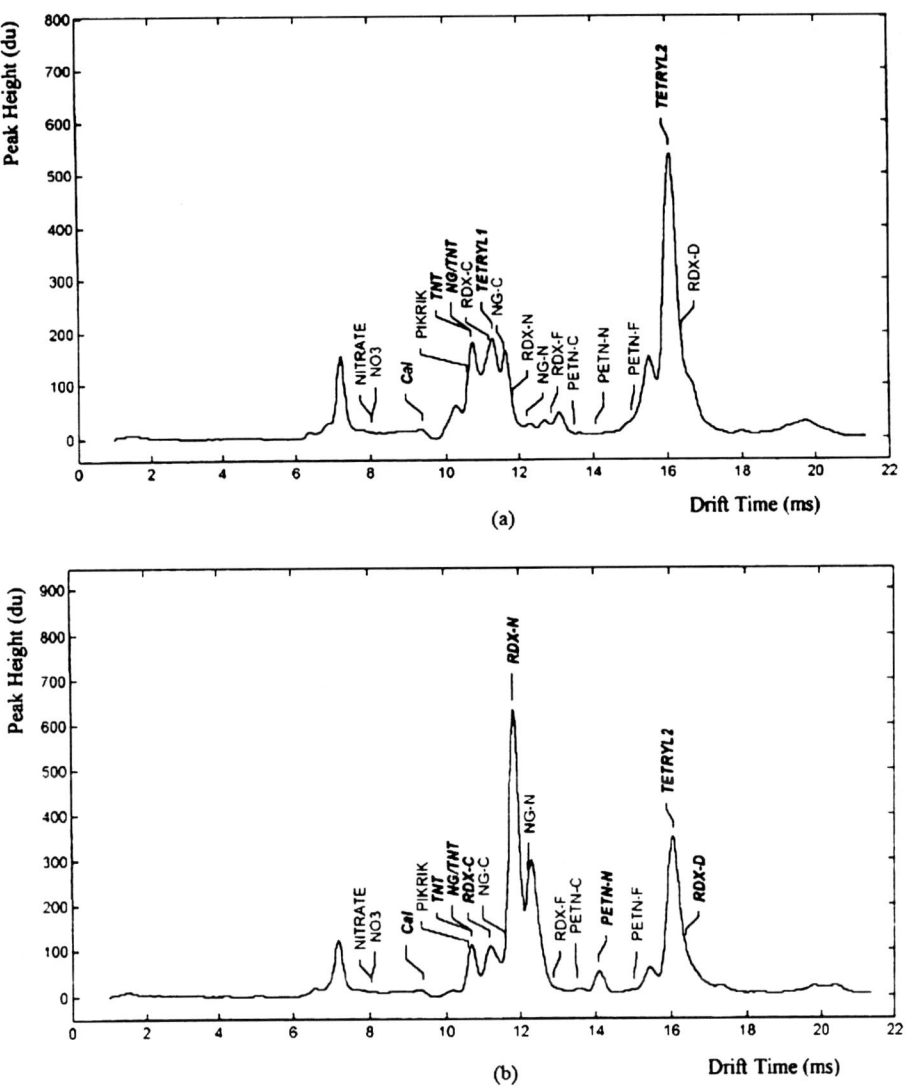

Fig. 24.11. IMS plasmagrams of a standard mixture (a) and a handswab sample (b). Reprinted with permission from H. Koyuncu *et al.*, Turk. J. Chem., 29 (2005) 255, Copyright 2005, TUBITAK.

24.8 CONCLUDING SUMMARY

Analysis of explosives from post-explosion debris and from soil and water can be carried out by GC/MS and LC/MS, with a distinct trend towards LC/MS. While chromatographic methods are based on retention times and therefore need comparison with standards, mass spectrometry is an identification method and provides a "fingerprint" of the investigated sample. The combination of chromatography and mass spectrometry incorporates both separation and identification capabilities and is therefore an excellent choice for the analysis of explosive residues in the forensic laboratory.

Tandem mass spectrometry (MS/MS) provides an additional dimension for improved selectivity and confirmation.

TLC and IMS are recommended for screening purposes only and need additional confirmation by an identification technique.

24.9 REFERENCES

1   J. Yinon; S. Zitrin, The analysis of explosives, Pergamon Press, Oxford, 1981.
2   J. Yinon. In: M.J. Bogusz (Ed.), Handbook of analytical separations, Vol. 2 – Forensic Science, Elsevier, Amsterdam, 2000, p. 604.
3   J. Yinon; S. Zitrin, Modern methods and applications in analysis of explosives, Wiley, Chichester, 1993.
4   J. Yinon, Forensic and environmental detection of explosives, Wiley, Chichester, 1999.
5   J.C. Oxley, J.L. Smith, K. Shinde and J. Moran, Propell. Explos. Pyrotech., 30 (2005) 127.
6   A.D. Beveridge, Forensic Sci. Rev., 4 (1992) 17.
7   K.G. Furton, L. Wu and J.R. Almirall, J. Forensic Sci., 45 (2000) 857.
8   T. Tamiri, S. Zitrin, S. Abramovich-Bar, Y. Bamberger and J. Sterling. In: J. Yinon (Ed.), Advances in analysis and detection of explosives, Kluwer, Dordrecht, 1993, p. 323.
9   S. Calderera, D. Gardebas and F. Martinez, Forensic Sci. Int., 137 (2003) 6.
10 R. Waddell, D.E. Dale, M. Monagle and S.A. Smith, J. Chromatogr. A, 1062 (2005) 125.
11 H.E. Cullum, C. McGavigan, C.Z. Uttley, M.A.M. Stroud and D.C. Warren, J. Forensic Sci., 49 (2004) 684.
12 C.A. Crowson, H.E. Cullum, R.W. Hiley and A.M. Lowe, J. Forensic Sci., 41 (1996) 980.
13 Restek Application Note # 59361, Restek Corporation, Bellefonte, PA 16823, USA, 2001.
14 T. Borch and R. Gerlach, J. Chromatogr. A, 1022 (2004) 83.
15 B. Paull, C. Roux, M. Dawson and P. Doble, J. Forensic Sci., 49 (2004) 1181.
16 E.M. Sigman and C.-Y. Ma, J. Forensic Sci., 46 (2001) 6.
17 J.G. McDonald, K.H. Mount, J.M. Egan, G.L. Egan and M.L. Miller, Proceedings of the 8th International Symposium on Analysis and Detection of Explosives, Ottawa, Canada, June 2004, p. 85.
18 D. Muller, A. Levy, R. Shelef, S. Abramovich-Bar, D. Sonenfeld and T. Tamiri, J. Forensic Sci., 49 (2004) 935.
19 G.L. Egan, M.E. Russ, M.L. Miller, K.H. Mount and J.G. McDonald, Proceedings of the 8th International Symposium on Analysis and Detection of Explosives, Ottawa, Canada, June 2004, p. 544.
20 J. Yinon, Forensic analysis of explosives by LC/MS. Forensic Sci. Rev., 13 (2001) 19.
21 M.L. Miller, R. Mothershead, J. Leibowitz, K. Mount and R. Martz, Proceedings of the 45th ASMS Conference on mass spectrometry and allied topics, Palm Springs, CA, 1997, p. 52.
22 X. Zhao and J. Yinon, J. Chromatogr. A, 977 (2002) 59.
23 H.R. Beller and K. Tiemeier, Environ. Sci. Technol., 36 (2002) 2060.
24 A. Gapeev, M. Sigman and J. Yinon, Rapid Commun. Mass Spectrom., 17 (2003) 943.
25 C.S. Evans, R. Sleeman, J. Luke and B.J. Keely, Rapid Commun. Mass Spectrom., 16 (2002) 1883.
26 J.A. Mathis and B.R. McCord, Rapid Commun. Mass Spectrom., 19 (2005) 99.
27 L. Widmer, S. Watson, K. Schlatter and A. Crowson, Analyst, 127 (2002) 1627.
28 T. Tamiri, Rapid Commun. Mass Spectrom., 19 (2005) 2094.
29 I. Cotte-rodriguez, Z. Takats, N. Talaty, H. Chen and R.G. Cooks, Anal. Chem., 77 (2005) 6755.
30 G.A. Eiceman and J.A. Stone, Anal. Chem., 76 (2004) 390A.
31 T.L. Buxton and P.D. Harrington, Appl. Spectrosc., 57 (2003) 223.
32 H. Koyuncu, E. Seven and A. Calimli, Turk. J. Chem., 29 (2005) 255.
33 G.A. Buttigieg, A.K. Knight, S. Denson, C. Pommier and M.B. Denton, Forensic Sci. Int., 135 (2003) 53.

M.J. Bogusz (Ed.). Forensic Science
Handbook of Analytical Separations, Vol. 6
© 2008 Published by Elsevier B.V.

CHAPTER 25

Chemical warfare agents

Paul A. D'Agostino

DRDC Suffield, P.O. Box 4000 Station Main, Medicine Hat, AB, Canada, T1A 8K6

25.1 INTRODUCTION

Chemical warfare agents [1–5] are toxic chemicals controlled by the Convention on the Prohibition of the Development, Production, Stockpiling, and Use of Chemical Weapons and their Destruction (commonly referred to as the Chemical Weapons Convention (CWC)). Poisonous or toxic compounds have been utilized in an effort to gain military superiority throughout history but it is only during the past century that chemical warfare agents have been produced and used on a large scale. Lachrymator (tear agents) grenades were used by the French and German armies at the outbreak of the First World War, but it was not until the German army used chlorine near Ypres in 1915 that the world entered the modern era of chemical warfare. Other chemical warfare agents including phosgene and mustard were weaponized during the First World War and were used by both sides throughout the conflict. Widespread use of these weapons resulted in more than a million chemical weapons casualties, with attributable deaths exceeding 100,000.

Chemical warfare agent development and use continued following the First World War despite the signing of the 1925 Geneva Protocol, which banned the first use of chemical weapons. Mustard was used by the Italian army against Ethiopia during the 1936–1937 war and chemical weapons were used in China by the Japanese armed forces during the Second World War. Germany discovered tabun and sarin, two highly toxic organophosphorus compounds, in the late 1930s and produced substantial stocks of the former during the Second World War. However, although all the major powers established extensive weapons stockpiles, aside from their use in China and a number of other isolated incidents, chemical weapons were not employed in that conflict. Following the Second World War most of the chemical warfare agent stocks were burned or dumped at sea, the latter method of disposal resulting in a continuing environmental hazard in the Baltic Sea and other areas.

Research during the 1950s resulted in the discovery of the V-agents, the most toxic of the standard nerve agents. VX was weaponized by the United States, while similar efforts in the former Soviet Union resulted in the discovery of a somewhat more toxic isomer of VX, which was also weaponized. Nerve and mustard agents were used repeatedly by Iraq in the Gulf War (1980–1988), with Iranian casualties reported to be in the tens of thousands. The willingness of the Iraqi regime to use chemical weapons in the war against Iran and against the indigenous Kurdish population was recognized by United Nations armed forces during the Gulf War (1990–1991) and considered a real threat to the coalition forces during the liberation of Kuwait. Following the Gulf War, the United Nations established a special commission (UNSCOM) to uncover and oversee the destruction of Iraqi chemical and biological stocks and related delivery systems. Most recently, sarin was released by the Aum Shinrikyo cult in the Tokyo underground transit system resulting in thousands seeking medical attention and 12 deaths.

Two years after the Tokyo transit system release, on April 29, 1997, the CWC collected sufficient signatures for the treaty to come into force. To date, 175 Member States have ratified the CWC and agreed not to develop, produce, stockpile, transfer, or use chemical weapons and agreed to destroy their own chemical weapons and production facilities. A strong compliance monitoring regime involving site inspections was built into the CWC to ensure that the treaty remains verifiable. The Organisation for the Prohibition of Chemical Weapons (OPCW) based in the Hague has responsibility for implementation of the treaty. Routine OPCW inspections have taken place at declared sites, including small-scale production, storage, and destruction sites, and challenge inspections may take place at sites suspected of non-compliance. Proliferation of chemical weapons and their use will hopefully decrease over the coming years as the OPCW proceeds towards its goal of worldwide chemical weapons destruction.

The al-Qaeda terrorist attacks of September 2001 and subsequent Capitol Hill anthrax letters heightened concern worldwide over the possible use of chemical or biological weapons by terrorist groups. Concerns over terrorist use of chemical weapons within the homeland security and defense communities and the requirements of a verifiable CWC have all driven the development and application of analytical methods for the detection and identification of chemical warfare agents [5]. Analytical techniques play an important role in this process, as sampling and analysis will be conducted to ensure treaty compliance, to investigate allegations of use, and to verify the use of these weapons for forensic purposes.

Chemical warfare agents have been classified into a number of categories based on their effect on humans. Categories include nerve, blister, choking, vomiting, blood, tear, and incapacitating, with the nerve and blister agent categories being most significant in terms of military utility and past use. For these reasons, the analytical methods developed for these compounds will be emphasized over the other groups. The choking, blood, and vomiting agents, generally considered obsolescent chemical agents, were employed during the First World War. The tear agents were used during the Vietnam War by the United States, but their primary use today is in riot control and training.

Table 25.1 lists the most common chemical warfare agents, with their Chemical Abstracts registry numbers. It has been estimated that more than 10,000 compounds are controlled under the CWC, although in practical terms the actual number of

TABLE 25.1
COMMON CHEMICAL WARFARE AGENTS

Full Name (Trivial Name(s))	CAS No.
(a) Nerve (reacts irreversibly with cholinesterase which results in acetylcholine accumulation, continual stimulation of the body's nervous system, and eventual death)	
O-Isopropyl methylphosphonofluoridate (sarin, GB)	107-44-8
O-Pinacolyl methylphosphonofluoridate (soman, GD)	96-64-0
O-Cyclohexyl methylphosphonofluoridate (GF)	329-99-7
O-Ethyl *N,N*-dimethylphosphoramidocyanidate (tabun, GA)	77-81-6
O-Ethyl *S*-2-diisopropylaminoethyl methylphosphonothiolate (VX)	50782-69-9
(b) Blister (affects the lungs, eyes, and produces skin blistering)	
Bis(2-chloroethyl)sulfide (mustard, H)	505-60-2
1,2-Bis(2-chloroethylthio)ethane (sesquimustard, Q)	3563-36-8
Bis(2-chloroethylthioethyl)ether (T)	63918-89-8
Tris(2-chloroethyl)amine (HN-3)	555-77-1
2-Chlorovinyldichloroarsine (lewisite, L)	541-25-3
(c) Choking (affects respiratory tract and lungs)	
Chlorine	7782-50-5
Phosgene (CG)	75-44-5
(d) Vomiting (causes acute pain, nausea, and vomiting in victims)	
Diphenylarsinous chloride (DA)	712-48-1
10-Chloro-5,10-dihydrophenarsazine (adamsite, DM)	578-94-9
Diphenylarsinous cyanide (DC)	23525-22-6
(e) Blood (prevents transfer of oxygen to the body's tissues)	
Hydrogen cyanide (HCN, AC)	74-90-8
(f) Tear (causes tearing and irritation of the skin)	
[(2-Chlorophenyl)methylene]propanedinitrile (CS)	2698-41-1
2-Chloro-1-phenylethanone (CN)	532-27-4
Dibenz[*b,f*][1,4]oxazepin (CR)	257-07-8
(g) Incapacitating (prevents normal activity by producing mental or physiological effects)	
3-Quinuclidinyl benzilate (BZ)	6581-06-2

chemical warfare agents, precursors, and degradation products that are contained in the OPCW database is in the hundreds. The structures of some of the more common nerve and blister agents and their hydrolysis products are illustrated in Fig. 25.1.

This chapter reviews field (on-site) and laboratory (off-site) based analytical methods used for the determination of chemical warfare agents, their characteristic degradation products, and related compounds, with an emphasis on analytical methods developed over the past five to seven years which could be used for forensic or other purposes.

25.2 CHEMICAL WARFARE AGENT ANALYSIS – OVERVIEW

Chemical warfare agents have often been referred to as warfare gases, but in fact most chemical warfare agents exist as liquids at ambient temperatures. The common

Nerve Agents

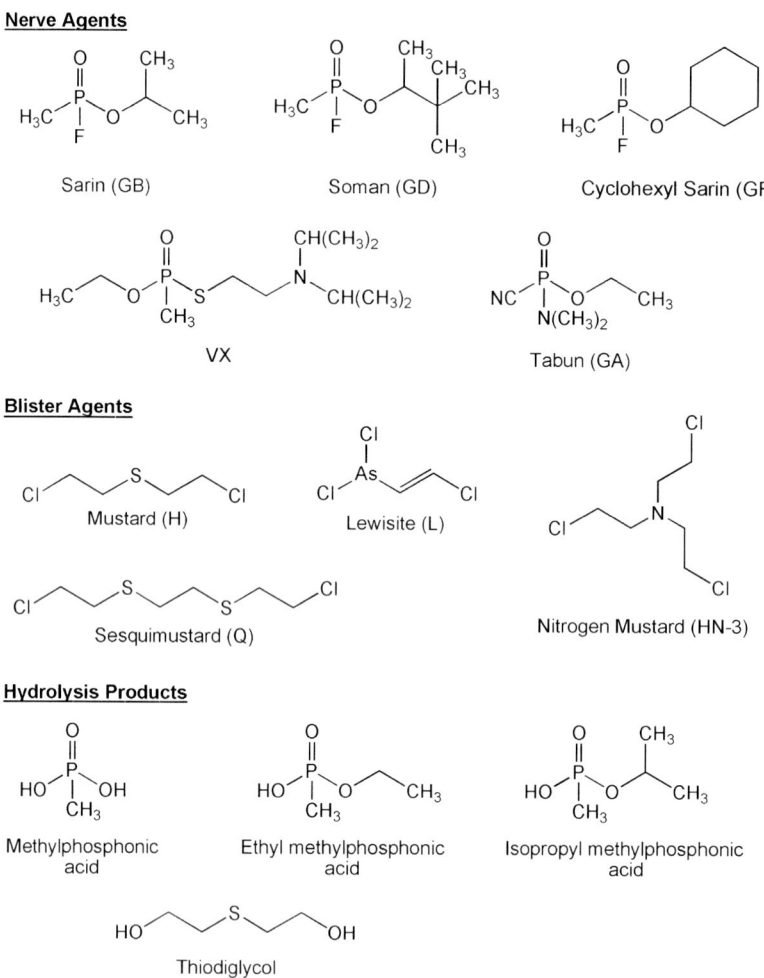

Sarin (GB)   Soman (GD)   Cyclohexyl Sarin (GF)

VX   Tabun (GA)

Blister Agents

Mustard (H)   Lewisite (L)

Sesquimustard (Q)   Nitrogen Mustard (HN-3)

Hydrolysis Products

Methylphosphonic   Ethyl methylphosphonic   Isopropyl methylphosphonic
acid   acid   acid

Thiodiglycol

Fig. 25.1. Common chemical warfare agents and related hydrolysis products.

chemical warfare agents have varying degrees of volatility and may pose both a vapor hazard as well as a liquid contact hazard. This physical characteristic has made the analysis of chemical warfare agents amenable to the analytical techniques commonly employed for most environmental analyses, namely gas chromatography (GC) and liquid chromatography (LC) with a variety of detectors including mass spectrometry (MS) [3,5–11]. Synthetic or relatively pure samples not requiring chromatographic separation are also frequently characterized by nuclear magnetic resonance (NMR) [12] or Fourier transform infrared (FTIR) spectroscopy [13].

The OPCW inspectorate, an important end user of analytical techniques for chemical warfare agents, requires the use of two or more spectrometric techniques and the availability of authentic reference standards for the unambiguous identification of controlled compounds. For this reason, the combined use of GC-FTIR [13] has received increased attention as newer technologies have led to detection limits approaching those routinely reported during GC-MS analysis. For analyses involving low levels of chemical warfare agents in the presence of high levels of interfering chemical background, tandem mass spectrometry (MS/MS) is often employed.

Samples containing chemical warfare agents typically contain multiple components that are best characterized following chromatographic separation. GC has been used extensively for the separation and identification of chemical warfare agents, with GC-MS being used frequently for the characterization of these compounds [6–10]. GC-MS analysis methods form the cornerstone of the OPCW recommended analytical procedures and have been used extensively during designated laboratory proficiency testing [14]. Electron impact mass spectrometric data and spectrometric or spectroscopic data from a second analytical technique (e.g. LC-MS, FTIR, or NMR) have typically been acquired to meet OPCW identification requirements. This "battery-of-tests" approach to compound confirmation often leads to identification of target compounds by multiple spectrometric or spectroscopic techniques [14–16].

As nerve and blister agents undergo hydrolysis in the environment [17,18], methods are required for the detection and confirmation of their hydrolysis products as well. The most commonly targeted degradation compounds include the alkyl methylphosphonic acids and methylphosphonic acid resulting from nerve agent hydrolysis and the primary hydrolysis product of mustard, thiodiglycol (Fig. 25.1). These compounds are significant as they would not be routinely detected in environmental samples and their identification strongly suggests the prior presence of chemical warfare agents.

Many degradation products of chemical warfare agents, especially those formed following hydrolysis of nerve agents, are much less volatile than the parent compounds and must be derivatized prior to GC analysis. A number of derivatization reagents, leading to the formation of pentafluorobenzyl, methyl, or silyl esters, have been investigated to allow GC-MS analysis of hydrolysis products of chemical warfare agents [9]. Increasingly researchers have developed atmospheric pressure ionization (API) based LC-MS methods (e.g. electrospray (ESI), ionspray (IS), and atmospheric pressure chemical ionization (APCI)) as complementary or replacement methods for the characterization of chemical warfare agents and/or their degradation products. A number of LC-MS methods have been reported for the confirmation of these compounds, with this technique being used most frequently during the analysis of aqueous samples or extracts [8,11,19].

25.3 SAMPLE HANDLING

Samples contaminated with chemical warfare agents generally fall into one of the following general categories: (a) munitions or munition fragments (e.g. neat liquid or artillery shell casing), (b) environmental (e.g. soil, water, vegetation, or air samples), (c) man-made materials (e.g. swabs, polymers, clothing, or painted surfaces), and (d) biomedical media (e.g. blood or urine). The ease of analysis depends on the amount of sample preparation required to obtain a suitable sample or extract for chromatographic analysis. In the simplest case where neat liquids can be obtained, the sample typically only requires dilution with a suitable solvent prior to analysis. Aqueous samples may be analyzed directly or following solvent extraction with an organic solvent, while biological fluids typically require extensive sample handling and/or derivatization prior to analysis [19].

Soil and other solid samples are often collected following a chemical warfare agent incident and these types of samples generally require, at a minimum, solvent extraction and concentration prior to analysis. Extraction of chemical warfare agents from soil samples may be accomplished using a number of solvents (e.g. hexane, dichloromethane or water), with dichloromethane being the most commonly employed extraction solvent for GC-MS determination of chemical warfare agents. A small portion of soil sample would typically be ultrasonically extracted with dichloromethane in screw-capped glass vial for ~10 min. If necessary, fine particles may be removed from the dichloromethane extract by centrifugation or filtration prior to concentration and analysis by GC-MS (or another analytical technique). Sample handling procedures developed for contaminated soils can often be extended to other solid samples with modifications to extraction solvent, pH, volume, or vessel [20]. These handling procedures vary with the methods described in the sections dealing with gas chromatography–mass spectrometry, liquid–chromatography mass spectrometry, and other analytical techniques. Detailed information on the sample handling procedures may be obtained from the individual papers reviewed in this chapter.

Between 1989 and 1994, five international round robin analytical exercises were hosted by the Finnish Institute for Verification of the Chemical Weapons Convention in an effort designed to establish recommended operating procedures for the analysis of chemical warfare agents in a variety of matrices. Sample handling procedures evaluated and adopted by the OPCW have recently been reviewed [20,21]. These recommended operating procedures, complete with flow diagrams, represent sound methods for sample handling and analysis of chemical warfare agents in a variety of matrices. The OPCW continues to develop and evaluate new methods for target compounds and different media. This is in part achieved through their on-going OPCW proficiency testing of designated laboratories that serves a dual purpose, the evaluation of a laboratory's capability as well as a means to refine existing recommended operating procedures. The requirements of the OPCW and the increased threat of terrorism have in large part driven the development of newer sample handling procedures [22–24], including those involving single drop microextraction [25] and supercritical or pressurized fluids [26–29] for the selective concentration of chemical warfare agents. Solid phase microextraction (SPME), a technique that is increasingly used for chemical warfare agent applications, is discussed in Section 25.7 (Field Analyses).

25.4 GAS CHROMATOGRAPHY-MASS SPECTROMETRY

Samples contaminated with chemical warfare agents typically contain multiple components that are best characterized following chromatographic separation. Capillary column GC remains the most frequently employed analytical separation method for the screening of samples suspected to contain chemical warfare agents [30]. Separation of chemical warfare agents may be achieved with many of the commercially available fused silica columns coated with polysiloxane or other films and retention index data relative to *n*-alkanes [31] and alkyl bis(trifluorome-thyl)phosphine sulfides (M-series) [32] have been reported for hundreds of chemical warfare agents and related compounds relevant to the CWC [30]. In general, the best separations have been achieved with 5% phenyl methylpolysiloxane or (86%)-di-methyl-(14%)-cyanopropylphenyl-polysiloxane films. Chiral stationary phases have also been developed for the resolution of stereoisomers of several chiral nerve agents, including sarin [33].

Most of the GC detectors commonly applied to pesticide residue analysis have also been applied to the screening of samples for chemical warfare agents with detection limits typically being in the nanogram to picogram range [30]. Flame ionization detection (FID) is routinely used for preliminary analyses, as this technique provides a good indication of the complexity of a sample extract. The need for higher specificity and sensitivity has led to the application of element specific detectors such as flame photometric detection (FPD) [14,15,34–38] , thermionic detection (TID) [36–38], atomic emission (AED) [15,16], and electron capture detection (ECD) [38,39]. Simultaneous acquisition of FID and element specific detection may be beneficial [14], particularly for more complex matrices. While data obtained with these detectors may provide strong collaborative evidence for the presence of chemical warfare agents, they cannot be used for full confirmation. Use of GC with one or more spectrometric techniques such as MS is required to confirm the presence of chemical warfare agents and would typically be used to support forensic investigations.

MS is the method of choice for the detection and characterization of chemical warfare agents, their precursors, degradation products, and related compounds. Extensive use has been made of GC-MS and the mass spectra of numerous chemical warfare agents and related compounds have been published, with the mass spectra of chemical warfare agents and related compounds being available in the OPCW Central Analytical Database, commercial, and defense community databases.

Most of these data were obtained under electron impact (EI) ionization conditions. However many of the chemical warfare agents, in particular the organophosphorus nerve agents and the longer chain blister agents related to mustard, do not provide molecular ion information under EI-MS. This hinders confirmation of these chemical warfare agents and makes identification of novel chemical warfare agents or related impurities difficult. Considerable effort has been devoted to the use of chemical ionization (CI) as a complementary ionization technique. This milder form of ionization generally affords molecular ion information for the chemical warfare agents and has been used extensively for the identification of related compounds or impurities in chemical warfare agent munitions samples and environmental sample extracts. The identity of these related compounds could be important during forensic investigations since this data may provide information on the origin of samples, synthetic procedure(s), or degree of degradation (weathering).

Isobutane, ethylene, and methane gases were initially demonstrated as suitable CI gases for the acquisition of organophosphorus nerve agent CI-MS data [40]. The efficacy of ammonia CI-MS for organophosphorus nerve agents and related compounds has been demonstrated [41] and many laboratories now employ this complementary confirmation technique [10,42,43]. Ammonia CI not only offers abundant molecular ion data but also affords a high degree of specificity, as less basic sample components are not ionized by the ammonium ion. Additional data may be obtained through the use of deuterated ammonia CI [44], as this technique provides useful hydrogen/deuterium exchange data that indicates the presence of exchangeable hydrogen(s) in CI fragmentation ions.

CI data can also be quite useful for differentiating compounds. VX and many VX-related compounds, including bis[2-(diisopropylamino)ethyl] disulfide, exhibit virtually indistinguishable EI data. These compounds lack a molecular ion and contain a base ion at m/z 114 due to $[CH_2N(iPr)_2]^+$ and additional ions related to the $SC_2H_4N(iPr)_2$ substituent. However, under ammonia CI conditions, mass spectra containing pseudo-molecular and CI fragmentation ions may be acquired, with this data being valuable to confirm molecular mass and to differentiate VX from the related compounds often found in VX-containing samples.

Capillary column GC-MS/MS offers the analyst the potential for highly specific, sensitive detection of chemical warfare agents as this technique significantly reduces the chemical noise associated with complex sample extracts. The specificity of product scanning with moderate sector resolution, as well as the specificity of ammonia CI, were first demonstrated with a hybrid tandem mass spectrometer during analysis of diesel exhaust samples [45,46], and this technique has been used recently during the analysis of more complex biomedical samples [47–52].

Both the nerve and blister agents undergo hydrolysis and methods are required for retrospective detection and confirmation of these compounds. The degradation products of the chemical warfare agents, in particular the nerve agents, are non-volatile hydrolysis products that must be derivatized prior to GC analysis. A variety of derivatization reagents [9], leading to the formation of pentafluorobenzyl, methyl, *tert*-butyldimethylsilyl, and trimethylsilyl ethers (or esters), have been investigated to allow GC analysis of these compounds.

GC-MS has become firmly entrenched as an important method for the identification of chemical warfare agents, their hydrolysis products, and related compounds in a variety of different sample types that could be important during a forensic investigation. Table 25.2 lists recently reported methods by sample media and includes the mode of ionization, GC column used for analysis, and the compounds analyzed. Analytical techniques for environmental samples such as air, water, and soil [42,53–60], the media considered most significant from a battlefield perspective during the Cold War, have declined in relative importance compared to techniques for biomedical samples, given the prospect of chemical warfare agent use

TABLE 25.2
GC-MS ANALYSES OF CHEMICALS WARFARE AGENTS AND RELATED COMPOUNDS

Media	Reference	Ionization	GC Column	Compounds Analyzed
Air	[53]	EI	HP-5MS or RTX-5 amine (30 m × 0.25 mm)	Nitrogen mustards (HN-1, HN-3)
Air	[54]	EI	DB-5MS (25 m × 0.25 mm)	Mustard
Air	[55]	EI	DB-5MS (25 m × 0.25 mm)	Mustard
Air				Lewisites (derivatives)
Water	[56]	EI	HP-5MS (30 m × 0.25 mm)	Thiodiglycol (derivative)
Water	[57]	EI PICI (methane)	DB-5MS (30 m × 0.25 mm)	Thiodiglycol (derivative)
Soil	[42]	PICI (ammonia)	DB-5 (30 m × 0.25 mm)	Alkyl methylphosphonic acids (derivatives)
Soil	[58]	EI	HP-5MS (30 m × 0.25 mm)	Alkyl methylphosphonic acids and methylphosphonic acid (derivatives)
Soil	[59]	EI	HP-5MS (30 m × 0.25 mm)	Alkyl methylphosphonic acids and methylphosphonic acid (derivatives)
Soil	[60]	EI	DB-35MS (15 m × 0.25 mm)	Thiodiglycol and long chain diols associated with munitions grade mustard degradation
Munition	[61]	EI	HP-5MS (30 m × 0.32 mm)	Mustard and numerous munitions related compounds
Diesel	[62]	EI	DB-1701 and DB-5 (15 m × 0.32 mm)	Tear gas agents (CR, CN, CS, CH)
Diesel	[63]	EI	DB-5MS (30 m × 0.25 mm)	Lewisites (derivatives)
Synthetic	[64]	EI PICI (methane, ammonia)	HP-5 (25 m × 0.32 mm)	Large series of alkyl methylphosphonofluoridates
Synthetic	[60]	EI	DB-35MS (15 m × 0.25 mm)	Tabun related compounds
Urine	[48]	EI	DB-5 (30 m × 0.32 mm)	Alkyl methylphosphonic acids (derivatives)
Urine	[49]	PICI (isobutene)	DB-5MS (30 m × 0.25 mm)	Thiodiglycol (derivative) and the mustard metabolite 1,1'-sulfonylbis[2-(methylyhio) ethane]
Urine	[50]	PICI (isobutene)	DB-5MS (30 m × 0.25 mm)	Mustard metabolite 1,1'-sulfonylbis[2-(methylyhio) ethane]
Urine	[51]	PICI (isobutane)	DB-5MS (30 m × 0.25 mm)	Alkyl methylphosphonic acids and methylphosphonic acid (derivatives)
Urine	[52]	NICI (methane)	RTX-5MS (30 m × 0.25 mm)	Alkyl methylphosphonic acids and methylphosphonic acid (derivatives)
Urine	[65]	NICI (isobutene)	DB-1 (30 m × 0.53 mm)	Alkyl methylphosphonic acids (derivatives)
Urine	[66]	EI	DB-1 (30 m × 025 mm)	2-Chlorovinyl arsonous acid (derivative)

TABLE 25.2
CONTINUED

Media	Reference	Ionization	GC Column	Compounds Analyzed
Urine	[67]	EI	HP-5MS (30 m × 0.25 mm)	Alkyl methylphosphonic acids and methylphosphonic acid (derivatives)
Urine	[68]	EI	CP-SIL 5 CB (50 m × 0.32 mm)	2-Chlorovinyl arsonous acid (derivative)
Saliva	[65]	NICI (isobutene)	DB-1 (30 m × 0.53 mm)	Alkyl methylphosphonic acids (derivatives)
Blood	[43]	PICI (ammonia)	HP-5MS or RTX-1701 (no column dimensions)	Sarin
Blood	[47]	NICI (methane)	BPX5 (25 m × 0.22 mm)	Valine and histidine adducts of mustard
Blood	[68]	EI	CP-SIL 5 CB (50 m × 0.32 mm)	2-Chlorovinyl arsonous acid (derivative)
Blood	[69]	EI	HP-5MS (30 m × 0.25 mm)	Isopropyl methyl phosphonic acid (derivative)
Blood	[70]	NICI (methane)	CP-SIL 5 CB (50 m × 0.32 mm)	Valine adducts of mustard
Serum	[65]	NICI (isobutene)	DB-1 (30 m × 0.53 mm)	Alkyl methylphosphonic acids (derivatives)
Plasma	[71]	PICI (ammonia)	CIP SIL 19 CB (50 m × 0.32 mm)	Tabun, sarin, soman, GF, VX
Tissue	[72]	EI	HP-5MS (30 m × 0.25 mm)	Isopropyl methylphosphonic acid (derivative)

by terrorist groups. Following a chemical event like the Tokyo transit system release, the general population would want to know if they were exposed and the
forensic community would want to be able to establish the fact. As a result, a
number of new methods for the hydrolysis products of chemical warfare agents or
specific metabolites have been reported for the analysis of these compounds in
biomedical samples [43,47–52,65–72].

Most GC separations continue to be performed on low to moderately polar capillary GC columns (e.g. DB-5 and DB-35), with many of the newer methods making
use of the low bleed "MS" designated columns designed for GC-MS applications. EI
remains the most significant ionization method for the chemical warfare agents [53–
55,60–62], with CI being frequently used to confirm molecular mass or to support
confirmation [43,71]. CI has been used in both positive ion (PI) and negative ion
(NI) mode and with a variety of reagent gases, with most researchers reporting
improved method detection limits for the derivatized hydrolysis products or metabolites in biomedical samples [47,49–52,65,70]. This improvement usually results
from reduced chemical interference, often the result of using more specific means of
ionization such as NICI or PICI with ammonia. Detection limits vary with compound and matrix but almost without exception the best detection limits for these
compounds have been reported during GC-MS or GC-MS/MS applications, either
during selected ion monitoring (SIM) or reaction ion monitoring (RIM) applications. Picogram sensitivity has become routine, with researchers reporting method
detection limits in biomedical samples [48–52,57,65] and groundwater [56] in the ng/
mL range or lower.

Figure 25.2 illustrates the total ion current chromatogram obtained during GC-
MS/MS analysis of number of pentafluorobenzyl esters of alkyl methylphosphonic
acids in urine at the 1 ng/mL level, about an order of magnitude above the reported
detection limit [52]. The derivatives were well resolved with an RTX-5MS capillary
column and ionized by NICI (methane) to take advantage of the pentafluorobenzyl
substitution [52].

25.5 LIQUID CHROMATOGRAPHY-MASS SPECTROMETRY

Recent LC separation methods have made use of UV [73,74] or FPD [75–77] detectors, but the majority of new methods have been based on mass spectrometric
detection, as this means of detection provides a higher level of confirmation. LC-MS
is being used increasingly for chemical warfare agent analyses, as mass spectrometric
data may be used to directly identify chemical warfare agents, their hydrolysis
products, and related compounds in collected aqueous samples or extracts during a
single analysis. Wils and Hulst were the first to demonstrate the use of LC-MS for
the direct analysis of nerve agent hydrolysis products [78] and VX [79], using
thermospray ionization, a technique that has now been superseded by API. Increasingly, researchers have developed API based LC-MS methods (e.g. electrospray
(ESI), ionspray (IS), and atmospheric pressure chemical ionization (APCI)) as
complementary or replacement methods for the characterization of chemical

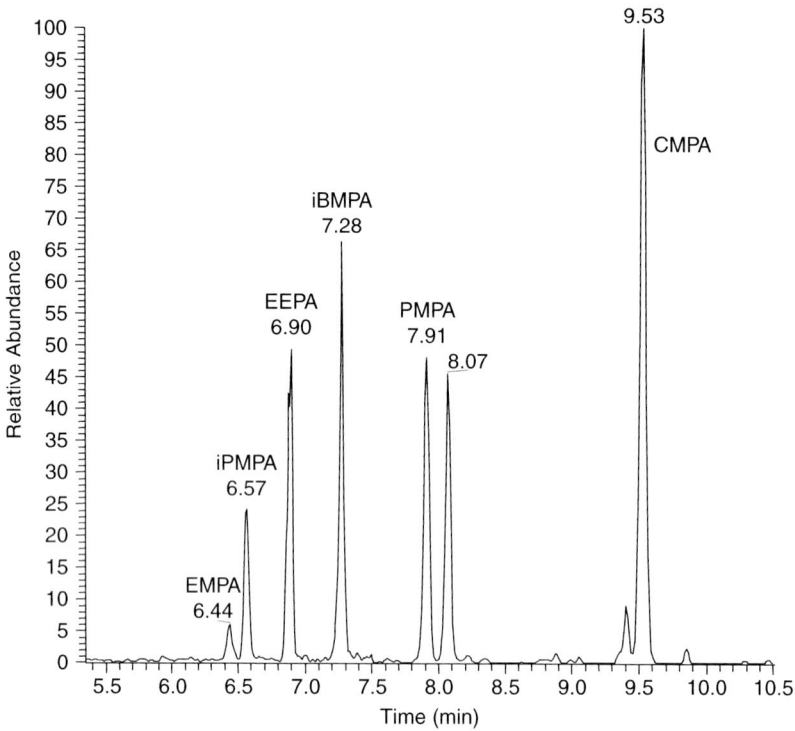

Fig. 25.2. Total ion current chromatogram obtained during GC-MS/MS (RIM) analysis of six alkyl methylphosphonic acids spiked into human urine at the 1 ng/mL level (EMPA: ethyl methylphosphonic acid, iPrMPA: isopropyl methylphosphonic acid, EEPA: ethyl ethylphosphonic acid, iBMPA: isobutyl methylphosphonic acid, PinMPA: pinacolyl methylphosphonic acid, CMPA: cyclohexyl methyl-phosphonic acid) [52] (Copyright © (2005) with permission from Elsevier).

warfare agents and/or their degradation products. A number of LC-MS methods have been reported for the confirmation of these compounds, with this technique being used most frequently during the analysis of aqueous samples or extracts [8,11,19]. In many cases LC-MS proved to be an attractive alternative to GC-MS for aqueous analyses, as both the organophosphorus chemical warfare agents and their hydrolysis products could be analyzed directly without the need for additional sample handling and derivatization steps associated with GC-MS analysis [9].

The majority of the LC-MS applications listed in Table 25.3 involve the analysis of degradation products of chemical warfare agents, as these compounds can be analyzed directly using API methods of ionization. Both PI and NI modes have been used, with advantages in selectivity [92] or sensitivity [88] generally being cited as the reasons for the choice. Screening of the wide range of possible chemical warfare agent degradation products would likely involve a PI screening procedure or one that targets a number of key compounds using both PI and NI modes depending on the compounds being targeted [88]. Figure 25.3 illustrates LC-APCI-MS selected ion

TABLE 25.3
LC-MS ANALYSES OF CHEMICALS WARFARE AGENTS AND RELATED COMPOUNDS

Media	Reference	Ionization	Chromatography	Compounds Analyzed
Water	[80]	ESI (PI)	Mixed C_8/C_{18} (250 × 2.1 mm) column MeOH/water (0.1% formic acid) gradient at 200 μL/min	Alkyl methylphosphonic acids, alkyl ethylphosphonic acids, alkyl alkylphosphonic acids, and dialkyl alkylphosphonates
Water	[81]	ESI (PI)	C_{18} (150 × 0.32 mm) column ACN/water (0.1% trifluoroacetic acid) gradient at 5 μL/min	Munitions grade mustard hydrolysis products including thiodiglycol and longer chain diols
Water	[82]	APCI (PI)	C_{18} (250 × 2.1 mm) column Water (0.001–0.05 M ammonium acetate) isocratic at 250 μL/min	Hydrolysis products of VX, lewisite and nitrogen mustard using post column derivatization
Water	[83]	ESI (PI)	C_{18} (150 × 0.32 mm) column ACN/water (0.1% trifluoroacetic acid) gradient at 5 μL/min	VX and numerous VX degradation products and related compounds
Water	[84]	ESI (PI)	C_{18} (150 × 0.32 mm) column ACN/water (0.1% trifluoroacetic acid) gradient at 5 μL/min	Nerve agents, sarin, soman, tabun, and cyclohexyl methylphosphonofluoridate
Water	[85]	Fast atom bombardment (PI)	C_{18} (150 × 1.5 mm) column ACN/water (0.005 M ammonium acetate) isocratic at 100 μL/min	Derivatized (*p*-bromophenacyl) alkyl phosphonic acids
Water	[86]	IS (NI)	PGC (150 × 2.1 mm) column ACN/water (trifluoroacetic acid) gradient/ isocratic at 200 μL/min	Alkylphosphonic acids, alkyl methylphosphonic acids, and alkyl ethylphosphonic acids
Water	[87]	IS (NI)	PGC (150 × 2.1 mm) column ACN/water (trifluoroacetic acid) gradient/ isocratic at 200 μL/min	Alkylphosphonic acids
Water	[88]	APCI (PI/NI)	C_{18} (250 × 2.0 mm) column MeOH/water (0.02 M ammonium formate) gradient at 200 μL/min	Hydrolysis products of ten nerve agents, mustard, nitrogen mustard, and quinuclidinyl benzilate (BZ)
Water	[89]	ESI (PI)	C_{18} (150 × 0.32 mm) column MeOH/water (0.2% formic acid) isocratic at 6 μL/min	Thiodiglycol and other hydrolysis products of sulfur mustards

TABLE 25.3
CONTINUED

Media	Reference	Ionization	Chromatography	Compounds Analyzed
Water	[90]	ESI (PI)	C_{18} (150 × 0.32 mm) column ACN/water (0.1% trifluoroacetic acid) gradient at 16 µL/min	Nerve agents, sarin, soman, tabun, and cyclohexyl methylphosphonofluoridate and their hydrolysis products
Water	[91]	ESI (PI)	C_{18} (150 × 0.32 mm) column ACN/water (0.1% trifluoroacetic acid) gradient at 16 µL/min	Sarin and its degradation products and numerous related compounds
Water	[92]	ESI (NI)	C_{18} (150 × 2.1 mm) column MeOH/water (0.01 M ammonium formate) gradient at 200 µL/min	Alkyl methylphosphonic acids
Water	[93]	ESI (PI) ICP	C_4 (150 × 2.1 mm) column ACN/water (nitric acid to pH = 2) gradient at 200 µL/min	Degradation products of organoarsenic chemical warfare agents (diphenylchlorophenylarsine)
Water	[94]	ESI (PI/NI)	C_{18} (150 × 0.5 mm) column ACN/water (0.005 M ammonium formate) gradient at 20 µL/min	Alkyl methylphosphonic acids
Soil	[89]	ESI (PI)	C_{18} (150 × 0.32 mm) column MeOH/water (0.2% formic acid) isocratic at 6 µL/min	Thiodiglycol and other hydrolysis products of sulfur mustards
Soil	[90,95]	ESI (PI)	C_{18} (150 × 0.32 mm) column ACN/water (0.1% trifluoroacetic acid) gradient at 16 µL/min	Nerve agents, sarin and soman and their hydrolysis products
Soil	[92]	ESI (NI)	C_{18} (150 × 2.1 mm) column MeOH/water (0.01 M ammonium formate) gradient at 200 µL/min	Alkyl methylphosphonic acids
Soil	[60,96]	ESI (PI)	C_{18} (150 × 0.32 mm) column ACN/water (0.1% trifluoroacetic acid) gradient at 10 µL/min	Thiodiglycol and longer chain diols associated with hydrolysis of munitions grade mustard
Munition	[16]	APCI (PI)	C_{18} (100 × 2.1 mm) column ACN/water (0.05 M ammonium acetate) gradient at 250 µL/min	Phosphorothioates and related compounds

Sample	Reference	Ionization	LC conditions	Analyte
Munition	[90]	ESI (PI)	C$_{18}$ (150 × 0.32 mm) column ACN/water (0.1% trifluoroacetic acid) gradient at 16 µL/min	Tabun, its hydrolysis product and numerous related compounds
Synthetic	[60]	ESI (PI)	C$_{18}$ (150 × 0.32 mm) column ACN/water (0.1% trifluoroacetic acid) gradient at 10 µL/min	Tabun and numerous tabun related compounds
Synthetic	[97]	ESI (PI)	C$_{18}$ (250 × 2.0 mm) column MeOH/water (0.02 M ammonium formate) gradient at 200 µL/min	Hydrolysis and oxidation products of two longer chain sulfur vesicants (Q and T)
Urine	[98]	ESI (PI)	C$_{18}$ (250 × 2.0 mm) column MeOH/water (0.02 M ammonium formate) gradient at 200 µL/min	Beta-lyase metabolites of sulfur mustard
Urine	[99]	ESI (PI/NI)	C$_{18}$ (150 × 2.0 mm) column ACN/water (0.05% formic acid) gradient at 200 µL/min	Sulfur mustard metabolite 1,1'-sulfonylbis[2-S-(N-acetylcysteinyl)ethane]
Serum	[100]	ESI (PI/NI)	PRP-X100 (200 × 0.32 mm) column ACN/water (0.5% formic acid) isocratic at 20 µL/min	Isopropyl methylphosphonic acid
Serum	[85]	Fast atom bombardment (PI)	C$_{18}$ (150 × 1.5 mm) column ACN/water (0.005 M ammonium acetate) isocratic at 100 µL/min	Derivatized (p-bromophenacyl) alkyl phosphonic acids
Serum	[101]	ESI (PI)	C$_{18}$ (150 × 0.30 mm) column ACN/water (0.2% formic acid) gradient at 6 µL/min	Albumin/sulfur mustard adducts
Plasma	[102]	APCI (PI)	OD-H (250 × 4.6 mm) column Hexane/isopropanol isocratic at 800 µL/min	Enantiomers of VX

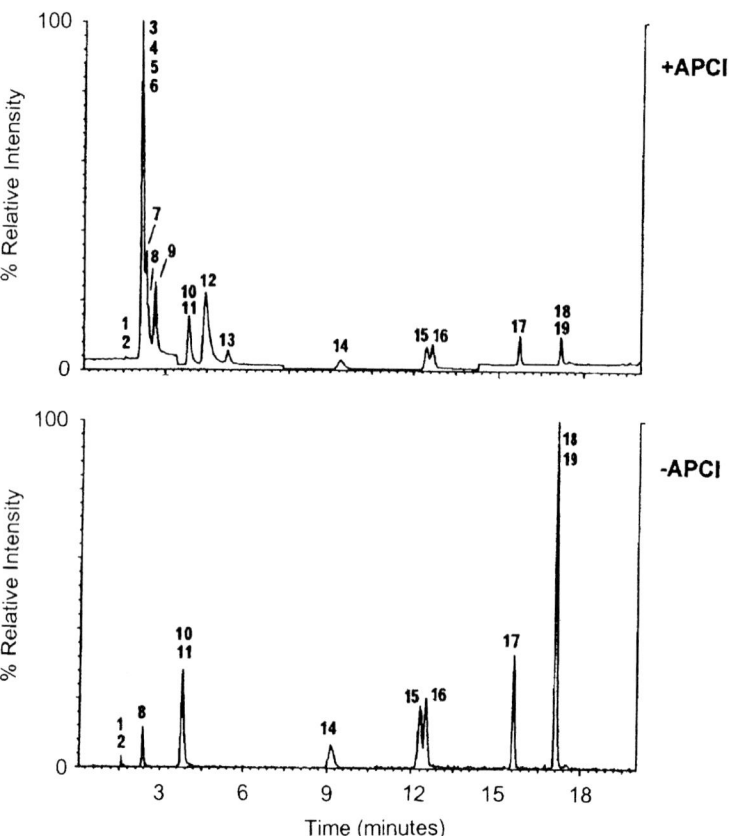

Fig. 25.3. LC-MS selected ion current chromatograms, PI APCI (upper) showing the detection of 19 degradation products (each 0.1 µg/mL in water), and NI APCI (lower) showing the selective detection of phosphonic acids and benzilic acid in the mixture. 1. methylphosphonic acid, 2. ethylphosphonic acid, 3. *N*-methyldiethanolamine, 4. thiodiglycolsulfoxide, 5. 3-quinuclidinol, 6. *N*-ethyldiethanolamine, 7. thiodiglycol sulfone, 8. ethyl methylphosphonic acid, 9. triethanolamine, 10. isopropyl methylphosphonic acid, 11. ethyl ethylphosphonic acid, 12. *N,N*-diisopropylaminoethanol, 13. thiodiglycol, 14. *sec*-butyl methylphosphonic acid, 15. isobutyl methylphosphonic acid, 16. *n*-butyl methylphosphonic acid, 17. cyclohexyl methylphosphonic acid, 18. benzilic acid, and 19. pinacolyl methylphosphonic acid [88] (Copyright © (1999) with permission from Elsevier).

monitoring (SIM) chromatograms in both PI and NI mode for 19 acidic, neutral, and basic CW agent degradation products in water using a screening procedure developed by Read and Black [88].

Recent publications have focused on the direct analysis of hydrolysis products related to the more common chemical warfare agents, including, thiodiglycol and longer chain diols that form following hydrolysis of mustard and munitions grade mustard formulations [60,81,88,89,96,97] and alkyl methylphosphonic acids and methylphosphonic acid that form following hydrolysis of the nerve agents [80,82,83,86–88,90–92,95,100]. Other degradation products related to nitrogen mustard, quinuclidinyl benzilate, and other compounds listed in the CWC schedules have been included in reports dealing with preparing for or actual OPCW proficiency testing [14,80,86,88].

D'Agostino, Hancock, and Chenier recently demonstrated the applicability of packed capillary LC-ESI-MS/MS for the identification of mustard hydrolysis products in aqueous extracts of soil. The method successfully characterized thiodiglycol and nine longer chain diols in soil samples taken from a former mustard storage site at different locations and depths as part of an ongoing environmental assessment. Figure 25.4 illustrates the ESI-MS/MS product mass spectra obtained for the $[M + H]^+$ ion of thiodiglycol, bis(hydroxyethyl)disulfide, 1,2-bis(2-chloroethylthio) ethane (sesquimustard, Q), and bis(2-chloroethylthioethyl)ether (T) at collision energies that resulted in the formation of characteristic product ions that could be used for identification purposes [96].

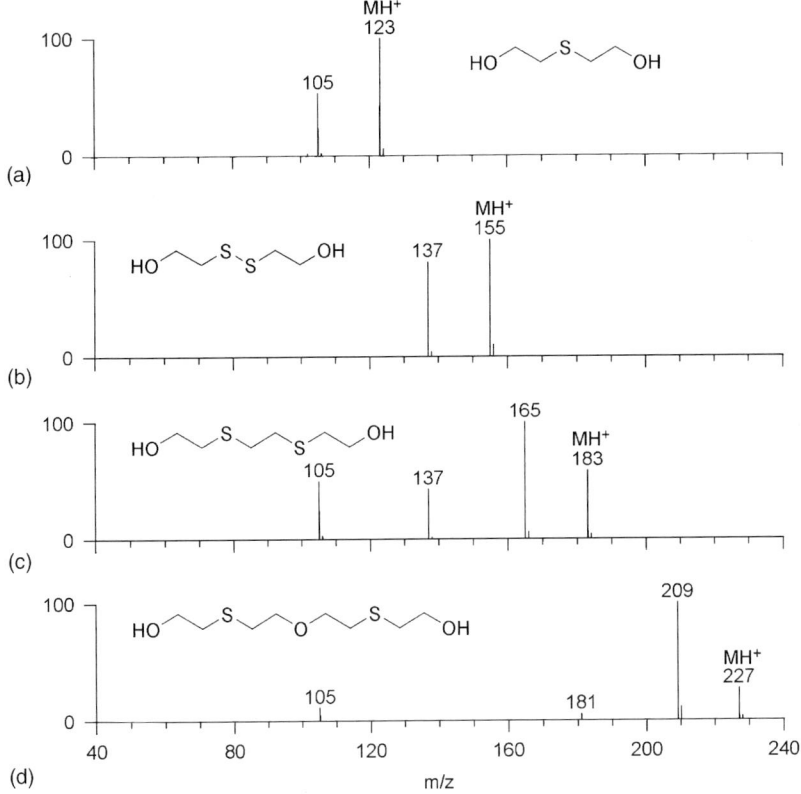

Fig. 25.4. ESI-MS/MS product mass spectra obtained for the MH^+ ion of (a) thiodiglycol (collision energy: 2 V), (b) bis(hydroxyethyl)disulfide (collision energy: 2 V), (c) 1,2-bis(2-chloroethylthio)ethane (sesquimustard, Q) (collision energy: 4 V), and (d) bis(2-chloroethylthioethyl)ether (T) (collision energy: 5 V) during LC-ESI-MS/MS analysis of an aqueous extract of a soil sample taken from a former mustard storage site [96] (Copyright © (2004) with permission from Elsevier).

An important advantage of LC-ESI-MS is that it may also be used for the identification of intact organophosphorus chemical warfare agents and related compounds
that are often also present in samples or sample extracts [60,83,84,90,91,95,102].
Figure 25.5 illustrates the ESI-MS data acquired for a snow sample contaminated
during the destruction of a chemical munition. LC-ESI-MS was used to determine
the presence of sarin, its hydrolysis products, and a number of related organophosphorus compounds. Full ESI mass spectra were acquired for 14 compounds in the
snow sample, with all compounds exhibiting $[M+H]^+$, $[M+H+ACN]^+$ ions, and/or

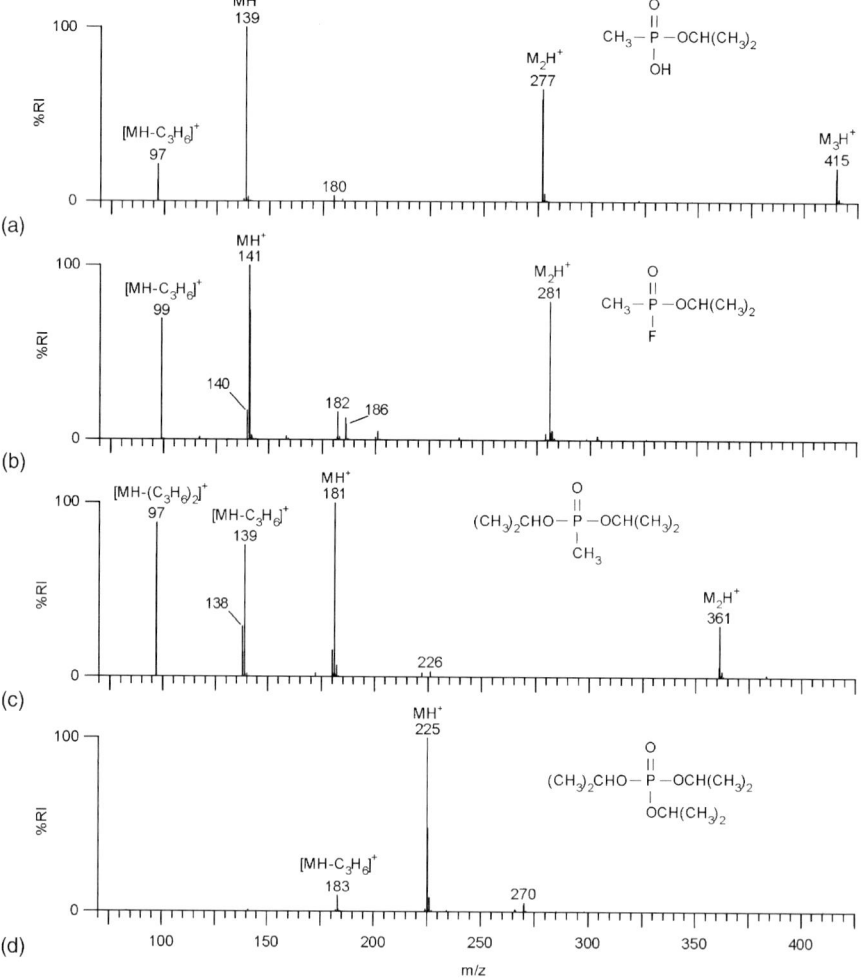

Fig. 25.5. ESI-MS data acquired for (a) isopropyl methylphosphonic acid (sampling cone voltage: 20 V),
(b) sarin (sampling cone voltage: 20 V), (c) diisopropyl methylphosphonate (sampling cone voltage: 30 V)
and (d) triisopropyl phosphate (sampling cone voltage: 20 V) during LC-ESI-MS of a melted snow sample
diluted 1:10 with distilled water [91] (Copyright © (2002) with permission from Elsevier).

protonated dimers that could be used to confirm molecular mass, as well as structurally significant product ions (sampling cone voltages in the 20–30 V range). This LC-ESI-MS application demonstrates an important advantage of LC-MS over GC-MS for aqueous analyses. Both the chemical warfare agent, sarin, and its hydrolysis products were analyzed during a single analysis, without the need for additional sample handling and derivatization steps associated with GC-MS analysis of sarin hydrolysis products. The identification and characterization of related organophosphorus sample components in the snow sample was significant as the identification of these compounds may provide source or synthetic clues during a forensic investigation. While this approach has value for organophosphorus chemical warfare agent identification, it should be noted that mustard does not ionize during conventional LC-ESI-MS. Mustard analyses should be performed by GC-MS [60] or by another spectrometric or spectroscopic technique.

The reported LC separations were usually performed with C_{18} LC columns and acetonitrile (ACN) or methanol (MeOH)/water gradients using trifluoroacetic acid, formic acid, or ammonium formate modifiers. Trifluoroacetic acid generally provides the best chromatographic resolution [88] and this modifier has been used frequently for lower flow rate analyses [90,91,96]. However, this acid has been associated with suppression of analyte signal at higher flow rates and may not be the best choice for NI analyses, as it would compete with the analyte during API-MS. Black and Read selected ammonium formate as a buffer compromise for a broad screening procedure where both PI and NI data were acquired. Gradient separations with C_{18} provided the flexibility to analyze a variety of analytes from small, polar hydrolysis products, such as thiodiglycol and methylphosphonic acid, to larger, less polar compounds such as VX and related compounds [83]. The principal limitation associated with use of C_{18} columns remains the poor retention for the smaller, polar hydrolysis products. This situation may be improved through the use of porous graphitic carbon (PGC) or other columns [86,87]. Chiral separations can also be quite valuable for toxicological studies, with Smith reporting the first chiral separation of the two enantiomers of VX using a Chiralcel OD-H column and a hexane/isopropanol mobile phase [102].

All the API methods have been utilized for LC-MS analysis of chemical warfare agents and their degradation products, with ESI being preferred for the lower flow rates (e.g. 10 μL/min) associated with packed capillary LC separation [89,90,96,]. IS and APCI methods have usually involved the use of larger bore LC columns and higher flow rates (e.g. 200 μL/min) and were found to be less susceptible to variable adduct ion formation associated with some ESI-MS analyses [88]. However, regardless of the API technique or instrument used, the acquired mass spectra remain similar in ion content, exhibiting adduct ions (e.g. $[M+H]^+$ or $[M+Na]^+$) that may be used to determine molecular mass and characteristic product ions (e.g. $[M+H-H_2O]^+$ or $[M+H-C_nH_{2n}]^+$) indicative of the compound's structure. Relative intensities of the observed ions will vary depending on instrumental conditions [90].

Newer LC-MS instruments have made it possible to routinely detect compounds in full scanning mode in the 0.5–5 ng range and at levels about two orders of magnitude better during SIM or RIM. Reported detection limits for the determination of chemical warfare agent hydrolysis products using LC-API-MS have been found to be quite compound dependent. In an earlier paper, Black and Read used SIM and detected neutral and basic compounds in the < 200 pg range (or < 10 ng/mL water), but some acidic compounds, like methylphosphonic acid and thiodiglycol, exhibited detection limits of up to 8 ng (or 400 ng/mL) [80]. Later improvements to the above method, including the use of both PI and NI, dropped detection limits by a factor of four at the higher end [88]. A similar method, using LC-ESI-MS (NI), resulted in SIM detection limits that ranged from 250 pg to 5 ng, with methylphosphonic acid being at the upper limit [92]. Figure 25.6 illustrates a typical LC-ESI-MS (NI) separation for the six alkyl methylphosphonic acids at the 5 μg/mL level.

D'Agostino *et al.* determined LC-ESI-MS detection limits for triethyl phosphate, a chemical warfare agent stimulant resistant to hydrolysis. Triethyl phosphate was detected at 50 pg under full scanning conditions [90]. Detection limits may be improved by employing larger volume injections with peak compression. Hooijschuur *et al.* reported full scanning detection limits of 500–800 ng/mL with this technique for longer chain diols associated with mustard hydrolysis [89]. Reducing the chemical background through the use of MS/MS resulted in some of the best reported detection limits. Noort *et al.* reported an absolute detection limit of 2 pg for isopropyl methylphosphonic acid [100] and 4 pg for a mustard adduct [101] during RIM.

Acquisition of high-resolution data for chemical warfare agents and related compounds was greatly aided by the introduction of instruments with time-of-flight (TOF) mass analyzers. These instruments, while lacking the dynamic range and sensitivity associated with triple quadrupole instruments, may be used to acquire high resolution, full scanning data (typically 5000–17,000 resolution, 50% valley) for unknown sample components without the signal losses typically associated with

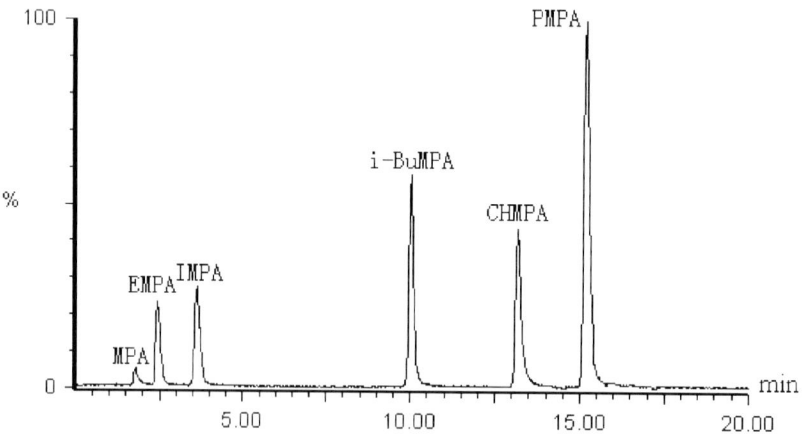

Fig. 25.6. LC-ESI-MS SIM chromatogram of six alkyl methylphosphonic acids (5 μg/mL, ions monitored: *m/z* 95, 123, 137, 151, 177, 179). 1. methylphosphonic acid, 2. ethyl methylphosphonic acid, 3. isopropyl methylphosphonic acid, 4. *i*-butyl methylphosphonic acid, 5. cyclohexyl methylphosphonic acid, and 6. pinacolyl methylphosphonic acid [92] (Copyright © (2004) with permission from Elsevier).

magnetic sector instrumentation. D'Agostino *et al.* demonstrated the utility of high resolution LC-ESI-MS and LC-ESI-MS/MS data during the identification of numerous tabun impurities in a synthetic sample [60], for the identification of sarin related compounds in snow [91], and for the determination of longer chain diols in soil samples collected from a former mustard storage site [96]. High resolution LC-ESI-MS in NI mode was also used by Liu *et al.* for the confirmation of spiked alkyl methylphosphonic acids in water and soil samples [92]. Errors associated with the mass measurements during these analyses were generally < 0.001 Da.

This approach was used during a recent investigation involving typical forensic media that might be collected at the scene of an indoor terrorist attack. A variety of indoor sample media, including flooring, wall surfaces, office fabrics, window coverings, and paper products or packaging, were spiked with chemical warfare agents to assess the applicability of aqueous extraction and LC-ESI-MS and LC-ESI-MS/MS analysis for the identification of chemical warfare agents [103]. The spiked chemical warfare agents were recovered and positively identified by ESI-MS with efficiencies that varied with media. Figure 25.7 illustrates LC-ESI-MS and LC-ESI-MS/MS chromatograms acquired during a single analysis of an aqueous extract of an office carpet sample spiked at the 0.5–5 µg/g level with a complex munitions grade tabun sample. The crude chemical warfare agent mixture was included in this study since being able to identify the related organophosphorus compounds in an extract may aid in establishing source information during a forensic investigation. Figure 25.8 illustrates the ESI-MS/MS data acquired for three of the identified compounds, including the chemical warfare agent, tabun. All ESI-MS/MS data acquired during this study were acquired using a QToF instrument under high-resolution conditions to enable identification of the tabun components (Table 25.4).

25.6 OTHER INSTRUMENTAL TECHNIQUES

The majority of published methods for the determination of chemical warfare agents, their degradation products, and related compounds involve GC or LC separation, usually with MS detection. Capillary electrophoresis (CE) has also been demonstrated as a viable separation technique, particularly for the chemical warfare agent degradation products. The first application of IS for the determination of chemical warfare agent degradation products was demonstrated during CE-MS analysis of methylphosphonic acid and alkyl methylphosphonic acids [104]. A more recent application involved the use of CE-IS-MS and MS/MS for the determination of methylphosphonic acids and alkyl methylphosphonic acids in tap water [105]. Other reported applications involving sensitive detection of chemical warfare agent degradation products make use of indirect laser induced fluorescence detection [106], derivatization to enable laser induced fluorescence detection [107,108], and conductivity or indirect ultraviolet detection [109–111]. CE separation, often influenced by the background matrix, was successfully employed for the determination of methylphosphonic acid and alkyl methylphosphonic acids in three soil types and four water samples at low µg/L levels [111]. Continuing efforts towards developing

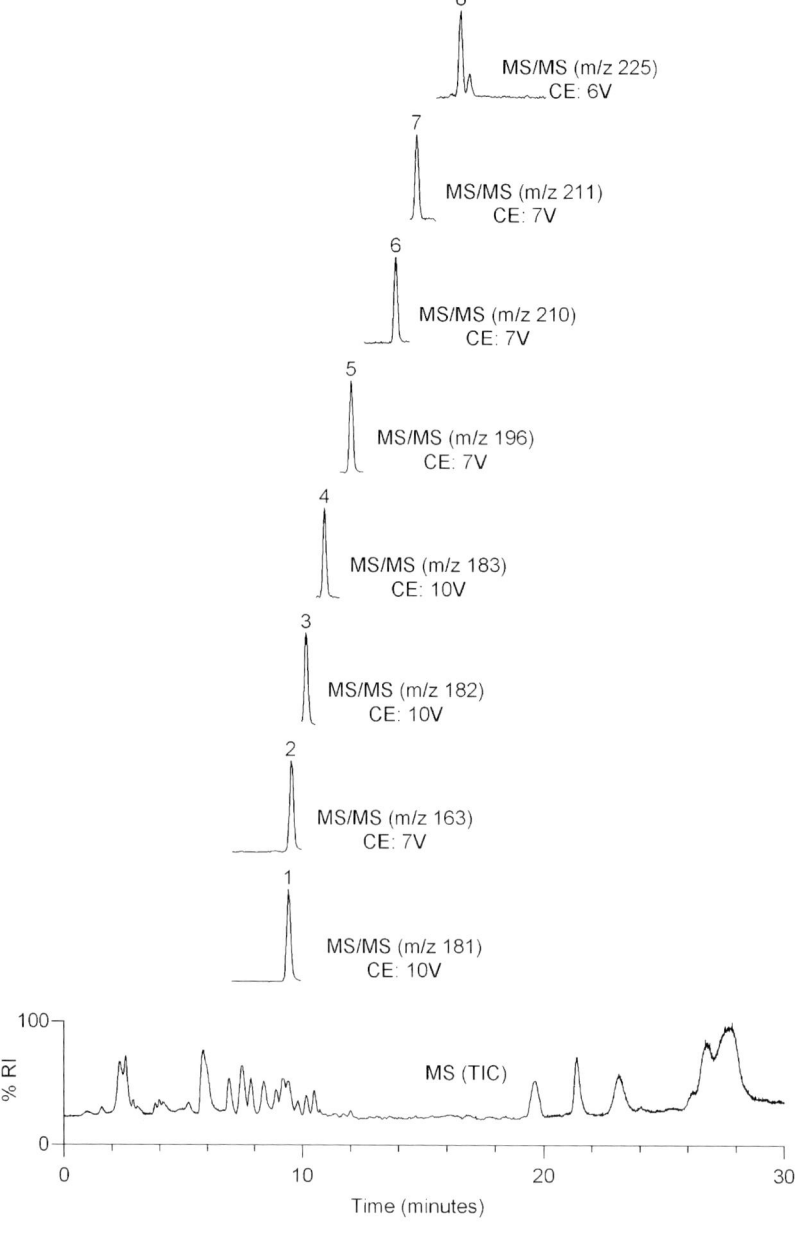

Fig. 25.7. LC-ESI-MS (lowest) and LC-ESI-MS/MS chromatograms (above) of an extract of an office carpet spiked with munitions grade tabun (0.5–5 µg/g per component). Components 1–8 identified in Table 25.4. (CE: Collision energy) [103] (Copyright © (2006) with permission from Elsevier).

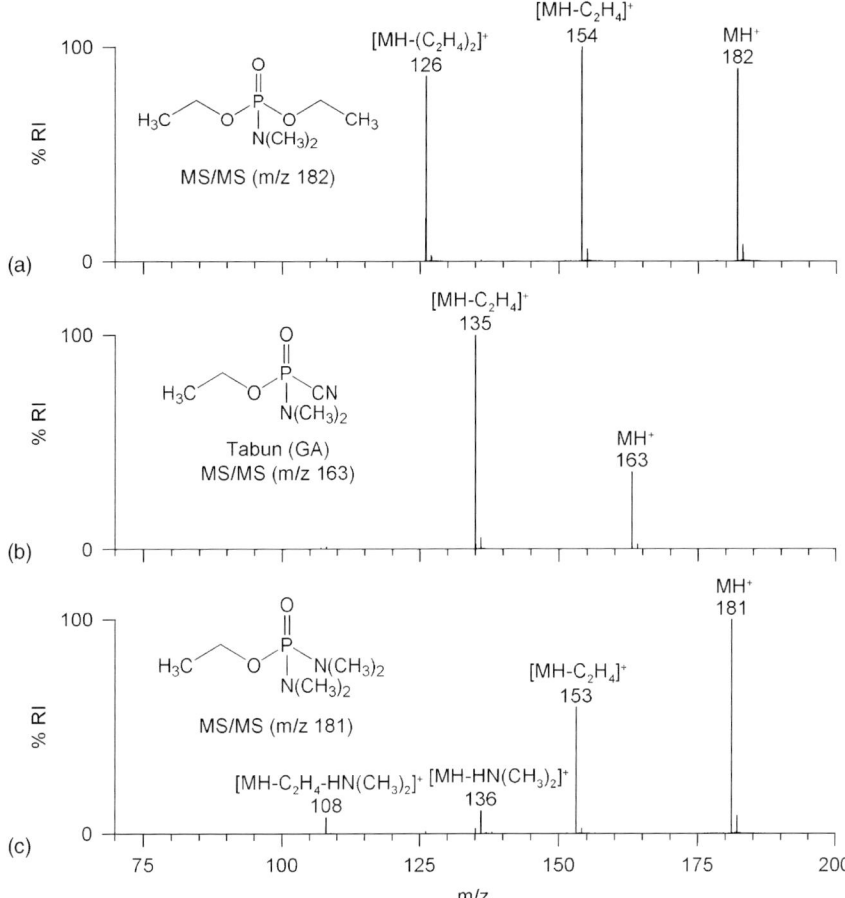

Fig. 25.8. Product ion mass spectra obtained for (a) diethyl dimethylphosphoramidate (*m/z* 182, Collision energy: 10 V), (b) GA (*m/z* 163, Collision energy: 7 V), and (c) ethyl tetramethylphosphoramidate (*m/z* 181, Collision energy: 10 V) during LC-ESI-MS/MS analysis of an office carpet sample spiked with munitions grade tabun (0.5–5 µg/g per component) [103]. (Copyright © (2006) with permission from Elsevier).

"laboratory-on-a-chip" technology have also resulted in several CE applications for chemical warfare agent degradation products [112,113].

Ion mobility spectrometry (IMS), a technology used in hand-held military detectors for chemical warfare agents, has shown promise for very rapid separations. This means of separation has been investigated for a number of chemical warfare agent simulants and degradation products with Hill *et al.* reporting a number of different instrumental methods [114–117], the most recent involving the analysis of aerosol samples [117]. An aspiration IMS method for VX degradation products and a multicapillary column GC-IMS method for chemical warfare agents [118] have also been recently reported.

References pp. 868–872

TABLE 25.4
ESI-MS/MS DATA ACQUIRED FOR SIGNIFICANT MUNITIONS GRADE TABUN COMPONENTS IDENTIFIED IN A SPIKED OFFICE CARPET EXTRACT

Peak Number[a]	Compound Name	Ion	Observed Mass (Da)[b]	Theoretical Mass (Da)	Error (Da)
1	Ethyl tetramethylphosphorodiamidate	MH^+	181.1108	181.1106	0.0002
		$[MH-C_2H_4]^+$	153.0795	153.0793	0.0002
		$[MH-HN(CH_3)_2]^+$	136.0533	136.0527	0.0006
		$[MH-C_2H_4-HN(CH_3)_2]^+$	108.0215	108.0214	0.0001
2	Ethyl dimethylphosphoramidocyanidate (Tabun, GA)	MH^+	163.0628	163.0636	0.0008
		$[MH-C_2H_4]^+$	135.0316	135.0323	0.0007
3	Diethyl dimethylphosphoramidate	MH^+	182.0950	182.0946	0.0024
		$[MH-C_2H_4]^+$	154.0637	154.0633	0.0004
		$[MH-(C_2H_4)_2]^+$	126.0322	126.0320	0.0002
4	Triethyl phosphate	MH^+	183.0805	183.0786	0.0019
		$[MH-C_2H_4]^+$	155.0470	155.0473	0.0003
		$[MH-(C_2H_4)_2]^+$	127.0153	127.0160	0.0007
		$[MH-(C_2H_4)_3]^+$	98.9836	98.9847	0.0011
5	Ethyl isopropyl dimethylphosphoramidate	MH^+	196.1109	196.1102	0.0007
		$[MH-C_3H_6]^+$	154.0630	154.0633	0.0003
		$[MH-C_3H_6-C_2H_4]^+$	126.0327	126.0320	0.0007
6	Diisopropyl dimethylphosphoramidate	MH^+	210.1282	210.1259	0.0023
		$[MH-C_3H_6]^+$	168.0790	168.0789	0.0001
		$[MH-(C_3H_6)_2]^+$	126.0316	126.0320	0.0004
7	Diisopropyl ethyl phosphate	MH^+	211.1109	211.1099	0.0010
		$[MH-C_3H_6]^+$	169.0647	169.0629	0.0018
		$[MH-(C_3H_6)_2]^+$	127.0172	127.0160	0.0012
8	Triisopropyl phosphate	MH^+	225.1273	225.1255	0.0018
		$[MH-C_3H_6]^+$	183.0791	183.0786	0.0005
		$[MH-(C_3H_6)_2]^+$	141.0324	141.0316	0.0008
		$[MH-(C_3H_6)_3]^+$	98.9842	98.9847	0.0005

[a]Refer to Fig. 7.
[b]Average of scans across the chromatographic peak (lock mass used).

NMR is an important technique for the structural analysis and characterization of chemical warfare agents, particularly for the authentication of synthetic reference materials or unknown chemical warfare agents and related compounds [12,119,120]. The presence of heteronucleii such as ^{31}P and ^{19}F in the nerve agents leads to diagnostic splitting patterns and coupling constants due to ^{1}H-^{31}P and ^{1}H-^{19}F spin–spin coupling. The utility of conventional NMR for analysis of complex sample mixtures or for trace analysis is somewhat limited, but a novel ^{31}P NMR technique using coaxial inserts has recently been demonstrated for purity determinations [120]. While often not the most significant technique for trace level confirmation, one- and two-dimensional NMR have often been successfully employed in the battery of analyses typically performed during OPCW exercises [12].

Condensed phase infrared (IR) data exists for many chemical warfare agents and related compounds, as this technique was routinely used prior to the advent of GC-MS. Capillary column GC-FTIR offers considerably more promise for the identification and characterization of chemical warfare agents in multiple component sample extracts and has been utilized as a complementary confirmation technique. Sensitivity is generally poorer than that obtained by MS but may be improved by using large volume (e.g. 50 µL) injections with peak compression onto an uncoated pre-column with lightpipe technology or through the use of cryodeposition [13].

Direct analysis of contaminated media would provide the analyst with a more rapid analysis without the need for sample preparation. Groenewold *et al.* and Gresham *et al.* have investigated the direct analysis of pinacolyl methylphosphonic acid, 2-chloroethyl sulfide, VX, nitrogen mustards, and mustard on surfaces, including soil and concrete, using static secondary ion mass spectrometry (SIMS) and MS/MS in the PI or NI mode [121–125]. Although not as precise as other analytical methods, this technique enables detection in a fraction of the time normally associated with conventional methods requiring extraction and analysis [125].

Rapid analysis in real-time developments have continued with the development of the direct analysis in real time (DART) source by Cody *et al.* [126] and the development and application of desorption electrospray ionization (DESI) by Cooks *et al.* for direct sample ionization and analysis [127]. Cody *et al.* used the DART source to directly analyze VX on concrete with a TOF instrument [126], while the DESI technique has been used to directly determine sarin and soman from contaminated SPME fibers [103].

25.7 FIELD ANALYSES

The development of field detection methods for chemical warfare agents was driven by specific military requirements [128–130], with a variety of detection devices and other chemical warfare agent defense equipment having been developed for military applications [128]. Most of the effort in this area resulted from the perceived threat during the Cold War era and although this threat has decreased dramatically, interest in chemical detection equipment persists because of worldwide chemical weapons proliferation. During the 1990–1991 Gulf War, chemical detection equipment was deployed in the operational theatre and similar equipment was used to support the United Nations Special Commission during the destruction of Iraqi chemical weapons. Equipment of this type has been used by the OPCW and has increasingly been put in the hands of first responders to counter the possibility of a chemical warfare agent terrorism incident.

Most equipment or devices used for field detection of chemical warfare agents by the military falls into one of several general categories. Tests making use of chemical reagents have been used extensively, with commonly used examples being 3-way detection paper, the Nerve Agent Immobilized-enzyme Alarm and Detector (NAIAD), Dräger tubes, and the M256A1 Chemical Agent Detection Kit.

The development of point detectors improved the response capability of military forces and most of the commonly deployed devices have been based on rapid, sensitive detection of chemical warfare agents at a particular location. The Chemical Agent Monitor (CAM), first fielded in the 1980s as a hand-held alarm, was adopted by many countries for military field detection. Improved IMS devices have been produced and put into service in a variety of detection roles. Newer IMS based devices include the GID-2 and GID-3, the Improved and Enhanced Chemical Agent Monitors (ICAM and ECAM), the RAID detectors, and the M90 DIC Chemical Agent Detector, employing both IMS and metal conductivity technologies. Competition to IMS based point detectors comes primarily from devices based on FPD, with the AP2C being widely used for chemical warfare agent detection due to its reliability [131]. Use of only one technology for field detection is generally not recommended and most military forces and first responders will approach detection with a suite of options, with many opting to employ two or more detection technologies (e.g. IMS and FPD backed up with a reagent based test(s)) to improve certainty of detection. Standoff detection, based on passive IR and other techniques, while desirable, has not been developed to a point of widespread adoption.

Identification of the chemical warfare agent in the field has become increasingly important and the equipment that can provide this information in real time or near real time has been developed. The MiniCAMS, based on solid adsorbent sampling, gas chromatographic separation, and FPD detection was developed for sensitive detection of chemical warfare agents at storage, demilitarization, and other sites. Increasingly, mass spectrometric-based instruments have been developed for identification purposes, with a number of military and environmental field portable instruments having been developed. Mass spectrometric instruments developed include the MM-1 Mobile Mass Spectrometer (deployed in the German "Fuchs" recon vehicle), the Block II Chemical and Biological Mass Spectrometer [132], and the Viking and Bruker portable GC-MS systems. Sampling and analysis in a field situation remains an integral part of the OPCW strategy and the scientists involved in verifying the CWC currently use specialized field portable GC-MS instrumentation [133–135]. Table 25.5 lists examples of selected chemical detection equipment by country and indicates the principle of detection and capabilities of each system [128,129].

Many nations concerned about possible terrorist attacks have equipped their First Responders with military detection equipment used for chemical warfare agent

TABLE 25.5

SELECTED MILITARY CHEMICAL WARFARE AGENT DETECTION DEVICES

Country	Device Name and Capabilities
Canada	Chemical agent detection system (CADS II)
	Early warning system that controls a network of chemical agent monitors (see UK) for the real time detection of nerve and blister agents
Finland	Chemical agent detection system, M90 DIC
	Alarm for the ion mobility spectrometric and metal conductivity detection of nerve, blister, and other agents
France	PROENGIN portable chemical contamination monitor AP2C
	Hand-held flame photometric detection of nerve and blister agents
	Also designs for fixed sites (AP2C-V and ADLIF)
Germany	MM-1 mobile mass spectrometer
	Quadrupole mass spectrometric detection of chemical warfare agents
	Rapid alarm and identification device – 1 (RAID-1)
	– Ion mobility spectrometric detection of nerve and blister agents
Switzerland	IMS 2000 CW agent detector
	Ion mobility spectrometric detection of nerve and blister agents
UK	Chemical agent monitor (CAM), GID-2/GID-3 detectors
	Ion mobility spectrometry based monitor for the detection of nerve and blister agents
	NAIAD
	Nerve agent immobilized enzyme detector and alarm
USA	ICAD miniature chemical agent detector
	Personal detector based on electro-chemical principals for the detection of nerve, blister, blood and choking agents
	MINICAMS
	Gas chromatographic detection of nerve and blister agents
	M21 remote sensing chemical agent alarm (RSCAAL)
	Passive infrared detection of chemical warfare agents
	Chemical agent detection kit, M256A1
	Wet chemistry detection of nerve, blister, choking and blood agents
	SAW MINICAD MK II
	Surface acoustic wave detection of nerve and blister agents

identification and taken steps to ensure that laboratory networks are available to support forensic, remediation, and other purposes. Detection networks utilizing IMS and other means of detection and transportable laboratories containing analytical equipment have been developed and these may be deployed to sites such as the Olympics or Heads-of-States meetings where there is a perceived risk.

Improved field detection and identification has been the focus of a number of researchers with several new approaches being investigated. Fast GC instruments have interfaced to FPD to improve speed of separation [136,137], new sampling devices have been developed for sample concentration [137,138], and methanol has been suggested as a possible CI-MS reagent gas since it does not require gas cylinders [139]. Use of commercial bench top GC-MS instrumentation has been demonstrated during chemical warfare agent applications [140] and MS has been incorporated into field laboratories to improve army medical capabilities [141].

Considerable effort has been expended on the development of field portable MS and GC-MS instruments, as this technique may be used to rapidly confirm chemical warfare agents under field situations. Concepts of use usually involve either bringing the instrument into the contaminated area for sample collection and analysis or collection of sample and transport of the sample to a GC-MS located in a clean area. The availability of SPME fibers [142], which may be thermally desorbed into an on-site GC-MS instrument, has resulted in increased use of SPME applications in chemical defense [143–147], including applications in field sampling and analysis [148–154]. SPME fibers may be taken into a contaminated area to directly sample air [152,154], the headspace above solutions or soil [145–148,151], or water [143,145]. Subsequent analysis may then be performed in a clean area [153]. This concept has been demonstrated with analysis of SPME fibers coated with polydimethylsiloxane/divinylbenzene (PDMS/DVB) using resistively heated GC columns (fast GC) with a field portable GC-MS. Figure 25.9 illustrates near real time resistively heated GC-MS

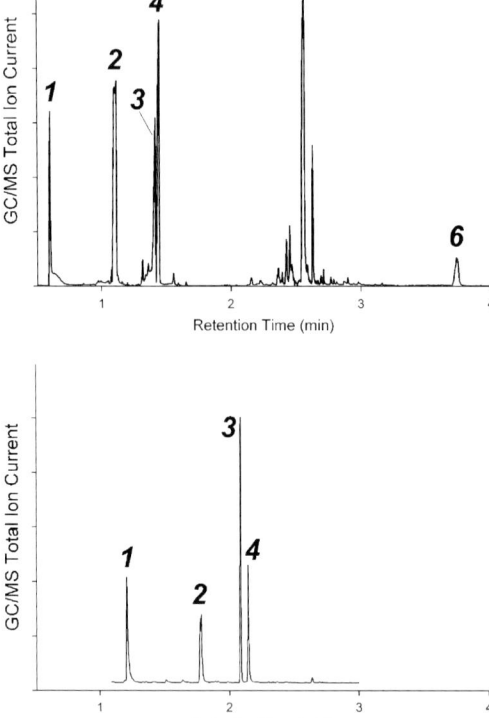

Fig. 25.9. Resistively heated GC-MS analysis of (a) direct SPME sampling of a water sample (5 min) and (b) SPME sampling of air (5 min). The water sample was spiked with six compounds: (1) sarin, (2) soman, (3) tabun, (4) mustard, (5) VX, and (6) T-2 toxin, while the air was spiked with compounds (1) to (4) [154] (Copyright © (2005) with permission from Elsevier).

analysis and identification of six chemical warfare agents, sarin, soman, tabun, mustard , VX, and T-2 toxin, following water sampling with a SPME fiber and identification of sarin, soman, tabun, and mustard following air sampling with a SPME fiber. Both analyses were completed in less than 4 min, a savings of ~20 min over conventional GC-MS analyses [154]. Design and interfacing of a smaller mass spectrometer to complement the fast GC component appears likely in the future.

LC-MS has not yet been used in a field portable role like GC-MS, but the versatile nature of this approach has found researchers interested in developing field portable instrumentation. A number of research efforts, including the development of a field portable chemical warfare agent analysis platform at DRDC Suffield for the Canadian NBC Company, are presently underway. It would be advantageous to utilize instrumentation based on API-MS since this ionization technique has the potential to rapidly detect and identify many chemical warfare agents, chemical warfare agent degradation products, toxic industrial chemicals, higher molecular mass toxins, and biological warfare agent biomarkers.

API-MS data, while less reliable for database searching, may still be useful for identification of chemical warfare agents. An ESI-MS library containing data for ~60 chemical warfare agents, their hydrolysis products, and related compounds has been created and is available on the Internet [155]. It contains both higher and lower sampling voltage mass spectra with the former containing more product ion information. During most ESI-MS analyses, the molecular mass will be evident, simplifying identification and limiting the number of possible matches in the mass spectral database. While ESI-MS data is not as amenable to database searching as EI-MS data, spectra obtained with different instruments generally exhibit the same ions, albeit with differences in their relative intensities. Fits may not be as good as EI-MS, but use and creation of API-MS databases will aid future analyses, particularly in cases where standards are unavailable.

25.8 SAFETY AND DISPOSAL

Chemical warfare agents are extremely hazardous and lethal compounds. They should only be used in designated laboratories by personnel trained in safe-handling and decontamination procedures and with immediate access to medical support. Safety and standard operating procedures must be developed and approved before any chemical warfare agents are handled. Chemical warfare agents can only be safely handled in laboratory chemical hoods with a minimum face velocity of 100 linear feet per minute equipped with emission control devices that limit exhaust concentration to below $0.0001 \, mg/m^3$. Personnel handling chemical warfare agents should wear rubber gloves, lab coats, and full-faceshields and keep a respirator (gas mask) within easy reach. Sufficient decontaminant to destroy all chemical warfare agents being handled must be on hand before commencing operations.

Blister and nerve agents can be destroyed using saturated methanolic solutions of sodium or potassium hydroxide. Decontaminated chemical warfare agents must be disposed of in an environmentally approved method according to local legislation.

References pp. 868–872

25.9 CONCLUSIONS

The current security environment and concern over terrorist use has been responsible for a renewed emphasis on the development of methods for the analysis of chemical warfare agents, their hydrolysis products, and related compounds in environmental and biomedical samples that might be collected in support of military, homeland security, or forensic requirements. Methods involving the use of GC, LC, CE, MS, NMR, FTIR, SIMS, and IMS have all been reported in the past few years. GC-MS remains the method of choice for many chemical defense applications. Newer applications involving SPME sampling and rapid GC-MS analysis show considerable promise and have been demonstrated for near real time field identification of chemical warfare agents. Increasingly, analysts have used LC-MS as a complementary or replacement technique for the confirmation of chemical warfare agents or their degradation products in aqueous samples or extracts and biomedical fluids. Advances in biomedical analyses could be extremely important in the future from both a forensic and public health perspective since assessment of chemical warfare agent exposure within a population could be a priority following a chemical terrorist incident.

25.10 ANNEX

During the review process, several additional papers were published. Pumera reviewed capillary ectrophoresis and "laboratory-on-a-chip" [156], Hill *et al.* investigated the use of thermal desorption ion mobility spectrometry for rapid detection of chemical warfare agent degradation products [157], and Hanaoka *et al.* conducted analyses on chemical munitions abandoned by the Japanese Imperial Forces at the end of the Second World War [158]. Sng *et al.* investigated LC-MS for the screening of nitrogen mustards and their degradation products [159] and Dubey *et al.* continued their sample preparation research with a new paper on hollow fiber-mediated liquid phase microextraction [160].

25.11 REFERENCES

1   S.M. Somani, Chemical warfare agents, Academic Press Inc., New York, 1992.
2   D.H. Ellison, Handbook of chemical and biological warfare agents, CRC Press, Washington, 2000.
3   O. Kostiainen, . In: M.J. Bogusz (Ed.), Handbook of analytical separations, Vol. 2. Forensic Science, Elsevier, Amsterdam, NL, 2000, p. 405.
4   L.K. Engman; A. Lindblad; A.-K. Tunemalm; O. Claesson; B. Lilliehöök, FOI briefing book on chemical weapons: threat, effects and protection, Edita Vastra Aros AB, Stockholm, 2002.
5   M. Mesilaakso, Chemical weapons convention chemical analysis: sample collection, preparation and analytical methods, Wiley, Chichester, UK, 2004.
6   Z. Witkiewicz, M. Mazurek and J. Szulc, J. Chromatogr., 503 (1990) 293.
7   Ch.E. Kientz, J. Chromatogr. A, 814 (1998) 1.
8   E.W.J. Hooijschuur, C.E. Kientz and U.A.T. Brinkman, J. Chromatogr. A., 982 (2002) 177.
9   R.M. Black and B. Muir, J. Chromatogr. A, 1000 (2003) 253.

10   E.R.J. Wils, . In: M. Mesilaakso (Ed.), Chemical weapons convention chemical analysis: sample collection, preparation and analytical methods, Wiley, Chichester, UK, 2004, p. 249.

11   R.M. Black, R.W. Read, . In: M. Mesilaakso (Ed.), Chemical weapons convention chemical analysis: sample collection, preparation and analytical methods, Wiley, Chichester, UK, 2004, p. 283.

12   M. Mesilaakso, A. Niederhauser, . In: M. Mesilaakso (Ed.), Chemical weapons convention chemical analysis: sample collection, preparation and analytical methods, Wiley, Chichester, UK, 2004, p. 321.

13   M.T. Söderström, . In: M. Mesilaakso (Ed.), Chemical weapons convention chemical analysis: sample collection, preparation and analytical methods, Wiley, Chichester, UK, 2004, p. 353.

14   E.W.J. Hooijschuur, A.G. Hulst, A.L. de Jong, L.P. de Reuver, S.H. van Krimpen, B.L.M. van Baar, E.R.J. Wils, C.E. Kientz and U.A.T. Brinkman, TRAC Trends Anal. Chem., 21 (2002) 116.

15   W.R. Creasy, J.R. Stuff, B. Williams, K. Morrissey, J. Mays, R. Duevel and H.D. Durst, J. Chromatogr. A, 774 (1997) 253.

16   M.D. Brickhouse, W.R. Creasy, B.R. Williams, K.M. Morrissey, R.J. O'Connor and H.D. Durst, J. Chromatogr. A, 883 (2000) 185.

17   A.F. Kingery and H.E. Allen, Toxicol. Environ. Chem., 47 (1995) 155.

18   N.B. Munro, S.S. Talmage, G.D. Griffin, L.C. Waters, A.P. Watson, J.F. King and V. Hauschild, Environ. Health Perspect., 107 (1999) 933.

19   D. Noort, H.P. Benschop and R.M. Black, Toxicol. Appl. Pharmacol., 184 (2002) 116.

20   M.-L. Kuitunen, . In: M. Mesilaakso (Ed.), Chemical weapons convention chemical analysis: sample collection, preparation and analytical methods, Wiley, Chichester, UK, 2004, p. 163.

21   J. Hendrikse, . In: M. Mesilaakso (Ed.), Chemical weapons convention chemical analysis: sample collection, preparation and analytical methods, Wiley, Chichester, UK, 2004, p. 89.

22   T.J. Reddy, U. Saradhi, S. Prabhakar and M. Vairamani, J. Chromatogr. A, 1038 (2004) 225.

23   C. Montauban, A. Begos and B. Bellier, Anal. Chem., 76 (2004) 2791.

24   D. Pardasani, M. Palit, A.K. Gupta, P. Shakya, K. Sekhar and D.K. Dubey, Anal. Chem., 77 (2005) 1172.

25   M. Palit, D. Pardasani, A.K. Gupta and D.K. Dubey, Anal. Chem., 77 (2005) 711.

26   X. Chaudot, A. Tambute and M. Caude, J. High Resolut. Chromatogr., 21 (1998) 457.

27   X. Chaudot, A. Tambute and M. Caude, J. Chromatogr. A, 866 (2000) 231.

28   X. Chaudot, A. Tambute and M. Caude, J. Chromatogr. A, 888 (2000) 327.

29   N.V. Beck, W.A. Carrick, D.B. Cooper and B. Muir, J. Chromatogr. A, 907 (2001) 221.

30   O. Kostiainen, . In: M. Mesilaakso (Ed.), Chemical weapons convention chemical analysis: sample collection, preparation and analytical methods, Wiley, Chichester, UK, 2004, p. 185.

31   P.A. D'Agostino and L.R. Provost, J. Chromatogr., 331 (1985) 47.

32   A. Manninen, M.-L. Kuitunen and L. Julin, J. Chromatogr., 394 (1979) 465.

33   H.E.T. Spruit, H.C. Trap, J.P. Langenberg and H.P. Benschop, J. Anal. Toxicol., 25 (2001) 57.

34   E.M. Jakubowski, L.S. Heykamp, H.D. Durst and S.A. Thomson, Anal. Lett., 34 (2001) 727.

35   H.J. O'Neill, K.L. Brubaker, J.F. Schneider, L.F. Sytsma and T.A. Kimmell, J. Chromatogr. A, 962 (2002) 183.

36   I.N. Stan'kov, A.A. Sergeeva, I.D. Derevyagina and K.V. Konovalov, J. Anal. Chem., 58 (2003) 160.

37   I.N. Stan'kov, A.A. Sergeeva, V.B. Sitnikov, I.D. Derevyagina and O.T. Morozova, J. Anal. Chem., 59 (2004) 260.

38   I.N. Stan'kov, A.A. Sergeeva, V.B. Sitnikov, I.D. Derevyagina, O.T. Morozova, S.N. Mylova and V.B. Forov, J. Anal. Chem., 59 (2004) 447.

39   R. Haas, Environ. Sci. Pollut. Res., 5 (1998) 2.

40   S. Sass and T.L. Fisher, Org. Mass Spectrom., 14 (1979) 257.

41   P.A. D'Agostino and L.R. Provost, Biomed. Environ. Mass Spectrom., 13 (1986) 231.

42   D.K. Rohrbaugh and E.W. Sarver, J. Chromatogr. A, 809 (1998) 141.

43   E.M. Jakubowski, J.M. Mcguire, R.A. Evans, J.L. Edwards, S.W. Hulet, B.J. Benton, J.S. Forster, D.C. Burnett, W.T. Muse, K. Matson, C.L. Crouse, R.J. Mioduszewski and S.A. Thomson, J. Anal. Toxicol., 28 (2004) 357.

44   P.A. D'Agostino, L.R. Provost and K.M. Looye, J. Chromatogr., 465 (1989) 271.
45   P.A. D'Agostino, L.R. Provost, J.F. Anacleto and P.W. Brooks, J. Chromatogr., 504 (1990) 259.
46   P.A. D'Agostino and L.R. Provost, J. Chromatogr., 541 (1991) 121.
47   R.M. Black, R.J. Clarke, J.M. Harrison and R.W. Read, Xenobiotica, 27 (1997) 499.
48   W.J. Driskell, M. Shih, L.L. Needham and D.B. Barr, J. Anal. Toxicol., 26 (2002) 6.
49   A.E. Boyer, D. Ash, D.B. Barr, C.L. Young, W.J. Driskell, R.D. Whitehead, M. Ospina, K.E.
      Preston, A.R. Woolfitt, R.A. Martinez, L.A. Silks and J.R. Barr, J. Anal. Toxicol., 28 (2004) 327.
50   C.L. Young, D. Ash, W.J. Driskell, A.E. Boyer, R.A. Martinez, L.A. Silks and J.R. Barr, J. Anal.
      Toxicol., 28 (2004) 339.
51   J.R. Barr, W.J. Driskell, L.S. Aston and R.A. Martinez, J. Anal. Toxicol., 28 (2004) 372.
52   J. Riches, I. Morton, R.W. Read and R.M. Black, J. Chromatogr. B, 816 (2005) 251.
53   J.R. Stuff, R.L. Cheicante, H.D. Durst and J.L. Ruth, J. Chromatogr. A, 849 (1999) 529.
54   W.A. Carrick, D.B. Cooper and B. Muir, J. Chromatogr. A, 925 (2001) 241.
55   B. Muir, S. Quick, B.J. Slater, D.B. Cooper, M.C. Moran, C.M. Timperley, W.A. Carrick and
      C.K. Burnell, J. Chromatogr. A, 1068 (2005) 315.
56   B.A. Tomkins and G.A. Sega, J. Chromatogr. A, 911 (2001) 85.
57   I. Ohsawa, M. Kanamori-Kataoka, K. Tsuge and Y. Seto, J. Chromatogr. A, 1061 (2004) 235.
58   M. Kataoka, K. Tsuge, H. Takesako, T. Hamazaki and Y. Seto, Environ. Sci. Technol., 35 (2001)
      1823.
59   M. Noami, M. Kataoka and Y. Seto, Anal. Chem., 74 (2002) 4709.
60   P.A. D'Agostino, J.R. Hancock and C.L. Chenier, Eur. J. Mass Spectrom., 9 (2003) 609.
61   M. Mazurek, Z. Witkiewicz, S. Popiel and M. Sliwakowski, J. Chromatogr. A, 919 (2001) 133.
62   P.A. D'Agostino and L.R. Provost, J. Chromatogr. A, 695 (1995) 65.
63   B. Muir, B.J. Slater, D.B. Cooper and C.M. Timperley, J. Chromatogr. A, 1028 (2004) 313.
64   H.D. Durst, J.R. Mays, J.L. Ruth, B.R. Williams and R.V. Duevel, Anal. Lett., 31 (1998) 1429.
65   A. Miki, M. Katagi, H. Tsuchihashi and M. Yamashita, J. Anal. Toxicol., 23 (1999) 86.
66   T.P. Logan, J.R. Smith, E.M. Jakubowski and R.E. Nielson, Toxicol. Methods, 9 (1999) 275.
67   M. Kataoka and Y. Seto, J. Chromatogr. B, 795 (2003) 123.
68   A. Fidder, D. Noort, A.G. Hulst, L.P.A. de Jong and H.P. Benschop, Arch. Toxicol., 74 (2000)
      207.
69   M. Nagao, T. Takatori, Y. Matsuda, M. Nakajima, H. Iwase and K. Iwadate, Toxicol. Appl.
      Pharmacol., 144 (1997) 198.
70   H.P. Benschop, G.P. van der Schans, D. Noort, A. Fidder, R.H. Mars-Groenendijk and L.P.A. de
      Jong, J. Anal. Toxicol., 21 (1997) 249.
71   M.J. Van Der Schans, M. Polhuijs, C. Van Dijk, Degenhardt Ceam, K. Pleijsier, J.P. Langenberg
      and H.P. Benschop, Arch. Toxicol., 78 (2004) 508.
72   Y. Matsuda, M. Nagao, T. Takatori, H. Niijima, M. Nakajima, H. Iwase, M. Kobayashi and K.
      Iwadate, Toxicol. Appl. Pharmacol., 150 (1998) 310.
73   W. Abu-Qare and B. Abou-Donia, Chromatographia, 53 (2001) 251.
74   L. Shanmao, L. Wei, Z. Boli, Y. Lijun and W. Qinpel, S. Afr. J. Chem., 58 (2005) 82.
75   E.W.J. Hooijschuur, C.E. Kientz and U.A.Th. Brinkman, J. Chromatogr. A, 849 (1999) 433.
76   E.W.J. Hooijschuur, C.E. Kientz and U.A.Th. Brinkman, J. Chromatogr. A, 907 (2001) 165.
77   E.W.J. Hooijschuur, C.E. Kientz and U.A.T. Brinkman, J. Chromatogr. A, 928 (2001) 187.
78   E.R.J. Wils and A.G. Hulst, J. Chromatogr., 454 (1988) 261.
79   E.R.J. Wils and A.G. Hulst, J. Chromatogr., 523 (1990) 151.
80   R.M. Black and R.W. Read, J. Chromatogr. A, 794 (1998) 233.
81   P.A. D'Agostino, L.R. Provost and J.R. Hancock, J. Chromatogr. A, 808 (1998) 177.
82   W.R. Creasy, J. Am. Soc. Mass Spectrom., 10 (1999) 440.
83   P.A. D'Agostino, J.R. Hancock and L.R. Provost, J. Chromatogr. A, 837 (1999) 93.
84   P.A. D'Agostino, J.R. Hancock and L.R. Provost, J. Chromatogr. A, 840 (1999) 289.
85   M. Katagi, M. Tatsuno, M. Nishikawa and H. Tsuchihashi, J. Chromatogr. A, 833 (1999) 169.
86   J.-P. Mercier, P. Morin and M. Dreux, Chimia, 53 (1999) 511.
87   J.-P. Mercier, Ph. Morin, M. Dreux and A. Tambute, J. Chromatogr. A, 849 (1999) 197.

88   R.W. Read and R.M. Black, J. Chromatogr. A, 862 (1999) 169.
89   E.W.J. Hooijschuur, C.E. Kientz, A.G. Hulst and U.A.T. Brinkman, Anal. Chem., 72 (2000) 1199.
90   P.A. D'Agostino, J.R. Hancock and L.R. Provost, Adv.Mass Spectrom., 15 (2001) 297.
91   P.A. D'Agostino, C.L. Chenier and J.R. Hancock, J. Chromatogr. A, 950 (2002) 149.
92   Q. Liu, X. Hu and J. Xie, Anal. Chim. Acta, 512 (2004) 93.
93   K. Kinoshita, Y. Shida, C. Sakuma, M. Ishizaki, K. Kiso, O. Shikino, H. Ito, M. Morita, T. Ochi and T. Kaise, Appl. Organomet. Chem., 19 (2005) 287.
94   M. Kanamori-Kataoka and Y. Seto, Jpn. J. Forensic Toxicol., 23 (2005) 21.
95   P.A. D'Agostino, J.R. Hancock and L.R. Provost, J. Chromatogr. A, 912 (2001) 291.
96   P.A. D'Agostino, J.R. Hancock and C.L. Chenier, J. Chromatogr. A, 1058 (2004) 97.
97   C.M. Timperley, R.M. Black, M. Bird, I. Holden, J.L. Mundy and R.W. Read, Phosphorus Sulfur Silicon, 178 (2003) 2027.
98   R.W. Read and R.M. Black, J. Anal. Toxicol., 28 (2004) 346.
99   R.W. Read and R.M. Black, J. Anal. Toxicol., 28 (2004) 352.
100   D. Noort, A.G. Hulst, D.H.J.M. Platenburg, M. Polhuijs and H.P. Benschop, Arch. Toxicol., 72 (1998) 671.
101   D. Noort, A. Fidder, A.G. Hulst, A.R. Woolfitt, D. Ash and J.R. Barr, J. Anal. Toxicol., 28 (2004) 333.
102   J.R. Smith, J. Anal. Toxicol., 28 (2004) 390.
103   P.A. D'Agostino, J.R. Hancock, C.L. Chenier and C.R.J. Lepage, J. Chromatogr. A, 1110 (2006) 86.
104   R. Kostiainen, A.P. Bruins and V.M.A. Hakkinen, J. Chromatogr., 634 (1993) 113.
105   J.-P. Mercier, P. Chaimbault, Ph. Morin, M. Dreux and A. Tambute, J. Chromatogr. A, 825 (1998) 71.
106   J.E. Melanson, C.A. Boulet and C.A. Lucy, Anal. Chem., 73 (2001) 1809.
107   J. Jiang and C.A. Lucy, J. Chromatogr. A, 966 (2002) 239.
108   C.L. Copper and G.E. Collins, Electrophoresis, 25 (2004) 897.
109   A.-E.F. Nasser, S.V. Lucas, W.R. Jones and L.D. Hoffland, Anal. Chem., 70 (1998) 1085.
110   A.-E.F. Nasser, S.V. Lucas, C.A. Myler, W.R. Jones, M. Campisano and L.D. Hoffland, Anal. Chem., 70 (1998) 3598.
111   A.-E.F. Nasser, S.V. Lucas and L.D. Hoffland, Anal. Chem., 71 (1999) 1285.
112   J. Wang, M. Pumera, G.E. Collins and A. Mulchandani, Anal. Chem., 74 (2002) 6121.
113   J. Wang, J. Zima, N.S. Lawrence, M.P. Chatrathi, A. Mulchandani and G.E. Collins, Anal. Chem., 76 (2004) 4721.
114   G.R. Asbury, C. Wu, W.F. Siems and H.H. Hill Jr., Anal. Chim. Acta, 404 (2000) 273.
115   W.E. Steiner, B.H. Clowers, L.M. Matz, W.F. Siems and H.H. Hill, Anal. Chem., 74 (2002) 4343.
116   W.E. Steiner, B.H. Clowers, P.E. Haigh and H.H. Hill, Anal. Chem., 75 (2003) 6068.
117   W.E. Steiner, S.J. Klopsch, W.A. English, B.H. Clowers and H.H. Hill, Anal. Chem., 77 (2005) 4792.
118   K. Tuovinen, H. Paakkanen and O. Hanninen, Anal. Chim. Acta, 440 (2001) 151.
119   C. Albaret, D. Loeillet, P. Auge and P.-L. Fortier, Anal. Chem., 69 (1997) 2694.
120   T.J. Henderson, Anal. Chem., 74 (2002) 191.
121   J.C. Ingram, A.D. Appelhans and G.S. Groenewold, Int. J. Mass Spectrom. Ion Proc., 175 (1998) 253.
122   G.S. Groenewold, A.D. Appelhans, J.C. Ingram, G.L. Gresham and A.K. Gianotto, Talanta, 47 (1998) 981.
123   G.S. Groenewold, A.D. Appelhans, G.L. Gresham, J.E. Olson, M. Jeffery and J.B. Wright, Anal. Chem., 71 (1999) 2318.
124   G.L. Gresham, G.S. Greonewold and J.E. Olson, J. Mass Spectrom., 35 (2000) 1460.
125   G.L. Gresham, G.S. Groenewold, A.D. Appelhans, J.E. Olson, M.T. Benson, M.T. Jeffery, B. Rowland and M.A. Weibel, Int. J. Mass Spectrom., 208 (2001) 135.
126   R.B. Cody, J.A. Laramee and H.D. Durst, Anal. Chem., 77 (2005) 2297.
127   Z. Takats, J.M. Wiseman, B. Gologan and R.G. Cooks, Science, 306 (2004) 471.

128   J. Eldridge, Jane's nuclear, biological and chemical defence, Jane's Information Group Limited, Coulsdon, UK, 2006.

129   G.M. Murray, D.S. Lawrence, . In: M. Mesilaakso (Ed.), Chemical weapons convention chemical analysis: sample collection, preparation and analytical methods, Wiley, Chichester, UK, 2004, p. 65.

130   B.A. Eckenrode, J. Am. Soc. Mass Spectrom., 12 (2001) 683.

131   S. Kendler, A. Zaltsman and G. Frishman, Instrum. Sci. Technol., 31 (2003) 357.

132   K.J. Hart, M.B. Wise, W.H. Griest and S.A. Lammert, Field Anal. Chem. Tech., 4 (2000) 93.

133   S. Mogl, . In: M. Mesilaakso (Ed.), Chemical weapons convention chemical analysis: sample collection, preparation and analytical methods, Wiley, Chichester, UK, 2004, p. 7.

134   S. Krüger, . In: M. Mesilaakso (Ed.), Chemical weapons convention chemical analysis: sample collection, preparation and analytical methods, Wiley, Chichester, UK, 2004, p. 33.

135   M. Sokolowski, . In: M. Mesilaakso (Ed.), Chemical weapons convention chemical analysis: sample collection, preparation and analytical methods, Wiley, Chichester, UK, 2004, p. 51.

136   G. Frishman and A. Amirav, Field Anal. Chem. Tech., 4 (2000) 170.

137   S. Kendler, A. Zifman, N. Gratziany, A. Zaltsman and G. Frishman, Anal. Chim. Acta, 548 (2005) 58.

138   A. Gordin and A. Amirav, J. Chromatogr. A, 903 (2000) 155.

139   D.K. Rohrbaugh, J. Chromatogr. A, 893 (2000) 393.

140   E. Davoli, L. Cappellini, R. Fanelli, M. Bonsignore and M. Gavinelli, Field Anal. Chem. Tech., 5 (2001) 313.

141   J.R. Smith, M.L. Shih, E.O. Price, G.E. Platoff and J.J. Schlager, J. Appl. Toxicol., 21 (2001) S35–S41.

142   J. Pawliszyn, Sampling and sample preparation for field and laboratory, Elsevier, Amsterdam, 2002.

143   H.-A. Lakso and W.F. Ng, Anal. Chem., 69 (1997) 1866.

144   B.A. Tomkins, G.A. Sega and C. Ho, J. Chromatogr. A, 909 (2001) 13.

145   J.F. Schneider, A.S. Boparai and L.L. Reed, J. Chromatogr. Sci., 39 (2001) 420.

146   S.D. Harvey, D.A. Nelson, B.W. Wright and J.W. Grate, J. Chromatogr. A, 954 (2002) 217.

147   J.V. Wooten, D.L. Ashley and A.M. Calafat, J. Chromatogr. B, 772 (2002) 147.

148   G.L. Kimm, G.L. Hook and P.A. Smith, J. Chromatogr. A, 971 (2002) 185.

149   G.L. Hook, G.L. Kimm, T. Hall and P.A. Smith, TRAC Trends Anal. Chem., 21 (2002) 534.

150   G.L. Hook, G. Kimm, G. Betsinger, P.B. Savage, A. Swift, T. Logan and P.A. Smith, J. Sep. Sci., 26 (2003) 1091.

151   G.L. Hook, G. Kimm, D. Koch, P.B. Savage, B.W. Ding and P.A. Smith, J. Chromatogr. A, 992 (2003) 1.

152   G.L. Hook, C. Jackson Lepage, S.I. Miller and P.A. Smith, J. Sep. Sci., 27 (2004) 1017.

153   P.A. Smith, C.R. Jackson Lepage, D. Koch, H.D.M. Wyatt, G.L. Hook, G. Betsinger, R.P. Erickson and B.A. Eckenrode, TRAC Trends Anal. Chem., 23 (2004) 296.

154   P.A. Smith, M.T. Sng, B.A. Eckenrode, S.Y. Leow, D. Koch, R.P. Erickson, C.R.J. Lepage and G.L. Hook, J. Chromatogr. A, 1067 (2005) 285.

155   P.A. D'Agostino. http://www.suffield.drdc-rddc.gc.ca/ResearchTech/Products/CB_PRODUCTS/index_e.html

156   M. Pumera, J. Chromatogr. A, 1113 (2006) 5.

157   A.B. Kanu, P.E. Haigh and H.H. Hill, Anal. Chim. Acta, 553 (2005) 148.

158   S. Hanaoka, K. Normura and T. Wada, J. Chromatogr. A, 1101 (2006) 268.

159   H.C. Chua, H.S. Lee and M.T. Sng, J. Chromatogr. A, 1102 (2006) 214.

160   D.K. Dubey, D. Pardasani, A.K. Gupta, M. Palit, P.K. Kanaujia and V. Tak, J. Chromatogr. A, 1107 (2006) 29.

M.J. Bogusz (Ed.). Forensic Science
Handbook of Analytical Separations, Vol. 6
© 2008 Published by Elsevier B.V.

CHAPTER 26

Forensic analysis of fire debris

Julia A. Dolan

*Bureau of Alcohol, Tobacco, Firearms and Explosives, Forensic Science Laboratory – Washington,
Ammendale, Maryland, USA*

26.1 INTRODUCTION

Forensic science, in its broadest sense, refers to the application of various scientific disciplines to matters that may have legal ramifications. In this sense, the application of forensic science to the investigation of fires and fire-related crimes includes truly all aspects of the cause and origin investigation of the fire scene. For origin and cause analysis is itself its own discipline of forensic science. By applying the scientific method to the scene of a fire of unknown cause, the investigator makes observations, collects data, and forms and tests hypotheses, so fire investigation is, in fact, a unique area of forensic science.

The application of forensic science to fire and arson investigation often refers to chemical and other laboratory tests of materials recovered from fire scenes, rather than the overall scene investigation. There are many types of physical evidence that can provide information to an investigator to help determine the cause of a fire. Most frequently, investigators submit samples of debris from fire scenes that they believe may have been set, to determine if an accelerant may have been used. It is this examination of fire debris for the presence of ignitable liquids that makes up the vast majority of forensic laboratory support to fire investigations. Other types of evidence are also subjected to examination by forensic science laboratories, often to search for trace evidence or latent fingerprints. The analysis of evidence recovered from fire scenes for ignitable liquid residues, trace evidence, or latent fingerprints can generally be conducted in most public sector forensic laboratories, and in private laboratories focused on fire investigation.

Another major area of physical evidence that often requires examination by a trained expert is that relating to electrical equipment. Electrical engineers are often called upon to provide forensic examination of electrical equipment recovered from fire scenes to determine if it may have caused the fire. This type of examination is often crucial in civil cases, when it has been determined that no criminal act has been committed, but in which it may be necessary to determine liability. Examination by a qualified electrical engineer may also be necessary in order to eliminate an electrical fire cause.

This chapter will focus primarily on the forensic examination of fire debris for the presence of ignitable liquid residues. Although there are many aspects of the application of forensic science to fire investigation, they cannot be adequately covered in this text. Therefore, the scope of this chapter will be limited, and will provide a detailed presentation of various issues relating to the chemical analysis of fire debris for the presence of ignitable liquids.

26.2 THE ARSON PROBLEM

Fires cause billion-dollar damages annually throughout the world, and even in the United States alone, direct dollar losses exceed $10 billion [1,2]. Because of the devastating effects of fire – whether intentionally set or not – ensuring that they are adequately and scientifically investigated is a major priority. Understanding the causes of accidental fires, and being able to prosecute those responsible for intentionally set fires are important steps in reducing the incidence of and damage due to uncontrolled fires.

26.2.1 Fire scene investigation

The field of fire investigation is unique in many aspects. Unlike many criminal investigations, the investigator must first determine if the fire was the result of a criminal act. This portion of the investigation is referred to as the origin and cause investigation. It is during this phase of the investigation that the fire scene is examined in an attempt to determine where the fire started (the origin) and why the fire started (the cause). This process may in some cases be relatively simple and straightforward, and in other cases may be extremely complex. Regardless of the complexity of the fire scene, each investigation of the origin and cause of a fire should be treated with a practical scientific methodology. In cases of incendiary fires – those that have been intentionally set under circumstances in which the person knows that the fire should not be ignited – there is in addition to the origin and cause investigation, a criminal investigation aimed at not only determining who set the fire, but also for obtaining evidence sufficient to prove the circumstances regarding the fire. If a fire cause is determined to be accidental, there may be additional inquiries relating to insurance or product liability issues; however, for the public fire investigator, the finding of an accidental cause is usually the end of their investigation.

Fire investigation has changed dramatically since its earliest days, in which the determination of cause was based more on assumptions rather than science. A major effort has been made to increase the foundation of scientific knowledge required for appropriate fire scene investigations. The publication of NFPA 921 Guide for Fire and Explosion Investigations was a driving force in this process. The first edition of NFPA 921 was published in 1992 and was developed by the Committee on Fire

Investigations for the purpose of "improving the fire investigation process and the quality of information on fires resulting from the investigative process" [3]. Since its initial publication, new editions have been published every 3 years and it is now in its fifth edition (2004) [4]. In addition to NFPA 921, the textbook "Kirk's Fire Investigation" by John DeHaan, now in its fifth edition, has also been an instrumental force in placing a greater emphasis on scientific methodology in the investigation of fires [5]. As a result of these efforts, numerous books have since been published that promulgate the necessity of integrating science into the investigation of the fire scene. John Lentini has been a staunch proponent of the need for scientific methodology in all aspects of fire investigation, and his text "Scientific Protocols for Fire Investigation" is a valuable reference, providing comprehensive coverage of fire dynamics, combustion science and how such phenomena will affect the behavior of a fire, and ultimately, the fire scene [6]. Additional texts focus on other aspects of fire investigation and may provide different perspectives on the application of the scientific method to examining a fire scene [7,8]. Consequently, fire investigators have much greater access to training materials that emphasize the scientific nature of fire scene analysis. As a result of these efforts, fire investigation today is a scientific endeavor that encompasses principles of engineering, fire dynamics, chemistry, and logic.

One aspect of fire investigation that has historically been recognized as a scientific discipline is that of recovering and identifying residues of ignitable liquid accelerants. Fire investigators have long sought the assistance of a forensic science laboratory to examine debris recovered from a fire scene to determine if materials consistent with accelerants were present. Identification of ignitable liquid residues from fire debris is often an important part of the fire investigation. The successful preservation of residues recovered from a scene is vital to the potential identification of these residues. The unique properties associated with this type of evidence offer several challenges to the investigator and scientist, which are discussed in the following section.

26.2.2 Challenges associated with the recovery of potential accelerants

Accelerants may be any material that can contribute to the ease of ignition or the speed or spread of a fire, although those most commonly encountered are volatile ignitable liquids [4]. Technically, an arsonist could use newspapers, fabric softener sheets, or even potato chips to aid in increasing the rate of fire spread. An accelerant is therefore defined not by what it is, but rather by how it is used. The examination of fire debris for potential accelerants does not generally include ordinary combustible materials that may have been intentionally misused for the purpose of spreading a fire. Most often, when fire debris is submitted to an analytical laboratory to determine if an accelerant is present, it is examined for the presence of volatile ignitable liquids.

Volatile ignitable liquids are particularly suitable for use as accelerants. They are readily available, relatively inexpensive, and effective. Volatility refers to a liquid's tendency to vaporize; a volatile liquid therefore will have a rich vapor phase. It is this quality that makes this type of liquid so effective as an accelerant. Although there are a few materials that will burn in the solid state, most materials must be in a gaseous or vapor state [9]. For this reason, volatile liquids are often easily ignited, and when burning, can provide sufficient energy to ignite surrounding materials that are not as readily ignited.

It is this very quality that makes these liquids such good accelerants, and that also makes their survivability and subsequent recovery more challenging. Their recovery can be affected in two basic ways. Because ignitable liquids have a tendency to vaporize, there is a greater likelihood that an ignitable liquid used as an accelerant in an intentionally set fire may evaporate prior to collection of the evidence. If not packaged appropriately – in an airtight container – evaporation and loss may even occur subsequent to collection [10]. The other major problem that affects the ability to recover ignitable liquid residues from a fire scene is the fact that they are good fuels. Because ignitable liquids tend to vaporize readily, they are very efficient fuels, and may be completely consumed in an intense fire.

There are a variety of conditions that may affect the ability to recover an ignitable liquid residue. First and foremost is the nature and amount of the ignitable liquid. A highly volatile ignitable liquid, such as some cigarette lighter fluid is more likely to be consumed in the fire, as compared with a similar quantity of a less volatile liquid, such as kerosene. Another factor to consider is the water miscibility of the accelerant used in the fire. Very often, petroleum-based liquids such as gasoline (petrol), which are immiscible in water, are used. This type of liquid offers an improved chance for recovery when compared with liquids that are miscible with water, such as some of the light oxygenated solvents. Because these solvents can mix with water, they are often lost or excessively diluted in the copious amounts of water that are invariably applied to the fire during the suppression process. In addition to these qualities of the ignitable liquids used, the amount applied will also have an effect. When a greater amount of ignitable liquid is used, the chances of its successful recovery are increased.

Factors other than the type and amount of ignitable liquid will also affect the survivability and therefore their potential for recovery. The intensity of the fire, response time, and suppression techniques will all contribute to the likelihood of recovery. In addition, the fact that ignitable liquids may be inherent to the fire scene environment can pose additional challenges, as can the contributions from the pyrolysis and partial combustion of the materials recovered from the fire scene [11–13].

26.3 IGNITABLE LIQUIDS

A variety of liquids may be used as accelerants to enhance an incendiary fire. The hydrocarbon-based ignitable liquids that are commonly used as accelerants in intentionally set fires have been examined by forensic examiners, and placed into general categories based on their chemical composition and how they were manufactured. Currently, the consensus-based standard published by ASTM for the application of gas chromatography-mass spectrometry (GC-MS) to fire debris analysis recognizes eight different classifications of ignitable liquids, whereas the gas chromatography standard recognizes nine [14,15]. These include six that are by definition, petroleum-derived, which include gasoline, petroleum distillates, isoparaffinic products, aromatic products, naphthenic paraffinic products, and *n*-alkane products. The current gas chromatography standard and the previous edition of the GC-MS standard include a classification system that also contains a separate category for dearomatized petroleum distillates [15,16]. This category was absorbed into the broader classification of petroleum distillates and is no longer a distinct category in the GC-MS standard [14]. Of the petroleum-derived products, all but gasoline can be further classified as light, medium, or heavy to describe the approximate boiling point range of the material. In addition, the ASTM standards recognize a classification for oxygenated products and a miscellaneous category. In order to have a clear understanding of the data analysis process, it is important to understand each of these types of products that could potentially be encountered as accelerants.

26.3.1 Petroleum and the refining process

Petroleum is a vast mixture of hydrocarbons and other chemical compounds, ranging from small gaseous compounds, such as methane up to compounds that are solids at ambient temperatures with 40 and even more carbons. As it occurs naturally, it is of little use. However, when it is processed for the removal of impurities, and is separated based on usable boiling point fractions, and further refined to meet desired properties, it can be transformed into a variety of useful products. Refining consists of many steps, which are beyond the scope of this chapter. The major steps can be summarized into three general processes: (1) separation based on boiling point by fractional distillation, (2) chemical conversions of compounds in order to enrich the product with desired compounds, and (3) clean-up procedures in which specific chemical compounds are removed. From these basic steps a wide variety of products are made for use as solvents, fuels, and other applications. Many of these commercially marketed products can be placed into the ignitable liquid classification schemes published by ASTM [14,15]. A brief description of each of the classifications follows.

26.3.2 Classification of ignitable liquids

26.3.2.1 Petroleum distillates

Petroleum distillates are the least refined of the petroleum-derived products, and as such, have compositions, which mirror the hydrocarbon composition of the fraction from which they were derived. Petroleum distillates are primarily aliphatic, but do have a distinct aromatic component, depending on their boiling point range. Petroleum distillates will have a strong presence of normal alkanes (*n*-paraffins), with isoalkanes (isoparaffins), cycloalkanes (naphthenes), and aromatic compounds. The aromatic contribution may be benzene-based compounds (mononuclear aromatics or MNAs), indane-based compounds ("indanes") or compounds based on two or more fused six-membered aromatic rings (polynuclear aromatics, or PNAs). Because of the distillation process, there is not a sharp delineation marking the beginning or ending boiling point of the product; rather the composition of petroleum distillates is indicated by a bell-shaped Gaussian distribution. The recently abolished classification of dearomatized distillates is now included as part of this class, and describes petroleum distillates that have been specifically treated for the purpose of removing aromatic compounds. The additional descriptor of light, medium or heavy refers to the approximate boiling point range or *n*-alkane range of the product. A light petroleum distillate (LPD) would encompass normal alkanes in the range of C_4–C_9; medium petroleum distillates (MPD) fall within the range of C_8–C_{13}, and heavy petroleum distillates (HPD) may start as low as C_8 and extend to C_{23} and beyond. Also, narrower range products that begin around C_{11} may also be considered as HPD. These criteria that describe the classification of distillates will be used in the data analysis process. Fig. 26.1 shows total ion chromatograms (TICs) of some typical petroleum distillates.

Petroleum distillates and their dearomatized counterparts are used in a wide variety of applications. LPDs are most commonly encountered in products such as pocket lighter fuels and camp stove fuels. Medium range products have historically been encountered as charcoal starter fluids, paint thinners, dry cleaning solvents, mineral spirits, and some automotive fuel treatments [17]. HPDs are most commonly seen as kerosene, diesel fuel, home heating fuel, and lamp oils, but are also used as specialty solvents and as jet fuel [18].

26.3.2.2 Isoparaffinic products

Isoparaffinic products are derived from petroleum and have undergone significant processing in order to remove unwanted compounds. Isoparaffinic products are entirely aliphatic and have virtually no aromatic compounds present. They consist of fairly narrow boiling point products, and are composed virtually entirely of branched-chain alkanes. Isoparaffinic products because of their lack of cyclic compounds tend to be clean-burning, with little smoke production, making them useful as fuels. Also due to their lack of aromatic compounds, they are very low-odor products, which makes them useful in food-related applications, such as packaging [19]. Depending on the application, light, medium, or heavy, isoparaffinic products may be used as solvents in a variety of applications [20]. In addition, light isoparaffinic products are used in some aviation fuels. Medium-range products may also be commonly seen in commercial products, such as charcoal lighter fluids, copier toners, and paint thinners.

26.3.2.3 Aromatic products

As with the isoparaffinic classification described above, aromatic products are highly refined to remove undesirable components. This classification consists of fairly narrow range products that are composed exclusively of aromatic compounds. There are virtually no aliphatic compounds present in these products. Aromatic products

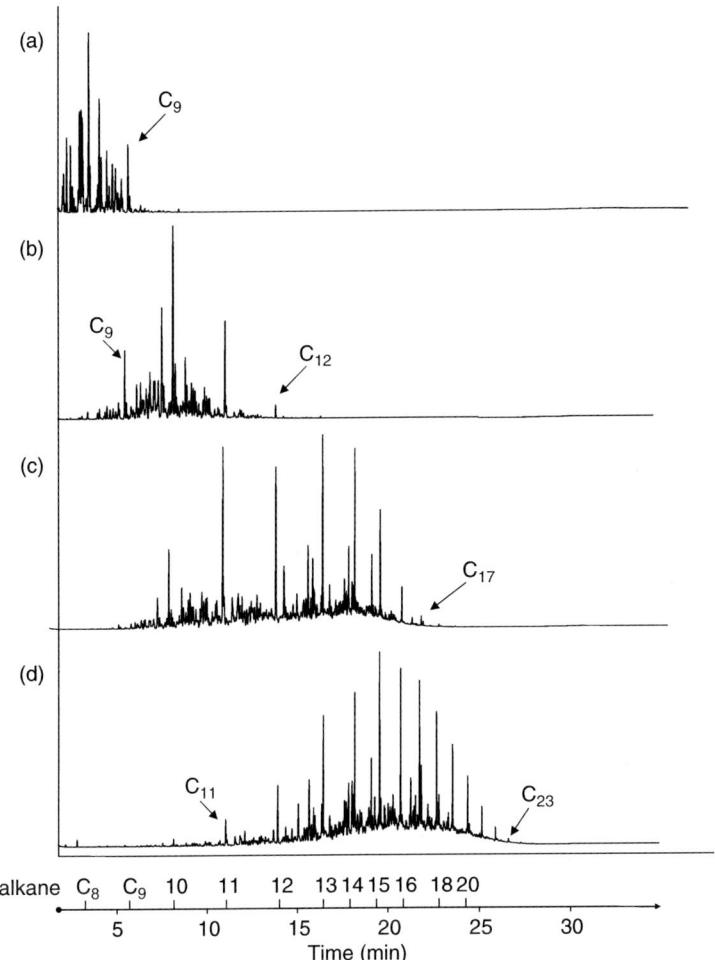

Fig. 26.1. Total ion chromatogram of selected petroleum distillates. (a) Cigarette lighter fluid (light petroleum distillate/LPD); (b) paint thinner (medium petroleum distillate/MPD); (c) kerosene (heavy petroleum distillate/HPD); and (d) diesel fuel (HPD). Each petroleum distillate is characterized by prominent *n*-alkanes and a bell-shaped distribution. Note that in the LPD, some of the left half of the curve (prior to $\approx C_6$) is not visible under common chromatographic conditions. Obtained in the authors' ATF Laboratory under the following conditions: instrument, HP 6890 Series GCMS with HP-1 column 22 m, 0.20 mm ID, film 0.5 μm. Split 20:1. Temperature program: 60°C for 3 min, ramp 5°C/min to 120°C, ramp 12°C/min to 300°C, 5 min at 300°C.

come in a variety of boiling point ranges, depending on the specifications for the intended use. Light aromatic products are based on blends of benzene-based aromatics, such as the alkylbenzenes. Medium-range products may contain alkylbenzenes, but will often also contain indane-based compounds. Aromatic products in the heavy range tend to be composed almost exclusively of PNA compounds, whether indane- or naphthalene-based.

References pp. 919–922

Aromatic compounds contain proportionally more carbon than their aliphatic counterparts and tend to have strong odors associated with them. Consequently, aromatic products tend to burn with significant smoke production, and give off odors when burning. In addition, many aromatic compounds have health risks associated with them. For these reasons, aromatic products are not often commercially marketed as fuels. The benefit provided by aromatic products is that of their solvating ability. The Kauri-butanol value is a measure of a hydrocarbon solvent's strength and tends to be very high for aromatic solvents, as compared with aliphatic solvents [21]. Because of their ability to dissolve a wide range of materials, aromatic products are often used as solvents or cleaning products.

26.3.2.4 Naphthenic paraffinic products

Naphthenic paraffinic products are petroleum-based products derived from the distillate class. Similar to distillate-type products, the naphthenic paraffinic products tend to be fairly broad in terms of boiling point range, and do not have distinct cut-off points. Rather, the distribution of naphthenic paraffinic products mirrors the distillates, in having a bell-shaped Gaussian-type distribution. What distinguishes naphthenic paraffinic compounds from distillates, is the fact that they have undergone additional refining. Naphthenic paraffinic products consist primarily of cycloalkanes and branched alkanes (naphthenes and isoparaffins). The name of the naphthenic paraffinic classification is sometimes considered a misnomer, as most of the normal paraffins have been removed, and that the paraffins present are primarily isoparaffins. Some in the field had promoted the idea of renaming the class "naphthenic isoparaffinic" in order to provide a clearer description of the composition, however, the idea did not garner widespread support. Unlike the less refined distillate classification, there are virtually no aromatic compounds present in naphthenic paraffinic products, and the relative amount of normal alkanes is greatly reduced, compared with distillates and crude oil. Naphthenic paraffinic products are more commonly encountered as commercial products in the medium– heavy range, although solvents based on cyclohexane and the alkyl-substituted cyclohexanes do exist, and are properly classified as light naphthenic paraffinic products. Products in the medium–heavy range include lamp oils, solvents for insecticides, some charcoal starter fluids and industrial solvents [18]. Many of the applications for which naphthenic paraffinic products are used are those that had previously used distillates or dearomatized distillates.

26.3.2.5 n-alkane products

Normal alkane products are very simple products, usually containing three to five consecutive normal alkanes. These products have had substantial refining so that there are virtually no aromatic, cyclic, or branched compounds present. The resulting product is a simple, clean-burning product with a fairly narrow boiling point range. Although there are products with broader ranges that can be categorized as *n*-alkane products, such as some paraffins, the majority of the ignitable liquids that comprise this classification scheme span from three to five *n*-alkanes. Because these products burn with little smoke or odor, they are commonly used as fuels for liquid candles, often with the addition of a dye and or fragrance. Normal alkane products are also used as solvents for inks in some no carbon required (NCR) papers that are commonly seen as multi-page forms [11]. This category also includes single normal alkane solvents, such as *n*-pentane, *n*-hexane, and *n*-heptane, although these types of solvents are seen much less frequently in common commercially available solvents or products.

26.3.2.6 Gasoline

Gasoline is a unique product, with one specific use: fuel for internal combustion engines such as those used in automobiles. Virtually all of the other categories are classifications that include products with similar chemical compositions, but that can have markedly different end uses. Gasoline, however, does not have other commercial uses; its sole purpose is as a fuel for the internal combustion engine. Much of the refining process is designed around maximizing the production of gasoline. It is the most economically important petroleum product that is produced [22]. In order to fulfill its purpose as an effective fuel, gasoline must have a volatility range that will allow sufficient vapor for ignition, and it must also have adequate anti-knock characteristics. Refining a product to meet the necessary vapor pressure specifications is not difficult. Older automotive fuels had compositions similar to LPDs, and met the needs of the engine as far as vapor pressure and volatility were concerned. However, engines improved over the years and they became more powerful. There were no major changes to the fundamental design of the four-stroke engine; rather, much of the additional power came from additional cylinders, and increased compression ratios. The use of increased compression ratios to gain more power from an engine had one problem associated with it. With increased compression ratios came an increased tendency toward pre-ignition, also known as knocking. In order to minimize knocking and attain the maximum potential power from high performance engines, the octane rating of gasoline had to be increased. Octane number is simply a measure of a fuel's tendency to resist knock. It is determined by using a standard engine, and comparing the performance of the fuel being tested to various combinations of standard fuels. There are two standard test engines that are used for determining octane rating. One simulates highway driving and the other city driving. Depending on the engine used, one obtains either a Research Octane Number (RON) or a Motor Octane Number (MON). The Posted Octane Number (PON), which is the number posted on gas pumps in the United States, is the arithmetic average of the RON and MON. In most other parts of the world, it is the RON that is used at the pump and for comparison purposes. The pure compounds isooctane and *n*-heptane are used as standards and have been assigned values of 100 and 0, respectively, because isooctane has good anti-knock properties whereas *n*-heptane performs very poorly in this regard. Therefore, a fuel that exhibits the same anti-knock characteristics as a mixture that is 89% isooctane and 11% *n*-heptane would be assigned an octane rating of 89.

Unfortunately, the composition of crude oil in the boiling point range that is needed for gasoline – roughly C_4–C_{11} – does not result in a high-octane fuel. A straightforward fractional distillation of crude oil would result in an LPD-type product, which would have the right boiling point range, but an inadequate octane rating. In order to meet the requirements for modern day automotive fuels, significant processing of crude oil is required to manufacture gasoline.

The composition of most petroleum is fairly rich in normal alkanes and cycloalkanes, with lesser amounts of isoparaffins and aromatics, although the relative amounts of these classes of compounds will vary depending on the source [23]. As a general rule, aromatic compounds have the highest octane ratings, with isoparaffins also being very good. Naphthenes and straight-chain alkanes are significantly worse. Therefore, in order to produce a gasoline it is necessary to perform additional steps in the refinery so as to increase the octane rating to an acceptable level. While there are many processes in the modern integrated refinery that are used to enhance gasoline, two of the most important processes are reforming and isomerization. In simplest terms, the primary goal of the reforming process is to increase the proportion of aromatics, primarily by the conversion of naphthenes into aromatic compounds. The goal of isomerization is to convert normal paraffins to isoparaffins [22]. The use of the reforming and isomerization processes provide blending stocks which are used to improve the octane rating of the final gasoline. Because of these processes the chemical composition of the resulting gasoline is very different from that of crude oil. Gasoline is much richer in aromatic and isoparaffinic compounds than the corresponding light straight run fraction. These methods for refining gasoline do not remove normal paraffins or naphthenes; however, because many of these components are converted to isoparaffins and aromatics, their relative amounts are significantly reduced. A typical gasoline will therefore be rich in aromatic compounds such as the mononuclear alkylbenzenes, as well as in branched-chain paraffins. Normal alkanes and naphthenes will also be present, although at much lower levels than is seen in the less refined distillate-type products. PNAs based on indane and naphthalene will also be present, but not as major components. Commercially marketed gasoline is therefore a unique petroleum product, with a composition that reflects in intended uses and required specifications.

26.3.2.7 Oxygenated solvents and miscellaneous products

In addition to the categories of ignitable liquids derived from petroleum, the ASTM standards also recognize a classification for oxygenated solvents, as well as a catch-all miscellaneous category [14,15]. Most oxygenated products are not derived from petroleum, although a blend of a petroleum-based product with an oxygenated compound would be classified according to the ASTM E 1618 criteria as an oxygenated solvent [14]. Products classified as oxygenated solvents include alcohols, ketones, and other compounds that contain oxygen as part of their chemical formula. In addition, blends of other products with these oxygen-containing compounds are also included within this classification [24]. Because this class of ignitable liquids does not originate from petroleum, there are no general statements that can be made regarding the types of compounds present, other than the necessity of the oxygenated compound, which defines the category.

The miscellaneous classification exists because it is recognized that no classification scheme will be able to completely categorize all ignitable liquids that can be potentially encountered. As with oxygenated solvents, there are no general rules for the types of chemical compounds that will be present in materials that are classified under the miscellaneous heading. This classification exists to address liquids that are ignitable, yet that cannot be adequately described by any of the major classifications. Examples of products that fall within this category are turpentine, limonene-based products, and blends of individual components.

A general understanding of the characteristics common to frequently encountered ignitable liquids is necessary in order to develop an appropriate analytical scheme. Table 26.1 provides a summary of the major categories of ignitable liquids and a brief description of their compositions. Knowledge of how ignitable liquids are refined from crude oil provides information as to the expected composition of the various commercial products that are used as accelerants. This understanding will be used in not only the data analysis portion of identifying ignitable liquid residues, but will also be of value in the selection of appropriate extraction techniques.

26.4 ANALYSIS OF FIRE DEBRIS

The analysis of fire debris for the presence of ignitable liquids is a useful tool for the investigator attempting to determine the cause of a fire. It is important, however, to remember that the results of laboratory examination of debris collected from a fire scene is only one piece of evidence in a complex investigation, and must be examined within the context of the entire investigation. There are many types of physical evidence and categories of laboratory examination that could potentially be useful to an origin and cause investigator. Often appliances and other equipment that carry or use electricity may be of interest to a fire investigator. The nature of this type of examination is a science unto itself, and will not be covered in this chapter. In addition to the examination of fire debris for potential accelerants, the fire investigator should consider evidence that could potentially link perpetrators to the scene of the fire, rather than simply support a particular fire cause of incendiary or accidental. This type of associative evidence, which is often emphasized in other types of crime scenes, is often overlooked in the investigation of fires due to the excessive damage caused by the fire.

The analysis of fire debris involves a series of steps, each of which must be carried out properly in order to ensure the most accurate results. ASTM International is a voluntary standards development organization that has developed and published numerous consensus standards in the various disciplines of forensic science and criminalistics. There are several ASTM standards relating to the analysis of fire debris. In addition to some of the standards that have a more general applicability to forensic science, there are published standards for the extraction and sampling of fire debris, as well as for the instrumental analysis of ignitable liquids and extracts recovered from fire debris, including data interpretation [14–15,25–31]. The published ASTM standards are accepted within the scientific community, and have been

TABLE 26.1
SUMMARY OF CHEMICAL COMPOSITION OF VARIOUS IGNITABLE LIQUIDS BY ASTM CLASSIFICATION

	Manufacture	Common Uses	Major Components	Minor Components
Petroleum distillates	Derived from crude oil; separation based on boiling point; minimal processing	Camp stove fuels, pocket lighter fuels; charcoal starter fluid, paint thinners, lamp oils; diesel fuel, home heating fuels, jet fuel	Predominantly aliphatic in composition; normal alkanes abundant; marked presence of isoalkanes and cycloalkanes	Aromatic compounds may be present; depending on boiling point range, benzene-based, indane-based and/or naphthalene-based
Isoparaffinic products	Derived from crude oil; separation based on boiling point; molecular sieve	Aviation fuels, charcoal starters, hand cleaners, copier toners; specialty solvents	Exclusively isoalkanes	Virtually no aromatic compounds; normal alkanes, if present are in very low abundance
Aromatic products	Derived from crude oil; separation based on boiling point; various methods of further refinement	Paint removers; automotive parts cleaners, insecticide solvents/diluents, fuel additives, specialty solvents	Exclusively aromatic compounds; depending on boiling point range, may be benzene-based, indane-based and/or naphthalene-based	Virtually no aliphatic compounds present
Naphthenic paraffinic products	Derived from crude oil; separation based on boiling point; reduction/ removal of aromatics and normal alkanes	Charcoal starter fluids; lamp oils, insecticide solvents/ diluents	Predominantly cycloalkanes and isoalkanes	Virtually no aromatic compounds; normal alkanes, if present are at a lower abundance
Normal alkane products	Derived from crude oil; separation based on boiling point; molecular sieve	Candle oils; solvent in carbonless forms; copier toners; paraffin oils; solvents	Exclusively normal alkanes	Virtually no aromatic or other aliphatic compounds
Gasoline	Derived from light fraction of crude oil; reforming and isomerization to increase aromatic and isoalkane content	Fuel for automotive engines; fuel for other internal combustion engines	Predominantly benzene-based aromatic compounds and isoalkanes	Also contains indane-based and naphthalene-based aromatic compounds; contains cycloalkanes and low levels of normal alkanes

	Alkanes	Aromatics	Indanes	Naphthenes/Olefins	PNAs		
Gasoline	▨	■	▨	▨	■	Varies; depends on product; solvent	
Distillates	■	▨	▨	▨	▨		
Isoparaffinic Products	■	□	□	□	□		
Aromatic Products	□	■	▨	□	■	Must contain at least one significant oxygen-containing compound	
Naphthenic/ Paraffinic Products	■	□	□	■	□		
n-Alkane Products	■	□	□	□	□		
De-Aromatized Distillates	■	□	□	▨	□	May contain other oxygen-containing or hydrocarbon-based compounds	

Oxygenated products — Varies; depends on product
Others- miscellaneous — Varies; depends on product

Varies; depends on product

Varies; depends on product

Varies; depends on product

Varies; depends on product

■ Present and Predominant pattern if within range

▨ Minor pattern present* if within range

□ Not present or virtually none

* a pattern may be present in the specified profile, however the peaks are not necessarily of that class validated. Each standard provides information regarding the appropriate use of each technique, including the scope and limitations of each technique.

26.4.1 Extraction methods

There is no single best extraction method for the analysis of fire debris. Selection of an extraction technique is dependent on a variety of factors, many of which are beyond the control of the examiner. The ideal extraction technique would be sensitive yet nondestructive – that is, it would not fundamentally alter the evidence in such a way that it could not be retested. It should allow for the extraction process to recover any ignitable liquid of interest from the debris, but should minimize the recovery of chemicals related to the substrate or its combustion and pyrolysis products. The composition of the extracted sample should be perfectly representative of the ignitable liquid residue in the debris, with no skewing. The ideal extraction method should recover a wide range of ignitable liquids equally well. Finally, the ideal technique would be inexpensive, simple, safe, and quick. Of course, no single method meets all of these criteria; therefore, there are several useful extraction techniques in widespread use today.

26.4.1.1 Solvent extraction

The technique of solvent extraction is based on one of the simplest and most well-known chemical principles: like dissolves like. The technique of solvent extraction requires that the analyst use an appropriate solvent to dissolve any ignitable liquids that may be present in the sample being tested. It is a simple, quick, inexpensive, and uncomplicated technique that requires little in the way of equipment and has long been applied to the analysis of fire debris for the recovery of accelerants [32–34]. Solvents are selected based on their ability to recover common ignitable liquids. Because the most frequently encountered accelerants are composed of hydrocarbons derived from crude oil, nonpolar solvents are most effective [35]. Commonly used solvents include pentane, carbon disulfide, diethyl ether, and chlorinated solvents [34,36–38]. A consensus-based standard method for performing the solvent extraction technique is published as ASTM E 1386 Standard Practice for Separation and Concentration of Ignitable Liquid Residues from Fire Debris by Solvent Extraction [28].

Solvent extraction offers several advantages to the fire debris analyst. It can take only minutes to perform, allowing the analyst to begin the instrumental portion of the analysis nearly immediately. It also offers the distinct advantage of minimal discrimination in extracting residue from the sample matrix. Because many of the commonly used extraction techniques rely on the collection of ignitable liquid vapors from a sample's headspace, they do not allow for a complete recovery of the less volatile components of ignitable liquids. In contrast, the solvent extraction method recovers all miscible components of the ignitable liquid residue more or less equally [39]. This allows for the extract to better represent the actual composition of the ignitable liquid residue present in the sample matrix. Solvent extraction offers some disadvantages as well. It can often require fairly substantial amounts of hazardous solvents, which can pose risks to health and safety. In addition, this technique, although sensitive, is less sensitive than some of the available techniques. It often requires concentration of the extract, which may result in the loss of the more volatile ignitable liquid components and the concentration of low-level impurities present in the solvent. In cases of very volatile ignitable liquids, especially single component oxygen-containing compounds, the ignitable liquid of interest may not be adequately separated from the solvent *via* chromatography, making its identification difficult or impossible. Another fairly significant disadvantage of this technique is that it often co-extracts substantial amounts of matrix-related components [38]. This results in the extract being fairly "dirty," which can complicate data interpretation and can also cause the instruments to require more frequent cleaning and maintenance. Finally, this technique is capable of fundamentally changing the evidence. Removal of essentially all ignitable liquid residues and the potential dissolution of the debris may result in spoliation of the item of evidence. The actual item of evidence can be substantially changed by this technique, such that it is not suitable for retesting by another competent expert. In addition, if a concentration step is used on the extract, the potential exists for the complete and irrecoverable loss of volatile compounds to the environment [40]. Because of this, the technique of solvent extraction may be considered a destructive technique when applied in certain circumstances. Because of this, it is recommended that a sample of the fire debris be removed for solvent extraction, rather than extracting the entire sample. While it is recognized that fire debris will not be homogenous, by preserving a portion of the evidence the analyst will minimize the likelihood that spoliation will be alleged. When selecting analytical methods to be applied to forensic evidence, it is often desirable to avoid the use of destructive techniques.

Solvent extraction offers both advantages and disadvantages, and may be the technique of choice depending on the type of sample to be extracted. It is the preferred technique in situations in which there is expected to be little contribution to the extract from the sample matrix, such as with containers and other nonporous substrates. Also, in cases in which there is a high concentration of ignitable liquid present – often indicated by a noticeable odor – solvent extraction may be the most appropriate technique. Other techniques, such as the charcoal-based adsorption techniques, may result in disproportionate recovery of ignitable liquids when they are present at high concentration [41]. The presence of a high concentration of ignitable liquid also makes the slightly lower sensitivity of the method not a significant disadvantage, and also makes the sample more suitable for removing a sample for extraction and preserving the remainder of the debris, thereby minimizing the potential for spoliation. Finally, in cases in which the examiner knows that the residue being collected is relatively nonvolatile, such as with kerosene, diesel fuel, and other fuel oil-type products, the use of the solvent extraction technique offers the best chance for optimal recovery of the ignitable liquid. Unfortunately, in most forensic cases, the examiner has little idea of the nature or concentration of the ignitable liquid that may be present; therefore, it is most often recommended that unless there is an obviously high concentration of ignitable liquid, as indicated by either a screening technique or a noticeable odor, that this technique only be used as a secondary technique following application of one of the available nondestructive extraction methods.

26.4.1.2 Headspace sampling

Like solvent extraction, the technique of headspace sampling is quick and simple, easy to perform, and requires minimal equipment. The technique of headspace sampling relies on the physical property of the volatility of ignitable liquids. Volatility refers to a liquid's tendency to have a rich vapor phase, and to evaporate readily. In headspace sampling, the sample of fire debris may be heated in order to enrich the vapor above the sample, or it may be done at ambient temperature. To perform the technique of headspace sampling, a portion of the vapors above the sample of fire debris is removed, and directly introduced to the instrument being used for analysis – generally, the gas chromatograph (GC) [42]. This technique is explained in ASTM E 1388 Standard Practice for Sampling of Headspace Vapors from Fire Debris Samples [27].

The headspace sampling method offers several advantages as a method for the recovery of ignitable liquids. Although the theory and dynamics of the phase distribution that govern headspace analysis may be quite complex, the application of the technique is rather simple; there is no special equipment needed, other than a suitable syringe for collecting and introducing the sample. No solvents are used, thereby offering two advantages: there are reduced health, safety, and environmental hazards, and there is no added solvent, which may interfere with the products being recovered from the sample matrix. Because this sampling method only removes a relatively small portion of the headspace vapors, the evidence is essentially unchanged and the technique is considered to be nondestructive. The disadvantage associated with the fact that only a small portion of the headspace vapors being sampled is that this technique offers no means of concentrating low levels of ignitable liquids; therefore, it is one of the least sensitive techniques still in modern use [43]. The headspace technique is also not suitable for the recovery of all types of ignitable liquids. Because it is based on the collection of vapors, this technique may be useful for lighter (more volatile) ignitable liquids, but will not efficiently recover all classes of ignitable liquids, such as those classified as heavy or medium [39,44]. Headspace sampling can provide important information about some types of samples; however, because of its inability to recover the full range of commonly encountered ignitable liquids, it should not be used as the sole sampling technique.

Headspace sampling can be a very useful first step in a thorough analytical scheme for the recovery of ignitable liquids. It is one of the most effective techniques for the recovery of very volatile water-miscible ignitable liquids, such as the lower molecular weight alcohols and ketones. Because it is one of the few techniques that does not require the addition of a solvent, it will not alter the sample, nor will it interfere with subsequent extraction techniques. Although there exist headspace autosampling devices that are routinely used in the field of toxicology, they have not been practically applied to the analysis of fire debris, due to the cumbersome nature and heterogeneity of the samples. Because of the challenges to automating headspace sampling in this discipline, many forensic science laboratories do not routinely use this technique within their overall analytical scheme. It is most commonly utilized in cases in which there is a reason to suspect that an ignitable liquid may be present which could potentially be difficult to identify if an extraction technique requiring the addition of a solvent was used. It is important to reiterate, however, that because this technique does not recover the full range of ignitable liquids, it is not suitable to use it as the sole method for sample recovery.

26.4.1.3 Passive adsorption (passive headspace concentration)

The most commonly used extraction method for the recovery of ignitable liquids from fire debris is that of passive adsorption according to reports from recent proficiency tests and a survey of practitioners [45–49]. This method involves concentrating vapors from the headspace of a sample within a closed system onto a suitable adsorbent, then desorbing the recovered species [50]. Activated charcoal, also referred to as active carbon, is the most commonly used adsorbent in the United States, and is generally desorbed *via* solvent [51]. Other adsorption-based methods exist including solid phase microextraction (SPME) and those that utilize an adsorbent, such as Tenax®TM TA, which is desorbed thermally. Passive headspace concentration goes by a variety of names, including passive adsorption, adsorption–elution, and the charcoal strip method (when the adsorbent used is in the form of a charcoal-impregnated polymer strip). To use this technique, adsorbent in an appropriate form is placed in the sample container along with the fire debris, and the container is usually heated for a period of time. This is the adsorption period in which vapors present in the sample container become adsorbed onto the activated charcoal. Following the adsorption period, the adsorbed compounds are eluted – or removed – from the charcoal with solvent. While many solvents have been researched for this application, the most important factors to consider are miscibility and the solvents affinity for adsorption sites [52]. Because of these factors, carbon disulfide is commonly used, although other solvents such as pentane, diethyl ether, and methylene chloride have also been used in this application [53,54]. The activated charcoal needs to be in such a form that it can be relatively easily manipulated so that it can be placed in and removed from the container. In addition, the active sites on the charcoal must be accessible to the headspace vapors in the can. Very often the form of charcoal used is that of the activated charcoal strip (ACS), which consists of robust polymer support for the charcoal, and is commercially marketed. Other ways of introducing the charcoal to the sample include "C-bags," which are constructed in a manner similar to the tea bags from which their name was derived, and charcoal wires [50,55]. These are usually made in-house, and consist of granular charcoal either contained within a permeable filter or coated onto a wire. Detailed information describing the use of this extraction technique is presented in the standard ASTM E 1412 Standard Practice for Separation of Ignitable Liquid Residues from Fire Debris by Passive Headspace Concentration with Activated Charcoal [29].

The popularity of the passive adsorption method of extraction is not without reason. The technique has broad appeal for the many advantages that it offers to the

References pp. 919–922

analyst. The main advantage offered by this technique is that it is suitable for recovering a fairly broad range of ignitable liquids. Extraction conditions can be altered depending on sample condition; time and temperature are often adjusted to more efficiently extract samples based on indications of ignitable liquid concentration or volatility [41]. Because of this factor, when only one extraction technique is to be used, this is most often the method of choice. In addition to the petroleum-based ignitable liquids, this method can also recover very volatile ignitable liquids, which can be detected and identified *via* GC-MS when instrument parameters are set to collect data prior to the elution of the solvent [56]. The other major advantage that passive adsorption offers is that it requires relatively little examiner time. Fire debris samples can be batched and extracted simultaneously, with the majority of the extraction time being unattended [57]. This technique allows for concentration of vapors onto the charcoal, and elution generally occurs with a minimal volume of solvent, so it is also one of the most sensitive techniques available. Studies have shown that in all but the most extreme situations, this technique can be repeated numerous times, with no significant change in the extracts obtained; therefore, it may be considered nondestructive in that it does not permanently alter the parent evidence [58,59]. All of these factors lead to the passive adsorption method being one of the most efficient of the available techniques.

As with every technique, however, there exist some disadvantages. Although a minor factor, when commercially prepared activated charcoal strips are used for the processing of evidence the cost per sample is relatively high compared with other sample preparation methods. While the cost factor can be minimized by the use of homemade charcoal devices such as the C-bag or charcoal wire, their use would require additional examiner time to prepare the sampling devices. Another disadvantage is that while this technique does a fair job of collecting a broad range of ignitable liquids, some skewing may result depending on sample condition. Samples that are extremely concentrated may saturate the available active sites, which leads to the phenomenon of displacement [41]. Displacement results when compounds with a greater adsorption affinity for charcoal displace those with a lesser affinity. Consequently, the sample recovered by the charcoal will not necessarily be representative of the ignitable liquid residue in the container; it will be skewed toward compounds that have a greater affinity for the adsorption sites – typically favoring large molecules over small, and aromatic compounds over aliphatic. Another potential for skewed samples arises from the fact that this technique is based on the recovery of vapors residing in the headspace above the sample. Because it is vapors that are collected, it is often not possible to adequately recover those components of ignitable liquid residues that are less volatile. This deficiency is most pronounced in situations in which the substrate from which the ignitable liquid residue is recovered is a complex or charred material that provides additional sites for adsorption. The adsorption sites present on the charred debris may retain ignitable liquid residues, thereby affecting the composition of the vapor phase, and consequently, what is available for adsorption. This concept is referred to as competitive adsorption and can significantly impact the composition of the adsorbed sample [60]. For example, products such as kerosene and diesel fuel are compositionally similar to one another. Both are HPD-type products, however, diesel fuel products have a significantly higher end-cut. Whereas a typical kerosene may range from C_8 to C_{16}, a diesel fuel will more typically range from C_9 to C_{23}. The effects of competitive adsorption from a complex substrate and the effects of a vapor-based recovery will combine to yield results from a sample of diesel fuel that are often indistinguishable from those of a kerosene product. This is due to the incomplete recovery of the compounds in the C_{16} and higher region. Consequently, the passive charcoal adsorption technique may not be able to provide a truly accurate representation of ignitable liquids with compounds in the C_{16} and higher range. Competitive adsorption may also negatively impact the sensitivity of this technique.

Although there are some limitations when using charcoal-based passive adsorption, knowledge of those limitations goes a long way in mitigating them. Understanding that parameters may need to be changed based on information gained in the preliminary examination of a sample, and that distinctions amongst various types of heavy products will generally not be possible, allows the examiner to properly interpret data gained from this type of extract. The technique of passive headspace concentration *via* activated charcoal offers so many advantages, that it will continue to remain one of the most commonly used methods for the forensic laboratory's extraction of fire debris for the presence of ignitable liquids.

26.4.1.4 Dynamic adsorption (dynamic headspace concentration)

The technique of dynamic adsorption relies on the same basic principles as the passive adsorption extraction method. Versions of a dynamic adsorption method have been applied in the petroleum industry and in environmental and industrial hygiene applications [61–63]. Dynamic adsorption was in common use for the extraction of ignitable liquid residues from fire debris for a long time prior to being almost completely replaced by the passive technique [63–66]. The dynamic technique is still considered an acceptable method for the extraction of fire debris for the recovery of ignitable liquid residues, and its use and application are described in ASTM E 1413 Standard Practice for Separation and Concentration of Ignitable Liquid Residues from Fire Debris by Dynamic Headspace Concentration [30]. (Use of a dynamic adsorption using Tenax TA, rather than activated charcoal is described in Section 26.4.1.6.) Like the passive technique, the dynamic method of extraction relies on the adsorption of vapors onto a suitable material, most commonly charcoal, and their subsequent elution *via* solvent. The critical difference between these two techniques is that while the passive method uses a closed system, the dynamic system uses a system in which air or an inert gas is forced through the sample container. This may be done by either negative or positive pressure. The gases and vapors that are forced through the sample container are recovered on a tube filled with activated charcoal, which serves as the collection device. This adsorption phase, which takes place at an elevated temperature, is followed by elution of the adsorbed species from the charcoal with a solvent, such as carbon disulfide. While the techniques of passive and dynamic adsorption are similar in general principle, the underlying differences are what led to the almost universal replacement of the dynamic method by the passive method.

Dynamic adsorption is a good technique for extracting ignitable liquid residues from fire debris for several reasons. As with the passive method, this technique offers the capability of recovering a broad range of ignitable liquid residues with very good sensitivity. Unlike its passive counterpart, however, a complete extraction – from adsorption to elution – can often be done in less than an hour. While the ability to conduct the extraction in less time does offer an advantage, it must be kept in mind that the time conducting this extraction is much more labor-intensive than the longer time spent completing the passive extraction. Whereas the passive extraction requires minimal examiner time during its approximately 16-h extraction process, and allows for sample batching, the dynamic extraction process requires the examiner to be present and attending the dynamic extraction. In addition, this method cannot be feasibly automated and is not amenable to batching numerous samples [66]. In order to batch samples, it is necessary to have numerous sets of equipment and even when such equipment is available, the number of samples being extracted coincidentally is limited by the examiners ability to monitor them. Because samples of fire debris may differ greatly from sample to sample in terms of the amount, density, and nature of the material present, a set of samples will not heat equally, nor approach equilibrium in uniform time. This results in the need to monitor individual samples. Based on these minor differences in the methodology, it would appear that the techniques of passive adsorption and dynamic adsorption seem equally well suited to the extraction of fire debris.

In considering the technique of dynamic adsorption, there are significant issues that affect its suitability for extracting ignitable liquids from fire debris. The same issues that result in skewing due to competitive adsorption when using the passive adsorption method arise with the dynamic technique as well; however, it becomes a greater problem in the dynamic system because it is not a closed system. Therefore, when compounds with greater affinity for the adsorbent displace those with less affinity, those with less affinity are removed from the charcoal, and are irrecoverably lost to the vacuum system. This is in contrast to the displacement that occurs in the closed (passive) system because in cases of a closed system, molecules that are displaced return to the sample container. Because of the airflow through the dynamic system, molecules that are displaced cannot be recovered; in this case, the phenomenon of displacement is referred to as breakthrough. This can be a significant disadvantage. An experienced examiner that is familiar with this phenomenon may take precautions to avoid breakthrough, however, the possibility of it occurring and the extent to which it can affect data cannot be fully controlled. Dynamic adsorption can fundamentally alter the composition of the sample, making the possibility of retesting it much less likely. Because the composition of the vapor phase is changing as the flow moves through it, equilibria are continuously being reestablished, or more accurately, a true equilibrium is never established. This can result in a more complete recovery of the ignitable liquid residues, making this method among the most sensitive extraction techniques available. The consequence of this complete recovery is that dynamic adsorption is more often than not considered to be destructive technique because the primary sample is fundamentally altered, and is generally not amenable to retesting [67,68]. Dynamic adsorption can be very susceptible to changes in conditions such as temperature and collection time; therefore, an examiner must exercise caution to minimize the likelihood of break-through and skewing. Because so many of the disadvantages of this technique are reduced by using the passive method, use of the dynamic method of extracting fire debris samples has been steadily declining.

26.4.1.5 Solid phase microextraction (SPME)

The most recent extraction technique to be added to the fire debris analyst's toolbox is that of SPME. It is a relatively new technique that has been used in a variety of other separation applications, as well as in the extraction of ignitable liquids from fire debris [69–71]. The first version of an ASTM standard describing the effectiveness of this technique to fire debris was published in 2001 as ASTM E 2154 Standard Practice for Separation and Concentration of Ignitable Liquid Residues from Fire Debris Samples by Passive Headspace Concentration with Solid Phase Microextraction (SPME) [31]. Extraction by SPME most commonly involves the exposure of a fiber bearing a adsorptive coating to the headspace above a sample, which may be heated or held at ambient temperatures. Less frequently, the SPME fiber may be immersed directly into an aqueous liquid. Following this adsorption step, the fiber is then thermally desorbed directly into the instrument being used for analysis. A variety of fiber coatings are commercially available, depending on the application; Table 26.2 shows some of the available fibers and their utility. For most fire debris applications, the analyst is looking primarily for nonpolar ignitable liquids. In this application, polydimethylsiloxane (PDMS) is most commonly used. Little research has been conducted on applications in which more polar analytes are of interest, such as oxygen-containing compounds such as acetone or alcohols; however, toxicological applications have shown that the carbowax/divinylbenzene (CW/DVB) and carboxen/polydimethylsiloxane (CAR/PDMS) are effective for recovering alcohol from biological fluids [72–74]. Almirall has reported success with the use of the CAR/PDMS fiber at low temperatures for the recovery of polar analytes from fire debris; however, further research focusing on optimizing parameters of volatile polar compounds encountered in fire debris is needed [75].

TABLE 26 2
AVAILABLE FIBERS AND THEIR UTILITY

Fiber Type	Analytes
Polydimethylsiloxane (PDMS)	Nonpolar and polar
Polydimethylsiloxane/divinlybenzene (PDMS/DVB)	Polar, especially amines
Polyacrylate	Very polar, i.e. phenols
Carboxen/polydimethylsiloxane (CAR/PDMS)	Gaseous/volatiles
Carbowax/divinylbenzene (CW/DVB)	Polar, especially alcohols
Carbowax/templated resin	Surfactants (HPLC)
Divinylbenzene/carboxenPDMS (DVB/CAR/PDMS)	Broad range of polarities

The application of SPME for the recovery of ignitable liquids from fire debris offers many advantages. Perhaps the most significant of these advantages is that it is both quick and easy to perform. Extraction times for SPME are typically in the range of 5–20 min, and examiner intervention is generally not necessary during the exposure period. The process is relatively simple, requiring that the examiner expose the SPME fiber to the samples headspace for the prescribed duration, then desorb the fiber directly into the instrument's injection port. Because of the quick and simple nature of this technique, instrumental analysis can begin with 20 min of beginning the examination, thereby allowing for very rapid results. The SPME fiber is small, and has relatively few adsorption sites. This means that the recovered sample is most often only a small portion of the ignitable liquid residue vapors present in the can, so the extraction may be conducted repeatedly, without significantly altering the sample. Because this technique will not fundamentally change the primary sample so that the sample can be subjected to retesting, it may be considered a nondestructive technique, which is an important advantage for any technique with forensic implications. Another advantage of this technique is that it does not require use of a solvent in the desorption phase. Lack of solvent offers several advantages. There is a significantly lower health and safety risk when no solvent is used, and there are no expenditures for the purchase or disposal of solvents. In terms of analytical advantages, the lack of solvent means that there is no concern with a solvent masking a component of an ignitable liquid that may co-elute with the solvent, or that may be lost within the solvent front. This allows SPME to be effectively applied to cases in which there may be very volatile ignitable liquid residues. A final benefit of not needing solvent is that there is no dilution of the adsorbed species. Whatever ignitable liquid residue is recovered onto the SPME fiber is introduced in its entirety to the instrument being used for analysis. As a result, this technique is also one of the most sensitive techniques. The disadvantage that accompanies analyzing the sample in its entirety is that there is no way to preserve a portion of the sample for later reanalysis, which may be problematic for forensic cases. The purchase of several SPME assemblies is necessary for the efficient processing of forensic casework, which requires an initial one-time expenditure. The fibers onto which the analyte is adsorbed, however, although reusable, have a limited lifespan, and must be replaced on a regular basis. Literature reports that the fibers may last for up to 50–100 analyses, suggesting that this a fairly cost efficient method of analysis [69,71]. Practical use has not always supported such longevity of the fibers, however. Consequently, the cost and fragility of the fibers may be considered to be a disadvantage. While the majority of the listed disadvantages associated with the application of SPME to fire debris analyses are merely minor nuisances, there is one serious concern that must be clearly understood before using this technique. As with other vapor adsorption techniques, the phenomenon of displacement will occur, resulting in skewed data. Components that have a greater affinity for the adsorbent will displace those that have a lesser affinity; this becomes more pronounced when there are few adsorption sites relative to the amount of vapors to be collected. While this presents a concern in the previously discussed adsorption methods in some cases – where there is a particularly strong sample – it consistently affects sample recovery when SPME is used. Because the SPME fiber is so small, and the number of adsorption sites so few, displacement seriously affects the composition of the sample being analyzed; it is virtually never representative of the ignitable liquid residue being collected, but rather is skewed to favor the less volatile portion of the sample. As a result, the sample being analyzed may have serious compositional differences from the actual ignitable liquid residue in the container due to the fact that the more volatile portions of a sample may not be adequately retained and represented in the adsorbed sample. For this reason, the application of this technique is limited to use as part of a comprehensive analytical scheme and it is not recommended that this sampling method be used as a standalone technique.

The use of SPME offers some unique benefits for the examiner analyzing fire debris. Because it is quick, easy, sensitive, and can recover a fairly broad range of volatilities, it may be particularly useful in screening samples. In this manner, it could provide information regarding the approximate concentration of any ignitable liquids present in the matrix, thereby allowing the examiner to optimize parameters for the primary extraction method. When used as a screening technique, SPME can also show the presence of very volatile components that could potentially be missed if only extraction techniques requiring solvents were used. It is important to remember, however, that the data obtained from this method does not provide an accurate representation of the ignitable liquid residue present in the sample; therefore, this method should not be used as a standalone method of sampling for general fire debris casework. SPME provides a great deal of information about a sample, and when used as part of a comprehensive analytical scheme can be a very powerful tool.

26.4.1.6 Tenax TA

While adsorption-based methods that use activated charcoal are most commonly used in North America and Australia, methods that use the adsorbent Tenax TA have garnered more widespread use in Europe. Tenax TA is a porous diphenylene oxide-based polymer developed for the purpose of collecting volatile compounds [76]. Tenax TA is capable of adsorbing a wide range of nonpolar compounds in the range of volatilities of interest to the fire debris analyst and has been applied to the analysis of fire debris for many years [77]. Because the fundamental principles effecting the application of Tenax TA to fire debris analysis are essentially the same as those effecting the other adsorption-based techniques described in Sections 26.4.1.3–26.4.1.5, many of the same advantages and disadvantages apply. Thus, Tenax TA offers the advantages of effectively extracting a wide range of compounds, relative simplicity and good sensitivity. As with other adsorption-based techniques, compounds having a lesser affinity for the adsorbent may be displaced by those having a greater affinity for the adsorbent in cases in which saturation of adsorption sites is approached or exceeded or in cases in which the extraction conditions are too severe. Saturation occurs when the concentration of ignitable liquid is high relative to the number of available adsorption sites; severity of conditions refers to combinations of temperature and exposure time. Because Tenax TA shows a greater preference for higher boiling compounds than does activated charcoal, loss of lighter compounds due to displacement will result in greater compositional differences of the sample collected when compared to the ignitable liquid in the primary sample.

Currently, there is not a published ASTM method for the use of Tenax TA for extracting fire debris. When Tenax TA is used, it is most commonly used as a dynamic extraction method, although recent work has shown that passive methods may be used with Tenax TA as the adsorbent [78]. Unlike the charcoal-based dynamic extraction, in which multiple volumes of sample headspace are passed through the adsorbent, usually only a relatively small volume of headspace is collected – on the order of 30–50 mL when Tenax TA is used [79]. As a result, extraction *via* Tenax TA tends to have properties intermediate to the passive and dynamic methods that rely on charcoal. The technique is easy to perform and only a portion of the headspace is sampled; therefore, it is expected that resampling would be effective. In this manner, it is similar to the passive headspace technique. However, the technique does not require long sampling times and breakthrough is a potential concern; in this manner it is similar to the dynamic headspace technique.

A major difference between the use of Tenax TA and charcoal-based adsorption techniques is that as with SPME, desorption is done thermally, rather than *via* solvent. Because the entire sample is introduced to the instrument for analysis, there is a relative increase in sensitivity when compared with a solvent-based desorptions. In addition, it is necessary to have instrumentation equipped with a thermal desorption device for analysis. These devices are capable of holding multiple sample tubes, thereby allowing for automation of the instrumental analysis. Extraction *via* Tenax TA therefore offers many advantages in terms of its ease of use and efficiency as well as its ability to effectively recover most components of commonly encountered ignitable liquids.

26.4.1.7 Developing an analytical scheme for sample preparation

Choice of extraction methods and the parameters for their use are highly sample dependent. As a result, it is not possible to develop a single overall analytical approach that is suitable for all samples. Based on the experience of the examiner, the nature of the sample, and the time and equipment available, the examiner will develop an analytical scheme appropriate for the given set of circumstances. It should favor nondestructive techniques, and should provide suitable recovery for a wide range of ignitable liquids that may be encountered. Some cases may require multiple extractions, whereas in other cases a single recovery method may be sufficient. Fig. 26.2 shows a comprehensive analytical scheme based on indications of the presence of potential ignitable liquids and the benefits of each of the commonly used extraction techniques. This scheme is based on conducting extractions in a sequence that begins with methods that are least invasive, and uses the most invasive methods only when dictated by the sample. Many laboratories do not employ such a scheme, however, and select the single technique that provides the most efficient means for recovering an unknown liquid, usually either the passive adsorption technique or the Tenax TA technique.

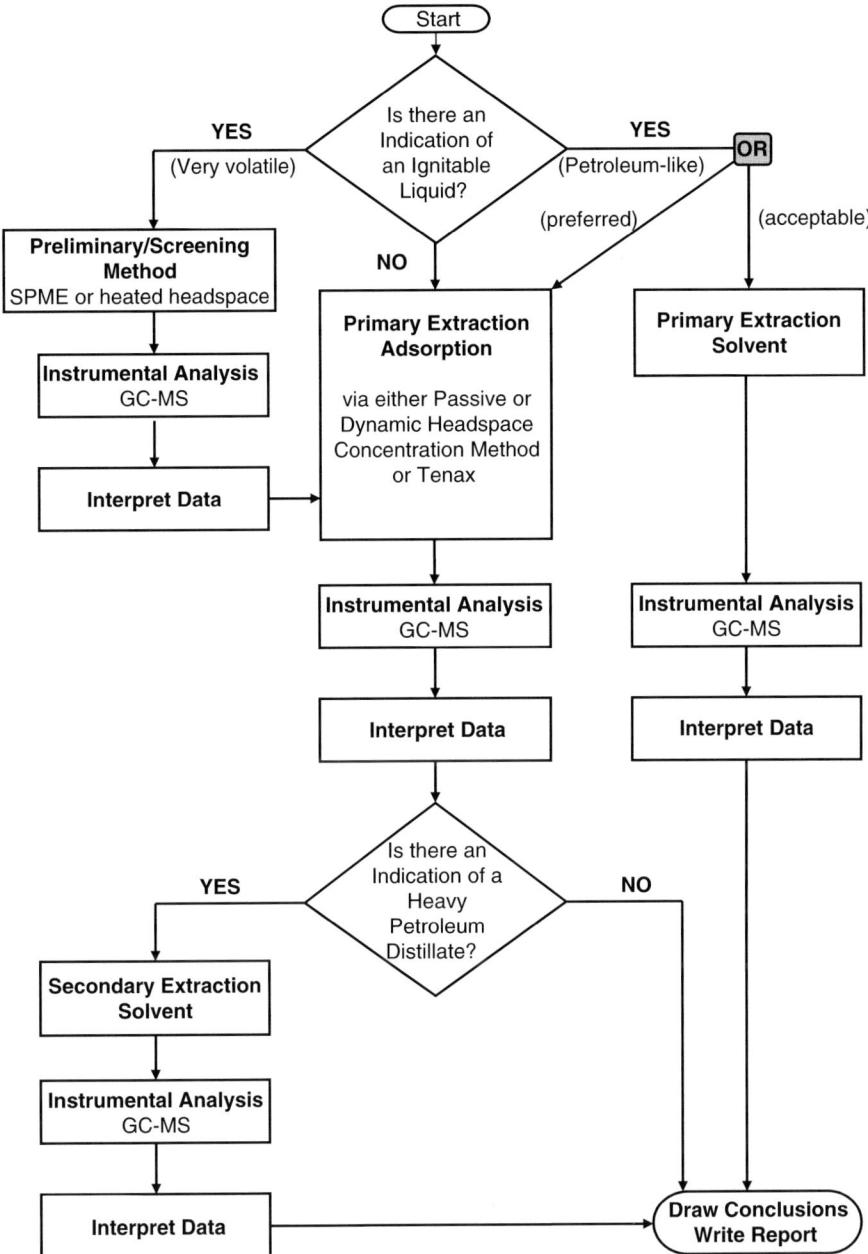

Fig. 26.2. Flow chart for sample preparation. Sample preparation methods should be selected based on the individual sample. This flow chart shows an analytical scheme that will recover a wide range of ignitable liquids using the least invasive techniques. Depending on the sample, there may only need to be one sample preparation technique used, or as many as three.

Another factor to consider in developing a protocol for sampling evidence for the presence of ignitable liquids is the ability to preserve a portion of the sample for subsequent retesting. In some cases, the primary sample may be suitable for retesting. This is generally possible when the original concentration of ignitable liquid in the sample was relatively high, and when simple headspace, passive adsorption, SPME, or Tenax TA is used for sample preparation. Because some samples may be stored for a relatively long period of time before reanalysis is attempted, there is the potential for loss of ignitable liquid residues, even when stored in an "airtight" container. For this reason, it is recommended that a portion of sample extracts be preserved when choice of extraction method makes it feasible. A portion of the sample extract can be easily preserved when passive adsorption, dynamic adsorption, or solvent extraction is used; however, extract preservation is not feasible for headspace, SPME, and Tenax TA extractions.

26.4.2 Detection and identification of ignitable compounds

Following appropriate sample preparation methods, the extract from the fire debris is then subjected to an instrumental analysis. Although there have been many different types of instruments researched to examine their applicability to the identification and classification of ignitable liquids and their residues, one instrument has clearly surpassed all others with regard to its effectiveness for this purpose: the GC [38,80–82]. The GC is designed to separate and detect components of complex mixtures, based on their physical and chemical properties. Although, gas chromatography is not generally considered an identification technique, because of the complex nature of the mixtures that comprise petroleum products, which are the vast majority of accelerants, it can, in most instances, be used as an independent means of identifying ignitable liquids. Although a variety of detectors exist that can be used in with the GC, the use of a mass spectrometer (MS) in conjunction with the GC provides a wealth of information, is widely accepted, and is considered by many to be the best analytical technique readily available for the analysis of ignitable liquids.

26.4.2.1 Gas chromatography

26.4.2.1.1 Overview
Chromatography refers to a physical method of separation in which the analytes are separated based on their distribution between two phases – a stationary phase and a mobile phase. In gas chromatography, the mobile phase is an inert gas, and the stationary phase is a nonvolatile liquid (gas–liquid chromatography). Separation takes place in a column that contains the stationary phase, and through which the mobile phase flows. A mixture is introduced to the system, usually *via* injection, and the separated components are detected and recorded following their separation. The data reflecting the separation is presented graphically as abundance versus time, and is referred to as a chromatogram.

There are a multitude of variables that can be changed in order to optimize a chromatographic separation. Some of these are system-based, such as the style of inlet and type of detector, which are not as easily changed as method-based variables, such as the temperature program. A discussion of some of the major components and parameters as they pertain to the analysis of ignitable liquids will follow.

26.4.2.1.2 Selection of system components

There are several major choices when selecting a GC system for any given application. Some of the most critical decisions must be made at the time of instrument purchase, such as whether to use packed column technology or open tubular (capillary) technology, inlet system capabilities, and what type of detector is most suitable. For the analysis of ignitable liquids, resolution is a major consideration. It is not unusual for a petroleum-based product to contain hundreds of individual components; therefore, the system must use an open tubular column so as to provide adequate separation. More discussion on column selection appears in Section 26.4.2.1.3.

Sample introduction systems that can be operated isothermally or by use of a temperature program have not been widely used, although they offer unique benefits. A temperature-programmable inlet is especially useful in cases in which the sample is susceptible to thermal degradation or when sample discrimination due to nonrepresentative vaporization is a concern. Currently, most fire debris analysis applications utilize inlet systems capable of only isothermal operation; however, it is anticipated that as the benefits of temperature-programmable inlets become more widely known, their frequency of use will increase.

There are many types of GC detectors available and the detector selection is entirely dependent on the types of analyses being conducted. Detectors can be broadly classified as either specific or universal. Specific detectors, which do not respond to all chemical compounds, tend to be more sensitive, but only for the particular types of chemicals to which they are sensitive. In contrast, universal detectors are less sensitive, but will respond to nearly all organic compounds. There are two universal detectors, which are appropriate for the analysis of fire debris for the presence of ignitable liquid residues: the flame ionization detector (FID) and the MS or mass selective detector (MSD). Both are universal in nature, and capable of detecting the hydrocarbon mixtures that comprise the vast majority of ignitable liquids encountered in fire investigations. The FID, which will provide a generic signal for unknown compounds, is not capable of providing identification of the components of the mixture. Consequently, the use of FIDs for ignitable liquid residue analysis has steadily declined as the use of the MS has increased, although it is still an acceptable means for identifying most ignitable liquids. The MS can provide critical structural information, which is often sufficient to identify the components of the mixture eluting from the GC column. Another benefit of the MS is that it offers flexibility in analysis, allowing the examiner to set MS parameters to mimic a less sensitive, but nearly universal detector or a more sensitive and selective detector. In this regard, the MS conditions can be adjusted to the needs of the specific analysis. Because of its immense importance in the analysis of fire debris, the MS and especially its use in conjunction with the GC is discussed in much greater detail in Section 26.4.2.2.

26.4.2.1.3 Selection of system parameters

The chromatographic column may in some regards be considered a component of the system, but it may also be considered a parameter due to the fact that it can be changed with relative ease. The column is the most important part of the chromatographic system. There are a variety of specifications that should be considered when deciding on the most appropriate column to use. Foremost should be the type of stationary phase, stationary phase thickness, column length, and column diameter. When considering options for the nature of the stationary phase, the application must be considered. In the analysis of ignitable liquids, the vast majority of the materials being analyzed will be complex mixtures that are primarily nonpolar, and will often be composed of hydrocarbons that are very similar to one another. In order to adequately separate these types of mixtures, the column must be nonpolar, so that the components of the hydrocarbon-based mixtures will interact sufficiently with the stationary phase, and not merely travel through the column along with the mobile phase. Because of the inherent similarity of the components of routinely encountered hydrocarbon mixtures, much of the separation will be based on boiling point of the components; however, that does not negate the importance of proper stationary phase selection. The stationary phase must be chemically similar to the materials to be analyzed; that is, it must be nonpolar. Stationary phases that have been shown to be suitable for this type of separation include those that are 100% PDMS or those with small percentages of phenyl. The thickness of the stationary phase is referred to as film thickness and will have an effect on retention and capacity. A thicker film allows for more time in the stationary phase, therefore, a longer retention time. A film that is too thin will not allow for sufficient separation, and the components of the mixture will elute quickly, with little separation. Conversely, if the film is too thick, it will result in excessively long retention times. A thick film can be useful in the analysis of the most volatile mixtures because it will allow their separation at ambient temperatures, making expensive and cumbersome cryogenic systems unnecessary. Use of a thinner film would be beneficial in moving less volatile materials through the system more quickly. In the same manner that the film thickness affects the volume of the stationary phase, the diameter of the column affects the volume of the mobile phase. Therefore, a larger diameter results in shorter retention times. Most fire debris applications use capillary columns, which are of narrow diameter, typically less than 0.53 mm. A final consideration in selecting a column selection is that of column length. As column length increases, so does the retention time. A long column can be particularly beneficial for the separation of gases, or for the separation of the most volatile portions of gasolines, which may be necessary for sample discrimination [83–85]. These columns, typically in the 60–100 m length can add considerably to the time required for analysis, and are often not practical for routine casework in which samples are screened for a wide variety of ignitable liquid residues. Column length must be adequate to achieve the desired resolution and separation of components and this can generally be achieved with columns in the range of 12–30 m in length. Columns may be changed with relative ease as applications dictate, however, a good, robust nonpolar column can be expected to serve a variety of applications.

Another major factor to consider in developing a method to test for ignitable liquid residues is that of the temperature parameters. Most commonly, inlets and detectors are set at a single temperature – they are run isothermally. The temperature for the inlet area must be sufficient to insure that the sample is rapidly and completely vaporized upon being introduced to the GC and the detector temperature must be high enough to avoid condensation. Because common ignitable liquids contain numerous components having a relatively wide boiling point range, there will be no single oven temperature that is appropriate for all of the individual chemical compounds that comprise a typical hydrocarbon-based ignitable liquid. The solution to the challenge of separating materials with a wide range of boiling points is to use a temperature program. A temperature program works by starting the oven and column at a relatively low temperature, at which the more volatile components can be separated. The temperature is then gradually and precisely increased, allowing for the elution of the higher boiling components of the mixture. The use of temperature program allows volatile components to be adequately separated at a relatively low temperature, and for the less volatile components to elute from the column with reasonable retention times and good peak shapes. All routine screening of fire debris analysis requires the use of a temperature program due to the nature of the materials being sought.

26.4.2.2 Gas chromatography-mass spectrometry

26.4.2.2.1 Overview
The application of the hyphenated technique of GC-MS to the field of forensic fire debris analysis has offered great advances. While GC with an FID can provide a great deal of information regarding the presence or absence of ignitable liquid residues within a sample, numerous circumstances will preclude definitive determinations due to complex contributions from sample materials as well as the inability to easily recognize simpler or single component ignitable liquids. The ability of the MS portion of the instrument to identify specific components can often provide critical information regarding the composition of an unknown sample. However, the greatest benefit of GC-MS to fire debris analysis lies not in its ability to identify a specific peak at a specific retention time, but in the synergy of the combination of techniques. Using gas chromatography in conjunction with mass spectrometry has been referred to as mass chromatography because of the way the information provided is not merely two sets of independent instrumental data, but combines the information in a way that allows the user to benefit from structural information along with basic GC patterns. The following sections will explain how this combined technique is used in the analysis of ignitable liquids and their residues.

26.4.2.2.2 Optimizing GC-MS data acquisition parameters
Instrument parameters for an MS can be set so as to be universal in which case virtually all compounds are detected, yet the sensitivity is not maximized. To achieve this goal, the MS is set to scan from a high mass to a low mass, so that virtually everything is detected. This means of operation is somewhat inefficient, yet is

capable of providing full mass spectra and is referred to as the scanning mode. An alternative way of using an MS is desirable when only specific compounds are of interest, and it is beneficial to maximize sensitivity. In such cases, the instrument may be set to collect data only for specific ions of interest. When the compounds of interest are known, the parameters can be set so that the majority of instrument time is spent collecting data of interest, rather than by scanning a wide range of m/z values. In this method, several specific ions are pre-selected, and only these selected ions are monitored; data for any other mass-to-charge value is not collected. This is referred to as selected ion monitoring (SIM). In determining which method of data collection to use, one must consider several factors. In cases of true unknowns, it is absolutely necessary to use the scanning method. In cases in which it is necessary or desirable to have full spectra, the scanning method must be used. SIM offers distinct advantages, but only in situations in which one is interested in only a limited number of known compounds. In these situations, the use of SIM can increase sensitivity due to the fact that fewer ions are being monitored, but also because there is less ex-traneous data collected. In its application to fire debris analysis, a technique of data analysis referred to as extracted ion profiling (EIP) offers the best of both of these techniques. Full scan data is acquired, therefore, the analyst has access to full spec-tra for any point in time. Data analysis mimics SIM data, and effectively screens out data not containing ions of interest. The use of EIP has been extensively used in fire debris analysis due to the complex nature of samples, and its utility will be further explained.

Selection of MS parameters must be based on the types of analytes to be en-countered. Ignitable liquids may be as light as methanol, with a molecular weight of 32 amu, with heavier ignitable liquids not extending much beyond the C_{23} normal alkane, which has a molecular weight of 324 amu. Although less volatile molecules with higher molecular weights may be encountered in some ignitable liquids, it is with much less frequency. In addition, these heavier compounds are not as easily collected by common extraction means. Instrument parameters may be customized to look for very volatile compounds in the early portion of the analysis so that it is not necessary to look for smaller, less diagnostic fragments throughout an entire analysis [56]. General data collection for screening purposes is aimed at detecting compounds in the hexane (C_6) to eicosane (C_{20}) range. It is recommended that scanning parameters not include m/z values below 33, as contributions from water and air will contribute excessively to the data and will not provide useful diagnostics. Scanning above m/z 330 is unlikely to provide much useful data. Therefore, selection of scan parameters must consider optimizing the collection of useful data, so that sensitivity is not needlessly lost. Scanning from m/z 33 to 300 will cover most major ignitable liquids, and minimizes collection of nondiagnostic or rarely encountered ions. These parameters are not appropriate however for low molecular weight ignitable liquids. In order to avoid missing the most volatile ignitable liquids such as the light alcohols and acetone, parameters must be adjusted. It has been proven effective to change parameters within an analysis so that data for the more volatile components can be collected prior to the solvent, and then optimized for general screening after the solvent [56]. This may be done by scanning from m/z 15 to 100, turning the MS off to allow the solvent to pass through undetected, and then scanning from m/z 33 to 300. Another major consideration regarding scan parameters that may affect both sensitivity and spectral quality is the speed of the scan. A rapid scan is desirable in order to ensure adequate chromatography; it is recommended that on the order of ten complete mass spectra be acquired during the time required for a peak to fully elute [86]. If scan parameters are set with excessive dwell times, resulting in a slow scan, skewing will result in both the chromatographic peak as well as in the mass spectra. If the dwell time is inadequate, however, and too little time is spent at each mass-to-charge value, then sensitivity is compromised. Sensitivity and spectral quality are therefore both dependent on scan range and scan speed.

There are other MS parameters that may be set by the user that will affect the quality of the data collected. Temperatures may be set in several different zones and calibration values may be set. An in-depth discussion of these parameters is beyond the scope of this chapter.

26.4.2.2.3 Ignitable liquid data analysis by GC-MS

When examining gas chromatography data, the analyst can be considered to be looking at two dimensions of data: abundance versus time; there is no structural information. Similarly, in the examination of mass spectral data, there are two dimensions of data: abundance and the mass-to-charge information that provides the basis of the structural identification, but there is no element of time. When GC-MS is used as an integrated technique, the examiner has access to three dimensions of data: time, abundance, and mass-to-charge data. By virtue of computers, the examiner sifts through the data in a variety of ways. The data can be examined in the two dimensions traditionally used in GC applications, by examining the total ion chromatogram (TIC), also referred to as a reconstructed ion chromatogram (RIC). This means of examining data is analogous to having no structural information, but having abundance and retention time data. Analogously, a single point in time within the data can be examined, and the ion fragment information and abundance information at that particular retention time can be examined. Again, this provides only two dimensions of data, allowing the examiner to look at a full mass spectrum for a particular peak, thereby providing a means for its identification, but not providing the benefit of how this particular component fits in with the mixture as a whole. EIP, however, allows the examiner to extract information from both the separation (GC) and the structural analysis (MS), and to use that information in an integrated format enabling the analyst to focus on compounds of interest and their proportions relative to one another. EIP, which relies on full scan data – it is not SIM – provides chromatographic profiles that minimize interferences from background materials. This technique may be applied by looking at individual m/z values, or by combining several ions into a single profile to represent a class of compounds [87]. It is also referred to as mass chromatography and extracted ion chromatography (EIC), and is the crux of GC-MS data analysis for the detection and identification of ignitable liquids in complex debris samples.

EIP relies on the fact that the chemical compounds comprising various chemical families will have structural similarities that are reflected in their spectra. Fragmentation

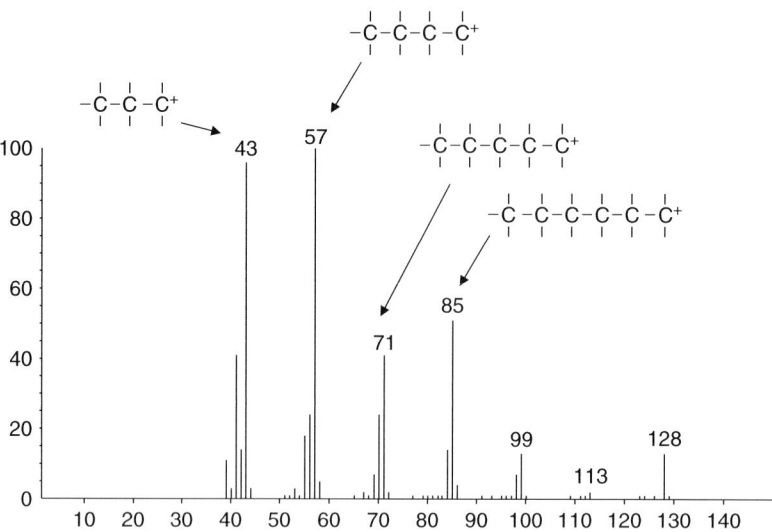

Fig. 26.3. Ion Fragments typically associated with the mass spectral analysis of alkanes (paraffins). Alkanes will fragment into an easily recognized and inherently sensible series of alkyl fragments of $m/z = 43, 57, 69...(+14)$. The differences in mass of 14 correspond to methylene groups. The spectrum shown is that of a branched nonane compound, with a molecular weight of 128 amu. Obtained in the authors' ATF Laboratory under the following conditions: instrument, HP 6890 Series GCMS with HP-1 column 22 m, 0.20 mm ID, film 0.5 μm. Split 20:1. Temperature program: 60°C for 3 min, ramp 5°C/min to 120°C, ramp 12°C/min to 300°C, 5 min at 300°C, Scanning, M/Z 33–330.

of alkanes, whether branched- or straight-chain, will result in the formation of alkyl groups of varying lengths as shown in Fig. 26.3. Therefore, the presence of ions with m/z values of 43, 57, 71, 85, etc. can be indicative of an alkane. Similarly, cycloalkanes will typically form a cyclohexyl ion during the fragmentation; therefore, the m/z 83 ion is often indicative of this class of compound. PNA compounds tend to show relatively little fragmentation due to their inherent stability. As a consequence, these compounds typically have a strong molecular ion. Table 26.3 shows ions that are typical of some of the types of compounds that are commonly found in ignitable liquids, or from debris-related interferences. EIP selectively displays data in chromatographic form, using structural indicators to extract pertinent data from the entire set of collected data. For example, Fig. 26.4a shows the TIC of a MPD. A typical MPD is expected to have aromatic content, however, it is not easily seen in the TIC. Fig. 26.4b shows an extracted ion profile of m/z 91, 105, 119, and 133. This mass chromatogram reveals every component that has one or more of the selected ions in its fragmentation pattern. Because the mass spectrum of alkylbenzenes have these ions present in significant abundances, this extracted ion profile is useful in revealing low levels of aromatic compounds that were not easily discernable amongst the much stronger aliphatic contributions. In this way, EIP allows the analyst to utilize structural information in combination with chromatography, so as to focus on components of interest.

TABLE 26.3
IONS AND THE TYPE OF COMPONENTS

Type of Compound	Diagnostic Ions
Alkanes, branched or straight-chain	43, 57, 71...
Alkenes	41, 55, 69...
Cycloalkanes	83
Indanes	117, 118, 131, 132, 145, 146
Simple aromatics (alkylbenzenes)	91, 105, 119, 133, 134
Polynuclear aromatics (alkylnaphthalenes)	128, 142, 156
Styrenes	104
Terpenes	93, 136

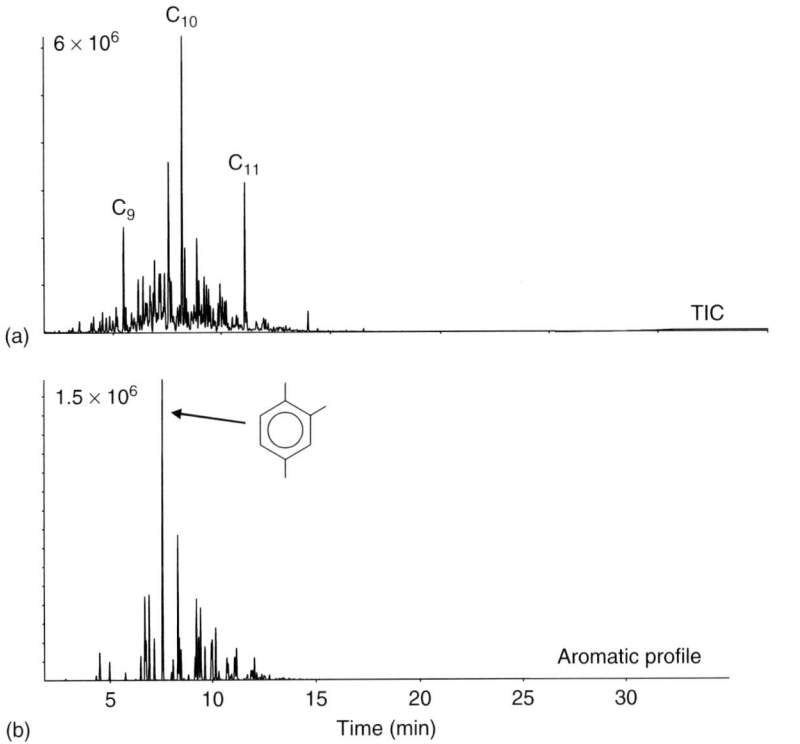

Fig. 26.4. GC-MS data from a typical medium petroleum distillate. (a) Total Ion Chromatogram (TIC). (b) Aromatic profile. The TIC of this MPD (a) shows all of the chemical compounds that were detected. However, because this liquid has a much greater proportion of aliphatic compounds, it is difficult to see the benzene-based aromatic compounds that are expected. By using the technique of extracted ion profiling (EIP) one can extract data for components that contain the ions 91, 105, 119, or 133, which are indicative of alkylbenzenes (benzene-based aromatics). Examination of chromatogram (b) shows the clear presence of aromatic compounds, which is consistent with the aromatic content of that boiling point range of petroleum. Obtained in the authors' ATF Laboratory under the following conditions: instrument, HP 6890 Series GCMS with HP-1 column 22 m, 0.20 mm ID, film 0.5 μm. Split 20:1. Temperature program: 60°C for 3 min, ramp 5°C/min to 120°C, ramp 12°C/min to 300°C, 5 min at 300°C.

EIP is widely accepted as a means of data analysis for ignitable liquid residues. Typically, an analyst will collect full scan data throughout the course of a temperature programmed chromatographic analysis. From this data, the analyst can examine the TIC, and the relevant mass chromatograms, or extracted ion profiles. Full mass spectral data is also available should it be necessary to identify a particular component of the mixture. Extracted ion profiles commonly used in ignitable liquid residue analysis include alkanes, aromatics (benzene-based), PNAs based on naphthalene-, indane- and indene-based aromatics, and alkenes and cycloalkanes. In this manner, a complex mixture can be separated based on not only chromatography, but also chemical characteristics.

26.4.2.2.4 Petroleum distillates, dearomatized distillates, and naphthenic paraffinic products

In examining the GC-MS data of petroleum distillates and dearomatized distillates, the TIC will show features that are characteristic of how it was refined. Refer back to Fig. 26.1, which shows various petroleum distillates in the light, medium, and heavy ranges. In each chromatogram, there is a distribution of peaks resembling a normal (Gaussian) distribution. This bell-shaped curve is an important feature of petroleum distillates, and is a result of the fractional distillation process. Another feature observed in the TIC that is common to all petroleum distillates is the prominent *n*-alkane pattern. The homologous series of *n*-alkanes is relatively abundant compared with the other compounds present in the distillate. This is due to their relative proportion in crude oil. The other peaks present between the *n*-alkanes represent the aromatics, cycloalkanes, and branched alkanes that are present in lesser quantities. For liquid reference standards such as those shown in Fig. 26.1, the class of product can be identified based solely on examination of a TIC.

The use of EIP becomes helpful when looking for compounds that are present in lesser abundance. For example, it is often not possible to distinguish between a traditional straight run petroleum distillate and a dearomatized distillate without the use of EIP. Figs. 26.4a and 26.5a show the TICs of two similar petroleum products. Each displays the features characteristic of a medium range distillate – a bell-shaped distribution with prominent *n*-alkanes in the C_8–C_{12} range, and less abundant peaks between the homologous *n*-alkanes. It cannot be definitively determined if aromatics are present in these products, because if they were present, they would be present at such low abundance that they would be overwhelmed by the more abundant aliphatic compounds. By using EIP, however, the presence or absence of aromatic compounds can easily be determined. Figs. 26.4b and 26.5b show the benzene-based aromatic profiles for these two products. This technique allows for the easy identification of the aromatic pattern typically found in petroleum-based products, as seen in Fig. 26.4b. Similarly, examination of the extracted ion profile shown in Fig. 26.5b shows that there are virtually no aromatic compounds present. The use of EIP in this instance demonstrates its utility in combining structural and chromatographic data to selectively focus on classes of compounds that would not be seen *via* a simple time versus abundance chromatogram due to their relatively low abundance within a complex mixture.

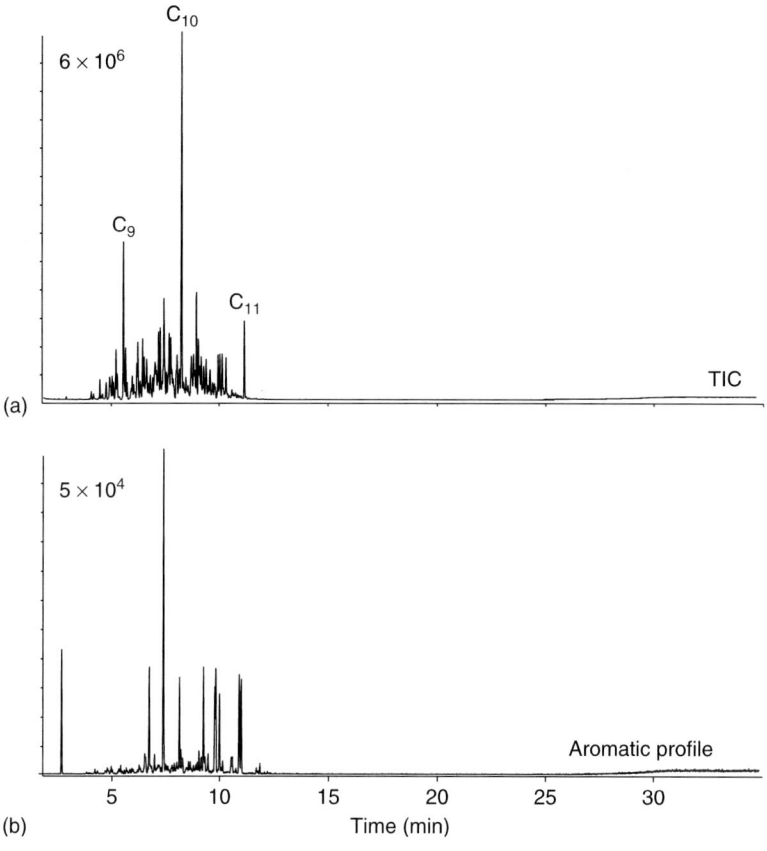

Fig. 26.5. GC-MS data from a typical dearomatized medium petroleum distillate. (a) Total Ion Chromatogram (TIC). (b) Aromatic profile. The TIC of this MPD (a) shows all of the chemical compounds that were detected. Because aromatic components if present, are expected to be at much lower levels than the aliphatic components, it cannot be easily determined from examining the TIC if this product contains aromatic compounds. By using the technique of extracted ion profiling (EIP) one can extract data for components that contain the ions 91, 105, 119, or 133, which are indicative of alkylbenzenes (benzene-based aromatics). Examination of chromatogram (b) easily shows the absence of aromatic compounds. The peaks that are present are at approximately 1/100 of the intensity of the TIC, and do not represent the aromatic content of that boiling point range of petroleum; there is no aromatic pattern. Obtained in the authors' ATF Laboratory under the following conditions: instrument, HP 6890 Series GCMS with HP-1 column 22 m, 0.20 mm ID, film 0.5 μm. Split 20:1. Temperature program: 60°C for 3 min, ramp 5°C/min to 120°C, ramp 12°C/min to 300°C, 5 min at 300°C.

Naphthenic paraffinic products are derived from petroleum distillates, and are modified by the removal of aromatic compounds, and by either removing or greatly reducing the amount of normal alkanes. As a result, the overall distribution of the peaks is similar to the bell-shaped curve seen in distillates. However, because of the marked reduction in normal alkanes, the pattern will not show the series of normal alkanes spiking above the rest of the pattern. The major series of homologous peaks observed in naphthenic paraffinic products is that of the alkylcyclohexanes. Normal alkanes may also be present, but will be more similar in abundance to the other peaks present in the chromatogram. Naphthenic/paraffinic products may be recognized with only a TIC; however, the pattern is not as readily recognized as that of most other common petroleum-based ignitable liquids. The use of mass chromatography can be helpful in emphasizing the alkylcyclohexanes and the lack of aromatic compounds.

Another highly useful application of mass chromatography is that of finding ignitable liquids in the presence of overwhelming amounts of interfering products. Many items of evidence that are submitted to the forensic science laboratory for ignitable liquid residue analysis are synthetic materials such as carpet or furnishings, which have been exposed to extreme fire conditions. The many chemical compounds created from the pyrolysis and partial combustion of the substrate materials will also be present in the sample extract, and therefore in the instrumental data. EIP allows the analyst to selectively examine the various classes of compounds typically present in petroleum-based ignitable liquids. If using traditional GC analysis, the presence of an ignitable liquid could easily be lost amongst the overwhelming presence of interfering products derived from the sample substrate. The clear benefit of using EIP in this situation is that it may allow for the detection and identification of ignitable liquids which otherwise would not have been possible.

26.4.2.2.5 Aromatic products, n-alkane products, and isoparaffinic products

The ignitable liquid classifications of aromatic products, *n*-alkane products, and isoparaffinic products have an important characteristic in common: they have undergone significant processing in order to make products that are composed of a single class of chemical compound. For these types of products, the use of EIP is not particularly useful when examining reference standards; the TIC will be sufficient. However, in cases in which there are significant contributions from the debris, EIP may help to find low levels of ignitable liquid residues in a complex chromatogram. Because isoparaffinic products and normal alkane products both consist exclusively of alkanes, when EIP is applied, the patterns seen in the alkane and alkene profiles will appear similar to that present in the TIC. Alkane peaks will appear in the alkane profile at a lower abundance than in the TIC, and will also appear in the alkene profile, at an even lower abundance. This is because although the ions that are selected to represent the alkenes class, typically m/z 55, 69, and 83, are well represented in the fragmentation of alkenes, and therefore are most indicative of alkenes, these fragment ions will also appear in the mass spectra of alkanes, although at a lesser abundance. Fig. 26.6 shows the mass spectrum of an alkane and an alkene. Examination of these spectra shows that while the 55, 69, and 83 series of ions will be good indicators of alkenes, these fragments also appear in alkane spectra. Therefore, when EIP uses these ions, both alkenes and alkanes will appear in the profile, although not in proportion to their abundances in the TIC – the alkene will have a greater relative abundance. Figs. 26.7 and 26.8 show the TIC and selected EIPs for an isoparaffinic product and a normal alkane product.

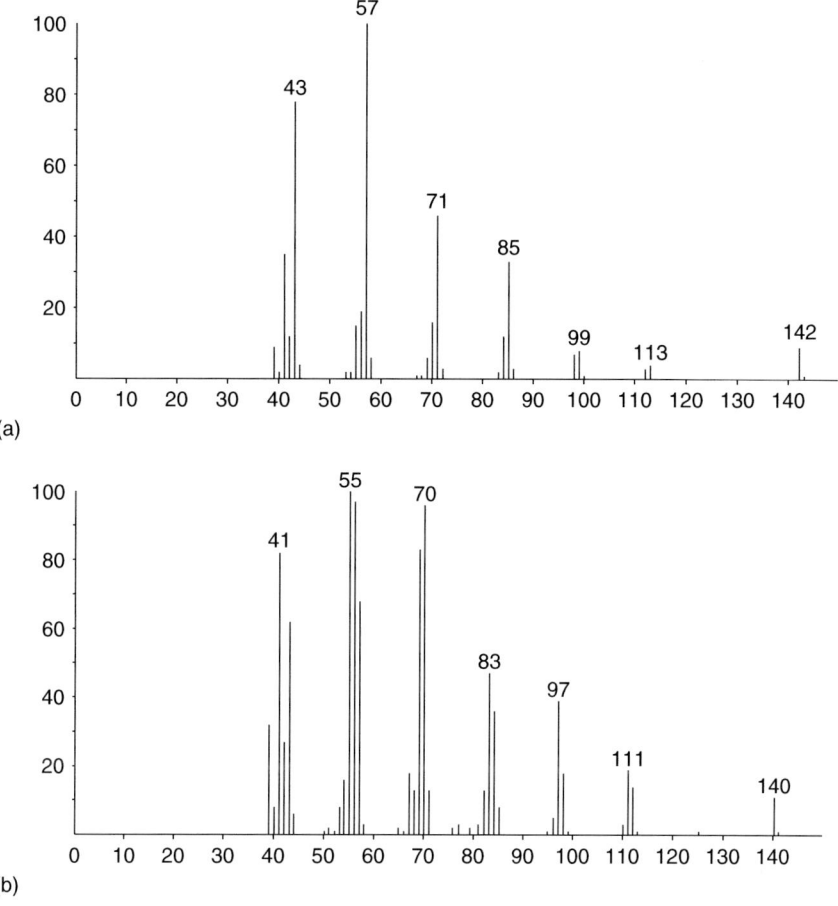

Fig. 26.6. Mass spectra of (a) an alkane (decane) and (b) an alkene (decene). The structural difference between an alkane and an alkene is the presence of a double bond, which results in an alkene having two fewer hydrogen atoms that the corresponding alkane. This is reflected in the molecular weight of an alkene being two less than that of the corresponding alkane (C_nH_{2n+2} for alkanes and C_nH_{2n} for alkenes). This difference is also represented in the mass spectral fragmentation pattern. Whereas spectrum (a) is dominated by the 43, 57, 71... series of ions associated with alkyl fragments, the predominant fragments in the alkene spectrum are at m/z values that are two less: 41, 55, 69.... Note that although 57 is better represented in the alkane spectrum, there is however a 57 ion in the alkene spectrum. Conversely, although 55 is better represented in the alkene spectrum, there is also a 55 ion in the alkene spectrum. Obtained in the authors' ATF Laboratory under the following conditions: instrument, HP 6890 Series GCMS with HP-1 column 22 m, 0.20 mm ID, film 0.5 µm. Split 20:1. Temperature program: 60°C for 3 min, ramp 5°C/min to 120°C, ramp 12°C/min to 300°C, 5 min at 300°C, Scanning, M/Z 33–300.

In the case of aromatic products, the aromatic compounds comprising the product may be benzene-based or polynuclear, depending on the boiling point range and end use of the product. Light and medium range aromatic products will contain benzene-based aromatic compounds, whereas PNA compounds will be found in the

References pp. 919–922

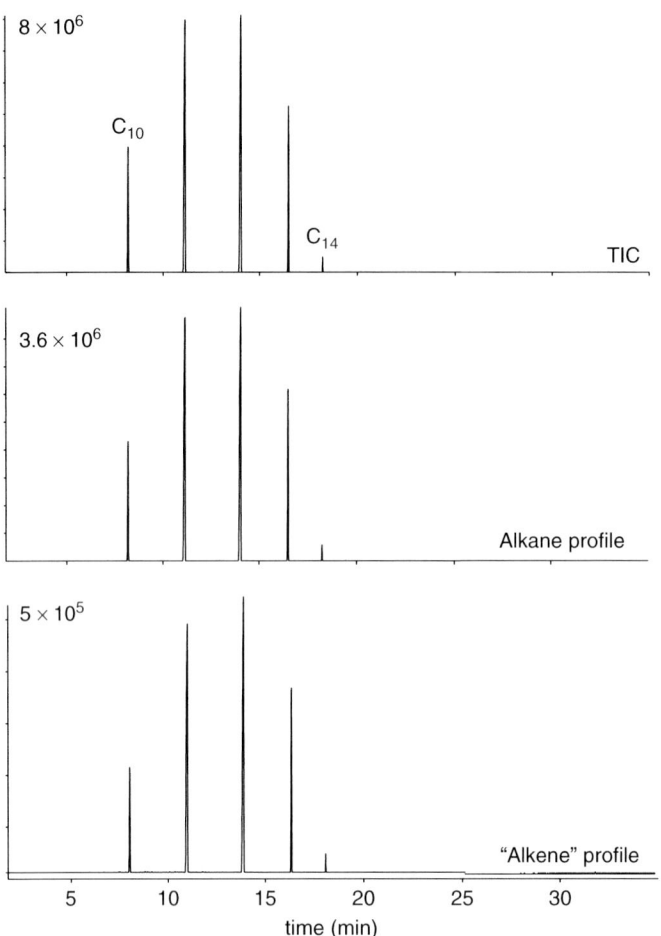

Fig. 26.7. Total ion chromatogram (TIC) and selected extracted ion profiles of normal alkane product. Note that the chromatograms observed in the alkane profile (sum of ions 57, 71, 85, and 99) and the alkene profile (sum of ions 55, 69, and 83) show the same pattern, with the only significant difference being the abundance. There is no significant pattern in any of the aromatic profiles. Obtained in the authors' ATF Laboratory under the following conditions: instrument, HP 6890 Series GCMS with HP-1 column 22 m, 0.20 mm ID, film 0.5 μm. Split 20:1. Temperature program: 60°C for 3 min, ramp 5°C/min to 120°C, ramp 12°C/min to 300°C, 5 min at 300°C.

medium and heavy range products. Because the best ions to represent aromatic compounds tend to be molecular ions or M-1 fragments, the pattern shown in a TIC may not appear in its entirety in any one profile. Components such as the indane- and indene-based compounds are naturally present in lower abundances than some of the benzene- and naphthalene-based aromatic compounds, so an indane profile may provide information not readily observed in the TIC. Fig. 26.9 shows an aromatic product with selected profiles.

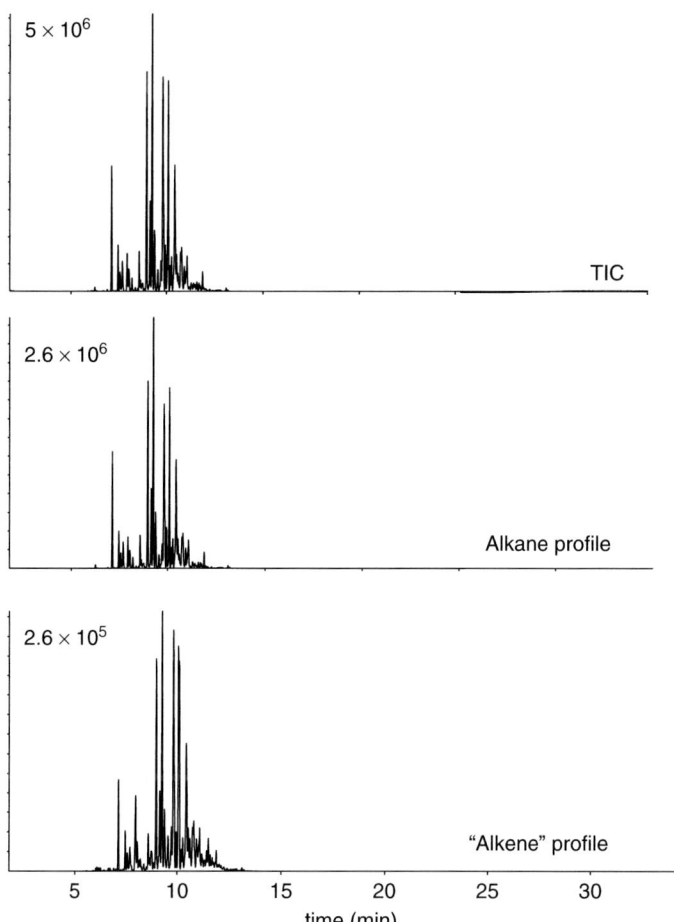

Fig. 26.8. Total ion chromatogram (TIC) and selected extracted ion profiles of isoparaffinic product. Note that the chromatograms observed in the alkane profile (sum of ions 57, 71, 85, and 99) and the alkene profile (sum of ions 55, 69, and 83) show the same pattern, with the only significant difference being the abundance. There is no significant pattern in any of the aromatic profiles. Obtained in the authors' ATF Laboratory under the following conditions: instrument, HP 6890 Series GCMS with HP-1 column 22 m, 0.20 mm ID, film 0.5 μm. Split 20:1. Temperature program: 60°C for 3 min, ramp 5°C/min to 120°C, ramp 12°C/min to 300°C, 5 min at 300°C.

26.4.2.6 Gasoline

Gasoline is composed of all of the major classes of hydrocarbons found in the appropriate boiling point range of crude oil. It contains normal alkanes, isoalkanes, cycloalkanes, and aromatic compounds. The benzene-based aromatic compounds are enriched as compared with the starting material, and are readily observable in a TIC. Similarly, in the boiling point range prior to toluene, the isoalkanes can easily be seen, as they are the dominant class. Because the alkanes and cycloalkanes are

References pp. 919–922

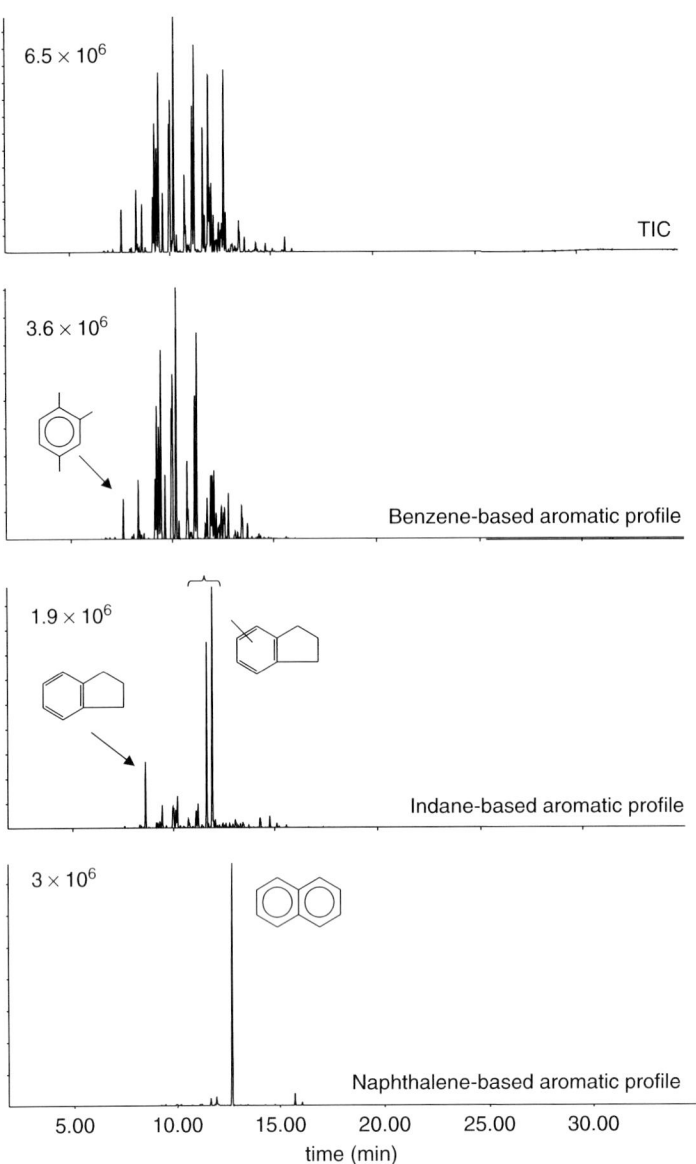

Fig. 26.9. Total ion chromatogram (TIC) and extracted ion profiles of an aromatic product. Peaks present in aromatic products may be benzene-, indane- or naphthalene-based, depending on the boiling point range of the product. There is no significant pattern in the aliphatic profiles. Obtained in the authors' ATF Laboratory under the following conditions: instrument, HP 6890 Series GCMS with HP-1 column 22 m, 0.20 mm ID, film 0.5 μm. Split 20:1. Temperature program: 60°C for 3 min, ramp 5°C/min to 120°C, ramp 12°C/min to 300°C, 5 min at 300°C.

present in fairly low amounts relative to the benzene-based aromatic compounds, the use of EIP can help the analyst to see these compounds more easily, as well as the indane- and indene-based compounds, which also may be overwhelmed by the alkylbenzenes. The use of EIP may be beneficial in the examination of gasoline data in order to show the many types of compounds that are present at relatively low levels. As with the other types of ignitable liquids, the technique of EIP is most helpful when identifying a low level of gasoline in a complex matrix.

26.4.2.2.7 Oxygenated solvents and others-miscellaneous

The classifications of oxygenated solvents and others-miscellaneous include products that are not necessarily derived from petroleum; consequently, traditional pattern recognition techniques cannot always be applied. Both of these classifications can include single component ignitable liquids, such as acetone or methyl isobutyl ketone. For these types of ignitable liquids, the best approach to data analysis is to identify the chemical by comparing its retention time and full mass spectrum with those of a known reference standard. For oxygenated solvents in which the oxygenated compound is present as part of a blended mixture, the oxygenated component should still be identified *via* retention time and mass spectrum along with any other major components comprising the blend. In cases in which an oxygenated component is blended with a petroleum-derived product, the petroleum portion of the pattern may be evaluated *via* traditional chromatographic pattern recognition and EIP techniques.

The others-miscellaneous classification is defined quite broadly. It may include other ignitable liquids that cannot be defined by the existing criteria for the other specific classifications, such as mixtures and products not derived from petroleum. Examples of ignitable liquids, which may be placed in the miscellaneous category include natural products such as turpentine, limonene, and essential oils, or blends of individual products, such as some specialty solvents. Data interpretation may present a challenge, especially in instances in which the examiner does not have a suitable reference ignitable liquid. For simple blends, individual component identification may be necessary. In cases in which the miscellaneous product can be described as a blend of existing classifications, it is recommended that the patterns be analyzed individually for conformance to the criteria defined for each classification. Data from natural products such as turpentine may be evaluated *via* comparison of patterns and EIPs, supplemented by individual peak identification. It is not possible to provide specific criteria for the products that fall within this classification. Depending on the nature of the data, the examiner will need to determine the most appropriate way to evaluate it.

26.4.2.2.8 Detection of ignitable liquids from fire debris samples

When ignitable liquids are recovered from actual samples of fire debris, many complications arise that will affect data interpretation. Extraction techniques may skew data so that the recovered sample is not truly representative of the ignitable liquid present in the item of evidence. Significant contributions from the material comprising the substrate can contribute significantly to the data [12]. Interferences from the debris may result from pyrolysis of the substrate material, contributions from materials naturally occurring in the substrate that are released during the extraction process, or from products of partial combustion [88]. The presence of these interfering products can make the detection and identification of an ignitable liquid residue much more difficult [89]. The use of EIP as described in previous sections reduces the complications due to the presence of interfering products in the sample, however, it will not completely eliminate their contribution.

Another means of data analysis that has been applied to fire debris analysis is that of target compound gas chromatography (TCGC)-MS. Several researchers have studied the utility of applying the concept of target compounds to the analysis of ignitable liquids [90–91]. In this method of data analysis, pre-selected target compounds are entered into the computer software, with sufficient information to tentatively identify the compounds, usually retention time, the base peak and two to three significant ions and their relative ratios. Target compounds are selected that will survive most fire conditions and that are not common interferences from substrate materials. The software then "looks" for the target ions, and extracts abundance information for the target compounds from the complex data. A simplified stick plot can be created that provides a visual representation of the target compounds present and their relative abundances. This method is very effective for identifying very low levels of ignitable liquid residue in the presence of relatively great amounts of interferences derived from the sample substrate. The concept of creating a stick plot to enable the analyst to examine data that simulates a chromatogram, rather than merely tabulated data was introduced by Keto and Wineman [90]. Because much of ignitable liquid analysis is dependent on patterns and the relative amounts of compounds, the ability to visualize target compounds as a pattern is beneficial. While target compound chromatography offers unique benefits for detecting low levels of ignitable liquids from complex data, it has not garnered the widespread use that EIP has.

Evaluation of GC-MS data for the identification of ignitable liquids requires an understanding of the composition of commonly encountered ignitable liquids, a basic foundation in GC-MS theory, and access to numerous reference standards. Unlike many other chemical analyses, the mere presence of a particular compound is not sufficient to identify an ignitable liquid or residue. Contributions from the partial combustion and pyrolysis of substrate materials can significantly complicate interpretation of data. The fact that compositions of ignitable liquids will change as a result of exposure to fire can further complicate the task of data analysis. In order to competently test evidence for the presence of ignitable liquids, an analyst must have a thorough understanding of a wide variety of interrelated topics, and undergo extensive hands-on training.

26.5 QUALITY ASSURANCE

One of the most critical aspects of any forensic science laboratory is the establishment of an appropriate quality system. Quality assurance and quality control go hand in hand to ensure that the requirements for quality are achieved. Quality assurance is defined by the American Society of Crime Laboratory Directors – Laboratory

Accreditation Board (ASCLD-LAB) as "Those planned and systematic actions necessary to provide sufficient confidence that a laboratory's product or service will satisfy given requirements for quality," whereas quality control is described as "internal activities, or activities conducted according to externally established standards, used to monitor the quality of analytical data and to ensure that it satisfies specified criteria" [92]. In forensic science laboratories, quality is often considered to have foundations built on three separate, but intimately related mainstays: the laboratory system and its policies; the individual examiners and their qualifications; and the methods of analyses that are used. Quality systems for each of these components exist in the form of laboratory accreditations; examiner certifications; and the use of validated methods.

26.5.1 Laboratory accreditation

The purpose of an accreditation program is to demonstrate that an entity or organization has procedures in place that allow it to function as intended, and to detect and correct the errors that will inevitably occur. Accreditation is a seal of approval and indicates that the organization meets the requirements set forth by the individual accrediting organization. For forensic science laboratories, there are two major organizations with established accreditation standards: the ASCLD-LAB and the International Organization for Standardization (ISO). Other organizations provide related services, such as the National Center for Forensic Science and Technology (NFSTC), an organization based in Florida that provides certification for DNA laboratories that they meet (US) national standards. Each of these types of external quality assessments may be valuable to forensic science laboratories.

ASCLD-LAB offers accreditation internationally, although the majority of ASCLD-LAB accredited laboratories are in the United States [93]. The accreditation program developed by ASCLD-LAB was designed exclusively for forensic science laboratories. As a result, the ASCLD-LAB program has elements that address the variety of specific issues encountered in forensic science laboratories, including chain of custody, method validation, and sample preservation. ISO offers two main standards that have been applied to forensic science laboratories. The 9000 group of standards is not specific to forensic science applications, or even to laboratories. Rather, it is a standard that addresses management systems to ensure that appropriate measures for achieving quality are in place. The ISO 17025 standard is a more narrowly focused standard that addresses issues specifically related to testing laboratories. The ASCLD-LAB program is a dynamic accreditation model that has adapted over the years to address new issues as they arise. Currently, the ASCLD-LAB program has adopted the ISO 17025 standard as a significant part of its International Accreditation program. This modification to the ASCLD-LAB accreditation program offers the benefits of a seemingly more external and therefore objective program than the one developed through an offshoot of an association of crime laboratory directors, while still maintaining its applicability to forensic science laboratories.

In examining the measures taken to ensure quality, it is important to consider the quality management system of a laboratory. Adherence to external standards, such as those promulgated by the ASCLD-LAB accreditation program and the ISO standards is an objective measure of a laboratory's dedication to a quality system.

26.5.2 Examiner qualifications

The examination of fire debris for the presence of ignitable liquids is, first and foremost, a chemical analysis. Although there are generally accepted methods for conducting these types of analyses, each step of the process requires decision making; therefore, the analyst conducting the tests must have a minimum level of training. Various professional associations and working groups have examined some of the issues regarding minimum qualifications, including training specific to the analysis of fire debris. In addition, the American Board of Criminalistics (ABC) has developed a certification program dedicated to providing an external measure of competency for those practicing forensic science.

The ABC is a voluntary, peer-based organization providing an external measure of competency for those practicing in the criminalistics field of forensic science. It offers different levels of certification to individuals who meet their peer-developed criteria for certification. The designation of Diplomat of the ABC provides an external measure that the analyst "is qualified to supervise multidisciplinary examinations of physical evidence" and the designation of Fellow indicates that the individual "is qualified to supervise multidisciplinary evidence examinations and to conduct examinations in the specialty area" [94]. Certification by the ABC is based on education, experience, both written and practical examinations, and continuing professional education.

The ABC certification program is one means of providing an external review of an analyst's capabilities. Other organizations have also provided input to the minimum requirements necessary for an analyst to practice in the field of fire debris analysis. The Technical Working Group for Fire and Explosions (TWGFEX) is a working group composed of investigators forensic examiners specializing in the analysis of fire- and explosive-related incidents. As part of TWGFEX's mission, the organization has published recommendations for core knowledge, skills, and abilities that are required for analysts performing fire debris analyses [95]. This document presents recommendations for providing training in the areas essential for an examiner analyzing ignitable liquids and their residues. Other organizations and independent laboratories will have their own criteria for minimum qualifications.

26.5.3 Valid methods

With the exception of DNA, the analysis of fire debris is probably one of the most standardized and best-documented disciplines of forensic science. One of the first major steps in developing consensus-based standards for the analysis of ignitable liquid residues from fire debris was the publication of a classification system developed by the Bureau of Alcohol, Tobacco and Firearms (ATF) and the National Bureau of Standards (NBS, currently the National Institute of Standards and Technology or NIST) in the Arson Analysis Newsletter [96]. From that initial classification scheme, came the resurgence of the ASTM E 30 Committee of Forensic Sciences, and the subsequent publication of numerous standard methods relevant to fire debris analysis. ASTM International is a voluntary consensus-based standards development organization that develops and publishes standards related to a wide variety of subject matters, of which forensic science is a very, very small part [97]. Currently, ASTM International publishes eight standards specific to the analysis of fire debris, including standard practices which describe sample preparation and extraction methods, and standard test methods which describe instrumental analysis and data interpretation criteria and procedures [14,15,27–31]. In addition, other more general forensic science standards also exist, related to issues such as evidence storage, labeling, and documentation [25,26]. The process of developing a standard from its initial proposal to published method includes many opportunities for input, and therefore is quite lengthy and time-consuming. Yet a standard is never truly "finished," because each standard must be reviewed on a regular basis to ensure that it still meets the needs of its users. The benefit to the laboratories conducting fire debris analysis is that valid methods exist which have met the requirements of general acceptance within the applicable scientific community. This allows the laboratory to merely verify the standard method in their laboratory, rather than go through the more extensive validation process, as would be necessary for a new method. In addition, the use of standard methods gives a laboratory greater credibility in court and accreditation processes.

26.6 THE FUTURE OF FIRE DEBRIS ANALYSIS

Forensic science is a dynamic field, with methods of analysis evolving as the result of research aimed at improving existing methods. The bulk of recent research has been focused on improving sensitivity and in increasing specificity of identifications, particularly in comparing a recovered ignitable liquid to one that can be associated with a suspect. Research in the area of fire debris analysis has examined sample preparation methods, but has primarily focused on instrumental methods of analysis [98].

While GC-MS will continue to be the primary method of analyzing fire debris extracts, there may be some utility to the use of tandem mass spectrometry. Preliminary research into the use of GC-MS/MS has shown promise in cases of some challenging samples [99,100]. In samples in which a very complex chromatogram results due to a low concentration of ignitable liquid residue in the presence of substantial interferences from the sample substrate, GC-MS/MS has demonstrated a superior ability to discern the components of interest necessary to identify an ignitable liquid. In this circumstance, lower limits of detection may be achievable through the use of GC-MS/MS. Although there has been limited work in this area, the benefits of GC-MS/MS may provide a solution to the challenge of identifying a low level of surviving ignitable liquid amongst various pyrolysis and decomposition products originating from the debris.

Interest in sourcing recovered accelerants will always be of interest to the fire investigator. Research conducted by Mann in the eighties demonstrated that variations in the most volatile portions of gasoline could be used as a tool for determining if different samples could have had a common source [83,84]. Expansions of this work showed that the technique could be applied to current formulations of gasoline through GC-MS analysis and semi-automated data analysis [101]. Further application of this technique showed that even in cases in which significant loss of the more volatile components of gasoline occurred due to fire exposure, that limited comparisons could still be conducted [85]. Because of the significant loss of volatiles from gasoline in fire situations, this research provides a technique that may be more widely applicable to evidence recovered from scenes with extensive fire damage. Another approach to the comparison and sourcing of petroleum-based ignitable liquids was born from methods used for sourcing oil spills. Work by Frysinger and Gaines showed that comprehensive two-dimensional gas chromatography (GC × GC) may provide the solution to more fully characterizing source-to-source variations in gasoline products [102]. Analysis of gasoline by comprehensive two-dimensional gas chromatography provides significant improvements in resolution that cannot be accomplished *via* a single chromatographic separation. Separations of potential ignitable liquids are better resolved as a result of a two-step chromatographic separation that uses a boiling point based separation followed by a separation based on polarity. This improved resolution has potential to be applied to the comparisons of ignitable liquids, and research in this area is continuing. Another novel approach to the comparisons of petroleum products involves the use of stable isotope ratios. While stable isotope ratios have been successfully used in other applications, research into the comparison of ignitable liquids to one another is still in its early stages [103]. Preliminary research has shown that gas chromatography-isotope ratio mass spectrometry (GC-IRMS) may be able to provide a high degree of specificity for the comparison of common accelerants; however, the practical value of the technique will be better established with further research.

The use of ultra-high resolution mass spectrometry has been applied to the analysis of ignitable liquids and has shown that identification may be accomplished without benefit of a chromatographic separation [104]. This application of electron ionization Fourier transform ion cyclotron resonance (EIFT-ICR) is based on the presence or absence of specific compounds that are identified based solely on molecular weight, thereby not allowing for discrimination amongst isomers. The technique has been successfully applied to only a small number of ignitable liquids, so additional work in this area is warranted. Unfortunately, the equipment required to conduct this type of analysis is extraordinarily expensive; therefore, it is likely that this technique will remain the purview of research laboratories, and not be widely used for routine analyses in forensic science laboratories.

Increasing interest is being placed on the detection of ignitable liquids that are not derived from petroleum. Examination of fire debris for the presence of animal or vegetable oils is often desired, particularly in cases in which spontaneous combustion is suspected [105,106]. Methods of analysis for these types of oils have long been applied in the food industry, but are not routinely applied to fire debris. Concern with spontaneous combustion and the potential use of nonpetroleum-based accelerants will result in a greater demand for this type of analysis. Research by Stauffer has shown that several options are available to the fire debris analyst desiring to include vegetable oil residue testing in their analytical scheme [107].

The ability to take the laboratory to the field has been a topic of interest. The use of accelerant detection canines in the field continues to be a valuable resource; however, canines are not infallible [108–110]. Research in this area has focused on rugged, field portable GC-MS systems [111]. While the desire to put analytical instrumentation in the field has been slightly tempered by challenges associated with sample preparation, efforts continue to improve field detection capabilities.

It is anticipated that research into the application of novel instrumental techniques to fire debris analysis will continue. Studies will continue to focus on increasing sensitivity and the ability to better separate materials of interest from interfering compounds contributed by sample materials. Investigators continue to desire a conclusive link between suspects and crime scenes, so until a DNA-like profile of ignitable liquids is developed, additional efforts will likely be aimed at providing additional individualization of recovered ignitable liquids. As improvements in current methods are achieved, it will still remain up to the fire investigator, with the help of the forensic analyst, to determine the meaning of any recovered ignitable liquids. Some things will never change.

26.7 REFERENCES

1   QuickStats: The Overall Fire Picture 2005, Retrieved August 24, 2006, from http://www.usfa.dhs.gov/statistics/quickstats
2   The Geneva Association,(2005) World fire statistics: Information bulletin of the world fire statistics. The Geneva Association, Geneva, Switzerland, 2005.
3   National Fire Protection Association, NFPA 921: Guide for fire and explosion investigations. NFPA, 1992.
4   National Fire Protection Association, NFPA 921: Guide for fire and explosion investigations. NFPA, 2004.
5   J.D. DeHaan, Kirk's fire investigation, (5th ed.). Pearson Education, Inc., Upper Saddle River, NJ, 2002.
6   J.J. Lentini, Scientific protocols for fire investigation, series: Protocols in forensic science, Vol. 3. CRC Press, Boca Raton, FL, 2006.
7   N. Nic Daéid, Fire investigation, CRC Press, Boca Raton, FL, 2004.
8   J.R. Almirall, K.G. Furton, Analysis and interpretation of fire scene evidence, CRC Press, Boca Raton, FL, 2004.
9   V. Babrauskas, Ignition handbook, Fire Science Publishers, Issaquah, WA, 2003.
10   W. Bertsch and Q. Ren. In: M.J. Bogusz (Ed.), Forensic science, series: Handbook of analytical separations, Vol. 2. Elsevier, Amsterdam, 2000.
11   J.J. Lentini, J.A. Dolan and C. Cherry, J. Forensic Sci., 45 (2000) 968–989.
12   E. Stauffer, Sci. Justice, 43 (2003) 29–40.
13   J.J. Lentini, National Fire and Arson Report 2:3 (1983) 3.
14   ASTM Method E 1618–06, Standard test method for ignitable liquid residues in extracts from fire debris samples by gas chromatography-mass spectrometry, in Annual Book of ASTM Standards, 2006. ASTM International, West Conshohocken, PA, 2006.

References pp. 919–922

15   ASTM Method E 1387–01, Standard test method for ignitable liquid residues in extracts from fire debris samples by gas chromatography, in Annual Book of ASTM Standards, 2006. ASTM International, West Conshohocken, PA, 2006.

16   ASTM Method E 1618–01, Standard test method for ignitable liquid residues in extracts from fire debris samples by gas chromatography-mass spectrometry, in Annual Book of ASTM Standards, 2001. ASTM International, West Conshohocken, PA, 2001.

17   J.A. Dolan and E. Stauffer, J. Forensic Sci., 49 (2004) 992–1004.

18   R. Newman, M. Gilbert, K. Lothridge, GC-MS guide to ignitable liquids, CRC Press, Boca Raton, FL, 1998.

19   Exxon Isopar$^{\text{®}}$ Product Information, Retrieved December 13, 2006, from http://www.exxon.com/USA-English/Lubes/PDS/Pds_Files/nause2indexisopar.pdf

20   J. DeHaan, Fire Arson Investigator, 52 (2002) 46–47.

21   ASTM Method D133, Test method for Kauri-butanol value of hydrocarbon solvents, in Annual Book of ASTM Standards, 2006. ASTM International, West Conshohocken, PA, 2006.

22   J.H. Gary, G.E. Handwerk, Petroleum refining technology and economics, Marcel Dekker, Inc., New York, 1994.

23   J.G. Speight, The chemistry and technology of petroleum, (3rd ed.). Marcel Dekker, Inc., New York, 1998.

24   E. Stauffer and J.J. Lentini, Forensic Sci. Int., 32 (2003) 63–67.

25   ASTM Method E 1459–92, Standard guide for physical evidence labeling and related documentation, in Annual Book of ASTM Standards, 2006. ASTM International, West Conshohocken, PA, 2006.

26   ASTM Method E 1492–05, Standard practice for receiving, documenting, storing and retrieving evidence in a forensic science laboratory, in Annual Book of ASTM Standards, 2006. ASTM International, West Conshohocken, PA, 2006.

27   ASTM Method E 1388–05, Standard practice for sampling headspace vapors from fire debris samples, in Annual Book of ASTM Standards, 2006. ASTM International, West Conshohocken, PA, 2006.

28   ASTM Method E 1386–00, Standard practice for separation and concentration of ignitable liquid residues from fire debris samples by solvent extraction, in Annual Book of ASTM Standards, 2006. ASTM International, West Conshohocken, PA, 2006.

29   ASTM Method E 1412–00, Standard practice for separation and concentration of ignitable liquid residues from fire debris samples by passive headspace concentration with activated charcoal, in Annual Book of ASTM Standards, 2006. ASTM International, West Conshohocken, PA, 2006.

30   ASTM Method E 1413–00, Standard practice for separation and concentration of ignitable liquid residues from fire debris samples by dynamic headspace concentration, in Annual Book of ASTM Standards, 2006. ASTM International, West Conshohocken, PA, 2006.

31   ASTM Method E 2154–01, Standard practice for separation and concentration of ignitable liquid residues from fire debris samples by passive headspace concentration with solid phase microextraction (SPME), in Annual Book of ASTM Standards, 2006. ASTM International, West Conshohocken, PA, 2006.

32   J.W. Brackett Jr., J. Crim. Law Criminol. Police Sci., 46 (1955) 554–561.

33   D.L. Adams, J. Crim. Law Criminol. Police Sci., 47 (1956) 593–596.

34   B.V. Ettling, J. Forensic Sci., 8 (1963) 261–267.

35   B.V. Ettling and M.F. Adams, J. Forensic Sci., 13 (1968) 76–89.

36   D. Willson, Forensic Sci., 10 (1977) 243–252.

37   R.A. Rouen and V.C. Reeve, J. Forensic Sci., 19 (1974) 607–617.

38   C.R. Midkiff, Arson Anal. Newsl., 2(6) (1978) 8–16.

39   R.N. Thaman, Arson Anal. Newsl., 31 (1979) 9–14.

40   M.J. Camp, Anal. Chem., 52 (1980) 423A–426A.

41   R.T. Newman, W.R. Dietz and K. Lothridge, J. Forensic Sci., 41 (1996) 361–370.

42   D.G. Kubler, D. Greene, C. Stackhouse and T. Stoudmeyer, Arson Anal. Newsl., 5 (1981) 64–79.

43   W. Bertsch and Q.W. Zhang, Anal. Chim. Acta, 236 (1990) 183–195.

44   C.R. Midkiff and W.D. Washington, J. AOAC Int., 55 (1972) 840–845.
45   Collaborative Testing Services, Forensic Testing Program, Report for Test 02–536 (2002).
46   Collaborative Testing Services, Forensic Testing Program, Report for Test 03–536 (2003).
47   Collaborative Testing Services, Forensic Testing Program, Report for Test 04–536 (2004).
48   Collaborative Testing Services, Forensic Testing Program, Report for Test 05–536 (2005).
49   S.P. Allen, S.W. Case and C. Frederick, Forensic Sci. Commun., 2(1) .
50   W.R. Dietz, J. Forensic Sci., 36 (1991) 111–121.
51   R. Newman. In: N. Nic Daéid (Ed.), Fire investigation, CRC Press, Boca Raton, FL, 2004, pp. 146–147.
52   J. A. Dolan and R. Newman, Solvent options for the desorption of activated charcoal in fire debris analysis. Proceedings of the American Academy of Forensic Sciences Annual Meeting, Seattle, WA, February 19–24, 2001; American Academy of Forensic Sciences (AAFS), Colorado Springs, CO, 2001.
53   J.J. Lentini and A.T. Armstrong, J. Forensic Sci., 42 (1997) 307–311.
54   G.D. Hicks, A.R. Pontbriand and J.M. Adams, Carbon disulfide vs. dichloromethane for use of desorbing ignitable liquid residues from activated charcoal strips. Proceedings of the American Academy of Forensic Sciences, Annual Meeting Chicago, IL, February 17–22, 2003; American Academy of Forensic Sciences (AAFS), Colorado Springs, CO, 2003.
55   D.J. Tranthim-Fryer, J. Forensic Sci., 35 (1990) 271–280.
56   J.L. Phelps, C.E. Chasteen and M.M. Render, J. Forensic Sci., 39 (1994) 194–206.
57   R. Newman. In: N. Nic Daéid (Ed.), Fire investigation, CRC Press, Boca Raton, FL, 2004, p. 147.
58   R.T. Newman, An evaluation of multiple extractions of fire debris by passive diffusion. Proceedings of the International Symposium on the Forensic Aspects of Arson Investigations, Fairfax, VA, July 31–August 4, 1995, U.S. Department of Justice Federal Bureau of Investigation, Washington, DC, 1995.
59   L.V. Waters and L.A. Palmer, J. Forensic Sci., 38 (1993) 165–183.
60   J.M. Twibell, J.M. Home and K.W. Smalldon, J. Forensic Sci. Soc., 22 (1982) 155–159.
61   P.V. Peurifoy, L.A. Woods and G.A. Martin, Anal. Chem., 40 (1968) 1002–1004.
62   J.P. Mieure and M.W. Dietrich, J. Chromatogr. Sci., 11 (1973) 559–570.
63   J.E. Chrostowski and R.N. Holmes, Arson Anal. Newsl., 3(5) (1979) 1–17.
64   M. Frenkel, S. Tsaroom, Z. Aizenshtat, S. Kraus and D. Daphna, J. Forensic Sci., 29 (1984) 723–731.
65   P.M.L. Sandercock, Can. Soc. Forensic Sci. J., 27 (1994) 179–201.
66   J.A. Juhala, Arson Anal. Newsl., 6(2) (1982) 32–40.
67   R. Newman. In: N. Nic Daéid (Ed.), Fire investigation, CRC Press, Boca Raton, FL, 2004, p. 149.
68   J.J. Lentini, Scientific protocols for fire investigation, series: protocols in forensic science, Vol. 3. CRC press, Boca Raton, FL, 2006, 139–147.
69   K.G. Furton, J. Bruna and J.R. Almirall, J. High Resolut. Chromatogr., 18 (1995) 625–630.
70   K.G. Furton, J.R. Almirall and J.C. Bruna, J. Forensic Sci., 41 (1996) 12–22.
71   J.R. Almirall, J. Bruna and K.G. Furton, Sci. Justice, 36 (1996) 283–287.
72   Z. Penton, Can. Soc. Forensic Sci. J., 30 (1997) 7–12.
73   O. Suzuki and K. Sato, Chromatographia, 43 (1996) 393–397.
74   Z. Lee, T. Kumazawa, K. Sato, H. Seno, A. Ishii and O. Suzuki, Chromatographia, 47 (1998) 593–596.
75   J.R. Almirall and J. Perr. In: J.R. Almirall, K.G. Furton (Eds.), Analysis and interpretation of fire scene evidence, CRC Press, Boca Raton, FL, 2004, p. 237.
76   Tenax^{TM} TA, Adsorbent resin physical properties, Retrieved December 8, 2006 from http://www.sisweb.com/index/referenc/tenaxtam.htm
77   L.W. Russell, J. Forensic Sci. Soc., 21 (1981) 317–326.
78   R. Borusiewicz and J. Zieba-Palus, J. Forensic Sci., 52 (2007) 70–74.
79   R. Newman. In: N. Nic Daéid (Ed.), Fire investigation, CRC Press, Boca Raton, FL, 2004, pp. 149–150.
80   K.L. Bryce, I.C. Stone and K.E. Daugherty, J. Forensic Sci., 26 (1981) 678–685.
81   J. Alexander, G. Mashak, N. Kapitan and J.A. Siegel, J. Forensic Sci., 32 (1987) 72–86.
82   L. Meal, Anal. Chem., 58 (1986) 834–836.

83   D.C. Mann, J. Forensic Sci., 32 (1987) 606–615.

84   D.C. Mann, J. Forensic Sci., 32 (1987) 616–628.

85   A.T. Barnes, J.A. Dolan, R.J. Kuk and J.A. Siegel, J. Forensic Sci., 49 (2004) 1015–1023.

86   J.T. Watson, Introduction to mass spectrometry,   (3rd ed.). Lippincott-Raven Publishers, Philadelphia, PA, 1997.

87   M.W. Gilbert, J. Forensic Sci., 43 (1998) 871–876.

88   E. Stauffer, Identification and characterization of interfering products in fire debris analysis. Master's thesis, Florida International Univeristy, FL, 2001.

89   W. Bertsch, J. Chromatogr. A, 674 (1994) 329–333.

90   R.O. Keto and P.L. Wineman, Anal. Chem., 63 (1991) 1964–1971.

91   C.J. Lennard, V.T. Rochaix and P. Margot, Sci. Justice, 35 (1995) 19–30.

92   American Society of Crime Laboratory Directors Laboratory Accreditation Board (ASCLD/ LAB), 2005 Manual, ASCLD/LAB, Garner, NC, 2005.

93   Laboratories accredited by ASCLD/LAB, Retrieved September 6, 2006 from http://www.ascld-lab. org/legacy/aslablegacylaboratories.html

94   Certification program overview, Retrieved September 19, 2006 from http://www.criminalistics. com/cert_ovw.cfm

95   TWGFEX Fire/Arson Training Committee, Training guidelines for the fire debris analyst. The Technical Working Group for Fires and Explosions (TWGFEX), Orlando, FL, 2000.

96   AAN Notes, Arson Anal. Newsl., 6 (1982) 57–59.

97   About ASTM International, Retrieved September 7, 2006 from http://www.astm.org/cgi-bin/SoftCart. exe/ABOUT/aboutASTM.html?L + mystore + ylcj6660 + 1157673833

98   A.D. Pert, M.G. Baron and J.W. Birkett, J. Forensic Sci., 51 (2006) 1033–1049.

99   D.A. Sutherland, Can. Soc. Forensic Sci., 30 (1997) 185–199.

100   B.J. deVos, M. Froneman and D.A. Sutherland, J. Forensic Sci., 47 (2002) 736–756.

101   J.A. Dolan and C.J. Ritacco, Gasoline comparisons by gas chromatography-mass spectrometry utilizing an automated approach to data analysis. Proceedings of the American Academy of Forensic Sciences Annual Meeting, Atlanta, GA, February 11–16, 2002; American Academy of Forensic Sciences (AAFS), Colorado Springs, CO, 2002.

102   G.S. Frysinger and R.B. Gaines, J. Forensic Sci., 47 (2002) 471–482.

103   J.T. Jasper, J.S. Edwards, L.C. Ford and R.A. Corry, Fire Arson Investigator, 51(2) (2002) 30–34.

104   R.P. Rodgers, E.N. Blumer, M.A. Freitas and A.G. Marshall, J. Forensic Sci., 46 (2001) 268–279.

105   R. Coulombe, J. Forensic Sci., 47 (2002) 195–201.

106   E. Stauffer, J. Forensic Sci., 50 (2005) 1091–1100.

107   E. Stauffer, J. Forensic Sci., 51 (2006) 1016–1032.

108   R.T. Newman and K. Lothridge, J. Forensic Sci., 40 (1995) 561–564.

109   D.J. Tranthim-Fryer and J.D. DeHaan, Sci. Justice, 37 (1997) 39–46.

110   M.E. Kurz, S. Schultz, J. Griffith, K. Broadus, J. Sparks, G. Dabdoub and J. Brock, J. Forensic Sci., 41 (1996) 868–873.

111   B. Eckenrode, Field Anal. Chem. Tech., 2 (1998) 3–20.

M.J. Bogusz (Ed.). Forensic Science
Handbook of Analytical Separations, Vol. 6
© 2008 Elsevier B.V. All rights reserved.

CHAPTER 27

Writing media and documents

Valery N. Aginsky

Riley, Welch & Aginsky, Forensic Documents Examinations, Inc., PO Box 80225, Lansing, MI 48908, USA

27.1 INTRODUCTION

This chapter focuses on the chromatographic procedures applied to the analysis and dating of ink on documents. It discusses most recent techniques covered in the scientific literature since the first edition of the Handbook of Analytical Separation was published in 2000 [1]. Some important older techniques and review papers will also appear in the reference list.

27.2 THIN-LAYER CHROMATOGRAPHY

Thin-layer chromatography (TLC) remains one of the principal analytical methods employed by most laboratories in forensic ink analysis. The reason for this is that this method is comparatively simple, rapid, and cost efficient. In most cases, this method is used to visually evaluate the qualitative and semi-quantitative composition of ink dye components separated on the TLC plate.

A review of Pagano *et al.* [2] considers TLC procedures used for the analysis of color jet inks, writing inks, and color toners. The sample preparation for the TLC analysis of ink is discussed in detail. It is recommended that color jet ink should be removed from the uppermost layer of paper using a scalpel. This method increases the ink-to-paper ratio and reduces potential contamination from ink on the reverse of the document. Samples must contain all of the imaging colors appearing on the document (e.g. YMCK), preferably in equal proportions. A sample of the ink analyzed is extracted with a solution of ethanol–water (1:1), applied on a TLC silica gel plate, and the plate is developed using ethyl acetate–ethanol–water (70:35:30), butyl acetate–butanol–water–acetic acid (10:41:17:32), or another mobile phase that gives adequate separation of ink colored components. For writing inks, approximately 1 cm of an ink line is removed from the document using a scalpel, blunted hypodermic needle, or sampling device (commercially available forensic document sampling punch), transferred to a vial, and the ink is extracted using either ethanol–water (1:1) for water-based and some solvent-based inks, or pyridine for ballpoint and most solvent-based inks. The extracts obtained are applied on a TLC silica gel plate, and the plate is developed using ethyl acetate–ethanol–water (70:35:30), butanol–ethanol–water (41:35:32), cyclohexane–chlorobenzene–ethanol (10:2:1), or another mobile phase that gives adequate separation of ink colored components. To remove a toner sample from the document, a clean scanning electron microscope aluminum stub is placed over the sampling area and heated with a soldering iron. The heat causes some of the toner to be transferred to the stub. Since this method removes only the toner, paper interactions are avoided. As with jet inks, all of the process colors should be sampled from the document in equal proportions. The toner is washed from the stub with chloroform into a sampling vial. The toner wash is applied on a TLC silica gel plate, and the plate is developed using ethyl acetate–ethanol–water (70:35:30), ethyl acetate–ethanol–chlorobenzene (1:1:5), or another mobile phase that gives adequate separation of ink colored components. The next step of the procedure is a comparison of the chromatogram obtained for the questioned jet ink, toner or writing ink with the chromatograms within the jet ink, toner or writing ink libraries, respectively. A list of possible matches is recorded, and the corresponding library specimens are sampled and analyzed, along with another sample taken from the questioned ink, using the same TLC procedure. At this stage of the analysis, a new TLC plate is used along with a solvent system (mobile phase) different from the one used for developing the first TLC plate. The match is confirmed by interpreting the TLC plate under both UV and visible light. These authors conclude that TLC is not an identification method unless used for comparison with a complete collection (library) of all inks manufactured throughout the world. To maintain such a library, it is necessary to obtain ink samples from all manufactures on a regular basis since deficiencies in a comparative library will weaken the forensic conclusion and increase the number of nonmatches. They emphasize that a "match becomes an identity only if the match is known to be unique and/or the library is complete."

LaPorte and Ramotowski [3] used TLC to examine the effects of a latent print development technique (ninhydrin, physical developer, and a bleach enhancer) on the chemical examination of documents produced from copiers and laser and inkjet printers. Three 5 mm hole punch specimens were removed from each of the printed samples both prior to and after these printed samples were treated for latent prints. The inkjet samples were extracted with an ethanol–water solution (1:1) and the toner samples were extracted with chloroform. The TLC plates were Whatman polyester plates coated with silica gel, and the solvent system utilized was ethyl acetate–ethanol–water (70:35:30). The results obtained show that latent print processing did not cause any significant changes to the TLC elution profile: all the bands of the colored components of the toners and inkjet printer inks present in the untreated samples were present, at the same R_f, in the processed samples treated for latent prints.

LaPorte *et al.* [4] used TLC for the chemical analysis of printing inks on documents produced with thermal transfer printers. Extractions using pyridine were performed on 81 different samples, which included a total of 54 printer samples

(43 photographic prints on paper and 11 plastic card samples) and 27 printer ribbons, all from various manufacturers. Five millimeter hole punches were removed from the paper samples, small cuttings measuring approximately 5×5 mm were cut from the ribbons, and scrapings, using a scalpel, were removed from the plastic cards to minimize the amount of plastic removed. The samples were dissolved in pyridine and applied on silica gel 60 precoated EM Science® glass plates. The plates were placed in an oven ($100°C$) for approximately 5 min. Once removed, the plates were immediately developed in hexane–methyl ethyl ketone–ethyl acetate (80:3:17) and then digitally imaged using the Foster and Freeman® Video Spectral Comparator 2000 High Resolution (VSC 2000 HR). By utilizing this procedure, 43 photographic-like prints from 21 different manufacturers were differentiated into 19 categories.

Thakur *et al.* [5] analyzed 37 samples of photocopy toners belonging to different batches of various brands using TLC. Samples of the toners were dissolved in acetone and applied on precoated silica gel TLC plates. The plates were developed in butanol–ethanol–water (90:15:10) and then examined in daylight and under UV light for the location of the spots of the toners' dye components. The same plates were developed in carbon tetrachloride–ethyl acetate (6:3) or dichloroethane–acetone (12:2.4), then sprayed with 3% $KMnO_4$ in concentrated sulfuric acid for the location of the spots of the toners' resin components. Most of the samples analyzed, including the toners of the same formulations but of different manufacturing batches, were discriminated using this TLC procedure.

Wilson *et al.* [6] used TLC for differentiating dye-based black gel inks. The samples to be analyzed were prepared by taking 1.3 mm punch holes from written ink lines. The samples were extracted with ethanol–water (1:1), applied on silica gel 60 precoated EM Science glass plates, and the plates were developed in ethyl acetate–ethanol–water (75:35:30). The VSC 2000 HR was used to examine and document the TLC plates under visible, UV, and infrared reflectance modes. Using this procedure, the nine dye-based black gel inks were separated into eight groups.

Aginsky [7] showed with examples that though TLC is an excellent method for discriminating between similarly colored inks that have been made using different dyes, this method has only limited capabilities for the identification of ink formula and for the determination of whether two or more entries were written with the ink of the same formula or batch. The main reason for this is that TLC allows an examiner to obtain only some partial information about the composition of ink, mainly, information regarding ink dye components, which, as known, constitute usually a smaller part of ink composition. Thus, for ballpoint inks, colorants (ink dyes) never represent more than a quarter of the ink composition. The remaining part of an ink composition consists mainly of noncolored viscous liquid and solid substances, such as solvents, resins, modifiers, by-products (impurities and micro impurities), etc., which constitute more than 75% of the mass of ink contained in a ballpoint pen cartridge and more than 50% of the mass of the "old" ink that has dried on paper. Fig. 27.1 shows the resulting thin-layer chromatogram obtained for nine red ballpoint inks, the dye components of which were separated on the high performance TLC silica gel 60-F_{254} (10×10 cm) precoated glass plates (Merck, Germany) using ethyl acetate–isopropanol–water–acetic acid (30:15:10:1) as an eluent.

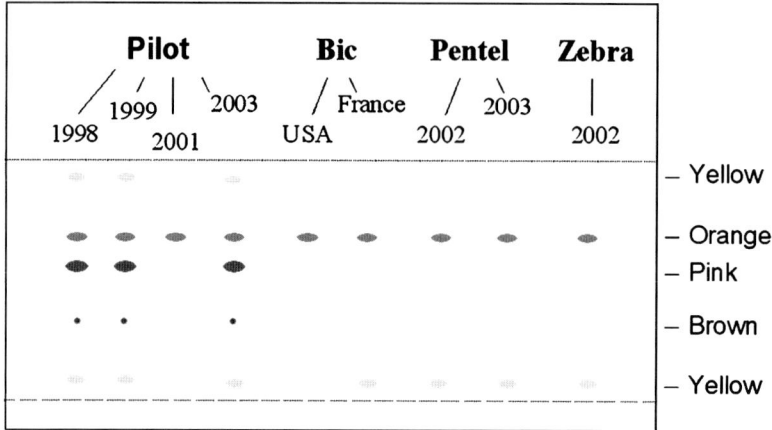

Fig. 27.1. The thin-layer chromatogram obtained for the nine red ballpoint inks analyzed shows that there are two groups of inks for which the brand cannot be discriminated by TLC. The amount of dye components in the analyzed inks varies from one orange dye in the Bic (USA) and Pilot-2001 up to five dye components of orange, pink, yellow (two dyes), and brown (minor component) colors in three Pilot inks manufactured in 1998, 1999, and 2003 respectively. (Reproduced from Ref. [7] with permission of the Journal of the American Society of Questioned Document Examiners.)

The chromatogram in Fig. 27.1 shows that the amount of dye components in the analyzed inks varies from one orange dye in the Bic (USA) and Pilot-2001 up to five dye components of orange, pink, yellow (two dyes), and brown (minor component) colors in three Pilot inks manufactured in 1998, 1999, and 2003, respectively. This chromatogram clearly illustrates the difficulties that any examiner can face when using TLC for ink comparison and formula identification, namely:

- The same ink manufacturer can use different dye mixtures for producing different ink formulations of similar color or different batches of the same ink formulation. For example, the Pilot red ballpoint ink manufactured in 2001 contains only one dye, while each of the Pilot red ballpoint inks manufactured in 1998, 1999, and 2003 contains a mixture of the same five dye components.
- Different ink manufacturers using exactly the same dyes may produce different ink formulations used in pens with different brand names. Among the nine inks analyzed, the brand of two groups of inks could not be discriminated by TLC. Given that, within each group, the inks contain the same dyes: one group comprises the Bic (USA) and Pilot-2001 inks that contain the same orange dye, and the other group includes the Pentel-2002, Pentel-2003, Bic (France), and Zebra-2002 inks that contain the same two dyes of orange and yellow colors.

It is a well-known fact that there are a small number of commercially available dyes that are often used by different ink manufacturers. For this reason, compared inks of different formulations that contain the same dyes may easily be misidentified as the inks of the same formulation. In this connection, it becomes evident that when an examiner tries to determine whether two or more entries were written with the ink of the same formula or identify a questioned ink formulation (and determine the year when this formulation first came into existence), the chemical analysis should not be restricted by analysis of ink dye components using TLC. Coupling TLC with gas chromatography-mass spectrometry (GC-MS) allows one to obtain significantly more information about the composition of the ink analyzed (see Section 27.3).

27.3 GAS CHROMATOGRAPHY-MASS SPECTROMETRY

GC-MS, being a well-proven analytical method of determining the qualitative composition of multi-component systems, is effective for the analysis of inks most of which are complex mixtures of organic chemical compounds.

Vu [8] used solid phase microextraction (SPME)/GC-MS to characterize the volatile components associated with U.S. currency, U.S. currency inks, and Canadian currency. A 1 cm, 50/30 µm divinylbenzene/carboxen/polydimethylsiloxane (PDMS) SPME fiber (Supelco) was chosen for all currency analyses. Data were acquired on a Hewlett Packard 5980/5970 GC/MSD using a 30 m × 0.25 mm × 0.25 µm Restek XTI-5 column and helium as the carrier gas. Headspace components were identified by comparing their mass spectra with those found in reference libraries (Wiley, 6th ed.; Pfleger, 2nd ed.; and NIST, 1998 ed.). Additionally, selected components were identified by both their retention times and mass spectra as generated from running standards, made from reagent grade chemicals. Over 100 compounds were identified in the headspace of currency-related samples. Compounds that can be attributed to the ink curing process include series of straight-chain aldehydes, alkenals, acids, alcohols, and ketones and a series of lactones and 2-alkyl furans. Solvent compounds include naphthenic and paraffinic hydrocarbons with a profile typical of petroleum products, alkyl cyclohexanes, various ethylene glycol alkyl ethers, and traces of chlorinated solvents. A comparison of the headspace profiles of Canadian and U.S. currency showed substantial similarities in the ink curing products region, indicating that it is these compounds that are important to canine detection of currency. Because it is still unknown which of the ink curing compounds or class of compounds plays a key role in canine detection of currency, it was suggested that currency-related training aids should have headspace profiles (in the ink curing region) closely matching that of authentic currency. The results of this study showed that authentic currency remains the best available training aid, provided that the notes are frequently exchanged so that contamination from handling is random and the only consistent odor picture is that arising from the inks themselves.

LaPorte *et al.* [9] used GC-MS for characterizing the vehicles (solvents and resins) found in ballpoint inks. The authors chose to investigate a single volatile compound, 2-phenoxyethanol (PE), that had been reported by the industry to be in many formulations of inks. They analyzed 633 ballpoint inks and identified PE in 85% (237 of 279) and 83% (293 of 354) black and blue ballpoint inks, respectively. They conclude that the identification of PE in over 80% of black and blue ballpoint ink formulations has shown that studies investigating PE as it relates to the aging of writing inks have been and continue to be significant.

References pp. 940–941

Wilson *et al.* [6] used GC-MS to differentiate pigment-based black gel inks and also to examine differences between gel and roller-ball inks by examining volatile organic compounds (VOCs) present in the ink samples. The inks on paper were sampled fresh approximately 1 h, 1 week, and 1-month-old. The samples were extracted with ethanol, and the ethanol extracts were analyzed on a Perkin Elmer Autosystem XL Gas Chromatograph-Turbomass Mass Spectrometer using an HP-5 column. Of the 15 analyzed pigment-based black gel inks, 10 contained glycerin. The glycerin peak was very abundant in the fresh sample, as well as in the week-old sample, and was detectable in samples aged 6 months. After differentiation by detection of a glycerin peak, other major peaks were used to separate the samples. These peaks corresponded to triethylene glycol, pentaethylene glycol, and triethanolamine. The five samples that did not contain any glycerin were separated into four groups based on the presence or absence of triethylene glycol and triethanolamine. The ten samples that contained glycerin were separated into three groups using pentaethylene glycol and triethanola-mine peaks. The time that these peaks remained in the sample was examined and determined to be less than that of glycerin. Triethanolamine and triethylene glycol were detected in some quantity in samples that were 3 months old. However, penta-ethylene glycol dissipated after approximately 2 weeks. Numerous smaller peaks were detected on the total ion chromatograms and may be used to differentiate the pig-mented inks further; however, the small peaks were usually detected only in fresh samples and could not be detected a few days later. These peaks would not be of interest in most casework since it is unlikely that documents are received for analysis in this time frame. Besides the gel inks, 20 roller-ball inks were analyzed using GC-MS to determine the volatile compounds in the inks. Several major peaks were detected in the roller-ball samples that were similar to those of the gel inks. It was found that the presence of hexagol, heptaethylene glycol, and pentaethylene glycol most likely in-dicated a gel pen as these compounds were not found in the roller-ball samples. Three major solvents detected and identified in the roller-ball samples were as follows: 2-pyrrolidinone, 2,2-oxybis-ethanol, and triethanolamine.

Bügler *et al.* [10] used thermal desorption followed by GC-MS for the charac-terization of binder resins, solvents, additives, and colorant compounds from ball-point pen ink samples on paper. They analyzed handwritten entries made by 121 ballpoint pen ink of different formulations and determined that a number of com-ponents of binder resins (characteristic monomers and oligomers), as well as solvents and additives used in the manufacturing of ink, can be detected and characterized in old ink samples on paper. Thermal desorption of ink samples on paper was per-formed in a thermal desorption unit TDS2 with auto-sampler (Gerstel, Mülheim, Germany). This unit was connected to a KAS2 injector (Gerstel) with glass wool liner equipped with a liquid nitrogen cooling unit. The injector is part of a GC-MS Agilent 6840N/5973N. Samples were placed in preconditioned glass thermal des-orption tubes (Gerstel). Tube and glass wool conditioning were performed by purg-ing with argon (100 mL/min) in a tube conditioner (Gerstel) at 250°C for 30 min. A sample of ballpoint pen ink on paper was cut out (length: 5–30 mm; width: 1–2 mm) and placed in a thermal desorption tube. The tube was heated in the TDS2 up to the desired temperature (at 12°C/min). Analytes were collected for 5 min by a flow of helium (40 mL/min) and trapped in the cooled injector at $-100°C$. Subsequently, the injector was heated (280°C, 60°C/min), the compounds were transferred onto the column and analyzed in the GC-MS (column: HP5-MS, 30 m $\times$ 0.25 mm $\times$ 0.25 µm; carrier gas He, constant flow: 1.2 mL/min; oven program: 45°C, 1 min; 45–100°C at 30°C/min; 100–190°C at 12°C/min; 190–270°C at 50°C/min; 270°C, 3 min). Integration of the PE signal was performed on the extracted ion chromatogram ($m/z = 138.10$). For the unambiguous assignment of characteristic components to each binder polymer, thermal desorption analysis of 24 pure binder polymers as used by ink manufacturers was performed and a routine data analysis procedure for the identification of the main binder resin(s) in a ballpoint pen ink formulation was established. Seventy out of the 121 analyzed ballpoint inks showed signals for acetophenone–formaldehyde resin only, 19 contained cyclohexanoneformaldehyde resin only, 11 had alkyd resins, 4 had phenolformaldehyde resin, 9 had combinations thereof, and 8 showed no signals attributable to a pure polymer. The 121 pens were further grouped in five solvent classes, depending on the type(s) of solvents found in the ink samples on paper. Together with the information about ink solvents and binder resins, the presence of certain additives was established in many ink samples. These additives are characteristic for an ink sample and can be used for further identification of inks on paper. Typical additives include softening agents, such as phthalates; corrosion inhibitors, such as organophosphates; antioxidants, such as butylated phenols; and surfactants, such as alkylamines. Additionally, in many ink compositions compounds originating from the colorants were found. Diphenylamine, for example, points to the presence of the yellow organic azo dye Tropaeoline G (Metanil Yellow), being a major dye in many black ink compositions. Many black ballpoint pen inks also contain nigrosines, the main structural unit of which is phenazine. Thus, the presence of phenazine in the chromatogram indicates that nigrosine may be present in the ink. In order to highlight the practical value of the proposed method, the authors analyzed a widely used group of 17 black ballpoint pen inks containing Metanil Yellow and Methyl Violet as colorants. The pens in this group (their market introduction ranged from 1991 to 2001) could not be differentiated by analysis of their colorants using TLC or HPLC. The analysis of binder resins and solvents by thermal desorption and GC-MS resulted in 13 subgroups for the 17 pens, thereby increasing the probability for finding an anachronism in the dating of questioned ink entries with the aforementioned colorant composition. Based on the results of this study, it was concluded that the conventional arrangement of ballpoint pen inks into colorant groups could further be refined by defining groups with respect to binder resins and solvents. These new groups enable more comprehensive discrimination of ballpoint pen inks being indistinguishable by common colorant analysis.

Aginsky [11] reported that GC-MS is capable of detecting and identifying PE (and other high boiling solvents used for manufacturing ballpoint inks) even in many-year-old ballpoint ink writings. Also, he showed [12] that GC-MS allows identification of many ink's volatile solid ingredients, such as nonreacted low molecular monomers or oligomers, reagents, by-products, and various proprietary additives. These ingredients, at temperatures typically used in the injector of a gas chromatograph (average 250°C), have structural stability in the vapor phase and have vapor pressure that is sufficient for their GC-MS analysis. Finally, Aginsky [7] showed that the GC-MS analysis of noncolored ink components generally provides more information on individualizing ballpoint ink composition than the TLC analysis of the ink's dye components. As a typical example, he compared nine red ballpoint inks produced either by different ink manufacturers (different ink formulations) or by the same manufacturer but in different years (different manufacturing batches of the same formulation). The inks sampled from paper were extracted with chloroform, and the extracts were analyzed by GC-MS (Agilent 6850 gas chromatograph interfaced with an Agilent 5973N mass selective detector; column: HP5-MS). All the inks analyzed, including all those that could not be discriminated by TLC, were reliably differentiated by GC-MS (see Fig. 27.2 that shows GC-MS chromatograms of two red ballpoint inks that cannot be discriminated by TLC).

As seen from Fig. 27.2, the Pilot-1998 and Pilot-1999 inks contain the same liquid components: the residue of benzyl alcohol and PE. However, two different solid components were detected and identified in these inks: the Pilot-1998 ink contains diphenoxyethane and the Pilot-1999 ink contains *N*-butyl-benzenesulfonamide. The data obtained indicates that these two inks represent either two manufacturing batches of the same ink formulation or two different formulations in which the same dyes and solvents were used (compare Figs. 27.1 and 27.2 above). This example shows that GC-MS provides analysts with valuable information about ink composition. This information, being coupled with that obtained by common colorant analysis, is very helpful both for ink comparison and for ink formula identification. Thus, for example, GC-MS, coupled with TLC, can determine that two inks being compared have the same unique composition that characterizes these inks as belonging to the same manufacturing batch. If an ink contains a sufficient number of chromatographically separable organic ingredients, they, in the aggregate, may form a unique combination – a complex mixture of colored and noncolored organic substances present in the ink in a certain, measurable relative proportion. If this "chemical fingerprint" is complex enough, then it is practically improbable that it can be coincidentally duplicated either by an ink of a different formulation or by the ink of the same formulation but of a different manufacturing batch. If one were to assume what is very unlikely to happen in practice, namely, that another batch of ink was manufactured using the very same major and minor ingredients, even then the "chemical fingerprints" of the inks representing these two expectedly identical batches should be *a priori* different. This takes place because no technology, at least in chemical industry, is ideal. As acknowledged by ink manufacturers, variations between batches should produce measurable differences. These differences between batches occur due to inevitable variations of various parameters of the technological process. Among these variations are: an amount of each ingredient added to the reaction mixture; the chemical purity of these ingredients (the presence of by-products and contaminants); the chemical composition of ink resins (including nonreacted chemical substances used for synthesizing the resin, molecular mass distributions of ink resins' components, and a composition of primer and minor proprietary additives to ink resins); and fluctuations of the temperature and the duration of various processes in the ink manufacturing. Depending on the

Abundance

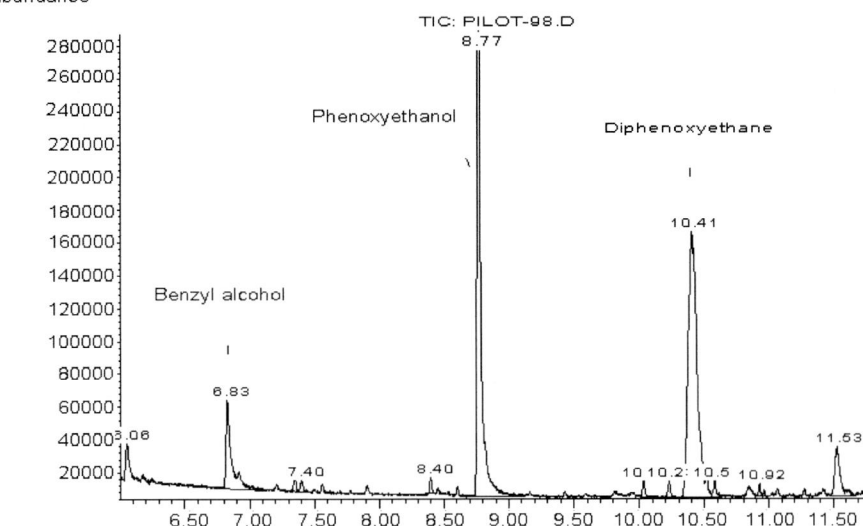

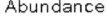
Time-->

Abundance

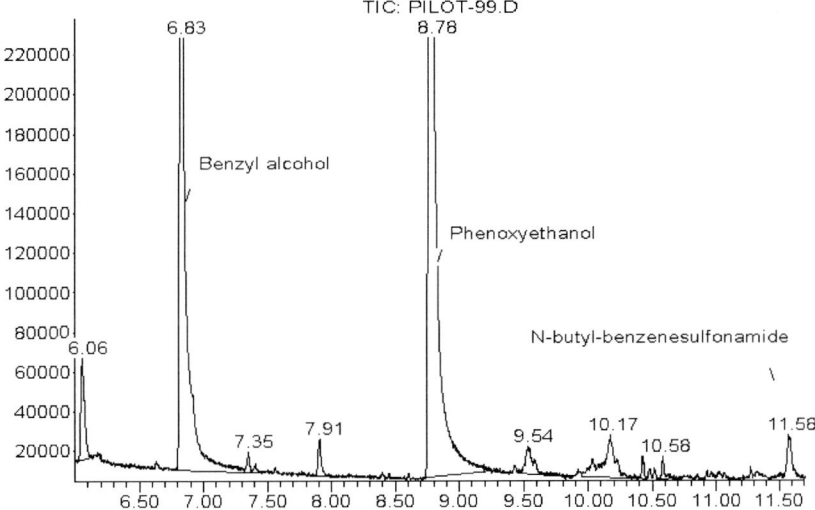

Time-->

Fig. 27.2. Total ion chromatograms of the Pilot red ballpoint inks manufactured in 1998 (upper chromatogram) and 1999 (lower chromatogram). (Reproduced from Ref. [7] with permission of the Journal of the American Society of Questioned Document Examiners.)

References pp. 940–941

composition of ink, batch variations are more or less pronounced to be detected chromatographically. There is no universal rule that could be applicable to any case, in which an examiner would need to determine whether the similarity found between the inks being compared is enough to come to a scientifically sound conclusion that these inks belong to the same manufacturing batch. In some cases, a key feature can be a unique component present in the inks being compared. In other cases, a complex composition of the ink ("chemical fingerprint") can show that such a composition is so unique that its duplication is practically impossible. There are inks, however, whose compositions are not complex and far from being unique. If this is the case, then it may be impossible not only to determine that the compared inks (matching at the applied level of analysis) came from the same manufacturing batch, but also to determine if these inks are of the same formulation.

27.4 HIGH PERFORMANCE LIQUID CHROMATOGRAPHY

Andrasko used the high performance liquid chromatography (HPLC) analysis with diode array detection to monitor changes in the chemical composition of ballpoint pen inks after exposure to light and under normal aging conditions [13]. He studied blue ballpoint pen inks (different batches manufactured by Ballograf Bic AB, Sweden) placed on ordinary writing paper. Changes in dye composition have been evaluated quantitatively using a Hewlett Packard series II Liquid Chromatograph connected to the HP's HPLC3D ChemStation. The instrument was equipped with an auto-sampler, an auto-injector, and a diode array detector from Hewlett Packard. HPLC separations were performed by using a 25 cm TSKgel ODS-120 T (4.6 mm ID, Tosohaas Bioseparation Specialists) stainless steel column. The mobile phase consisted of two solvents. Solvent A was a mixture of 20% acetonitrile and 80% water containing 10 mM KClO4, pH adjusted to approximately 3 with hydrochloric acid. Solvent B was 100% acetonitrile. The gradient was linear from solvent A to solvent B in 20 min at a flow rate of 1.0 mL/min and operated at room temperature. The diode array detector was used to acquire full spectra on all significant peaks from 190 to 600 nm. Both Crystal Violet and Methyl Violet were present in every blue ballpoint ink analyzed. It was found that when a text written by a ballpoint pen is exposed to daylight, Crystal Violet, Methyl Violet, and other dyes decompose by successively losing their methyl groups, which are substituted by hydrogen atoms. Crystal Violet thus decomposes into Methyl Violet, which subsequently decomposes into other, structurally similar compounds, by successive loss of methyl groups. Considerable quantitative changes in composition of the ink were detected already after several hours' storage at indoor daylight (the ink was not exposed to sunshine). For comparison, no changes in composition were noted when the same ink was stored in darkness for 3 weeks. The author concludes that it is not always straightforward to state that two ink entries on different documents are different. If the differences between the ink entries (on different documents or even on the same document, if the entries appear on the front and back) are not of qualitative character, storage of the documents containing the same ink at different light conditions might have caused the observed differences. Both optical properties (color shade, IR luminescence) and chemical composition will change, and the changes occur already after rather short exposure to daylight (a few hours at daylight). The presence of various amounts of decomposition products indicates that the documents may have a different history of storage.

Kher *et al.* [14] studied the feasibility of the coupling of HPLC and the principal components analysis (PCA) for the forensic characterization of ballpoint pen inks. Inks from seven black and eight blue ballpoint pens were separated by a HPLC method utilizing a photodiode array detection. A classifier flowchart was designed for the chromatographic data based on the presence or absence of certain peaks at different wavelengths to qualitatively discriminate between the inks. The same data were quantitatively classified by PCA to estimate the separation between a pair of classes of ink samples. It was found that the black ballpoint pen inks were discriminated satisfactorily utilizing two-dimensional data of the peak areas and retention times at the optimum wavelengths. The blue pens (seven out of eight) were discriminated by analyzing the chromatographic data at four different wavelengths simultaneously with a cross-validated PCA.

Andrasko and Kunicki [15] used HPLC to study the inhomogeneity and aging of ballpoint pen inks inside of cartridges, particularly near the tip of a pen. No aging of dyes in the ink bulk material was discovered when the results of analyses of fresh ink entries produced by the same pen at widely separated occasions (more than 6 years apart) were compared. Inhomogeneities in terms of the ratio between volatile solvents, such as PE, and dyes were observed in many of the pens examined. Inhomogeneities in terms of the composition of the dye mixture were not observed in pens frequently used for writing. When pen cartridges not used for a considerable period of time (in many cases for more than 10 years) were investigated, aging of ink at the tip of the cartridge was detected. Two processes were investigated – the evaporation of PE from ink and changes in the composition of cationic dyes. When starting the writing with these pens, the concentration of PE and Crystal Violet was sometimes considerably lower than that in bulk material. For almost all the pens investigated, this aging and evaporation process was only observed for approximately three first written asterisks (each asterisk corresponds to a line of approximately 1 cm). The consequence of this finding for ink examination is that when comparing inks in, for example, two signatures, taking samples from the beginning of the writing should be avoided and several ink entries from different positions should be analyzed. This is also important for relative ink dating, when two ink entries, written actually simultaneously, might be found different in the way normally caused by aging.

27.5 CAPILLARY ELECTROPHORESIS

Vogt *et al.* [16] analyzed blue and black ballpoint pen inks using capillary electrophoresis (CE). Their study was focused on the optimization of the separation of ink extracts from paper material of commercially available inks with respect to resolution and analysis time. During the method development process different buffers, organic modifiers, and surfactants were tested. To determine how much ink should be sampled from paper for analysis, up to 20 plugs with a diameter of 1.5 mm were cut out from a writing line of ink. The plugs were transferred to a micro vial and 25 mL buffer (50 mM borate pH 9.0, 50% acetonitrile, 30 mM SDS) as well as 25 mL acetonitrile was added. After 10 min ultrasonication the supernatant of the extract was pipetted to a fresh microvial and injected directly into the separation unit. Extracts obtained from 20 plugs were scanned for the three dye peaks (Methyl Violet) with the highest peak area. These peaks were used as markers for the assessment of the extracts obtained from less than 20 plugs. Even one plug of the writing line was sufficient to generate all three dye peaks with areas high above the detection limit, so that quantification was still possible. Nevertheless, only with 10 and more extracted plugs (1.5–2.0 cm writing line) electropherograms have been obtained which showed other components in addition to the dyes. Electropherograms of 20 inks of different formulations examined in this study showed patterns that were in most cases distinctly different from each other.

Egan *et al.* [17] proposed two CE methods to differentiate and identify black and blue ballpoint inks and colored inks for ink jet printers. Two separate buffers and separation conditions were utilized for CE analysis of ballpoint pen inks. One buffer was designed to adequately separate cationic dyes (general class of basic dyes), and another buffer was designed to separate anionic dyes (general class of acidic dyes). These buffers were successful in investigating a number of black and blue ballpoint inks and identified specific dye compounds by comparison of electrophoretic mobility and absorbance spectra to standard chemical references. CE separations of colored inks for ink jet printers were performed with the anionic buffer because of the prevalence of anionic dyes encountered in inkjet formulations. Unlike the ballpoint ink samples that were extracted with methanol, CE injections were performed with inkjet cartridge samples diluted in the CE analysis buffer. The analysis of ballpoint inks showed that the identification of five ink components was sufficient enough to distinguish among the inks investigated. Most of these ink components were ink dyes such as Crystal Violet, Methyl Violet, Victoria Blue B, Victoria Pure Blue BO, Metanil Yellow, and sulfonated copper phthalocyanines. Some of the ballpoint inks analyzed contained diarylguanidines. These substances are known to be used by the dye industry to form salts with acidic dyes or pigments that would be otherwise water insoluble. Each blue ballpoint ink that was observed to contain diarylguanidine(s) also possessed sulfonated copper phthalocyanines. The copper phthalocyanine sulfonation process leads to multiple sulfonate groups at different ring locations, which explains the observation of multiple peaks corresponding to the phthalocyanine dye components present in some ballpoint inks. Several examined inks contained more than one diarylguanidine compound, which aided in formula differentiation. The amount of different diarylguanidines used in a dye formulation appears to be manufacturer specific. The type and quantity of diarylguanidines in an ink sample provides another aspect of formula differentiation, which could be particularly useful if comparing inks of the same formulations manufactured from different raw dye components.

27.6 CHROMATOGRAPHIC METHODS USED FOR DATING INK ON DOCUMENTS

Earlier researches on ink aging that involved TLC were reviewed in the first edition of the Handbook of Analytical Separation [1]. For the last several years, various approaches that used GC-MS and HPLC for studying the aging of ink on documents have been reported and applied to cases. Most of these studies used GC-MS and were aimed to further develop a dynamic approach to ink dating by analyzing ink volatile components (IVCs) reported by Aginsky [12]. This author described the ink drying process and summarized that, at a certain stage of aging of ink on paper, a process of the diffusion and evaporation of IVCs stops completely, and the remaining IVCs become trapped inside the ink on paper "for a practically infinite period of time" until they are freed by heating or solvent extraction [11]. In addition to the diffusion and evaporation of IVCs from the aging ballpoint ink on paper, another very important age-transforming process is the hardening of ballpoint ink resin. Aginsky [18] reported that ballpoint ink resins harden as a result of some physical and chemical processes that slow down gradually to the zero asymptote during a period of time comprising up to 2 years after the ink has been placed on paper. During this period of time, the physical properties of ballpoint ink resins are changing. As far as the dating of ink on documents is concerned, two most important physical properties of ballpoint ink resins that are changing with the ink age (within up to 2 years after an ink being placed on paper) are as follows:

- the solubility of ink resins in so-called weak solvents gradually decreases. As ink resin (binder resin) is a matrix that contains all other ingredients of the ink, the decrease of the extractability (from paper) of this hardening matrix causes a corresponding decrease of the extractability of the other ink ingredients, including ink dye and volatile components;
- the ability of an ink resin to retain IVCs (PE and similar ink solvents) inside the body of the aging ink on paper gradually increases in proportion as this hardening resin is becoming a "tighter" matrix. Therefore, if an ink on paper undergoes heating at a moderately elevated temperature, the effect of the heating should be significantly greater for a recently made writing than for an older one: "fresh" ink should loose a significantly larger part of its solvent (contained in the ink before the ink was heated) than "old" ink.

To evaluate the age of an ink by measuring the ink's ability to "free" (loose) its IVCs when the ink on paper undergoes heating at moderately elevated temperatures, it is necessary to compare the content of IVCs in the ink prior to and after the heating. This approach was introduced by Aginsky in 1996 [12]. He used the following procedure: 20 micro discs (1 mm diameter) of the ink on paper are removed using a boring device and divided on two parts, 10 discs/each (samples 1 and 2). As the technique is not mass invariant, the sampling procedure must minimize sample variation, that is, the difference between the amount of ink contained in samples 1 and 2. To meet this requirement, each pair of samples to be included in set 1 (one sample) and set 2 (the other sample), respectively, must be taken from ink lines that are similar in thickness (degree of pressure), appearance, and arrangement (distribution) of ink, particularly within the area of the ink line from which a pair of samples will be taken. Sample 2 is heated for 1 h at 70°C and sample 1 is used as it is. Then the amounts of PE (or another ink solvent) in sample 1 (P) and sample 2 (P_T) are determined using GC-MS, and the value of $R(\%)$ characterizing the "solvent loss" rate is calculated as follows:

$$R(\%) = \left(\frac{P - P_T}{P} \right) \times 100$$

If the value of $R(\%)$ is larger than 20%, it shows (on condition that the content of the analyzed ink's solvent is not too small) that the natural aging of the ink analyzed is still in progress.

Gaudreau and Brazeau [19] have developed this method further (they have called it a "Solvent Loss Ratio method") and presented their findings on an extensive research with experimental data (obtained for numerous ballpoint inks) on how PE levels change over time following an ink entry placed on paper. They discuss solvent loss process that takes place as ink ages on paper and conclude that the "…phenoxyethanol in [ballpoint] ink evaporates at a high rate during the first six to eight months following its application on paper. The rate of evaporation stabilizes over a period of six to eighteen months. This process is no longer significant after a period of about two years." They determined that the following two "broad time thresholds" can be used for evaluating the actual age of ink on paper: if the value of $R(\%)$ is larger than 50%, then the age of the ink is less than 5 months, and if the value of $R(\%)$ is larger than 25%, then the ink on paper is less than 1 year old.

In 2003, Andrasko offered a method for distinguishing between fresh and old ballpoint pen ink entries [20]. This method is a modification of the solvent loss ratio method. Only one ink sample is removed from the suspect document. Using SPME followed by GC-MS analysis a concentration of PE in vapor phase in equilibrium with the ink sample is determined. Afterwards, the bulk concentration of this compound in the same ink sample is measured by solvent extraction, using methanol. The ratio between the concentration of PE in headspace (vapor phase) to that in the whole ink was found to decrease with age of the ink. This method does not need to determine either the ink formulations or the amount of ink used for the analysis. The method is used to reveal if a suspect ink entry is fresh – up to approximately 4–6 months old. It cannot distinguish between older ink entries, for example, between 1- and 5-year old entries. Best results are achieved with inks containing large amounts of PE. A serious problem encountered is a possibility of contamination of the suspect document by close contact with a fresh writing. The analysis of paper background near the suspect text must be carried out to check this source of error. Also airing of the questioned text for several days may reveal such a contamination.

Cochran *et al.* [21] used SPME and gas chromatography (GC)–time-of-flight mass spectrometry (TOFMS) to characterize ballpoint ink for VOCs and determine those remaining in ballpoint ink from aged writings on paper. Five millimeter plugs of ink on paper were placed into 4 mL vials that were capped with polytetrafluoroethylene (PTFE)-silicone septa for SPME sampling. Five plugs were used for the characterization experiment and eight were used for the ink aging work. The vial was sampled by headspace SPME either at 22°C (characterization) or 70°C (aging). For the aging work, the vial was equilibrated at 70°C for 10 min prior to SPME. A 10 min SPME sampling time was used to collect ink volatiles. The fiber (Supelco) was desorbed into a 0.75 mm injection sleeve (Supelco) in a split/splitless injector at 270°C either at a 20:1 split ratio (characterization) or splitless with a valve time of 60 sec (aging). The GC-TOFMS parameters were as follows: column: 20 m × 0.25 mm × 0.71 μm Rtx-TNT (Restek); carrier: helium at 2 mL/min, constant flow; oven program: 40°C (1 min), 40°C/min to 340°C; MS: LECO Pegasus TOFMS; ionization: electron ionization at 70 eV; source temperature: 200°C; stored mass range: 35–450 amu; acquisition rate: 10 spectra/sec; and data processing: LECO ChromaTOF® software with automated peak find and spectral deconvolution was used to locate all eluted ballpoint ink VOCs. Based on the results obtained, it was concluded that documents could not be aged past 8 months for using SPME GC-TOFMS and PE concentrations. It was theorized that using a less volatile component, one that might stay in the ink on paper longer than PE, could be used to age ballpoint pen ink documents. It was shown that diphenylamine (a "less volatile" solid compound detected in some inks analyzed) gave similar results to PE and that the ratio of less volatile to more volatile compounds, as analyzed by SPME GC-TOFMS, was no help in determining ballpoint pen ink age. The authors concluded that the SPME GC-TOFMS method needs further investigation as an ink dating technique for samples more than 8 months old.

Lociciro *et al.* [22] modified the evaporating compound-to-stable compound ratio ("solvent-to-solid component ratio") approach reported by Aginsky [12] in which the ratio of the contents of an IVC (PE) and a chosen nonvolatile solid component (detected and identified by GC-MS) is measured as a function of the ink age. They applied this modified method to study the aging of two inks – Bic and Staedtler Sick 400 M (GB) blue ballpoint inks. Some of the basic modifications of the previous technique were as follows: a nonpolar substance (a solid hydrocarbon nonadecane, $C_{19}H_{40}$) was used as an internal standard for the quantitative determination of ballpoint ink solvents, which are all polar chemical compounds; "one cm of the ink strokes were scraped off [from paper] with a scalpel and then cut into smaller pieces" (this sampling procedure unpredictably changes the specific surface of ink samples taken from paper for analysis); PE was extracted from ink samples and transformed into a derivative (for its subsequent quantification by GC-MS) using a mixture of chloroform/pyridine/MSTFA in proportion 5:5:1 (the authors assumed that the yield of the derivatization reaction was 100%); 2 μL of the extracts obtained were injected into a GC (splitless, 220°C, injection port pressure: 12 psi, carrier gas: helium, 1 mL/min); two substances of unknown chemical formulas (the GC-MS failed to identify these substances in the Bic and Staedtler ballpoint inks, respectively) were chosen as the "stable" compounds. These "stable" compounds were quantified and used to calculate the evaporating compound-to-stable compound ratios. No correlation was found between these ratios obtained for the Bic and Staedtler ballpoint inks and the age of the inks. The authors based their research on the premise that if the "ratios of the evaporating compound to a stable compound of ink according to time" are taken, the "results [ratios] obtained are independent of a quantity of ink sample [taken] for analysis." However, Aginsky [23] has recently shown that if two samples taken from an ink on paper have the same mass of the ink (and thus the same mass of any stable component containing in the ink) but significantly different specific surface, then these ink samples should *a priori* contain different amounts (masses) of the ink's volatile components. If one calculates "ratios of the evaporating compound to a stable compound of ink according to time" for two ink samples with substantially different specific surface, then these ratios should be significantly different. This explains why two ink samples that are to be compared (using an ink dating technique) must be taken only from those ink lines that are similar in thickness (degree of pressure), appearance, and arrangement (distribution) of ink, particularly within the area of the ink line from which a pair of samples will be taken.

Wang *et al.* [24] described a method for evaluating the age of blue ballpoint ink on paper that they apply to cases. They determined aging curves for 74 blue ballpoint inks of different formulations by using an ink aging parameter the ratio of the content of IVCs – benzyl alcohol and PE (measured by GC), and the content of ink dyes (measured by UV-visible spectrophotometry). This method is a further development of the "vehicle solvent-to-dye ratio" approach first reported by Bezhanishvili *et al.* [25] and then modified by Aginsky [11].

Bügler *et al.* [26] used thermal desorption and GC-MS to determine the age of ballpoint ink on paper. Their extensive research further developed the solvent loss ratio approach. First they randomly took 230 ballpoint pens from the collection of more than 4500 samples of inks maintained by the Forensic Sciences Institute of the Bavarian State Bureau of Investigation and determined that more than 99% of these inks contain PE as major solvent. Then they studied the "outgassing of PE from ink samples on paper vs. time" and determined that "more than 95% of the initial amount of PE in ballpoint inks is lost during the first 3 days after writing. Thereafter, the amount of PE decreases slightly and steadily and stays constant within the accuracy of the analytical method after a few weeks. This remaining amount of the ink solvent PE is trapped in the matrix ink resin/paper and can be detected in significant quantities even in samples as old as 50 years". Further, the authors describe a procedure for age determination of ballpoint inks on paper that includes two thermal desorption steps. First, a sample of ballpoint ink is cut out of the questioned document and is directly heated in the thermal desorption unit at a moderately elevated temperature (e.g. 70°C). The amount of outgassing PE is quantified using the GC-MS. Then, the same sample is further heated at a high temperature (e.g. 200°C) and the total amount of PE in the sample is quantified by GC-MS. The sampling procedure and the thermal desorption and GC-MS parameters were the same as described previously [10] (see Section 27.3). Finally, the authors calculated the ratio V of the amount of PE obtained at the moderately elevated temperature *vs.* the total amount of PE obtained in both steps as a direct measure for ballpoint ink age. (The ratio V has the same physical meaning as the ratio $R(\%)$ introduced by Aginsky [12]. Both V and $R(\%)$ show what part (in percent) of PE was lost/evaporated from the ink on paper as a result of the ink being heated at a certain moderately elevated temperature during a certain period of time). The authors applied this procedure to 60 ballpoint inks of different age on paper and they were able to determine the ratio V for 29 of these inks. They determined that the following three threshold values can be used for evaluating the actual age of ink on paper: if the value of V is higher than 20%, then the age of the ink is less than 3 months, and the values of V "above 15% and 10% indicate that the ink is not older than 9 month and 15 month, respectively". If the ratio V is below 10%, then "no conclusion is drawn, and it is stated that the method is not applicable in this case". The authors concluded that their method is independent of sample size, does not depend on the type of paper and the writing pressure, and it is applicable to ballpoint inks with an age of up to 1.5 years. They also noted that as PE migrates from a ballpoint pen stroke into the surrounding paper, and as PE may be absorbed by paper from contacting ballpoint strokes on other sheets in a pile, a paper blank measurement has to be conducted when the method is applied to date ink on document.

Chromatographic methods other than GC-MS have also been used to study ink aging. Aginsky [11] used TLC to study changes in the nonvolatile and colorless ink components (such as ballpoint ink resins) taking place as inks age on paper. He showed that, in some ballpoint inks, these components have different stability with age. As an example, it was shown that a ratio of two resin's components found in a Parker blue ballpoint ink changed as the ink aged on paper, and this ratio leveled off at approximately 3 years after the ink had been placed on paper. This author has also studied the aging of Crystal Violet and Methyl Violet – two dyes widely used for manufacturing blue, violet, and black writing inks [27]. It is a well-known fact that these dyes are not lightfast (being exposed to light), they decompose (fade) at a fast rate forming the products of their successive demethylation. It was determined that when ink on paper is stored in darkness, Crystal Violet, and Methyl Violet mainly undergo oxidative destruction with the formation of the derivatives of diphenyl-methane (Michler's ketone) and phenol. No correlation, however, was found between the "concentration" of Michler's ketone in the inks analyzed and the age of these inks [27,28].

Andrasko [29] used HPLC to study the relative age of ink entries written by the same ballpoint pen on documents stored in darkness. He had prepared ink entries repeatedly (once a month for approximately 2 years), in the form of asterisks, on different types of paper, and stored them in various places and conditions. One of the documents involved in this project was an ordinary notebook containing lined paper. Blue ballpoint inks from several different ink manufacturers were used for writing asterisks in this notebook. Each asterisk corresponded to approximately 1 cm ink in length. The notebook was kept at room temperature and closed all the time. For the analysis, single asterisks were cut out from the paper in the notebook, placed in glass vials and extracted with methanol. During the extraction procedure, the ink and the extract were protected from exposure to intense light. The extracts were analyzed by HPLC (as described in Section 27.4). The results of the analyses showed that changes in the chemical composition of the ballpoint inks on paper aged in darkness were similar to those observed when the inks were exposed to light or heat. These changes included decomposition (successive demethylation) of Crystal Violet and Methyl Violet. For inks investigated in this study, stored in darkness and with age range from fresh to approximately 2 years old, the amount of the decomposition products increased linearly with the age of ink. The author noted that (1) compared with compositional changes on exposure to light, the observed changes due to aging in darkness were much slower and (2) the kind of ink, the thickness of ink layer on paper, and the kind of paper are all parameters that can influence the rate of compositional changes. In his experiments, for ink entries inside a notebook, differences in the composition of the same ink were detectable for ink entries separated at least 3–4 months in time. The author had used this method in casework for the relative dating of ink entries in diaries, notebooks, pads of paper, etc., that is, in documents where several ink entries were written by the same ink.

Hofer [30] used HPLC for the relative dating of the handwritten entries by the quantification of the ballpoint pen ink dyes and their degradation products. Samples of ballpoint pen ink were taken from the entries in such a manner as to build up a chronological database. Each sample was extracted with methanol, and the extracts were analyzed by HPLC (HP 1090, stationary phase: RP-18 CC 150/4.6 Nucleosil 100–5 column, mobile phase: potassium perchloride buffer pH 3.0/methanol). The results of the quantification of the ballpoint pen ink dyes and their degradation products using HPLC show that, if there are enough number of samples written by the same ink and stored at the same environmental conditions, and enough time space between the samples, their relative age can be determined within a relatively short time scale.

27.7 REFERENCES

1   V.N. Aginsky. In: Handbook of analytical separations, M.J. Bogusz (Ed.), Forensic science, Vol. 2. Elsevier, Amsterdam, 2000, p. 679.
2   L.W. Pagano, M.J. Surrency, A.A. Cantu. In: I.D. Wilson (Ed.), Encyclopedia of separation science, Academic Press, New York, 2000, p. 3101.
3   G.M. LaPorte and R.S. Ramotowski, J. Forensic Sci., 48 (2003) 658.
4   G.M. LaPorte, J.D. Wilson, S.A. Mancke, J.A. Payne, R.S. Ramotowski and S.L. Fortunato, J. Forensic Sci., 48 (2003) 1163–1171.
5   V. Thakur, O.P. Jasuja and A.K. Singla, J. Forensic Identif., 54 (2004) 53.
6   J.D. Wilson, G.M. LaPorte and A.A. Cantu, J. Forensic Sci., 49 (2004) 364–370.
7   V.N. Aginsky, J. Am. Soc. Questioned Doc. Examiners, 9 (2006) 19.
8   D.T. Vu, J. Forensic Sci., 48 (2003) 754.
9   G.M. LaPorte, J.D. Wilson, A.A. Cantu, S.A. Mancke and S.L. Fortunato, J. Forensic Sci., 49 (2004) 155.
10   J.H. Bügler, H. Buchner and A. Dallmayer, J. Forensic Sci., 50 (2005) 1209–1214.
11   V.N. Aginsky, J. Forensic Sci., 38 (1993) 1134.
12   V.N. Aginsky, Int. J. Forensic Doc. Examiners, 2 (1996) 103.
13   J. Andrasko, J. Forensic Sci., 46 (2001) 21.
14   A.A. Kher, E.V. Green and M.I. Mulholland, J. Forensic Sci., 46 (2001) 878.
15   J. Andrasko and M. Kunicki, J. Forensic Sci., 50 (2005) 542–547.
16   C. Vogt, A. Becker and J. Vogt, J. Forensic Sci., 44 (1999) 819.
17   J.M. Egan, J.D. Brewer, K.A. Hagan and C.L. Strelko, J. Am. Soc. Questioned Doc. Examiners, 9 (2006) 37.

18   V.N. Aginsky, Serving justice through science. Proceedings of the 60th Annual Conference of the American Society of Questioned Document Examiners, Aug 14–18, San Diego, CA, 2002.

19   M. Gaudreau and L. Brazeau, Serving justice through science. Proceedings of the 60th Annual Conference of the American Society of Questioned Document Examiners, Aug 14–18, San Diego, CA, 2002.

20   J. Andrasko, Forensic Sci. Int., 136 (2003) 80.

21   J. Cochran, F. Glisson and G. LaPorte, PittConn, Chicago, IL, March 8–11, 2004.

22   S. Lociciro, W. Mazzella, L. Dujourdy, E. Lock and P. Margot, Sci. Justice, 44 (2004) 165.

23   V.N. Aginsky, Int. J. Forensic Doc. Examiners, 4 (1998) 214.

24   Y. Wang, L. Yao, P. Zhao, J. Wang and Y. Wang, Chin. J. Chromatogr., 23 (2005) 202.

25   G.S. Bezhanishvily, E.A. Trosman, P.B. Dallakian and G.P. Voskerchian, Ballpoint ink age - a new approach. Proceedings of the 12th International Forensic Scientists Symposium, Adelaide, Australia, Oct. 15–19, 1990.

26   J.H. Bügler, H. Buchner and A. Dallmayer, Towards a measurable science. Proceedings of the 63rd Annual Conference of the American Society of Questioned Document Examiners, Montreal, Quebec, Canada, Aug., 11–16, 2005.

27   V.N. Aginsky. In: B. Jacob, W. Bonte (Eds.), Advances in forensic sciences, Vol. 3. Verlag Dr. Köster, Berlin, 1995, p. 320.

28   V.N. Aginsky, J. Forensic Sci., 40 (1995) 475.

29   J. Andrasko, J. Forensic Sci., 47 (2002) 324.

30   R. Hofer, J. Forensic Sci., 49 (2004) 1353.

PART 5:
FORENSIC IDENTIFICATION OF INDIVIDUALS AND BIOLOGICAL TRACES

M.J. Bogusz (Ed.). Forensic Science
Handbook of Analytical Separations, Vol. 6
945

CHAPTER 28

Forensic DNA typing technologies: a review

Angel Carracedo[1,2], Beatriz Sobrino[2] and María Victoria Lareu[1]

[1]Unidad de Xenética, Instituto de Medicina Legal, Facultad de Medicina, Universidad de Santiago de Compostela, 15782, Santiago de Compostela, Spain
[2]Centro Nacional de Xenotipado (CeGen), Hospital Clínico Universitario, 15706; Galicia, Spain

28.1 INTRODUCTION

DNA profiling as it is now known, was first described in 1985 by Alec Jeffreys and coworkers [1] and it has had a tremendous impact in forensic genetics. Before that, all the forensic genetic casework (paternity testing, criminal casework, individual identification) was performed using classical serological genetic markers. Blood groups, HLA, and polymorphic protein and enzymes were used for solving forensic genetic casework using immunological and electrophoretic methodologies. These genetic markers were nevertheless limited when it was necessary to analyze minimal or degraded material, which is commonly involved in forensic cases. It was, in addition, difficult to analyze biological material other than blood, and therefore the information obtained from hair, saliva, and even semen in rape cases was rather limited.

Due to the fact that the polymorphic proteins and enzymes were infrequent, it was necessary to obtain as much information as possible. For this reason, sophisticated electrophoretic methods such as isoelectric focusing, immobilines, or hybrid isoelectric focusing were developed and applied. In spite of this, the information that the forensic geneticists were able to report in many cases was clearly insufficient.

Since the discovery of polymorphisms in repetitive DNA by Jeffreys et al. [1], highly informative and robust DNA typing systems have been developed which are quite powerful for the individualization of biological material of human origin.

DNA typing has advantages over traditional protein assays, first of all because it is more informative and can be analyzed in minute or degraded material since DNA is physically much more resistant to degradation than proteins. In addition, the same DNA genotype can be obtained from any tissue (i.e. blood, saliva, semen, hair, skin, bones) whereas the analysis of protein markers is restricted to cells where these proteins are expressed.

References pp. 955–957

Nowadays, DNA analysis has become the standard method in forensic genetics as it is currently used by most of the labs for the majority of forensic genetic expertise and especially in criminal forensic casework (stain analysis and hairs) and identification.

From the beginning of the introduction of DNA technology it was realized that DNA databases would provide the criminal justice systems with an efficient way of crime solving, and consequently some local databases were created. It was not until the introduction of the amplification technology linked to the analysis of short tandem repeats (STRs) that a sufficiently sensitive and robust system was available for the formation of efficient and effective DNA databases. Comprehensive legislation enacted in the UK in 1995 enabled forensic scientists to set up the first national DNA database that would hold both personal DNA profiles together with results obtained from crime scenes. Other countries quickly followed but in some the legislation has severely restricted the amount and type of data that can be retained and, therefore, the effectiveness of the databases is limited. The widespread use of commercially produced multiplex kits has produced a situation in which nearly all laboratories around the world are using compatible systems. However, the exchange of results between countries is hampered by the various legislations that currently exist [2]. At not in this moment most of Western countries have national DNA criminal databases with a variety of legislations from very restrictive ones (i.e. France) to more universal (i.e. U.K.) with a variety of requirements for data entry and protection, removal of data, or inclusion or not of DNA samples versus DNA profiles.

28.2 DNA POLYMORPHISMS

Hidden in the $\sim$3 billion base pairs of DNA of the haploid human genome are an estimated number of 30,000 genes [3]. All human genes are encoded in $\sim$5% of the human genome. Thus, the great majority (more than 90%) of the human genome represents "non-coding" parts of the genome as they do not contain genetic information directly relevant for protein synthesis. Genetic variation is rather limited in coding DNA with the exception of the HLA region. This is due to the fact that expressed genes are subjected to selection pressure during evolution to maintain their specific function. In contrast, the non-coding part of the genome is not mainly controlled by selection pressure, and thus mutations in these regions are usually kept and transmitted to the offspring leading to a tremendous increase in genetic variability. Therefore, these regions are very appropriate for forensic genetics because they are very informative, and at the same time not useful for drawing conclusions about the individual other than for identification purposes.

An important percentage of the non-coding DNA (50%) consists of repetitive sequences that can be divided into two classes: tandemly repetitive sequences and interspersed elements. The majority of forensic typing systems in current use are based on genetic loci with tandem repetitive DNA sequences.

Tandemly repeated sequences can be found in satellite DNA, but from the forensic point of view regions of repetitive DNA much shorter than satellite DNA are much more interesting. These regions can be classified into minisatellites [1] and microsatellites or STRs [4,5]. Minisatellites, otherwise known as variable number of tandem repeats (VNTRs) loci [6] are composed of sequence motifs ranging from around 15 to 50 bp in length, reiterated tandemly for a total length of 500 bp–20 kb. STRs are much shorter. The repeat unit ranges from 2 to 6 bp for a total length between 50 bp and 500 bp.

In addition, minisatellites and STRs have differences in their distribution in the human genome and probably in their biological function. Thus, minisatellites are more common in subtelomeric regions, whilst STRs are widely distributed throughout the human genome, occurring with a frequency of 1 locus every 6–10 kb [7]. The origin of the variability seems to be different as well. Whilst unequal crossing over and even gene conversion are involved in the variability of minisatellites, replication slippage is mainly involved in the origin of the variability in microsatellites [8].

The genetic variation between individuals in these minisatellites and STR systems is mainly based on the number of tandemly arranged core repeat elements; however, it is also based on differences in the DNA sequence itself, since the repeats can have slight differences in the sequence.

After the efforts of the human genome project, another source of variation has attracted the interest of forensic geneticists: the single nucleotide polymorphisms or SNPs. These represent the most simple type of polymorphisms (a simple variation in a single base) and they are usually biallelic markers. More than 11 million SNPs have been identified so far and the HapMap project [9] has provided information about the distribution of these SNPs in haplotype blocks in the human genome.

28.3 DNA TYPING METHODS

Technologies used for DNA typing for forensic purposes differ in their ability to differentiate two individuals and in the speed and sensitivity with which results can be obtained. The speed of analysis has dramatically improved for forensic DNA analysis. DNA testing that previously took more than 1 week can now be performed in a few hours.

Southern blotting with single-locus (SLP) DNA probes has been used for paternity testing and forensic stain typing but they are rarely used nowadays except in complicated paternity cases.

However until the introduction of STR analysis by polymerase chain reaction (PCR), minisatellite analysis with single locus probes was very popular in forensic laboratories.

The main advantage of SLP analysis is the enormous variability of some of the minisatellites. The main disadvantages are the time needed for the analysis and especially the need for the relatively large amount of non-degraded DNA required for SLP typing. Since DNA extracted from forensic specimens is often degraded due to environmental conditions, these techniques have often failed to produce reliable results. The PCR has overcome these difficulties and it has strongly enhanced the usefulness of DNA profiling techniques in forensic science.

The PCR is a technique for the *in-vitro* amplification of specific DNA sequences by the simultaneous primer extension of complementary strands of DNA. The PCR method was devised and named by Mullis and colleagues at the Cetus Corporation [10], although the principle had been described in detail by Khorana *et al.* [11] over a decade earlier. The use of PCR was limited until heat-stable DNA polymerases became available. DNA polymerases carry out the synthesis off a complementary strand of DNA in the 5′ to 3′ direction using a single-stranded template, but starting from a double-stranded region. The PCR uses the same principle, but employs two primers, each complementary to opposite strands of the region of DNA that have been denatured by heating. The primers are arranged so that each primer extension reaction directs the synthesis of DNA towards the other. This results in the novo synthesis of the region flanked by the two primers.

Automated temperature cyclers (usually called thermocyclers) allow the exact control of successive steps of denaturation, annealing of the primers, and extension (when the DNA polymerase extends the primer by using a complementary strand as a template). All these three steps constitute a cycle, and a normal PCR reaction consists of 20–25 cycles, allowing the formation of 2^{20-25} molecules from a single molecule of template DNA.

Most PCR-based typing systems allow alleles to be identified as discrete entities, thus avoiding most of the statistical issues that arise in matching and binning SLPs bands and making standardization easier. Also, apart from the increased sensitivity inherent in any PCR technique, it is more likely to be successful in analyzing old or badly degraded material mainly because of the small size of some of the DNA polymorphisms (SNPs and STRs) susceptible to analysis by PCR [12,13].

Once PCR has been used to generate a large number of copies of a DNA segment of interest, different approaches may be taken to detect genetic variation within the segment amplified. Because 10^6 or more copies of the target sequence can be produced, it is possible to use nonisotopic methods of detection. A number of imaginative methods for PCR product detection have been described.

The first one was the use of sequence specific oligonucleotide (SSO) probes [14] to detect variation in HLA-Class II genes, especially in the HLA DQA1 system [15]. The AmpliType PM PCR amplification kit (Perkin-Elmer, Foster City, CA) was very popular in forensic labs some years ago. With this kit, the loci HLA DQA1, LDLR, GYPA, HBGG, D7S8, and GC are amplified in a multiplex fashion. The last five loci listed are typed simultaneously in a single reverse dot-blot strip containing ASO probes; HLADQA1 must be typed in a separate strip.

The efforts of forensic scientists have mainly addressed the amplification of fragment length polymorphisms. The minisatellite D1S80 (pMCT118) was the first one to be applied to the forensic routine [16] but all these polymorphisms have been substituted by STRs. Analysis of STRs by PCR is the method of choice for forensic identification nowadays. Dinucleotide STRs are the most common STRs in the human genome and are the genetic markers most commonly used for linkage analysis, although they are not being used in forensic science. The reason is that analysis of these STRs is affected by strand slippage during amplification, producing artifactual stutter bands [17]. Nevertheless, tetra and pentanucleotide repeats appear to be less prone to slippage and are more suitable for forensic purposes. The percent of stutters is very interesting to identify and select ideal STRs for forensic purposes, since having a low percentage of stutters is critical for the analysis of mix stains. Some tetranucleotide STRs (such as TH01) are known to have a good behavior regarding these characteristics but specially pentanucleotides are ideal systems for analyzing mixtures.

According to its structure, STRs range from the extremely complex STRs to the most simple [18]. Complex STRs have the advantages of hypervariability. Simple STRs have the advantages of easy standardization and low mutation rates. Mutation events are more frequent in the male germ line and the rates of different loci can differ by several orders of magnitude, the structure and length being the most influential factors in the rate [19]. In addition to the characteristics already mentioned, the selection of ideal STRs for forensic purposes include the analysis of other artifactual bands, the robustness, and the size. In general, short sizes are desirable since the size of the amplified product is critical in degraded samples and small fragments can be amplified when larger fragments failed [20].

Another important fact is the possibility of amplifying multiple STR loci in a single multiplex reaction. This, coupled with the direct detection of amplified products to polyacrylamide gels, makes STR DNA profiling amenable to automation. For this reason the ability to be included in multiplexes is another characteristic that should be analyzed for the selection of good STRs for forensic purposes.

STRs were firstly analyzed in manual electrophoretic systems. Denaturing polyacrylamide gels are recommended for standardization purposes, given that with native gels sequence variation can also be detected making the typing prone to errors. STR electrophoretic mobility under native and denaturing conditions should also be checked since some STRs (especially AT rich ones) have been shown to have anomalous mobility in polyacrylamide gels [21].

The introduction of fluorescent-based technology and the use of DNA sequencers have revolutionized the field, allowing the typing of large multiplexes (including up to ten systems) as well as the automation of the typing procedure. Commercially available STR multiplexes for manual electrophoretic systems are available, but the major advantages of the use of sequencers is automation and the possibility of using intelligent systems of interpretation. The use of sequence reference allelic ladders is essential for STR typing. In general, the reference allelic ladders comprise most of the alleles of the system, but intermediate alleles are always possible even in the most simple STRs. Interpretation guidelines have been produced [21,22] to distinguish these intermediate alleles, and can be easily implemented in automatic sequencers. There are many multiplexes commercially available, the most popular being the Poweplex16 (Promega Corp) and the Identifier (Applied Biosystems) that include 15 STRs plus the amelogenine (for gender identification).

In general, the combined discrimination power of STRs is enormous and the probabilities of two unrelated individuals matching by chance (pM) are lower (10^{-15}) for some of these large multiplexes.

The difficulties in analyzing degraded material due to the size of the fragments have opened a new tendency to use miniSTRs [23]. These are usually obtained from classical STRs with new primers designed to produce shorter amplicons. They have demonstrated clear advantage over traditional STRs in typing degraded material.

References pp. 955–957

MiniSTRs have also demonstrated to be very efficient in low copy number cases. These are cases where the amount of DNA is critical (DNA from a few cells). It has been demonstrated [24] that by increasing the PCR amplification regime to 34 cycles, it is possible routinely to analyze 100 pg DNA. However the success rate was not improved (without impairing quality) by increasing cycle number further. Compared to the amplification of 1 ng DNA at 28 cycles, it was shown that increased imbalance of heterozygotes occurred, along with an increase in the size (peak area) of stutters. The analysis of mixtures by peak area measurement becomes increasingly difficult as the sample size is reduced and rules have been produced to guide interpretation [25].

28.4 STRS IN SEXUAL CHROMOSOMES AND MITOCHONDRIAL DNA

Y chromosome specific polymorphisms have proved to be especially useful in forensic work. The applications of Y polymorphisms range from deficiency paternity testing when a male offspring is in question to different applications in criminal casework. Y polymorphisms are especially interesting for the analysis of male DNA fraction in stains involving male/female mixtures, the most common biological material available in sexual crimes. Especially important is the use of these markers in cases where preferential sperm DNA extractions fails (this is estimated to occur in 5–15% of forensic cases) and also in rapes committed by azoospermic individuals. Although the variation in the Y chromosome is low, the non-pseudoautosomal region still bears different kinds of polymorphisms including biallelic markers, STRs, and minisatellites. SNPs and STRs are the most interesting. The most used Y-STRs are the trinucleotide repeat DYS392, and the tetranucleotide repeats DYS19, DYS385, DYS389-I, DYS389-II, DYS390, DYS391, and DYS393. These STRs comprise the so-called minimum Y-STR haplotype [26], but new STRs have recently been described [27,28] and extended haplotypes are currently used in forensic labs. Also, commercially validated Y-STR plexes are available.

As for mtDNA, statistical interpretation in cases of match is more complicated and appropriate corrections taking into account population substructure and sampling errors need to be performed. Population compilations are therefore very important and many efforts have been done regarding this [29,30].

STRs in the X chromosome have actually been introduced [31] and they are of interest for some deficiency paternity testing cases.

Analysis of the mtDNA Control Region is an efficient method for the study and comparison of bones, old and degraded DNA, and, especially, the analysis of telogenic hairs.

In these cases, samples of mtDNA variation can be analyzed using a variety of strategies. The combination of PCR amplification with direct DNA sequencing is usually the ultimate choice for identification and it has proved to be a reliable and reproducible method in forensic casework [32].

Analysis of mtDNA is a valid method to be applied in forensic genetics and it is accepted in courts all over the world. However, problems such as mutation rate, heteroplasmy, or the statistical approach, make sometimes the interpetation difficult

[33]. ISFG DNA Commission recommendations and European DNA profiling group (EDNAP) recommendations on the use of mtDNA, including nomenclature, prevention of contamination (aspect that it is crucial in mtDNA analysis), and statistical interpretation have been recently published [34,35].

Sexual chromosome polymorphisms and mtDNA are described in depth in the following chapters.

28.5 SNP TYPING

SNPs have a number of characteristics that make them ideal markers for human identification. First, they have lower mutation rates than the STR and VNTR loci typically used for relationship analysis in paternity and immigration testing. Second, SNPs can be analyzed after PCR amplification of very short DNA regions surrounding the substitution site, making SNPs preferable for anthropological and crime case investigations where the DNA is often degraded. Third, SNPs can be genotyped with a growing range of high-throughput technologies; an important factor in the implementation of large criminal DNA databases. Finally, SNPs, as binary polymorphisms, are comparatively easy to validate, because precise allele frequency estimates, required for the accurate interpretation of forensic genotyping data, can be obtained by analysing fewer samples compared to those needed for allele frequencies estimates of STRs and VNTRs. Seeking to match the discriminatory power of the 10–15 multiple allele STRs routinely used in forensic investigations, a set of ~50 polymorphic SNP markers are predicted to be required [36]. Furthermore, it has been suggested that 50 unlinked SNP loci with high overall heterozygosity should be sufficient to adjust for population stratification in population-based associations studies [37]. SNPs that are polymorphic in one population may be almost or completely monomorphic in another population [38], while others are known to be polymorphic in all major population groups. Thus, it should be possible to select SNPs that are useful for human identification purposes in the majority of populations, and to supplement these with SNPs showing highly contrasting allele frequency distributions in particular populations. These latter SNPs can provide valuable information for population admixture detection, in addition to the estimation of biogeographical ancestry.

It is now clear that SNP typing on a large scale is and will be of prime importance in human genetics and particularly valuable in the identification of genes that predispose individuals to common, multifactorial disorders by using linkage disequilibrium mapping. In addition, there are crucial markers for pharmacogenetics and pharmacogenomics. These potential applications of SNP typing, together with progress in identifying large sets of SNPs are the driving forces behind intense efforts to establish the technology for large-scale analysis of SNPs.

A great variety of chemistries and detection platforms have been proposed for SNP typing.

There is not a single ideal method for typing SNPs and the choice depends on both the need and the field of application, but for most of the applications (forensics included) the choice of the method must be a high-throughput technique that can be easily applied in molecular labs.

SNP genotyping technologies have been developed rapidly in the last few years. As a result, a great variety of different SNP typing protocols have become available for researchers; however there is no single protocol that meets all research needs. Different aspects should be taken into account to determine which technology is the most suitable for forensic purposes, such as the sensitivity, the reproducibility, the accuracy, the capability of multiplexing, and the level of throughput. It is also important to have in mind the flexibility of the technology, the time-consumption, and the cost, considering both the equipments required and the cost per genotype.

The level of throughput required depends on each application. Some applications use few SNP markers but a large sample size, other applications require large number of SNPs in a few samples, and, finally, there are other applications that need large number of both SNPs and samples. For forensic purposes, a medium throughput is required for paternity testing and criminal casework, but a high throughput is necessary to implement criminal DNA databases.

An important limiting step for forensic genetics in all these technologies is the amount of DNA required per genotype. There are some techniques that interrogate the SNP directly on genomic DNA, without a previous PCR, such as the SNPlex, Illumina, and the Invader assay [39]. In these cases, the minimum amount of DNA required for the analysis should be determined, but it is always higher than when a previous PCR is performed. For the other technologies requiring a PCR before the allelic discrimination reaction, the development of multiplex PCR is essential, not only for the small amount of DNA available to be analyzed in the majority of the criminal casework but also from the throughput point of view. In critical cases, the use of whole genome amplification (WGA) should be explored [40] prior to starting the analysis of the selected markers to avoid the problem of the lack of enough amount of DNA.

Another important issue in forensic genetics is the analysis of mixtures. Due to the biallelic nature of most of the SNPs, it will be more difficult to detect the presence of a mixture in a sample using these markers. Therefore, the possibility of the quantification of each allele in a sample can help in the determination of the contribution of each component in a mixed profile. Some technologies such as mass spectrometry and pyrosequencing allow the possibility of some quantification. This feature is routinely used for estimating allele frequencies in pooled samples, but could be an advantageous feature useful in forensic genetics.

In technologies based on homogenous hybridization with FRET detection such as the LightCycler, TaqMan, and Molecular Beacon (Bea Review), PCR and allelic discrimination reactions are performed in the same reaction. This advantage avoids further manipulation steps, favoring the automation and throughput of the process, especially when the high-throughput equipments for TaqMan assays are used. The main drawback is the limited multiplexing capability. As a consequence, these technologies are a good option for validating candidate SNPs and for building criminal DNA databases, but not for being used as routine technology in forensic casework.

The minisequencing technologies are at this moment the most popular methods in forensic laboratories, especially the SnaPshot [31] because the detection is performed on an automatic capillary electrophoresis instrument that is also used for STRs analysis. Several studies have been carried out in different forensic laboratories related to not with the analysis of SNPs with this technology [40–44]. The multiplex capability is feasible for forensic requirements, but a lot of work is needed to optimize the design and concentration of the primers for PCR and minisequencing reaction of each set of SNPs. The other detection methods such as microarrays and MALDI-TOF seem to be suitable for forensic purposes, and there are also some studies in those directions [45–46]. However, the main drawback of these technologies is that specific equipment is required, in contrast with SnaPshot. Furthermore, the multiplex capability of MALDI-TOF is lower than the other minisequencing technologies, although it is rapidly increasing and actually it is possible to analyze 30 plexes or even more SNPs at a time using this technology. The main problem of classical microarrays detection [47] is that the reproducibility and validation is more difficult but they have proved to be also an efficient method for analyzing SNP variation. Even the attractive electronic microarrays (Nanogen) have made extraordinary progresses in last couple of years [48].

Pyrosequencing has the limitation of the multiplex capability, which is very low, and the automation, because several steps need to be performed before the detection. The main advantage is the possibility of quantification of the contribution of each allele, a very useful feature in the analysis of mixture profiles.

The multiplex capability of several technologies based on OLA is very high, even above the forensic requirements. The main problem of these technologies is that the allelic discrimination reaction is performed directly on genomic DNA. This is a big advantage when there is enough amount of DNA for the analysis but not for criminal casework. However new strategies OLA-PCR have not has been recently described that look very promising for solving most of the problems related with forensic SNP typing technologies [49].

The technologies that reach the criteria required for forensic purposes need to be explored to determine their limitations and the possibility of being used in forensic genetics. It is difficult to find one that fits all the requirements, and therefore it is not an easy task to define which technology is the best for SNP typing in forensic genetics. Furthermore, there is a rapid technological progress, due to biotechnology companies that are making a big effort in developing new strategies for SNP typing, making it more difficult to choose the appropriate method for specific applications. Probably, different technologies will be used for routine casework, where the DNA amount and integrity are much more critical, than for paternity testing or for creating criminal DNA databases. Furthermore, it is possible to make differences between autosomal SNPs and Y chromosome and mitochondrial SNPs regarding the number of markers that is necessary to analyze per sample and the strategy of the analysis. Therefore the multiplex requirements are not the same for both applications and also the appropriate technology could be different for different types of SNPs.

In Europe the SNP for ID consortium has made an effort on the identification and forensic validation of SNP panels useful for forensic purposes. Thus, a 52 plex SNP set has been validated [50] and it has shown to be useful for the solution of complicated paternity and criminal casework cases. In addition the group has developed specific panels for Y-chromosome typing [51], mtDNA [52], population specific SNPs, and non-binary SNPs [53].

The introduction of SNP typing has opened new perspectives to forensic genetics further than the classical applications in identification or paternity testing. The identification of geographical origin of sample, the investigation of some causes of death (forensic molecular pathology), toxicogenetics, or the investigation of physical traits are examples of new areas of interest of this exciting branch of forensic sciences.

28.6 STANDARDIZATION EFFORTS AND THE VALUE OF THE DNA EVIDENCE

If DNA analysis is nowadays accepted in countries all over the world, it is in part due to the progress made in standardization.

Standardization of forensic DNA analysis has made enormous progress in the last few years and this innovation in standardization is comparable to the introduction of DNA technology itself.

Standards are crucial for forensic geneticists. This is due to the fact that only with an agreement about standards is it possible to develop quality control and quality assurance programs. In other words, standards are the only way to guarantee to the judges, juries, and the public that the tests performed and laboratory efficiency are reliable in any specific case.

In addition, standards are necessary to allow for second opinions for the exchange of data between labs and for the creation of uniform searching procedures in cross-border crime.

Two types of standards need to be addressed: technical and procedural. Technical standards include matters such as the genetic systems to be used (including type, nomenclature, and methodology), the statistical methods for evaluating the evidence, and the communication of the final report. Procedural standards encompass matters of operation such as laboratory accreditation, laboratory performance, accreditation and licensing of personnel, record keeping, and proficiency testing.

In the United States and in some European countries the development of procedural standards for forensic genetic labs has made considerable progress in the last few years. In some of these countries laboratories have agreed on the requirements necessary for organization and management, personnel, facilities and security, evidence control, validation, analytical procedures, equipment calibration and maintenance, proficiency testing, corrective actions, and audits. Proficiency testing programs for DNA analysis are established in some countries, and external and internal controls have been set up by most of the labs in western countries. Progress in accreditation has been effective in many countries in the last few years.

Even more advances have been made in attaining common technical standards. Agreement on genetic systems, types, and nomenclature is widespread.

Establishing common standards in Forensic DNA analysis is not easy because very different legal systems exist as well as a variety of laboratories performing forensic genetic analysis. The success of the forensic geneticist in achieving common standards (at least compared with other aspects of forensic science and genetics) has been greatly facilitated because a great number of them are members of the International Society of Forensic Genetics (ISFG) (www.isfg.org), which has many national and international working groups actively involved in establishing common standards, particularly in Europe – ENFSI (European Network of Forensic Science Institutes) and EDNAP. The Interpol Working Party on DNA (DNA MEG) has also contributed to the common aim to achieve common standards. In the States SWGDAM (Scientific Working Group on DNA Analysis Methods) has intensively worked in the field in connection with other regulatory bodies.

In addition, the existence of commercially available kits for DNA typing, a shared aim of geneticists in the search for common standards, and the influence of some leading groups, are other reasons for this success.

Efforts in standardization should be encouraged at national and international level, but these efforts must be coordinated in some way. If not, a constellation of different standards will emerge with more confusion than benefits. In Europe, EDNAP and ENFSI have defined precisely their respective roles and standardization efforts are well coordinated, although coordination with the local working groups is necessary.

Strategies for the coordination of standardization efforts are therefore needed. The DNA commission of the ISFG has an important role in resolving scientific conflicts between standardization groups and producing recommendations for the use of DNA polymorphisms in forensic casework (www.isfg.org).

However, efforts in standardization should continue. Progress in common procedural standards and particularly progress in similar requirements between countries for accreditation are necessary. Concerning technical standards other priorities include the harmonization of criminal databases, the coordination and compilation of population databases (especially for mtDNA and Y STRs), and a continuation of the progress initiated over the last few years on statistical evaluation and communication of the value of evidence provided by DNA analysis.

The latter one is a priority field of standardization. The question is that in most cases, but not always, the value of the evidence provided by DNA analysis is enormous. However, uncertainty always exists. As scientists, we must measure this uncertainty and for this we use a standard: the probability. Likelihood ratios are nowadays used for weighing the value of the evidence and for communicating this value to the courtroom, and the Bayesian approach to inference provides a coherent framework for interpretation.

28.7 REFERENCES

1   A.J. Jeffreys, V. Wilson and S.L. Thein, Nature, 314 (1985) 67–73.
2   P.D. Martin, H. Schmitter and P.M. Schneider, Forensic Sci. Int., 119 (2001) 225–231.
3   M. Litt and J.A. Luty, Am. J. Hum. Genet., 44 (1989) 397–401.
4   D. Tautz, Nucleic Acids Res., 17 (1989) 6463–6471.
5   Y. Nakamura, M. Leppert, P. O' Connell, R. Wolff, T. Holm, M. Culver, C. Martin, E. Fujimoto, M. Hoff, E. Kumlin and R. White, Science, 235 (1987) 1616–1622.
6   J.S. Beckman and J.L. Weber, Genomics, 12 (1992) 627–631.

7   A.J. Jeffreys, K. Tamaki, A. MacLeod, D.G. Monckton, D.L. Neil and J.A.L. Armour, Nat. Genet., 6 (1994) 136–145.

8   A. Di Rienzo, A.C. Peterson, J.C. Garza, A.M. Valdes, M. Slatkin and N.B. Freimer, Proc. Natl. Acad. Sci. USA, 91 (1994) 3166–3170.

9   The International Human Genome Mapping Consortium, Nature, 409 (2001) 934.

10 K. Mullis and F. Faloona, Specific synthesis of DNA in vitro via polymerase-catalyzed chain reaction. In: R. Wu (Ed.), Methods in enzymology, Academic Press, New York, 1987, pp. 335–350.

11 K. Kleppe, E. Ohstuka, R. Kleppe, L. Molineux and H.G. Khorana, J. Mol. Biol., 56 (1971) 341–361.

12 E. Hagelberg, B. Sykes and R. Hedges, Nature, 342 (1989) 485.

13 A. Alvarez-García, I. Muñoz, C. Pestoni, M.V. Lareu, M.S. Rodríguez-Calvo and A. Carracedo, Int. J. Legal Med., 109 (1996) 125–129.

14 B.J. Conner, A.A. Reyes, C. Morin, K. Itakura, R.L. Teplitz and R.B. Wallace, Proc. Natl. Acad. Sci. USA, 80 (1983) 278–282.

15 R. Saiki, T.L. Bugawan, T.G. Horn, K.B. Mullis and H.A. Erlich, Nature, 324 (1986) 163.

16 B. Budowle, A.M. Giusti and R.C. Allen, Analysis of PCR products (pMCT118) by polyacrylamide gel electrophoresis. In: H.F. Polesky, W.R. Mayr (Eds.), Advances in forensic haemogenetics, Springer, Berlin, 1990, pp. 148–150.

17 X.Y. Hauge and M. Litt, Hum. Mol. Genet., 2 (1993) 411–415.

18 A. Urquhart, C. Kimpton, T.J. Downes and P. Gill, Int. J. Legal Med., 107 (1994) 13–20.

19 B. Brinkmann, M. Klintschar, F. Neuhuber, J. Hühne and B. Rolf, Am. J. Hum. Genet., 62 (1998) 1408–1415.

20 M.V. Lareu, C. Pestoni, C. Phillips, F. Barros, D. SynderCombe-Court, P. Lincoln and A. Carracedo, Electrophoresis, 19 (1998) 1566–1573.

21 P. Gill, R. Sparkes and C. Kimpton, Forensic Sci. Int., 89 (1997) 185–197.

22 P. Gill, J. Whitaker, C. Flaxman, N. Brown and J. Buckleton, Forensic Sci. Int., 112 (2000) 17–40.

23 J.M. Butler, Y. Shen and B.R. McCord, J. Forensic Sci., 48 (2003) 1054–1064.

24 P. Gill, J. Whitaker, C. Flaxman, N. Brown and J. Buckleton, Forensic Sci. Int., 112 (2000) 17–40.

25 P. Gill, C.H. Brenner, J.S. Buckleton, A. Carracedo, M. Krawczak, W.R. Mayr, N. Morling, M. Prinz, P.M. Schneider and B.S. Weir, Forensic Sci. Int., 160 (2006) 90–101.

26 M. Kayser, A. Caglià, D. Corach, N. Fretwell, C. Gehrig, G. Graziosi, F. Heidorn and S. Herrmann, Int. J. Legal Med., 110 (1997) 125–133.

27 P.S. White, O.L. Tatum, L.L. Deaven and J.L. Longmire, Genomics, 57 (1999) 433–437.

28 Q. Ayub, A. Mohyuddin, R. Qamar, K. Mazhar, T. Zerjal, S. Mehdi and C. Tyler-Smith, Nucleic Acids Res., 28 (2000) e8.

29 L. Roewer, M. Krawczak, S. Willuweit, M. Nagy, C. Alves, A. Amorim, K. Anslinger, C. Agustin, A. Betz, E. Bosch, A. Caglià, A. Carracedo, D. Corach, A. Dekairelle, T. Dobosz, B.M. Dupuy, S. Furedi, C. Gehrig, L. Gusmao, J. Henke, L. Henke, M. Hiddinng, C. Hohoff, B. Hoste, M.A. Jobling, H.J. Kargel, P. de Knijff, R. Lessig, E. Liebeherr, M. Lorente, B. Martínez-Jarreta, P. Nievas, M. Nowak, W. Parson, V.L. Pascali, G. Penacino, R. Ploski, B. Rolf, A. Sala, U. Schmidt, C. Schmitt, P.M. Schneider, R. Szibor, J. Teifel-Gredding and M. Kayser, Forensic Sci. Int., 118 (2001) 106–113.

30 L. Roewer, Forensic Sci. Int., 118 (2001) 105.

31 S. Hering and R. Szibor, J. Forensic Sci., 45 (2000) 929–931.

32 A. Carracedo, E. D'Aloja, B. Dupuy, A. Jangblad, M. Karjalainen, C. Lambert, W. Parson, H. Pfeiffer, H. Pfitzinger, M. Sabatier, D. Syndercombe Court and C. Vide, Forensic Sci. Int., 97 (1998) 165–170.

33 M.M. Holland and T.J. Parsons, Forensic Sci. Rev., 11 (1999) 1–25.

34 A. Carracedo, W. Bär, P.J. Lincoln, W. Mayr, N. Morling, B. Olaisen, P. Schneider, B. Budwole, B. Brinkmann, P. Gill, M. Holland, G. Tully and M. Wilson, Forensic Sci. Int., 110 (2000) 79–85.

35 G. Tully, W. Bär, B. Brinkmann, A. Carracedo, P. Gill, N. Morling, W. Parson and P. Schneider, Forensic Sci. Int., 124 (2001) 83–91.

36 A. Amorim and L. Pereira, Forensic Sci. Int., 150 (2005) 17–21.

37 K. Hao, C. Li, C. Rosenow and W.H. Wong, Eur. J. Hum. Genet., 12 (2004) 1001–1006.

38 M. Shriver, M. Shriver, G. Kennedy, E. Parra and H. Lawson, Hum. Genomics, 1 (2004) 274–286.

39 B. Sobrino, M. Brion and A. Cariacedo, Forensic Sci. Int., 154 (2005) 181–194.

40 P.M. Schneider, K. Balogh, N. Naveran, M. Bogus, K. Bender, M. Lareu, A. Carracedo, Progress in forensic genetics (10th Ed.). Elsevier, 2004, pp. 24–26.

41 J. Sanchez, C. Børsting, C. Hallenberg, A. Buchard, A. Hernandez and N. Morling, Forensic Sci. Int., 137 (2003) 74–84.

42 B. Quintans, V. Alvarez-Iglesias, A. Salas, C. Phillips, M.V. Lareu and A. Carracedo, Forensic Sci. Int., 140 (2004) 251–257.

43 M. Brion, B. Sobrino, A. Blanco-Verea, M.V. Lareu and A. Carracedo, Int. J. Legal Med., 119 (2004) 10–15.

44 S. Inagaki, Y. Yamamoto, Y. Doi, T. Takata, T. Ishikawa, K. Imabayashi, K. Yoshitome, S. Miyaishi and H. Ishizu, Forensic Sci. Int., 144 (2004) 45–57.

45 B. Sobrino, M. Lareu, M. Brion and A. Carracedo, Progress in forensic genetics (10th Ed.). Elsevier, 2006, pp. 331–333.

46 J. Mengel-Jorgensen, J. J. Sanchez, C. Borsting, F. Kirpekar and N. Morling, Progress in forensic genetics (10th Ed.). Elsevier, 2006, pp. 15–17.

47 A.M. Divne and M. Allen, Forensic Sci. Int., 154 (2005) 111–121.

48 M. Bogusz, B. Sobrino, K. Bender, A. Carracedo and P.M. Schneider, Prog. Forensic Genet., 11 (2006) 37–40.

49 C. Phillips, R. Fang, D. Ballard, M. Fondevila, C. Harrison, F. Hyland, E. Musgrave- Brown, C. Proff, E. Ramos-Luis, B. Sobrino, A. Carracedo, M.R. Furtado, D. Syndercombe Court and P.M. Schneider, Forensic Sci. Genet. 1 (2007) 180–185.

50 J.J. Sanchez, C. Phillips, C. Borsting, K. Balogh, M. Bogus, M. Fondevila, C.D. Harrison, E. Musgrave-Brown, A. Salas, D. Syndercombe-Court, P.M. Schneider, A. Carracedo and N. Morling, Electrophoresis, 27 (2006) 1713–1724.

51 M. Brion, J.J. Sanchez, K. Balogh, C. Thacker, A. Blanco-Verea, C. Borsting, B. Stradmann-Bellinghausen, M. Bogus, D. Syndercombe-Court, P.M. Schneider, A. Carracedo and N. Morling, Electrophoresis, 26 (2005) 4411–4420.

52 A. Brandstatter, A. Salas, H. Niederstatter, C. Gassner, A. Carracedo and W. Parson, Electrophoresis, 27 (2006) 2541–2550.

53 E. Musgrave-Brown, Anwar, K. Elliot, C. Phillips, Syndercombe-Court, A. Carracedo, N. Morling, P. Schneider and B. McKeown, Prog. Forensic Genet., 11 (2006) 34–37.

M.J. Bogusz (Ed.). Forensic Science
Handbook of Analytical Separations, Vol. 6
© 2008 Elsevier B.V. All rights reserved.

CHAPTER 29

Mitochondrial DNA: future challenges in forensic genetics

Antonio Salas[1,2], Vanesa Álvarez-Iglesias[1], María Cerezo[1], Ana Mosquera[1],
Nuria Naverán[1], Chris Phillips[2], María Victoria Lareu[1] and
Ángel Carracedo[1,2]

[1]*Unidad de Xenética, Instituto de Medicina Legal, Facultad de Medicina, Universidad de Santiago de Compostela, 15782, Santiago de Compostela, Spain*
[2]*Grupo de Medicina Xenómica, Hospital Clínico Universitario, 15706; Galicia, Spain*

29.1 INTRODUCTION

Often, the nuclear DNA (nDNA) obtained from evidential samples is highly de-graded or cannot be recovered in sufficient quantity from forensic material. In such circumstances the recovery and analysis of mitochondrial DNA (mtDNA) molecule can become the only option available for forensic DNA analysis. This is mainly because there is a high number of mtDNA molecules per cell, $\sim$1,000–10,000 [1,2], but only two (or one; X and Y chromosomes in males) copies per cell for each nuclear chromosome. Thus, the study of the mtDNA molecule is highly efficient when analyzing bone samples, old and degraded DNA, as well as telogen hairs [3,4]. The mtDNA is maternally inherited [1] and therefore it does not go through re-combination at each generation. Some recent studies have questioned this dogma [5–7]. However, these attempts do not explain, for instance, why there are no signals of recombination in the worldwide mtDNA phylogeny [8,9] or why heteroplasmies occur generally at just one or two positions in a single profile instead of multiple positions, a fact that is incompatible with the expected multiple heteroplasmic state under a model of patrilineal contribution. On the other hand, the techniques em-ployed by these authors are very prone to errors (author's unpublished data); this is not uncommon if we take into account the high rate of mtDNA errors in the scientific literature and public databases (see below). Maternal inheritance of the mtDNA therefore allows the study of familiar relationships between members shar-ing the same maternal lineage.

References pp. 965–967

MtDNA variation can be analyzed using a variety of strategies but PCR amplification coupled with direct DNA sequencing is the most common approach using globally standardized procedures. The use of the mtDNA test in forensic casework began more than 10 years ago and although technical improvements have been achieved during this decade, serious problems still remain to be resolved; here we briefly review the future challenges related to the forensic use of the mtDNA test.

29.2 AVERAGE AND SITE-SPECIFIC MUTATION RATES

It is commonly said that the evolutionary substitution rate of the mtDNA molecule is about five to ten times higher than the average mutation rate in nDNA. This may be due in part to the activity derived from the mitochondrial oxidative metabolic pathways that provide most of the energy to the cell. The mtDNA is prone to oxidative damage mainly because: (a) the mitochondria consume $>90\%$ of the oxygen that enters the cell, the resulting accumulation of free oxygen radicals may thus preferentially damage mtDNA; (b) the lack of protective histones in the mtDNA molecule; and (c) the mitochondrial repair mechanisms are less efficient than the nucleus in correcting DNA damage and replication errors [10]. The mechanism by which mtDNA mutation arises and becomes fixed in mammalian maternal lineages [11] is relevant not only for the investigation of mtDNA diseases, but also for the analysis of human populations (where, e.g. estimates of mtDNA sequence diversity are used to date demographic events) and for forensic genetics in identification and criminal cases.

Two main different approaches have been followed for the investigation of the mutation rate of the mtDNA molecule: pedigree and phylogenetic studies. Unfortunately, there exists a large disparity in the mutation rates estimated for mtDNA from different authors (see, e.g. [12]) and depending on the approach employed. Although estimation of the average mutation rate is important, from a practical point of view, it is much more relevant to gather information about site-specific mutation rates. Let us imagine a common case in forensic casework where a maternal relative is used as a reference sample for the identification of missing family members. The fact that the mtDNA molecule is highly heterogeneous in terms of site-specific mutation rates, it should be an important part of the interpretive framework in which to evaluate the importance of a single mismatch. How can site-specific mutation rates be accurately estimated? We envision two potential approaches: The first approach would be to analyze several thousand inter-generational events (e.g. mother–son) and obtain an approximation of the pedigree mutation rate estimated for each site; this option is evidently unrealistic if we take into account the limitations of current laboratory capacity and the logistic and economical efforts required for such a study. The second (much cheaper) option would be to extrapolate phylogenetic substitution rates estimates from the observed genealogical mutations. These two estimates are not exactly the same thing but it can be stated with certain reservations that the theoretical phylogenetic estimates can act as a proxy for use in forensic interpretation (author's unpublished data).

29.3 THE IMPORTANCE OF CODING REGION SNPS IN FORENSIC GENETICS

During the last decade many different techniques developed in forensic, clinical, and population genetics laboratories have been applied for screening the mtDNA variation. Among the most well established analytical approaches for the determination of point or length polymorphisms is restriction fragment length polymorphism (RFLP) analysis [13,14]. Other less popular analysis methods include hybridization with allele specific probes [15], oligonucleotide ligation assay, and oligonucleotide ligation solid phase minisequencing. In addition, some other alternative strategies to DNA sequencing have been developed in the past and have since became popular among medical and forensic analysts; these techniques allow for the identification of variation through the observation of different electrophoretic patterns in DNA tested samples in comparison to controls. Use of Single Strand Conformation Polymorphisms (SSCPs) was one of the most popular methods in the field of medical research and forensic genetics [16]. SSCPs analysis and accompanying modifications (e.g. [17]) or closely related techniques (e.g. heteroduplex analysis) are based on the different conformations that single stranded DNA adopts when the samples are denatured and run in native gels. Using this method, a large number of samples can be analyzed in with a rapid and straightforward assay. However, the advantages of this method are counterbalanced by the fact that standardization is difficult. In addition SSCP fails to detect an important amount of the variation and this approach does not provide specific information about which variant(s) is/are responsible for the observed conformational electrophoretic patterns. As such these represent, first and foremost, a simple system for screening samples.

Today, the most popular method for screening the variability at the coding region or for increasing the discrimination power of the mtDNA test is the technique of minisequencing [18–21]. It is now possible to interrogate several dozen SNP sites in one multiplexed reaction. Another advantage of minisequencing is that this technique performs well with highly degraded samples because it involves amplification of short amplicons. Lastly, this technique can be less prone to the artifacts that afflict many of the other techniques (including to some extent sequencing).

29.4 HIGH-DENSITY DNA ARRAYS: CHIPS FOR MTDNA

In 1996, Chee *et al.* [22] described the analysis of the entire human mitochondrial genome using DNA arrays containing up to 135,000 probes complementary to the 16.6 kb of the human mitochondrial genome. According to these authors, these arrays, generated by light-directed chemical synthesis, have the resolution of a single-base and the results are obtained in a few minutes. More recently alternative attempts, primarily aimed at analyzing a substantial number of mtDNA SNPs rather than the whole genome, have been published by Sigurdsson *et al.* [23]. The authors interrogated 150 SNPs in the coding region that were tested using a sample set of 265 mtDNAs from different worldwide populations. According to the authors, the discrimination power provided is even higher than analyzing 500 bp of the hypervariable region, although it is likely that this is lower than typing the whole control region.

However the great expectations generated by the publication of Chee's results, have since diminished for several reasons: (a) the microarrays are expensive and therefore not readily applicable to the throughput of forensic laboratories, and (b) the microarrays have not been standardized in forensic labs (an obligatory step in this field prior to the adoption of new techniques for real casework); furthermore the prospects of securing a standardized procedure for this technique in the near future are not good. Similarly the use of chips for "complete genome sequencing" is limited, but for other reasons. For instance, the set of probes implemented in these arrays does not cover 100% of the variability seen in all populations; because: (a) the databases used as "templates of variability" for designing probes just cover populations representing small parts of the worldwide phylogeny; and (b) these databases are too small to adequately represent even small geographical regions or ethnic groups.

Molecular anthropologists or population geneticists feel uncomfortable with techniques that exclude a small but unknown (unchecked) proportion of total variability when the main aim is to cover the full variation of the mtDNA genome. In clinical analysis, with the purpose of examining the mtDNA of a patient in search of a causal variant, it is also undesirable to leave out a percentage of the molecule unchecked, especially in cases where the disease of interest is a complex trait (and the causal variant SNP is of low penetrance). The main application of microarrays in medical genetics remains concerned with certain applications related to the routine work of molecular diagnostics. Additionally some commercial kits have been designed for complete genome analysis. One example is the mitoSEQr^{TM} resequencing system (Applied Biosystems, Foster City, CA, USA) for reasons not immediately apparent that fails to cover the whole mtDNA sequence. The percentage of targeted regions covered by the mitoSEQr^{TM} probe sets amounts to 98.7% and it should be noted that this value does not relate directly to the percentage of variation coverage by the kit which is probably less than 98.7%.

Finally, some of the most recent attempts at screening the total mtDNA variation in an individual using whole molecule microarrays are least so error-prone that current sequencing methods (author's unpublished data, but see also Sigurdsson *et al.* [23]) in comparison gave fully reliable results. We can tentatively say that mtDNA microarrays will need many years of further development in order to occupy a relevant niche in the forensic field. In addition, microarrays have yet to be tested with problematic samples, such as those containing degraded DNA; the performance of the technique is likely to be poor in comparison to standard sequencing based approaches. Other disadvantages can be anticipated: cost, lack of flexibility of the technique under the continuously changing circumstances of analyzing evidential samples, ability to interpret unexpected profiles (length, variation, etc.), reliably detecting heteroplasmy, etc.

29.5 USING PHYLOGENETIC APPROACHES TO CORRECT ERRORS IN MTDNA DATASETS

The high incidence of errors in mtDNA testing have been the subject of many alerting publications in the forensic literature [9,24–27]. These errors are universal and occur as often in other fields of research, such as medical and population genetics studies [8,27–33]. The typical problems observed in mtDNA datasets can be classified in the following categories [24,27]:

- Type I or base shift
- Type II or reference bias
- Type III or phantom mutations
- Type IV or base mis-scoring
- Type V or artifactual recombination

Most of these errors can be detected using a phylogenetic approach (see [26] for a review) but this method can only detect a limited number of errors in databases; this number depends mainly on the haplotypic composition of the population under study because the worldwide phylogeny contains parts that are better defined than others.

No other field of genetic research is the beneficiary of such a posteriori phylogenetic analysis of datasets that are able to reveal the presence of important deficiencies related to data management (e.g. documentation errors) and genotyping accuracy. Unfortunately, forensic laboratories have made insufficient efforts to add this powerful tool to their current mtDNA database management practices. This is all the more paradoxical when taking into account the high demands of accuracy and precision made by forensic laboratories for all forms of DNA testing. Furthermore the lack of impetus for adopting phylogenetic data checks is all the more surprising if we consider that a single mistake can completely alter a mtDNA test result; the latter situation can change the course of a case in court from innocence to culpability. Therefore it remains all the more mysterious that forensic scientists are still so reluctant to use the phylogenetic approach [26].

29.6 TROUBLESOME MTDNA DATABASES

A database is an absolute necessity in order to interpret the weight of the DNA evidence in a criminal or identification context. There are three main problems concerning mtDNA databases.

Firstly, the weight of evidence is dependent on the frequency of the profile in the reference population but, since sample sizes of most current databases are very low in relation to the large amount of variability in populations and most mitochondrial haplotypes are rare, there is a large degree of uncertainty concerning the estimation of profile frequencies. Statistical interpretation of the mtDNA test should seek to incorporate the uncertainty created by the small sample sizes of most databases, yet this remains a rare practice among forensic laboratories [26,34]. Therefore, reliable estimates of haplotype frequencies require large databases [26,34]. Today, the largest public mtDNA database is the one from the Scientific Working Group on DNA Analysis Methods (SWGDAM) and consists of less than 5,000 profiles. The SWG-DAM database is however divided into smaller datasets that range from ~1,200 to a few hundred profiles. In contrast to SWGDAM, the EMPOP database (http://www.empop.org) has been designed to avoid the addition of erroneous profiles as much as possible with the main aim of covering samples from worldwide populations. Although the total number of profiles is 4,527 (January 31, 2007), as with SWGDAM these are grouped into several small datasets that limit their use for haplotype estimation. The growth of EMPOP seems assured since a growing number of laboratories are continuously submitting profiles to the database. Most of the profiles (3,830) were collected from West Eurasia, including several country-specific samples (Austria, Belgium, Germany, Spain, etc.). At the current time EMPOP represents the most promising tool for mtDNA analysis in forensic laboratories.

Secondly, on several occasions the literature has alerted users to the high prevalence of errors in datasets produced in forensic labs, as well as laboratories from other disciplines. However little progress has been made in resolving this important drawback. Despite such errors being problematic *per se*, it is important to address the primary causes of the errors caused in the first place. Given the priority for the majority of forensic laboratories in providing effective analysis for the bulk of cases analyzed it is perhaps not surprising that database checking and management is not the foremost priority. However the continuing high incidence of errors in forensic databases has not been taken seriously by most practicing laboratories. Even worse, some influential laboratories have eluded the responsibility of proper correction of their databases. A notable example is one of the most widely used databases employed by forensic practitioners: SWGDAM [25–27,31,35].

Thirdly, the use of a database in a particular criminal case implicitly assumes that the database is representative of the populations found in the region. Generally, this corresponds to a geographic location (neighborhood, city, etc.) or an ethnic group. Without a guarantee that the database is representative for the forensic case under study, any frequency estimate based on this database will be inherently biased [26,34]. The idealized assumption that a database can adequately represent the geographic context where the crime was committed is however questionable; one example of a database that shows questionable representativeness is SWGDAM [26,34]. The structure of SWGDAM begs a number of questions in this respect: Which population is represented by the "Hispanic" dataset? How is population stratification, known to exist in many different regions or cities in US, properly addressed? What is the current process to properly evaluate these potential problems [26,34]?

29.7 FINAL REMARKS: VALIDATION OF MTDNA IN FORENSIC GENETICS AND STATISTICAL APPROACH

Today the analysis of mtDNA is used by several hundred laboratories around the world working in criminal or identification casework. In the last few years, some effort has been made to validate mtDNA in forensic genetics laboratories. An experimental validation was first carried out by Wilson *et al.* [3] analyzing the effect of chemical contaminants on DNA extracted from blood and semen. This study also aimed to evaluate the effect of typing DNA extracted from body-fluid samples deposited on different types of substrates. The results obtained confirmed that mtDNA typing using PCR and direct automated sequencing was a valid and reliable means of forensic identification. In Europe, the European DNA Profiling Group (EDNAP) has also contributed to the validation of the mtDNA test [36]. In general, mtDNA analysis is a valid method for forensic genetics and in fact, it has been included in the proficiency testing programs of several organizations, such as the Spanish and Portuguese Group of the International Society for Forensic Haemogenetics or GEP-ISFG [37–40]. The implementation of mtDNA analysis in forensic laboratories is however not exempt from serious problems (a selection of these have been discussed in this paper) and unfortunately little or no progress has been made in the field to properly address these. In addition to these numerous challenges mention must be made of the statistical interpretation of mtDNA data; at the moment this issue is also deeply deficient, in part due to the very problems highlighted in this paper. The nature of mtDNA variation is highly complex and our limited knowledge about inter-generational and inter-tissue mutation rates, heteroplasmy, etc., is hindering progress in developing an appropriate interpretative framework for assessing mtDNA results.

Analysis of mtDNA is still the only realistic option for dealing with challenging forensic evidential material meaning it forms an essential component of the forensic casework armory. Nevertheless, there remains ample opportunity for misinterpretation of mtDNA data, with the inevitable conclusion that forensic analysts must bring into play the full range of checks, balances, and statistical adjustments in the routine use of the mtDNA test.

29.8 ACKNOWLEDGMENTS

The "Ramón y Cajal" Spanish programme from the Ministerio de Educación y Ciencia (RYC2005-3), the grant of the Xunta de Galicia (PGIDIT06PXIB208079PR), and a grant from the Fundación de Investigación Médica Mutua Madrileña, awarded to AS, partially supported this project; as well as the projects from the Ministerio de Educación y Ciencia (BIO2006-06178) and Xunta de Galicia (PGIDT06P-XIB228195PR) given to MVL.

29.9 REFERENCES

1   R.E. Giles, H. Blanc, H.M. Cann and D.C. Wallace, Proc. Natl. Acad. Sci. USA, 77 (1980) 6715–6719.
2   W.M. Brown, M. George Jr. and A.C. Wilson, Proc. Natl. Acad. Sci. USA, 76 (1979) 1967–1971.
3   M.R. Wilson, J.A. DiZinno, D. Polanskey, J. Replogle and B. Budowle, Int. J. Legal Med., 108 (1995) 68–74.
4   A. Salas, M.V. Lareu and A. Carracedo, Int. J. Legal Med., 114 (2001) 186–190.
5   Y. Kraytsberg, M. Schwartz, T.A. Brown, K. Ebralidse, W.S. Kunz, D.A. Clayton, J. Vissing and K. Khrapko, Science, 304 (2004) 981.

6   G. Zsurka, K.G. Hampel, T. Kudina, C. Kornblum, Y. Kraytsberg, C.E. Elger, K. Khrapko and W.S. Kunz, Am. J. Hum. Genet., 80 (2007) 298–305.

7   G. Zsurka, Y. Kraytsberg, T. Kudina, C. Kornblum, C.E. Elger, K. Khrapko and W.S. Kunz, Nat. Genet., 37 (2005) 873–877.

8   H.-J. Bandelt, Q.-P. Kong, W. Parson and A. Salas, J. Med. Genet., 42 (2005) 957–960.

9   H.-J. Bandelt, A. Salas and S. Lutz-Bonengel, Int. J. Legal Med., 118 (2004) 267–273.

10   D.A. Clayton, Ann. Rev. Biochem., 53 (1984) 573–594.

11   K.E. Bendall, V.A. Macaulay, J.R. Baker and B.C. Sykes, Am. J. Hum. Genet., 59 (1996) 1276–1287.

12   T.J. Parsons, D.S. Muniec, K. Sullivan, N. Woodyatt, R. Alliston-Greiner, M.R. Wilson, D.L. Berry, K.A. Holland, V.W. Weedn, P. Gill and M.M. Holland, Nat. Genet., 15 (1997) 363–368.

13   S. Horai, T. Gojobori and E. Matsunaga, Hum. Genet., 68 (1984) 324–332.

14   A. Torroni, H.-J. Bandelt, L. D'Urbano, P. Lahermo, P. Moral, D. Sellitto, C. Rengo, P. Forster, M.-L. Savantaus, B. Bonné-Tamir and R. Scozzari, Am. J. Hum. Genet., 62 (1998) 1137–1152.

15   M. Stoneking, D. Hedgecock, R.G. Higuchi, L. Vigilant and H.A. Erlich, Am. J. Hum. Genet., 48 (1991) 370–382.

16   F. Barros, M.V. Lareu, A. Salas and A. Carracedo, Electrophoresis, 18 (1997) 52–54.

17   A. Salas, E.M. Rasmussen, M.V. Lareu, N. Morling and A. Carracedo, Forensic Sci. Int., 124 (2001) 97–103.

18   A. Brandstätter, A. Salas, H. Niederstätter, C. Gassner, Á. Carracedo and W. Parson, Electrophoresis, 27 (2006) 2541–2550.

19   V. Álvarez-Iglesias, J.C. Jaime, Á. Carracedo and A. Salas, Forensic Sci. Int. (2006) submitted.

20   B. Quintáns, V. Álvarez-Iglesias, A. Salas, C. Phillips, M.V. Lareu and Á. Carracedo, Forensic Sci. Int., 140 (2004) 251–257.

21   P.M. Vallone, R.S. Just, M.D. Coble, J.M. Butler and T.J. Parsons, Int. J. Legal Med., 118 (2004) 147–157.

22   M. Chee, R. Yang, E. Hubbell, A. Berno, X.C. Huang, D. Stern, J. Winkler, D.J. Lockhart, M.S. Morris and S.P. Fodor, Science, 274 (1996) 610–614.

23   S. Sigurdsson, M. Hedman, P. Sistonen, A. Sajantila and A.C. Syvanen, Genomics, 87 (2006) 534–542.

24   H.-J. Bandelt, P. Lahermo, M. Richards and V. Macaulay, Int. J. Legal Med., 115 (2001) 64–69.

25   H.-J. Bandelt, A. Salas and C.M. Bravi, Problems in FBI mtDNA database. Science, 305 (2004) 1402–1404.

26   A. Salas, H.J. Bandelt, V. Macaulay and M.B. Richards, Forensic Sci. Int., 168 (2007) 1–13.

27   A. Salas, Á. Carracedo, V. Macaulay, M. Richards and H.-J. Bandelt, Biochem. Biophys. Res. Commun., 335 (2005) 891–899.

28   A. Salas, Y.-G. Yao, V. Macaulay, A. Vega, Á. Carracedo and H.-J. Bandelt, PLoS Med., 2 (2005) e296.

29   H.J. Bandelt, A. Salas and C.M. Bravi, J. Hum. Genet., 51 (2006) 1073–1082.

30   H.-J. Bandelt, A. Achilli, Q.-P. Kong, A. Salas, S. Lutz-Bonengel, C. Sun, Y.-P. Zhang, A. Torroni and Y.-G. Yao, Biochem. Biophys. Res. Commun., 333 (2005) 122–130.

31   H.-J. Bandelt, L. Quintana-Murci, A. Salas and V. Macaulay, Am. J. Hum. Genet., 71 (2002) 1150–1160.

32   Y.G. Yao, A. Salas, C.M. Bravi and H.J. Bandelt, Hum. Genet., 119 (2006) 505–515.

33   Q.-P. Kong, H.-J. Bandelt, C. Sun, Y.-G. Yao, A. Salas, A. Achilli, C.Y. Wang, L. Zhong, C.L. Zhu, S.F. Wu, A. Torroni and Y.-P. Zhang, Hum. Mol. Genet., 15 (2006) 2076–2086.

34   T. Egeland, G.O. Storvik and A. Salas, Statistical considerations for haploid databases, Springer Verlag, Berlin, 2007, in press.

35   A. Brandstätter, T. Sänger, S. Lutz-Bonengel, W. Parson, E. Béraud-Colomb, B. Wen, Q.P. Kong, C.M. Bravi and H.J. Bandelt, Electrophoresis, 26 (2005) 3414–3429.

36   A. Carracedo, E. D'Aloja, B. Dupuy, A. Jangblad, M. Karjalainen, C. Lambert, W. Parson, H. Pfeiffer, H. Pfitzinger, M. Sabatier, D. Syndercombe Court and C. Vide, Forensic Sci. Int., 97 (1998) 165–170.

37   A. Alonso, A. Salas, C. Albarrán, E. Arroyo, A. Castro, M. Crespillo, A.M. di Lonardo, M.V. Lareu, C.L. Cubria, M.L. Soto, J.A. Lorente, M.M. Semper, A. Palacio, M. Paredes, L. Pereira, A.P. Lezaun, J.P. Brito, A. Sala, M.C. Vide, M. Whittle, J.J. Yunis and J. Gómez, Forensic Sci. Int., 125 (2002) 1–7.

38   M. Crespillo, M.R. Paredes, L. Prieto, M. Montesino, A. Salas, C. Albarrán, V. Álvarez-Iglesias, A. Amorin, G. Berniell-Lee, A. Brehm, J.C. Carril, D. Corach, N. Cuevas, A.M. Di Lonardo, C. Doutremepuich, R.M. Espinheira, M. Espinoza, F. Gómez, A. González, A. Hernández, M. Hidalgo, M. Jimenez, F.P. Leite, A.M. López, M. López-Soto, J.A. Lorente, S. Pagano, A.M. Palacio, J.J. Pestano, M.F. Pinheiro, E. Raimondi, M.M. Ramón, F. Tovar, L. Vidal-Rioja, M.C. Vide, M.R. Whittle, J.J. Yunis and J. García-Hirschfel, Forensic Sci. Int. (2007) in press.

39   L. Prieto, M. Montesino, A. Salas, A. Alonso, C. Albarrán, S. Álvarez, M. Crespillo, A.M. Di Lonardo, C. Doutremepuich, I. Fernández-Fernández, A.G. de la Vega, L. Gusmão, C.M. López, M. López-Soto, J.A. Lorente, M. Malaghini, C.A. Martínez, N.M. Modesti, A.M. Palacio, M. Paredes, S.D. Pena, A. Pérez-Lezaun, J.J. Pestano, J. Puente, A. Sala, M. Vide, M.R. Whittle, J.J. Yunis and J. Gómez, Forensic Sci. Int., 134 (2003) 46–53.

40   A. Salas, L. Prieto, M. Montesino, C. Albarrán, E. Arroyo, M.R. Paredes-Herrera, A.M. Di Lonardo, C. Doutremepuich, I. Fernández-Fernández, A.G. de la Vega, C. Alves, C.M. López, M. López-Soto, J.A. Lorente, A. Picornell, R.M. Espinheira, A. Hernández, A.M. Palacio, M. Espinoza, J.J. Yunis, A. Pérez-Lezaun, J.J. Pestano, J.C. Carril, D. Corach, M.C. Vide, V. Álvarez-Iglesias, M.F. Pinheiro, M.R. Whittle, A. Brehm and J. Gómez, Forensic Sci. Int., 148 (2005) 191–198.

M.J. Bogusz (Ed.). Forensic Science
Handbook of Analytical Separations, Vol. 6

969

CHAPTER 30

The human Y chromosome male-specific polymorphisms and forensic genetics

Leonor Gusmão[1], María Brión[2] and Iva Gomes[1,2]

[1]*IPATIMUP, Institute of Molecular Pathology and Immunology of the University of Porto, 4200-465
Porto, Portugal*
[2]*Grupo de Medicina Xenómica, CeGen-Institute of Legal Medicine, University of Santiago de Compostela,
15782 Santiago de Compostela, Spain*

30.1 INTRODUCTION

30.1.1 Y-chromosome structure

The Y chromosome is one of the smallest human chromosomes, with an estimated average size of 60 million base pairs (Mb) (Fig. 30.1). During male meiosis recombination only takes place in the pseudoautosomal regions at the tips of both arms of Y and X chromosomes (PAR1, with 2.6 Mb, and PAR 2, with 0.32 Mb). Along ~95% of its length the Y chromosome is male-specific and effectively haploid, since it is exempt from meiotic recombination. Therefore, this Y-chromosome segment where X-Y crossing over is absent has been designated as the non-recombining region of the Y chromosome or NRY. Because of the high non-homologous recombination occurring within this Y chromosome specific region, a more appropriately name of male-specific region or MSY is nowadays used to designate it [1].

The MSY is a mosaic of heterochromatic and euchromatic regions. Besides the centromeric heterocromatin, a large heterochromatic region is located on the distal long arm of the Y chromosome (Yq) and constitutes more than half of the chromosome in some normal males, but is virtually undetectable in others [2]. A third heterochromatic region was recently discovered by Skaletsky *et al.* [1], interrupting the euchromatic sequences of proximal Yq (see Fig. 30.1). These regions are composed of highly repeated sequences of non-functional DNA: DYZ1, DYZ2, DYZ3, DYZ17, DYZ18, and DYZ19.

References pp. 992–1000

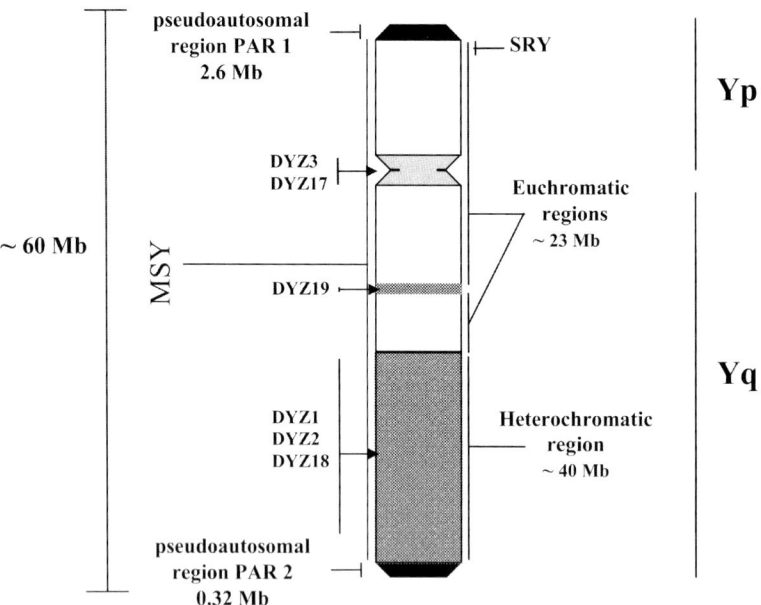

Fig. 30.1. Y-Chromosome structure.

The euchromatin is a constant size region and includes sequences homologous to the X chromosome, Y-specific repetitive sequences, and all the genes identified in the Y chromosome, which include the now identified 27 distinct protein-coding genes or gene families. Near-complete sequence of the Y-chromosome euchromatin has been recently revealed by Skaletsky *et al.* [1] that classifies the euchromatic sequences into three categories. First, the X-transposed, consisting in a stretch recently transposed from the X chromosome – ~3–4 million years ago, that still presents 99% homology to their X-chromosome counterparts. Second, the X-degenerate, consisting of a class of sequences more distantly related to the X chromosome – remnants of ancient autosomal sequences from which the modern X and Y derive. And at last, the ampliconic class composed largely of sequences that exhibit as much as 99.9% identity to other sequences in the MSY, maintained by frequent Y–Y gene conversion events. These sequences are located in seven segments scattered across the long and proximal short arms, and the most striking structural feature are eight massive palindromes located in the ampliconic regions of Yq, six of which carry testis genes.

30.1.2 The evolution of sex chromosomes

The similarities between the X and Y chromosome sequences are consistent with the hypothesis of a common origin. The mammalian advanced sex chromosome systems originated 300 million years ago from systems in which the X and Y were initially largely genetically homologous [3,4]. The evolution of sex chromosomes involved

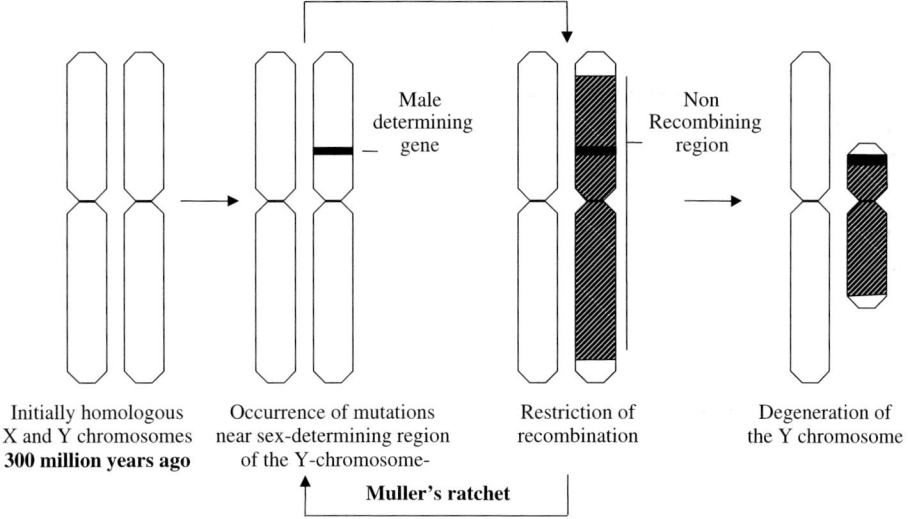

| Initially homologous
X and Y chromosomes
300 million years ago | Occurrence of mutations
near sex-determining region
of the Y-chromosome- | Restriction of
recombination | Degeneration of
the Y chromosome |

Muller's ratchet

Fig. 30.2. Differentiation of the initially morphologically homogeneous X and Y chromosomes.

mechanisms of restriction of gene recombination, transposition, and translocation. The sequence of events that induced the morphological and genetic differentiation of the X and Y chromosomes and the genetic inactivation of the Y-chromosome genes is still not completely understood. The presently accepted explanation of the differentiation of the initially morphologically homogeneous X and Y chromosomes invokes successive processes where alternated steps of mutation and restriction of recombination were involved (Fig. 30.2).

In time, the Y chromosome comes to carry genes that are beneficial to the male but not to the female sex. If linked to the sex-determining region of the Y chromosome, those genes, favored in males and selected negatively in females, will tend to spread through the population. In order to keep this genetic heterogeneity between X and Y chromosomes, restriction of recombination involves sex determination genes and loci controlling secondary sexual characteristics being promoted by selection mechanisms. In a process referred to as "Muller ratchet," the lack of exchange through all or part of the originally homologous X and Y chromosomes will accumulate deleterious recessive mutations, since they are not restricted by selection. If there is no recombination, some mutations are more liable to be lost from the population and spread of a favorable Y-linked mutant allele through a population that would allow for the fixation of deleterious alleles at other loci. The accumulation of recessive deleterious alleles on the Y chromosome favors a selection for increased activity of the homologous loci on the X chromosome. On the other hand, with the reduction of Y chromosome genetic activity, there will be weak selection against insertions into the Y chromosome. In the absence of gene exchange and selective pressures, transposable elements and tandem repeat sequences are expected to accumulate, leading to a step-by-step reduction of the Y activity.

Sequencing of the MSY provided the opportunity to reexamine the model of evolution of the human sex chromosomes, showing that it is a consequence of two opposed evolutionary dynamics acting on the Y chromosome: gene decay versus gene acquisition and conservation [1,5].

Because of the presence of MSY gene pairs in the ampliconic sequences of the euchromatin which are subject to frequent gene conversion [1,5], and of the little or no X-degenerate gene loss or decay observed during the last 6 million years of human evolution [6], the predictions that the Y chromosome would be vanished completely in the next 10 million years seem to no longer have support.

30.1.3 The Y-chromosome inheritance

As a result of the evolutionary process, exchange between X and Y chromosomes is limited to two small regions of the X-Y pair and, consequently, to a great extent, the Y chromosome is paternally inherited and haploid. Along generations, the MSY is transmitted from father to son unchanged unless a mutational event takes place. For this reason, the Y chromosome contains a record of all the mutational events that occurred among his ancestors, reflecting the history of paternal lineage. Therefore, all modern Y chromosomes have a single paternal ancestor, on their male-specific region.

30.1.4 Y-chromosome-specific polymorphisms

In 1985, Casanova *et al.* [7] undertook the first search for Y-linked restriction fragment length polymorphisms (RFLPs) in humans, with the report of two Y-specific polymorphisms. This and latter surveys on Y-specific markers by RFLP studies [8–10] and sequence analysis [11,12] emphasized the low level of polymorphism of this chromosome, compared with other chromosomes [13]. The attempt to identify new Y-specific polymorphisms in different population samples, mainly in Caucasians [8,9] and Africans [10], showed that the Y chromosome is apparently devoid of polymorphic genetic markers. Jakubiczka *et al.* [8] estimated a frequency of less than 1 point mutation in 18,000 nucleotides, and Malaspina *et al.* [9] calculated less than 1 per 46,515 nucleotides. Spurdle and Jenkins [10] screened a 20,808 bp segment for Y-specific RFLPs and did not detect any new polymorphism.

The low variation found in the Y chromosome was unexpected in view of its origin and is best explained, simply, by its presence at one quarter of the frequency of the autosomes, in diploid populations. Therefore, it is especially subject to drift that will be reflected in a corresponding reduced diversity [14]. The effective population size of the Y chromosome can also be reduced by a particular pattern of mating behavior found in specific populations. The lack of recombination also explains the low Y chromosome variation found, due to the effect of selective pressure in which a whole haplotype is involved instead of a specific allele [15].

Despite the scarcity of polymorphisms, with the availability of the complete sequence of the Y chromosome and with the improvement and applications of new techniques, a high number of Y-specific polymorphisms are now well characterized and available to population, evolutionary, and forensic genetics e.g. [16,17].

30.2 Y CHROMOSOME IN FORENSIC APPLICATIONS

Y-specific polymorphisms have been proven to be especially useful in routine forensic casework. The male specificity and especial inheritance features, distinct from autosomal, X-chromosomal, and mitochondrial markers transmission, determined the relevance of the Y-specific polymorphic markers, first in evolutionary and population genetics and then, based on the population studies, in forensic field applications e.g. [18–22].

30.2.1 Paternity testing

The pattern of inheritance along the male lineage makes the Y-STR polymorphisms suitable in paternity testing. However, the applicability of these markers is limited to approximately one half of paternity cases (those in which a male descendency is in question). It must be considered that a result based exclusively on Y-chromosome STRs does not exclude as father any male relative in the same patrilineage. Whenever possible, autosomal (AS) markers should be also used in order to avoid or reduce this possibility.

The probability of excluding in paternity for Y-chromosome markers is equal to the probability of having two different alleles in the population, which means that it is identical to the probability of discrimination in criminal identification. Thus, Y-linked polymorphisms have a much higher exclusion power than equally variable autosomal loci [19].

The possibility of using these markers is especially important in deficiency cases. In these situations, namely when the alleged father is deceased, it is possible to access his complete Y-chromosome information using male relatives in patrilineage.

30.2.2 Stain analysis

Y-chromosome STRs are useful in the discrimination of stains in forensic investigation when a male suspect is involved. This is the case of most violent crimes including sexual offences [18].

Mixtures of body fluids from different individuals are frequent in forensic casework. Y-chromosome analysis can be particularly helpful to detect male DNA fraction in stains involving male/female mixtures [23,24], the most common biological material available in sexual assaults.

Although a differential lysis can be applied in stains where semen from the offender is mixed with cells from the victim, this is laborious and is not always successful in achieving complete separation of the two cellular fractions. Differential extraction can fail in very small and degraded samples or when the fraction of semen is much reduced, leading to loss of sperm DNA [25,26]. Differential lysis is supposed to fail in 10–20% of forensic cases although Y-specific amelogenin can be detected in most of them. In other mixtures of body fluids from different individuals, such as blood–blood or blood–saliva, differential extraction cannot be applied in order to separate the DNA fraction from male and female sources. In such cases, when AS markers are used, preferential amplification of the major component of the mixture (usually female DNA from the victim) can mask the genetic profile of the assailant. By typing Y markers, even a minor male DNA component in a mixed male/female stain of a sexual assault yields a male-specific profile that can be compared with the DNA of suspects [27]. Prinz *et al.* [23] identified Y-specific STR alleles in male/female DNA mixture in a ratio of 1:2,000, compared with a 1:50 maximum ratio to detect a minor component for an autosomal locus. Another application is in rape cases committed by azoospermic individuals. They represent around 1–2% of all rapes but this percentage is increasing due to the fact that vasectomy is now more frequent. With Y-specific analysis only the male component is detected, and this allows a direct determination of the Y haplotype.

Y-specific STRs can also be useful in the determination of the number of semen contributors in multiple rape cases and as a screening method for linking rape series or for excluding suspects.

30.2.3 Y's counterpart: the X chromosome in forensic genetics

The X chromosome has a particular place in forensic genetics, regardless of its limitations in identity testing, when compared with the Y chromosome and the autosomes. The unique properties of genetic inheritance of this chromosome, consequence of the sex chromosomes evolution, are responsible for its role in forensics. In males, one single copy and the absence of recombination allows direct haplotyping; therefore the single male copy is entirely transmitted to female descendants. This characteristic itself can lead to paternity exclusion if the alleged father is unavailable and two sisters or half sisters are under investigation, since they will share the same paternal alleles [28]. This is also true in kinship analysis when accessing an alleged paternal grandmother/granddaughter relationship. X-linked genetic markers singly handed or supplementing the analysis of Y chromosomal or AS markers, are mainly useful in complex cases of kinship analysis, with female offspring, representing one of the major advantages of the use of X-chromosome markers [28–34]. In fact, in paternity deficiency cases where paternal relatives must be studied, X-STRs can occasionally resolve these cases more efficiently than common AS loci. When evaluating the forensic efficiency of X and AS loci in trios involving a daughter, X-chromosome markers are more efficient than AS markers, since the mean exclusion chance (MEC) is higher for these specific markers [28]. On the other hand, the X chromosome recombines in females providing an interesting system of genetic variation in each generation. The combination of these features, recombination in females and direct haplotyping in males, also makes the X chromosome an interesting system for historical studies [35].

At the present time, studies using X-STRs are of great interest in the fields of population genetics, as well as in forensics and kinship testing and, several X-STRs have been recently validated. Large PCR multiplexes for X-linked genetic markers make population studies and databasing more efficient and need to be designed and optimized. Several X-STR multiplexes have been described in the literature e.g. [36–38]; however, reference to amplifications in one single PCR reaction containing a high number of STR loci has not so far been common. Nevertheless, further studies are still needed on allele frequency distributions in different populations, mutation rates, and linkage disequilibrium, in order to establish reference population databases [36,38,39]. A forensic chromosomal X-STR database compiling population data previously published on allele frequencies and forensic parameters is also available [40]. In the case of X-STRs it is particularly important to study simultaneous transmission of different markers in order to screen possible associations between loci (this can be especially important in admixed and possibly substructured populations). It is expected that linkage disequilibrium intervals are longer on the X chromosome because recombination occurs only in females; therefore only two-thirds of the X chromosome recombines in each generation [35]. Although the lack of association between X-STRs contributes to increase the power of discrimination of a set of markers, haplotype analysis has been demonstrated to be a valuable tool in pedigree-based-kinship testing [29–31,37]. Reliable estimates of mutation rates are also important in kinship analysis; but few X-STRs segregation studies have been reported for these markers. Even so, after combining data previously reported [28,41–44], a total of 18 mutations were compiled in 8,698 allele transmissions, with an average mutation rate of 2.07×10^{-3} (95% CI $1.23–3.27 \times 10^{-3}$) per locus/meiosis. This value is inside the 95% CI of other mutation rates estimates published for Y-chromosomal and autosomal STRs e.g. [45–48].

The common application of AS, Y-chromosomal markers, and the mtDNA genome in forensics exceed the use of X chromosome genetic markers in this field. However, the X chromosome can be an important additional tool in human identification, unravelling some of the challenges presented in forensic genetics. A promising future for its application in this field can only be expected.

30.3 Y-CHROMOSOME MARKERS IN FORENSICS

The non-pseudoautosomal region of a Y chromosome contains different kinds of polymorphisms. These different loci have different mutation rates and consequently it is possible to select appropriate Y polymorphisms to study evolution events over different time scales [15]. STRs seem to be the most suitable Y-chromosome markers in forensic genetics due to their levels of diversity when compared to the biallelic polymorphisms. Taking into account the polymorphic levels, the minisatellite MSY1

TABLE 30.1
CONTINUED

Haplogroup	SNP[a]
D2a1a	P42
D2a2	M151
D3	P47
E	SRY$_{4064}$ (SRY8299 or M40), M96, P29
E1	M33, M132
E1a	M44
E2	M75
E2a	M41
E2b	M54, M90, M98, M85
E2b1	M85
E2b1a	M200
E3	P2, DYS391p
E3a	M2 (SY81 orDYS271), M180, P1,
E3a1	M58
E3a2	M116.2
E3a3	M149
E3a4	M154
E3a5	M155
E3a6	M10, M66, M156, M195
E3a7	M191, U186, U247
E3a7a	U174
E3a8	U175, U209
E3a8a	U290
E3a8a1	U181
E3b	M215
E3b1	M35
E3b1a	M78
E3b1a1	V12
E3b1a1a	M224
E3b1a1b	V32
E3b1a2	V13
E3b1a2a	V27
E3b1a3	V22
E3b1a3a	M148
E3b1a3b	V19
E3b1b	M81
E3b1b1	M107
E3b1b2	M183, M165
E3b1c	M123
E3b1c1	M34
E3b1c1a	M136, M84
E3b1c1b	M290
E3b1d	M281
E3b1e	V6
E3c	M329
F	P14, M89, M213, M235
F1	P91, P104
F2	M427, M428
G	M201

TABLE 30.1
CONTINUED

Haplogroup	SNP[a]
G1	M285, M342
G1a	P20
G2	P15
G2a	P16
G2a1	P17, P18
G2b	M286
G3	M287
G4	M377
H	M69
H1	M52
H1a	M82
H1a1	M36, M197
H1a2	M97
H1a3	M39, M138
H1b	M370
H2	Apt
I	M170, M258, P19, U179
I1	P38
I1a	M253, M307, P30, P40
I1a1	M227
I1a2	M21
I1a3	M72
I1b	S31
I1b1	P37.2
I1b1a	P41.2 (M359.2)
I1b1b	M26
I1b1b1	M161
I1b2	S23, S30, S32, S33
I1b2a	M223, S24
I1b2a1	M284
I1b2a2	M379
I1b2a3	P78
I1b2a4	P95
J	12f2.1, M304
J1	M267
J1a	M62
J1b	M365
J1c	M367, M368
J1d	M369
J1e	M390
J2	M172
J2a	M410
J2a1	DYS413 $\leqslant$ 18
J2a1a	M47, M322
J2a1b	M67 (S51)
J2a1b1	M92, M260
J2a1b1a	M327
J2a1b2	M163, M166
J2a1c	M68

References pp. 992–1000

TABLE 30.1
CONTINUED

Haplogroup	SNP[a]
J2a1d	M137
J2a1e	M158
J2a1f	M289
J2a1g	M318
J2a1h	M319
J2a1i	M339
J2a1j	M419
J2a2	M340
J2b	M12, M314, M221
J2b1	M102
J2b1a	M241
J2b1a1	M99
J2b1a2	M280
J2b1a3	M321
J2b1b	M205
K	M9
K1	M353, M387
K1a	SRY9138 (M177)
K2	M70, M184, M193, M272
K2a	M320
K3	M147
K4	M230
K4a	M254
K4a1	M226
K5	P117
L	M11, M20, M22, M61, M185, M295
L1	M27, M76
L2	M317
L2a	M274
L2b	M349
L3	M357
L3a	PK3
M	M4 (M296), M5, M106, M186, M189, P35
M1	P34
M1a	P51
M2	P87
M2a	M104 (P22)
M2a1	M16
M2a2	M83
NO	M214
N	LLY22g, M231
N1	M128
N2	P43
N3	Tat (M46)
N3a	M178
N3a1	P21
O	M175
O1	MSY2.2
O1a	M119

TABLE 30.1
CONTINUED

Haplogroup	SNP[a]
O1a1	M101
O1a2	M50, M103, M110
O2	P31, M268
O2a	M95
O2a1	M88, M111
O2a1a	PK4
O2a2	M297
O2b	SRY465 (M176)
O2b1	P49
O2b1a	47z
O3	M122
O3a	M324
O3a1	M121, DYS257
O3a2	M164
O3a3	LINE1, M159
O3a4	M7
O3a4a	M113, M188, M209
O3a4a1	N4
O3a4a2	N5
O3a5	M134
O3a5a	M117, M133
O3a5a1	M162
O3a6	M300
O3a7	M333
P	92R7, M45, M74 (N12), P27
Q	MEH2, M242, P36
Q1	M120, N14 (M265)
Q1a	M378
Q2	M25, M143
Q3	M3
Q3a	M19
Q3b	M194
Q3c	M199
Q4	M323
Q5	M346
R	M207 (UTY2)
R1	M173
R1a	$SRY_{10831.2}$ (SRY1532)
R1a1	M17, M198
R1a1a	M56
R1a1b	M157
R1a1c	M64.2, M87, M204
R1b	M343
R1b1	P25
R1b1a	M18
R1b1b	M73
R1b1c	M269, S3, S10, S13, S17
R1b1c1	M37
R1b1c2	M65

References pp. 992–1000

TABLE 30.1
CONTINUED

Haplogroup		SNP[a]
R1b1c3	M126	
R1b1c4	M153	
R1b1c5	M160	
R1b1c6	SRY2627 (M167)	
R1b1c7	M222	
R1b1c8	U152	
R1b1c9	U106	
R1b1c9a	U198	
R1b1d	M335	
R2	M124	

[a]Includes data from Jobling and Tyler-Smith [51]; Mohyuddin *et al.* [138]; Shen *et al.* [139]; Cruciani *et al.* [57,58,67]; Underhill *et al.* [53]; Sengupta *et al.* [71]; Hammer *et al.* [66]; Sims *et al.* [56]; Semino *et al.* [140]; Cinnioglu *et al.* [69]; Rootsi *et al.* [72]; Karlsson *et al.* [141]; Scheinfeldt *et al.* [142]; Di Giacomo *et al.* [73]; HYPERLINK "http://isogg.org/tree/ISOGG_HapgrpK07.html" Kayser *et al.* [63,143]; Regueiro *et al.* [144]; Shi *et al.* [77]; Deng *et al.* [145]; and the Ethnoancestry (Genealogy-DNA Rootsweb List, 11 January 2006).

Usually, there is a strong correlation between the Y-chromosome variation and the geography [51]. Some of the haplogroups are confined to specific areas or continents, like haplogroups A and B, confined to Africa. While other clades show an extensive geographic representation. This condition has been used to explore patterns of migrations, population substructure, and mixture among diverse populations [58–60]. In addition, this geographical distribution can be used to make inferences on the possible geographic origin of any sample of interest. In a forensic scenario this may serve as a tool to predict the origin of the paternal lineage of any stain contributor.

30.3.1.1 Global distribution of Y haplogroups

Y DNA haplogroup A represents the oldest branch of the Y-chromosome phylogeny. Like haplogroup B, it only appears in Africa, with the highest frequency among the hunter-gatherer groups in Ethiopia and Sudan [58,61]. A3b1 is a Khoisan exclusive haplogroup.

Clade C was found in Central Asia, South Asia, and East Asia [53,62–65]. C1 lineage is found exclusively in Japan. C2 is found in New Guinea, Melanesia, and Polynesia. C3 lineage is believed to have originated in Southeast or Central Asia, spreading from there into northern Asia, the Americas and Central Europe. C4 appears to be restricted among aboriginal Australians and is dominant in that population. C5 has a significant presence in India.

Haplogroup D appears in Central Asia, Southeast Asia, and in Japan [62,76] showing the highest frequencies in Tibet and Japan (50% and 35%, respectively).

Haplogroup E is one of the most branched, with many subhaplogroups described [57,58,67]. E1 and E2 were described in Northeast Africa, and E3 shows a wide geographic distribution, with two main clades: E3a, present all around Africa and among African-Americans; and E3b, present in Western Europe, North Africa, and the Near East.

Haplogroup F is the parent of haplogroups from G to R; however excluding these common haplogroups, the minor clades F*, F1, and F2, seem to appear in the Indian continent [68].

The highest frequencies of haplogroup G appear in the Caucasus region; however it also shows significant frequencies in the Mediterranean areas and the Middle East [69,70].

Until now, haplogroup H has not been well studied, members of this haplogroup were mainly found in the Indian continent [68,71].

Haplogroup I is a clear European haplogroup; it is one of the most frequent haplogroups among northwestern European populations [72].

It is generally agreed that haplogroup J was dispersed by the westward movement of people from the Middle East to North Africa, Europe, Central Asia, Pakistan, and India [54,71]. However, Di Giacomo *et al.* [73] also consider it as a signature of the expansion of the Greek world, with an accompanying novel quota of genetic variation produced during its demographic growth.

Haplogroup K is the ancestral haplogroup of major groups L to R, but, in addition, also includes the minor K* and K1 to K5 haplogroups, which are present at low frequencies in dispersed geographic regions all around the world [63].

Haplogroup L is found mainly in India and Pakistan, as well as in the Middle East and, very occasionally, in Europe, particularly in Mediterranean countries [68,71].

The highest frequencies of haplogroup M are shown in Melanesia, being restricted to the geographical distribution of Papuan languages [74–76].

The Y-DNA haplogroup N has a wide distribution, primarily in northern Eurasia, and is often associated (but not necessarily) with current and earlier Uralic speakers [72].

Lineage O represents nearly 60% of chromosomes in East Asia. The O3 haplogroup has the highest frequency, being absent outside East Asia. The O1 and O2 haplogroups appear in Malaysia, Vietnam, Indonesia, South China, Japan, and Korea [66,76,77].

The P clade is the parent of haplogroups Q and R, and is rarely found. It has been detected at low frequencies in the Caucasus and India [52,70].

Haplogroup Q is found in Asia, the Americas, Europe, and the Middle East. One of its sub-clades, group Q3 is almost exclusively associated with the Native Americans [78,79].

Finally, the last clade of the Y-chromosome tree is the extensive haplogroup R, which is mainly represented by two lineages – R1a and R1b [64,69,80,81]. The members of R1b are believed to be the descendants of the first modern humans who entered Europe, and is now the most common Y haplogroup in Europe. More than half of men of European descent belong to R1b. Haplogroup R1a is currently found in central and western Asia, India, and in Slavic populations of Eastern Europe.

30.3.1.2 Y-SNPs typing technologies

A number of different technologies have been described for the analysis of SNP variation; however the decision on the appropriate method depends mainly on the number of SNPs to be analyzed.

Most of the multiplex reactions described during these last years are based on minisequencing or base extension reactions, followed by capillary electrophoresis. Some examples are the 6 PCR multiplex reactions used by Onofri *et al.* [82] for typing 37 Y-chromosome SNPs, or the 29-plex reaction described by the SNP for ID European working group (Fig. 30.3) [81].

An alternative approach is the use of MALDI-TOF MS (matrix assisted laser desorption ionization time-of-light mass spectrometry) for detection of minisequencing products [83]. Other authors have also used microarrays for Y-SNP typing [84,85].

Finally, there are several kits commercially available based on the Luminex platform (The Signet^{TM} Y-SNP Genotyping Kits, Marligen Biosciences, Inc). These kits enable the detection of up to 97 SNPs, distributed in 12 multiplexes [86].

30.3.2 Minisatellite (MSY1-DYF155S1)

The most variable Y-specific marker is the minisatellite MSY1. An extremely high degree of structural variation is observed in this minisatellite using an MVR-PCR strategy, with a virtual heterozygosity of 99.9% [87–89]. From the observed diversity, a high mutation rate is expected in human minisatellites [90,91]. Based on the high degree of structural variation found in this locus, Jobling *et al.* [88] estimates the mutation rate to be between 2 and 11% per generation.

Despite the high informative potential of this minisatellite, the MVR-PCR method is technically complex and therefore the analysis of a large number of samples in population surveys is difficult [92]. Thus, the use of this minisatellite in population or in forensic genetics was not widespread.

MSY2 (DYS440) was the second minisatellite described on the Y chromosome [93]; however compared with the MSY1, it shows a very low diversity, with only 3 types of long repeat units.

30.3.3 Alphoid satellite DNA

Alphoid satellite DNA sequences are tandemly repeated arrays present in the centromere region. In the Y chromosome the alphoid DNA (Yα1, DYZ3) seems to be a functional part of the centromere [94]. A large number of alphoid patterns can be distinguished by the combination of restriction enzymes and allows the identification of haplogroup types that contribute to the definition of paternal lineages [95–98].

The variation found in alphoid haplotypes is mainly due to point mutations and insertion/deletion events. It can be detected as RFLPs and analyzed using different

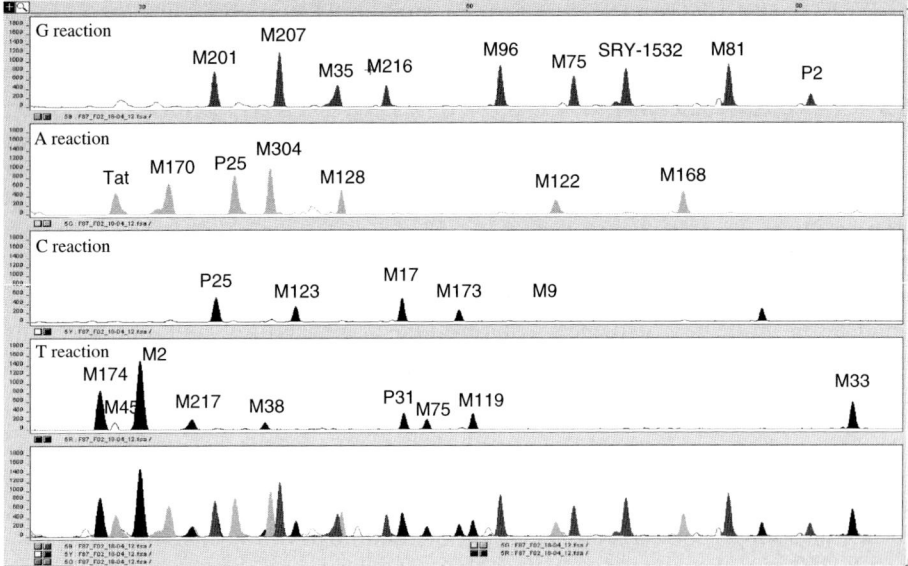

Fig. 30.3. Y-chromosome SNP 29-plex reaction described by the SNP for ID European working group.

techniques, namely conventional gel electrophoresis, pulse-field gel electrophoresis [95], or even by the identification of heteroduplex patterns of PCR-amplified fragments [99].

Although a high variability is present in these sequences, the methods usually used are not suitable for the analysis of forensic DNA samples, except those described by Santos *et al.* [95] using PCR amplification and heteroduplex analysis. Despite the importance of the information disclosed by PCR variant alphoid satellite DNA subunits in the study of human populations, this methodology can be considered quite complex for forensic purposes when compared with the existing ones for the study of other type of markers, namely the microsatellites.

30.3.4 STRs (short tandem repeats) or microsatellites

Although different kinds of polymorphisms were described on the haploid male-specific portion of the human Y chromosome, due to their levels of diversity and typing simplicity, the STRs are the most used markers in the forensic field. All the advantages already emphasized in the forensic application of autosomal STRs when compared with the study of other kind of markers are also applied to the Y-chromosome-specific STRs. In particular, they can be studied using very simple and reliable PCR techniques and, moreover, most PCR-STRs produce small amplicons with the advantage that they can be studied in degraded samples.

Since the first report of a Y-specific tetranucleotide repeat polymorphism, DYS19 (Y27H39), by Roewer and Epplen [100], over 200 Y-STR polymorphisms have been described [19,101–106], and Y-chromosome-specific STR analysis has been extensively applied to human migrations and evolution as well as to forensics e.g. [18,107].

Regarding the Y chromosome polymorphic STRs described until now, DYS19, DYS385, DYS389I, DYS389II, DYS390, DYS391, DYS392, DYS393, DYS438, and DYS439 have been selected by The Scientific Working Group on DNA Analysis Methods (SWGDAM) as the core set for forensic DNA analysis in the U.S. This same set of Y-STRs is included in the Y-STR Haplotype Reference Database (YHRD; http://www.yhrd.org), known as extended haplotype (ExtHt). With the help of collaborative efforts made in the construction of this database, these markers are the best studied, concerning amplification performance and specificity, multiplex amplification strategies, sequence structure and nomenclature, mutation rates, as well as worldwide allele frequencies distribution. Therefore, this group of markers has been, until now, the most used in population and forensic genetics.

30.3.4.1 Guidelines on the use of Y-STRs in forensic analysis

The use of Y-STRs in the forensic field has been greatly improved by coordinating efforts concerning, simultaneously, typing methodologies, nomenclature, and databasing. The use of a consensus nomenclature is crucial to allow for second opinions, proficiency testing, exchange of data, and databasing. Sequence data on Y-STRs are important in the establishment of allele nomenclature and, although for some STRs, with simple repeat structure, it is easy to find a consensus nomenclature; in others, with a complex structure, it becomes more difficult [108]. For example, for the first described STRs, nomenclature changes were made for DYS19, DYS390, and DYS389 in order to include non-repetitive motives or motives that were found to be variable with the accumulation of new sequence data. In order to clarify some confusion that still exists in the field, mainly as a consequence of the large number of new markers that have been introduced in recent years, the DNA Commission of International Society for Forensic Genetics (ISFG) recently provided recommendations on the nomenclature of Y-STR loci and alleles, as well as on its use in forensic genetics [109]. Some Y-STR markers for which sequence information is available, and a nomenclature based on the recommendations of the DNA Commission of the ISFG has already been published, are listed in Table 30.2.

30.3.4.2 Y-STR typing strategies

Multiplex PCR amplification techniques can be used in order to increase the information content of the Y-STR haplotype typing approach, and also to reduce sample quantity in forensic cases, since working with minimal amounts of material is common in casework. A great effort has been done in order to develop STR multiplex systems including a large number of markers, which will greatly improve the power of discrimination between unrelated individuals, as well as minimize costs and labor.

Nowadays, there are many different PCR multiplexing strategies described for the amplification of a large number of Y-STRs e.g. [110–114]. In addition widely used

TABLE 30.2
Y-STRS REPEAT STRUCTURE AND NOMENCLATURE

GDB Locus Name	STR Reference	Repeat Structure	Nomenclature Reference
YCAII [MC]	Mathias et al. [96]	$(CA)_n$	Schmidt et al. [129]
YCAIII [MC]	Mathias et al. [96]	$(CA)_n$	Kayser et al. [19] De Knijff et al. [101]
DYS19/DYS394	Roewer and Epplen [101]	$(TAGA)_3 tagg(TAGA)_n$	Kayser et al. [19] De Knijff et al. [101]
DYS385	Kayser et al. [19] De Knijff et al. [101]	$(aagg)_{6-7}(GAAA)_n$	Kayser et al. [19] De Knijff et al. [101]
DYS388	Kayser et al. [19] De Knijff et al. [101]	$(ATT)_n$	Butler et al. [111]
DYS389 I	Kayser et al. [19] De Knijff et al. [101]	$(TCTG)_3(TCTA)_n$	Kayser et al. [19] De Knijff et al. [101]
DYS389 II	Kayser et al. [19] De Knijff et al. [101]	$(TCTG)_n(TCTA)_n N_{28}(TCTG)_3(TCTA)_n$	Kayser et al. [19] De Knijff et al. [101]
DYS390	Kayser et al. [19] De Knijff et al. [101]	$(tcta)_2(TCTG)_n(TCTA)_n(TCTG)_n(TCTA)_n tca(tcta)_2$	Kayser et al. [19] De Knijff et al. [101]
DYS391	Kayser et al. [19] De Knijff et al. [101]	$(tctg)_3(TCTA)_n$	Kayser et al. [19] De Knijff et al. [101]
DYS392	Kayser et al. [19] De Knijff et al. [101]	$(TAT)_n$	Kayser et al. [19] De Knijff et al. [101]
DYS393/DYS395	Kayser et al. [19] De Knijff et al. [101]	$(AGAT)_n$	Kayser et al. [19] De Knijff et al. [101]
DYS426	Jobling et al. [106]	$(GTT)_n$	Butler et al. [111]
DYS434	Ayub et al. [103]	$(TAAT)_{1-2}(CTAT)_n$	Gusmão et al. [108]
DYS435	Ayub et al. [103]	$(TGGA)_n$	Gusmão et al. [108]
DYS436	Ayub et al. [103]	$(GTT)_n$	Gusmão et al. [108]
DYS437	Ayub et al. [103]	$(TCTA)_n(TCTG)_{1-3}(TCTA)_4$	Gusmão et al. [108]
DYS438	Ayub et al. [103]	$(TTTTC)_1(TTTTA)_{0-1}(TTTTC)_n$	Gusmão et al. [108]
DYS439 (GATA A4)	Ayub et al. [103]	$(GATA)_n$	Gusmão et al. [108]
DYS441	Iida et al. [104]	$(TTCC)_n$	[a]
DYS442	Iida et al. [104]	$(TATC)_2 (TGTC)_3 (TATC)_n$	[a]

TABLE 30.2
CONTINUED

GDB Locus Name	STR Reference	Repeat Structure	Nomenclature Reference
DYS443	Iida *et al.* [104]	$(TTCC)_n$	Iida *et al.* [104]
DYS444	Iida *et al.* [104]	$(ATAG)_n$	[a]
DYS445	Iida *et al.* [104]	$(TTTA)_n$	Iida *et al.* [104]
DYS446	Redd *et al.* [105]	$(TCTCT)_n$	Redd *et al.* [105]
DYS447	Redd *et al.* [105]	$(TAATA)_n(TAAAA)_1(TAATA)_n(TAAAA)_1(TAATA)_n$	Redd *et al.* [105]
DYS448	Redd *et al.* [105]	$(AGAGAT)_n\ N_{42}(AGAGAT)_n$	Redd *et al.* [105]
DYS449	Redd *et al.* [105]	$(TTTC)_n\ N_{50}(TTTC)_n$	Redd *et al.* [105]
DYS450	Redd *et al.* [105]	$(TTTTA)_n$	Redd *et al.* [105]
DYS452	Redd *et al.* [105]	$(TATAC)_2(TGTAC)_2(TATAC)_n(CATAC)_1(TATAC)_1(CATAC)_1$ $(TATAC)_{3-4}(CATAC)_{0-2}(TATAC)_{0-3}(CATAC)_1(TATAC)_3$	Redd *et al.* [105]
DYS453	Redd *et al.* [105]	$(AAAT)_n$	Redd *et al.* [105]
DYS454	Redd *et al.* [105]	$(AAAT)_n$	Redd *et al.* [105]
DYS455	Redd *et al.* [105]	$(AAAT)_n$	Redd *et al.* [105]
DYS456	Redd *et al.* [105]	$(AGAT)_n$	Redd *et al.* [105]
DYS458	Redd *et al.* [105]	$(GAAA)_n$	Redd *et al.* [105]
DYS459 [MC]	Redd *et al.* [105]	$(TAAA)_n$	Redd *et al.* [105]
DYS460 (GATA A7.1)	White *et al.* [102]	$(ATAG)_n$	Gusmão *et al.* [108]
DYS461 (GATA A7.2)	White *et al.* [102]	$(TAGA)_n(CAGA)$	Gusmão *et al.* [108]
DYS462 (G09411)	Bosch *et al.* [110]	$(TATG)_n$	Bosch *et al.* [110]
DYS463	Redd *et al.* [105]	$(AAAGG)_n(AAGGA)_2$	Redd *et al.* [105]
DYS464 [MC]	Redd *et al.* [105]	$(CCTT)_n$	Redd *et al.* [105]
DYS485	Kayser *et al.* [17]	$(TTA)_n$	Butler *et al.* [130]
DYS490	Kayser *et al.* [17]	$(TTA)_n$	Butler *et al.* [130]
DYS495	Kayser *et al.* [17]	$(AAT)_n$	Butler *et al.* [130]
DYS504	Kayser *et al.* [17]	$(TCCT)_n$	Butler *et al.* [130]
DYS505	Kayser *et al.* [17]	$(TCCT)_n$	Butler *et al.* [130]
DYS508	Kayser *et al.* [17]	$(TATC)_n$	Butler *et al.* [130]
DYS510	Kayser *et al.* [17]	$(TAGA)_3(TACA)(TAGA)(TACA)(TAGA)_n$	Dai *et al.* [131]
DYS513	Kayser *et al.* [17]	$(TATC)_n$	Dai *et al.* [131]
DYS520	Kayser *et al.* [17]	$(ATAG)_n\ (ATAC)_n$	Butler *et al.* [130]
DYS522	Kayser *et al.* [17]	$(GATA)_n$	Butler *et al.* [130]

Marker	Reference	Repeat motif	Reference
DYS525	Kayser et al. [17]	$(TAGA)_n$	Butler et al. [130]
DYS532	Kayser et al. [17]	$(CTTT)_n$	Butler et al. [130]
DYS533	Kayser et al. [17]	$(ATCT)_n$	Butler et al. [130]
DYS534	Kayser et al. [17]	$(CTTT)_n$	Butler et al. [130]
DYS540	Kayser et al. [17]	$(TTAT)_n$	Butler et al. [130]
DYS542	Kayser et al. [17]	$(ATAG)_2$ ATAA $(ATAG)_n$	Butler et al. [130]
DYS544	Kayser et al. [17]	$(GATA)_3$ GATG $(GATA)_n$	Dai et al. [131]
DYS552	Kayser et al. [17]	$(TCTA)_3$ TCTG $(TCTA)_n$ N_{40} $(TCTA)_n$	Dai et al. [131]
DYS556	Kayser et al. [17]	$(AATA)_n$	Butler et al. [130]
DYS557	Kayser et al. [17]	$(TTTC)_n$	Butler et al. [130]
DYS561	Kayser et al. [17]	$(GATA)_n$ $(GACA)_4$	Dai et al. [131]
DYS570	Kayser et al. [17]	$(TTTC)_n$	Butler et al. [130]
DYS575	Kayser et al. [17]	$(AAAT)_n$	Butler et al. [130]
DYS576	Kayser et al. [17]	$(AAAG)_n$	Butler et al. [130]
DYS587	Kayser et al. [17]	$(ATACA)_n$ $[(GTACA)(ATACA)]_3$	Dai et al. [131]
DYS593	Kayser et al. [17]	$(AAAAC)_2$ AAAAT $(AAAAC)_4$ $(AAAAT)_n$	Dai et al. [131]
DYS594	Kayser et al. [17]	$(TAAAA)_n$	Butler et al. [130]
DYS632	Kayser et al. [17]	$(CATT)_n$	Butler et al. [130]
DYS635 (GATA-C4)	White et al. [102]	$(TCTA)_4(TGTA)_2(TCTA)_2(TGTA)_2(TCTA)_2(TGTA)_{0.2}(TCTA)_n$	Gusmão et al. [108]
DYS641	Kayser et al. [17]	$(TAAA)_n$	Butler et al. [130]
DYS643	Kayser et al. [17]	$(CTTTT)_n$	Butler et al. [130]
GATA-A10	White et al. [102]	$(TCCA)_2(TATC)_n$	Gusmão et al. [108]
GATA-H4	White et al. [102]	$(AGAT)_4$ CTAT$(AGAT)_2(AGGT)_3(AGAT)_n$ $N_{24}(ATAG)_4(ATAC)_1(ATAG)_2$	Gusmão et al. [108]
GATA-H4.1	White et al. [102]	$(AGAT)_4$ CTAT$(AGAT)_2(AGGT)_3(AGAT)_n$	Gusmão et al. [108]

Note: [MC] Multi-copy Y-STR.

[a]Modified in order to observe the ISFG recommendations.

commercial kits are also available allowing the simultaneous typing of as much as 12 (PowerPlex® Y System [115]; Promega Corporation) and even 17 markers (AmpFlSTR YFiler Amplification Kit [116]; AB Applied Biosystems).

30.3.4.3 Population genetics and databasing

The determination of Y-STR haplotype frequencies in different populations is a main point for the correct interpretation of the genetic profile matches in kinship analysis and forensic casework.

The use of Y-STRs as inclusion evidence involves population genetic profile definition, with the elaboration of a large number of databases. Because of the lack of recombination between Y-chromosome-specific markers, they are transmitted as haplotypes in the same way as single locus alleles, and the construction of Y-specific STR databases seems to be more complex than for unlinked AS markers, since the whole haplotype must be typed for each sample. First, the informative content of Y-specific STRs, results from the possibility of constructing highly discriminative haplotypes. The suitability of the Y-STR databases for practical use will be greatly increased with the typing of each individual to as many loci as possible, instead of typing a great number of individuals for a small number of Y-STRs. On the other hand, population substructuring seems to be more severe in the case of Y chromosome than for unlinked AS markers. Due to the lower effective number of Y chromosomes in a population, Y haplotypes/haplogroups tend to present a higher proportion of variation between populations than that observed for other markers located on autosomes or X chromosome. The interpopulational variability on Y-profiles makes the definition of local databases crucial for the application of Y-specific markers in practice. DNA Commission of the International Society of Forensic Genetics (ISFG) recommended the use of regional Y-STR haplotype databases, and pooling data from different regions is only valid after verifying that no population substructure exists [109].

A large amount of Y-haplotype data is, therefore, essential for two main reasons: (a) in match probability calculation, since it is not valid to multiply the allele frequency of each locus, a large number of haplotypes are needed to allow reliable frequency estimates; (b) population substructure analysis is highly dependent on the amount of data available.

Due to their importance in forensics, Y-STR haplotype distributions in populations worldwide have been made available not only through classical publications but also, more recently, through large-scale forensic databases. The YHRD is the most extensive survey available on line and the development of these databases is important not only for haplotype frequency estimation and subsequent application for match probability calculations in forensic studies, but also for performing comparative population analysis.

30.3.4.4 Y-STR mutation rates

According to data on autosomal STRs, the frequency of mutation events in the male germ line is higher than in the female germ line [45,117]. In a survey on Y and

TABLE 30.3
Y-STR MUTATION RATES

Locus	Nr. Mutations	Total[a]		
		Allele Transmissions	Frequency ($\times 10^{-3}$)	95% CI $\times 10^{-3}$
DYS19	13	7314	1.7774	0.947–3.038
DYS389 I	11	5518	1.9935	0.996–3.564
DYS389 II	12	5505	2.1798	1.127–3.805
DYS390	15	6796	2.2072	1.236–3.638
DYS391	23	6744	3.4104	2.163–5.113
DYS392	4	6710	0.5961	0.162–1.526
DYS393	4	5498	0.7275	0.198–1.862
DYS385	24	10207	2.3513	1.507–3.497
DYS437	5	2437	2.0517	0.667–4.781
DYS438	1	2476	0.4039	0.010–2.248
DYS439	12	2451	4.8960	2.532–8.537
GATA A10	4	946	4.228	1.153–10.971
DYS460	5	1109	4.509	1.465–10.490
DYS461	0	873	–	0.000–4.217
DYS635	3	873	3.436	0.709–10.010
GATA H4	3	1036	2.896	0.598–8.439
Total	139	66493	2.090	1.758–2.468

[a]Includes data from Heyer *et al.* [132]; Bianchi *et al.* [133]; Kayser *et al.* [46]; Dupuy *et al.* [47]; Kurihara *et al.* [125]; Góes *et al.* [134]; Budowle *et al.* [135]; Ballard *et al.* [136]; Gusmão *et al.* [48]; Turrina *et al.* [43]; and Domingues *et al.* [137].

X-linked loci, Scozzari *et al.* [118] report an overall higher diversity for the Y-linked loci and suggest a higher rate of accumulation of variants on this chromosome. These results can be explained by the higher number of divisions involved in male gametogenesis when compared with the female one, and should be reflected in a higher mutation rate on Y-STRs.

A large amount of data is necessary to estimate reliable allele specific mutation rates, essential for consistent dating of Y-SNP defined lineages (haplogroups) and data interpretation in kinship analysis. Although studies on Y-STR mutation rates are still scarce and have only considered a restricted number of markers, no significant differences were found between the average Y-STR mutation rates (see Table 30.3) and those found in autosomal STRs [45,119,120]. Data on Y-STR mutations also support that slippage is the mechanism involved, and that there is an agreement with the generally accepted single-step mutation model for microsatellites in which the alleles are known to mutate primarily through the gain and loss of single repeat units [117,121,122].

STR mutation rates present not only inter- but also intra-locus variation, depending on the locus structure and allele length e.g. [45,123]. When compared to the average value, a significantly lower mutation rate was observed at DYS392 and higher at DYS391 and DYS439 loci [48]. For TAGA repeats, a trend for higher mutability was confirmed for longer alleles.

References pp. 992–1000

In Y-STRs, repeat gains were found to be twice more frequent than losses [46–48,124,125], as expected for young microsatellites [126]. When comparing repeat gains and losses in different allele classes, Gusmão *et al.* [48] found no evidence for an excess of repeat losses at longer alleles supporting that this mechanism is biased toward microsatellite expansion e.g. [45,127,128] until a certain repeat length at which the rates of expansion and contraction mutations are equal [126].

30.4 ACKNOWLEDGMENTS

IPATIMUP is partially supported by Fundação para a Ciência e a Tecnologia (through POCI, Programa Operacional Ciência e Inovação). Iva Gomes is a recipient of a grant from Fundação para a Ciência e a Tecnologia (SFRH/BD/21647/2005). CeGen (National Genotyping Center of Spain) is funded by "Fundacion Genoma España."

30.5 REFERENCES

1   H. Skaletsky, T. Kuroda-Kawaguchi, P.J. Minx, H.S. Cordum, L. Hillier, L.G. Brown, S. Repping, T. Pyntikova, J. Ali, T. Bieri, A. Chinwalla, A. Delehaunty, K. Delehaunty, H. Du, G. Fewell, L. Fulton, R. Fulton, T. Graves, S.F. Hou, P. Latrielle, S. Leonard, E. Mardis, R. Maupin, J. McPherson, T. Miner, W. Nash, C. Nguyen, P. Ozersky, K. Pepin, S. Rock, T. Rohlfing, K. Scott, B. Schultz, C. Strong, A. Tin-Wollam, S.P. Yang, R.H. Waterston, R.K. Wilson, S. Rozen and D.C. Page, The male-specific region of the human Y chromosome is a mosaic of discrete sequence classes. Nature, 423 (2003) 825–837.
2   S. Foote, D. Vollrath, A. Hilton and D.C. Page, The human Y chromosome: overlapping DNA clones spanning the euchromatic region. Science, 258 (1992) 60–66.
3   B. Charlesworth, The evolution of sex chromosomes. Science, 251 (1991) 1030–1033.
4   B.T. Lahn and D.C. Page, Four evolutionary strata on the human X chromosome. Science, 286 (1999) 964–967.
5   S. Rozen, H. Skaletsky, J.D. Marszalek, P.J. Minx, H.S. Cordum, R.H. Waterston, R.K. Wilson and D.C. Page, Abundant gene conversion between arms of palindromes in human and ape Y chromosomes. Nature, 423 (2003) 873–876.
6   J.F. Hughes, H. Skaletsky, T. Pyntikova, P.J. Minx, T. Graves, S. Rozen, R.K. Wilson and D.C. Page, Conservation of Y-linked genes during human evolution revealed by comparative sequencing in chimpanzee. Nature, 437 (2005) 100–103.
7   M. Casanova, P. Leroy, C. Boucekkine, J. Weissenbach, C. Bishop, M. Fellous, M. Purrello, G. Fiori and M. Siniscalco, A human Y-linked DNA polymorphism and its potential for estimating genetic and evolutionary distance. Science, 230 (1985) 1403–1406.
8   S. Jakubiczka, J. Arnemann, H.J. Cooke, M. Krawczak and J. Schmidtke, A search for restriction fragment length polymorphism on the human Y chromosome. Hum. Genet., 84 (1989) 86–88.
9   P. Malaspina, F. Persichetti, A. Novelletto, C. Iodice, L. Terrenato, J. Wolfe, M. Ferraro and G. Prantera, The human Y chromosome shows a low level of DNA polymorphism. Ann. Hum. Genet., 54 (1990) 297–305.
10 A. Spurdle and T. Jenkins, The search for Y chromosome polymorphism is extended to negroids. Hum. Mol. Genet., 1 (1992) 169–170.
11 R.L. Dorit, H. Akashi and W. Gilbert, Absence of polymorphism at the ZFY locus on the human Y chromosome. Science, 268 (1995) 1183–1185.
12 L.S. Whitfield, J.E. Sulston and P.N. Goodfellow, Sequence variation of the human Y chromosome. Nature, 378 (1995) 379–380.
13 M.A. Jobling, A survey of long-range DNA polymorphisms on the human Y chromosome. Hum. Mol. Genet., 3 (1994) 107–114.

14   A. Sajantila, A.H. Salem, P. Savolainen, K. Bauer, C. Gierig and S. Pääbo, Paternal and maternal DNA lineages reveal a bottleneck in the founding of the Finnish population. Proc. Natl. Acad. Sci., 93 (1996) 12035–12039.

15   M.A. Jobling and C. Tyler-Smith, Fathers and sons: the Y chromosome and human evolution. Trends Genet., 11 (1995) 449–456.

16   P.A. Underhill, P. Shen, A.A. Lin, L. Jin, G. Passarino, W.H. Yang, E. Kauffman, B. Bonne-Tamir, J. Bertranpetit, P. Francalacci, M. Ibrahim, T. Jenkins, J.R. Kidd, S.Q. Mehdi, M.T. Seielstad, R.S. Wells, A. Piazza, R.W. Davis, M.W. Feldman, L.L. Cavalli-Sforza and P.J. Oefner, Y chromosome sequence variation and the history of human populations. Nat. Genet., 26 (2000) 358–361.

17   M. Kayser, R. Kittler, A. Erler, M. Hedman, A.C. Lee, A. Mohyuddin, S.Q. Mehdi, Z. Rosser, M. Stoneking, M.A. Jobling, A. Sajantila and C. Tyler-Smith, A comprehensive survey of human Y-chromosomal microsatellites. Am. J. Hum. Genet., 74 (2004) 1183–1197.

18   M.A. Jobling, A. Pandya and C. Tyler-Smith, The Y chromosome in forensic analysis and paternity testing. Int. J. Legal Med., 110 (1997) 118–124.

19   M. Kayser, A. Caglià, D. Corach, N. Fretwell, C. Gehrig, G. Graziosi, F. Heidorn, S. Herrmann, B. Herzog, M. Hidding, K. Honda, M. Jobling, M. Krawczak, K. Leim, S. Meuser, E. Meyer, W. Oesterreich, A. Pandya, W. Parson, G. Penacino, A. Perez-Lezaun, A. Piccinini, M. Prinz, C. Schmitt, P.M. Schneider, R. Szibor, J. Teifel-Greding, G. Weichhold, P. de Knijff and L. Roewer, Evaluation of Y-chromosomal STRs: a multicenter study. Int. J. Legal Med., 110 (1997) 125–149.

20   M. Prinz, A. Ishii, A. Coleman, H.J. Baum and R.C. Shaler, Validation and casework application of a Y chromosome specific STR multiplex. Forensic Sci. Int., 120 (2001) 177–188.

21   L. Roewer, M. Krawczak, S. Willuweit, M. Nagy, C. Alves, A. Amorim, K. Anslinger, C. Augustin, A. Betz, E. Bosch, A. Caglia, A. Carracedo, D. Corach, A.F. Dekairelle, T. Dobosz, B.M. Dupuy, S. Furedi, C. Gehrig, L. Gusmão, J. Henke, L. Henke, M. Hidding, C. Hohoff, B. Hoste, M.A. Jobling, H.J. Kargel, P. de Knijff, R. Lessig, E. Liebeherr, M. Lorente, B. Martinez-Jarreta, P. Nievas, M. Nowak, W. Parson, V.L. Pascali, G. Penacino, R. Ploski, B. Rolf, A. Sala, U. Schmidt, C. Schmitt, P.M. Schneider, R. Szibor, J. Teifel-Greding and M. Kayser, Online reference database of European Y-chromosomal short tandem repeat (STR) haplotypes. Forensic Sci. Int., 118 (2001) 106–113.

22   A. Dettlaff-Kakol and R. Pawlowski, First Polish DNA "manhunt": an application of Y-chromosome STRs. Int. J. Legal Med., 116 (2002) 289–291.

23   M. Prinz, K. Boll, H. Baum and B. Shaler, Multiplexing of Y chromosome specific STRs and performance for mixed samples. Forensic Sci. Int., 85 (1997) 209–218.

24   R. Zehner and U. Bohrer, DYS19 and amelogenin in artificial blood stains with defined amounts of male and female cells. Int. J. Legal Med., 111 (1998) 340–342.

25   U.B. Gyllensten, A. Josefsson, K. Schemschat, T. Saldeen and U. Petterson, DNA typing of forensic material with mixed genotypes using allele-specific enzymatic amplification (polymerase chain reaction). Forensic Sci. Int., 52 (1992) 149–160.

26   J. Kreike and A. Lehner, Sex determination and DNA competition in the analysis of forensic mixed stains by PCR. Int. J. Legal Med., 107 (1995) 235–238.

27   A. Betz, G. Babler, G. Dietl, X. Steil, G. Weyermann and W. Pflug, DYS STR analysis with epithelial cells in a rape case. Forensic Sci. Int., 118 (2001) 126–130.

28   R. Szibor, M. Krawczak, S. Hering, J. Edelmann, E. Kuhlisch and D. Krause, Use of X-linked markers for forensic purposes. Int. J. Legal Med., 117 (2003) 67–74.

29   R. Szibor, I. Plate, J. Edelmann, S. Hering, E. Kuhlisch, M. Michael and D. Krause, Chromosome X haplotyping in deficiency paternity testing principles and case report. Int. Congr. Ser., 1239 (2003) 815–820.

30   R. Szibor, S. Hering, E. Kuhlisch, I. Plate, S. Demberger, M. Krawczak and J. Edelmann, Haplotyping of STR cluster DXS6801-DXS6809-DXS6789 on Xq21 provides a powerful tool for kinship testing. Int. J. Legal Med., 119 (2005) 363–369.

31   S. Hering, C. Augustin, J. Edelmann, M. Heidel, J. Dressler, H. Rodig, E. Kuhlisch and R. Szibor, DXS10079, DXS10074 and DXS10075 are STRs located within a 280 kb region of Xq 12 and provide stable haplotypes useful for complex kinship cases. Int. J. Legal Med., 120 (2006) 337–345.

32 C. Toni, S. Presciuttini, I. Spinetti and R. Domenici, Population data of four X-chromosome markers in Tuscany, and their use in a deficiency paternity case. Forensic Sci. Int., 137 (2003) 215–216.

33 S. Turrina, R. Atzei and D. De Leo, Polymorphism of four X-chromosomal STRs: DXS7423, DXS7424, DXS8378 and DXS6809 in a North Italian population sample and their use in kinship testing. Forensic Sci. Int., 168 (2007) 241–243.

34 J. Edelmann, R. Lessig, M. Klintschar and R. Szibor, Advantages of X-chromosomal microsatellites in deficiency paternity testing: presentation of cases. Int. Congr. Ser., 1261 (2004) 257–259.

35 S.F. Schaffner, The X chromosome in population genetics. Nat. Rev. Genet., 5 (2004) 43–51.

36 C. Bini, S. Ceccardi, G. Ferri, S. Pelotti, M. Alu', E. Rocaglia, G. Beduschi, L. Caenazzo, E. Ponzano, P. Tasinato, C. Turchi, M. Mazzanti, A. Tagliabracci, C. Toni, I. Spinetti, R. Domenico and S. Presciuttini, Development of a heptaplex PCR system to analyse X-chromosome STR loci from five Italian population samples: a collaborative study. Forensic Sci. Int., 153 (2005) 231–236.

37 C. Robino, A. Giolitti, S. Gino and C. Torre, Development of a heptaplex PCR system to analyse X-chromosome STR loci from five Italian population samples: a collaborative study. Int. J. Legal Med., 120 (2006) 315–318.

38 I. Gomes, M. Prinz, P. Pereira, C. Meyers, R.S. Mikulasovich, A. Amorim, A. Carracedo and L. Gusmão, Genetic analysis of three US population groups using an X-chromosomal STR decaplex. Int. J. Legal Med., 121 (2007) 198–203.

39 M.A. Moreno, J.J. Builes, P. Jaramillo, C. Espinal, D. Aguirre, M.M. Pancorbo, L. Gusmão and M.L.J. Bravo, Validation of five X-chromosomal STRs DXS6800, DXS6807, DXS6798, DXS8377 and DXS7423 in an Antioquian population sample. Int. Congr. Ser., 1288 (2006) 295–297.

40 R. Szibor, S. Hering and J. Edelmann, A new Web site compiling forensic chromosome X research is now online. Int. J. Legal Med., 120 (2006) 252–254.

41 M.T. Zarrabeitia, T. Amigo, C. Sanudo, A. Zarrabeitia, D. González-Lamuño and J.A. Riancho, A new pentaplex system to study short tandem repeat markers of forensic interest on X chromosome. Forensic Sci. Int., 129 (2002) 85–89.

42 K.J. Shin, B.K. Kwon, S.S. Lee, J.E. Yoo, M.J. Park, U. Chung, H.Y. Lee, G.R. Han, J.H. Choi and C.Y. Kim, Five highly informative X-chromosomal STRs in Koreans. Int. J. Legal Med., 118 (2004) 37–40.

43 S. Turrina and D. De Leo, Population genetic comparisons of three X-chromosomal STRs (DXS7132, DXS7133 and GATA172D05) in North and South Italy. Int. Congr. Ser., 1261 (2004) 302–304.

44 M. Poetsch, H. Petersmann, A. Repenning and E. Lignitz, Development of two pentaplex systems with X-chromosomal STR loci and their allele frequencies in a northeast German population. Forensic Sci. Int., 155 (2005) 71–76.

45 B. Brinkmann, M. Klintschar, F. Neuhuber, J.B. Hühne and B. Rolf, Mutation rate in human microsatellites: influence of the structure and length of the tandem repeat. Am. J. Hum. Genet., 62 (1998) 1408–1415.

46 M. Kayser, L. Roewer, M. Hedman, L. Henke, J. Henke, S. Brauer, C. Krüger, M. Krawczak, M. Nagy, T. Dobosz, R. Szibor, P. de Knijff, M. Stoneking and A. Sajantila, Characteristics and frequency of germline mutations at microsatellite loci from the human Y chromosome as revealed by direct observation in father/son pairs. Am. J. Hum. Genet., 66 (2000) 1580–1588.

47 B.M. Dupuy, M. Stenersen, T. Egeland and B. Olaisen, Y-chromosomal microsatellite mutation rates: differences in mutation rate between and within loci. Hum. Mutat., 23 (2004) 117–124.

48 L. Gusmão, P. Sánchez-Diz, F. Calafell, P. Martín, C.A. Alonso, F. Álvarez-Fernández, C. Alves, L. Borjas-Fajardo, W.R. Bozzo, M.L. Bravo, J.J. Builes, J. Capilla, M. Carvalho, C. Castillo, C.I. Catanesi, D. Corach, A.M. Di Lonardo, R. Espinheira, E. Fagundes de Carvalho, M.J. Farfán, H.P. Figueiredo, I. Gomes, M.M. Lojo, M. Marino, M.F. Pinheiro, M.L. Pontes, V. Prieto, E. Ramos-Luis, J.A. Riancho, A.C. Souza Go'es, O.A. Santapa, D.R. Sumita, G. Vallejo, L. Vidal Rioja, M.C. Vide, C.I. Vieira da Silva, M.R. Whittle, W. Zabala, M.T. Zarrabeitia, A. Alonso, A. Carracedo and A. Amorim, Mutation rates at Y chromosome specific microsatellites. Hum. Mutat., 26 (2005) 520–528.

49 V.L. Pascali, M. Dobosz and B. Brinkmann, Coordinating Y-chromosomal STR research for the Courts. Int. J. Legal Med., 112 (1999) 1.

50 M.E. Hurles, C. Irven, J. Nicholson, P.G. Taylor, F.R. Santos, J. Loughlin, M.A. Jobling and B.C. Sykes, European Y-chromosomal lineages in Polynesians: a contrast to the population structure revealed by mtDNA. Am. J. Hum. Genet., 63 (1998) 1793–1806.

51 M.A. Jobling and C. Tyler-Smith, The human Y chromosome: an evolutionary marker comes of age. Nat. Rev. Genet., 4 (2003) 598–612.

52 P.A. Underhill, P. Shen, A.A. Lin, L. Jin, G. Passarino, W.H. Yang, E. Kauffman, B. Bonne-Tamir, J. Bertranpetit, P. Francalacci, M. Ibrahim, T. Jenkins, J.R. Kidd, S.Q. Mehdi, M.T. Seielstad, R.S. Wells, A. Piazza, R.W. Davis, M.W. Feldman, L.L. Cavalli-Sforza and P.J. Oefner, Y chromosome sequence variation and the history of human populations. Nat. Genet., 26 (2000) 358–361.

53 P.A. Underhill, G. Passarino, A.A. Lin, P. Shen, M. Mirazon Lahr, R.A. Foley, P.J. Oefner and L.L. Cavalli-Sforza, The phylogeography of Y chromosome binary haplotypes and the origins of modern human populations. Ann. Hum. Genet., 65 (2001) 43–62.

54 L. Quintana-Murci, C. Krausz, T. Zerjal, S.H. Sayar, M.F. Hammer, S.Q. Mehdi, Q. Ayub, R. Qamar, A. Mohyuddin, U. Radhakrishna, M.A. Jobling, C. Tyler-Smith and K. McElreavey, Y-chromosome lineages trace diffusion of people and languages in southwestern Asia. Am. J. Hum. Genet., 6 (2001) 537–542.

55 O. Semino, G. Passarino, P.J. Oefner, A.A. Lin, S. Arbuzova, L.E. Beckman, G. De Benedictis, P. Francalacci, A. Kouvatsi, S. Limborska, M. Marcikiae, A. Mika, B. Mika, D. Primorac, A.S. Santachiara-Benerecetti, L.L. Cavalli-Sforza and P.A. Underhill, The genetic legacy of paleolithic *Homo sapiens sapiens* in extant Europeans: a Y chromosome perspective. Science, 290 (2000) 1155–1159.

56 L.M. Sims, D. Garvey and J. Ballantyne, Sub-populations within the major European and African derived haplogroups R1b3 and E3a are differentiated by previously phylogenetically undefined Y-SNPs. Hum. Mutat., 28 (2007) 97.

57 F. Cruciani, R. La Fratta, A. Torroni, P.A. Underhill and R. Scozzari, Molecular dissection of the Y chromosome haplogroup E-M78 (E3b1a): a posteriori evaluation of a microsatellite-network-based approach through six new biallelic markers. Hum. Mutat., 27 (2006) 831–832.

58 F. Cruciani, P. Santolamazza, P. Shen, V. Macaulay, P. Moral, A. Olckers, D. Modiano, S. Holmes, G. Destro-Bisol, V. Coia, D.C. Wallace, P.J. Oefner, A. Torroni, L.L. Cavalli-Sforza, R. Scozzari and P.A. Underhill, A back migration from Asia to sub-Saharan Africa is supported by high-resolution analysis of human Y-chromosome haplotypes. Am. J. Hum. Genet., 70 (2002) 1197–1214.

59 M. Brion, B. Quintans, M. Zarrabeitia, A. Gonzalez-Neira, A. Salas, V. Lareu, C. Tyler-Smith and A. Carracedo, Micro-geographical differentiation in Northern Iberia revealed by Y-chromosomal DNA analysis. Gene, 329 (2004) 17–25.

60 M.F. Hammer, V.F. Chamberlain, V.F. Kearney, D. Stover, G. Zhang, T. Karafet, B. Walsh and A.J. Redd, Population structure of Y chromosome SNP haplogroups in the United States and forensic implications for constructing Y chromosome STR databases. Forensic Sci. Int., 164 (2006) 45–55.

61 O. Semino, A.S. Santachiara-Benerecetti, F. Falaschi, L.L. Cavalli-Sforza and P.A. Underhill, Ethiopians and Khoisan share the deepest clades of the human Y-chromosome phylogeny. Am. J. Hum. Genet., 70 (2002) 265–268.

62 T.M. Karafet, L. Xu, R. Du, W. Wang, S. Feng, R.S. Wells, A.J. Redd, S.L. Zegura and M.F. Hammer, Paternal population history of East Asia: sources, patterns, and microevolutionary processes. Am. J. Hum. Genet., 69 (2001) 615–628.

63 M. Kayser, S. Brauer, G. Weiss, W. Schiefenhövel, P. Underhill, P. Shen, P. Oefner, M. Tommaseo-Ponzetta and M. Stoneking, Reduced Y-chromosome, but not mitochondrial DNA, diversity in human populations from West New Guinea. Am. J. Hum. Genet., 72 (2003) 281–302.

64 R.S. Wells, N. Yuldasheva, R. Ruzibakiev, P.A. Underhill, I. Evseeva, J. Blue-Smith, L. Jin, B. Su, R. Pitchappan, S. Shanmugalakshmi, K. Balakrishnan, M. Read, N.M. Pearson, T. Zerjal, M.T. Webster, I. Zholoshvili, E. Jamarjashvili, S. Gambarov, B. Nikbin, A. Dostiev, O. Aknazarov, P. Zalloua, I. Tsoy, M. Kitaev, M. Mirrakhimov, A. Chariev and W.F. Bodmer, The Eurasian heartland: a continental perspective on Y-chromosome diversity. Proc. Natl. Acad. Sci., 98 (2001) 10244–10249.

65   T. Zerjal, R.S. Wells, N. Yuldasheva, R. Ruzibakiev and C. Tyler-Smith, A genetic landscape
      reshaped by recent events: Y-chromosomal insights into central Asia. Am. J. Hum. Genet., 71
      (2002) 466–482.
66   M.F. Hammer, T.M. Karafet, H. Park, K. Omoto, S. Harihara, M. Stoneking and S. Horai, Dual
      origins of the Japanese: common ground for hunter-gatherer and farmer Y chromosomes. J. Hum.
      Genet., 51 (2006) 47–58.
67   F. Cruciani, R. La Fratta, P. Santolamazza, D. Sellitto, R. Pascone, P. Moral, E. Watson, V.
      Guida, E.B. Colomb, B. Zaharova, J. Lavinha, G. Vona, R. Aman, F. Cali, N. Akar, M. Richards,
      A. Torroni, A. Novelletto and R. Scozzari, Phylogeographic analysis of haplogroup E3b (E-M215)
      Y chromosomes reveals multiple migratory events within and out of Africa. Am. J. Hum. Genet.,
      74 (2004) 1014–1022.
68   T. Kivisild, S. Rootsi, M. Metspalu, S. Mastana, K. Kaldma, J. Parik, E. Metspalu, M. Adojaan,
      H.V. Tolk, V. Stepanov, M. Golge, E. Usanga, S.S. Papiha, C. Cinnioglu, R. King, L. Cavalli-
      Sforza, P.A. Underhill and R. Villems, The genetic heritage of the earliest settlers persists both in
      Indian tribal and caste populations. Am. J. Hum. Genet., 72 (2003) 313–332.
69   C. Cinnioglu, R. King, T. Kivisild, E. Kalfoglu, S. Atasoy, G.L. Cavalleri, A.S. Lillie, C.C.
      Roseman, A.A. Lin, K. Prince, P.J. Oefner, P. Shen, O. Semino, L.L. Cavalli-Sforza and P.A.
      Underhill, Excavating Y-chromosome haplotype strata in Anatolia. Hum. Genet., 114 (2004)
      127–148.
70   I. Nasidze, D. Quinque, I. Dupanloup, S. Rychkov, O. Naumova, O. Zhukova and M. Stoneking,
      Genetic evidence concerning the origins of South and North Ossetians. Ann. Hum. Genet., 68
      (2004) 588–599.
71   S. Sengupta, L.A. Zhivotovsky, R. King, S.Q. Mehdi, C.A. Edmonds, C.E. Chow, A.A. Lin, M.
      Mitra, S.K. Sil, A. Ramesh, M.V. Usha Rani, C.M. Thakur, L.L. Cavalli-Sforza, P.P. Majumder
      and P.A. Underhill, Polarity and temporality of high-resolution Y-chromosome distributions in
      India identify both indigenous and exogenous expansions and reveal minor genetic influence of
      Central Asian pastoralists. Am. J. Hum. Genet., 78 (2006) 202–221.
72   S. Rootsi, C. Magri, T. Kivisild, G. Benuzzi, H. Help, M. Bermisheva, I. Kutuev, L. Barac, M.
      Pericic, O. Balanovsky, A. Pshenichnov, D. Dion, M. Grobei, L.A. Zhivotovsky, V. Battaglia, A.
      Achilli, N. Al-Zahery, J. Parik, R. King, C. Cinnioglu, E. Khusnutdinova, P. Rudan, E. Ba-
      lanovska, W. Scheffrahn, M. Simonescu, A. Brehm, R. Goncalves, A. Rosa, J.P. Moisan, A.
      Chaventre, V. Ferak, S. Furedi, P.J. Oefner, P. Shen, L. Beckman, I. Mikerezi, R. Terzic, D.
      Primorac, A. Cambon-Thomsen, A. Krumina, A. Torroni, P.A. Underhill, A.S. Santachiara-
      Benerecetti, R. Villems and O. Semino, Phylogeography of Y-chromosome haplogroup I
      reveals distinct domains of prehistoric gene flow in Europe. Am. J. Hum. Genet., 75 (2004)
      128–137.
73   F. Di Giacomo, F. Luca, L.O. Popa, N. Akar, N. Anagnou, J. Banyko, R. Brdicka, G. Barbujani,
      F. Papola, G. Ciavarella, F. Cucci, L. Di Stasi, L. Gavrila, M.G. Kerimova, D. Kovatchev, A.I.
      Kozlov, A. Loutradis, V. Mandarino, C. Mammi, E.N. Michalodimitrakis, G. Paoli, K.I. Pappa,
      G. Pedicini, L. Terrenato, S. Tofanelli, P. Malaspina and A. Novelletto, Y chromosomal haplo-
      group J as a signature of the post-neolithic colonization of Europe. Hum. Genet., 115 (2004)
      357–371.
74   C. Capelli, J.F. Wilson, M. Richards, M.P.H. Stumpf, F. Gratrix, S. Oppenheimer, P. Underhill,
      V.L. Pascali, T.M. Ko and D.B. Goldstein, A predominantly indigenous paternal heritage for the
      Austronesian-speaking peoples of insular Southeast Asia and Oceania. Am. J. Hum. Genet., 68
      (2001) 432–443.
75   M.E. Hurles, J. Nicholson, E. Bosch, C. Renfrew, B.C. Sykes and M.A. Jobling, Y chro-
      mosomal evidence for the origins of Oceanic-speaking peoples. Genetics, 160 (2002)
      289–303.
76   M.P. Cox and M.M. Lahr, Y-chromosome diversity is inversely associated with language affil-
      iation in paired Austronesian- and Papuan-speaking communities from Solomon Islands. Am. J.
      Hum. Biol., 18 (2006) 35–50.

77   H. Shi, Y.L. Dong, B. Wen, C.J. Xiao, P.A. Underhill, P.D. Shen, R. Chakraborty, L. Jin and B. Su, Y-chromosome evidence of southern origin of the East Asian-specific haplogroup O3-M122. Am. J. Hum. Genet., 77 (2005) 408–419.

78   M.C. Bortolini, F.M. Salzano, M.G. Thomas, S. Stuart, S.P. Nasanen, C.H. Bau, M.H. Hutz, Z. Layrisse, M.L. Petzl-Erler, L.T. Tsuneto, K. Hill, A.M. Hurtado, D. Castro-de-Guerra, M.M. Torres, H. Groot, R. Michalski, P. Nymadawa, G. Bedoya, N. Bradman, D. Labuda and A. Ruiz-Linares, Y-chromosome evidence for differing ancient demographic histories in the Americas. Am. J. Hum. Genet., 73 (2003) 524–539.

79   S.L. Zegura, T.M. Karafet, L.A. Zhivotovsky and M.F. Hammer, High-resolution SNPs and microsatellite haplotypes point to a single, recent entry of Native American Y chromosomes into the Americas. Mol. Biol. Evol., 21 (2004) 164–175.

80   C. Capelli, N. Redhead, V. Romano, F. Cali, G. Lefranc, V. Delague, A. Megarbane, A.E. Felice, V.L. Pascali, P.I. Neophytou, Z. Poulli, A. Novelletto, P. Malaspina, L. Terrenato, A. Berebbi, M. Fellous, M.G. Thomas and D.B. Goldstein, Population structure in the Mediterranean basin: a Y chromosome perspective. Ann. Hum. Genet., 70 (2006) 207–225.

81   M. Brion, J.J. Sanchez, K. Balogh, C. Thacker, A. Blanco-Verea, C. Borsting, B. Stradmann-Bellinghausen, M. Bogus, D. Syndercombe-Court, P.M. Schneider, A. Carracedo and N. Morling, Introduction of a single nucleotide polymorphism-based "Major Y-chromosome haplogroup typing kit" suitable for predicting the geographical origin of male lineages. Electrophoresis, 26 (2005) 4411–4420.

82   V. Onofri, F. Alessandrini, C. Turchi, M. Pesaresi, L. Buscemi and A. Tagliabracci, Development of multiplex PCRs for evolutionary and forensic applications of 37 human Y chromosome SNPs. Forensic Sci. Int., 157 (2006) 23–35.

83   S. Paracchini, B. Arredi, R. Chalk and C. Tyler-Smith, Hierarchical high-throughput SNP genotyping of the human Y chromosome using MALDI-TOF mass spectrometry. Nucleic Acids Res., 30 (2002) e27.

84   M. Raitio, K. Lindroos, M. Laukkanen, T. Pastinen, P. Sistonen, A. Sajantila and A.C. Syvanen, Y-chromosomal SNPs in Finno-Ugric-speaking populations analyzed by minisequencing on microarrays. Genome Res., 11 (2001) 471–482.

85   M. Lareu, B. Sobrino, C. Phillips, M. Brión and A. Carracedo, Typing Y-chromosome single nucleotide polymorphisms with DNA microarray technology. Prog. Forensic Genet., 9 (2003) 21–25.

86   J.H. Wetton, K.W. Tsang and H. Khan, Inferring the population of origin of DNA evidence within the UK by allele-specific hybridization of Y-SNPs. Forensic Sci. Int., 152 (2005) 45–53.

87   A.J. Jeffreys, A. MacLeod, K. Tamaki, D.L. Neil and D.G. Monckton, Minisatellite repeat coding as a digital approach to DNA typing. Nature, 354 (1991) 204–209.

88   M.A. Jobling, N. Bouzekri and P.G. Taylor, Hypervariable digital DNA codes for human paternal lineages: MVR-PCR at the Y-specific minisatellite, MSY1 (DYF155S1). Hum. Mol. Genet., 7 (1998) 643–653.

89   N. Bouzekri, P.G. Taylor, M.F. Hammer and M.A. Jobling, Novel mutation processes in the evolution of a haploid minisatellite, MSY1: array homogenization without homogenization. Hum. Mol. Genet., 7 (1998) 655–659.

90   A.J. Jeffreys, K. Tamaki, A. MacLeod, D.G. Monckton, D.L. Neil and J.A. Armour, Complex gene conversion events in germline mutation at human minisatellites. Nat. Genet., 6 (1994) 136–145.

91   A.J. Jeffreys, M.J. Allen, J.A.L. Armour, A. Collick, Y. Dubrova, N. Fretwell, T. Guram, M. Jobling, C.A. May, D.L. Neil and R. Neumann, Mutation processes at human minisatellites. Electrophoresis, 16 (1995) 1577–1585.

92   M. Brion, R. Cao, A. Salas, M.V. Lareu and A. Carracedo, New method to measure minisatellite variant repeat variation in population genetic studies. Am. J. Hum. Biol., 14 (2002) 421–428.

93   W. Bao, S. Zhu, A. Pandya, T. Zerjal, J. Xu, Q. Shu, R. Du, H. Yang and C. Tyler-Smith, MSY2: a slowly evolving minisatellite on the human Y chromosome which provides a useful polymorphic marker in Chinese populations. Gene, 244 (2000) 29–33.

References pp. 992–1000

94   C. Tyler-Smith, R.J. Oakey, Z. Larin, R.B. Fisher, M. Crocker, N.A. Affara, M.A. Ferguson-Smith, M. Muenke, O. Zuffardi and M.A. Jobling, Localization of DNA sequences required for human centromere function through an analysis of rearranged Y chromosomes. Nat. Genet., 5 (1993) 368–375.

95   R. Oakey and C. Tyler-Smith, Y chromosome DNA haplotyping suggests that most European and Asian men are descended from one of two males. Genomics, 7 (1990) 325–330.

96   N. Mathias, M. Bayes and C. Tyler-Smith, Highly informative compound haplotypes for the human Y chromosome. Hum. Mol. Genet., 3 (1994) 115–123.

97   N.O. Bianchi, G. Bailliet, C.M. Bravi, R.F. Carnese, F. Rothhammer, V.L. Martinez-Marignac and S.D. Pena, Origin of Amerindian Y-chromosomes as inferred by the analysis of six polymorphic markers. Am. J. Phys. Anthropol., 102 (1997) 79–89.

98   F.R. Santos, A. Pandya, C. Tyler-Smith, S.D. Pena, M. Schanfield, W.R. Leonard, L. Osipova, M.H. Crawford and R.J. Mitchell, The central Siberian origin for native American Y chromosomes. Am. J. Hum. Genet., 64 (1999) 619–628.

99   F.R. Santos, S.D. Pena and C. Tyler-Smith, PCR haplotypes for the human Y chromosome based on alphoid satellite DNA variants and heteroduplex analysis. Gene, 165 (1995) 191–198.

100   L. Roewer and J.T. Epplen, Rapid and sensitive typing of forensic stains by PCR amplification of polymorphic simple repeat sequences in case work. Forensic Sci. Int., 53 (1992) 163–171.

101   P. de Knijff, M. Kayser, A. Caglià, D. Corach, N. Fretwell, C. Gehrig, G. Graziosi, F. Heidorn, S. Herrmann, B. Herzog, M. Hidding, K. Honda, M. Jobling, M. Krawczak, K. Leim, S. Meuser, E. Meyer, W. Oesterreich, A. Pandya, W. Parson, G. Penacino, A. Perez-Lezaun, A. Piccinini, M. Prinz, C. Schmitt, P.M. Schneider, R. Szibor, J. Teifel-Greding, G. Weichhold and L. Roewer, Chromosome Y microsatellites: population genetic and evolutionary aspects. Int. J. Legal Med., 110 (1997) 134–149.

102   P.S. White, O.L. Tatum, L.L. Deaven and J.L. Longmire, New, male-specific microsatellite markers from the human Y chromosome. Genomics, 57 (1999) 433–437.

103   Q. Ayub, A. Mohyuddin, R. Qamar, K. Mazhar, T. Zerjal and C. Tyler-Smith, Identification and characterisation of novel human Y-chromosomal microsatellites from sequence database information. Nucleic Acids Res., 28 (2000) e8.

104   R. Iida and K. Kishi, Identification, characterization and forensic application of novel Y-STRs. Legal Med., 7 (2005) 255–258 (Tokyo).

105   A.J. Redd, A.B. Agellon, V.A. Kearney, V.A. Contreras, T. Karafet, H. Park, P. de Knijff, J.M. Butler and M.F. Hammer, Forensic value of 14 novel STRs on the human Y chromosome. Forensic Sci. Int., 130 (2002) 97–111.

106   M.A. Jobling, V. Samara, A. Pandya, N. Fretwell, B. Bernasconi, R.J. Mitchell, T. Gerelsaikhan, B. Dashnyam, A. Sajantila, P.J. Salo, Y. Nakahori, C.M. Disteche, K. Thangaraj, L. Singh, M.H. Crawford and C. Tyler-Smith, Recurrent duplication and deletion polymorphisms on the long arm of the Y chromosome in normal males. Hum. Mol. Genet., 5 (1996) 1767–1775.

107   L.A. Zhivotovsky, P.A. Underhill, C. Cinnioglu, M. Kayser, B. Morar, T. Kivisild, R. Scozzari, F. Cruciani, G. Destro-Bisol, G. Spedini, G.K. Chambers, R.J. Herrera, K.K. Yong, D. Gresham, I. Tournev, M.W. Feldman and L. Kalaydjieva, The effective mutation rate at Y chromosome short tandem repeats, with application to human population-divergence time. Am. J. Hum. Genet., 74 (2004) 50–61.

108   L. Gusmão, A. Gonzalez-Neira, C. Alves, M. Lareu, S. Costa, A. Amorim and A. Carracedo, Chimpanzee homologous of human Y specific STRs: a comparative study and a proposal for nomenclature. Forensic Sci. Int., 126 (2002) 129–136.

109   L. Gusmão, J.M. Butler, A. Carracedo, P. Gill, M. Kayser, W.R. Mayr, N. Morling, M. Prinz, L. Roewer, C. Tyler-Smith and P.M. Schneider, DNA Commission of the International Society of Forensic Genetics (ISFG): an update of the recommendations on the use of Y-STRs in forensic analysis. Int. J. Legal Med., 120 (2006) 191–200.

110   E. Bosch, A.C. Lee, F. Calafell, E. Arroyo, P. Henneman, P. de Knijff and M.A. Jobling, High resolution Y chromosome typing: 19 STRs amplified in three multiplex reactions. Forensic Sci. Int., 125 (2002) 42–51.

111   J.M. Butler, R. Schoske, P.M. Vallone, M.C. Kline, A.J. Redd and M.F. Hammer, A novel multiplex for simultaneous amplification of 20 Y chromosome STR markers. Forensic Sci. Int., 129 (2002) 10–24.

112   J.G. Shewale, H. Nasir, E. Schneida, A.M. Gross, B. Budowle and S.K. Sinha, Y-chromosome STR system, Y-PLEX 12, for forensic casework: development and validation. J. Forensic Sci., 49 (2004) 1278–1290.

113   A. Hall and J. Ballantyne, The development of an 18-locus Y-STR system for forensic casework. Anal. Bioanal. Chem., 376 (2003) 1234–1246.

114   S. Beleza, C. Alves, A. González-Neira, M. Lareu, A. Amorim, A. Carracedo and L. Gusmão, Extending STR markers in Y chromosome haplotypes. Int. J. Legal Med., 117 (2003) 27–33.

115   B.E. Krenke, L. Viculis, M.L. Richard, M. Prinz, S.C. Milne, C. Ladd, A.M. Gross, T. Gornall, J.R. Frappier, A.J. Eisenberg, C. Barna, X.G. Aranda, M.S. Adamowicz and B. Budowle, Validation of male-specific, 12-locus fluorescent short tandem repeat (STR) multiplex. Forensic Sci. Int., 151 (2005) 111–124.

116   J.J. Mulero, C.W. Chang, L.M. Calandro, R.L. Green, Y. Li, C.L. Johnson and L.K. Hennessy, Development and validation of the AmpFlSTR Yfiler PCR amplification kit: a male specific, single amplification 17 Y-STR multiplex system. J. Forensic Sci., 51 (2006) 64–75.

117   J.L. Weber and C. Wong, Mutation of human short tandem repeats. Hum. Mol. Genet., 2 (1993) 1123–1128.

118   R. Scozzari, F. Cruciani, P. Malaspina, P. Santolamazza, B.M. Ciminelli, A. Torroni, D. Modiano, D.C. Wallace, K.K. Kidd, A. Olckers, P. Moral, L. Terrenato, N. Akar, R. Qamar, A. Mansoor, S.Q. Mehdi, G. Meloni, G. Vona, D.E. Cole, W. Cai and A. Novelletto, Differential structuring of human populations for homologous X and Y microsatellite loci. Am. J. Hum. Genet., 61 (1997) 719–733.

119   L. Henke and J. Henke, Mutation rate in human microsatellites. Am. J. Hum. Genet., 64 (1999) 1473–1474.

120   A. Sajantila, M. Lukka and A.C. Syvanen, Experimentally observed germline mutations at human micro- and minisatellite loci. Eur. J. Hum. Genet., 7 (1999) 263–266.

121   A. Di Rienzo, A.C. Peterson, J.C. Garza, A.M. Valdes, M. Slatkin and N.B. Freimer, Mutational processes of simple-sequence repeat loci in human populations. Proc. Natl. Acad. Sci., 91 (1994) 3166–3170.

122   L.A. Zhivotovsky and M.W. Feldman, Microsatellite variability and genetic distances. Proc. Natl. Acad. Sci., 92 (1995) 11549–11552.

123   A. Di Rienzo, P. Donnelly, C. Toomajian, B. Sisk, A. Hill, M.L. Petzl-Erler, G.K. Haines and D.H. Barch, Heterogeneity of microsatellite mutations within and between loci, and implications for human demographic histories. Genetics, 148 (1998) 1269–1284.

124   G. Cooper, N.J. Burroughs, D.A. Rand, D.C. Rubinsztein and W. Amos, Markov chain Monte Carlo analysis of human Y-chromosome microsatellites provides evidence of biased mutation. Proc. Natl. Acad. Sci., 96 (1999) 11916–11921.

125   R. Kurihara, T. Yamamoto, R. Uchihi, S.L. Li, T. Yoshimoto, H. Ohtaki, K. Kamiyama and Y. Katsumata, Mutations in 14 Y-STR loci among Japanese father–son haplotypes. Int. J. Legal Med., 118 (2004) 125–131.

126   X. Xu, M. Peng and Z. Fang, The direction of microsatellite mutations is dependent upon allele length. Nat. Genet., 24 (2000) 396–399.

127   C.R. Primmer, N. Saino, A.P. Moller and H. Ellegren, Directional evolution in germline microsatellite mutations. Nat. Genet., 13 (1996) 391–393.

128   W. Amos, S.J. Sawcer, R.W. Feakes and D.C. Rubinsztein, Microsatellites show mutational bias and heterozygote instability. Nat. Genet., 13 (1996) 390–391.

129   U. Schmidt, N. Meier and S. Lutz, Y-chromosomal STR haplotypes in a population sample from southwest Germany (Freiburg area). Int. J. Legal Med., 117 (2003) 211–217.

130   J.M. Butler, A.E. Decker, P.M. Vallone and M.C. Kline, Allele frequencies for 27 Y-STR loci with U.S. Caucasian, African American, and Hispanic samples. Forensic Sci. Int., 156 (2006) 250–260.

References pp. 992–1000

131   H.L. Dai, X.D. Wang, Y.B. Li, J. Wu, J. Zhang, H.J. Zhang, J.G. Dong and Y.P. Hou, Characterization and haplotype analysis of 10 novel Y-STR loci in Chinese Han population. Forensic Sci. Int., 145 (2004) 47–55.

132   E. Heyer, J. Puymirat, P. Dieltjes, E. Bakker and P. de Knijff, Estimating Y chromosome specific microsatellite mutation frequencies using deep rooting pedigrees. Hum. Mol. Genet., 6 (1997) 799–803.

133   N.O. Bianchi, C.I. Catanesi, G. Bailliet, V.L. Martinez-Marignac, C.M. Bravi, L.B. Vidal-Rioja, R.J. Herrera and J.S. López-Camelo, Characterization of ancestral and derived Y-chromosome haplotypes of New World native populations. Am. J. Hum. Genet., 63 (1998) 1862–1871.

134   A.C.S. Góes, E.F. Carvalho, I. Gomes, D.A. da Silva, E.H.F. Gil, A. Amorim and L. Gusmão, Population and mutation analysis of 17 Y-STR loci from Rio de Janeiro (Brazil). Int. J. Legal Med., 119 (2005) 70–76.

135   B. Budowle, M. Adamowicz, X. Aranda, C. Barna, R. Chakraborty, D. Cheswick, B. Dafoe, A. Eisenberg, R. Frappier, A. Gross, L.C. Ladd, H. Lee, S.C. Milne, C. Meyers, M. Prinz, M.L. Richard, G. Saldanha, A.A. Tierney, L. Viculis and B.E. Krenke, Twelve short tandem repeat loci Y chromosome haplotypes: genetic analysis on populations residing in North America. Forensic Sci. Int., 150 (2005) 1–15.

136   D.J. Ballard, C. Phillips, C.R. Thacker, C. Robson, A.P. Revoir and D. Syndercombe Court, A study of mutation rates and the characterisation of intermediate, null and duplicated alleles for 13 Y chromosome STRs. Forensic Sci. Int., 155 (2005) 65–70.

137   P.M. Domingues, L. Gusmão, D.A. da Silva, A. Amorim, R.W. Pereira and E.F. de Carvalho, Sub-Saharan Africa descendents in Rio de Janeiro (Brazil): population and mutational data for 12 Y-STR loci. Int. J. Legal Med., 121 (2007) 238–241.

138   A. Mohyuddin, Q. Ayub, P.A. Underhill, C. Tyler-Smith and S.Q. Mehdi, Detection of novel Y SNPs provides further insights into Y chromosomal variation in Pakistan. J. Hum. Genet., 51 (2006) 375–378.

139   P. Shen, T. Lavi, T. Kivisild, V. Chou, D. Sengun, D. Gefel, I. Shpirer, E. Woolf, J. Hillel, M.W. Feldman and P.J. Oefner, Reconstruction of patrilineages and matrilineages of Samaritans and other Israeli populations from Y-chromosome and mitochondrial DNA sequence variation. Hum. Mutat., 24 (2004) 248–260.

140   O. Semino, C. Magri, G. Benuzzi, A.A. Lin, N. Al-Zahery, V. Battaglia, L. Maccioni, C. Triantaphyllidis, P. Shen, P.J. Oefner, L.A. Zhivotovsky, R. King, A. Torroni, L.L. Cavalli-Sforza, P.A. Underhill and A.S. Santachiara-Benerecett, Origin, diffusion, and differentiation of Y-chromosome haplogroups E and J: inferences on the neolithization of Europe and later migratory events in the Mediterranean area. Am. J. Hum. Genet., 74 (2004) 1023–1034.

141   A.O. Karlsson, T. Wallerström, A. Götherström and G. Holmlund, Y-chromosome diversity in Sweden: a long-time perspective. Eur. J. Hum. Genet., 14 (2006) 963–970.

142   L. Scheinfeldt, F. Friedlaender, J. Friedlaender, K. Latham, G. Koki, T. Karafet, M. Hammer and J. Lorenz, Unexpected NRY chromosome variation in Northern Island Melanesia. Mol. Biol. Evol., 23 (2006) 1628–1641.

143   M. Kayser, S. Brauer, R. Cordaux, A. Casto, O. Lao, L.A. Zhivotovsky, C. Moyse-Faurie, R.B. Rutledge, W. Schiefenhoevel, D. Gil, A.A. Lin, P.A. Underhill, P.J. Oefner, R.J. Trent and M. Stoneking, Melanesian and Asian origins of Polynesians: mtDNA and Y chromosome gradients across the Pacific. Mol. Biol. Evol., 23 (2006) 2234–2244.

144   M. Regueiro, A.M. Cadenas, T. Gayden, P.A. Underhill and T.J. Herrera, Iran: tricontinental nexus for Y-chromosome driven migration. Hum. Hered., 61 (2006) 132–143.

145   W. Deng, B. Shi, X. He, Z. Zhang, J. Xu, B. Li, J. Yang, L. Ling, C. Dai, B. Qiang, Y. Shen and R.E. Chen, Evolution and migration history of the Chinese population inferred from Chinese Y-chromosome evidence. J. Hum. Genet., 49 (2004) 339–348.

Subject index